Microarray Analysis

MICROARRAY ANALYSIS

Mark Schena, Ph.D.

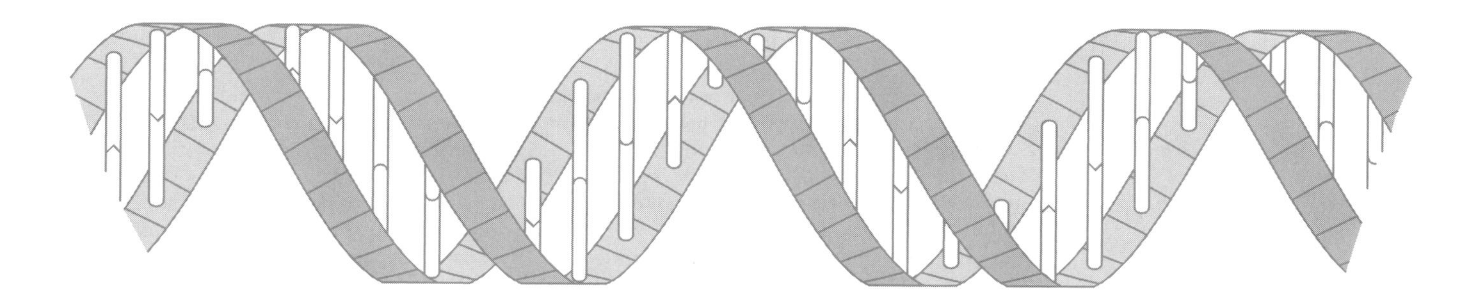

WILEY-LISS
A John Wiley & Sons, Inc., Publication

Published by John Wiley & Sons, Inc., Hoboken, New Jersey.

Published simultaneously in Canada.

Library of Congress Cataloging-in-Publication Data

Schena, Mark.
  Microarray analysis / Mark Schena.
    p. ; cm.
  Includes bibliographical references and index.
  ISBN 0-471-41443-3 (cloth : alk. paper)
  1. DNA microarrays. 2. Protein microarrays. 3. Biochips.
  [DNLM: 1. Oligonucleotide Array Sequence Analysis. 2. Gene Expression Profiling.
3. Microchemistry—methods. QH 441 S324m 2002] I. Title.
QP624.5.D726 S34 2002
572.8′636 dc21 2002011157

Printed in the United States of America.
10 9 8 7 6 5 4 3

To my wife, family, friends, mentors, and humankind.

Contents

Preface

> Some painters transform the sun into a yellow spot; others transform a yellow spot into the sun. —*Pablo Picasso*

> What does not destroy me, makes me strong.
> —*Friedrich Nietzsche*

Why are we here? What is the meaning of life? What happens when we die? Does God exist? This textbook is unlikely to provide immediate answers to these fundamental questions, but it does promise a broad introduction to microarray analysis, which may actually seem more like a punishment than a primer once you study its contents. But seriously...please understand that a scientist is no more equipped to spend 18 months (1.9 millions) writing a textbook than a writer would be if he or she were required to spend 18 months in a laboratory. This preface endeavors to direct, though admittedly it also conveys some of the delirium generated at the hands of a most challenging and demanding project.

I was warned amply about Ron Davis before departing the University of California at San Francisco for Stanford, and I heeded those warnings for several years before falling. But what started as biology, ended up technology, and it is the latter that is the main focus of this textbook. In a brief conversation in a Stanford café in 1990, Ron said in no uncertain terms (and with all the prescience of the French astronomer Nostradamus) that he thought I should develop a "major new technology," and I suppose that is exactly what happened, albeit several years after his suggestion and not necessarily the technology that either of us (certainly not I) had envisioned. Ron and I began our odyssey together exploring the function of transcription factors in the flowering plant *Arabidopsis thaliana* and, through this endeavor, came to realize that a new gene-expression technology based on solid-surface assays would revolutionize the medical and agricultural sciences. I'm not sure either of us understood the magnitude of what we were embarking on, and perhaps neither of us (again, certainly not I) can anticipate the full scope of discoveries that will be made with this technological advance.

There is always a "eureka" moment in research, a term that is even more appropriate in this context because *Eureka* is the motto of California. During a discussion in Ron's office, he told me about "a company in Santa Clara" that was developing high-density arrays of oligonucleotides for gene sequencing

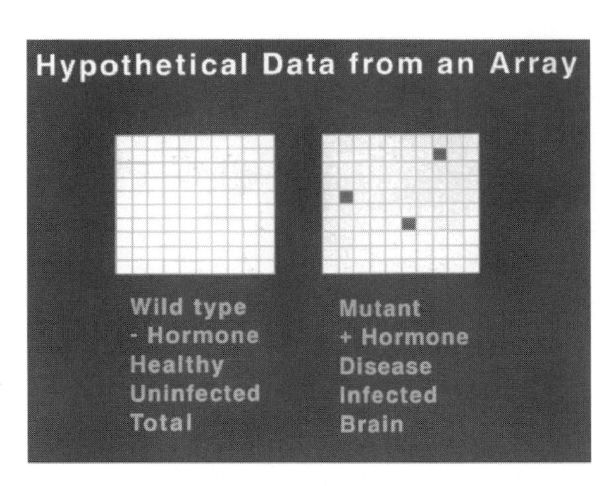

FIGURE P.1. Microarray analysis concept. Microarrays can be used to profile gene expression patterns in labeled mixtures prepared from reference (left) *and test* (right) *samples, yielding information on genotype, hormone action, disease state, infection, and tissue-specific gene expression. (Reprinted with permission from M. Schena, 1994. Data courtesy of M. Schena and R. Davis, Stanford University.)*

applications. After seeing some unpublished Affymetrix data, the idea of using microarrays for gene expression studies hit me with the force of a thunderbolt. Instead of using short oligonucleotide arrays for sequencing by hybridization, it seemed much more powerful to use microarrays of long oligonucleotides for gene expression analysis. Eureka!

We began our work on microarrays at a time when there was no evidence that biological experiments could be performed on glass chips. The microarray concept was presented in Holland in the summer of 1994 amid (I kid you not) howling laughter from the audience; and it was on this occasion that I outlined the basic experimental approach that would come to be known as microarray analysis (Fig. P.1). During that presentation, I also introduced the first microarray enzymatic labeling procedure, demonstrating the feasibility of preparing fluorescent probes from yeast and plant messenger RNA (Fig. P.2).

As a proof of principle study, we manufactured the first microarrays in collaboration with Affymetrix and used fluorescent oligonucleotide probes to demonstrate the specificity of the microarray assay (Fig. P.3). Hybridization experiments with complex gene expression mixtures yielded promising data, although the hybridization specificity was somewhat compromised owing to a suboptimal probe preparation procedure (Fig. P.4). Improved hybridization

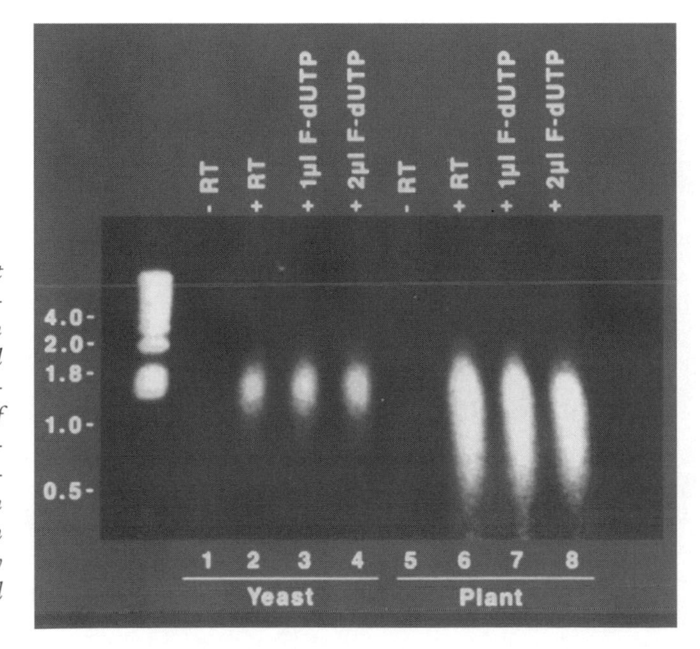

FIGURE P.2. Data from the first microarray labeling procedure. Total messenger RNA samples from yeast and Arabidopsis *were primed with oligo-dT, and labeled with reverse transcriptase in the presence of fluorescein-dUTP. The labeled products were separated by agarose gel electrophoresis and stained with ethidium bromide. (Reprinted with permission from M. Schena, 1994. Data courtesy of M. Schena and R. Davis, Stanford University.)*

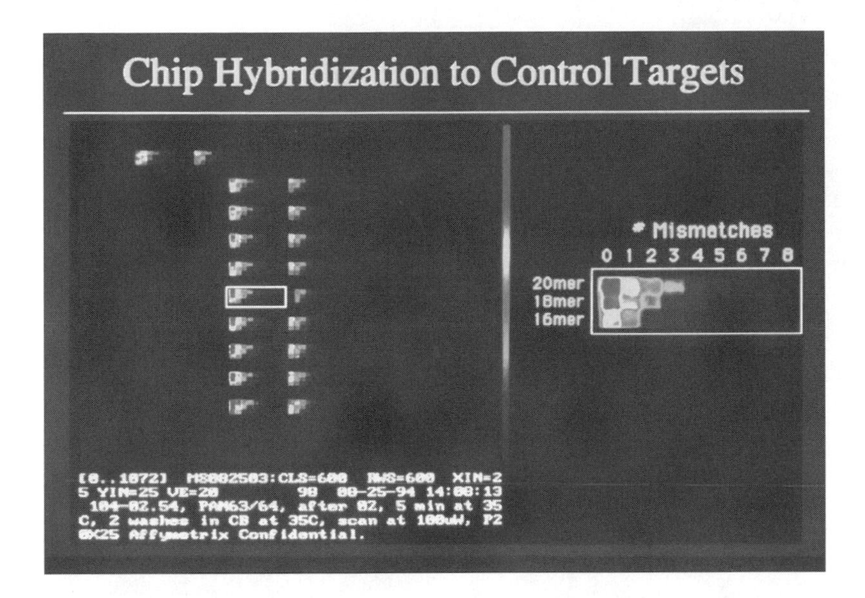

FIGURE P.3. *Data from the first manufactured microarray, which was by Affymetrix and used very large scale immobilized polymer synthesis (VLSIPS) technology. It was hybridized with two fluorescent oligonucleotides (PAN63 and PAN64) to assess assay specificity and detectivity. A small microarray region (white box) is enlarged for ease of viewing (right). (Reprinted with permission from M. Schena, 1994. Data courtesy of M. Schena and R. Davis, Stanford University.)*

specificity was obtained using complementary DNA microarrays in a second collaboration, this time with Dari Shalon and Patrick Brown at Stanford (Fig. P.5). A microarray robot was used for complementary DNA microarray manufacture and, as expected, these microarrays yielded unambiguous gene expression information for several different lines of *Arabidopsis*. The data from these experiments were presented at the University of Wisconsin at Madison in the summer of 1995 (Fig. P.6). Printed microarrays were also employed for the first human microarray analysis experiments, in a collaboration with Synteni (Palo Alto, CA). This work was undertaken in spite of warnings from some luminaries in the field who noted that repetitive sequences in the human genome would prevent the use of microarray assays for human studies (huh?). The first human microarray data were presented at the Stanford Sierra Retreat in October 1995 (Fig. P.7).

The field has blossomed immensely since the first paper on microarrays and microarray analysis appeared in *Science* magazine in the fall of 1995 (Schena

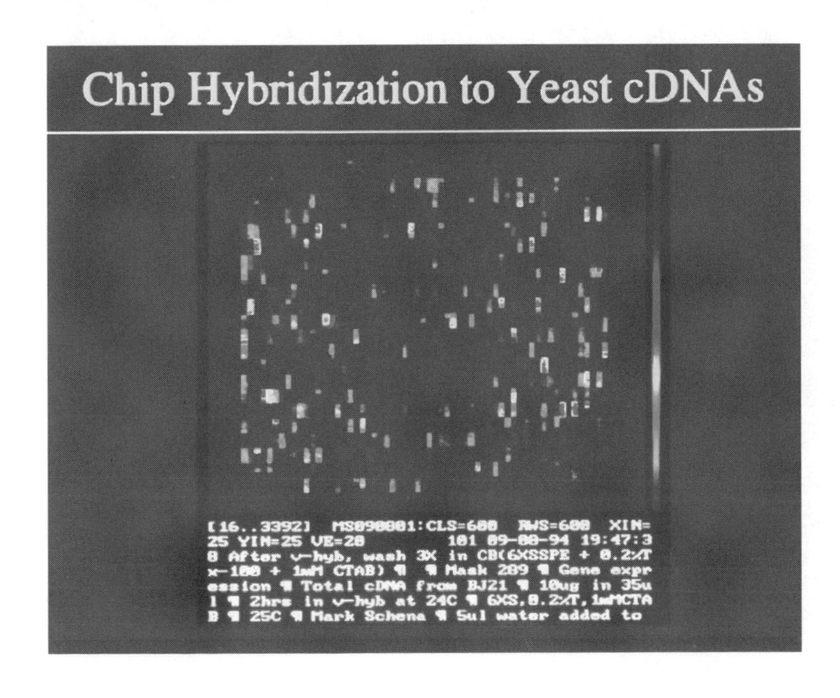

FIGURE P.4. *First microarray analysis experiment. Shown are data from the first microarray analysis experiment. The yeast oligonucleotide microarray was manufactured at Affymetrix (Santa Clara, CA) using VLSIPS technology, and hybridized under a cover slip using a fluorescent probe mixture prepared by reverse transcription of yeast total mRNA. Data were generated by Mark Schena on September 9, 1994, and presented on October 5, 1994 at the Stanford Sierra Camp. Data were provided courtesy of Mark Schena and Ron Davis (Stanford University).*

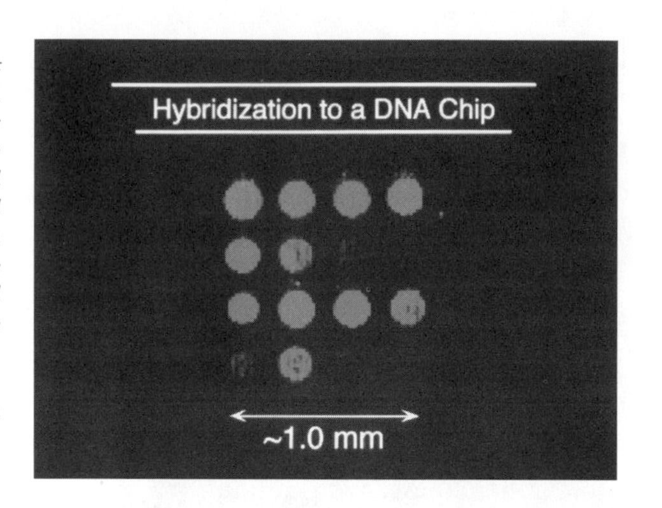

FIGURE P.5. *Data from the first printed complementary DNA (cDNA) microarray, which was manufactured in the laboratory of P. Brown using a tweezer-based printing robot that deposited small droplets of amplified* Arabidopsis *cDNA onto a treated glass microscope slide. The microarray was hybridized with fluorescent cDNA to assess the specificity and detectivity of the assay. (Reprinted with permission from M. Schena and D. Shalon, 1995. Data courtesy of M. Schena and R. Davis, Stanford University.)*

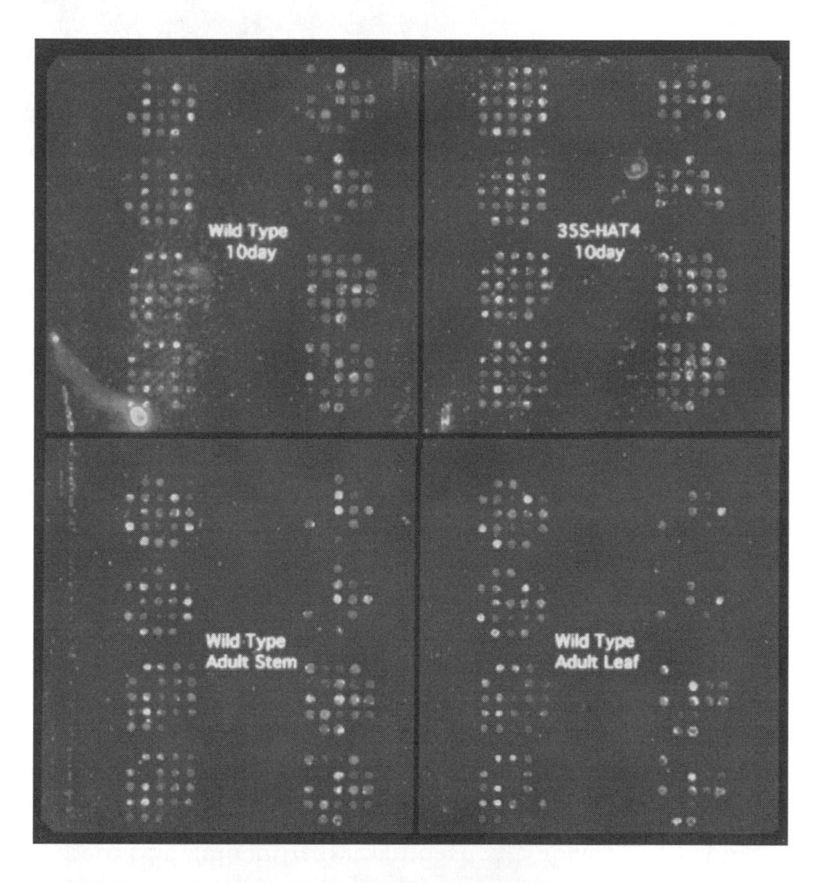

FIGURE P.6. *Data from the first complementary DNA (cDNA) microarray analysis experiment. An* Arabidopsis *cDNA microarray was hybridized with fluorescent cDNA mixtures prepared from four different sources of* Arabidopsis *messenger RNA. The experiments were performed to profile gene expression patterns in a wild-type line, a transgenic line overexpressing the* HAT4 *cDNA, wild-type stem tissue, and wild-type leaf tissue. (Reprinted with permission from M. Schena and D. Shalon, 1995. Data courtesy of M. Schena and R. Davis, Stanford University.)*

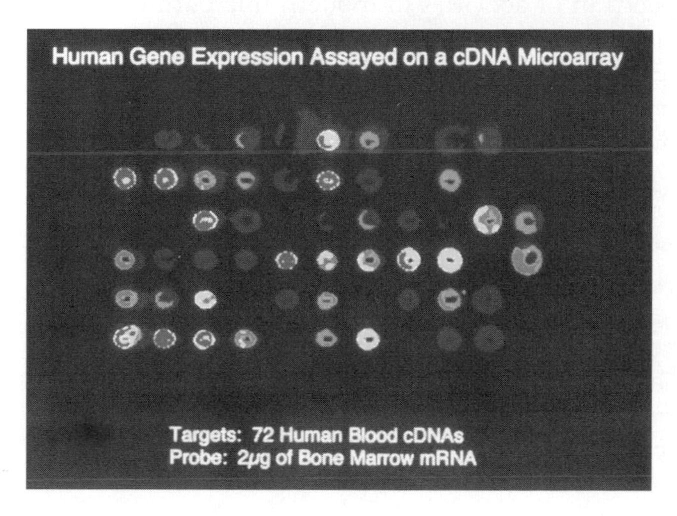

Human Gene Expression Assayed on a cDNA Microarray
Targets: 72 Human Blood cDNAs
Probe: 2µg of Bone Marrow mRNA

FIGURE P.7. *Data from the first human microarray manufactured by printing 72 amplified complementary DNA (cDNA) inserts from a human blood cDNA library onto a glass substrate. Microarray analysis was performed by hybridizing a fluorescent cDNA mixture labeled from bone-marrow messenger RNA and reading the fluorescent signals, which were coded to a rainbow palette. (Reprinted with permission from M. Schena 1995. Data courtesy of M. Schena and R. Davis, Stanford University.)*

et al., 1995). Assays now encompass a remarkable breadth of organisms, including human, yeast, mouse, rat, chimp, gorilla, fruit fly, worm, corn, rice, and bacteria, and enjoys an equally impressive breadth of applications, including genotyping, tissue analysis, and protein studies. More than 3000 scientific publications showcase a broadening repertoire of microarray assays (see References), and this wealth of published work emphasizes the richness of the technology, the collegiality of the field, and the abundant availability of commercial resources.

The creation of *Microarray Analysis* was motivated by the explosive proliferation of the technology, the highly technical nature of the field, and the flood of requests from young scientists for a foundational compendium. The textbook was also motivated by a clear need to assemble a resource aimed at complementing the specialty books and primary scientific literature in the field, and by my good friend Tom Tisone. I have been compensated enormously over the years for my technical expertise, and the notion of giving something back to the scientific community at this juncture seemed entirely reasonable, logistically possible, and probably a bit overdue.

And so when Luna Han, life and medical sciences editor at John Wiley & Sons, contacted my publicist Paul "Hollywood" Haje to initiate *Microarray Analysis,* I jumped at the opportunity and an agreement was hammered out quickly. I slated six months to write the textbook, and the project took almost exactly three times that long, a remarkably poor estimate for someone with a Ph.D. in biochemistry. In my defense, let me assure you that no amount of grooming by the best mentors at the top biochemistry departments in the country could prepare a biochemist for writing a 1500-page science manuscript. I joked several times with Luna that I came to consider myself a POW (Prisoner of Wiley), but in retrospect that title turned out to be less than accurate because real POWs actually enjoy somewhat better treatment than textbook writers! With all kidding aside, the project would not have succeeded without Luna's constant support and guidance, and dearest thanks are due to Luna and the entire staff at Wiley for making *Microarray Analysis* a success.

Microarray Analysis is written for a broad audience of undergraduates, graduate students, postdoctoral fellows, faculty members, and deans as well as researchers, clinicians, investors, lawyers, and businesspeople from universities, companies, hospitals, government agencies, and nonprofit organizations. The book is intended to provide a conceptual, experimental, and methodological foundation for the full spectrum of activities in modern microarray analysis. Chapter 1 introduces the field, and Chapters 2–4 cover basic concepts in chemistry, biochemistry, and genomics pertinent to microarray analysis. Chapters 5–9 explore microarray surfaces, targets and probes, manufacturing, detection, and data analysis and modeling. Chapters 10–13 explain microarray methodology, cleanroom technology, gene expression profiling, and genotyping and diagnostics. Separate chapters are devoted to novel technologies (Chapter 14), commercial opportunities (Chapter 15), and future trends (Chapter 16). Each chapter also contains a set of questions intended for use as a teaching aid, so that university courses, technical workshops, and adult-education programs can be designed around *Microarray Analysis.* The continued expansion and success of the field is predicated on having highly educated students and scientists, and it is my hope that *Microarray Analysis* will assist in teaching a new generation of microarray researchers, including our youngest and most precocious students (Fig. P.8).

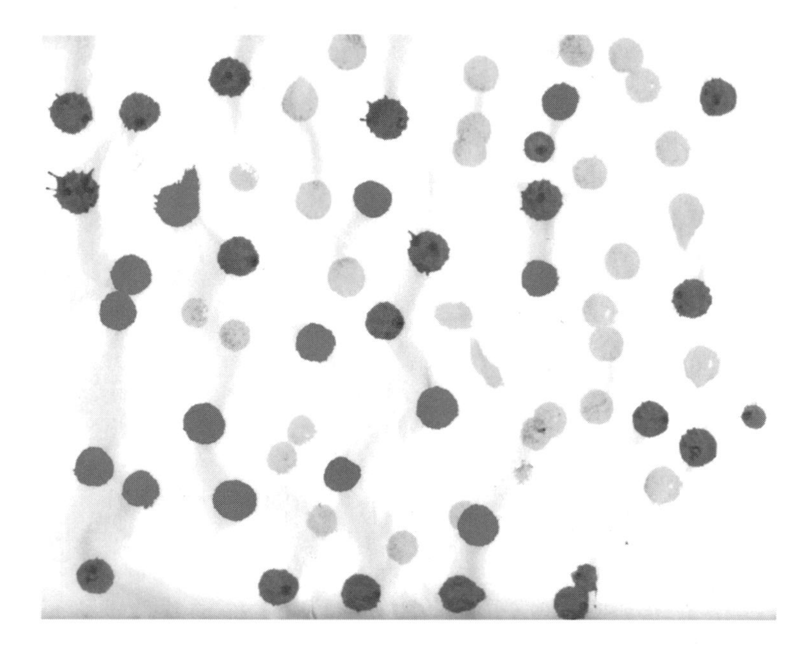

FIGURE P.8. *A first attempt at microarray manufacture from the youngest known microarray researcher. (Data courtesy of Julia Kaplan, age 5, San Carlos, CA.)*

Have we reached a turning point in our civilization? Since our earliest days as humans on the plains of Africa, genetic and infectious diseases have ravaged our communities. What is disease? Disease is little more than the disruption of normal cellular gene function, and microarrays can be used to rapidly understand disease states by enabling the analysis of gene expression patterns, sequence variation, and other biochemical processes. Fifty years from now, and long after human disease has been eradicated, we will look back incredulously at the start of this millennium and wonder how we ever endured cancer, heart disease, AIDS, and the thousands of other illnesses that compromise our well-being. But that will be then, and this is now. Let's get to work!

Mark Schena

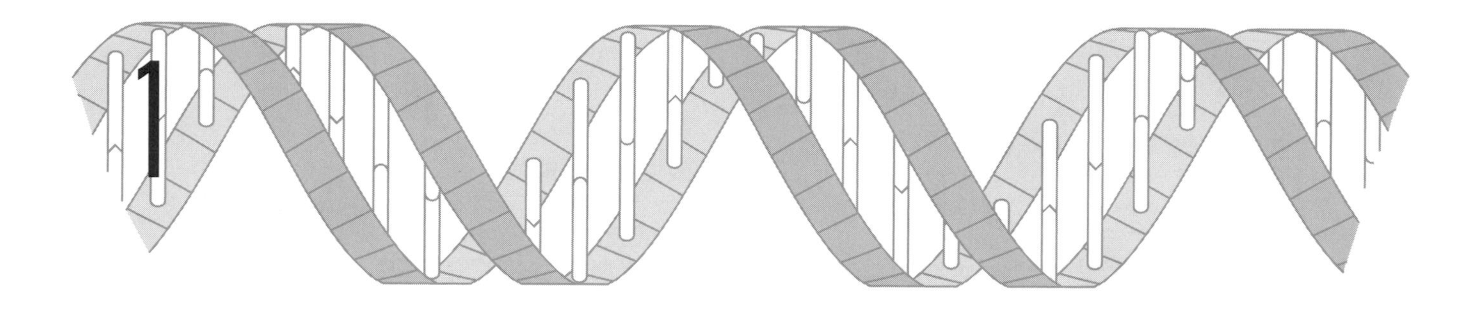

Introduction to Microarray Analysis

Any sufficiently advanced technology is indistinguishable from magic.
—*Arthur C. Clarke*

The magic of **microarray analysis** is sweeping through the agricultural and medical sciences, replacing traditional biological assays based on gels, filters, and purification columns with small glass chips containing tens of thousands of DNA and protein sequences. Microarrays function like biological microprocessors, enabling the rapid and quantitative analysis of **gene expression** patterns, patient genotypes, drug mechanisms, and disease onset and progression on a genomic scale. This chapter provides an introduction to microarray analysis, including definitions, origins, and historical underpinnings of the field along with experimental tips and discussions of how this technology is revolutionizing research.

Microarray analysis. *Five-step experimental procedure that uses microarrays to explore the biological, chemical, and physical world.*

Gene expression. *The cellular process by which genetic information flows from gene to messenger RNA to protein.*

WHAT IS A MICROARRAY?

A **microarray** is a small analytical device that allows genomic exploration with speed and precision unprecedented in the history of biology. Glass chips containing tens of thousands of genes are used to examine fluorescent samples prepared by labeling **messenger RNA** (mRNA) from cells, tissues, and other biological sources (Fig. 1.1). Molecules in the fluorescent sample react with cognate sequences on the chip, causing each **spot** to glow with an intensity proportional to the activity of the expressed **gene** (Fig. 1.2). The enormous capacity of these miniature devices allows the analysis of the entire human

Microarray. *An ordered array of microscopic elements on a planar substrate that allows the specific binding of genes or gene products.*

Messenger RNA. *The class of cellular RNA that undergoes extensive editing, contains the protein-coding sequences of genes, and functions as an informational intermediate between DNA and protein.*

Spot. *A microarray element or feature that contains target molecules.*

Gene. *Segment of genomic DNA that encodes a specific cellular mRNA and protein.*

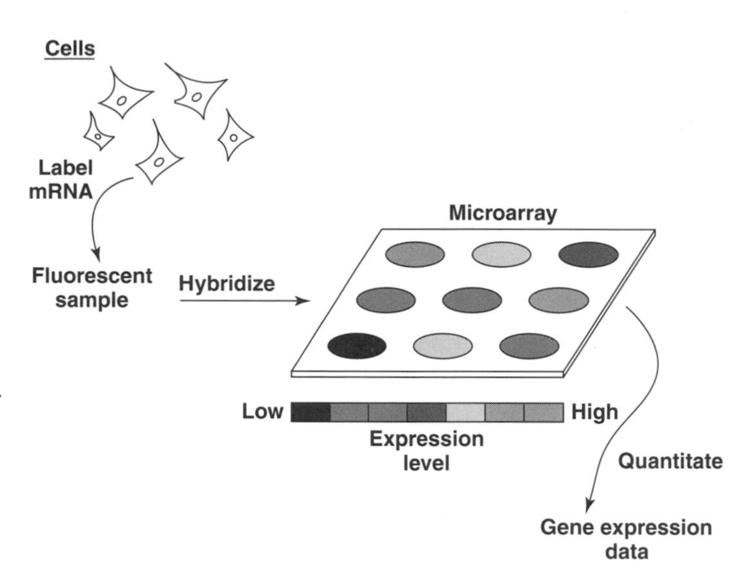

FIGURE 1.1. For microarray analysis, mRNA is isolated from cells, labeled with fluorescent tags, and hybridized to a glass chip containing a microarray of gene sequences. Each spot on the chip binds specifically to fluorescent molecules in solution, causing the spots to glow with intensities proportional to the expression level of each gene. Intensities can be coded to a color palette (horizontal bar). *Quantitative gene expression data are obtained by determining the fluorescence intensity at each microarray location.*

Genome. *Entire DNA content of a cell, including the nucleotides, genes, and chromosomes.*

Biochemistry. *The field of study that endeavors to understand the chemical basis of life by focusing on the study of DNA, RNA, proteins, and other biomolecules.*

Transcription. *First step in gene expression in which messenger RNA is synthesized from a DNA template.*

genome in a single experiment. Because patterns of gene expression correlate strongly with function, microarrays are providing unprecedented information on human disease, aging, drug and hormone action, mental illness, diet, and many other clinical matters. Microarrays can also be used to find alterations in gene sequences, paving the way for a new era of genetic screening, testing, and diagnostics. Tissue and protein microarrays are miniaturizing traditional histological and biochemical assays, speeding the analysis of tumor specimens, protein–protein interactions, and enzymes. The capacity to explore the genomes of bacteria, viruses, worms, fruit flies, plants, cows, chickens, mice, rats, and primates renders microarrays the Noah's Ark of **biochemistry**.

WHAT ARE THE ORIGINS OF MICROARRAYS?

Microarrays were developed at Stanford University by Schena and co-workers in early 1990s. The idea arose during a conversation with Davis in which we discussed the feasibility of developing a revolutionary new technology to study plant gene expression. We had spent considerable time isolating and characterizing plant **transcription** factors during the early 1990s (Lloyd et al., 1994; Schena et al., 1992, 1993), but the lack of sufficient analytical tools in plant research made the study of these proteins slow and arduous. Davis supported the outlandish proposal of developing glass chips to study plant transcription factors, and we moved ahead aggressively with the project in spite of the formidable technical hurdles that challenged the success of the work. Glass chips had never yielded biological information on any experimental system, let alone gene expression data from a mustard plant that contains minute quantities of mRNA. How would we manufacture the chips? How would the chips be read? Was it possible to hybridize complementary DNA (cDNA) mixtures on glass? How would we label the probes? Would we obtain sufficient signals from complex probe mixtures? These and other formidable questions were addressed and solved by drawing on the rich tradition of the Davis Laboratory and the Stanford biochemical department. The work would not have succeeded without strong support for risky, cutting-edge technological research and a thorough appreciation of nucleic acid biochemistry.

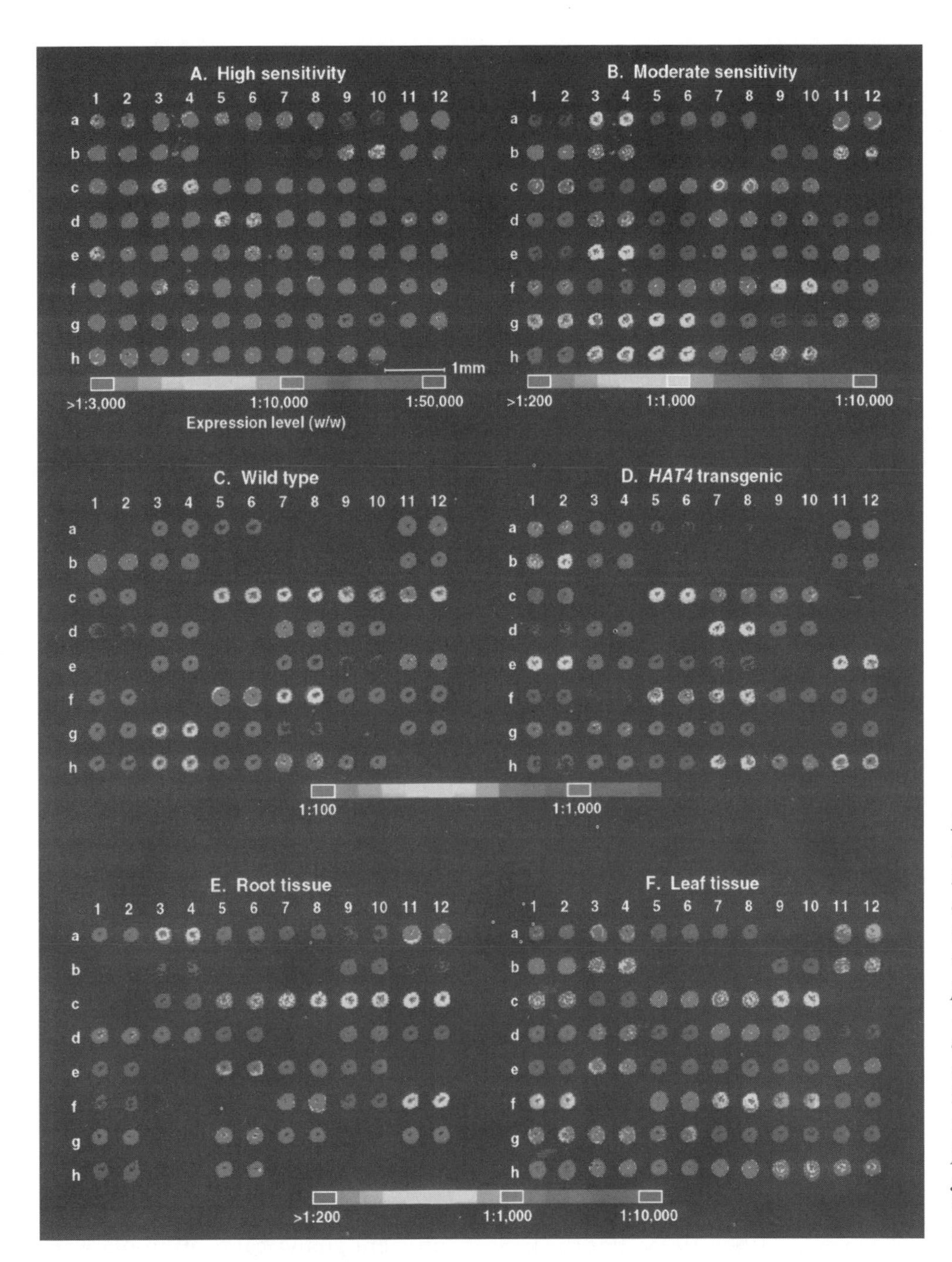

FIGURE 1.2. Scanned images from the first published microarray experiments. Plant cDNAs were printed onto treated microscope slides and then denatured, hybridized with fluorescent cDNA mixtures prepared from plant mRNA samples, and scanned. The fluorescent images were coded to a rainbow color palette which was calibrated using known quantities of human mRNA. Gene expression patterns were examined in the whole plant (A, B), wild type (C) and transgenic plant lines (D), and root (E) and leaf tissue (F). (Reprinted with permission from Science.)

Our research strategy was to manufacture microscopic arrays (microarrays) containing plant gene sequences attached to a glass substrate and use the manufactured microarrays to measure plant gene expression in hybridization experiments with fluorescent samples prepared by the enzymatic labeling of plant mRNA. Under the proper experimental conditions, the fluorescent signals on the surface would provide a quantitative measure of each plant gene represented in the microarray. Three advanced approaches (photolithography, ink jetting, and contact printing) were envisioned for microarray manufacture (Lemieux et al., 1998; Schena et al., 1998), and these three technologies are the main strategies used currently (see Chapter 7).

The approach leveraged Stanford's rich tradition of nucleic acid biochemistry, including the knowledge of DNA polymerase, recombinant DNA, filter array methods, and gene expression technologies developed between 1950 and

1990 (see below). The technology required collective contributions from many different disciplines and drew from early experiments on solid surfaces (see below). Our early work in gene expression (Heller et al., 1997; Schena et al. 1995, 1996a, 1996b) paved the way for the explosive proliferation of microarray methods used at present (see References).

HOW ARE MICROARRAYS DEFINED?

A *microarray* is an ordered array of microscopic elements on a planar substrate that allows the specific binding of genes or gene products (Figs 1.1–1.3). *Microarray* is a new scientific word derived from the Greek word *mikro* (small) and the French word *arayer* (arranged). Microarrays (also known as biochips, DNA chips, and gene chips) contain collections of small elements or spots arranged in rows and columns. To qualify as a microarray, the analytical device must be (1) ordered, (2) microscopic, (3) planar, and (4) specific. Devices that fulfill only a subset of these criteria do not afford the advantages of microarrays, do not qualify as microarrays, and should not be considered as such. A brief discussion of each of these criteria assists in defining the basic attributes of a microarray and highlight how microarrays differ from earlier devices. One obvious exception to the definition of a microarray pertains to non-genetic microarray applications. Microscopic arrays of organic chemicals, semiconductor materials, minerals, and many other types of substances are finding increasing use in chemistry, physics geology, and material science. Devices of this sort qualify as microarrays providing they satisfy three of four criteria (ordered, microscopic, and planar).

Ordered Array

Ordered. *A collection of microarray elements configured in rows and columns.*

An ***ordered*** array is any collection of analytical elements configured in rows and columns (Fig. 1.4). Analytical elements are also known as spots or features in the microarray literature. Each row of elements must form a straight line horizontally across the substrate, and each column of elements must form a straight line vertically across the substrate, in a manner perpendicular to the rows. Ordered elements must have a uniform size and spacing and a unique

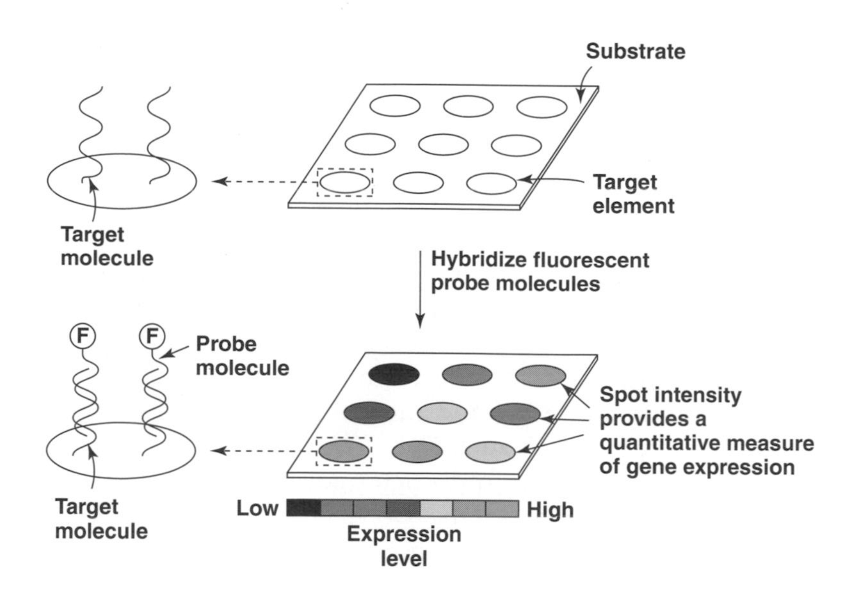

FIGURE 1.3. *Target elements, spots, or features* (circles) *contain target DNA molecules attached to a glass substrate* (top inset). *The single-stranded target molecules hybridize to fluorescent probe molecules in solution* (bottom inset), *causing the spots to glow with intensities proportional to the expression level of each gene. Signal intensities are coded to a color palette* (horizontal bar) *for ease of interpretation, and quantitative gene expression information is obtained by measuring the fluorescence intensity at each microarray location.*

Rows & columns

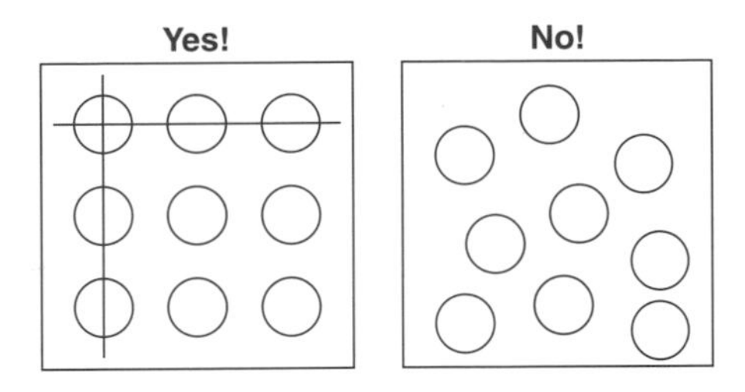

Uniform size and spacing

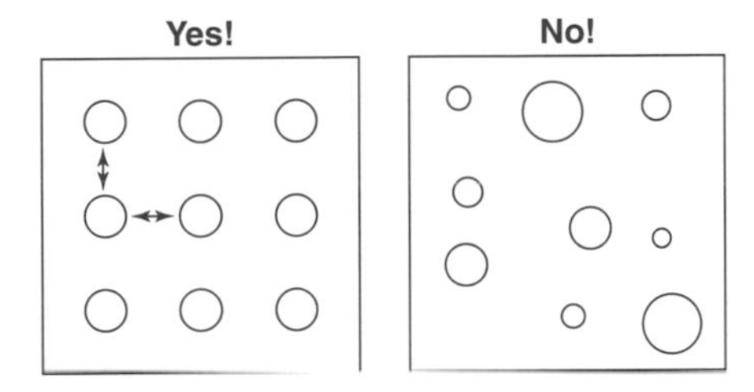

Unique address

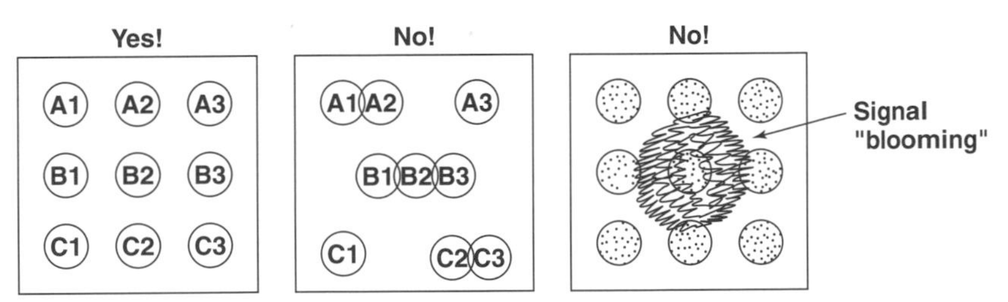

FIGURE 1.4. *Microarray elements are configured in rows and columns rather than in an irregular pattern. Elements should have a uniform size and spacing, rather than nonuniform sizes and irregular spacing. Each element should reside at a unique address, rather than at ambiguous locations. Labeling and detection schemes that cause the signal to spread, or bloom into adjacent locations complicate data analysis and should be avoided.*

location on the microarray substrate. The basic criteria for an ordered array are, therefore, rows and columns, uniform size and spacing, and unique address. Arrays that do not conform to these basic criteria are not ordered and thus do not meet the first of the four standards required to qualify as a microarray. Multiple microarrays can be combined on a single planar substrate to provide larger microarrays, and this provision is useful in that in enables microarray manufacture using multiple printing implements such as pins and ink-jet nozzles (see Chapter 7).

What is the value of having the microarray elements arranged in rows and columns? Elements configured in rows and columns (Fig. 1.4) are an essential and enormously valuable component of microarray analysis, because this design allows rapid manufacturing, detection, and quantification. Rows and columns allow the use of standard motion control technologies, including linear actuators and encoders for microarray manufacture and detection (see Chapters 7 and 8), keeping equipment costs low and quality high. Microarray printers, scanners, and other devices can read and write ordered microarrays in a rapid and highly automated fashion, enabling a speed and precision that would not be possible with irregular formats. Software quantitation tools adapted from the computer industry can be used to extract rows and columns of microarray data (see Chapter 9), a step that would be hindered with an irregular format. Any device containing an arrangement other than rows and columns would require extensive customization of current hardware and software, and the expense and complexity of such designs have prevented their proliferation.

What is the value having the microarray elements spaced uniformly on the microarray substrate? The requirement of uniform element spacing (Fig. 1.4) is an extension of the importance of rows and columns, namely it simplifies and speeds microarray manufacture, detection, and analysis. Uniform spacing allows the use of standard offsets (e.g., 125 μm) in controller software, which simplifies the routines used to move photomasks, printing pins, and other manufacturing devices (see Chapter 7). Irregular spacing or bizarre element geometry would be extremely difficult to implement in a laboratory or manufacturing environment. Uniform spacing simplifies detection and data analysis by enabling the use of standard quantification templates (see Chapter 9) that define each location like squares on a checkerboard and rely on equal spacing between the elements to extract the data.

What is the value of having microarray elements of uniform size? The microarray format requires uniform spot size for ease of quantification and assay precision. Elements of uniform size (Fig. 1.4) simplify the use of quantification templates, which define the location and size of each element and calculate average signal intensities based on the number of pixels contained at each location. Formats with elements of different sizes would hamper signal intensity calculations. Uniform spot size also improves assay precision by ensuring a uniform target density at each address (see Chapter 5). Target density and signal strength are related to spot size (see Appendix H) in that a given number of molecules configured in a large spot would have a lower density than the same number of molecules deposited in a smaller region. Uniform element size ensures that the same number of molecules are present at each address on the microarray substrate, and the intrinsic importance of uniform sized elements is underscored by the absence of any successful microarray format with random or variable feature size.

What is the value of having each microarray element located at a unique address on the substrate? Unique spot location or address (Fig. 1.4) ensures the accurate quantification of signal intensities and their unambiguous assignment to the correct target sequence. If three spots containing gene sequences A, B, and C at unique substrate locations produce signal intensities of 50,000, 5,000, and 500 counts, respectively, a unique address for each feature ensures that the data will be quantified and assigned correctly to genes A–C. Quantified data can be saved in delimited formats (see Chapter 9), as long as each target element is present at a unique substrate location. The capacity to exploit existing delimited formats and other software tools greatly speeds data analysis, mining, and archiving. Spots that overlap, have random spacing, or have nonunique locations would hinder accurate microarray data analysis, and such formats should be avoided.

A related aspect of unique addresses is the importance of using labeling and detection schemes that spatially restrict spot signals (Fig. 1.4). The unwanted phenomenon of signal blooming, in which an intense signal from one feature contaminates adjacent features, confounds microarray analysis and must be avoided under all circumstances. Fluorescent labeling and detection schemes (see Chapter 8) maintain the spatial separation of signals and are desirable for this reason, though various nonfluorescent approaches also restrict signals in a spatial manner (see Chapter 14). Radioactive signals, such as those emitted by phosphorus 32 (P^{32}), spread aggressively in three dimensions and impair the capacity to quantitate adjacent features accurately. Radioisotopes and other diffusible labeling and detection schemes should not be used in microarray analysis.

The requirement for ordered arrays guides the implementation of many different enabling technologies, including photolithography, contact printing, ink jetting, scanning, and imaging. Methodological principles (see Chapter 10) describe what needs to be done and why, rather than suggesting how the aforementioned will be accomplished. The epistemological distinction between methodology and method is significant, and students should endeavor to understand it. Methodology maintains the integrity of microarray analysis without stifling technology innovation. Microarray methods and tools have evolved immensely since the first microarray paper was published (Schena et al., 1995), and this evolution is likely to continue at a brisk pace as long as the toolmakers adhere to the general criteria put forth for microarray formats.

Microscopic Elements

Microscopic is defined as any object that cannot be seen clearly without the use of a microscope, which is nominally anything smaller than about 1 mm (1000 μm). Microarrays made by photolithography and the other semiconductor-based strategies typically produce 15- to 30-μm features, and printed microarray spot size is generally 50–350 μm. Most tissue microarrays contain spots of 200–600 μm, though the tissue microarray format is likely to shrink as the technology improves. The use of submillimeter microscopic elements is a clear departure from the earlier glass- and filter-based methods (see below), which used large printed elements > 1 mm in diameter. To qualify as a microarray element, the element must be smaller than 1.0 mm.

Microscopic. *An object that cannot be seen clearly without the use of a microscope and is measured typically in microns (1000 microns = 1 mm).*

Microarray elements are collections of **target** molecules that allow specific binding of **probe** molecules including genes and gene products, and a typical printed DNA spot contains approximately 1 billion (10^9) molecules attached to the glass substrate (see Appendix H). Microarray target material can be derived from whole genes or parts of genes, and may include genomic DNA, cDNA, mRNA, protein, small molecules, tissues, or any other type of molecule that allows quantitative gene analysis. Target molecules include natural and synthetic derivatives obtained from a variety of sources, such as cells, enzymatic reactions, and machines that carry out chemical synthesis. Synthetic oligonucleotides, short single-stranded molecules made by chemical synthesis, provide an excellent source of target material (see Chapter 6).

Target. *Molecule tethered to a microarray substrate that reacts with a complementary probe molecule in solution*

Probe. *Labeled molecule in solution that reacts with a complementary target molecule on the substrate*

What is the advantage of having microscopic elements? Small features or spots enable high density (>5000 elements/cm^2), rapid reaction kinetics and the analysis of entire genomes on a single chip. Experiments that examine all of the genes in the genome provide a comprehensive of a biological phenomenon (see Chapter 12) that is not possible with technologies limited to gene subsets. Microscopic spots enable miniaturization and automation, two key features of microarrays and microprocessors (see below). Filter arrays and other nonmicroarray formats made with large elements prevent miniaturization and automation and do not allow whole genome analysis in a miniature format.

Planar Substrate

A ***planar*** substrate is parallel and unbending support—such as glass, plastic, or silicon—onto which a microarray is configured. Glass is the most widely used substrate material owing to the many advantages offered by silicon dioxide (see Chapter 5), though other planar materials also work well. It is important to note that the requirement for a planar substrate is somewhat more stringent than the specification of a solid support. Glass, plastic, and silicon are solids but so

Planar. *Evaluative criterion for a microarray substrate that refers to the parallelism of the surface over the entire substrate.*

are the nitrocellulose and nylon filters used in nonmicroarray assays developed in the 1970s and 1980s (see below). To qualify as a microarray, the substrate must be *planar*. All planar materials are solids, but not all solids are planar.

What is the value of having a planar substrate for the microarray support? Planar materials are flat over the entire surface. Flat supports are amenable to automated manufacture, providing an accurate distance from photomasks, pins, ink-jet nozzles, and other manufacturing implements, and ensuring high quality of manufactured microarrays (see Chapter 7). Planar materials also allow accurate scanning and imaging, which rely on a uniform detection distance between the optical element and the microarray surface (see Chapter 8). Planar materials tend to be impermeable to liquids, allowing small feature size and low reaction volumes. Automation, increased precision in manufacture and detection, and impermeability are afforded by planar substrates but not by the solid nitrocellulose and nylon supports used in the earlier filter-based formats. Some planar supports are better than others, but there is no compelling reason per se to specify glass as the material for microarray substrates.

Specific Binding

Specific binding refers to unique biochemical interactions between probe molecules in solution and their cognate target molecules on the microarray. Binding specificity allows a gene or gene product to be analyzed quantitatively with a single microarray target element.

Quantitative. *Any microarray assay that provides a precise measure of the number, amount, or concentration of the molecules present in a sample.*

Quantitative analysis is the process of measuring the number, amount, or concentration of the molecules present in a sample. It is distinct from qualitative analysis, which provides an approximate "yes or no" answer or estimate. Specific binding is an immensely important aspect of microarray analysis and a criterion that imparts much of the value of this new science. Knowing the quantitative levels of gene transcripts or their ratios in two or more samples tells much about gene function and the physiological basis of cell signaling, disease, and environmental responsiveness (see Chapter 12). Numerical microarray outputs enable quantitative analysis of microarray data, and the application of a complete spectrum of statistical measurements, including reproducibility, precision, confidence levels, variation, and correlation (see Chapter 9). Accurate patient genotyping (see Chapter 13) requires methods that can distinguish homozygotes and heterozygotes that differ by only 50% in gene concentration, and such methods rely entirely on binding specificity between target and probe molecules.

Each microarray spot (target) should bind essentially to a single species in the labeled probe mixture (Fig. 1.5) to provide the most accurate measure of genes or gene products. Microarray assays exploit a "one target per probe molecule" paradigm, and assay precision can be enhanced using multiple microarray elements per gene. In most hybridization experiments, 15- to 25-nucleotide target sequences define the minimal target length required to achieve single gene specificity. Many gene expression formats use 50- to 5000-nucleotide targets, which provide extensive target–probe complementarity and intense fluorescent signals.

Rigid. *Physically inflexible or unbending, such as glass and other materials used for microarray manufacture.*

Microarrays differ fundamentally from some of the early oligonucleotide array formats involving hybridization to short oligonucleotides on **rigid** supports, in that these formats did not provide the hybridization specificity

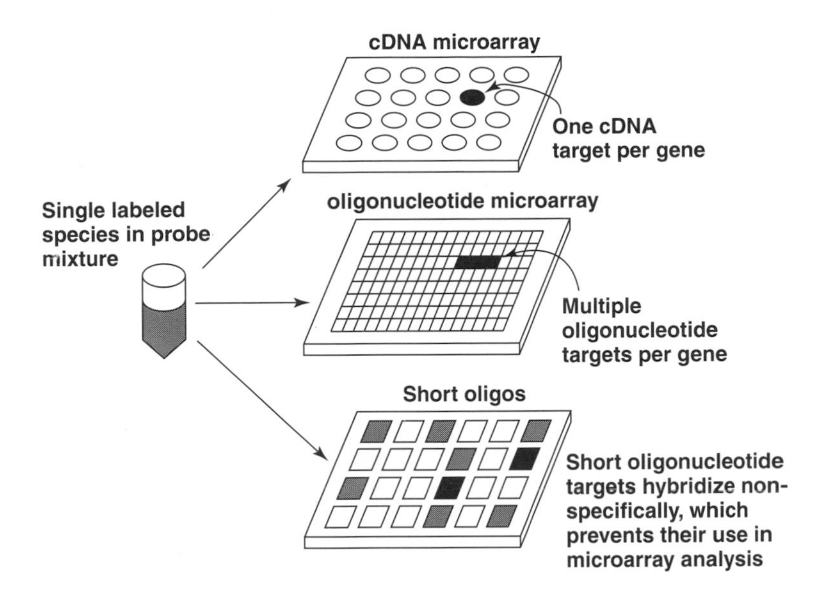

FIGURE 1.5. A probe mixture containing a single labeled mRNA species should hybridize to a single cDNA target element in a cDNA microarray or to a related set of oligonucleotide targets in an oligonucleotide microarray. An array of short oligonucleotides produces a confusing and complex hybridization pattern when reacted with a single probe species.

required for quantitative analysis. The suggestion that short oligonucleotide arrays could be used to assay gene expression based on hybridization pattern is undermined by the fact that the pattern obtained by hybridization of a complex mixture (e.g., labeled mRNA from a cell) would be too complex to provide quantitative information (Fig. 1.5). High specificity between target and probe is a key criterion of microarray analysis, and formats that do not provide specific binding and quantitative data do not qualify as microarrays.

WHAT ARE THE HISTORICAL UNDERPINNINGS OF MICROARRAYS?

Microarray analysis is unique in the history of biology because no other technology has ever involved so much technological complexity, combined expertise from so many different disciplines, and provided a quantitative and systematic view of a biological system. The development of microarray technology in the early 1990s at Stanford University (Schena et al., 1995) drew heavily from six major disciplines: biology, chemistry, physics, engineering, mathematics, and computer science. Reviewing all of the contributions from these major disciplines would require an entire textbook on its own, but a brief summary of some of the foundational discoveries from biology highlights the complexity of microarray science and places microarrays into historical context.

The correlation between gene mutation, altered protein, and disease was made first by Pauling and co-workers (1949). Pauling showed that hemoglobin from sickle cell patients differs from hemoglobin in healthy individuals in that it migrates aberrantly in gel electrophoresis assays, a finding the authors attributed correctly to a change in the surface charge of hemoglobin. By examining normal individuals, carriers, and patients with sickle cell disease, Pauling and co-workers concluded that changes in the hemoglobin gene were responsible for the altered protein, which was verified later in gene sequencing studies. This remarkable paper paved the way for the molecular genetic analysis of human disease, and provided a conceptual foundation for the use of microarrays in genetic screening, testing, and diagnostics (see Chapter 13). The hemoglobin paper in *Science* was a landmark publication.

Watson and Crick (1953) predicted the chemical structure of DNA in their remarkable letter to *Nature*. Using structural, chemical and modeling data, the authors correctly speculated that DNA contains two strands that run in opposite directions, held together by hydrogen bonds between the DNA bases. Watson and Crick suggested specific base pairing interactions (A-T and G-C), and a structure in which the phosphate groups reside on the exterior of the **double helix**. Subsequent biochemical and structural studies verified these predictions, and Watson, Crick, and Wilkins shared the Nobel Prize in 1962 "for their discoveries concerning the molecular structure of nuclear acids and its significance for information transfer in living material." The discovery of the double helix was one of the most important breakthroughs in twentieth-century science, and among its many contributions is the chemical basis of microarray hybridization reactions.

Double helix. *The two-stranded configuration of native DNA, wherein the complementary strands are interwoven around a center axis like a spiral staircase.*

DNA and RNA polymerase enzymes link nucleotide building blocks together into DNA and RNA chains. Kornberg (2001) and co-workers at Washington University in St. Louis discovered **DNA polymerase** in the 1950s, motivated by earlier work on glycogen phosphorylase in the Cori Laboratory. **RNA polymerase** activity was discovered independently by Ochoa, another of Cori's students, and Kornberg and Ochoa shared the Nobel Prize in 1959 "for their discovery of the mechanisms in the biological synthesis of ribonucleic acid and deoxiribonucleic acid," and rightly so. Polymerases have proven to have numerous practical applications, including their roles as key enzymes in recombinant DNA, the **polymerase chain reaction** (PCR), and microarray analysis, and their discovery ranks among the most significant in the history of biology.

DNA polymerase. *Cellular enzyme that synthesizes an exact DNA copy from a single-stranded DNA template.*

RNA polymerase. *Enzyme that synthesizes RNA molecules from a DNA template.*

Polymerase chain reaction (PCR). *Revolutionary technology developed during the 1980s that allows massive amplification of any gene sequence of interest.*

A specialized DNA polymerase known as **reverse transcriptase** (RT) catalyzes DNA synthesis from an RNA template, an enzymatic activity discovered by Baltimore (1970) and Temin and Mizutani (1970). The authors used DNA polymerase assays and radioisotope incorporation experiments to demonstrate the presence of reverse transcriptase in preparations of *Rous sarcoma virus* and other RNA viruses. The ribonuclease sensitivity of the RT activity and the requirement for a viral preparation suggested an RNA template from a viral source, both of which were proven correct in subsequent experiments. The discovery of RT was unexpected and dramatic in that it seemingly reversed the flow of genetic information thought previously to occur from DNA into RNA, but not vice versa (see Chapter 3). Baltimore, Temin, and Dulbecco shared the Nobel Prize in 1972 "for their discoveries concerning the interaction between tumour viruses and the genetic material of the cell." The discovery of RT has had a myriad of fundamental and practical implications, including its use as the labeling enzyme in the first microarray experiments (Schena et al., 1995).

Reverse transcriptase. *Enzyme that synthesizes complementary DNA from a messenger RNA template.*

A myriad of nitrocellulose and nylon filter methods were developed at Stanford during the 1970s, and such methods provided important foundational principles for the development of microarray assays 20 years later. Grunstein and Hogness (1975) at Stanford appear to have published the first paper describing a DNA array. The authors used a nitrocellulose filter array of bacterial colonies to isolate fruit fly genes, establishing, among other important principles, the value of rows and columns in DNA hybridization experiments. Davis and co-workers at Stanford developed nitrocellulose filter assays to examine bacterial plaques, and such methods were used to identify the first differentially expressed genes from higher organisms (Benton and Davis 1977, St. John and Davis 1979).

DNA sequencing. *Experimental process of determining the primary nucleotide sequence of a DNA molecule.*

Maxam and Gilbert (1977) at Harvard and Sanger and co-workers (1977) at the MRC developed **DNA sequencing** technology independently. Gilbert and

Sanger shared the Nobel Prize in 1980 "for their contributions concerning the determination of base sequences in nucleic acids." Sanger chemistry was used to sequence the human genome and has provided much of the sequence database information used to manufacture DNA microarrays (see Chapter 7).

The Nobel Prize in chemistry is 1980 was awarded to Berg at Stanford University "for his fundamental studies of the biochemistry of nucleic acids, with particular regard to recombinant-DNA." The development of **recombinant DNA** technology by Berg and co-workers (Jackson et al., 1972) was one of the most important advances of the twentieth century, and has shown many practical applications, including its extensive use in the preparation of cloned libraries for microarray analysis.

Recombinant DNA. *Revolutionary technology developed in the 1970s that allows genes from different organisms to be spliced together.*

A revolutionary offshoot of the discovery of DNA polymerase was the development of PCR by Mullis and co-workers at Cetus Corporation in the early 1980s (Mullis, 1990; Sakai et al., 1985). PCR allows millions of copies of DNA to be produced from a small amount of genetic material (see Chapter 4), enabling molecular analysis of any gene sequence from any organism. PCR is used extensively in microarray manufacture and in the diagnostic applications of microarrays (see Chapter 13).

Fluorescent dyes have been used for decades to examine biological membranes, including early studies in the 1970s by Waggoner and Stryer (1970). Subsequent work in the early 1990s demonstrated the usefulness of cyanine dyes in the enzymatic preparation of DNA probes (Yu et al., 1994). Pinkel and co-workers developed two-color labeling and detection strategies for chromosome analysis in the 1980s and early 1990s (Weier et al., 1991). This and other pioneering work in fluorescence and fluorescence microscopy laid the foundation for the fluorescent labeling and detection techniques used in microarray analysis (see Chapter 8).

The first **hybridization** experiments on glass were performed in the late 1980s and early 1990s by Mirzabekov and co-workers in Moscow (Khrapko et al., 1989), Fodor and co-workers (1991) at Affymax, Maskos and Southern (1992) at the University of Oxford, Eggers and co-workers at Baylor (Lamture et al., 1994), Smith and co-workers at the University of Wisconsin (Guo et al., 1994), and others. These early experiments established the feasibility of glass-based hybridization, combinatorial oligonucleotide synthesis, linker and surface chemistry, contact printing based on capillary action, and early detection technologies, and many of these principles were used to develop the first microarray assays (Schena et al., 1995).

Hybridization. *The chemical process by which two complementary DNA or RNA strands zipper up to form a double-stranded molecule.*

Hans Lehrach and co-workers at the Imperial Cancer Research Fund (ICRF) pioneered the use of robots for high-speed array manufacture in the late 1980s (Nizetic et al., 1991). The authors used solid pins to prepare large nylon filter arrays of human genomic DNA clones. Despite the large format, this foundational work demonstrated the usefulness of robots for array manufacture, and high-precision motion control systems are used extensively in photolithography, contact printing, and ink jetting (see Chapter 7).

WHAT IS THE SIMILARITY BETWEEN MICROPROCESSORS AND MICROARRAYS?

The genesis of microarray technology at Stanford University in the early 1990s bears a fascinating resemblance to birth of the computer chip industry in the same geographical region 35 years earlier. The same conceptual principles account for the enormous processing power of microarrays and microprocessors, both devices experienced rapid technological advance in a short period of time, and the history of both fields are defined by colorful personalities and tumult. A brief discussion of computer chips sheds light on the similarities shared by microarray processors and microarrays and reveals how the latter may provide a second technological revolution in Silicon Valley.

No discussion of computer chips would be justified without mention of Shockley, co-creator of the transistor and a member of the 1956 Nobel Prize–winning team in physics recognized "for their researches on semiconductors and their discovery of the transistor effect." Shockley is by most accounts the "Father of Silicon Valley" and his pioneering Palo Alto organization, Shockley Semiconductor Laboratories (founded in 1955), helped transform northern California from a sleepy pastoral community into a world center for technological innovation.

Shockley's company made significant progress in transistor development during the late 1950s, but his inexplicable aversion to silicon and his abrasive style generated dissatisfaction within the organization. A group of disgruntled employees known as the "Traitorous Eight" left Shockley's company in 1957 to found Fairchild Semiconductor, a few miles down the road from Palo Alto in Mountain View. The Fairchild Semiconductor founders included Noyce and Moore, and the new company moved quickly to explore silicon-based fabrication methods to streamline computer chip manufacture. In 1959, researchers at Fairchild Semiconductor and Texas Instruments independently developed the *integrated circuit,* and these silicon-based microprocessors defined what is now known as Silicon Valley and ushered in the modern computer industry.

The 1960s were punctuated by landmark discoveries in the computer chip field, including the first commercial integrated circuits from Fairchild Semiconductor in 1961, the invention of the computer mouse by Engelbart at Stanford Research Institute in 1962, and the formulation of Moore's Law by Moore in 1965. Moore, the head of research and development at Fairchild, noticed that transistor density and computing power were doubling every 12–18 months, and he predicted presciently that this trend would continue for several decades according to a specific law. Moore's Law predicted accurately the increasing density and speed of microprocessors over time, and a similar technological trend has occurred in the microarray field, albeit at slightly faster rate (see below).

Noyce and Moore left Fairchild Semiconductor in 1968 to found Integrated Electronics (Intel) in Santa Clara. Intel released the first commercial microprocessor in 1971, and the 4004 chip contained 2,300 transistors capable of performing approximately 100,000 calculations per second (108 KHz). As Moore predicted, modern chips would vastly outperform the 4004 chip over time, and accordingly the Pentium IV chip released in 2000 contains 42,000,000 transistors capable of carrying out 1,500,000,000 calculations per second (1.5 GHz) with 0.18-μm circuit lines. Microarrays have similarly shown an increase in analytical power and decrease in feature size. The first plant microarrays printed in 1995 contained 96 genes with 200-μm features, compared to the highest density microarrays manufactured in 2001, which contain 30,000 genes with 16-μm features. Microarray gene content has increased by more than 300-fold in 6 years, doubling once every 8 months during the 6 year period.

Increasing microarray gene content and decreasing feature size are not the only similarities between microarrays and microprocessors. Parallelism, miniaturization, and automation, the three conceptual cornerstones of the computer chip industry, also provide the foundation of the microarray industry. Parallelism is the process of performing a large number of calculations simultaneously, miniaturization refers to the microscopic size of the microprocessors, and automation denotes the fact that computer chips are manufactured with robots and other automated devices. Microarrays perform gene analyses in a parallel and miniature format, and employ automated devices (e.g., arrayers and scanners) in their manufacture and use. Given the extraordinary conceptual similarities between microprocessors and microarrays, it is little surprise that both technologies were spawned in Silicon Valley.

WHAT IS MICROARRAY ANALYSIS?

Microarray analysis is the process of using microarrays for scientific exploration. The field has undergone massive technological innovation and expansion since the early 1990s, but the general strategies and approaches remain the same. This section covers basic aspects of microarray analysis, including types of microarrays, experimental design, and tips for experimental success.

Types of Microarrays

The first microarray experiments were performed with cDNA microarrays (Fig. 1.2). A cDNA is a nucleic acid molecule derived from mRNA (see Chapter 4), and cDNA microarrays continue to find wide use in gene expression assays (see Chapter 12). The length of a cDNAs is typically 500–2500 base pairs, and microarrays containing such molecules provide intense hybridization signals (see Chapters 6–9) because of their extensive complementarity to fluorescent probe molecules in solution. Database analysis of 2000 microarray citations (arrayit.com/e-library) reveals that cDNA microarrays account for approximately 65% of all microarray publications (Figs. 1.6 and 1.7).

Oligonucleotide microarrays are another commonly used microarray, finding wide use in a variety of applications, including gene expression profiling (see Chapter 12) and genotyping (see Chapter 13). Oligonucleotides are single-stranded 15- to 70-nucleotide molecules made by chemical synthesis (see Chapter 6), and these synthetic targets produce high specificity and good signal

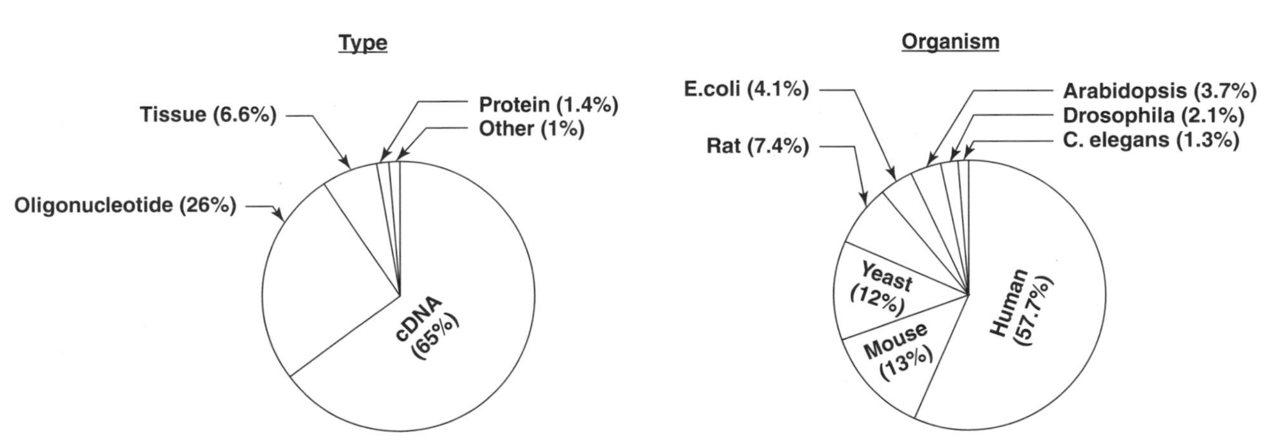

FIGURE 1.6. Microarray papers published since 1995, categorized according to target type and organism. (Data obtained by analysis of the Microarray Electronic Library, arrayit.com/e-library.*)*

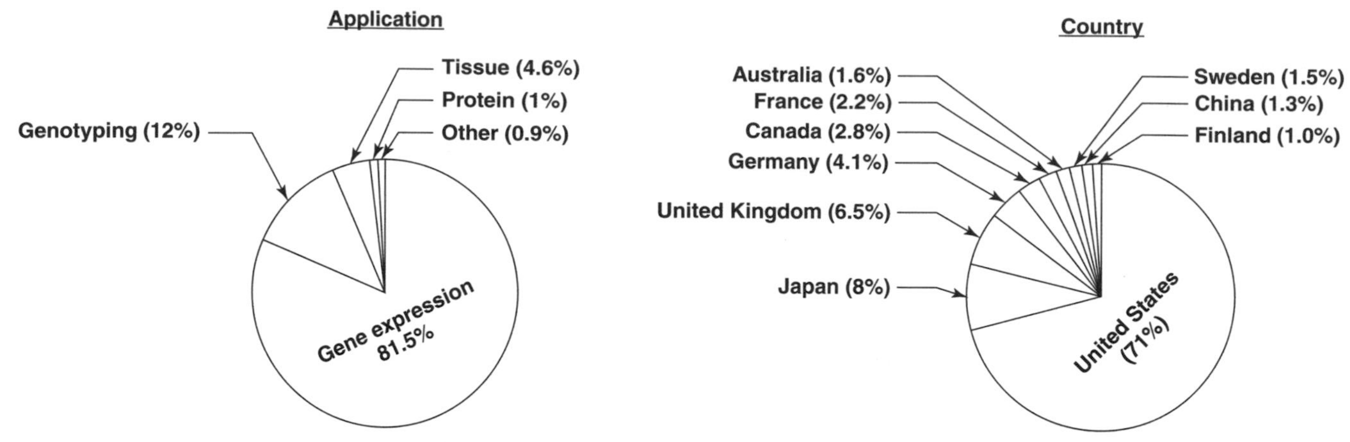

FIGURE 1.7. Microarray papers published since 1995, categorized according to research application and country of origin. (Data were obtained by analysis of the Microarray Electronic Library, arrayit.com/e-library.*)*

strength in hybridization reactions. More than one quarter of all microarray publications to date use oligonucleotides are the target molecules (Fig. 1.6). Complementary DNA and oligonucleotide microarrays both exploit the chemical process of hybridization (see Chapter 3) to generate microarray signals. Oligonucleotide and cDNA microarrays fall into a broader category known as ***nucleic acid microarrays,*** which encompasses microarrays containing any type of DNA or RNA as the target material.

Nucleic acid microarray. *A broad category of microarray that includes any microarray that contains DNA or RNA target elements.*

Tissue and protein microarrays are more recent than nucleic acid microarrays, but these two types of microarrays are being used with increasing frequency, combining for nearly 10% of the scientific publications to date (Fig. 1.6). Tissue microarrays contain sections from human tumor specimens and other tissues of interest, and protein microarrays contain pure proteins or cell extracts at each microarray location. These new microarray formats are replacing many of the traditional histological and biochemical assays because the parallelism, miniaturization, and automation of microarray assays afford a precision, speed, and information content unattainable with the antecedent technologies.

Though microarray studies of humans account for more than half of the scientific publications to date, three other organisms—mouse (13%), yeast (12%) and rat (7.4%)—account for nearly one third of the microarray citations (Fig. 1.6). The common laboratory bacterium *Escherichia coli,* mustard plant (*Arabidopsis thaliana*), fruit fly (*Drosophila melanogaster*) and worm (*Caenorhabditis elegans*) account for a combined 11.2% of the total publications, further underscoring the diversity of organisms that are being studies by microarray. Molecules from any organism in the biosphere can be configured into a microarray, and the broadness of microarray analysis is key attribute of this new technology.

Experimental Design

Microarray analysis differs from traditional research in a number of striking ways, one of which is the relationship between the amount of experimental time required and the amount of data obtained (Fig. 1.8). Traditional experimental approaches based on gels and filter blots required a relatively large amount of experimental time to obtain a small volume of data, whereas microarray

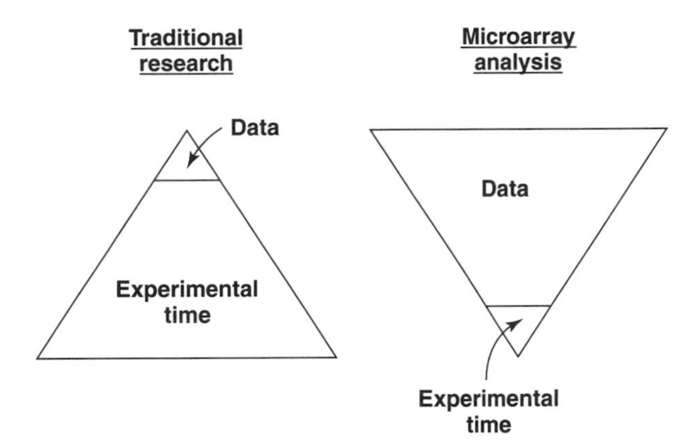

FIGURE 1.8. The relationship between experimental time and data using traditional research methods and microarray analysis.

analysis affords vast quantities of data with relatively little experimental time. Microarrays purchased commercially provide an extreme example, allowing a single researcher to generate millions of datum points in a few weeks. This paradigm shift and upside down relationship between experimentation and data output places tremendous importance on sound experimental design in microarray analysis. Properly designed experiments that include the right experimental components and controls enable researchers to avoid the data avalanche that can quickly bury the uninitiated.

Every microarray experiment should contain a positive control, a negative control, and an experimental component. A ***positive control*** is a microarray element or substrate that provides a readable signal, irrespective of the results obtained from the experimental component of the assay. Readable results from the positive controls greatly improve the capacity to evaluate the experimental data, particularly if negative results are obtained from the experimental components. Intense signals from the positive controls exclude trivial explanations for a failed experiment, such as a defect in hybridization, washing, scanning, or data analysis. No formal conclusions can be drawn from a negative result in a microarray experiment unless the positive controls produce readable signals.

Positive control. *A microarray element or substrate that provides a readable signal, irrespective of the results obtained from the experimental component of the assay.*

A ***negative control*** is a microarray element or substrate that provides little or no readable signal, irrespective of the results obtained from the experimental component of the assay. Negative results add confidence to the experimental data by excluding or reducing the possibility that a nonspecific biochemical event (e.g., staining or cross-hybridization) is producing the signals at the experimental locations. It is risky to draw conclusions concerning the experimental components of a microarray assay if the assay does not include one or more negative controls.

Negative control. *A microarray element or substrate that provides little or no readable signal, irrespective of the results obtained from the experimental component of the assay.*

The *experimental component* of a microarray assay corresponds to the new information that is sought in a given experiment. The experimental data contain information regarding gene expression patterns, genotypes, and other biological processes or pathways. Microarray assays that contain positive controls, negative controls, and an experimental component yield reliable experimental data that can be quantified, mined, and modeled using an increasing powerful collection of software tools (see Chapter 9).

Experimental design can be simplified by understanding the five basic steps in the microarray analysis cycle: a biological question, sample preparation, a biochemical reaction, detection, and data analysis and modeling (Fig. 1.9). A biological question should be formulated before embarking on a microarray experiment. If the goal of the work is to understand the patterns of gene

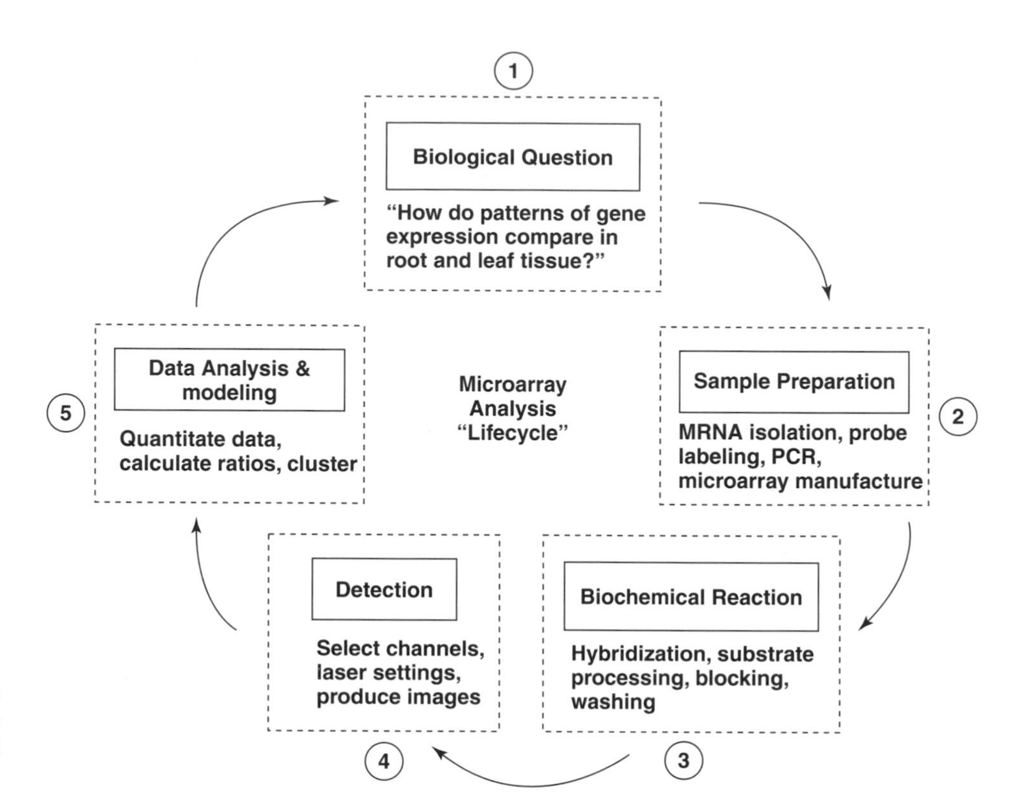

FIGURE 1.9. *The five steps of the microarray analysis cycle, with specific examples of some of the experimental activities performed at each step.*

expression in different plant tissues, the researcher might begin by asking "How do patterns of gene expression compare in plant leaf and root?" The process of formulating a question before every experiment seems mechanical, but it is a habit worth forming because it focuses the research, identifies potential pitfalls, assists in the selection of controls, and streamlines data analysis and modeling. Microarray analysis need not begin with a hypothesis (e.g., "I believe that 20% of all plant genes will exhibit expression differences in leaf and root"), but it is extremely valuable to formulate a question on the upstream side of microarray analysis. A more formal treatment of the microarray analysis **life cycle** is provided in Chapter 10.

Life cycle. *Experimental cycle of microarray analysis that contains the five components of biological question, sample preparation, biochemical reaction, detection, and data analysis and modeling.*

Sample preparation (Fig. 1.9) includes DNA and RNA isolation and purification, target synthesis, probe amplification and preparation, and microarray manufacture (see Chapters 6 and 7). The biochemical reaction involves the incubation of the fluorescent sample with the microarray to allow productive biochemical interactions to occur between target and probe molecules. DNA microarrays exploit hybridization for this step, and protein microarrays use protein–protein interactions. Detection, the forth step in the microarray life cycle, involves capturing an image from the microarray using a scanning or imaging instrument. Captured images are analyzed and modeled to complete the five-step procedure.

It is important to emphasize that the microarray life cycle is a methodological procedure, rather than a series of methods based on a particular enabling technology (see Chapter 10). Microarray manufacture, for example, can be achieved using many different methods, including photolithography, contact printing, and ink jetting, and the same is true for each phase of the life cycle. The microarray life cycle provides a guide to *what* must be done and *why*, without specifying *how* to perform a given procedure. The methodological nature of the experimental cycle allows the implementation of new tools and procedures as innovation moves the field forward. There are currently several thousand commercial products that enable one or more steps of the microarray life cycle (see Chapter 15).

Ten Tips

Microarray analysis is a relatively new science, distinguished from earlier biological technologies in some key respects, including the use of microscopic formats, solid surfaces, small volumes, intensive automation, and highly interdisciplinary approaches. In conjunction with the experimental design guidelines provided above, the following 10 tips will help ensure success in microarray analysis. Additional information regarding the practical aspects of microarray analysis is available online (arrayit.com/wiley) and on CD.

1. *Follow the protocol.* Scientists tend to have a tinkering mentality, which is good because tinkering drives discovery. But microarray analysis encompasses tens of thousands of different protocols developed through the collective efforts of university and company researchers worldwide. In many cases, these experimental recipes have been optimized within the context of a specific set of reagents, surfaces, fluorescent labels, tools, methods, and techniques. When embarking on a microarray procedure for the first time, it makes sense to follow the protocol exactly as it is written, and this applies equally to protocols obtained from the scientific literature, Internet, or commercial products. Once positive results are obtained, the protocol can be optimized and streamlined to improve the procedure.

2. *Read the manual.* Few occasions in the laboratory are more intoxicating than receiving a new scanner or printing robot, but the rush of excitement can have devastating consequences. Owing to the newness of microarray science and the relative inexperience of many users, manufacturers estimate that > 50% of the damage that occurs to expensive microarray instruments is incurred during the first 24 h of use. Scanners, printing robots, and liquid handling systems are extremely durable if used properly, but improper use can damage the motors, linear actuators, and electrical components (see Chapter 7) that control these instruments. Before using a new microarray instrument, it is prudent to read the instruction manual.

3. *Think small.* Microarrays differ considerably from older assays that use large reagent volumes and porous filters, as illustrated in the following comparisons. The nucleic acid concentration in a microarray hybridization reaction containing 1,000 ng of fluorescent probe in 5 μL (200 ng/μL) is 40,000 times greater than a filter hybridization containing 100 ng of probe in 20 mL (0.005 ng/μL). The staggering 40,000-fold increase in nucleic acid concentration, achieved by the use of microscopic elements and a rigid substrate, allows faster hybridization kinetics and greater detectivity. These attributes, coupled with the massively parallel format, allow much higher throughput than traditional assays. A researcher can obtain quantitative gene expression data for 10,000 genes in a 5-min scan (2000 genes/min) using a microarray, compared to the 2 weeks required to measure a single gene (0.000025 genes/min) using a filter blot. The 80 million-fold increase in throughput of a microarray assay relative to a filter blot highlights one of the many advantages of microarray analysis over traditional methods. When throughput is required, think small.

4. *Keep it clean.* The use of small (1–100 μL) reaction volumes and small feature size (16–600 μm), render microarray assays much more sensitive to contaminants than traditional methods. A 5.0-μL microarray hybridization reaction contaminated with a single human fingerprint (~0.25 μL) would contain 5% hand oil by volume, and this extremely high level of contamination would alter the microarray reaction and skew the data. Hand oils can also tarnish and impair the function of high-precision implements (such as microarray printing devices), elevate fluorescent background, and create spurious and uneven scanning artifacts. Many airborne particles are 25–250 μm in diameter, and a single dust particle that comes to rest on a microarray substrate can ruin the entire printed microarray. Because of the miniature format of microarrays, extra care is required to minimize the detrimental effects of biological, chemical, and airborne contaminants. Protective gloves should be worn at all times when working with microarrays, and environmental enclosures and clean rooms are recommended for exacting procedures (see Chapter 11). When performing microarray analysis, keep it clean.

5. *Keep it warm, keep it hydrated.* One of the most common causes of poor microarray data is elevated background fluorescence, a phenomenon usually caused by the precipitation of fluorescent probe molecules on the microarray substrate, followed by irreversible attachment of the fluorescent molecules to the surface. Fluorescent molecules are more prone to precipitation than native nucleic acids owing to the presence of the fluorescent organic dye molecules attached to the nucleotide labels (see Chapter 6). Organic dyes are oily molecules that reduce the solubility of probe molecules in aqueous buffers. Background fluorescence can be minimized using elevated reaction temperatures and proper hydration; thus, low temperatures and desiccation should be avoided whenever possible. Desiccation is a particular challenge in the microarray format because water evaporates at ~0.1 μL/min at ambient conditions, and a low volume microarray reaction can lose a significant percentage of its volume quickly if steps are not taken to minimize evaporation. Use of the proper hardware and environmental settings maintains the proper temperature and hydration levels and minimizes fluorescent background. When working with microarrays, keep it warm and hydrated for best results.

6. *Think globally.* Traditional biological research focuses on single genes and proteins, but biological systems are complex biochemical "machines" that derive their functionality from global interactions between genes and proteins (see Chapter 13). Researchers using microarrays to examine entire genomes must modify their thinking to accommodate the global view of the cell afforded by the microarray data. Drugs, hormones, environmental signals, and a myriad other physiological cues exert global effects on gene expression, and these genome-wide biochemical changes require a holistic view of biological systems to be interpreted accurately.

7. *Do the small experiment first.* Microarray analysis can produce spectacular data for tens of thousands of genes in a single experiment, but the rush to discover can be costly if flaws exist in the experimental technique or design. It is always prudent carry out a pilot study on a small number of samples or genes before investing the time and expensive involved in doing a large experiment. Once high-quality results are obtained from a small experiment, the procedure can be scaled to accommodate more ambitious research plans, and scale-up is much easier if all of the buffers, surfaces, and protocols have been optimized beforehand. For peace of mind (and for the sake of your supervisor's pocketbook), do the small experiment first.

8. Confirm as you go. Microarrays are being used extensively for gene expression analysis, and this discovery-oriented research is providing remarkable new biological insights that are transforming our understanding of biological systems (see Chapter 12). But the complexity of microarray manufacture, clone identity, database information, sample tracking, and data analysis can lead to occasional errors in gene identification or the accuracy of results, even if great care is taken to ensure the integrity of a microarray assay. It is recommended that the researcher confirm the identity of a small number of genes by microarray or by an independent means before he or she examines a large number of precious samples or achieves gigabyte quantities of data. Gene identity can be confirmed using oligonucleotide probes specific to a small number of target elements on the microarray, and expression levels can be checked using filter blot or PCR methods. Confirming microarray data early in the microarray analysis pathway provides peace of mind and prevents the frustration and expense associated with inaccurate data.

9. Look early. Cells are exquisitely sensitive to myriad biochemical signals, and cell signaling pathways often contain primary, secondary, and tertiary response genes that are regulated sequentially in response to a biochemical signal (see Chapter 4). The hierarchical structure of regulatory pathways is such that the long-term treatment of cells with drugs, hormones, and other chemical and environmental elicitors can lead to changes in the expression of thousands of cellular genes, obscuring the identity of the genes involved in the primary response. The best way to combat the complexity of cell signaling pathways is look early (1–4 h) after stimulation, which will maximize the chances of identifying the primary response genes and will yield a gene fingerprint specific to a particular response.

10. Don't panic. One invariable consequence of microarray analysis is the occasional panic that ensues after an experiment yields too many promising genes. The complexity of the data from an experiment that yields 500 candidate genes, for example, can be overwhelming if efforts are made to understand all the genes within the context of the primary scientific literature (e.g. >50,000 publications). One way to combat the microarray data flood is to use data quantification, mining, modeling tools (see Chapter 9), and focused experimentation to narrow the list of candidate genes before embarking on a detailed study of each gene. Biology is complex, but microarray methods are extremely powerful. Don't panic at the site of vast quantities of data.

WHAT ARE SOME APPLICATIONS OF MICROARRAYS?

Two trends in microarray research are the diversification of the assays and the worldwide spread of the technology. Gene expression applications account for 81% of the scientific publications to date, but microarrays are being used for many other purposes, including genotyping, tissue analysis, and protein studies (Fig. 1.7). Researchers in the United States have contributed 71% of the microarray publications to date, but scientists from nine other nations (Japan, United Kingdom, Germany, Canada, France, Australia, Sweden, China, and Finland) have provided nearly one third of the citations. The diversification of the microarray platform and the collective efforts of scientists worldwide have added breadth and depth to the field and built collegiality among researchers. This section provides a brief look at some of the many applications of microarrays that are now in use.

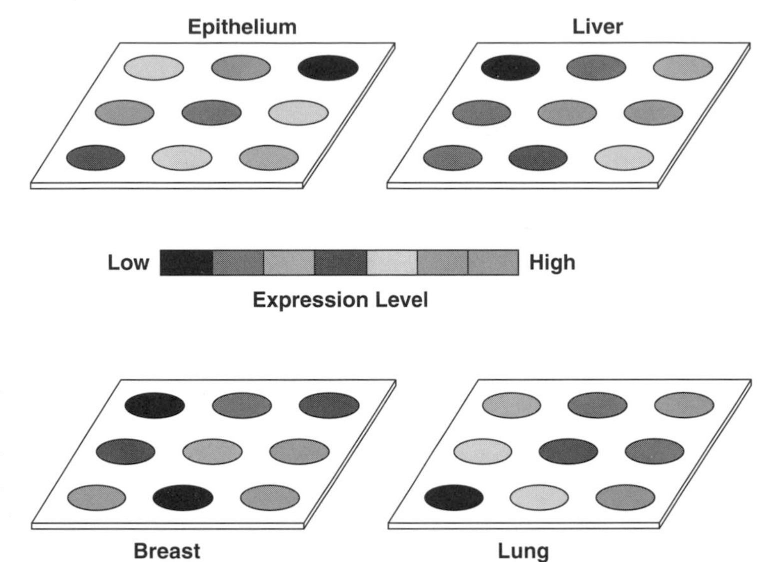

FIGURE 1.10. Microarrays can be used to examine patterns of gene expression in different human tissues. Signal intensities are coded to a rainbow palette so that genes expressed at a high level appear red; genes expressed at low level appear black; and genes expressed at intermediate levels are coded to violet, blue, green, yellow, and orange.

Development

Essentially every cell in a multicellular organism contains an identical copy of the genetic (DNA) blueprint, but cells differ dramatically in terms of shape, size, and function. Developmental identity is achieved, in part, through genetic programming that alters the repertoire of genes expressed in each cell type. By examining gene expression patterns on a genomic scale, microarrays can be used to build a database of gene expression levels as a function of cell and tissue type (Fig. 1.10). Such databases are extremely valuable because they provide a deeper understanding of the basic mechanisms that control multicellular development and shed light on pathological cellular events, including the onset and progression of human disease. The human brain has been the most actively studied human tissue to date, accounting for 19% of the all citations, but nine other tissues (liver, breast, prostate, lung, colon, kidney, heart, bladder, and skin) account for at least 81% of the publications (Fig. 1.11). Building gene expression fate maps for each cell type in the human and other organisms is shedding remarkable new light on development.

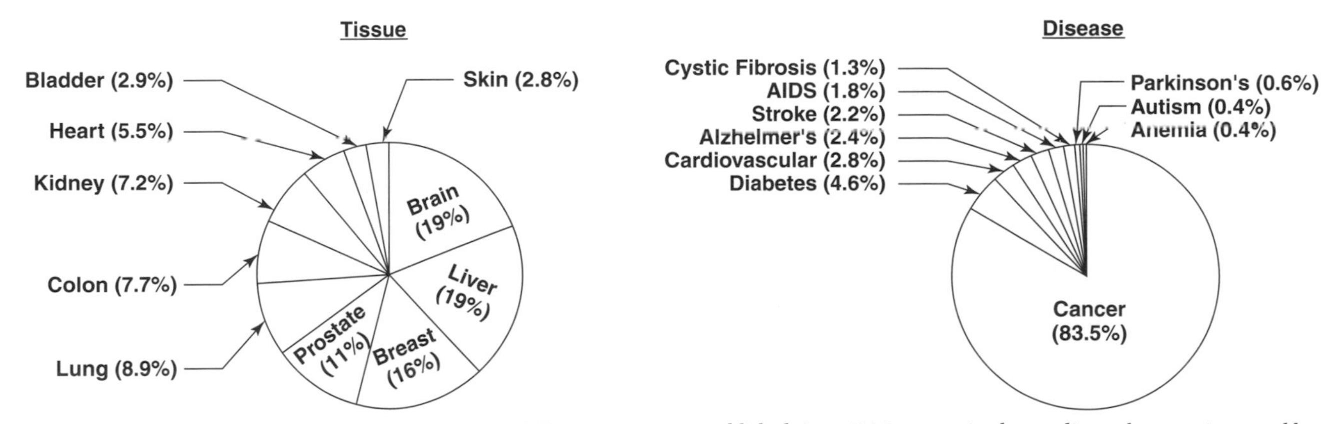

FIGURE 1.11. Microarray papers published since 1995, categorized according to human tissue and human disease. (Data obtained by analysis of the Microarray Electronic Library, arrayit.com/e-library.*)*

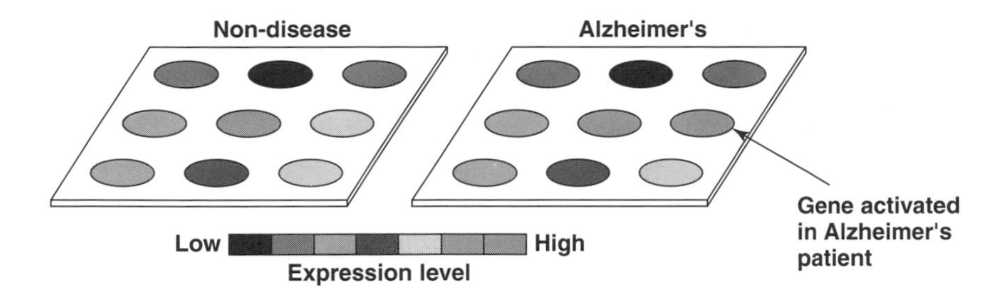

FIGURE 1.12. Quantitative expression analysis of a normal individual and an Alzheimer disease patient reveals an up-regulated gene in the patient (red spot) *that may contribute to the disease. Signal intensities are coded to a rainbow color palette for ease of viewing.*

Human Disease

The onset and progression of human disease are determined by a complex set of factors, including genetics, diet, the environment, and the presence of infectious agents, and microarray analysis is unique in its ability to detect each of these contributing factors. Cancer has accounted for a remarkable 83.5% of the microarray publications on human disease to date, but diabetes, cardiovascular disease, Alzheimer, stroke, AIDS, cystic fibrosis, Parkinson, autism, and anemia are under intensive investigation by microarray analysis. By comparing gene expression patterns in brain tissue from normal individuals and Alzheimer patients, for example, it should be possible to elucidate the genetic basis of this devastating illness (Fig. 1.12). Knowing the molecular basis of Alzheimer and other diseases enhances our ability to understand genetic predisposition, onset, and regression and expedites the development of safer and more effective treatments. All human illness can be studied by microarray analysis, and the ultimate goal of this work is to develop treatments or cures for every human disease by 2050.

Drug Discovery

Many drugs impart their therapeutic activity by binding to specific cellular targets, inhibiting protein function, and altering the expression of cellular genes. It is possible in principle to use microarrays for drug discovery and clinical trials by generating gene expression profiles in patients undergoing disease progression or drug treatment (Fig. 1.13). Many illnesses result in specific changes in gene expression, and drugs that reverse these changes are expected to ameliorate the disease. Expression profiling may also be useful in identifying drugs

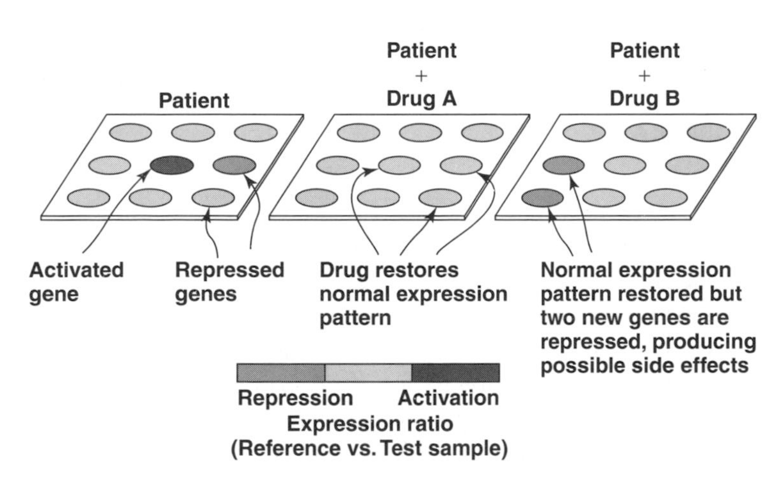

FIGURE 1.13. Disease progression results in the activation of one cellular gene (green circle) *and the repression of two cellular genes* (red circles) *in a patient. These changes in gene expression can be reversed by treating the patient with drug A, which restores the normal expression pattern. Treatment of the patient with drug B restores normal expression levels but represses two other genes* (red circles)*, producing potentially deleterious side effects.*

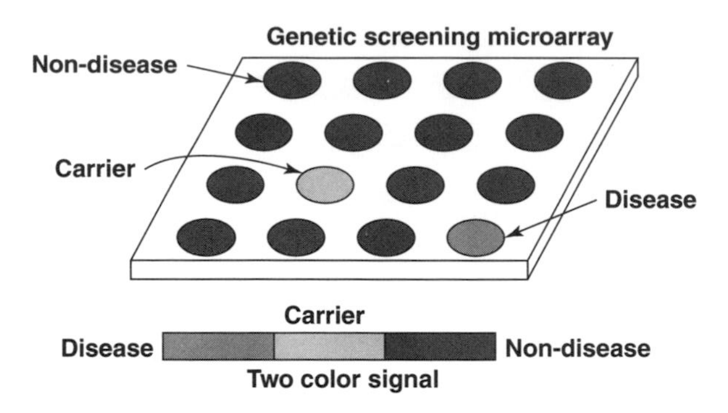

FIGURE 1.14. Patient samples are amplified, printed into a microarray, and hybridized with fluorescent oligonucleotides bearing green and red labels. Genotypes are determined by examining the color at each microarray location, allowing a carrier (yellow circle) *and a patient* (red circle) *to be distinguished easily from the larger nondisease population* (green circles).

that alter the expression of nontarget genes, as a means of identifying drugs with potential side effects. This intelligent approach to drug discovery may reduce the cost of drug development and produce safer medicines with fewer side effects. Microarray analysis can also be used for patient genotyping, and the capacity to partition the population into drug responders and nonresponders based on genotype may enable more personalized medicine.

Genetic Screening and Diagnostics

Small errors in the genetic code can lead to the synthesis of a defective cellular protein, and proteins incapable of performing their normal cellular functions can cause human disease. Thousands of disease-causing sequence variants are known, and affordable microarray screens for these diseases are of tremendous scientific and commercial interest. In one microarray screening procedure (see Chapter 13), patient samples are amplified by PCR, printed into microarrays, and hybridized with synthetic oligonucleotides. The microarrays are then scanned for fluorescence emission, and the data are represented in a two-color composite image (Fig. 1.14). This microarray screening procedure allows normal, carrier, and disease genotypes to be easily distinguished; and acquiring such information for treatable and curable genetic disease, particularly at an early stage, will improve the quality of health care and reduce its cost. One could also envision genetic testing kits in local pharmacies that would allow patients to sample their genetic material, and obtain their genotype from a testing center equipped to perform microarray-based screens. Consumer kits for the common inherited disorders, including cystic fibrosis, sickle cell anemia, Tay Sachs disease, and breast cancer, are likely to empower the public by providing confidential access to their genomic information.

SUMMARY

Microàrray analysis is a revolutionary new science developed by me and my co-workers at Stanford University in the early 1990s. The approach uses microscopic glass arrays (microarrays) for the quantitative analysis of genes and gene products. Traditional research methods based on gels, filters, and purification columns are yielding to biological chips, which offer an enormous increase in speed and precision. Microarrays can be used to analyze the entire human genome in a single step, generating quantitative gene expression information for 30,000 human genes in $<$ 10 min. Microarray analysis is unique in the history of biology because no other technology has used so much technological sophistication, combined expertise from so many different disciplines, and provided such a detailed view of the cell.

Microarray analysis has its roots in advances made between 1950 and 1990, including the discovery of DNA polymerase, recombinant DNA, and PCR. Microarray technology exploits parallelism, miniaturization, and automation, the same three cornerstones of the computer chip industry. Experiments with microarrays are facilitated by sound experimental design and a five-step microarray analysis life cycle. Nucleic acid, protein, antibody, and tissue microarrays are expanding the repertoire of exciting microarray applications, including studies of development and human disease, drug discovery, and genetic screening and diagnostics. Microarray assays for genetic and infectious diseases may improve health care by providing rapid and affordable genotyping data for treatable and curable illnesses.

SELECTED READING

Baltimore, D. RNA-dependent DNA polymerase in virions of RNA tumour viruses. *Nature* 226:1209–1211, 1970.

Benton, W. D., and Davis, R. W. Screening lambdagt recombinant clones by hybridization to single plaques in situ. *Science* 196:180–182, 1977.

Ceruzzi, P. E. *A History of Modern Computing*. MIT Press, Cambridge, MA, 1998.

Fodor, S. P., Read, J. L., Pirrung, M. C., Stryer, L., Lu, A. T., and Solas, D. Light-directed, spatially addressable parallel chemical synthesis. *Science* 251:767–773, 1991.

Grunstein, M., and Hogness, D. S. Colony hybridization: a method for the isolation of cloned DNAs that contain a specific gene. *Proc Natl Acad Sci U S A* 72:3961–3965, 1975.

Guo, Z., Guilfoyle, R. A., Thiel, A. J., Wang, R., and Smith, L. M. Direct fluorescence analysis of genetic polymorphisms by hybridization with oligonucleotide arrays on glass supports. *Nucleic Acids Res* 22:5456–5465, 1994.

Heller, R. A., Schena, M., Chai, A., Shalon, D., Bedilion, T., Gilmore, J., Woolley, D. E., and Davis, R. W. Discovery and analysis of inflammatory disease-related genes using cDNA microarrays. *Proc Natl Acad Sci U S A* 94:2150–2155, 1997.

Jackson, D. A., Symons, R. H., and Berg, P. Biochemical method for inserting new genetic information into DNA of simian virus 40: circular SV40 DNA molecules containing lambda phage genes and the galactose operon of *Escherichia coli*. *Proc Natl Acad Sci U S A* 69:2904–2909, 1972.

Khrapko, K. R., Lysov, Y. P., Khorlyn, A. A., Shick, V. V., Florentiev, V. L., and Mirzabekov, A. D. An oligonucleotide hybridization approach to DNA sequencing. *FEBS Lett* 256:118–122, 1989.

Kornberg, A. Remembering our teachers. *J Biol Chem* 276:3–11, 2001.

Lamture, J. B., Beattie, K. L., Burke, B. E., Eggers, M. D., Ehrlich, D. J., Fowler, R., Hollis, M. A., Kosicki, B. B., et al. Direct detection of nucleic acid hybridization on the surface of a charge coupled device. *Nucleic Acids Res* 22:2121–2125, 1994.

Lemieux, B., Aharoni, A., and Schena, M. Overview of DNA chip technology. *Molec Breeding* 4, 277–289, 1998.

Maskos, U., and Southern, E. M. Oligonucleotide hybridizations on glass supports: A novel linker for oligonucleotide synthesis and hybridization properties of oligonucleotides synthesised in situ. *Nucleic Acids Res* 20:1679–1684, 1992.

Maxam, A. M., and Gilbert, W. A new method for sequencing DNA. *Proc Natl Acad Sci U S A* 74:560–564, 1977.

Mullis, K. B. The unusual origin of the polymerase chain reaction. *Sci Am* 262:56–61, 64–65, 1990.

Nizetic, D., Zehetner, G., Monaco, A. P., Gellen, L., Young, B. D., and Lehrach, H. Construction, arraying, and high-density screening of large insert libraries of human chromosomes X and 21: Their potential use as reference libraries. *Proc Natl Acad Sci U S A* 88:3233–3237, 1991.

Pauling, L., Itano, H. A., Singer, S. J., and Wells, I. C. Sickle cell anemia: a molecular disease. *Science* 110:543–548, 1949.

Saiki, R. K., Scharf, S., Faloona, F., Mullis, K. B., Horn, G. T., Erlich, H. A., and Arnheim, N. Enzymatic amplification of beta-globin genomic sequences and restriction site analysis for diagnosis of sickle cell anemia. *Science* 230:1350–1354, 1985.

Sanger, F., Nicklen, S., and Coulson, A. R. DNA sequencing with chain-terminating inhibitors. *Proc Natl Acad Sci U S A* 74:5463–5467, 1977.

Schena, M., ed. *DNA Microarrays: A Practical Approach,* 2nd ed. Oxford University Press, Oxford, UK, 2000.

Schena, M. Genome analysis with gene expression microarrays. *BioEssays* 18:427–431, 1996.

Schena, M., ed. *Microarray Biochip Technology*. Eaton, Natick, MA, 2000.

Schena, M., Heller, R. A., Theriault, T. P., Konrad, K., Lachenmeier, E., and Davis, R. W. Microarrays: Biotechnology's discovery platform for functional genomics. *Trends Biotechnol* 16:301–306, 1998.

Schena, M., Shalon, D., Davis, R. W., and Brown, P. O. Quantitative monitoring of gene expression patterns with a complementary DNA microarray. *Science* 270:467–470, 1995.

Schena, M., Shalon, D., Heller, R., Chai, A., Brown, P. O., and Davis, R. W. Parallel human genome analysis: Microarray-based expression monitoring of 1,000 genes. *Proc Natl Acad Sci U S A* 93:10614–10619, 1996.

Singer, M., and Berg, P. *Genes and Genomes*. University Science Books, Herdon, VA, 1991.

St. John, T. P., and Davis, R. W. Isolation of galactose-inducible DNA sequences from *Saccharomyces cerevisiae* by differential plaque filter hybridization. *Cell* 16:443–452, 1979.

Streitweiser, A., and Heathcock, C. H. *Introduction to Organic Chemistry,* 2nd ed. Macmillan, New York, 1981.

Stryer, L. *Biochemistry,* 4th ed. Freeman, New York, 1995.

Temin, H. M., and Mizutani, S. RNA-dependent DNA polymerase in virions of *Rous sarcoma* virus. *Nature* 226:1211–1213, 1970.

Waggoner, A. S., and Stryer, L. Fluorescent probes of biological membranes. *Proc Natl Acad Sci U S A* 67:579–589, 1970.

Watson, J. D., and Crick, F. H. C. Molecular structure of nucleic acid. A structure for deoxyribose nucleic acid. *Nature* 171:737–738, 1953.

Weier, H. U., Lucas, J. N., Poggensee, M., Segraves, R., Pinkel, D., and Gray, J. W. Two-color hybridization with high complexity chromosome-specific probes and a degenerate alpha satellite probe DNA allows unambiguous discrimination between symmetrical and asymmetrical translocations. *Chromosoma* 100:371–376, 1991.

Yu, H., Chao, J., Patek, D., Mujumdar, R., Mujumdar, S., and Waggoner, A. S. Cyanine dye dUTP analogs for enzymatic labeling of DNA probes. *Nucleic Acids Res* 22:3226–3232, 1994.

REVIEW QUESTIONS

1. A microarray is a contraction of which two words of English?
2. A researcher uses a traditional filter blot with 5-mm DNA spots and mRNA labeled with radioactive phosporus-32 to measure gene expression. Identify three ways that this experimental approach differs from microarray analysis and explain each briefly.
3. The four basic criteria for a microarray are (1) ordered, (2) microscopic (3), planar, and (4) specific. Explain the four criteria briefly and describe one advantage of each in a microarray experiment.
4. Which of the following materials would meet the planarity requirement for a microarray substrate? Explain briefly. (a) 1.0-mm-thick glass slide, (b) 1.0-mm-thick plastic slide, (c) a 50-μm-thick plastic slide, (d) a 50-μm-thick nitrocellulose membrane.
5. High binding specificity between targets on the substrate and probe molecules in solution is a key aspect of microarray analysis. What does high binding specificity enable? A one-word answer will suffice.
6. Microarray analysis is unique in science history because it combines (1) a high degree of technological complexity, (2) contributions from many different disciplines, and (3) a quantitative and systematic view of the cell. Provide one specific example for each of the three categories.
7. The history of science is important because it reminds us that scientific discoveries build on work from previous decades. Name one key scientific

discovery from the 1950s, 1960s, 1970s, and 1980s that contributed to the development of microarray analysis.

8. Microprocessors and microarrays use the same three fundamental principles. Name them. One word for each will suffice.
9. Based on the scientific literature, what is the most commonly studied organism in microarray analysis?
10. In terms of experimental design, every microarray experiment should contain three components. Name them. A few words for each will suffice.
11. Name the five steps in the microarray analysis life cycle, providing one concrete example in each phase.
12. The chapter provides 10 tips to experimental success. Name 3 of them and describe the importance of each briefly.
13. Three patients are admitted to a hospital of the future with an unexplained illness characterized by dizziness and nausea. Detailed medical records are available for all three patients. The physician withdraws samples of blood from each patient and uses microarray analysis to identify the genes that are activated and repressed in the three patients. The data reveal two main findings: (1) all three patients share an unusual subset of activated and repressed genes and (2) each patient has a few additional changes in expression not shared by the other two patients. What conclusions might the doctor draw from the data?
14. Microarray analysis is used to determine the genotypes of 1000 patients for a genetic disease. Each amplified patient sample is printed once on a microarray substrate according to the method described in Figure 1.14. Data analysis reveals 997 green spots, 2 red spots, and one yellow spot. What are the genotypes of the 1000 patients with respect to nondisease, carrier, and disease?
15. A microarray image is color-coded using a rainbow palette as shown in Figure 1.1. Which spot color, yellow or blue, corresponds to the gene with a greater level of gene expression?

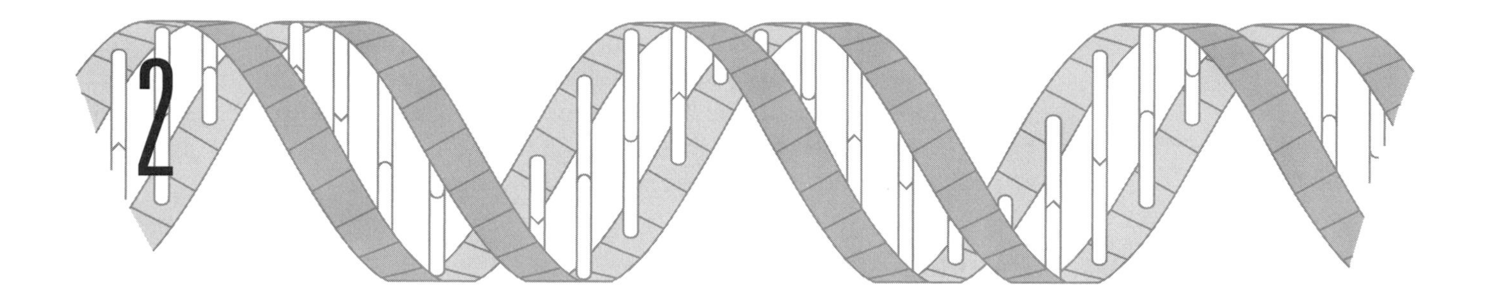

Introduction to Chemistry

Biochemistry is organic chemistry on a grand scale. —*Carl Ward*

A childhood mentor of mine (Carl Ward) once told me that "biochemistry is organic chemistry on a grand scale," a quip that I have held dear for 25 years and one that has gained increasing relevance as we advance the forefronts of microarray analysis. It was never clear to me whether this was Ward's original line or whether it was something he picked up at Stanford, but either way it not only has retained its value since the 1970s but has provided a nice return on investment. Indeed, microarray analysis is organic chemistry on a grand scale, and students are served well by equipping themselves with a basic knowledge of chemistry before embarking on microarray studies. This chapter provides an introduction to chemistry, focusing on atoms and molecules, chemical bonding, properties of matter, and organic compounds.

ATOMS AND MOLECULES

This section provides an overview of some basic concepts in chemistry, including atomic theory, the periodic table, the structure of atoms, and chemical formulae.

Atomic Theory

The term ***atom,*** coined by the Greek philosopher Democritus in about 400 B.C.E., derives from the Greek word *atomos* or "indivisible". Democritus

Atom. *The indivisible building block of all molecules, including solids, liquids, and gases.*

Matter. *Any substance that has mass and occupies space.*

hypothesized correctly that **matter** is composed of atoms, which are the indivisible building blocks of all molecules, including solids, liquids, and gases. The concept of the atom was bolstered in the early 1800s by the experimental work of John Dalton, who proposed his *atomic theory* based on experimental observations. Dalton's key concepts were that *elements* are composed of atoms, atoms of the same element share the same size and mass, atoms of different elements have a different size and mass, compounds are formed by joining atoms of different elements, and atoms combine in compounds in integer proportions. Though the atomic theory proposed by Dalton proved overly simplistic, the basic tenets of the theory remain valid today.

Periodic Table

Element. *Any of a set of the 118 fundamental buildings block of all matter. Alternatively, a printed spot or feature in a microarray.*

Every known substance on earth contains one or more of 118 possible chemical building blocks or elements, which make up the chemical alphabet. An ***element*** is the fundamental building block of all matter, analogous to the four nucleotides found in DNA and the 20 amino acids used to make proteins (see Chapters 3 and 4). Elements are atoms that cannot be broken down further by ordinary chemical changes. Water (H_2O) can be broken down into hydrogen and oxygen but no further, and two atoms of hydrogen (H) and one atom of oxygen (O) combine to form water (Fig. 2.1). Hydrogen and oxygen are two of the 118 known elements, and these atoms figure centrally in biochemistry.

Periodic table. *An organizational chart that contains the names and atomic properties of the 118 known elements.*

Atomic number. *Corresponds to the number of protons in the nucleus of the atom and provides the basis by which the 118 known elements are arranged in the periodic table.*

Each element is identified using a full name and a single letter abbreviation. Carbon (C), nitrogen (N), oxygen (O), phosphorus (P), and sulfur (S) are important elements found in biomolecules. The 118 elements are configured into a graphical representation known as the ***periodic table.*** The periodic table lists each of the elements consecutively, in order of increasing atomic complexity. Each element has an ***atomic number*** (1–118), which corresponds to the number of protons in the nucleus of the atom (see below). Elements are arranged in the periodic table in order of an increasing number of protons (Fig. 2.2). Hydrogen, carbon, nitrogen, oxygen, phosphorus, and sulfur contain 1, 6, 7, 8, 15, and 16

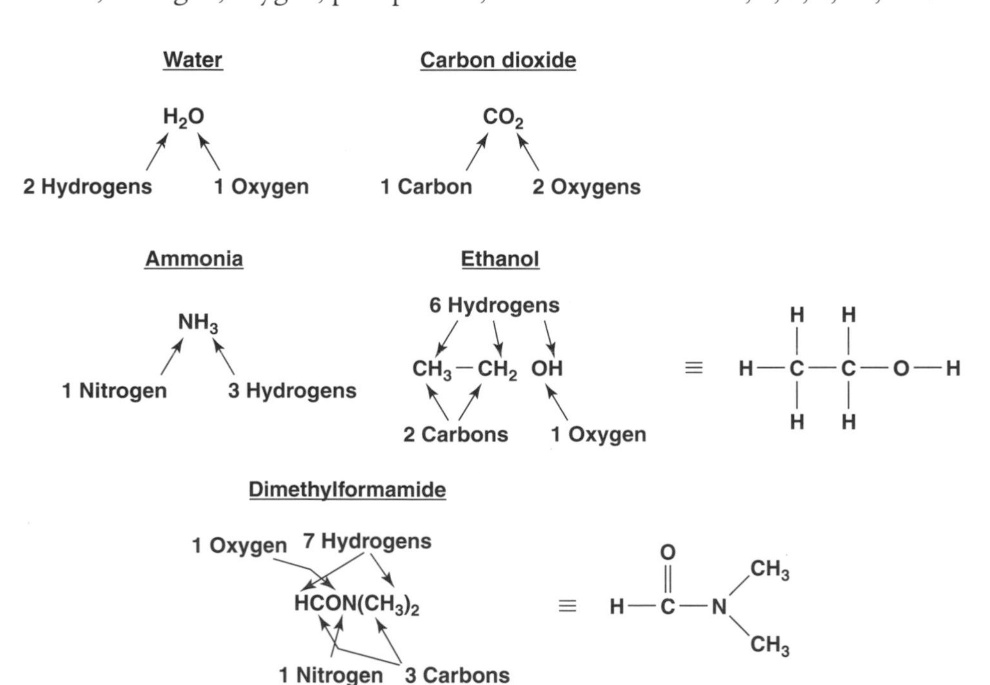

FIGURE 2.1. The chemical formulae for some simple molecules. Atom type and number are provided (arrows) for each molecule. Molecular drawings are provided for ethanol and dimethylformamide.

1 H Hydrogen 1.0																	2 He Helium 4
3 Li Lithium 6.9	4 Be Beryllium 9.0											5 B Boron 10.8	6 C Carbon 12.0	7 N Nitrogen 14	8 O Oxygen 16	9 F Fluorine 19	10 Ne Neon 20.2
11 Na Sodium 23	12 Mg Magnesium 24.3											13 Al Aluminium 27	14 Si Silicon 28.1	15 P Phosphorus 31	16 S Sulfur 32.1	17 Cl Chlorine	18 Ar Argon 40
19 K Potassium 39	20 Ca Calcium 40.1	21 Sc Scandium 45	22 Ti Titanium 47.9	23 V Vanadium 50.9	24 Cr Chromium 52	25 Mn Manganese 54.9	26 Fe Iron 55.8	27 Co Cobalt 58.9	28 Ni Nickel 58.7	29 Cu Copper 63.5	30 Zn Zinc 65.4	31 Ga Gallium 69.7	32 Ge Germanium 72.6	33 As Arsenic 74.9	34 Se Selenium 79	35 Br Bromine 80	36 Kr Krypton 83.8
37 Rb Rubidium 85.5	38 Sr Strontium 87.6	39 Y Yttrium 88.9	40 Zr Zirconium 91.2	41 Nb Niobium 92.9	42 Mo Molybden. 95.9	43 Tc Technetium (98)	44 Ru Rutherium 101.1	45 Rh Rhodium 102.9	46 Pd Palladium 106.4	47 Ag Silver 107.9	48 Cd Cadmium 112.4	49 In Indium 114.8	50 Sn Tin 118.7	51 Sb Antimony 121.8	52 Te Tellurium 128	53 I Iodine 127	54 Xe Xenon 131.3
55 Cs Cesium 132.9	56 Ba Barium 137.3	57–71	72 Hf Hafnium 178.5	73 Ta Tantulum 180.9	74 W Tungsten 183.8	75 Re Rhenium 186.2	76 Os Osmium 190.2	77 Ir Iridium 192.2	78 Pt Platinum 195	79 Au Gold 197	80 Hg Mercury 200.6	81 Tl Thallium 204.3	82 Pb Lead 207.2	83 Bi Bismuth 209	84 Po Polonium (209)	85 At Astatine (210)	86 Rn Radon (222)
87 Fr Francium (223)	88 Ra Radium (226)	89–103	104 Rf Rutherford. (261)	105 Db Dubnium (262)	106 Sg Seaborgium (263)	107 Bh Bohrium (262)	108 Hs Hassium (265)	109 Mt Meitnerium (266)	110 Uun Ununnilium (269)	111 Uuu Unununium (272)	112 Uub Unumbium (277)	113	114 Unq Ununquad. (289)	115	116 Uuh Ununhexium (289)	117	118 Uuo Ununoctium (293)

57–71	57 La Lanthanum 139	58 Ce Cerium 140.1	59 Pr Praseodyn. 140.9	60 Nd Neodymium 144.2	61 Pm Promethium (145)	62 Sm Samarium 150.4	63 Eu Europium 152	64 Gd Gadolinium 157.2	65 Tb Terbium 159	66 Dy Dysprosium 162.5	67 Ho Holmium 164.9	68 Er Erbium 167.3	69 Tm Thulium 168.9	70 Yb Yterrbium 173.0	71 Lu Lutetium 175
89–103	89 Ac Actinum (227)	90 Th Thorium 232	91 Pa Protactin. 231.0	92 U Uranium 238	93 Np Neptunium (237)	94 Pu Plutonium (244)	95 Am Americium (243)	96 Cm Curium (247)	97 Bk Berkelium (247)	98 Cf Californium (251)	99 Es Einsteinium (252)	100 Fm Fermium (257)	101 Md Mendelevium (258)	102 No Nobelium (259)	103 Lr Lawrencium (262)

FIGURE 2.2. *The periodic table.*

protons, respectively; and they assume the corresponding positions (1, 6, 7, 8, 15, and 16) in the periodic table. A neutral atom is one that contains an equivalent number of protons and electrons (see below), and charged atoms or *ions* are molecules that contain a proton–electron imbalance.

The mass of a carbon atom is approximately 1.9×10^{-23} g, highlighting the extraordinarily small mass of a single atom. Carbon provides a reference standard known as the ***atomic mass unit,*** defined as one twelfth the mass of a carbon atom. The carbon atom is divided by 12 because it contains 6 protons and 6 neutrons (see below). One atomic mass unit equals approximately 1.6×10^{-24} g. Each of the 118 elements in the periodic table is assigned an ***atomic mass,*** based on the atomic mass units it contains relative to the carbon reference (Fig. 2.2). Hydrogen, carbon, nitrogen, oxygen, phosphorus, and sulfur have atomic masses of approximately 1, 12, 14, 16, 31, and 32, respectively. The atomic mass is equal to the sum of the protons and neutrons in each atom or element, and this value is useful for determining the number of atoms present in a given mass of a substance (see below).

Atomic mass unit. *One-twelfth (1/12) the mass of a carbon atom.*

Atomic mass. *The weight of an atom given in atomic mass units.*

Structure of Atoms

Atoms range in diameter from 1 to 5 Å (1 Å = 10^{-10} m), with the smallest atom being hydrogen. Atoms are made up of protons, neutrons, and electrons, and these ***subatomic particles*** derive their name because they are smaller than the intact atom. Protons, neutrons, and electrons are identical in every atom, and different atoms derive their chemical uniqueness by possessing a different number of these subatomic particles. Hydrogen contains one proton and one electron and is chemically distinct from carbon, which contains six protons, six neutrons, and six electrons (Fig. 2.2).

Subatomic particle. *Any of a family of quantum mechanical entities, including protons, neutrons, and electrons, that make up atoms.*

Atoms have a dense positively charged core, or ***nucleus,*** surrounded by an electronic cloud consisting of orbiting electrons (Fig. 2.3). The approximate diameter of a nucleus, proton, and electron are believed to be approximately 0.0001 Å (1×10^{-14} m), 0.00001 Å (1×10^{-15} m), and 0.00000001 Å (1×10^{-18} m), respectively. To put this into perspective, if an atom had the diameter of a baseball stadium, a proton would be the size of a baseball and an electron would be smaller than a grain of sand. It should be obvious from this analogy that atoms consist mainly of empty space and derive their size predominantly from the nuclear forces that attract protons and electrons.

Nucleus. *The dense, positively charged core of an atom. Alternatively, the organelle inside a cell that contains the genetic material.*

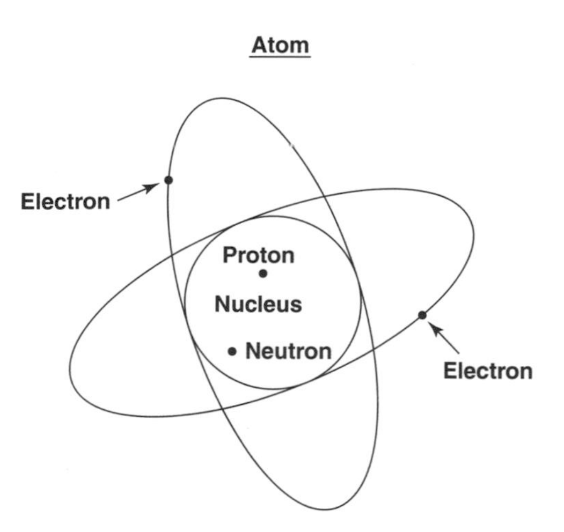

FIGURE 2.3. *The structure of an atom.*

Protons reside in the nucleus and carry a charge of +1, ***neutrons*** reside in the nucleus and carry no charge, and ***electrons*** orbit around the nucleus and have a charge of −1. Neutral atoms have an equivalent number of protons (+1) and electrons (−1) to balance the positive and negative charge, and the electromagnetic force between protons and electrons provides the atomic "glue" that holds atoms together. Charged atoms or ions have an excess of positive or negative charge.

Proton. *A subatomic particle that resides in the nucleus of an atom and carries a charge of +1 and a mass of approximately 1 atomic mass unit.*

Neutron. *A subatomic particle that resides in the nucleus of an atom, carries no charge, and has a mass of approximately 1 atomic mass unit.*

Protons have a mass of approximately 1 atomic mass unit (1.673×10^{-24} g), neutrons have a mass of approximately 1 atomic mass unit (1.675×10^{-24} g), and electrons have a mass of approximately 0.0005 atomic mass units (9.1×10^{-28} g). By analogy, if a proton had the mass of a grand piano, an electron would weigh about as much as the musical score. It should be clear from this comparison that an electron has an extraordinarily small mass, and that the nucleus accounts for nearly the entire mass of an atom. The atomic properties of all 118 elements can be understood largely in terms of protons, neutrons, and electrons, and students are served well by understanding the basic properties of these subatomic particles.

Electron. *A subatomic particle that resides in an orbit around the nucleus of an atom, and carries a charge of −1 and a mass of approximately 0.0005 atomic mass units.*

Electrons do not orbit randomly around the nucleus but rather assume discrete positions defined by principle energy levels and sublevels. The electrons present in each atom can be represented using an ***orbital diagram,*** which defines the electrons in each principle energy level and sublevel (Fig. 2.4). Orbitals are represented by boxes that accommodate one pair of electrons, and each electron pair is represented by arrows pointing in opposite directions to denote that fact that electrons pairs share opposite ***spin states.***

Orbital diagram. *An illustration of the energy levels, occupancy, and spin states of the electrons in atom.*

Spin state. *The intrinsic angular momentum of an electron characterized by the quantum number $\frac{1}{2}$; depicted with an arrow in orbital diagrams.*

The first 18 elements in the periodic table fill successive orbitals in the first three principle energy levels and sublevels known as 1s, 2s, 2p, 3s, and 3p. The s and p sublevels accommodate two and six electrons, respectively (Fig. 2.4). A completely filled sublevel is energetically favored, and atoms seek to acquire filled sublevels by bonding with other atoms. The number of ***valence electrons*** in the outer shell plays a major role in determining the chemical reactivity and bonding pattern of an atom. Carbon, nitrogen, and oxygen require four, three, and two electrons, respectively, to fill the outer p level, and these atoms form bonds with four, three, and two atoms, respectively. The formation of methane (CH_4), ammonia (NH_3), and water (H_2O) results in filling the 2p level of carbon, nitrogen, and oxygen, respectively. Filling the outer orbital reduces the energy

Valence electron. *An electron that occupies the outer energy shell of an atom and plays a major role in chemical reactivity and bonding.*

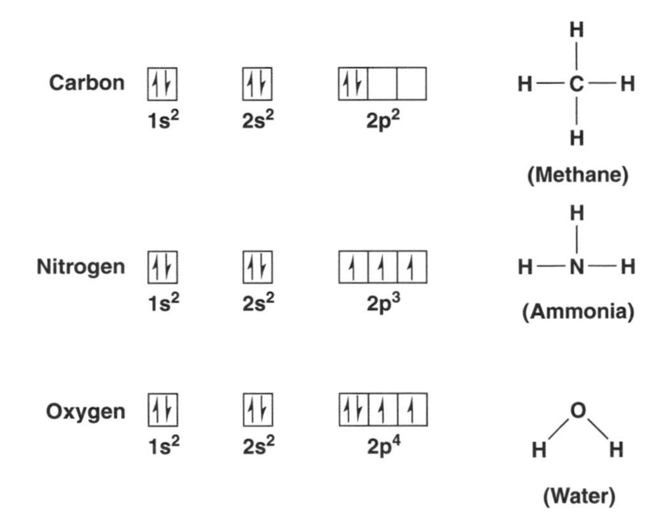

FIGURE 2.4. The pairs of electrons present in carbon, nitrogen, and oxygen along with their spin states (arrows) *in each sublevel* (boxes). *The carbon, nitrogen, and oxygen atoms in methane, ammonia, and water, respectively, have completely filled outer shells.*

level of an atom and is a main driver in chemical bonding (see below). The second principle energy level contains two electrons in 2s and six electrons in 2p, and the so-called octet rule accounts for all eight electrons and provides a useful guide for drawing chemical structures (see below).

Chemical Formulae

Molecule. *A substance that contains two or more atoms.*

Chemical formula. *A shorthand notation that describes the type and number of atoms present in the molecule.*

Molecular weight. *The sum of the atomic masses of all the atoms in a molecule.*

Avagadro's number. *The constant 6.022 × 10^{23}, which is equivalent to the number of atoms in 12 grams of carbon.*

Molarity. *The number of moles of a substance present in per liter of liquid.*

Any combination of two or more atoms is known as a ***molecule.*** Water (H_2O), carbon dioxide (CO_2), and ammonia (NH_3) are examples of simple molecules (Fig. 2.1), and nucleotides and amino acids are more elaborate biological molecules (see Chapter 3). A molecule can be written in shorthand notation, using a ***chemical formula*** that describes the type and number of atoms present in the molecule. A water molecule contains two hydrogen atoms and one oxygen atom, and a molecule of carbon dioxide has one carbon and two oxygens. Slightly more complex molecules such as ethanol (CH_3CH_2OH) use an extended notation to describe the types and numbers of atoms present, and their relationship to each another in the molecule. Parentheses are used if a molecule contains a chemical group present more than one time in the molecule, such as the two methyl groups (CH_3) in dimethyformamide: $HCON(CH_3)_2$.

Chemical formulae are useful for determining the molecular weight of a molecule. The ***molecular weight*** is the number of grams of a molecule in 1 mole (6.022×10^{23} molecules) of that substance. The constant 6.022×10^{23} is known as ***Avagadro's number,*** and it is equivalent to the number of atoms in 12 g of carbon. The molecular weight (MW) of a molecule is simply the sum of the molecular weights all of the atoms present in the molecule:

(2.1) Molecular weight = Sum of the atomic mass of all atoms

The molecular weight of water (H_2O) is 18 g per mole (18 g/mol) because water contains two hydrogen atoms and one oxygen atom [(1)2 + (16)1 = 18]. The molecular weight of carbon dioxide (CO_2) is 44 g/mol [(12)1 + (16)2 = 44], and ammonia (NH_3) has a molecular weight of 17 g/mol [(14)1 + (1)3 = 17].

The molecular weight can be used to prepare a solution of known concentration. The standard unit of liquid concentration is ***molarity*** (M), defined as the number of moles of a substance present per liter of solvent (typically water).

$$\text{(2.2)} \qquad \text{Concentration (M)} = \frac{\text{Number of moles}}{\text{Number of liters}}$$

A solution containing 2.5 mol (146.25 g) of sodium chloride in 0.5 mL of H_2O would have a concentration of 5.0 molar (2.5/0.5 = 5.0). In the laboratory, it is useful to know the amount of solid required to make 1 L of solution of a given concentration, which is the product of the molecular weight of the solid, times the molarity (mol/L) of the solution.

(2.3) Amount of solid (g) = MW (g/mol) × molarity (mol/L)

The amount of NaCl (MW = 58.5 g/mol) required to make 1 L of 5M NaCl solution is 292.5 g (58.4 × 5 = 292.5). A solution of 5 M NaCl would be prepared by dissolving 292.5 g of sodium chloride in a volume of water sufficient to make 1.0 L of solution. Note that the NaCl is not added to 1.0 L of H_2O because sodium atoms take up space, and a solution of NaCl prepared in this manner would have a final volume > 1.0 L (and a concentration < 5.0 M). Solutions are always prepared by bringing the final volume up to the desired volume to compensate for the volume contributed by the solid.

In addition to molarity, ***percent concentration*** is a unit of concentration used commonly in the laboratory, and it is calculated simply as the number of grams of solid per 100 mL of total liquid, with water being the most common solvent:

Percent concentration. *The number of grams of solid per 100 mL liquid.*

(2.4) Percent concentration (%) = Number of grams per 100 mL

A 2% solution of sodium dodecyl sulfate (SDS) would be made by dissolving 2.0 g of solid SDS in a volume of water sufficient to produce 100 mL. A 2% solution of NaCl would be made by dissolving 2.0 g of solid NaCl in 100 mL. Note that percent concentration does not use the molecular weight of the substance, but rather the mass of the substance per 100 mL. A 2% solution of SDS would have a different molar concentration than a 2% solution of NaCl, because SDS (288.4 g/mol) and NaCl (58.5 g/mol) have different molecular weights.

CHEMICAL BONDING

An attractive force between two atoms is known as a ***chemical bond.*** A bond that occurs between two atoms in the same molecule is an ***intramolecular bond,*** and one that occurs between two atoms in different molecules is an ***intermolecular bond.*** The four main types of bonds are covalent, electrostatic, hydrogen, and van der Waals. Covalent bonds are usually intramolecular bonds, and electrostatic, hydrogen, and van der Waals bonds can be either intramolecular or intermolecular. A closer look at chemical bonding helps define the forces that dictate the biochemistry of nucleic acids, proteins, carbohydrates, and other biomolecules used in microarray analysis.

Chemical bond. *An attractive force between two atoms.*

Intramolecular bond. *An attractive chemical force between two atoms in the same molecule.*

Intermolecular bond. *An attractive chemical force between two atoms in different molecules.*

Covalent Bond

A ***covalent bond*** is a chemical bond formed by the sharing of electrons between two atoms. Atoms share electrons with other atoms and form covalent bonds to fill the outer shell of electrons, which provides greater atomic stability than an outer shell that is partially filled (see above). Atoms and molecules tend toward the lowest possible energy state, and this tendency drives chemical reactions. The outer shell or valence electron capacity of most biologically important atoms (e.g., carbon, oxygen, nitrogen) is eight, and the propensity to obtain a complete set of eight valence electrons is known as the ***octet rule.*** The octet rule is helpful because it provides a simple guide for configuring the chemical bonds in nucleotides, amino acids, and other biomolecules.

Covalent bond. *An attractive chemical force characterized by the sharing of electrons between two atoms.*

Octet rule. *The chemical propensity of an atom to obtain a complete set of eight valence electrons.*

A covalent bond involves the sharing of two electrons and is drawn as straight line between the two participating atoms. Carbon can share one, two, or three chemical bonds with another carbon atom as shown for ethane, ethene, and ethyne (Fig. 2.5). The single, double, and triple bonds present in these molecules contain two, four, and six shared electrons, respectively; and the remaining electrons required to fill the outer shell (eight) are provided by bonds to hydrogen. Ethane, ethene, and ethyne obey the octet rule, though the three molecules use different bonding schemes. The hydrogen atom is an exception to the octet rule; due to its small size and 1s orbital, hydrogen requires only two electrons to fill its outer electron shell.

The octet rule can be satisfied by bond formation or by the presence of nonbonding ***lone pair*** electrons, both of which provide two electrons to the electronic octet. The nitrogen atom in ammonia (NH_3) forms three single bonds with hydrogen (six electrons), and the remaining two electrons required for the

Lone pair. *Two electrons that occupy the outer energy level of an atom but do not participate in covalent bond formation.*

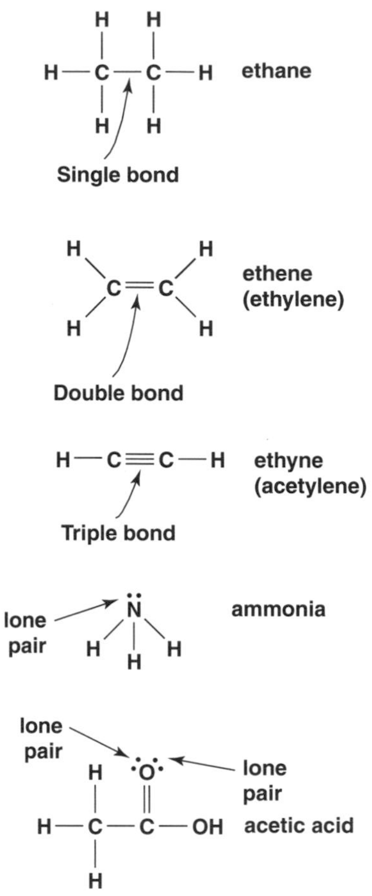

FIGURE 2.5. The octet rule requires eight outer shell electrons in carbon, oxygen, nitrogen, and other bonded atoms and can be satisfied by the formation of single (ethane), double (ethene) or triple (ethyne) bonds or by the presence of lone pair electrons (double dots), as in ammonia and acetic acid.

octet are provided by a lone pair (Fig. 2.5). The carboxylate carbon in acetic acid fulfills the octet rule by forming a single bond with OH (two electrons), a single bond with carbon (two electrons), and a double bond with oxygen (four electrons). The double-bonded oxygen in acetic acid acquires its eight electrons via a double bond with carbon (four electrons) and by the presence of two lone pairs on oxygen (four electrons). Lone pair electrons are important biologically because they participate in hydrogen bond formation and in acid–base reactions (see below).

A chemical bond can be broken by adding an increasing amount of heat to the two bonded atoms, which causes the atoms to vibrate more and more rapidly until the bond breaks. The amount of heat required to break the chemical bonds in 1 mole of a substance is known as the ***dissociation energy.*** The dissociation energy for a typical covalent bond is 100 kilocalories per mole (100 kcal/mol). Covalent bonds, the strongest chemical bonds in biochemistry, are up to 100 times stronger than noncovalent bonds (see below), and covalent bonds between carbon, hydrogen, nitrogen, oxygen, phosphorus, and other atoms are used to build nucleotides, amino acids, and other important biomolecules (see Chapter 3). The noncovalent electrostatic, hydrogen, and van der Waals bonds are much weaker than covalent bonds, but noncovalent interactions are extremely important in biology, particularly when they occur in large numbers.

Dissociation energy. *The amount of heat required to break the chemical bonds in 1 mole of a substance, expressed typically in kilocalories per mole.*

Electrostatic bond. *A noncovalent chemical interaction formed by the association of two molecules of opposite charge.*

Noncovalent. *Any of several types of attractive chemical forces between atoms possessing opposite charge character.*

Electrostatic Bond

An ***electrostatic bond*** is a **noncovalent** interaction formed by the association of two molecules of opposite charge. Biomolecules contain many positively and

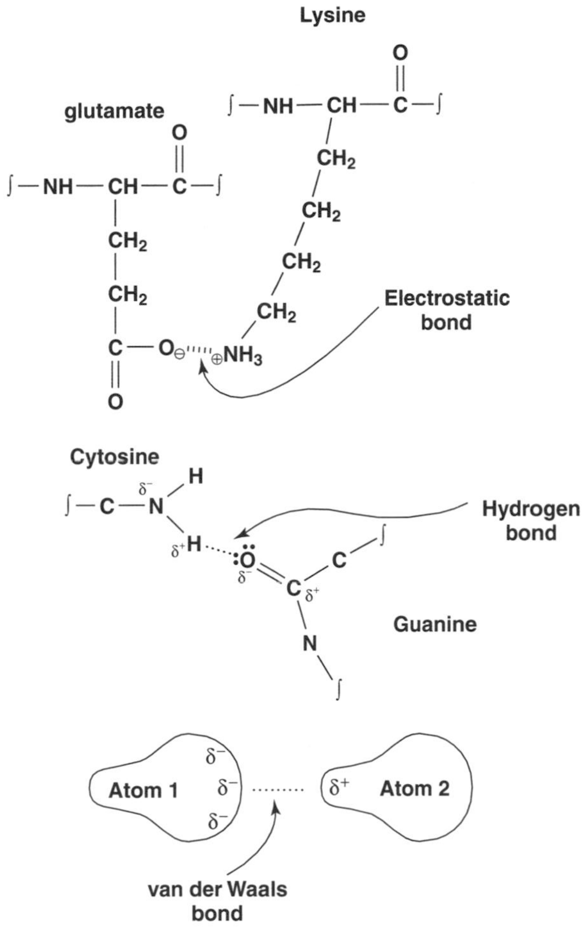

FIGURE 2.6. *Three types of noncovalent bonds are important in biology: electrostatic* (top), *hydrogen* (center), *and van der Waals* (bottom).

negatively charged groups, and these opposite charges interact with each other in important ways both in intramolecular and in intermolecular contexts. Electrostatic bonds are found extensively in intramolecular and intermolecular protein interactions. The negatively charged amino acid side chains of glutamate can bond electrostatically with the positively charged lysine side chain (Fig. 2.6). Electrostatic bonds help maintain the folded state of proteins and mediate protein–protein interactions. The dissociation energy for a typical electrostatic bond is 30 kcal/mol, about a third of the strength of an average covalent bond. Electrostatic bonds are also known as salt bridges.

Hydrogen Bond

Hydrogen bond. *A noncovalent chemical interaction formed by the sharing of a hydrogen atom between two molecules.*

Hydrogen donor. *An atom bonded to hydrogen that creates a partial positive charge on the hydrogen atom and facilitates the formation of a hydrogen bond.*

Hydrogen acceptor. *An electron-rich atom that has an electron lone pair that interacts with the partially positively charged hydrogen atom in a hydrogen bond.*

A ***hydrogen bond*** is a noncovalent interaction formed by the sharing of a hydrogen atom between two molecules. A ***hydrogen donor*** is an atom (typically oxygen or nitrogen) bonded to hydrogen that creates a partial positive charge on hydrogen, and "donates" the hydrogen atom to the hydrogen bond. A ***hydrogen acceptor*** is an electron-rich atom (typically oxygen or nitrogen) that has one or more lone pairs, and this partial negative charge accepts the partially positively charged hydrogen atom to form the hydrogen bond (Fig. 2.6).

Hydrogen bonds form between a myriad hydrogen donors and acceptors, including the DNA bases A-T and G-C (Fig. 2.6). Hydrogen bonds have a strength of approximately 5 kcal/mol, providing only about one twentieth the strength of a covalent bond, and about one fifth the strength of an electrostatic bond. The relatively weak bonding force of hydrogen bonds can be compensated by the fact that H-bonds tend to occur in large numbers in biologically important molecules. A 1000-base pair DNA molecule (see Chapter 4), for example, would form 2000–3000 hydrogen bonds, depending on the nucleotide sequence, and the sum of the 5 kcal/mol interactions would provide enormous avidity between the complementary DNA chains. A 1000-base pair DNA molecule requires near boiling temperatures (100°C) to disrupt the complementary DNA strands held together tightly by hydrogen bonds.

Hydrogen bonding also accounts for many of the striking properties of water, including its unexpectedly high boiling point (100°C) and prominent surface tension. Surface tension in microarray droplets is an important determinant of spot diameter in microarray printing procedures (see Chapter 7).

van der Waals Bond

van der Waals bond. *A noncovalent interaction between two molecules, created by transient fluctuations in electron density.*

Dipole. *A transient or permanent asymmetry in electron density producing a positive or negative bias around an atom.*

Contact distance. *The atomic separation between two atoms that maximizes the strength of a van der Waals bond.*

A ***van der Waals bond*** is a noncovalent interaction between two molecules created by transient changes in electron density. Electrons orbiting around the nucleus of an atom can create transient changes in the electron cloud and a momentary positive and negative bias known as a ***dipole.*** The creation of transient dipoles allows the formation of van der Waals bonds between adjacent atoms. All atoms exhibit van der Waals interactions if they are brought to within 3–4 Å of each other, an atomic distance known as the van der Waals ***contact distance.*** At 1 kcal/mol, van der Waals bonds have one hundredth the strength of a covalent bond, one thirtieth the strength of an electrostatic bond, and one fifth the attractive force of a hydrogen bond, making the van der Waals bond the least avid of any of the main types of chemical bonds. Despite their relatively weak attractive force, van der Waals bonds are important in biology and in microarray analysis. The formation of van der Waals bonds between microarray surfaces and target molecules is believed to play an important role in target attachment (see Chapter 5).

PROPERTIES OF MATTER

Matter. *Any substance that has mass and occupies space.*

Matter is any substance has mass and occupies space. This section provides a brief discussion of some of the main aspects of matter, including physical states, solubility, surface tension, and capillarity.

Physical State

Solid. *The physical state of matter characterized by a defined shape and volume.*

Liquid. *The physical state of matter characterized by an indefinite shape and a defined volume.*

Gas. *The physical state of matter characterized by an indefinite shape and an indefinite volume.*

Amorphous solid. *A substance such as glass having a defined shape and volume but lacking the regular pattern of molecules found in true solids.*

The three main physical states are solid, liquid, and gas. A ***solid*** has a defined shape and volume, a ***liquid*** has a defined volume and an indefinite shape, and a ***gas*** has an indefinite shape and an indefinite volume.

Most solids, including metals, salts, and other substances, have a regular pattern of atoms and are known as crystalline solids. Glass is a fairly unusual solid in that it has a defined shape and volume, but silicon dioxide atoms lack a regular pattern and glass is therefore known as an ***amorphous solid*** (see Chapter 5). Glass is used widely as a substrate in microarray manufacture, and understanding the basics of glass chemistry is important in microarray research.

The biological samples and buffers used in microarray analysis are liquids and as such assume the form of the containers in which they carried.

Microplates, microfuge tubes, and a variety of printing devices (including pins and ink jets) are some of the solid implements used to contain microarray liquids. Gases, including water in the vapor phase (i.e., humidity), play a role in the environmental aspects of microarray reactions and manufacture. Gas molecules diffuse quickly in three dimensions, and gaseous organic emissions are an important consideration in cleanroom manufacturing processes (see Chapter 11).

Substances can be converted from one physical state to another, in a process known as a ***change of state.*** The transitions from solid to liquid, liquid to gas and solid to gas are known as ***melting, evaporation,*** and ***sublimation,*** respectively. One common change of state in microarray reactions is evaporation, or more specifically the conversion of liquid water into gaseous water during microarray sample preparation, printing, and hybridization. The loss of a small volume of water from a small sample size can have a large impact on microarray manufacture and use, and care must be taken to minimize evaporation during microarray experimentation. Evaporation during printing can be minimized by maintaining reasonable experimental temperatures (< 25°C) and by minimizing air currents and turbulence. Microplates and other storage containers should be covered when not in use, to minimize evaporation.

Change of state. *A chemical transition characterized by the conversion of one physical state to another.*

Melting. *The chemical transition from a solid to a liquid.*

Evaporation. *The chemical transition from a liquid to a gas.*

Sublimation. *The chemical transition from a solid to a gas.*

Solubility

The amount of a solid that can be dissolved in a defined volume is known as ***solubility.*** As a general rule, solubility increases with increasing temperature and solvent compatibility. A polar substance (e.g., NaCl) dissolves more readily in warm solvent than in cool solvent and more readily in a polar solvent such as water than in a less polar solvent like ethanol. Polar solids are weakly soluble in nonpolar solvents such as benzene. Most of the biochemical reactions encountered in microarray analysis use polar solids and solvents at relatively low temperatures (25–65°C), and microarray researchers should understand the relationship between solubility, solvent temperature, and solvent polarity.

Solubility. *The amount of solid that can be dissolved in a defined volume of liquid.*

Surface Tension

The attractive force between molecules in a liquid is known as ***surface tension.*** Surface tension accounts for several important properties of water, including the concave appearance that droplets assume when placed on hydrophobic surfaces such as chemically modified glass. The beading of liquid droplets occurs because the water molecules interact tightly with each other to minimize the repulsive energy that occurs when highly hydrophilic water molecules are deposited on a hydrophobic surface. The energy state of the droplet is minimized by minimizing the surface area of the droplet, a force that accounts for the beading of water molecules. The greater the hydrophobicity of the surface, the greater the repulsive energy and the more pronounced the concave appearance of the droplet. The contact angle between the droplet and the surface provides a semiquantitative measure of surface hydrophobicity (see Chapter 6).

Surface tension is extremely strong in pure water because water molecules exhibit strong hydrogen bonding interactions that exert powerful attractive forces. A ***surfactant*** is an agent that reduces the surface tension of a liquid, and surfactants reduce the surface tension of water by disrupting the hydrogen bonding interactions between water molecules. Several types of molecules, including detergents and organic solvents, reduce the surface tension of

Surface tension. *Cohesive force, such as hydrogen bonding exerted between liquid molecules in a droplet, contributing to meniscus formation and other properties of liquids.*

Surfactant. *A chemical agent that reduces the surface tension of a liquid, typically by disrupting hydrogen bonding interactions between water molecules.*

microarray samples, causing an increase in the diameter of printed spots. Contaminants such as SDS, triton, ethanol, and dimethylsulfoxide (DMSO) reduce surface tension and are common contributors to variability in spot size. High-quality microarrays require uniform spot diameter, a parameter that can be controlled by controlling the surface tension of the printed samples (see Chapter 7). Researchers are served well by understanding the positive correlation between strong surface tension and small spot size and correspondingly reduced surface tension and larger spot diameter.

Capillary Action

Capillary action. *The spontaneous loading of a liquid into an elongated tube or channel.*

Capillary action is the spontaneous loading of a liquid into an elongated tube or channel. The phenomenon was discovered by the Egyptians about 3000 years ago and was exploited, in conjunction with reeds and other hollow objects, in early writing implements. Attractive forces between the liquid and the loading implement drive capillary action. If the attractive forces between the loading implement and the liquid exceed the liquid–liquid attractive forces, capillary action drives the filling of the capillary device until the forces are equalized. In highly elongate capillaries, attractive forces often cause the liquid to rise to a higher level along the walls of the device than in the center, leading to the formation of curved liquid surface known as a ***meniscus.*** One of the early microarray printing technologies, based on tweezers, formed a meniscus that required a tapping force to expel the sample from the microarray printing device (see Chapter 7). A more robust contact printing technology loads by capillary action but does not form a meniscus, obviating the need for tapping during printing and extending the durability of the printing implement (see Chapter 7).

Meniscus. *The curved surface of a liquid produced by the interaction between the liquid and the walls of a capillary device or container.*

ORGANIC COMPOUNDS

Organic compound. *A molecule that contains one or more carbon-to-hydrogen bonds.*

Inorganic compound. *A molecule that does not contain carbon-to-hydrogen bonds.*

An ***organic compound*** is a molecule that contains one or more carbon–hydrogen bonds. Organic compounds are distinct from ***inorganic compounds*** such as salts, minerals, and gases that do not contain these bonds. Some carbon-containing compounds such as carbon dioxide (CO_2), contain carbon but not hydrogen and are considered inorganic. Biological molecules, including DNA, RNA, and protein, contain carbon–hydrogen bonds and are classified as complex organic compounds. This section provides a short summary of organic compounds, focusing on functional groups, oxidation and reduction, and acids and bases. These concepts are important for understanding the chemical basis of microarray analysis.

Functional Groups

Functional group. *An group of atoms that possesses a characteristic chemical property.*

A ***functional group*** is an atom or group of atoms that possesses a characteristic chemical property. The presence of a functional group provides clues to the properties of the molecule that contains the group, placing considerable importance on the understanding the functional groups present in DNA, RNA, proteins, and other biological molecules. The functional groups most pertinent to biochemistry and microarray analysis include alcohol, aldehyde, amide, amine, carboxylic acid, disulfide, ester, ether, ketone, and thiol.

Alcohol (OH), aldehyde (CHO), and carboxylic acid (COOH) are related functional groups (Fig. 2.7) that mix well with water and exhibit hydrogen

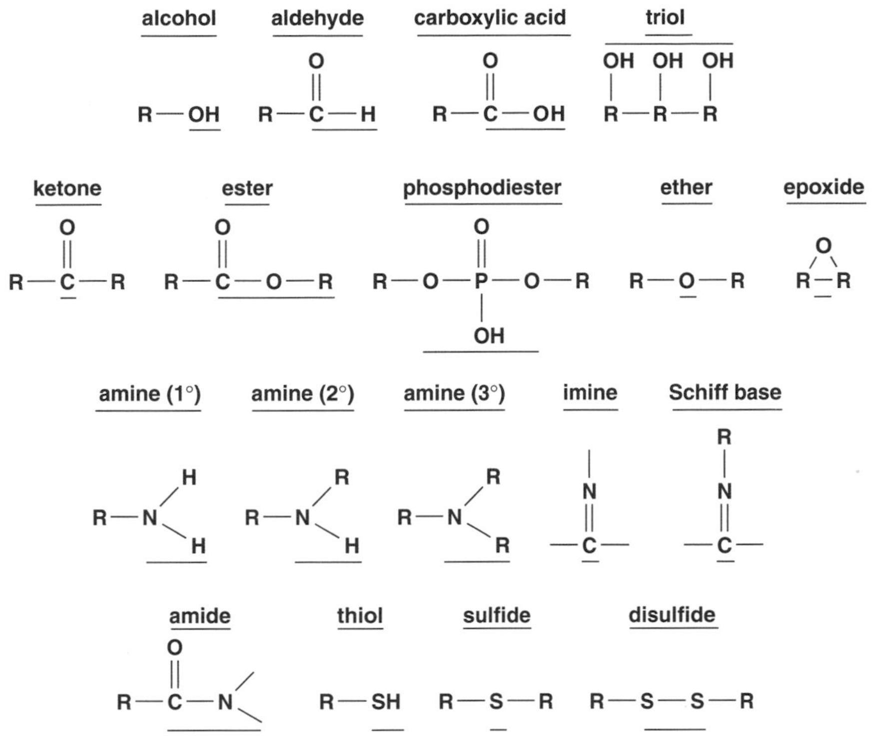

FIGURE 2.7. *Functional groups pertinent to biochemistry and microarray analysis.* R, *alkyl groups.*

bonding capabilities, the latter quality conferring a low ***vapor pressure*** and high boiling point to molecules that contain these groups. Alcohol (hydroxyl) groups are present on the deoxyribose and ribose sugars present in DNA and RNA, three of the amino acid side chains (serine, threonine, and tyrosine), carbohydrates, and other important biomolecules. Some of the solvents used in microarray analysis, including water and ethanol, also contain the OH group. Hydroxyls can function as nucleophiles in chemical reactions (see below). Specialized alcohols known as triols contain three alcohol groups on adjacent atoms, and the common triol glycerol is used in some protein and enzyme buffers to stabilize protein structure.

Vapor pressure. *The propensity of a liquid to enter the gas phase.*

Aldehydes (Fig. 2.7) are used to attach nucleic acids and proteins to microarray surfaces during microarray manufacture (see Chapter 5). The carbon group in the aldehyde is electrophilic (see below), bonding to primary amines and allowing covalent attachment of nucleic acids and proteins via Schiff base formation (see Chapter 5). The carboxylic acid group is found on all 20 free amino acids (see Chapter 3), and this functional group deprotonates readily to carry a negative charge at neutral pH. The negatively charged carboxylic acid forms electrostatic bonds with positively charged functional groups present in lysine, arginine, and other molecules.

The phosphodiester group (Fig. 2.7), similar to the carboxylic acid, is a highly soluble functional group with hydrogen bonding and ***ionization*** capacity. Phosphodiester bonds link deoxyribonucleotides and ribonucleotides together in DNA and RNA chains (see Chapter 4). The negative charge of phosphodiester groups imparts DNA and RNA their characteristic negative charge.

Ionization. *The chemical process whereby a molecule acquires a positive or negative charge.*

Ketone, ether, ester, and epoxide functional groups (Fig. 2.7), contain carbon–oxygen bonds, but not carbon–hydrogen bonds. These functional groups are ***polar*** but do not form hydrogen bonds, and small molecules containing these groups have lower boiling temperatures and high vapor pressures than alcohols, aldehydes, and carboxylic acids due to the lack of hydrogen bonding. Some DNA and RNA extraction protocols use ethers (e.g., diethyl

Polar. *A chemical bond or molecule that exhibits an unequal distribution of electron density.*

ether) to remove trace quantities of phenol and chloroform before ethanol precipitation. Because of their high vapor pressure, ethers are removed easily from DNA preparations by mild vacuum centrifugation and thus are a convenient extraction solvent. Ketones such as acetone are strong solvents that are used in a number of manufacturing procedures, including substrate production. Esters are volatile compounds that give plant products (e.g., bananas, strawberries) their fruity aroma. Epoxides are ringed structures that provide a common functional group for microarray surface chemistry. Due to the high energy state (~25 kcal/mol) produced by ring strain, epoxides are highly reactive electrophiles that form covalent bonds with DNA and protein molecules containing nucleophilic primary amines (see below). Deoxyribose and ribose sugars contain ringed epoxide functional groups, though not much strain energy is involved with these relatively large (five atom) ring structures (see Chapter 3).

Ketone groups (C=O) are present in four of the five DNA and RNA bases (guanine, cytosine, uracil, and thymine). Ether groups are present in DNA and RNA sugars (deoxyribose and ribose) and in carbohydrates (see Chapter 3). In a biological context, some esters are formed by specialized enzymes known as proteases, which break peptide bonds through the formation of ester linkages.

Amine and imine are important nitrogen-containing functional groups found in many biomolecules. Primary, secondary, and tertiary amines contain one, two, and three nonhydrogen atoms (usually carbon), respectively, attached to the amine group (Fig. 2.7). Primary amine groups (see Chapter 3) occur in three of the five nucleic acid bases (adenine, guanine, and cytosine), and two amino acids (lysine and arginine). Secondary amines are found in all five DNA bases (adenine, guanine, cytosine, uracil, and thymine) and in three amino acids (tryptophan, arginine, and histidine). Primary amines are more nucleophilic than secondary amines, which are better nucleophiles than tertiary amines (see below).

Aliphatic. *A functional group or molecule containing a straight chemical chain.*

Aromatic. *A functional group or molecule containing a ring structure and conjugated double bonds.*

Amines are also classified as aliphatic or aromatic, depending on the nature of the substituent attached to the amine group. ***Aliphatic*** and ***aromatic*** amines contain straight chain and aromatic ringed groups, respectively, attached to the nitrogen atom. Primary aliphatic amines are the strongest nucleophiles of any of the amines (see below) and as such are particularly useful as linker arms for attaching oligonucleotides and cDNAs to microarray surfaces. Primary aliphatic amines can be linked to synthetic oligonucleotides during chemical synthesis, providing excellent nucleophiles for oligonucleotide attachment to aldehyde and epoxide microarray surfaces (see Chapter 5). Two of the amino acids (lysine and arginine) contain primary aliphatic amines as their side chains, which explains why proteins attach tightly to aldehyde and epoxide surfaces (see Chapter 5).

Schiff base. *A specialized imine consisting of a primary amine that forms a double bond between carbon and nitrogen.*

An imine is functional group containing a carbon–nitrogen double bond (Fig. 2.7). A specialized imine known as a ***Schiff base,*** contains a primary amine as the double-bonded nitrogen. Imines are important in microarray research because they represent the intermediates that form during covalent attachment of DNA and protein to aldehyde surfaces (see Chapter 5). The double bond in the Schiff base can be reduced (see below) with sodium borohydride to produce a secondary amine, which is more stable chemically than the imine (see Chapter 5).

Amide is the functional group that links amino acids together into proteins (see Chapter 3). Amide or peptide bonds contain a hydrogen donor (N-H) and hydrogen acceptor (C=O), and therefore exhibit the strong hydrogen bonding properties that, in part, account for the rich secondary and tertiary structures of folded proteins. Amide groups are also present in four of the five nucleotide bases (guanine, cytosine, uracil and thymine), and in the side chains of two amino acids (asparagine and glutamine).

Several biologically important functional groups contain a sulfur atom (S), including thiol, sulfide, and disulfide (Fig. 2.7). Thiol groups (SH) are chemically similar to alcohols (OH), though sulfur holds its electrons less tightly than oxygen, and the less electronegative sulfur atom in the SH group renders thiols more prone to oxidation. Thiol oxidation presents a main challenge in maintaining the stability of proteins prepared in the laboratory. Thiols are also weaker hydrogen donors than alcohols and have weaker hydrogen bonding characteristics than hydroxyl-containing compounds, leading to a lower boiling point and greater vapor pressure. The human nose is particularly sensitive to thiols, capable of detecting 1 part per 100 billion in air. Thiol compounds are responsible for some of familiar stinky odors encountered in daily life, including the foul odor produced in the urine after the consumption of an asparagus-containing meal (more of a problem in California than elsewhere!). The thiol group is present in the cysteine side chain (see Chapter 3).

Sulfide and disulfide groups are also found in amino acids and proteins. Methionine contains a sulfide functional group (Fig. 2.7), characterized by the presence of two alkyl groups bonded to sulfur (R-S-R). Bond formation between two oxidized thiol groups produces a disulfide group or ***disulfide bond.*** Disulfide bond formation between cysteine residues in proteins helps maintain the elaborate secondary and tertiary structure found in folded cellular proteins (Fig. 2.8). Disulfide bond formation is also used to link polypeptides together to form proteins that contain a multisubunit structure.

Disulfide bond. *Covalent chemical interaction between two sulfur atoms.*

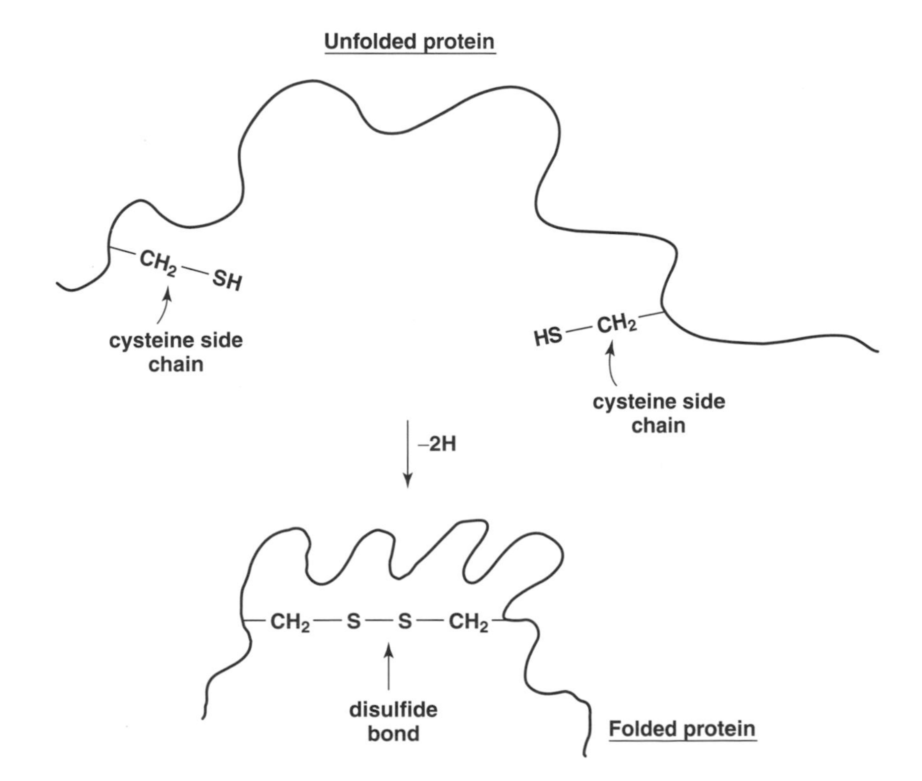

FIGURE 2.8. Some newly synthesize proteins in the unfolded state (top) *contain cysteine residues (CH_2-SH) that can undergo oxidation (-2H) and disulfide bond (CH_2-S-S- CH_2) formation, which stabilizes the structure of the folded protein* (bottom).

Oxidation and Reduction

Oxidation. *The chemical process involving the loss of electrons.*

Reduction. *The chemical process involving the acquisition of electrons.*

The process by which electrons are lost by an atom is known as ***oxidation,*** and the reciprocal process by which atoms are gained by an atom is known as ***reduction.*** Oxidation and reduction reactions (known collectively as redox reactions) are important in several microarray research contexts, and students should endeavor to understand the chemical basis of these two important reactions. The generalized oxidation reaction is provided in Equation 2.5, whereby an atom (A) loses an electron (e^-) and acquires a positive charge (A^+):

$$A \rightarrow A^+ + e^- \tag{2.5}$$

The generalized reduction reaction is provided in Equation 2.6, whereby an atom (A) gains an electron (e^-) and acquires a negative charge (A^-):

$$A + e^- \rightarrow A^- \tag{2.6}$$

Electron affinity. *The propensity of an atom to acquire or lose electrons.*

Electronegative. *The chemical property of an atom or functional group, such as a nucleophile, characterized by a propensity for negative charge.*

Electropositive. *The chemical property of an atom or functional group, such as an electrophile, characterized by a propensity for positive charge.*

The propensity of an atom to gain or lose electrons is defined by its ***electron affinity.*** Atoms with a high electron affinity are termed ***electronegative,*** and atoms with a low electron affinity are said to be ***electropositive.*** Electron affinity increases from left to right across the periodic table (Fig. 2.2), so that elements on the right side of the periodic table (e.g., oxygen and chlorine) are highly electronegative, and elements on the left side of the periodic table (e.g., hydrogen and sodium) are highly electropositive. The propensity of oxygen to gain electrons and "rip" them away from other atoms is the chemical basis of a specialized oxidation reaction known as corrosion (see below). Extremes in electron affinity also account for the capacity of certain elements to exist as ions. The highly electropositive sodium atom occurs readily as a cation (Na^+), and the highly electronegative chlorine often exists as an anion (Cl^-). The two ionization reactions for sodium and chlorine are as follows:

$$Na \rightarrow Na^+ + e^- \tag{2.7}$$

$$Cl + e^- \rightarrow Cl^- \tag{2.8}$$

Reducing agent. *A chemical compound that contributes electrons to a chemical reaction.*

Several microarray-processing protocols use sodium borohydride ($NaBH_4$) as a ***reducing agent,*** to convert imines formed in Schiff base coupling reactions to the more stable amine configuration (see Chapter 5). Sodium borohydride dissolves readily into sodium ions and borohydride ions in aqueous buffers as shown:

$$NaBH_4 \rightarrow Na^+ + BH_4^- \tag{2.9}$$

The borohydride anion (BH_4^-) then reduces the imine bond into a secondary amine as shown:

$$\text{R-N=CH-} + BH_4^- + H^+ \rightarrow \text{R-HN-CH}_2\text{-} + BH_3 \tag{2.10}$$

In addition to reducing imine bonds between the microarray surface and bound DNA molecules, borohydride anion also reduces unreacted aldehyde groups on the microarray surface into nonreactive alcohols:

$$\text{R-HC=O} + BH_4^- + H^+ \rightarrow \text{R-H}_2\text{C-OH} + BH_3 \tag{2.11}$$

Several common functional groups can be interconverted via oxidation and reduction. Successive reduction of a carboxylic acid group (R-COOH) produces an aldehyde (R-CHO), and then an alcohol ($R\text{-}CH_2\text{-}OH$). Sequential oxidation of an alcohol ($R\text{-}CH_2\text{-}OH$) produces an aldehyde (R-CHO), then a carboxylic acid (R–COOH). Students should endeavor to learn the relationships of related functional groups vis-à-vis oxidation and reduction reactions.

One undesired oxidation reaction can sometimes complicate microarray manufacture and other aspects of microarray analysis. The oxidation of metals such as iron (Fe) is known as ***corrosion.*** Corrosion results in the loss of electrons from iron and the formation of ferrous (Fe^{2+}) and ferric (Fe^{3+}) ions as shown:

Corrosion. *The oxidation of iron and other metals, producing rust and other oxidative surface changes.*

(2.12) $$Fe \rightarrow Fe^{2+} + 2e^-$$

(2.13) $$Fe^{2+} \rightarrow Fe^{3+} + e^-$$

The main **oxidizing agent** in these reactions is oxygen gas (O_2), which is present at a relatively high concentration (21%) in the atmosphere. The rate of the oxidation reaction is accelerated by the presence of moisture (H_2O) and salt (e.g., NaCl), the latter facilitating the flow of electrons required for the oxidation process. The generalized equation for the formation of iron oxide ("rust") via corrosion is written as follows:

Oxidizing agent *A chemical compound that removes electrons in a chemical reaction.*

(2.14) $$4Fe^{2+} + O_2 + 6H_2O \rightarrow 2Fe_2O_3(H_2O)_n + 8H^+$$

The presence of atmospheric oxygen, moisture and salt in microarray manufacturing environments can lead to the corrosion of printing pins and other metal hardware. Researchers can minimize corrosion by understanding its chemical basis and minimizing the exposure of microarray hardware to moisture and salt.

Acids and Bases

An ***acid*** is a proton donor and a ***base*** is a proton acceptor. The generalized reactions of an acid and base are as follows:

Acid. *A functional group or molecule that donates a proton in a chemical reaction.*

(2.15) $$HA \rightarrow H^+ + A^-$$

(2.16) $$B^- + H^+ \rightarrow BH$$

Base. *A functional group or molecule that acquires a proton in a chemical reaction. Alternatively, any one of five nitrogenous molecules (adenine, cytosine, guanine, thymine, and uracil) contained in the nucleotides that make up cellular DNA and RNA.*

Proton loss and gain are commonplace in biochemistry and many of the molecules, buffers, and reactions used in microarray research involve acids and bases. The acidity or basicity of a solution is measured by pH, the negative logarithm of the H^+ concentration of the solution expressed in units of moles per liter (M):

(2.17) $$pH = -\log[H^+]$$

Neutral water has an H^+ concentration of 1×10^{-7} M and a pH value of 7.0 ($-\log 1 \times 10^{-7} = 7.0$). Acidic solutions contain a high concentration of H^+ ions and exhibit pH values < 7; basic solutions have a low H^+ concentration, and their pH values are > 7. Strong acids (e.g., HCl) have pH values < 1 and weaker acids are in the range of 2–5. Stomach acid, tomato sauce, and coffee have pH values of approximately 2, 4, and 5, respectively. Human blood is slightly basic with a pH of 7.5, and baking soda and ammonia provide slightly more basic solutions with pH values of 9 and 11, respectively. Household bleach is a fairly strong base with a pH of 13 and strong bases (e.g., NaOH) have pH readings of 14. Most buffers used in microarray research have pH values in the range of 5–9, as required to maintain the integrity of biological molecules. Acidic and basic extremes generally compromise the structure and integrity of biomolecules, though certain methods use high pH to selectively degrade mRNA molecules in RNA–DNA hybrids (see Chapter 3).

Proton donors (acids) can be understood in terms of the electron affinity of the donor atoms that carry the negative charge after proton loss. Acids have atoms with a relatively high electron affinity and are happy to accept the additional electron that results from deprotonation. The side chain of the amino acid aspartate (see Chapter 3), for example, contains an acidic carboxy group that donates a proton readily because the electrophilic oxygen atoms can accommodate the negative charge:

$$\text{R-CH}_2\text{-COOH} \rightarrow \text{R-CH}_2\text{-COO}^- + \text{H}^+ \quad (2.18)$$

Most aspartate groups are negatively charged at neutral pH, enabling the formation of intramolecular and intermolecular electrostatic bonds with positively charged side chains (see above). Phosphate atoms readily carry a negative charge as well, donating protons efficiently and accounting for the negative charge present in the phosphate backbones of DNA and RNA (see Chapter 3).

Proton acceptors (bases) can be understood in terms of how readily they remove protons (H^+) from solution. Strong bases remove protons from solution more readily than weak bases. Most of the acids and bases in biological molecules are relatively weak compared to chemical acids and bases, but the acidity and basicity of biomolecules is considerable and enormously important structurally and catalytically. The strength of weak acids and bases, such as those present in biomolecules, is often expressed as a pK value, which is approximately equal to the pH at which the acid or base is 50% deprotonated. Acidic molecules have small pK values and protonated basic molecules have large pK values. The acidic side chains of aspartate and glutamate have pK values of 3.9 and 4.3, compared to the pK values of 10.8 and 13.2 for the basic side chains of lysine and arginine.

The side chain of lysine (pK = 10.8), though weakly basic compared to NaOH, contains an electron lone pair that binds H^+ ions and confers a positive charge at neutral pH:

$$\text{R-(CH}_2\text{)}_4\text{-NH}_2 + \text{H}^+ \rightarrow \text{R-(CH}_2\text{)}_4\text{-NH}_3^+ \quad (2.19)$$

Positively charged lysine side chains form electrostatic bonds with negatively charged side chains of aspartate and glutamate, and such noncovalent interactions are important in protein folding and catalysis (see Chapter 4). The arginine side chain (pK = 13.2) also contains an electron lone pair on nitrogen, and arginine like lysine is positively charged at neutral pH.

A widely used microarray surface contains primary amine groups attached to the glass substrate. Each primary amine contains an electron lone pair that has basic properties, and, therefore, binds protons in solution and confers the glass surface with a positive charge. This positively charged surface binds to negatively charged molecules including DNA, RNA, and proteins via electrostatic bond formation (see Chapter 5). The acidic phosphate groups on DNA and RNA and the acidic amino acid side chains in aspartate and glutamate confer the negative charge required for electrostatic binding to the amine substrate. Electrostatic binding, plus some covalent attachment added during the coupling procedure, provides a highly stable microarray that can be used to study gene expression, protein–protein interactions, and other biochemical reactions.

CHEMICAL REACTIONS

The process whereby chemical bonds are formed and broken is known as a ***chemical reaction.*** Chemical reactions include the formation and breakage of covalent and noncovalent bonds. Both types of chemical reactions occur in microarray analysis and students will do well to understand the basics of chemical reactivity. Attachment of organic molecules such as reactive amines and aldehydes to a glass substrate involves the formation of covalent bonds. Hybridization of a DNA double helix and many protein–protein interactions involve the formation of noncovalent hydrogen and electrostatic bonds (see above). This section provides an introduction to chemical reactions, including chemical equations, types of reactions, chemical polarity, nucleophiles and electrophiles, and photochemistry.

Chemical reaction. *The process whereby chemical bonds are formed and broken.*

Chemical Equations

Chemists and biochemists use a shorthand notation known as a ***chemical equation*** to describe a chemical reaction. In the context of chemistry, the guidelines for writing chemical equations are rather strict, and each equation must include all reactants and products, as well as the physical state of all substances, and the reaction conditions. A ***reactant*** is any substance consumed in a chemical reaction, and a ***product*** is any substance formed in a chemical reaction. The chemical state (solid, liquid, or gas) of reactants and products should be denoted, along with the solvents and thermal conditions (e.g., heat) used. The treatment of methane (CH_4) with heat (Δ) leads to the formation of acetylene (C_2H_2) and hydrogen (H_2) gas (g), and the chemical equation is written as follows:

Chemical equation. *A shorthand notation used by chemists and biochemists to describe a chemical reaction.*

Reactant. *A substance consumed in a chemical reaction.*

Product. *A compound or chemical produced a chemical reaction.*

$$2\ CH_4\ (g) \xrightarrow{\Delta} C_2H_2\ (g) + 3\ H_2\ (g) \qquad (2.20)$$

Note that two molecules of methane ($2\ CH_4$) produce one molecule of acetylene (C_2H_2) and three molecules of hydrogen (H_2) gas. Every chemical equation must be ***balanced*** for reactants and products, so that the same number of atoms occur on both sides of the chemical equation. Equation 2.20 is balanced by placing a two in front of methane to provide two carbon atoms and eight hydrogen atoms as reactants, and two carbons and eight hydrogens as products. All three substances are gases (g), as noted in the equation.

Balanced. *Refers to a chemical equation written so that the same number of atoms occur on both sides of the equation.*

Biochemical reactions are sometimes written in a more descriptive manner, to obviate the vastly complex equations that would otherwise be involved with writing the chemical structures of large molecular weight biomolecules such as DNA and protein. A biochemist might write a biochemical reaction for DNA methylation as shown in Equation 2.21, describing an important biochemical reaction without bogging down the equation with an exact description of the the reactants and products.

$$\text{DNA} \xrightarrow{\text{methylase}} \text{DNA-Me} \qquad (2.21)$$

This reaction is carried out enzymatically by DNA methylase, an enzyme that attaches methyl groups to DNA in a covalent manner. Many enzymatic and biochemical reactions are written using the shorthand notation presented in Equation 2.21.

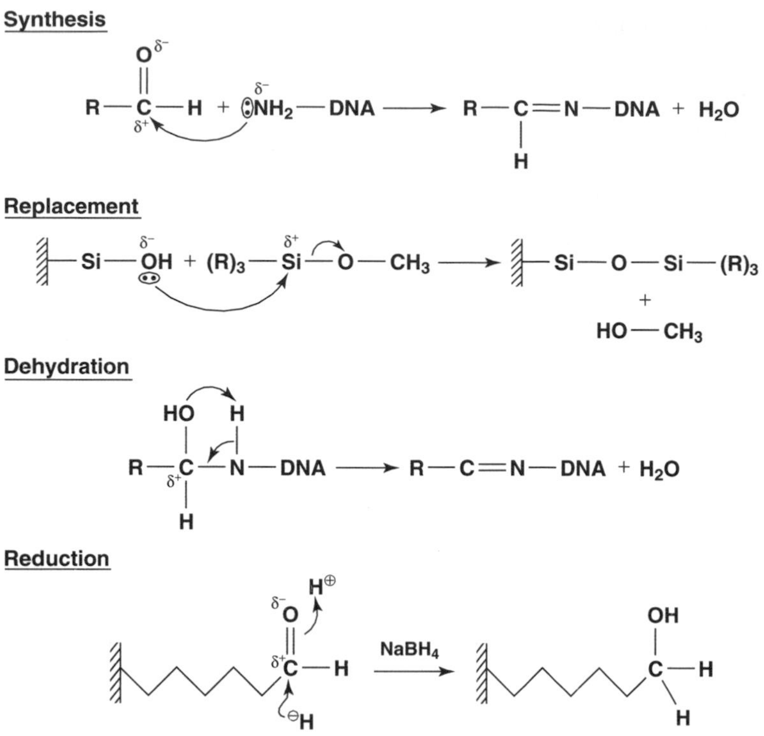

Synthesis

$$R-\overset{\overset{O^{\delta-}}{\|}}{\underset{\delta+}{C}}-H + \overset{\delta-}{:NH_2}-DNA \longrightarrow R-\underset{\underset{H}{|}}{C}=N-DNA + H_2O$$

Replacement

$$-Si-\overset{\delta-}{OH} + (R)_3-\overset{\delta+}{Si}-O-CH_3 \longrightarrow -Si-O-Si-(R)_3 + HO-CH_3$$

Dehydration

$$R-\overset{\overset{HO}{|}}{\underset{\underset{H}{|}}{C}}-\overset{\overset{H}{|}}{N}-DNA \longrightarrow R-C=N-DNA + H_2O$$

Reduction

$$\text{R}-\overset{\overset{O^{\delta-}}{\|}}{C^{\delta+}}-H \xrightarrow{NaBH_4} \text{R}-\overset{\overset{OH}{|}}{\underset{\underset{H}{|}}{C}}-H$$

FIGURE 2.9. Types of chemical reactions. Lone pair electrons (double dots)*, moving electrons* (arrows) *and positively (δ^+) and negatively (δ^-) charged centers are indicated.*

Types of Reactions

Chemical reactions can be classified using a number of different criteria, including the pattern of atomic rearrangements that occur, the chemical processes involved, and the reaction mechanisms. Some of the chemical reactions pertinent to microarray analysis include synthesis, replacement, dehydration, reduction, and oxidation. Each of these reactions occur in a number of different contexts. Microarray surface chemistry provides a good illustration of synthesis, replacement, dehydration, and reduction reactions (Fig. 2.9).

Synthesis reaction. *A chemical reaction in which two or more compounds are combined into a more complex compound.*

A ***synthesis reaction*** is a chemical reaction that combines two or more compounds into a more complex compound:

$$A + X \rightarrow A - X \quad (2.22)$$

The coupling of an amino-linked DNA molecule to an aldehyde surface is a synthesis reaction. A ***replacement reaction*** is one in which a more reactive molecule takes the place of a less reactive molecule to form a new substance:

Replacement reaction. *A chemical reaction in which a more reactive molecule takes the place of a less reactive molecule during the formation of a new substance.*

$$A + BX \rightarrow A - X + B \quad (2.23)$$

The reaction of glass with organosilane is a replacement reaction (Fig. 2.9). A ***dehydration reaction*** involves the loss of water (H_2O) during a chemical reaction:

Dehydration reaction. *A chemical reaction involving the loss of water.*

$$A \rightarrow B + H_2O \quad (2.24)$$

Dehydration is an important chemical reaction in microarray coupling processes. A reduction reaction involves the gain of electrons (see above), an example of which involves the reduction of an aldehyde to an alcohol using sodium borohydride (Fig. 2.9).

Chemical Polarity

Chemical polarity is defined as the unequal sharing of electrons in a chemical bond. Polar bonds form if the bonded atoms differ in electron affinity. Chemical polarity can be predicted for a given chemical bond by determining the position of the bonded atoms within the periodic table. Atoms on the left side of the table are relatively electropositive and atoms to the right of the table are relatively electronegative. A bond formed between an electropositive atom and an electronegative atom exhibits chemical polarity, resulting in electron deficiency (positive character) around the electropositive atom, and electron abundance (negative character) around the electronegative atom. An abundance and shortage of electron density defines nucleophiles and electrophiles, respectively (see below).

Chemical polarity. *The unequal sharing of electrons in a chemical bond.*

Oxygen (O) and nitrogen (N) have relatively high electron affinities and hold their electrons tightly, whereas hydrogen (H), carbon (C) and silicon (Si) have lesser electron affinities and hold their electrons more loosely. Chemical polarity is observed in H-O, H-N, C-O, C-N, and Si-O bonds, and these polar bonds are important in biochemistry and microarray analysis.

Solvents are also assessed as polar and nonpolar, depending on whether the solvent bonds exhibit chemical polarity or the polar bonds are equalized across the solvent molecule. A solvent that contains polar bonds or uneven charge distribution is known as a ***polar solvent,*** and a solvent that contains nonpolar bonds or uniform charge distribution is known as a ***nonpolar solvent.*** Water (H_2O) is an extremely polar solvent (Fig. 2.10) and is known as the ***universal solvent*** because of its capacity to dissolve a wide range of different compounds—salts, acids, and bases and a full spectrum of biomolecules, including DNA, RNA, protein, and carbohydrate (see Chapter 3). An ***aqueous*** reaction is one that contains water as the solvent, and many microarray reactions are performed using aqueous buffers.

Polar solvent. *A solvent that contains polar bonds or uneven distribution of charge.*

Nonpolar solvent. *A solvent that contains nonpolar bonds or a uniform distribution of charge.*

Universal solvent. *Water, so called because of its capacity to dissolve a wide range of different solids.*

Aqueous. *A buffer or chemical reaction that is composed off or occurs in water.*

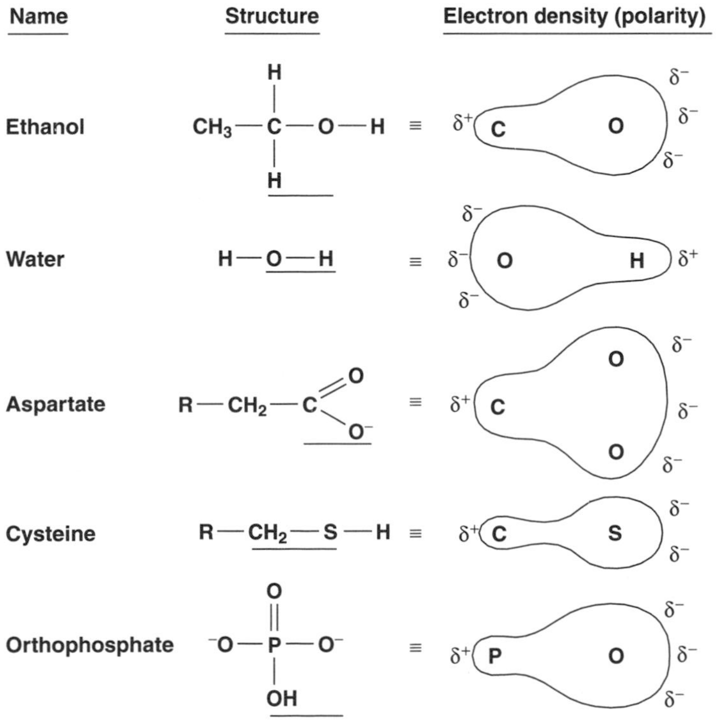

FIGURE 2.10. Polar bonds form between atoms that share electrons unequally. The electron-rich (δ^-) and electron-deficient (δ^+) centers are indicated for each molecule; the amino acid backbone is abbreviated (R) for aspartate and cysteine.

Carbon–oxygen bonds are less polar than hydrogen–oxygen bonds (Fig. 2.10), and consequently methanol and ethanol are less polar solvents than water. Methanol and ethanol are widely used as solvents in manufacturing because they are relatively inexpensive and dissolve many organic compounds that will not dissolve as easily in water. The reduced solubility of highly polar DNA and RNA molecules in alcohol is exploited extensively in nucleic acid purification schemes that use buffers containing high concentrations of alcohol (>70%) to precipitate the nucleic acids. Centrifugation and membrane-based purification schemes use buffers containing ethanol, isopropanol, and other alcohols.

There are many different types of carbon–oxygen bonds, including alcohol, ketone, aldehyde, epoxide, and carboxylic acid functions (see above), and each exhibits chemical polarity. Carbon–oxygen bonds are found in a wide spectrum of important biomolecules, including DNA, RNA, and carbohydrate, and side chains of seven amino acids (serine, threonine, aspartate, glutamate, asparagine, glutamine, and tyrosine) contain polar carbon–oxygen bonds.

Carbon–nitrogen, phosphorus–oxygen, and carbon–sulfur bonds are also important in biochemistry, and in each case, the electron density is concentrated around the more electronegative atom, causing a partial positive charge to form around the more electropositive atom. Electron density is concentrated around nitrogen, oxygen, and sulfur, respectfully, in N–O, P–O and C–S bonds. The SH-containing side chain of cysteine is polar, albeit less so than the OH-containing side chains, because sulfur is less electronegative than oxygen. The phosphate-oxygen bond in orthophosphate (Fig. 2.10) is a polar bond exploited in cells to achieve DNA and RNA synthesis (see Chapter 4). During polynucleotide synthesis, oxygen functions as a nucleophile, reacting with the electropositive phosphorus atom (see below).

Nonpolar solvents such as benzene (C_6H_6) and toluene (C_6H_5–CH_3) are used frequently in chemistry but much less so in biochemistry and microarray analysis. Proteins and other biomolecules denature rapidly in nonpolar solvents, and the damage to protein molecules is irreversible in most cases. Nonpolar solvents solubilize the side chains of the aromatic amino acids (phenylalanine, tyrosine, and tryptophan), which are usually located on the interior of folded proteins. Solubilization of the protein interior leads to rapid denaturation and loss of protein function. Nonpolar amino acids are folded into the protein interior to shield these nonpolar groups from the polar environment of the cell (mostly water). Understanding the interplay of bond and solvent polarity and biomolecular function is helpful in a number of areas of microarray analysis, including the rapidly emerging field of protein microarrays.

Nucleophiles and Electrophiles

Nucleophile. *A molecule or functional group that contains a free pair of electrons or a strongly negative charge bias, and an affinity for an electrophile.*

Electrophile. *A molecule or functional group that has strongly positive charge bias and an affinity for a nucleophile.*

An important extension of chemical polarity is chemical reactivity, which is determined largely by the types of bonds present the chemical reactants. Molecules with polar bonds function as nucleophiles and electrophiles in chemical reactions. A ***nucleophile*** is molecule, functional group or atom that contains a free pair of electrons or a strong negative charge bias. Nucleophile derives from "nucleus" and the Greek word *philos,* or "love," underscoring the fact that nucleophiles are attracted strongly to positively charged nuclei. An ***electrophile*** is a molecule, functional group or atom that bears a positive charge or strong positive charge bias, and these electron-seeking centers bond avidly with nucleophiles. Reactions between nucleophiles and electrophiles are central to organic chemistry, biochemistry, and microarray analysis. Equation 2.25 shows a generalized reaction between a nucleophile (N) and an electrophile (E), whereby the nucleophile attacks the electrophile and pushes out the leaving group (X).

$$\text{H–N:} + \text{E–X} \rightarrow \text{N–E} + \text{XH} \tag{2.25}$$

There are hundreds of complex chemical reactions involving nucleophiles and electrophiles, and students can get lost quickly in an "alphabet soup" of names and formulae. But organic chemistry and biochemistry can be simplified immensely by developing an *intuitive understanding* of chemical reactivity. The basic steps in the process are (1) identify the positive and negative centers (e.g., electrophiles and nucleophiles), (2) move the electrons from the negative centers to the positive centers, and (3) construct the new molecule based on the structure of the reactants and the movement of electrons. Nearly all chemical and biochemical reactions can be predicted using this simple three-step process.

Positive and negative centers are located by identifying the polar bonds in the reactants. Once the reactive centers are identified, bond formation and breakage is simply a matter of pushing electrons around to form the new molecule. Nucleophiles are essentially electron-dense portions of a polar bond, and electrophiles can be viewed as electron deficient components of a polar bond. The electron-rich (nucleophilic) and electron-poor (electrophilic) centers in organic molecules can be labeled by placing partial negative (δ^-) and positive (δ^+) charge symbols at these locations. Once the reactive centers are identified, electron pushing and bond formation and breakage flow naturally with the movement of electrons.

A closer look at the aldehyde and epoxide coupling chemistries used in microarray analysis provide a clear example of how a nucleophile (NH_2–DNA) can react readily with an electrophilic center to form a new chemical compound (Fig. 2.11). The nitrogen atom in amino-linked DNA (NH_2–DNA) is nucleophilic, owing to the presence of an electron lone pair. The lone pair seeks the electrophilic carbon atom in the aldehyde group, which contains a partial positive charge due to the double-bonded oxygen atom, which pulls electron density away from carbon. Nucleophilic attack "kicks out" one of the two carbon bonds, forming a hydroxyl group, which then dehydrates to form an imine.

The reaction of amino-linked DNA with an epoxide surface is analogous to the aldehyde surface, though the details are different (Fig. 2.11). The nucleophilic lone pair on nitrogen attacks the positive carbon center, so rendered by the oxygen-containing epoxide group, which pulls electron density away from carbon and creates an electrophile. Nucleophilic attack kicks out the epoxide, forming an alcohol group and a secondary amine. Dehydration is not a major reaction in epoxide coupling chemistry.

Hundreds of different organic and biochemical reactions can be drawn simply by charting the natural flow of electrons from nucleophiles to electrophiles. Most chemical reactions are little more than "negative seeking positive," and students who understand this simple concept are well along their way to being able to predict the vast majority of organic and biochemical reactions without memorizing hundreds of baroque names and mechanisms. Nucleophiles and electrophiles are nature's "hook-and-loop tape," attaching two molecules together by chemical bond formation. Once formed, chemical bonds (similar to

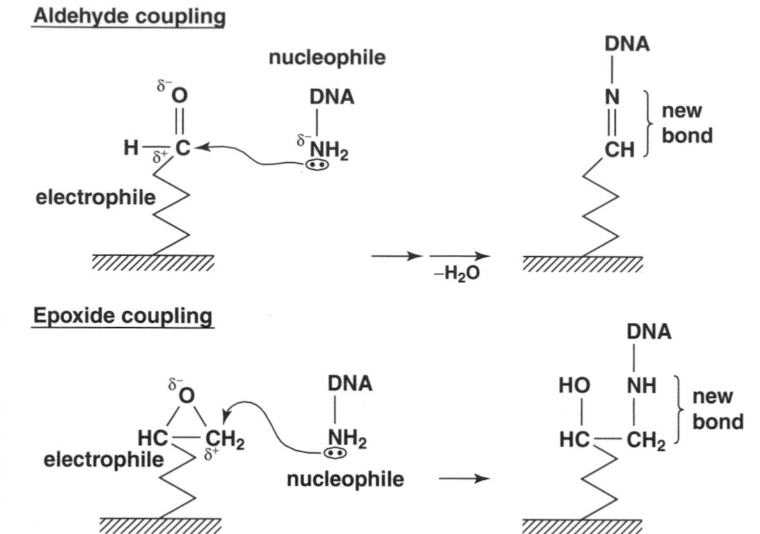

FIGURE 2.11. Molecules that bear an extra pair of electrons (nucleophiles) react efficiently with molecules that have positively charged centers (electrophiles), as demonstrated by the attachment of amino-linked DNA to aldehyde (top) *and epoxide* (bottom) *surfaces.*

Velcro strips) require considerable energy to separate. Noncovalent bonds such as electrostatic, hydrogen, and van der Waals bonds (see above) use the similar "opposites attract" concept of covalent bonds, though noncovalent bonds do not have nearly the attractive force of covalent bonds.

Photochemistry

Photochemistry. *A chemical reaction involving the interaction of organic molecules with light.*

A chemical reaction involving the interaction of organic molecules and light is known as ***photochemistry.*** Photochemical reactions are relevant to microarray analysis in a number of important respects, including the fact that fluorescence (see Chapter 8) and photochemical cross-linking (see Chapter 5) both involve photochemistry. Fluorescence results from the absorption of light ($h\nu$) by a fluorescent molecule (F) leading to electronic excitation (F*), followed by the release of a photon ($h\nu'$) during electronic relaxation (see Chapter 8). The generalized reaction for fluorescence is written as follows:

$$F \xrightarrow{h\nu} F^* \rightarrow F + h\nu' \quad (2.26)$$

The emitted photon ($h\nu'$) has slightly less energy and therefore slightly longer wavelength than the absorbed photon ($h\nu$), due to energy dissipation before relaxation. The longer wavelength of fluorescence emission relative to the excitation light is known as the Stokes shift, a phenomenon that plays an important role in microarray detection (see Chapter 8).

Fluorescence is not the only outcome of light absorption. Many organic molecules absorb light, but only a small fraction of those molecules emit fluorescence. Another outcome of light absorption, particularly with higher energy excitation wavelengths (e.g., ultraviolet light), is bond cleavage or ***photolysis:***

Photolysis. *Chemical bond breakage caused by the absorption of light.*

$$P\text{–}P \xrightarrow{h\nu} P\text{–}P^* \rightarrow P\cdot + P\cdot \quad (2.27)$$

The photochemical molecule (P–P) absorbs a photon leading to electronic excitation (P–P*), followed by bond cleavage and the formation of two free radicals (P·). A ***free radical*** is an atom with a single, unpaired electron. Free radicals are extremely reactive, and atoms harboring unpaired electrons seek the much

Free radical. *An atom with a single, unpaired electron.*

more stable double electron configuration, achieved usually by the formation of a chemical bond. Free radicals in the DNA bases, generated by the exposure of printed DNA samples to ultraviolet light, provide a nonspecific means of cross-linking DNA to a microarray surface (see Chapter 5).

Fluorescence and photochemical bond breakage are related phenomena in that both require photon absorption by the organic molecule. What chemical attribute in an organic molecule accounts for photon absorption? The common chemical feature in most fluorescent and photoreactive molecules, as well as nearly all pigments and dye molecules, is a system of alternating single and double bonds known as a conjugated system. A ***conjugated system*** contains a series of π (pronounced *pie*) electrons that reside above and below the plane of the molecule. Electrons are free to move within the conjugated system and are said to be ***delocalized*** because of their indefinite position within the π system. Electron delocalization reduces orbital energy, making it easier for molecules containing delocalized electrons to absorb light. As a general rule, a molecule with a greater number of double bonds holds its electrons more loosely and absorbs less energetic (longer wavelength) light, than a molecule with a lesser number of double bonds.

Conjugated system. *A series of alternating single and double bonds that contains electrons residing above and below the plane of the molecule.*

Delocalized. *The quantum mechanical state of electrons characterized by a freedom to move within and maintain indefinite positions within the conjugated electron system.*

The positive correlation between double bond number and absorption wavelength is evident in a comparison of tyrosine, cyanine 3 (Cy3), and cyanine 5 (Cy5), three organic molecules that contain 3, 9, and 10 double bonds, respectively, and absorb light at increasingly longer wavelengths of 275, 550, and 649 nm (Fig. 2.12). All three molecules contain delocalized electrons in a conjugated π system, but the photon energy required for excitation decreases as the number of double bonds increases, owing the greater delocalization in Cy5 compared to Cy3, and Cy3 compared to tyrosine. Cy3 and Cy5 absorb light in the visible range, and these fluorophores are used widely in microarray detection (see Chapter 8). The absorption of ultraviolet light (280 nm) by tyrosine and the other aromatic amino acids is not useful in microarray detection, but it does provide a way to assess protein concentration using analytical equipment

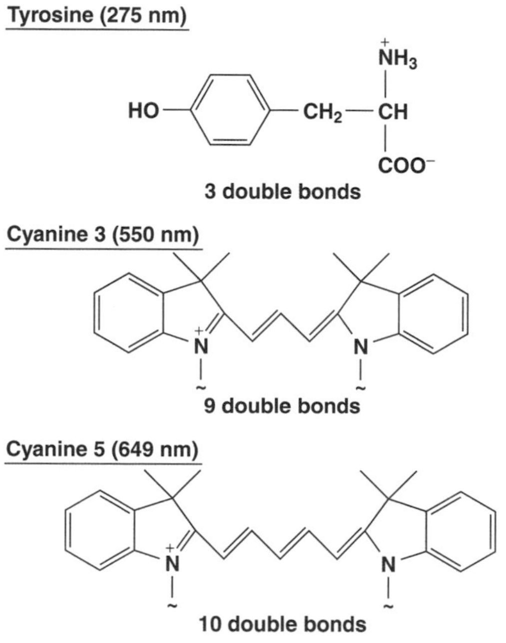

Tyrosine (275 nm)

$HO-C_6H_4-CH_2-CH(\overset{+}{N}H_3)-COO^-$

3 double bonds

Cyanine 3 (550 nm)

9 double bonds

Cyanine 5 (649 nm)

10 double bonds

FIGURE 2.12. The alternating single and double (conjugated) bonds of organic molecules contain delocalized π electrons that absorb light readily, and the absorption wavelength depends on the extent of conjugation.

set at an excitation wavelength of 280 nm. The delocalized electrons in the nucleic acid bases allow DNA and RNA to be measured similarly, by absorbance at 260 nm.

SUMMARY

Organic and inorganic substances are composed of 118 elements, which are listed in the periodic table. Elements are atoms composed of subatomic particles known as protons, neutrons, and electrons. The diameter of a hydrogen atom is approximately 1 Å. Atoms combine to form molecules and are held together by electron sharing in covalent chemical bonds. Noncovalent electrostatic, hydrogen, and van der Waals bonds are also important in biology, though these attractive forces are much weaker than covalent bonds. Numerous hydrogen bonds can combine forces to hold together important biomolecules, including the DNA chains in a double helix. Matter, any substance that has mass and occupies space, exists as a solid, liquid, or gas. Solubility, surface tension, and capillary action are physical aspects of matter pertinent to microarray analysis.

An organic compound contains one or more carbon–hydrogen bonds, including functional groups that possess characteristic chemical properties. Chemical groups can undergo important reactions, including oxidation and reduction, and function as acids and bases. Many biomolecules contain acidic and basic groups that are important determinants of structure and function. Chemical reactions involve the formation and breakage of chemical bonds, and chemical equations describe such reactions. Synthesis, replacement, and dehydration are important reaction types. Reactivity is determined by the unequal sharing of electrons and chemical polarity. Polar bonds form between atoms of different electron affinity, and nucleophiles and electrophiles define reactive sites. Photochemistry, the interplay of organic molecules with light, includes fluorescence and photolysis. Microarray analysis is organic chemistry on a grand scale, and students should endeavor to learn the basics of chemistry before embarking on microarray studies.

SELECTED READING

Blackburn, G. M., and Gait, M. J., eds. *Nucleic Acids in Chemistry and Biology,* 2nd ed. Oxford University Press, Oxford, UK, 1996.

Bloomfield, V. A., Crothers, D. M., and Tinoco, I. *Nucleic Acids: Structure*. University Science Books, Herndon, VA, 2000.

Brook, M. A. *Silicon in Organic, Organometallic, and Polymer Chemistry*. Wiley, New York, 2000.

Fersht, A. *Structure and Mechanism in Protein Science: A Guide to Enzyme Catalysis and Protein Folding*. Freeman, New York, 1999.

Hanson, J. R., and Abel, E. *Functional Group Chemistry*. Wiley, New York, 2002.

Hecht, S. M. *Bioorganic Chemistry: Nucleic Acids*. Oxford University Press, Oxford, UK, 1996.

Hein, M., Best, L. R., Pattison, S., and Arena, S. *Introduction to General, Organic, and Biochemistry,* 6th ed. Brooks/Cole, Pacific Grove, CA, 1997.

Lakowicz, J. R. *Principles of Fluorescence Spectroscopy,* 2nd ed. Kluwer Academic/Plenum, New York, 1999.

Lehninger, A. L., Nelson, D. C., and Cox, M. M. *Principles of Biochemistry,* 3rd ed. Freeman, New York, 2000.

Matthews, H. R., Freedland, R. A., and Miesfeld, R. L. *Biochemistry: A Short Course*. Wiley, New York, 1997.

Rost, F. W. D. *Fluorescence Microscopy*. Cambridge University Press, Cambridge, UK, 1995.

Schena, M., ed. *DNA Microarrays: A Practical Approach,* 2nd ed. Oxford University Press, Oxford, UK, 2000.

Schena, M., ed. *Microarray Biochip Technology*. Eaton, Natick, MA, 2000.

Singer, M., and Berg, P. *Genes and Genomes*. University Science Books, Herdon, VA, 1991.

Smith, MB., and March, J. *March's Advanced Organic Chemistry: Reactions, Mechanisms, and Structure,* 5th ed. Wiley, New York, 2000.

Streitweiser, A., and Heathcock, C. H. *Introduction to Organic Chemistry*. 2nd ed. Macmillan, New York, 1981.

Stryer, L. *Biochemistry,* 4th ed. Freeman, New York, 1995.

Weeks, D. P. *Pushing Electrons: A Guide for Students of Organic Chemistry,* 3rd ed. International Thomson, Cambridge, MA, 1998.

REVIEW QUESTIONS

1. The 118 chemical elements, the building blocks of all known substances, are configured in a table known as what?
2. The single-letter abbreviations S, P, O, and N correspond to which four elements.
3. Rank the following in order of smallest diameter to largest diameter: atom, golf ball, electron, nucleus, baseball stadium, and proton.
4. Identify the type and number of atoms in each molecule: HCl, CH_3OH, HCN, C_6H_6.
5. A p orbital accommodates a total of how many electrons? A pair of single-headed arrows pointing in opposite directions symbolizes what?
6. One gram of sodium chloride with a molecular weight of 58.4 g/mole contains how many molecules?
7. A 2% solution of DNA contains how many grams of DNA in 100 mL?
8. Rank the following chemical bond types in order of weakest to strongest: electrostatic, hydrogen, covalent, van der Waals.
9. The two strands of a DNA double helix are held together primarily by what type of chemical interactions?
10. Which of the following functional groups contain oxygen: ether, ester, primary amine, thiol, aldehyde?
11. Identify the atom in each pair of bonded molecules that contains the greater electron density: C-O, O-H, P-O, C-H, C-S.
12. The molecules on the left and right of an arrow in a chemical equation are known as what?
13. Water (H_2O) is a polar or nonpolar solvent?
14. A researcher prints protein onto two different glass microarray substrates—one containing reactive aldehyde groups and one, reactive epoxide groups—and finds that the protein binds more efficiently to the epoxide surface. Is the protein an electrophile or nucleophile in these two coupling reactions? What chemical property of the epoxide group accounts for its greater reactivity with protein?
15. Three large organic molecules are vastly different in chemical structure, but all three molecules appear bright purple to the naked eye. What common chemical feature in the three molecules probably accounts for their colored appearance?

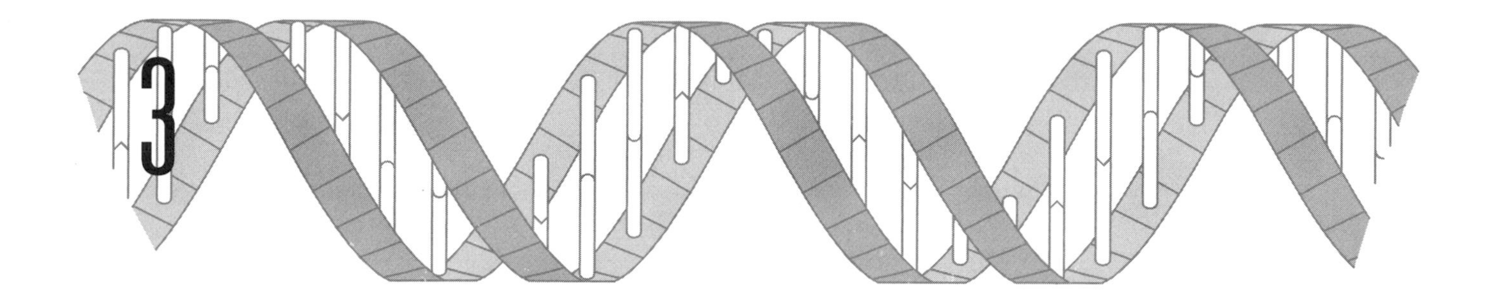

3 Basics of Biochemistry

There was never a genius without a tincture of madness. —*Aristotle*

Biochemistry is, intellectually speaking, a truly luxurious discipline. The field is populated with marvelously ingenious (and somewhat mad) practitioners, a quality that provides a continuous stream of discoveries and nary a dull moment. Biochemistry also examines biological problems at the "correct level of resolution," avoiding the narrowness of chemistry and the broadness of some of the other biological sciences. As a practical matter, the extraordinary academic popularity of biochemistry, and the enormous commercial value of its discoveries, render the field hectic and pressure filled. This chapter provides an introduction to biochemistry, with brief discussions of history of the field and an overview of some biochemical building blocks (DNA, RNA, proteins, and carbohydrates).

HISTORY AND DEFINITIONS

Biochemistry is the field of study that endeavors to understand the chemical basis of life, focusing on DNA, RNA, proteins, and other biomolecules. It is perhaps of little surprise that microarray analysis was developed in a biochemistry department (at Stanford University). Biochemists have a keen appreciation for genes and genomes, and they use experimental approaches that draw heavily from mechanistic studies and technology. Training in biochemistry provides a blend of chemistry and biology that is well suited for major technical

Biochemistry. *The field of study that endeavors to understand the chemical basis of life by focusing on the study of DNA, RNA, proteins, and other biomolecules.*

innovation, and a brief summary of the history of the biochemistry brings some of these innovations and discoveries into view.

Biochemistry as a scientific discipline is approximately 200 years old, though traditional peoples have been practicing biochemistry in a nonsystematic manner for thousands of years. Some of the key advances in biochemistry in the nineteenth century included the first demonstration of enzymatic catalysis by Berzelius, early evolutionary work by Darwin, Mendel's research on pea genetics, and fermentation studies by Pasteur. Substantial progress was made in the first half of the twentieth century, including Fisher's work on sugar and purine synthesis; Buchner's discovery of cell-free fermentation; Hunt's work on heredity; and basic advances in enzyme purification and crystallization by Sumner, Northrop, and Stanley.

Between 1950 and 1975, the premodern era of biochemistry expanded rapidly with an impressive list of discoveries and advances, including the mechanism of DNA and RNA synthesis by Ochoa and Kornberg, the double-helical structure of DNA by Watson and Crick, genetic control circuits by Jacob and Monod, and the discovery of reverse transcriptase by Baltimore and Temin. Since the mid-1970s, the modern era of biochemistry has seen exponential growth with respect to scientific progress and the number of researchers in the field. Approximately 100 times as many biochemists now compete for essentially the same number of academic awards as 50 years ago, a dynamic that has rendered modern biochemistry one of science's most competitive and exciting fields. Key modern era discoveries in biochemistry include restriction enzymes by Nathans and Smith, cancer genes by Bishop and Varmus, recombinant DNA by Berg, DNA sequencing by Gilbert and Sanger, site-directed mutagenesis by Smith, and the polymerase chain reaction by Mullis. Advances in modern biochemistry, coupled with key technological breakthroughs, paved the way for the development of microarray analysis.

Biopolymer. *Biological molecules, including DNA, RNA, protein, and carbohydrate, characterized by many small chemical building blocks linked together into a highly repeated structure.*

Monomer. *A nucleotide, amino acid, sugar, or other biochemical building block that makes up a biopolymer.*

Nucleotide. *A biochemical building block that makes up cellular DNA and RNA, consisting of a base, sugar, and phosphate group.*

Base. *A functional group or molecule that acquires a proton in a chemical reaction. Alternatively, any one of five nitrogenous molecules (adenine, cytosine, guanine, thymine, and uracil) contained in the nucleotides that make up cellular DNA and RNA.*

Sugar. *Any of a family of biomolecules, including deoxyribose and ribose in DNA and RNA, that contains carbon, hydrogen, and oxygen in a ratio of approximately 1:2:1.*

Phosphate (PO_4^{3-}). *The negatively charged component of DNA and RNA.*

BIOCHEMICAL BUILDING BLOCKS

Many of the important biological molecules in cells including DNA, RNA, protein, and carbohydrate occur as elongated structures known as ***biopolymers,*** which are analogous to a "beads on a string" in that they are composed of a small molecules or ***monomers*** linked together to form the repeated structures. Monomers that connect together into biopolymers are sometimes known as biochemical "building blocks", because these small biomolecules are used to "build" DNA, RNA, protein, carbohydrate, and other biopolymers using cellular enzymes to form monomer–monomer covalent bonds. This section describes some of the important types of biochemical building blocks and explains briefly how these compounds are joined together to form biopolymers.

Nucleotides

The biochemical building blocks that make up cellular DNA and RNA are known as ***nucleotides***. Because DNA and RNA harbor the genetic information of essentially every known life form, nucleotides are arguably the most important biochemical building blocks in the cell. Nucleotides contain three different biochemical components: a ***base, sugar,*** and ***phosphate***. The base, sugar, and

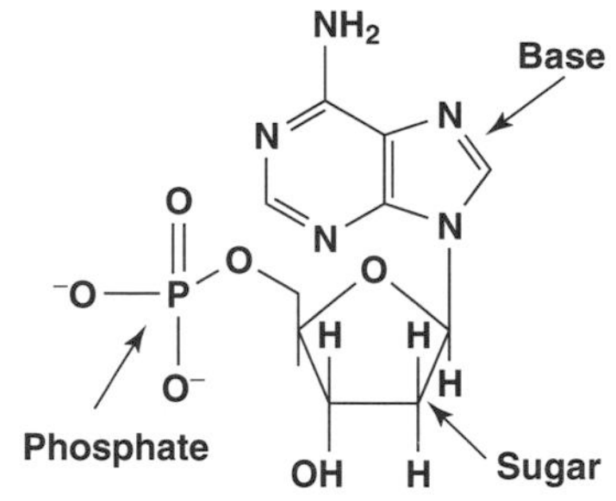

FIGURE 3.1. The chemical structure of the nucleotide deoxyadenosine 5′-monophosphate, including the base (adenine), sugar, and phosphate groups.

phosphate are connected by covalent bonds to form a nucleotide as shown in Figure 3.1.

DNA nucleotides contain one of four different bases on each nucleotide (Fig. 3.2), and the four bases are known as ***adenine*** (A), ***guanine*** (G), ***cytosine*** (C), and ***thymine*** (T). RNA nucleotides also contain one of four bases which are adenine, guanine, and cytosine, and *uracil* (U). The DNA and RNA bases are identical except that DNA contains A, G, C, and T, and RNA contains A, G, C, and U. The T and U bases are identical chemically, except that T contains a methyl group and U contains a hydrogen atom at one of the ring positions. The five bases are divided into two chemical groups known as ***purines*** (A and G) and ***pyrimidines*** (C, T, and U). The purines (A and G) contain two chemical rings, and the pyrimidines contain a single ring structure. During the formation of a DNA double helix, base pairs are formed by interactions between purines and pyrimidines in a base-specific manner (see below).

Adenine, guanine, cytosine, thymine, and uracil are known as *bases* because each of these molecules exhibits basic (as opposed to acidic) properties in solution. A, G, C, T, and U contain multiple nitrogen (N) atoms in the purine and pyrimidine rings and bear the name ***nitrogenous base*** because nitrogen is present in these heterocycles. Each nitrogen atom contains an electron lone pair that binds a proton in solution and elevates the pH above 7.0 (neutral) when the bases are dissolved in water. The A and G bases contain five nitrogen atoms, C contains three nitrogen atoms, and T and U contain two nitrogen atoms (Fig. 3.2). The dissociation constants (pKa values) for the DNA and RNA bases are in the range of 9–11.

The numbering system for the bases begins with one and proceeds in a sequential manner until each ring atom has been numbered with a successive integer (Fig. 3.3). Pyrimidines contain six ring atoms, and purines contain nine ring atoms, with base positions 1 and 9 corresponding to the respective attachment

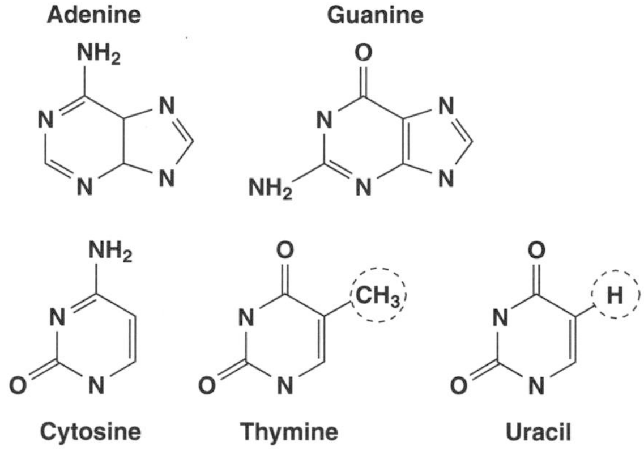

Adenine. *One of the four nitrogen-containing bases of DNA and RNA.*

Guanine. *One of the four nitrogen-containing bases of DNA and RNA.*

Cytosine. *One of the four nitrogen-containing bases of DNA and RNA.*

Thymine. *One of the four nitrogen-containing bases of DNA and RNA.*

Purine. *A two-ringed chemical structure that includes adenine and guanine.*

Pyrimidine. *A single-ringed chemical structure that includes cytosine, thymine, and uracil.*

Nitrogenous base. *A heterocyclic chemical structure that contains multiple nitrogen atoms, and includes the five bases (adenine, guanine, cytosine, thymine, and uracil) that make up DNA and RNA.*

FIGURE 3.2. Chemical structures for the nucleotide bases adenine (A), guanine (G), cytosine (C), thymine (T), and uracil (U). Thymine and uracil are identical except at position 5 (circled), which contains a methyl group in T and a hydrogen atom in U.

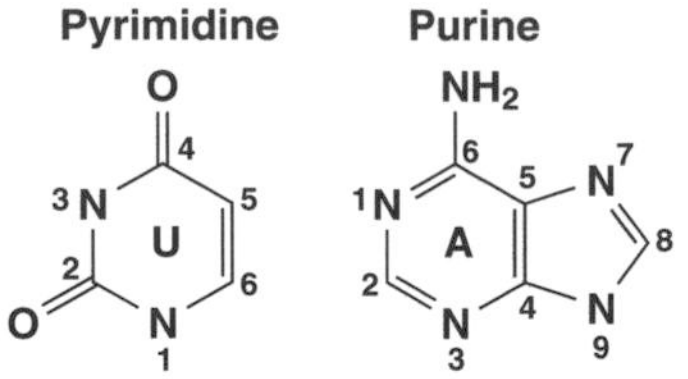

FIGURE 3.3. *Pyrimidine (uracil) and purine (adenine) rings are numbered with successive positive integers assigned to each ring atom. Pyrimidine and purine contain five and nine ring atoms, respectively, and the attachment position of the sugar moiety is 1 for pyrimidine and 9 for purine. The same numbering scheme applies to the other bases (G, C, and T).*

positions for the sugar moieties (see below). The base numbering system is useful because it allows unambiguous naming of modified nucleotides, such those that contain fluorescent organic molecules. Fluorescent nucleotides are prepared by attaching fluorescent tags to specific carbon atoms on the DNA bases.

Pentose. *Any five-carbon sugar including deoxyribose and ribose.*

Nucleoside. *A biomolecule consisting of a base and a sugar.*

In addition to a nitrogenous base, each nucleotide contains a five-carbon sugar moiety known as a ***pentose*** molecule, which derives its name from the presence of five (penta) carbon atoms in the ring structure (Fig. 3.4). The sugar is attached to the base via a covalent bond between a carbon atom on the sugar and ring atom number 1 (pyrimidines) or 9 (purines). A molecule containing a base and a sugar is known as a ***nucleoside,*** which is distinct from a nucleotide, which contains a base, a sugar, and a phosphate.

Deoxyribose. *The five-carbon DNA sugar that contains an hydrogen atom at the 2′ position instead of the hydroxyl group, which is found in the ribose RNA sugar.*

Ribose. *The five-carbon RNA sugar that contains a hydroxyl group at the 2′ position, instead of the hydrogen atom found in the deoxyribose DNA sugar.*

Sugar atoms are numbered in a clockwise manner from 1′ to 5′, and the 1′ carbon is the attachment position for the nitrogenous base (Fig. 3.5). The sugar ring atoms are designated with a "prime" to distinguish them from the ring atoms in the bases. The DNA and RNA sugars are identical chemically, except at the 2′ position, which contains a hydrogen atom (H) in DNA and a hydroxyl group (OH) in RNA. The DNA sugar is known as ***deoxyribose*** because it lacks an oxygen atom (deoxy) at the position that contains the OH group in the ***ribose*** RNA sugar. The chemical difference between deoxyribose and ribose is subtle, yet this subtle chemical difference has a profound effect on the chemical and enzymatic stability of DNA versus RNA (see below).

Nucleotide monophosphate. *A nucleotide in DNA and RNA that contains a single phosphate group.*

Nucleotide triphosphate. *A nucleotide in DNA and RNA that contains three phosphate groups.*

Nucleotides contain one, two, or three phosphate groups (PO_4^{2-}) on the 5′ carbon (Fig. 3.6), depending on the biochemical context in which the nucleotide is used. A nucleotide present in DNA and RNA chains contains a single phosphate group and is known as a ***nucleotide monophosphate,*** whereas a nucleotide used to synthesize DNA or RNA enzymatically contains three phosphates and is known as a ***nucleotide triphosphate***. Monophosphate and triphosphate

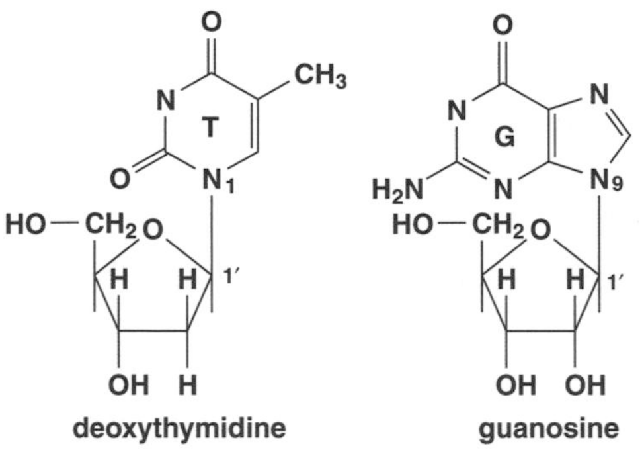

FIGURE 3.4. *The chemical structures of the nucleosides deoxythymidine and guanosine, which contain a deoxyribose sugar and a T base, and a ribose sugar and a G base, respectively. The sugar base covalent bond forms between the 1′ position of deoxyribose and 1 position of thymine* (left), *and between the 1′ position on ribose and position 9 of guanine* (right).

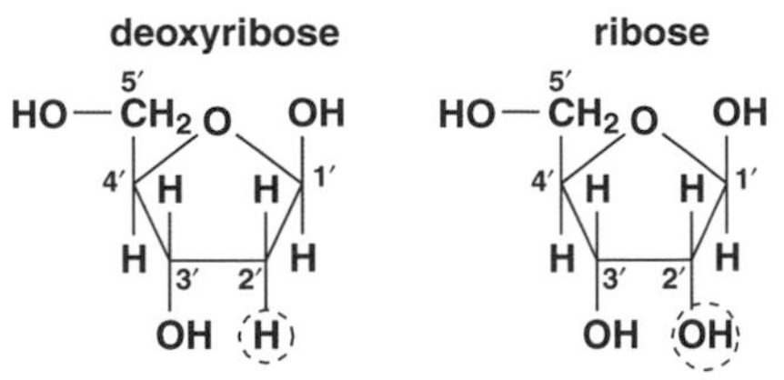

FIGURE 3.5. The molecular structures of deoxyribose (left) *and ribose* (right)*, the sugar molecules found in DNA and RNA, respectively. The ring atoms are numbered using successive integers, beginning with the atom that attaches to the base (1′) and proceeding in a clockwise manner until all five atoms have been numbered. The prime is used to prevent confusion between the sugar atoms and the atoms of the nitrogenous bases (see Fig. 3.3).*

nucleotides are often written in shorthand notation, such as GMP for guanosine monophosphate or dCTP for deoxycytidine triphosphate.

A complete DNA nucleotide contains a phosphate molecule, a deoxyribose molecule, and one of the four DNA bases (A, G, C, or T). A complete RNA nucleotide consists of a phosphate, a ribose, and one of the four RNA bases (A, G, C, or U). The nucleotide phosphate group loses its two protons in neutral solution, carries a net negative charge at neutral pH, and exhibits chemical acidity. The acidic nature of nucleotides, coupled with the fact that DNA and RNA are located in the nucleus, account for the term ***nucleic acid,*** which is used as a general descriptor for DNA and RNA.

Nucleic acid. *Any of a family of negatively charged polymeric biomolecules, including DNA and RNA, that contain a repeating series of nucleotide building blocks.*

Amino Acids

A biochemical building block used to make cellular proteins is known as an ***amino acid***. There are 20 different commonly-occurring amino acids, and each contains a common core structure as well as a ***side chain*** unique to it (Fig. 3.7). The core structure contains an amino group and a carboxylic acid group, forming the basis of the name amino acid. At neutral pH, the amino group acquires

Amino acid. *Any of a family of 20 biochemical building blocks that make up cellular proteins.*

Side chain. *A chemical group on an amino acid that that determines its chemical and biochemical properties.*

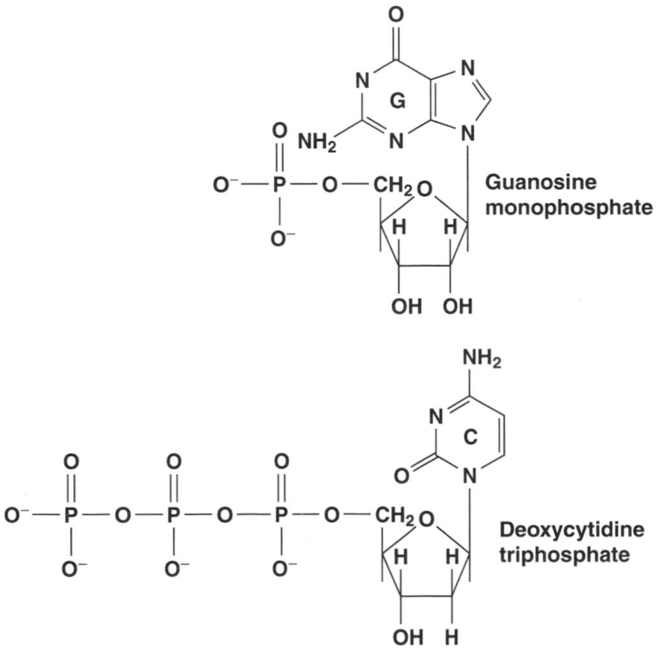

FIGURE 3.6. Molecular structures for guanosine monophosphate and deoxycytidine triphosphate. Nucleotide monophosphates make up cellular DNA and RNA, and nucleotide triphosphates are the energetically activated derivatives used to synthesize nucleic acid chains. Triphosphates are converted into monophosphates during enzymatic synthesis, through the release of a diphosphate molecule that drives the synthesis of DNA and RNA.

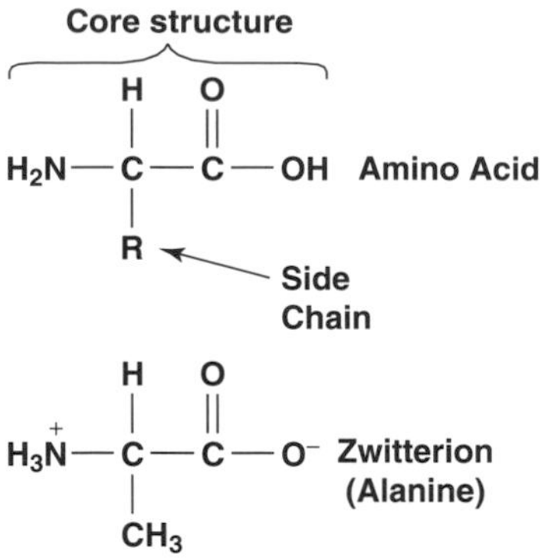

Core structure

$$H_2N-\underset{\underset{R}{|}}{\overset{\overset{H}{|}}{C}}-\overset{\overset{O}{\|}}{C}-OH \quad \text{Amino Acid}$$

R ← Side Chain

$$\overset{+}{H_3N}-\underset{\underset{CH_3}{|}}{\overset{\overset{H}{|}}{C}}-\overset{\overset{O}{\|}}{C}-O^- \quad \text{Zwitterion (Alanine)}$$

FIGURE 3.7. The structure of an amino acid showing the amino and carboxyl groups of the amino acid core (bracket) *and the position of the side chain* (R). *At physiologic (neutral) pH, the amino and carboxyl groups of amino acids acquire a positive and negative charge, respectively; molecules that carry a positive and negative charge simultaneously are known as zwitterions. The zwitterion for alanine (R = CH_3) is shown.*

Zwitterion. *Molecule that bears a both positive and a negative charge.*

a proton and becomes positively charged and the carboxylic acid loses a proton and becomes negatively charged. A molecule that bears a both positive and negative charge simultaneously is known as a ***zwitterion***.

The 20 amino acids are classified by several different schemes, all according to the chemical characteristics of the side chains. One classification scheme groups the amino acids into nonpolar, polar, and charged categories. The nonpolar category contains amino acids with relatively hydrophobic side chains, including alanine (ala), valine (val), leucine (leu), isoleucine (ile), methionine (met), phenylalanine (phe), proline (pro), and tryptophan (trp) (Fig. 3.8). The polar amino acids have relatively hydrophilic side chains and include glycine (gly), serine (ser), threonine (thr), cysteine (cys), asparagine (asn), glutamine (gln), and tyrosine (tyr) (Fig. 3.9). The charged amino acids are hydrophilic and have side chains that carry a positive or negative charge at neutral pH and include the negatively charged aspartate (asp) and glutamate (glu), and the positively charged lysine (lys), arginine (arg), and histidine (his) (Fig. 3.10). As a general rule, the hydrophobic side chains are located in the "oily" protein

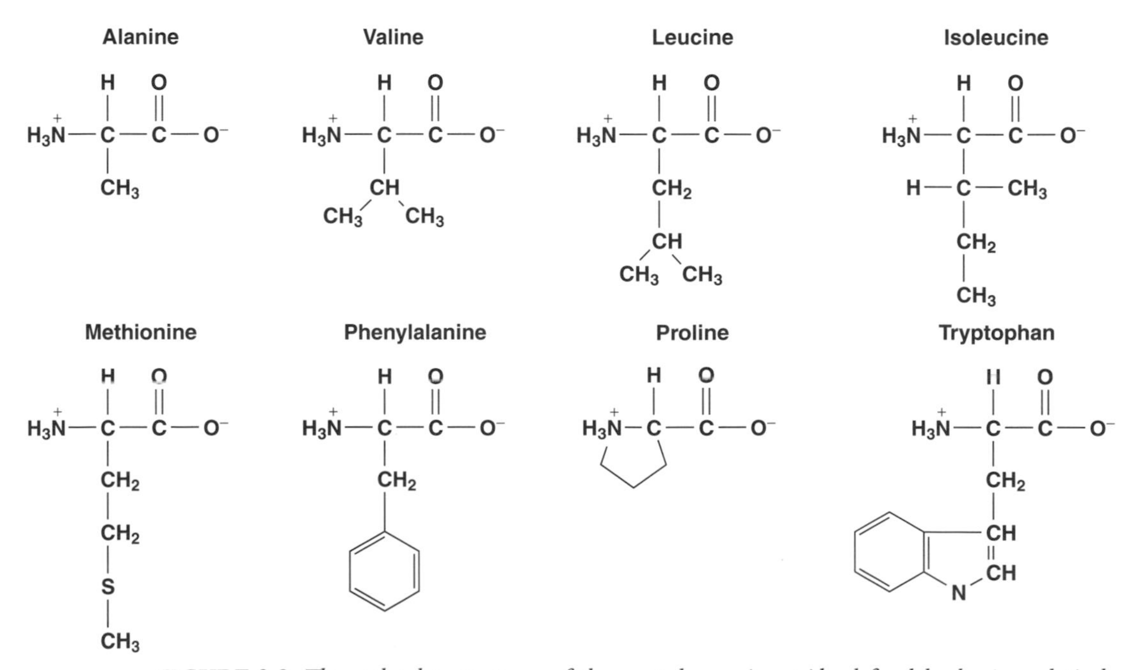

FIGURE 3.8. The molecular structures of the nonpolar amino acids, defined by having relatively hydrophobic side chains.

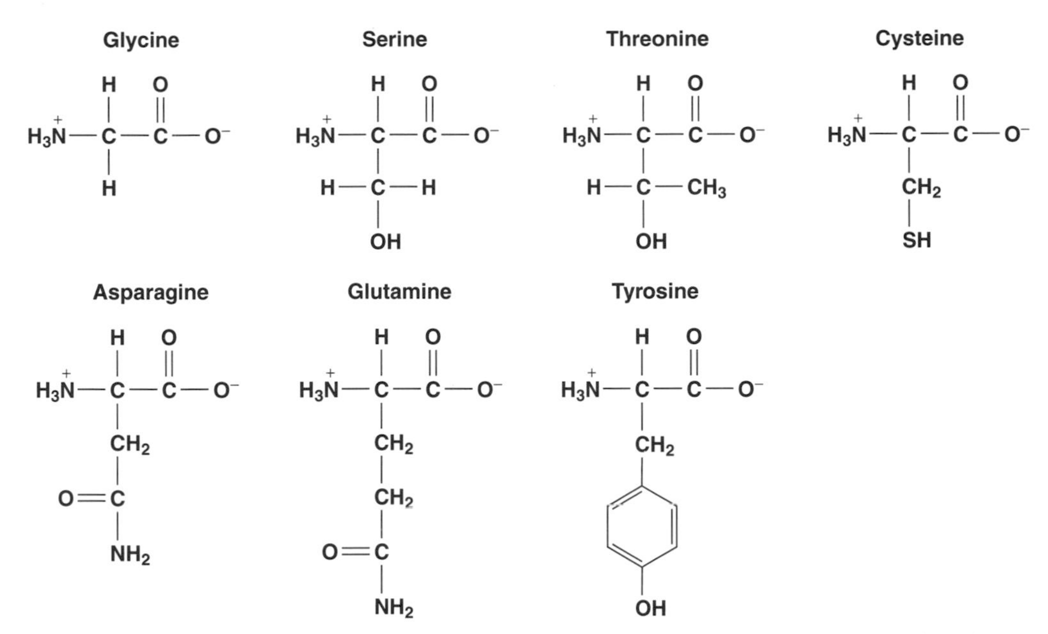

FIGURE 3.9. *The molecular structures of the polar amino acids, defined by having relatively hydrophilic side chains.*

interior, and the hydrophilic and charged amino acids cluster on the protein exterior, where they interact favorably with water molecules (Table 3.1).

The amino acid nomenclature includes a full chemical name, a three-letter abbreviation, and a one-letter abbreviation (Table 3.1). Serine, for example, is also known as *ser* and *S* and a similar nomenclature is used for each amino

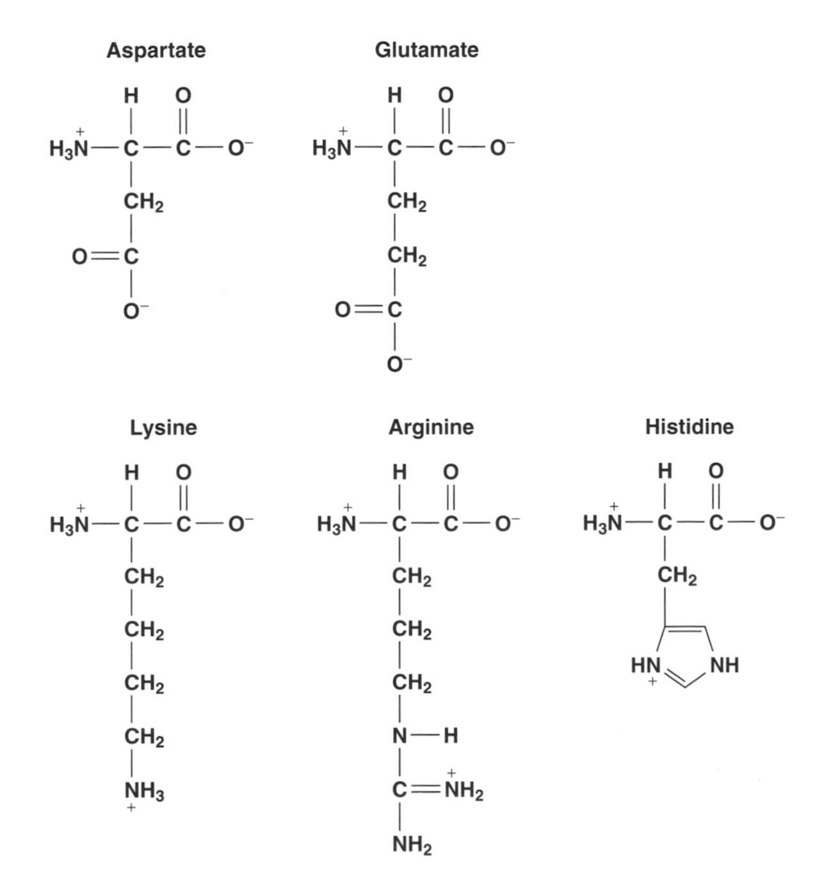

FIGURE 3.10. *The molecular structures of the charged amino acids, defined by having side chains that carry a negative* (top) *or positive* (bottom) *charge at neutral pH.*

TABLE 3.1. The Amino Acids

Amino Acid	Three-Letter Abbreviation	Single-Letter Abbreviation	Hydrophobicity[a] (kcal/mole)
Alanine	ala	A	0.100
Arginine	arg	R	1.910
Asparagine	asn	N	0.480
Aspartate	asp	D	0.780
Cysteine	cys	C	−1.420
Glutamine	gln	Q	0.950
Glutamate	glu	E	0.830
Glycine	gly	G	0.330
Histidine	his	H	−0.500
Isoleucine	ile	I	−1.130
Leucine	leu	L	−1.180
Lysine	lys	K	1.400
Methionine	met	M	−1.590
Phenylalanine	phe	F	−2.120
Proline	pro	P	0.730
Serine	ser	S	0.520
Threonine	thr	T	0.070
Tryptophan	trp	W	−0.510
Tyrosine	tyr	Y	−0.210
Valine	val	V	−1.270

[a] A measure of the free energy required to transfer the amino acid from a polar solvent (water) into a nonpolar solvent (octanol). Reprinted with permission from Guy et al. (1985).

acid. Because of the central functional role of proteins in the cell, microarray researchers should take some time to learn the names, abbreviations, and chemical properties of all 20 amino acids.

Monosaccharides

Monosaccharide. *A highly soluble chemical building block of carbohydrates containing carbon, hydrogen, and oxygen in a ratio of about 1:2:1.*

Disaccharide. *Any of a family of sugar molecules containing two monomer building blocks.*

A highly soluble biochemical compound containing carbon, hydrogen, and oxygen in a ratio of about 1:2:1 is known as a ***monosaccharide***. Monosaccharides, also known as sugars, make up a diverse family of molecules, including glucose, fructose, mannose, and galactose (Fig. 3.11). The *-ose* suffix is associated with nearly all monosaccharides, including those found in DNA (deoxyribose) and RNA (ribose). Monosaccharides are single molecules that can be joined together in cells to form successively larger molecules, such as the ***disaccharide***

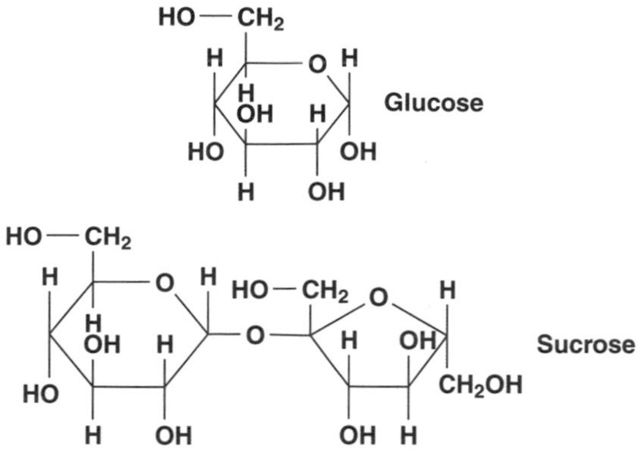

HO—CH2
H O H
H OH H
Glucose
HO OH
H OH

HO—CH2
H O H HO—CH2 O H
H OH H O H OH
Sucrose
HO CH2OH
H OH OH H

FIGURE 3.11. The molecular structures of the monosaccharide glucose and the discaccharide sucrose. Monosaccharides are joined together in cells to form cellulose, glycogen, starch, and other members of the carbohydrate family.

sucrose, which contains two monosaccharides, and larger molecules, which contain three (trisaccharide) and four (tetrasaccharide) monosaccharide molecules. Chains of monosaccharides containing about 10 or more sugar molecules are known as carbohydrates and, analogous to nucleic acids and proteins that are made by joining nucleotides and amino acids, respectively, monosaccharides are joined like beads on a string to form the carbohydrate family (see below).

DNA

The genetic blueprint of virtually every organism in the biosphere is stored in the biopolymeric molecule known as ***deoxyribonucleic acid*** (DNA). DNA is composed of a long string of nucleotides, each of which contains one of four bases (A, G, C, or T), a deoxyribose sugar, and a phosphate group. This section provides a brief overview of the structure of DNA, its biochemical properties, and the mechanism by which genetic information is stored in this essential biopolymer.

Deoxyribonucleic acid. *The biopolymeric molecule that constitutes the genetic blueprint of virtually every organism in the biosphere.*

Structure of DNA

Nucleotides are joined together in a covalent manner in the cell to build linear DNA sequences. Nucleotide bonds are formed by the enzymatic joining of nucleotide triphosphates, whereby the 3′ hydroxyl group of one nucleotide is attached covalently to the 5′ phosphate group of another nucleotide (Fig. 3.12). Two nucleotides joined in this manner form a molecule known as a ***dinucleotide,*** though nearly all DNA chains contain a rather large number of nucleotides joined in succession. Short synthetic DNA chains contain 10–100 nucleotides

Dinucleotide. *A molecule consisting of two nucleotides joined in succession.*

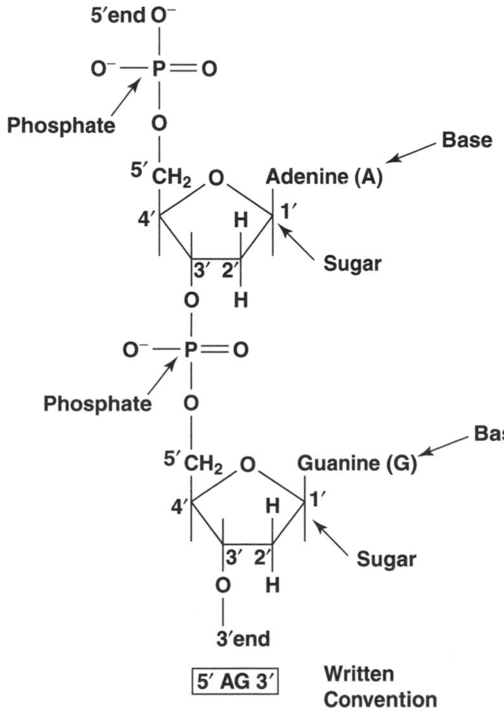

FIGURE 3.12. A dinucleotide DNA chain, including the phosphate, sugar, and base moieties. The 5′ end of the chain contains a terminal phosphate group attached to the 5′ carbon (circle) *of deoxyribose, and the 3′ end bears a 3′ hydroxyl group attached to the 3′ carbon* (circle) *of deoxyribose. In shorthand, DNA sequences are written from the 5′ to the 3′ end and from left to right, as shown.*

(see below), and a typical human gene and chromosome contain approximately 20,000 and 100,000,000 nucleotides, respectively (see Chapter 4).

Polarity. *A term that refers to the chemical nonsymmetry of DNA and RNA chains, specified as 5′ and 3′ ends.*

DNA chains exhibit chemical ***polarity,*** which describes the fact that a DNA chain has different chemical groups on each end of the molecule (Fig. 3.12). The "top" end of a DNA chain contains a terminal phosphate group located on the 5′ carbon atom of deoxyribose, and is known as the 5′ (five prime) end. The "bottom" end of a DNA chain contains a terminal hydroxyl group located on the 3′ carbon atom of deoxyribose, and this end is known as the 3′ (three prime) end of the chain. The nucleotide sequence of a DNA chain is written from the 5′ to the 3′ end, from left to right, using the single-letter abbreviations for the bases (A, G, C, T) as a shorthand notation. A DNA sequence containing A, G, T, and T in succession, for example, would be written 5′ AGTT 3′. DNA sequences in the scientific literature and in electronic databases are written left to right and top to bottom in a 5′ to 3′ manner, and researchers should become familiar with this convention, which is based on the underlying chemical polarity of DNA.

DNA synthesizer. *An automated machine used to manufacture synthetic oligonucleotides.*

Synthetic oligonucleotide. *A short chain of synthetic DNA manufactured using a DNA synthesizer.*

DNA chains can be synthesized chemically by joining phosphoramidite derivatives of the four nucleotides (see Chapter 6), using a machine known as a ***DNA synthesizer***. Single-stranded chains made with a DNA synthesizer are known as ***synthetic oligonucleotides***. Synthetic oligonucleotides are typically 10–100 nucleotides in length.

Double-stranded. *The form of naturally occurring DNA consisting of two complementary strands held together by hydrogen bonding.*

Antiparallel. *The configuration of the two strands in double-stranded DNA, wherein one strand runs in the 5′ to 3′ direction and the other strand runs in the 3′ to 5′ direction.*

Genes, chromosomes, and other naturally occurring DNA molecules are ***double-stranded*** molecules in which the two stands are held together by hydrogen bonds between the bases. Hydrogen-bonded DNA chains are configured in an ***antiparallel*** manner, such that one DNA strand runs in the 5′ to 3′ direction and the other strand runs in the 3′ to 5′ direction (Fig. 3.13). Antiparallel, double-stranded DNA is held together by specific interactions between A and T and between G and C bases (Fig. 3.14). Base pairing occurs exclusively between A and T and G and C, and not between other combinations of bases. Adenine,

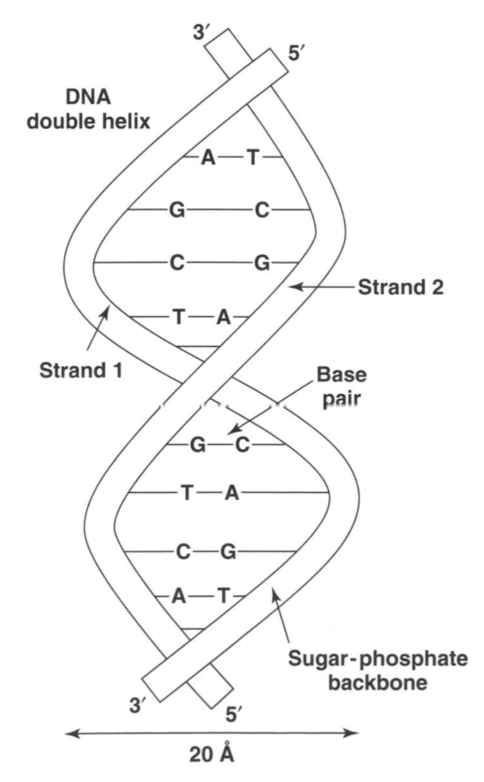

FIGURE 3.13. A DNA double helix shown with the molecule flattened and unwound slightly for ease of viewing. Hydrogen bonds (horizontal lines) *between complementary bases hold the two helices, containing the sugar and phosphate moieties* (ribbons), *together tightly. One strand runs from 5′ to 3′, and the complementary strand runs from 3′ to 5′. Each turn of the double helix corresponds to 10 base pairs and the width of the double-stranded molecule is 20 Å* (double arrow).

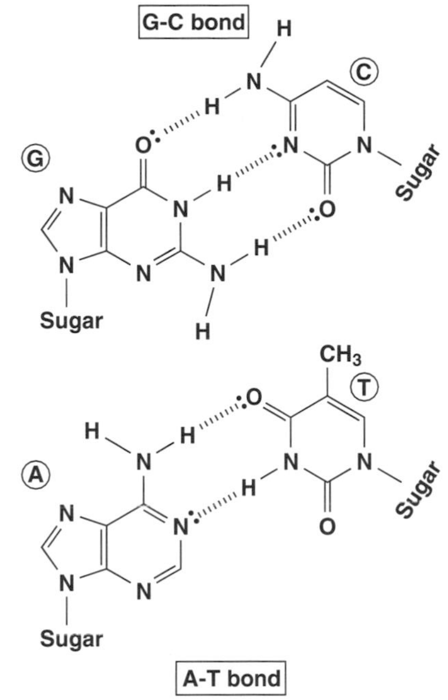

FIGURE 3.14. Complementary DNA bases form hydrogen bonds with each other (dashed lines), *such that G pairs exclusively with C, and A pairs exclusively with T.*

for example, does not form a base pair with guanine or cytosine. The molecular basis of A-T and G-C specificity should be clear from the geometry of bases. Hydrogen bonds are directional and angle dependent, and other combinations of bases simply do not support robust hydrogen bonding. The G-C base pairs are chemically more avid than the A-T base pairs, because three hydrogen bonds are formed for G-C pairings and only two for A-T pairings (Fig. 3.14).

DNA chains that bond to each other through A-T and G-C interactions are known as ***complementary strands***. The chemical process by which complementary strands "zipper up" into a double-stranded molecule is known as hybridization. Complementary strands of DNA form a spiral molecule or double helix, whereby the two interwoven chains coil around a center axis like a spiral staircase. By this analogy, the complementary bases extend toward the center of the staircase forming the "steps," and the sugar–phosphate backbone provides the staircase supports (Fig. 3.13). The width of a double helix is approximately 20 Å, and the helical molecule makes a complete revolution every 10 nucleotides. A DNA double helix can also be viewed as a "railroad track," in which the sugar–phosphate backbone forms the rails and the base pairs serve as the railroad ties. In the railroad model of DNA, one complementary strand runs north, the other strand runs south, and polymerase serves as the engine that travels along the rails and performs DNA and RNA synthesis (see Chapter 4).

Complementary strand. *A DNA strand that has the complementary nucleotide sequence and opposite chemical polarity of a second DNA strand.*

A double helix with perfect A-T and G-C base pairs is a highly stable molecule that requires considerable energy (e.g., heat) to "melt" apart. By contrast, a double helix that contains one or more ***mismatches*** in the base pairing scheme, zippers up less efficiently than a DNA molecule in which the two strands are fully complementary. Mismatches weaken a double helix, and the hydrogen bonding disparity between a perfectly matched double helix and one that contains mismatches provides the specificity required for hybridization-based genotyping assays (see Chapter 13). The strength of a double helix increases as the number of mismatches decreases and vice versa, and researchers should understand the relationship between base pairing and the avidity of the DNA chains for one another.

Mismatch. *A location in a DNA double helix wherein the bases on opposite strands do not share complementary sequences and therefore do not form productive adenine–thymine or guanine–cytosine base pairs.*

The number of different DNA sequences that can be built with four nucleotide building blocks is 4^n, where n equals the number of nucleotides in the DNA chain:

$$\text{Number of DNA sequences} = 4^n \tag{3.1}$$

A total of 16 (4^2) different dinucleotide DNA sequences can be built with the four bases, and greater than one trillion (1×10^{12}) different sequences are possible if the DNA chain contains 20 nucleotides (4^{20}). An enormously powerful aspect of the DNA structure is that cells can construct a virtually infinite number of different DNA sequences and genes using a simple set of four nucleotides, a biochemical fact that enables enormous diversity in the biosphere.

Properties of DNA

DNA is a highly soluble, negatively charged, stable molecule that is easy to isolate, purify, and manipulate in the laboratory. The aqueous solubility of DNA derives from the hydrophilic phosphate and sugar moieties present in the DNA backbone. DNA occurs at a very high concentration in the nucleus of the cell; and high solubility is required to achieve efficient enzymatic reactivity required for DNA replication, transcription, and other processes (see Chapter 4).

The phosphate groups, in addition to conferring solubility, account for the enormous negative charge of DNA. At neutral pH, the phosphate groups lose their proteins, causing the oxygen atoms to harbor negative charge. The abundance of negative charge in the DNA backbone would cause prohibitive strand repulsion during base pairing if it were not for the presence of cations (e.g. Na^+), which bind to and neutralize the phosphate groups. Cations prevent electrostatic repulsion of the DNA strands and enable hybridization, which explains why cells and microarray hybridization buffers contain high cation concentrations. The negative charge on DNA conferred by the phosphates, is often exploited in laboratory assays that use positively charged surfaces for DNA binding. One popular type of microarray substrate contains positively charged amine groups ($R\text{-}NH_3^+$), which bind to phosphates on the DNA backbone and enable stable attachment of DNA to the substrate (see Chapter 5).

The absence of a hydroxyl group (OH) on the 2′ carbon of deoxyribose (Fig. 3.5) accounts for the greater chemical and enzymatic stability of DNA relative to RNA. Deoxyribose lacks a 2′ hydroxyl group and is, therefore, resistant to cleavage by nucleophilic attack of the 5′ carbon atom with the 2′ hydroxyl. The absence of a 2′ hydroxyl renders DNA highly resistant to alkali (e.g., NaOH) and enzymatic digestion (e.g., ribonuclease), two treatments that degrade RNA readily owing the presence of a 2′ hydroxyl on ribose (see below). Researchers often exploit the alkali and ribonuclease resistance of DNA as a means of degrading RNA selectively in samples that contain a mixture of RNA and DNA (see Chapter 6).

Codon. *Any one of 64 three-nucleotide sequences or triplets in messenger RNA that specify one of the 20 amino acids used for protein synthesis.*

Triplet. *Any one of 64 three-nucleotide sequences or triplets in messenger RNA that specify one of the 20 amino acids during protein synthesis.*

Genetic Code

How does a linear sequence of nucleotides encode functional cellular proteins? Genetic information is stored in a fundamental unit of genetic information known as a ***codon***. A codon, or ***triplet,*** contains three successive nucleotides that are read by the cellular machinery in a 5′ to 3′ manner to specify an amino acid. There are 64 possible combinations of the four nucleotides (4^3), and all

TABLE 3.2. The Genetic Code

First Position (5′)	Second Position (Middle) A	G	C	T	Third Position (3′)
A	lys	arg	thr	lle	A
	lys	arg	thr	met	G
	asn	ser	thr	lle	C
	asn	ser	thr	lle	T
G	glu	gly	ala	val	A
	glu	gly	ala	val	G
	asp	gly	ala	val	C
	asp	gly	ala	val	T
C	gln	arg	pro	leu	A
	gln	arg	pro	leu	G
	his	arg	pro	leu	C
	his	arg	pro	leu	T
T	stop	stop	ser	leu	A
	stop	trp	ser	leu	G
	tyr	cys	ser	phe	C
	try	cys	ser	phe	T

64 codons are used in the cell. The cellular "conversion table" between codon and amino acid is known as the ***genetic code*** (Table 3.2). Of the possible combinations, 61 codons specify the 20 amino acids; and the remaining 3 combinations are stop codons, the genetic signals in the code that signal protein termination (see below). Except for methionine (ATG) and tryptophan (TGG), each amino acid is designated by multiple codons; thus the genetic code is termed ***degenerate***. For example, 4 codons specify leucine (CTT, CTC, CTA and CTG); 2, glutamate (GAA and GAG); and 2, lysine (AAA and AAG).

Genetic code. *The cellular alphabet that specifies one of the 20 amino acids or a stop codon from the 64 triplets in messenger RNA.*

Degenerate. *Term used to describe the fact that many of the 20 amino acids are encoded by more than one codon.*

In many cases, a set of related codons are identical at the first two positions but differ at the third position, as in the case of four codons that specify valine (GT*T*, GT*C*, GT*A*, GTG). The degeneracy of the genetic code allows different nucleic acid sequences to encode identical proteins, which has a number of important evolutionary implications, including the fact that gene sequences can accumulate changes or ***diverge*** over time, without altering protein sequence or function (see Chapter 4).

Diverge. *To accumulate changes in gene and protein sequences over evolutionary time.*

RNA

The DNA encodes the genetic blueprint of an organism, and the blueprint is converted into protein information using ***ribonucleic acid*** (RNA) as a molecular intermediary. Certain classes of RNA also play a role in protein synthesis. RNA is similar to DNA, but possesses some novel structural, physical, and functional properties. This section provides a brief overview of RNA, including RNA structure, biochemical properties, and the different types of RNA found in cells.

Ribonucleic acid. *Any of a broad family of single stranded, negatively charged nucleic acids composed of ribonucleotides.*

Structure of RNA

RNA is formed by the linear assembly of ribonucleotides, in the same manner as deoxyribonucleotides are used to make DNA. Ribonucleotides are identical to deoxyribonucleotides, except that the RNA building blocks contain ***uracil***

Uracil. *The nitrogen-containing base found in RNA but not in DNA.*

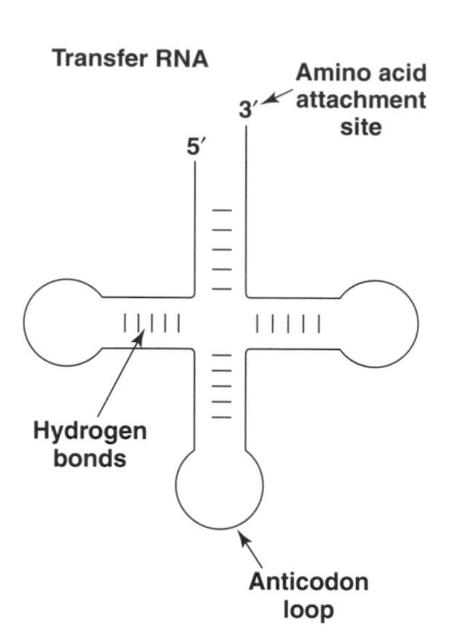

FIGURE 3.15. The structure of tRNA, which contains a single RNA chain of approximately 75 ribonucleotides, folded so that the 5′ and 3′ ends are adjacent spatially (top) *in the functional molecule. Folding occurs via intramolecular hydrogen bonds* (dashed lines) *between bases in a single strand, enabling the formation of three loops, one of which (the anticodon loop) pairs with codons during protein synthesis.*

Secondary structure. *Two- and three-dimensional configurations of proteins and nucleic acids that originate from intramolecular interactions of primary linear sequences.*

Intramolecular. *Biochemical interactions that occur between functional groups within a single molecule of DNA, RNA, or protein.*

Anticodon loop. *The portion of a transfer RNA molecule that interacts with cognate codons in the messenger RNA.*

(U) instead of thymine (T) and ribose instead of deoxyribose (see above). RNA chains are generally much shorter than DNA chains; most RNA molecules contain 70–10,000 ribonucleotides. Unlike DNA, which forms a double-stranded double helix, nearly all RNA molecules are single stranded. RNA molecules have the capacity, however, to form ***secondary structure,*** which allows RNA to fold into intricate structures through the formation of hydrogen bonds between bases in the same strand. So-called ***intramolecular*** hydrogen bonds between bases of transfer RNA (tRNA) molecules (see below) allow RNA to form double-helical regions even though such regions contain a single RNA chain (Fig. 3.15). Transfer RNA molecules also contain three "loop" structures, one of which is known as the ***anticodon loop***. The anticodon loop interacts with codons through base pairing and directs the formation of proteins during the cellular translation process (see Chapter 4). A second class of RNA, known as ribosomal RNA (rRNA; see below), also forms secondary structure.

Properties of RNA

One of the extraordinary aspects of biochemistry is the extent to which subtle biochemical differences between molecules can exert a profound functional effect. One nucleotide change in the 3 billion-base human genome can produce a fatal disease. Analogously, RNA is nearly identical chemically to DNA but is dramatically less stable, owing to the presence of a 2′ hydroxyl group on ribose (Fig. 3.5). The 2′ hydroxyl group can function as a nucleophile in chemical and enzymatic reactions, attacking the 5′ carbon atom and breaking the RNA strand (Fig. 3.16). At high pH (e.g., sodium hydroxide) or upon treatment with enzymes (e.g., ribonuclease), the proton is removed from the OH group, allowing oxygen to function as a nucleophile and catalyze strand breakage. DNA lacks a 2′ hydroxyl group and is, therefore, resistant to high pH and ribonuclease treatment.

The enzymatic instability of RNA serves a critical cellular function, allowing the rapid turnover of RNA molecules and dynamic changes in gene expression. Cells must respond rapidly to cellular signals such as hormones, nutrients, and environmental stimuli; and RNA instability allows cells to adjust their RNA levels quickly in response to physiologic changes. If RNA were as stable as DNA, organisms would be preprogrammed chemically and as such would not have

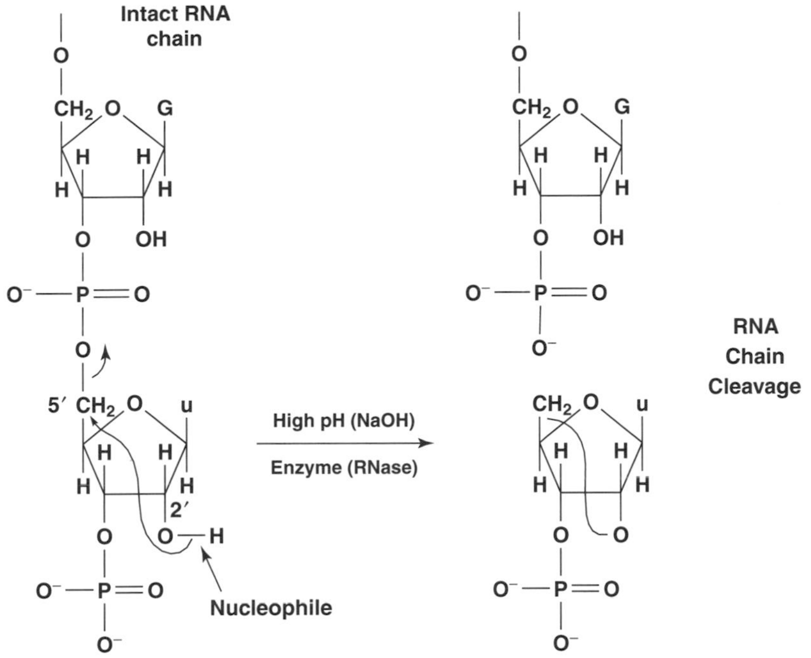

FIGURE 3.16. The ribose moiety of RNA (left) *contains a 2′ hydroxyl group (OH), which can function as a nucleophile, attacking the 5′ carbon atom and causing breakage of the RNA strand* (right). *Nucleophilic attack requires the removal of a proton (H) from the OH group, a reaction catalyzed by high pH or ribonuclease (RNase).*

the capacity to fine-tune their physiology to meet the needs of a dynamic environment.

The relative instability of RNA has some practical consequences, one of which is that great care must be taken when isolating RNA from biological samples. Cells, tissues, and human secretions (e.g., hand oils and perspiration) contain large quantities of ribonuclease, which will degrade RNA samples if care is not taken during RNA purification. Researchers should use protocols that inactivate the endogenous ribonucleases and should wear gloves at all times when handling RNA preparations. RNA is used extensively in gene expression experiments (see Chapters 6 and 12), and microarray researchers are served well by understanding the chemical basis of RNA instability and how RNA degradation can be avoided.

Except for the chemical and enzymatic instability differences, many of the other properties of RNA and DNA are quite similar. RNA is a highly soluble, negatively charged molecule that is purified readily by a number of different means, including alcohol precipitation, filter binding, chromatography, and centrifugation.

Types of RNA

The genome encodes the three main types of cellular RNA known as messenger RNA (mRNA), ***ribosomal RNA*** (rRNA) and ***transfer RNA*** (tRNA). Each type is single stranded, though tRNA and rRNA form double-stranded regions via intramolecular hydrogen bonding, as described above. Messenger RNA is an informational intermediate between gene (DNA) and protein, carrying the genetic information from the nucleus (location of the DNA) to the cytoplasm (location of protein synthesis). Genetic information flows from DNA into RNA and from RNA into protein in a molecular phenomenon known as the ***fundamental dogma of molecular biology*** (Fig. 3.17). Each messenger RNA sequence corresponds to a specific gene contained in the DNA. There are approximately 35,000 genes

Ribosomal RNA. *Specialized class of cellular RNA, located in ribosomes, that plays structural and catalytic roles during protein synthesis.*

Transfer RNA. *Specialized class of cellular RNA that binds specific amino acids and facilitates protein synthesis by mediating codon recognition.*

Fundamental dogma of molecular biology. *Doctrine that specifies that genetic information flows from DNA into RNA into protein.*

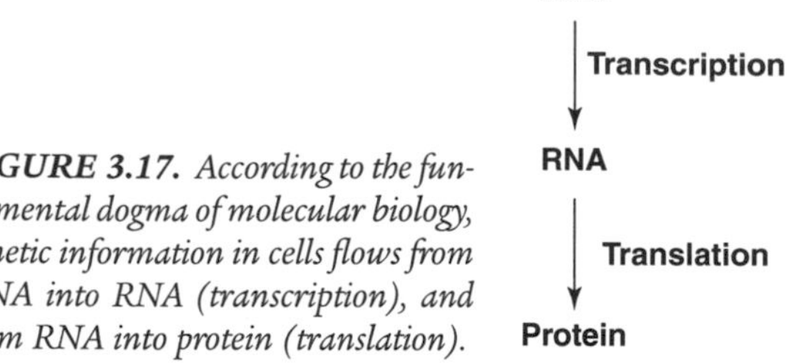

FIGURE 3.17. According to the fundamental dogma of molecular biology, genetic information in cells flows from DNA into RNA (transcription), and from RNA into protein (translation).

in the human genome and correspondingly 35,000 different mRNA sequences, each containing a specific string of condons specified by the DNA. Messenger RNA is 1,000–10,000 ribonucleotides in length and is read by tRNA to make protein (see Chapter 4). Cellular mRNA contains the genetic instructions of the cell and is used as the source of RNA for gene expression microarray experiments (see Chapter 12). Cellular mRNA concentrations are relatively low, accounting for 1–5% of the total RNA found in cells; tRNA and rRNA contribute the remaining 95–99%.

The two other classes of RNA (tRNA and rRNA) do not carry the genetic instructions of the cell but rather play a fundamental role in the cellular process of protein synthesis (see below). Transfer RNA molecules bind to specific amino acids and mRNA codons, enabling protein synthesis from mRNA templates (see Chapter 4). Mammalian cells contain several dozen different tRNA molecules, each of which contains about 75 ribonucleotides (Fig. 3.15). The set of cellular tRNA molecules accounts for 10–15% of the total cellular RNA.

Ribosome. *Large cytoplasmic structure that facilitates protein synthesis.*

Ribosomal RNA, the third major class of cellular RNA, constitutes an important component of the protein synthesis molecule known as the ***ribosome*** (see Chapter 4). Molecules of rRNA coordinate protein synthesis by providing part of the molecular scaffold that binds mRNA and tRNA to the ribosome. Mammalian cells contain four major rRNA molecules, each of which is 100–5,000 ribonucleotides in length. Ribosomal RNA molecules are extremely abundant in cells, accounting for 75–85% of the total cellular RNA.

PROTEINS

The cellular genome (DNA) makes up the genetic blueprint, but proteins actually carry out the functional instructions encoded by the genes. This section provides a brief overview of protein structure and function, describing some of the different types of proteins found in cells.

Peptide bond. *An amide linkage between two amino acids formed between the carboxyl group of the first amino acid to the amino group of the second.*

Protein. *Any member of the major family of cellular biomolecules encoded by a unique cellular gene and consisting of a repeating series of amino acids linked together by peptide bonds.*

Polypeptide *A naturally occurring chain of amino acids.*

Structure of Proteins

Proteins are composed of amino acids linked together into protein chains, analogous to the manner in which nucleotides are joined like beads on a string to make DNA and RNA. Amino acids are linked in a covalent manner by a ***peptide bond,*** which connects the carboxyl group of the first amino acid to the amino group of the second amino acid (Fig. 3.18). Through a repeating series of peptide bonds, hundreds or thousands of amino acids can be connected in a linear sequence to form a ***protein***. The term ***polypeptide*** is also used to describe proteins, because many (poly) peptide bonds are used in their synthesis.

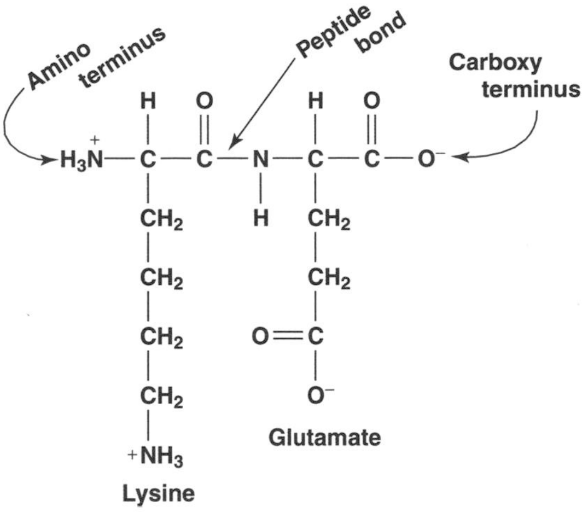

FIGURE 3.18. Amino acids are joined together by peptide bonds, which connect the carboxyl group of one amino acid to the amino group of another. Peptide bond formation between amino acids lysine (left) *and glutamate* (right), *results in the formation of a two-amino acid polypeptide containing an amino terminus (NH_3^+) and a carboxy terminus (COO^-).*

A total of 20 chemically diverse amino acids (see above) are used to make proteins, compared to the four nucleotides making up DNA and RNA, a chemical fact that imparts proteins with much more functional diversity than nucleic acids. The number of different proteins that can be built with the 20 amino acids, is calculated as 20 to the *n*th power, where *n* is the number of amino acids in the protein chain.

$$\text{Number of different proteins} = 20^n \tag{3.2}$$

Thus 400 dipeptides (20^2) can be formed from the 20 amino acids, and more than 10 trillion (1×10^{13}) different proteins are possible if the polypeptide contains 10 amino acids. It should be obvious that the cell can generate enormous protein diversity using the simple set of 20 amino acids.

The cellular process of protein synthesis reads mRNA codons in a successive and nonoverlapping manner, such that a protein sequence is ***co-linear*** with the DNA and mRNA (Fig. 3.19). A DNA sequence containing three codons (5′ TTT CAC GGT 3′), would specify an mRNA containing three codons (5′ UUU CAC GGU 3′), and a three amino acid protein (NH_3^+) phe his gly COO^-). The two ends of a protein molecule, known as the amino (NH_3^+) and carboxy (COO^-) termini, contain an amino and carboxy group, respectively. The co-linearity of the genetic code is such that the 5′ and 3′ ends of the DNA and mRNA, correspond to the amino and carboxy termini of the protein, respectively.

Co-linear. *Any case in which two molecules share genetic information along the primary sequence, such as the codons in messenger RNA that specify a cognate series of amino acids.*

The co-linearity of the genetic code imparts an evolutionary advantage, allowing the cellular machinery to "read" the codons in an efficient and linear manner, rather than by a more complicated de-coding scheme that would slow the transfer of genetic information from gene to protein. The co-linearity of the code also has the practical consequence of allowing the researcher to

5′ – TTT CAC CGT – 3′ DNA
5′ – UUU CAC CGU – 3′ mRNA
$H_3\overset{+}{N}$ – Phe His Arg – COO^- Protein

FIGURE 3.19. The genetic code is co-linear, meaning that the order and identity of amino acids in proteins are the same as the codons specified in the mRNA and DNA. The DNA coding sequence 5′ TTT CAC CGT 3′ produces an mRNA containing 5′ UUU CAC GGU 3′, which specifies the three-amino acid protein NH_3^+ phe his gly COO^-.

deduce unambiguously a protein sequence from a DNA or mRNA sequence, a property of the genetic code that allows the "virtual translation" of proteins from the nucleic acid sequence alone. Virtual translation of proteins by computer provides clues to protein function before obtaining bona fide biochemical data. The degeneracy of the genetic code does not, however, allow the deduction of the nucleic acid sequence from an given amino acid sequence. Nearly every amino acid is specified by multiple codons, making it impossible to deduce a gene or mRNA sèquence unambiguously from amino acid data alone.

Types of Proteins

Proteins possess enormous functional diversity and make up a large family of molecules, including enzymes, antibodies, transcription factors, receptors, hormones, and transporters. A specific cellular gene encodes each protein; therefore, the approximately 35,000 genes in the human genome encode about 35,000 different proteins.

Enzyme. *A protein that carries out a biochemical reaction in the cell.*

An ***enzyme*** is a protein that carries out a biochemical reaction in the cell. Enzymes are used in a vast array of cellular processes, including DNA synthesis (e.g., DNA polymerase I), complementary DNA synthesis (e.g., reverse transcriptase), RNA synthesis (e.g., RNA polymerase II), protein synthesis (e.g., aminoacyl-tRNA synthetase), cellular metabolism (e.g., hexokinase), DNA degradation (e.g., deoxyribonuclease I), RNA degradation (e.g., ribonuclease H), protein degradation (e.g., proteinase K), cell wall degradation (e.g., cellulase), and DNA joining (e.g., DNA ligase). The names of enzymes often include a description of their biochemical function, followed by a common suffix *-ase* (pronounced "ace"). An enzyme that polymerizes DNA is a polymerase, one that breaks down a cell wall is a cellulase, and so forth. ***Restriction enzymes,*** an extremely important class of enzymes used in genetic engineering (see Chapter 4), provide an exception to the enzyme-naming system. Restriction enzymes are named according to the bacteria from which there were purified (e.g., *Eco*RI, *Bam*HI, *Sac*I).

Restriction enzyme. *Any of a family of specialized enzymes used in genetic engineering that cleave DNA at discrete locations.*

An enzyme uses its amino acid side chains to hold molecules in position, and speeds biochemical reactions facilitating the breakage and formation of chemical bonds. Chemical bonds are formed and broken with the assistance of functional groups (e.g. acids and bases) present on the amino acid side chains. Because of the richness of their chemistry, arginine, lysine, glutamate, aspartate, and other amino acids are often located in the active site of the enzyme, the location of the biochemical reaction. The precursors and products of enzymatic reactions are known as reactants and products, respectively. An enzyme is a ***catalyst*** because it increases the rate of biochemical reaction, but does not change the energy of the reactants and products (Fig. 3.20). A catalyst reduces the ***activation energy*** of a biochemical reaction, which is the "hump" in the energy profile that must be overcome for a reaction to take place. Once the enzyme reduces the activation energy, reactions occur very rapidly, and many enzymes catalyze hundreds or thousands of biochemical reactions per second. It is not uncommon for an enzyme to increase the rate of a biochemical reaction by more than a millionfold, and the exquisite rate enhancement imparted by enzymes allows essential cellular processes to occur at rates compatible with the functional requirements of cells.

Catalyst. *Enzyme, chemical, or other agent that increases the rate of a biochemical or chemical reaction but does not change the energy of the reactants and products.*

Activation energy. *Barrier in a reaction profile that must be overcome for a chemical or biochemical reaction to occur.*

Some cellular proteins possess enzymatic activity, though a large number of important proteins including antibodies, transcription factors, receptors,

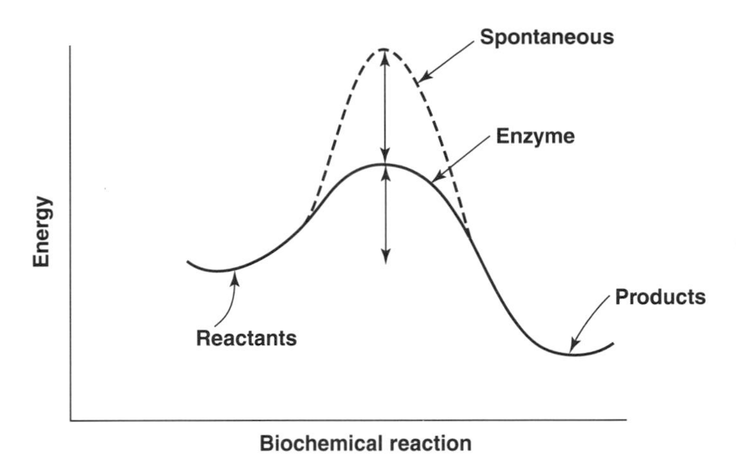

FIGURE 3.20. Energy required for a biological reaction to occur spontaneously (dotted line) *and with the assistance of an enzyme.* (solid-line). *Enzymes do not change the balance between reactants and products; they reduce the energy required to activate reactions* (double arrows).

hormones, and transporters do not carry out enzymatic reactions. Antibodies bind to invading viruses and bacteria and protect cells from these foreign intruders by assisting in their removal from the body. Transcription factors are proteins that bind near promoter sequences, enabling the cellular process of transcription (see Chapter 4). Receptors are a class of proteins that bind with high specificity to signaling molecules and thereupon mediate cell signaling. Certain types of polypeptide hormones, including insulin and endorphin, are themselves proteins. Cellular proteins also assist in transporting critical molecules, such as oxygen, throughout the body, a role performed by the protein hemoglobin. Each cellular gene or protein can be viewed as a unique member in the orchestra of life or as a mechanical component in the cellular machine (see Chapter 10). The analogies of cells as orchestras and as mechanical devices are intended to convey the notion that the genome contains thousands of different functional components, corresponding to myriad cellular genes and proteins, each with unique biochemical characteristics.

CARBOHYDRATES

A molecule composed of repeating sugar molecules is known as a ***carbohydrate***. Sugar molecules are the biochemical building blocks of cellular carbohydrates and thus are analogous to nucleotides in DNA and RNA and to amino acids in proteins. Some familiar members of the carbohydrate family are cellulose, glycogen, and starch. ***Cellulose*** is the main structural component of plants, providing the biopolymeric material that plants use to maintain their form and rigidity. Animal cells make large quantities of ***glycogen*** and ***starch*** and use these carbohydrates as sources of stored metabolic energy. Carbohydrates are also linked to the surface of proteins and cells and function in numerous cell-signaling activities. Carbohydrates do not encode any genetic information but rather play nongenetic roles in cell structure, energy storage, and cell signaling.

Carbohydrate. *Any of a large family of cellular biomolecules composed of repeating sugar monomers.*

Cellulose. *Plant carbohydrate that is the main structural component of the cell wall.*

Glycogen. *Highly branched glucose-based carbohydrate used to maintain blood sugar levels in animal cells.*

Starch. *Plant and animal carbohydrate that provides a cellular reservoir of stored energy.*

SUMMARY

Biochemistry is the scientific field of study focused on DNA, RNA, proteins, and other molecules of biological importance. The field has a rich tradition of discovery and innovation; since the mid-1970s, we have witnessed the development of recombinant DNA technology, polymerase chain reaction (PCR), and microarray analysis. Nucleic acids, proteins, and carbohydrates are composed of nucleotide, amino acid, and monosaccharide building blocks, respectively. DNA is made up of four different nucleotides, which are linked together into long chains that form a double-stranded double helix. The genome, which is made of DNA, contains the genetic blueprint for the cell. Cells contain three types of RNA, known as messenger, transfer, and ribosomal RNA. Messenger RNA carries the genetic instructions between gene and protein, using the three-letter genetic code to assemble amino acids from codon information. Proteins—a diverse family of molecules made by linking amino acids into polypeptides—include enzymes, antibodies, transcription factors, receptors, hormones, and transporters. Carbohydrates are made of sugar molecules and play roles in cellular structure, energy storage, and signaling.

SELECTED READING

Blackburn, G. M., and Gait, M. J., eds. *Nucleic Acids in Chemistry and Biology,* 2nd ed. Oxford University Press, Oxford, UK, 1996.

Bloomfield, V. A., Crothers, D. M., and Tinoco, I. *Nucleic Acids: Structure*. University Science Books, Herndon, VA, 2000.

Devlin, T. H., ed. *Textbook of Biochemistry with Clinical Correlations,* 5th ed. Wiley, New York, 2002.

Fersht, A. *Structure and Mechanism in Protein Science: A Guide to Enzyme Catalysis and Protein Folding*. Freeman, New York, 1999.

Gilbert, H. F., ed. *Basic Concepts in Biochemistry: A Student's Survival Guide*. McGraw-Hill Medical, New York, 2000.

Guy, H. R. Amino acid side-chain partition energies and distribution of residues in soluble proteins. *Biophys. J.* 47:61–70, 1985.

Hein, M., Best, L. R., Pattison, S., and Arena, S. *Introduction to General, Organic, and Biochemistry,* 6th ed. Brooks/Cole, Pacific Grove, CA, 1997.

Lehninger, A. L., Nelson, D. C., and Cox, M. M. *Principles of Biochemistry,* 3rd ed. Freeman, New York, 2000.

Lewin, B. *Genes VII*. Oxford University Press, Oxford, UK, 2000.

Matthews, H. R., Freedland, R. A., and Miesfeld, R. L. *Biochemistry: A Short Course*. Wiley, New York, 1997.

Schena, M., ed. *DNA Microarrays: A Practical Approach,* 2nd ed. Oxford University Press, Oxford, UK, 2000.

Schena, M., ed. *Microarray Biochip Technology*. Eaton, Natick, MA, 2000.

Singer, M., and Berg, P. *Genes and Genomes*. University Science Books, Herdon, VA, 1991.

Streitweiser, A., and Heathcock, C. H. *Introduction to Organic Chemistry,* 2nd ed. Macmillan, New York, 1981.

Stryer, L. *Biochemistry,* 4th ed. Freeman, New York, 1995.

REVIEW QUESTIONS

1. The field of study focused on understanding the chemical basis of life is known as what?
2. DNA, RNA, protein, and carbohydrate share what common repeating chemical characteristic?
3. Name the three chemical components of nucleotides.
4. Name the four nitrogenous bases found in the DNA nucleotides.
5. DNA and RNA differ chemically in several respects. Describe three.
6. The letters A, E, G, F, and K are abbreviations for which 20 amino acids?
7. A nucleotide is to DNA as a sugar is to what biomolecule?
8. A biochemist obtains structural data indicating that a protein contains two tripeptides with amino acid sequences Lys-Arg-Glu and Phe-Tyr-Ile. Which tripeptide is likely to reside on the interior of the protein and why?

9. Write the complementary sequence to a DNA molecule containing the following nucleotides: 5′ ACGTAATTCAGG 3′. Be sure to label and 5′ and 3′ ends of the complementary 12mer.
10. How many different tetrapeptide (4-amino acid) sequences can be synthesized from the 20 amino acids?
11. A DNA microarray is likely to be more stable than an RNA microarray. Explain your answer in a few sentences.
12. Describe the protein information specified by the following three codons 5′ ATG-CCC-TAA 3′.
13. A microarray researcher identifies a peptide that binds to a specific protein on a microarray. The eight amino acid peptide has the sequence NH_2-KKAVLPRD-COOH. The researcher's supervisor asks her to design an oligonucleotide that will match the gene sequence that encodes the peptide exactly. After studying the protein sequence, the researcher concludes that it is not possible to specify a single 24-mer. What aspect of the genetic code makes this task impossible? How many different 24-mers would be necessary to ensure that one of the oligonucleotides would match with the gene sequence exactly?
14. A microarray researcher is given three unlabeled glass substrates containing a DNA microarray, an RNA microarray, and a protein microarray and three enzymes that the researcher is told can be used to establish the identity of the three microarrays. Name the three enzymes and describe a simple procedure that could be used to identify the three microarrays.
15. An unknown double-stranded nucleic acid molecule is degraded into simpler components and chemical analysis reveals the presence of phosphate, ribose, thymine, adenine, guanine, cytosine, uracil, and deoxyribose. What general structure of double helix would contain these components? One sentence will suffice.

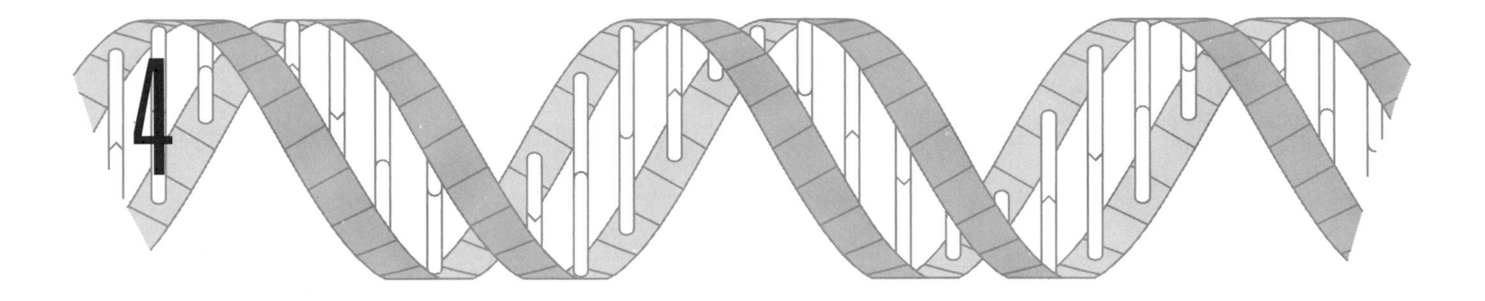

Genes and Genomes

> The essence of life is statistical improbability on a colossal scale. —*Richard Dawkins*

The genetic blueprint is carried in the genome, an improbable assembly of DNA bases, genes, and chromosomes. Cells pass an exact copy of the genome to other cells during cell division, and the blueprint is inherited during reproduction. Human genes are structurally complex, and minor changes in the sequence can produce disease. The tools of recombinant DNA and the polymerase chain reaction (PCR) enable the isolation, cloning and amplification of genes. Each cell in the human body contains the same genomic sequence as every other cell, but gene expression varies greatly from cell to cell. Transcription and translation convert gene information into protein, and DNA replication synthesizes exact copies the genomic sequence. DNA sequencing technology has recently afforded the sequence of the entire human genome, as well as sequences of dozens of other organisms, including viruses, bacteria, yeast, worms, insects, plants, and rodents. This chapter provides a survey of genes and genomes, and offers a glimpse into the exciting and fast-moving field of genomics.

GENES

The basic unit of genetic function is the gene. Genes are continuous segments of genomic DNA constructed from four nucleotide building blocks (see Chapter 3). Each gene encodes a specific mRNA and protein, the latter of which imparts biological function in the cell. This section provides a brief overview of gene structure, sequence variants, gene manipulations, and evolution.

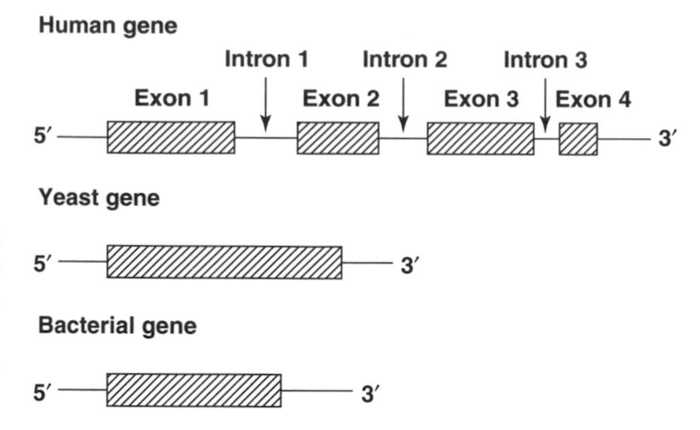

FIGURE 4.1. Three hypothetical genes, drawn to scale. The four exons (hatched boxes) *in the human gene are separated by three introns* (lines), *and the yeast and bacterial genes each contain a single exon.*

Gene Structure

Base pair. *Unit of measure of double-stranded DNA comprising one set of complementary nucleotides, and abbreviated bp.*

Higher eukaryote. *Category of biological organisms that includes humans, primates, rodents, insects, plants and other multicellular life forms.*

Exon. *Segment of a gene retained in the mRNA after processing, and often containing the codons represented as amino acids in proteins.*

Intron. *Segment of a gene removed from the mRNA during processing, and not represented in proteins.*

Noncoding. *Any gene sequence that does not contain protein coding information.*

A gene is composed of deoxyribonucleic acid (DNA), and an average human gene contains approximately 10,000 DNA building blocks or nucleotides (nt), each containing one of the four bases A, G, C, or T (see Chapter 3). Genes are composed of double-stranded DNA, and gene size is measured in a unit known as the ***base pair***, corresponding to one nucleotide of double-stranded DNA. Genes in ***higher eukaryotes*** contain exons and introns (Fig. 4.1). An ***exon*** is gene segment that is copied into mRNA and maintained after mRNA processing (see below), and an ***intron*** is a gene segment that is copied into mRNA, but removed from the mature mRNA before protein synthesis. Genes in lower eukaryotes, such as yeast, are essentially devoid of introns, and bacteria do not contain any introns at all. The presence of introns in complex organisms, but not in simple systems, suggests an evolutionary role for these ***noncoding*** gene sequences (see below).

Genes are illustrated schematically from left to right, as a linear arrangement of boxes and lines corresponding to exons and introns, respectively (Fig. 4.1). The left end of a gene is the 5′ end, and the right end of a gene is the 3′ end. The 5′ and 3′ nomenclature is a bit confusing for gene sequences, because genes are double-stranded molecules and as such contain 5′ and 3′ termini on both ends of the double-stranded gene (see Chapter 3). The nomenclature is based on the fact that the coding (top) strand is written 5′ to 3′, with the first codon on the left side of the drawing and the last codon on the right. In common usage, the 5′ (left) end of a gene corresponds to the beginning of a coding sequence and the 3′ (right) end refers to the end of coding sequence. Researchers should commit this nomenclature to memory because it is used commonly, despite its being somewhat confusing.

Kilobase. *Segment of DNA equal to 1,000 base pairs, and abbreviated kb.*

A typical human gene contains 6–8 exons and introns, with an average length of 100–200 and 1000 base pairs, respectively, for the two elements. A gene segment containing 1000 base pairs equals one ***kilobase*** (kb) of DNA. An average human gene covers 10 kb, and the exons and introns are 0.2 kb and 1.0 kb, respectively. The structure and size of human genes vary considerably, and the cystic fibrosis (CF) gene, spans 250 kb and contains 25 exons. The 250-kb CF gene encodes a 1480 amino acid protein, and mutations in the CF gene causing cystic fibrosis disease occur in newborns at a rate of ~1:3000. The largest known human gene, involved in Duchenne muscular dystrophy, contains 75 exons and covers 2.4 million base pairs (2400 kb).

```
             Normal :  5′-TGATTGACCT-3′
                SNP :  5′-TGATCGACCT-3′
                                ↑
                          T Changed to C
          Insertion :  5′-TGATTGGACCT-3′
                                 ^
                               G inserted
           Deletion :  5′-TGAACCT-3′
                             △
                            TTG deleted
             Normal :  5′-TGAT[T]GACCT-3′
    (double-stranded)  3′-ACTA[A]CTGGA-5′
              SNP   :  5′-TGAT[C]GACCT-3′
    (double-stranded)  3′-ACTA[G]CTGGA-5′
```

FIGURE 4.2. A 10-nucleotide hypothetical DNA sequence (normal), and corresponding variants. Double-stranded versions of the normal and SNP sequences (bottom) *illustrate the fact that sequence variants* (boxes) *alter both DNA strands in a duplex.*

Sequence Variants

A change in the primary nucleotide sequence of DNA is known as a ***sequence variant.*** A sequence variant that occurs within a gene, produces a gene variant known as an ***allele.*** A human gene with 15 different sequence variants would contain 15 different alleles of that gene. The most common types of sequence variant are the ***single nucleotide polymorphism*** (SNP), ***mutation, insertion,*** and ***deletion.*** All four types of variants produce different alleles (Fig. 4.2). A single nucleotide polymorphism refers typically to a single nucleotide change within a population of organisms (e.g., the human population). SNPs, inherited and passed through the generations during DNA replication and reproduction (see below), occur at a rate of approximately 1:1,000 base pairs in the human genome. SNP analysis provides a powerful means of determining the genetic makeup or ***genotype*** of an organism, and each human in the population possesses a unique repertoire of SNPs.

A mutation is a sequence variant that is acquired during the lifespan of an organism. A chemical agent or ***mutagen*** can alter the primary DNA sequence and produce mutations that lead to uncontrolled cell growth and cancer. A mutagen that causes cancer is known as a ***carcinogen.*** Organic molecules containing multiple aromatic rings, nitroso groups, and other chemical functions fall into the carcinogen class. The term mutation is also used more broadly to describe any change in a DNA sequence. An insertion results in the addition of one or more nucleotides to a DNA sequence, and a deletion results in the removal of one or more nucleotides. Genes are double-stranded molecules; therefore, every sequence variant changes the nucleotide sequence of both DNA strands, a practical consequence of which is that either strand of DNA can be used for genotyping assays (see Chapter 13).

Sequence variant. *Any minor change in the primary nucleotide information in a gene.*

Allele. *Any sequence variant of a gene.*

Single nucleotide polymorphism. *Common sequence variant containing a one-base-pair change relative to the normal gene.*

Mutation. *Any change in a DNA sequence, but typically acquired during the life span of an organism.*

Insertion. *Mutation that results in the addition of one or more nucleotides to a DNA sequence.*

Deletion. *Mutation that results in the removal of one or more nucleotides from a DNA sequence.*

Genotype. *The genetic makeup of an organism.*

Mutagen. *Chemical agent that alters the primary nucleotide sequence of DNA.*

Carcinogen. *Chemical, biological, radiological or environmental agent that causes cancer.*

Gene Manipulations

Genes can be manipulated in variety of ways experimentally to expedite genetic analysis. The two most powerful methods of gene manipulation, developed in the 1970s and 1980s, respectively, are recombinant DNA technology and the polymerase chain reaction. Recombinant DNA technology, also known as gene splicing or genetic engineering, uses enzymes to cut and paste DNA into "recombinant" molecules. The capacity to recombine human and bacterial sequences, for example, allows large quantities of human genes to be grown in

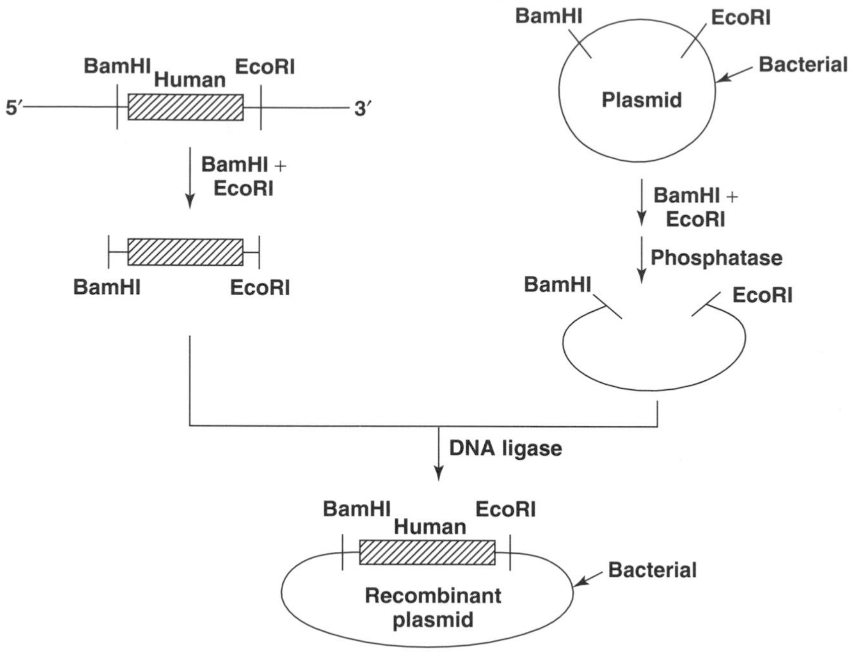

FIGURE 4.3. A segment of human DNA (left) *is digested with two restriction enzymes* Bam*HI and* Eco*RI, yielding a double-stranded DNA molecule containing* Bam*HI and* Eco*RI sticky ends. The bacterial plasmid* (right) *is digested with the same enzymes and treated with phosphatase to prevent plasmid reclosure during ligation. The human and plasmid DNAs are mixed together and joined with DNA ligase, creating covalent bonds between the sticky DNA ends to yield a recombinant DNA molecule* (bottom).

Nuclear transfer. *Technology used in cloning that allows the introduction of a somatic cell nucleus into an egg lacking a nucleus.*

Plasmid. *Any circular DNA molecule that has the capacity to replicate independently in the cell.*

DNA ligase. *Enzyme used in gene splicing that creates phosphodiester bonds between free ends of double-stranded DNA.*

Recognition sequence. *Nucleotide target site at which a restriction enzyme cleaves the DNA.*

bacterial cells. Recombinant DNA technology is also known as cloning, because the recombinant molecules spliced or engineered in the laboratory are identical clones of each other. Cloning in the context of recombinant DNA should not be confused with the cloning of sheep and other organisms using ***nuclear transfer*** techniques (see below).

One popular recombinant strategy involves the insertion of a human gene sequence into a replicating bacterial molecule known as a ***plasmid.*** The human sequence is cut with restriction enzymes (e.g., *Bam*HI and *Eco*RI) that recognize specific nucleotides in the human DNA and cleave the two strands at precise locations, leaving overhanging sticky ends that can be inserted into the bacterial plasmid (Fig. 4.3). In preparation for DNA insertion, the bacterial plasmid is cut with *Bam*HI and *Eco*RI, treated with phosphatase to prevent plasmid reclosure, and incubated with ***DNA ligase*** to generate a recombinant molecule containing the human sequence inserted into the plasmid DNA at the restriction sites. The availability of hundreds of different restriction enzymes each containing a unique ***recognition sequence,*** allows insertion of virtually any DNA sequence into a plasmid, and its rapid propagation in the bacterium known as *E. coli*. Recombinant DNA molecules including cDNAs, expressed sequence tags (ESTs), and bacterial artificial chromosomes (BACs) (see below) are used widely in microarray manufacture (see Chapter 7).

PCR, developed by Mullis and co-workers at Cetus (Emeryville, CA) in the early 1980s, is a revolutionary method that allows selective amplification of any nucleic acid sequence from small quantities of starting material. The PCR process reacts a target DNA with two synthetic oligonucleotides and polymerase, resulting in the amplification of the target in successive heating and cooling cycles (Fig. 4.4). The synthetic oligonucleotides hybridize to the target DNA, providing a start site for polymerase, which synthesizes exact copies the target DNA between the two primers. A single round of PCR makes two templates, 2 rounds makes four templates, and so forth, such that 30 rounds of PCR can provide a 1 million-fold amplification of a target sequence. The exponential PCR

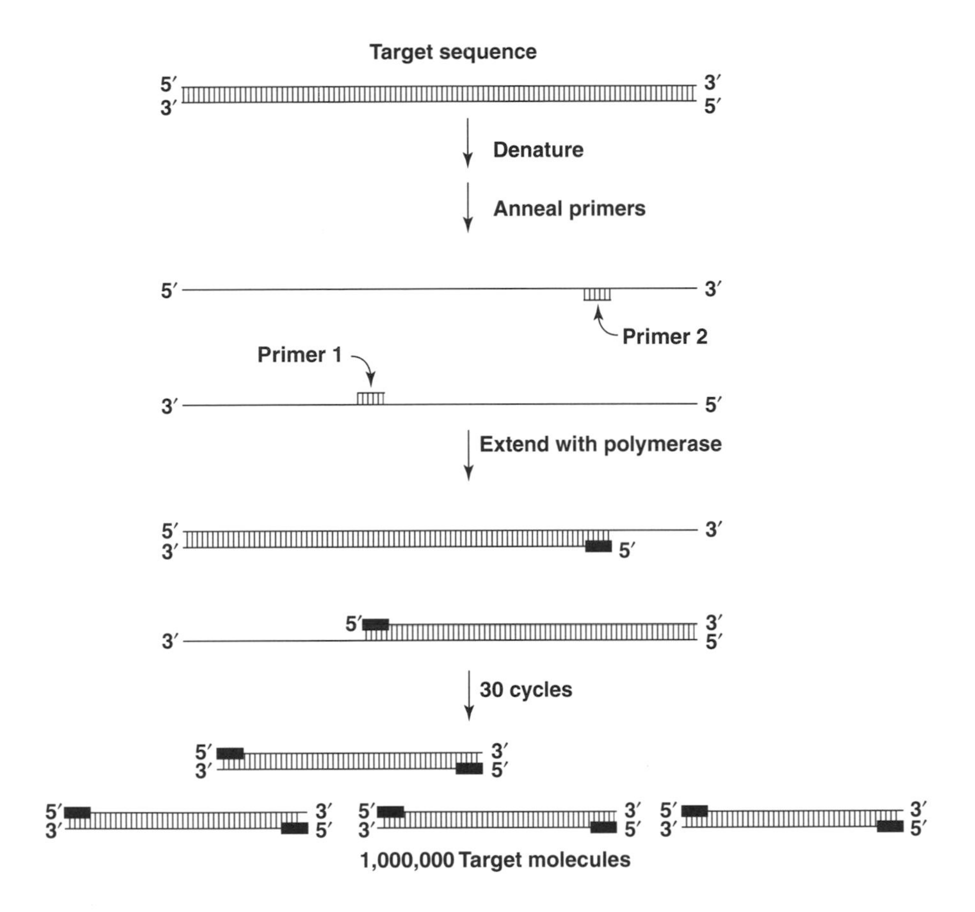

FIGURE 4.4. PCR allows selective amplification of any DNA sequence. A target sequence is denatured, annealed with two synthetic oligonucleotide primers, and extended with polymerase enzyme to produce two amplified segments. This process is repeated 30 times to produce a 1 million-fold amplification of the target sequence.

process allows synthesis of any gene fragment of interest from minute quantities of cells or tissue specimens, and has found wide use in forensics, diagnostics, and research. PCR is also used widely in microarray analysis for the amplification of target DNA and is gaining use in several emerging probe amplification strategies.

Cloned DNA

Several types of cloned DNA are used commonly in microarray analysis, including cDNA, EST, and BAC. ***Complementary DNA*** (cDNA) is a double-stranded molecule of ~0.2–5 kb in length that is an exact replica or "complement" of an mRNA molecule. Collections of cDNAs are extremely useful for gene expression analysis, because cDNAs lack introns and therefore contain only the exon content of cellular genes . Microarrays of cDNAs allow profiling of mRNA levels in hybridization-based assays, such that the fluorescence intensity at each cDNA location provides a quantitative measure of the corresponding mRNA. Steady-state mRNA levels provide an excellent approximation of gene expression levels, and cDNA microarrays allow expression measurements of many genes in parallel.

Complementary DNA. *DNA version of messenger RNA.*

The DNA polymerase known as reverse transcriptase is used to synthesize cDNA molecules from mRNA. Reverse transcriptase is an ***RNA-dependent DNA polymerase,*** because the enzyme depends on an RNA template for DNA synthesis. Any DNA molecule made from mRNA is known as a cDNA, whether the cDNA is single or double stranded, and microarray researchers should be aware of this somewhat ambiguous usage. Most microarray assays use

RNA-dependent DNA polymerase. *Specialized family of enzymes, including reverse transcriptase, that synthesize DNA from a ribonucleic acid template.*

double-stranded cDNAs as target elements and mixtures of single-stranded cDNA molecules as labeled probe mixtures, though beginning experimentalists should exercise caution when reading the published literature, microarray protocols, and the Internet, to avoid confusion over whether a given cDNA is a single-stranded probe or a double-stranded target.

Expressed sequence tag. *Specialized complementary DNA molecule that has been subjected to a single pass of DNA sequencing.*

An ***expressed sequence tag*** is a cDNA molecule that has been subjected to a single pass of DNA sequencing (see below), providing primary nucleotide information for all or part of one cDNA strand (Claverie et al., 1993; White et al., 1993). All ESTs are cDNAs, but not every cDNA is an EST, with the distinguishing criterion being the availability of DNA sequence information. Single-pass sequence data can be used to query sequence databases (see Chapter 9), providing functional clues to the cellular gene corresponding to the EST. Microarrays of ESTs, similar to cDNA microarrays, find wide use in gene expression analysis. The public sequence database from the National Center for Biotechnology Information (Bethesda, MD) contains currently >10 million ESTs from nearly 400 different organisms, providing a rich source of target sequences for microarray manufacture.

Bacterial artificial chromosome. *Recombinant plasmid that replicates autonomously in a prokaryotic host and contains 50–250 kilobase segments of inserted DNA.*

Bacterial artificial chromosome DNA contains 50- to 250-kb segments of genomic DNA, human or another organism of interest, inserted into replicating bacterial plasmids. BACs are constructed using the tools of genetic engineering and are therefore artificial chromosomes. The capacity to propagate relatively large fragments of human DNA in bacteria assists in the physical analysis of human chromosomes, including gene mapping, gene identification, and evolutionary analysis. BAC clones are finding increasing use in microarray assays, because they allow the representation of a large amount of genetic information on a single chip. The entire 3 billion bases of the human genome could be configured in a single microarray containing 30,000 BAC clones with nonoverlapping 100-kb inserts. Microarrays bearing the entire human genome have yet to be manufactured, but this technical feat is likely to be accomplished in the near future.

Gene Evolution

The most widely held model of evolution posits that life began with the creation of a primordial cell (e.g., bacterium) that evolved over billions of years to give rise to humans and other complex organisms. The nucleotide and amino acid similarity of human and bacterial genes supports this view. A striking experiment in evolutionary conservation, made in the mid-1980s, revealed that a mammalian steroid receptor functions when expressed in yeast cells (Metzger et al., 1988; Schena and Yamamoto, 1988), even though approximately 1 billion years of evolutionary time separates mammals and fungi. The activity of mammalian steroid receptors in yeast demonstrates the conservation of gene sequence and function, as well as the intricate protein–protein contacts required for transcriptional activation (see below). These and other observations argue that a basic set of cellular genes was used to build complex organisms, but human and bacterial genes are quite different in terms of size and structure. How did relatively simple cellular genes evolve into the more complex genes found in higher organisms?

Several observations point to a potential role of introns in gene evolution. Comparison of human and yeast genes reveals that human genes are generally larger and contain multiple introns and exons; yeast genes are smaller and are

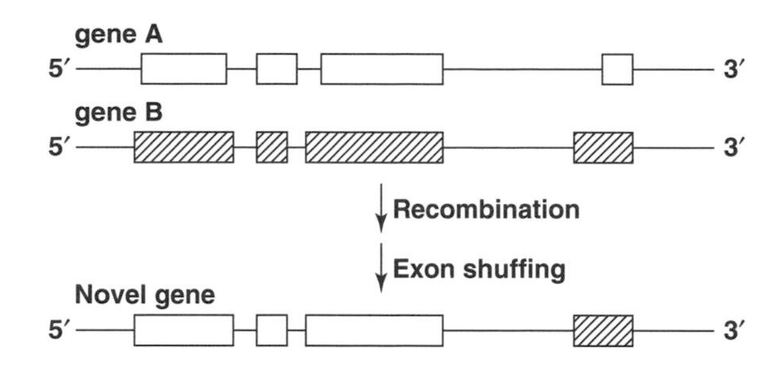

FIGURE 4.5. Exon shuffling postulates the creation of novel genes (bottom) *through genetic recombination between exon (rectangles) and intron (lines) sequences of related genes.*

basically exon free. Inspection of exon sequences reveals that the functional domains of human proteins tend to localize to discrete exons. One hypothesis is that genes in complex organisms evolved by ***exon shuffling,*** whereby exons from different genes were mixed and matched, or shuffled, during evolution to create proteins with novel function (Maki et al. 1980). Exon shuffling postulates that new genes are created by ***genetic recombination*** between related sequences, using introns to shuffle in the new exons, thereby maintaining the integrity of the coding sequence (Fig. 4.5). Novel genes and proteins allow evolution to test the fitness of new sequences, selecting and evolving the organisms that bear advantageous functionality. The exon shuffling model postulates that genes in humans and other complex organisms maintain their introns to allow exon shuffling, explaining the apparent paradox of why an organism would maintain gene sequences that serve no role in protein coding. Exon shuffling tends to conserve protein domains within exons, to maintain the presence of introns, and to drive the expansion of gene size, all of which are seen in comparisons of human and yeast genes.

Exon shuffling. *Evolutionary model that postulates that novel genes are formed by the mixing and matching of coding sequences through genetic recombination.*

Genetic recombination. *Genetic crossing over between two genomic segments that share sequence similarity.*

GENE EXPRESSION

The process by which genetic information at the DNA level is converted into functional proteins is known as gene expression. Because cells express their genes only when they are required for a cellular process, patterns of gene expression recorded under different physiological conditions provide important clues to gene function. This section discusses transcription, mRNA processing, and translation, three key steps in the gene expression process.

Transcription

The synthesis of mRNA from a DNA template is termed ***transcription.*** Transcription occurs at discrete genomic sites (genes), which are contiguous linear DNA sequences composed of coding sequences, noncoding sequences, and regulatory elements. Coding sequences or exons specify protein information, and noncoding sequences (e.g., introns) are either removed from the transcript by splicing, or do not contain coding information (see below). ***Regulatory elements*** are short (10–100 base pairs) DNA sequences that control the expression of genes (Fig. 4.6). There are several different types of regulatory elements, including promoters and enhancers, and all regulatory elements are located near the genes they regulate. A ***promoter*** is a regulatory element that determines the start site for RNA polymerase, the enzyme that makes mRNA from the DNA template. Many promoters in prokaryotes and eukaryotes contain a conserved

Transcription. *First step in gene expression in which messenger RNA is synthesized from a DNA template.*

Regulatory element. *Short segment of DNA near a promoter that binds one or more cellular proteins and modulates gene expression.*

Promoter. *Genomic location upstream of a cellular gene that determines the start site for RNA polymerase.*

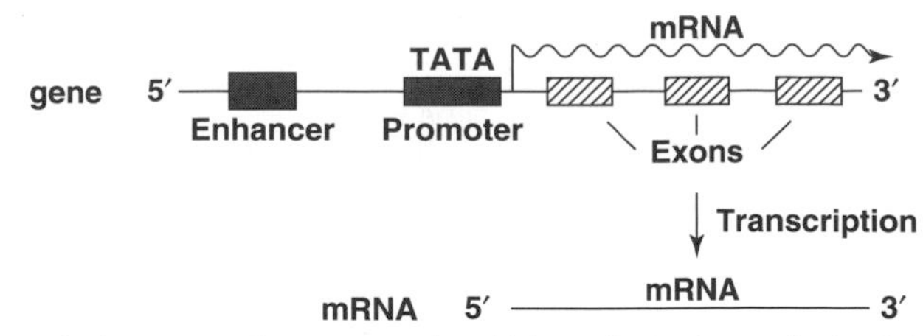

FIGURE 4.6. The transcription process (wavy arrow), *which results in the synthesis of single-stranded cellular mRNA, is mediated by specific DNA sequences or regulatory elements located adjacent to cellular genes. An enhancer* (solid rectangle) *modulates the efficiency of transcription, and a promoter* (thick line) *provides a start site for the RNA polymerase enzyme.*

Enhancer. *Regulatory element that alters promoter efficiency by increasing or decreasing the rate of transcription.*

Activation. *An increase in the rate of transcription of a cellular gene.*

Repression. *A decrease in the rate of transcription of a cellular gene.*

Basal expression. *Low level of gene transcription that occurs in the absence of activation.*

Transcription factor. *Cellular protein that binds to a promoter, enhancer, or other regulatory element and modulates expression of a cellular gene.*

Activator. *Cellular protein that binds to a regulatory element and increases the rate of transcription.*

Repressor. *Cellular protein that binds to a regulatory element and decreases the rate of transcription.*

AT-rich promoter sequence, or TATA box (pronounced *tata box*), located approximately 25 base pairs upstream of the start site in eukaryotic promoters. The TATA box is believed to play a role in positioning RNA polymerase for accurate transcription initiation.

An ***enhancer*** is a regulatory element that alters promoter efficiency by increasing or decreasing the rate of transcription (Fig. 4.6). ***Activation,*** up-regulation, or enhancement corresponds to an increase in the rate of transcription, and ***repression*** or down-regulation corresponds to a transcriptional rate decrease. Genes that are activated produce more mRNA relative to the ***basal expression*** level, and genes that are repressed produce less mRNA. Microarrays allow the quantitative assessment of mRNA levels and, therefore, the identification of activated and repressed genes. Cellular genes undergo activation and repression in response to thousands of different hormonal and environmental cues, and systematically recording the changes in ~35,000 human genes under myriad physiological conditions is a major technical challenge in modern microarray analysis.

The activity of promoters, enhancers, and other types of gene regulatory elements is mediated by cellular proteins known as ***transcription factors,*** which bind to specific nucleotide sequences within regulatory elements and modulate transcription by a variety of mechanisms (Fig. 4.7). A transcription factor

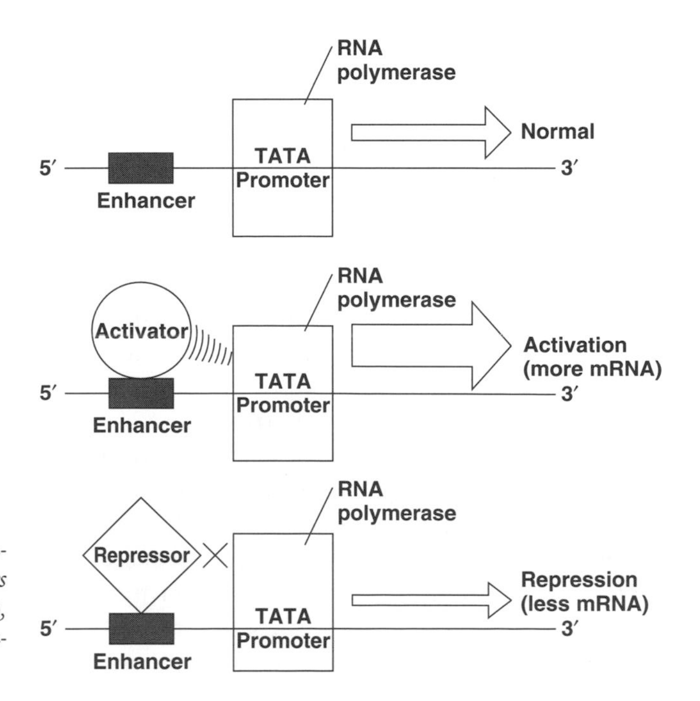

FIGURE 4.7. Transcription of cellular genes is regulated by activators (oval) *and repressors* (diamond), *functioning through enhancer regulatory elements* (solid rectangles).

that increases the rate of transcription is known as an ***activator,*** and a transcription factor that decreases the rate of transcription is termed a ***repressor.*** Activators facilitate transcription initiation by RNA polymerase, and repressors inhibit initiation by polymerase, leading to concomitant increases or decreases in the rate of mRNA synthesis. Enhancers, working in conjunction with transcription factors, are striking in their capacity to function at considerable distances (e.g., up to 100 kb) from the promoter and in either physical orientation.

Transcription activators and repressors often control the expression of subsets of cellular genes by ***coordinate regulation.*** Cells use coordinate regulation as a means of regulating efficiently the expression of gene subsets that share a common function. ***Primary response genes*** are cellular genes regulated directly by an activator or repressor. Certain primary response genes can, in turn, regulate a second tier of genes known as ***secondary response genes,*** and so forth. Groups of genes configured in regulatory tiers constitute a ***regulatory cascade*** or regulatory hierarchy. Cells use coordinate regulation and regulatory cascades to trigger complex cellular responses with a relatively small set of activators and repressors. The steroid receptors, an important family of hormone-dependent transcription factors in humans, regulate gene expression in response to changes in steroid hormone levels (Yamamoto, 1985). Steroid responsive genes are configured in massive regulatory cascades, allowing activation and repression of approximately 1000 cellular genes in response to changes in the concentration of a single soluble hormone.

Coordinate regulation. *Process by which the expression of multiple cellular genes is controlled by a common mechanism.*

Primary response gene. *Cellular gene under direct and immediate regulation in a regulatory cascade.*

Secondary response gene. *Cellular gene under indirect regulation in a regulatory cascade.*

Regulatory cascade. *Group of cellular genes configured in a complex gene expression hierarchy.*

mRNA processing. *Cellular mRNA editing steps that include capping, polyadenylation and splicing.*

mRNA Processing

Messenger RNA molecules are synthesized from DNA templates during the transcription of cellular genes. The transcription initiation site defines the first nucleotide in the mRNA chain, followed by the linear sequence of exons and introns, and the 3′ untranslated region (Fig. 4.6). Following transcription, mRNAs undergo a series of editing steps known as ***mRNA processing,*** which include the addition of a cap structure on the 5′ end, a poly A sequence onto the 3′ end, and the removal the introns from the coding sequence (Fig. 4.8). A mature processed mRNA contains a 5′ cap (a single G nucleotide), a 3′ ***poly A sequence*** (100–200 nt), one or more exons, and a 3′ untranslated region. The processes of cap addition, poly A addition and intron removal are known as ***capping, polyadenylation,*** and ***splicing,*** respectively. Capping and polyadenylation increase the

Poly A sequence. *Sequence of 10–100 adenine residues added to the 3′ end of eukaryotic messenger RNAs.*

Capping. *Enzymatic processing of messenger RNA whereby a modified G nucleotide is added to the 5′ end of the transcript. Alternatively, the step in the chemical synthesis of oligonucleotides and other synthetic molecules in which an unreacted group such as the 5′ hydroxyl is blocked by acetylation or some other means to prevent incorrect additions to the growing polymer chain.*

Polyadenylation. *Enzymatic processing of mRNA whereby a series of adenine nucleotides are added to the 3′ end of the transcript.*

Splicing. *Enzymatic processing of messenger RNA whereby the introns are removed from the nascent messenger RNA.*

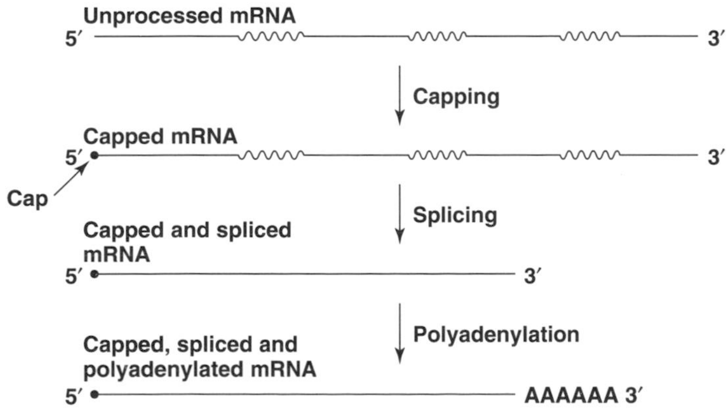

FIGURE 4.8. *In mRNA processing, an unprocessed mRNA (top) with exons* (straight lines) *and introns* (wavy lines) *undergoes capping, which adds a single G residue* (solid circle) *to the 5′ end and increases mRNA stability. Splicing removes introns from the mRNA to create a functional coding sequence. Polyadenylation results in the addition of a poly A tail to the 3′ end, which increases mRNA stability and the efficiency of protein synthesis (translation).*

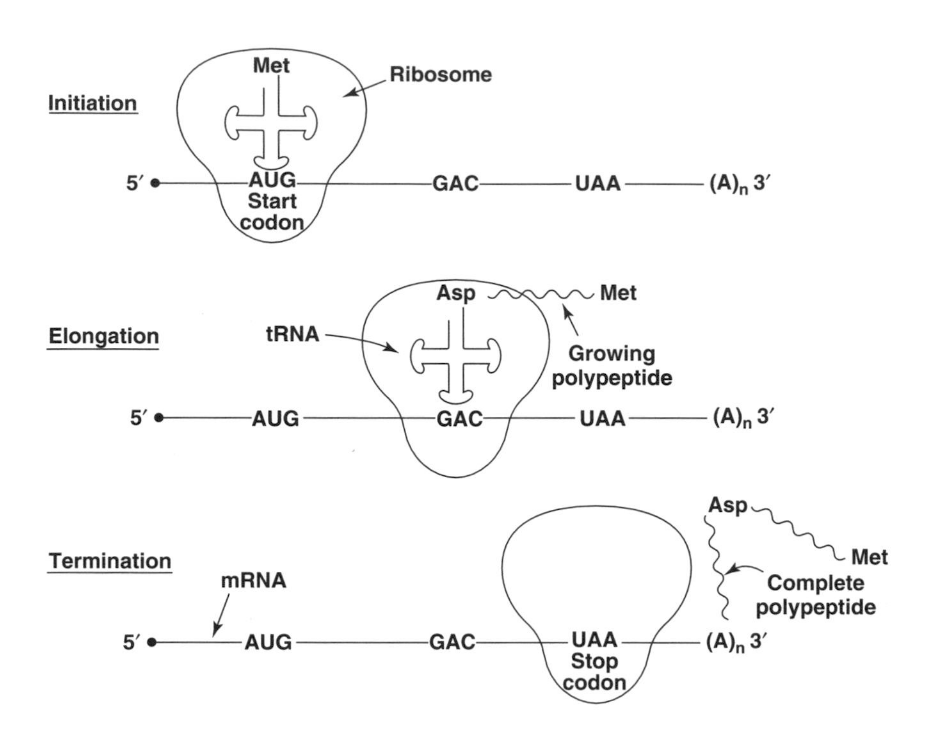

FIGURE 4.9. The ribosome initiates translation, coordinating interactions between the start codon (AUG) in the mRNA, and the tRNA that contains the methionine (Met) amino acid. Translation proceeds in a 5′ to 3′ manner along the mRNA during the elongation phase, adding amino acids (e.g., Asp) to the growing polypeptide chain. Translation is terminated when the ribosome encounters a stop codon (UAA) for which there is no tRNA, causing the release of a fully synthesized polypeptide (wavy line).

stability of mRNAs and their export from the nucleus, and splicing removes the introns to afford a continuous coding sequence. mRNA processing also improves the efficiency of protein synthesis, a cellular process that occurs in the ***cytoplasm.***

Cytoplasm. *Portion of a cell located outside the nucleus.*

Translation

Translation. *Synthesis of a polypeptide chain from a processed mRNA molecule.*

Aminoacyl-tRNA. *Cellular molecule that binds a specific amino acid and facilitates protein synthesis by codon recognition.*

The synthesis of a polypeptide chain from processed mRNA is known as ***translation.*** Transcription and translation together constitute the cellular process of gene expression. Translation of mRNA occurs on large cytoplasmic structures known as ribosomes. The ribosome binds to the capped (5′) end of the mRNA and moves in a 5′ to 3′ manner until it reaches a start codon (AUG), whereupon translation begins and methionine (Met) is added as the first amino acid in the polypeptide chain (Fig. 4.9). Translation proceeds codon by codon, and the ribosome coordinates interactions between the mRNA and ***aminoacyl-tRNA*** molecules, the latter of which contain bound amino acids. One amino acid is added per codon to the growing polypeptide chain, and amino acids are joined like beads on a string via peptide bonds. Translation continues until a stop codon (UAA, UGA, or UAG) is encountered in the mRNA sequence, which triggers translation termination. Nascent proteins fold into precise three-dimensional structures and carry out the instructions of their genes, functioning as enzymes, antibodies, transcription factors, transporters, and the like.

GENOMES

The entire DNA content of a cell, including the nucleotides, genes, and chromosomes is known as the genome. The genome of an organism contains the genetic blueprint, as demonstrated strikingly by recent advances in cloning, which show that adult sheep and other fertile animals can be grown from the genomic

DNA of a single cell using nuclear transfer technology (Campbell et al. 1996). Cloning demonstrates unambiguously that the genome is both necessary and sufficient to direct the creation of life, placing tremendous importance on continued genomic analysis of human and other organisms. This section provides a glimpse into the exciting field of ***genomics,*** focusing on the basics of heredity, DNA replication, DNA sequencing, and genome structure.

Genomics. *Field of study focused on the analysis of genomes.*

Heredity. *Transmission of genetic traits from one generation to the next.*

Diploid. *Cell that possesses two copies of each chromosome.*

Chromosome. *Large segment of genomic DNA that replicates autonomously in the cell and segregates during cell division.*

Heredity

The transmission of genetic traits from one generation to the next is termed ***heredity.*** Heredity is determined almost entirely by the primary nucleotide sequence of the DNA obtained during reproduction, with exceptions due to mutations acquired during the life span of an organism. Nearly all eukaryotic organisms–including humans and other primates, rodents, plants, and fungi–are ***diploid,*** meaning they possess two copies of each cellular gene in every cell except the reproductive cells (see below). Genes are carried on large pieces of replicating DNA termed ***chromosomes,*** and higher organisms, such as humans, obtain one set of chromosomes from the mother and one set from the father during ***fertilization,*** the fusion of an egg and sperm. Human eggs and sperm are known as ***gametes*** or ***germ cells,*** because unlike ***somatic cells,*** gametes are ***haploid*** cells, which contain only one copy of each chromosome. Human gametes have 23 chromosomes, and fertilization produces a diploid human cell with 46 chromosomes. Every cell in the human body is a somatic cell, except for the gametes.

Fertilization. *The cellular union of egg and sperm to create an embryo.*

Gamete. *Cell such as an egg or sperm that carries a single copy of each chromosome.*

Germ cell. *Cell such as an egg or sperm that carries a single copy of each chromosome.*

Somatic cell. *Diploid cell of an organism.*

Gametes are formed during ***meiosis,*** the process whereby the chromosomes are partitioned into haploid germ cells. The genetic constitution or genotype of a gamete is determined by the genotype of the diploid somatic cells that undergo meiosis. For each gene and ***trait,*** somatic cells are either ***homozygous*** or ***heterozygous,*** depending on whether the two genes are identical or different at the nucleotide level. Homozygous traits are represented in 100% of the gametes, and heterozygous traits produce a 50:50 mixture of gametes, representing the two gene variants or alleles. A organism that is heterozygous for a disease trait is known as a disease ***carrier.*** For recessive disorders, carriers of human diseases do not manifest the disease ***phenotype,*** but can pass on the disease to their children if the other parent is also a carrier for the same disease. For simple genetic disorders, the progeny of two carriers have a 25% chance of acquiring the disease, a frequency that derives from the fact that 50% of the gametes in each parent bear the disease allele, leading to a 25% chance (0.5×0.5) of a child inheriting two copies of the defective gene during fertilization. The risk associated with being a carrier of human disease provides a compelling argument for comprehensive genetic screening in adults (see Chapter 13), particularly in the case of common genetic disorders such as cystic fibrosis, which afflicts nearly 0.1% of all newborns.

Haploid. *Cell that possesses a single copy of each chromosome.*

Meiosis. *Process of cell division resulting in the production of gametes.*

Trait. *Phenotypic quality produced by a gene.*

Homozygous. *Diploid cell or organism that contains two identical copies of a given gene.*

Heterozygous. *Diploid cell or organism that contains two different variants of a given gene.*

Carrier. *An individual with a normal phenotype that is heterozygous for a disease gene and capable of passing on the disease gene to its progeny.*

A mutated allele that does not change the phenotype in the presence of a normal allele is said to be genetically ***recessive.*** Most alleles are recessive because the normal allele produces enough functional protein to satisfy the needs of the cell. Alleles that change the phenotype in the presence of a normal copy of the gene are said to be genetically ***dominant.*** Though dominance has several different mechanisms, a common one is that the "defective" protein produced by the dominant locus interferes with the function of the normal protein to produce a visible phenotype.

Phenotype. *The physical manifestation of genotype.*

Recessive. *An allele that does not manifest a phenotype in the presence of a normal copy of the gene.*

Dominant. *An allele that manifests a phenotype in the presence of a normal copy of the gene.*

Mitosis. *Cellular process whereby two identical cells are produced from one dividing cell.*

An organism is formed during the process of ***mitosis,*** successive divisions of the fertilized egg that produce two identical diploid cells per cell division. Mitosis does not alter the genotype of the resultant cells, but rather increases the number of cells in the growing organism. Mitosis ensures that every cell in the human body is genetically identical to every other cell, such that specialized cell types (e.g., nerve, skin, and brain) arise from changes in gene expression rather than DNA sequence. Gene expression does not alter the sequence of the genes, but rather alters the amounts of the gene products (e.g., mRNAs and proteins) produced in each cell type. Mitosis, in conjunction with gene expression, provides an efficient means of obtaining a complex organism from a fertilized egg.

One practical consequence of mitosis is that the genotype of a patient can be determined by examining the genomic DNA from any somatic cell type, including blood, skin cells, tissue biopsies, and the like. Genotypes can also be determined from semen because sperm (gametes) contain every allele in the genome in a ratio of 1:1. The genome of an organism is synthesized during mitosis, by a complex interplay of cellular enzymes.

DNA Replication

Replication. *Cellular process by which DNA is copied from a DNA template to produce an exact copy of the genome.*

Fidelity. *Measure of the precision of DNA replication, represented typically as the number of errors per nucleotide copied.*

DNA ***replication*** is the cellular process by which exact copies of the genome are synthesized by copying the double-stranded DNA template. Similar to transcription and translation, DNA replication is a complex cellular process involving multiple steps and enzymes. One striking aspect of DNA replication is the precision or ***fidelity*** of the process. The replication error rate in human cells is approximately one mistake in 10^9–10^{10} nucleotides, allowing the cellular replication machinery to duplicate the entire human genome without making more than about one mistake per 3 billion base pairs. The extraordinary fidelity of DNA replication ensures that the each cell in the human body contains exactly the same genomic sequence, which maintains the integrity of the genetic blueprint.

Nucleosome. *Higher order nuclear structure in which DNA is wound up in three dimensions.*

The highly elongated structure of DNA is such that the chromosomes that make up the human genome are much longer than the size of the nucleus in which they are contained. At a length of approximately 3.3 Å per base pair, the 257,000,000 base pairs of human chromosome 1 has a physical length of 8.4 cm, which is >1000 times the width of a 5-μm nucleus. To accommodate the highly elongated genome inside the nucleus, DNA is compacted into a structure akin to a ball of yarn; higher order structures, including ***nucleosomes,*** wind up the DNA in three dimensions. The knotted configuration of nuclear DNA means it is imperative to unwind and separate the DNA strands before replication, tasks that are provided by the cellular replication machinery.

Replication origin. *Discrete chromosomal location at which DNA replication occurs.*

Cell cycle. *Biochemical process that divides mitosis into four discrete phases known as presynthetic gap, synthesis, postsynthetic gap, and mitosis.*

S phase. *Second step in the cell cycle, during which DNA synthesis occurs.*

Once the strands are unwound and separated, the cellular enzyme DNA polymerase uses the single-stranded DNA templates to replicate the DNA (Fig. 4.10). Polymerase uses the DNA strands as templates to create exact copies of the existing chains, laying down nucleotides in a 5′ to 3′ manner. After both DNA strands have been copied by polymerase, a second enzyme known as DNA ligase closes the breaks in the sequence, forming covalent bonds between the newly synthesized polynucleotide chains. DNA replication occurs at discrete chromosomal sites known as ***replication origins,*** during a portion of the ***cell cycle*** known as synthesis, or the ***S phase.*** The entire human genome is replicated

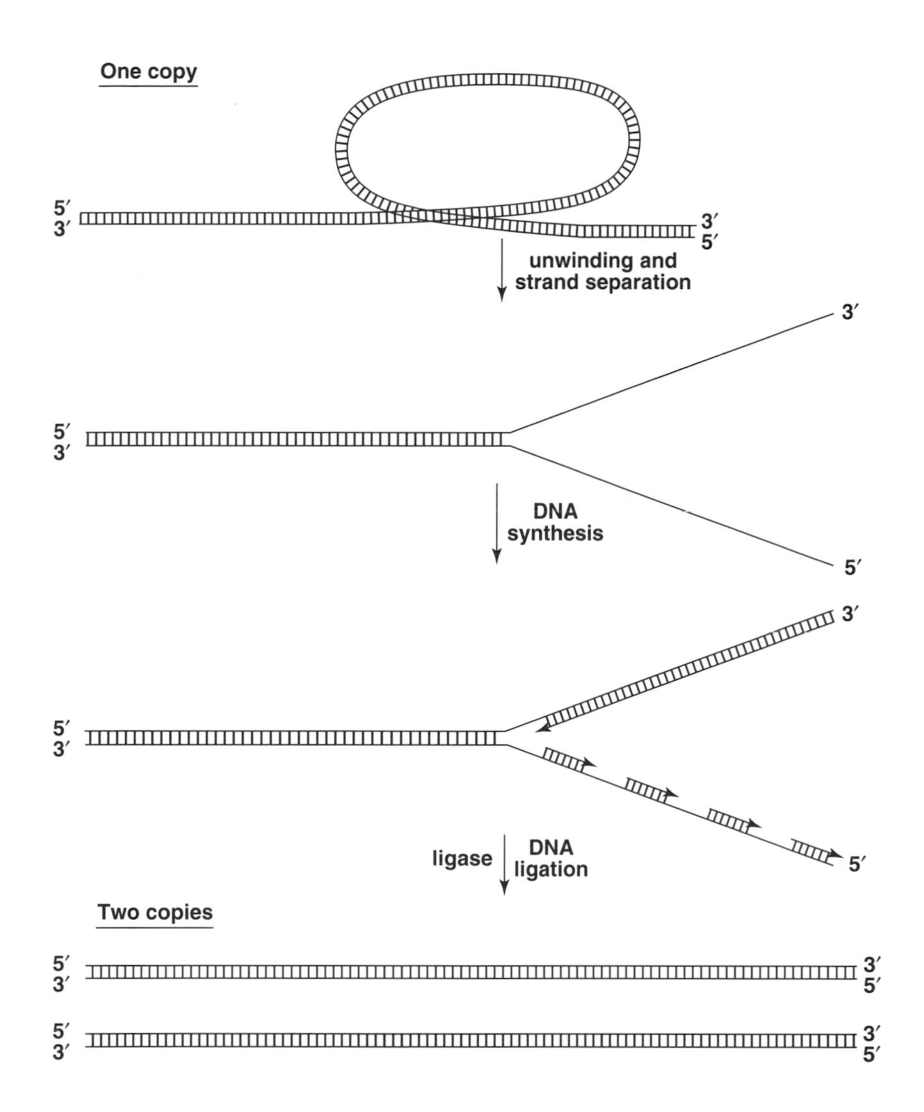

FIGURE 4.10. Highly compacted chromosomal DNA (top) *is unwound and the strands are separated to provide single-stranded templates for replication* (forked structure). *DNA polymerase copies both strands, laying down polynucleotide chains* (arrows) *in a 5′ to 3′ manner. DNA ligase closes the breaks in the newly formed chains, yielding two identical copies of DNA* (bottom).

once during each S phase of the cell cycle, ensuring that the cells produced by mitosis receive one complete copy of the genome per cell division.

DNA Sequencing

The process of determining the primary nucleotide sequence of a unknown DNA molecule is known as DNA sequencing. DNA sequencing has undergone significant technological advance since the late 1970s, though the fundamental chemistry has remained unchanged since its inception by Sanger and coworkers (1977). ***Dideoxy sequencing*** uses modified deoxyribose sugars that lack a hydroxyl group on the 3′ carbon, rendering them incapable of extending a growing DNA chain (Fig. 4.11). The Sanger method uses a trace concentration of the four dideoxynucleotides, together with a high concentration of the four deoxynucleotides, to produce a ladder of terminated DNA chains, each containing a dideoxynucleotide derivative at the 3′ end. Each dideoxynucleotide contains a unique fluorescent tag (Smith et al., 1985), allowing their unambiguous identification in the sequencing process. By reading the fluorescent tags on the 3′ end of each terminated sequencing product, it is possible to deduce the sequence of any DNA molecule of interest because the synthesized strand is an

Dideoxy sequencing. *Method of DNA sequencing that relies on chain termination by modified nucleotides that lack 2′ and 3′ hydroxyl groups.*

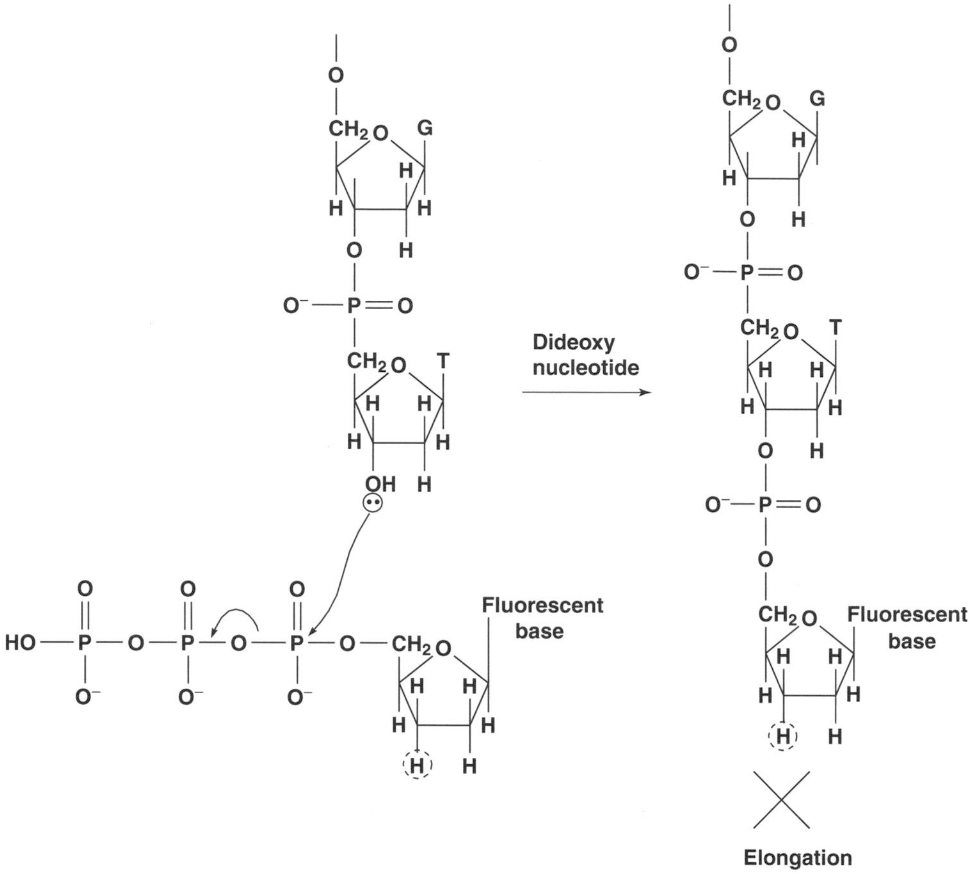

FIGURE 4.11. The dideoxy sequencing method uses fluorescent dideoxy nucleotides (bottom), *which lack a hydroxyl group on the 3′ carbon (dashed circle), preventing elongation of the DNA chain* (X). *The dideoxy nucleotide is incorporated into the growing chain by nucleophilic attack* (arrow) *between the 3′ OH group and the 5′ phosphate of the dideoxy nucleotide* (left). *Chains that contain a dideoxy nucleotide cannot undergo further elongation* (right). *Each of the four dideoxy bases bears a different fluorescent tag, allowing their identification on sequencing gels.*

Capillary electrophoresis. *Method of chromatography used in DNA sequencing in which the DNA molecules are separated by movement through thin glass tubes.*

Shotgun sequencing. *Method of determining an unknown sequence by breaking the sequence into small pieces, analyzing the fragments at random, and compiling the overlapping fragments to generate a complete sequence.*

Sequence trace. *Data from a DNA sequencing machine corresponding to the nucleotide sequence of one DNA sample.*

Coverage. *Term from DNA sequencing that refers to the number of times a given genomic segment of genome is represented in the sequencing data, represented typically as fold coverage.*

Finishing. *Late phase of the DNA sequencing process in which small gaps are filled and sequences aligned.*

exact complement of the unknown target sequence. Dideoxy sequencing is used extensively in microarray analysis to determine the nucleotide composition of target DNA.

The development of ***capillary electrophoresis*** is a recent advance that sped human genome sequencing. The original Sanger method used large vertical slab gels to separate the DNA chains, a rather slow process that was limited by the difficulty in dissipating the heat generated during slab gel electrophoresis. Capillary electrophoresis uses fine (e.g., 100 μm) glass tubes to separate the DNA chains, a format that allows rapid heat dissipation and enables much faster electrophoresis than the original format. The capillary format is highly compact and uses low reagent volumes, which conserves laboratory space and reduces cost.

Shotgun sequencing is a method that breaks genomic DNA into small pieces, sequences the DNA templates at random, and then assembles the primary nucleotide sequence of the genome by compiling the overlapping sequences obtained from the DNA fragments. The genome is broken into small pieces (0.2–10 kb) using enzymes or mechanical shearing, and the genomic fragments are introduced at random into a cloning vector and propagated in bacteria. The ***sequence traces*** are stored in a large sequence database, and computer programs are used to compile a complete genomic sequence by assembling the overlapping fragments into a continuous linear sequence. A 5- to 10-fold ***coverage*** is usually sufficient to afford a nearly complete genomic sequence for most organisms, though a tedious ***finishing*** process is required to fill in the large number of tiny gaps that are produced using the shotgun method. Capillary electrophoresis, in conjunction with fluorescent Sanger chemistry and the shotgun strategy, was used to determine the nucleotide sequence of the human genome (see below).

Genome Structure

What is the structure of a genome? The answer is that each organism contains a unique genomic sequence with a unique structure. Complete and partial sequences are available for approximately 800 organisms, providing an unprecedented view of how genomes are configured in different life forms. Because the technical difficulty of genome sequencing increases as the size of the genome increases, most complete genome sequences have been obtained from relatively simple organisms such as viruses and bacteria. Nearly complete sequences are available for a small number of complex organisms including yeast, *Drosophila* (fruit fly), *Caenorhabditis elegans* (worm), *Arabidopsis* (mustard plant), mouse, and human (Table 4.1). Several observations emerge from this wealth of sequence information.

The genomes of many viruses and bacteria consist of a single circular chromosome, whereas higher organisms including humans have multiple linear chromosomes (Table 4.1). Viral and bacterial genomes are highly compact with respect to gene density, and genes in these organisms are often overlapping or use both strands of the DNA, in cases where genes are expressed in opposing directions. Genes in higher organisms are spaced relatively far apart on the chromosomes, rarely overlap, and typically contain one or more introns in the coding sequence. The yeast genome is organizationally intermediate between bacteria and multicellular organisms, with a high gene density and few overlapping genes and introns. Genomic complexity generally tracks with biological complexity, such that complex organisms contain larger genomes and a greater gene and chromosome number than simple organisms. The human genome contains 3,000 ***megabases*** and 35,000 genes distributed across 23 chromosomes, compared to the 140-megbase fruit fly genome that contains 14,000 genes and 5 chromosomes, and the 125-megabase mustard plant genome with its 25,000 genes and 5 chromosomes. The human and mouse genomes appear to be similar in size and gene number, though the mouse has 3 fewer chromosomes than the human.

Megabase. *Length of DNA containing 1 million base pairs.*

TABLE 4.1. Genome Specifications for Some Well-Studied Organisms

Organism	Type	Chromosome Number	Gene Number[a]	Genome Size (bp)[a]
Hepatitis B	Virus	1 (circular)	4	3,215
Escherichia coli	Bacterium	1 (circular)	4,394	4,639,221
Saccharomyces cerevisiae	Yeast	16 (linear)	6,183	12,000,000
Drosophila melanogaster	Fruit fly	5 (linear)	14,000	140,000,000
Caenorhabditis elegans	Worm	6 (linear)	19,000	90,000,000
Arabidopsis thaliana	Flowering plant	5 (linear)	25,000	125,000,000
Mus musculus	Mouse	20 (linear)	35,000	3,000,000,000
Homo sapiens	Human	23 (linear)	35,000	3,000,000,000

[a] Approximations based on current data.

The recent completion of the human genome sequence (Lander et al., 2001; Venter et al., 2001) has provided an unprecedented view of human physiology, behavior, aging, and disease at the DNA level. The human genetic blueprint provides a powerful entry point for elucidating gene function and sequence differences that affect human health. Before the completion of the human genome, gene number predictions ranged from 50,000 to 150,000, far greater than the number of genes actually present in humans (~35,000). Early gene number predictions were based on the assumption that humans are much more complex than simple organisms and therefore must possess a much greater number of genes, an argument that is flawed if one considers the combinatorial aspect of genes and gene products. If one assumes that a cellular gene can exist in either of two expressed states (on or off), and that each gene product can alter the functional state of an organism, the number of different functional states in an organism is 2^n, where 2 is the number of states (on or off) and n is the number of genes in the organism.

$$\text{Number of functional states} = 2^n \quad (4.1)$$

It should be obvious from Equation 4.1 that a relatively small increase in gene number can produce a massive increase in the number of functional states, potentially enabling a significant increase in biological complexity with only a modest increase in gene number. An organism with 100 more genes than another organism, for example, could execute 2^{100} or 1×10^{30} additional functional states, demonstrating that humans need not have a vastly greater gene number than simpler organisms to execute much more complex genetic programs.

SUMMARY

The gene is the genetic unit of life, and an average human gene contains approximately 20,000 nucleotides divided into exons and introns. Sequence variants, including SNPs and mutations, can alter protein function and produce disease. Genes can be easily manipulated in the laboratory using the tools of recombinant DNA technology and PCR. Cloned and amplified cDNAs, ESTs, and BACs find use as microarray targets. Exon shuffling provides a molecular model for gene evolution and natural selection.

Genetic information in the DNA is converted into mRNA (transcription) and protein (translation) during the two-step gene expression process. Transcription uses activators and repressors to increase and decrease the rate of transcription by altering the efficiency of promoter usage. Newly transcribed mRNA molecules undergo extensive processing, including capping, polyadenylation, and splicing. Processed mRNAs are exported from the nucleus into the cytoplasm, to undergo ribosome-directed protein synthesis. The genetic code is co-linear, such that the codons specify a corresponding sequence of amino acids.

Heredity is imparted through genomic inheritance, and the genome is copied during DNA replication. A fluorescent version of dideoxy DNA sequencing, coupled with the shotgun method, was used to sequence the 3 billion base pair human genome. Complete genome sequences of viruses, bacteria, fungi, worms, insects, plants, rodents, and mammals provide an unprecedented view of the genetic basis of life. The exciting field of genomics, together with microarray analysis, speeds advancing forefronts in molecular medicine, including genetic screening, testing, and diagnostics.

SELECTED READING

Baxevanis, A. D., and Ouellette, B. F. F., eds. *Bioinformatics: A Practical Guide to the Analysis of Genes and Proteins,* 2nd ed. Wiley-Interscience, New York, 2001.

Blackburn, G. M., and Gait, M. J., eds. *Nucleic Acids in Chemistry and Biology,* 2nd ed. Oxford University Press, Oxford, UK, 1996.

Bloomfield, V. A., Crothers, D. M., and Tinoco, I. *Nucleic Acids: Structure.* University Science Books, Herndon, VA, 2000

Campbell, K. H., McWhir, J., Ritchie, W. A., and Wilmut, I. Sheep cloned by nuclear transfer from a cultured cell line. *Nature* 380:64–66, 1996.

Claverie, J. M., Hardelin, J. P., Legouis, R., Levilliers, J., Bougueleret, L., Mattei, M. G., Petit, C. Characterization and chromosomal assignment of a human cDNA encoding a protein related to the murine 102-kDa cadherin-associated protein (alpha-catenin). *Genomics* 15:13–20, 1993.

Devlin, T. M. *Textbook of Biochemistry with clinical correlations,* 5th ed. Wiley, New York, 2002.

Innis, M. A., Gelfand, D. H., and Sninsky, J. J., eds. *PCR Applications: Protocols for Functional Genomics.* Academic Press, San Diego, CA, 1999.

Kornberg, A., and Baker, T. A. *DNA Replication,* 2nd ed. Freeman, New York, 1991.

Lander, E. S., Linton, L. M., Birren, B., Nusbaum, C., Zody, M. C., Baldwin, J., Devon, K., Dewar, K., and co-workers. Initial sequencing and analysis of the human genome. *Nature* 409:860–921, 2001.

Lehninger, A. L., Nelson, D. C., and Cox, M. M. *Principles of Biochemistry,* 3rd ed. Freeman, New York, 2000.

Lewin, B. *Genes VII.* Oxford University Press, Oxford, UK, 2000.

Maki, R., Traunecker, A., Sakano, H., Roeder, W., and Tonegawa, S. Exon shuffling generates an immunoglobulin heavy chain gene. *Proc Natl Acad Sci U S A* 77:2138–2142, 1980.

Matthews, H. R., Freedland, R. A., and Miesfeld, R. L. *Biochemistry: A Short Course.* Wiley, New York, 1997.

Metzger, D., White, J. H., and Chambon, P. The human oestrogen receptor functions in yeast. *Nature* 334:31–36, 1988.

Sanger, F., Nicklen, S., and Coulson, A. R. DNA sequencing with chain-terminating inhibitors. *Proc Natl Acad Sci U S A* 74:5463–5467, 1977.

Schena, M., ed. *DNA Microarrays: A Practical Approach,* 2nd ed. Oxford University Press, Oxford, UK, 2000.

Schena, M., ed. *Microarray Biochip Technology.* Eaton Natick, MA, 2000.

Schena, M., and Yamamoto, K. R. Mammalian glucocorticoid receptor derivatives enhance transcription in yeast. *Science* 241:965–967, 1988.

Singer, M., and Berg, P. *Genes and Genomes.* University Science Books, Herdon, VA, 1991.

Smith, L. M., Fung, S., Hunkapiller, M. W., Hunkapiller, T. J., and Hood, L. E. The synthesis of oligonucleotides containing an aliphatic amino group at the 5′ terminus: Synthesis of fluorescent DNA primers for use in DNA sequence analysis. *Nucleic Acids Res* 13:2399–2412, 1985.

Streitweiser, A., and Heathcock, C. H. *Introduction to Organic Chemistry,* 2nd ed. Macmillan, New York, 1981.

Stryer, L. *Biochemistry,* 4th edition. Freeman, New York, 1995.

Venter, J. C., Adams, M. D., Myers, E. W., Li, P. W., Mural, R. J., Sutton, G. G., Smith, H. O., Yandell, M., and co-workers. The sequence of the human genome. *Science* 291:1304–1351, 2001.

White, O., Dunning, T., Sutton, G., Adams, M., Venter, J. C., and Fields, C. A quality control algorithm for DNA sequencing projects. *Nucleic Acids Res* 21:3829–3838, 1993.

Yamamoto, K. R. Steroid receptor regulated transcription of specific genes and gene networks. *Annu Rev Genet* 19:209–252, 1985.

REVIEW QUESTIONS

1. A segment of genomic DNA that encodes a specific mRNA and protein is known as a what? One word will suffice.
2. Is protein-coding information specified by introns, exons or both?

3. Small changes in the primary DNA sequence are known as sequence variants. Name three types of sequence variants.
4. Recombinant DNA and the polymerase chain reaction are two revolutionary technologies that enable gene studies. Describe each technology briefly.
5. Name two types of cloned DNA molecules and describe each briefly.
6. The flow of genetic information from DNA into mRNA, and mRNA into protein is known as what and what? One word for each will suffice.
7. A gene expression microarray experiment identifies 10 cellular genes that are activated by a steroid hormone. Sequencing studies reveal that all 10 cellular genes share an identical eight-nucleotide sequence immediately upstream of the TATA box. Describe a molecular mechanism to explain the microarray and sequencing data.
8. Molecules of mRNA undergo extensive processing. Name three processing reactions and describe each briefly.
9. Cellular ribosomes are macromolecular structures that coordinate protein synthesis. Name two types of biomolecules that interact with the ribosome during protein synthesis.
10. Which of the following is *not* contained in genomic DNA? Exon, intron, gene, polyA sequence, enhancer, or TATA box.
11. Essentially every cell in the human body contains an exact copy of the genome, but cells adopt different fates and perform different functions (e.g., nerve cells, liver cells, and skin cells.). What cellular process allows different functions to be achieved from the same DNA blueprint? Explain your answer in a few sentences.
12. A microarray genotyping assay reveals that the DNA from a patient contains a 1:1 ratio of normal and mutated allele for a recessive genetic disease locus. Will the patient manifest the illness associated with the locus? Explain your answer in molecular terms.
13. The out of Africa hypothesis suggests that all humans evolved from a common African descendant >250,000 years ago. If the genetic blueprint were maintained perfectly during human evolution, all humans would share an identical DNA sequence, which is different from the observation that DNA in modern humans differs at about 1 nucleotide per 1000 bases from person to person. Name two contributors to DNA sequence variation during evolution.
14. Dideoxynucleotides prevent the elongation of DNA chains in Sanger sequencing chemistry. Name the functional group missing on the dideoxynucleotide compared to the deoxynucleotide.
15. Rank the following organisms in terms of increasing gene number and genome size: mouse, bacterium, yeast, fruit fly, and virus.
16. A researcher analyzes the genomes of a newly sequenced insect and mammal and finds that the mammal contains only 2.5 times as many genes as the insect. He concludes that the mammalian sequence must be incomplete because mammals are vastly larger and more complex than insects, and at least 10 times as many genes must be present to perform all of the functions in the mammal. Is this rationale correct or incorrect? Use your knowledge of gene expression to answer the question.

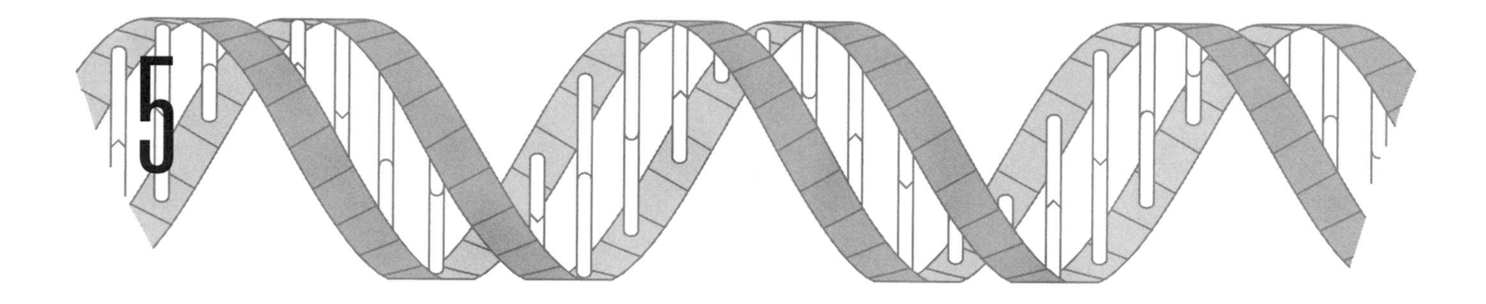

5

Microarray Surfaces

Only torture will bring out the truth.
—*C. S. Lewis*

Robust microarray analysis requires high-quality surfaces, the preparation of which can be a demanding (and sometimes torturous) process. The surface plays an important role in determining how well the molecules attach to it as well as the efficiency of the biochemical reactions, the precision of the detection steps, and the quality of the resultant data. This chapter provides an overview of microarray surfaces, including optimal properties, reaction kinetics, substrate materials, and surface chemistry.

OPTIMAL SURFACE PROPERTIES

In many respects, a microarray experiment is only as good as the surface that is used to create it. The first microarray experiments were actually performed on inexpensive microscope slides coated with polylysine. As microarray analysis has evolved and improved since the mid-1990s, the quality of microarray surfaces and the availability of sophisticated commercial products have grown steadily. An ideal microarray surface should be ***dimensional, flat, planar, uniform, durable, inert, efficient,*** *and* ***accessible.*** Though materials and surface chemistries vary considerably and will continue to diversify, all good microarray surface should endeavor to possess as many of these qualities as possible. A brief description of each of these surface properties follows.

Dimensional. *Evaluative criterion for a microarray substrate that refers to the physical size of the substrate.*

Flat. *Evaluative criterion for a microarray substrate that refers to the smoothness of the surface over a small area.*

Planar. *Evaluative criterion for a microarray substrate that refers to the parallelism of the surface over the entire substrate.*

Uniform. *Evaluative criterion for a microarray substrate that refers to the regularity of the spacing of the chemical groups on the surface.*

Durable. *Evaluative criterion for a microarray substrate that refers to the stability of the chemical groups, treatment or coating on the surface.*

Inert. *Evaluative criterion for a microarray substrate that refers to the non-contribution of the substrate to the assay signal.*

Dimensional

Efficient. *Evaluative criterion for a microarray substrate that refers to the efficacy by which the surface attaches printed target molecules.*

Accessible. *Evaluative criterion for a microarray substrate that refers to the efficacy by which the surface enables productive interactions between target and probe molecules.*

Dimensionality defines the physical size of the microarray substrate. The most commonly used substrate at present shares the dimensions of the traditional 1 × 3-in. microscrope slide (25 × 76 × 0.94 mm). Another common format developed by Affymetrix (Santa Clara, CA) is 0.5 × 0.5 in. Constant physical dimensionality of the substrate is desirable because it adds precision to the manufacturing and detection steps in microarray analysis. Particularly for cases in which instruments use the substrate edges or surfaces for location, printing, or detection purposes, dimensionality improves microarray analysis. Common benchmarks in the x and y (width and length) and z (thickness) axes are $\pm$ 300 μm and $\pm$10 μm, respectively.

Flat

Flatness refers to the evenness or two-dimensional quality of the microarray surface over a local region. An ideal microarray surface is flat to a precision of $\pm$ 10 μm or better in the "up and down," or z axis, relative to the horizontal plane (x and y axes). Deviations from flatness impair microarray analysis in two serious ways. First, such surfaces can compromise the quality of microarray printing. This is particularly the case for noncontact methods, such as ink-jet printing, that rely on a precise distance from the ink-jet nozzle to the surface to create spots of uniform diameter. Flat surfaces produce superior microarrays relative to those that vary over local regions (Fig. 5.1), though surface unevenness can be compensated by some of the contact printing methods that use "floating pins" (see Chapter 6). Deviations from flatness can also impair microarray analysis by changing the distance of the microarray surface relative to the detection device. Because many detectors have a relatively small depth of focus (e.g., 20–30 μm), deviations from flatness can lead to unevenness in the detected signal (Fig. 5.2). One means of achieving flat microarray substrates is to polish glass to optical-quality specifications (discussed later in this chapter).

Planar

The criterion of planarity refers to the evenness or two-dimensional quality of the entire microarray substrate. For the standard microarray substrate used in most laboratories (25 × 76 mm), a planar substrate is one that is even or level over the entire 25 mm in the x axis and 76 mm in the y axis. Because the

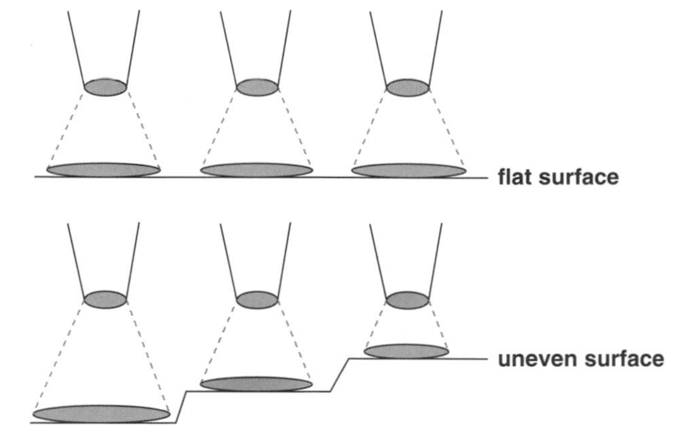

FIGURE 5.1. *Because the distance between an ink-jet nozzle and the microarray surface determines the sample spray diameter, flat surfaces produce microarrays with uniform spot diameters, whereas uneven surfaces produce microarrays with variable spot diameters.*

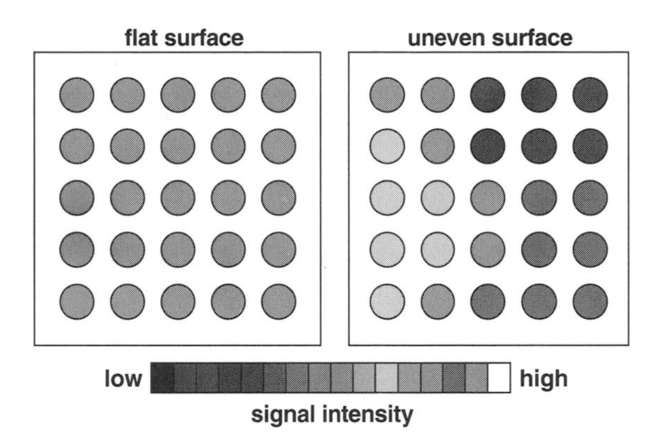

FIGURE 5.2. Because microarray detectors have a limited depth of focus, a hypothetical microarray of identical samples formed on a flat surface will produce a uniform image, whereas the same microarray formed on an uneven surface will produce an image that has variations in signal intensity. Figure also appears in Color Figure section.

same considerations for flatness hold for planarity, a substrate level to within $\pm$ 10 μm is considered planar with respect to printing and detection. Substrates that are tilted or twisted along either x or y will produce inaccurate readings in microarray experiments. Because glass manufacturing is highly advanced, nonplanarity is rarely a problem with glass substrates. Nonplanarity can be a consideration with experimental materials, such as injection-molded plastics, or when the glass microarray is enclosed in a plastic cassette. In the latter case, care must be taken to ensure that the microarray cassette does not exert any twisting or torsional force on the glass chip that it is holding.

Uniform

Uniformity refers to the atomic or molecular regularity of chemically reactive groups on a microarray surface. Uniform microarray surfaces are those that have an equal number of reactive groups per unit area (i.e., equal density); a common standard is $\pm$ 25% variation over the entire microarray substrate. Uniformity applies to both surface treatments and surface coatings. A ***surface treatment*** is defined as a chemically reactive monolayer formed by the covalent linkage of reactive groups directly to substrate molecules. Organosilanes (see below) are a class of chemical reagents commonly used in the preparation of surface treatments. A ***surface coating*** is defined as a thin film placed on top of the microarray substrate in a noncovalent manner. Acrylamide, polylysine, and nitrocellulose (see below) are common surface coatings. Because the printed target molecules attach to the substrate via the reactive groups in the surface treatment or surface coating, the uniformity of the reactive groups determines the resultant uniformity of the target molecules. Because microarray assays involve direct biochemical interactions between target molecules attached to the chip and probe molecules in the sample, uniform surface treatments and surface coatings produce more uniform and hence more accurate signals than non-uniform surfaces (Fig. 5.3).

Surface treatment. *Chemically reactive monolayer formed by the covalent linkage of reactive groups to the microarray substrate.*

Surface coating. *Thin film placed in a noncovalent manner on a microarray substrate.*

Durable

The durability of the microarray surface refers to the stability of the substrate and surface treatment or surface coating over the duration of the microarray assay. A typical benchmark for durability is that a processed microarray should loose fewer than 10% of the target molecules over the assay duration. Microarray analysis protocols vary considerably, but many involve a range of processing and

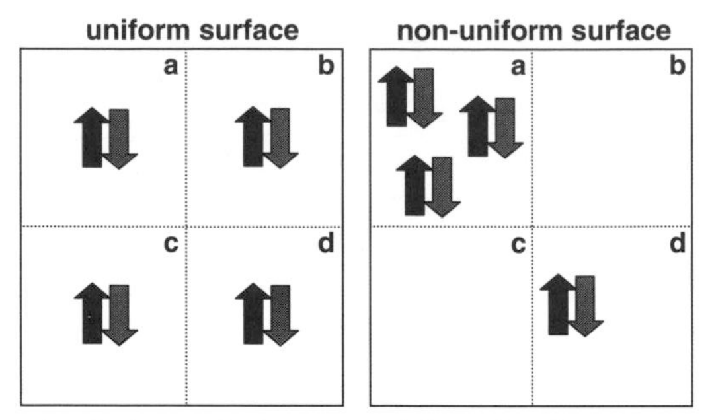

FIGURE 5.3. Target molecules (up-pointing arrows) *bound to the microarray substrate via reactive groups interact directly with probe molecules* (down-pointing arrows) *in the sample. The microarray formed on the chemically uniform surface produces uniform signals throughout the microarray, whereas one formed on the chemically non-uniform surface gives variable signal intensities.*

reaction conditions, including the use of elevated temperatures (e.g., boiling water to denature double-stranded DNA) and organic reagents (e.g., formamide in the hybridization buffer). Given that the target molecules confer binding of the probe molecules and hence determine the strength of microarray signals, robust microarray assays require durable substrates, surface treatments, and surface coatings. The loss of bound target molecules during the assay will reduce microarray signals (Fig. 5.4), and variable loss of surface treatments or coatings can greatly impair diagnostic applications of microarrays. Given the inherent stability of glass and the strength of the covalent bonds used in surface treatments, durability is rarely an issue with glass substrates and organic silane treatments. Durability can become a problem with substrates such as silicon or plastics that exhibit less physical stability than glass under certain conditions. Surface coatings, which are attached in a noncovalent manner to the substrate, are also the subject of durability concerns under certain conditions.

Inert

A microarray surface is described as inert if the surface binds target molecules in a stable manner but does not contribute any gain or loss of signal in the microarray assay. Similar to the criterion of durability, inertness pertains to

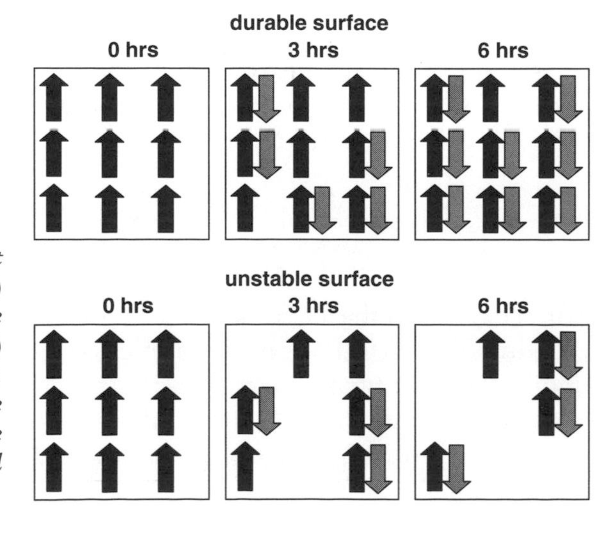

FIGURE 5.4. Stably bound target molecules (up-pointing arrows) *bind progressively (0–6 h) to probe molecules* (down-pointing arrows) *over a 6-h microarray experiment. Signal strength is greater on the durable surface than on the unstable surface because the latter looses bound target molecules over time.*

both the substrate and the surface treatment or surface coating. Common benchmarks for the substrate and surface treatment or surface coating are that each contributes less than a twofold additional signal relative to the signal obtained from a detector without a loaded microarray. According to these standards, the absolute detected signals would be 250 counts, ≤ 500 counts, and $\leq 1,000$ counts for the empty detector, the detector plus substrate, and the detector plus substrate with a surface treatment or surface coating, respectively. Given the inherent transparent property of glass, contributions to the microarray signal are rarely a concern with this inert substrate material. More common deviations from inertness are seen with surface treatments and surface coatings, which can complicate microarray signals by elevating or reducing true microarray readings. Common sources of erroneously elevated readings are background fluorescence caused by the chemicals used to prepare the surface treatments and surface coatings, light scattering attributable to the surface coating, and light scattering due to dust and other particulate contaminants obtained during microarray manufacture or use. Erroneously low readings can be observed in cases in which enzymatic (e.g., ribonucleases) or chemical agents (e.g., oxidants) reduce true microarray signals by damaging target or probe molecules during the microarray assay.

Efficient

Efficiency pertains to the potency or efficacy of the surface treatment or surface coating to bind printed target molecules. The common benchmark for target-binding efficiency is in the range of 10–30% for a target mixture that contains an oligonucleotide at 30 μM concentration or a polymerase chain reaction (PCR) product that is 0.3 $\mu g/\mu L$. Most surface treatments and surface coatings provide efficiencies within this range. Microarray surfaces with efficiencies $< 10\%$ require greater concentrations of target solution and, therefore, are undesirable because they increase the consumption of expensive reagents, leading to elevated costs in microarray manufacture. Binding efficiencies of $> 30\%$ can be achieved by preparing surfaces with an excessive density of reactive groups, though excessive reactive group density can lead to elevated background reactivity and exaggerated hydrophobicity, both of which can diminished microarray performance (see below). Binding efficiencies need to be measured and adjusted empirically for each surface treatment or surface coating.

Accessible

An accessible microarray surface is one that permits as many productive biochemical interactions (e.g., hybridization) as possible between target and probe molecules. A common accessibility benchmark is that a target-optimized microarray surface should provide productive interactions for 50% of all target molecules under conditions of probe excess (see below). Accessibility is rarely as issue with planar surfaces, such as glass, that have organic monolayer surface treatments (e.g., organoamine or organoaldehyde); the targets on these surfaces tend to be readily accessible, and the flat glass surface provides little steric interference. Accessibility can be an issue with some surface coatings (e.g., acrylamide and nitrocellulose) that require the diffusion of probe molecules through the coating matrix, though reduced accessibility of certain surface coatings can be compensated by an increase in target density for these three-dimensional surfaces (see below).

SURFACE INTERACTIONS AND REACTION KINETICS

One common misconception in the rapidly evolving landscape of microarray science, is the notion that "more is better" with respect to the density of bound target molecules on the microarray surface and to the concentration of probe molecules in the labeled sample. In fact, it is really an optimal density of target molecules and an optimal concentration of probe molecules that produce the best results in microarray experiments. A brief consideration of microarray reaction kinetics will help define and explain optimal target density and optimal probe concentration.

Optimal Target Density

A simple demonstration that more is not always better can be seen in a target molecule titration experiment in which microarray signals are measured as a function of target molecule concentration in an oligonucleotide-to-oligonucleotide hybridization experiment. In this experiment, a 15-base oligonucleotide (15-mer) was printed on a microarray substrate at a range of different concentrations from 1 to 100 μm. The printed microarray was then processed to remove unbound oligonucleotide and hybridized with a probe solution containing a fluorescent 15-mer complementary to the target sequence. The microarray was scanned and the fluorescent signals were measured at each location on the microarray corresponding to six different target molecule concentrations: 1, 3, 10, 30, 50, and 100 μM. The fluorescent intensities were then plotted as a function of target concentration (Fig. 5.5).

Examination of the data reveals that the fluorescent intensity increases steadily in the range of 1–10 μM target and reaches a peak intensity at 30 μM oligonucleotide, at which point the signals levels off and decreases significantly as the target concentration approaches 100 μM (Fig. 5.5). The target concentration of 30 μM gives the strongest signal in the microarray hybridization experiment; thus this concentration defines the ***optimal target concentration.*** Given the optimal target concentration, it is straightforward to calculate the number of target molecules bound to the microarray surface per unit area (Appendix H). Assuming that 30% of the printed oligonucleotide couples to the substrate and that a typical printed droplet is 300 pL, a 30-μM solution of oligonucleotide gives 2.6×10^5 oligonucleotide molecules per square micron of substrate. Because this number produces the strongest fluorescent signals in a hybridization experiment, the value 2.6×10^5 molecules/μm^2 is known as the ***optimal target density.*** Additional area conversions reveal that 2.6×10^5 molecules/μm^2

Optimal target concentration. *Number of target molecules per unit volume of printed sample that provides the strongest signal in a microarray assay.*

Optimal target density. *Number of target molecules per unit area on a microarray substrate that provides the strongest signal in a microarray assay.*

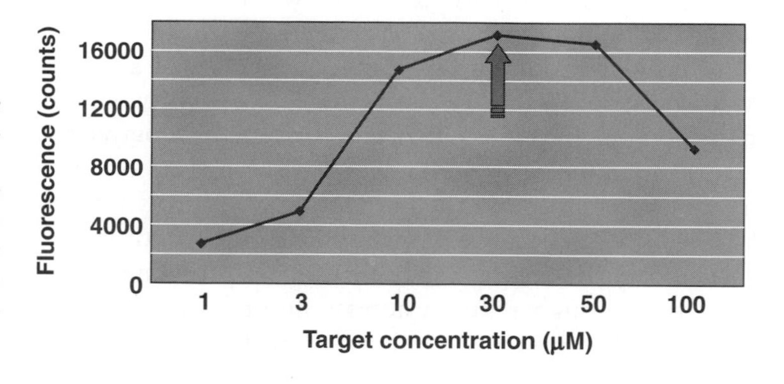

FIGURE 5.5. *Scanned fluorescence intensities were plotted as a function of target sample concentration. The target concentration that gives the strongest fluorescent signal* (arrow) *defines the optimal target density. The conversion from target concentration to target density is provided in Appendix I.*

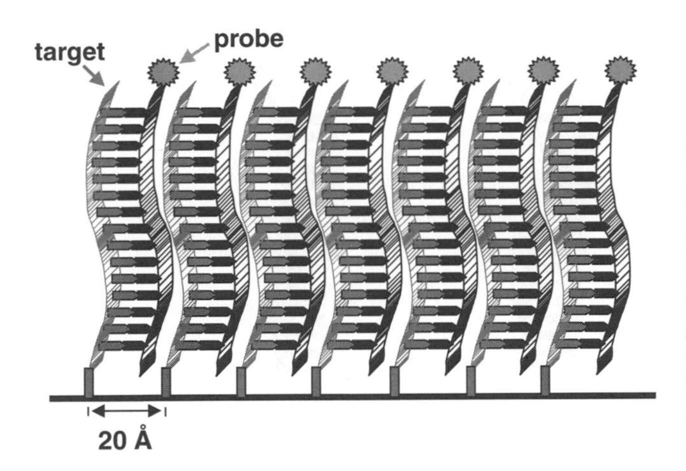

FIGURE 5.6. Target DNA molecules are attached by linkers to a microarray substrate (horizontal bar) *and hybridized with probe molecules that bear fluorescent tags. A target density of 1 target molecule per 20 Å provides the tightest possible packing of target–probe duplexes (optimal target density) and the strongest fluorescent signal. Data from Figure 5.5; see Appendix H for conversions.*

corresponds to 1 oligonucleotide per 400 Å^2, or 1 target molecule per 20 Å in a single dimension. Why does 1 molecule every 20 Å (1 per 400 Å^2) define the optimal target density, while lesser and greater target densities give weaker fluorescent signals?

The answer to understanding optimal target density lies in the physical size of DNA and in the physical or steric considerations that play out when target and probe DNA molecules hybridize to each other on the microarray surface. A single DNA strand (e.g., the target) is approximately 12 Å in diameter, and the hybridized target–probe duplexes are approximately 24 Å in diameter. Because DNA has both a major and minor groove, which allow for some additional packing on the microarray surface, the effective diameter of hybridized target–probe molecules is approximately 20 Å. A spacing of 1 target per 20 Å (1 per 400 Å^2) defines the optimal target density because it allows for tightest possible packing of target–probe duplexes, without causing physical interference (i.e., steric hindrance) during the hybridization (Fig. 5.6). Prohibitive target concentration produces too many target molecules on the microarray surface, and steric hindrance by target molecules prevents incoming probe molecules from hybridizing, thereby reducing the fluorescent signal (Fig. 5.7). Insufficient target density corresponds to a suboptimal number of target–probe duplexes, fewer fluorescent probe molecules per unit area, and a weaker fluorescent signal (Fig. 5.8). Insufficient target density can also lead to a saturation of target molecules during the hybridization reaction and assay nonlinearity (see below).

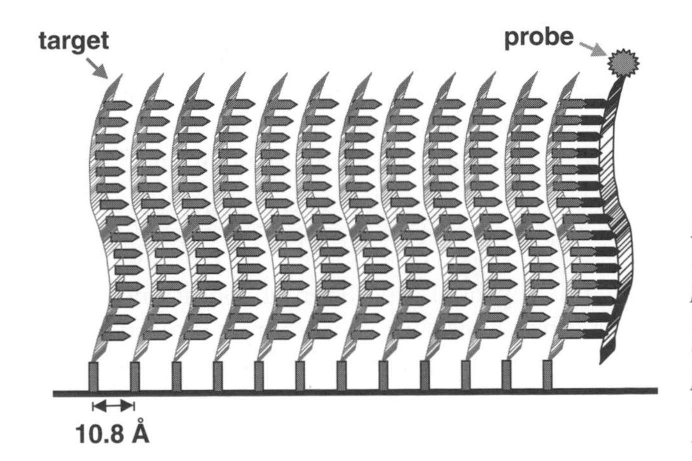

FIGURE 5.7. A target density of 1 target molecule per 10.8 Å is prohibitive, causing physical occlusion (steric hindrance) and preventing the efficient formation of target–probe duplexes, resulting in a reduced fluorescent signal. Data from Figure 5.5; see Appendix H for conversions.

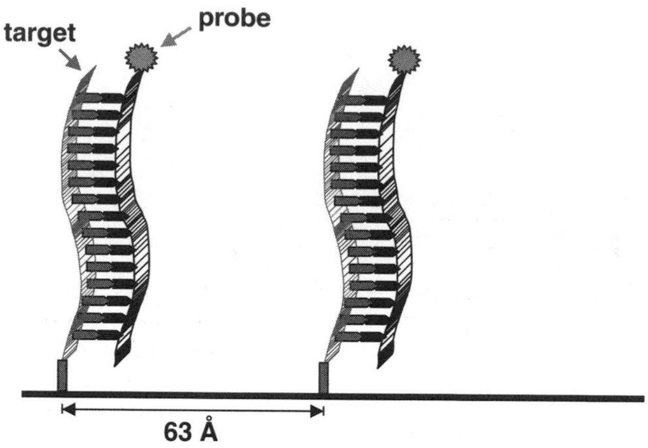

FIGURE 5.8. A target density of 1 molecule per 63 Å provides insufficient density, resulting in fewer target–probe duplexes and a reduced fluorescent signal. Data from Figure 5.5; see Appendix H for conversions.

Optimal Probe Concentration

Optimal probe concentration. *Number of probe molecules per unit volume of sample that provides the strongest signal in a microarray assay.*

Similar to target concentration and density, greater probe concentrations produce stronger signals in microarray experiments, though care must be taken not to use too much probe. The probe concentration that produces the strongest microarray signals without compromising microarray quantitation is known as the ***optimal probe concentration.*** Probe concentrations greater than the optimal concentration are useful under certain circumstances (see below), though extra care must taken when interpreting microarray experiments that use excessive probe concentrations. A closer look at microarray reaction kinetics helps explain the concept of optimal probe concentration.

Target (T) molecules on the microarray surface form productive interactions with probe (P) molecules in the solution to form target–probe (T–P) pairs. In the case of an oligonucleotide hybridization reaction, for example, the target oligonucleotide bound to the microarray surface forms a duplex with a labeled oligonucleotide in the probe mixture. The generalized biochemical reaction for target–probe binding can be represented as follows:

$$\text{T} + \text{P} \rightarrow \text{T–P} \tag{5.1}$$

The rate of formation of target–probe products depends on the concentration of the two reactants and can be expressed as the product of the concentration of T and P times a proportionality constant (k):

$$\text{Rate} = k[\text{T}][\text{P}] \tag{5.2}$$

Second order. *Biochemical reaction, such as hybridization, in which the rate depends on the concentration of two reactants.*

Rate constant. *Number in a kinetic equation that has a fixed value.*

Because the microarray reaction involves two reactants (target and probe), it is known kinetically as a ***second-order*** biochemical reaction. The constant k is known as the ***rate constant.*** During the course of a microarray reaction, the concentrations of T and P decrease and the concentration of T–P increases. These rates can be expressed by calculus as follows:

$$\text{Rate} = -\frac{d[\text{T}]}{d\text{T}} = -\frac{d[\text{P}]}{d\text{T}} = \frac{d[\text{T–P}]}{d\text{T}} = k[\text{T}][\text{P}] \tag{5.3}$$

Target excess. *Kinetic conditions in a microarray assay in which the concentration of the target molecules on the surface exceeds the concentration of the probe molecules in solution.*

Under optimal experimental conditions, the printed microarray will contain a much larger number of target molecules than are required to form T–P pairs during the course of the reaction. When the number of free target molecules is much greater than the number of molecules that form productive T–P pairs, the kinetic condition is known as ***target excess.*** Under microarray conditions of target excess, the effective change in concentration of T is

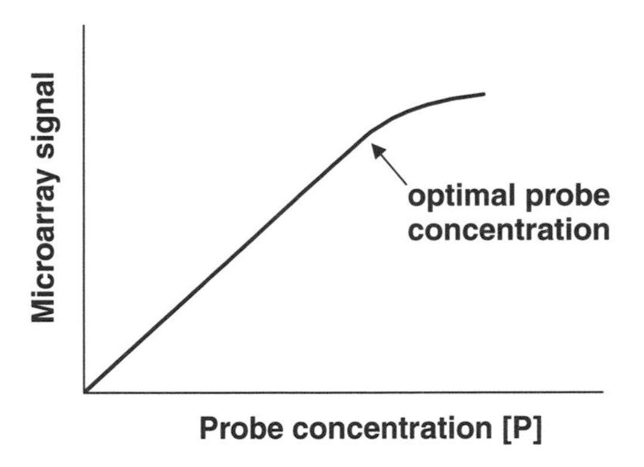

FIGURE 5.9. At optimal target concentration and at low probe concentration, microarray reactions follow pseudo-first-order kinetics whereby the fluorescent signal doubles with every doubling of probe concentration until the target becomes limiting (rounded portion of curve). The greatest probe concentration that gives the strongest signal (arrow) without diminishing quantitation in the assay is known as the optimal probe concentration.

relatively small over the course of the reaction. Under target excess conditions, [T] is relatively constant and can therefore be considered part of the constant (k term), in which case the rate equation can be simplified to:

$$\text{Rate} = k'[\text{P}] \tag{5.4}$$

k' denotes the fact that the constant target concentration has become part of this term. Equation 5.4 means that the rate at which productive target–probe pairs are formed depends solely on one reaction component, the concentration of the probe solution. Because Equation 5.4 is an approximation of a first-order reaction, the reaction conditions as written are known as ***pseudo-first-order*** kinetics. It follows from this equation that doubling the concentration of a microarray probe solution will double the rate of the reaction. Because faster rates result in more target-probe pairs per unit time and greater [T–P] means greater signal, it is desirable to use as much probe material as possible in any given microarray experiment as long as the performance of the assay is not compromised. The probe concentration that gives the strongest microarray signals without compromising quantitation is known as the optimal probe concentration. A plot of microarray signal as a function of probe concentration depicts pseudo-first-order kinetics and the optimal probe concentration (Fig. 5.9). The linear portion of this plot is known as the ***linear range*** of the assay.

Pseudo first order. *Second-order kinetic reaction that can be approximated as a first-order reaction because the concentration of one of the two reactants remains nearly constant over the duration of the reaction.*

Linear range. *Portion of microarray data in which an increase in probe concentration produces a concomitant increase in assay signal.*

If the concentration of the probe becomes too high, the microarray signal tails off, such that a given increase in P no longer creates a corresponding increase in microarray signal (Fig. 5.9). Tailing off of the microarray signal occurs because the availability of free target sites diminishes and eventually becomes ***saturated*** over the duration of the reaction (Fig. 5.10). The conditions known

Saturated. *Microarray target element in which most or all of the target molecules contain bound probe molecules.*

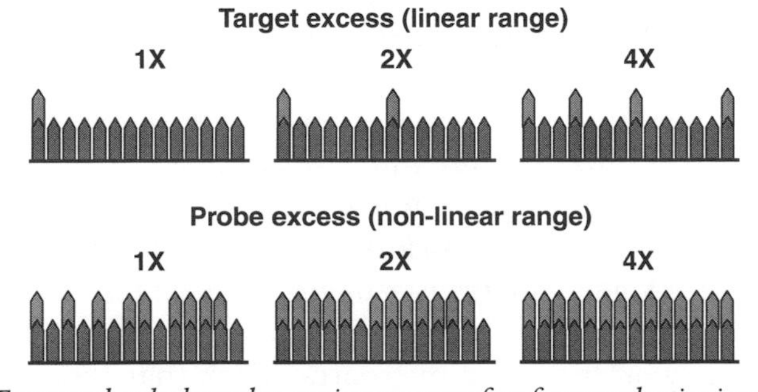

FIGURE 5.10. Target molecules bound to a microarray surface form productive interactions with probe molecules from the solution. ***Top,*** *Under conditions of* target excess, *successive twofold increases in probe concentration produce twofold increases in the reaction rate and microarray signal.* ***Bottom,*** *Under conditions of* probe excess, *successive twofold increases in probe concentration produce less than twofold increases in reaction rate and microarray signal.*

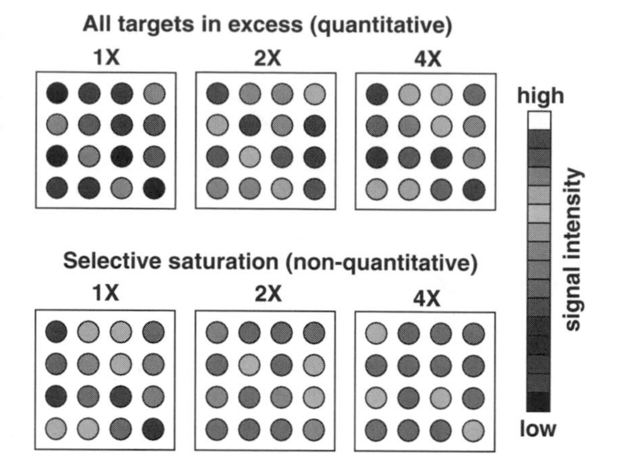

FIGURE 5.11. ***Top,*** *When all the targets are present in excess of probe, successive twofold increases in probe concentration produce successive twofold increases in signal intensity at each microarray location.* ***Bottom,*** *Under conditions of probe excess, there is a selective saturation of targets corresponding to abundant species, so that successive twofold increases in probe concentration produce twofold increases in signal intensity at only some microarray locations. Figure also appears in Color Figure section.*

Probe excess. *Microarray assay condition in which the probe concentration results in target saturation and a loss of quantiation.*

Selective target saturation. *Microarray assay condition in which a subset of the target elements become largely or fully bound leading to a loss of quantiation.*

Signal compression. *Microarray assay condition in which the fluorescent readings underestimate the number of molecules present on the target element or in the probe mixture, leading to a loss of assay quantitation.*

as ***probe excess,*** which result in a saturation of free target sites, are undesirable because they impair quantitation in microarrays assays.

Probe excess is particularly detrimental in reactions of complex probe mixtures such as those derived from total mRNA. In complex probe mixtures, the concentration of a given species (i.e., a given labeled mRNA species) can vary tremendously from gene to gene. Abundant mRNAs may represent as much as 1% of the total mixture, whereas rare transcripts may represent < 0.001%. Because a given species in the probe mixture reacts with a particular target on the microarray, excessive probe concentrations can result in a selective saturation of microarray target sites, such that the sites representing abundant transcripts become saturated and the sites representing rare transcripts remain in the linear range. Under conditions of ***selective target saturation,*** microarray data not only are nonquantitative but are selectively nonquantitative for the abundant transcripts (Fig. 5.11). This effect, known as ***signal compression,*** can lead to erroneous readings at the high end of the concentration range, and these conditions should be avoided in all but a few experimental situations (see below).

Probe Excess

There are a few situations in which the use of excessive probe concentrations can be useful in microarray analysis. First, probe excess conditions can be exploited to determine the absolute target density on a microarray surface. In this type of analysis, a vast excess of probe is used to drive the formation of productive target–probe interactions so that all of the free target on the surface is converted to target–probe pairs. By reading the fluorescent signal on the microarray and comparing signals to known concentrations of control dyes, it is possible to calculate the absolute number of target molecules per unit area of microarray surface (e.g., to determine the target density). Second, probe excess conditions can be used to measure the concentration of rare species in complex mixtures, such as rare transcripts in gene expression experiments. In this approach, a high concentration of fluorescent probe prepared from an mRNA mixture is hybridized to a microarray such that the abundant species saturate their respective targets, but the rare species hybridize in the linear concentration range. Excess probe can be used to "drive" the gene expression assay so that the weak fluorescent signals at rare transcript target locations can be read more easily than if the experiment were conducted under optimal probe conditions.

GLASS SUBSTRATES

Though several different types of materials have been used for microarray analysis, glass is by far the most common material. This section provides a historical overview of glass, describes different glass types, and explains the properties of glass that make it an appealing substrate for microarray analysis.

Structure of Glass

The principle component of glass is ***silicon dioxide*** (SiO_2), or silica, which is the main component of beach sand. Because of the high melting temperature of sand (> 1500°C) and technical limitations, glass manufacturing first occurred rather late in human history. About 5000 years ago, the Egyptians noticed shiny deposits around the perimeter of campfires built on sandy shores, leading to the observation that addition of ash from wood fires (potash, or K_2O) lowered the melting point of sand (see below). Further improvements were made by the Romans who developed the first commercial glass products. The first eyeglasses and telescope lenses were developed by the Italians during the sixteenth century, and the Germans are generally credited with manufacturing the first high-end optical components in the 1800s.

Silicon dioxide (SiO_2). *Main component of glass.*

Modern-day glass manufacturing includes the preparation a large family of related materials that solidify from the molten state without forming crystals; thus, by definition, glass is neither a true solid nor a true liquid. Because of this unique property, glasses are sometimes referred to as supercooled liquids or amorphous solids.

The basic chemical unit of glass is a four-sided (tetrahedron) molecule of silicon and oxygen atoms, such that each silicon atom is bonded to four oxygen atoms and each oxygen atom is bonded to two silicon atoms (Fig. 5.12). Because silicon is highly electropositive and oxygen is highly electronegative, the covalent bonds between silicon and oxygen are quite strong and account for the high melting temperature of this inorganic polymer. The angles around the oxygen-silicon-oxygen (O–Si–O) bonds are highly regular and actually show little variation from the 109° angle expected for a tetrahedron. At the level of the tetrahedral SiO_4 groups, glass is actually a regular substance and thus is said to have regular short range order (SRO). The amorphous property of glass

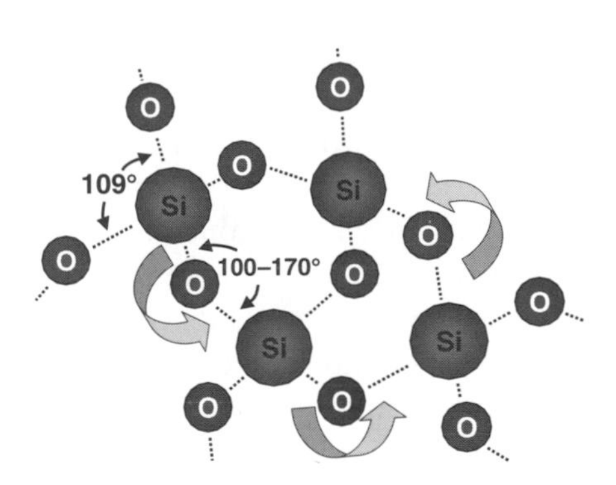

FIGURE 5.12. The main component of glass (SiO_2) consists of tetrahedral atoms of silicon (Si) and oxygen (O). The tetrahedral O-Si-O bond angles are highly regular (109°), whereas the Si-O-Si bond angles show a broad distribution of 100–180° and are freely rotational (arrows). *The SiO_4 tetrahedra link together to form the extensive networks that make up this amorphous inorganic polymer.*

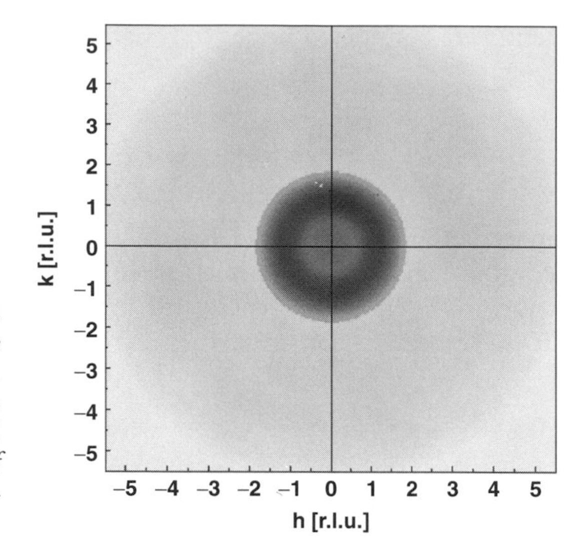

FIGURE 5.13. An x-ray diffraction pattern of a silicon atom, one of the two main atomic components of glass. Electron densities are represented in a gray scale. (Data courtesy of T. Proffen and R. B. Neder, University of Wurzburg, Germany.) Figure also appears in Color Figure section.

derives from variability at the level of tetrahedron–tetrahedron interactions and above. Unlike the highly regular O-Si-O bonds, the angles around the silicon-oxygen-silicon (Si–O–Si) bonds can vary from 100 to 180°, and rotation of the bonds around these axes is largely unencumbered. The result is a disordering of tetrahedrons and a lack of regularity at the levels of intermediate range order (IRO) and long range order (LRO). The unique property of a high degree of SRO with little or no IRO or LRO renders glass an amorphous material that lacks the crystalline properties of a true solid.

A number of physical techniques make it possible to examine silicon and oxygen atoms at the atomic scale. To visualize atoms, it is necessary to use a wavelength of electromagnetic radiation that is approximately one half the size of the atoms of interest. For silicon and oxygen atoms, which have atomic radii of 1.1 and 0.7 Å, respectively, wavelengths for atomic visualization would have to be in the range of ~0.5 Å. One technique that meets the wavelength criterion is X-ray diffraction. In an X-ray diffraction experiment, very short wavelength electromagnetic radiation (X rays) is beamed at the material of interest, and the electrons around the atoms cause the X rays to change direction (diffract). The diffraction pattern is collected and converted into a map that shows the size and shape of a given atom, such as silicon (Fig. 5.13). Another useful technique, known as small-angle neutron scattering (SANS), uses the scattering of a short wavelength neutron beam (< 1.0 Å) to obtain information about the size and shape of molecules. Neutron scattering has been used to visualize SiO_2 molecules (Fig. 5.14).

Types of Glass

With the exception of fused silica, one or more metal oxides are added to all commercial glass to reduce the melting temperature of the silicon dioxide and to impart other desirable characteristics (Table 5.1). Metal oxide additives are sometimes known as fluxes or stabilizers. Common glass types that have been used for microarray substrates or are amenable to use include aluminosilicate, borosilicates, silicas, soda lime, and zinc titania. Each glass type differs in terms of its composition, thermal coefficient of expansion and softening point.

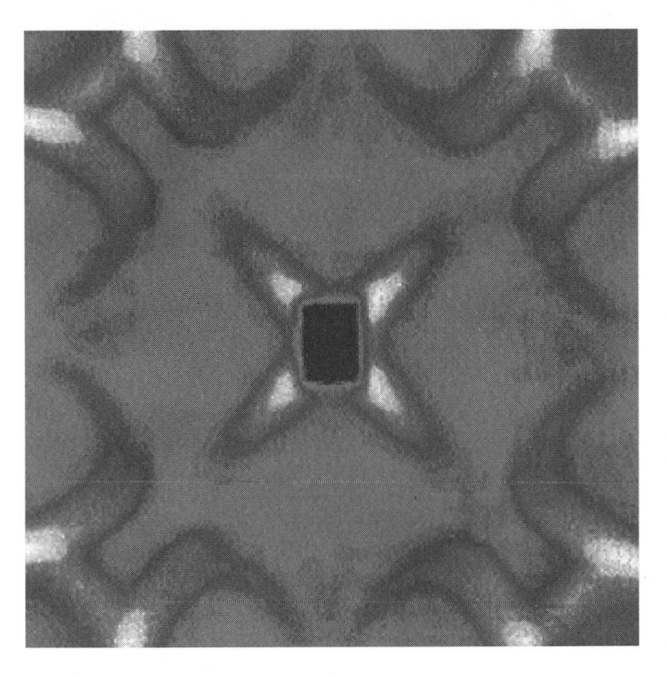

FIGURE 5.14. SANS of silicon dioxide (SiO_2), the principle component of glass. A single crystal of silicon was heated for 500 h at 600°C, and trace oxygen (~ 30 ppm) diffused to form SiO_2 precipitates. (Data courtesy of ILL, France.) Figure also appears in Color Figure section.

There are several manufacturing and postmanufacturing steps that are commonly used to strengthen glass. Annealed glass is made in an annealing oven, which allows gradual cooling and a reduction in the internal stresses that are seen if the surface of molten glass cools much more quickly than the interior. Heat-strengthened glass, which is about twice as strong as annealed glass, is made by bringing manufactured annealed glass to the softening point and allowing it too cool slowly. Tempered glass, which is up to five times stronger than annealed glass, is made by bringing manufactured annealed glass to the softening point and allowing the surface to cool rapidly. The use of hardened glasses for microarray analysis is in the experimental phase.

Surface irregularities (e.g., scratches, pits, and other blemishes) can impair the quality of a microarray substrate, leading to compromised printing, chemical and biochemical reactivity, and detection. The postmanufacturing

TABLE 5.1. Composition and Properties of Glass[a]

Type	SiO_2	Al_2O_3	Ba_2O_3	CaO	Na_2O	TiO_2	ZnO	CE[b] ($\times 10^7$ cm/cm/°C)	SP[c] (°C)
Aluminosilicate	57	20	4	5	1			88	915
Borosilicate (high expansion)	70	6	10	1	9			62	740
Borosilicate (low expansion)	80.6	2.3	13		4			32.5	821
Fused silica	100							5.5	1500–1670
High silica	96.4	0.5	3					7.5	1530
Soda-lime	73			12	14			90	700
Zinc titania	64	3	9		7	3		74	720

[a] The chemical composition (most components) by weight.
[b] Coefficient of expansion.
[c] Softening point, defined as the temperature at which a thin glass rod bends under its own weight.
Data from About.com.

TABLE 5.2. Scratch/Dig Specifications for Polished Glass

Scratch or Dig Specification	Maximum Scratch Width (μm)	Maximum Dig Diameter (μm)	Dig Separation Distance (mm)
120	120	1200	20
80	80	800	20
60	60	600	20
50	50	500	20
40	40	400	20
30	30	300	20
20	20	200	20
15	15	150	20
10	10	100	1
05	5	50	1
03	3	30	1

Data adapted from military specification MIL-O-13830 and Sinclair Manufacturing Company, Chartley, MA.

Scratch/Dig. *Glass polishing specification used as a benchmark for microarray substrates.*

Atomic force microscopy (AFM). *Analytical technique that is used to generate high-resolution, three-dimensional plots of microarray surfaces.*

step of glass polishing can be used to remove surface irregularities, providing optical quality substrates for exacting microarray applications. Manufactured glass is polished with mechanical devices that rub fine abrasives over the glass surface in a rotary fashion. The smoothness of the resultant polished surface is measured by a criterion known as the ***scratch/dig*** specification (Table 5.2). Scratch is defined as any mark or tear on the glass surface; dig refers to any small rough spot or pit. Scratch and dig specifications denote the maximum allowable deformation over the specified separation distance as measured with an optical comparator. High-quality microarray substrates available at present use a 10/5 scratch/dig specification, which is well beyond the quality of the standard 80/60 microscope slide.

The smoothness of glass surfaces can be accessed at high resolution using ***atomic force microscopy*** (AFM). The AFM technique employs a fine silicon tip that traces back and forth across the surface, detecting and recording surface irregularities as it moves. Typical scans provide three-dimensional images and corresponding roughness calculations over an area of several square microns (Fig. 5.15). AFM analysis of an optically flat microarray substrate indicates a maximal roughness (R_{max}) of 5.3 nm (53 Å) over a 4 μm^2 area, corresponding to a distance of approximately 40 silicon–oxygen bonds or about twice the diameter of duplex DNA. Given that most microarray detection systems allow a

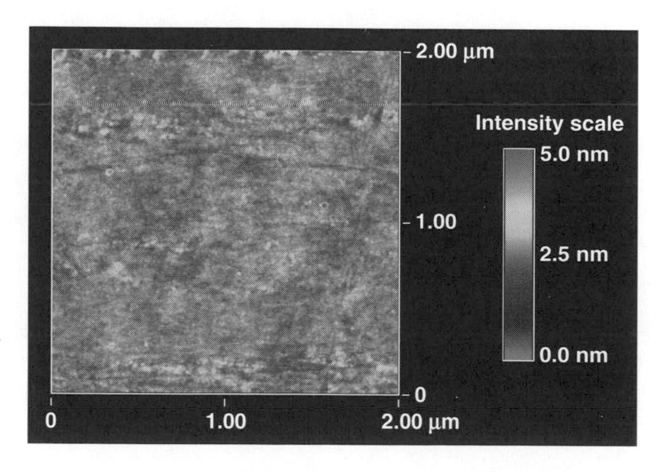

FIGURE 5.15. An optically flat glass substrate was examined by AFM over an area of 4 μm^2. The roughness image is displayed in a rainbow intensity scale, and the mean flatness is ~2.1 nm. (Data courtesy of TeleChem/ArrayIt.com, Sunnyvale, CA.) Figure also appears in Color Figure section.

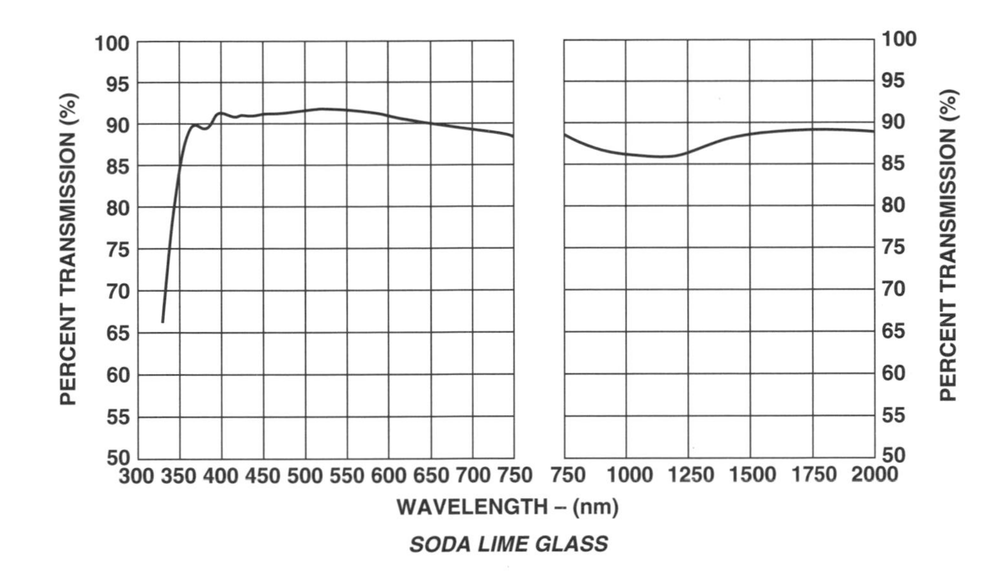

FIGURE 5.16. Transmission (%) for soda-lime glass is plotted against wavelength (nm) for the entire visible range (380–780 nm), part of the ultraviolet range (200–380 nm), and part of the infrared range (780–3000 nm) of light. (Data courtesy of Sinclair Manufacturing Co., Chartley, MA.)

$\pm$ 10 μm ($\pm$ 100, 000 Å) deviation from flatness, a surface that is flat to an $R_{\max}$ of 53 Å would be orders of magnitude flatter than the detector requirement. Because most microarray reactions are performed in volumes corresponding to 2 $\mu L/cm^2$, which gives a 20-μm-thick (200,000 Å) sample layer, a surface with an $R_{\max}$ of 53 Å would change the thickness of the sample layer by $< 0.03\%$ thereby contributing little error to the measurement.

Another type of postmanufacturing modification is known as ***etching.*** One common etching compound is the highly reactive substance hydrofluoric (HF) acid. HF acid reacts with silicon atoms on the glass surface according to the following reaction:

Etching. *Chemical process used to score glass surfaces for the purpose of labeling or identification.*

$$SiO_2 + 4\ \text{HF acid} \rightarrow SiF_4 + 2H_2O \tag{5.5}$$

Treatment of glass with HF acid results in the selective dissolving of SiO_2 atoms on the glass surface, producing an etched or frosted appearance that can be used to demarcate regions of interest on the microarray substrate.

All of the glasses described in this section share a number of properties that make these materials well suited as microarray substrates, including a high degree of chemical and physical stability, low coefficients of thermal expansion, low intrinsic fluorescence, low reflectivity, chemical uniformity, amenable to polishing, and efficient transmission throughout the visible range. The latter point is important because microarray detection often involves the use of selective excitation and emission wavelengths of light. Efficient transmission by the substrate during excitation ensures relative "optical inertness" of the substrate during the microarray detection process (see Chapter 7). Transmission for the visible wavelengths is in the 90% range for most high-quality glass substrates (Fig. 5.16).

AMINE AND ALDEHYDE SURFACES

Of the many different types of surfaces used for microarray analysis, the most common chemistries at present are treatments that provide chemically reactive amine or aldehyde groups. These widely used surface treatments are prepared

Silane reagent. *Any member of the specialized class of chemical compounds containing a silicon atom that is used to add organic groups to glass microarray surfaces.*

by exploiting the selective reactivity of glass with a special class of chemical compounds known as ***silane reagents.*** This section describes silane chemistry, amine and aldehyde surface treatments, and the attachment of molecules to these surfaces.

Silane Chemistry

Compounds that contain an organic group (R) bonded directly to a metal (M) are known as organometallic compounds (R-M). These compounds are named by using the organic substituents as the prefix and the metal as the suffix followed by *-ane.* The simplest silicon-containing member is silane (SiH_4). Silane is actually a gas at room temperature and most of the interesting reactivity with silane compounds is achieved using derivatives that contain organic functional groups in place of the hydrogen atoms.

Organosilane. *Any member of the specialized class of chemical compounds that contains silicon-to-carbon bonds.*

Compounds that contain a silicon-to-carbon bond are known as ***organosilanes.*** The chemistry of organosilanes is generally similar to that of carbon-based molecules, except that silicon is more electropositive than carbon, resulting in more facile nucleophilic attack at silicon position and greater reactivity. Derivatives of silane that have electronegative substituents create polar bonds, such that the silicon atom has an electropositive character and the R group is electronegative:

(5.6)
$$\overset{\delta^-}{\text{R}}\text{–}\overset{\delta^+}{\text{Si}}$$

Silicon ester. *Any member of the specialized class of chemical compounds that contains an oxygen atom between the silicon-to-carbon bond.*

Silicon esters make up a subfamily of organosilane compounds and contain an oxygen atom in between silicon and carbon:

(5.7)
$$\text{Si–O–R}$$

Because the central silicon atom allows bonding to as many as four organic substituents, silicon esters constitute a large and diverse family of compounds, including triethylmethoxylsilane $(CH_3CH_2)_3SiOCH_3$. The electropositive silicon atom is susceptible to nucleophilic attack by hydroxyl groups present on a glass surface, allowing organic groups of various types to be bonded directly to the microarray substrate.

Amine Surface

A wide spectrum of amine-containing silane derivatives (aminosilane compounds) have been prepared by organic synthesis (Table 5.3), and each has different properties and reactivity, depending on the nature of the amine substituent and on the other three organic groups bonded to the central silicon atom. Several of these compounds have been used to prepare amine surfaces for microarray analysis and many derivatives of these compounds are currently under testing.

The reaction of glass with 3-aminopropyltrimethoxysilane provides a general view of the mechanism of these reactions (Fig. 5.17). Hydroxyl groups present on the glass surface act as nucleophiles that attack the silicon atoms, resulting in covalent bond formation between the hydroxyl group and the silicon atom, creating a surface that contains reactive amine groups. In the case shown in Figure 5.17, the amine nitrogen contains a single organic group joined to

TABLE 5.3. Some Organosilane compounds That Contain Amino Groups

Number	Compound Name
1	3-aminopropylmethyldiethoxysilane
2	aminopropylsilanetriol
3	allylaminotrimethylsilane
4	3-(2-aminoethylamino)propyltrimethoxysilane
5	4-aminobutyltriethoxysilane
6	3-aminopropylpentamethyldisiloxane
7	(aminoethylaminomethyl)phenethyltrimethoxysilane
8	*n*-(2-aminoethyl)-3-aminopropylmethyldimethoxysilane
9	3-aminopropyltrimethoxysilane
10	*n*-(6-aminohexyl)aminopropyltrimethoxysilane
11	aminomethyltrimethylsilane
12	3-(m-aminophenoxy)propyltrimethoxysilane

it and is hence termed a ***primary amine.*** Because the primary (propyl) organic group is a straight-chain hydrocarbon, it is termed an ***aliphatic amine.*** The nonbonding electron pair on the aliphatic amine nitrogen group acts as a base that attracts protons in the solution and thus carries a net positive charge at neutral pH. Aliphatic amines are generally more basic than their aromatic counterparts and so are favored for amine microarray substrates that use the positive charges to attach molecules of interest. Treatment of glass with aminosilane causes changes in the surface topology that can be visualized by AFM (Fig. 5.18).

Primary amine. *Amino group in which the nitrogen atom is bonded to a single organic group.*

Aliphatic amine. *Chemical compound in which the amino group is attached to the end of a straight-chain hydrocarbon.*

The overall positive charge of amine microarray surfaces allows attachment of printed biomolecules that carry negative charges. Attachment occurs primarily via ***electrostatic interactions*** or attractive forces between positive charges on the amine groups and negative charges on biomolecules, such as nucleic acids. Attachment of nucleic acids to an amine surface occurs via interactions between negatively charged phosphate groups on the DNA backbone and positively charged amine groups (Fig. 5.19). Because attachment occurs by general positive–negative electrostatic interactions, the process is termed ***adsorption.*** Adsorption occurs without regard to the sequence composition or spatial alignment of the target molecules; therefore, attachment on amine surfaces is said to be nonspecific. Because it is convenient and easy to implement, the amine surface is gaining wide use among microarray scientists.

Electrostatic interaction. *Noncovalent attractive force between positive and negative charges such as an amine surface and the phosphate groups on DNA.*

Adsorption. *Nonspecific attachment of a molecule to a surface.*

Amine surfaces tend to perform well over a wide range of experimental conditions (Table 5.4) and generally manifest both low intrinsic fluorescence (i.e., before hybridization) and low background fluorescence (i.e., after hybridization). The low intrinsic fluorescence generally derives from the fact that the

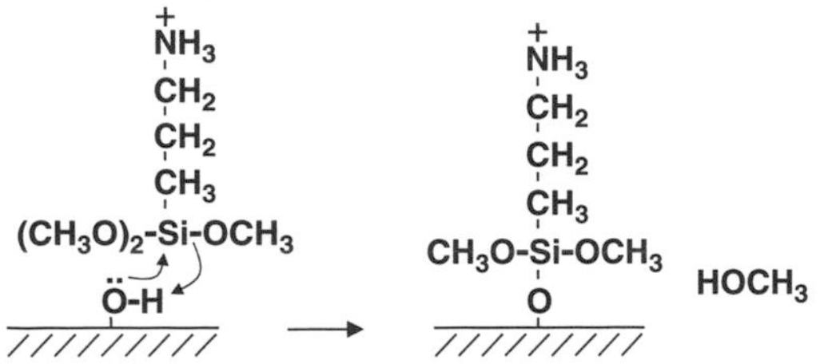

FIGURE 5.17. *Lone pair electrons from a hydroxyl group on the glass surface act as a nucleophile, attacking the electropositive silicon atom in 3-aminopropyltrimethoxysilane and displacing methanol. The result is the covalent attachment of a reactive primary amine group to the glass surface.*

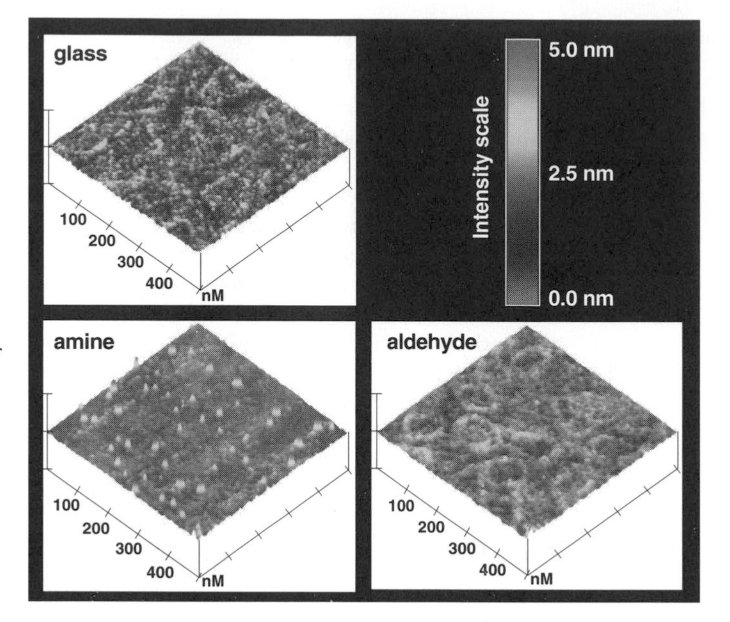

FIGURE 5.18. AFM was used to examine three surfaces over an area of 0.25 μm^2. Data for optically flat glass, optically flat glass with reactive amine groups, and optically flat glass with reactive aldehyde groups are represented in a rainbow intensity scale. (Data courtesy of TeleChem/ArrayIt.com, Sunnyvale, CA.) Figure also appears in Color Figure section.

Hydrophilic. *Measure of the extent to which a microarray surface or some other substance or molecule dissolves water.*

Hydrophobic. *Measure of the extent to which a microarray surface or some other substance or molecule repels water.*

surface treatment can be implemented without the creation of conjugated or unsaturated ring structures, which tend to fluoresce in the visible wavelength range and impair the detectivity of the microarray assay (see Chapter 7). The low background fluorescence is generally the result of the ***hydrophilic*** amine surface, which energetically repels ***hydrophobic*** molecules such as some of the dyes (e.g., cyanine) commonly used in direct labeling strategies for microarray analysis.

In experiments with fairly large molecules such as PCR products, cDNAs, and proteins (see Chapter 13), few problems are encountered with amine surfaces. In experiments with short oligonucleotides (5–30 nucleotides) or peptides (5–20 residues), however, the nonspecific attachment mechanism presents some steric or spatial limitations. Because the molecules are adsorbed onto the surface, attachment occurs nonspecifically along the entire length of the molecule and can therefore hinder the formation of productive interactions with incoming target molecules (Table 5.4). Particularly in single nucleotide polymorphism (SNP) experiments, amine surfaces can result in a loss of hybridization specificity, and for this reason, end-attachment schemes (see below) are generally preferred. Another minor limitation of amine surfaces is that, owing to the hydrophilicity, aqueous solutions of molecules tend to spread out on the amine surface upon printing, thereby creating slightly larger spots than on hydrophobic surfaces.

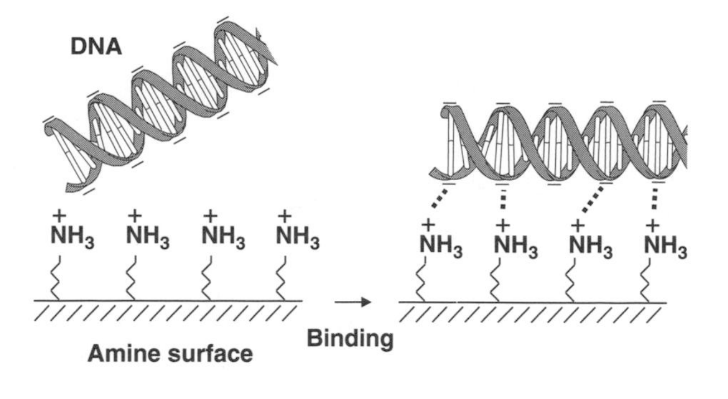

FIGURE 5.19. Electrostatic binding of DNA to a reactive amine substrate. Positive charges on the primary amine groups form electrostatic interactions with the negative charges on the phosphate backbone of DNA. (Adapted from D. Leja, National Human Genome Research Institute, Bethesda, MD.)

TABLE 5.4. Amine and Aldehyde Surface Comparisons

Criterion	Amine Surface	Aldehyde Surface
Surface charge	positive	neutral
Intrinsic fluorescence	very low	very low
Surface character	slightly hydrophilic	hydrophobic
Printed sample spreading	slight	very slight
Attachment mechanism	adsorption	covalent
Linkers required	no	primary amine
Attachment chemistry	electrostatic (positive - negative)	dehydration reaction (Schiff base)
Attachment geometry	nonspecific	specific (end attachment)
Attachment processing (nucleic acids)	drying oven or cross-linking with ultraviolet light	ambient dehydration
Denaturation (nucleic acids)	boiling water	boiling water
Assay sensitivity (cDNAs or proteins)	excellent	excellent
Assay sensitivity (oligos or peptides	good	excellent
Assay background fluorescence	very low	low
Ease of use	excellent	very good

Once the printed DNA is absorbed onto the amine surface, the printed substrates are often placed in a drying oven at elevated temperatures (e.g., 60 min at 80°C) to drive off any remaining water molecules, and strengthen the attachment. Attachment can also be strengthened by the use of ultraviolet light, which cross-links the DNA to the surface. If the printed DNA is double-stranded (e.g., PCR products or cDNAs), the double-stranded molecules on the surface must be ***denatured*** to create single-stranded DNA strands before hybridization (Fig. 5.20). The denaturation step is commonly performed by placing the substrates in boiling water for a period of 2–3 min, during which time the elevated temperature and absence of salt results in a rapid unzipping of the DNA.

Denature. *Process of converting DNA into single-stranded form or, more generally, any process that renders a molecule unstructured.*

The denatured DNA on the surface is available for hybridization to molecules in the probe solution. Hybridization reactions are typically performed in the presence of salt and at elevated temperatures. The addition of salt (e.g., 0.5–1.0 M NaCl) helps shield the negatively charged phosphate groups on adjacent strands, thereby reducing ***electrostatic repulsion*** when the two DNA strands hybridize to each other (Fig. 5.20). Hybridization occurs more rapidly at elevated temperatures because heating increases molecular motion and thus is used to drive collisions between target molecules on the surface and probe molecules in solution. Signal intensities increase with increasing temperature, though care must be taken not to use excessive temperature, because melting of the duplex molecules will occur as temperatures approach the melting temperature. Most microarray hybridization reactions are performed ~10°C below the melting temperature; hybridizations with cDNAs and PCR products generally use 65°C, whereas temperatures for 15mer oligonucleotides tend to be in the range of 37–42°C. Detergents and other additives are often used to reduce background hybridization and nonspecific hybridization. A typical

Electrostatic repulsion. *Repellant force that occurs when molecules with the same electrostatic charge are brought into close proximity.*

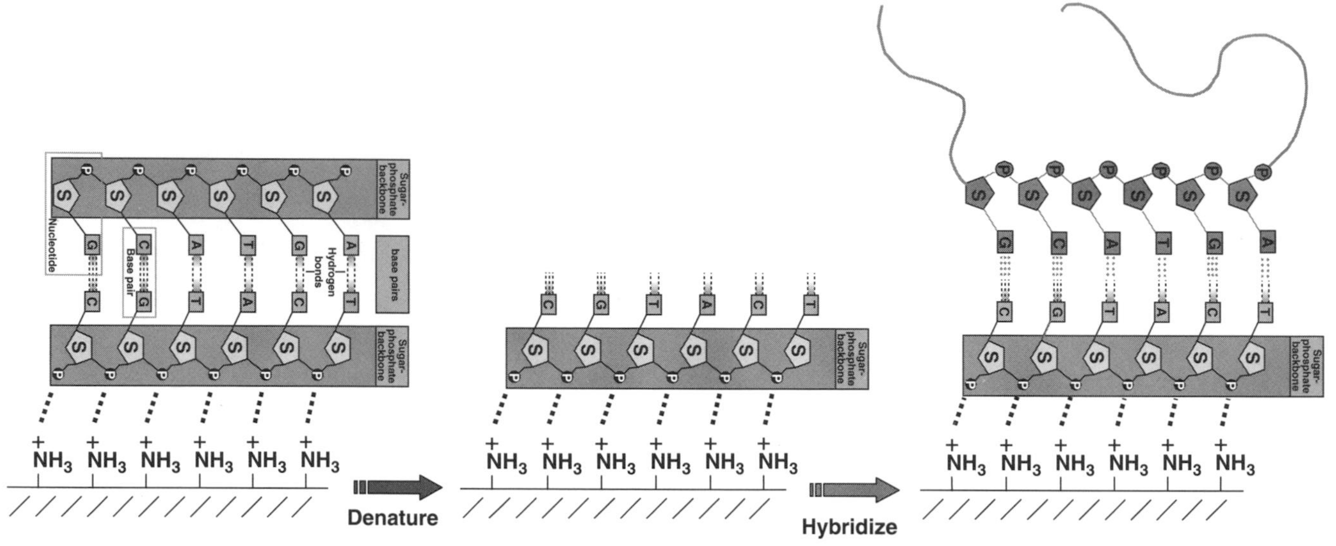

FIGURE 5.20. A printed substrate containing a double-stranded DNA target molecule is attached by electrostatic interactions to the surface. The substrate is treated with boiling water to denature the target DNA and then hybridized with a probe solution so a fluorescent molecule hybridizes to its complementary target on the surface. (Adapted from D. Leja, National Human Genome Research Institute, Bethesda, MD.)

human gene expression experiment with 1024 genes reveals the strong signals and low background observed with amine surfaces (Fig. 5.21).

Aldehyde Surface

Spacer arm. *Chemical group such as an aliphatic hydrocarbon used to create distance between a functional group and a microarray substrate.*

Amino linker. *Chemical group consisting of a primary amine that is used to attach DNA to an amine-reactive microarray substrate.*

Another widely used surface in microarray analysis contains reactive aldehyde groups. Aldehyde surface preparation, similar to amine surface chemistry, uses silane reagents to form covalent bonds between silicon atoms on the microarray substrate and organic reactive groups. An aldehyde surface typically contains an aliphatic group or ***spacer arm*** of approximately 10 carbon atoms at the end of which is a reactive aldehyde group (Fig. 5.22).

Attachment to the aldehyde surface occurs by a different mechanism from that of attachment to an amine surface. For DNA attachment to an aldehyde surface, a primary amine group must first be added to the DNA of interest. The 5′ amine group or ***amino linker*** contains a primary amine attached via a 6-12 carbon atom spacer arm to the phosphate group of the 5′-most nucleotide in the DNA molecule. Amino linkers can be added to double-stranded DNA during PCR amplification using oligonucleotides that contain a 5′ amino

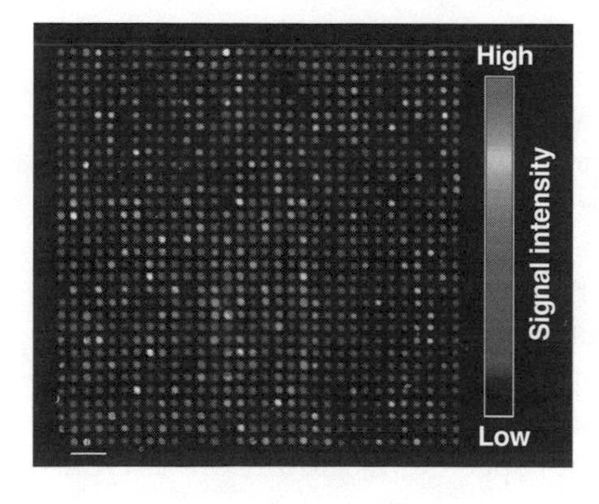

FIGURE 5.21. Target cDNAs corresponding to 1024 human genes were printed on a reactive amine surface and hybridized to fluorescent probes prepared from human liver RNA. The scanned data are represented in a rainbow color palette. Space bar = 500μm. (Data courtesy of Y. Li, Fudan University, Shanghai, People's Republic of China.) Figure also appears in Color Figure section.

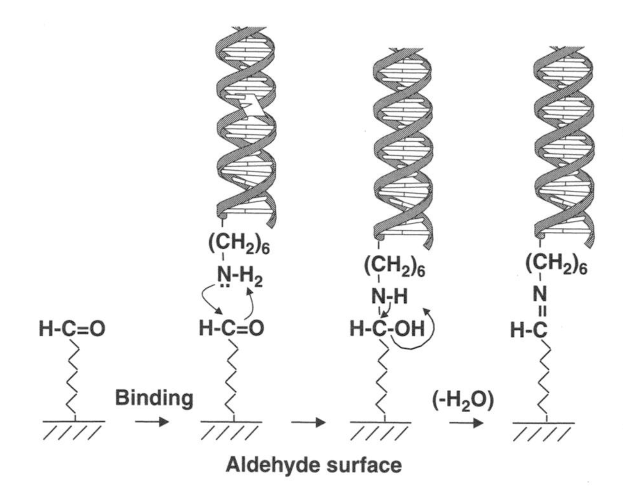

FIGURE 5.22. A microarray surface containing a reactive aldehyde group reacts with a DNA molecule that contains an aliphatic amine linker. The lone pair electrons on the amine nitrogen act as a nucleophile, attacking the electropositive aldehyde carbon atom. A dehydration reaction results in the formation of a covalent imine bond (Schiff base) between the DNA and the microarray surface. (Adapted from D. Leja, National Human Genome Research Institute, Bethesda, MD.)

modification. Oligonucleotides themselves can be linked directly to an aldehyde surface through the use of the same 5′ amino linkers. Binding occurs when the nonbonding electron pair on the amine linker acts as a nucleophile that attacks the electropositive carbon atom of the aldehyde group (Fig. 5.22). A subsequent dehydration step, which occurs as a natural part of most microarray printing processes (i.e., at $< 40\%$ humidity), affords a covalent bond between the amino-linked DNA molecule and the microarray surface. The resultant linkage creates a ***substituted imine*** that is commonly known as a Schiff base.

Substituted imine. *Chemical function that contains a carbon atom double bonded to the nitrogen atom of a primary, secondary, or tertiery amine.*

Because attachment to an aldehyde surface involves covalent bond formation, binding to this surface is termed ***covalent coupling.*** Different from attachment to amine surfaces, molecules couple to an aldehyde surface in a directional manner such that the end of the molecule containing the amino linker bonds to the microarray surface. This end-attachment process is selective for DNA molecules that contain 5′ amino linkers and is thus termed ***specific coupling.*** The aldehyde surface is widely used in the microarray community; and though it is a bit more difficult to implement than amine substrates, it is extremely stable and performs well under diverse reaction conditions.

Covalent coupling. *Attachment scheme that involves electron sharing between target molecules and the microarray substrate.*

Specific coupling. *Any covalent attachment scheme including a Schiff base, in which target attachment to the microarray surface occurs with defined chemistry.*

Aldehyde surfaces generally manifest slightly higher intrinsic fluorescence and background fluorescence than do amine surfaces, but the performance of this surface is extremely high (Table 5.4). The slightly elevated intrinsic fluorescence generally derives from the fact that the preparation of the surface often leads to the creation of a slight amount of fluorescent by-product in the form of chemical side reactions. Though background fluorescence is generally low with this surface, slightly elevated nonspecific fluorescence can be observed under certain experimental conditions. Elevated background fluorescence is almost always observed with probe mixtures that contain nucleic acids labeled directly with hydrophobic dyes such as cyanine-3 (Cy3) and cyanine-5 (Cy5). Because the aldehyde surface is hydrophobic, care must be taken to prevent drying of the fluorescent sample during the hybridization reaction. Sample dessication leads to the reduced solubility of the dye molecules, which can precipitate on the hydrophobic aldehyde surface and cause elevated background fluorescence.

Proper reaction conditions and blocking agents all but eliminate background fluorescence with aldehyde surfaces. Heating the probe solution and substrate to 55°C before initiating the hybridization reaction is a common procedure used to essentially eliminate background fluorescence. Raising the temperature increases molecular motion and probe solubility, both of which reduce

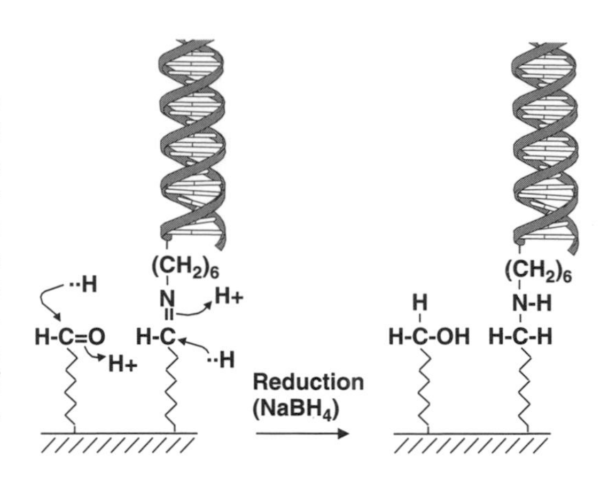

FIGURE 5.23. An aldehyde surface containing a DNA molecule attached by a Schiff base is treated with the reducing agent sodium borohybride ($NaBH_4$). Hydride ions (H··) attack unreacted aldehydes and the imine bond of the Schiff base, resulting in the formation of a nonreactive alcohol group and a stable secondary amine. Protons in solution (H+) bond with displaced electrons to form alcohol (OH) groups.

nonspecific interactions between fluorescent probe molecules and the aldehyde surface.

Because DNA molecules must be modified with amino linkers before attachment to aldehyde substrates, sample preparation for nucleic acids is a bit more difficult than with amine surfaces. Aldehyde coupling chemistry is nonetheless widely used for gene expression analysis and is also gaining popularity in SNP experiments; the latter application takes advantage of the fact that the targets of interest are attached to the surface by the ends of the molecules. End attachment is highly desirable for single nucleotide detection experiments because maximal ***steric availability*** of the target improves the specificity of the hybridization reaction. Genomic DNAs, bacterial artificial chromosomes (BACs), yeast artificial chromosomes (YACs), and oligonucleotides as short as five bases have all been successfully hybridized on aldehyde surfaces.

Steric availability. *Desirable spatial configuration, such as end attachment, that maximizes the physical accessibility of target molecules to incoming probe molecules.*

In addition to nucleic acid binding, the Schiff base chemistry allows an extremely wide range of other types of molecules to be coupled to reactive aldehydes, and all of the different biochemical reactions can be performed without any significant concern for the steric issues that can complicate the use of the electrostatic amine substrates. Owing to the natural occurrence of amino linkers in the form of surface lysine and arginine residues, it is possible to attach proteins, cell extracts, and intact cells to aldehyde surfaces without the addition of synthetic amine groups. The versatility of the aldehyde surface is causing an explosion in the development of novel microarray assays (see Chapter 13). Because aldehyde surfaces are hydrophobic, printed droplets from diverse sample types exhibit little spreading and thus spot sizes tend to be smaller on aldehyde surfaces than on amine surfaces. Because spot diameter determines printing density, aldehyde substrates afford greater printing densities than their amine counterparts.

Once the amino-linked DNA is printed onto the aldehyde surface, the substrates are allowed to dry to allow the dehydration coupling reaction to proceed to completion (Fig. 5.22). Simple overnight treatment at ambient humidity (30–40%) is typically sufficient for complete coupling, though a drying oven (e.g., 60 min at 80°C) can also be used if the ambient conditions do not afford sufficiently low humidity. To further strengthen the coupling linkages, printed substrates are often treated with a reducing agent, such as sodium borohydride ($NaBH_4$), to reduce the imine linkage into the more stable secondary amine (Fig. 5.23). Reduction of the Schiff base prevents hydrolysis of the imine linkage (i.e., the reverse reaction) from occurring during the hybridization reaction. Treatment of an aldehyde surface with sodium borohydride also reduces

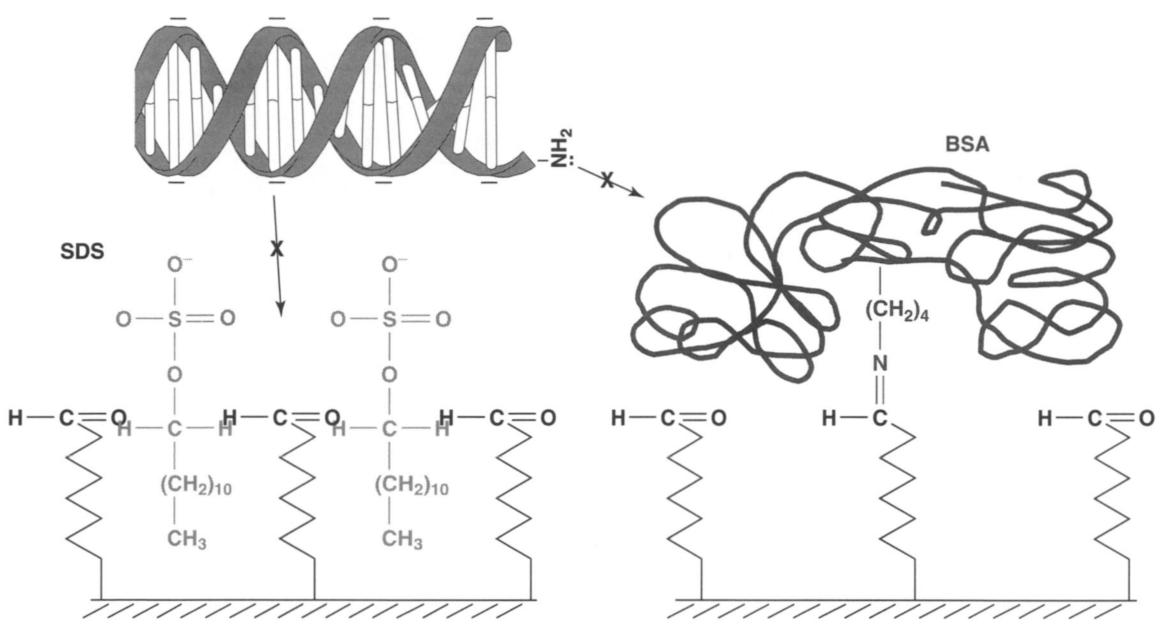

FIGURE 5.24. SDS and BSA are commonly added to microarray hybridization reactions to reduce nonspecific binding of DNA (arrows). *The sulfate groups of SDS provide negative charges that shield the microarray surface from the negatively charged phosphate backbone. Lysine residues of BSA form Schiff base bonds with reactive aldehyde groups, thereby preventing nonspecific interactions between the surface and internal amine residues present on the DNA bases.*

reactive aldehyde groups into their nonreactive alcohol counterparts. The conversion of aldehydes to alcohols reduces nonspecific attachment of molecules to the microarray surface and, therefore, reduces background fluorescence that can be acquired during the hybridization step. Because sodium borohybride treatment reduces nonspecific attachment, solutions containing $NaBH_4$ are often known as ***blocking agents.***

Blocking agent. *Chemical or biochemical agent such as borohydride or bovine serum albumin used to inactivate reactive groups on a microarray substrate to prevent nonspecific reactivity.*

As shown for amine surfaces (Fig. 5.20), double-stranded DNA attached to aldehyde surfaces must be denatured before hybridization. Denaturation for 3 min in boiling water is typically sufficient to convert the targets into single-stranded DNA. Hybridization reactions are performed in the presence of salt (0.5–1.0 M NaCl) and at a hybridization temperature (4–65°C) that is optimal for the length of the target–probe interaction. Microarrays of cDNAs generally use 65°C, whereas hybridizations with 15mers often use 42°C. A good guideline is to increase or decrease the hybridization temperature by 1°C for each nucleotide that is added or subtracted, respectively, from the length of the hybridized duplex. The detergents sodium lauroyl sarcosine or sodium dodecyl sulfate (SDS) and the protein blocking agent bovine serum albumin (BSA) are commonly added to aldehyde surface hybridization reactions to decrease background fluorescence (Fig. 5.24). A time course hybridization experiment with oligonucleotide microrarrays reveals the specificity and linearity of the fluorescent signals on an aldehyde surface (Fig. 5.25).

Preheating the microarray substrate immediately before adding the fluorescent probe solution reduces background fluorescence, in some cases, by preventing nonspecific attachment of probe to the surface during the time that it takes the microarray to reach the hybridization temperature (Fig. 5.26). Some researchers combine preheating with one or more blocking agents to produce extremely low background and increased detectivity.

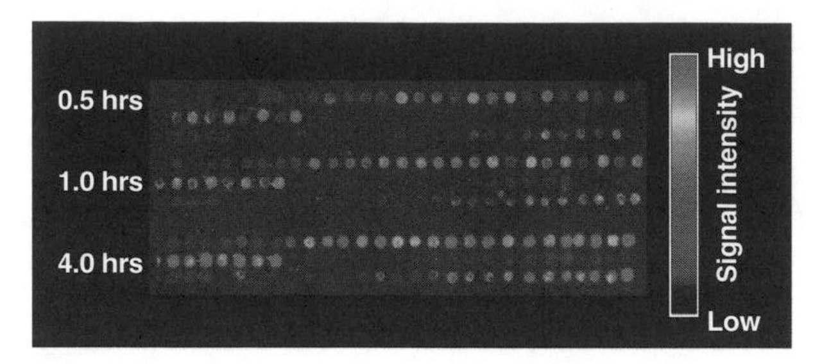

FIGURE 5.25. *Oligonucleotides (15mers) containing amino linkers were attached to an aldehyde surface and hybridized for different lengths of time (0.5–4 h) with a mixture of fluorescent oligonucleotides (15mers) at a concentration of 2 μM for each microarray. After hybridization, the chips were washed to remove unbound probe and scanned for Cy3 emission. Fluorescence intensities are represented in a rainbow scale. Because the hybridization was carried out under conditions of target excess, signal intensities increased as a function of time. Figure also appears in Color Figure section.*

SUMMARY

High-quality microarray surfaces are critical for optimal assay performance. Substrates should be dimensional, flat, planar, uniform, durable, inert, efficient, and accessible. An appreciation for the biophysical issues involved in microarray analysis—including reaction kinetics, steric interactions, and optimal probe and target concentrations—are crucial to understanding the biochemical interactions that take place on microarray surfaces. Because of its transparency, flatness, and affordability, glass is the most commonly used material for microarray substrates. There are many different types and qualities of silicon dioxide, and most types of glass contain metal oxide additives, which provide desired characteristics. Optical-quality glasses, obtained by surface polishing, increase assay precision by maximizing flatness and minimizing error during the reaction and detection steps. Analytical techniques including X-ray diffraction and AFM can be used to examine microarray surfaces at the atomic level. Reactive organic molecules are attached to the glass surface by treating the glass with organic compounds known as organosilanes. The most common types of surfaces contain reactive amine and reactive aldehyde groups, which allow attachment

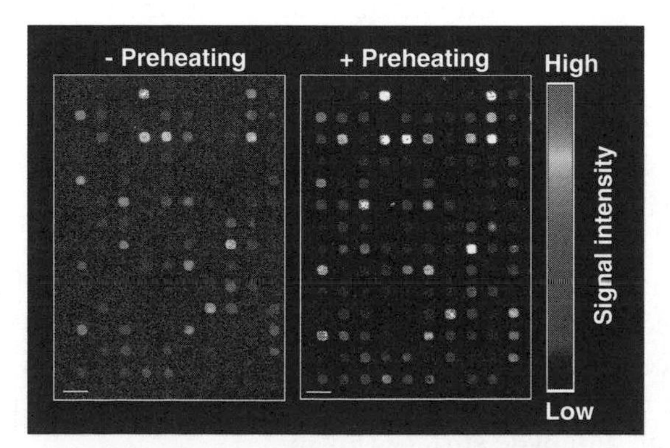

FIGURE 5.26. *Two microarrays were hybridized with fluorescent probes prepared from 3 μg of total* Arabidopsis *RNA and scanned for Cy3 emission. Image data are displayed in rainbow palette. Hybridizations were initiated by adding room-temperature probe to a room-temperature aldehyde microarray (−Preheating) or a 55°C probe to a 55°C aldehyde microarray (+Preheating). Space bar = 250 μm. The materials used include a Submicro Expression Array detection kit (Genisphere, Montvale, NJ), GlassHyb buffer (Clontech, Palo Alto, CA), and SuperAldehyde substrates (TeleChem/ArrayIt.com, Sunnyvale, CA). (Data courtesy of S. Ruuska, Michigan State University, East Lansing, MI.) Figure also appears in Color Figure section.*

of biomolecules of interest by electrostatic interactions and Schiff base covalent bonds, respectively. Detergents, blocking agents, and probe preheating are commonly used to reduce background fluorescence. Continued technological advances in microarray surface technology, including the use of epoxide surface chemistry, will improve the quality of microarray analysis and promote the development of new microarray assays.

SELECTED READING

Blackburn, G. M., and Gait, M. J., eds. *Nucleic Acids in Chemistry and Biology,* 2nd ed. Oxford University Press, Oxford, UK, 1996.

Bloomfield, V. A., Crothers, D. M., and Tinoco, I. *Nucleic Acids: Structure*. University Science Books, Herndon, VA, 808 pp. 2000.

Brook, M. A. *Silicon in Organic, Organometallic, and Polymer Chemistry*. Wiley, New York, 2000.

Creighton, T. E. *Proteins: Structures and Molecular Properties,* 2nd ed. Freeman, New York, 1992.

Fersht, A. *Structure and Mechanism in Protein Science: A Guide to Enzyme Catalysis and Protein Folding*. Freeman, New York, 1999.

Hecht, S. M. *Bioorganic Chemistry: Nucleic Acids*. Oxford University Press, Oxford, UK, 1996.

Hein, M., Best, L. R., Pattison, S., and Arena, S. *Introduction to General, Organic, and Biochemistry,* 6th ed. Brooks/Cole, Pacific Grove, CA, 1997.

Lehninger, A. L., Nelson, D. C., and Cox, M. M. *Principles of Biochemistry,* 3rd ed. Freeman, New York, 2000.

Matthews, H. R., Freedland, R. A., and Miesfeld, R. L. *Biochemistry: A Short Course,* Wiley, New York, 1997.

Rousseau, J. *Basic Crystallography*. Wiley, New York, 1998.

Schena, M., ed. *DNA Microarrays: A Practical Approach,* 2nd ed. Oxford University Press, Oxford, UK, 2000.

Schena, M., ed. *Microarray Biochip Technology*. Eaton, Natick, MA, 2000.

Streitweiser, A., and Heathcock, C. H. *Introduction to Organic Chemistry,* 2nd ed. Macmillan, New York, 1981.

Stryer, L. *Biochemistry,* 4th ed. Freeman, New York, 1995.

Sun, S. F. *Physical Chemistry of Macromolecules: Basic Principles and Issues*. Wiley, New York, 1994.

Van Holde, K. E., Johnson, W. C., and Ho, P. S. *Principles of Physical Biochemistry*. Prentice-Hall, Upper Saddle River, 1998.

REVIEW QUESTIONS

1. Eight basic criteria are used to evaluate microarray substrates. List three of the eight and explain what each one means and its importance briefly.
2. A researcher prints a 1000-element microarray using an ink-jet printer loaded with a fluorescent oligonucleotide. Although all 1000 spots were printed with the same fluorescent sample, not all of the spots emit the same signal. What surface deficiency would explain these results? Explain your answer in a few sentences.
3. A microarray researcher prints a microarray using a range of oligonucleotide concentrations, and notices that the sample with the highest concentration actually produces a *weaker* signal in a hybridization experiment than several samples of lower concentration. Explain these observations within the context of oligonucleotide target density.
4. Under conditions of target excess, which probe concentration will give the greatest fluorescent signal in a hybridization experiment: 1 μM, 10 μM, 100 nM, 100 pM?
5. What is the principle chemical component of glass and what is its chemical formula? Is glass a crystalline solid? Explain your answer.
6. Atomic force microscopy (AFM) is used to examine three glass substrates, and the maximal roughness (R_{max}) values are reported as follows: 2.3 nm,

47 Å , and 0.0025 μm. Which R_{max} parameter corresponds to the smoothest surface?

7. Microarray substrates containing reactive amine and reactive aldehyde groups are used widely in microarray analysis. Describe the basic chemical properties of these two surfaces and explain their mechanisms of DNA and protein binding. A few sentences for each will suffice.
8. A researcher reacts an amine substrate with a fluorescent reagent that binds specifically to reactive amine groups on the glass surface, and after washing and detection, the substrate image shows uneven fluorescence. If the fluorescent reagent reacted with equal efficiency at all surface locations, what can be concluded about the reactive amine groups on the surface?
9. A researcher amplifies a set of target DNAs using PCR, prints the PCR products into a microarray, washes the microarray, and hybridizes it with a fluorescent mixture derived by labeling mRNA. Even though the PCR products correspond to abundant transcripts in the mRNA population, fluorescent scanning reveals weak fluorescence at all microarray locations. What basic step did the researcher probably forget to perform that produced the weak signals? Explain your answer.
10. A scientist prints amino-modified DNA onto reactive aldehyde substrates at 70% humidity and leaves the printed substrates in the humidity enclosure overnight with the humidity setting on 70%. The next day, she processes and hybridizes the substrates only to find they produce weak fluorescent signals. With respect to aldehyde coupling chemistry, identify the experimental problem and propose an easy solution.
11. The addition of metal oxides reduces what basic physical property of glass?
12. A single nucleotide polymorphism (SNP) experiment produces greater discrimination between a perfectly matched duplex and a single-base mismatch when performed on an aldehyde surface compared to an amine surface. With respect to coupling chemistry and DNA topology, explain this observation.

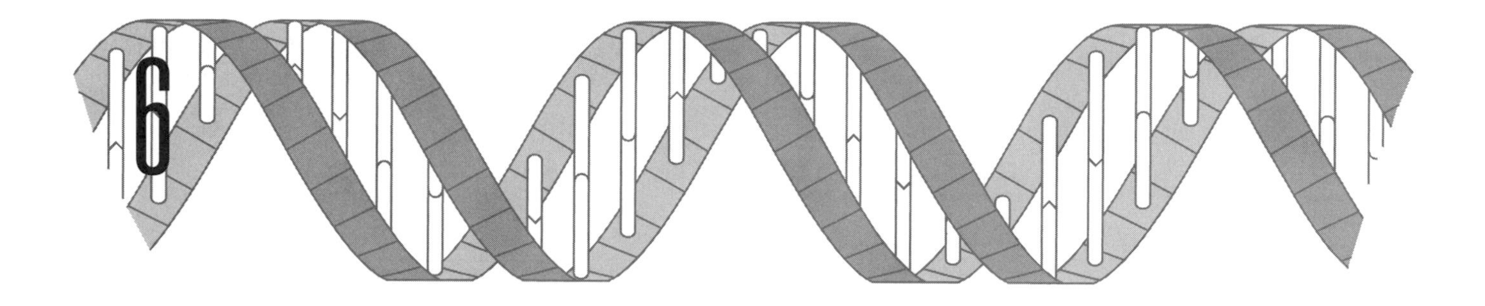

Targets and Probes

Adventure is worthwhile in itself.
—*Amelia Earhart*

Success in microarray analysis relies on a thorough understanding of both the target elements and the labeled probes used in each experiment. The preparation of target elements leverages many existing methods, including the polymerase chain reaction (PCR), traditional phosphoramidite synthesis, and other several more recently developed approaches. Likewise, probe preparation exploits existing enzymatic methods, including the use of reverse transcriptase and RNA polymerase, as well as a host of new strategies for making fluorescent mixtures from biological samples that can be reacted with target elements on the microarray. This chapter provides an adventure into the preparation of microarray targets and probes.

NOMENCLATURE

The original publication describing microarray analysis (Schena et al., 1995) uses the term ***target*** to describe the DNA (PCR products) attached to the microarray substrate, and ***probe*** to describe the sample (mRNA) that is labeled with fluorescent tags. Since the original publication in *Science,* there has been some ambiguity in the scientific literature about the use of two terms, and some authors have even suggested the opposite usages. The term *probe* was chosen by Schena and co-workers to maintain a convention that has been used by the overwhelming majority of the scientific community for 25 years. *Probe* was actually

Target. *Molecule tethered to a microarray substrate that reacts with a complementary probe molecule in solution.*

Probe. *Labeled molecule in solution that reacts with a complementary target molecule on the substrate.*

established during the development of recombinant DNA and blotting procedures developed at Stanford University in the mid-1970s by P Berg, R Davis, G Stark and their colleagues. In this context, probes were labeled with radioactive tags and hybridized to target sequences immobilized on filters. Given the conceptual and technical similarities of these early techniques with modern microarray analysis and in deference to the pioneering efforts of these great scientists, we will maintain the conventional use of the terms *target* and *probe*.

TARGETS

The quality of each target element on a microarray is an important consideration in microarray analysis. Though most microarrays to date have used nucleic acid targets, in principle microarrays can be constructed of virtually any type of molecule of interest, including antibodies, enzymes, carbohydrates, lipids, small molecules, cell extracts, intact cells, inorganic compounds, and semiconductor materials. This section provides an overview of modern microarray target preparation, focusing on synthesis approaches, comparative evaluations, enzymology, specificity, and reactivity.

Types of Approaches

Delivery. *Method of microarray manufacture in which the target molecules are made offline and delivered to the microarray surface by contact printing or some other means.*

Synthesis. *Method of microarray manufacture in which the target molecules are manufactured directly on the microarray surface using photolithography or some other chemical means.*

There are two approaches to microarray target preparation known as ***delivery*** and ***synthesis***. The delivery approaches prepare the target molecules off-line by a technique such as PCR, and they subsequently deliver the target molecules onto the microarray surface by contact printing or an equivalent means (Fig. 6.1). The synthesis methods create target elements directly on the microarray surface by joining monomer building blocks together via a combinatorial strategy such as photolithography. Both approaches suffice to create target elements of high quality, though there are strengths and weaknesses to these two general categories of approach (Table 6.1).

The delivery approaches excel in a number of respects, including ease of implementation, affordability, macromolecular diversity, and quality. Because of the widespread commercial availability of printing robots and a large number of target synthesis schemes, any researcher with reasonable training can prepare microarrays of oligonucleotides, PCR products, proteins, and virtually any other molecule of interest. At low to medium target complexity, the delivery approaches are also quite affordable; a single 100-nanomole scale oligonucleotide synthesis provides material sufficient for >10,000 microarrays. The diversity

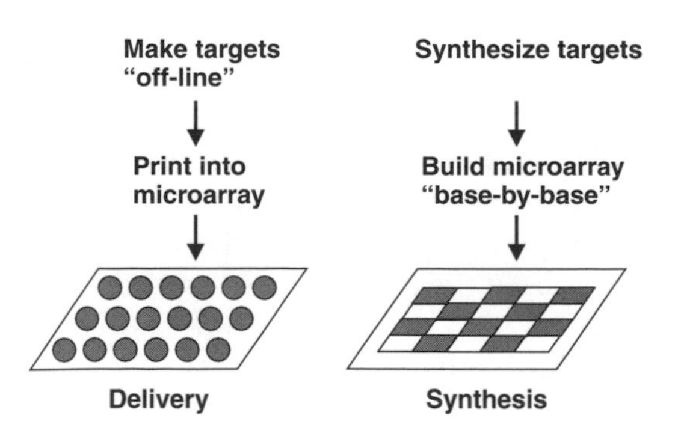

FIGURE 6.1. *The delivery strategy for target synthesis uses targets made by an independent means such as PCR. The products are formed into a microarray by contact or noncontact printing. The synthesis approach builds the targets on the surface of the chip in a step-by-step manner using chemical building blocks.*

TABLE 6.1. Comparative Analysis of Delivery and Synthesis Strategies

Item	Delivery	Synthesis
Method of target preparation	independent means; targets made off-line	built on the chip, monomer by monomer
Ease of implementation	easy to moderate	moderate to difficult
Main enabling technologies	contact and noncontact printing	photolithography; micromirror; noncontact printing
Target density	low and medium; 100–10,000 elements/cm^2	low, medium or high; 100–500,000 elements/cm^2
Target complexity	low to medium	high
Oligonucleotide length	up to 120 bases	up to 25 bases
Target diversity	any molecule that can be made and attached	limited mainly to oligonucleotides and peptides
Oligonucleotide coupling efficiency per base	> 99% (phosphoramidite)	95–99%, depending on the chemistry used
Cost for low complexity (100–10,000 elements)	\$50–5,000 per chip	\$50–10,000 per chip
Cost for high complexity (10,000–500,000 elements)	\$1,000–50,000 per chip	\$100–10,000 per chip
Sample tracking	moderate to difficult	easy

of microarrays made by delivery approaches is limited only by the types of molecules that can be created off-line and by the attachment chemistries available. Using the existing aldehyde and amine surfaces as well as recent advances in new attachment schemes, such as streptavidin–biotin (see Chapter 15), it may soon be possible to manufacture microarrays of literally thousands of different types of molecules. The quality of oligonucleotide targets made off-line is extremely high, owing to the fact that coupling efficiencies for phosphoramidite synthesis are often > 99% per synthesis cycle (see below).

There are some disadvantages inherent to all delivery approaches, however, including uncertainties about sample identification and elevated cost at high target complexity (Table 6.1). The delivery approaches require the preparation and storage of individual target samples, and this is typically carried out in 96- or 384-well microplates. Liquid handling, pipeting, and purification steps with multiple microplates can lead to sample tracking errors and targets incorrectly assigned to their presumptive microarray locations. Microarrays containing targets at incorrect locations can cause massive errors in data analysis and interpretation, "contaminating" functional databases and confounding the researcher. Bar coding, DNA sequencing, mass spectrometry, and hybridization tests are all used to minimize sample tracking errors, and together these techniques can almost completely remove uncertainty about sample identification in microarrays made by target delivery. There is also significant cost associated with high-complexity microarrays made with delivered targets. A microarray containing 100,000 25mers synthesized at \$0.25 per base would cost \$625,000 for the oligonucleotides alone. A generalized equation for the cost of the targets required for a microarray made by one of the delivery approaches is simply the product of the number of targets times the price per target:

$$\text{Target cost} = \text{Number of targets} \times \text{Price per target} \tag{6.1}$$

Microarrays of low to medium complexity (< 10,000 elements) are probably better suited to one of the delivery strategies, whereas microarrays of high target complexity (>10,000 elements) are perhaps better suited to the synthesis approaches.

The synthesis approaches to target preparation excel in a number of key ways, including high density, cost effectiveness at high target complexity, and extremely reliable target identification (Table 6.1). In cases in which the targets are synthesized from chemical building blocks using semiconductor technology, microarrays containing >100,000 elements/cm^2 are readily achieved, paving the way for relatively inexpensive generic chips representing whole genomes for potentially hundreds or thousands of different organisms. The solid-phase synthesis and flow cell technology used for the synthesis strategies are such that the same volume of reagent is consumed whether the microarray contains 1,000 elements or 100,000 elements. Whereas the cost of the microarray increases linearly with target number using the delivery approaches, the synthesis strategies offer high-complexity microarrays at little or no additional cost. The synthesis strategies also obviate the need for sample handling, thereby minimize uncertainties concerning target identification. With the semiconductor-based synthesis strategies, the resultant microarrays are guaranteed to have the correct targets as long as the correct photomasks or binary code is used to implement the synthesis.

There are some disadvantages intrinsic to the synthesis approaches, including relatively high cost of implementation and restricted chemistry (Table 6.1). In the context of oligonucleotide synthesis, the use of light-activated DNA bases also tends to be somewhat less efficient (95–99%) than traditional phosphoramidite synthesis (> 99%). The target yield for a synthesis process is calculated by raising the coupling efficiency (%) to the *n*th power, where *n* is the number of coupling steps:

$$\text{Target yield} = (\text{Synthesis efficiency})^n \quad (6.2)$$

Oligonucleotide targets containing 30 bases coupled at 99.4% efficiency would contain 83% pure 25mer (0.994^{30}) compared to only a 21% yield for a synthesis strategy providing a 95% coupling efficiency (0.95^{30}). The effect of reduced coupled efficiency is even more pronounced for longer target sequences (Table 6.2). Target elements containing mixtures of shorter oligonucleotides can complicate hybridization specificity and degrade gene expression and genotyping data.

TABLE 6.2. Target Synthesis Yields (percent) as a Function of Target Length and Coupling Efficiency

Target Length (nucleotides)	Coupling Efficiency (%)					
	90	92	94	96	98	99.4
10	35	43	54	66	82	94
20	12	19	29	44	67	89
30	4.2	8.2	16	29	55	83
40	1.4	3.6	8.4	20	45	79
50	0.52	1.5	4.5	13	36	74
60	0.18	0.67	2.4	8.6	30	70
70	0.063	0.29	1.3	5.7	24	66

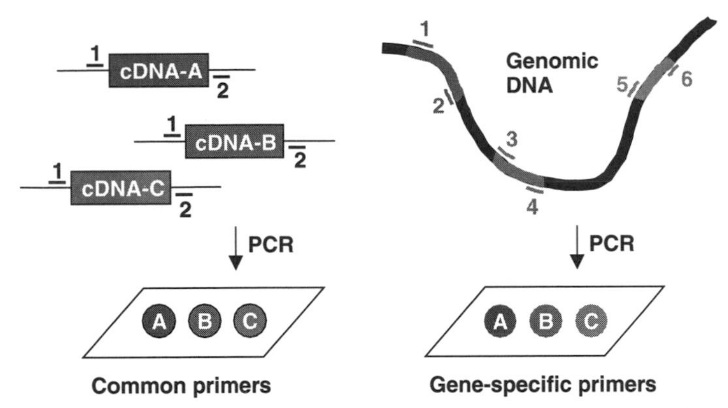

FIGURE 6.2. PCR in conjunction with a single pair of common primers (1, 2) *can be used to amplify target sequences for each cDNA that is bordered by the same vector sequence. Targets* (A–C) *obtained using common primers each contain a small amount of contaminating vector sequence* (outline). *Gene-specific primers* (1–6), *used to obtain target sequences from genomic DNA, produce targets that are free of contaminating sequences but require a separate primer pair to produce each amplicon* (A–C).

PCR Products

The first experiments in microarray analysis employed DNA targets prepared by PCR. The PCR process provides microgram quantities of target material for any DNA segment of interest and from any organism, including bacteria, fungi, plants, and animals. PCR products for microarrays are generated using either ***common primers*** or ***gene-specific primers*** and are formed into microarrays by contact printing or an equivalent delivery technology. Targets made with common primers are amplified using vector sequences that border each target of interest (Fig. 6.2). With a single PCR mixture containing a single pair of common primers, it is possible to amplify thousands of different target elements for microarray manufacture. This approach is extremely useful in a research setting because it allows scientists to amplify cDNAs or expressed sequence tags (ESTs) from any available library, thus creating microarrays of gene segments from any tissue or organism. A common primer pair allows diverse sequences to be amplified with nearly the same efficiency, because the primers are optimized to bind with high affinity to the vector sequences, which are common to every clone. One minor drawback of the common primer approach is that each target element contains a small amount of unwanted vector sequence, though careful selection of primers minimizes the presence of this "contaminating" DNA in the target element.

Common primers. *Two oligonucleotides used for polymerase chain reaction amplification that bind to vector sequences bordering each target sequence and allow amplification of all of the targets using a single oligonucleotide pair.*

Gene-specific primers. *Two oligonucleotides used for polymerase chain reaction amplification that bind uniquely to a given target sequence and allow amplification of a unique target from a complex mixture of complementary or genomic DNA.*

Gene-specific primers are used to obtain unique targets from total genomic DNA or to amplify targets containing no contaminating vector sequence. Gene-specific primers have been used to amplify all 6,000 open reading frames (ORFs) from the yeast *Saccharomyces cerevisiae*. Because yeast has few introns, segments of genomic DNA corresponding to most yeast genes are virtually identical to cDNAs. One advantage of gene-specific primers is that the amplified material contains little or no contaminating vector sequence, owing to the use of a specific pair of primers for each gene. One disadvantage is that each target element requires two primers, and so, for example, 12,000 different primers are required to amplify the 6,000 ORFs from yeast, which adds significant cost to this approach. Common primers and gene-specific primers both work well for making microarrays of genomic and cDNA sequences, and both are widely used in microarray manufacturing.

The use of PCR products for hybridization-based studies has a number of advantages over oligonucleotides, a main one being that no gene sequence information is required before microarray analysis. A researcher can simply amplify cDNAs from a library of interest by PCR, print microarrays with the amplicons, hybridize samples of interest, identify genes that are regulated in a positive and negative manner, and sequence the corresponding genes on the downstream side of the analysis. This "poor person's microarray" as it was called in the early days of the technology is enormously powerful because it allows immediate entry into the genomics of any organism, including organisms for which there is no formal sequencing program, including fungi, insects, agricultural species, and less-studied animals. Another advantage of PCR products is that their relatively large size (500–5000 base pairs) provides extensive complementarity for hybridization, generating intense fluorescent signals in virtually every experiment. PCR is also a widely used technique that is easily implemented in a conventional laboratory setting, providing target material for microarrays in the absence of more expensive and complicated technologies, such as oligonucleotide synthesis.

One disadvantage of PCR products is that these double-stranded nucleic acid targets require denaturation before use, adding some time and complexity to microarray processing. The presence of both the coding and noncoding strands in the target DNA means that PCR products will not hybridize to incoming single-stranded probe molecules unless they are "melted" before hybridization (Fig. 6.3), which is accomplished typically by treating the microarrays with boiling water for several minutes. The presence of complementary target strands in the PCR products also means that reannealing or snapping back can occur during the course of the hybridization. Targets that reanneal cannot hybridize to probe molecules, leading to a loss in fluorescent signal intensity. Reannealing varies with the coupling chemistry, denaturation, and processing; in most cases, however, it is a rather minor concern in terms of reduced signal.

Cross-hybridization. *Undesired hybridization between a target molecule and probe molecules that share sequence similarity.*

A second disadvantage of PCR-generated targets is that their relatively large size provides extensive sequences for hybridization but can also cause unwanted hybridization or ***cross-hybridization*** when studying a family of genes that share

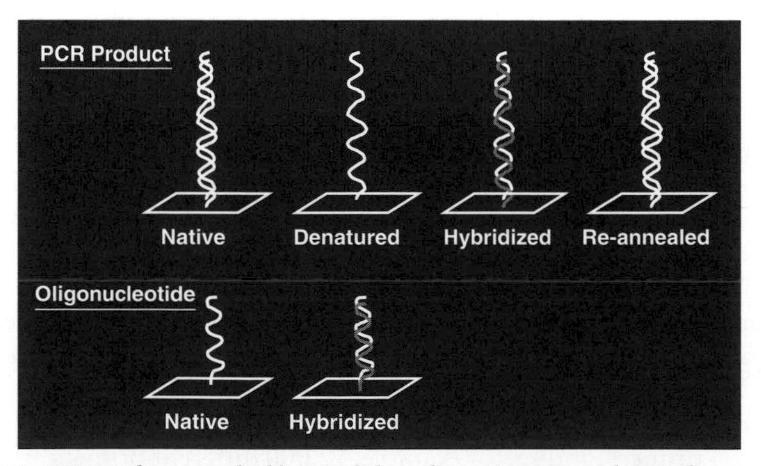

FIGURE 6.3. *PCR products are double-stranded in the native state and must be converted into the single-stranded (denatured) form to allow hybridization with complementary fluorescent probe molecules. The two strands of PCR products can also reanneal to each other, preventing productive hybridization events. Oligonucleotides are single-stranded molecules, allowing hybridization to complementary probe molecules without denaturation or concerns about reannealing.*

considerable sequence identity. Cross-hybridization occurs when a given target shares significant sequence identity (>70% over ~1000 nucleotides) with more than one species in the probe mixture, leading to a loss of gene specificity in microarray experiments. Cross-hybridization erroneously elevates some signals and impairs the capacity of the assay to identify changes in gene expression when comparing two different samples.

Cross-hybridization with PCR-generated targets occurs most commonly when studying highly conserved gene families or gene families that contain a large number of members. The *hsp90* gene from humans, for example, represents a large family of highly conserved heat-shock protein (hsp) genes that protect cells against elevated temperatures and prevent proteins from unfolding during this stressful physiological condition. The complication of using a 2912-nucleotide cDNA sequence from *hsp90* is illustrated by examining the entire human genome for the presence of sequences that are similar or ***homologous*** to *hsp90*. Using the basic local alignment search tool (BLAST) (see Chapter 9) to search the entire human genome (Fig. 6.4) reveals the existence of 15 *hsp90*-related sequences that share strong homology (BLAST score > 80) and 6 *hsp90*-related sequences that share weaker homology (BLAST score 50–80). Though none of these sequences would hybridize as strongly to the *hsp90* target as *hsp90* itself, one or more of these *hsp90*-related sequences might cross-hybridize to the 2.9 kb *hsp90* target, complicating the interpretation of the fluorescent signal in a microarray experiment. BLAST analysis with a 70-nucleotide portion of *hsp90* reveals only a single hit in the human genome and just 3 *hsp90*-related sequences that show weaker homology. In the context of large or highly conserved gene families, the use of long oligonucleotides (70-mers) has an advantage over cDNA targets with respect to reducing the potential for cross-hybridization (see below).

Homologous. *Two or more biological sequences that display sequence similarity.*

The cross-hybridization complication when using PCR-generated targets, cDNAs or ESTs should not be overemphasized, however. For most human genes, large targets are highly specific in microarray hybridization experiments. BLAST analysis of the human interleukin 21 (*IL21*) gene, for example, reveals a single hit in the entire human genome on chromosome 4, irrespective of

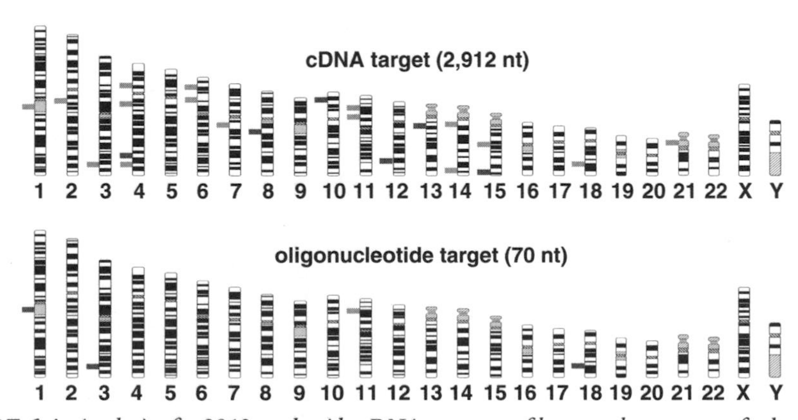

FIGURE 6.4. *Analysis of a 2912-nucleotide cDNA sequence of human chromosomes for* hsp90 *gene produced 16 highly homologous sequences* (light bars), *with BLAST scores > 80, and 6 regions of weaker homology* (dark bars), *with BLAST scores of 50–80. The same BLAST analysis with a 70-nucleotide oligonucleotide target from* hsp90 *produced only a single highly homologous region* (light bar) *and 3 regions of weaker homology* (dark bars). *Map locations are approximate. (Chromosome diagrams courtesy of the Human Genome Organization, Baltimore, MD; homology data courtesy of the National Center for Biotechnology Information, Bethesda, MD.)*

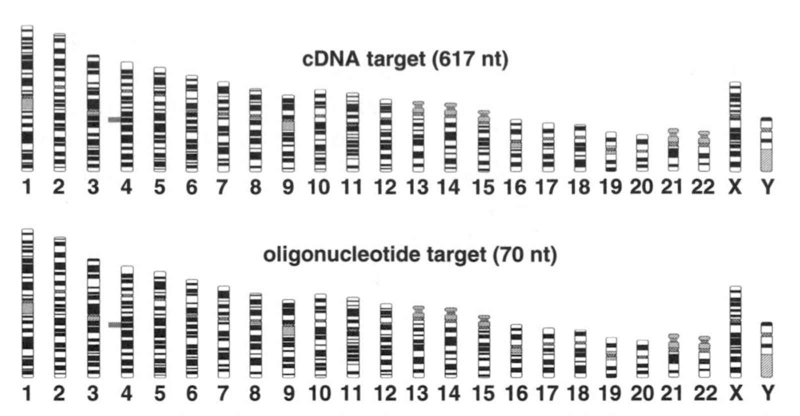

FIGURE 6.5. BLAST analysis of a 617-nucleotide cDNA sequence of the human genome for IL21 *produced a single hit* (bar) *on chromosome 4 as did the same search with a 70-nucleotide oligonucleotide target from* IL21. *Map locations are approximate. (Chromosome diagrams courtesy of the Human Genome Organization, Baltimore, MD; homology data courtesy of the National Center for Biotechnology Information, Bethesda, MD.)*

whether the homology search is performed with a 617-nucleotide cDNA or a 70mer oligonucleotide of *IL21* (Fig. 6.5). A high degree of gene specificity would therefore be expected using either a PCR-generated target or a long oligonucleotide to *IL21,* and the same is true for most human genes. That *IL21* is believed to play a critical role in the proliferation of T cells and immune responsiveness in humans underscores the importance of specific microarray targets for this gene, a criterion satisfied both by cDNAs and oligonucleotides.

Oligonucleotides

Oligonucleotide. *Short chain of single-stranded DNA or RNA.*

Single-stranded ***oligonucleotides*** provide another common source of target sequences for nucleic acid microarrays. Microarrays of oligonucleotides can be prepared using either delivery or synthesis strategies (Fig. 6.6), with conventional phosphoramidite synthesis (see below) or one of the in situ approaches, such as photolithography, providing the microarray targets, respectively. In the delivery strategies, oligonucleotides made off line are prepared using standard phosphoramidite synthesis, suspended in a suitable printing buffer and formed into a microarray using a contact or noncontact printing technology. In the synthesis approaches, oligonucleotides are made in situ one base at a time,

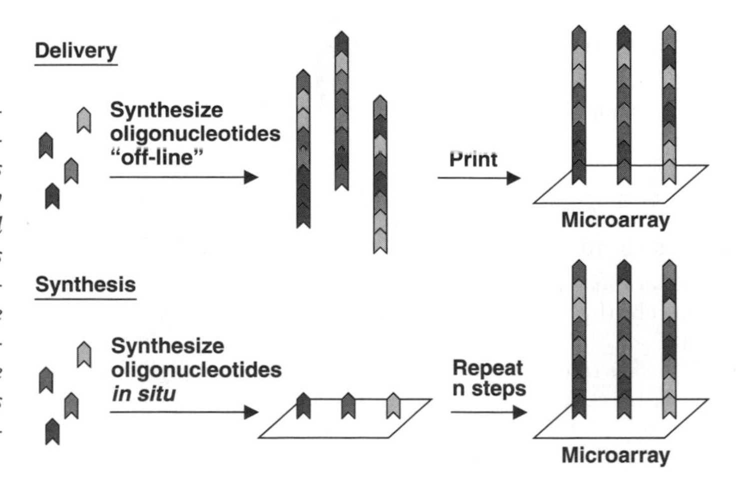

FIGURE 6.6. In the delivery approaches to microarrays, oligonucleotides containing the four DNA bases (bullets) *are prepared off-line by phosphoramidite synthesis and printed into a microarray. In the synthesis approaches, oligonucleotides are synthesized in situ using one of the semiconductor or noncontact printing technologies, and microarrays are formed base by base, thus many cycles* (n *steps) are repeated until the microarrays are complete.*

and many synthesis cycles are used until the microarrays are complete. Because of slightly reduced coupling efficiency and extended synthesis time, oligonucleotides made in situ are generally restricted to 5–25 nucleotides (nt) in length, whereas the delivery strategies allow oligonucleotides as long as can be made by phosphoramidite synthesis (5–120 nt). Both approaches are used widely and both provide high-quality oligonucleotide targets. Oligonucleotide microarrays have a number of strengths and weakness compared to long targets such PCR products.

The main advantages of oligonucleotide targets are increased specificity and the capacity to work directly from sequence database information. The increased specificity of oligonucleotides derives from their small size (<120 nt), allowing unique targets to be made for large gene families, highly homologous sequences, and mRNAs from the same gene that differ solely on the basis of mRNA splicing or processing (see Chapter 4). In each of these cases, oligonucleotide targets allow unique identification of a given gene or transcript and PCR products would not (Fig. 6.7). For genotyping applications involving single base mismatch detection (see Chapter 13), oligonucleotides are essential.

The expanding electronic databases of DNA sequence information, including the recent availability of a draft of the entire human genome (www.ncbi.nlm.nih.gov/), provide a nearly unlimited wealth of sequence information for oligonucleotide manufacture. Oligonucleotides can be selected in an electronic manner from sequence databases (Fig. 6.8), obviating the requirement of preparing and storing cDNA clones and conferring an advantage over PCR-generated products in this respect.

Two disadvantages of oligonucleotide targets are the requirement for sequence information prior to manufacture and the loss of signal when using certain types of fluorescent probes. The requirement for sequence information is less of a problem than it used to be, owing to the frantic pace of genome sequencing internationally. Nonetheless, for obscure organisms, it is convenient to manufacture microarrays in the absence of any sequence information, a capacity afforded by the PCR approach but not by synthetic oligonucleotides. Care must also be taken when using oligonucleotides to make sure that the fluorescent probe mixture is complementary to the target sequences on the chip.

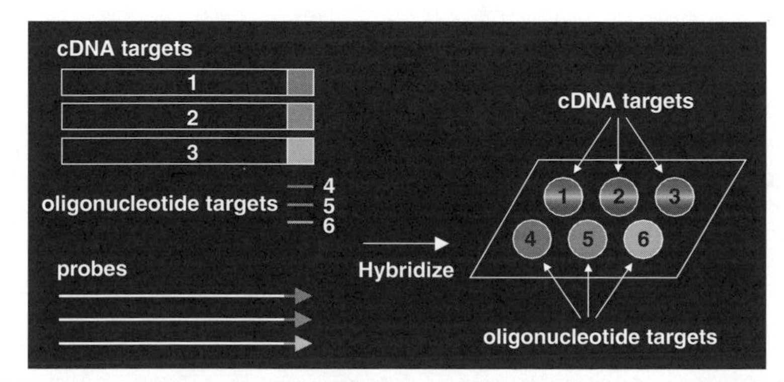

FIGURE 6.7. *Three cDNA targets with highly conserved 5′ ends* (white rectangles) *and divergent 3′ ends* (gray regions) *are printed into a microarray, together with three oligonucleotide targets* (gray lines) *corresponding to the divergent regions of the genes. The microarray targets are hybridized with a fluorescent mixture* (probes) *complementary to the three cDNA targets. The PCR-generated cDNA targets (1–3) will produce cross-hybridization* (rainbow circles) *because each contains the highly conserved 5′ region; the oligonucleotide targets (4–6) mediate specific hybridization on the microarray* (single = color circles) *owing to the presence of specific 3′ target sequences.*

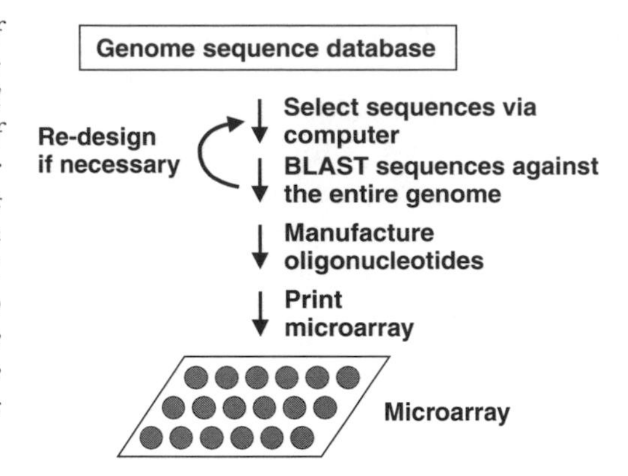

FIGURE 6.8. *The availability of complete genome sequences allows oligonucleotide targets to be selected electronically. Target sequences of interest are selected via computer and examined for uniqueness against the entire genome sequence using a computational tool such as BLAST. Sequences that show homology to more than one region in the genome are redesigned, and those that are unique are printed into microarrays for use in genome analysis.*

Complementarity is rarely a concern when using oligonucleotides corresponding to 3′ gene sequences, but can be a problem if 5′ sequences are chosen, particularly when used in conjunction with labeling by reverse transcription (see below).

Phosphoramidite Synthesis

Phosphoramidite. *Any of the set of modified oligonucleotides used in the manufacture of synthetic oligonucleotides.*

The central importance of oligonucleotide synthesis in microarray target preparation justifies a closer look at this revolutionary chemistry. There are several methods in use presently, but the predominant one uses ***phosphoramidite*** chemistry developed by Caruthers and co-workers in the early 1980s. Phosphoramidite-based oligonucleotide synthesis underlies most of the synthetic DNA market, which today includes more than 75 commercial vendors worldwide and annual revenues totaling hundreds of millions of dollars. Building on concepts developed in the 1960s and 1970s, the Caruthers approach uses a sophisticated process that allows oligonucleotides of any sequence to be synthesized from the four DNA building blocks. The four DNA bases used most commonly are known as cyanoethyl (CE) phosphoramidites (Fig. 6.9). Each base is identical to its natural counterpart except for the presence of several

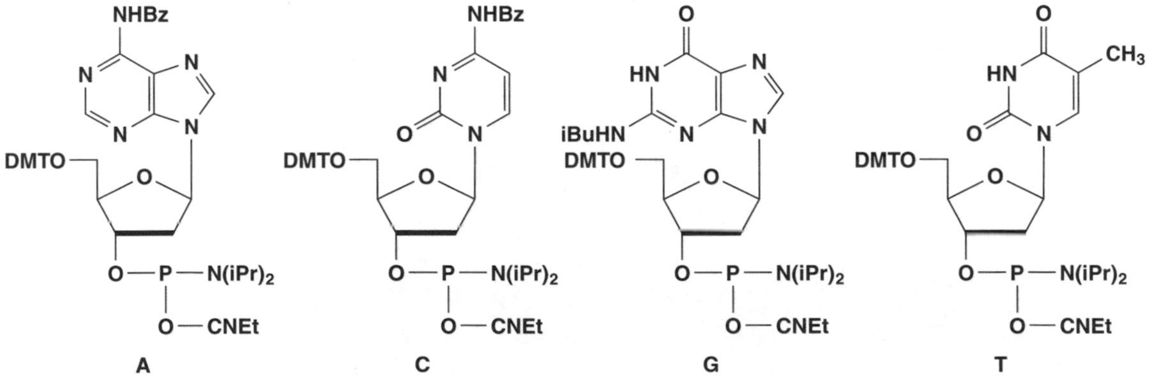

FIGURE 6.9. *Each CE phosphoramidite base contains a cyanoethyl (CNEt) group and a diisopropyl ($N(iPr)_2$) group on the 3′ phosphate and a dimethoxytrityl (DMT) group on the 5′ hydroxyl. In addition, A and C contain a benzoyl (Bz) group and G contains an isobutyryl (iBu) group on the primary amine position in the bases. The T derivative lacks a primary amine and thus does not require a benzoyl or isobutyryl protecting group. (Adapted from Glen Research, Sterling, VA.)*

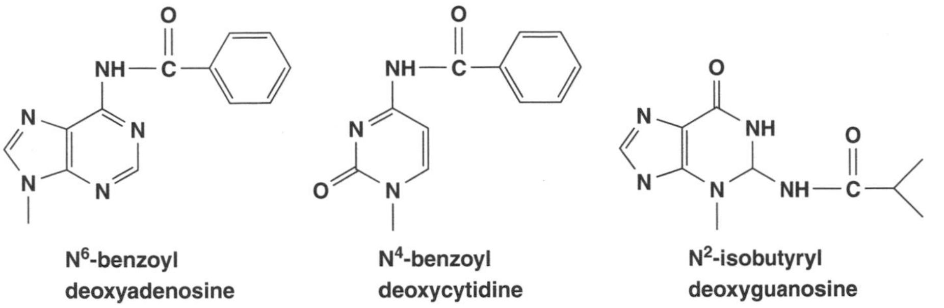

FIGURE 6.10. To protect the A, C, and G phosphoramidite bases, a benzoyl or isobutyryl group is attached to the 6, 4, and 2 position, respectively. These groups are removed at the end of the oligonucleotide synthesis process by treatment with ammonium hydroxide. (Adapted from Interactiva, Ulm, Germany.)

chemical substituents that protect the phosphoramidites during synthesis and activate the 3′ phosphate for chemical coupling.

Three of the phosphoramidite bases (A, C, and G) contain a reactive primary amine on the purine or pyrimidine ring and, therefore, require a protecting group on the amine to avoid damaging this position during the synthesis process (Fig. 6.10). A benzoyl protecting group is typically used for bases A and C, whereas an isobutyryl group is usually employed on G. The fourth base (T) does not contain a primary amine on the pyrimidine ring and thus does not require a protecting group. All four phosphoramidite bases also contain a dimethoxytrityl (DMT) group on the 5′ hydroxyl, which blocks the 5′ hydroxyl from chemical coupling until it is intentionally deprotected during the synthesis process (Fig. 6.11). Selective deprotection allows synthesis to proceed in a stepwise manner. The 3′ phosphate is protected against side reaction and activated for nucleophilic attack by the presence of β-cyanoethyl and diisopropyl groups, respectively. The protecting groups are removed at the end of the synthesis, yielding an oligonucleotide that is identical to native DNA. The synthesis process proceeds in a 3′ to 5′ direction as follows.

The initial step in oligonucleotide synthesis involves coupling the first base to the solid support. Oligonucleotides can be synthesized on a variety of different supports, but the most common matrix is ***controlled pore glass*** (CPG), so named because these glass beads contain pores of defined diameters, inside of which synthesis occurs. The porous nature of the material increases the surface

Controlled pore glass. *Porous silicon dioxide matrix used in the manufacture of synthetic oligonucleotides.*

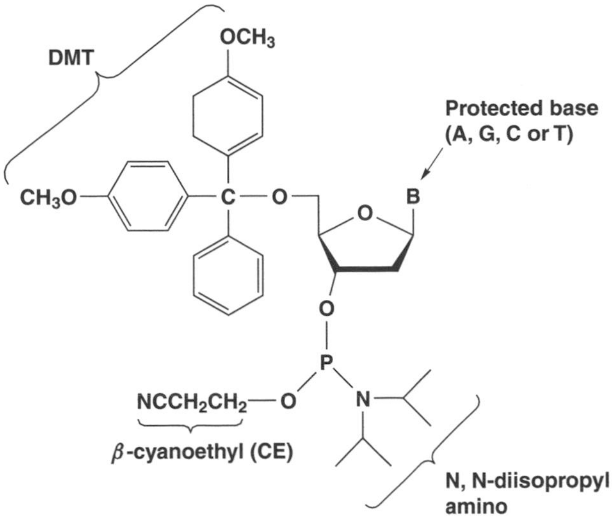

FIGURE 6.11. The locations of the protected bases and DMT on the deoxyribose ring (center). Removal of the DMT group during synthesis exposes the 5′ hydroxyl group, which acts as a nucleophile, attacking the 3′ phosphate position activated by CE and N,N- diisopropylamino groups. (Adapted from Interactiva, Ulm, Germany.)

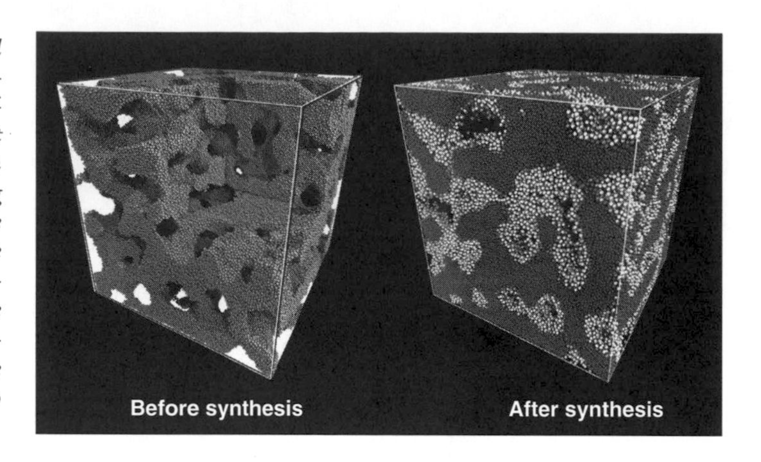

FIGURE 6.12. Computer-generated models of CPG simulating the oligonucleotide synthesis matrix before and after oligonucleotide synthesis. The first DNA base is bound directly to silicon dioxide atoms (gray matrix). *During the synthesis process, phosphoramidite reagents* (white spheres) *enter the porous CPG and become incorporated into the growing oligonucleotide chains.(Courtesy of Dr. L. Gelb, Department of Chemistry, Florida State University, Tallahassee.) Figure also appears in Color Figure section.*

area of the synthesis matrix, which improves the efficiency of the process and reduces reagent consumption. Computer-generated models of CPG illustrate the properties of this material (Fig. 6.12). Pores sizes for CPG are 50–100 nm for most oligonucleotide synthesis applications. The first base is coupled to CPG via a chemical linker arm that attaches at the 3′ hydroxyl group on the deoxyribose ring (Fig. 6.13). Once the first base is attached to the CPG, oligonucleotide synthesis proceeds in a stepwise fashion until the last base is coupled to the growing chain. The steps are performed in the liquid phase, and most current synthesizers are capable of manufacturing oligonucleotides in a 96-well or 384-well format.

Deprotection. *Step in oligonucleotide synthesis in which a protecting group such as dimethoxytrityl is removed from the 5′ hydroxyl group to allow a subsequent round of coupling.*

Detritylation. *Step in oligonucleotide synthesis in which the dimethytrityl group is removed from the 5′ hydroxyl group to allow a subsequent round of coupling.*

Coupling. *Step in the chemical synthesis of oligonucleotides and other synthetic molecules in which a monomer is added to the growing polymer chain.*

Capped. *Inactivated functional group prevented from undergoing chemical reactivity by the addition of a blocking group.*

To couple the first base to the second, the CPG matrix is treated with a weak acid, such as tetrazole, to remove the DMT group from the 5′ hydroxy (Fig. 6.14). This ***deprotection*** or ***detritylation*** step allows the 5′ hydroxyl to act as a nucleophile, attacking the 3′ activated phosphate group of the second base, which is added to the activated CPG matrix by coupling to the first base. The result is dinucleotide bond formation in the 3′ to 5′ direction. After the ***coupling*** step, unreacted 5′ hydroxyl groups are inactivated or ***capped*** by acetylation to prevent these bases from reacting with phosphoramidites in subsequent coupling steps. Capping prevents the formation of frame shift oligonucleotides that are missing one or more bases compared to the full-length product, a process that occurs if unreacted 5′ hydroxyls are not capped before the next coupling cycle. After capping, the phosphite triester of the newly formed dinucleotide is oxidized to the phosphate form to stabilize the phosphate linkage.

The four-step process of de-protection, coupling, capping, and oxidation is the basis of phosphoramidite synthesis (Fig. 6.14). Repeating the cycle allows oligonucleotides of any sequence of interest to be synthesized efficiently using

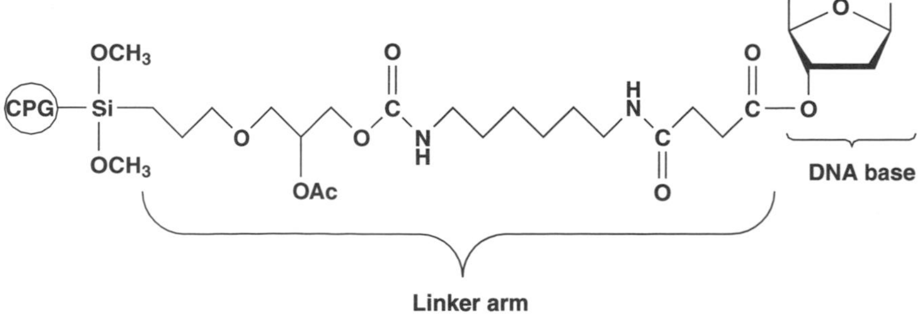

FIGURE 6.13. A CPG bead containing a chemical linker arm is attached to a phosphoramidite DNA base containing a DMT protecting group. Treatment of the coupled CPG with a weak acid removes the DMT group and allows the first base to be coupled to the second. (Adapted from Interactiva, Ulm, Germany.)

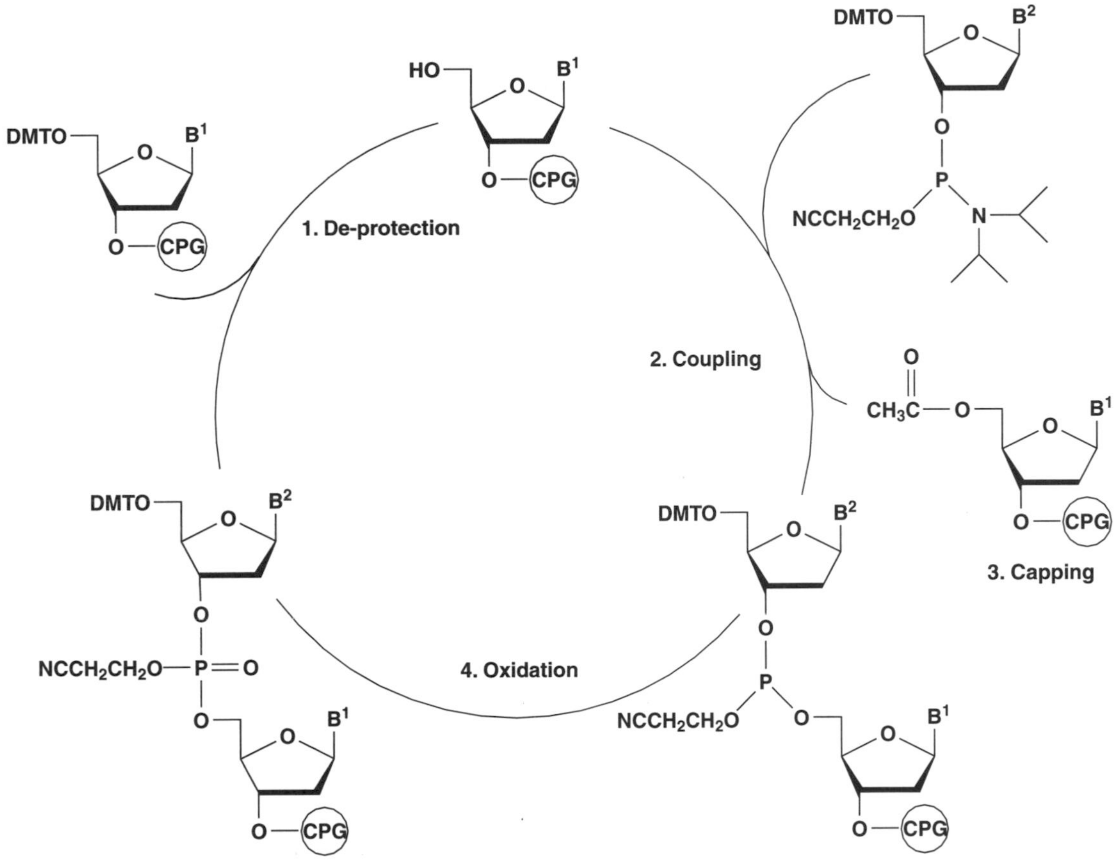

FIGURE 6.14. *The four-step process of the oligonucleotide synthesis cycle on CPG. (Adapted from Interactiva, Ulm, Germany.)*

the phosphoramidite bases and supporting reagents. Each four-step cycle takes 5–7 min, enabling synthesis of a synthetic 70-mer in < 8 h. Following synthesis, the nascent oligonucleotides are treated overnight with ammonium hydroxide to remove the protecting groups from the bases and phosphate groups and to cleave the oligonucleotides from the CPG support. With coupling efficiencies exceeding 99% per cycle, a synthetic 70-mer preparation would contain more than 60% full-length product (Table 6.2). Full-length oligonucleotides can be purified away from shorter products using polyacrylamide gel electrophoresis (PAGE) or high-pressure liquid chromatography (HPLC).

Proteins

Protein biochemistry has a glorious past; but generally speaking, the techniques for studying protein–protein interactions, catalysis, structure, and the like are rather conventional. There is an opportunity to leverage microarray technology into the protein area and in so doing expand greatly the capacity to study protein function in a parallel manner. Owing to their naturally advantageous chemistry, proteins attach with high efficiency to microarray surfaces. This section provides a brief look at how protein samples are prepared and formed into microarrays.

Protein microarrays are prepared from diverse sources such as purified preparations, synthetic peptides, and native cellular extracts. Proteins can be overexpressed in a number of different systems including *Escherichia coli* (*E. coli*), yeast, and insect and mammalian cells. The use of recombinant vector systems such as those based on baculovirus provide microgram quantities of protein from a few milliliters of culture. Recombinant proteins produced in a 96- or 384-well microplate format provide throughput sufficient to prepare microarrays containing all of the proteins (5,000–30,000) expressed in a given organism. Fusion proteins containing a successive string of histidine residues (e.g., His tag) bind with high affinity to nickel columns, allowing the rapid purification of the recombinant proteins for use as microarray targets. Other types of convenient molecular tags, or ***epitopes,*** such as glutathione-*S*-transferase (GST), c-Myc, influenza virus haemagglutinin (HA), and green fluorescent protein (GFP), provide a multitude of different strategies for isolating protein targets expressed from recombinant vectors.

Epitope. *Protein moiety used in antibody binding or to stimulate an immune response.*

For energetic reasons, folded proteins tend to direct the amino acids containing charged side chains onto the surface of the protein. The result for most proteins is a surface studded with both the positively charged amino acids (His, Arg, and Lys) and the negatively charged residues (Glu and Asp). The side chains of arginine and lysine are primary amines that can attach to microarray substrates by Schiff base interactions (see Chapter 5) with reactive aldehyde groups (Fig. 6.15). The negatively charged glutamic and aspartic residues bind to microarray substrates via ionic interactions (see Chapter 5) with reactive amine surfaces. The same surfaces that allow attachment of nucleic acid targets, allow attachment of proteins. Though the details of the reactive groups are different for nucleic acids and proteins, the underlying attachment chemistries (Schiff base and ionic) are the same for the two types of molecules.

With respect to stability, there are some special issues that apply to proteins but not to nucleic acids, among them is the fact that protein binding is sensitive to ***denaturation*** and oxidation. When removed from their cellular milieu, proteins can loose their native shape, and the undesired process of denaturation diminishes the capacity of the protein to interact with other proteins, perform catalysis, and so forth. The traditional view is that proteins are inherently unstable molecules, though this view is based on experimental protocols that

Denaturation. *Process of denaturing.*

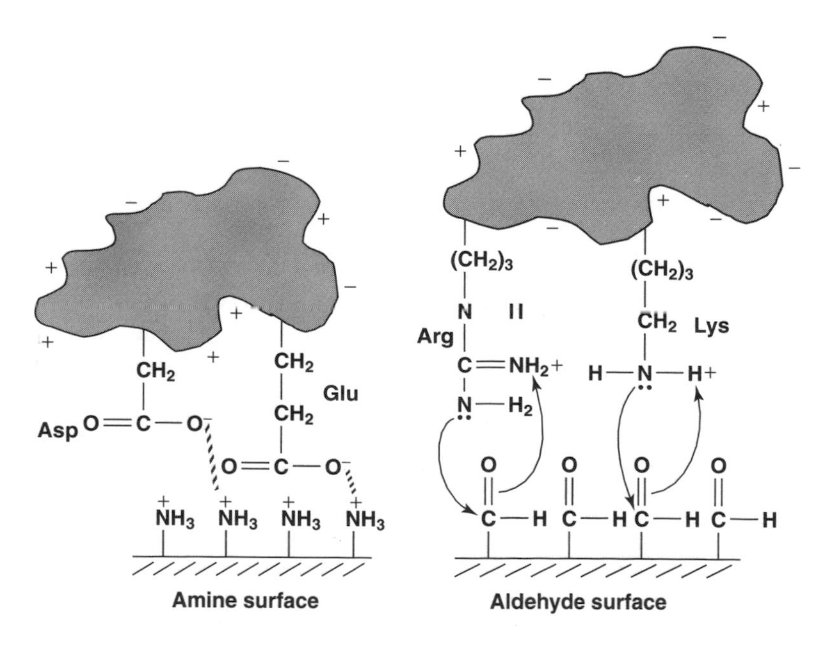

FIGURE 6.15. **A,** *The negatively charged Asp and Glu residues allow attachment of proteins (gray) via ionic bonds to microarray surfaces containing reactive amine groups. B, The positively charged Arg and Lys residues enable Schiff base attachment of proteins to microarray surfaces containing reactive aldehyde groups.*

often require days or even weeks of manipulation and often involve large (e.g., 1000-fold) dilutions of the proteins. The arduous nature of traditional protein biochemistry and the massive dilution of these macromolecules may in fact be responsible for most of the presumed instability of proteins. The microarray format might therefore offer a much better setting in which to study protein function. Microarray manufacture is rapid, reactions with probe solutions are brief (2–4 h), and the concentration of the printed targets is extremely high (10 mg/mL). All of these factors contribute to stabilizing protein structure in microarray formats. Printing buffers tailored to protein microarrays (see below) also aid in maintaining native protein conformation.

Printing Buffers

Target solutions of PCR products, oligonucleotides, proteins, and other molecules require suspension in an appropriate buffer before microarray printing. ***Printing buffers*** of various types have been used, and all attempt to accomplish one or more of the following: increase printing efficiency, reduce sample spreading, promote uniform delivery, stabilize the target molecules, denature the target molecules, reduce sample drying, and increase the visibility of the printed spots.

Printing buffer. *Chemical mixture used in microarray manufacture to stabilize target molecules and improve sample spreading and attachment.*

Printing efficiency (P_e) is simply a measure of the number of spots that are printed (S_p) divided by the total number of spots that are attempted (S_a)

$$P_e = (S_p/S_a) \times 100 \tag{6.3}$$

The printing efficiency for a microarray that is intended to have 1000 spots but has only 995 would be 99.5%. This efficiency would suffice for many research applications, but not for commercial chip manufacturing, for which a 100% printing efficiency is required for every manufactured microarray. Given the premium placed on 100% printing efficiency, buffers that improve the printing process are extremely valuable. A brief summary of the forces acting on microarray droplets during contact printing helps explain how printing buffers can improve printing efficiency.

During contact printing, sample is held in the pin by the combined forces of ***adhesion*** and surface tension. Adhesion is the force exerted by the printing pin (e.g., stainless steel) on the sample, and surface tension is the cohesive force (e.g., hydrogen bonding) exerted on the liquid by molecules within the liquid at the droplet surface. Microarray target solutions adhere to the tip of a microspotting pin because the pin pulls upward on the sample and the sample attracts itself within the liquid layer (Fig. 6.16). There is also a minor downward pull on the droplet owing to gravity.

Adhesion. *Attractive force exerted by a solid surface such as a printing pin on a liquid sample, contributing to meniscus formation and other phenomena.*

When the droplet touches the printing surface, the substrate (e.g., chemically treated glass) exerts an adhesive force on it. If the combined force of substrate–liquid adhesion and gravity exceeds the forces of pin–liquid adhesion and sample surface tension, the sample adheres to the printing surface and a microarray droplet is deposited (Fig. 6.16). If substrate–liquid adhesion and gravity are less than the forces of pin–liquid adhesion and surface tension, the sample adheres to the pin on the upstroke, and a microarray spot is not formed. The latter scenario is a major source of printing failure in microarray manufacture and is accentuated on hydrophobic surfaces (e.g., aldehyde), which exert less adhesive force on aqueous samples than do hydrophilic surfaces (e.g., amine). Printing failures tend to be more frequent with solid pin technologies

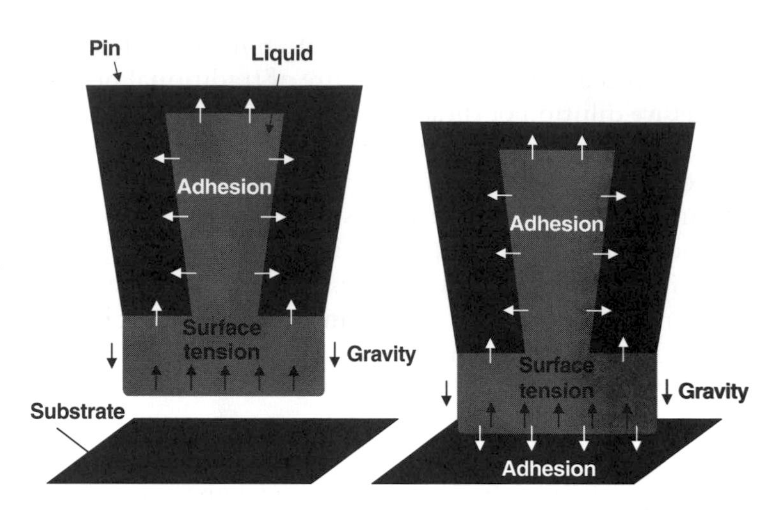

FIGURE 6.16. The sample in a printing pin is held in place by adhesion (white arrows) *and surface tension* (upward black arrows). *When the sample touches the printing surface, the substrate exerts adhesive forces on the liquid. If the sum of the forces of substrate–liquid adhesion and gravity exceeds the forces of pin–liquid adhesion and liquid surface tension, a sample droplet is deposited, forming a microarray spot.*

than with microspotting pins, because the adhesive forces are larger for solid pins, owing to the larger surface area of the solid pin tip.

Viscosity. *Property of a liquid corresponding to its resistance to flowing or spreading.*

One way to improve the efficiency of contact printing is to increase sample ***viscosity.*** The stickiness, or viscosity, of PCR products and genomic DNA tends to be fairly high, so simple salt solutions (e.g., 3× standard saline citrate or 3× SSPE) usually suffice as printing buffers for these molecules. Oligonucleotides, which have lower viscosity than PCR products and genomic DNA, often require the used of buffers that contain thickening agents to ensure robust printing efficiency. Viscosity increases the adhesive force exerted on the sample by the substrate; thus more viscous liquids print better than less viscous ones. There is a trade-off, however, between viscosity and spot-to-spot contamination in that viscous solutions are generally more difficult to remove from the pins during the wash steps and tend to pose a greater threat in terms of spot-to-spot carry over than less viscous solutions. Proper pin cleaning allows samples of nearly any viscosity to be printed without sample carry over, but extra care should be taken when using sticky samples.

Printing buffers are also formulated to improve the visibility of printed spots, a capacity that is extremely useful because printing efficiency is often assessed by visual inspection of microarrays with a dissecting microscope. Upon drying, additives such as salts and polymers form translucent crystals on the surface of printed spots, allowing optical detection with moderate (10–50×) magnification. Dried spots can also be seen by scanning or imaging the microarray, though it should be emphasized that the detected signal in this case corresponds to light that is reflected off the surface of the salt crystals rather than from a true fluorescent signal. Most printing buffers produce spots that are discernible in scanners and imagers, though in most cases neither the sample nor the printing buffer is appreciably fluorescent.

Wettable. *Extent to which a solid surface allows spreading of a liquid.*

Surface free energy. *Thermodynamic description of a liquid spreading across a solid surface.*

In addition to controlling sample viscosity, printing buffers are also designed to minimize sample spreading, which reduces spot diameter and increases printing density. The effect of printing buffers on this criterion is best understood in the context of ***wettability,*** which describes how a liquid droplet spreads across a surface once it is delivered by a printing device. Spreading of the droplet is determined mainly by the opposing forces of cohesion and adhesion or ***surface free energy,*** the balance of which determines how readily a liquid coats the microarray substrate (Fig. 6.17). For aqueous solutions whose

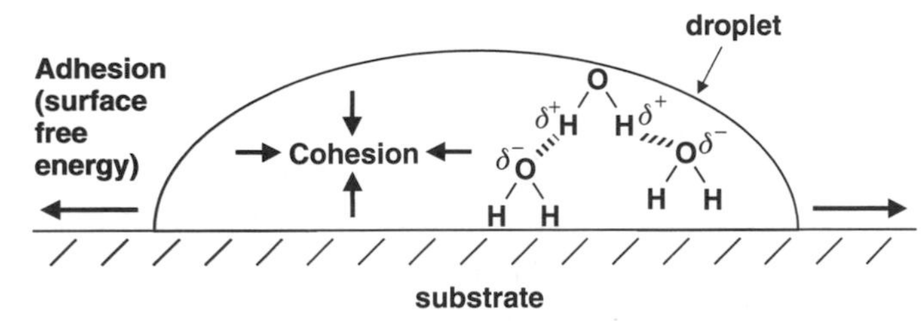

FIGURE 6.17. A sample droplet deposited on a microarray substrate spreads across the surface until it reaches a balance or equilibrium between the outward forces of adhesion or surface free energy and the inward forces of cohesion. For aqueous samples and printing buffers, the predominant cohesive force is hydrogen bonding (dashed lines) *between water molecules in the droplet.*

main ingredient is water, the main cohesive force is hydrogen bonding, and so printing buffers with strong hydrogen bonding properties produce smaller spots than those with weaker hydrogen bonding. Samples containing bases (sodium hydroxide, potassium hydroxide), alcohols (ethanol or methanol) or solvents (dimethylsulfoxide, formamide) have weaker hydrogen bonding properties than aqueous salt solutions; thus printing buffers containing these agents produce larger spots than their aqueous counterparts. The unwanted phenomenon of spot merging, whereby printed spots overlap on the microarray, is commonly due to contaminants that disrupt hydrogen bonding in the target samples.

Printing solutions act in conjunction with surfaces to determine the final spot diameter; hydrophobic surfaces (e.g., aldehyde) produce smaller spots than do hydrophilic surfaces (e.g., amine) for a given sample. A droplet will always seek the ***lowest total energy*** on a surface, such that the spreading of an aqueous droplet across aldehyde groups is less favorable than spreading of the same droplet across amine groups. The wettability of a given solution on a microarray substrate can be determined optically by measuring the ***contact angle,*** which is the tangent formed between the printed droplet and the substrate 1 s after the droplet is delivered to the microarray surface (Fig. 6.18). The contact angle for an aqueous sample on an aldehyde substrate is markedly larger than on an amine surface because droplets tend to bead up on the oily surface, forming relatively tight droplets. Contact angle measurements can provide quantitative information about the properties of printing buffers and microarray surfaces.

Lowest total energy. *Minimum energy state sought by a liquid spreading across a solid surface.*

Contact angle. *Tangent formed between a printed droplet and a microarray substrate, measured typically 1 s after the droplet is deposited onto the surface.*

Hydrophilic solutions such as those containing oligonucleotides can pose a printing problem, particularly on hydrophobic surfaces, because the sample tends to wet unevenly across the substrate. Uneven sample wetting can cause punctate fluorescence within a given spot, which can impair quantitation in a

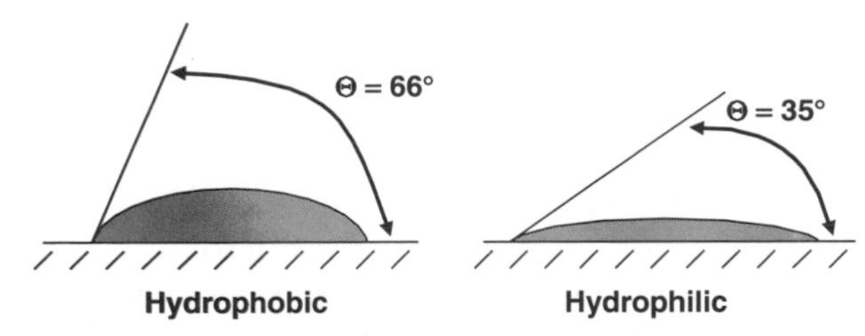

FIGURE 6.18. An aqueous printing buffer, such as a saline solution, beads up on a hydrophobic surface and produces a larger contact angle than the same solution on a more hydrophilic surface. The contact angle (θ) measures the wettability of a surface for a given sample.

given assay. Reducing the cohesive forces of the liquid with low concentrations of a detergent or adding reagents that provide a sieving capacity can promote uniform sample spreading.

Printing buffers can also be formulated to increase the stability of the target molecules, a capacity more important for proteins than for nucleic acids, given the relative instability of polypeptides compared to DNA. Glycerol and other additives that are used to stabilize proteins in conventional purification schemes are used in microarray printing buffers to stabilize the proteins once they are printed on the substrate. Glycerol also prevents complete drying of protein samples on the microarray surface, which minimizes protein unfolding. Dimethylsulfoxide (DMSO) has been used in nucleic acid printing buffers to prevent dehydration of samples from microplates during extended print runs. DMSO works well to slow dehydration; but it has the trade-off of exaggerating environmental influences on printing quality, particularly the effects of small changes in humidity on spot diameter. Heat, mild base, and DMSO have also been used in printing buffers to denature double-stranded DNA before printing, though most users find that denaturation postprinting is a better way to obtain single-stranded targets.

PROBES

The quality of the probe solution is a key determinant of the quality of the microarray data. Fluorescence is the dominant probe label, and so most of this section focuses on labeling methods that employ fluorescent tags, though other types of labeling and detection schemes have been used for microarray analysis (see Chapter 14). Probes can be derived from many different biological and chemical sources, although nucleic acid probes are the most common at present. This section presents an overview of microarray probe preparation, focusing on synthesis approaches, labeling methods, amplification schemes, and comparative analyses.

Types of Approaches

Direct labeling. *Probe preparation scheme in which the fluorescent tags are attached in a covalent manner directly to the probe molecule using an enzymatic or chemical means.*

Indirect labeling. *Probe preparation scheme in which the fluorescent tags are attached in a noncovalent and indirect manner to the probe molecules using dendrimers, antibodies, or some other reagent.*

There are two approaches to probe preparation: ***direct labeling*** and ***indirect labeling***. The direct labeling schemes attach fluorescent tags in a covalent manner to the probe molecules by an enzymatic or chemical means, whereas the indirect labeling approaches attach fluorescent tags in a noncovalent manner to the probe molecules via a bridge molecule (e.g., biotin, oligonucleotide), which, in turn, is attached to a fluorescent reagent. Both approaches provide high-quality probes, and there are pros and cons to these two general strategies (Table 6.3).

Reverse Transcription

Reverse transcription. *Enzymatic process in which genetic information in cellular RNA is converted into DNA.*

Building on the Nobel prize–winning work of Temin and Baltimore for studies on the enzyme reverse transcriptase (RT) in the early 1970s, ***reverse transcription*** is the direct labeling method used for the first microarray experiments. In the most basic reverse transcription approach, samples of mRNA are hybridized with oligo-dT, and fluorescent nucleotides are incorporated into cDNA using reverse transcriptase, which initiates synthesis on the primed templates (Fig. 6.19). Following this direct labeling reaction, the templates are treated with

TABLE 6.3. Direct and Indirect Probe-Labeling Schemes

Criterion	Reverse Transcription	RNA Polymerase	Eberwine Procedure	TSA[a]	Dendrimer[b]
Type	direct	direct or indirect	direct	indirect	indirect
Biological template	RNA	double-stranded DNA containing a T3 or T7 promoter	double-stranded DNA containing a T7 RNA polymerase promoter	RNA or DNA labeled with a small molecule tag	RNA or DNA containing a dendrimer binding sequence
Labeling tag	modified nucleotide	modified nucleotide	modified nucleotide	modified nucleotide	oligonucleotide
Fluorescent reagent	modified nucleotide	modified nucleotide, antibody conjugate, TSA or dendrimer	modified nucleotide	tyramide cyanine analogs	oligonucleotide–dye complex
Molecular interaction	hybridization	hybridization or small molecule-antibody	hybridization	small molecule-antibody	Hybridization
Amount of signal amplification	none	10- to 100-fold	100- to 1,000,000-fold	100-fold	10- to 350-fold
Type of signal amplification	none	none, enzymatic or passive	passive by increasing the amount of RNA	enzymatic	passive
Fluorescent dyes	cyanine, Alexa[c]	cyanine, Alexa,[c] Phycoerythrin	any	cyanine	cyanine, Alexa,[c] BODIPY[c]
Posthybridization processing	none	none to 3 h	None, but RNA amplification takes several days	3 h	1 h

[a]Tyramide signal amplification (Life Science Products, Boston).
[b]Provided by Genisphere (Datascope, Montvale, NJ).
[c]From Molecular Probes (Eugene, Oregon OR).

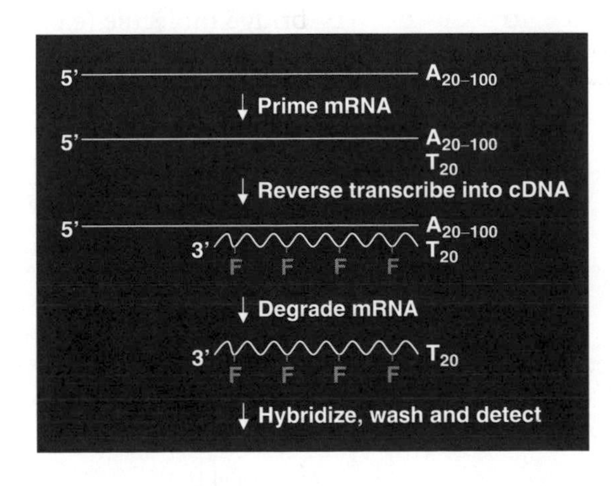

FIGURE 6.19. mRNA containing the typical polyA sequence at the 3′ end (A_{20-100}) is primed with oligo-dT (T_{20}) and reverse transcribed into cDNA in the presence of fluorescent nucleotides (F). The mRNA is degraded with sodium hydroxide to provide a single-stranded, fluorescent cDNA molecule that can be hybridized to a microarray.

sodium hydroxide to degraded the mRNA, providing a single-stranded cDNA mixture that is subsequently purified to remove unincorporated nucleotides, salts, enzyme, and other unwanted components. The RT-generated fluorescent probe is hybridized to a microarray, and the chip is detected after several brief washes to remove unhybridized probe molecules.

Since the original microarray publication in 1995, there have been dozens of modifications of this direct labeling method. Some protocols use total cellular RNA instead of mRNA, which provides little loss of labeling efficiency but a large savings in time because total RNA is much easier to obtain than mRNA. Some protocols use random 9mer primers instead of 20mer oligo-dT primers to allow labeling of mRNAs that do not contain polyA sequences, such as those from bacterial sources. A number of different reverse transcriptase enzymes have been used and it appears that some RT enzymes provide better labeling than others, perhaps due to an improved capacity to use fluorescent nucleotide analogs. A variety of different probe purification schemes have also been described, including ethanol precipitation, size exclusion, and membrane-based purification. Probe solutions free of unincorporated fluorescent nucleotides and other contaminants always yield superior microarray results, owing largely to reduced background fluorescence.

Aminoallyl. *Aliphatic primary amine containing a double bond that allows direct labeling of microarray probe molecules by nucleophilic attack of fluorescent reagents by the modified bases.*

Another adaptation of direct labeling with RT has been the development of a protocol that incorporates ***aminoallyl*** nucleotide analogs (Fig. 6.20), which allows direct labeling of cDNA molecules by reacting the aminoallyl groups with reactive fluorescent dyes. Aminoallyl is a primary amine that acts as a nucleophile, attacking fluorescent dye molecules, such as cyanine and Alexa derivatives, that contain reactive groups and coupling the dye directly to the nascent DNA or mRNA via the aminoallyl function. The main advantage of this approach is that, owing to the small size of the aminoallyl group, aminoallyl nucleotide analogs incorporate much more easily into DNA or mRNA than the bulky fluorescent bases. More efficient incorporation enables more uniform labeling and, in some cases, probes that are more brightly fluorescent.

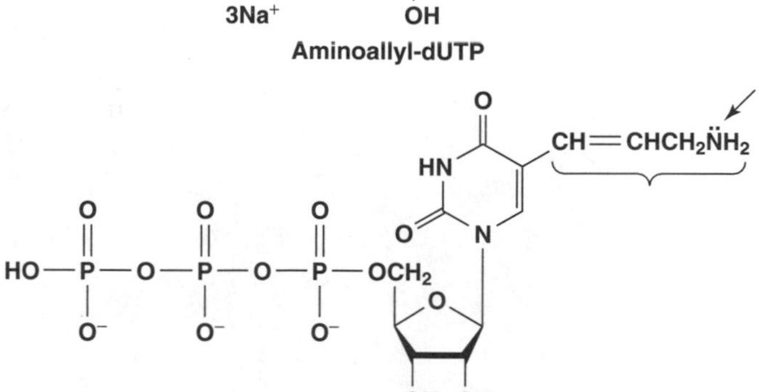

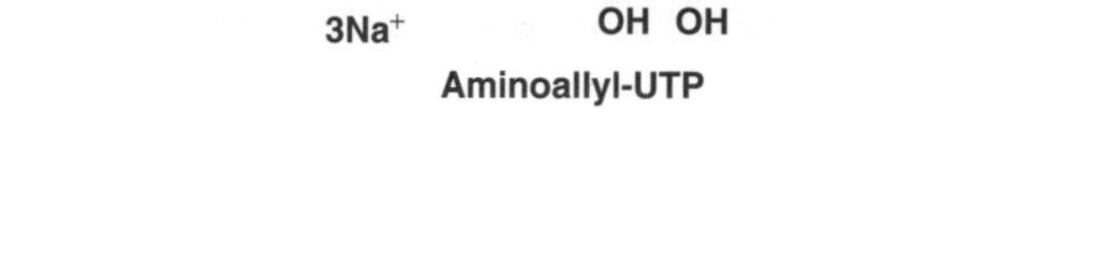

Aminoallyl-dUTP

Aminoallyl-UTP

FIGURE 6.20. *Aminoallyl-2′-deoxyuridine-5′-triphosphate (dUTP) and aminoallyl-uridine triphosphate (UTP) are identical, except that aminoallyl-UTP has a hydroxyl group (OH) at the 2′ position on the ribose ring. The aminoallyl group* (brace) *is added chemically to the 5 position of the uracil ring and the lone pair of electrons* (arrow) *acts as a nucleophile, attacking fluorescent dye molecules that contain reactive groups, coupling the dye directly to the nascent DNA (aminoallyl-dUTP) or RNA (aminoallyl-UTP). Both derivatives are supplied commercially as trisodium salts (3 Na^+). (Adapted from Molecular Probes, Eugene, OR.)*

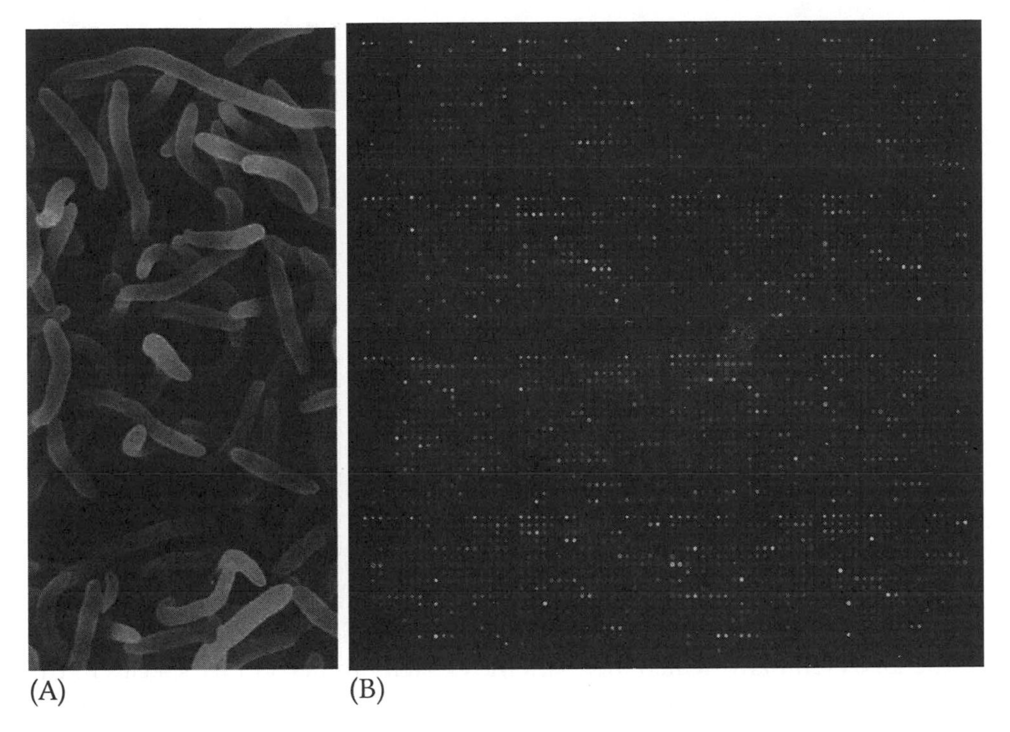
(A) (B)

FIGURE 6.21. **A,** *Photomicrograph of the bacterium* V. cholera. *(Courtesy of Dennis Kunkel Microscopy, Kailua, HI.)* **B,** *Fluorescent image of a microarray containing the entire genome of* V. cholera. *(Courtesy of K. Chong, Schoolnik Laboratory at Stanford University, Stanford, CA.) Total RNA from* V. cholera *was labeled with reverse transcriptase in the presence of Alexa546-dUTP, and the direct-labeled probe was hybridized to the microarray and scanned with a ScanArray 5000 (Packard Biochip Technologies, Billerica, MA). The fluorescent TIF data are coded to a monochromatic green look-up table. (Photomicrograph was provided). Figure also appears in Color Figure section.*

The main advantage of direct labeling with RT is the simplicity of the approach. Probe molecules that are labeled directly obviate the need for posthybridization reactions, which can be arduous and time-consuming. The RT methods are also highly versatile, allowing the use of many different fluorescent dyes—including fluorescein, Lissamine, cyanine, Alexa, and BODIPY—as well as the aminoallyl derivatives; and each has been shown to work well in the microarray format. Microarrays containing genes representing the entire genome of *Vibrio cholera,* (a bacterial agent that causes a serious class of food poisoning) highlight the brightness of the signals obtained by direct labeling with an Alexa dye (Fig. 6.21). The main disadvantage of direct labeling by reverse transcription is its reduced detectivity compared to the indirect approaches that provide signal amplification (Table 6.3).

RNA Polymerase

RNA polymerase. *Enzyme that synthesizes RNA molecules from a DNA template.*

In addition to the use of reverse transcriptase (see above), ***RNA polymerase*** is an enzyme used widely for microarray probe synthesis. The RNA polymerases are actually a family of enzymes from bacterial viruses (e.g., T3 and T7) that catalyze the synthesis of RNA from double-stranded DNA templates containing specific bacteriophage promoter sequences (see Chapter 4). Catalysis by T3 or T7 RNA polymerase is extremely robust, with tens of micrograms of RNA being produced from less than one microgram of DNA template. RNA polymerase is used in number of capacities to synthesize RNA for microarray applications: synthesis in the presence of "cold" ribonucleotide triphosphates (rNTPs) to make large quantities of RNA for subsequent labeling reactions, RNA amplification in the context of the Eberwine procedure (see below), and synthesis in the presence of fluorescent or nonfluorescent nucleotide analogs for direct and indirect labeling procedures. The incorporation of biotin-labeled rNTPs for use in a subsequent indirect labeling approach illustrates the enzymology of T7 RNA polymerase (Fig. 6.22). Conceptually similar reactions underpin all of the RNA polymerase-based probe preparation methods.

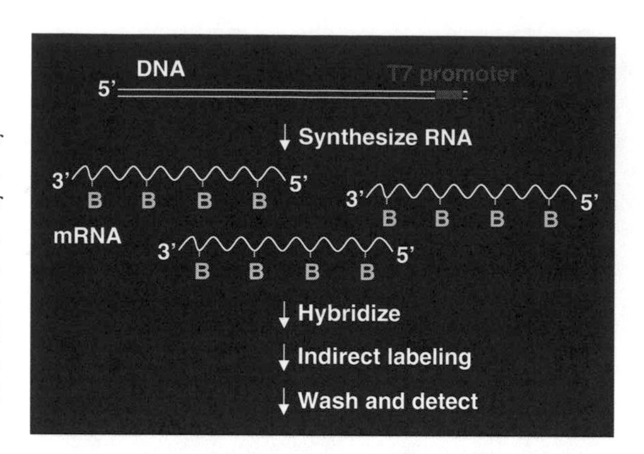

FIGURE 6.22. In the presence of biotin-labeled rNTPs, T7 RNA polymerase synthesizes multiple copies of RNA containing biotin tags (B) *from templates that contain the T7 promoter* (rectangle). *The mixture is hybridized to a microarray and labeled by an indirect means. Finally, the fluorescent microarray is washed and detected.*

There are a number of advantages to using RNA polymerase to generate probe molecules. The synthesis products are RNA instead of DNA, making it much easier to fragment the RNA into smaller pieces for hybridization to oligonucleotide microarrays (see below). Owing to the robustness of the RNA polymerase reaction, there is also a ~100-fold amplification of the RNA relative to the DNA template, which increases the capacity to detect rare species in the mixture. The enzyme is active on any double-stranded template that contains the corresponding bacteriophage promoter, making it possible to synthesize RNA templates from double-stranded DNA mixtures (e.g., cDNA) or cloned sequences (e.g., cDNA libraries). It is also possible to generate a large amount of in vitro transcript corresponding to a specific gene of interest, and these in vitro transcripts provide excellent controls for mRNA labeling experiments.

The main disadvantage of RNA polymerase-generated products is that the resultant RNA molecules are susceptible to chemical and enzymatic digestion to a much greater extent than cDNA mixtures made with RT. To avoid degradation of synthesized RNA, chips must be free of ribonuclease contamination. In this context, gloves must be worn at all times to prevent the transfer of ribonucleases onto the microarray surface, and careful cleanroom practices using ribonuclease-free reagents and buffers are also extremely helpful.

Eberwine Procedure

Van Gelder and Eberwine developed an ingenious method for mRNA amplification in the late 1980s based on T7 RNA polymerase. The so-called Eberwine procedure converts mRNA into cDNA and then into amplified RNA (aRNA), in a manner that allows up to 10^6-fold amplification of the starting material. The method relies on the ***linear amplification*** of the cloned sequences by RNA polymerase, thereby minimizing misrepresentation of the amplified RNA relative to the starting material. The procedure has been used to construct cDNA libraries from single human cells and to examine gene expression patterns in a few hundred isolated neurons by microarray. The amplification procedure works as follows.

Linear amplification. *Experimental procedure in which multiple copies of nucleic acids are obtained by a nonexponential process.*

Messenger RNA (mRNA) is isolated from a small number of cells using a standard mRNA purification procedure. Cells can be small aliquots from cell culture populations, single cells isolated from tissue sections using a technique known as laser capture microdissection (see Chapter 4), needle biopsies, and other sources. The isolated mRNA is hybridized with an oligo-dT primer that contains additional sequences corresponding one strand of the T7 RNA

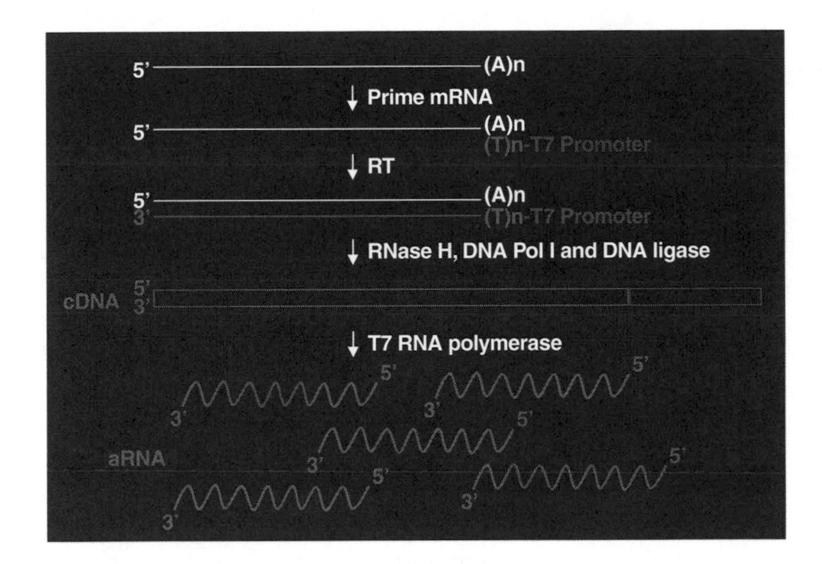

FIGURE 6.23. For the Eberwine procedure, mRNA is hybridized with an oligo-dT primer that contains T7 promoter sequences (blue and red primer). *Reverse transcriptase is used to produce a single-stranded cDNA* (blue line) *from the primed RNA template. Ribonuclease H* (RNase H), *DNA polymerase I* (DNA pol I), *and DNA ligase are used to synthesize double-stranded cDNA* (blue rectangle) *containing the T7 promoter* (red rectangle). *T7 RNA polymerase is then used to synthesize many copies of aRNA* (blue wavy lines) *from the double-stranded template.*

polymerase promoter (Fig. 6.23). The oligo-dT-primed mRNA is converted into single-stranded cDNA with reverse transcriptase and then into double-stranded cDNA with DNA polymerase. Because each double-stranded cDNA molecule contains a T7 promoter, large quantities of amplified RNA can be synthesized from the cDNA mixture with RNA polymerase. Each round of the Eberwine procedure produces about a 100-fold amplification of the starting material, and three rounds of Eberwine yields a 1 million-fold amplification (100 × 100 × 100) of the input mRNA. Given that a human cell contains about 0.5 pg of mRNA, a 10^6-fold amplification produces about 0.5 μg aRNA from a single cell, material sufficient for the preparation of fluorescent probes for microarray analysis. RNA amplification by the Eberwine procedure can be used to study biological questions that are effectively impossible by any other current means.

There are several important advantages of the Eberwine procedure in the context of microarrays. First, the procedure allows physical amplification of the RNA, enabling fluorescent probes to be made directly from the amplified material. Second, the procedure allows the largest amplification factor (10^6) of any of the probe preparation procedures, sufficient to allow gene expression analysis of single human cells. The procedure is also rather quantitative owing to the fact that amplification occurs in a linear manner from double-stranded templates, each of which contains an identical T7 promoter. The quantitative aspect of the Eberwine procedure contrasts somewhat with PCR, in which PCR-amplified material can be skewed relative to the input material because of the exponential PCR process that amplifies short templates more efficiently than long templates. The central importance of quantitation in gene expression studies renders the Eberwine procedure well suited for mRNA-based microarray analysis.

The main disadvantages of RNA amplification using the Eberwine procedure are that it is arduous and time-consuming. Three rounds of Eberwine amplification requires 2–3 days of steady experimentation; and most of the steps involve manipulations of nucleic acid that are "invisible" to the experimentalist. Loss of the amplified material, inactive reagents, or ribonuclease contamination over the several-day procedure can lead to less than desirable results. Modifications to the procedure, such as the use of affinity matrices containing oligo-dT sequences fused to the T7 promoter, are sorely needed and would serve to simplify RNA amplification by this method and proliferate its routine use in microarray analysis.

Tyramide Signal Amplification

Tyramide signal amplification. *Microarray labeling procedure in which fluorescent signals are enhanced by the enzymatic deposition of fluorescent reagents.*

Tyramide signal amplification (TSA) technology has its roots in assays developed by Du Pont (Wilmington, DE) scientists in the late 1980s; it was later used in a microarray context by researchers at NEN Life Science Products (Boston). This sophisticated method uses the enzymatic activity of horseradish peroxidase (HRP) to catalyze the deposition of fluorescent tyramide molecules on microarray spots that are hybridized with molecules containing tags for the antibodies conjugated to HRP. TSA can produce 100-fold more intense fluorescence than direct labeling procedures, without any increase in probe concentration. The method works as follows.

Reverse transcriptase is used to incorporate biotin, dinitrophenol (DNP) or some other small molecule into cDNA (Fig. 6.24). The tagged cDNA is then hybridized to a microarray and incubated with an antibody conjugated to HRP. The antibody binds to tags in the synthesized cDNA, bringing conjugated HRP into close proximity to the microarray surface. The chip is then incubated with hydrogen peroxide (H_2O_2), which the HRP enzyme uses to oxidize fluorescent tyramide molecules (e.g., Cy3-tyramide). Oxidized tyramide is highly reactive and couples rapidly to the microarray surface. This enzymatic signal amplification procedure increases fluorescent signals by 100-fold under optimal conditions.

There are a number of advantages to the TSA labeling scheme, which provides the only method of enzymatic signal amplification for microarray analysis. Because reactive tyramide molecules have a short half-life, deposition of the fluorescent tyramide species occurs on or near each microarray spot, providing a high degree of spatial resolution. There is no spreading or blooming of the signal, which makes TSA well suited to the high-density microarray format. Tyramide derivatives are available for fluorescein, cyanine 3, cyanine 5, and other fluorescent dyes, enabling experimental schemes that exploit multiple colors. The signal is amplified instead of the number of probe molecules, preventing the saturation of target molecules on the microarray surface and maintaining linearity over many orders of magnitude. The technology allows the use of small amounts of total RNA (e.g., 2–5 μg), which minimizes consumption of precious biological material and provides access into blood samples and other

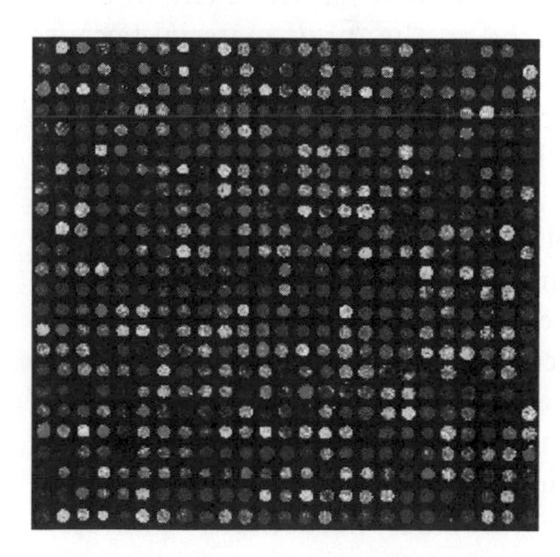

FIGURE 6.24. *Human total RNA (4 μg) was labeled and hybridized to a microarray containing full-length human cDNAs prepared using Full-Length Expressed Gene (FLEX) technology (AlphaGene, Woburn, MA). The hybridized microarray was then stained with TSA reagents from a MICROMAX system (NEN Life Science Products, Boston), and fluorescent gene expression signals were displayed in a rainbow palette. Figure also appears in Color Figure section.*

tissue sources of diagnostic value. The 100-fold signal amplification allows detection of human transcripts that are impossible to measure with direct labeling schemes (Fig. 6.24). The main disadvantages of TSA are that the protocol is somewhat time-consuming and arduous relative to direct labeling, but the richness of the resultant data almost always justifies the extra effort required with this approach.

Dendrimer Technology

Dendrimer technology is an enormously promising labeling approach, the foundation of which was established at a number of different centers, including Dow Chemical (Midland, MI), in the early 1990s and later in a nucleic acid context by Cantor and co-workers at Boston University. Fluorescent dendrimer technology for microarray applications was developed by Stears and Gullans at Harvard Medical School (Boston) in collaboration with Genisphere (Montvale, NJ). Dendrimers provide up to 300-fold brighter signals than direct labeling approaches, without the use of enzymatic catalysis or any increase in the number of probe molecules required. This revolutionary approach is likely to have a major impact on microarray research.

Dendrimer. *Large, highly fluorescent, branchlike microarray labeling reagent built by the sequential hybridization and cross-linking of fluorescent oligonucleotides.*

The term dendrimer derives from the Greek words *dendron,* meaning tree, and *meros,* meaning "part." Nucleic acid dendrimers are highly branched molecules, composed of oligonucleotide building blocks that are hybridized and then cross-linked into large macromolecular structures. Dendrimers are built in a step-wise manner, beginning with the formation of a monomer, which contains a central annealed portion and four single-stranded "arms" (Fig. 6.25). Four subsequent rounds of annealing and chemical cross-linking produce dendrimers that contain 12, 36, 108 and 324 single-stranded arms, respectively, for a final molecular weight of nearly 12,000 kd. A 12,000-kd dendrimer contains about 36,000 bases of DNA and has a diameter of approximately 0.2 μm. Synthesis with oligonucleotides that contain fluorescent dye molecules (Fig. 6.26) produces dendrimers with 300 fluors per molecule.

Dendrimer labeling requires the attachment of dendrimer-binding sequences to probe molecules. For gene expression studies, oligo-dT primers that contain dendrimer-binding sequences located on the 5′ end of the oligonu-

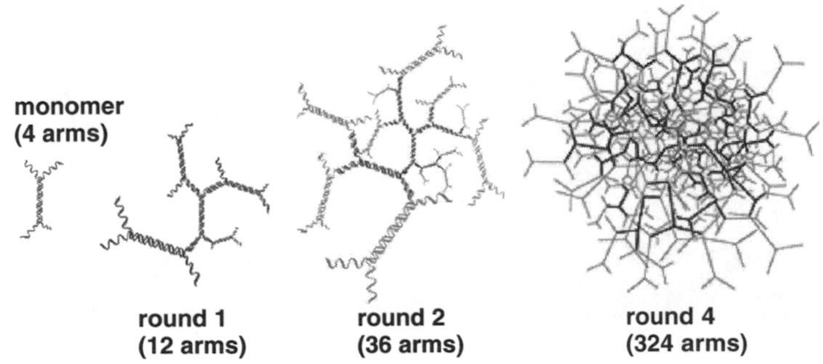

FIGURE 6.25. For the dendrimer synthesis process, two oligonucleotides are annealed to produce a molecule (monomer) that has a double-stranded central portion and four single-stranded arms. Monomers are then annealed and cross-linked in a stepwise manner; four successive rounds of annealing and cross-linking produce dendrimers with 324 arms. If the synthesis process is performed with fluorescent oligonucleotides, each dendrimer contains as many as 324 fluorescent dye molecules. (Courtesy of Dr. R. C. Getts, Genisphere, Montvale, NJ.) Figure also appears in Color Figure section.

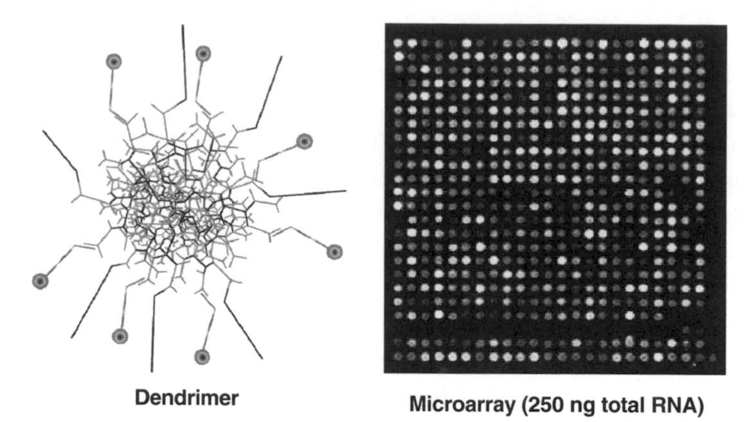

FIGURE 6.26. A dendrimer reagent was used to label a microarray indirectly by hybridizing the single-stranded capture sequences (black lines) *on the dendrimers to a microarray hybridized with a probe mixture prepared from 250 ng mouse total RNA. The intense fluorescent signals are due to passive signal enhancement by the dye molecules* (circles) *attached to the dendrimers. (Dendrimer courtesy of Dr. R. C. Getts, Genisphere, Montvale, NJ; microarray data courtesy of University of California at San Francisco Cancer Center.) Figure also appears in Color Figure section.*

cleotides are used to prime messenger RNA (Fig. 6.27). The primed mRNA is then extended with reverse transcriptase, incorporating dendrimer-binding sequences into the 5′ end of each cDNA molecule. The tagged cDNA probe molecules are then hybridized to a microarray and stained with fluorescent dendrimers, allowing the dendrimer-capture sequences to hybridize to dendrimer-binding sequences present on each cDNA. Microarray spots that contain a large number of bound probe molecules bind a greater number of dendrimers and produce more intense fluorescence than microarray spots that contain fewer bound probe molecules. The capacity to detect gene expression with only

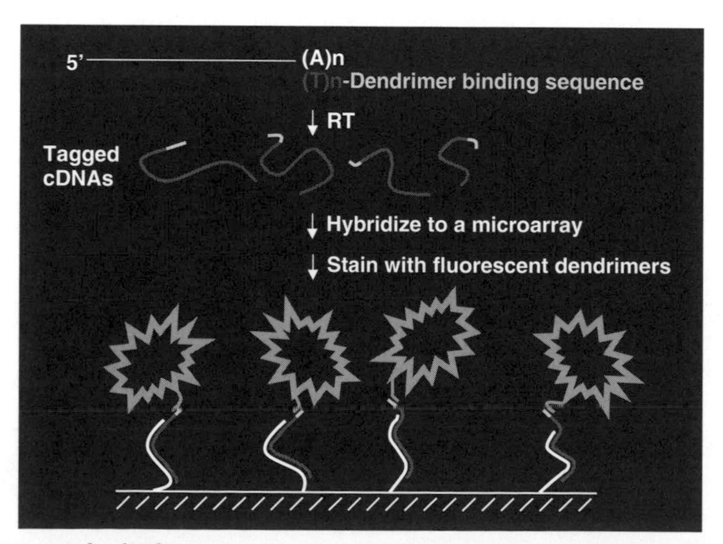

FIGURE 6.27. For the dendrimer labeling procedure, mRNA is primed with oligo-dT that contains dendrimer-binding sequences, and reverse transcriptase extends the primed mRNA, producing tagged cDNAs (gray lines) *that each contain a dendrimer-binding sequence located on the 5′ end of the molecule. The tagged cDNAs are hybridized to a microarray, which is stained with fluorescent dendrimers containing capture sequences that are complementary to the dendrimer-binding sequences. The intensely fluorescent dendrimers* (starbursts) *produce up to 300-fold signal enhancement relative to direct labeling procedures.*

250 ng mouse total RNA illustrates the extent to which dendrimer labeling improves microarray detectivity (Fig. 6.26).

In a second dendrimer labeling approach, oligonucleotides (instead of cDNAs) containing dendrimer-binding sequences are hybridized to amplified patient samples that contain disease alleles of interest. The hybridized amplicons are then stained with dendrimers that contain capture sequences, and the fluorescence intensity allows single nucleotide discrimination, a strategy that has shown great promise in genetic screening and genotyping applications (see Chapter 13).

There are several advantages to using dendrimers for signal enhancement. Dendrimers are branched nucleic acids and are, therefore, highly soluble in most hybridization buffers, producing in low background fluorescence on most microarray surfaces. The relatively small size (0.2 μm) of these particles combats precipitation and allows dendrimers to diffuse rapidly at modest hybridization temperatures (42–65°C). High solubility and rapid diffusion minimizes the amount of time (1–2 h) required to hybridize dendrimers to the microarray. The dendrimer geometry provides spacing between adjacent dye molecules, which prevents energy transfer and other types of quenching interactions that can reduce fluorescent signal. Dendrimers with 300 dye molecules provide 300-fold signal enhancement, compared to the lower efficiencies that are observed with some of the other labeling methods. Probe mixtures can be synthesized using unmodified dNTPs, which improves synthesis efficiency and reduces the cost of probe synthesis. The capacity to build dendrimers using any of a large family of fluorescent dyes extends the versatility of the dendrimer approach, allowing labeling with cyanine, Alexa, BODIPY, and other may other fluors.

The 0.2 μm-diameter dendrimers are so intensely fluorescent that single dendrimer molecules can be observed directly if microarrays hybridized with fluorescent dendrimers are detected at high resolution (Fig. 6.28). The capacity to distinguish rare and abundant transcripts by counting the number of dendrimer molecules bound to a given microarray spot demonstrates that single fluorescent dendrimers can be detected on microarray surfaces, enabling extraordinary precision in microarray analysis because measurements can be taken by counting individual bound dendrimers. Performing an 'end-point' assay, in which each probe molecule bound at a given spot is bound by a dendrimer, allows quantitative assessment of the actual number of probe molecules bound at a given microarray location. This capability also opens the

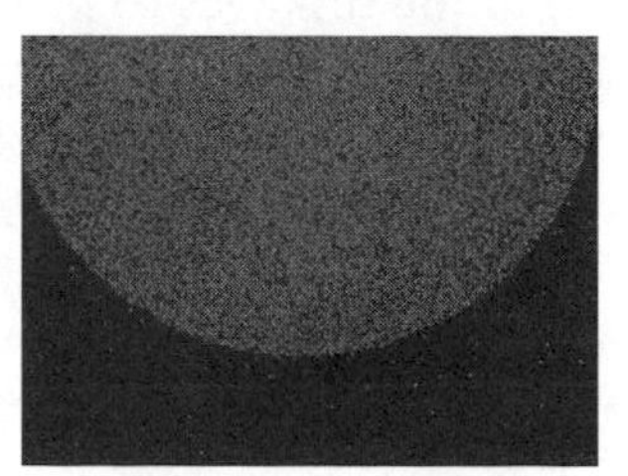

FIGURE 6.28. *The 0.2-μm-diameter dendrimer molecules are so intensely fluorescent that high-resolution (60 ×) microarray images allow visualization of single macromolecules. A microarray spot that binds cDNA probe molecules corresponding to a rare transcript contains a relatively small number of dendrimer molecules, whereas a target corresponding to an abundant transcript contains a large number of dendrimer molecules. (Data courtesy of Dr. R. C. Getts, Genisphere, Montvale, NJ.)*

door to a number of other types of studies, including kinetic and biophysical analysis.

Proteins

Protein probes have not been as widely used as nucleic acid probes because protein microarray technology has developed more slowly than nucleic acid microarrays. Nonetheless, several key reports indicate that fluorescent protein probes can be prepared and reacted with protein microarrays to study a host of fundamental problems in protein biochemistry. The abundance of charged amino acid side chains on the protein surface render proteins well suited for direct labeling with fluorescent dyes. Most of the protocols link fluorescent dyes to proteins in a covalent manner via reactive lysine and arginine residues on the protein surface. These approaches have been used for many years in the context of fluorescent antibody preparation.

Fluorescein isothiocyanate (FITC) contains a reactive isothiocyanate group (see Chapter 8) that is susceptible to nucleophilic attack by primary amine groups on the surface of proteins, resulting in covalent attachment of this fluorescent dye to any protein that has reactive amines on the surface. Because lysine and arginine residues make up nearly 10% of the amino acids in a given protein, purified proteins and cellular extracts can be labeled with high efficiency with FITC. The abundance of lysine and arginine residues on the surface of proteins is conserved among plants, animals, and microorganisms, suggesting that fluorescent labeling can be used to label protein mixtures from any organism. A labeling reaction containing 50 μg FITC and 1.0 mg protein usually provides brightly fluorescent protein mixtures.

Cyanine dyes are available as amine-reactive derivatives, allowing the cyanine dyes to be conjugated to proteins via nucleophilic attack by lysine and arginine residues in a manner analogous to the FITC labeling reaction. Amersham Pharmacia Biotech (Piscataway, NJ) offers Cy dyes as reactive NHS esters, allowing rapid coupling of these fluorescent derivatives to proteins of interest.

The fluorescent protein phycoerythrin (PE) is widely used in microarray probe labeling; and, with 25 fluors per molecule, this protein provides an intensely fluorescent label for any molecule to which it is attached. The protocols used to attach PE to proteins are somewhat more elaborate than the procedures required for the FITC and cyanine dyes, but the basic chemistry is the same. Phycoerthrin is first reacted with a succinimide reagent to provide a maleimide derivative of PE. The protein sample is treated with dithiothreitol (DTT) to reduce protein disulfide linkages to provide reactive sulfhydryl groups. The activated protein is then reacted with maleimide-PE such that the free sulfhydryl groups on the proteins act as nucleophiles, attacking the maleimide group on the phycoerythrin and coupling fluorescent PE to the protein of interest.

TARGET AND PROBE INTERACTIONS

The specificity and affinity of interactions between target molecules bound to the microarray substrate and probe molecules in solution largely determine the quality of microarray assays. This section provides a brief description of nucleic acid and protein reactions and the factors that influence them.

Hybridization

The first experiments in microarray analysis used target–probe hybridization interactions, and this common bimolecular reaction continues to be the predominant one used in microarray analysis. Hybridization reactions between single-stranded target and probe molecules occur by hydrogen bond formation between the bases of complementary nucleic acid sequences (see Chapter 3). Sequence composition, target and probe length, hybridization temperature, secondary structure, degree of homology, salt concentration, pH, and a number of other factors influence hybridization efficiency and the strength of the duplexes (Table 6.4). A brief discussion of these factors is helpful with respect to the theoretical and practical issues that underlie DNA microarrays.

Base pairs that form between guanine and cytosine residues (G:C base pairs) contain three hydrogen bonds and are thus more avid than A:T base pairs, which contain only two hydrogen bonds (Table 6.4). Sequence composition is extremely important in hybridization reactions involving short targets and probes (e.g., oligonucleotides), because a given heteroduplex can consist entirely of G:C bonds or A:T bonds (Fig. 6.29) and thus oligonucleotides of the same length can differ enormously in their hybridization efficiency, depending on sequence composition. An 11mer with all G:C base pairs would form a much stronger duplex than an 11mer with all A:T base pairs, and a duplex of mixed sequence composition would provide an intermediate hybridization affinity. Hybridization affinity and signal intensity correlate directly in microarray assays, with the G:C sequences producing extremely intense fluorescence, the A:T sequences producing much weaker fluorescence, and ***mixed sequences*** exhibiting intermediate fluorescent signals. Sequence composition is a minor consideration if the targets and probes are long (e.g., cDNAs) because hybridization affinity is determined by the average sequence composition along the heteroduplex, a fact that minimizes the effect of local G:C or A:T richness.

Mixed sequence. *Nucleic acid containing a heterogenous composition of adenine, cytosine, guanine, and thymine bases.*

In addition to sequence composition, the number of hybridized base pairs is an important determinant of hybridization affinity. Other factors being equal, long heteroduplexes have a much greater affinity than short heteroduplexes because of the greater number of hydrogen bonds present in long base pair regions. For gene expression measurements in which gene specificity is paramount, heteroduplexes of 25–2,500 base pairs are commonly used. For genotyping applications, single base discrimination is the key and so shorter target–probe interactions (e.g., 12–20 base pairs) are employed most often.

The extent of complementarity between probe and target is another important determinant of hybridization affinity. Double-stranded regions that share 100% complementarity between target and probe molecules are known as ***perfect heteroduplexes***. For a perfect heteroduplex sharing 15 base pairs, an average hybridization temperature in a standard hybridization buffer is 42°C, which is approximately 10°C below the melting temperature. For each additional base pair added to the length of the heteroduplex, the hybridization temperature must be elevated by 1–2°C to maintain sufficient hybridization specificity. Hybridizations involving 25mer heteroduplexes typically require hybridization temperatures of ~60°C. Decreasing the number of base pairs requires a concomitant decrease in hybridization temperature of 1–2°C per base pair subtracted. Most researchers use 30–37°C when hybridizing 9mer oligonucleotide probes to perfectly complementary target sequences.

Perfect heteroduplex. *Target and probe hybrid that shares 100% complementarity.*

TABLE 6.4. Hybridization Parameters

Parameter	Effects	Comments
G:C base pair	stronger hybridization affinity than A:T base pairs because of three hydrogen bonds	G:C and A:T differences can be minimized by the use of buffers containing TMAC[a]
A:T base pair	weaker hybridization affinity than G:C base pairs because of only two hydrogen bonds	G:C and A:T differences can be minimized by the use of with buffers that contain TMAC[a]
Perfect homology	100% homology provides the best hybridization efficiency	Hybrids can be DNA–DNA, DNA–RNA or RNA–RNA
Imperfect homology	< 100% homology reduces hybridization efficiency due to fewer hydrogen bonds	< 70% homology produces little to no signal under stringent conditions
Elevated temperature	improves hybridization efficiency by increasing thermal energy, which speeds diffusion and reduces secondary structure	optimal hybridization temperature is ~10°C below the melting temperature of the heteroduplex
Insufficient temperature	reduces hybridization efficiency by reducing diffusion and increasing secondary structure	> 10°C below the melting temperature is insufficient in most cases
Excessive temperature	reduces hybridization efficiency by melting the heteroduplexes	temperatures near or above the melting temperature are too high
Salt	improves hybridization efficiency by shielding negatively charged phosphate groups and minimizing electrostatic repulsion	most hybridization buffers use Na^+ concentrations in the range of 0.4–1.0 molar
pH	neutral pH favors hybridization by promoting hydrogen bond formation between the bases	optimal pH is in the range of 5.5–8.5
Low	Can reduce hybridization efficiency by damaging the DNA bases through depurination	depurination results from N-glycosidic bond breakage between the sugar and the base (A or G)
High	Reduces hybridization efficiency by weakening hydrogen bonding between the bases	deprotonation of the bases impairs hydrogen bond formation
Reducing the size of target molecules	can improve hybridization specificity by reducing or eliminating cross-hybridization	can be important in gene expression experiments involving highly conserved gene families
Reducing the size of probe molecules	can improve hybridization efficiency by speeding diffusion and reducing secondary structure	accomplished by mechanical shearing, enzymatic digestion, or chemical cleavage
Ribonuclease contamination	can reduce hybridization efficiency by degrading RNA targets and probes	A concern only for RNA targets and probes
Formamide addition	reduces the melting temperature and allows hybridization reaction to be performed at a lower temperature	addition of 50% formamide reduces the melting temperature by ~25°C, and often reduces background

[a]Tetramethylammonium chloride, TMAC

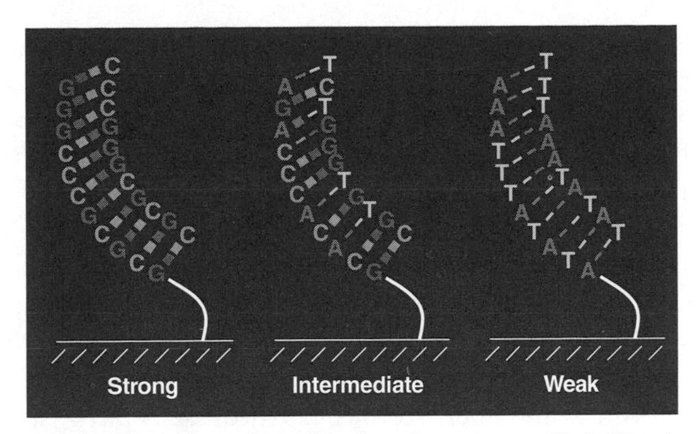

FIGURE 6.29. Sequence composition is an important determinant of hybridization affinity, particularly with short heteroduplexes such as those that form between oligonucleotide targets and probes. G:C base pairs contain three hydrogen bonds (thick lines) *and are stronger than A:T base pairs, which contain only two hydrogen bonds* (thin lines). *Sequences that consist of only G:C base pairs form strong heteroduplexes, whereas sequences that contain only A:T base pairs form much weaker heteroduplexes. Mixed sequences hybridize with an intermediate affinity.*

As long as the hybridization temperature is 5–10°C below the melting temperature of the heteroduplex, it is advantageous to use as high a temperature as possible because thermal energy increases the rate of diffusion and melts secondary structure in the probe molecules (Table 6.4), both of which promote hybridization in microarray assays. Increasing the rate of diffusion by elevating the temperature increases collisions between targets and probes and thus expedites the formation of productive hybrids. High temperature also removes secondary structure or knots from both DNA and RNA, which increases the hybridization efficiency of both types of molecules by minimizing conformational impediments to hybridization. RNA has a much greater propensity for secondary structure than DNA, so reducing the size of the RNA probe molecules by shearing them into smaller pieces improves hybridization efficiency by preventing secondary structure formation. Smaller probe molecules also diffuse more readily than larger ones, and thus there is often an improvement in fluorescent signal if the molecules are broken into smaller pieces before hybridization.

DNA molecules can be fragmented into smaller pieces by passing them through a small-diameter orifice at high pressure in a process known as ***mechanical shearing,*** or by subjecting the DNA to digestion with the enzyme DNase I (Table 6.4). RNA molecules can be broken into smaller pieces by mechanical shearing, digestion with ribonucleases, or by a chemical degradation with magnesium at elevated temperature.

Mechanical shearing. *Experimental procedure that uses high pressure, sonic waves, or some other means to break nucleic acid molecules into smaller fragments.*

In most gene expression studies, homology is 100% because target sequences hybridize to probe molecules derived from their cognate genes. In this case, the heteroduplexes are extremely strong, and the hybridization reaction can be performed at high stringency. Hybridization is not an all-or-none process, however, and sequences with less than 100% homology can form productive heteroduplexes even though there is some ***mispairing*** of the bases between the target–probe hybrid. Imperfect hybrids are an important consideration in several respects. Gene families that are highly conserved or have a large number of members will hybridize to noncognate targets, and this cross-hybridization can complicate gene expression measurements. Cross-hybridization is observed if the sequence identity between target and probe is $\geq$ 70% but is generally not

Mispairing. *Juxtaposition of a noncomplementary base in a hybridized nucleic molecule.*

a factor for sequences share < 70% nucleotide identity. It is important to emphasize that the relevant criterion is nucleic acid identity and not amino acid identity. Owing to the degeneracy of the genetic code (see Chapters 3 and 4), two proteins can have nearly identical amino acid sequences but < 70% identity at the DNA level.

The capacity to hybridize similar but nonidentical targets and probes can be exploited in evolutionary studies, whereby a human chip can be reacted with probes derived from gorilla, chimpanzee, orangutan, and other primates. The fluorescent signals observed at each human gene location can provide extremely important information about sequence conservation between humans and primates on a genomic scale. Such studies can provide rapid assessment of the evolutionary events that unfolded during prehuman evolution. Analogous experiments allow comparisons of different strains or variants of yeast, mouse, worm, fruit fly, plant, or any other set of organisms for which comparative analysis is informative.

The capacity to distinguish single base differences between targets and probes forms the basis of genotyping applications of microarrays. Most human diseases result from a single base change in a critical region of the coding sequence, which leads to the synthesis of a defective protein. Most patients afflicted with a genetic condition can be distinguished from the normal population via probes that hybridize with high affinity to the **wild type** locus and with lesser affinity to the disease locus (see Chapter 13). To maximize the capacity to distinguish single base differences, the heteroduplexes must be relatively short (e.g., 15 base pairs) and the mutation located in the center position of the heteroduplex. If longer heteroduplexes are used or if the mismatch resides near the ends of the hybrid, single base discrimination will be diminished and the genotyping assay will be impaired.

Wild type. *Genetics term that refers to a gene or organism that represents the most commonly found allele or strain.*

There are a number of chemical contributors to microarray hybridization efficiency, including salt, detergent, and destabilizing reagents. The addition of cations such as Na^+ stabilizes heteroduplex formation by binding to negatively charged phosphate groups and reducing electrostatic repulsion between the target and the probe sequences. All microarray hybridization buffers contain salt as an important ingredient as well as low concentrations of detergents, which improve the efficiency of hybridization by decreasing the surface tension forces of the hybridization buffer. Reagents that destabilize base pairing, including formamide and tetramethylammonium chloride (TMAC), are used in some microarray hybridization reactions. Formamide reduces the melting temperature of hybridized sequences in a global manner, and some users have noted that lower hybridization temperatures with formamide-containing buffers improves detectivity by reducing background fluorescence. TMAC selectively reduces the strength of G:C bonds relative to A:T bonds and thus minimizes binding differences between G:C-rich and A:T-rich sequences, a property useful for genotyping applications in which similar signals for each heteroduplex irrespective of the sequence composition, improves the quality of the assay.

Protein Reactions

Proteins carry out the genetic instructions of all cells and understanding their interactions on a genome-wide scale is one of the most challenging and important problems in modern biology. Microarray analysis provides a technology platform for massive, parallel analysis of protein–protein interactions that is

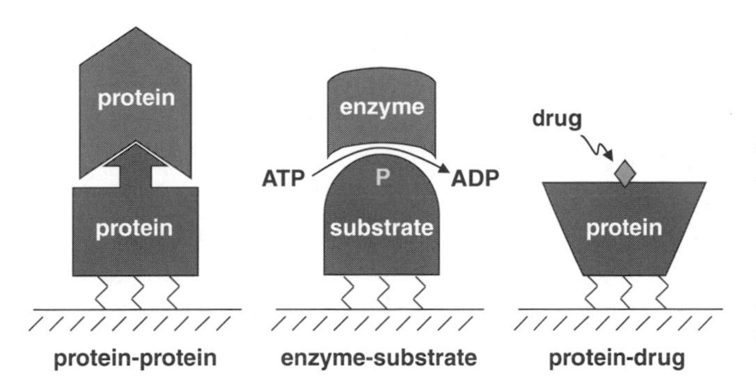

FIGURE 6.30. Proteins or substrates are attached in a covalent manner to a microarray surface (black line) *by electrostatic or Schiff base or epoxy mechanisms; and interactions between protein, enzyme, or drug occur with probe solutions that contain these molecules.*

both quantitative and comprehensive. Building on the concepts and protocols developed for nucleic acids, protein microarray technology is witnessing explosive growth. Because the spectrum of problems in protein biochemistry is much larger than in nucleic acid biochemistry, protein microarrays may be ultimately more widely used than DNA microarrays. This section explores some of the issues surrounding protein biochemistry in a microarray format.

Pioneering studies by MacBeath and Schrieber (2000) at Harvard University and Joos and co-workers (2000) at the Natural and Medical Sciences Institute (Germany) reveal that all of the key reactions of proteins can be studied in a microarray format, including protein–protein interactions, protein–drug binding, and enzymatic catalysis (Fig. 6.30). The capacities to analyze protein interactions on a genomic scale and to compare binding constants in a parallel format suggest that in the near future protein chips may replace many of the traditional methods in protein biochemistry.

Proteins bind to other proteins via several different atomic interactions, including electrostatic bonds, hydrogen bonds, and van der Waals forces (see Chapter 3). The same interactions that mediate protein–protein binding facilitate the binding of proteins to small molecules (e.g., drugs), substrates, carbohydrates, and myriad other cellular molecules. The basic concept of target–probe binding established for nucleic acids holds for proteins, though many of the details are different. Nucleic acids hybridize by zipping interactions between complementary strands, whereas proteins bind to each other using lock-and-key-type mechanisms between globular molecules (Fig. 6.31). The complex three-dimensional structure of proteins enables binding interactions of extraordinary specificity and affinity, though three-dimensionality poses both

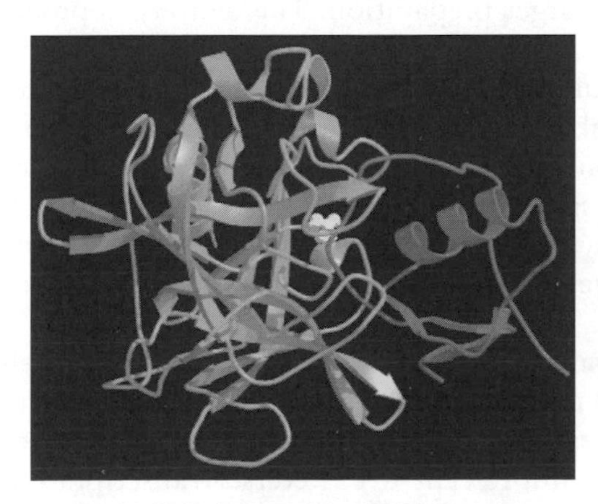

Protein-protein complex

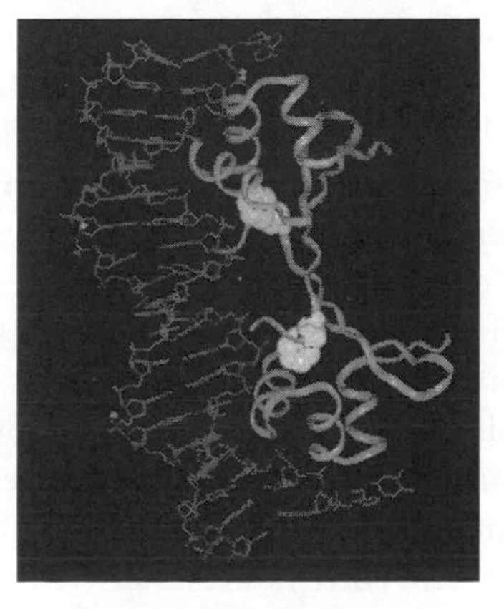

Protein-DNA complex

FIGURE 6.31. Crystal structures of a protein–protein binding reaction between the elastase enzyme (light blue) *and the ovomucoid substrate* (red), *and a protein–DNA binding reaction between the Cro-repressor protein* (green) *and double-stranded DNA* (blue and magenta). *(Protein-protein complex courtesy of Dr. K. P. Murphy, Department of Biochemistry, University of Iowa at Iowa City; protein–DNA complex courtesy of Dr. M. C. Mossing, University of Notre Dame, Notre Dame, IN.) Figure also appears in Color Figure section.*

theoretical and practical challenges to protein microarray assay development that are not a concern with nucleic acid chips.

One theoretical challenge to protein microarray assay development is that, because protein structure cannot be predicted from the primary amino acid sequence, it is impossible to determine a priori which proteins will bind to each other. This is in marked contrast to nucleic acids, in which binding can be predicted in advance simply by comparing the nucleotide sequence of two strands in question. The lack of predictability of protein interactions encourages discovery-based approaches, one of the strengths of microarray technology. Protein microarray analysis uncovers novel interactions in virtually every experiment conducted, due simply to the fact that little is understood about how amino acid sequence dictates protein folding and function.

A practical challenge posed by proteins derives from their relatively delicate tertiary structure, which is susceptible to unfolding due to oxidation, dessication, and other insults that can accompany samples during microarray printing. Protein denaturation can lead to a loss of binding affinity and specificity and thus impairs microarray analysis. Several approaches have been used to minimize the denaturation of protein targets, including the use of stabilizing agents such as dithiothreitol, glycerol, and printing buffer additives. Another approach uses relatively high protein concentrations during printing, which tends to stabilize protein structure. It will be interesting to explore the extent to which protein folding can be recapitulated in a microarray format. Probe solutions containing caperonins and other folding enzymes may facilitate the refolding of peptides and native proteins that have been denatured on the microarray substrate. The capacity to study protein folding in a massively parallel format may help define the rules that govern protein folding in a physiologically relevant setting.

SUMMARY

Microarray analysis requires a thorough understanding of the targets and probes used in each experiment. Nucleic acids represent the most common type of target molecule, but microarrays can be constructed of enzymes, antibodies, small molecules, carbohydrates, lipids, viruses, bacteria, plant and animal cells, protein extracts, inorganic compounds, semiconductor materials, and any other substance that can be attached to a solid surface. Delivery and synthesis represent the two approaches to target preparation. The delivery approaches prepare the target molecules off-line, and the synthesis methods create target elements by joining molecular building blocks on the chip. Phosphoramidite synthesis using the recently available human genome sequence allows microarray targets to be made directly from sequence databases. Both contact and noncontact printing require an understanding of adhesion, surface tension, surface free energy, and wettability. The synthesis approaches excel in the high-density applications, providing whole genome chips in an affordable and reliable manner.

Probe preparation takes advantage of both direct and indirect labeling schemes. Direct labeling attaches fluorescent tags directly to probe molecules, whereas indirect labeling attaches fluorescent labels via a bridge molecule, which in turn is attached to the fluorescent dye. Reverse transcriptase and RNA polymerase are used widely in probe preparation. Probe amplification and signal enhancement schemes, including the Eberwine procedure, tyramide signal amplification, and dendrimer technology, provide improved detectivity. Proteins can also be labeled directly with fluorescent tags via covalent linkage to surface lysine and arginine residues.

The specificity and affinity of target–probe interactions determine the usefulness of microarray assays. Factors such as sequence composition, target and probe length, homology, salt, pH, temperature, and secondary structure all influence the efficiency of hybridization. The important protein-binding reactions, including protein–protein, enzyme–substrate, and protein–drug interactions, have been recently recapitulated as microarray assays. Protein microarrays may supersede many of the traditional methods in protein biochemistry in the near future.

SELECTED READING

Baxevanis, A. D., and Ouellette, B. F. F., eds. *Bioinformatics: A Practical Guide to the Analysis of Genes and Proteins,* 2nd ed. Wiley-Interscience, New York, 2001.

Benton, W. D., and Davis, R. W. Screening λgt recombinant clones by hybridization to single plaques in situ. *Science* 196:180–182, 1977.

Caruthers, M. H. Gene synthesis machines: DNA chemistry and its uses. *Science* 230:281–285, 1985.

Higgins, D., and Taylor, W., eds. *Bioinformatics: Sequence, Structure, and Databanks: A Practical Approach.* Oxford University Press, Oxford, UK, 2000.

Innis, M. A., Gelfand, D. H., and Sninsky, J. J., eds. *PCR Applications: Protocols for Functional Genomics.* Academic Press, San Diego, CA, 1999.

Joos T. O., Schrenk, M., Hopfl, P., Kroger, K., Chowdhury, U., Stoll, D., Schorner, D., Durr, M., Herick, K., Rupp, S., Sohn, K., and Hammerle, H. A microarray enzyme-linked immunosorbent assay for autoimmune diagnostics. *Electrophoresis* 21:2641–2650, 2000.

Kramer, R. A., Cameron, J. R., and Davis, R. W. Isolation of bacteriophage lambda containing yeast ribosomal RNA genes: Screening by in situ RNA hybridization to plaques. *Cell* 8:227–232, 1976.

Lewin, B. *Genes VII.* Oxford University Press, Oxford, UK, 2000.

Macbeath, G., and Schrieber, S. L. Printing proteins as microarrays for high-throughput function determination. *Science* 289:1760–1763, 2000.

Reed, S. I , Stark, G. R., and Alwine, J. C. Autoregulation of simian virus 40 gene A by T antigen. *Proc Natl Acad Sci U S A* 73:3083–3087, 1976.

Rigby, P. W., Dieckmann, M., Rhodes, C., and Berg, P. Labeling deoxyribonucleic acid to high specific activity in vitro by nick translation with DNA polymerase I. *J Mol Biol* 113:237–251, 1977.

Schena, M., Shalon, D., Davis, R. W., and Brown, P. O. Quantitative monitoring of gene expression patterns with a complementary DNA microarray. *Science* 270:467–470, 1995.

Schena, M., ed. *DNA Microarrays: A Practical Approach,* 2nd ed. Oxford University Press, Oxford, UK, 2000.

Schena, M., ed. *Microarray Biochip Technology.* Eaton, Natick, MA, 2000.

Stears, R. L., Getts, R. C., and Gullans, S. R. A novel, sensitive detection system for high-density microarrays using dendrimer technology. *Physiol Genomics* 3:93–99, 2000.

Streitweiser, A., and Heathcock, C. H. *Introduction to Organic Chemistry,* 2nd ed. Macmillan, New York, 1981.

Stryer, L. *Biochemistry,* 4th ed. Freeman, New York, 1995.

REVIEW QUESTIONS

1. In the conventional microarray literature, is the target or the probe attached covalently to the microarray substrate?
2. Which approach, synthesis or delivery, builds microarray target elements on the glass substrate in a base-by-base manner?
3. A researcher uses PCR products and 70mer oligonucleotides corresponding to a homologous family of cellular genes and finds that the PCR

products produce more intense hybridization signals but the specificity of the hybridization reactions appear to be less than with the 70mers. Provide an explanation for the results.

4. The basic local alignment search tool (BLAST) is used to examine the entire human genome sequence, using a 500-nucleotide target as the query sequence. BLAST identifies 5 hits in the search, though none shows perfect homology. In view of the evolution of biological systems, a 500-nucleotide target from which of the following organisms is likely to produce the 5 hits with the greatest homology to the human database: yeast, bacterium, mouse, fly, worm, or mustard plant?
5. A supervisor asks his student to manufacture a microarray containing gene sequences from a moss isolated during a recent camping trip. Briefly describe an experimental approach to microarray manufacture that would not require the knowledge of any sequence information from the moss.
6. Phosphoramidite bases differ chemical from natural nucleotides in several respects. Name three.
7. A researcher uses a porous glass substrate for microarray manufacture and finds that target DNA attaches efficiently to the surface, but the background fluorescence is high after hybridization, washing, and imaging. With respect to surface topology, provide an explanation for these results.
8. Protein targets attach efficiently to glass substrates containing reactive epoxide groups. Write an organic chemical mechanism for protein attachment to an epoxy surface.
9. Four aqueous microarray printing buffers are printed on an aldehyde surface and their spot diameters are compared. Which of the following printing buffer additives is likely to produce spots with the largest diameter: (a) 0.1% SDS, (b) 10 mM NaCl, (c) 1 mM NaCl, or (d) 0.01% SDS? Explain your answer.
10. Reverse transcriptase and oligo-dT are used to prepare a fluorescent probe mixture by labeling mRNA, and the fluorescent probe is hybridized to four different 70mer target sequences. Rank the following 70mers targets, in order of weakest to strongest signal produced, when hybridized to the fluorescent probe mixture: (a) AT-rich 70mer complementary to the 3′ end of the mRNA, (b) GC-rich 70mer complementary to the 5′ end of the mRNA, (c) AT-rich 70mer complementary to the 5′ end of the mRNA, and (d) GC-rich 70mer complementary to the 3′ end of the mRNA.
11. An in situ synthesis method has an average coupling efficiency of 96%. After 25 coupling cycles, what percentage of microarray targets will contain 25 nucleotides?
12. A researcher generates a 1-kb cDNA molecule from purified mRNA using reverse transcriptase and a 1:1 mixture of Cy3-dCTP and unmodified dCTP. Microarray analysis indicates, on average, 10 fluorescent nucleotides per 1 kb cDNA, even though the cDNA contains approximately 250 cytosine residues. How many fluorescent nucleotides would be present if Cy3-dCTP and dCTP incorporated at equal efficiency? Suggest a reason for the reduced incorporation of Cy3-dCTP.
13. Which of the following procedures does not use an enzyme: (a) reverse transcription, (b) dendrimer staining, (c) Eberwine procedure, or (d) tyramide signal amplification?
14. Which of the following would be expected to decrease the melting temperature (T_m) of a hybridization reaction involving a perfectly matched

heteroduplex: (a) sequence mismatch, (b) reduced salt concentration, (c) formamide addition, and (d) shorter heteroduplex?

15. A traditional protein biochemist is certain that a purified kinase has five cellular targets, but a protein microarray experiment with 10,000 cellular proteins reveals intense fluorescence at five locations and additional fluorescence at 117 other locations. Name two advantages of the microarray format over traditional biochemical methods for enzymatic studies.

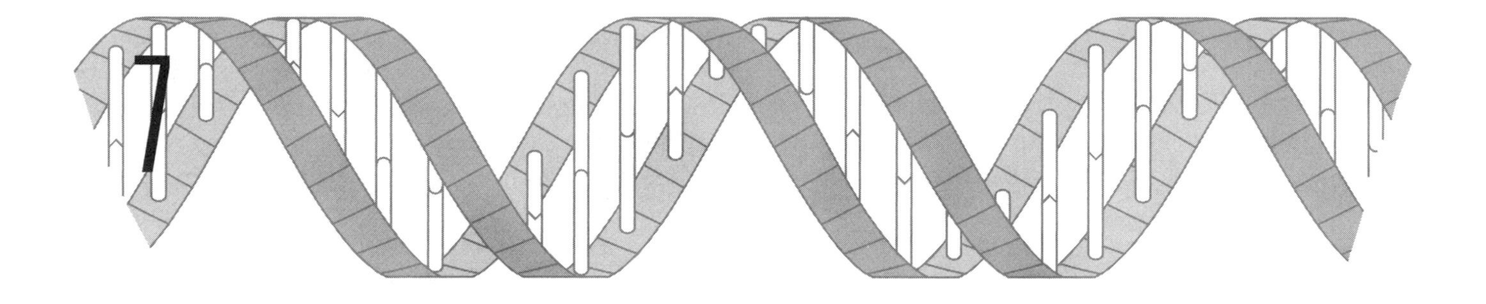

Microarray Manufacturing

Imagination is more important than knowledge.
—*Albert Einstein*

Microarray manufacturing techniques are as imaginative as they are useful. At the center of microarray manufacturing is a creative fusion of biology and engineering. Tools, gadgets, materials, and motion-control systems developed for applications ranging from aerospace science to the computer chip and video projection industries have been adapted for use in microarray manufacturing. Advanced contact printing methods, ink-jet technology, and semiconductor techniques are all used to manufacture microarrays at present. This chapter provides an overview of basic manufacturing requirements and motion-control technology, explains how each microarray manufacturing method works, and compares the different technologies vis-à-vis diverse applications.

MANUFACTURING CRITERIA

Modern microarray manufacturing techniques use diverse technologies, but each endeavors to produce microarrays that are both affordable and high in quality. Because data mining constitutes a tremendously time-consuming component of microarray analysis, care must be taken to ensure that the chips that produce the data are worthy of the many hours that will be spent in data analysis. Microarray manufacturing methods and the resultant microarrays are evaluated on the basis of a large number of criteria, including affordability, content, density, feature size, feature purity, feature reactivity, regularity, ease of implementation, and throughput. This section provides a brief overview of these evaluative criteria.

Affordability

Affordability. *Criterion of microarray manufacture that pertains to cost of a given technology.*

Because microarrays are widely used in basic research, the cost or ***affordability*** of the manufacturing tools is an important consideration. The pervasiveness of a given technology in the research community is to a certain extent cost driven, particularly in university settings where research budgets are limited. The cost of microarray manufacturing tools ranges from several hundred dollars for simple hand-held devices, to tens of millions of dollars for state-of-the-art semiconductor facilities in commercial settings. A typical user making microarrays in a research laboratory will probably spend ~\$50,000 for an automated printing robot. Continued cost reduction through expanding sales and miniaturization will provide increasingly affordable microarray manufacturing tools. Economical manufacturing tools will, in turn, speed the proliferation of microarray analysis.

As the availability of commercial microarrays increases, affordability of the chips is an important consideration. The current cost of preprinted microarrays is ~\$100 for chips containing a few hundred genes to > \$20,000 for microarrays containing tens of thousands of proprietary genes, or gene sets representing entire genomes. With respect to commercial microarrays, both the absolute cost per chip and the cost per gene or per data point are commonly used for cost comparisons. A \$1,000 microarray that contains 5000 genes is five times more expensive than a \$200 microarray that contains 100 genes, but the per gene cost (\$0.20) of the "more expensive" chip is actually ten times less than the cost per gene (\$2.00) of the "less expensive" chip. At the outset microarrays sound a bit pricey, until one considers that the cost of a traditional gene expression measurement by northern blot is in the range of \$20 per gene. If evaluated on the basis of cost per gene or per data point, microarrays represent the most cost-effective tool in the history of biology!

Content

Content. *Criterion of microarray manufacture that pertains to the amount of biological or chemical information on a microarray.*

The term ***content*** refers to the total amount or value the of the genetic, biochemical, or chemical material present on a given microarray or deliverable by a given manufacturing technology. A chip with 10,000 cDNAs would be said to contain superior content compared to a similar microarray with a 1,000-cDNA subset of the first chip. Analogously, a chip with 10,000 unique chemical compounds printed in triplicate (30,000 spots) would be superior content-wise to a microarray that contains 10,000 unique chemical compounds printed in duplicate (20,000 spots). If two chips each contain 5,000 different gene fragments or proteins, the content might be evaluated based on which chip contains cDNAs derived from rare transcripts or proteins that are difficult to overexpress or purify.

Content is also used in reference to a given manufacturing method. A manufacturing technology that allows the preparation of microarrays of oligonucleotides of 20 bases or more is said to deliver superior content compared to one that is able to prepare microarrays of oligonucleotides of only < 10 bases. Content of the chips and deliverable content of the manufacturing platform are important considerations because it is the content that determines the value of the resultant microarray data. Chips that contain a greater number of valuable targets deliver more valuable data than chips that contain a lesser number of less valuable targets. In terms of the legal issues surrounding microarray content, it will be interesting to see whether legal battles arise over microarrays that contain patented genes, gene products, sequences, or chemical compounds (see Chapter 14).

Content and cost are interrelated in the sense that chips with greater content are generally more expensive to manufacture than chips with lesser content. More content is generally better than less content, except for the cost-to-manufacture aspect and the fact that mining and storing the information from high-content chips can be arduous and expensive. If a pathway involves a 100-gene subset of the genome, it is generally unnecessary and perhaps undesirable to examine the entire genome. In this case, dedicated chips containing the 100 genes of interest might actually be more desirable to the researcher than whole genome chips, even though the former has less content.

Density

The criterion of ***density*** refers to the number of target spots or features per unit area of microarray substrate. A chip that contains 1000 cDNA spots in a 0.9×0.9-cm area (0.81 cm^2) would be said to have a density of 1235 spots $(1{,}000/0.81)/cm^2$. Density can be calculated for any microarray by knowing or measuring the distance between the centers of two adjacent spots, a distance that is known as the ***center-to-center spacing***. Microarray density (D) in features per centimeter squared can be calculated from the center-to-center spacing (CTC) measured in microns using the following formula:

Density. *Criterion of microarray manufacture equal to the number of elements or features per unit area of microarray substrate, expressed typically as spots per square centimeter.*

Center-to-center spacing. *Physical distance between the centers of two printed microarray spots, expressed typically in microns.*

$$D = 100 \times (1000/CTC)^2 \qquad (7.1)$$

A microarray with spots set at 140 μm center-to-center spacing would have a density of $100 \times (1000/140)^2$ or 5102 spots/cm^2. Because density goes as a square of center-to-center spacing, relatively modest changes in center-to-center spacing can have a large effect on density. A microarray with spot spacing of 120 μm has a density of 6944 spots/cm^2, whereas a microarray with 140 μm center-to-center spacing supports 5102 spots/cm^2.

Microarray density is sometimes designated in units of a thousand, such that a chip with 25,000 spots is referred to as a *25K chip*. Density is an important consideration because, other factors being equal, it determines the total amount of data that will be obtained per unit area of microarray. Because smaller microarrays use lower volumes of expensive reagents and are faster to scan than larger microarrays, higher density is better than lower density in most cases. Greater density can become a problem if the microarray features are so small so as to exceed the detection limit of the scanning device (see feature size). Most contact and noncontact printing devices deliver spot densities in the range of 1,000–10,000 features per centimeter squared, and densities in this range are readable by all of the popular detection technologies. The semiconductor-based technologies that use photomasks and micromirrors (see below) can reach densities $> 250{,}000$ feature/cm^2, and chips of this density require detection systems with greater scanning density capabilities.

Density and content are interrelated in the sense that two chips with the same 2500 spot/cm^2 density can have different analytical value. A 2.5K microarray with 1250 cDNAs printed in duplicate is generally perceived to have a greater analytical value than a 2.5K microarray with 25 cDNAs printed 100 times each.

Feature Size

Feature size. *Criterion of microarray manufacture that pertains to the physical size of a printed microarray element or feature, expressed typically as the diameter in microns.*

Element. *Any of a set of the 118 fundamental buildings block of all matter. Alternatively, a printed spot or feature in a microarray.*

The term ***feature size*** refers to the size of the spots or ***elements*** in a microarray, usually expressed as an average spot diameter. If the features are square, as in the case of microarrays prepared with photomasks or micromirrors (see below), feature size is measured as the width of each element. Most contact and noncontact printing devices deliver feature sizes in the range of 75–300 μm, whereas the semiconductor technologies offer feature sizes in the range of 10–40 μm. Feature size is an important criterion because it determines the density of the microarray (see above). For printed microarrays in which spots are deposited at discrete locations, the center-to-center spacing can be calculated from the feature size by multiplying by a factor of 1.2–1.4. A feature size of 110 μm, for example, will allow spots to be placed at a center-to-center distance of 132–154 μm. For microarrays prepared by photolithography or micromirrors in which the features are essentially continuous with one another, feature size and center-to-center spacing are equivalent.

Because smaller features allow greater density and greater density provides more information per unit area of microarray, manufacturing technologies that deliver smaller features are generally more valuable than technologies that deliver larger features. Since the late 1990s, manufacturing and detection technologies have evolved together so that increases in density on the manufacturing side have been met by improved resolution of the detection devices (see Chapter 8). A suitable detection device should provide approximately 10 pixels per feature diameter (see Chapter 8).

Feature Purity

Feature purity. *Criterion of microarray manufacture that pertains to the chemical or biochemical homogeneity of the target molecules in a given feature, expressed typically as a percentage of the desired sequence or of the total target molecules present.*

Feature purity refers to the homogeneity of the molecules present in a given spot or feature on the microarray; a common feature purity benchmark is 99% for a given microarray location. Deviations from 100% feature purity are caused both by contaminants in the sample and by errors in a given manufacturing process. For printed microarrays, feature purity is usually compromised either by sample contaminants such as polymerase chain reaction (PCR) primers and nucleotides or by carry over from sample to sample caused by the incomplete washing of the printing device during each printing cycle in the manufacturing process. Contaminants in the sample can usually be removed by the use of purification methods such as those with microplate-based filter binding procedures. Contamination during the printing process can be virtually eliminated by extensive washing of the pins or ink-jet nozzles between each round of printing. For oligonucleotide microarrays, particularly those made by combinatorial approaches such as photolithography and micromirrors, feature purity is usually compromised by incomplete synthesis products that contaminate the full-length products at a given microarray location. Care to fully deprotect each site and cap the reactants with each round of synthesis greatly minimize feature contamination in the combinatorial approaches.

Specificity. *Uniqueness of the biochemical reaction between a target and probe molecule.*

Microarray surfaces such as the aldehyde surface that work by a specific coupling process enable an additional level of feature purity by binding selectively to target molecules of interest. Feature purity for all types of microarrays is important because it determines the ***specificity*** of the biochemical reaction at each microarray location. Aberrant signals arising from interactions between probe molecules and contaminants on the microarray surface will complicate analysis of the microarray data and should be avoided to any extent possible.

Feature Reactivity

The activity or efficacy of the molecules at a given microarray location is known as the ***feature reactivity***. An ideal microarray would contain features with 100% of the molecules in the reactive state, though a common benchmark is 50% for most microarrays. Feature reactivity can be compromised either during the manufacturing process or during the microarray processing and blocking steps. On the manufacturing side, inactivation of the target molecules is generally a minimal problem, though there have been reports of molecular shearing caused by various ink-jet devices and some photodamage occurring during manufacturing processes that use wavelengths of light in the ultraviolet range. Most compromises to feature reactivity are caused by dessication of samples on the microarray surface or by thermal, chemical or enzymatic damage introduced during the processing steps. Because DNA is a relatively hardy molecule, most microarrays of cDNAs and oligonucleotides approach the 50% feature reactivity benchmark. Microarrays of proteins and RNAs, being more sensitive to sample drying and enzymatic digestion, present a greater challenge in terms of the integrity of the target molecules at a given microarray location.

Feature reactivity. *Criterion of microarray manufacture that pertains to the chemical or biochemical activity of the target molecules in a given feature, expressed typically as a percentage of the total target molecules present.*

Regularity

Microarray ***regularity*** pertains to the evenness of the rows and columns of spots across a given microarray. Regularity is generally measured by determining deviations in the center-to-center spacing of each feature, with a common regularity benchmark allowing for 10% error from the center of one feature to then center of an adjacent feature. According to this benchmark, a microarray with a nominal center-to-center spacing of 140 μm, would allow the center of an adjacent spot to reside within 126–154 μm of the first spot. The regularity requirement (R) for any two spots in a microarray can be calculated using the center-to-center distance (CTC) according to the formula:

Regularity. *Criterion of microarray manufacture that pertains to the evenness of the spacing of rows and columns in a printed microarray, expressed typically as a coefficient of variation percentage of the theoretical center-to-center distance.*

$$R = CTC \pm 0.1(CTC) \quad (7.2)$$

The allowable deviation from regularity between two spots in a microarray with a center-to-center spacing of 175 μm would be 175 $\pm$ 17.5 μm. For printed microarrays that use robotic motion control systems, repeatability specifications of $\pm$ 1–10 μm ensure adequate microarray regularity. Manufacturing technologies that use semiconductor approaches and photomasks or micromirrors produce extremely regular microarrays.

Deviations from microarray regularity are attributable to a number of different sources, including the surface, the manufacturing device and the manufacturing environment. The most common causes of irregularity owing to the substrate, derive from local variations in the surface properties (e.g., hydrophobicity) or surface topology. Poor quality surfaces exert local forces on a printed droplet sufficient to cause migration of the droplet away from the original deposition location. Surfaces with uniform surface treatments and coatings and atomically flat substrates provide a significant advantage in terms of microarray regularity. The most common sources of irregularity contributed by the manufacturing device usually pertain to imprecision in the movement of pins, ink-jet nozzles, photomasks, micromirrors, and the like. Movements of the printing substrate caused by vibrations or poor substrate location can also lead to deviations from regularity. Environmental forces typically contribute little to microarray irregularity, though air currents, severe humidity gradients and static electricity have been observed to cause uneven sample deposition and drying.

Regularity is an important criterion because it determines the ease with which the data can be extracted from a given microarray image. Most automated approaches to data analysis use a checkerboard-type grid to demarcate the features in each row or column, thereby allowing average intensity values to be calculated at each microarray position (see Chapter 9). Microarrays with irregular feature spacing impede data analysis by slowing the rate at which a quantiation grid can be placed over each target location. In extreme cases, irregularity can actually result in one feature overlapping an adjacent feature such that spot merging causes the mixing of two different samples, making accurate data analysis difficult.

Ease of Implementation

Ease of implementation. *Criterion of microarray manufacture that pertains to the simplicity by which a given technology can be used in a laboratory setting.*

The manufacturing criterion ***ease of implementation*** refers to the degree to which a given technology can be assimilated and used in a laboratory setting. For robotic printing devices, an ideal system is one that allows a reasonably skilled researcher to successfully manufacture microarrays during the first day of implementation. Ease of implementation pertains not only to the speed with which a technology is implemented but also to the durability of a system over time. The short- and long-term utility of a given technology is determined by a combination of design, hardware, and software considerations. Printing devices that require frequent replacement, ink-jet nozzles that clog, and software that contains baroque icons or commands all impede ease of implementation and should be avoided.

Ease of implementation from the perspective of the researcher does not apply to microarray manufacturing at companies, but it does apply to the commercial microarrays themselves. In this regard, commercial microarrays should be easy to use along every step of the microarray experimental process. Though ease of implementation of commercial manufacturing technology does not affect individual researchers, it is an important internal consideration for commercial entities (see Chapter 15).

Throughput. *Criterion of microarray manufacture that pertains to the number of microarrays that can be manufactured per unit time, typically expressed as the number of chips per day. Alternatively, the microarray detection system criterion that refers to the number of microarrays of a defined size that can be scanned or imaged at a defined resolution per unit time, typically expressed as microarrays per hour.*

Throughput

Throughput refers primarily to the rate at which a given device can manufacture microarrays. For contact printing and ink-jet systems, throughput is largely determined by the rate at which samples from microplates can be transferred onto the printing substrates. Manufacturing throughput with robotic systems is determined by the number of printing pins or nozzles in the printing device, and by the speed at which the robot moves the printing device from substrate to substrate. Throughput is usually expressed as the number of samples that can be deposited on set of 50 substrates per 8-h session per robot, and specifications vary from 100–10,000, depending on number of printing implements used, the speed of the motion control system and how many replicates of each sample are deposited per substrate. Because most motion control systems move at similar speeds, the largest increases in throughput are usually achieved by increasing the number of printing implements or by increasing the number of printing robots.

For combinatorial manufacturing technologies in which the targets are built on the microarray surface in a base-by-base fashion, throughput is determined by the time required to complete each round of synthesis (i.e., cycle time) and by the number of individual substrates or chips that can be addressed and manufactured in a given round of synthesis. Cycle times for photolithographic, micromirror, and ink-jet approaches vary from 5–60 min; an average technology is capable of manufacturing dozens of chips containing 20mers on a given synthesis station per day.

MOTION-CONTROL TECHNOLOGY

Computer-controlled robots provide the technological basis of virtually all microarray manufacture devices. Motion-control technology allows precise, rapid movement in three-dimensional space (Fig. 7.1). Three-dimensional movement is required for many different types of microarray manufacturing technologies, including those that use printing pins, ink-jet devices, and photomasks. This section describes the basic components of motion-control systems.

Electricity and Magnetism

To understand how robots work, it is useful to review the basic phenomena of electricity and magnetism. Electricity is the flow of electrons in a given substance. The flow of electrons or ***current*** (I) is directly proportional to ***voltage*** (V)

Current. *Flow of electrons per unit time.*

Voltage. *Difference in electric potential between two terminals in an electrical source.*

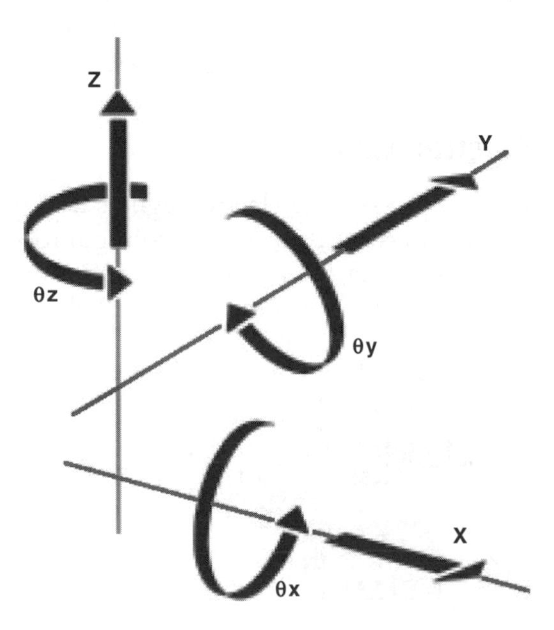

FIGURE 7.1. *The movements of motion-control systems are defined in three dimensions, with left to right as x, front to back as y, and up and down as z. Each axis also has a corresponding rotational axis (θx, θy, and θz). (Courtesy of Newport Corp., Irvine, CA).*

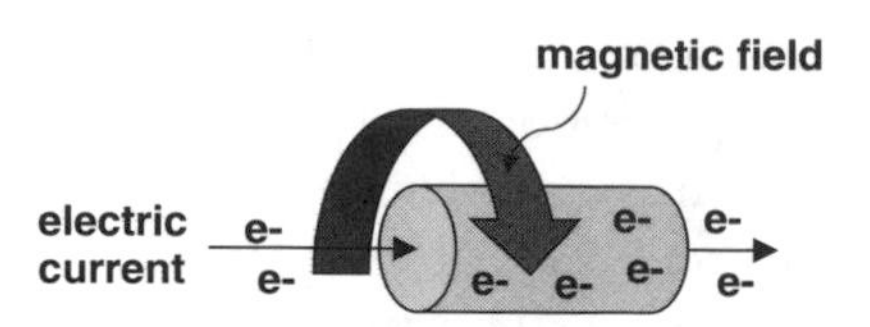

FIGURE 7.2. When electrons (e^-) flow along a wire (cylinder), *a magnetic field* (arrow) *is created with a force that is perpendicular to the direction of the current.*

and indirectly proportional to ***resistance*** (R) according to ***Ohm's Law:***

Resistance. *Propensity of a material to impede the flow of electricity.*

Ohm's Law. *Mathematical relationship between current, voltage and resistance, given as* $I = V/R$.

Conductor. *Material that supports the flow of electrons.*

Insulator. *Material that inhibits the flow of electrons.*

Potential difference. *Discrepancy in the capacity to support the flow of electricity between two terminals of a conducting substance, measured in volts.*

Direct current. *Electricity that flows in a unidirectional manner and at a constant rate.*

Alternating current. *Electricity that flows in a bidirectional manner and at a defined frequency.*

Magnetic field. *Force created by the flow of electrons through a conductive material that is perpendicular to the direction of the current.*

$$I = V/R \quad (7.3)$$

Materials such as copper wire allow electrons to flow easily and are known as ***conductors,*** whereas materials such as plastic coatings allow little charge to pass and are termed ***insulators***. Semiconductors are so named because such silicon-based materials conduct electricity with moderate ("semi") efficiency.

Electrical connections in robots are achieved by the use of wires (conductors) wrapped in protective coatings (insulators). If a ***potential difference*** is created across a conducting substance such as a wire, the flow of electrons is initiated and an electrical current is established. There are two common types of current known as ***direct current*** (DC) and ***alternating current*** (AC), which are typified by disposable batteries and normal laboratory electrical outlets, respectively.

Direct current flows in a unidirectional manner and at a constant rate, whereas AC flows in a bidirectional manner with a defined periodicity or frequency. The standard electrical outlets in domestic laboratories provide alternating currents in the range of 50–60 cycles per second (50–60 Hz) and 110–120 V. Voltages in other countries vary and microarray robots manufactured in the United States need to include the appropriate converters.

When current flows in a piece of wire, a force known as a ***magnetic field*** is created around the wire (Fig. 7.2). A current-carrying wire that creates a magnetic field is known as an electromagnet. The force of the circulating magnetic field in the electromagnet is perpendicular to the flow of electrons, and its direction is determined by the direction of the electron flow. If the conducting wire is brought in the vicinity of a permanent magnet, the wire exerts a force on the magnet, and these forces can used to create mechanical movement (Fig. 7.3). The process by which mechanical movement is achieved by a current-carrying wire in a magnetic field provides the basis for all of the electric motors used in microarray manufacturing devices.

Electric Motors

Rotor. *Permanent magnet that rotates in a circular manner in an electric motor.*

Stator. *Fixed electromagnetic that deflects a rotor in an electric motor.*

Electric motors use electromagnets and magnetic fields to create mechanical movement. In the most basic configuration, the mechanical movement occurs in a circular fashion, whereby a permanent magnet or ***rotor*** is deflected in a circular manner by a series of fixed electromagnets or ***stators***. The polarity of

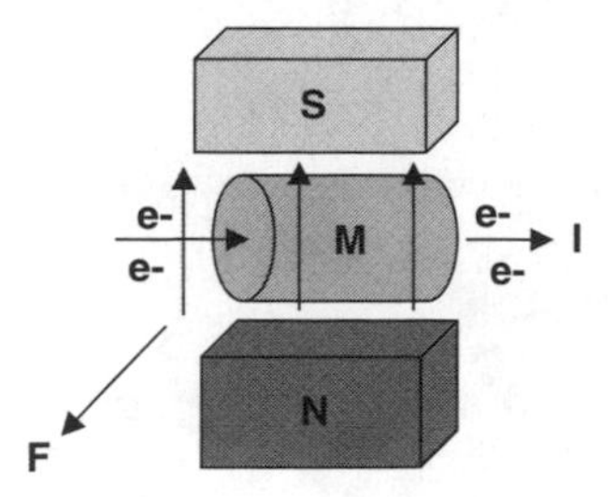

FIGURE 7.3. Two magnets of unlike poles (S and N) *create a magnetic field* (M) *that exerts a force* (F) *on a wire (teal) carrying an electric current* (I). *The force can be used to achieve mechanical movement.*

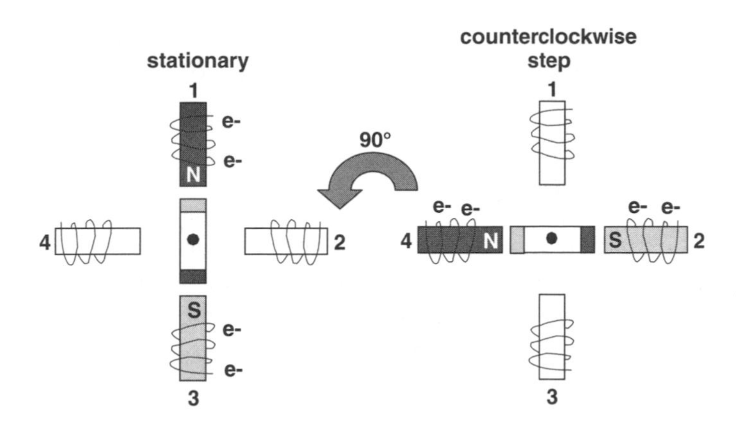

FIGURE 7.4. A generic stepper motor contains a permanent magnet as a rotor (center) and four stators (1–4), which can be polarized. Voltage applied to stators 1 and 3 causes the rotor to align and assume a stationary position. When voltage is applied to stators 2 and 4, a 90° movement of the rotor (counterclockwise step) occurs. (Adapted from T. Mercer, NewTech Education Resources, Victoria, Australia.)

the stators determines the resting position and movements of the rotor (Fig. 7.4). Circular movement of the rotor is achieved by applying voltages to the stators to create a given polarity. The rotor can be moved in precise clockwise and counterclockwise movements, or ***steps,*** by adding, removing, or reversing the polarity of the stators. Electrical motors that use rotors and stators to achieve stepwise mechanical movement are known as ***stepper motors*** (Fig. 7.5*A*). A related motor type is the known as the ***servomotor*** (Fig. 7.5*B*). Stepper motors and servomotors are used extensively in microarray manufacturing.

Step. *Precise clockwise or counterclosewise movement of an electric motor.*

Stepper motor. *Electric motor that uses rotors and stators to achieve stepwise mechanical movement.*

Servomotor. *Electric motor that achieves precise mechanical motion by the use of control circuitry.*

Stepper and servomotors offer precise motion-control capabilities. The main difference is that the precision of stepper motors depends on the specifications of the rotor, whereas precision in servomotors is determined primarily by the stability of the electronic feedback control systems. Stepper motors have several advantages over servomotors: greater affordability, no regular maintenance, and a high degree of safety owing to the fact that most errors cause stepper motors to stop moving. Some disadvantages of stepper motors are that they are relatively noisy at high speeds and generate a large amount of heat at standstill because continuous current needs to be applied to hold the motor in a stationary position (Fig. 7.4). The latter is a consideration, particularly in a controlled environment such as a microarray cleanroom (see Chapter 11), where thermal emissions are an important consideration in terms of maintaining sufficiently cool manufacturing temperatures.

Servomotors are available in brush and brushless models. Though extremely affordable, brush servomotors have the disadvantage that brush wear creates dust, which can be a source of contamination in a cleanroom environment. Brushless servomotors have the advantage of reduced maintenance, little or no dust, extremely smooth operation, and moderate thermal emission. The primary disadvantage of brushless servos is that they can be rather costly. The AC servomotor is the most common type of brushless model used

*FIGURE 7.5. **Left,** A stepper motor. (Courtesy of Electronics Information Online.) **Right,** A servomotor. (Courtesy of HVW Technologies, Alberta, Canada.)*

Linear actuator. *Device used in microarray manufacturing robots that converts circular motion into linear motion.*
Ball screw. *Linear actuator that uses ball bearings and a corkscrew-style track to convert circular motion to linear motion.*
Belt drive. *Linear actuator that uses flat rubber or plastic cords and pulleys to convert circular motion to linear motion.*
Round rail. *Linear actuator component that uses linear bushings and a round stainless-steel shaft as the guide mechanism.*
Square rail. *Linear actuator component that uses linear bushings and a square stainless-steel shaft as the guide mechanism.*
Ball spline. *Linear actuator component that uses a bearing and a stainless-steel shaft with uniformly spaced ridges as the guide mechanism.*
Speed. *Magnitude of travel per unit time of an electric motor or microarray robot, expressed typically in centimeters per second. Alternatively, a microarray detection system criterion that refers to the area of microarray substrate that can be captured per unit time at a given resolution, expressed typically as square centimeters per minute.*
Resolution. *Smallest detectable movement increment of an electric motor or microarray robot, typically expressed in microns. Alternatively, a microarray detection system criterion that refers to the pixel size in a microarray image, expressed in microns.*
Accuracy. *Maximal observed discrepancy between the desired and observed movement of an electric motor or microarray robot, expressed typically in microns.*

for microarray applications. Stepper motors and servomotors are both used widely in microarray manufacturing to position microplates, printing devices, and wafers.

Linear Actuators

The circular movements of stepper motors and servomotors need to be coupled to devices that convert circular motion into linear motion. Such devices are known as ***linear actuators***. Linear actuators or linear motion guides are available in dozens of different designs, with their classification generally based on the mechanism that is used to drive the actuator or on the type of guides that are used to support the movement of the actuator. Common drive mechanisms include ***ball screws*** (Fig. 7.6) and ***belt drives***. The ball screw mechanism uses ball bearings that roll in a corkscrew-style track, whereas belt drives use flat rubber or plastic cords and pulleys to transmit motor torque. Common types of supports include the ***round rail,*** which uses linear bushings that travel along a round stainless-steel shaft. The ***square rail*** and ***ball spline*** are other common support types.

Amount of thermal contamination present in a microarray cleanroom.

Linear actuators and motion-control systems are rated according to set of precise criteria, with ***speed, resolution, accuracy, precision, repeatability, backlash, durability,*** and ***load*** being some of the primary considerations for microarray applications. These terms are defined as follows:

Speed: the magnitude of travel per unit time (cm/sec).
Resolution: the smallest movement increment that can be detected (μm).
Accuracy: the maximal observed discrepancy between the desired and observed movement (μm).
Precision: the range of deviations that occur for 95% of the movements for a given input ($\pm$ μm).
Repeatability: the reliability of achieving a specified position over repeated attempts ($\pm$ μm).
Backlash: the magnitude of the input that produces no movement when the actuator is moved in the opposite direction along a given axis (μm).
Durability: the total length of time that a system can be used without visible deterioration in performance (machine hours).
Load: the maximal weight (g or kg) that can be moved without significant deterioration of the performance specifications.

Ball screw systems allow movements at moderate speeds (10–200 cm/s), with repeatability in the $\pm$ 10μm range, negligible backlash, durability exceeding several thousand machine hours, and load specifications in 10–30 kg range. Belt drives provide similar specifications although they generally allow

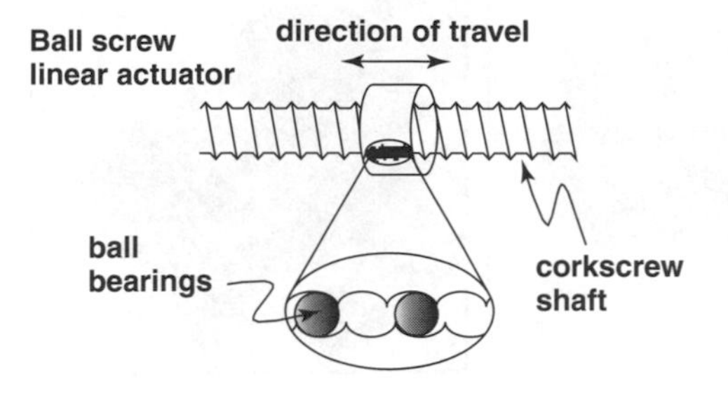

FIGURE 7.6. Bidirection travel (arrow) *is achieved with a stage* (cylinder) *that travels via ball bearings* (arrow) *along a corkscrew shaft. (Adapted from Motion Systems, Eatontown, NJ.)*

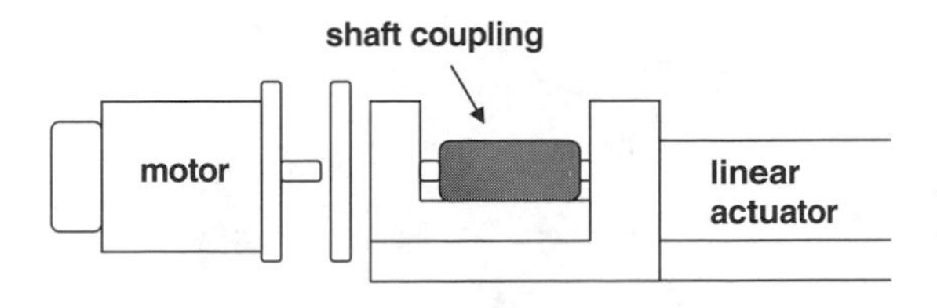

FIGURE 7.7. Motors are linked to linear actuators via a shaft coupling. (Adapted from THK Co., Tokyo.)

faster movements (100–1000 cm/s) and have somewhat lower repeatability (± 20–100μm).

Linear actuators are connected to motors by the use of linking device such as a ***shaft coupling*** (Fig. 7.7). A ***linear drive system*** consisting of a ball screw drive, double rectangular guides, and a servomotor is one configuration that provides particularly accurate actuator function.

Linear Encoders

As microarray applications become increasingly exacting, repeatability in the sub-20 μm range becomes important. The repeatability of linear actuators can be improved in a variety of ways. One means used in conjunction with stepper motors employs ***microstepping***, which allows incremental positions between steps to be achieved. Microstepping and ball screw devices allow repeatability in the ± 10 μm range. Another common approach employs ***linear encoders***, which are measuring devices that are applied along an axis of travel to allow increased precision (Fig. 7.8). Linear encoders are often provided in the form of an adhesive strip that contains a bar-code-like pattern of reflective gold to measure steps. The adhesive strips are read by a light source, such as a light-emitting diode (LED), which projects reflected light back into a photoreceptor (Fig. 7.9). Using this "microscopic ruler" as a measuring device, the encoder is able to convert linear position information into electrical output signals. Linear encoders that use a beam of light in the measuring device are known as ***optical encoders***. With measuring steps in the 0.001–1 μm range, many linear encoders provide precision of ± 1 μm without a major sacrifice in microarray manufacturing speed.

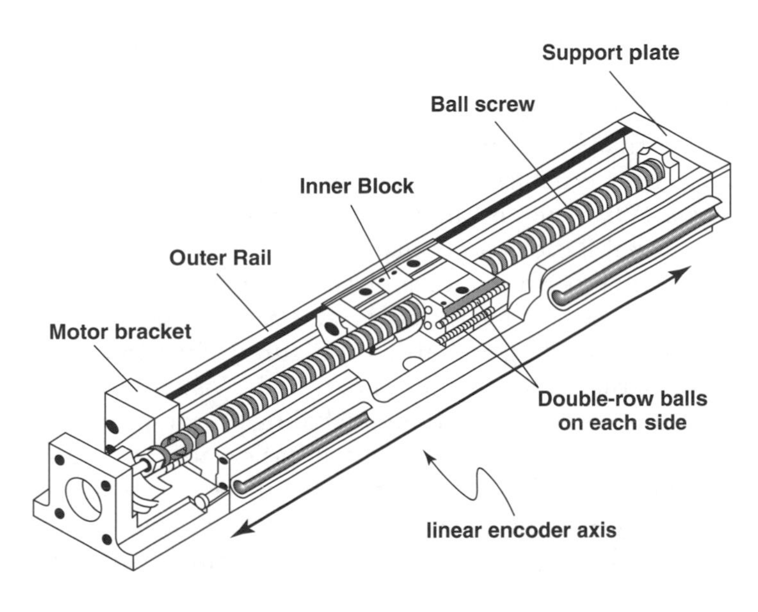

FIGURE 7.8. Linear encoders are applied to the travel axis (double arrow) *of linear actuators to increase repeatability, accuracy, and precision. (Adapted from THK Co., Tokyo.)*

Precision. *Range of deviations that occur for 95% of the movements of an electric motor or microarray robot for a given input, typically expressed in microns. Alternatively, a microarray detection system criterion that refers to the consistency of microarray images acquired during different detection sessions.*

Repeatability. *Reliability of achieving a specified position of an electric motor or microarray robot over repeated attempts, typically expressed in microns. Alternatively, a microarray detection system criterion that refers to the line-to-line accuracy of a scanning device.*

Backlash. *Magnitude of the input that produces no movement of an electric motor or microarray robot when the actuator is moved in the opposite direction along a given axis, expressed typically in microns.*

Durability. *Cleanroom evaluative criterion that refers to the extent to which the physical structure, environmental control, and other components of a microarray cleanroom maintain their specifications over time. Alternatively, the total length of time that an electric motor or microarray robot can be used without visible deterioration in performance, expressed typically in machine hours.*

Load. *Maximal weight that can be moved without a significant deterioration in the performance specifications of an electric motor or microarray robot, expressed typically in grams or kilograms.*

Shaft coupling. *Linking device in a microarray robot that connects a motor to a linear actuator.*

Linear drive system. *Microarray robot component consisting of a motor, linear actuator, and other components required to convert circular motion into precise linear motion.*

Microstepping. *Stepper motor technique that allows increased motion control precise by moving in fractions of steps.*

Linear encoder. *Measuring device on the linear actuator of a microarray robot consisting of a bar-code-like tape with microscopic hash marks and a reading device that allows increased precision in motion control.*

Optical encoder. *Measuring device on the linear actuator of a microarray robot consisting of a bar-code-like tape with microscopic hash marks and an optical reading device, such as a laser, that allows increased precision in motion control.*

FIGURE 7.9. *Linear encoders that use gold-electroplated steel tape and a noncontact optical readout for high-resolution linear registration. (Courtesy of Renishaw, Gloucestershire, UK.)*

Motion-Control Software

Software. *Programs, routines, and commands that a computer uses to control hardware devices.*

Virtual instrument. *Software module in LabVIEW software that can be constructed and used to control robots and other instruments.*

The programs, routines, and commands that a computer uses to control hardware devices are known as ***software***. For motion-control applications, a number of different software languages are commonly used, including LabVIEW, Visual Basic, and C++. All three languages run on Microsoft Windows operating systems. LabVIEW is a graphical programming language that provides an intuitive way to control the movement of robots and is easy to implement. With LabVIEW, the user specifies functionality by assembling block diagrams (Fig. 7.10) rather than by writing programs. LabVIEW uses a programming model known as G, which enables the user to assemble software modules called ***virtual instruments*** (VIs). The VIs can be used to control robots and other instruments.

Visual Basic is a popular programming language that allows the user to create high-performance applications quickly by using standard controls such as Web browsers, file list boxes, and data-bound grids. Reports and forms can be created by drag-and-drop procedures, which speeds programming by eliminating the need to write primary code. Visual basic is ideal for a number of motion-control applications, including prototype development, which places a premium on functionality. Visual Basic native code applications leverage the compiler technology used in C++. Run windows allow the user to initiative movements of the robot with ease, and dialogue boxes provide feedback concerning the position and status of robot movement (Fig. 7.11).

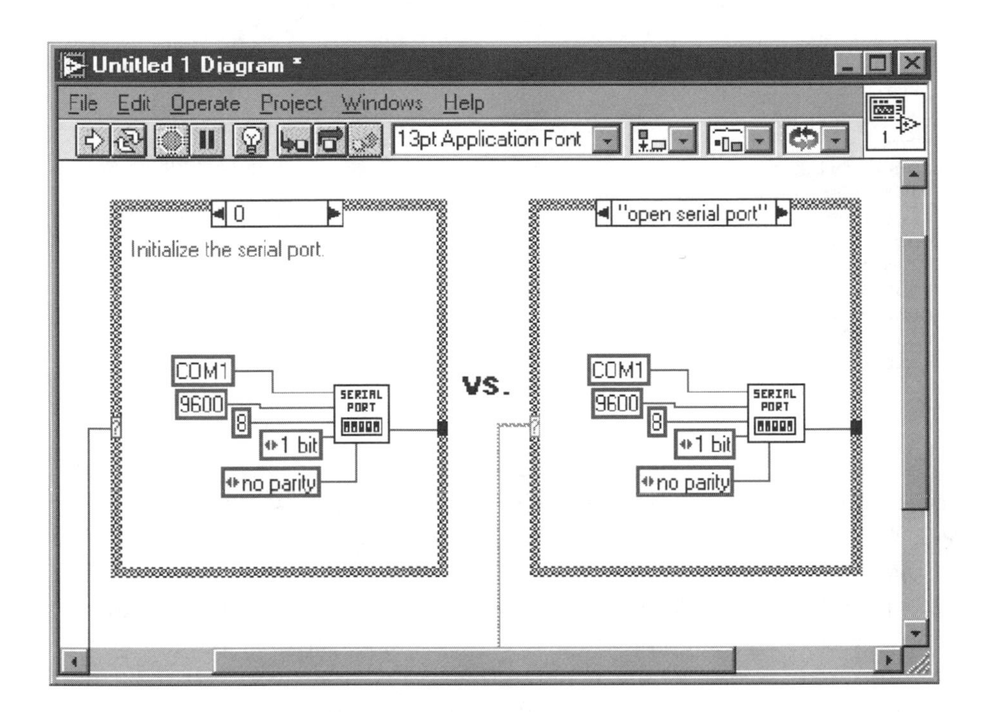

FIGURE 7.10. The graphical design of LabVIEW. *(Courtesy of R. Johnson, National Instruments, Austin, TX.)*

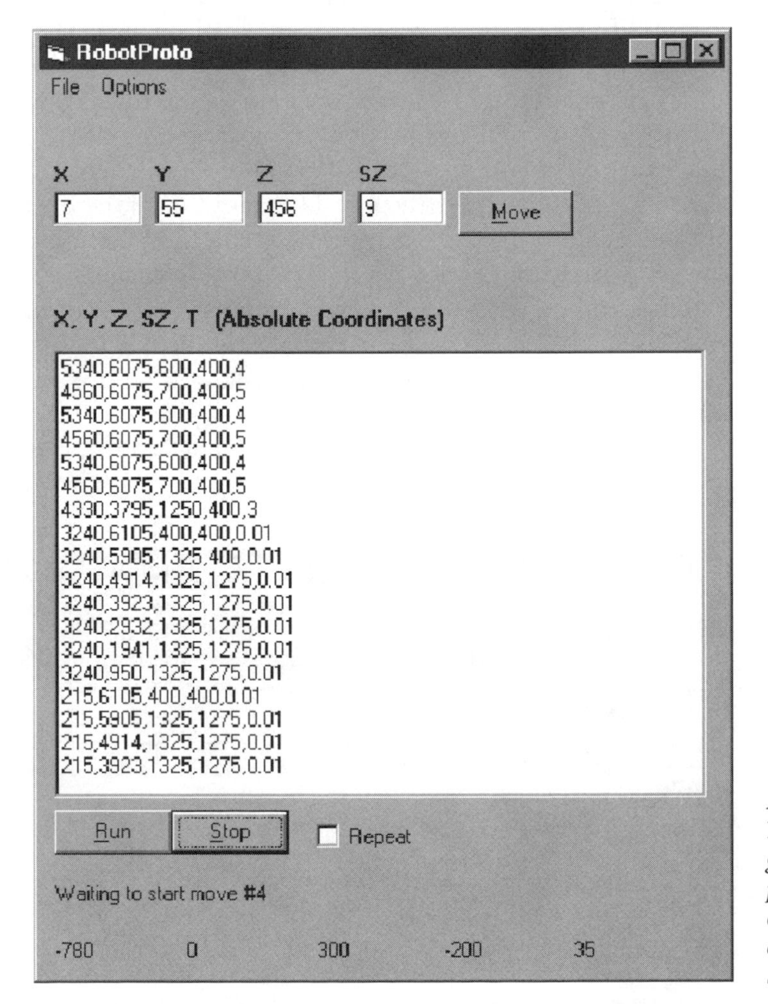

FIGURE 7.11. Microsoft Visual Basic *allows rapid computer programming for motion-control applications, in part through the use of intuitive windows that simplify coding. (Courtesy of K at Korteks, San Diego, CA.)*

```
{
    Vector v1, v2, v3(0.0,0.0);
    v1=Vector(1.1,2.2);
    v2=Vector(1.1,2.2);
    v3=v1+v2;
    cout << "v1 is ";
    v1.PrintOn (cout);
    cout << endl;
    cout << "v2 is ";
    v2.PrintOn (cout);
    cout << endl;
    cout << "v3 is ";
    v3.PrintOn (cout);
    cout << endl;
}
```

FIGURE 7.12. *Software code written in the C++ programming language. (Courtesy of S. Reddy, Vanderbilt University, and G. Bowden Wise, Rensselaer Polytechnic Institute, Vanderbilt, Nashville, TN RPI, Troy, NY.)*

C++ is an object-oriented, high-end programming language built on the procedural language C (Fig. 7.12). C++ is basically a robust version of C that frees the programmer from constant access to the primary code while maintaining access to C when necessary. C++ is widely used by programming experts and is ideal for the development of professional motion-control packages. The stability of the language provides robust, reliable performance for the most exacting microarray manufacturing applications. Microsoft Visual C++ offers real-time programming tips, and syntax assistance enables the programmer to write code quickly. The availability of software wizards in Visual C++ supports complex functionality and technology development. The efficiency, simplicity, and scalability of C++ make it ideal for large-scale software development.

Scripting language. *Software routines used in microarray motion control programs that run on top of the underlying programming language and simplify coding.*

Macro. *Higher-level routine in a scripting language used to achieve sophisticated motion control without requiring the user to code in the primary programming language.*

Microarray motion-control programs often use ***scripting languages*** that run on top of an underlying programming language such as C++. Scripting languages are often composed of a series of higher-level routines known as ***macros.*** Macro scripts are extremely useful for rapid motion-control programming because they allow the user to implement common commands using calls, loops, and pauses without coding in the primary language. A series of macros linked together can be used to implement sophisticated motion-control operations (Fig. 7.13).

All of the traditional programming languages are based on classical logic, which defines code in binary terms whereby a command or movement is based on either 0 or 1. A more recent development is a value system that allows for

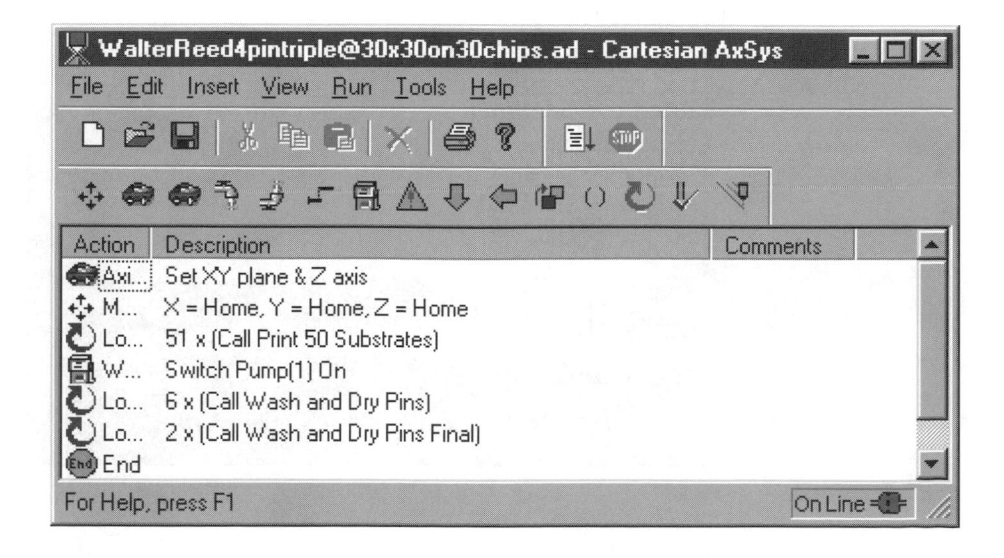

FIGURE 7.13. *Macros can be used to implement sophisticated motion-control routines, while shielding the user from the primary code. (Courtesy of TeleChem/Arrayit.com, Sunnyvale, CA; scripting language AxSys courtesy of Cartesian Technologies, Irvine, CA.)*

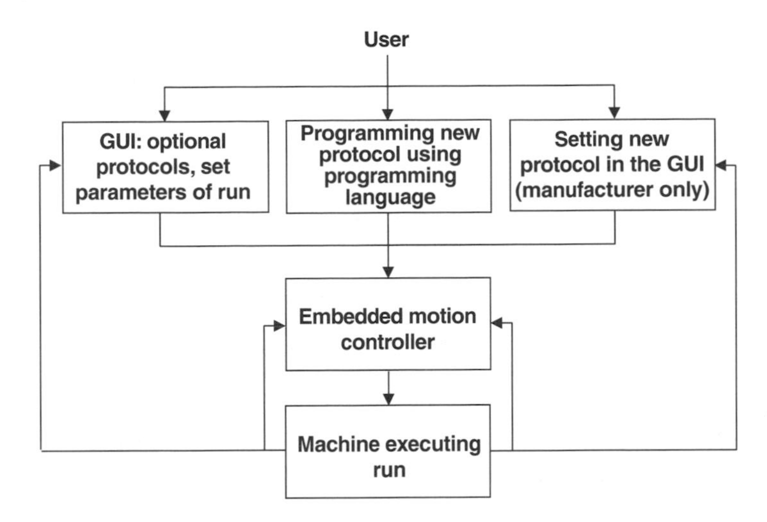

FIGURE 7.14. The GUI and programming language for a motion controller and microarray robot. (Courtesy of Dr. A. Goldenberg, University of Toronto, Toronto.)

intermediate values between 0 and 1 by the consideration of multiple variables. A programming system that allows logical conclusions to be drawn from inexact or complex relationships is known as ***fuzzy logic***. The use of fuzzy logic in microarray applications provides extremely precise and sophisticated software–hardware interactions (see below). Robots that learn by executing a series of commands may eventually provide motion control based on ***artificial intelligence***.

Fuzzy logic. *Programming system that allows logical conclusions to drawn from inexact, complex, or nonbinary relationships.*

Artificial intelligence. *Capacity of a computer to think and learn in a manner analogous to the human mind.*

Motion-Control Systems

Motion-control systems are integrated, computer-controlled devices that combine software and hardware tools to implement precise movement in three-dimensional space. The microarray user interacts with the robot via a computer screen and a set of software windows known as the ***graphical user interface*** (GUI), pronounced "gooey." In many cases, the GUI allows the user to select standard motion-control protocols as well as to generate custom protocols for unique motion-control routines and user-defined microarrays (Fig. 7.14). Figure 7.15 illustrates the sophistication of such devices for microarray manufacture.

Graphical user interface. *Intuitive software package that allows a user to efficiently control a robotic device such as a microarray robot or scanner via computer.*

Integrated motion-control systems contain linear actuators along all three axes (x, y, and z). Stages connected to the actuators allow attachment of microarray devices of interest, including pin tools, ink-jet dispensers, pipeting devices, glass printing substrates, glass and silicon wafers, and microplates. There are two main types of integrated motion control systems: ***moving platen*** and ***overhead gantry***. Most moving platen systems contain two actuators that move along the y and z axes above the base of the robot, and a third actuator that moves along the x axis (Fig. 7.16). The stage on the third actuator (x axis) holds the printing surface, which is known as a ***platen***. Other moving platen systems move in x and z above base, with the platen moving in the y dimension. The resulting bipartite configuration of moving platen systems accomplishes three-dimensional movement by moving the top portion of the robot in two dimensions (either y and z or x and z) and the bottom portion (platen) of the robot in one dimension (x or y).

Moving platen. *Microarray robot design that achieves movement along the x or y axis by movement of the platform that holds the printing substrates.*

Overhead gantry. *Microarray robot design that achieves movement along all three axes, using a suspended motion-control system that moves over a stationary platen.*

Platen. *Bed or base of a microarray robot that holds the microarray substrates, microplates, and other components and accessories.*

Most moving platen printing systems hold 1–64 pins or ink-jet dispensers (see below), 20–120 substrates, and 1–10 microplates. The addition of a microplate "hotel" allows up to 100 microplates to be loaded and unloaded from

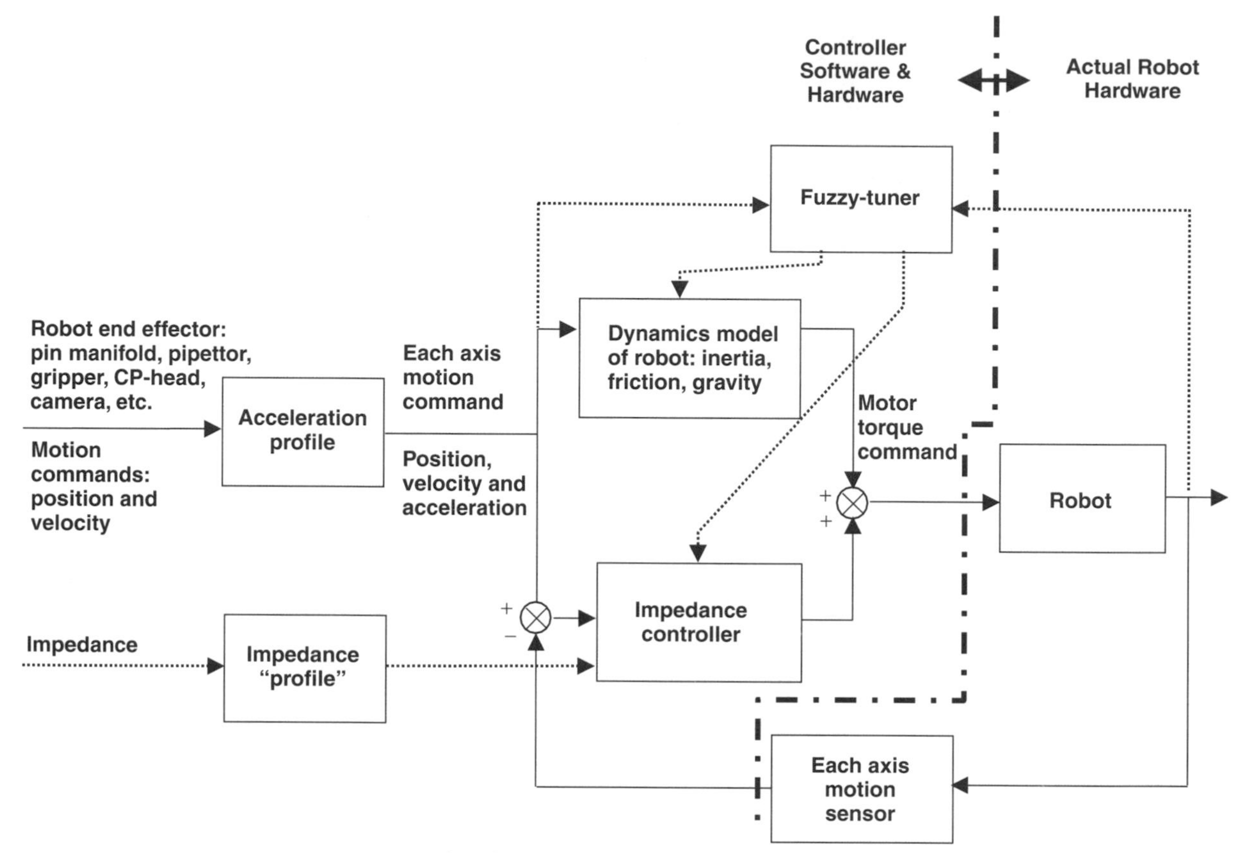

FIGURE 7.15. The software and hardware components of a microarray motion-control system. Courtesy of Dr. A. Goldenberg, University of Toronto, Toronto.)

the moving platen system in an unattended fashion (Fig. 7.17). A standard moving platen system allows the manufacture of chips containing 10,000 different samples in a single 8-h printing run, with precision in the $\pm$ 10 μm range (Fig. 7.18).

Overhead gantry systems represent the second main design of microarray motion control systems. The overhead gantry systems contain three linear actuators moving in the *x, y,* and *z* dimensions above a stationary platen. Some overhead gantry systems use ball screws, servomotors, and optical encoders to achieve precise ($\pm$ 1 μm) movement (Fig. 7.19). Microarrays prepared with this type of design are highly regular (Fig. 7.20). Other overhead gantry systems use ball screws and stepper motors or belt-driven actuators and encoders to enable extremely accurate three-dimensional movement (Fig. 7.21).

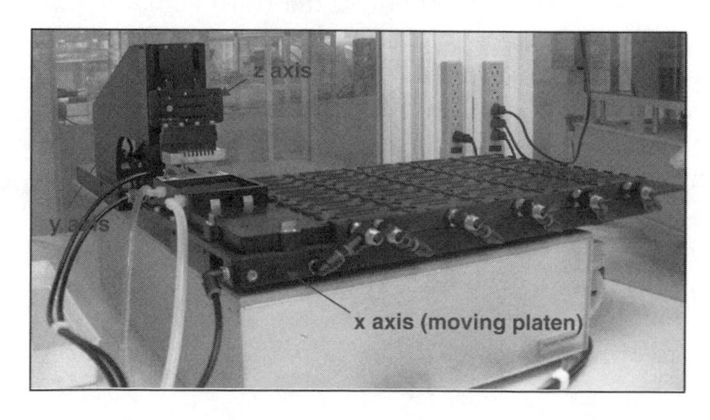

FIGURE 7.16. Stepper motors and ball screws are two linear actuators set above the platen (y and z axes); a third actuator is below the platen (x axis). Shown is a PixSys 5500 robot (Cartesian Technologies, Irvine, CA) configured in a class 100 microarray cleanroom. (Courtesy TeleChem/Arrayit.com, Sunnyvale, CA.)

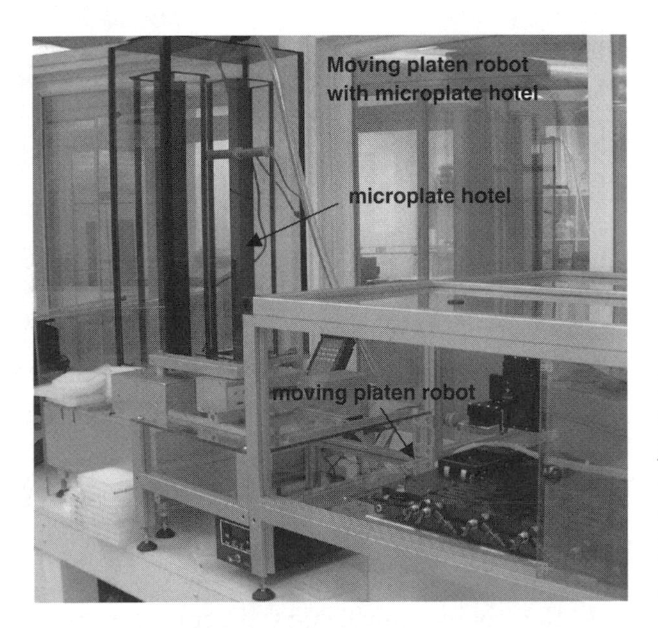

FIGURE 7.17. A moving platen robot with a microplate hotel configured in a class 100 microarray cleanroom. Shown is a PixSys 5500 XL robot (Cartesian Technologies, Irvine, CA). (Courtesy of TeleChem/Arrayit.com, Sunnyvale, CA.)

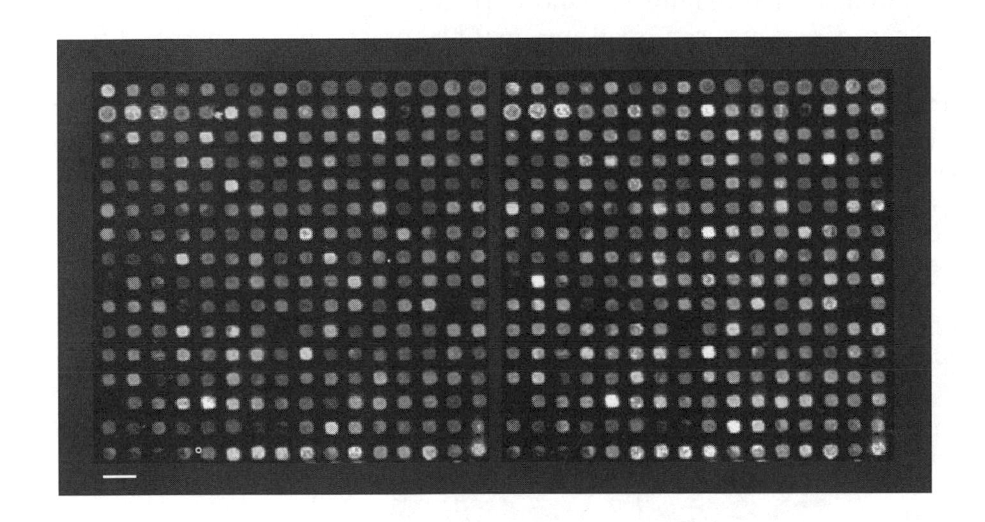

FIGURE 7.18. Microarray printed with a moving platen robot; the data are represented in a rainbow palette. Space bar $= 200\mu m$. (Courtesy of B. McIntosh, Cartesian Technologies, Irvine, CA.)

FIGURE 7.19. Three linear actuators with servomotors, ball screws, and optical linear encoders define the x, y, and z axes above the stationary platen. Shown is a ChipWriter Pro robot. (Courtesy of T. Bond, Virtek Vision, Waterloo Ontario, Canada.)

FIGURE 7.20. A microarray printed with the overhead gantry system shown in Figure 7.19. The chip contains 8500 human EST targets spotted in duplicate for a total of 19,000 spots. The probe for this experiment was obtained by converting 10 μg total human RNA into Cyanine-3-labeled cDNA. The scanned image is represented in a rainbow palette. The space bar = 500 μm. (Data courtesy of University Health Network Microarray Centre, Toronto.)

Microrobotics involving stepper motors and ball screw linear actuators can be combined to afford miniature motion control systems (Fig. 7.22). These overhead gantry systems use a small amount of laboratory bench space (~1 ft^3) and are extremely affordable. The total weight of the microrobots is < 6 kg, and the attachments on the gantry are generally < 50 g. This lightweight design provides some key advantages in terms of precision and durability that can be understood by considering basic Newtonian mechanics. As given in Equation 7.4, Newton's laws dictate that momentum (p) is proportional to mass (m) times

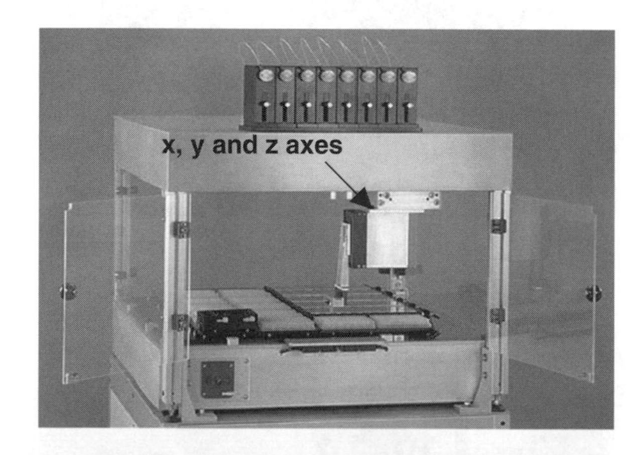

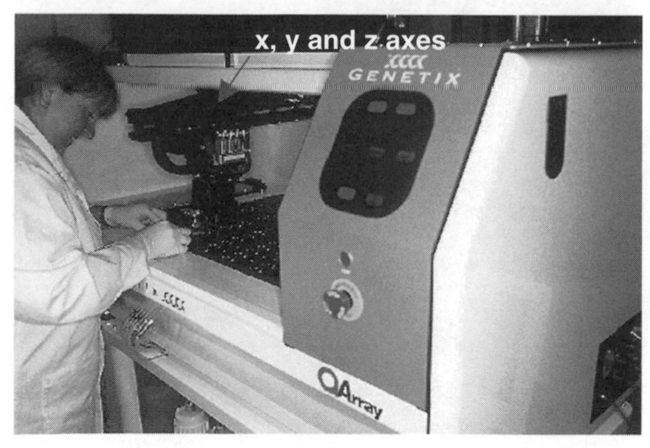

FIGURE 7.21. Microarray robots with three-axis (x, y, and z) movement above a stationary platen. Shown are **Top** the ProSys 5510B PA Workstation (courtesy of Cartesian Technologies, Irvine, CA) and **Bottom** the QArray (courtesy of Dr. S. Stephens and Dr. S. Richards, Genetix, Hampshire, UK).

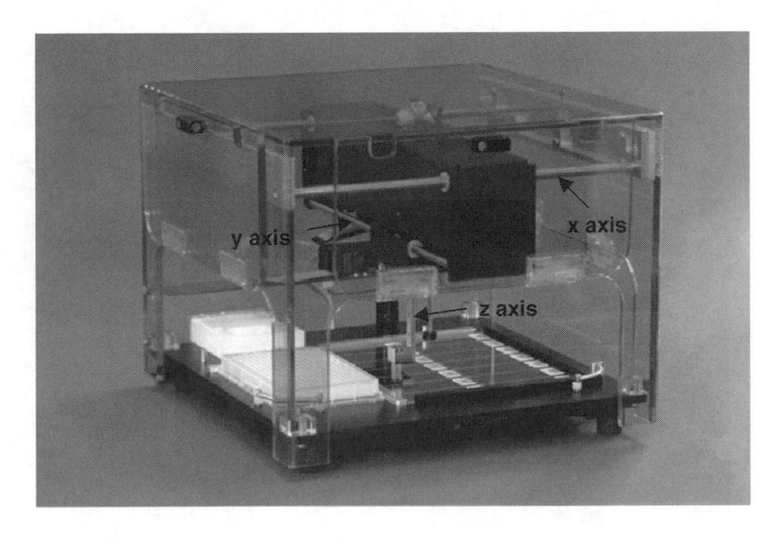

FIGURE 7.22. Stepper motors, ball screws, and round rail guides make up the three linear actuators that define the x, y, and z axes. Shown is the Spot-Bot microrobot. (Courtesy of Jennifer Bohlken, TeleChem/Arrayit.com, Sunnyvale, CA.)

velocity (v):

$$p = m \times v \tag{7.4}$$

Newton's laws also dictate that force (F) is proportional to the rate of change of momentum ($\Delta p/\Delta t$):

$$F = \Delta p/\Delta t \tag{7.5}$$

It follows from Equations 7.4 and 7.5 that a smaller mass produces a smaller momentum, and reducing momentum reduces the forces generated during movement. Because smaller forces improve precision and durability, the lightweight design of microrobotic systems provides repeatability in the $\pm$ 10–20 μm range and > 2000 machine-h durability, even though the motion-control components are relatively inexpensive compared to the larger systems.

MICROMACHINING

Motion-controls systems allow precise three-dimensional movement, but the microarrays manufactured with such systems are only as good as the precision of the printing implements and other components attached to them. For this reason, high-precision machining technologies, known as ***micromachining***, play an extremely important role in microarray manufacture. This section provides a brief overview of some of the software and hardware tools used in modern micromachining approaches.

Micromachining. *High-precision machining technology used to manufacture printing pins and other fine implements used in microarray analysis.*

Computer-Assisted Design

The first step in making an accessory for a microarray motion-control system is to create a drawing of the device. Drawings in an electronic format are preferable to those made by traditional drafting procedures, because the former can be saved in suitable formats and downloaded directly into computers that control the micromachining devices (see below). Electronic file formats are also preferable because they allow rapid redesign once prototypes are made and tested. Because electronic drawings are often subjected to dozens of modifications during

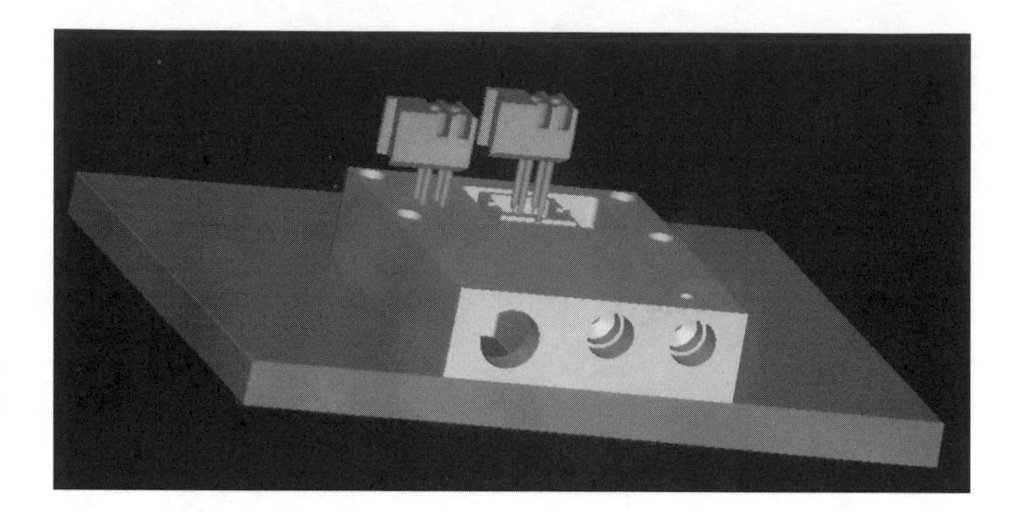

FIGURE 7.23. *A microarray wash station drawn as a 3D object using* ProE. *(Courtesy of R. Martinsky, TeleChem/Arrayit.com, Sunnyvale, CA.)*

the prototyping process, the ability to make and save changes on the computer greatly speeds the engineering process.

Computer assisted design. *Sophisticated electronic approach to generating two- and three-dimensional representations of objects used in micromachining and other manufacturing approaches.*

Solid modeling. *Computer-assisted design approach that allows the rendering of three-dimensional objects on a computer screen.*

Sophisticated electronic drawings are most commonly prepared by the use of ***computer-assisted design*** (CAD) software programs. CAD programs allow the engineer to convert an engineering concept or physical prototype into an electronic file. The pioneering CAD programs such as AutoCAD (AutoDesk, San Rafael, CA), allow the designer to prepare and visualize two-dimensional (2D) renderings of a device of interest. The more recent advent of ***solid modeling*** allows the design engineer to create and examine three-dimensional (3D) renderings on the computer screen. Because it is often difficult to transpose 2D objects into 3D ones, solid modeling technologies offer an advantage over two-dimensional CAD programs in terms of design speed and ease. Popular solid modeling programs include *Pro/ENGINEER* (*ProE*) (Parametric Technologies, Needham, MA) and SolidWorks (SolidWorks, Concord, MA). Figure 7.23 demonstrates the utility of solid modeling CAD programs in designing motion-control accessories.

Computer Numerical Control

Computer numerical control. *Sophisticated machining approach that uses computer drawings, stepper motors, and other technologies to prepare high-precision parts and implements used in microarray analysis.*

Tolerance. *Amount of error that can be tolerated in the manufacture of a component by micromachining, typically expressed as ± 0.0001″.*

Once suitable electronic files are created of an apparatus of interest, the part must be machined to exacting specifications. Precise micromachining using computer-generated drawings is commonly carried out using ***computer numerical control*** (CNC) devices. CNC machines, similar to microarray motion-control devices, use stepper motors and linear actuators to move machining tools in three-dimensional space so that a part of interest can be prepared using CAD files. The CNC machinist interfaces with the part via a simple GUI that runs on a personal computer (Fig. 7.24). The GUI allows the machinist to control laser cutters, routers, drills, mills, and lathes (Fig. 7.25) with a precision approaching a few microns. CNC devices provide microarray implements of extraordinary quality. In machine shop parlance, exactness or ***tolerance*** is often expressed in units of a mil; 1.0 mil equals one thousandth of an inch (0.001 in.). High-end CNC machines are capable of holding tolerances down to one tenth of a mil (0.0001 in.), or 2.5 μm. Some microarray printing devices (see below) are machined to tolerances in the 0.1 mil range.

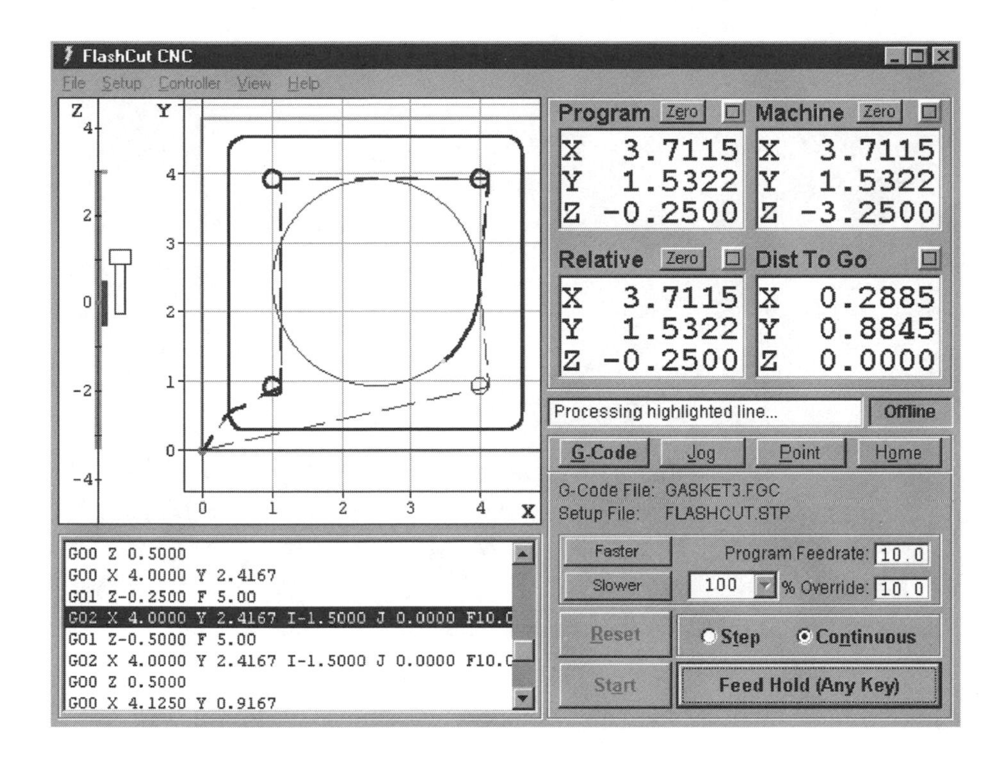

FIGURE 7.24. The GUI for a CNC machine. (Courtesy of FlashCut CNC, Menlo Park, CA.)

Electrical Discharge Machining

Another widely used engineering procedure is ***electrical discharge machining*** (EDM). The term EDM refers to any manufacturing procedure that employs a solid electric conductor (electrode) to cut metal using the sparks generated between the two surfaces. In the EDM process, the energy generated during the discharge of electricity causes the material to melt at the interface of the electrode and the part (Fig. 7.25). EDM devices use electricity to cut metal at very high (0.1 mil) tolerances and with speeds approaching 10 mm/s. Because a great deal of heat is generated during the EDM process, cutting is often performed in a water bath to dispel the heat.

Electrical discharge machining. *Manufacturing procedure that uses a solid electric conductor to cut metal by generating sparks between the two surfaces.*

CONTACT PRINTING

The methods of microarray manufacture known as ***contact printing*** include any technology that involves direct contact between the printing implement

Contact printing. *Microarray manufacturing technology that involves direct interaction between the printing implement or the sample and the microarray substrate.*

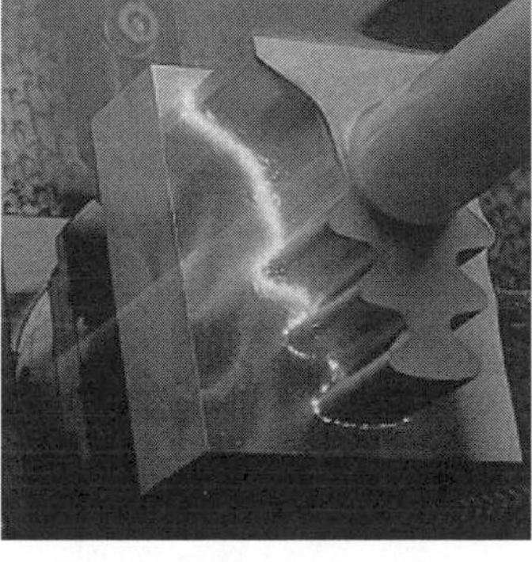

FIGURE 7.25. ***A,*** *CNC machining.* ***B,*** *The EDM process. (Courtesy of Mercatech, Dallas.)*

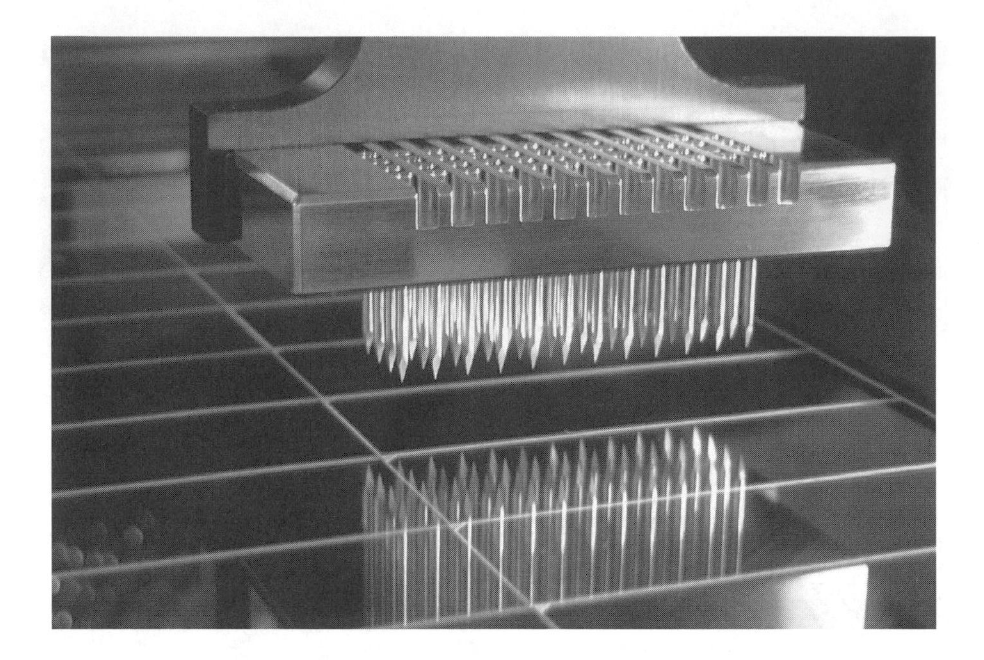

FIGURE 7.26. *A microspotting device consisting of a brass printhead and 48 pins. (Courtesy of T. Martinsky, TeleChem International, Sunnyvale, CA.)*

or the sample contained in the printing implement and the microarray substrate. Contact printing devices include a broad family of delivery implements such as microspotting pins, tweezers and split pins, capillary tubes, and ***Pin & Ring*** (PAR) devices. Of the three main categories of microarray manufacturing technology, contact printing is probably the most widely used by the research community. This section provides an overview of the various contact printing technologies, emphasizing design features, delivery mechanisms, performance specifications, and comparative discussions.

Pin and Ring. *Contact printing device that uses a circular loop to load sample by capillary action and a solid pin that moves up and down through the liquid in the loop to enable printing by direct contact with the microarray substrate.*

Micro Spotting. *Contact printing method that uses printing pins that contain a defined uptake channel and a flat or horizontally level tip; allows microarray printing without tapping the printing implement on the surface.*

Printhead. *Printing device that guides the movement of pins, ink jets, and other printing implements with high precision.*

Micro Spotting

Micro Spotting technology, the brainchild of Martinsky (TeleChem, Sunnyvale, CA), is a widely used microarray manufacturing method (Martinsky, 2000). This contact printing technology uses a motion-control system fitted with a holder or ***printhead*** and one or more Micro Spotting pins (Fig. 7.26) to prepare microarrays containing virtually any chemical or biological sample of interest. A digital movie showing the complete Micro Spotting process is available (http://arrayit.com/wiley/microspottingmovie). The apertures in the printhead that hold each pin are machined to tight tolerances, so that each pin rides against an air cushion that forms between the pin and the printhead bushing (Fig. 7.27). The air cushion provides a nearly frictionless environment that essentially eliminates pin wear and obviates the need for lubricants, the latter

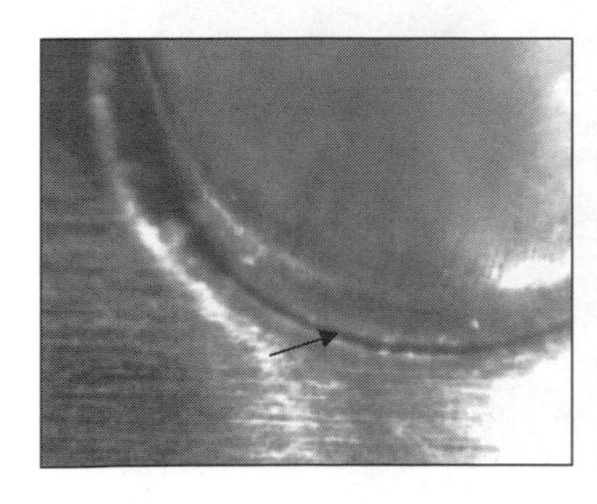

FIGURE 7.27. *A Micro Spotting pin bushing **(A)** and collar **(B)**. The arrow in A denotes the space ("air bushing") between the printhead and pin. (Courtesy of TeleChem International, Sunnyvale, CA.)*

TABLE 7.1. Stainless-Steel Compositions (% by weight) (Data adapted from The Hendrix Group, Houston.)

Alloy	C	Mn	P	S	Si	Cr	Ni	Mo	Other
201	0.15	6.5	0.06	0.03	1.0	17	4.5	—	0.25 N
303	0.15	2.0	0.2	0.15	1.0	18	9.0	0.6	—
409	0.08	1.0	0.045	0.045	1.0	11.5	35.5	—	0.48 Ti
416Se	0.15	1.25	0.06	0.06	1.0	13.0	—	—	0.15 Se
440C	1.0	1.0	0.04	0.03	1.0	17.0	—	—	—
2205	0.03	2.0	0.03	0.02	1.0	22	5.5	3.0	0.15 N

being an important advantage for high-end manufacturing environments such as cleanrooms. Square collars at the end of each pin ride up and down in a rectangular channel in the printhead, preventing rotation of the pins during the printing process and greatly improving printing precision. The printhead and pins are machined from brass and series-400 stainless steel, with the two materials providing superior gliding properties and durability, respectively. Grades of stainless steel vary considerably in terms of their chemical composition (Table 7.1) and properties; toughness and corrosion resistance are among the most important properties for Micro Spotting pins.

Micro Spotting pins are made by the use of EDM, and as such the pin tips hold tolerances to within a few tenths of a mil ($\leq$ 0.0002 in.). Tight pin tip tolerances ensure that the size of the printed spots and the amount of sample delivered by each pin exhibit a small coefficient of variation (C_V) from pin to pin. The coefficient of variation, a common statistical formula for comparing multiple independent trials, is expressed as a percentage by dividing the standard deviation (σ) by the mean ($\overline{X}$) as follows:

$$\frac{C_V = \sigma \times 100\%}{\overline{X}} \tag{7.6}$$

Typical C_V values for microspotting pins are in the 5–10% range, ensuring that microarrays manufactured with robots equipped with multiple pins show little variation from one region of the microarray to another. The C_V values for a single pin are well below 5% (Fig. 7.28).

The tip of each Micro Spotting pin contains a sample channel that holds a predetermined volume of sample. Because the tips are made by EDM, the sample channel can be customized to hold a specific amount of liquid; two common pins have uptake volumes of 0.2 and 0.6 μL, respectively (Fig. 7.29). A key design element is that each pin has a flat (horizontally level) tip (Fig. 7.30), which causes a thin layer of sample to form at the end of the tip upon sample loading. Because sample is present at the end of the tip, micro spotting pins work by an ink-stamping mechanism, whereby the sample at the end of the tip is pulled off the end of the pin by surface tension between the substrate and the sample (Fig. 7.31). This novel printing mechanism has the following important

FIGURE 7.28. A microarray printed with a single Micro Spotting pin. Scanned data are represented in a rainbow palette. The space bar = 125μm. (Data courtesy of M. Schena and R. W. Davis, Stanford University, Stanford, CA.)

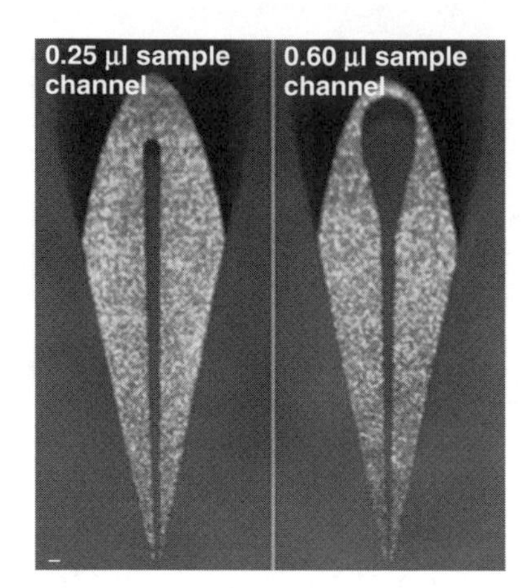

FIGURE 7.29. *Microspotting pin tips with uptake volumes of 0.25 µL and 0.6 µL. (Courtesy of TeleChem International, Sunnyvale, CA.)*

advantages over many other contact printing technologies: (1) the spot diameter is largely determined by the dimensions of the pin tip, which can be precisely controlled by EDM; (2) high-quality substrates produce precise spot diameters largely independent of the quality of the motion control system; (3) little or no contact between the pin and the surface is required for sample delivery, minimizing or eliminating impact forces and increasing pin longevity to millions of cycles; (4) the pins are capable of printing in any orientation, including horizontal and upside down, enabling custom manufacturing applications; and (5) pin movement in the downward ($+z$) direction is controlled by gravity instead of spring-loading, which minimizes surfaces forces and allows printing on delicate surfaces (e.g., acrylamide gel layers and silicon wafers).

Tweezers and Split Pins

Tweezer. *Early microarray printing implement that forms a meniscus on loading and requires a tapping force to break the meniscus and expel the sample onto the microarray printing surface.*

Another type of contact printing device, developed by Brown and co-workers at Stanford, uses ***tweezers*** (Fig. 7.32) as the sample-delivery mechanism (Brown and Shalon, 1998). The tweezers have two elongate members or prongs with pointed tips. Samples of interest are loaded into the tweezer by capillary action, and the sample forms a meniscus within the printing device. Because the sample is contained inside the tweezers, a tapping force is required to expel the sample out and onto the substrate. The tapping force is achieved by impacting the tweezers on the glass substrate during the printing process. Tweezers have the advantage of being inexpensive and relatively easy to implement in a research setting; they provided some of the first printed microarrays in the mid-1990s. The primary disadvantage of tweezers is that the tapping force required to expel the sample often leads to a blunting of the tweezers' points after several thousand tapping cycles, which causes the spot diameter to increase over time and necessitates resharpening at frequent intervals. The open capillary design

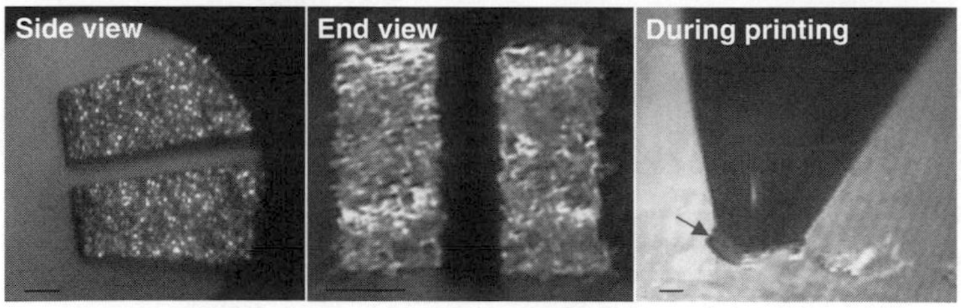

FIGURE 7.30. *Microspotting pins. The sample is between the end of the pin and the substrate* (arrow). *Space bar 50 µm. (Courtesy of TeleChem International, Sunnyvale, CA.)*

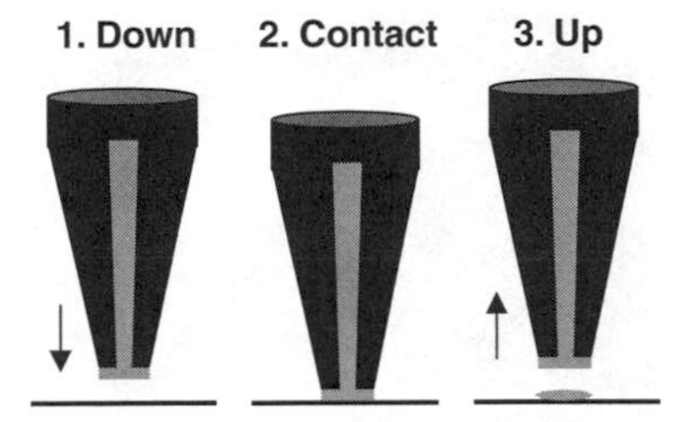

FIGURE 7.31. The ink-stamping mechanism consists of three movements dictated by the motion-control system: down, contact, and up. (Courtesy of TeleChem International, Sunnyvale, CA.)

also loads a variable sample volume, which depends on the viscosity and other properties of the solution; thus tweezers are difficult to implement and maintain in a serious manufacturing environment.

A mechanistically related device to tweezers is known as the ***split pin***. Split pins operate in a similar manner as tweezers except that split pins generally have shorter prongs. Split pins are somewhat easier to implement than tweezers and are relatively affordable. Their the main disadvantage is a lack of durability owing to the tapping force required for printing. Split pins enjoy some use in the research community, though the absence of flat tips compromises their effectiveness.

Split pin. *Tweezerlike printing implement that loads a volume of sample between two uneven points and enables microarray printing by tapping the implement on the substrate to expel the sample.*

Capillary Tubes

Capillary tubes represent the earliest microarray printing implement to use capillary action for sample loading. First used to print samples on substrates for chloramphenicol acetyl transferase (CAT) assays in the mid-1980s, capillary tubes are hollow cylinders that load samples by capillary action. Capillary tubes were first used in a microarray format in the early 1990s by Mirzabekov and co-workers at the Engelhardt Institute of Molecular Biology (Moscow). In this work, capillary tubes were used to transfer solutions of DNA onto acrylamide-coated glass surfaces. In the mid-1990s, capillary tube technology was used to print directly onto glass substrates in work by researchers at Molecular Dynamics, (Sunnyvale, CA) (Fig. 7.33). Capillary tubes are relatively easy to manufacture and are rather durable. One disadvantage of capillaries is that they are somewhat difficult to clean between printing cycles owing to the closed geometry of their structure, although advanced wash procedures largely overcome this problem. Capillary tube printing is currently used by dozens of laboratories, and the microarrays printed with such devices are of high quality (Fig. 7.34).

Capillary tube. *Enclosed, circular microarray printing implement that loads a sample by capillary action and delivers the sample onto the microarray substrate by direct contact.*

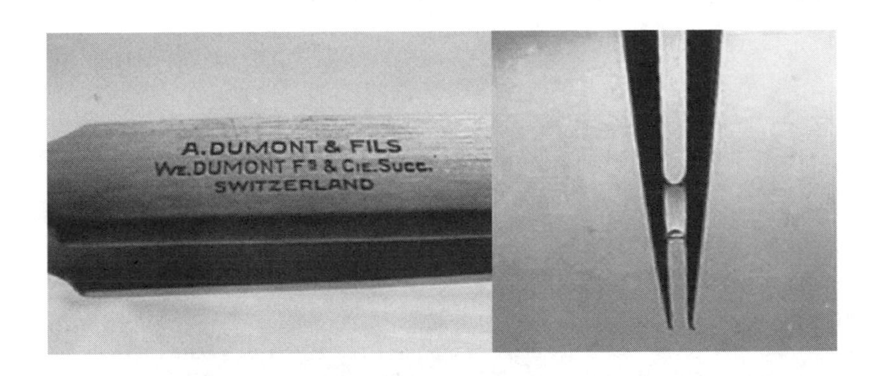

*FIGURE 7.32. Laboratory tweezers **(A)** loaded with a DNA sample **(B)**. (Courtesy of TeleChem International, Sunnyvale, CA.)*

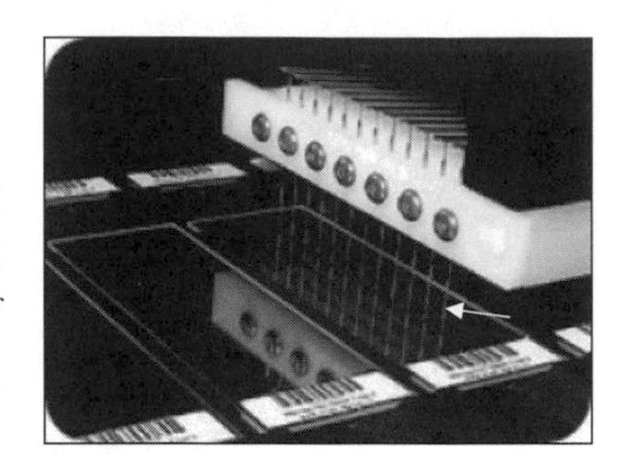

FIGURE 7.33. A capillary tube dispenser with 12 pens (arrow) fitted to a printhead and attached to a motion-control system. (Courtesy of Molecular Dynamics, Sunnyvale, CA, and Amersham Pharmacia Biotech Piscataway, NJ.)

Solid Pins

Solid pin. *Microarray printing implement containing a solid shaft and flat tip that holds sample on the end of the pin by adsorption and delivers sample onto the microarray substrate by direct contact.*

Array printing with ***solid pins*** was developed by Lehrach and co-workers at the Imperial Cancer Research Fund (London) in the mid-1980s. Solid pin printing uses steel pins that are dipped into a solution of interest and the sample that adheres to the end of the pin is transferred onto a printing membrane. Solid pins have also been used to print onto glass surfaces with moderate success. The main advantage of solid pins is that they are easy and inexpensive to manufacture and implement. The main disadvantage is that no more than one or a few spots can be printed with a single loading of sample, in contrast to microspotting pins that allow hundreds or thousands of spots to be printed from one loading. The large surface area of the solid pin also makes printing on glass somewhat unreliable, owing to the fact that the flat tip has a tendency to remove sample from the center of the spot during the up stroke. Solid pins enjoy moderate use for microarray applications.

Pin & Ring

PAR technology was developed by Montagu and co-workers at Genetic MicroSystems (Woburn, MA) in the mid-1990s. This ingenious microarray manufacturing method couples capillary rings and solid pins for sample uptake and delivery. The PAR ring is dipped into a solution of interest, and the sample loads into the ring by capillary action (Fig. 7.35). A solid pin is then propelled through the sample layer held in the ring, and the pendant droplet that adheres to the end of the pin is transferred onto the microarray printing surface. PAR has the advantage of allowing hundreds to thousands of spots to be printed with a single loading of the ring. One disadvantage of PAR is that the solid pins tend to produce somewhat irregular transfer volumes on solid surfaces, owing to the large surface area of the solid pin. PAR technology is used by dozens of laboratories worldwide and the quality of the microarrays produced is high.

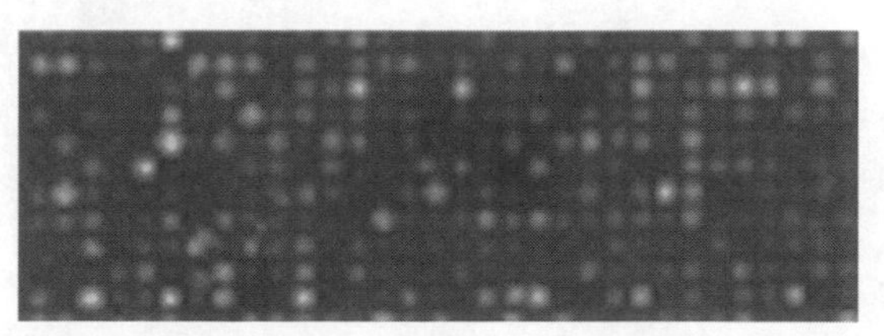

FIGURE 7.34. A microarray printed with a capillary tube dispenser. Data are represented in gray scale; the two channels correspond to two different samples (see Chapter 6). (Data courtesy of Molecular Dynamics, Sunnyvale, CA, and Amersham Pharmacia Biotech Piscataway, NJ.)

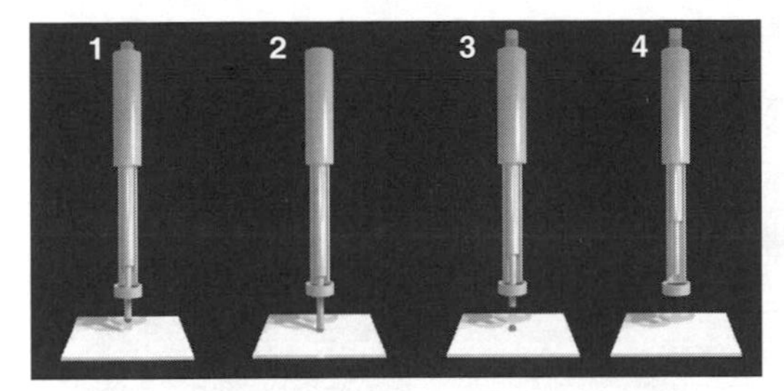

FIGURE 7.35. *For PAR, the sample of interest is loaded into the ring by capillary action and a pin is propelled through the sample, resulting in the attachment of a small pendant droplet to the end of the pin (1). The pin or the sample is then touched on the printing surface (2) and transferred by surface tension (3). The pin is then moved upward through the ring to initiate another round of printing (4). (Reprinted with permission from Eaton Publishing, Natick, MA.)*

NONCONTACT PRINTING

The microarray manufacturing methods known as ***noncontact*** printing include any technology that enables microarray printing without direct contact between the printing implement and the microarray substrate. Noncontact printing devices include ink-jets, which use piezoelectric crystals; microsolenoid-based systems; and thermal bubble jet delivery. Of the three main categories of microarray manufacturing technology, noncontact printing offers the most rapid delivery of a small number of samples onto solid surfaces. This section provides an overview of these technologies.

Noncontact printing. *Microarray manufacturing technology, including ink-jet printing, that does not require direct interaction between the printing implement or the sample and the microarray substrate.*

Piezoelectric

Piezoelectric dispensing systems take advantage of the unique property of certain materials, such as ceramics, to deform when subjected to an electric pulse. The deformation of the crystal, an effect discovered in the late 1800s by the Curie brothers, is caused by the displacement of ions in the crystal lattice. An element that confers the piezoelectric effect is known as a piezoelectric ***transducer***. If a piezoelectric transducer is fitted around a flexible capillary tube filled with a solution of interest, electric pulses can be used to create transient pressure waves inside the capillary tube, each of which results in the expulsion of a small volume of liquid out the nozzle. Because electrical pulses are used to actuate the transducer, piezoelectric dispensers can be used to fire > 60,000 droplets/s. Piezoelectric dispensers are used widely to propel ink onto paper surfaces and thus are commonly known as ***ink-jet*** dispensers. Popular color printers use ink-jet technology. A rapidly advancing technology forefront is providing extremely sophisticated ink-jet delivery systems (Fig. 7.36).

Piezoelectric. *Noncontact liquid-dispensing device that uses electrical pulses to deform a transducer material such as ceramic, and expels droplets in a controlled manner to achieve microarray printing.*

Transducer. *An element such as a ceramic coupling that confers the piezoelectric effect.*

Ink jet. *Noncontact liquid dispensing device that uses piezoelectric, microsolenoid, and other types of actuators to expel droplets in a controlled manner for microarray printing.*

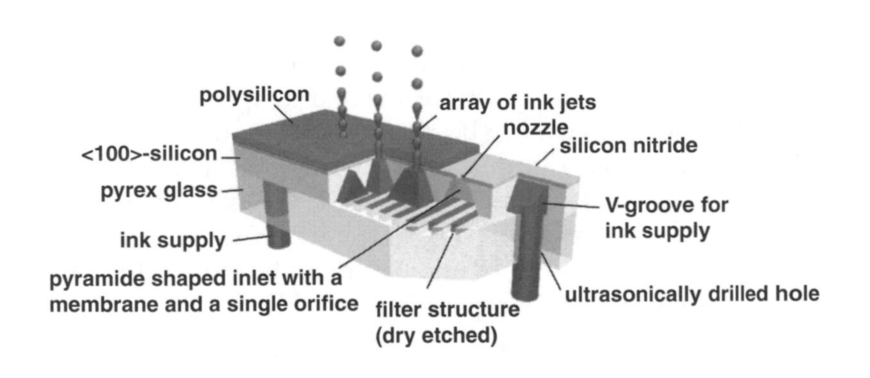

FIGURE 7.36. *A sophisticated ink-jet delivery system made by micromachining. (Courtesy of E. Obermeier and R. Jehring, Technology University of Berlin, Berlin.)*

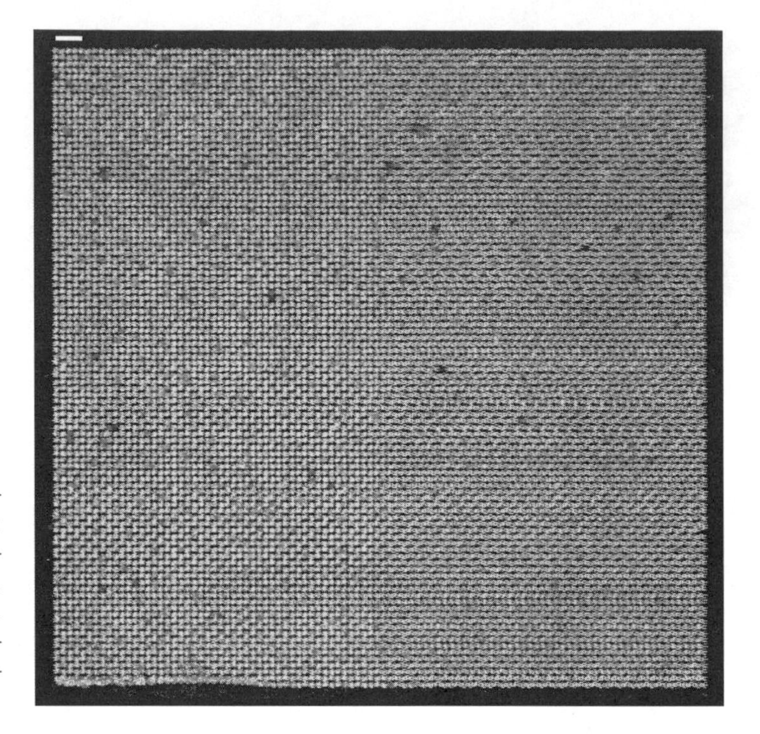

FIGURE 7.37. A microarray containing 10,000 DNA spots in a 100 × 100 configuration made with an ink-jet dispenser. The space bar = 500 μm. (Data courtesy of M. Schena and R. W. Davis, Stanford University, Stanford, CA, and T. P. Theriault, Combion, Pasadena, CA.)

The use of ink-jet technology for microarray manufacture (Theriault et. al. 1999, Rose 2000, Englert 2000) was adapted by several groups including Theriault and co-workers at Combion (Pasadena, CA) and Englert and co-workers at Packard Instrument (Meriden, CT). Piezo jets typically dispel sample volumes in the 300–500 pL range. Because the piezo jets print without contacting the microarray surface directly, suitable motion-control systems enable on-the-fly printing whereby a series of droplets can be transferred onto the surface without stopping and starting along the x–y axis. Printing with continuous movement enables microarrays containing thousands of spots of the same sample (Fig.7.37) to be manufactured in < 1 min with an ink-jet device. Ink-jet devices provide rapid droplet emission of one or a small number of samples and are extremely well suited for making chips that contain a small number of different target sequences.

Ink-jet approaches are also useful for combinatorial approaches (see Chapter 6) in which target elements are synthesized on the chip using the four DNA building blocks. In the combinatorial approaches, each DNA base is loaded into one of four ink-jet dispensers and target elements of the desired sequence are built on the chip by spraying one of the four bases at each microarray location. Because the chemically reactive DNA building blocks are relatively inexpensive, ink-jet approaches using the four DNA bases provide an affordable way to make oligonucleotide microarrays of any sequence of interest.

Because a suitable clearance (200–500 μm) is required to prevent the dispensers from hitting the printing surface, microarray spot sizes made with current ink jets (250–300 μm) tend to be somewhat larger those made with pins, because the sample spreads out in a circular fashion as it exits the piezoelectric nozzle. The recent preparation of ink-jet nozzles by micromachining has provided orifices as small as 5 μm (Fig. 7.38), which should ultimately provide extremely small feature sizes for microarray applications. The main advantage of ink-jet systems is rapid droplet delivery without surface contact. One

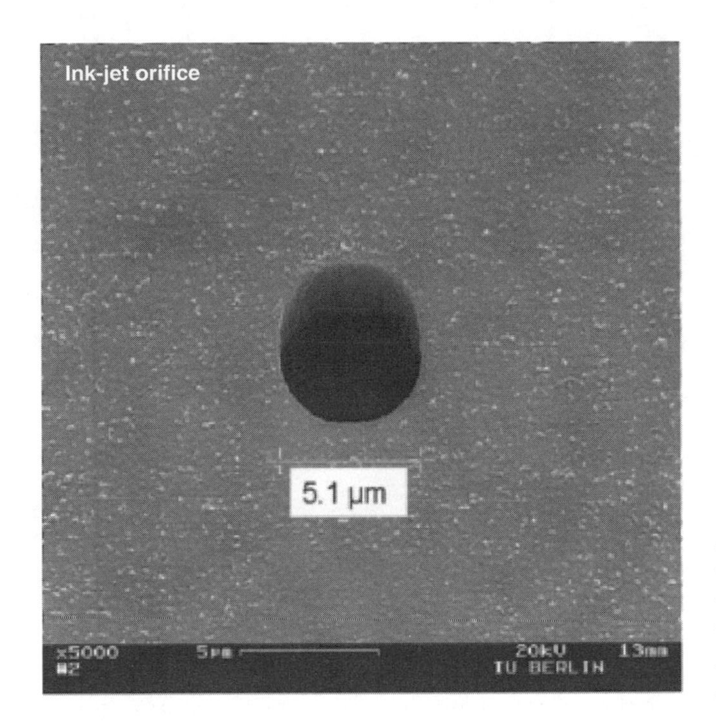

FIGURE 7.38. *A scanning electron micrograph of a 5-μm ink-jet orifice made by micromachining a silicon wafer. (Courtesy of E. Obermeier and R. Jehring, Technology University of Berlin, Berlin.)*

disadvantage of ink-jet systems is that they can be somewhat prone to clogging, though this problem can be largely overcome by using carefully filtered samples.

Microsolenoid

The key conceptual feature of all noncontact printing devices is the capacity to dispense small fluid volumes from a nozzle in a controlled manner. An ingenious ink-jet device using microsolenoid valves for fluid dispensing was developed by Tisone and co-workers (Rose, 2000) at Cartesian Technologies (Irvine, CA). The microsolenoid-based system has two key components (Fig. 7.39). A syringe pump is used to feed pressurized sample into an ink-jet nozzle fitted with a microsolenoid valve. When the microsolenoid valve is actuated by electrical pulses, the valve opens transiently and a small droplet of pressurized sample is released from the ink-jet orifice. By coupling the microsolenoid valve and syringe pump to a high-precision motion-control system, it is possible to move the ink-jet dispenser rapidly in three dimensions (Fig. 7.40). Sample volumes

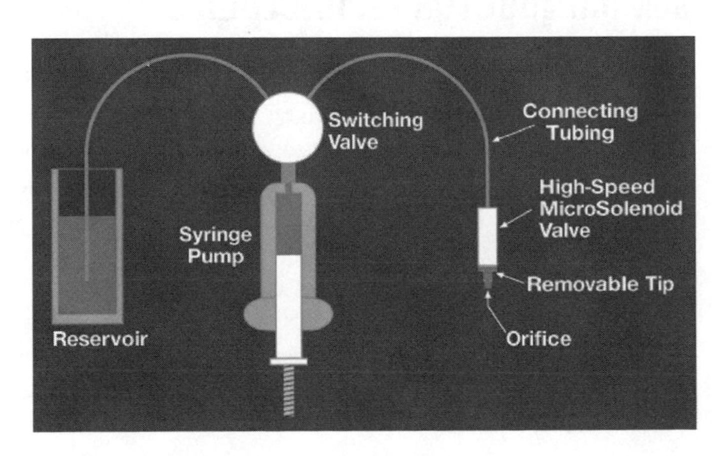

FIGURE 7.39. *An ink-jet device consisting of a microsolenoid valve and a syringe pump. (Courtesy of Cartesian Technologies, Irvine, CA.)*

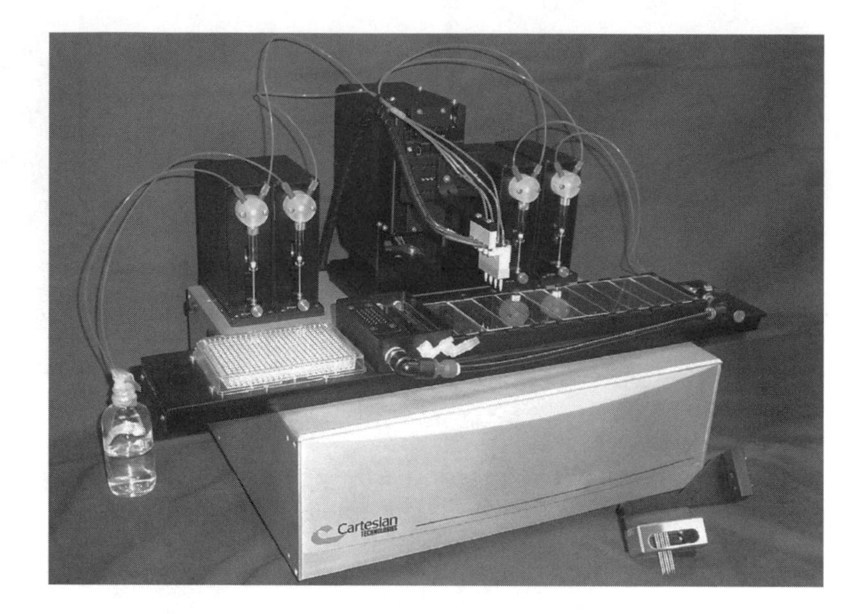

FIGURE 7.40. The MicroSys 5100, a microsolenoid-based ink-jet device attached to a motion-control system to allow movement in three dimensions. (Courtesy of Cartesian Technologies, Irvine, CA.)

for the microsolenoid-based systems are in the 1–10 nL range, providing microarray spots of ~300 μm in diameter. These systems are affordable, highly reproducible, and are used extensively in microarray and diagnostics settings.

Thermal Bubble Jet

Ink-jet systems based on thermal bubble-jet delivery were perfected at Canon (Tokyo, Japan) based on research carried out in the United States in the early 1970s. The basic principle of bubble-jet printing is that the viscosity of fluids can be made to vary with temperature. Pressurized fluid can, therefore, be expelled from delivery nozzles in a selective manner by raising the temperature of the sample, resulting in a decrease in viscosity. Thermal-based ink-jet devices use electrical currents to heat the ink in a transient manner. Because many biomolecules are heat sensitive, the use of bubble-jet printers for biological applications remains relatively unexplored, though use with phosphoramidite-based synthesis shows some success.

SEMICONDUCTOR TECHNOLOGIES

Semiconductor technology. *Any of the microarray manufacturing approaches, including photolithography, that uses technology developed in the computer chip industry for microarray manufacture.*

Combinatorial method. *Chemical synthesis approach that allows the generation of enormous chemical diversity by stepwise coupling of synthetic building blocks in a spatially addressable manner.*

Microarray manufacturing technologies that use miniaturization strategies from the computer chip industry are known as ***semiconductor technologies***. Though the microarrays themselves are not microprocessors, the strategies used to make the microarrays are the same as those used in the microelectronics industry. All of the semiconductor approaches use solid-phase synthesis to build the microarray on the chip in a stepwise manner using the four DNA bases. Because masks and bases can be combined in many different ways to achieve extensive synthesis diversity, the semiconductor approaches are commonly known as ***combinatorial methods***. This section describes the semiconductor microarray technologies, with a focus on manufacturing strategies and synthesis chemistry.

Photolithography

An ingenious adaptation from the microelectronics industry by Fodor and coworkers at Affymax (Santa Clara, CA), ***photolithography*** uses ultraviolet light and solid-phase chemical synthesis to manufacture microarrays (Fodor et. al. 1991). This light-based synthesis approach, developed in the early 1990s, has rapidly become one of the most widely used microarray manufacturing technologies. Affymetrix, which uses photolithography to make nucleic acid microarrays, sold > 200,000 chips made in this manner in the year 2000.

photolithography. *Method of microarray manufacture that uses a combination of modified phosphoramidite nucleotides, ultraviolet light, and photomasks to achieve solid phase chemical synthesis.*

Photolithography begins with a glass substrate modified with silane reagents to provide a surface containing reactive amine groups. The reactive amine groups are modified with a second reagent that contains a specific chemical group known as methylnitropiperonyloxycarbonyl (MeNPOC). The MeNPOC group is stable to a variety of chemical reagents but can be removed selectively by shining ultraviolet light on the chip for a period of about 30 s. MeNPOC groups prevents chemical reactions from taking place in the absence of ultraviolet light and are hence known as ***photoprotecting groups***. When the photoprotecting groups have been removed, the deprotected regions on the surface can react efficiently with a special family of DNA bases. Chemical bonds between the molecules on the chip surface and the DNA bases occur at the 3′ position of the deoxyribose, which contains a reactive phosphoramidite group (Fig. 7.41).

Photoprotecting group. *Chemical protecting group, such as methylnitropiperonyloxycarbonyl, used in microarray manufacture by photolithography that requires exposure to ultraviolet light for removal.*

When a DNA base attaches to the glass surface, the phenomenon is known as coupling. Each base that has undergone coupling has a MeNPOC photoprotecting group on its 5′ hydroxyl position (Fig. 7.41). The MeNPOC group on the coupled base can be removed by exposure to ultraviolet light, allowing

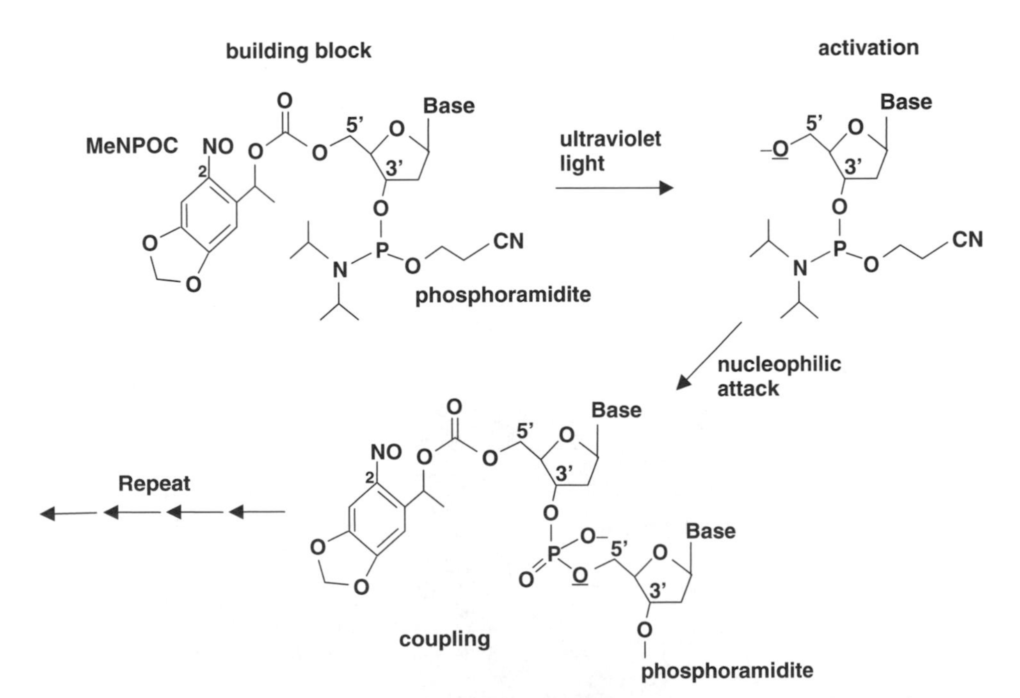

FIGURE 7.41. Photolithography uses DNA building blocks that contain an MeNPOC protecting group at the 5′ position and a phosphoramidite group at the 3′ position. Exposure to ultraviolet light results in the removal of the protecting group and activation of the base. The activated base contains a reactive oxygen anion (underlined) *that acts as a nucleophile, attacking the phosphoramidite position of a second base and coupling the first DNA base to the second. Repeated steps afford oligonucleotide microarrays.*

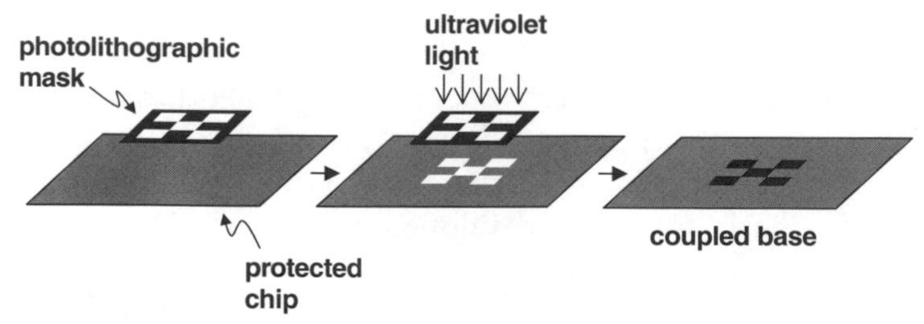

FIGURE 7.42. A glass substrate modified with photoprotecting groups is aligned with a checkerboard photomask that contains a chrome coating (black) *interspersed with clear regions. If the mask is exposed to ultraviolet light* (arrows), *select portions of the chip become deprotected and active for chemical coupling. The activated chip is flooded with a reactive DNA base, allowing coupling to selective regions* (solid) *defined by the photomask.*

coupling to a second base. A repeated series of deprotection and coupling steps allows oligonucleotides of any sequence to be synthesized on the glass surface in a stepwise manner.

Oligonucleotide synthesis can be directed to discrete locations on the microarray surface using photomasks, which allow the selective deprotection of certain regions of the chip. Photomasks are miniature chrome checkboards that are used in the semiconductor industry to manufacture microprocessors. The photomasks contain chrome-plated glass with interspersed clear regions (Fig. 7.42). Chrome prevents the passage of ultraviolet light, while the clear regions allow light to pass and impinge on the chip surface. Because the masks can be manufactured to contain any pattern of chrome and clear regions, it is possible direct light activation to any region of the chip in any desired order. A series of photomasks can thus be used to direct the stepwise synthesis of oligonucleotide microarrays containing any sequence of interest, with each mask directing the synthesis of one DNA base at each location on the substrate. The chrome checkerboards can be highly miniaturized, producing microarray features sizes in the 20–50 μm range. Affymetrix currently manufactures microarrays containing > 250,000 features/cm^2(Fig. 7.43). The 1.28-cm^2 chip is

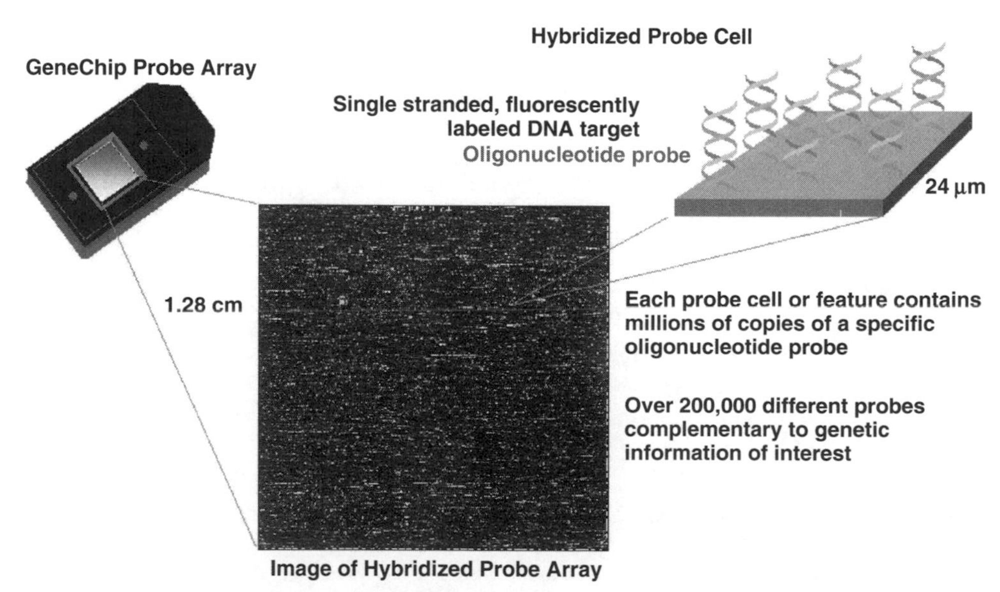

FIGURE 7.43. Aspects of photolithography, including a scanned microarray image displayed in a rainbow palette (center), *a hybridization graphic revealing the 24-µm feature size* (right), *and the hybridization enclosure that protects the chip* (left). *Microarray courtesy of Affymetrix, Santa Clara, CA.)*

FIGURE 7.44. The GeneChip (Affymetrix, Santa Clara, CA) includes a microarray enclosure that prevents the chip from being damaged and provides a convenient hybridization chamber.

enclosed in a plastic cassette that prevents it from being damaged and provides an extremely convenient hybridization chamber (Fig. 7.44).

The main advantage of photolithography over the contact and noncontact printing methods is that microarrays of any sequence can be constructed from a set of four nucleotide building blocks (A, G, C, and T). The capacity to use a simple set of reagents instead of preparing and storing an individual sample for each microarray element, is a big advantage, particularly when complex arrays are involved. The main disadvantage of photolithography is that microarray composition is restricted to relatively short oligonucleotides (< 30 nucleotides). The photomasks and chips are also rather expensive to manufacture, though photolithography probably provides the most economical means of obtaining large numbers of chips representing whole genomes.

Micromirrors

Micromirror. *Microscopic reflective device configured in a large array that is used to selectively direct light onto a microarray substrate during the digital light processing manufacturing approach.*

The use of microscopic mirrors or ***micromirrors*** for biological applications is a sensational adaptation of digital light processing (DLP) technology developed by Texas Instruments (Dallas) and Raytheon Digital Display Group (Plano, TX). DLP technology uses a digital micromirror device (DMD), which is a silicon chip that contains a vast array of microscopic aluminum mirrors mounted on a standard memory substrate (Fig. 7.45). Each mirror, positioned on a hinge and seated on top of an individual memory bit, can be deflected into either of two positions (off or on) by electrostatic forces that cause a $-10°$ or $+10°$ tilt of each mirror relative to the substrate (Fig. 7.46). Electrostatic deflections are determined by the binary state of each memory bit (0 or 1), which in turn is computer controlled. Micromirrors in the $+10°$ orientation reflect incident light into the light path, and mirrors in the $-10°$ position do not (Fig. 7.47). By using a DMD that contains as many as 2,000,000 individual micromirrors, appropriate software and hardware configurations produce extremely high-resolution projected images.

The adaptation of micromirror technology from video projectors to microarray manufacturing was achieved by researchers at the University of Wisconsin at Madison (Singh-Gasson et. al 1999). Maskless array synthesizer (MAS) technology uses micromirrors to direct an ultraviolet light beam to

FIGURE 7.45. *A DMD used for DLP. (Courtesy of Texas Instruments Dallas.)*

discrete locations on a glass substrate. The location of a given micromirror on the DMD corresponds to a single microarray feature on the glass surface. The glass substrate is activated for DNA synthesis using binary information sent from the computer, which directs each mirror to initiate synthesis at a discrete microarray location. Analogous to the photolithography approach, it is possible to use the maskless array synthesizer to build high-density microarrays of any desired oligonucleotide sequence. Though still at an early stage of development, the micromirror strategy is appealing because it obviates the need for photomasks, which are expensive and time consuming to manufacture. In a test study, the Wisconsin group was able to manufacture an oligonucleotide microarray containing 76,000 discrete microarray elements, each measuring 16 μm wide for a microarray density of $\sim$390,000 features/cm^2. In principle, the micromirror approach would allow a researcher to create any microarray of interest, simply by downloading sequence data from the Internet and using the sequence information to send instructions via computer to the micromirrors and DNA synthesizer. The recent availability of the complete human genome sequence makes the micromirror approach an exciting prospect.

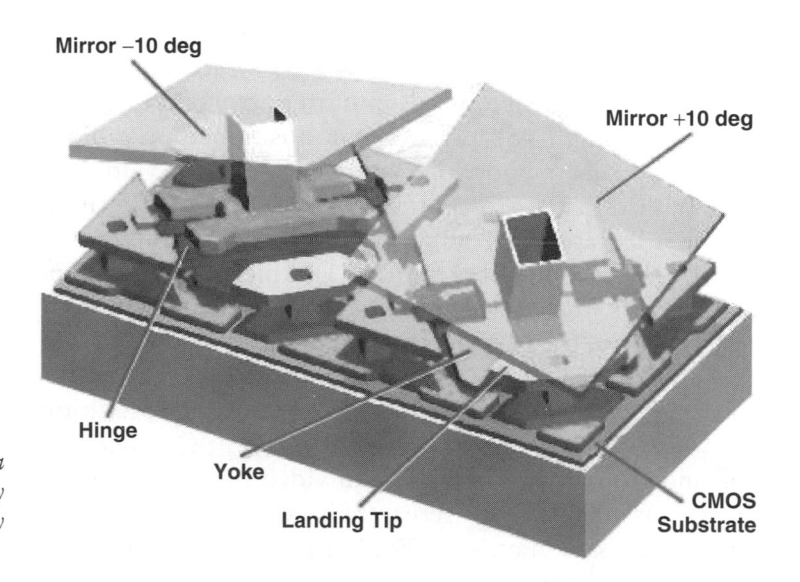

FIGURE 7.46. *Two pixels of a DMD.* CMOS, *complementary metal oxide semiconductor. (Courtesy of Texas Instruments, Dallas.)*

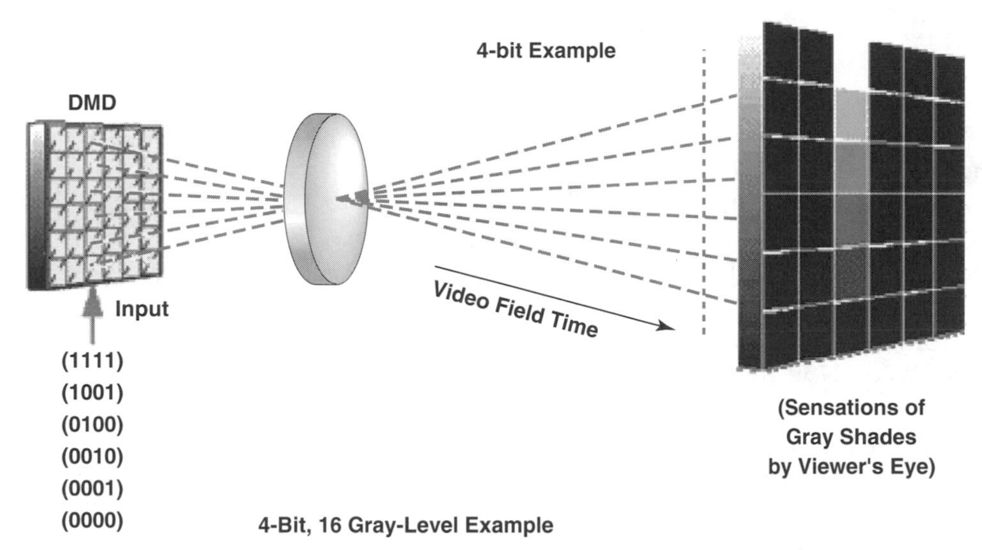

FIGURE 7.47. Binary signals sent by the computer determine the positions of the mirrors within the DMD, causing a specific pattern of reflected light. Filters in the light path allow the projection of a black-and-white image. (Courtesy of Raytheon Digital Display Group, Plano, TX.)

SUMMARY

Modern microarray manufacturing interfaces biology and engineering. Different methods can be evaluated with a set of criteria that includes affordability, content, density, feature size, feature purity, feature reactivity, regularity, ease of implementation, and throughput. Motion-control technology consisting of computer-controlled robots that allow 3D movement, is at the center of microarray manufacturing. Stepper motors and servomotors, coupled with linear actuators, convert electromagnetic forces into linear movement. Motion-control systems are evaluated on the basis of speed, resolution, accuracy, precision, repeatability, backlash, durability, and load capabilities. Linear encoders and optical encoders can be used to provide additional repeatability and precision during motion control. Software programs based on LabVIEW, Visual Basic, C++, and other programming languages connect the user to the robot via a GUI. Fuzzy logic and artificial intelligence provide exciting horizons on the software front. Moving platen and overhead gantries are the most common robot configurations, with microrobotic systems emerging because of their affordability. Accessories for manufacturing are made by micromachining approaches that use CAD, CNC, and EDM, to provide parts of remarkable quality.

The three main microarray manufacturing methods are contact printing, noncontact printing, and the semiconductor technologies. Contact printing uses direct contact between the printing implement or the sample contained in the implement and the microarray substrate for microarray manufacture. Microspotting pins, tweezers, split pins, capillary tubes, solid pins and PAR are the most common contact printing devices. Noncontact printing methods, which allow on-the-fly printing, include ink-jet technologies based on piezoelectric, microsolenoid technology, and thermal bubble jet dispensers. The semiconductor technologies are based on strategies used widely in the microelectronics industry. Photolithographic masks and micromirrors allow the stepwise manufacture of oligonucleotide microarrays at densities approaching 400,000 features/cm^2.

SELECTED READING

Benson, H. *University Physics,* rev. ed. Wiley, New York, 1996.

Brown, P. O., Shalon, T. D. Methods of fabricating microarrays of biological samples. United States Patent, 5, 807, 522, 1998.

Cawsey, A. *The Essence of Artificial Intelligence*. Prentice Hall, Upper Saddle River, NJ, 1998.

Ehrlich, R., Tuszynski, J., Roelofs, L., and Stoner, R. *Electricity and Magnetism Simulations*. Wiley, New York, 1995.

Englert, D. Production of microarrays on porous substrates using noncontact piezoelectric dispensing. In microarray Biochip Technology (M. Schena, ed), Eaton Publishing, Natick, MA, pp. 231–246, 2000.

Fodor, S. P., Read, J. L., Pirrung, M. C., Stryer, L., Lu, A. T., Solas, D. Light-directed, spatially addressable parallel chemical synthesis. Science 251, 767–773, 1991.

Giancoli, D. C. *Physics for Scientists and Engineers,* 3rd ed. Prentice Hall, Upper Saddle River, NJ, 2000.

Guitrau, E. B. *The EDM Handbook*. Hanser Gardner, Cincinnati, OH, 1997.

Halliday, D., Resnick, R., and Walker, J. *Fundamentals of Physics,* 6th ed. Wiley, New York, 2001.

Hearn, D., and Baker, M. P. *Computer Graphics, C Version,* 2nd ed. Prentice Hall, Upper Saddle River, NJ, 1996.

Hughes, A. *Electric Motors and Drives: Fundamentals, Types and Applications*. Butterworth-Heinemann, Woburn, MA, 1993.

Kaiser, J. *Electrical Power: Motors, Controls, Generators, Transformers*. Goodheart-Wilcox, Tinley Park, IL, 1998.

Khrapko, K. R., Lysov, Yu. P., Khorlin, A. A., Ivanov, I. B., Yershov, G. M., Vasilenko, S. K., Florentiev, V. L., Mirzabekov, A. D. A method for DNA sequencing by hybridization with oligonucleotide matrix. DNA Seq. 1:375–388, 1991.

Martin, F. G. *Robotic Explorations: A Hands-on Introduction to Engineering*. Prentice Hall, Upper Saddle River, NJ, 2001.

Martinsky, R. S. Microarray printing device including printing pins with flat tips and exterior channel and method of manufacture. United States Patent 6, 101, 946, 2000.

Morrison, R. *Solving Interference Problems in Electronics*. Wiley, New York, 1995.

Nishi, Y., and Doering, R., eds. *Handbook of Semiconductor Manufacturing Technology*. Marcel Dekker, New York, 2000.

Purcell, E. M. *Electricity and Magnetism,* Vol. 2, 2nd ed. McGraw-Hill, New York, 1985.

Razavi, B. *Design of Analog CMOS Integrated Circuits*. McGraw-Hill, New York, 2000.

Rose, D. Microfluidic technologies and instrumentation for printing DNA microarrays. In microarray Biochip Technology (M. Schena, ed.), Eaton Publishing, Natick, MA, pp. 19–38, 2000.

Schena, M., ed. *DNA Microarrays: A Practical Approach,* 2nd ed. Oxford University Press, Oxford, UK, 2000.

Schena, M., ed. *Microarray Biochip Technology*. Eaton, Natick, MA, 2000.

Singh-Easson, S., Green, R. D., Yue, Y., Nelson, C., Blattner, F., Sussman, M. R., Cerrina, F. Maskless fabrication of light-directed oligonucleotide microarrays using a digital micromirror array. Nat. Biotechnol. 17, 974–978, 1999.

Stroustrup, B. *The C++ Programming Language,* 3rd ed. Addison-Wesley, Glenview, IL, 2000.

Takashi, K., and Sugawara, A. *Stepping Motors and Their Microprocessor Controls*. Oxford University Press, Oxford, UK, 1994.

Theriault, T. P., Winder, S. C., Gample, R. C. Application of ink-jet printing technology to the manufacture of molecular arrays. In DNA microarrays: A Practical Approach, Oxford University Press, Oxford, UK and, pp. 101–120, 1999.

Valentino, J. V., Goldenberg, J. *Introduction to Computer Numerical Control (CNC),* 2nd ed. Prentice Hall, Upper Saddle River, NJ, 1999.

Wenckebach, W. T. *Essentials of Semiconductor Physics*. Wiley, New York, 1999.

Worley, J., Bechtol, K., Penn, S., Roach, D., Hanzel, D., Trounstine, M., Barker, D. A Systems Approach to Fabricating and Analyzing DNA Microarrays. In Microarray Biochip Technology, M. Schena (editor), Eaton Publishing, Natick, MA, pp. 65–85, 2000.

Yen, J., and Langari, R. *Fuzzy Logic: Intelligence, Control, and Information*. Prentice Hall, Upper Saddle River, NJ, 1999.

REVIEW QUESTIONS

1. Which of the following technologies have been used for microarray manufacture: (a) pins, (b) photolithography, (c) micromirrors, (d) ink jets, (e) pin and ring, and (f) tweezers.

2. Which chip contains the greatest amount of biochemical content: (a) 75,000 features representing one gene, (b) 7,500 features representing 10 genes, (c) 100 features representing 100 genes, (d) 300 features representing 100 genes?
3. A 26K microarray contains 26,450 features in a 9.7 cm^2 area. Calculate the spot density.
4. A 100-μm microarray spot was printed with 400 picoliters (pL) of a 60mer oligonucleotide at 50 μM concentration. Assuming a 30% coupling efficiency, calculate the approximate target density in oligonucleotides per μm^2 on the substrate.
5. Which of the following can impair the regularity of a printed microarray: (a) uneven surface properties of the substrate, (b) imprecise motion control of the printing robot, and (c) static electricity in the printing environment.
6. Mechanical movement in a stepper motor is achieved with what fundamental force? One word will suffice.
7. A linear actuator converts circular motion into what type of motion? One word will suffice.
8. A researcher notices that her printed microarrays have uneven rows and columns of printed spots. Which of the following microarrayer components could account for the diminished performance: (a) defective printhead, (b) linear encoder malfunction, (c) substrate movement on the platen during printing, or (d) use of human cDNAs instead of rat cDNAs?
9. An overhead gantry robot moves along which of the following axes: (a) x axis only, (b) y and z axes, (c) x, y, and z axes?
10. A computer is used in which of the following engineering and microarray procedures: (a) CAD, (b) CNC, and (c) ink-jet printing?
11. Which of the following would be expected to affect the printing efficiency of a microspotting pin: (a) pin collar composition, (b) pin tip flatness, (c) sample buffer composition, and (d) substrate hydrophobicity?
12. Which of the following are used in noncontact microarray printing methods: (a) solid pin (b) bubble jet, (c) pin and ring, (d) microsolenoid valve, and (e) piezoelectric transducer?
13. The MeNPOC protecting group in photolithography coupling chemistry absorbs light in the ultraviolet range. Based on the structure of the MeNPOC group and using your knowledge of photon absorption, explain why MeNPOC absorbs ultraviolet light and not visible light.
14. Which of the following are used in microarray motion control software: (a) fuzzy logic, (b) LabVIEW, (c) C^{++}, (d) GUI, and (e) Visual Basic?
15. Rank the following dimensions from smallest to largest: (a) 1.0 μm, (b) 1.0 mil, (c) 2,000 nm, and (d) 0.002″.

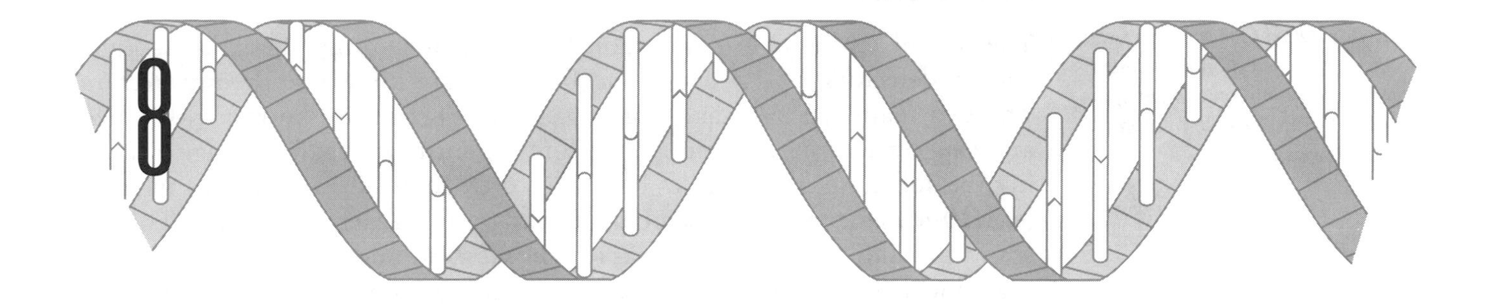

Microarray Detection

> Nothing tends so much to the advancement of knowledge as the application of a new instrument. —*Sir Humphrey Davy*

Microarray detection instruments are revolutionizing biology. These ingenious devices allow microarrays to be read in an automated fashion, providing an amount of data in a few minutes that would have taken months or even years to acquire with antecedent technologies. Most of these instruments use fluorescent detection schemes, which obviates the need for radioactivity, allows multiple biological samples to be examined on a single microarray, and provides a level of detectivity that enables the researcher to study even the rarest human transcripts. This chapter covers all of the main topics in modern microarray detection technology, including the evaluative criteria used to benchmark microarray systems, signal and noise, the physics of electromagnetic radiation, fluorescence and fluorescent labels, optical filters and lasers, detectors, and scanning and imaging architectures.

EVALUATIVE CRITERIA

Detection is a key step in microarray analysis. In this critical phase of the experimental process, probe molecules bound to target molecules at each microarray location must be discerned and quantified across the entire microarray. The detection step results in the procuring of a microarray image, which is essentially a photograph of the microarray. The microarray image provides a snapshot of the types and quantities of molecules that have reacted with the microarray and

Architecture. *Microarray detection system criterion that refers to the overall layout of the filters, lasers, light source, detectors, optics, stage, and other hardware components.*

Scanner. *One of the two categories of microarray detection instrument that relies on rapid movement of the optics, substrate, or both to capture a microarray image.*

Imager. *One of the two categories of microarray detection instrument that relies on stationary optics and substrate to capture large microarray images in single imaging steps.*

Moving substrate. *Microarray detection instrument design that employs physical translocation of the microarray substrate or slide to acquire an image.*

Moving optics. *Microarray detection instrument design that employs physical translocation of the laser path, lens, filters, or other optical components to acquire a microarray image.*

hence are present in the biological sample. This section explores the key criteria used to evaluate microarray detection systems.

Architecture

The overall layout of the detection hardware, including filters, lasers, light source, detectors, optics, stage, and footprint are known as the ***architecture.*** The two main types of detection systems at present are known as ***scanners*** and ***imagers.*** Microarray scanners capture the entire microarray image by moving the substrate or the optics back and forth across the microarray in small (e.g., 10-μm) increments, assembling the image from each line of data. The two scanner architectures are known as ***moving substrate*** and ***moving optics,*** depending on which component of the scanner is in motion relative to the stationary instrument. Most scanning systems use lasers as a source of excitation light and appropriate filter sets and photomultiplier tubes as detectors. Microarray imagers collect microarray data by taking snapshots of relatively large portions of the microarray (e.g., 1 cm^2), without moving either the substrate or the optics during the imaging step. The individual images are then combined into a single large composite image to provide complete data for an entire microarray. The most common hardware configuration in microarray imaging systems includes a white light source for excitation, appropriate filter sets, and a charge-coupled device (CCD) or complementary metal oxide semiconductor (CMOS) as the detector. At present, scanners outnumber imagers in the field by about 10:1.

Speed. *Magnitude of travel per unit time of an electric motor or microarray robot, expressed typically in centimeters per second. Alternatively, a microarray detection system criterion that refers to the area of microarray substrate that can be captured per unit time at a given resolution, expressed typically as square centimeters per minute.*

Speed

The ***speed*** of a detection system is a measure of the amount of microarray area captured per unit time at a given resolution. Early prototype scanning devices acquired microarray data at a rate of about 50 mm^2/min at 10 μm resolution. Current scanners are capable of detection speeds approaching 500 mm^2/min at 10 μm resolution, with an average system running at about 300 mm^2/min. At a scanning speed of 300 mm^2/min, an entire 20 $\times$ 72-mm microarray can be scanned in about 5 min. Most imaging devices are able to capture a 1.0 cm^2 frame of microarray data in about 10 s, with some additional time needed to move between frames and consolidate the image. Typical imaging devices, similar to scanners, produce final detection speeds in range of 100–500 mm^2/min at 10 μm resolution. Because most detection systems can provide information for >10,000 genes in less than 5 min, detection speeds are rarely seen as a limiting factor in microarray analysis.

Throughput. *Criterion of microarray manufacture that pertains to the number of microarrays that can be manufactured per unit time, typically expressed as the number of chips per day. Alternatively, the microarray detection system criterion that refers to the number of microarrays of a defined size that can be scanned or imaged at a defined resolution per unit time, typically expressed as microarrays per hour.*

Throughput

The criterion of ***throughput*** refers to the number of microarrays that can be scanned or imaged per unit time for a defined area and resolution. For microarrays of 1.0 cm^2 with a desired resolution of 10 μm, most detection systems provide a throughput in the range of 25–50 chips/h. Automatic substrate loaders, which load and unload the microarrays from the detection system in an unattended fashion, can be used to achieve an even higher throughput. In cases of large scan areas and multiple detection wavelengths, autoloading devices can be particularly helpful. Though throughput is not a major consideration for most research applications, drug screening and diagnostics applications require large numbers of chips and thus provide a driving force for greater detection throughput.

Precision

Microarray detection systems must be able to provide reliable and consistent images from minute to minute and day to day, a criterion known as ***precision.*** An ideal detection system is one that would provide an identical image for a single microarray detected in two successive trials. In practice, a common precision benchmark is $\pm$ 10% variation in absolute signal intensity for a single chip detected on two separate occasions, a standard pretty easily met by most current detection systems. Common contributors to the loss of precision include variations in the power of the excitation source, fluctuations in the efficiency of the detectors, and lack of consistent focusing owing to mechanical imprecision of the substrate relative to the collecting lens.

Precision. *Range of deviations that occur for 95% of the movements of an electric motor or microarray robot for a given input, typically expressed in microns. Alternatively, a microarray detection system criterion that refers to the consistency of microarray images acquired during different detection sessions.*

Repeatability

Repeatability is a measure of the line-to-line accuracy of a scanning device; the common benchmark is in the 1–2% range. The criterion of repeatability determines the sharpness of a microarray image, with repeatability deviations manifesting as jagged microarray spots in a phenomenon known as pixel jitter. If the repeatability of one line to another exceeds one pixel (e.g., $> 10 \mu m$), the microarray features take on a poorly defined appearance that can complicate data quantitation and analysis. The most common contributors to repeatability problems include poor substrate registration relative to the light path, inaccuracies in moving stages and improperly secured substrates. Sophisticated fiducials, suitable x–y stages, and advanced substrate holders all but eliminate repeatability problems. Because imaging systems use both a stationary substrate and stationary optics to capture a given microarray data frame, repeatability is not a significant issue with the imaging architecture.

Repeatability. *Reliability of achieving a specified position of an electric motor or microarray robot over repeated attempts, typically expressed in microns. Alternatively, a microarray detection system criterion that refers to the line-to-line accuracy of a scanning device.*

Resolution

A microarray image is a two-dimensional numerical representation in which intensity values are stored in a digital format in individual bins known as picture elements or ***pixels.*** A pixel is a square area representation of an image, easily viewed if the image is enlarged on the computer screen. The ***resolution*** of a microarray detection system or of the corresponding image is defined as the pixel size as measured in microns. The greater the number of pixels, the greater the image resolution or clarity. Early microarray detection systems had resolution settings in the 10–20 μm range, with more recent instruments boasting 3–5 μm resolution. A detection system with 10 μm resolution would mean that a given pixel would have a width of 10 μm, corresponding to 10,000 pixels (100 $\times$ 100) per mm^2. A common benchmark is that a given microarray spot should be represented by no fewer than 8 pixels per dimension or 64 total pixels per spot. As the resolution approaches the spot diameter, the spots appear grainy and pixilated, and the quantitation of the microarray data is impaired. Resolution can be estimated for a given detection system if the spot size is known, simply

Pixel. *Numerical bin or picture element used to store data in a graphics computer file.*

Resolution. *Smallest detectable movement increment of an electric motor or microarray robot, typically expressed in microns. Alternatively, a microarray detection system criterion that refers to the pixel size in a microarray image, expressed in microns.*

by enlarging the microarray image on the computer screen and counting the number of pixels across the spot diameter. For imaging technologies in which the detector has a defined number of pixels, the resolution defines the imaging area. A CCD device with a 1,000,000-pixel (1 megapixel) imaging chip reading at 10 μm resolution would produce an image measuring 1,000 $\times$ 1,000 pixels, which would correspond to a microarray area of 10,000 $\times$ 10,000 μm (1.0 cm^2).

Dynamic Range

Dynamic range. *Microarray detection system criterion that refers to the full gamet of signal intensities that can be discerned with a given detection instrument, typically 100,000-fold or more.*

The ***dynamic range*** of a microarray detection system is defined as the full gamut of signal intensities that can be discerned with a given detection instrument. Dynamic range is usually represented as the quotient of the brightest and dimmest signals that can detected over background; a 1,000-fold dynamic range within a given image is a common benchmark for most systems. A microarray image with a 1,000-fold dynamic range would have signal intensity values spanning three orders of magnitude, which for 16-bit data (see below) would correspond to absolute numbers ranging from 66 to 65,536. It should be clear that reducing the background signal can greatly extend the dynamic range, with each 2-fold reduction in background adding an additional 2-fold to the dynamic range. In the example above, reducing the background by 2-fold would allow the detection of signal values ranging from 33 to 65, 536 and a 2,000-fold dynamic range.

Because the actual signal intensities on a microarray often span four or five orders of magnitude, it is desirable to obtain a dynamic range beyond the three orders typically provided in a given image. Gene expression experiments place a particular premium on the value of dynamic range because abundant and rare transcripts often differ by as much as 100,000-fold in concentration. Additional dynamic range can be obtained by capturing two or more images of the same chip using different instrument settings. Microarray detection instruments enable the user to capture images of different exposures by adjusting the laser power, light intensity, detector settings, number of scans, and the like. By capturing two or more images of the same microarray at different settings, most contemporary detection systems provide a dynamic range approaching an impressive 100,000-fold, or five orders of magnitude.

Sensitivity

Sensitivity. *Microarray detection system criterion that refers to the efficiency by which a microarray detection device, such as a photomultiplier tube, converts photons from a fluorescent microarray into an electronic signal.*

Gain. *Amplification factor produced by a photomultiplier tube or other detector in a microaray detection instrument.*

Sensitivity refers to the efficiency by which a microarray detection device converts photons from a fluorescent microarray into electronic signal. Often confused with detectivity, the sensitivity of a detection system is defined as the output signal (i.e., electrons) divided by the input signal (i.e., photons) at a given wavelength. The sensitivity of an instrument provides a measure of instrument ***gain*** and can be expressed as an absolute value, such as the minimum number of detectable photoelectrons. Systems that use photomultiplier tubes often allow the user to adjust the sensitivity or gain of the instrument through the software that controls the scanner. Because microarray assays often involve a wide range of fluorescent signal intensities, the capacity to adjust the scanner sensitivity allows the user to capture the full gamut of signals from given microarray, even in cases in which spot intensities vary by more than five orders of magnitude.

Detectivity

Sometimes incorrectly referred to as sensitivity, ***detectivity*** is the extremely important criterion that defines the dimmest signal that can be detected over the total noise of the system (see below). Because fluorescence is the most common detection medium for microarray assays, detectivity is often expressed as the number of fluorescent molecules (fluors) per square micron of microarray substrate that can be detected above the background noise. Early microarray detection instruments provided detectivity in the range of 1 fluor/μm^2, whereas the most advanced systems at present provide detectivity in the range of 0.1–0.01 fluors/μm^2. An instrument that provides a detectivity of 0.1 fluors/μm^2 for a microarray spot measuring 90 μm in diameter (6,278 μm^2) would allow the experimentalist to detect ~630 flours/90 μm spot (Appendix G). If a 1000 nucleotide cDNA contains 10 fluorescent labels on average, a detectivity limit of 0.1 fluors/μm^2 would allow the detection of ~60 hybridized molecules in a 90-μm microarray spot.

Detectivity. *Microarray detection system criterion that refers to the dimmest signal that can be detected over the total noise of the system, typically expressed in fluors per square micron.*

Because light detection devices such as photomultiplier tubes allow detection of a single photon, the detectivity limit is rarely determined by the instrument, but rather by the amount of total noise generated by the microarray substrate and the sample. Because the substrate and sample almost always provide the major sources of noise and thus determine the detectivity limit, tremendous effort has been made over the past few years to reduce the background fluorescence of substrates and the nonspecific reactivity of fluorescent probes with surfaces. Microarray experimentalists routinely see 10-fold greater detectivity presently than in years past, owing mainly to improvements in surface chemistry, probe preparation and reaction chemistry. A 10-fold increase in detectivity is significant in that it allows the user to detect transcripts and other biomolecules of interest that have a 10-fold lower concentration. Because many important human genes are expressed at low levels, greater detectivity facilitates exploration of the human genome.

Uniformity

The ***uniformity*** of a detection system is a measure of the consistency of the recorded signal at different locations across a perfect substrate, with a common benchmark being ±10%. Though a perfect substrate is not obtainable, the atomically flat microarray substrates provide good approximations. A simple test for uniformity is achieved by recoding the background intensity across a 10 × 25-mm portion of an atomically flat microarray substrate. Average intensity values within each 5 × 5-mm portion should vary by < 10%, and many advanced detection systems provide uniformity in the 2–10% range.

Uniformity. *Microarray detection system criterion that refers to the consistency of the recorded fluorescent signal at different locations across a perfect substrate, typically measured as a percentage. Alternatively, a microarray cleanroom environmental control parameter that refers to the extent to which the temperature, humidity, and lighting are maintained evenly in the cleanroom space.*

A telltale sign of nonuniformity is a systematic gradient of background intensity along the x or y axis in the case of scanners or in a concentric pattern in the case of imagers. The so-called 180° test provides a simple way to access nonuniformity in scanning systems and is performed as follows. Insert a microarray substrate into the scanning device and record the background intensity over a 10-mm-long × 22-mm-wide portion of the substrate. If an intensity gradient is observed, remove the substrate from the instrument, rotate the substrate by 180°, reinsert it into the scanner, record the background intensity over the same area, and compare the images obtained from the two scans. If the gradient takes the same form in the two scans (left to right or top to bottom), then the nonuniformity is due to the scanner. If the gradient takes an opposite form in the two experiments, then the nonuniformity is attributable to the substrate.

The most common cause of nonuniformity for scanning systems is an uneven distance between the substrate and the detection lens, usually attributable either to a poorly constructed substrate holder or to low-quality substrates. This is a particularly pressing issue for confocal designs (see below) in which the depth of focus is often in the range of 10–30 μm, placing a premium on advanced engineering and atomically flat surfaces. The most common source of nonuniformity among imaging systems is due to the uneven power density of excitation light, generally owing to the fact that some flood illumination systems fail to provide an equal light intensity to all portions of the detection field (see below). Proper adjustments in the light path can correct nonuniformity problems in imaging systems that use flood illumination. Uniformity is an important criterion in all types of detection systems because it ensures that variations in signal readings are due to microarray signals and not to complications involving detection hardware or substrates.

Number of Channels

Number of channels. *Microarray detection system criterion that refers to the number of different wavelengths of light that can be detected with a given instrument.*

An important criterion for detection instruments is the number of different labels that can be read by a given device often, expressed as the ***number of channels*** supported by the instrument. Because the number of channels determines the number of different samples that can be examined on a given microarray, this criterion is an important feature among detection instruments. Some early home-built detection devices offered only a single channel, hence allowing analysis of only one sample at a time. Most of the next generation devices offered two channels, allowing a control sample and a test sample to be examined simultaneously on a single microarray. Recent systems offer five or six different channels, encompassing genotyping applications in which a different channel is devoted to each of the four DNA bases. For most microarray applications, three different channels are suitable for comparing two samples; the third channel providing a control for target molecule density.

Because nearly all microarray systems use fluorescence, the number of channels that can be supported by a given detection instrument is largely determined by number of separable wavelengths of light that can be obtained within the visible spectrum (400–700 nm). Both the Stoke shift and the tailing off of all emission spectra (see below) necessitate a ~50 nm emission frequency difference between two channels to obtain distinct signals. As the difference in emission wavelengths of two signals decreases below 50 nm, cross-talk between the channels (see below) begins to dominate the detection process.

Cross-Talk

Cross-talk. *Microarray detection system criterion that refers to the unwanted phenomenon in which signal from one channel is detected in another channel, measured typically as a percentage.*

The unwanted phenomenon in which the signal intended for one channel spills over into another channel is known as ***cross-talk.*** Because fluorescence emission is used in most microarray detection systems, the term optical cross-talk is sometimes used to describe this undesired carryover. Cross-talk can occur in experiments in which multiple labels are reacted with a single chip, and the researcher wants to measure the signal from each label using separate channels in the detection device. Because the emission wavelength for a given dye is longer than the excitation wavelength and because the emission spectra often extend over 50–100 nm, emitted light or signal from one channel can be erroneously detected in a second channel. A common benchmark for cross-talk is $\leq$1% from channel to channel, with $\leq$ 10% absolutely essential for any detection system designed to read multiple labels.

Cross-talk can be minimized using a number of strategies, including prudent choice of lasers, excitation filters, emission filters, and dye sets. For imaging systems that use continuous white light sources for excitation, cross-talk can be particularly problematic though avoided with appropriate light path designs. The importance of minimizing cross-talk can not be overstated, particularly in gene expression experiments in which the ratio of a given species in two separate samples can vary by several orders of magnitude. Detection instruments that suffer from excessive cross-talk will yield erroneous readings in ratiometric experiments, leading to falsely elevated signals for species whose concentration falls below the cross-talk threshold.

Sample Degradation

Microarrays reacted with samples or probe solutions often contain labels whose molecules can be damaged during the detection step; moreover, some two-step labeling procedures (see Chapter 6) include the use of labels with limited stability. The undesired process of damaging sample or probe molecules is known as ***sample degradation.*** Because fluorescence dominants microarray detection, the most common type of sample degradation is ***photobleaching.*** The process of photobleaching (see below) occurs when excessive illumination damages the fluorescent labels such that the absolute signal decreases in successive readings of a microarray. A typical benchmark for photobleaching is $\leq$10% from reading to reading, with <1% being desirable.

Sample degradation. *Unwanted phenomenon in which probe molecules are damaged by photobleaching or some other means during the detection process, typically measured as a percentage of the total molecules.*

Photobleaching. *Specialized form of sample degradation in which fluorescent molecules are damaged by exposure to excessive excitation light during the detection process, typically measured as a percentage.*

In early microarray detection systems, photobleaching was as high as 200–300% from one reading to another. In modern systems, photobleaching is generally in the 10% range or lower, depending on the illumination intensity and dye type. Because photobleaching leads to erroneously low signals in the microarray detection step, systems should be designed to minimize photobleaching as much as possible. Photobleaching is a particular killer in ratiometric experiments, because different dyes often exhibit differential photostabilities depending on the details of their molecular structures. Photobleaching can thus lead to false readings in one- and two-color applications. To minimize the negative consequences of photobleaching, users should use as little illumination possible for a given experiment, minimize the number of readings taken on a given microarray and avoid reading one portion of a chip more times than another portion.

Field Size

Field size pertains both to the size of the physical area that can be read in a given detection step and to the overall area that can be read on a standard microarray substrate using a detection instrument. Most scanners can read most or all of a standard 25 × 76-mm substrate in a single detection step, providing a microarray image that contains signal output for the entire substrate. Earlier scanners were generally limited to scanning physical areas of 15–22 mm wide and 40–60 mm long. Most imaging systems, particularly those that use CCD- or CMOS-based cameras, are generally restricted to reading a smaller area such

Field size. *Microarray detection system criterion that refers to the size of the physical area that can be read in a given detection step or the overall area that can be read on a standard microarray substrate using a given detection instrument, measured typically in millimeters.*

as 1–4 cm^2, though most imaging systems can be adjusted between readings to collect signal from different portions of the substrate. In the latter case, data clips from multiple images can be assembled by software to provide a composite image corresponding to the entire substrate.

Because the physical format of the printed microarray can vary considerably from lab to lab and vendor to vendor, detection systems that provide larger field sizes are preferred over systems that provide smaller field sizes. With the Affymetrix-style chips, the relatively small 1.6 cm^2 microarray area is read with a proprietary scanner designed to read the Affymetrix cassette.

Footprint

Footprint. *Physical width, height, and length of a microarray instrument, typically measured in centimeters.*

The ***footprint*** of a detection system refers to the physical size of the device as measured in three dimensions. Some of the early microarray breadboard systems measured 180 × 90 × 30 cm (width, depth, height) and had a total weight in excess of 200 kg! This massive footprint essentially precluded these devices from being used in most laboratory settings, and the mass was such that portability was nearly impossible. More recent systems have footprints closer to 30 × 60 × 30 cm, with weights in the range of 15–40 kg. Because bench space is limited and portability is an important feature, smaller is always better with respect to microarray detection devices, as long as there is no compromise with respect to performance.

Format

Format. *Microarray detection system criterion that refers to the physical dimensions of the microarray substrate that can read with a given detection instrument.*

The ***format*** that a detection instrument supports refers to the physical dimensions of the microarray substrate that it can read. The most common research format for microarrays is that of the microscope slide, which measures approximately 25 mm wide × 76 mm long × 1 mm thick. This format is supported by >90% of all currently used detection instruments. Because the microscope slide format can vary in physical size by 1 mm in length and width and by as much as 0.2 mm in thickness, depending on the manufacturer, the substrate holder supplied with detection devices that support this format should allow for slight variation in the substrate dimensions. Another common format is the Affymetrix GeneChip, which is a proprietary cassette supported by the Affymetrix detection station. The GeneChip is a square plastic cassette that houses the microarray and measures approximately 30 mm wide × 50 mm long × 10 mm thick. Formats are an important consideration because they provide standards that help members of the growing microarray community share and exchange instruments, chips, data, and ideas.

SIGNAL AND NOISE

Signal. *Numerical output from a microarray assay acquired by the detection instrument, and corresponding to the true experimental data.*

Noise. *Numerical output acquired by the detection instrument, corresponding to background fluorescence, dark current, shot noise and other nondata components of the assay.*

Signal in a microarray experiment is defined as the desired output in the assay, whereas ***noise*** is defined as the sum of unwanted contributions to the instrument readings. Because most microarray assays are based on fluorescent signals, the signal in nearly all microarray experiments derives from emission of fluorescent light from tags attached to the probe molecules. Noise, on the other hand, has many different origins (see below). The quotient of signal and total noise is known as the ***signal-to-noise ratio.*** Because the information in microarray assays in contained within the signal, one goal in all microarray experiments is to maximize the ratio of signal to noise.

Signal-to-noise ratio. *Quotient of signal and total noise in the microarray detection process.*

Signal Determinants

Intrinsic. *One of the three signal determinants in the microarray detection process, corresponding to the signal inherent to the labels attached to the probe molecules.*

Though the source of signal is much simpler to understand than the vast number of contributors to noise, total signal in a microarray analysis actually has three different determinants: ***intrinsic, extrinsic,*** and ***quantity.*** The intrinsic signal determinants are those that are inherent to the labels used on the probe molecules. Because nearly all microarray assays use fluorescent labels, intrinsic signal determinants include the physical properties of the fluorescent dyes, such as molar extinction coefficient and quantum yield (see below). To maximize microarray signals, prudent selection of dyes and other types of labels with superior intrinsic properties is key.

Extrinsic. *One of the three signal determinants in the microarray detection process, corresponding to the light source, excitation wavelength, pH, and other instrument and environmental contributions to signal.*

Extrinsic signal determinants are external contributors to signal, the most pertinent in microarray assays being instrument determinants and environmental determinants. With fluorescent detection instruments, some of the key instrument determinants of signal include the power of the light source, excitation wavelength, detection dwell time, and efficiency of the light path, detector, and analog-to-digital converter (see below). The main environmental determinants include the polarity of the solvent, pH of the buffer, presence of quenching species, and extent of energy transfer between adjacent dye molecules. Because most microarray substrates are detected in the dry state, solvent polarity, pH and presence of quenching species are fairly minor environmental signal determinants. Energy transfer or ***self-quenching,*** on the other hand, can exert a rather major effect on signal. Because energy transfer increases with the increasing proximity of dye molecules to each other, there can be a nonlinear relationship between probe quantity (see below) and signal output. To obtain maximum microarray signals, all of the main extrinsic signal determinants, including instrument and environmental contributions, need to be optimized.

Quantity. *One of the three signal determinants in the microarray detection process, corresponding to the number of probe molecules and labels present at a given microarray location.*

Self-quenching. *Unwanted detection phenomenon in which emission signal is lost by energy transfer between adjacent fluorescent molecules, resulting in a diminished signal.*

Microarray signal is also determined by the quantity of the label present at a given position on the microarray. Quantity determinants include the number of probe molecules bound and the number of labels present per bound probe molecule. The number of labels attached to a given probe molecule is sometimes referred to as the ***specific activity*** of the probe. Because the specific activity can vary greatly depending on the labeling scheme (see Chapter 6), there is not a 1:1 relationship between the number of molecules bound and the number of labels present at a given microarray location. Under conditions of target excess (see Chapter 5), a greater probe concentration results in a larger number of probe molecules binding to the surface. Other considerations being equal, stronger microarray signals are always observed with greater probe concentrations and higher specific activities and the researcher should always endeavor to maximize the quantity determinants to achieve maximum signal. With fluorescent labeling schemes, nonlinearity can be observed at high dye concentrations due to energy transfer between adjacent dye molecules (see below).

Specific activity. *Number of labels or fluorescent tags attached per probe molecule.*

Total Noise

Total noise. *Sum of all of the nondata numerical outputs acquired by the detection instrument, including background fluorescence, dark current, shot noise, and other unwanted sources.*

Total noise in microarray detection is defined as the sum of all unwanted contributions to the instrument readings. There are many different sources of noise, with the two types being instrument noise and microarray noise. Instrument noise includes dark current, electronic noise, shot noise, and optical noise. Microarray noise includes substrate noise and sample noise. Each of these sources of noise is explained below.

Dark current. *Instrument noise that originates in the absence of light, deriving mainly from thermal emissions and from current leaks, and measured typically in electrons per pixel per second at a given temperature.*

Dark count. *Numerical value in a microarray detector produced by dark current.*

Dark Current

Dark current, as the term implies, is instrument noise that originates in the absence of light. Dark current, or ***dark count,*** originates from the instrument detector and derives mainly from thermal emissions and from current leaks, usually measured in electrons (e^-) per pixel per second at a given temperature. All instrument detectors including photomultiplier tubes (PMTs), CCD cameras, and CMOS cameras exhibit measurable dark current. PMTs and cameras that have very low dark current ratings are obviously the devices of choice for microarray detection systems, which endeavor to provide the greatest detectivity possible. Because scanning systems acquire data over a given area very rapidly, dark current is a minor consideration for most modern scanners and can be nearly negated through the proper choice of PMTs. Dark current can be a major source of noise for imaging systems, which require up to 60 s to read a given area. For this reason, most CCD- and CMOS-based imaging systems used cameras cooled down to as low as −50°C to reduce dark current noise, with high-quality cooled cameras providing ratings in the range of 0.5–2.0 e^-/pixel/s (see below).

Electronic noise. *Unwanted contribution to detection instrument readings arising from the nondetector electrical components of the detection system notably the amplifiers, circuitry and analog-to-digital converter.*

Electronic Noise

A second source of instrument noise is ***electronic noise,*** which arises from the nondetector electrical components of the detection system, notably the amplifiers, circuitry, and analog-to-digital converter. For most microarray detection systems, electronic noise contributes less instrument noise than dark current.

Shot noise. *Unwanted contribution to detection instrument readings arising from the fundamental process of current flow, which corresponds to the discrete movement of electrons rather than a continuous flow process.*

Shot Noise

Shot noise is unwanted signal that derives from the fundamental process of electrical current flow, which corresponds to the discrete movement of electrons rather than a continuous flow process. Because microarray detection systems are light based, shot noise more precisely derives from the fact that electrical flow is determined by the emission of photons from fluorescent sources, which fundamentally consist of particles rather than continuous beams. Shot noise is generally modeled as increasing as the square root of signal intensity and as such can be minimized by running the instruments at fairly high sensitivity settings.

Optical noise. *Unwanted contribution to detection instrument readings, arising from reflected light from the substrate holder, spurious reflections from instrument enclosures, light leaks impinging on the detectors, cosmic rays, and other light-dependent sources.*

Optical Noise

Optical noise refers to all components of instrument noise that require light, excluding shot noise. The most common sources of optical noise include reflected light from the substrate holder, spurious reflections from instrument enclosures, light leaks impinging on the detectors, and cosmic rays. In properly designed systems, optical noise can be greatly minimized but not eliminated.

Microarray noise. *Unwanted contribution to detection instrument readings arising from the substrate, cross-reactivity of the sample, and any other chip-based source.*

Microarray Noise

The second main contributor to the total noise in a detection system is ***microarray noise,*** which includes all of the noninstrument noise components of the system. Microarray noise consists of the chip-based sources that include substrate noise and sample noise. In some of the early detection instruments, instrument noise was a major source of noise in the system. In most modern systems, the chief component of total noise derives from microarray noise, placing great importance on high-quality surfaces and reaction chemistries.

Substrate Noise

One of the two components of microarray noise is known as ***substrate noise.*** Noise from substrates derives either from the substrate material itself or from the surface treatment or surface coating that is applied to the substrate. Because most microarray substrates are made of glass, the inherent properties of transparency and low intrinsic fluorescence render glass substrates minimal contributors to substrate noise. Other substrate materials, including plastics and reflective metals, may present a considerable source of substrate noise in the system.

Substrate noise. *Unwanted contribution to detection instrument readings arising from surface reflection, organic treatments, surface coatings, or any other surface-dependent source.*

The noise contributed by surface treatments and surface coatings are generally the main source of substrate noise. The intrinsic fluorescence of different organic treatments can vary by more than three orders of magnitude, resulting in major research and developments efforts to reduce the noise contributed by the surface treatment. The high-quality organoamine and organoaldehyde surfaces (see Chapter 5) generally contribute less than twofold noise above and beyond the contribution of glass itself and therefore enable remarkable detectivity when used with an appropriate detection instrument. Some of the older organic surface treatments produced noise levels that were 1000-fold greater than glass and thus reduced detectivity greatly. Gel and nitrocellulose coatings (see Chapter 15) generally produce somewhat greater noise than the organic surface treatments, though microarray analysis with these surfaces can be successfully implemented by making adjustments to the instrument settings.

Sample Noise

Sample noise represents the second component of microarray noise. Noise from samples is introduced by the targets, probes, or solutions used to dissolve these components. Because most target molecules and target buffers are nonfluorescent, microarray targets generally contribute little to the sample noise. The main component of sample noise far and away, is attributable to the fluorescent probe molecules. Labeled probe solutions can react in a nonspecific manner with the surface. This nonspecific sticking of probe molecules to the surface can mask the productive interactions between targets and probes, obscuring the microarray signal. The noise attributed to nonspecific interactions between probe molecules and the microarray surface is known commonly as ***background.***

Sample noise. *Unwanted contribution to detection instrument readings arising from nonspecific attachment of probe molecules, probe reflection, salt crystals, or any other probe- or buffer-dependent source.*

Background. *Unwanted contribution to detection instrument readings arising from nonspecific interactions between probe molecules and the microarray surface.*

Background fluorescence reduces the signal-to-noise ratio by elevating the noise and, therefore, compromises microarray assay detectivity. Of all of the sources of noise in microarray systems, background noise contributed by nonspecific probe molecule interactions with the surface generally constitutes the main component. Vast resources have been devoted over the past few years to reducing background noise, with major successes in new blocking schemes, labeling procedures, and reaction and wash chemistries. At present, it is possible to perform microarray analyses with instrument controls adjusted to the highest settings of lasers and detectors. At these instrument settings, it is possible to detect a few dozen molecules bound to a single microarray spot.

FLUORESCENCE

Fluorescence is not the only detection scheme that can be used for microarray analysis, but it is by far the most widely employed; and most of the current commercial microarray detectors use fluorescence as the signal source. This section provides an overview of fluorescence, focusing on the physical phenomenon, advantages, types of molecules and labels used, photobleaching, and example data.

Phenomenon of Fluorescence

The first recorded observations of fluorescence were made by the German priest Athanasius Kircher in the 1640s, when he noticed that aqueous extracts of the wood *Lignum nephriticum* produced an unexplained blue color under certain illumination conditions. Kircher's observations were extended and fortified by Isaac Newton in the 1670s, as part of Newton's spectacular descriptions of the spectral properties of white light. In the 1850s, Cambridge professor George Stokes discovered that the fluorescence emission wavelength is longer than the excitation wavelength. The first fluorescence microscopes were of German origin and appeared in the early 1900s.

Any discussion of fluorescence must begin with a physical understanding of visible light. Visible light is actually one component of a broad spectrum of electromagnetic radiation whose wavelengths span 15 orders of magnitude (Fig. 8.1). The visible light spectrum occupies wavelengths from 400 to 700 nm and is located between the ultraviolet and infrared frequencies (Fig. 8.2). There are six easily recognized colors in the visible spectrum, with violet, blue, green, yellow, orange, and red representing increasing wavelengths from 400 to 700 nm. Light and all other forms of electromagnetic radiation are commonly modeled in two ways: as waves and as particles. When discussing beams of light, such as those emitted from lasers, the wave model provides a useful representation (Fig. 8.3). Light of a single wavelength is commonly known as monochromatic light. When discussing the function of photomultiplier tubes and other low-light detectors, it is often useful to think of light in the form of individual particles or ***photons.***

Photon. *Particle form of light.*

The wave and particle models of light have a number of useful associated formulae that help us understand the relationships between frequency, wavelength, and energy. The frequency of a light beam (ν) measured in cycles/s, or Hz, is equal to the speed of light in a vacuum ($c = 3.0 \times 10^{17}$ nm/s) divided by the wavelength (λ) measured in nanometers:

$$\nu = \frac{c}{\lambda} \tag{8.1}$$

As can be seen in Equation 8.1, the number of oscillations per second (ν) of a beam of light increase as the wavelength decreases, and vice versa. The

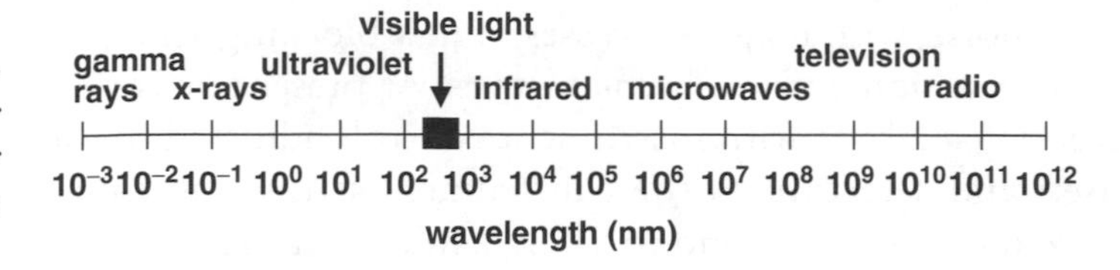

FIGURE 8.1. Shorter electromagnetic radiation wavelengths have higher energy and frequency than longer wavelengths, which have lower energy and frequency.

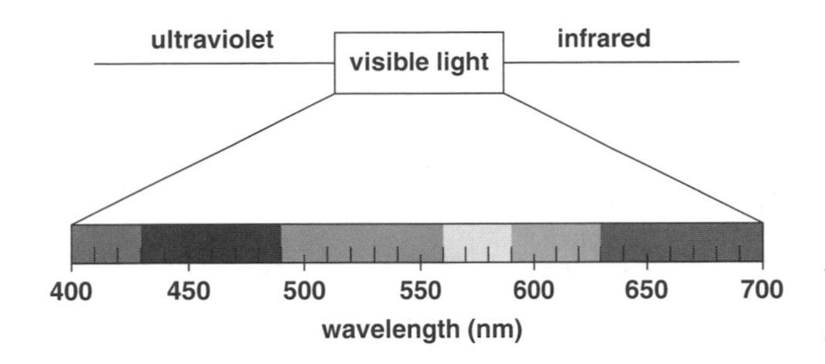

FIGURE 8.2. *The six colors of the visible light spectrum that are easily discernible to the human eye.*

energy of a given source of photons (E) is equal to Planck's constant ($h = 6.6 \times 10^{-34}$ J s) times the frequency (ν):

$$E = h \times \nu \tag{8.2}$$

One consequence of Equation 8.2 is that a photon with greater frequency possesses greater energy than a photon of lesser frequency. By combining Equations 8.1 and 8.2 and solving for λ, the wavelength of a photon is equal to Planck's constant (h) times the velocity of light (c) divided by the photon's energy (E):

$$\lambda = \frac{h \times c}{E} \tag{8.3}$$

Equation 8.3 shows that photons of lesser energy have a longer wavelength than photons with greater energy, a fact that underlies the Stokes shift (see below).

Certain molecules, known as ***fluorophores,*** are capable of absorbing light because of their special configuration of electrons (see Chapter 2). The process by which light is captured by a molecule is known as ***absorption.*** The absorption process can be studied in solution using an equation known as ***Beer's Law,*** which defines absorption (A) as the product of the molar extinction coefficient (ε) per mole per centimeter times the path length (b) in centimeters, times the concentration of the fluorescent molecule (C) in moles:

$$A = \varepsilon \times b \times C \tag{8.4}$$

Equation 8.4 is pertinent to microarray experiments in that it emphasizes that absorption by fluorescent molecules increases with increasing dye concentration and with larger molar extinction coefficients. The ε values for fluorescent dyes can vary considerably, ranging from ~1,000 to 250,000 M^{-1} cm^{-1}, and dyes with large ε values are highly desirable for microarray applications. The famous cyanine-3 (Cy3) and cyanine-5 (Cy5) microarray dyes from Amersham Pharmacia Biotech (Piscataway, NJ) have ε values of 150,000 M^{-1} cm^{-1} and 250,000 M^{-1} cm^{-1}, respectively.

Fluorophore. *Specialized class of organic molecules capable of light absorption and fluorescence emission owing to their conjugated configuration of π electrons.*

Absorption. *Process by which light is captured by a molecule.*

Beer's law. *Mathematical equation that represents light absorption as the product of the molar extinction coefficient times the path length times the concentration of the fluorescent molecule: $A = \varepsilon \times b \times C$.*

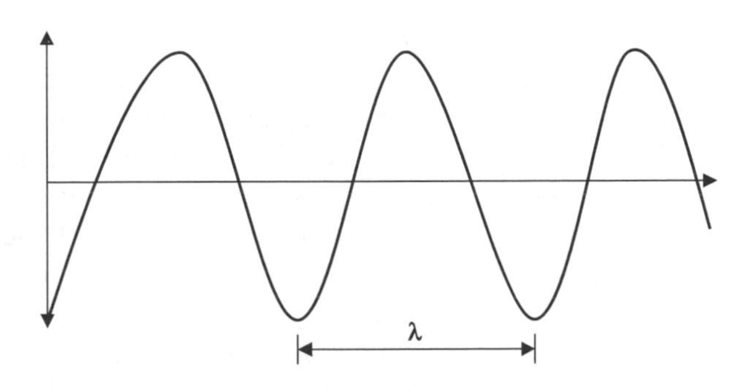

FIGURE 8.3. *Model for a beam of light having a single wavelength; the direction of propagation (x axis) and the amplitude (y axis) are indicated. The distance between successive peaks or valleys defines the wavelength (λ).*

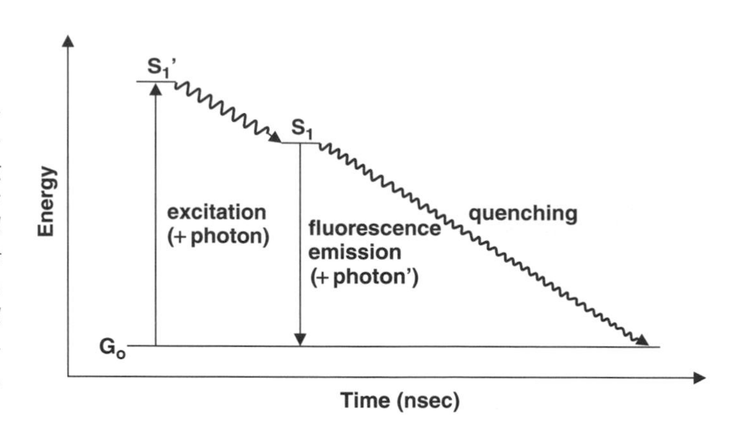

FIGURE 8.4. The Jablonski diagram depicts the absorption of light (excitation), fluorescence, and quenching as a function of time in nanoseconds (ns). The ground state (G_0), excited singlet state (S_1'), and relaxed singlet state (S_1) are shown for an electron in a molecule. The energy of an emitted photon (+photon') is less than the energy of excitation photon (+photon), accounting for the Stokes shift.

Excitation. *Light absorption process resulting in the conversion of electrons in a fluorophore from the ground state into a higher energy state.*

Fluorescence. *Light emission process in which a fluorophore in the excited state releases a photon of slightly longer wavelength than the excitation light.*

Jablonski diagram. *Schematic representation of the fluorescence process in which energy is plotted as a function of time, including excitation, emission, and quenching.*

Stokes shift. *Difference in wavelength between the excitation light and the emitted light in a fluorescent assay, typically expressed in nanometers.*

Incident light. *Electromagnetic radiation that impinges on a microarray substrate, mirror, or other component during microarray detection.*

Spectra. *Plural of spectrum.*

Absorption maximum. *Wavelength of light that is captured most efficiently by a fluorescent molecule.*

Emission maximum. *Wavelength of light that is emitted most efficiently by a fluorescent molecule.*

During the absorption process, photons captured by the fluorophore can result in the conversion of that molecule (M) into an excited state (M*). When absorption results in this conversion, the process is known as **excitation:**

$$\text{M} + \text{photon} \rightarrow \text{M}^* \tag{8.5}$$

For most fluorescent species, the length of time that a molecule spends in the excited state is short, on the order of 10^{-8}–10^{-9} s. During this brief period, a molecule in the excited state can undergo a number of different transitions; the desired one for microarray analysis is the release of a photon in a process known as ***fluorescence.***

$$\text{M}^* \rightarrow \text{M} + \text{photon}' \tag{8.6}$$

The process of fluorescence involves three discrete steps, which are illustrated by the common schematic known as a ***Jablonski diagram*** (Fig. 8.4). In the first step, photon absorption excites an electron from the ground state (G_0) into the excited singlet state (S_1'). The electron then looses a small amount of energy and moves to an excited singlet state of slightly lesser energy (S_1). In the third step, the molecule emits a photon, and the electron returns to the ground state. Because the emitted light (photon′) has slightly lesser energy that the excitation light (photon), the wavelength of fluorescent light is always longer than the wavelength of the excitation light. The wavelength shift between excitation and emission is known as the ***Stokes shift.***

The Stokes shift is a convenient aspect of fluorescence, because it allows optical discrimination between the emitted light (signal) and the excitation or ***incident light.*** The Stokes shift is measured by determining the correlation between the absorption and the emission for a given molecule as a function of wavelength. The values are then plotted on a graph to provide the molecule's ***spectra*** (Fig. 8.5). The wavelengths that correspond to the largest absorption and emission values, respectively, for a given fluorescent molecule are known as the ***absorption maximum*** and ***emission maximum.*** The Stokes shift is measured by subtracting the absorption maximum from the emission maximum. Most of the dyes used for microarray analysis have Stokes shift values in the range of 20–60 nm. Fluorescein isothiocyanate (FITC) has absorption and emission maxima of 494 and 518 nm, respectively, providing a Stokes shift of 24 nm. The Cy3 and Cy5 dyes have maxima of 550 and 570 and 649 and 670, respectively, providing Shokes shift values of 20 nm for Cy3 and 21 nm for Cy5. The greater the distance between the absorption and emission spectra, the easier it is to separate the fluorescent signal from the excitation light. For this reason, dyes with large Stokes shift values are extremely valuable for microarray applications.

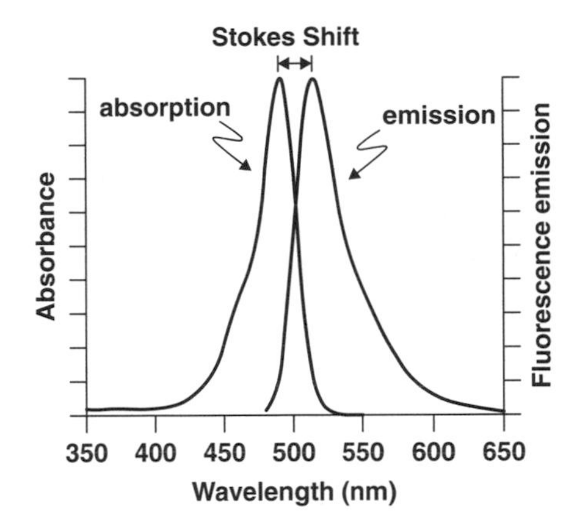

FIGURE 8.5. Absorption and emission spectra for the fluorescent dye FITC. The emission maximum is 518 nm and the absorption maximum is 494 nm; thus the Stokes shift is 24 nm. (Courtesy of Molecular Probes, Eugene, OR.)

In addition to fluorescence emission, excited molecules can undergo a number of nonproductive fates, including quenching, energy transfer, and intersystem crossing, all of which result in no fluorescence emission from an excited molecule. These nonproductive fates (Q) are given by the general formula:

(8.7) $$M^* + Q \rightarrow M + Q^*$$

The Q^* designation denotes the fact that excitation energy has been acquired by the quenching species. Fluorescence emission for a given dye that has absorbed light depends on the efficiency of photon release versus the combined undesired effects of the nonfluorescent pathways. The ratio of emitted photons ($h\nu'$) versus absorbed photons ($h\nu$), where h is Planck's constant and ν is the frequency, is known as the ***quantum yield*** (Q):

Quantum yield. *Efficiency of photon release by a fluorescent molecule, defined as the ratio of emitted photons divided by absorbed photons.*

(8.8) $$Q = \frac{h\nu'}{h\nu}$$

The quantum yield for dyes used in microarray experiments ranges from 0.1 to 0.5, with the commonly used labeling reagent Cy3-deoxycytidine triphosphate having a quantum yield of 0.15.

For a fixed detection instrument and a specified set of environmental contributions, the fluorescence intensity of a given dye (I_d) depends primarily on the molar extinction coefficient (ε), the quantum yield (Q), and the concentration of the dye (C) and is proportional to the product of those values:

(8.9) $$I_d \propto \varepsilon \times Q \times C$$

Microarray assays that use dyes with large values for ε, Q, and C produce more intense fluorescent signals than those that have lower values for one or more of the three variables.

Photobleaching

Using low to moderate levels of excitation light, a single fluorescent molecule can be excited and detected repeatedly. This cyclical excitation and emission property of fluorescent dyes enables a high level of assay sensitivity, because a single dye can emit hundreds of signal photons. If excessive excitation light is used however, the dye can become damaged physically in a process known as photobleaching. The chemistry of photobleaching is complex, but appears to result primarily through excited dye–dye interactions that cause irreversible changes in the dye structure and a loss of the photobleached molecules for

Intensity. *Amount of light used for microarray excitation equivalent to the power of the light source divided by the illumination area.*

Dwell time. *Duration of a pause in microarray printing, detection, or some other process, measured typically in milliseconds.*

Time illumination. *Quantity in microarray detection equal to the product of the light intensity times the dwell time.*

successive rounds of excitation and detection. Photobleaching is detrimental to microarray analysis because it reduces the absolute signals observed in the assay.

Photobleaching depends on the ***intensity*** of the light source and on the duration, or ***dwell time,*** of the illumination. For microarray detection systems, the ***time illumination*** (T_i) is equal to the intensity of the light source at a given wavelength (I_λ) in Watts per centimeter squared times the dwell time (T_d) in seconds:

$$T_i = I_\lambda \times T_d \tag{8.10}$$

A typical microarray scanner (see below) might use a 10-mW laser that produces an intensity (Eq. 8.14) of 10,000 W/cm^2 and a dwell time of 16 μs (16×10^{-6} s) on a microarray surface, for a time illumination of 0.16 W cm^{-2} s^{-1}. At this time illumination, photobleaching is rarely a concern even with dyes such as fluorescein that are prone to photobleaching. Photobleaching is rarely a concern with most contemporary microarray detection systems, though this detrimental process was an issue with some of the early home-built systems that used time illumination values in the range of 1–20 W cm^{-2}s^{-1}. Photobleaching can be a concern with microarray imaging systems that use flood illumination schemes and white light sources capable of T_i values >500 W cm^{-2}s^{-1}.

Photobleaching can be minimized or eliminated through the use of low to moderate intensity light sources, detectors set at high sensitivity levels, photostable dyes, large numerical aperture lenses, optimized exciter and emitter filter sets, and antifade reagents. A simple test for photobleaching is to perform a sequential series of detection steps on a single microarray, recording the images after each round of detection. By comparing the images for changes in the absolute signal intensity over 5 or 10 detection steps, photobleaching is easily observed if it is occurring. If the fluorescence intensity remains constant over the 5–10 recordings, photobleaching is not occurring. If, on the other hand, signal intensities decrease as a result of successive detection steps, the dyes are being degraded by photobleaching. A typical benchmark for detection systems is ≤1% loss of signal intensity per reading, with ≤10% being absolutely essential for functionality.

Advantages

Fluorescence has been established as the standard detection scheme for microarray analysis, offering many advantages including detectivity, speed, safety, high spatial resolution, durability, enzymatic compatibility, availability of different colors, and ease of use. These combined attributes suggest that fluorescence will be the standard detection label for the foreseeable future, though other interesting technologies on the horizon offer alternatives to fluorescence (see Chapter 14).

The high detectivity with fluorescence derives from the combined benefits of multiple photon emission from a single dye, high efficiency of photon detectors, and low background of glass substrates. Modern detection instruments easily provide detectivity in the range of a few hundred molecules per microarray spot. The high speed of fluorescence detection technology derives from the fact that data are collected in an automated fashion by scanners and imagers. In gene expression experiments, for example, a 10 second-scan provides a level of detectivity that would require an exposure time of several weeks with radioactive labels and filters.

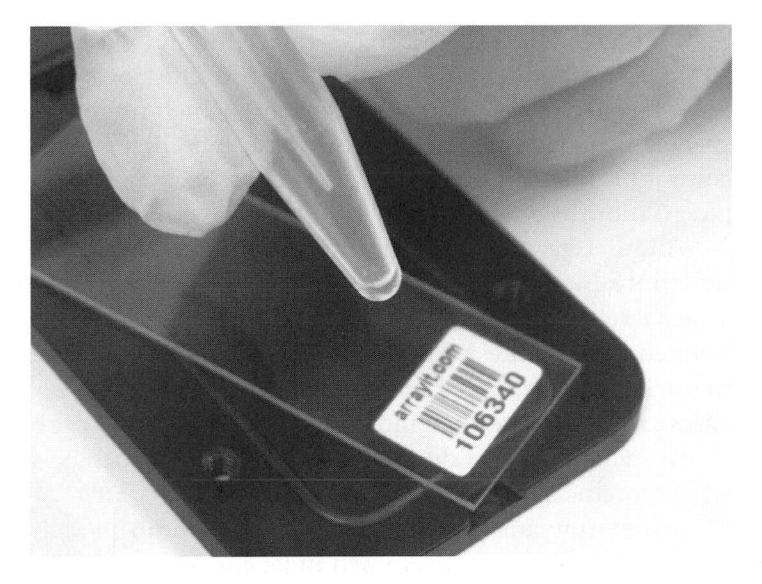

FIGURE 8.6. *Sample. A 10-μL oligonucleotide mixture labeled with the fluorescent Cy3.*

Detectivity and speed are two main advantages of fluorescence over radioactivity, and another is increased safety. Fluorescent labels are nonradioactive and rather nontoxic, both of which maximize user safety and eliminate cumbersome waste disposal issues. Unlike radioisotope labels, fluorescent dyes are also visible to the unaided eye and therefore allow easy tracking during labeling experiments. Most microarray samples produce colored solutions clearly visible to the naked eye (Fig. 8.6).

High spatial resolution is another major benefit of fluorescence, allowing microarray spots with strong and weak signals to exist at adjacent array positions without the concern of signal spreading from one spot to another (Fig. 8.7). Filter-based assays that use radioactivity often battle signal spreading, which can complicate measurements involving a large dynamic range. High spatial resolution enables extremely high spot densities, while maintaining accurate signal readings at each microarray location.

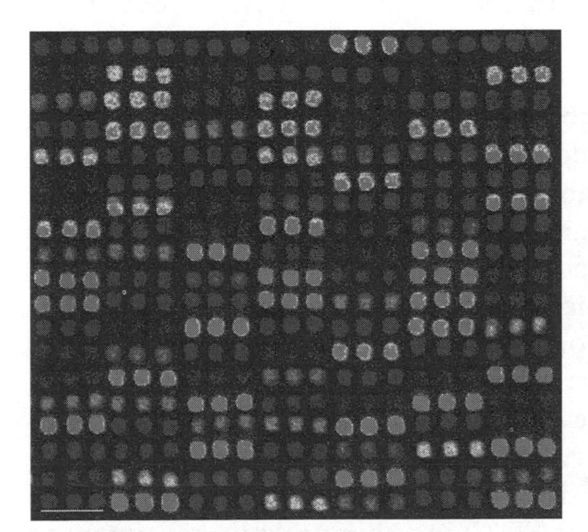

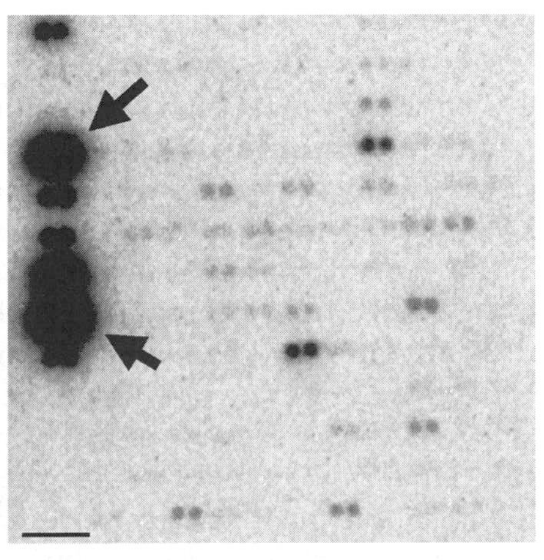

FIGURE 8.7. *A fluorescent microarray (**A**) and a P^{32} radioactive filter array (**B**). Fluorescent signals are spatially confined to each microarray spot, whereas signal spreading* (arrows) *occurs with assays that use radioisotopes. Scale bar = 500 μm (**A**) and 1 cm (**B**). (Data courtesy of Dr. T. C. Islam and C. I. Edvard Smith, Department of Biosciences, Novum, Karolinska Institute, Stockholm, Sweden.)*

Fluorescent labels are also extremely durable, with low temperature (−20 to −80°C) storage maintaining the integrity of the labels for months or years. This is in sharp contrast to radioactive labels, which undergo a twofold decrease in specific activity every few days or weeks. The half life of P^{32}, an isotope used commonly in traditional assays, is 14 days; thus <1% of the useful label remains after 3 months of storage.

Another key criterion for microarray labels is their capacity to be incorporated into biochemical preparations. Incorporation requires that the label be attachable to a biochemical building block (e.g., nucleotide) and easily incorporated into synthesized material (e.g., cDNA). Fluorescent dyes are attached to nucleotides and other molecular species with relative ease, and the modified bases can be incorporated into nucleic acid and other probe material using suitable enzymes. Certain dyes, such as fluorescein isothiocyanate, can be used to label proteins directly (see Chapter 7).

Microarray analyses often make use of multiple labels, which allow two or more samples to be examined independently on a single microarray. Dual-labeling and multilabeling schemes increase the precision of comparative analyses by removing the uncertainty involved in comparing two or more different microarrays. Simultaneous analysis of multiple samples on the same microarray requires the use of a different label for each sample (e.g., mRNA). The labels need to allow independent identification by the detection instrument. Fluorescent labels, with their different excitation and emission spectra, are well suited for multiple labeling schemes. A multitude of different fluorescent labels are available to microarray scientists (Table 8.1).

Fluorescent Dyes

Fluorescent dyes make up an interesting and important family of compounds that provide a large variety of labels for use in microarray analysis. At present, there are hundreds of different labeling reagents available commercially. All fluorescent dyes share a number of common features, the most important structural similarity being the presence of double bonds that occur on every other carbon atom. These ***conjugated bonds*** contain electrons that absorb and emit light in the visible range, accounting for the intense fluorescence of these compounds.

Conjugated bond. *Double bond that occurs on every other carbon atom containing π electrons; often associated with fluorescence.*

One of the most widely used dyes in microarray analysis is FITC. Fluorescein was synthesized by German chemist Adolph Von Baeyer in 1880. The molecule contains nine conjugated double bonds distributed over four ring structures, allowing the molecule to absorb light in the visible range (Fig. 8.8). Because the multiple ring structures are similar to benzene, organic molecules of this type are commonly known as polycyclic benzenoid compounds. The location of hydroxyl, carbonyl, carboxylic, and isothiocyanate substituents confer solubility on this otherwise highly nonpolar compound. Fluorescein absorbs and emits at 494 and 518 nm, respectively (Fig. 8.5), and microarray samples labeled with this dye appear to the eye as light green or yellow. The 494 nm absorption maximum of FITC coincides nicely with the 488 nm line of an argon-ion laser, making FITC a convenient dye for laser-based scanners (see below). There are several negative attributes of FITC, including relative sensitivity to photobleaching and a fluorescent signal that is sensitive to changes in pH. For these reasons, much effort has been spent developing FITC-related dyes with improved characteristics for fluorescence studies.

TABLE 8.1. Fluorescent Dyes and Proteins[a]

Dye[b]	Absorption	Emission	Quantum Yield	ε	Comments
DAPI	350	450	0.83	120,000	
Cy2	489	506		150,000	
FITC	490	520	0.71	67,000	fluorescein isothiocyanate
Alexa 488	495	519			
JOE	522	550			DNA sequencing dye
Rhodamine 6G	525	555	0.9	85,000	
B-phycoerythrin	546	575	0.98	2,410,000	protein of 240 kD, contains 25 fluors per protein
R-phycoerythrin	546	578	0.98	1,960,000	protein of 240 kD, contains 25 fluors per protein
Tetramethylrhodamine	555	580		176,000	
Cy3	550	570	0.14	150,000	
Lissamine	570	590			
Alexa 568	578	603			
ROX	580	605			DNA sequencing dye
Texas Red	596	620	0.51	85,000	
BODIPY 630/650	625	640			oligonucleotide labeling
Cy5	649	670	0.15	250,000	
BODIPY 650/665	646	660			oligonucleotide labeling
Cy5.5	675	694		250,000	
Cy7	743	767		250,000	

[a] Data courtesy of molecular probes, Eugene, OR, and Instituto Ricerche Farmacologiche, Milano, Italy.
[b] DAPI, diamidino-2-phenylindone; JOE, 6-carboxy-4′, 5′-dichloro-2′, 7′-dimethoxy fluorescein; ROX, 5-carboxy-x-rhodamine.

FITC-related dyes have been developed through sophisticated organic synthesis modifications of the fluorescein parent molecule, creating a large collection of dyes that are highly desirable for microarray experiments (Fig. 8.9). Some of the popular FITC derivatives include the Alexa series, JOE, Oregon Green, Lissamine, and the Rhodamine dyes. The derivatives contain additions of polar, aliphatic, and heterocyclic groups onto the parent FITC molecule. Each dye has distinct absorption and emission spectra (Table 8.1), which enables two or more labels to be used simultaneously in microarray analyses. In addition to altering the absorption and emission spectra, the addition of functional groups provides greater solubility, quantum yield, and photostability

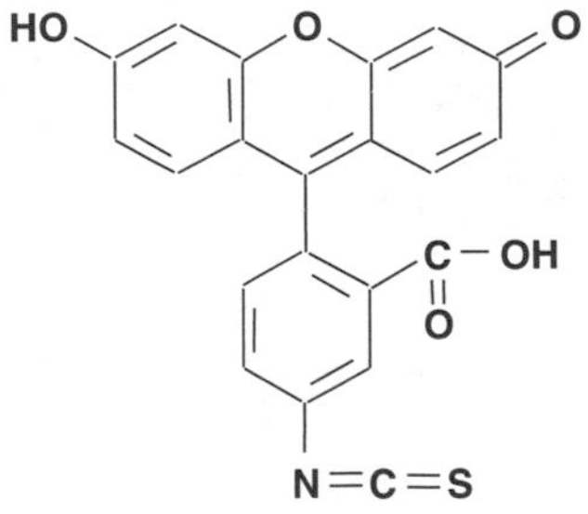

FIGURE 8.8. *The molecular structure of FITC.*

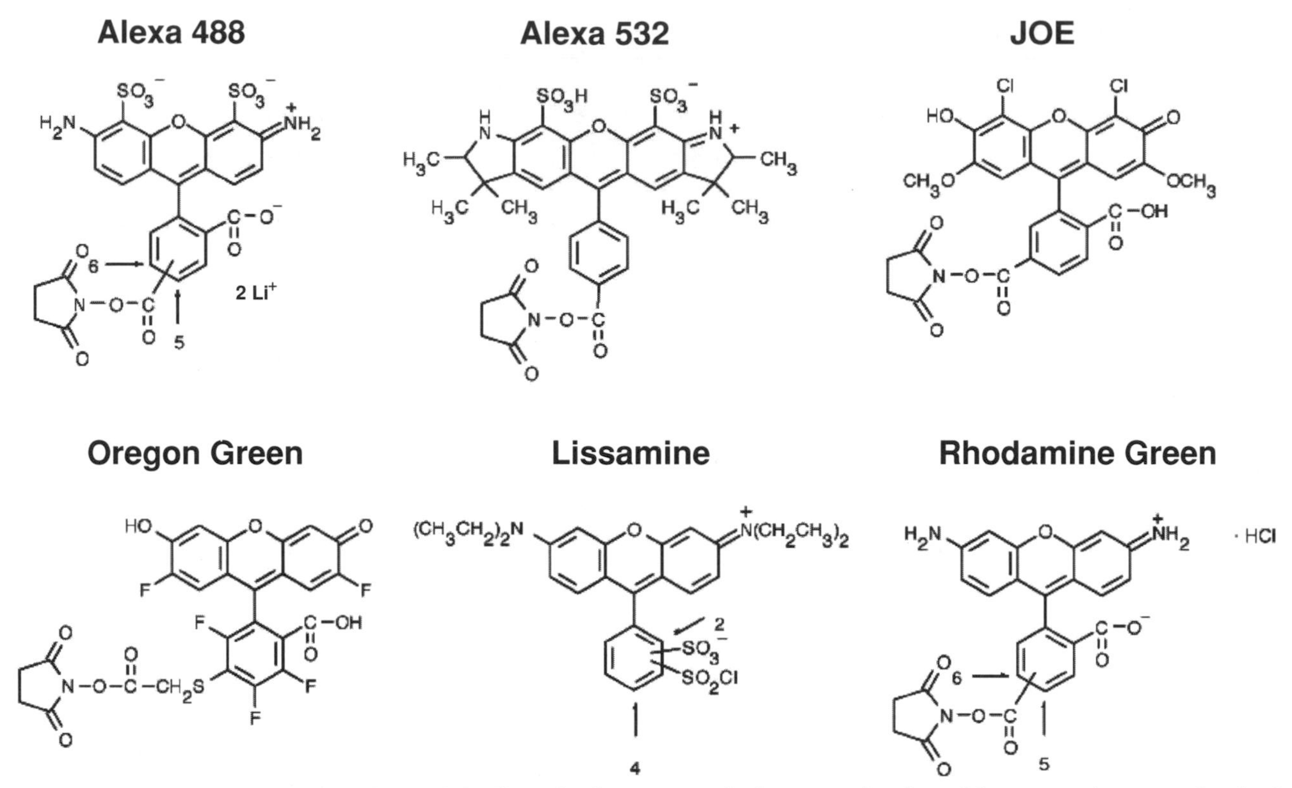

FIGURE 8.9. The molecular structures for fluorescent dyes derived from FITC. (Courtesy of Molecular Probes, Eugene, OR.)

as compared to FITC. The latter fact is evident in photobleaching studies with FITC and the derivative Alexa 488, which are rather similar in molecular structure but differ markedly in their photostabilities (Fig. 8.10). Further technology development in the form of chemical modifications will probably provide FITC-based dyes that are even better than those in current use.

Another very widely used fluorescent dye family, developed by Wagonner and co-workers at Carnegie Mellon University, is based on the cyanine molecule (Mujumdar 1989). The cyanine dyes are bright, stable to photobleaching, and easily added to nucleotides and oligonucleotides; plus they absorb and emit light at convenient wavelengths. Extensive photochemical studies have been

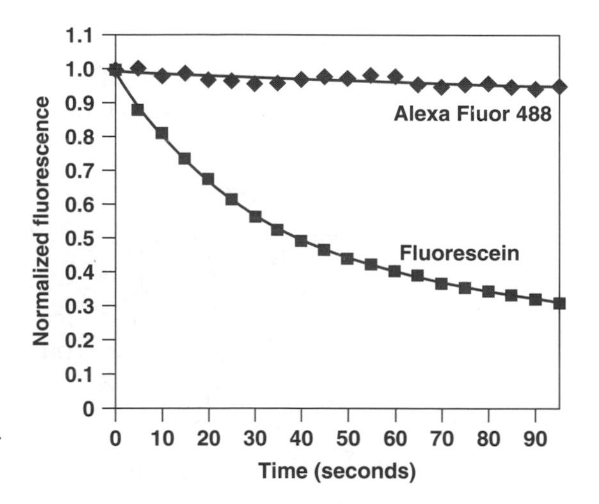

FIGURE 8.10. Photobleaching data for fluorescein and Alexa 488. Both dyes were exposed to a constant light source, and fluorescence was measured at 5-s intervals. (Data courtesy of Molecular Probes, Eugene, OR.)

indocarbocyanine (C3)

$I^{\ominus}$ $\oplus$ Et N N Et

indodicarbocyanine (C5)

$I^{\ominus}$ $\oplus$ Et N N Et

FIGURE 8.11. The molecular structures of indocarbocyanine and indodicarbocyanine. Each nitrogen atom contains an ethyl functional group (Et). (Adapted from the Oregon Medical Laser Center, Portland; original data courtesy of Dr. J. S. Lindsey and co-workers, Rockefeller University, New York.)

performed on many of these molecules, including indocarbocyanine (C3) and indodicarbocyanine (C5) (Fig. 8.11), which are closely related to the commercial Cy dyes. Spectral data for C3 and C5 demonstrate large molar extinction coefficients (Fig. 8.12) and convenient emissions of approximately 570 and 670 nm, respectively (Fig. 8.13).

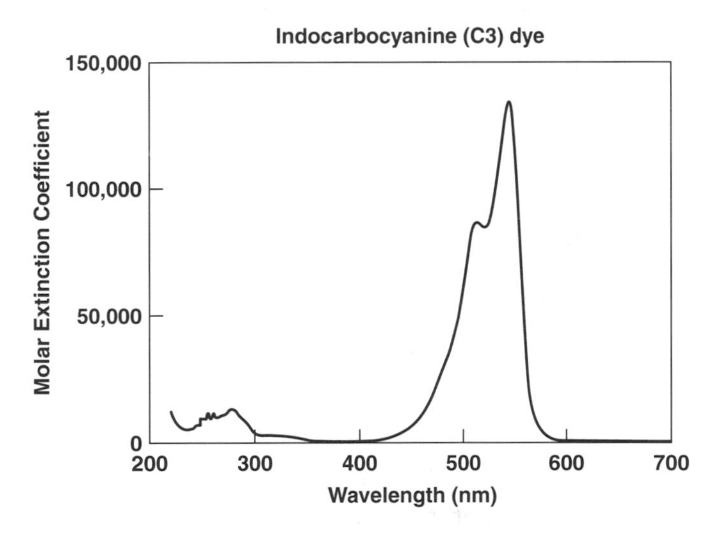

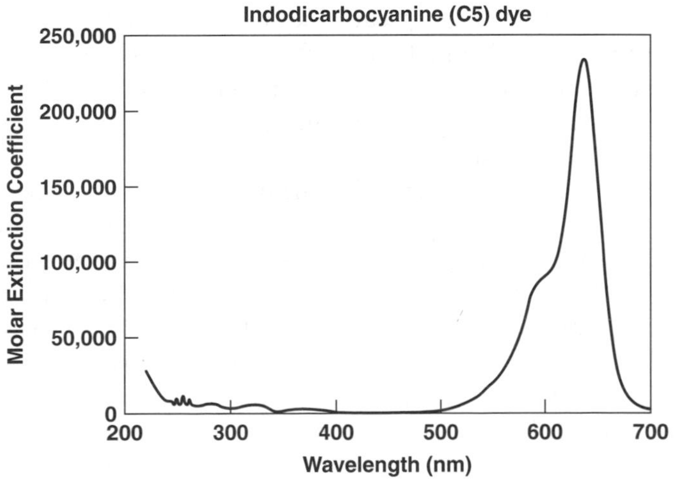

FIGURE 8.12. Molar extinction coefficient data for C3 and C5. (Adapted from the Oregon Medical Laser Center, Portland; original data courtesy of Dr. J. S. Lindsey and co-workers, Rockefeller University, New York.)

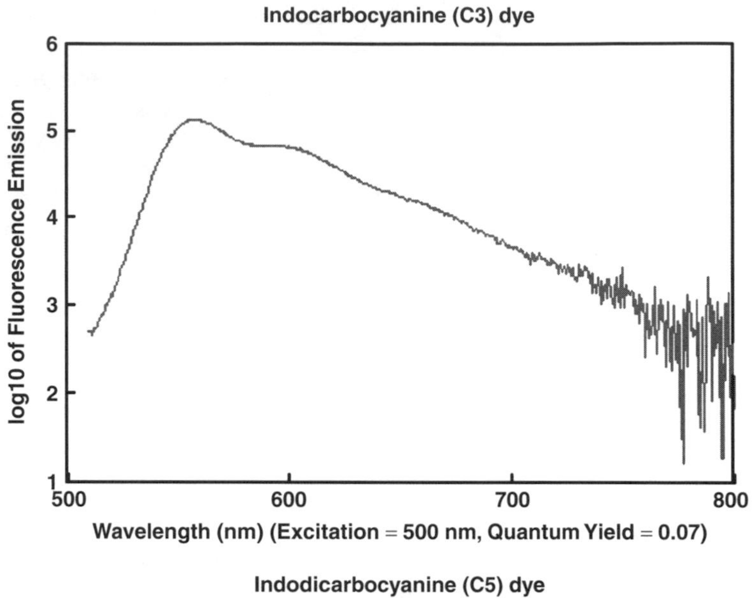

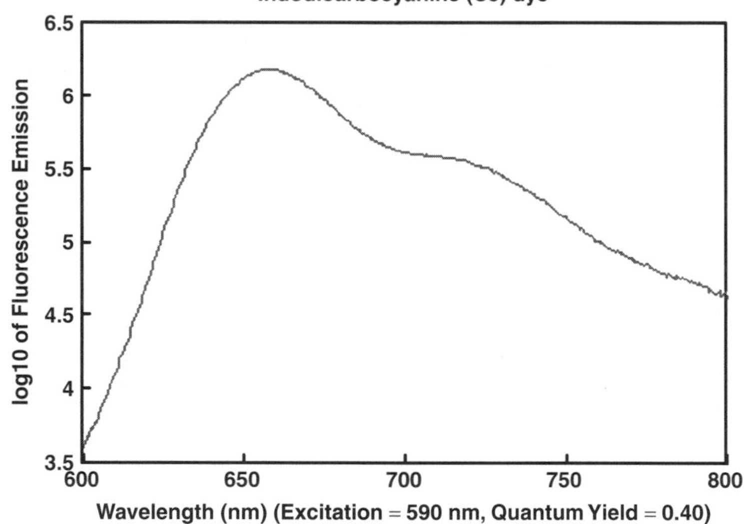

FIGURE 8.13. *Fluorescence emission data for the C3 and C5. For C3, excitation* = 500 *nm; quantum yield* = 0.07. *For C5, excitation* = 590 *nm; quantum yield* = 0.40. *(Adapted from the Oregon Medical Laser Center, Portland; original data courtesy of Dr. J. S. Lindsey and co-workers, Rockefeller University, New York.)*

The commercial cyanine dyes Cy3 and Cy5 (Fig. 8.14) are identical to C3 and C5, except that the Cy dyes contain two sulfate groups per dye and therefore have greater solubility than C3 and C5. The increased solubility of Cy3 and Cy5 render these dyes well suited for enzymatic labeling reactions, which are carried out in aqueous buffers. Cy3 and Cy5 are also available as phosphoramidite derivatives (Fig. 8.15) and can thus be added in a synthetic manner to oligonucleotides and used directly as hybridization probes in microarray experiments. The full chemical names of the Cy3 and Cy5 phosphoramidite derivatives are indocarbocyanine 3-1-O-(2-cyanoethyl)-(*N*,*N*-diisopropyl)-phosphoramidite and indodicarbocyanine 5-1-O-(2-cyanoethyl)-(*N*,*N*-diisopropyl)-phosphoramidite, respectively. Because the phosphoramidite versions of Cy3 and Cy5 lack sulfate groups, oligonucleotides bearing these labels sometimes produce slightly weaker signals in aqueous hybridization reactions owing to their reduced solubility.

In addition to the fluorescein and cyanine families of fluorescent dyes, a third dye family used in microarray analysis is the BODIPY series (Fig. 8.16).

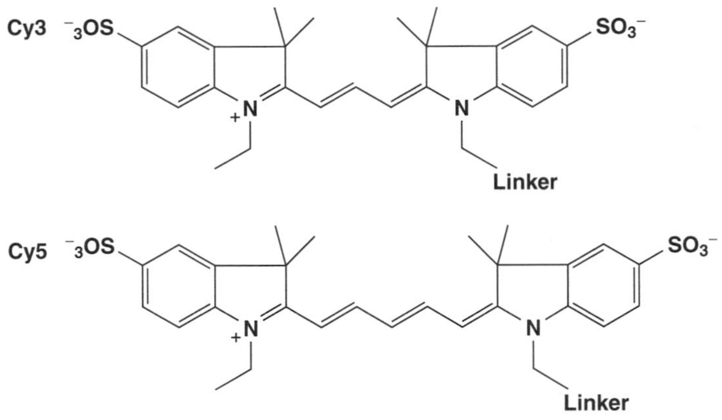

FIGURE 8.14. *The molecular structures of cyanine Cy3 and Cy5. The two molecules are identical except for an additional double bond in Cy5 between the two indole-like structures. (Adapted from Amersham Pharmacia Biotech, Piscataway, NJ.)*

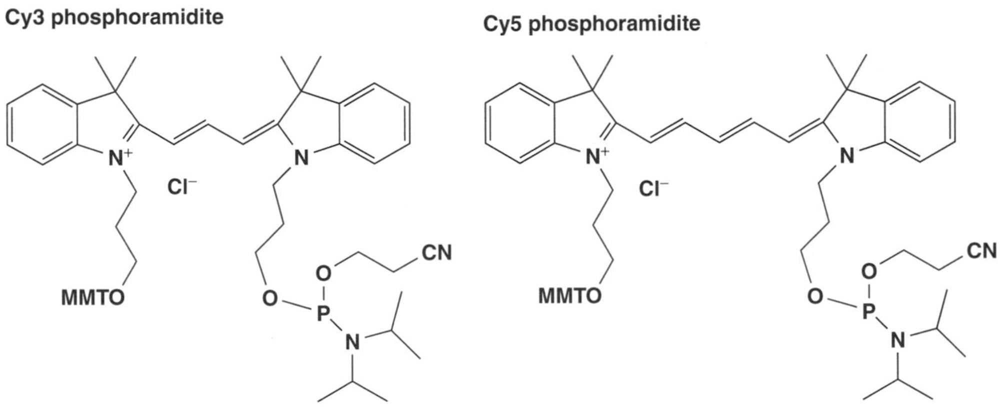

FIGURE 8.15. *The molecular structures of the cyanine Cy3 and Cy5 dyes as phosphoramidite derivatives. These molecules lack the two sulfate groups present on the Cy derivatives (Fig. 8.14) used in enzymatic labeling reactions. (Courtesy of Glen Research, Sterling, VA.)*

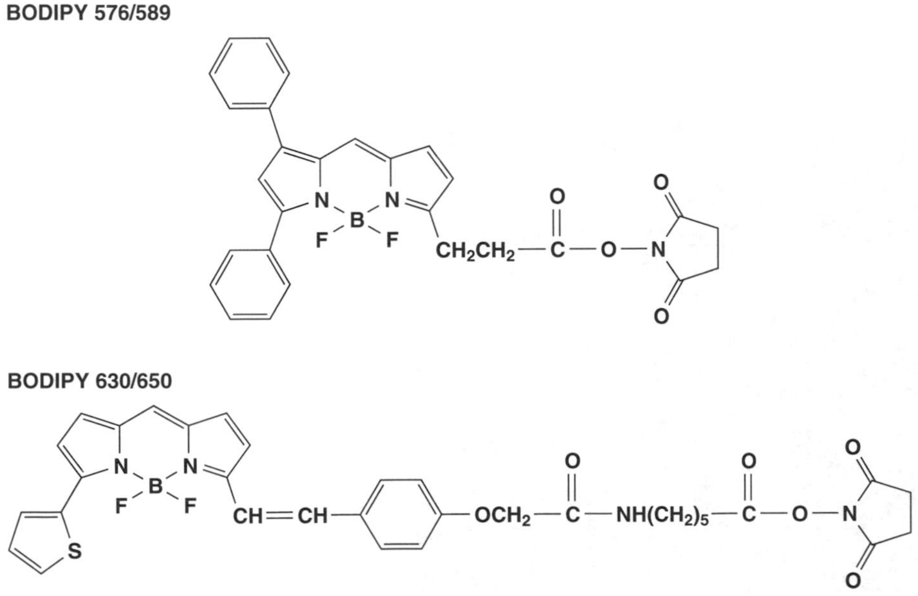

FIGURE 8.16. *The molecular structures of BODIPY 576/589 and BODIPY 630/650. The names derive from excitation and emission maxima. (Courtesy of Molecular Probes, Eugene, OR.)*

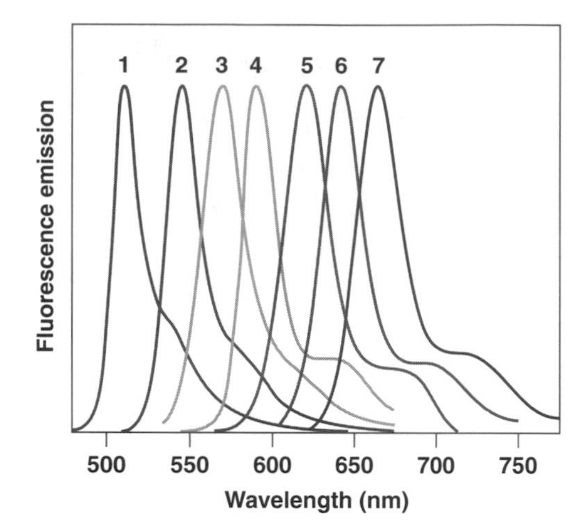

FIGURE 8.17. Normalized emission spectra of seven BODIPY dyes; each spectrum is coded according to the color of light required for excitation. (Courtesy of Molecular Probes, Eugene, OR.)

The BODIPY dyes have good spectral characteristics, with absorption and emission wavelengths spanning the entire visible region (Fig. 8.17).

The fluorescein, cyanine, and BODIPY dyes are used in labeling reactions and are incorporated into biochemical preparations by either enzymatic or synthetic means. There are also a number of fluorescent dyes that are used to stain DNA and RNA, allowing fluorescent detection without the requirement of labeling. The SYBR, PicoGreen, OliGreen, and RiboGreen dyes developed at Molecular Probes (Eugene, OR) have been used for microarray staining with good success (Fig. 8.18).

SCANNERS

Of the two main designs for microarray detection instruments, scanners represent the most widely used at present. The term scanner derives from the fact that detection is carried out by moving back and forth, or scanning, across the surface of the microarray. In essence, a microarray scanner can be viewed as a confocal microscope that moves quickly in the x and y dimensions to allow

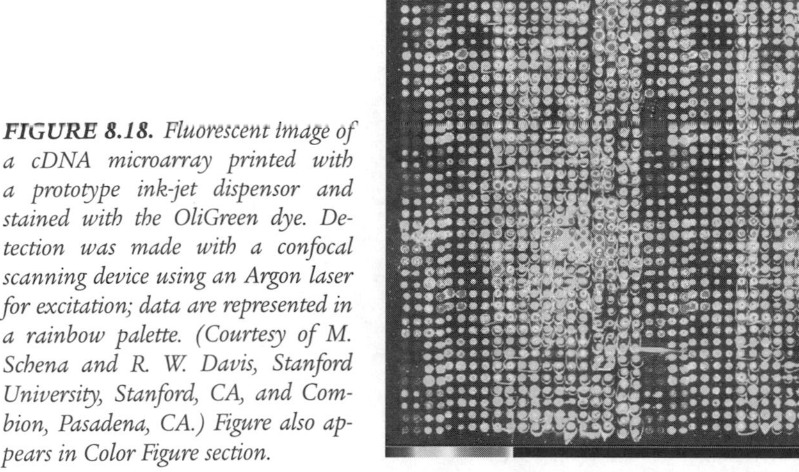

FIGURE 8.18. Fluorescent image of a cDNA microarray printed with a prototype ink-jet dispensor and stained with the OliGreen dye. Detection was made with a confocal scanning device using an Argon laser for excitation; data are represented in a rainbow palette. (Courtesy of M. Schena and R. W. Davis, Stanford University, Stanford, CA, and Combion, Pasadena, CA.) Figure also appears in Color Figure section.

large fluorescent images to be acquired. Some of the early scanners were built in the mid-1980s. Over the next decade, laser scanning systems were tailored for use in microarray applications. This section provides an overview of contemporary microarray scanning technology, including instrument architecture, scanning approaches, laser technology, beam splitters and optical filters, photomultiplier tubes, charge-coupled devices, analog-to-digital converters, and data file formats.

Instrument Architecture

Microarray scanners are sophisticated instruments, and specific design components can be used to achieve unique performance specifications. Though the details vary for each of the currently available commercial scanners, each instrument has most or all of the following components: laser, excitation filter, beam splitter, objective lens, stage, reflective mirror, emission filter, pinhole, photomultiplier tube, analog-to-digital converter, and personal computer. The basic architecture of a microarray scanner is shown in Figure 8.19.

In the basic design, laser light is directed through an excitation filter that allows passage of the excitation wavelength of interest (Fig. 8.19). The wavelength of interest (e.g., 488 nm) should correspond to the excitation maximum of the dye that is present on the microarray. The laser light is reflected through a microscope objective by a beam splitter, which focuses the excitation light onto the microarray surface. Excitation of the dye molecules results in the emission of fluorescent light, which passes back through the objective in a sporadic pattern and is converted into a parallel beam in a process known as collimation. A turn mirror reflects the collimated fluorescent beam through an emission filter, which passes the desired wavelength but rejects unwanted wavelengths including the original excitation light. The collimated emission beam is then refocused by a detector lens, and passes through a pinhole into the photomultiplier tube (PMT). Analogue signals from the PMT are converted into digital signals by an analog-to-digital (A/D) converter, and data are transferred onto a personal computer in the form of tag image file format (TIFF) files.

Microarray scanners acquire data one pixel at a time and therefore capturing a large image requires physical movement of the substrate or optics during the detection process. There are two types of microarray scanning architectures

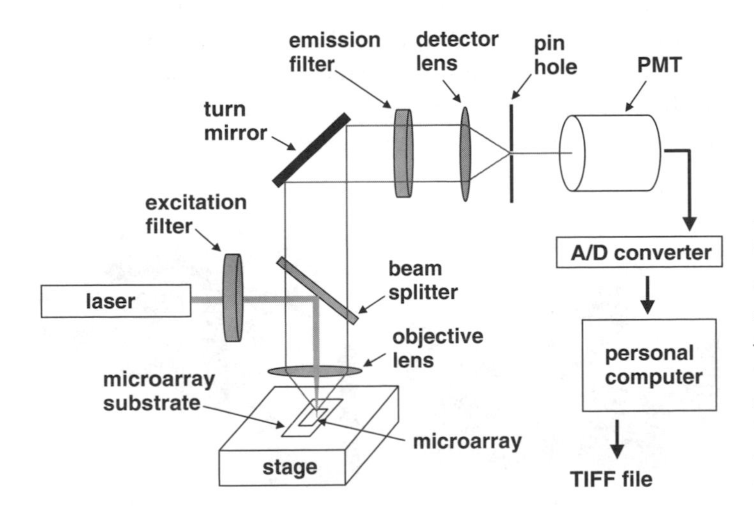

FIGURE 8.19. A generic microarray scanner. Laser light (thick line) *passes through a series of optical components and onto the microarray surface. Emitted fluorescent light* (thin line) *travels back through an optical path and into a PMT. Analog electrical signals are converted into digital signals, and the TIFF files are saved on a personal computer.*

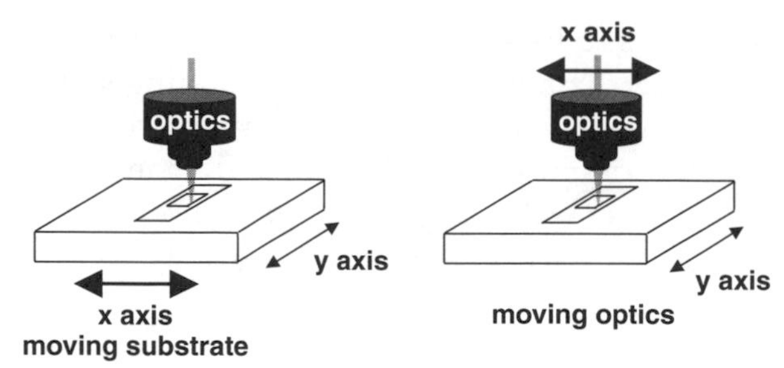

FIGURE 8.20. The moving substrate scanner design holds the optics stationary and moves the substrate quickly along the x axis (thick arrow) *and slowly along the y axis* (thin arrow). *The moving optics design moves one or more components of the optics quickly along the x axis* (thick arrow) *and moves the substrate slowly along the y axis* (thin arrow). *Both designs allow laser light to be directed over a large portion of the microarray substrate.*

known as moving substrate and moving optics designs. A moving substrate scanner translates the microarray slide in both the *x* and *y* dimensions, allowing fluorescent light to be captured by a stationary optical path. A moving optics scanner translates one or more components of the optics rapidly along the *x* axis and moves the substrate slowly along the *y* axis for image acquisition. The two types of configurations are shown in Figure 8.20.

Because the moving substrate design is the most easily implemented, many of early breadboard scanners used this design. The ScanArray instruments by Packard Biochip Technologies (Billerica, MA) are highly popular and use the moving substrate configuration (Fig. 8.21). Moving substrate instruments translate the microarray substrate rapidly along the *x* axis and slowly along the *y* axis until the desired portion of the microarray is detected. Rapid translation along the *x* axis is usually mediated by a slider controlled by a ***position transducer,*** such as a ***galvonometer,*** whereas a stepper motor carries out the slower movements in the *y* dimension. The rapid back and forth movements of these devices along the *x* axis is sometimes called a ***raster pattern.*** The instrument maintains a 90° alignment of the optics with the substrate during the scanning process, in a ***rectilinear*** scanning profile.

Position transducer. *Scanner component such as a galvonometer that translates rapidly along the x axis and enables the capture of a complete microarray image from the substrate.*

Galvonometer. *Position transducer in a microarray scanner that translates rapidly across the x axis and enables microarray image capture.*

Raster pattern. *Rapid translation of a microarray scanner back and forth across the x axis of a substrate.*

Rectilinear. *Microarray scanning profile characterized by a series of straight lines achieved by a 90° alignment of the optics with the substrate.*

One key advantage of the moving substrate systems is that the stationary optical path, which consists of delicate and complicated components, provides a high degree of instrument stability and durability. There are currently hundreds of commercial moving substrate scanners in microarray laboratories worldwide, some with more than 4000 h of machine time. Performance specifications for the ScanArray 5000 are provided in Table 8.2.

Moving optics scanners are also widely used in the microarray field, including several different designs that combine innovative ways to move one or more of the optical components. All of the moving optics scanners move

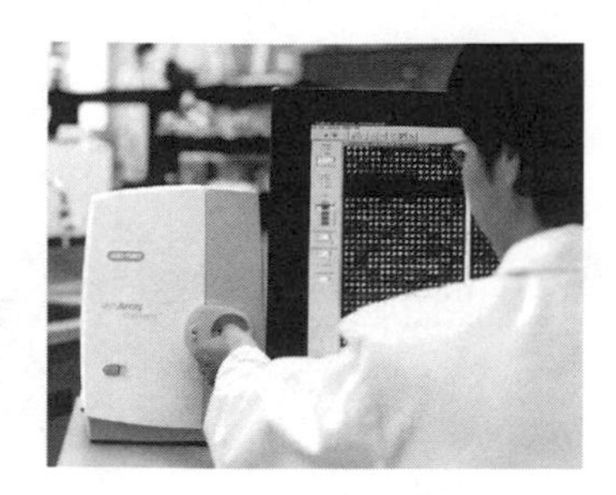

FIGURE 8.21. ***A,*** *The ScanArray 5000 from Packard Biochip Technologies (Billerica, MA) uses a moving substrate design.* ***B,*** *The VersArray ChipReader from Bio-Rad Laboratories (Hercules, CA) uses a moving optics design.*

TABLE 8.2. Performance Specifications for Two Scanners

Specifications	ScanArray 5000	ChipReader
Scanning design	moving substrate	moving optics
Substrate format (1 mm thick)	25 × 76 mm (1 × 3 in.)	25 × 76 mm (1 × 3 in.)
Maximum scan area	22 × 73 mm	22 × 65 mm
Scan time (10 μm resolution, 1 cm^2 area)	19 s	15 s
Scan resolution (pixel size)	5, 10, 20, 30, and 50 μm	3, 5, 10, 20, and 30 μm
Multiple channel scan sequence	sequential	sequential or simultaneous
Scanning pattern	raster	raster
Scanning profile	rectilinear	rectilinear
Number of lasers	5	2 (standard)
Excitation wavelengths	488–633 nm (11 settings)	532 and 635 nm
Laser type	gas	solid state
Maximum laser power	10 mW	10 mW
Optical design	confocal	confocal
Detectors	photomultiplier tubes	photomultiplier tubes
Bar code reader	yes	no
Maximum detectivity	< 0.1 fluors/μm^2	< 0.1 fluors/μm^2
Autoloader	yes (20 substrates)	no
Size (L × W × H)	97 × 39 × 41 cm	28 × 23 × 30 cm
Weight	75 kg	18 kg
Electrical	100–240 V	100–240 V
Digitization (per pixel)	16 bit	16 bit
File formats	TIFF, BMP, or raw	TIFF, BMP, or raw
Operating system	*Windows* NT 4.0	*Windows* NT 4.0
Computer processor	Pentium II	Pentium
Cost	$115,000 (U.S.)	$58,000 (U.S.)

quickly along the x axis and translate the substrate with a moving stage slowly along the y axis. Moving optics scanners that move the laser beam are known as ***preobjective*** designs because the dynamic component (the laser beam) occurs upstream (pre-) of the microscope objective in the light path. Moving optics scanners that move the lens are known as ***flying objective*** scanners because the lens flys back and forth rapidly over the microarray surface. Both preobjective and flying objective designs are available commercially.

Preobjective. *Moving optics microarray scanner design in which the laser beam, rather than the lens, is the dynamic component.*

Flying objective. *Moving optics microarray scanner design in which the objective, rather than the laser beam, is the dynamic component.*

The Array Scanner from Amersham Pharmacia Biotech (Piscataway, NJ) is an example of a preobjective moving optics scanner. It uses a galvonometer-controlled mirror to reflect the laser light back and forth across the microarray surface as the mirror oscillates rapidly about the x axis (Fig. 8.22). A sophisticated wide-angle lens maintains a collimated laser beam and a

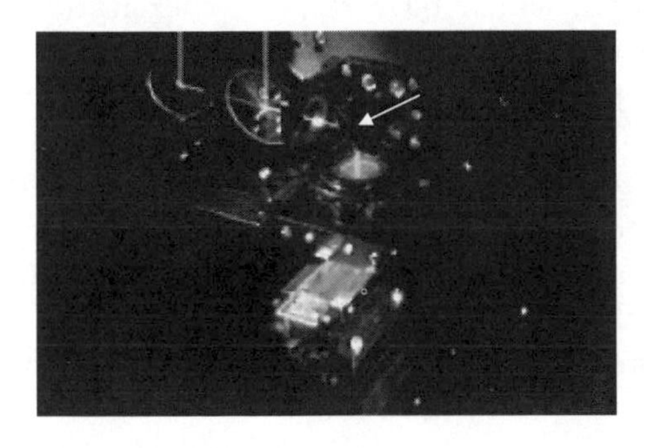

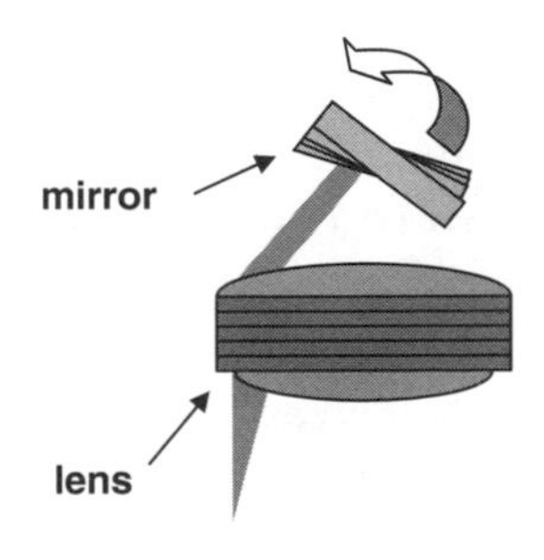

FIGURE 8.22. **A,** *The Array Scanner's light path.* Arrow points to *the galvonometer-controlled mirror.* **B,** *The galvonometer, lens and laser beam. (Courtesy of Amersham Pharacia Biotech, Piscataway, NJ.)*

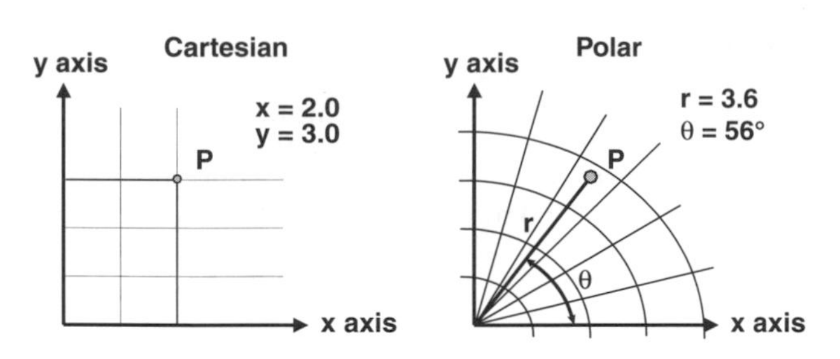

FIGURE 8.23. *Microarray scanners can provide planar data points either as x and y coordinates (Cartesian) or as coordinates defined by a radius (r) and a θ angle (polar). The Cartesian coordinates shown here are x = 2 and y = 3; the polar coordinates are r = 3.6 and θ = 56°.*

rectilinear scanning profile as the galvo deflects the mirror during the scanning process.

The flying objective hybrid scanners use one of several different mechanisms to translate the objective rapidly along the *x* axis. The GenePix microarray scanner from Axon Instruments (Union City, CA) uses a mechanism known as a ***voice coil*** to achieve rapid movements of the objective along the *x* axis. A voice coil exploits the fact that a magnet exerts a force on a current-carrying wire (coil), and this force can be used to rapidly move a slider in the *x* dimension to achieve microarray scanning. The ChipReader scanner from Virtek Vision (Ontario, Canada) uses a galvonometer to transduce the objective rapidly across the *x* axis (Table 8.2). Both the GenePix and the ChipReader systems maintain a perpendicular or rectilinear scanning profile and thus provide image data in which all the values can be described with simple *x* and *y* coordinates. Such *x* and *y* coordinates are commonly known as ***Cartesian coordinates*** (after the French scientist René Descarte), and images that can be described fully by Cartesian coordinates are known as ***Cartesian images.*** All rectilinear microarray scanners provide Cartesian images that correspond, on a 1:1 basis, to the fluorescent microarray.

Voice coil. *Scanner component consisting of a magnet and a current carrying wire that allows the rapid translation of the scanner objective along the x axis.*

Cartesian coordinate. *Mapping system used for most microarray printing robots and scanners consisting of x, y, and z values, each of which are perpendicular to each other.*

Cartesian image. *Microarray data file that can be described in terms of x and y values.*

Genetic MicroSystems (Woburn, MA) developed a variation of the flying objective moving optics scanner known as the ***rotary flying objective*** design. The rotary flying objective uses a periscope-like implement to move the objective rapidly along the *x* axis. Because the mass of the lens (<1 g) is small in this design, a galvonometer is able to deliver highly reliable rotary movements approaching 60 cycles per second (60 Hz). The minimal load on the galvo also means that the system is highly durable and the images are very good. One computational issue with the rotary flying objective design is that the microarray is scanned in a nonrectilinear arc pattern, and thus the data are collected in ***polar coordinates*** that must be deconvoluted into Cartesian coordinates by the host computer (Fig. 8.23).

Rotary flying objective. *Moving optics microarray scanner design in which the objective moves in a circular arc relative to the microarray substrate.*

Polar coordinate. *Mapping system used in rotary flying objective scanners, consisting of a radius and a θ angle instead of the traditional x and y Cartesian coordinates.*

All three scanner architectures lend themselves to detection at multiple wavelengths. Multiple wavelength detection is generally achieved through the use of separate lasers, filter sets, and photomultiplier tubes set to specific wavelengths. Detection of Cy3 and Cy5, which have respective excitation wavelengths of 550 nm and 649 nm, is usually carried out with two different lasers, filter sets, and PMTs, which are optimized to excite each dye and acquire the resultant fluorescence emissions. Current scanners allow as many as 11 different dyes to be detected over the 488–652 nm range of the visible spectrum. Data for multiple dyes can be captured one at a time or together. Scanners that capture multiple channels at the same time are known as ***simultaneous*** designs (Fig. 8.24), whereas scanners that that capture each different channel in successive scans are said to use a ***sequential*** scan sequence. A movie of the scanning process is available (http://arrayit.com/wiley/scanning).

Simultaneous. *Microarray scanner design that captures multiple channels at the same time.*

Sequential. *Microarray scanner design that captures multiple channels one at a time.*

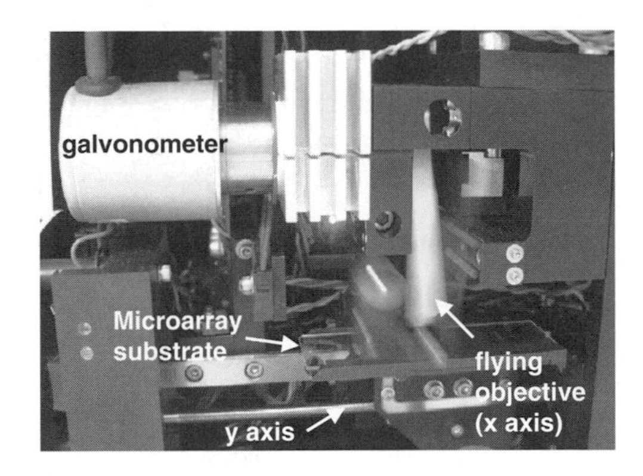

FIGURE 8.24. The ChipReader microarray scanner is able to scan simultaneously for Cy3 and Cy5 with green (532 nm) and red (635 nm) lasers, respectively.

To detect microarrays that contain more than one type of dye, the instrument design must prevent the signal from one channel from spilling over into another channel. Because the Stokes shift (Fig. 8.5) and emission spectra (Fig. 8.13) always result in emission wavelengths that are longer than the excitation wavelengths, channel-to-channel cross-talk can be a problem from shorter to longer wavelengths, but rarely vice versa. Cy3 emission can contaminate the Cy5 channel, for example, but Cy5 contamination of the Cy3 channel is not a significant concern. For dyes that have close or overlapping emission spectra, great care must be taken in the selection of lasers and filter sets to ensure that the signal from each channel is recorded accurately. Minimizing optical cross-talk is particularly important in gene expression experiments in which the readout for two different samples can vary by many orders of magnitude.

Lasers

Laser technology represents a large and complex field, with literally hundreds of millions of lasers in current use today, many of which are used in modern compact disc (CD) players. Despite the complexity of the technology, all lasers share common theoretical underpinnings and functional components. This section provides a simplified view of this complex technology area.

Laser. *Any of a family of devices that emits an intense beam of collimated, monochromatic light.*

Stimulated emission. *Phenomenon in which excited electrons exposed to a light source emit two photons of the same energy and direction, providing the physical basis of a laser.*

Photoelectric effect. *Interplay of photons and electron flow in which light impinging on a metal surface can produce an electrical current.*

The term ***laser*** is an acronym for light amplification by stimulated emission of radiation. As the name implies, laser technology has its origins in a process known as ***stimulated emission,*** which was first described by Albert Einstein in 1917. An extension of Einstein's Noble Prize–winning work on the ***photoelectric effect,*** the process of stimulated emission of light can be understood as follows. Electrons in the outer shell of an atom usually exist in the ground state (G), but can be excited by light or other stimuli into an excited state (E), which corresponds to a higher energy level (Fig. 8.25). Once in the excited state, an electron can return to the ground state by emitting a photon in a process known as fluorescence (see above). If an electron in the excited state absorbs a photon while in the excited state, two photons are emitted from the electron, both of which have the same energy and direction. This process is known as stimulated emission because two photons are emitted even though only a single photon is absorbed.

Because the outer-shell electrons in most atoms in nature exist in the ground state, stimulated emission is a fairly infrequent natural event. But building on Einstein's work, Schalow and Townes (1958) reasoned that stimulated emission could be made highly efficient if the starting atoms contained a large

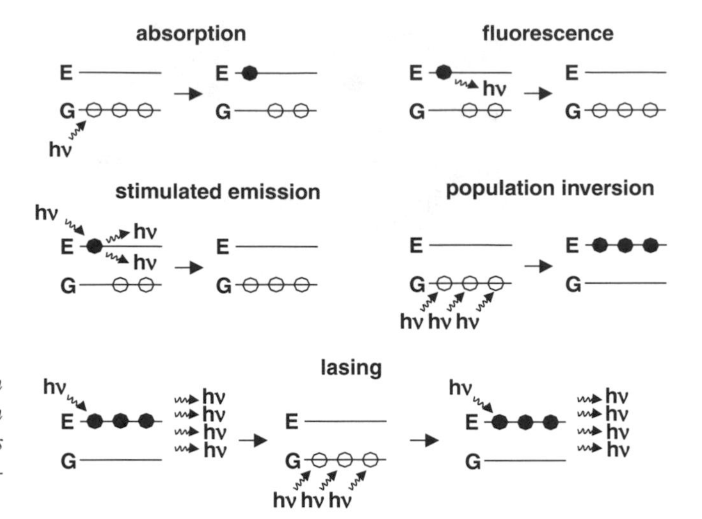

FIGURE 8.25. The electrons in an atom can exist in the ground (G) or in the excited (E) state, and transitions between these states occur by the acquisition or release of photons (hv).

Population inversion. *Phenomenon in which the majority of electrons in an atomic mixture exist in the excited state, providing a highly efficient source of electrons for stimulated emission.*

Pumping *Process in the operation of a laser in which an energy source is used to excite the atoms in the gain medium.*

Monochromatic light. *Electromagnetic radiation of a single or narrow wavelength, such as that emitted by a laser.*

Coherent. *Property of laser light in which all of the photons in the beam share the same wavelength.*

Collimated. *Property of laser light in which all of the photons in the beam are parallel to one another.*

number of electrons preexisting in the excited state (Schawlow, 1961). Mixtures of atoms that have undergone ***population inversion,*** so that the majority of outer-shell electrons are already in an excited state, provide a highly efficient source of electrons for stimulated emission. The phenomenon whereby an intense photon stream is produced from a chain reaction of excited atoms releasing photons by stimulated emission is known lasing (Fig. 8.25).

Though modern laser technology is complex and diverse, most lasers share three common components: an energy source, a gain medium, and a pair of reflectors (Fig. 8.26). The energy source excites the atoms in the gain medium in a process known as ***pumping.*** Atoms that are pumped into the excited state of population inversion, begin to release a stream of photons. Photons released in the laser cavity begin to bounce back and forth between the mirrors, leading to additional amplification by stimulated emission. By using one total mirror and one partial mirror, the amplified light is emitted as an intense laser beam from one end of the device.

Laser light is distinct from all other sources of light in that it is ***monochromatic, coherent*** and ***collimated.*** The monochromatic or single-color property of laser light occurs because excited atoms release light of a defined wavelength, and each type of atom used in a laser releases a characteristic wavelength. The beam is coherent such that the photons in the beam share the same phase. Laser light is also highly parallel or collimated. These unique properties of laser light provide a beam that is extremely intense, moves efficiently over great distances, and can be reflected in a precise manner through an optical path. Microarray scanners exploit all of these properties to direct laser light of desired wavelengths from the lasers onto the microarray substrate (Fig. 8.19).

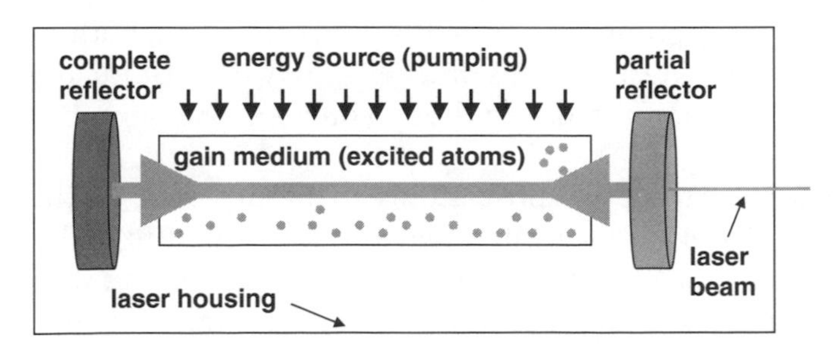

FIGURE 8.26. In a simple laser, an energy source pumps (arrows) *atoms in the gain medium into an excited electronic state* (solid circles). *Photons generated by stimulated emission bounce back and forth between two mirrors (or reflectors), forming a coherent beam of light that escapes through the partial reflector.*

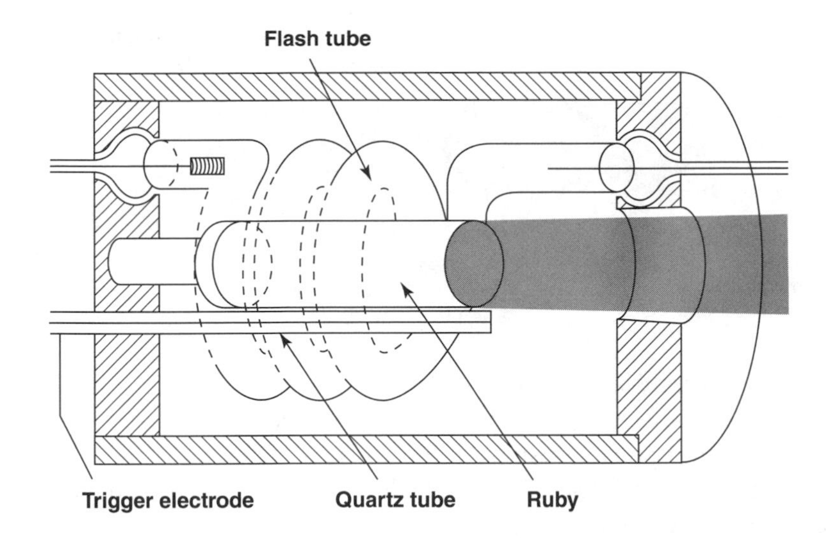

FIGURE 8.27. The world's first laser. (Reprinted with permission from Nature, *London.)*

The first laser was developed by Maiman at Hughes Research Laboratories (Malibu, CA) in 1960 (Fig. 8.27). Maiman used a synthetic ruby as the gain medium, a xenon flash tube as the energy source, and evaporated silver on each end of the ruby rod as the reflectors. Pulses from the flash tube caused population inversion of the chromium ion electrons in the ruby crystal, leading to stimulated emission and the release of an intense beam of red laser light. Because the ruby lasers use light as the energy source, they are sometimes referred to ***optical lasers.*** Building on Maiman's work, researchers have created an explosion in laser technology.

Optical laser. *Type of laser, such as a ruby laser, that uses a light source for pumping.*

Similar to the ruby laser, modern-day lasers are named according to the source of atoms used in the gain medium. Gain media consisting of solid, liquid, and gaseous materials are all used with success in lasers; gas lasers and solid-state lasers are the most widely used for microarray scanners. The most common gas lasers include those that use a mixture of helium and neon gas (HeNe lasers) and positively charged ions such as krypton and argon (Kr–Ar lasers). Popular solid-state lasers include those that use yttrium-argon-garnet (YAG), gallium arsenide, and other exotic semiconductor materials.

The HeNe (pronounced *hee nee*) lasers are widely used in microarray detection instruments (Table 8.3). They are low-cost lasers that have excellent optical properties; small size (10–30 cm); good durability (>10,000 h); and convenient wavelengths in the green (543 nm), yellow (594 nm), orange (611 nm), and red (633 nm) portions of the visible spectrum. The green and red HeNe wavelengths correspond closely to the excitation maxima of Cy3 (550 nm) and Cy5 (649 nm), respectively, providing extremely efficient excitation sources for

TABLE 8.3. Specifications of Lasers Used for Microarray Scanners[a]

Feature	HeNe	Ion	Diode	DPSS[b]
Type	gas	gas	solid state	solid state
Wavelengths (nm)	543, 594, 611 and 633	488, 514, 568, 647, 752	635, 650, 780	532
Beam	excellent	very good	good	very good
Cooling	convection	forced air	convection	convection
Efficiency	good	moderate	excellent	very good

[a] Data used with permission from M. Griot (Irvine, CA).
[b] Diode-pumped solid state.

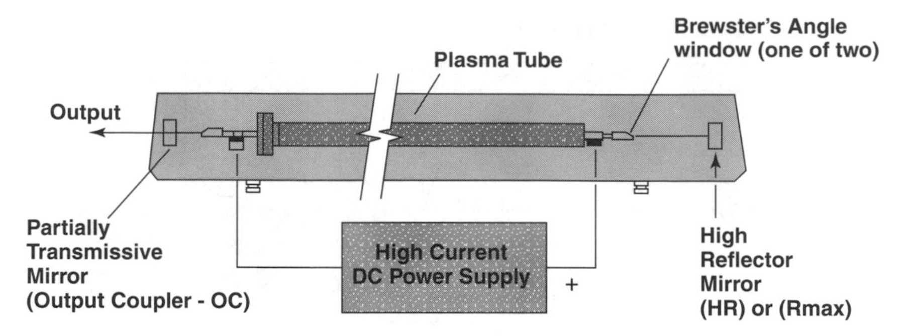

FIGURE 8.28. *An argon ion gas laser. (Courtesy of Rockwell Laser Industries, Cincinnati, OH.)*

these widely used fluorescent dyes. The lower-power HeNe lasers (5–20 mW) can be cooled passively by convection, which obviates the need for elaborate water-based cooling mechanisms and enables compact and simplified scanner systems to be built with these lasers. HeNe lasers typically use an electrical discharge in the gaseous gain media as a source of atomic excitation.

Mixed gas. *Type of laser, such as a krypton–argon laser, used in some microarray scanners that contains two or more gases as the gain medium.*

The second type of gas laser used for microarray applications are the ion lasers that commonly employ either argon or krypton gas (Table 8.3). The so-called ***mixed gas*** Kr–Ar ion lasers use a mixture of the two gases. Similar to the HeNe lasers, ion lasers have excellent optical properties, good durability, and convenient wavelengths. The strongest lines for the argon ion lasers are blue (488 nm) and green (514 nm), and the intense lines for krypton are yellow (568 nm) and red (647 nm). Mixed gas lasers provide output wavelengths of 454–752 nm, which essentially covers the important wavelengths in the visible spectrum. The blue line of argon at 488 nm is used commonly to excite the fluorescein dyes in microarray scanners. Ion lasers are generally somewhat larger and less efficient than the HeNe lasers and often require cooling by water or forced air to dissipate the heat that is generated during their operation (Fig. 8.28).

Diode laser. *Compact laser used in many microarray scanners that contains semiconductor materials as the gain medium.*

Diode-pumped solid-state laser. *Type of laser that uses the beam of a diode laser to excite the semiconductor gain medium in the second laser.*

Solid-state lasers are also widely used in microarray detection instruments (Table 8.3) and represent a rapidly advancing technology area. The most common solid-state lasers found in microarray scanners include the ***diode lasers*** (or laser diodes) and the ***diode-pumped solid-state*** (DPSS) ***lasers.*** The diode lasers, such as the red lasers used for laser pointers, are extremely compact (2–10 cm), highly efficient, and provide an intense beam at an affordable price. The gain media for all diode lasers is made from semiconductor materials such as gallium arsenide, and different combinations of molecules provide different emission wavelengths. Stimulated emission from the semiconductor material is achieved by applying an electric current across the gain material.

Because of the high efficiency of diode lasers and the naturally reflective properties of the semiconductor materials, reflective mirrors are not required to produce output beams of high intensity (10 mW). Diode lasers, to a somewhat greater extent than gas lasers, tend to show more laser-to-laser variation in terms of the output wavelength and optical quality. Diode laser performance can also vary with temperature, though appropriate feedback devices in microarray detection instruments virtually eliminate complications arising from different operating temperatures. Diode lasers provide output wavelengths of 630 nm and greater, with the 635 nm red line being commonly used to excite the Cy5 dyes. Their small size has allowed a high degree of miniaturization in microarray scanners.

The DPSS lasers are somewhat more complicated than the diode lasers. They use the beam from a diode laser (e.g., 820 nm) to excite a small chip of semiconductor material that usually consists of yttrium-argon-garnet doped with neodymium ions (Nd:YAG) and acts as the gain medium. DPSS lasers offer intense outputs in the blue (457 nm), green (532 nm), and infrared portions of the spectrum, with the green line often being used to excite the Cy3 dyes.

No discussion of laser technology would be complete without mention of the hazards of working with lasers. Lasers are classified on a scale of I to IV, with class I lasers posing no hazard to the human eye and class IV lasers being extremely dangerous. Most lasers used in microarray applications are class II or III; a 10-mW laser source being roughly equivalent in intensity to bright sunlight. Looking directly into a laser beam or pointing a laser of any kind including laser pointers at another person is *extremely dangerous* and may lead to permanent retina damage or blindness! The lasers in microarray scanners are enclosed in protective casings, providing a high degree of user safety. *Extreme caution* should be exercised at all times when working with lasers outside the scope of commercial instruments operating under their specified parameters!

Mirror. *Optical component used in microarray detection instruments that reflects light in a highly precise manner.*

Beam splitter. *Optical filter in a microarray detection instrument that divides excitation or emission light into reflected and transmitted components.*

Filter. *Round optical component of a microarray detection instrument that excludes specific wavelengths of light.*

Optical Mirrors and Filters

Laser beams provide intense, monochromatic light sources that are invaluable for exciting fluorescent dyes. But the usefulness of laser light in microarray scanners depends on the ability to direct, attenuate, and filter the desired excitation and emission wavelengths. Optical ***mirrors, beam splitters,*** and ***filters*** are used for this purpose.

A mirror is an optical component that reflects light in a highly precise manner. The basic components of a mirror include a substrate, such as polished glass, and a reflective coating (Fig. 8.29). A mirror reflects the incoming or incident light because the surface of the mirror does not absorb the energy of the electromagnetic radiation. When laser light shines on an optical mirror, the beam is reflected so that the angle between the mirror and the light beam, known as the ***angle of incidence,*** equals the ***angle of reflection.*** Laser mirrors are used widely in microarray scanners to direct laser light into optical filters and lenses. The ability to reflect a laser beam in a nonlinear manner enables a highly compact scanner design.

Angle of incidence (θ_1). *Arc formed between a mirror and the incoming light beam in a microarray detection instrument.*

Angle of reflection (θ_2). *Arc formed between a mirror and the outgoing light beam in a microarray detection instrument.*

Dielectric coating. *Special multiplayer material containing aluminum and other materials, used in the preparation of mirrors and optical components in microarray detection instruments.*

Because a laser beam is a highly intense light source, a special multilayer reflective ***dielectric coating*** is often used for scanner applications. One popular dielectric coating contains aluminum as the principle component. The glass substrates are polished to optical flatness at a scratch/dig specification of 10/10 or better. A typical mirror used to reflect a red laser line (e.g., 635 nm) might

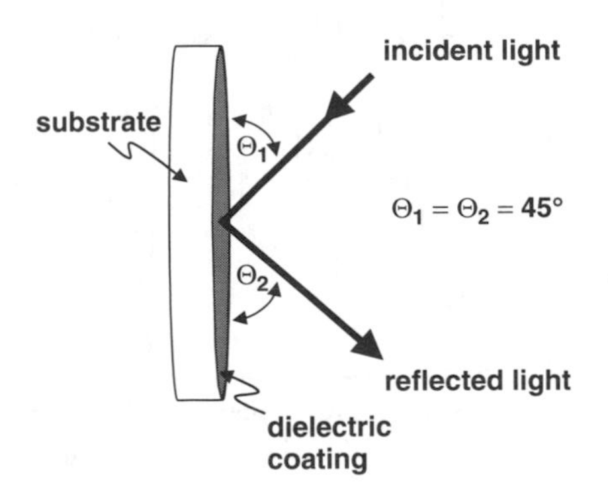

FIGURE 8.29. A glass substrate containing a dielectric coating (mirror) provides a highly reflective and durable surface for incoming (incident) light. The angle of incidence (θ_1) equals the angle of reflection (θ_2).

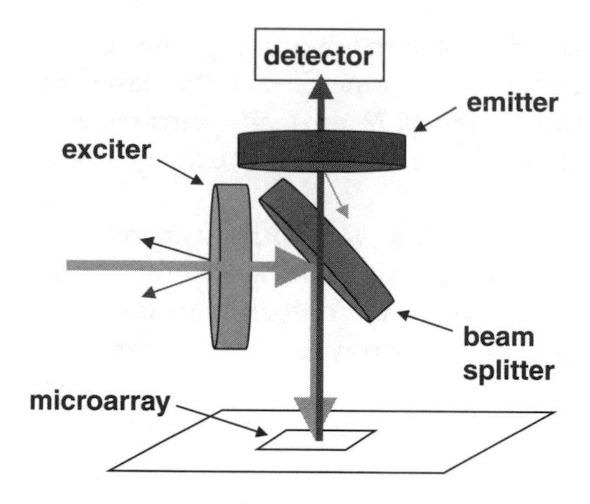

FIGURE 8.30. *How an excitation filter* (exciter), *a dichroic beam splitter, and an emission filter* (emitter) *are used to separate excitation laser light from the emitted fluorescent signal from a microarray. Only the excitation light is able to pass through to the detector.*

Optical filter. *Optical component in a microarray detection instrument that transmits light over a very narrow bandwidth.*

Filter set. *Pair of filters in a microarray detection instrument, typically consisting of an exciter and emitter.*

Filter wheel. *Rotary device in a microarray detection instrument that holds a plurality of optical filters.*

Exciter. *Optical filter in a microarray detection instrument that transmits the excitation light.*

Emitter. *Optical filter in a microarray detection instrument that allows passage of the fluorescence emission.*

have a reflectance rating of 99% or better in the 600–800 nm range and physical dimensions of about 25 mm wide × 5 mm thick.

Optical filters are also widely used in fluorescence detection instruments. Filters are optical components that transmit light over a narrow (20–50 nm) bandwidth, allowing specific excitation and emission wavelengths to be transmitted and detected during microarray scanning. Filters are often used in pairs known as ***filter sets,*** and the carousel-like devices that hold the filters are termed ***filter wheels.*** The two filters in a filter set are the ***exciter*** and the ***emitter,*** corresponding to the filter that transmits the excitation light and the one that allows passage of the fluorescence emission, respectively. Exciters and emitters are often used in conjunction with another optical component known as a beam splitter. As the name implies, beam splitters are optical devices that divide, or split, the laser light into two separate components. Beam splitters are available with different ratios of reflected and transmitted (R/T) light percentages; 50%R/50%T is a common configuration. A tandem configuration of exciter, beam splitter, and emitter provides a clean emitted signal from a fluorescent microarray (Fig. 8.30).

Transmission spectra for a filter set and beam splitter demonstrate the wavelengths that are transmitted when detecting a fluorescein dye; the filters are chosen so that the transmitted wavelength of the beam splitter resides between the transmission maxima of the exciter and emitter (Fig. 8.31). In most microarray experiments, the intensity of the fluorescence emission is often up to 1 million times weaker than the intensity of the excitation laser light, placing

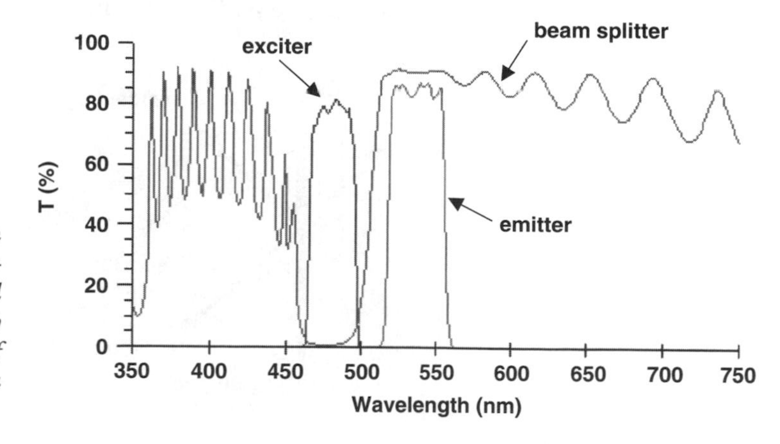

FIGURE 8.31. *Light transmission spectra for an excitation, beam splitter, and emission filter set designed to detect emission from FITC, which has an absorption maximum of 490 nm. (Data courtesy of Chroma Technology, Brattleboro, VT.)*

a premium on the quality of the optical components that transmit the fluorescent signal and reject the excitation light. The task is particularly challenging because a typical glass microarray surface can reflect as much as 1% or more of the laser light back into the collection path.

In constructing a microarray scanner, it is sometimes desirable to reduce, or ***attenuate,*** the intensity of a laser beam. Filters that attenuate a laser beam over a broad spectrum of wavelengths are known as ***neutral density filters.*** In contrast to exciters, emitters, and beam splitters, neutral density filters reduce laser power with roughly equal efficiency for all visible wavelengths. Neutral density filters attenuate the laser beam either by absorbing or reflecting the nontransmitted light, though care must be taken in using the latter type to avoid contaminating the light path with reflected laser light. Applications of neutral density filters include intensity reduction of lasers with excessive power and intensity matching of two lasers that have desired emission wavelengths but slightly different power ratings.

Attenuate. *Process of reducing the intensity of a laser beam.*

Neutral density filter. *Optical component in a microarray detection instrument that reduces the power of a laser beam.*

Lenses

Some microarray scanners make use of one or more lenses to carry out fluorescence detection. The lens is an optical component that focuses the excitation light or fluorescence emission into a discrete point at a fixed distance from the lens (Fig. 8.32). Focusing the excitation beam serves to increase the intensity of the laser light by focusing or ***converging*** the beam on the microarray surface. The input laser beam in most scanners is rather wide (0.5–2 mm); but in systems that use an objective lens, the focused beam width is about the size of a single pixel (5–10 μm). A lens that focuses a 5 mW laser beam into a 10 × 10-μm area produces a final illumination intensity of ~5,000 W/cm^2, an illumination intensity that would readily photobleach most dyes if it were not for the short (e.g., 10–20 μs) dwell time of the laser due to the rapid nature of the scanning process.

In confocal systems, focusing the emission beam can also be useful in that it allows the fluorescent signal to be passed through a pinhole, reducing the amount of noise that impinges on the detector (Fig. 8.19). In multiple lens microarray scanners, the ***objective lens*** focuses laser light on the microarray and the ***detector lens*** focuses the emitted light onto the detector. Most confocal systems are configured such that both the excitation and the emission beams pass through a single objective lens, a design feature known as ***epifluorescence.***

Converge. *Optical process of focusing a beam of light into a point source, such as is achieved with a lens.*

Objective lens. *Optical component in a microarray detection instrument that focuses the laser light on the microarray.*

Detector lens. *Optical component in a microarray detection instrument that focuses the emitted light onto the photomultiplier tube, charge-coupled device, or other light-sensing device.*

Epifluorescence. *Confocal microarray scanning architecture in which the excitation and emission beams pass through a single objective lens.*

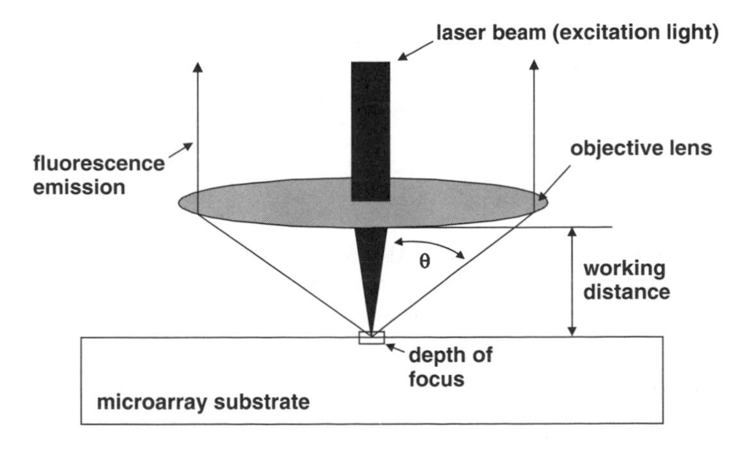

FIGURE 8.32. The laser beam (thick line) *is focused by the objective lens onto the microarray surface with a working distance* (double-headed arrow) *and depth of focus determined by the physical properties of the lens. Fluorescence emission* (arrows) *is captured by the objective lens as a cone of light, the angle of which* (θ) *defines the numerical aperture.*

TABLE 8.4. Technical Specifications for some Standard Microscope Objectives[a]

Power	Numerical Aperture	Working Distance (mm)
4×	0.10	16.70
10×	0.25	4.40
20×	0.40	3.30
40×	0.65	0.57
60×	0.85	0.28

[a] Data adapted from Edmund Scientific, Tonawanda, NY.

Working distance. *Measure of the physical distance between the end of the objective in a microarray detection instrument and the microarray surface when the lens is in focus.*

Depth of focus. *Distance between the closest and farthest images that remain in focus during microarray detection.*

Numerical aperture. *Measure of the angle or cone of light that is captured by an objective lens during microarray scanning or imaging; the larger the NA value the light more that is efficiently captured and the smaller the depth of focus.*

Efficiency. *Light-capturing ability of a lens in a microarray detection instrument.*

Several objective lens specifications are most useful in the context of microarray detection: ***working distance, depth of focus,*** and ***numerical aperture*** (Fig. 8.32). The working distance of an objective is a measure of the physical distance between the end of the objective and the microarray surface when the lens is in focus. A typical working distance for microarray objectives is 0.5–5 mm. The depth of focus refers to the distance between the closest and farthest images that remain in focus during microarray detection, with typical values being in the range of 10–30 microns. Numerical aperture (*NA*) is a measure of the angle or cone of light that is captured by an objective lens during microarray scanning, and is calculated simply by taking the sine (sin) of the maximum angle of light (θ) that can be captured by the lens:

$$NA = \sin\theta \tag{8.11}$$

Most microarray lenses have numerical aperture values of 0.5–0.8, with a value of 1.0 being the maximum that can be obtained using air (instead of oil) as the transmitting medium. A lens with a large numerical aperture can capture a larger cone of light than a lens with a small numerical aperture, and thus the light-capturing ability, or ***efficiency,*** of a lens increases as the numerical aperture increases. Efficiency increases exponentially with respect to numerical aperture, and so a small increase in *NA* can have a large increase in collection efficiency. Fluorescence emission from microarrays is a relatively weak light source, and thus large numerical aperture lenses are preferable in microarray scanning applications. There is an important trade-off, however, among numerical aperture, working distance, and depth of field, in that both of the latter parameters decrease as the numerical aperture increases (Table 8.4). A lens with a large numerical aperture will have a much smaller working distance and depth of field than a lens with a smaller numerical aperture, placing greater importance the quality of the substrate and scanner stage in instruments that use large *NA* lenses. Microarray spots that are outside the depth of focus will be rejected in the confocal configuration, making it imperative to have an extremely flat surface and substrate stage if large numerical aperture lenses are used.

Detector. *Photomultiplier tube, charge-coupled device, or other component of a microarray detection instrument that converts photons into electrical current.*

Photomultiplier tube. *Detector in a microarray scanner that converts photons into an amplified electrical signal.*

Photomultiplier Tubes

Microarray detection instruments convert the light emitted from fluorescent microarrays into an electrical current, which can then be measured and stored as a microarray image. The conversion of photons into current is carried out using a device known as a ***detector.*** The most common type of detector is the ***photomultiplier tube*** (PMT) located inside the scanner. A photomultiplier tube multiplies the signal from a light source, such as a fluorescent microarray, exploiting

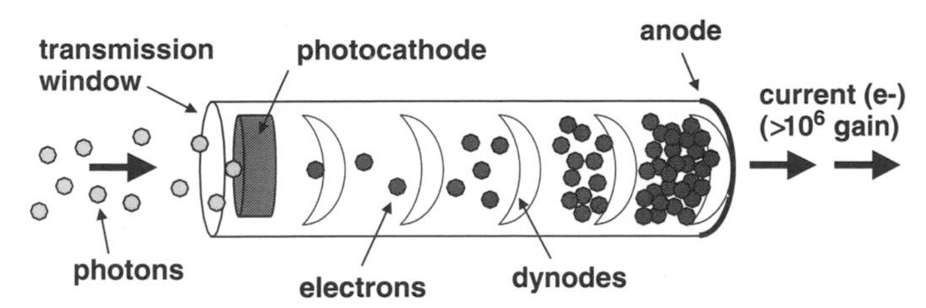

FIGURE 8.33. Photons (light circles) *emitted from a fluorescent microarray pass through a transmission window and impinge on a photocathode* (gray disk)*, resulting in the release of electrons* (dark circles) *by the photoelectric effect. The electrons are amplified through a series of dynodes* (crescents) *and gather at the anode. The current that flows out of the PMT can have an amplification or gain of more than a millionfold relative to the input signal.*

a process known as the photoelectric effect described by Einstein in the early 1900s. The photoelectric effect is the process whereby a beam of light (photons) shining on a metal surface results in the release of electrons from that surface. Photons of sufficient energy, such as those emitted from a fluorescent microarray, instigate the flow of electrons (electrical current) by exploiting the photoelectric effect.

Photocathode. *Light-sensitive device in a photomultiplier tube that captures the incident fluorescent photons and converts them into electrons.*

Dynode. *Charged electrode in a photomultiplier tube that amplifies the flow of electrons produced by the photocathode.*

Anode. *Component in a photomultiplier tube that collects the amplified electron stream from the dynodes.*

A photomultiplier tube contains several main components including a transmission window, a ***photocathode,*** a ***dynode,*** and an ***anode*** (Fig. 8.33). Photons emitted from the microarray pass through the transmission window and impinge on a light-sensitive device known as the photocathode, resulting in the release of electrons from the metal surface. Electrons dislodged from the photocathode jump onto a charged electrode called the dynode. Each electron that hits the dynode causes multiple electrons to be released, and because the dynodes are configured in a series of 10 or more elements, a huge amplification of the signal is achieved across the PMT. The amplified flow of electrons across the dynodes is collected at the anode and sent out of the PMT as an electrical current. The PMT thus converts a relatively small number of photons from a fluorescent microarray into a relatively large stream of flowing electrons (i.e., an electrical current) that is easily measured and recorded.

Photomultiplier tubes are well suited for microarray detection for a number of reasons, including uniform sensitivity over most of the visible spectrum, low dark current, rapid response time, and large dynamic range. The uniform sensitivity of the PMT to different wavelengths of light means a single type of tube can be used to capture fluorescent signals from many different dyes. The low dark current (i.e., low PMT signal in the absence of light) allows dim fluorescent signals to be recorded, including those that correspond to human genes expressed at a low level. Because the PMT signal relies on electronic excitation and electron flow, the tubes have rapid (nanosecond time scale) response times and are thus amenable to rapid microarray scanning, which occurs on the microsecond time scale. In scanning systems that allow detection of two or more dyes, multiple PMTs can be used to capture each signal. The use of a separate detector for each color minimizes the optical cross-talk between different channels.

The large dynamic range of PMTs derives from the low dark current and the fact that the current generated by the tubes is proportional to the photon flux over many orders of magnitude. The amplification factor, or gain, of a PMT, which corresponds to the output signal divided by the input signal, can be adjusted by changing the voltage applied to the dynodes (Fig. 8.34). PMTs that have a larger number of dynodes, or ***stages,*** generally allow higher gain at a

Stage. *Corresponds to the number of dynodes in a photomultiplier tube. Alternatively, the methodological term corresponding to the overall physical appearance of a system or organism in the methodology.*

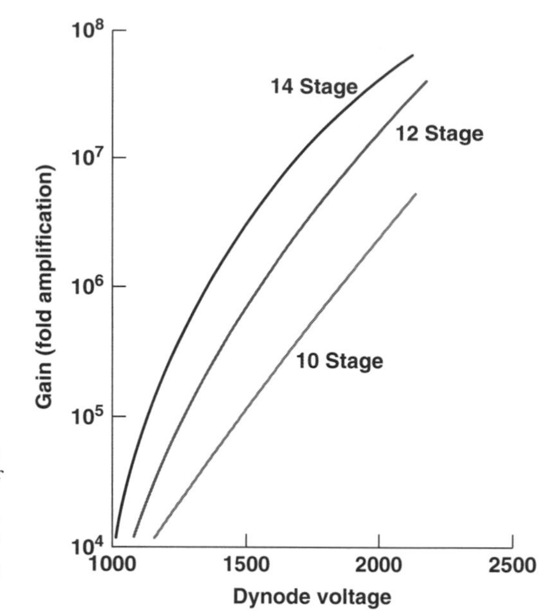

FIGURE 8.34. *Fold amplification (gain) measurements as a function of voltage for three PMTs with 10-, 12-, and 14-stage dynodes. (Data courtesy of M. W. Davidson, Florida State University, Tallahasse.)*

given voltage. The capacity to adjust the gain of the PMT (generally via scanner software) allows the microarray user to scan at a wide range of sensitivity settings, a capacity that is convenient for microarray analysis because microarray signals can vary more than a millionfold. If the sensitivity of the scanner is set at a high level, the scan is performed at what is known as a high gain. Note once again that sensitivity refers to the efficiency by which a microarray detection device converts photons into electronic signal, and detectivity is the dimmest fluorescent signal that can be read above background. The two terms are both important, but they are not interchangeable.

One limitation of a PMT, versus for example a charged-coupled device (see below), is that the PMT functions as a single pixel detector and thus does not allow capture of more than one pixel at a time. This is in contrast to charge-coupled devices that contain an array of detectors (e.g., 1000 × 1000) and so allow a relatively large area of a microarray (e.g., 1 cm^2) to be captured without moving the optics during detection. The somewhat cumbersome requirement of scanning with PMTs is more than offset, however, by the many advantages of a PMT, including low dark current, high gain, and large dynamic range. Several other types of detectors have also been used in microarray scanners, including ***avalanche photodiodes,*** which are semiconductor-based light sensing devices. Photodiodes generally have somewhat lesser gain capabilities and higher dark current ratings than photomultiplier tubes and are thus not used nearly as often as PMTs in microarray scanning applications.

Avalanche photodiode. *Semiconductor-based light-sensing device used in microarray detection instruments.*

Analog-to-Digital Conversion

A light detector, such as a PMT, converts a beam of photons into an electrical signal. The electrical signal sent from the anode of the photomultiplier tube consists of a continuously varying flow of electrons and is thus termed an ***analog*** signal because it is "analogous" to the physical flow of electrons. To make use of analog signal information, analog signals must be converted into a form that can be read by a computer. Computers understand data in a ***digital*** form,

Analog. *Electrical signal from the anode of a photomultiplier tube or other source that consists of a continuously varying flow of electrons.*

Digital. *Electrical signal or other data source represented as a series of binary (0 or 1) values that can be read with a computer.*

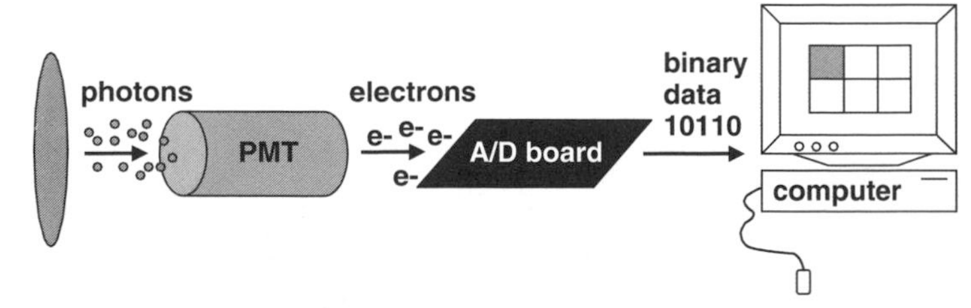

FIGURE 8.35. A fluorescent microarray spot (oval) *emits fluorescent light as photons* (solid circles), *which are collected by a PMT and converted into an electrical current (e^-). The analog electrical current is then converted into binary code by an A/D converter. The binary values are sent to a computer and appear on the screen as pixels with intensities proportional to the photon flux from a designated portion of the microarray.*

or more specifically as a ***binary code,*** which specifies all values in terms of 0 or 1. The analog signals that flow from the PMT of a microarray scanner are converted into digital information by a solid-state device known as an ***analog-to-digital converter.*** The A/D converter receives electronic signals from the PMT, converts the voltage information into discrete digital values, and sends binary information to the personal computer as an output (Fig. 8.35). Each pixel that appears on the computer screen corresponds to a single measurement taken by the A/D converter corresponding precisely to the number of photons emitted from a single region on the microarray.

Binary code. *Nomenclature system used in computers that represents all specified values in terms of 0 or 1.*

Analog-to-digital converter. *Solid-state device that converts a continuous flow of electrons into binary signals that can be read and manipulated with a computer.*

By way of example, a PMT might have a current flow corresponding to a voltage range of 0–100 such that no fluorescent signal from a microarray gives a reading of 0 V and a maximum fluorescent signal produces 100 V. The A/D converter will assign a digital number to each voltage reading at precise moments in time, with the numbers being proportional to the voltage sent from the anode of the photomultiplier tube. The number of values assigned to the voltage range is determined by the binary digit or ***bit*** rating of the A/D board. Most microarray scanners use A/D boards that provide digital conversions in a 16-bit format, meaning that output numbers can have 2^{16} (65,536) possible values. Owing to the use of either 0 or 1 in the binary code, the total range of values provided by any A/D converter is obtained by raising the number of bits (n) to the power of 2 as follows:

Bit. *Shortened form of binary digit that refers to the number of values assigned to the full spectrum of analog values.*

$$\text{Number of values} = 2^n \tag{8.12}$$

It follows from Equation 8.12 that 12-bit and 8-bit converters would provide 4,096 and 256 values, respectively, compared to the 65,536 different possible values provided by the 16-bit converter.

Analog-to-digital converters that provide 16 bits of information are preferred to those with lower specifications because they provide better accuracy and enable a larger dynamic range in microarray applications. The superior accuracy occurs because the current from the photomultiplier tube is divided into smaller bins with a 16-bit converter than with lower-bit converters. For the example given above, a 100-V range divided into 65,536 bins would assign a unique number to different voltages at 1.52 mV (100/65,536) intervals, versus 24.4 mV and 391 mV, respectively, for the 12- and 8-bit converters. Smaller bins mean that a smaller change in fluorescent intensity can distinguished, improving the accuracy of the data in terms of converting photons into digital values. Dynamic range is also improved with 16-bit converters, allowing more than four orders of magnitude dynamic range (65,536:1) compared to the

8-bit converter, which allows just a little more than two orders of magnitude (256:1).

Microarray Images

File. *Collection of digital microarray data stored on a computer.*

Personal computer. *Any of a wide spectrum of binary hardware devices used to control the robots employed in microarray manufacture and detection and to analyze and model microarray data.*

File name. *Text identifier for a microarray computer file, which in the DOS system corresponds to the letters to the left of the decimal point.*

Extension. *Text identifier in a microarray computer file that allows file recognition by the operating system, which in DOS corresponds to the three letters to the right of the decimal point.*

Microarray detection instruments examine microarrays by exciting fluorescent dyes on the surface and gathering the fluorescence emission by converting a stream of photons into digital values that can be stored on a computer. A collection of digital microarray data stored on a computer is known as a ***file.*** Computers that run most of the home-built and commercial microarray scanners are known as ***personal computers,*** and nearly all of the personal computers (PCs) use the Windows operating system developed by MicroSoft (Redmond, WA). The early microarray scanners used MicroSoft Windows 95 or 98, whereas most of the contemporary scanners employ Windows 2000, NT 4.0 or Millennium.

Files in the Windows operating system are identified on the basis of a ***file name*** and an ***extension,*** separated by a period in the character string. The file name provides the user a convenient means to identify a particular file and the extension allows file recognition by the operating system. One popular naming system for scanned microarray files employs the user's initials, the date, and scan as the file name, followed by a period and the file extension. For example, a scanned file generated by Jane Smith on April 17, 2001, and stored in the standard TIFF format (see below) would have the file name JS041701a.tif according to the following convention: user's initials (JS), month (04), day (17), year (01), scan number (a), period to separate the file name and extension (.), and file format (tif). The early versions of Windows required a maximum of eight characters to the left of the period and a maximum of three characters to the right of the period for the extension. More recent versions of Windows allow more than eight characters for the file name, but retain the three character extension on the right side of the period. In choosing file names for Windows computers, the use of spaces and punctuation marks (except to separate the file name and extension) should be avoided.

Bit map. *Digital file corresponding to a microarray surface.*

Graphics file. *Any of several different file types consisting of ordered rows and columns of digital values that provides a 1:1 numerical representation of the fluorescent values acquired from a microarray by a detection instrument.*

Microarray spots are fluorescent regions located on the surface of the microarray substrate, whereas digital microarray images are a numerical representation of the fluorescent spots. The digital file corresponding to a microarray surface is known as a ***bit map,*** derived from binary digit map. The bit map provides a 1:1 numerical representation of the fluorescent values acquired from the microarray and is also known as a ***graphics file.*** Graphics files contain ordered rows and columns of digital values, each derived from its corresponding position on the microarray. The standard graphics file format for microarray scanners is TIFF. TIFF, developed in the mid-1980s at Aldus is a graphics format designed for storing and retrieving high-quality images. Microarray TIFF files carry the .tif extension and provide the most accurate graphical representation of the true microarray data. TIFF files containing the standard 16-bit data are essentially two-dimensional bit map files wherein all fluorescent intensities are stored as a number between 1 and 65,536.

Byte. *Amount of memory on a personal computer required to store eight bits of information.*

Graphical microarray TIFF files are stored on a computer, and the size of each file occupies an amount of computer memory known as a ***byte.*** A byte of memory on a PC corresponds to the amount of memory required to store eight bits of information. A microarray image obtained from scanning at 10-μm resolution (see below) has a file size of approximately 1,900,000 bytes (kilobytes, KB) or 1.9 million bytes (megabytes, MB). A 1.9 MB TIFF file can be reduced

TABLE 8.5. Common Graphical File Formats Used for Storing Microarray Images

File Type	Extension	Size (KB)	Compression (fold)
TIFF	.tif	1900	none
GIF	.gif	52	27
JPEG	.jpg	28	68

in size or ***compressed*** to reduce the amount of memory required to store the file (Table 8.5). Compressed graphics files are convenient for display over the Internet because they load much more quickly than full-size TIFF files. The most common types of compressed graphical files used to store microarray images are the graphical interchange format (GIF) and joint photographic experts group (JPEG). Conversion of TIFF images into GIF and JPEG formats provides respective 27-fold and 68-fold file size reduction. TIFF, GIF, and JPEG files are extremely convenient in that all three formats can be used on many different operating systems, including Windows and Macintosh (Apple Computer, Cupertino, CA).

Compress. *Act of reducing the size of a computer file.*

To be stored as a graphics file, the microarray image is broken into discrete pieces or numerical bins, known as picture elements or pixels. The information stored in each pixel is generating by taking the average fluorescent signal at each location over the entire two-dimensional microarray surface. The number of pixels used to represent a given microarray image is known as the resolution, with most commercial scanners and imagers designed so that the user can select from among several different resolution settings. The convention in microarray detection dictates a resolution pixel size no greater than one tenth the diameter of the average spot present in the microarray. A microarray consisting of 100-μm-diameter spots should be scanned at a pixel size of no greater than 10 μm, and this scanner setting would be known as "10-μm resolution."

A microarray containing spots of 125 μm in diameter is represented nicely with a 10-μm scan, but the same microarray scanned at 50-μm resolution is nearly unrecognizable (Fig. 8.36). The grainy appearance of the 50-μm resolution scan results from the fact that too few pixels are used to represent the microarray image. High-quality images are critical for obtaining accurate information from the scanned microarray (see Chapter 9).

Most contemporary microarray scanners provide the experimentalist with a variety of resolution settings from 3 to 100 μm in pixel size. The choice of scanning resolution is an important consideration because a smaller pixel size

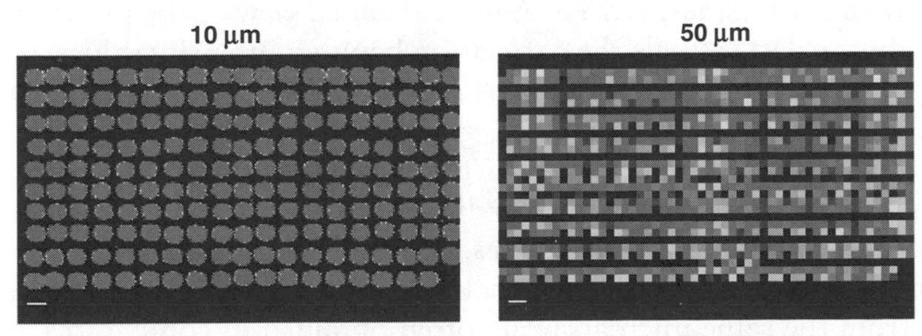

FIGURE 8.36. *Two microarray images obtained from the same fluorescent microarray, scanned at a pixel size of 10 μm and 50 μm. The TIFF data were obtained using a ScanArray 3000 (Packard Biochip, Billerica, MA) and are presented in a rainbow palette from* NIH Image 1.62 *software developed by Rosband (National Institutes of Health, Bethesda, MD). Space bars = 150 μm. Figure also appears in Color Figure section.*

TABLE 8.6. Relationship among Pixel Size, File Size, and Scan Time for a Microarray with an Area of 1.0 cm^2[a]

Pixel Size (μm)	File Size (MB)	Scan Time (s)
3	21.4	177
6	5.3	88
9	2.4	59
12	1.3	44
15	0.96	35
18	0.60	29
21	0.44	25
24	0.34	22

[a]Data obtained with a ChipReader scanner set to the Cy3 channel and a speed of 25 Hz.

improves the quality of the microarray data and image but slows the scanning speed and creates files that require greater storage space (Table 8.6). Because graphical files are two-dimensional representations of microarrays, a twofold decrease in pixel size results in a fourfold increase in file size. Scanning an entire microarray substrate (22 × 74 mm) at 3μm-resolution would generate a TIFF file of 348 MB, and 30 microarray files of this size would fill an entire 10-GB computer hard drive. Care should be taken to choose a scan resolution that is optimal for a given microarray feature size; the optimal pixel size (in microns) is obtained by dividing the microarray spot diameter (in microns) by 10:

$$\text{Optimal scan resolution} = \frac{\text{Spot diameter}}{10} \tag{8.13}$$

Color palette. *A look-up table used to index a graphical microarray image, producing red, green, and rainbow representations of microarray data.*

Each pixel in a microarray image contains a discrete value corresponding to the average fluorescent signal at the corresponding position on the chip. A microarray image at the level of computer files is a two-dimensional array of bytes. For 16-bit TIFF files, the values assigned to each pixel are integers from 1 to 65,536. Microarray images can be displayed on computer screens by using the pixel values to index a look-up table (LUT), also known as a ***color palette.*** Colorizing microarray data aids in the analysis of the results and makes the data easier to understand. Indexing is achieved by assigning a color from a palette to each pixel based on the intensity value of that pixel. The most common palettes used to display microarray image files are black and white, monochrome (e.g., green) and rainbow. In the examples shown in Figure 8.37, a high level of fluorescence intensity corresponds to white, bright green, and red on the three palettes, respectively. A more detailed view of the pixels displayed in a rainbow palette reveals the correlation between intensity values and color as it appears on the computer screen (Fig. 8.38). Many microarray scanners allow the creation of custom LUTs, giving the user the freedom to depict microarray images in virtually any palette of interest (Fig. 8.39).

Two-color overlay. *Graphical combination of two separate microarray images created by subtracting the values in the two images to obtain a composite image, and used to depict gene expression and genotyping data from two samples in a single image.*

It is also possible to combine two different images from the same microarray recorded in different color palettes, with each color palette corresponding to data obtained from one of two channels in the scanning instrument. Two images from the same microarray are often obtained in comparative gene expression or genotyping experiments wherein separate dyes (e.g., red and green) are used to label the two samples. A graphical combination of two separate images is known as a ***two-color overlay*** and is created by subtracting the values in the two images to obtain a composite image (Fig. 8.40). The composite image

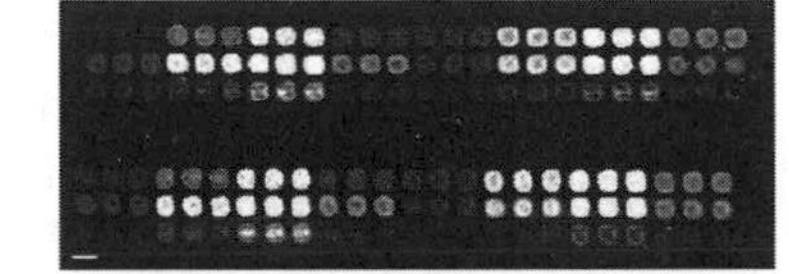

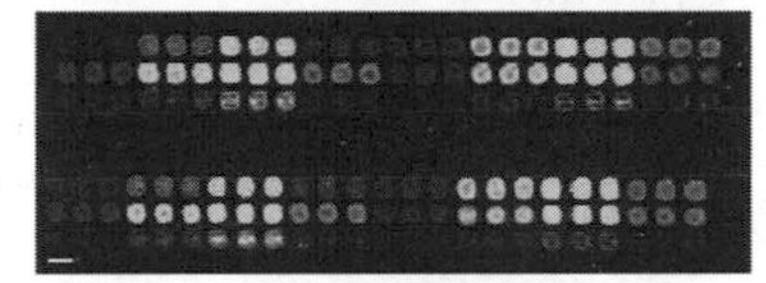

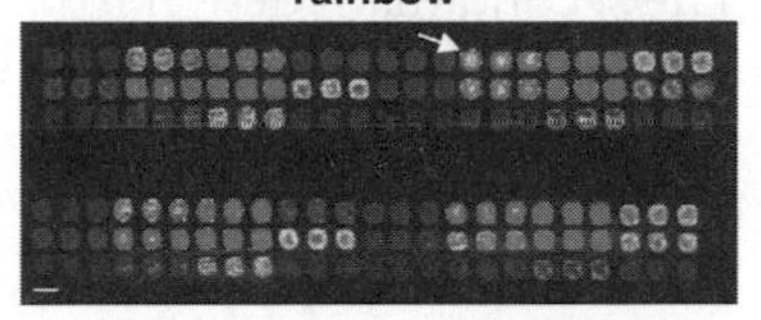

FIGURE 8.37. *A single microarray image with the values from the TIFF file used to index a LUT composed of a black and white, green, or rainbow scale. The brightest fluorescent signals are represented as white, bright green, and red in the three respective palettes. Space bars* = 150 *μm.* Arrow, *the portion of the image shown in Figure 8.38. (Data generated with a ScanArray 3000 instrument set to the Cy3 channel.) Figure also appears in Color Figure section.*

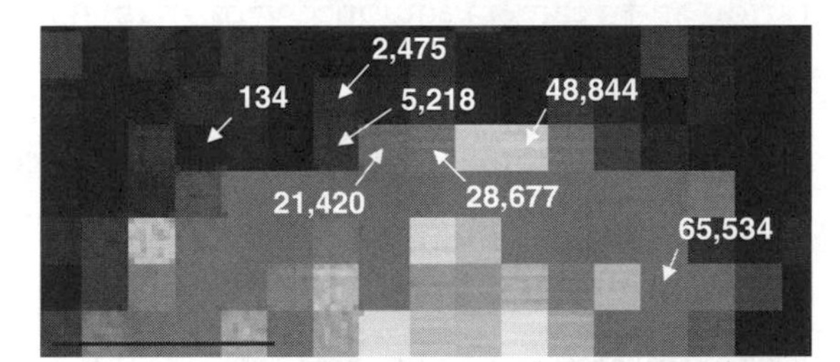

FIGURE 8.38. *An enlarged image of a microarray (from Fig. 8.37) with intensity values given for seven different pixels. The custom rainbow palette was indexed with a 16-bit TIFF file. Space bar* = 50 *μm. Figure also appears in Color Figure section.*

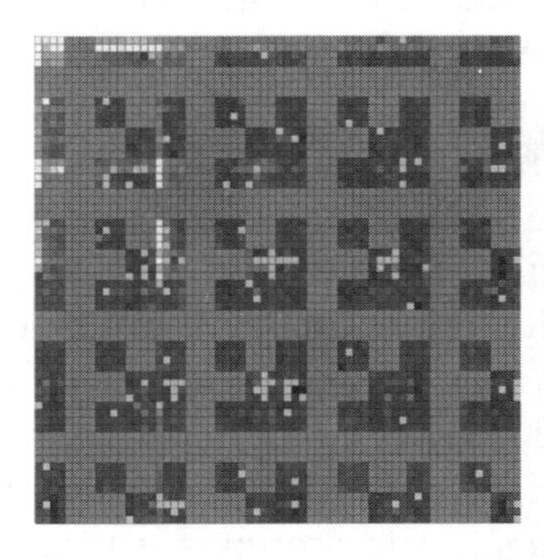

FIGURE 8.39. ***A,*** *Oligonucleotide hybridization data displayed via a custom color palette. (Reprinted with permission from* Nucleic Acids Research.*)* ***B,*** *Corning Microarray Technology slides and gene expression data displayed in a custom color palette. (Data courtesy of Corning, Corning NY.) Figure also appears in Color Figure section.*

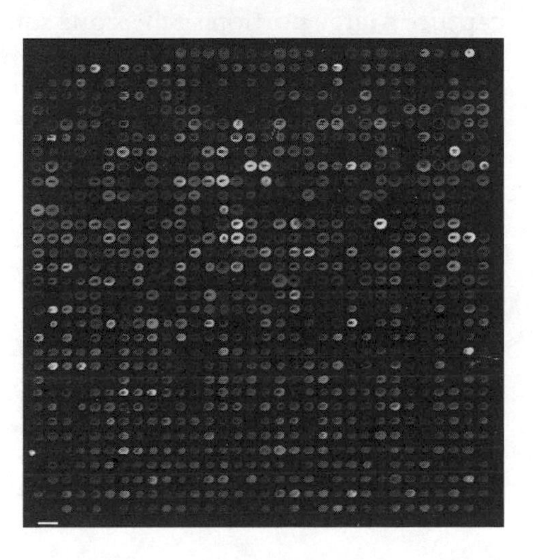

FIGURE 8.40. *A green–red overlay is created by subtracting image values obtained from separate channels of a microarray scanner and superimposing the images to create a composite. The samples were prepared by labeling mRNA from human cells grown at 37 and 42°C, respectively. (Courtesy of Dr. M. Schena and Dr. R. W. Davis, Stanford University, Stanford, CA, and by Dr. D. Shalon, Synteni, Palo Alto, CA.) Figure also appears in Color Figure section.*

Test sample. *Population of molecules such as messenger RNA and complementary DNA from which experimental data are derived.*

Reference sample. *Population of molecules, such as messenger RNA and complementary DNA, used as a control to facilitate comparison of one or more test samples.*

Green–red overlay. *Two-color composite microarray image generated by superimposing a green image and a red image.*

provides a convenient way of identifying genes or gene transcripts present in greater abundance in a ***test sample*** compared to a ***reference sample.*** In a ***green–red overlay,*** green spots correspond to sequences present in greater concentration in the test sample and red spots correspond to sequences more abundant in the reference sample. The two-color overlay representation allows intuitive identification of genes involved, for example, in the heat shock response in human cells.

IMAGERS

The second main design for microarray detection instruments is the imager. A microarray imager is a device that records fluorescent microarray data by generating a photograph or image of a microarray. Different from a scanner, a microarray imager does not move back and forth quickly across the microarray, but rather captures a relatively large portion of the microarray (e.g., 1 cm^2) in a single detection step. Imagers generally use a white light source for flood excitation and a camera equipped with an imaging chip to capture the fluorescence emission. This section provides an overview of current microarray imaging technology, including instrument architecture, white light sources, charged coupled devices, and complementary metal oxide silicon detectors.

Instrument Architecture

The architecture for a microarray imager is conceptually similar to a microarray scanner, though many of the technical details are different. Microarray imagers contain the following basic components: light source, optical filters, stage, camera, detector, A/D converter, and computer (Fig. 8.41). White light is directed through an optical filter to obtain a monochromatic light source, leading to the excitation of a fluorescent dye on the microarray surface. Fluorescence emission then passes through an emission filter to remove unwanted excitation light, and the signal is captured on a light sensor fitted to the camera. Similar to a photomultiplier tube, the detector in a microarray imager converts a stream of photons into an analog electrical current which, in turn, is converted into a digital signal that can be stored as a TIFF file on a personal computer.

The key difference between imagers and scanners is that imaging devices capture a large portion such as one square centimeter in a single detection step. To capture a large microarray snapshot, microarray imagers employ detectors

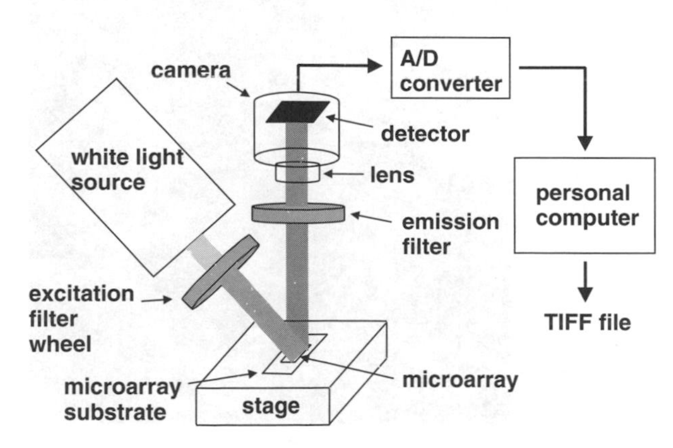

FIGURE 8.41. *For a microarray imaging device, a white light source* (light thick line) *passes through a filter wheel to select a wavelength of interest* (thick dark line). *The excitation light shines on the surface of the microarray, causing excitation emission* (gray line), *which passes into a camera and is recorded on a detector. The analog signal is converted into a digital output that can be stored and displayed on a PC as a TIFF file.*

known as ***area detectors.*** The use of an area detector simplifies instrument design by avoiding the need to move quickly back and forth across the substrate to capture an image. Imagers are therefore less expensive to build and easier to maintain. The first microarray imagers built at Stanford University in the mid-1990s actually used black cardboard as the microarray stage! More recent imagers employ somewhat more sophisticated substrate stages. In all imagers, microarrays larger than ~1 cm^2 are captured as a series of successive images and combined into a single image using software tools. A key consideration in software tools for reconstruction is that a given algorithm must stitch the images together accurately.

Area detector. *Microarray light-sensing implement, such as a charge-coupled device, that captures data from a large area of a microarray substrate in a single detection step.*

At present, commercial scanners outnumber imagers by about 10:1 though further improvements in microarray imager technology will help proliferate this platform. In addition to simplicity and affordability, the potential versatility of imagers is enormously appealing. Owing to the use of a white light source, a single microarray imager can be used to detect virtually any fluorescent dye of interest simply by dialing in a filter set tailored to that particular dye. A white light source has all possible excitation wavelengths, and the detector can record the full gamut of emission wavelengths, making it possible to read any fluorescent microarray by using filter sets designed for that dye. This is in sharp contrast to scanners, which employ specific laser lines for excitation and are thus restricted to a given dye set a priori. One popular microarray imager, developed by Applied Precision, (Issaquah, WA), uses a fiberoptic technology to direct the excitation light onto the microarray surface (Fig. 8.42).

White Light Sources

The source of excitation light used for microarray imagers contains the entire gamut of wavelengths in the visible spectrum and thus has a bright white appearance. Different from lasers, which have a defined monochromatic emission, white light sources contain many colors as a simultaneous emission and are thus said to be ***polychromatic.***

Polychromatic. *Any light source, such as a xenon lamp, that produces electromagnetic radiation of many different wavelengths.*

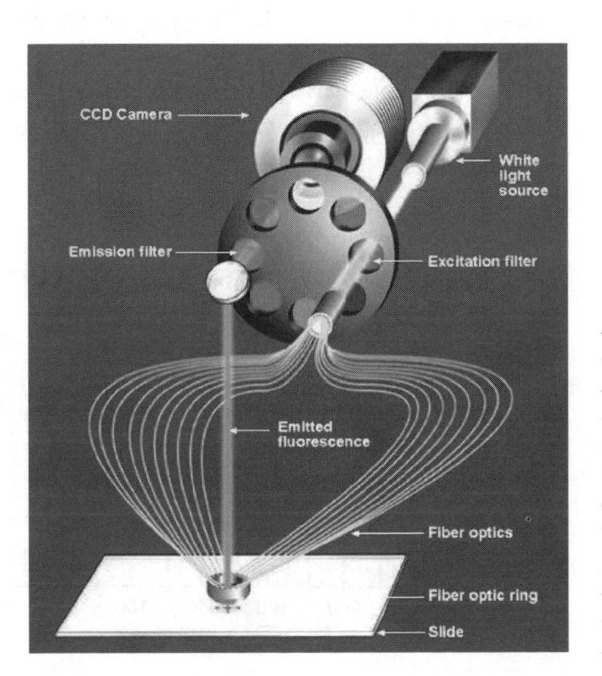

FIGURE 8.42. *The arrayWoRx microarray imaging instrument from Applied Precision. A white light source is directed through an excitation filter to select a wavelength of interest* (green lines), *which is directed by a fiberoptic bundle onto the microarray surface. The fluorescence emission* (red line) *is passed through an emission filter and detected with a CCD camera. (Courtesy of Applied Precision, Issaquah, WA.)*

Xenon arc lamp. *Polychromatic light source used for excitation in microarray imaging devices.*

Arc. *Discharge of current produced by applying a large voltage across two electrodes.*

Xenon ion. *Negatively charged atom produced when electrons are lost from xenon in an arc lamp or other device.*

Continuous spectrum. *Light source produced by an arc lamp or a related device characterized by the presence of many different wavelengths owing to the fact that the excited electrons occupy poorly defined orbits around the atom.*

Color temperature range. *Kelvin scale for color emission derived from the correlation between temperature and the color of light emitted by a hot substance.*

The white light output from a ***xenon arc lamp*** provides a convenient source of excitation light for microarray imagers. A xenon arc lamp contains a pair of electrodes separated by several millimeters. The electrodes are enclosed in a quartz bulb filled with xenon gas pressurized to 10–20 atm. A large voltage is used to create a discharge of current or ***arc*** between the electrodes. Xenon atoms collide with electrons in the arc, leading to a loss of electrons from the outer shell and to the creation of ***xenon ions,*** which are xenon atoms missing one or more electrons in their normally complete outer shell. Xenon ions are highly unstable and reacquire one or more excited electrons to complete their outer shell. When the excited outer shell electrons return to the ground state, photons are released from the xenon atoms, providing an intense source of light. Because of the pressurized state of the xenon gas in the xenon lamp, neighboring xenon atoms exert electronic influence on the outer shell electrons of adjacent xenon atoms, resulting in poorly defined excited electronic orbits. The indefinite orbits result in poorly defined wavelength emission, producing a large number of different emission wavelengths known collectively as a ***continuous spectrum.*** Xenon atoms releasing a continuous spectrum of light gives a xenon arc lamp its white appearance.

The repertoire of emission wavelengths from a white light source is commonly expressed in degrees Kelvin (K), which is a temperature scale obtained by adding 273 degrees to the Centigrade (C) scale. The Kelvin scale for color emission is also known as the ***color temperature range*** because it is derived from the correlation between temperature and the color of light emitted from hot substances. A simple demonstration of the color–temperature relationship can be obtained by heating a piece of metal with a natural gas flame from a conventional household stove. As the metal is heated, the surface will acquire a reddish glow that eventually transitions from red into orange and finally into white if heated sufficiently. The surface temperature of the metal determines the color of the emitted light, with red, orange, yellow and white corresponding to approximately 3500, 4500, 5000, and 6000 K. The surface temperature of earth's sun is 5000–6000 K, which gives sunlight its yellow–white appearance. The xenon arc lamps used in microarray imagers have Kelvin ratings of approximately 6000, and as such closely approximate sunlight in terms of their white light emission. The emission spectrum for a xenon source is relatively uniform across the visible wavelengths (Fig. 8.43). The presence of significant quantities of ultraviolet and infrared radiation mandate the use of special filtering to

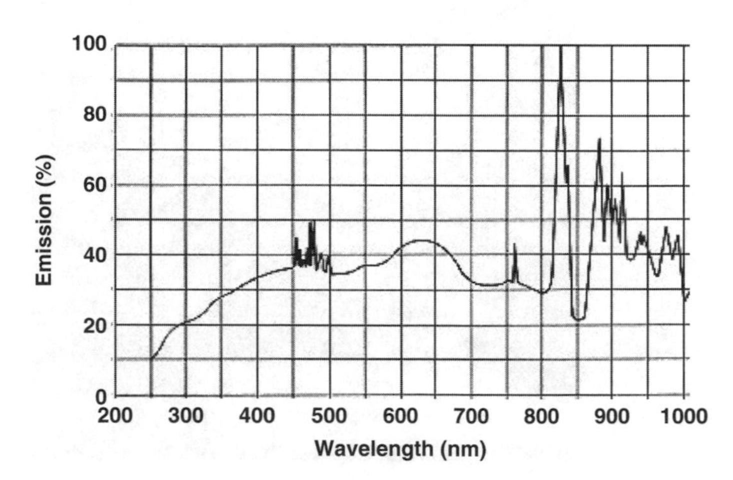

FIGURE 8.43. Light emission data from a xenon arc lamp. (Data courtesy of Photon Technology International, Lawrenceville, NJ.)

remove unwanted damaging radiation and heat, respectively, from xenon light sources.

The presence of all possible excitation wavelengths from white light sources provides a convenient excitation source for microarray imaging devices in that any fluorescent dye can be excited with a xenon lamp. As the availability of different dyes used in microarray analysis increases, white light sources have an advantage over laser-based systems in terms of versatility. One drawback of white light is that it is much more difficult to obtain clean excitation and emission wavelengths than with laser systems. Microarray imagers therefore require sophisticated engineering to minimize optical cross-talk between different imager channels when a xenon source is used for excitation. Xenon arc lamps are available with power ratings >1000 W, which is sufficient energy to excite any fluorescent dye even if the light output is focused over a 1-cm^2 area.

Despite the fact that xenon lamps have much greater power ratings than the lasers used in microarray scanners, the two sources have a similar ***intensity.*** The intensity (I), or ***power density,*** of a microarray light source is the power (P) measured in Watts divided by the area of the substrate in square centimeters that is illuminated:

Intensity. *Amount of light, abbreviated I, used for microarray excitation equivalent to the power of the light source divided by the illumination area.*

Power density. *Intensity of a light source used for microarray excitation.*

$$I = \frac{P}{A} \tag{8.14}$$

The xenon lamp (1,000 W) has a much greater power rating than the laser (0.010 W) but is focused into a much larger area (1.0 cm^2) than the laser (10^{-6} cm^2), and so the laser actually produces a 10-fold greater intensity (10,000 W/cm^2) than the xenon source (1,000 W/cm^2). But the *true* intensity of the xenon source is actually ~100-fold lower than the 1,000 W/cm^2 value because only a small portion (e.g., 1%) of the emitted spectrum is used to excite a given dye. On the other hand, the dwell time for imaging (1 s) is ~10,000-fold longer than that of a scanner (10–100 μs). In terms of photobleaching, the relevant value is the time illumination (see Eq. 8.10). Because the user can often increase the dwell time or exposure time of a microarray imager, care must be taken to avoid photobleaching when using the flood illumination schemes.

Similar to laser light sources, xenon lamps and other white light emitters represent a significant safety hazard. Both the intensity of the light from a xenon lamp and the presence of ultraviolet wavelengths pose a threat to the human retina. The pressurized gas and large voltages required also represent potential explosion and electric shock dangers. All safety precautions should be followed when handling or using powerful white light sources!

Charge-Coupled Device

A key component of imagers is the light detection device or detector located inside the camera. Detectors are solid-state devices consisting of a checkerboard matrix of light sensing pixels. The most common detector used in microarray imagers is the charge-coupled device (CCD). Each pixel in a CCD corresponds to an individual photosensor that converts photons from a given region of a fluorescent microarray into an electrical charge that can be measured, recorded, and displayed as a digital TIFF file. As the name implies, each pixel in a CCD is coupled to an adjacent pixel, which allows the charge that is accumulated at each photosensor to be transferred across the pixel matrix for

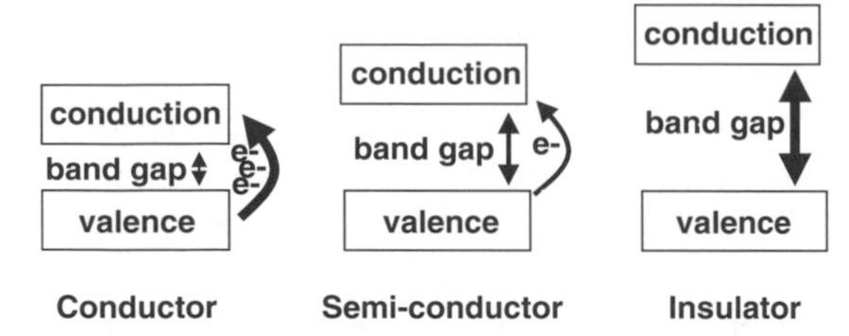

FIGURE 8.44. The amount of energy required to excite electrons (e^-) from the valence band into the conduction band is known as the band gap (double-headed arrows). *Conductors have little or no band gap and insulators have a large band gap. The band gap in semiconductors has an intermediate size.*

amplification and output. A brief consideration of semiconductor chemistry provides a good starting point for understanding how a charge-coupled device converts photons into electrical signals.

Semiconductors derive their name from the fact that these materials are partial conductors of electron flow. The conductivity of a substance is largely determined by the amount of energy that is required to move electrons from the bonded state into the conductive state, the latter allowing electrons to flow freely in the material and conduct electricity. Electrons that participate in chemical bonding are located in the ***valence band,*** and those participating in electrical flow are located in the ***conduction band*** (Fig. 8.44). The energy separating the valence band and the conduction band is known as the ***band gap.*** Metals such as copper have little or no band gap and thus conduct electricity readily, whereas nonmetals have a large band gap and are poor conductors but good insulators. Semiconductors such as silicon (Si) have a larger band gap than conductors and a smaller band gap than insulators, and are thus partial conductors (semiconductors) of electricity. If a semiconductor material is exposed to an energy source, such as photons from a fluorescent microarray, electrons can be dislodged from the valence band into the conduction band, allowing them to flow freely in the material.

Valence band. *Location of electrons that participate in chemical bonding in a conductor, semiconductor or insulator.*

Conduction band. *Location of electrons that participate in electrical flow in a conductor, semiconductor, or insulator.*

Band gap. *Energy separating the valence band and the conduction band in a conductor, semiconductor, or insulator.*

The flowing electrons are used to build up charge in each CCD pixel, and the amount of charge that accumulates is proportional to the number of photons that impinge on the surface of the CCD. Each pixel stores charge in a temporary manner and thus functions as a ***capacitor.*** A semiconductor capacitor is an electric circuit element that is built by creating a sandwich of silicon materials that differ with respect to their conductivity. Pure silicon can be made more or less conductive by substituting trace amounts of other atoms into the crystal lattice.

Capacitor. *Charge-coupled device pixel or other device that stores electrical charge in a temporary manner.*

Silicon has four valence electrons, and in pure silicon, each of the four electrons is shared with a neighboring silicon atom such that all four valence electrons form covalent bonds (Fig. 8.45). The conductivity of silicon can be decreased or increased by ***doping*** the silicon with atoms that contain a fewer or greater number of valence electrons. Positively charged or ***p-type*** silicon has slightly less conductivity than pure silicon and is made by substituting a trace amount of boron, which contains three valence electrons instead of four. Negatively charged or ***n-type*** silicon has slightly greater conductivity than pure silicon and is made by doping with trace quantities of phosphorus, which contains five electrons instead of four. P-type silicon contains unpaired electrons or holes and n-type silicon contains extra electrons with respect to pure silicon. The three types of silicon are used in conjunction to make the pixel capacitors in CCD cameras.

Doping. *Process of adding trace quantities of atoms that contain a lower or higher number of valence electrons to a pure substance, such as silicon, to increase or decrease its electrical conductivity.*

p-type. *Positively charged silicon made by substituting a trace amount of boron to provide a doped material that has slightly less conductivity than pure silicon.*

n-type. *Negatively charged silicon made by substituting a trace amount of phorphorus to provide a doped material that has slightly greater conductivity than pure silicon.*

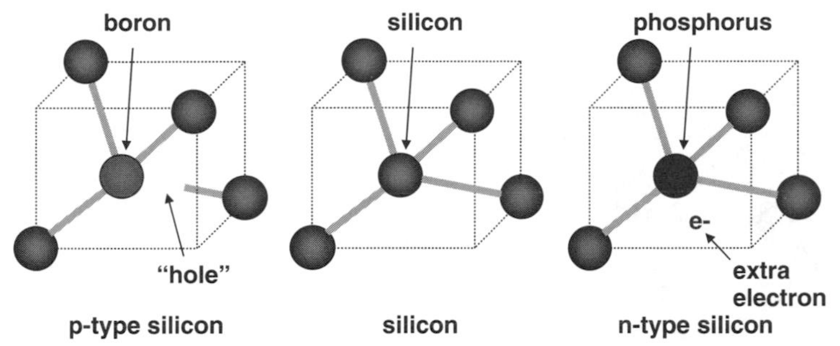

FIGURE 8.45. Tetrahedral models of the three forms of silicon used to make photodiodes for CCD cameras. Atoms of silicon (gray circles) *contain four valence electrons and form four covalent bonds* (lines) *with neighboring silicon atoms. Positively charged silicon is doped with boron, which has only three valence electrons and thus contains a hole corresponding to the missing electron. Negatively charged silicon contains phosphorus, which has five valence electrons and contains an extra electron.*

A closer look at a CCD pixel reveals a metal anode, an oxide coating, a p-type silicon layer, a depletion region, an n-type silicon layer, and a cathode (Fig. 8.46). Metal oxide semiconductor (MOS) detectors are present in all CCD cameras used in microarray imagers. The anode and cathode of the MOS (pronounced *moss*) allow external voltage to be applied to each pixel and allow the accumulated charge to flow during the read-out process. The oxide coating (e.g., SiO_2) protects the silicon layers and prevents reflection of the incident light. The p-type layer carries a positive charge, the depletion layer is neutral, and the n-type layer carries a negative charge. The three layers work in concert to create an electric field that moves newly created charges to their proper location.

In a simplified view, photons from a fluorescent microarray impinge on silicon atoms in the depletion layer, exciting electrons from the valence band into the conduction band thereby creating free electrons and holes. The electric field causes the holes to move into the p-layer and electrons to move into the n-layer (Fig. 8.46). Electrons and holes are formed at a steady rate, charging each pixel in a manner proportional to the photon stream. Charge accumulates in each CCD pixel rather than flowing out of the device because the p-side carries a

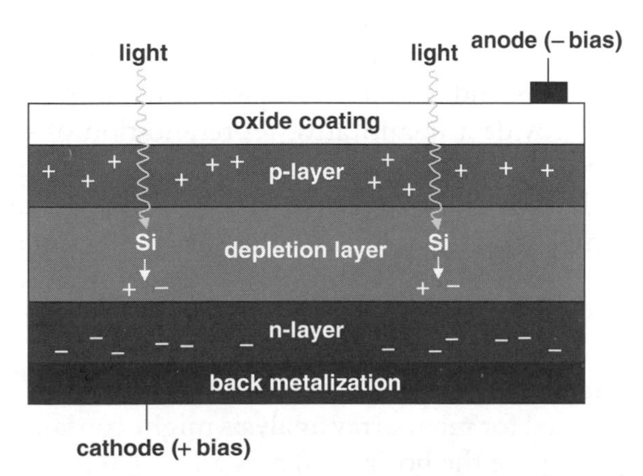

FIGURE 8.46. In a CCD pixel, an oxide coating protects the silicon surface and a metal surface coats the backside of the device. Light excites silicon atoms in the depletion layer, creating free electrons (−) and holes (+) that migrate into the n-layer silicon and p-layer silicon, respectively. The reverse bias on the anode and cathode allows charge to build up while the pixel is exposed to light. The accumulated charge is then allowed to flow, and the number of electrons or holes are counted and converted into a digital value.

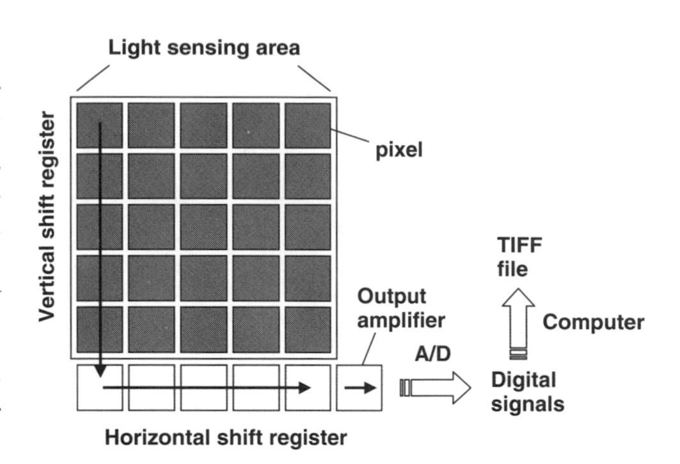

FIGURE 8.47. *A CCD detector showing how microarray data are gathered, transferred, and stored. A lightsensing area composed of a pixel array converts photons into stored charge. Charge stored in each pixel is transferred out of the CCD detector in a stepwise fashion using a vertical shift register and a horizontal shift register. Each pixel signal is amplified at an output node, processed using A/D converter, and stored on a computer as a TIFF file.*

Reverse bias. *Electrical configuration maintained by applying an external voltage to a charge-coupled device pixel so that the p-side carries a negative charge and the n-side carries a positive charge; thus allows charge to accumulate in the pixel equivalent to the number of photons emitted at a given position in the microarray.*

Vertical shift register. *Device in a charge-coupled device camera that reads the charge accumulated at each pixel by transferring the charge from top to bottom.*

Horizontal shift register. *Device in a charge-coupled device camera that reads the charge accumulated at each pixel by transferring the charge from left to right.*

Read time. *Number of seconds or minutes required to capture a microarray image with a microarray imaging instrument.*

negative charge and the n-side carries a positive charge in a configuration known as ***reverse bias.*** The reverse bias configuration is maintained by applying an external voltage to each pixel, which allows charge to accumulate for a designated amount of time, after which time the reverse bias is removed, allowing charge to flow out of each pixel and recorded as a digital value.

The amount of charge that accumulates over time is proportional to the amount of light that impinges on a given pixel. Because only a single electron is dislodged per captured photon, a CCD detector (unlike a PMT) does not amplify the light signal. The inherent lack of gain in a CCD renders this type of light detector intrinsically less sensitive than a photomultiplier tube, though the lesser sensitivity can be overcome simply by taking a longer exposure during the imaging process (see below).

A macroscopic view of a CCD chip reveals a regular array of light sensing pixels (Fig. 8.47), each of which captures incoming photons from its corresponding region on the fluorescent microarray. A CCD camera reads the charge accumulated at each pixel by transferring the charge downward along the ***vertical shift register*** and then from left to right along the ***horizontal shift register.*** Each charge is transferred in a serial manner to an output amplifier, where it is converted from an analog signal into a digital signal and stored as a TIFF file on a personal computer. A CCD detector therefore captures microarray signals with a two-dimensional array of light sensing pixels, reads the accumulated charge in a serial fashion, and remaps the digital values into a two-dimensional TIFF file image to provide a quantitative representation of the intensity values across the microarray surface. The amount of time it takes to capture a microarray image, known as the ***read time,*** is the sum of the time required for CCD exposure, charge transfer, digitization of the analog signals, transfer into computer memory, and display on a suitable computer monitor. Other factors being equal, faster CCD cameras are better than slower ones because they speed the rate at which microarray images can be acquired. A typical CCD camera used in a microarray imaging device can capture an image in about 60 s.

A CCD camera used for microarray analysis might contain a sensing region that has 1000 pixels along the horizontal axis and 1000 pixels along the vertical axis for a total of 1 million (1000 × 1000) pixels. This type of chip is often known either a "1K by 1K chip" to denote that it contains 1000 pixels in each dimension or as a "1 megapixel" camera because it contains a total of 1 million pixels. Most cameras used in microarray imagers contain physical pixel sizes in the range of 7–12 μm, which is an important consideration because the size of the pixel determines the total amount of charge that can stored by a given pixel. Pixel storage capacity can be estimated by taking the square of the pixel size (Px) in microns and multiplying by 1000:

$$\text{Storage capacity} = 1000 \times (Px)^2 \tag{8.15}$$

A CCD chip with 9-μm pixels allows approximately 81,000 stored charges whereas a corresponding camera with 7-μm pixels has a 49,000-charge storage capacity. Because each stored charge is converted into one count, the storage capacity sets the upper limit for the dynamic range of the detector. Using the example above, 16-bit data (65,536 counts) could be acquired with the 9-μm-pixel camera (81,000 count capacity) but not with the 7-μm-pixel (49,000 count capacity) one.

There is a trade-off between pixel size and resolution, however, with smaller pixels providing greater resolution than larger pixels for CCD detectors of equal physical size. A 1.0 cm^2 CCD chip with 7-μm pixels could contain as many as 2.0 megapixels compared to the 1.2-megapixel limit of the chip with 9-μm pixels. Images with a greater number of pixels are superior to those with a lesser number of pixels, because the physical image is represented by greater number of bins. The maximal number of pixels obtainable on a CCD chip can be approximated by taking the square of the quotient of the width of the CCD chip (in microns) divided by the pixel size (in microns):

$$\text{Number of pixels} = (\text{Chip size} \div \text{Pixel size})^2 \tag{8.16}$$

Dynamic range in a CCD detector is reduced by using smaller pixels but can also be affected negatively by the presence of noise in the system, which is typically specified in electrons (e^-). The extent of noise in the system is extremely important because it reduces the dynamic range of the CCD by "clipping" the detection range at low signal readings. A good approximation of the true dynamic range of a CCD detector in the imaging process can be calculated by taking the quotient of the storage capacity (counts) divided by the total noise in the system (electrons):

$$\text{True dynamic range} = \frac{\text{Storage capacity}}{\text{Total noise}} \tag{8.17}$$

A CCD with a storage capacity of 81,000 counts and a total noise of 100 electrons would have a true dynamic range of 8,100 (81,000/100) counts or just under 13 bits (8,192). The same CCD operating with a total noise of 1 electron would have a true dynamic range of 81,000 counts or greater than 16 bits (65,536). Given the enormous negative impact of total noise on dynamic range, great effort is taken to reduce the noise present in CCD detectors. Most CCD devices used in microarray imagers produce about 10 electrons of total noise.

One source of noise intrinsic to all CCD systems is dark count, which is the unwanted process by which pixels accumulate charge in the absence of light. Charge is accumulated in the absence of light because the silicon atoms present in the detector pixels undergo thermal excitation, producing free electrons and holes in the absence of a photon flux. Dark count is a thermal process and can therefore be reduced by decreasing the operating temperature of the CCD such that a 5–6°C drop in temperature reduces dark count by roughly a factor of 2. Cooling the operating temperature of a CCD from the ambient temperature (25°C) to −40°C would result in a 25-fold decrease in the dark count, or an

Blooming. *Unwanted phenomenon produced by a charge-coupled device in which charge from a completely filled pixel transfers to an adjacent, partially filled pixel to produce a blurry image.*

Antiblooming gate. *Charge-coupled device chip design feature that increases the spacing between detector pixels to reduce pixel-to-pixel charge transfer and blooming.*

Point defect. *Single defective pixel in a charge-coupled device detector.*

Cluster. *Small group of adjacent pixels that are partially or fully defective in a charged-coupled device.*

Hot pixel. *Defect in a charge-coupled device detector corresponding to a picture element that produces too much signal relative to a normally functioning picture element.*

Dead pixel. *Defect in a charge-coupled device detector corresponding to a picture element that produces no signal.*

Dark pixel. *Defect in a charge-coupled device detector corresponding to a picture element that produces too little signal relative to a normally functioning picture element.*

increase of nearly 5 bits of dynamic range. Because large improvements in CCD performance are obtained by cooling, most CCDs in microarray imagers are operated at well below zero. The so-called cooled CCD, which employs an operating temperature in the −20 to −40°C range, is the detector of choice for microarray imaging.

At high light intensities, one or more pixels in the CCD detector can fill completely with charge. Charge from completely filled pixels can sometimes transfer to adjacent, partially filled pixels in a process known as ***blooming.*** The phenomenon of blooming is detrimental because it degrades the quality of the image. Blooming can be observed with microarrays that contain spots with intense fluorescence or with microarrays that have a large dynamic range of signals, the latter requiring long exposures to capture the weakly fluorescent microarray features. In microarray imaging, blooming will manifest as a loss of spot sharpness and can complicate analysis if the features are closely spaced. Some CCD chips use special designs to prevent blooming. An ***antiblooming gate,*** which effectively increases the spacing between detector pixels, greatly reduces pixel-to-pixel charge transfer thus minimizing blooming. Antiblooming designs are not without a downside however; a loss of pixel charge capacity is a typical compromise with CCD chips that use antiblooming gates. Some early CCD-based microarray imagers suffered from blooming problems (Fig. 8.48), though most of the current systems have largely eliminated this complicating factor.

Because CCD chips contain a large number of pixels, manufacturing devices with one million or more perfectly functioning pixels is difficult and expensive. Most CCD chips contain a small number of defective pixels known as ***point defects*** and ***clusters.*** A point defect is a single defective pixel, whereas a cluster is a small group of adjacent pixels that are partially or fully defective. The three types of pixel defects are commonly known as ***hot pixels, dead pixels,*** and ***dark pixels,*** corresponding to pixels that produce too much, none, or too little signal, respectively, compared to the majority of pixels in the detector. Pixel defects are identified easily by enlarging a digital image on a computer screen and searching for erroneous signal intensities. A cluster of hot pixels is identified

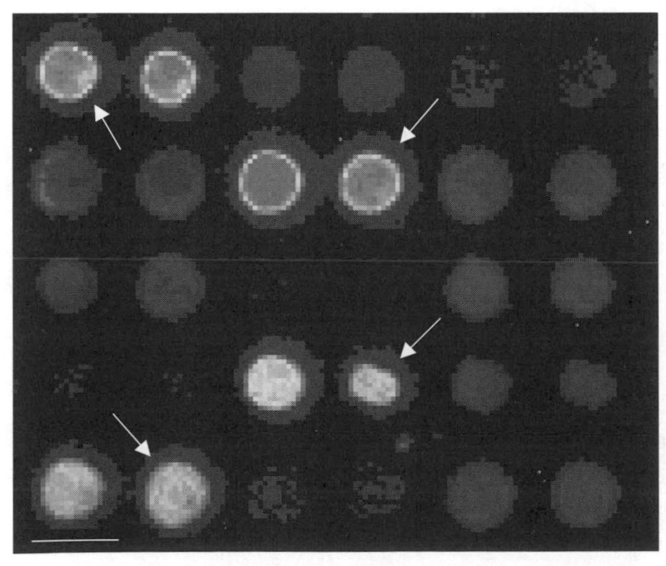

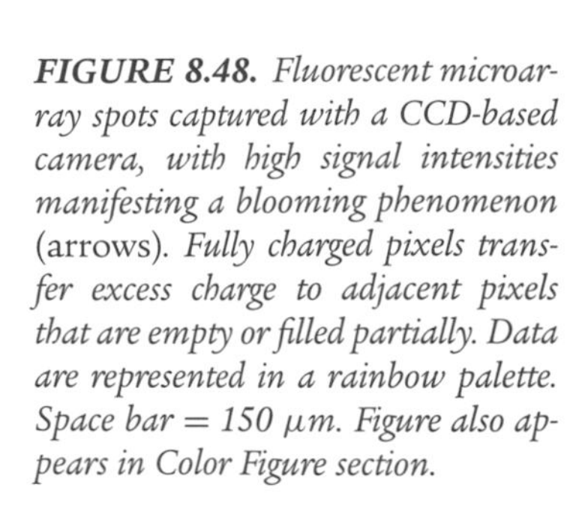

FIGURE 8.48. Fluorescent microarray spots captured with a CCD-based camera, with high signal intensities manifesting a blooming phenomenon (arrows). *Fully charged pixels transfer excess charge to adjacent pixels that are empty or filled partially. Data are represented in a rainbow palette. Space bar = 150 μm. Figure also appears in Color Figure section.*

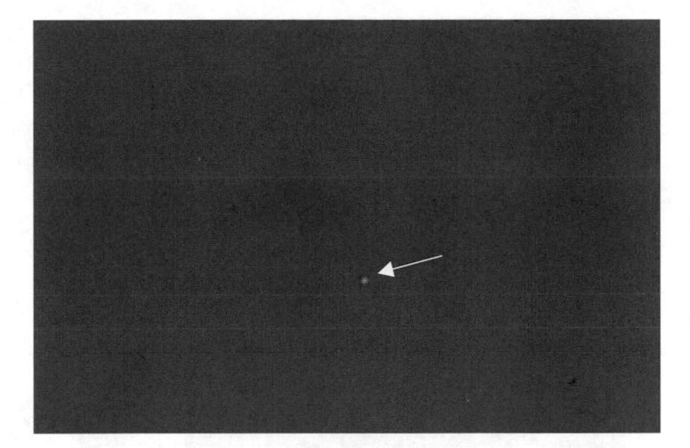

FIGURE 8.49. *A digital image taken with a CCD camera in a low-light environment. A small cluster of pixels gives a stronger signal than all of the surrounding pixels and thus represents a hot cluster, which has an elevated dark count with respect to the other sensors on the CCD.*

with a digital image captured in a low light environment (Fig. 8.49). Different grades of CCD chips are available commercially, and a Grade 0 chip corresponds to one that has no defective pixels. Better grades of CCD chips are preferable in microarray applications because defective pixels will produce erroneous microarray signal intensities.

CCD detectors offer many desirable qualities in the context of microarray imaging. The fact that these area detectors capture 1 million or more pixels of data in a single imaging step offers significant speed and precision in a microarray detection instrument. Though the dynamic range and detectivity of these devices are often viewed as less than that of photomultiplier tubes, proper cooling, design, and use afford CCD-based microarray data of exceptional quality (Fig. 8.50). Exposure times of 10–60 s with a high-quality CCD camera suffice to capture fluorescent signals from microarrays with signal intensities well below 0.1 fluors/μm^2, allowing detection of extremely rare human transcripts. For example, gene expression data captured with a CCD camera allow the identification of human genes induced and repressed by the chemotherapy drug camptothecin.

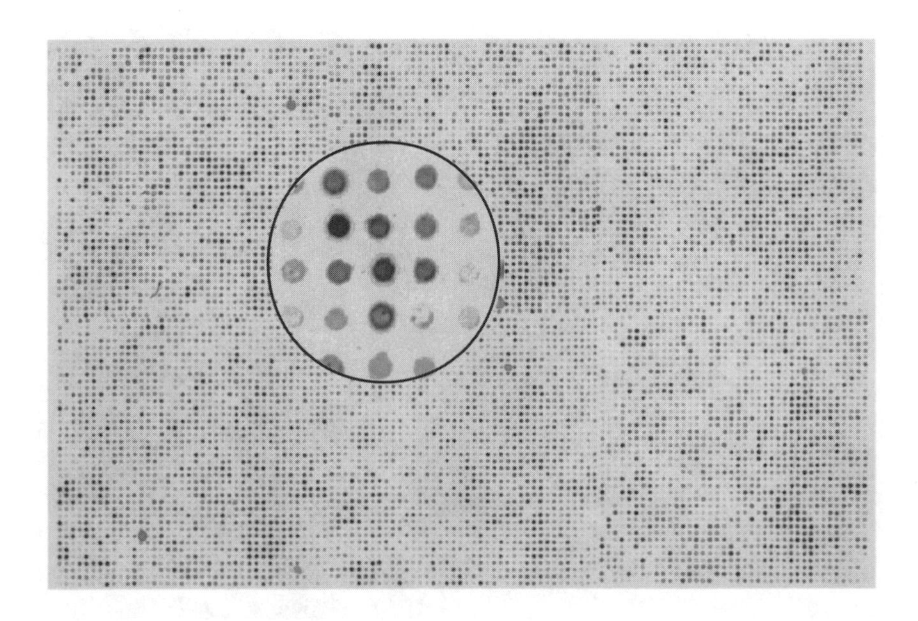

FIGURE 8.50. *Human oral carcinoma cells were treated with the chemotherapy drug camptothecin and compared to untreated cells. Two-color data are represented in a composite image so that dark and light circles correspond to genes that are activated and repressed, respectively, by the drug. A portion of the array* (circle) *is enlarged for ease of viewing and the data were obtained by Dr. Konan Peck. (Reprinted with permission from Academic Press, San Diego, CA.) Figure also appears in Color Figure section.*

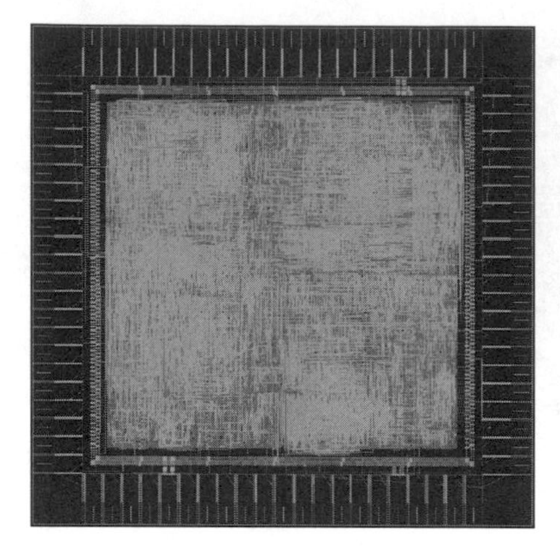

FIGURE 8.51. ***A,*** *A CMOS wiring diagram.* ***B,*** *A CMOS device. (Courtesy of Dr. K. von der Heide, Computer Science Department, University of Hamburg, Hamburg, Germany.)*

Complementary Metal Oxide Semiconductor

Complementary metal oxide semiconductor. *Relatively new type of microarray detector that has an opposite polarity charge relative to the metal oxide semiconductor switches used in charge-coupled device cameras.*

The CCD is the most commonly used detector in microarray imaging instruments, though recent advances have been made in developing a new type of detector known as a ***complementary metal oxide semiconductor*** (CMOS). CMOS (pronounced *sea moss*) devices operate in an manner opposite to the MOS switches used in CCD cameras. In a MOS device, high voltage closes the metal gates, resulting in the flow of current, whereas in a CMOS device, high voltage opens the metal gates, and current flow ceases. CMOS devices thus have the opposite polarity charge relative to standard MOS switches and derive their name because they are complementary to MOS circuits. Photographs of CMOS chips reveal the sophistication of these devices (Fig. 8.51).

Though CMOS detectors are generally viewed as less sensitive than CCDs and as having an inferior dynamic range, improvements in the technology may move CMOS-based devices on par with CCD cameras with respect to these important parameters. If so, CMOS chips show great promise for microarray detection by possessing a number of key advantages over CCD devices.

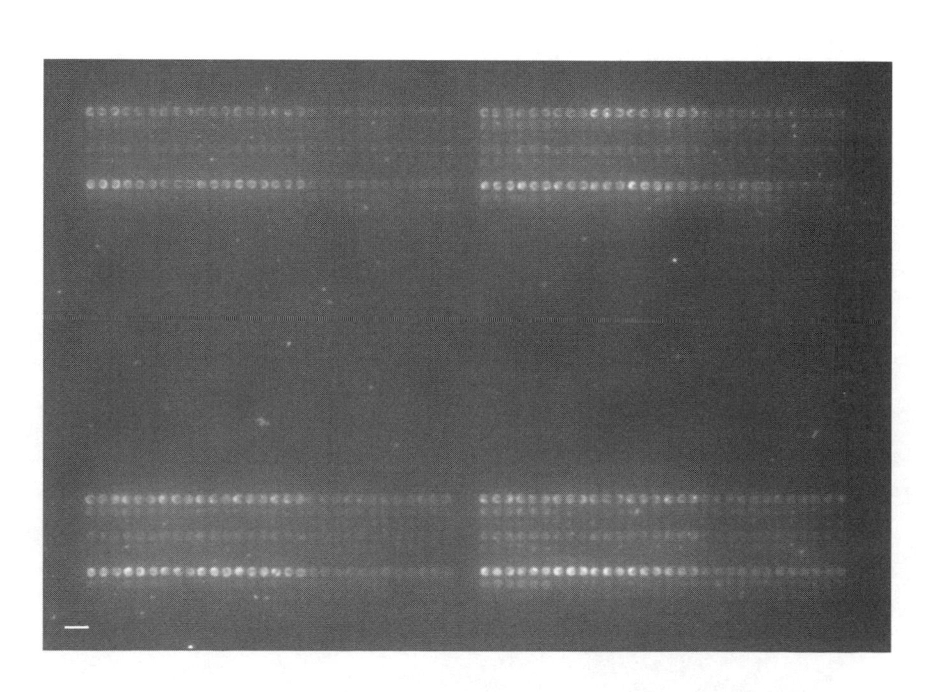

FIGURE 8.52. A microarray image captured with a camera equipped with a CMOS detector. Signals from a fluorescent microarray with spot sizes of approximately 100 μm were captured using a 60-s exposure of the imaging instrument prototype. Data are presented in a black and white scale. Space bar = 200 μm.

CMOS chips are generally smaller in size and less expensive to manufacture than CCD detectors. CMOS chips also use less energy than CCD devices owing to the fact that current flows in a CMOS device only when the circuits are switched. Some CMOS detectors have A/D converters built into the device, which simplifies design issues. CMOS detectors also avoid blooming, owing to the fact that there is no pixel-to-pixel charge transfer in these devices. Prototype CMOS detectors have been used to produce functional microarray images (Fig. 8.52).

SUMMARY

Microarray detection instruments are revolutionizing biology because they allow scientists to acquire data of extraordinary quality rapidly and automatically. Modern detection technology encompasses a wide range of issues, including evaluative criteria used to benchmark microarray scanners and imagers, instrument architecture, signal and noise considerations, optics, semiconductor technology, fluorescent dye chemistry, and laser and detector technology. Microarray scanners include moving substrate and moving optics designs and rapidly capture microarray images using laser scanning and detection with PMTs. Microarray imagers capture relatively large images in a single step, using white light sources for excitation and CCD or CMOS devices for detection. Excitation and emission filters, beam splitters, objective lenses, stages, reflective mirrors, and A/D converters are additional key components in scanners and imagers. Microarray data are stored on PCs in a digital format as graphical TIFF files and can be compressed into GIF or JPEG formats to converse space; they are displayed in wide range of different color formats. Fluorescence is the predominant detection technology, with highly conjugated dye molecules such as FITC, Cy, Alexa and BODIPY being widely used in microarray analysis. Fluorescence provides some key advantages over radioactive detection methods, including speed, safety, a diversity of different colors, and a high degree of spatial resolution. Continued advances in microarray scanning and imaging technology will provide detection instruments with increasing resolution, precision, accuracy, dynamic range, sensitivity, and detectivity.

SELECTED READING

Benson, H. *University Physics,* rev. ed. Wiley, New York, 1996.

Giancoli, D. C. *Physics for Scientists and Engineers,* 3rd ed. Prentice Hall, Upper Saddle River, NJ, 2000.

Hein, M., Best, L. R., Pattison, S., and Arena, S. *Introduction to General, Organic, and Biochemistry,* 6th ed. Brooks/Cole Publishing, Pacific Grove, CA, 1997.

Janesick, J. R. *Scientific Charge-Coupled Devices.* International Society for Optical Engineering, Bellingham, WA, 2001.

Kapon, E. *Semiconductor Lasers II : Materials and Structures.* Academic Press, San Diego, CA, 1999.

Lakowicz, J. R. *Principles of Fluorescence Spectroscopy,* 2nd ed. Kluwer Academic/Plenum, New York, 1999.

Lawrence, C. L. *The Laser Book—A New Technology of Light.* Prentice Hall, New York, 1986.

Mujumdar, R. D., Ernst, L. A., Mujumdar, S. R., Waggoner, A. S. Cyanine dye labeling reagents containing isothiocyanate groups. Cytometry 10, 11–19, 1989.

Myler, H. R., and Weeks, A. R. *The Pocket Handbook of Imaging Processing Algorithms in C.* Prentice Hall, Upper Saddle River, NJ, 1993.

Nishi, Y., and Doering, R., eds. *Handbook of Semiconductor Manufacturing Technology.* Marcel Dekker, New York, 2000.

Rost, F. W. D. *Fluorescence microscopy*. Cambridge University Press, Cambridge, UK, 1995.

Schawlow, A. L. Optical masers. Scientific American, June, 1961.

Schena, M., ed. *DNA Microarrays: A Practical Approach*, 2nd ed. Oxford University Press, Oxford, UK, 2000.

Schena, M., ed. *Microarray Biochip Technology*. Eaton, Natick, MA, 2000.

Siegman, A. E. *Lasers*. University Science Books, Herndon, VA, 1986.

Streitweiser, A., and Heathcock, C. H. *Introduction to Organic Chemistry*, 2nd ed. Macmillan, New York, 1981.

Tilley, D. E., and Thumm, W. *Physics for College Students*. Cummings, Menlo Park, CA, 1974.

Wenckebach, W. T. *Essentials of Semiconductor Physics*. Wiley, New York, 1999.

REVIEW QUESTIONS

1. The overall layout of instrument detection hardware including filters, lasers, and the like is known as what? One word will suffice.
2. A charge-coupled device is abbreviated how? (a) CDC, (b) CAD, (c) CMOS, or (d) CCD?
3. How many pixels would be contained in an image corresponding to a 1.0 cm^2 microarray captured at 10 μm scanning resolution: (a) 1,000,000, (b) 100,000, (c) 1,000, or (d) 10,000?
4. What instrument dynamic range would be required to detect a microarray with signal intensities spanning 325 to 63,234 counts: (a) 19.5-fold, (b) 195-fold, (c) 97-fold, or (d) 9.8-fold?
5. An 80-μm microarray spot containing 19,750 fluors would contain how many fluors per μm^2?
6. Which of the following extrinsic determinants would be expected to affect a fluorescent microarray signal: (a) surface dwell time, (b) light source power, (c) detector, (d) excitation wavelength, (e) dye molar extinction coefficient, and (f) dye quantum yield?
7. A microarray researcher compares the signal intensity of a microarray hybridized with a 100 ng/ml probe to a microarray hybridized with a 500 ng/ml probe and observes less than the expected fivefold increase in signal intensity. Provide two possible explanations for the nonlinearity in this microarray experiment.
8. Rank the energy of the photons in the following visible light colors in order of least to most energetic: (a) violet, (b) red, (c) green, (d) blue, (e) yellow, and (f) orange.
9. Would a smaller or larger Stokes shift facilitate the separation of excitation light and fluorescence emission in a microarray detection instrument?
10. Which of the following is not represented in a typical Jablonski diagram: (a) photon absorption, (b) photobleaching, (c) Stokes shift, (d) fluorescence emission, or (e) quenching?
11. One structural difference between the cyanine dyes and Cy dyes is the presence of sulfate groups on the aromatic rings of the latter. What fundamental property of the Cy dyes is likely to be improved by the presence of sulfate groups? One sentence will suffice.
12. A researcher labels two identical samples of mRNA by direct incorporation of Cy dye and Alexa dye and finds that the two samples produce slightly different emission patterns when hybridized to a 10-K microarray, even though no differences are observed when comparing two different Cy dyes or two different Alexa dyes labeling the same identical samples of mRNA. Provide an explanation.

13. Which of the following components might be found in a laser-based microarray scanner: (a) excitation filter, (b) photomultiplier tube, (c) A/D converter, (d) emission filter, (e) piezoelectric transducer, and (f) beam splitter.
14. Which of the following is *not* a typical property of gas and solid-state laser light: (a) intense beam, (b) monochromatic, (c) collimated, or (d) scattered?
15. Name one advantage and one disadvantage of having a large numerical aperture (NA) lens in a confocal scanner.
16. Order the following dyes with respect to maximal absorption wavelength from shortest to longest wavelength: (a) FITC, (b) Cy3, (c) lissamine, and (d) Cy5.
17. A photomultiplier tube (PMT) in a microarray scanner converts: (a) electrons to photons, (b) photons to electrons, (c) protons to electrons, or (d) electrons to neutrons.
18. Standard microarray images store fluorescent data in 16 bits, providing intensity values ranging from 1 to: (a) 6,536, (b) 16,536, (c) 65,536, or (d) 655,360?
19. Which of the following graphical file formats provides the truest representation of microarray data: (a) JPEG, (b) GIF, (c) TIFF, or (d) PICT?
20. A researcher scans a fluorescent microarray and acquires an image that appears blurry and highly pixilated. Her supervisor tells her that changing one setting on the microarray scanner will produce a much better image. Identify the instrument setting in two words.
21. Which one of the following is *not* likely to be found in a microarray imager: (a) galvonometer, (b) lens, (c) white light source, (d) excitation filter, (e) emission filter, or (f) CCD?
22. Which of the following accurately describe a charge-coupled device: (a) detects a relatively large area per exposure, (b) converts photons to stored electrical charge, (c) utilizes semiconductor materials, and (d) amplifies the light signal?

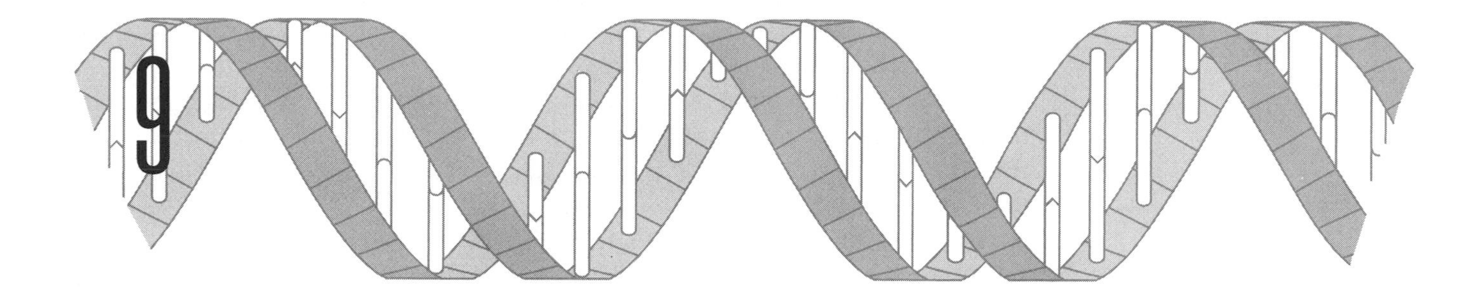

Microarray Informatics

I think there's a world market for about five computers. —*Thomas Watson*

The computer plays a role in biology that is more extensive than anyone, save for a handful of visionaries (e.g., Abarbanel et al., 1984), could have reasonably predicted 20 years ago. Personal computers have taken up residence on the desks of virtually every scientist in the world, and larger workstations are common equipment in most research departments. Microarray science provides an extreme case, with computer-based operations directing nearly every aspect of research. This chapter explores the myriad different facets of microarray informatics, including the electronic resources and biological databases; sequence and design informatics; and data quantitation, mining, and modeling.

ELECTRONIC RESOURCES

The linking of microarray scientists worldwide via the Internet is speeding technological advance like never before. The Internet, which supports the World Wide Web (WWW) and electronic mail (e-mail), allows scientists to share microarray protocols and data, obtain commercial products, and send electronic messages quickly and economically. A host of electronic support services, including an electronic library of microarray citations and PubMed, is assisting scientists in keeping apace with the rapidly expanding scientific literature.

Internet

Internet. *International network of computers that allows global communication in the form of the World Wide Web, electronic mail, and other applications.*

The international network of computers that allows global communication in the form of the Web, e-mail, and other applications is known as the ***Internet***. The meteoric popularity of this revolutionary communication tool and its key role in modern microarray technology justifies a brief look at the historical events that shaped its development.

The inventions of the telegraph and of the telephone and the laying of the transatlantic cable were the three events in the 1800s that laid the groundwork for the modern Internet. The telegraph demonstrated the concept of communicating at a distance using words encrypted in a code. Although modern computers use a different code, the telegraph and binary computer share conceptual common ground. Telephone lines, which allow dial-up access, provide the infrastructure of the Internet. The transatlantic cable, first laid down in the 1850s, established the feasibility of overseas communication, a key concept in worldwide communication via the Internet.

Nearly 70 years would pass from the time Alexander Graham Bell invented the telephone until another key event in the shaping of the Internet. By launching *Sputnik,* the world's first satellite, the former USSR demonstrated the concept of the artificial satellite, a concept that is used today in wireless communication. As a response to *Sputnik,* Eisenhower formed the Advanced Research Projects Agency (ARPA) to address the need for expedited advanced technologies. Scientists funded by ARPA, working under the direction of the Department of Defense, promoted the concept of an interlinked network (ARPANET) of computers that would allow computer-based communication at a distance. In 1969, the first ARPANET message was sent between the University of California at Los Angeles (UCLA) and Stanford Research Institute (SRI). The involvement of Stanford in the first Internet communication and in the first microarray analysis experiments links these important technologies.

The Internet pioneers moved the project forward slowly but steadily in the 1970s, establishing gateways, the file transfer protocol (FTP), and the first international communications. Key advances were also made in the 1980s, including the because it's time network (BITNET), transmission control protocol (TCP), and internet protocol (IP). UUNET, the first commercial Internet service, was established in 1987. The early 1990s marked key events, including the development the Web, and the first graphical user interface, known as Mosaic, in 1993 by Mosaic Communications (Mountain View, CA), founded by Jim Clark and Marc Andreessen, which became Netscape Communications in 1994.

By 1995, the same year that the first microarray paper was published (Schena et al., 1995), the Internet was growing in leaps and bounds, fueled by the availability of the user-friendly *Netscape* browser and commercial dial-up services, including Prodigy, America Online (AOL), and Compuserve. In 1996, the year that the first human microarray paper was published (Schena et al., 1996), a browser war began between *Netscape* and Microsoft (Redmond, WA). In 2000, Microsoft was found guilty of antitrust violations by U.S. District Judge Thomas Penfield Jackson, for the practice of bundling their Internet browser (Microsoft *Explorer*) with the Microsoft operating system. The case is now being reconsidered by the Bush administration.

There are currently >100 million Internet domains and more than 1 billion users worldwide, and these numbers have increased nearly exponentially since 1995 (Fig. 9.1). The modern Internet plays a key role in many different aspects of microarray research, including providing access to sequence databases and

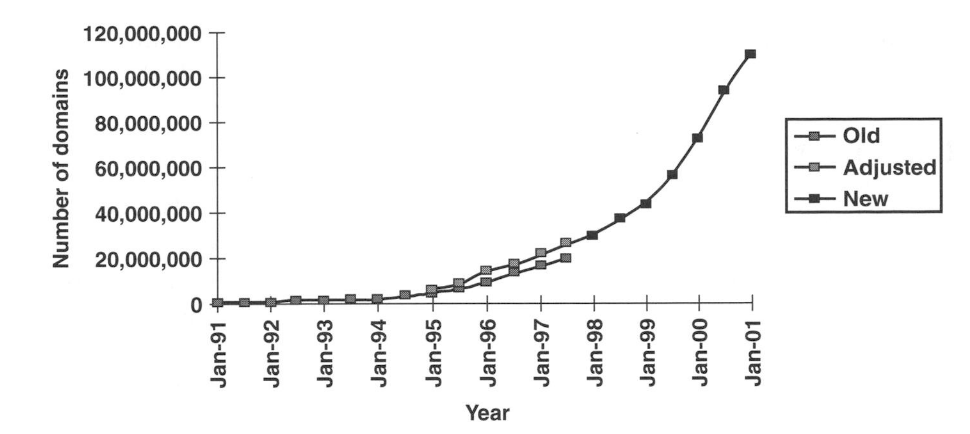

FIGURE 9.1. The number of domain names recorded during each January from 1991 to 2001. Between 1998 and 2001, the number of domains world wide increased from ~30 million to >100 million. (Data courtesy of Internet Software Consortium, Redwood City, CA.)

scientific publications, mediating communication between microarray scientists by e-mail, and allowing primary data to be posted and shared over the Web (see below). As microarray analysis moves into clinical use, the Internet will also support the results of genetic tests and diagnostic data, enabling the public to access their confidential results from any computer in the world.

Each computer connected to the Internet has a unique IP number, or ***IP address,*** consisting of four sets of 8-bit numbers (0–255) separated by periods (e.g. 171.65.21.118). The IP address is analogous to an international telephone number in that provides a unique identifier that allows computers to communicate globally over the Internet. If a user is unsure of the IP address for his or her computer, it can be obtained automatically by logging on to a Web site (www.ip-address.com/).

IP address. *Four sets of eight-bit numbers separated by periods (e.g., 171.65.21.118) that provide a unique identifier, allowing computers to communicate globally over the Internet.*

IP addresses are powerful identifiers for computers, but they are cumbersome and nonintuitive for human users. The ***domain name*** (e.g., genechips.com) provides an optional identifier on the Internet that is more intuitive than IP addresses and easier to remember, because the domain name often describes the nature of the academic institution or business that holds the domain. The affymetrix.com domain name is held by Affymetrix (Santa Clara, CA), the world's leading manufacturer of genechips. The suffix of most domain names allows the user to know whether the site involves an academic institution (edu), a nonprofit organization (org), a commercial entity (com), and so forth (Table 9.1). Domain names held by countries overseas–Germany (de), Australia

Domain name. *Intuitive Internet identifier (e.g., arrayit.com) that corresponds to the type of institution or business that holds the domain.*

TABLE 9.1. Common Domain Name Extensions

Domain Name Extension	Meaning	Example
edu	educational institution	genome-www4.stanford.edu
org	nonprofit organization	tigr.org
gov	governmental institution	nciarray.nci.nih.gov
com	commercial organization	affymetrix.com
net	major network center	he.net
int	international organization	europa.eu.int
mil	military agency	defenselink.mil
de	Germany	mpiz-koeln.mpg.de
uk	United Kingdom	ep.ebi.ac.uk
fr	France	proteigene.fr
au	Australia	advancedlabs.com.au
se	Sweden	labvision.se
ca	Canada	uhnres.utoronto.ca

Domain name system. *Internet function that interconverts Internet protocol addresses and domain names.*

(au), and many others—also include unambiguous identifiers. IP addresses and domain names are interconverted by the ***domain name system*** (DNS).

Electronic Mail

Electronic mail. *Messages containing text, graphics, images, and other forms of data sent back and forth between computers.*

Messages containing text, graphics, images, and other forms of data sent back and forth between computers is known as e-mail. Second only to the Web, e-mail has become the most widely used application of the Internet. ***Electronic mail*** is used on a daily basis by every microarray scientist in the world, supporting written communication and data exchange, custom microarray services, newsgroups and other important resources (see below). A brief look at the history and functionality of the Internet provides a foundation for understanding this powerful communication tool.

According to Business Communications (Norwalk, CT), nearly 80 trillion e-mail messages were sent worldwide in 1998 and, based on the previous rate of increase, the number of e-mail messages projected for 2001 tops 500 trillion. By most accounts, the first e-mail messages were sent in 1971 by Ray Tomlinson, a computer engineer working for Bolt Beranek and Newman (BBN), a defense contractor commissioned by ARPA to develop the Internet. Tomlinson apparently chose the at sign (@) in e-mail addresses to designate that the user was "at" another location, a piece of e-mail nomenclature that has been preserved to the present.

Similar to the Internet, e-mail developed slowly for several decades and did not enjoy wide popularity until the genesis of user-friendly graphical interfaces that simplified the tasks of sending, composing, and receiving electronic messages. The first graphical e-mail package, *Eudora,* was developed at the University of Illinois and was provided commercially by Qualcomm (San Diego, CA). *Eudora* was originally developed for the Macintosh operating system (OS), but is currently available for both Mac OS and Microsoft *Windows* and works with an ***Internet service provider*** (ISP) that uses standard e-mail protocols (see below). Microsoft *Outlook* is another widely used e-mail software package, and widespread public access to e-mail is provided by several national ISPs, including AOL, Prodigy, and earthlink, as well as by smaller local commercial organizations. The recent availability of the ***digital subscriber line*** (DSL) provides accelerated access to the World Wide Web and e-mail, transferring information much more quickly than standard dial-up modems.

Internet service provider. *Company or institution that provides Internet access to users.*

Digital subscriber line. *New technology that provides accelerated access and data transfer through the Internet.*

Local area network. *Group of interconnected computers that provide speedy and economical computer connectivity for a set of users that share common interests.*

Simple mail transfer protocol. *Internet procedure that allows messages to be sent from remote mail servers.*

Post office protocol. *Internet procedure that allows messages to be received from remote mail servers.*

Electronic mail is sent between users by e-mail servers, which are high-storage-capacity computers that receive, deliver, and store e-mail messages and function as intermediaries between sender and recipient. Some universities and office complexes use a ***local area network*** (LAN), which is a network of interconnected computers that provide speedy and economical computer connectivity for a group of users that share common interests. A ***simple mail transfer protocol*** (SMTP) and ***post office protocol*** (POP) are used to send and receive messages, respectively, from remote mail servers. Sending and receiving messages requires an e-mail address, which includes the user's name, the at symbol, and the computer domain name (e.g. todd@arrayit.com). A user checks for new e-mail by polling the POP server to retrieve messages that are being stored on the portion of the server disk designated for that person.

E-mail offers several advantages that are responsible for the meteoric rise of this communication tool. Electronic mail is rapid, easy to use, inexpensive, and highly versatile. Microarray users can send text messages, seminar presentations (e.g., *Powerpoint* talks), primary data (e.g., tif files), graphical images (e.g., jpg files), scientific manuscripts, audio and video clips, and other important types information. The ***asynchronous*** aspect of e-mail provides tremendous flexibility by allowing senders and recipients to compose, read, and respond to e-mail messages at their leisure. E-mail is also helpful in that it eliminates the "human element" in communication, allowing users to exchange information without requiring either party to look or speak in a certain manner. Electronic mail has totally revolutionized interpersonal communication.

Asynchronous. *Aspect of e-mail that provides tremendous flexibility by allowing senders and recipients to compose, read, and respond to e-mail messages without direct communication between the sender and recipient.*

One potential disadvantage of e-mail is that, unlike the telephone, it provides a permanent hard copy of each communication, and the content of e-mail messages provides legally relevant material in legal disputes. Because all electronic messages are stored on mail servers, the confidentiality of e-mail messages is also limited. System administrators, hackers, and other third parties may gain access to e-mail messages via the server and abscond with the contents of a putative confidential communication. Unlike the telephone and direct person-to-person communication, e-mail is also lacking in that it does not provide much context for the words and images it transports. Messages that are intended to be sincere can sometimes be misconstrued as sarcastic, and vice versa.

Internet etiquette (*netiquette*) is an informal set of manners that govern the personal conduct of e-mail users and the content of messages. Abbreviations such as BTW (by the way) and symbols that include a wink ;) and a smile :) find common use in e-mail messages and can provide endearing shorthand if used sparingly. The use of all capital letters (e.g., GIVE ME A BREAK) denotes shouting and should be used only under dire circumstances. Suggestive and profane language has little place in e-mail communication and should be avoided if at all possible.

Electronic Library

The microarray field is witnessing explosive expansion and with it a rapidly growing collection of scientific papers, books, and patents. There have been >2,000 citations on microarray analysis since the first paper appeared in *Science* magazine in 1995, and the total number of manuscripts is expected to top 10,000 within two years. Keeping up with the scientific literature now and in the near future presents a major challenge that is best addressed using computational tools and databases.

One successful approach is the creation of a virtual repository of microarray citations in relational database form, complete with publication titles, journal information, authors, addresses, summaries, and electronic links to complete manuscripts (Fig. 9.2). The e-library is a powerful electronic resource and educational tool provided free of charge to microarray scientists worldwide. Scientific papers contained in the database encompass the full gamut of subject areas, including basic science, clinical medicine, agriculture, and diagnostics, with descriptions of cutting-edge discoveries from thousands of scientists working in the field. The entire e-library, updated daily, uses enterprise-level informatics tools for data mining, retrieving, and archiving. The e-library supports thousands of visitors each month, enabling global access to the entire microarray literature via the Internet.

The contents of the electronic library can be queried using any keyword of interest (Fig. 9.3) or in alphabetical order according to the last name of the first author (Fig. 9.4). Keyword or alphabetical searches drill down into the second level of the database, which contains each citation in abbreviated form; a

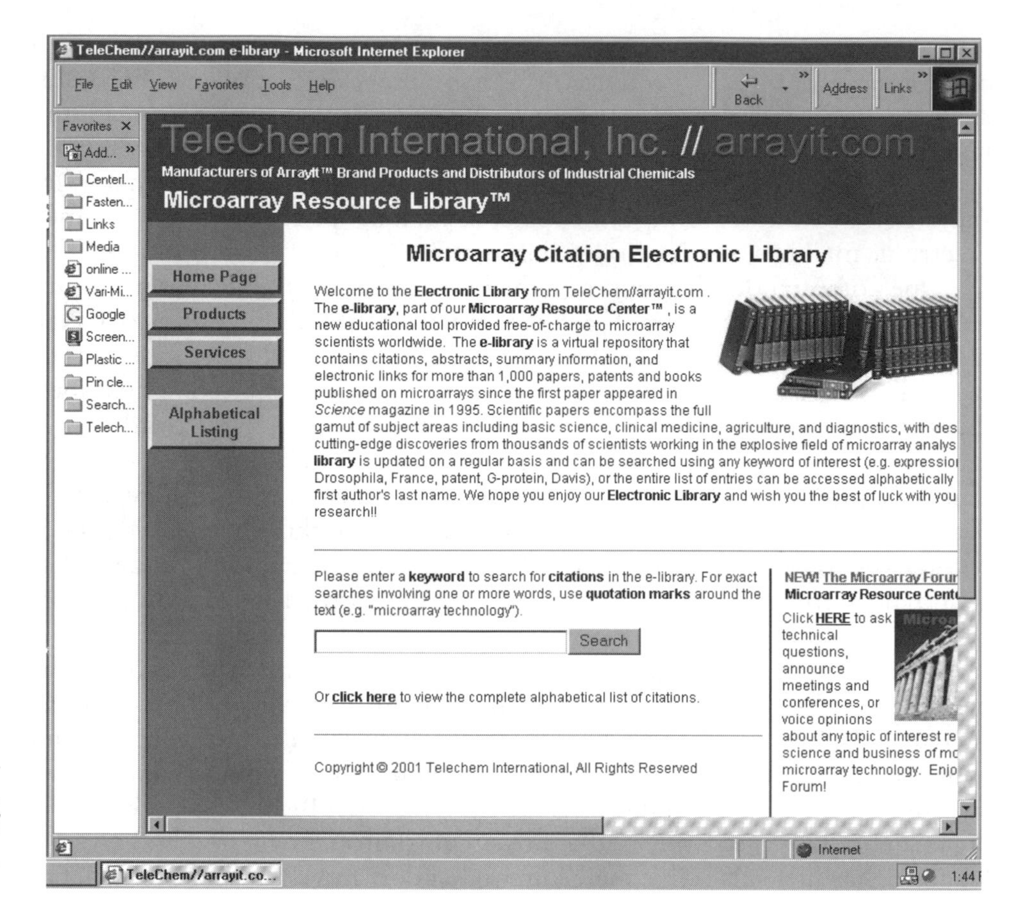

FIGURE 9.2. The Microarray Resource Library (arrayit.com/e-library). (Courtesy of Mark Shevetone and Kim Schena, TeleChem/arrayit.com, Sunnyvale, CA.)

Microarray Citation Electronic Library

Welcome to the **Electronic Library** from TeleChem//arrayit.com . The **e-library**, part of our **Microarray Resource Center**™ , is a new educational tool provided free-of-charge to microarray scientists worldwide. The **e-library** is a virtual repository that contains citations, abstracts, summary information, and electronic links for more than 1,000 papers, patents and books published on microarrays since the first paper appeared in *Science* magazine in 1995. Scientific papers encompass the full gamut of subject areas including basic science, clinical medicine, agriculture, and diagnostics, with descriptions of cutting-edge discoveries from thousands of scientists working in the explosive field of microarray analysis. The entire **e-library** is updated on a regular basis and can be searched using any keyword of interest (e.g. expression profiling, Drosophila, France, patent, G-protein, Davis), or the entire list of entries can be accessed alphabetically according to the first author's last name. We hope you enjoy our **Electronic Library** and wish you the best of luck with your microarray research!!

Please enter a **keyword** to search for **citations** in the e-library. For exact searches involving one or more words, use **quotation marks** around the text (e.g. "microarray technology").

Drosophila | Search

Or **click here** to view the complete alphabetical list of citations.

Copyright © 2001 Telechem International, All Rights Reserved

NEW! The Microarray Forum, part of our **Microarray Resource Center**™

Microarray Forum

Click **HERE** to ask technical questions, announce meetings and conferences, or voice opinions about any topic of interest related to the science and business of modern microarray technology. Enjoy your trip to the Forum!

FIGURE 9.3. The entire contents of the microarray e-library can be queried electronically using any keyword of interest, such as Drosophila (arrow). *(Courtesy of Mark Shevetone and Kim Schena, TeleChem/arrayit.com, Sunnyvale, CA.)*

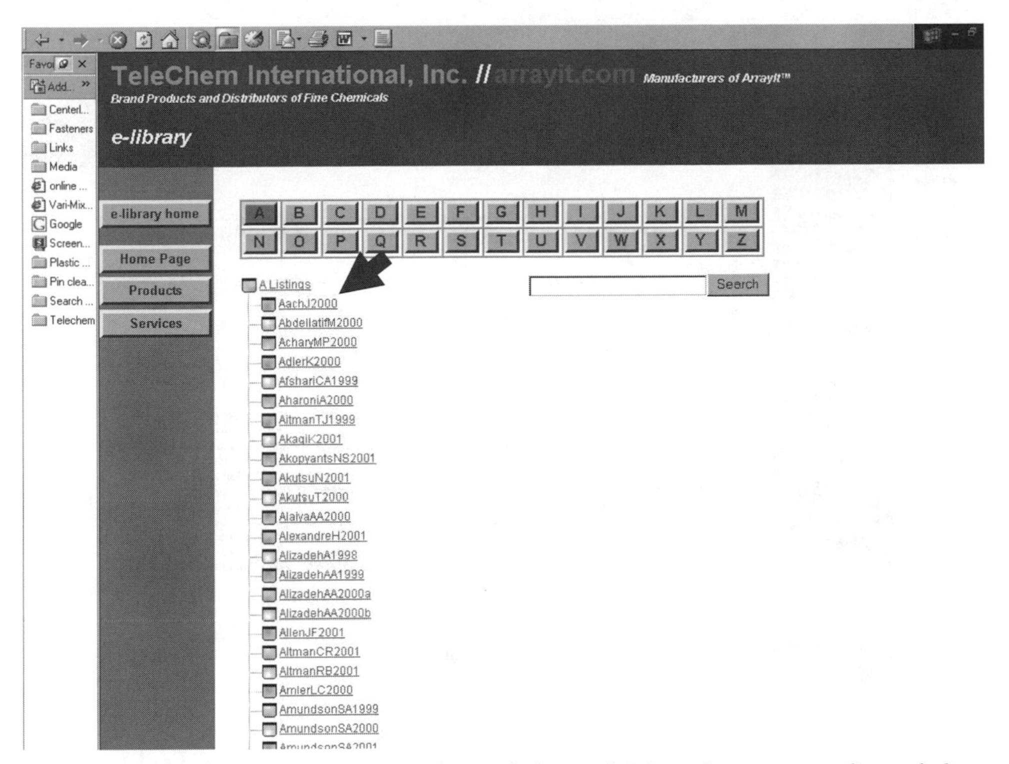

FIGURE 9.4. The microarray e-library can be searched in an alphabetical manner, according to the last name of the first author. Clicking on a citation (arrow) *provides the user with the full citation, summary information, electronic links, and other information. (Courtesy of Mark Shevetone and Kim Schena, TeleChem/arrayit.com, Sunnyvale, CA.)*

mouse click on any citation moves the user into the third tier of the e-library, which contains the full citation, summary information, and a link to an electronic manuscript (Fig. 9.5). Links to electronic manuscripts allow the user to navigate to sites provided by each scientific journal, patent information provider, or other agency. Some scientific publications are available in complete electronic form (Fig. 9.6), providing an enormously convenient resource to all microarray scientists.

PubMed

Microarray analysis often requires publication queries beyond the papers that involve microarrays. Such searches can provide the researcher with a wealth of biological, agricultural, and clinical information required to formulate experimental questions and interpret results. A wonderful electronic resource known as ***PubMed*** is provided the National Library of Medicine, which is part of the National Institutes of Health (Bethesda, MD). PubMed provides electronic access to $>$10 million scientific papers from nearly 5000 journals published in 70 countries since the mid-1960s. This near comprehensive scientific library is available via the Internet and can be accessed with any standard browser (Fig. 9.7). The entire contents of PubMed can be searched with any keyword of interest such as author name. The second level in PubMed contains all of the hits for a given search in the form of complete citations to each paper that contains the query term (Fig. 9.8). PubMed, which contains all of the contents of ***MEDLINE*** plus some additional content, allows researchers access to all but the most obscure medical publications.

PubMed. *Electronic resource from the National Library of Medicine that provides electronic access to $>$10 million scientific papers from nearly 5000 journals published in 70 countries since the mid-1960s.*

MEDLINE. *Main source of content for PubMed containing a comprehensive electronic library of medically oriented scientific publications.*

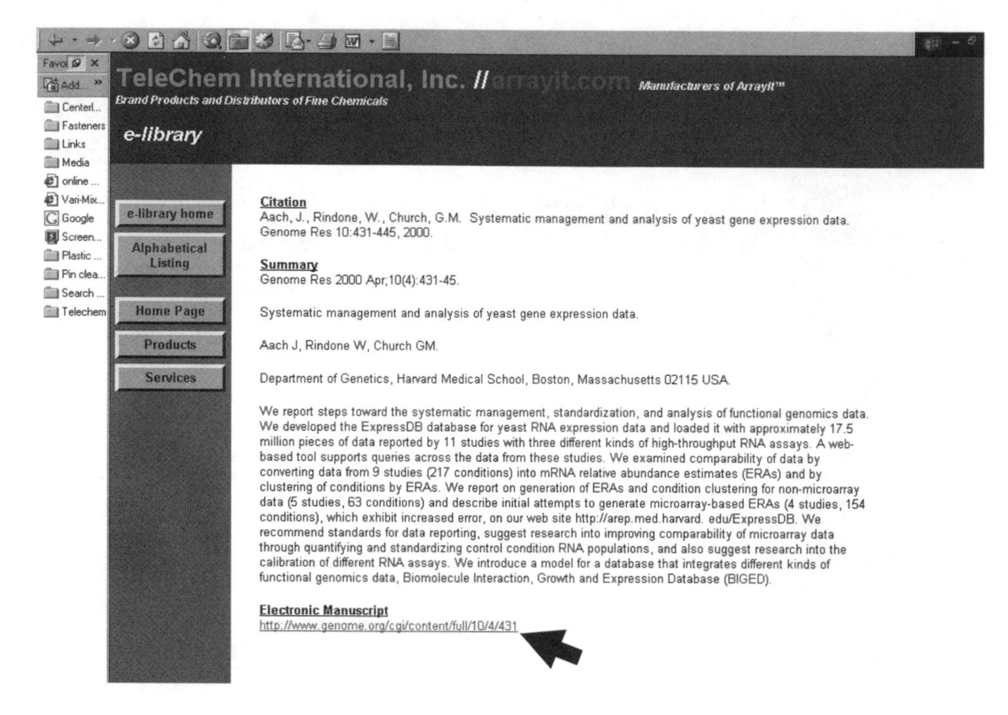

FIGURE 9.5. *Complete citation information, a summary, and a link to an electronic manuscript if available* (arrow). *(Courtesy of Mark Shevetone and Kim Schena, TeleChem/arrayit.com, Sunnyvale, CA.)*

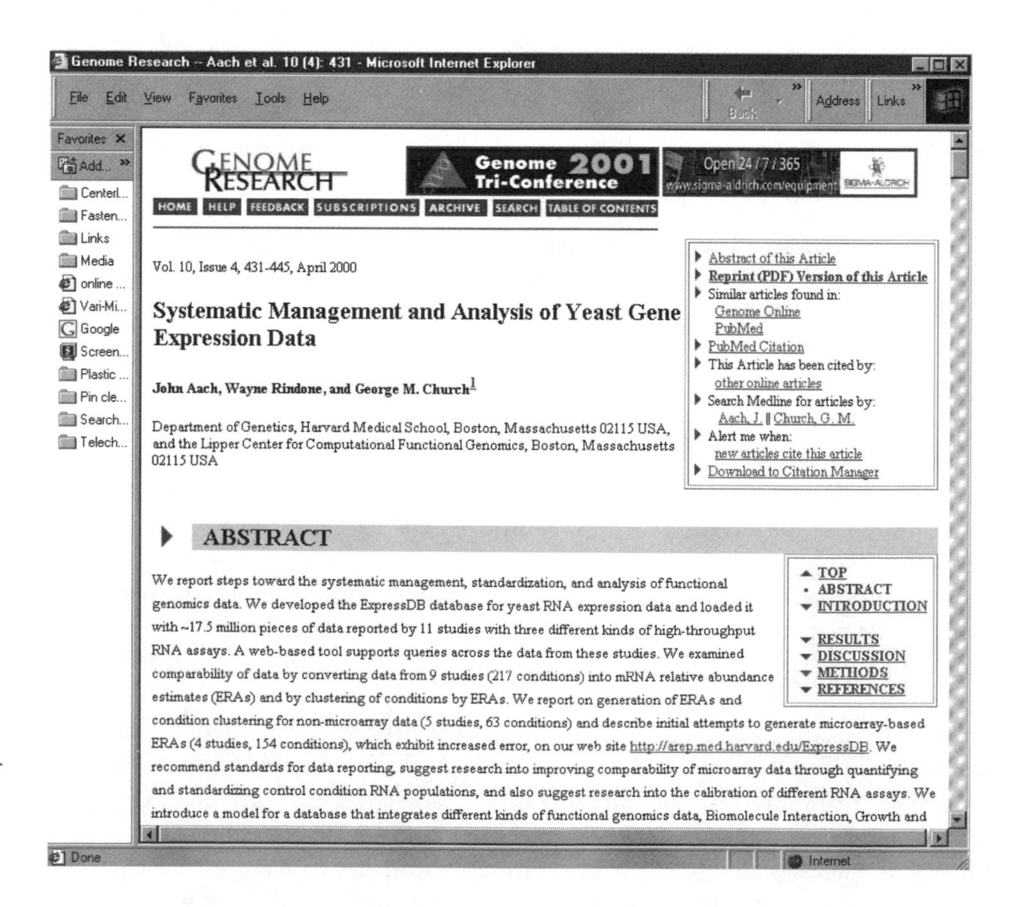

FIGURE 9.6. *Many files in the e-library contain links to journals, such as* Genome Research, *which provide complete manuscripts in an electronic format. (Courtesy of Mark Shevetone and Kim Schena, TeleChem/arrayit.com, Sunnyvale, CA.)*

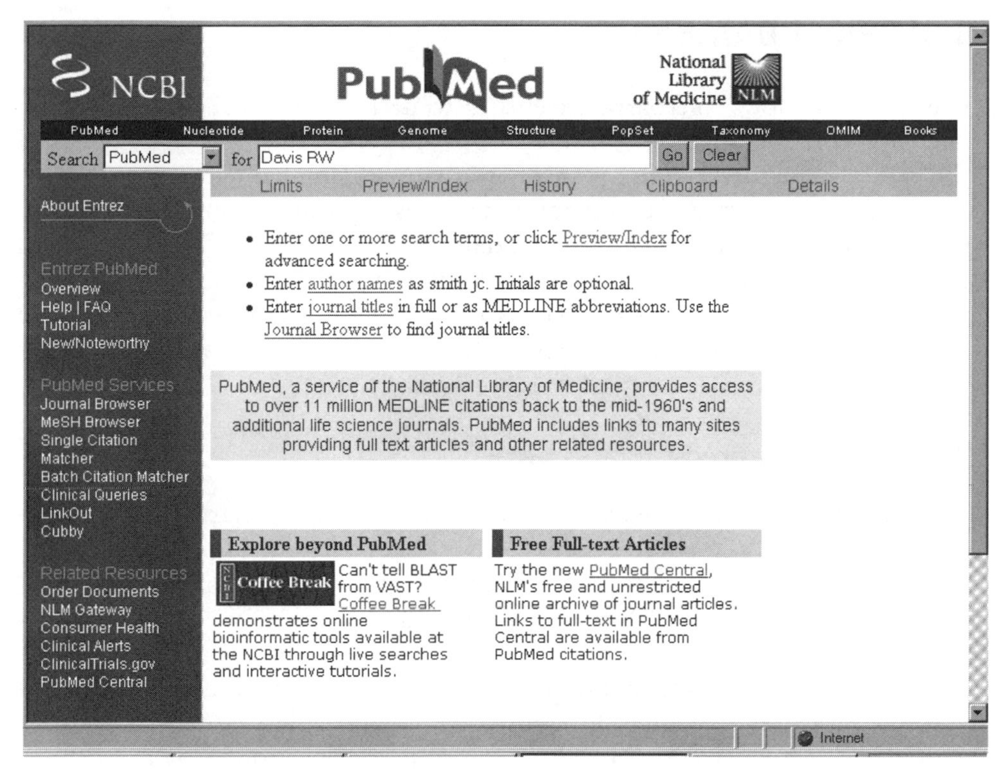

FIGURE 9.7. Search screen for PubMed (www.ncbi.nlm.nih.gov/PubMed/).

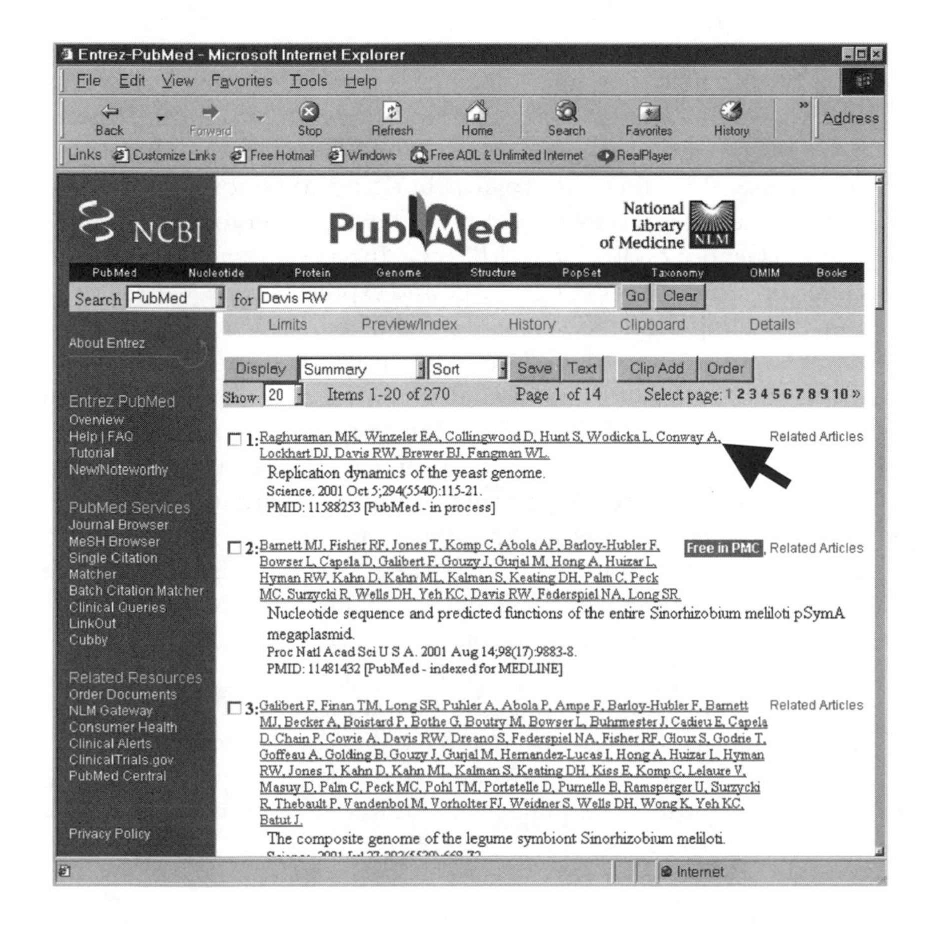

FIGURE 9.8. PubMed search results for keyword "Davis RW" revealed 270 hits.

BIOLOGICAL DATABASES

The Internet allows microarray scientists to access biological databases with extensive content, including information on genes and genomes, proteins and crystal structures, and sequence variants (e.g., single nucleotide polymorphisms), and to carry out comparative analyses, including homology searches and the construction of evolutionary relationships. These databases are useful in selecting sequences for microarray manufacture and in interpreting the results of microarray experiments. This section describes some of the biological databases available to microarray researchers, focusing on the practical aspects of how to use them via standard Web browsers.

Genes and Genomes

Gene and genome sequence databases are essential for microarray analysis. Sequence databases allow researchers to select sequences of interest to prepare oligonucleotides for microarray manufacture and to obtain primers for polymerase chain reaction (PCR) amplification of genes and cDNAs. Electronic access and search features allow the selected sequences to be checked against the database, to avoid redundancy, repeat elements, and gene families. Gene and genome databases for many different organisms enable the construction of "zoo chips" for microarray-based exploration of evolutionary relationships. Complete genomic sequences are preferable to partial sequences for an organism, because they provide the entire genetic blueprint for that organism, but much use can be made of partial gene sequence collections.

Microarray scientists commonly use three types of nucleotide sequence databases. Many laboratories, companies and universities have small in-house databases of select content, which is available on a limited basis. These sequence databases are useful for focused projects, but tend to be somewhat limited in terms of the amount of content. Several large biotechnology companies, including Celera Genomics (Rockville, MD) and Incyte Genomics (Palo Alto, CA), possess large private databases of genomic and expressed sequence tag (EST) information, providing the data on a paid basis. The Celera and Incyte databases provide extensive collections of high-quality sequence information for microarray analysis, but access can be somewhat expensive and restrictive, depending on the volume and intended use of the information.

GenBank. *Public sequence database maintained by the National Center for Biotechnology Information, consisting of an annotated collection of all publicly available sequences from the United States, Japan, and Europe; includes >13 billion bases of DNA from approximately 12 million submissions.*

BankIt. *Graphical form used to submit sequences to GenBank via the World Wide Web.*

Accession number. *Unique identification number attached to each sequence submitted to GenBank.*

The third source of sequence information for microarray experimentation is found in ***GenBank,*** a public sequence database maintained by the NCBI (Bethesda, MD). GenBank®is an annotated collection of all publicly available sequences collected from the United States, the DNA DataBank of Japan (DDBJ), and the European Molecular Biology Laboratory (EMBL), and it has seen explosive growth since its inception in 1982 (Fig. 9.9). As of August 2001, GenBank contained >13 billion bases of DNA sequence information from approximately 12 million submissions. The 7.3 billion bases of DNA information added during the year 2000 exceeded the 3.8 billion nucleotides submitted during the previous 18 years of Genbank combined. Regular sequence submissions comprise a diverse group of organisms, including human, mouse, fish, worm, parasite, bacteria, viruses, and others (Fig. 9.10).

Sequences can be submitted to GenBank via the Web using a graphical form known as ***BankIt***. Each sequence submitted to the database receives a unique identification number known as an ***accession number***. GenBank accession

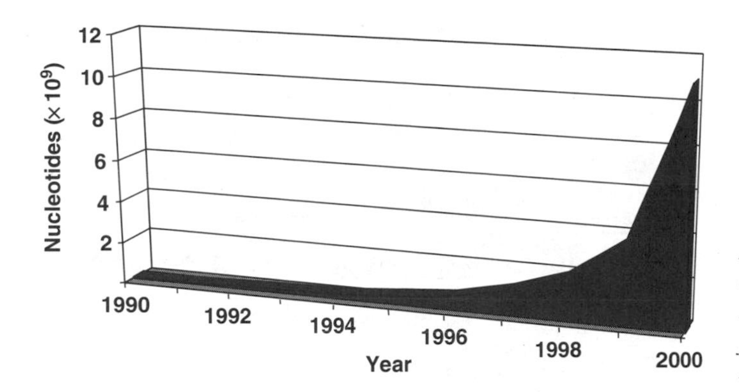

FIGURE 9.9. The number of nucleotides in billions of bases (10^9) submitted to the GenBank database from 1990 to 2000. (Data courtesy of NCBI).

numbers are used in publications describing the sequence and in microarray content maps (see below) that contain part of all of the sequence as a microarray target element. The GenBank accession number for the human galactose-1-phosphate uridyltransferase (GALT) coding sequence (NM_000155), for example, is distinct from the accession number for the equivalent sequence in the mouse (NM_016658). Accession numbers provide invaluable identifiers for microarray applications, particularly in cases when the chips contain thousands of different target sequences.

The contents of GenBank can be queried using a standard Web browser and any search term of interest (Fig. 9.11) with the search and retrieval system known as ***Entrez***. Queries are entered into a text box, and a mouse click sends the user into the second tier of the database, which contains summaries of all of the hits obtained with the specific query. Selecting a specific entry sends the user to a page that contains the sequence file, accession number, gene name, organism, submitter, and other essential information (Fig. 9.12). Scrolling down through the GALT sequence file provides the fully annotated cDNA sequence (Fig. 9.13), which can be used to generate PCR primers, oligonucleotide targets, and other important sequences. The sequence can be cut and pasted into e-mail messages and sent via the Internet, inserted into the body of text documents (e.g., Microsoft *Word*), or into text boxes for database searches (see below).

Entrez. *Search and retrieval system used to query the contents of GenBank via a standard World Wide Web browser.*

Complete or near complete genome sequences are available from the NCBI for dozens of organisms, including human; *Arabidopsis;* mouse; *Caenorhabditus elegans; Drosophila melanogaster;* and many bacteria, viruses, and organelles. All of the genome sequence information is accessible via the Internet with a standard

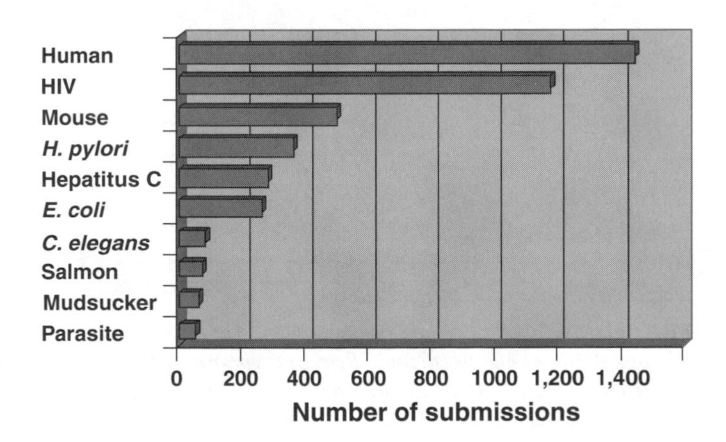

FIGURE 9.10. The 10 organisms that received the greatest number of GenBank submissions for August 2000. HIV, *human immunodeficiency virus type 1;* H. pylori, *Helicobacter pylori;* E. coli, *Escherichia coli;* C. elegans, *Caenorhabditus elegans;* parasite, *malaria. (Data courtesy of NCBI).*

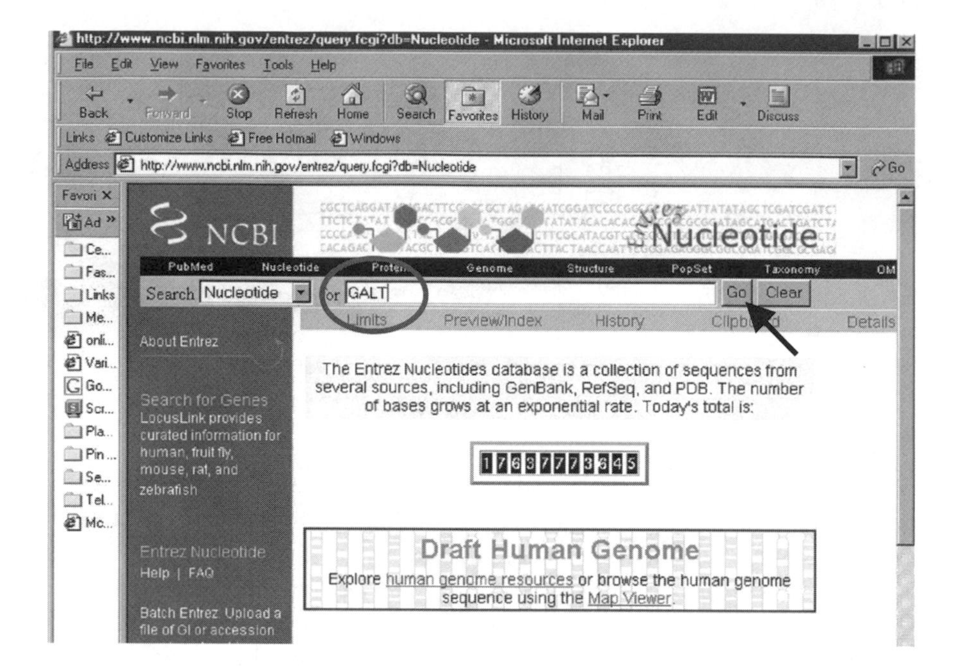

FIGURE 9.11. The search screen for the GenBank nucleotide database (www.ncbi.nlm.nih.gov:80/entrez/query.fcgi?db=Nucleotide). Microarray users enter a query term such as "GALT" (circle), and click on "Go" (arrow) to initiate the database search.

Web browser (www.ncbi.nlm.nih.gov:80/entrez/query.fcgi?db=Genome). Queries are made by entering a search term into the text box, and numerous useful links are available to microarray researchers on a point-and-click basis (Fig. 9.14). The availability and affordability of genomic databases, with the NCBI as curator, are invaluable to the microarray community.

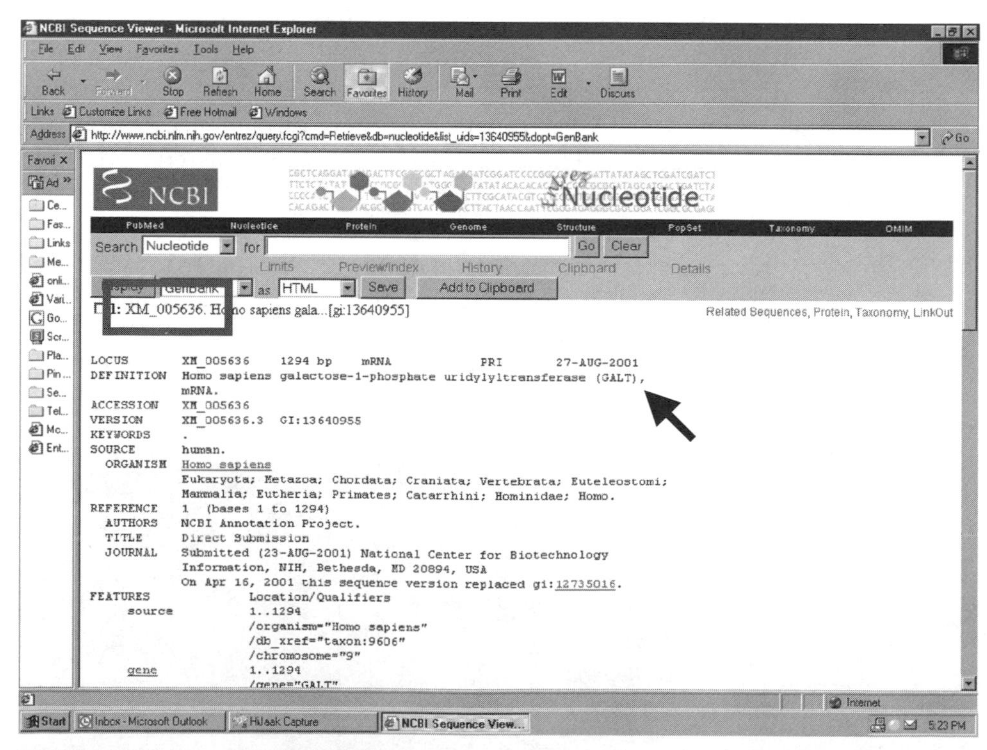

FIGURE 9.12. Users can find specific sequences of interest in the GenBank database (www.ncbi.nlm.nih.gov:80/entrez/query.fcgi?cmd=Retrieve&db=nucleotide&list_uids=13640955&dopt=GenBank). The human GALT mRNA (arrow) *contains a unique identification number (XM_005636;* box) *that can be used for archiving, retrieval, and generating microarray content files.*

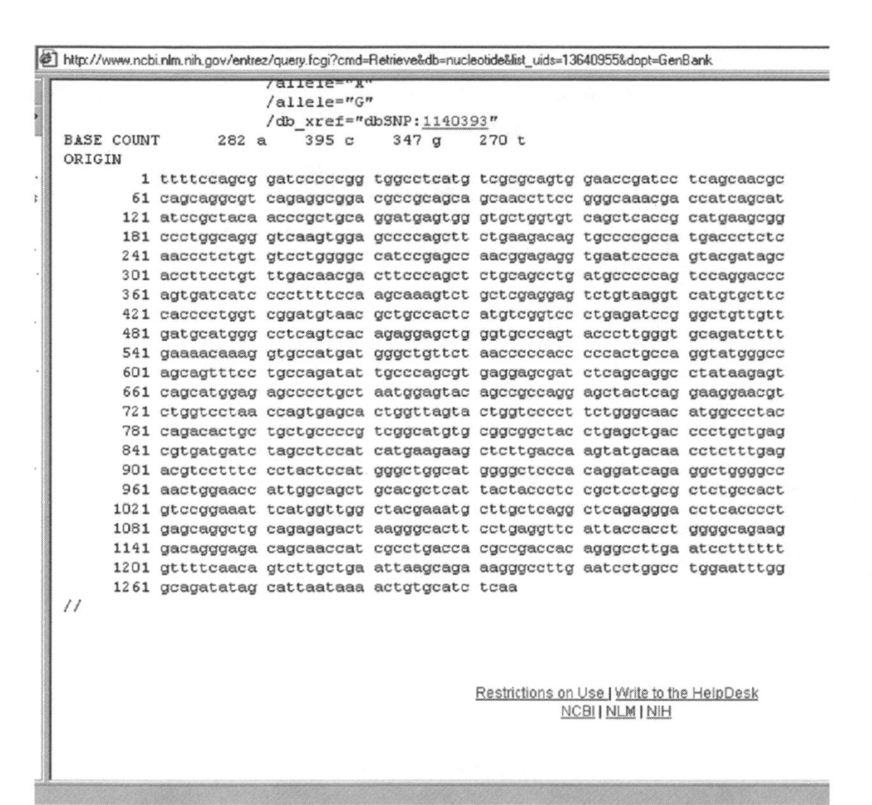

http://www.ncbi.nlm.nih.gov/entrez/query.fcgi?cmd=Retrieve&db=nucleotide&list_uids=13640955&dopt=GenBank
/allele="G"
/db_xref="dbSNP:1140393"
BASE COUNT 282 a 395 c 347 g 270 t
ORIGIN
1 ttttccagcg gatcccccgg tggcctcatg tcgcgcagtg gaaccgatcc tcagcaacgc
61 cagcaggcgt cagaggcgga cgccgcagca gcaaccttcc gggcaaacga ccatcagcat
121 atccgctaca acccgctgca ggatgagtgg gtgctggtgt cagctcaccg catgaagcgg
181 ccctggcagg gtcaagtgga gccccagctt ctgaagacag tgccccgcca tgaccctctc
241 aaccctctgt gtcctggggc catccgagcc aacggagagg tgaatcccca gtacgatagc
301 accttcctgt ttgacaacga cttcccagct ctgcagcctg atgcccccag tccaggaccc
361 agtgatcatc ccctttcca agcaaagtct gctcgaggag tctgtaaggt catgtgcttc
421 cacccctggt cggatgtaac gctgccactc atgtcggtcc ctgagatccg ggctgttgtt
481 gatgcatggg cctcagtcac agaggagctg ggtgcccagt acccttgggt gcagatcttt
541 gaaaacaaag gtgccatgat gggctgttct aacccccacc cccactgcca ggtatgggcc
601 agcagtttcc tgccagatat tgcccagcgt gaggagcgat ctcagcaggc ctataagagt
661 cagcatggag agcccctgct aatggagtac agccgccagg agctactcag gaaggaacgt
721 ctggtcctaa ccagtgagca ctggttagta ctggtcccct tctgggcaac atggccctac
781 cagacactgc tgctgccccg tcggcatgtg cggcggctac ctgagctgac ccctgctgag
841 cgtgatgatc tagcctccat catgaagaag ctcttgacca agtatgacaa cctctttgag
901 acgtcctttc cctactccat gggctggcat ggggctccca caggatcaga ggctggggcc
961 aactggaacc attggcagct gcacgctcat tactaccctc cgctcctgcg ctctgccact
1021 gtccggaaat tcatggttgg ctacgaaatg cttgctcagg ctcagaggga cctcacccct
1081 gagcaggctg cagagagact aagggcactt cctgaggttc attaccacct ggggcagaag
1141 gacagggaga cagcaaccat cgcctgacca cgccgaccac agggccttga atcctttttt
1201 gttttcaaca gtcttgctga attaagcaga aagggccttg aatcctggcc tggaatttgg
1261 gcagatatag cattaataaa actgtgcatc tcaa
//
Restrictions on Use | Write to the HelpDesk
NCBI | NLM | NIH

FIGURE 9.13. *Sequence files in Gen-Bank are annotated to provide a convenient format for designing primers, oligonucleotide targets, and synthetic probes (www.ncbi.nlm.nih.gov:80/entrez/query.fcgi?cmd=Retrieve&db=nucleotide&list_uids=13640955&dopt=GenBank).*

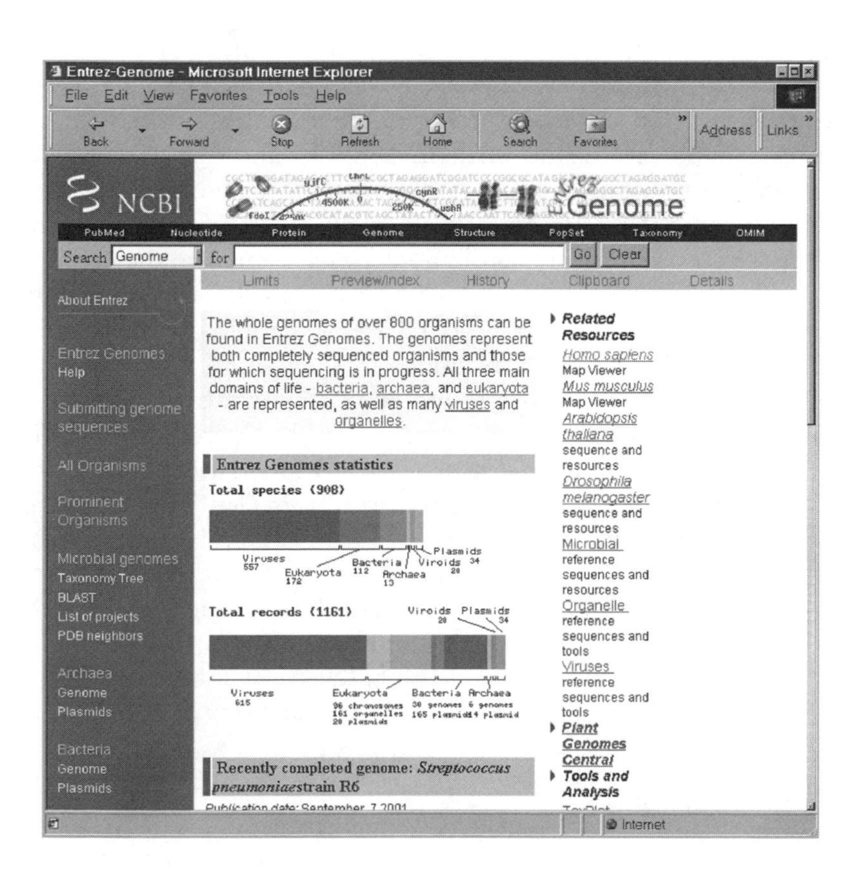

FIGURE 9.14. *Microarray researchers can access an extensive suite of genome sequence resources via Entrez Genome (www.ncbi.nlm.nih.gov:80/entrez/query.fcgi?db=Genome).*

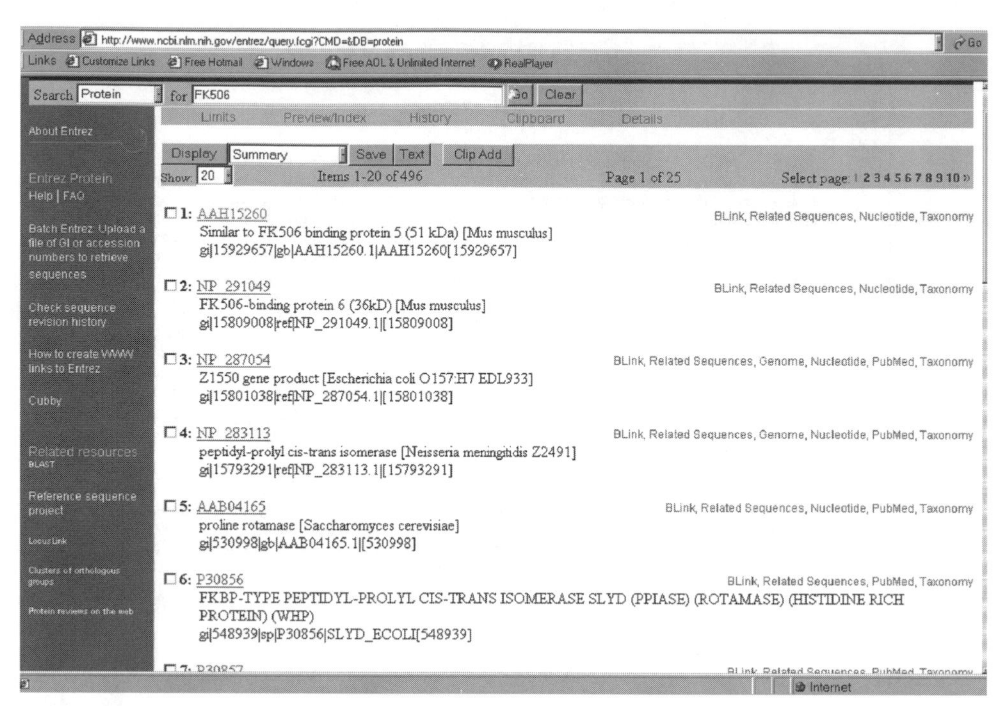

FIGURE 9.15. *Microarray researchers can access an extensive database of protein sequence information from the NCBI (www.ncbi.nlm.nih.gov:80/entrez/query.fcgi?db=Protein).*

Proteins

The rapid proliferation of protein microarrays is creating a greater and greater need for protein informatics. Microarray scientists require several different kinds of protein information, including primary amino acid sequences, post-translational modification sites, structural motifs, and crystallographic information. Protein informatics is essential for evaluating myriad different aspects of protein microarray experiments, specifically protein–protein, protein–drug, enzyme–substrate and other types of interactions. Similar to the nucleotide sequence databases, researchers draw commonly from three types of databases: in-house dedicated sources, private databases assembled by companies, and public databases provided by the NCBI.

NCBI's search-and-retrieval system allows the user to obtain protein information from a variety of sources, including the Protein Information Resource (PIR) database (Washington, DC), the Protein Research Foundation (PRF) database (Osaka, Japan), and translations from GenBank nucleotide coding regions. Protein databases can be queried using a standard Internet browser and a search term of interest such as FK506 (Fig. 9.15), an immunosuppressive drug that binds the FK506-binding protein. Each hit in the protein database is reported in abbreviated form, and additional information can be obtained by clicking on the link of interest (Fig. 9.16). Extensive crystallographic information is also available to microarray researchers online (Fig. 9.17).

Homology search. *Computational process in which a query sequence is aligned with every subject sequence in a gene or protein database.*

Homology Searches

The process by which a query sequence is aligned with every subject sequence in a database is known as a ***homology search***. Query sequences may be nucleic acid or protein, though most homology searches are performed at the nucleotide

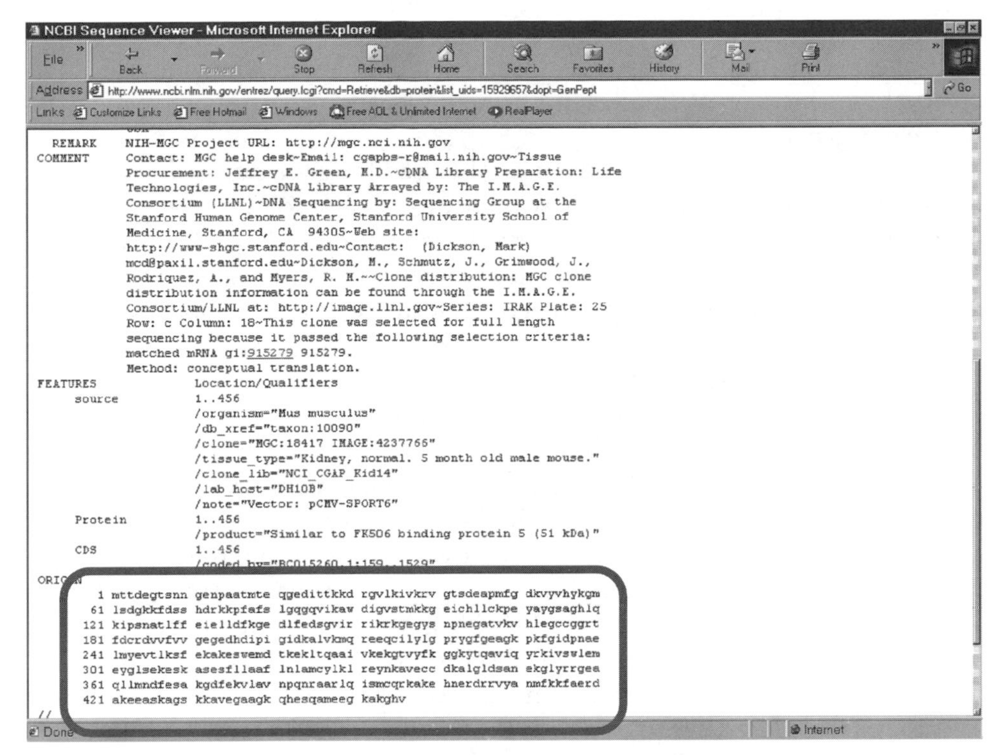

FIGURE 9.16. *The primary amino acid sequence* (rectangle) *of the mouse FK506-binding protein accessed from the NCBI.*

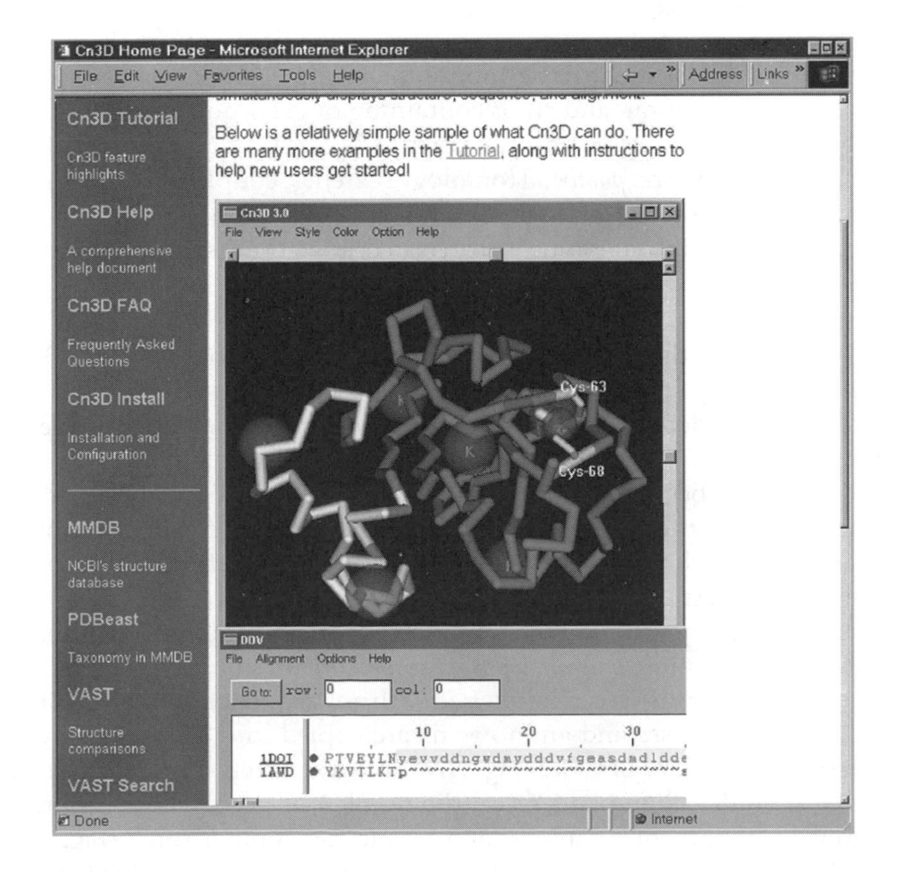

FIGURE 9.17. *The crystal structures of thousands of proteins including ferredoxin (shown) are available online from the NCBI.*

level. Microarray researchers employ homology searches as a regular component of experimental work because the process allows electronic identification of novel sequences, gene families in the genome, and evolutionary relatedness between organisms.

Microarrays of unknown cDNAs can be used for gene expression analysis, but such microarrays require clone sequencing and homology searching to interpret the experimental results. The first human microarray experiments used 1000 cDNAs of unknown sequence, followed by select clone sequencing and homology searches to identify human genes modulated by heat shock treatment and phorbol ester exposure (Schena et al., 1996).

Redundant. *Two or more target sequences in a microarray that share perfect or near-perfect sequence identity.*

Homology searches are also used extensively to design sequences for microarray manufacture, particularly long oligonucleotide and cDNA target elements. Potential oligonucleotide and cDNA targets are often run against the database to determine whether identical or similar sequences occur at multiple genomic locations, and if so, whether these sequences are expressed as mRNA. The use of ***redundant*** target sequences can complicate microarray analysis by hybridizing to more than one species in the probe mixture, producing an undesired effect known as cross-hybridization. Target elements subject to cross-hybridization produce erroneously elevated signals and can diminish differential expression measurements in ratio studies. Homology searches allow the microarray researcher to establish the uniqueness of a target sequence before experimental work, thereby avoiding cross-hybridization.

Homolog. *Gene or protein that shares sequence similarity with another gene or protein.*

Paralog. *Related gene or protein present in the same organism.*

Ortholog. *Single gene or protein that occurs in two different organisms.*

Two or more genes that share sequence similarity are known as ***homologs***. Related genes present in the same organism (e.g., members of a gene family) are known as ***paralogs,*** and related genes that occur in different organisms are known as ***orthologs***. Orthologous genes enable microarray analysis of evolutionary relatedness, such that a mouse microarray can be used to examine probes prepared from rat, and chips containing target sequences from baker's yeast (*Saccharomyces cerevisiae*) can be used to hybridize samples from fission yeast (*Schizosaccharomyces pombe*). Homology searches, coupled with microarray hybridization or protein reactions, facilitate interpretation of interspecies microarray analysis and allow evolutionary relatedness to be assessed on a genomic scale.

Homology searches are performed electronically using a number of different algorithms including the basic local alignment search tool (BLAST). BLAST facilitates extremely sensitive and robust searches because it compares similarity across short nucleotide and protein regions rather than complete genes and proteins. The BLAST tool is particularly well-suited for identifying distantly-related orthologs, because highly homologous regions tend to occur in local sequence regions rather than uniformly across the entire gene or protein.

BLAST-based homology searches can be performed using a standard Internet browser and data maintained by the NCBI (www.ncbi.nlm.nih.gov/BLAST/). Complete or near-complete genomic sequences of human, mouse, *Arabidopsis, Drosophila,* scores of bacteria, and many other genes and organisms can be queried with any sequence of interest. Strings of DNA and RNA bases and amino acids are copied into a text window and queried from a standard browser (Fig. 9.18), and a search is completed typically within 20 s. BLAST alignments reveal the number of positive scores obtained and provide alignment graphics and sequence comparisons (Fig. 9.19). A BLAST search using the nucleotide sequence of the rat glucocorticoid receptor (GR) gene, for example, produces a strongly homologous sequence

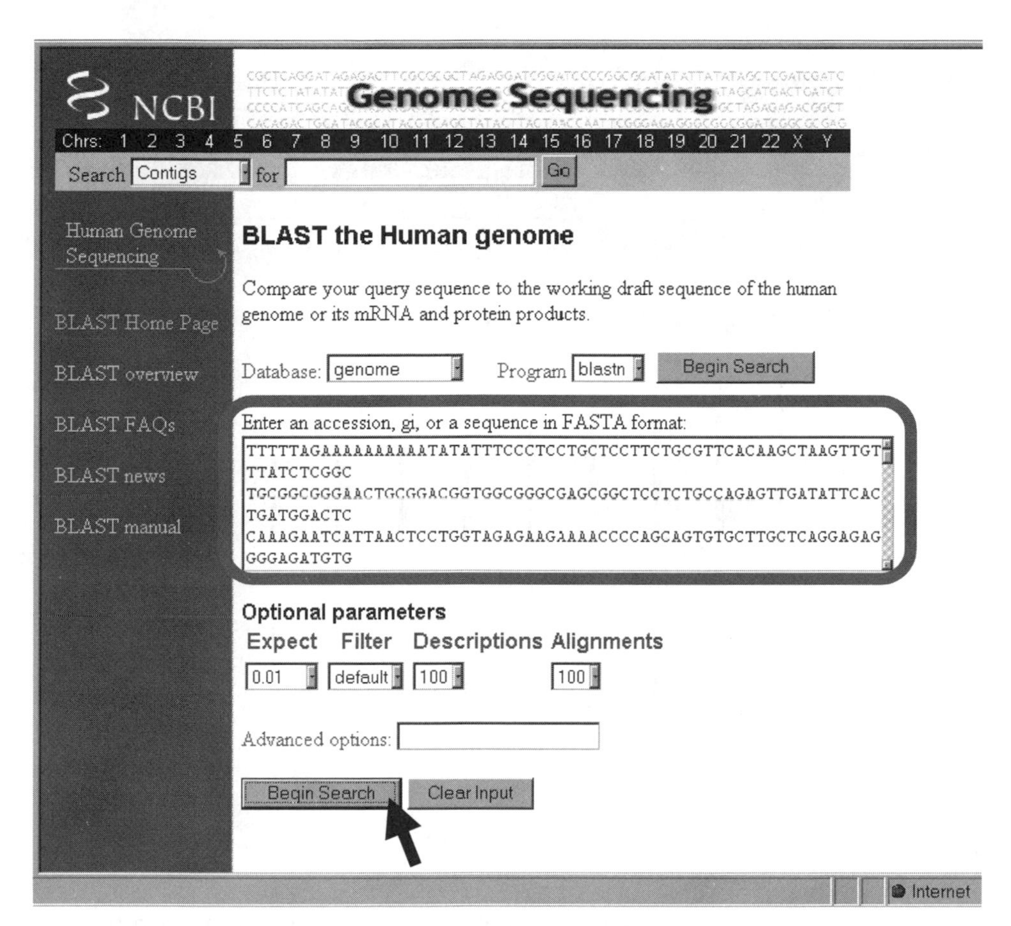

FIGURE 9.18. *Gene and genome databases at the NCBI (Bethesda, MD) can be queried with any sequence of interest via the Internet.*

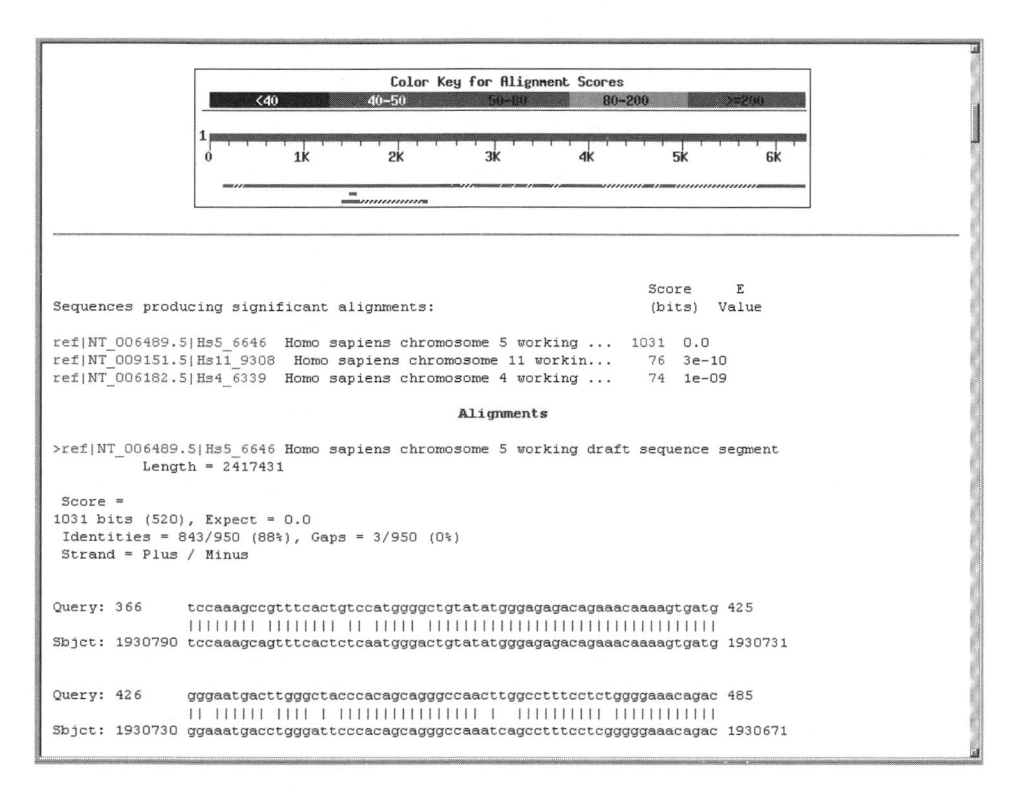

FIGURE 9.19. *A query using the rat glucocorticoid receptor gene against the entire human genome draft sequence produces a single BLAST hit, indicating the presence of one ortholog of the rat gene in human. BLAST results can be viewed in graphical form (top) or as a primary sequence alignment (bottom).*

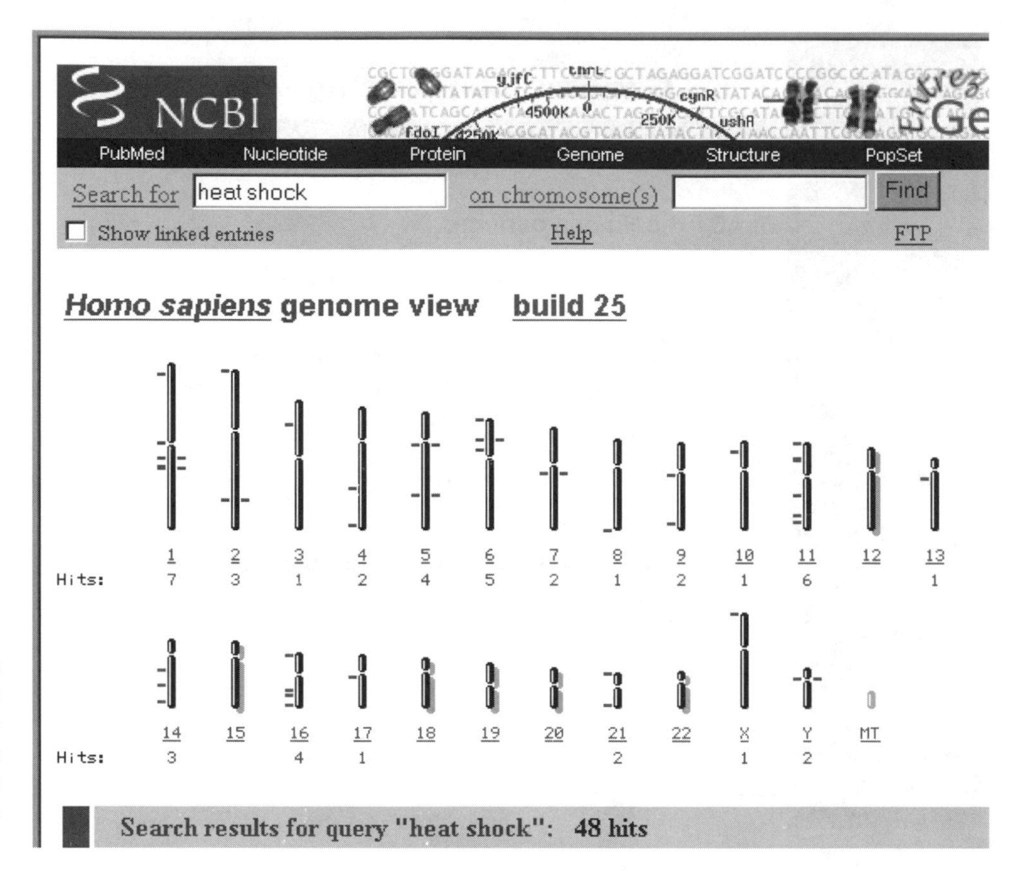

FIGURE 9.20. *Text queries, such as* heat shock, *can be used to search the entire human genome sequence; the results are represented in graphical form on the 22 human autosomes, two sex chromosomes (X and Y), and the mitochondrial sequence (MT).*

corresponding to the human ortholog of the rat GR, enabling comparative studies between rat and human.

Sequence and text queries can also be made and displayed as a function of all of the human chromosomes, with homologous regions denoted along the 22 autosomes, two sex chromosomes, and the mitochondrial sequence (Fig. 9.20). A BLAST search with the term *heat shock,* for example, produces 45 hits in the human genome, and greater and fewer positive scores are obtained using query sequences that occur more or less frequently in the genome. Homology searches on the genomic level provide a comprehensive view of the genetic blueprint of an organism vis-à-vis any microarray target sequence.

Sequence Variants

Single nucleotide polymorphisms (SNPs), insertions, deletions, and other genetic alterations in a gene, cDNA, mRNA, or protein are known as sequence variants. Changes in the primary sequences of genes and proteins can lead to disease phenotypes, and microarray assays that allow sequence checking, resequencing, and mutation detection are gaining wide use (see Chapter 13). To expedite microarray analysis of sequence variants, it is useful to have informatics tools that allow the search and retrieval of such sequences.

One straightforward approach is to use PubMed (see above) and the scientific literature to identify mutations and other sequence variants described in experimental studies. The PubMed approach is manual and somewhat laborious, but it offers an inexpensive means by which to collate results from many laboratories and manufacture the corresponding microarrays for genetic screening and diagnostic applications.

TABLE 9.2. Summary of the dbSNP[a]

Organism	Type	Number of SNPs
Human	primate	3,052,574
Mouse	rodent	483
Arabidopsis thaliana	mustard plant	184
Plasmodium	malaria parasite	4
Chimpanzee	primate	2
Rat	rodent	1

[a] From NCBI (www. ncbi.nlm.nih.gov/SNP).

Another approach is use the single nucleotide polymorphism database (dbSNP) provided by the NCBI (*www.ncbi.nlm.nih.gov/SNP/*), which is available free of charge to microarray researchers worldwide. The dbSNP contains the sequences of SNPs, small insertion and deletion mutations, and microsatellite repeats and therefore provides access to a broad spectrum of minor sequence variants. The database contains >3 million SNPs for human and a small number for other organisms (Table 9.2). The dbSNP can be queried using a keyword of interest such as "beta-globin" (Fig. 9.21), and the results of searches provide the location of mutations within each database sequence, listed as a function of the GenBank accession number (Fig. 9.22). Electronic access to sequence variant databases expedites disease testing, screening, diagnostic, and other applications of microarrays that focus on DNA information.

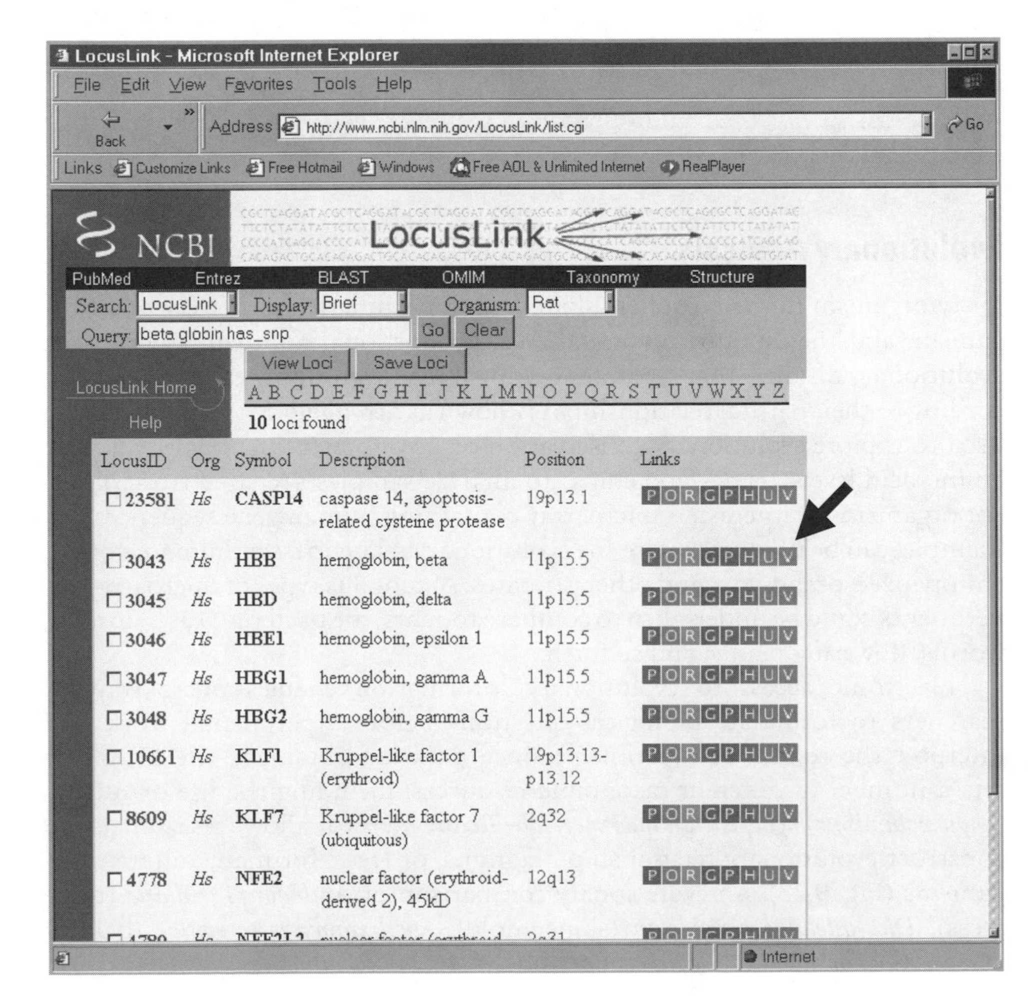

FIGURE 9.21. *The dbSNP can be queried using a gene name, such as beta globin; the browser displays the positive matches. Specific information on SNPs is accessed via a mouse click* (arrow).

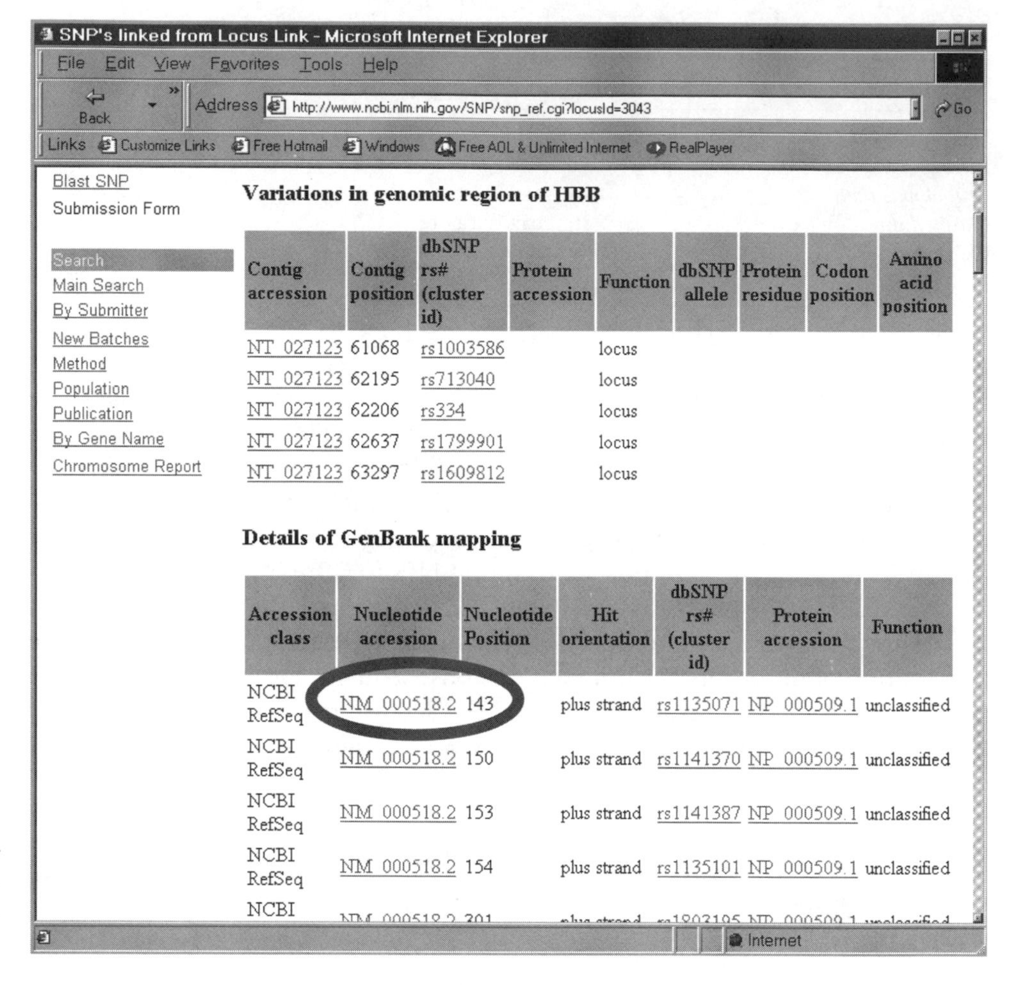

FIGURE 9.22. *Search results obtained using the query term* beta globin *in the dbSNP. The database provides the nucleotide position and accession number* (oval) *of each sequence variation, expediting the identification and use of SNP information in microarray assays.*

Evolutionary Analysis

Taxonomy. *Grouping of organisms according to their natural relationship.*

Every organism in the biosphere shares genetic similarity with every other organism, and the amount of similarity is proportional approximately to the evolutionary distance that separates organisms. The grouping of organisms according to their natural relationships is known as ***taxonomy***. Microarrays can be used to explore evolutionary conservation or taxonomy at the nucleotide and amino acid levels, using "zoo chips" to analyze samples prepared from different organisms of interest. A microarray containing human gene sequences, for example, can be used to explore the evolutionary relatedness of human, gorilla, chimpanzee, orangutan, and other primates. Analogous types of microarray experiments could be undertaken to compare rodents, the bacteria, yeasts, insects, worms, flowering plants, and so forth.

Electronic access to evolutionary information enables microarray researchers to formulate biological questions before experimental work and interpret the results of experiments once data are obtained. The NCBI offers a number of different taxonomic resources, including the use of BLAST (*www.ncbi.nlm.nih.gov/entrez/query.fcgi?db=Taxonomy*) to allow researchers to construct evolutionary relationship diagrams, or trees, for many different organisms (Fig. 9.23). An evolutionary comparison of *Arabidopsis thaliana* (thale cress), *Dictyostelium discoideum* (slime mold), *Saccharomyces cerevisiae* (baker's yeast), *Caenorhabditis elegans* (nematode), *Schizosaccharomyces pombe* (fission

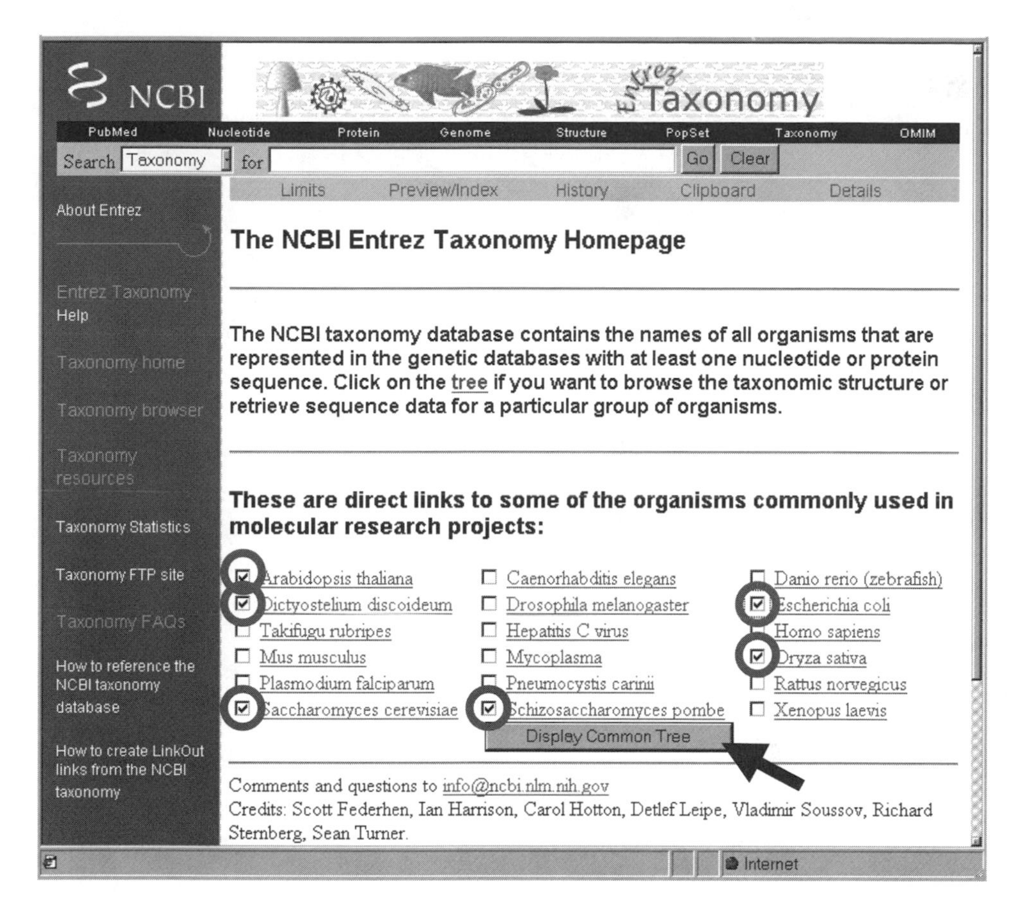

FIGURE 9.23. *The NCBI allows users to compare organisms of interest by checking the subject organisms* (circles) *of interest and selecting the "Display Common Tree" button in the browser* (arrow).

yeast), *Escherichia coli* (bacterium), and *Oryza sativa* (rice) using this NCBI resource produces an evolutionary tree that groups the organisms according to their sequence similarity (Fig. 9.24). The computational basis for evolutionary tree construction is described in further detail below.

Sequence Informatics

The rapid transformation of biology from a gene science into a genome science has placed a premium on computational tools that are capable of examining and manipulating large amounts of biological sequence information. Microarray analysis is facilitated by the availability of a wide assortment of sequence informatics tools, including software that provides automated melting temperature and sequence alignment information as well as optimal oligonucleotide targets and PCR primers. This section explores several important topics in sequence informatics.

Melting Temperature

Single-stranded nucleic acids of complementary sequence, such as the targets and probes used in microarray assays, zipper up to form double-stranded molecules in a chemical reaction known as hybridization. Hybridization reactions are enabled primarily by the formation of hydrogen bonds between DNA bases on the two complementary strands. The ***melting temperature*** (T_m) defines the thermal point at which 50% of the complementary molecules hybridize and 50% dissociate or "melt" away from their cognate strands (Fig. 9.25). Most

Melting temperature. *Thermal point at which 50% of the complementary molecules hybridize and 50% dissociate from their cognate strands.*

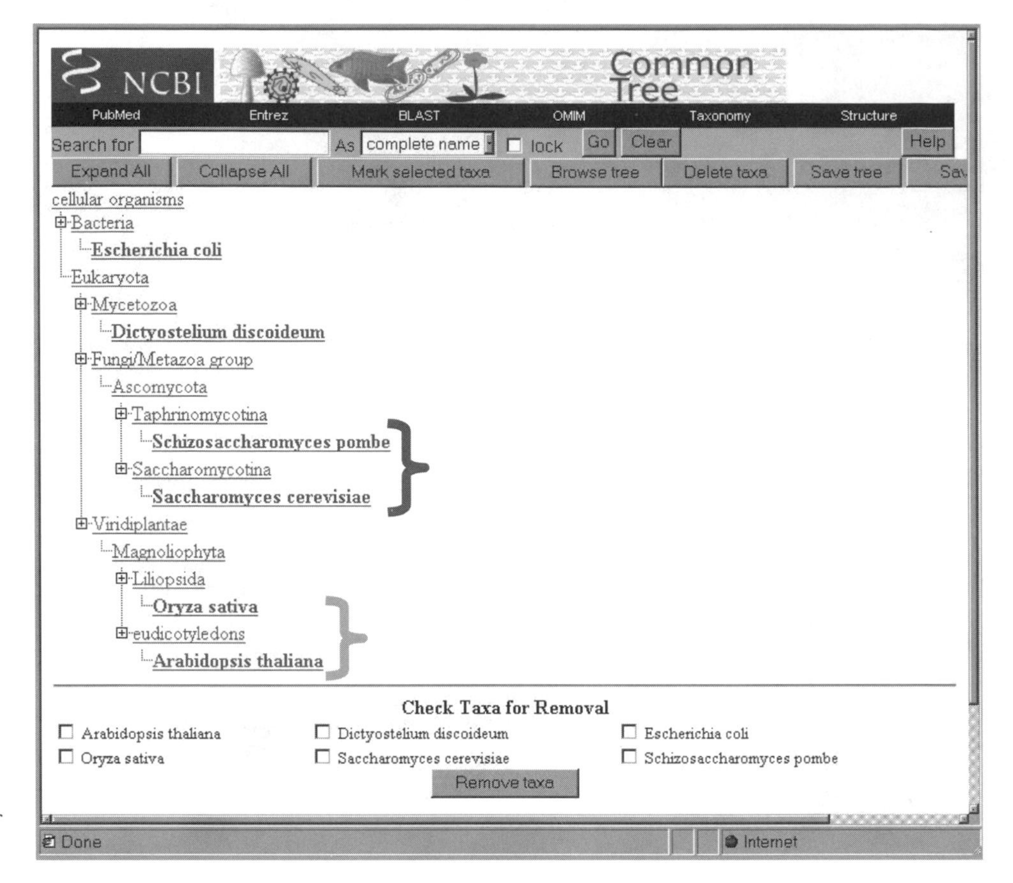

FIGURE 9.24. *The NCBI taxonomy site allows users to construct an evolutionary relationship diagram (dendrogram) using a standard web browser. The tree reveals taxonomic similarity among the organisms selected, including the yeasts* (top brace) *and plants* (bottom brace). *Additional information can be obtained by clicking on an organism of interest.*

melting temperature data are derived from "solution studies," in which hybridization occurs with both strands in the liquid phase, rather than one in solution and one on a solid support as in microarray assays. If the microarray target sequence is tethered with a suitable linker arm and if robust surface chemistries and buffers are used the solution data provide a good estimate for the T_m values observed in microarray assays. The capacity to estimate melting temperatures is important because it helps the researcher select hybridization and wash temperatures and enables the development of hybridization-based assays involving novel sequences of unknown melting temperatures.

Melting temperature is affected by a number of variables, including duplex length, sequence composition, percent complementarity, salt concentration, solvent, and pH. Other factors being equal, melting temperature increases with increasing oligonucleotide length, such that an 18mer of sequence

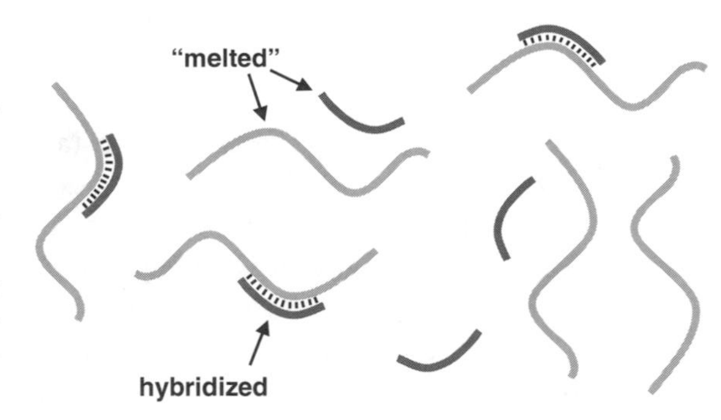

FIGURE 9.25. *The melting temperature defines the thermal point at which half of the probe molecules in a mixture hybridize* (hashed lines) *and half dissociate or "melt". The strength of the interactions between target* (long lines) *and probe* (short lines) *molecules depends on sequence composition, salt concentration, solvent, pH, and other factors.*

TABLE 9.3. Oligonucleotide Melting Temperatures[a]

Sequence	Length (nt)	Number of Hydrogen Bonds	Melting Temperature (°C)
5′-GGG	3	9	12
5′-GGGGGG	6	18	24
5′-GGGGGGGGG	9	27	36
5′-GGGGGGGGGGGG	12	36	48
5′-GGGGGGGGGGGGGGG	15	45	61
5′-GGGGGGGGGGGGGGGGGG	18	54	69
5′-GGGGGGGGGGGGGGGGGGGGG	21	63	74
5′-CGA	3	8	10
5′-CGACGA	6	16	20
5′-CGACGACGA	9	24	30
5′-CGACGACGACGA	12	32	40
5′-CGACGACGACGACGA	15	40	47
5′-CGACGACGACGACGACGA	18	48	55
5′-CGACGACGACGACGACGACGA	21	56	60
5′-AAA	3	6	6
5′-AAAAAA	6	12	12
5′-AAAAAAAAA	9	18	18
5′-AAAAAAAAAAAA	12	24	20
5′-AAAAAAAAAAAAAAA	15	30	24
5′-AAAAAAAAAAAAAAAAAA	18	36	28
5′-AAAAAAAAAAAAAAAAAAAAA	21	42	33

[a] Assuming 100% complementarity and a salt concentration of 0.195M sodium (1 × SSC). Values determined using the Oligo Calculator (Molecular Biology Care Facilities, Dana-Farber Cancer Institute (Boston, MA).

5′-CGACGACGACGACGACGA has a T_m of 55°C compared to 40°C for a 12mer 5′-CGACGACGACGA (Table 9.3). The 18mer forms a total of 48 hydrogen bonds and is more difficult to "melt away" from its complementary sequence than is the 12mer, which forms only 32 hydrogen bonds and therefore dissociates at a lower temperature. Microarray assays that use long oligonucleotide targets generally use higher hybridization and wash temperatures than assays that use shorter oligonucleotide targets, and it is important for researchers to understand the positive correlation between oligonucleotide length (i.e., the number of hydrogen bonds), melting temperature, and hybridization and wash conditions.

Most microarray assays use oligonucleotide sequences of diverse base composition, such that the number of GC base pairs and the number of AT base pairs is approximately equal. The approximately equal occurrence of the four bases in oligonucleotide sequences derives from the natural fact that the genes and genomes of most organisms contain a fairly equal representation of A, T, G, and C on the kilobase or megabase scale. But in local regions spanning 10–20 nucleotides, many genes and genomes exhibit local extremes in sequence composition, and oligonucleotide targets or probes corresponding to these regions may contain an abundance of GC or AT nucleotides. The human breast and ovarian cancer susceptibility gene (*BRCA1*), for example, contains an AT-rich 14mer (ATAAAGAAAAAAAA) and a GC-rich 14mer (CCCCAGATCCCCCA) within the 5711 nucleotide coding sequence (Fig. 9.26). The local bias in these two sequences produces dramatically different melting temperature values of 20 and 46°C, respectively, for the two 14mers even though the global sequence of the 5.7 kb *BRCA*1 mRNA is 58% GC and 42% AT. Local differences in nucleotide composition can effect the choice of PCR primers,

20°C

5' ...TGCAAATTGATAGTTGTTCTAGCAGTGAAGAGATAAAGAAAAAAAAGTACAACCAAATGCCA
GTCAGGCACAGCAGAAACCTACAACTCATGGAAGGTAAAGAACCTGCAACTGGAGCCAAGAAG
AGTAACAAGCCAAATGAACAGACAAGTAAAAGACATGACAGCGATACTTTCCCAGAGCTGAAGT
TAACAAATGCACCTGGTTCTTTTACTAAGTGTTCAAATACCAGTGAACTTAAAGAATTTGTCAAT...

...GTGTCCACCCAATTGTGGTTGTGCAGCCAGATGCCTGGACAGAGGACAATGGCTTCCATGCAA
TTGGGCAGATGTGTGAGGCACCTGTGGTGACCCGAGAGTGGGTGTTGGACAGTGTAGCACTCT
ACCAGTGCCAGGAGCTGGACACCTACCTGATACCCCAGATCCCCCACAGCCACTACTGA 3'

46°C

FIGURE 9.26. *Two portions of the 5711-nucleotide sequence (GenBank accession number U14680) of the human breast and ovarian cancer susceptibility* (BRCA1) *gene, spanning nucleotides 2035–2280 (top) and 5527–5711 (bottom). The two 14mers hybridizing to the sequences shown* (underlined) *would have predicted melting temperatures of 20 and 46°C, respectively.*

microarray targets, and probes (see below), and microarray researchers should be cognizant of the existence and consequence of local deviations in sequence composition.

Entirely GC- or AT-containing sequences exhibit extreme differences in melting temperature (Table 9.3), such that a 15mer that contains only G residues would have a predicted melting temperature of 61°C, much higher than the 24°C melting temperature of a 15mer containing only A bases. A 15mer of mixed composition that forms 10 GC bonds and 5 AT bonds has a predicted melting temperature of 47°C, in between the melting temperature extremes of the all-G 15mer and the all-A 15mer.

Longer oligonucleotides have greater melting temperature values than shorter oligonucleotides due to the formation of a greater number of hydrogen bonds on hybridization, and sequence-dependent differences in melting temperature are likewise attributable to differences in the number of hydrogen bonds formed between different pairs of complementary sequences. Base pairing interactions between G and C residues result in the formation three hydrogen bonds per base compared to the two hydrogen bonds per base that form between A and T, and GC-rich sequences display a greater melting temperature than AT-rich sequences because it is more difficult to melt apart GC than AT (Table 9.3). Stated differently, oligonucleotides that have an abundance of GC residues zipper up more tightly than AT-rich sequences and therefore require a greater temperature to denature the complementary strands. A good approximation of melting temperature takes the sum of the G and C bases times four and the A and T bases time two, according to the following equation:

$$T_m = 4(G + C) + 2(A + T) \tag{9.1}$$

The melting temperature of a 15mer oligonucleotide with the sequence AGTTCCTCACAGGAC would be calculated as follows: T_m (AGTTCCTCACAGGAC) = 4 (8) + 2 (7) = 46°C. Melting temperatures predicted by Equation 9.1, the so-called Wallace equation (Wallace et al., 1979), provide a good estimate for relatively short oligonucleotides (5–20 nt) hybridized in aqueous buffers containing low salt (<100 mM Na^+). Longer oligonucleotides and more exotic hybridization buffers are better estimated using more complex equations. Microarray users can obtain predicted melting temperatures for any oligonucleotide via the Internet, simply by entering the nucleotide sequence into a text window and clicking on "calculate" (Fig. 9.27). Internet-based melting temperature calculators are free of charge and provide information concerning oligonucleotide length, GC content, and molecular weight.

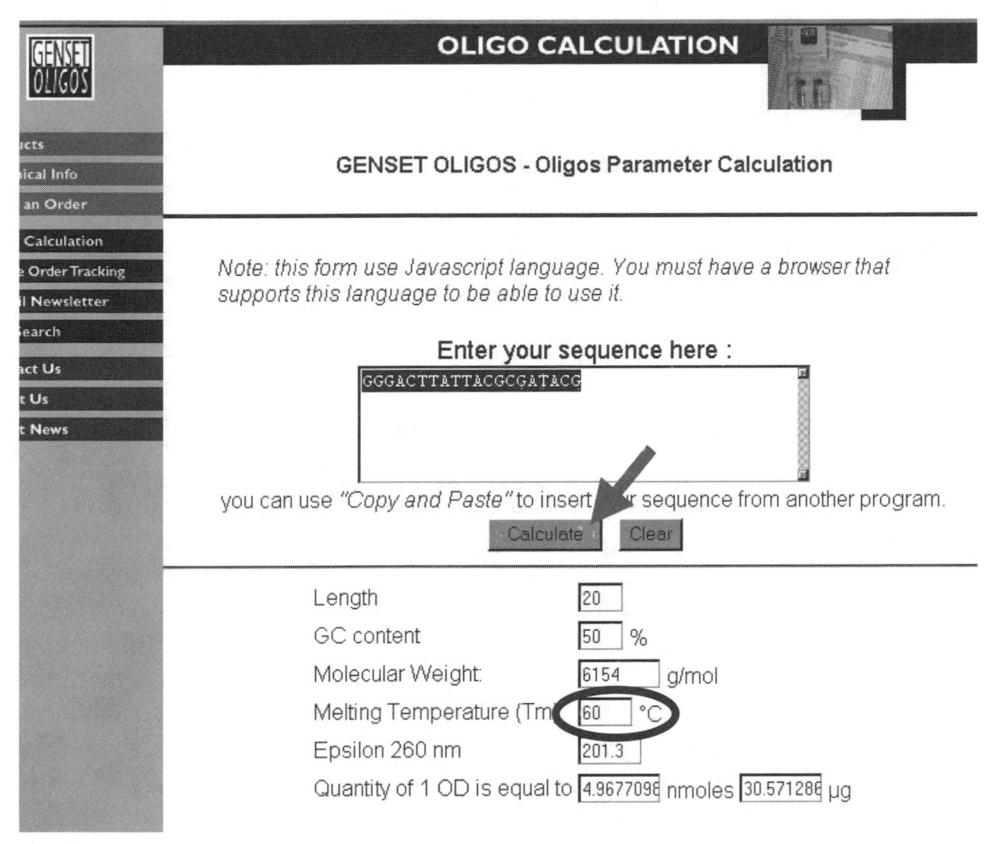

FIGURE 9.27. Melting temperature calculator (www.gensetoligos.com/Calculation/calculation.html). (Data courtesy of Genset Oligos, Paris.)

In addition to oligonucleotide length and sequence composition, the extent of complementarity between the microarray target sequence and the oligonucleotide probe effects the melting temperature, such that perfectly complementary sequences display greater melting temperature values than complementary sequences with one or more mismatches. A mismatch in a hybridized sequence corresponds to a nucleotide position that fails to form a productive (Watson–Crick) base pairing interaction because the bases that are juxtaposed lack the chemical geometry required for hydrogen bond formation. Of the 16 possible pairings (2^4) between the four DNA bases, only 4 (A-T, T-A, G-C and C-G) provide productive base pairing interactions (Table 9.4). The 12 remaining combinations do not allow the formation of hydrogen bonds, and their occurrence destabilizes any hybrid that contains them in a hybridized oligonucleotide. Nonproductive base pairs reduce the melting temperature of

TABLE 9.4. Productive and Nonproductive Base Pairing

	Strand 1			
Strand 2	A	G	C	T
A	–	–	–	+
G	–	–	+	–
C	–	+	–	–
T	+	–	–	–

+; productive hydrogen bond formation.

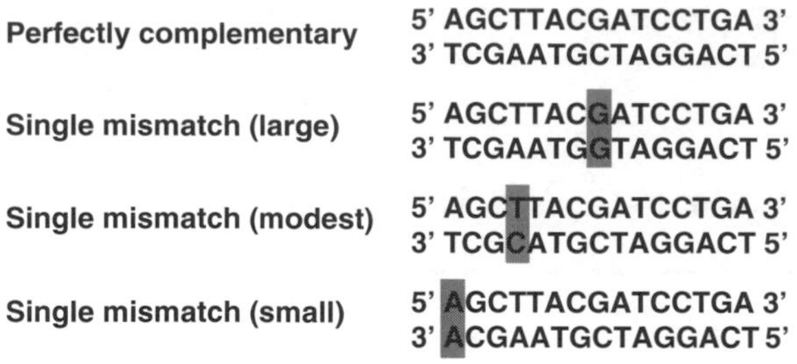

Perfectly complementary	**5' AGCTTACGATCCTGA 3'** **3' TCGAATGCTAGGACT 5'**
Single mismatch (large)	**5' AGCTTACGATCCTGA 3'** **3' TCGAATGGTAGGACT 5'**
Single mismatch (modest)	**5' AGCTTACGATCCTGA 3'** **3' TCGCATGCTAGGACT 5'**
Single mismatch (small)	**5' AGCTTACGATCCTGA 3'** **3' ACGAATGCTAGGACT 5'**

FIGURE 9.28. *Base-pairing interactions between four different sets of complementary oligonucleotides.*

a hybrid by weakening the zippering interactions between the complementary strands.

The extent to which a single base mismatch reduces the melting temperature of a hybrid depends on the location of the mismatch within the duplex. Though essentially every single base mismatch destabilizes a duplex relative to two perfectly complementary sequences (Fig. 9.28), mismatches in the center of the hybrid exert a much larger destabilizing effect than those located near the ends of the duplex. Figure 9.29 uses a hybridized 15mer as an example; a single base mismatch at the center nucleotide position (0) would be much more detrimental than a mismatch at position number -7 or $+7$, with mismatch locations -6 to -1 and $+6$ to $+1$ exerting increasingly greater deleterious effects on stability. Mismatches located in the center of a hybrid destabilize binding to a greater extent than mismatches on the ends of the duplex because nonproductive base pairing in the center maximally disrupts the double-stranded molecule by effectively breaking it into shorter oligonucleotides, each of which has a much lower melting temperature than the perfectly matched duplex. Center mismatches also exert "neighbor effects" on adjacent base pairs disrupting the hydrogen bonding capacity of adjacent nucleotides to a much greater extent than mismatches located on the ends of the duplex.

Oligonucleotides of greater length, GC content, and complementarity form more stable hybrids and therefore exhibit greater melting temperatures than shorter oligonucleotides than are AT-rich and possess one or more destabilizing mismatches. In addition to these factors, elevated salt (e.g., Na^+) concentration stabilizes base pairs and elevates melting temperatures by binding to phosphate groups on each single-stranded, complementary molecule and reducing the repulsive effects that would otherwise ensue for hybridization in a low salt environment. To speed the rate of hybridization and stabilize target–probe duplexes, most microarray hybridization buffers contain a high salt

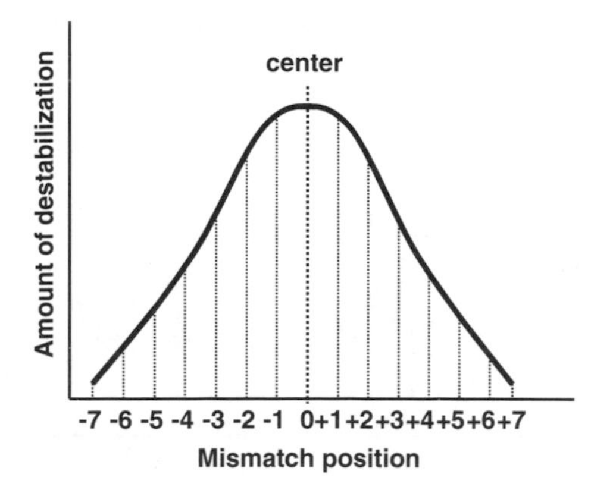

FIGURE 9.29. *How a single base mismatch destabilizes the hybridization of a 15mer oligonucleotide to a target sequence as a function of the position of the mismatch in the duplex.*

(e.g., 1.0 M) concentration. Neutral or near neutral pH (6–8) also stabilizes hybridization and elevates melting temperatures relative to high pH buffers. At very high pH (>12), double-stranded DNA denatures rapidly and sodium hydroxide (a strong base) is a common denaturant for this reason. Low pH buffers damage the DNA bases by encouraging ***depurination,*** the chemical cleavage and loss of the purine bases (A and G), rendering low pH buffers rather unsuited for most microarray hybridization reactions.

Depurination. *Chemical cleavage and loss of the purine bases adenine and guanine that occurs by treatment of DNA with acids and other low pH agents.*

Certain organic solvents (e.g., formamide) compete with the bases for hydrogen bonding interactions and reduce the melting temperature for this reason. Despite its negative effect on melting temperature, some microarray hybridization buffers contain formamide as a means of reducing nonspecific hybridization and background fluorescence. The organic reagent tetramethylammonium chloride (TMAC) is also used in some buffers to equalize the difference between AT and GC base pairs. Buffers containing TMAC allow melting temperature values to be estimated largely on the basis of oligonucleotide length alone, rather than length and sequence composition.

Microarray researchers can predict melting temperatures using affordable commercial software packages that take into account all of the main factors that effect melting temperature (length, GC-content, complementarity, salt, pH, solvent), and most commercial software provides several different algorithms to generate melting temperature values (Fig. 9.30). *OLIGO* 6 from Molecular Biology Insights (Cascade, CO; www.oligo.net), for example, provides sophisticated data, graphical depictions, and convenient outputs for oligonucleotide-based microarray assays.

A set of oligonucleotides with 35–65% GC sequence composition will exhibit fairly similar melting temperature values, and matched values should be employed in microarray experiments whenever possible to maximize the precision of the assay. Reactions with pools of synthetic oligonucleotides, particularly sequences in the 10- to 25-nt range, benefit from well-matched melting temperature values because a single hybridization temperature can be used to obtain similar fluorescent signals for the entire pool of different sequences, facilitating data analysis. Longer oligonucleotides (~50 nt), such as those used in gene expression assays, are less susceptible to local extremes in sequence composition, because GC- and AT-rich regions tend to span relatively short regions (<10 nt), which are averaged out across 50mers.

The relationship between melting temperature and hybridization and wash conditions is known ***stringency***. All other factors being equal, a hybridization temperature of 50°C would be more stringent than if the same assay were run at 42°C. More stringent conditions are desired for most applications, because greater stringency means greater specificity and less cross-hybridization. A good rule of thumb is that hybridization reactions should be run 5–10°C below the melting temperature to provide stringency sufficient to avoid cross-hybridization (see below). Reaction temperatures approaching or exceeding the melting temperature result in reduced signal because complementary strands melt apart under conditions of excessive stringency. Reaction temperatures that are far below the melting temperature lead to nonspecific hybridization (cross-hybridization) and a loss of assay specificity.

Stringency. *Sum total of the external factors that affect hybridization efficiency, including temperature, salt, and pH, with greater stringency corresponding to excessive temperature, low salt, high pH, and other conditions that reduce hybridization efficiency.*

Reduced stringency can be used selectively to examine sequences from organisms that share <100% sequence identity, such as in zoo chip experiments in which microarrays are used to compare related plants and animals. Reduced hybridization and wash temperature, higher salt, and a lower concentration of

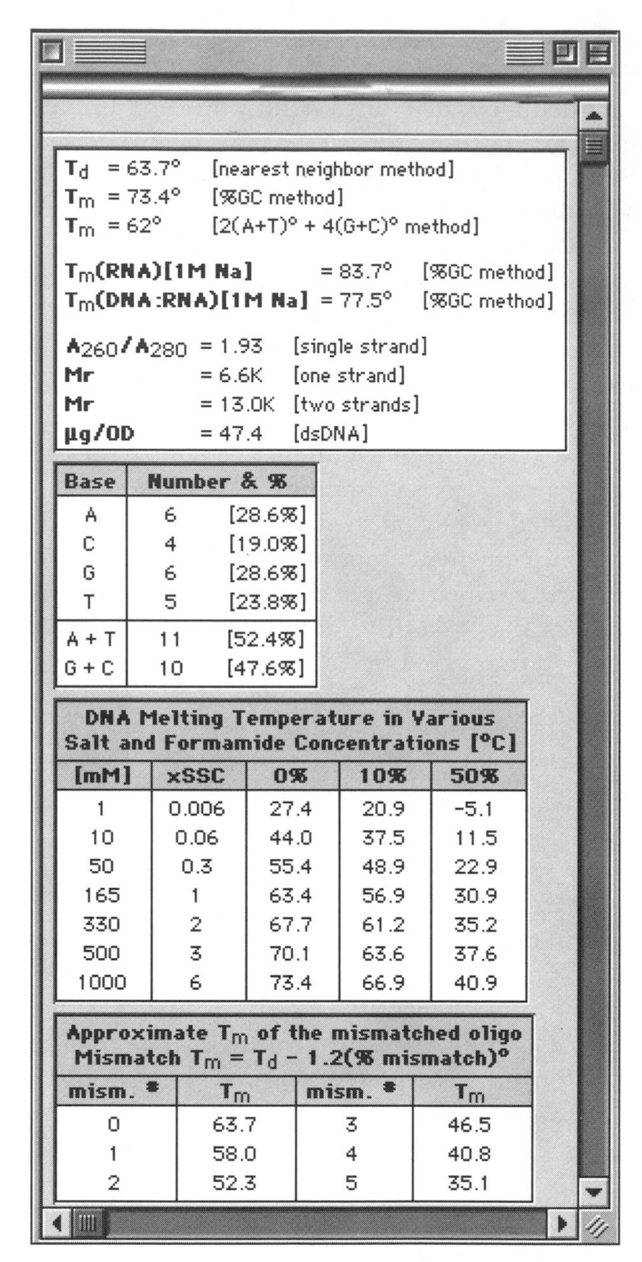

T_d = 63.7° [nearest neighbor method]
T_m = 73.4° [%GC method]
T_m = 62° [2(A+T)° + 4(G+C)° method]

T_m(RNA)[1M Na] = 83.7° [%GC method]
T_m(DNA:RNA)[1M Na] = 77.5° [%GC method]

A_{260}/A_{280} = 1.93 [single strand]
Mr = 6.6K [one strand]
Mr = 13.0K [two strands]
µg/OD = 47.4 [dsDNA]

Base	Number	& %
A	6	[28.6%]
C	4	[19.0%]
G	6	[28.6%]
T	5	[23.8%]
A + T	11	[52.4%]
G + C	10	[47.6%]

DNA Melting Temperature in Various Salt and Formamide Concentrations [°C]

[mM]	xSSC	0%	10%	50%
1	0.006	27.4	20.9	-5.1
10	0.06	44.0	37.5	11.5
50	0.3	55.4	48.9	22.9
165	1	63.4	56.9	30.9
330	2	67.7	61.2	35.2
500	3	70.1	63.6	37.6
1000	6	73.4	66.9	40.9

Approximate T_m of the mismatched oligo
Mismatch $T_m = T_d - 1.2$(% mismatch)°

mism. #	T_m	mism. #	T_m
0	63.7	3	46.5
1	58.0	4	40.8
2	52.3	5	35.1

FIGURE 9.30. OLIGO *6 software generating melting temperature data for the 21mer oligonucleotide 5′-ACCATGATTACGCCAAGCGTG-3′, including melting temperature predictions as a function of length, sequence composition, presence of mismatches, and salt and solvent concentrations and using several different algorithms. (Data courtesy of Molecular Biology Insights, Cascade, CO.)*

organic solvents all suffice to reduce hybridization stringency, allowing sequences with reduced complementarity to form target–probe hybrids.

Sequence Alignment

Sequence alignment.
Computational juxtaposition of two or more linear strings of nucleotides or amino acids.

The computational juxtaposition of a two or more linear strings of nucleotides or amino acids is known as a ***sequence alignment***. Aligning gene and protein sequences is useful to microarray researchers in several different ways. Computer-based sequence alignments identify minor sequence variants (single nucleotide polymorphisms, insertions, deletions) in full-length genes and proteins, an exercise that is performed easily by computer but is nearly impossible by hand. Minor sequence alterations serve as the basis for genetic screening, diagnostics tests, and genotyping assays (see Chapter 13). Aligning multiple sequences also facilitates the identification of unique target sequences among closely related genes, a capability that is essential for selecting high-quality target sequences for gene expression assays (see below). Sequence alignments provide a measure of sequence identity, which is key when examining closely related members of gene families or in microarray assays that use one kind of chip (e.g., human) to examine samples from a related organism (e.g., chimpanzee).

The practice of aligning sequences of nucleic acids and proteins is based on the assumption that linear similarity maps with functional similarity, an assumption that follows correctly from the fact that gene information is converted into protein information in a linear manner according to the genetic code (see Chapter 4). Proteins that share similar amino acid sequences fold similarly and therefore function in a similar manner. Nucleotide sequences that display sequence similarity are said to be homologous. Because homology at the nucleotide and amino acid levels tracks with function, there is great value in being able to align biological sequences quickly, accurately, and affordably.

The massive number of entries in contemporary sequence databases, coupled with the inherent mathematical complexity of sequence alignment, requires computers for this task. Any sequence of computational steps performed by computer is known as an ***algorithm,*** and dozens of different algorithms have been brought to bear on the computational challenge of sequence alignment, starting with the pioneering efforts of several laboratories in the 1970s (Needleman and Wunsch, 1970; Smith and Waterman, 1981). An algorithm that provides an approximate solution to an extremely complex problem is known as a ***heuristic*** algorithm. Heuristics can be applied in cases where "correct" solutions are impractical to obtain by computer.

Algorithm. *Any sequence of computational steps performed by computer.*

Heuristic. *Algorithm that provides an approximate solution to an extremely complex problem.*

The Needleman-Wunsch and Smith-Waterman algorithms employ ***dynamic programming*** to find optimal sequence alignments. Dynamic programming is a mathematical strategy that essentially tackles the alignment problem backwards and finds the best alignment by making decisions with a series of subalignments scored one after another. Though both the Needleman-Wunsch and Smith-Waterman algorithms employ dynamic programming, the two differ as to the amount of each sequence that is used to find the optimal alignment. Needleman-Wunsch uses a ***global alignment*** strategy in which the entire sequences of the genes or proteins are compared, whereas Smith-Waterman employs a ***local alignment*** method, allowing for comparisons and alignments based on small regions of sequence similarity (Fig. 9.31). Due to the computationally intensive requirements of comparing entire sequences and the inherent modularity of genes and proteins, local alignment strategies tend to compute much more quickly than global alignment methods and provide greater resolution in terms of aligning genes and proteins that may share extensive homology across a small region. Global and local approaches both produce valuable alignment data.

Dynamic programming. *Mathematical strategy that tackles the alignment problem backwards to find the best alignment by making decisions with a series of subalignments scored one after another.*

Global alignment. *Computational strategy in which the entire sequence of the genes or proteins is compared.*

Local alignment. *Computational strategy in which small regions of sequence similarity between genes or proteins is compared.*

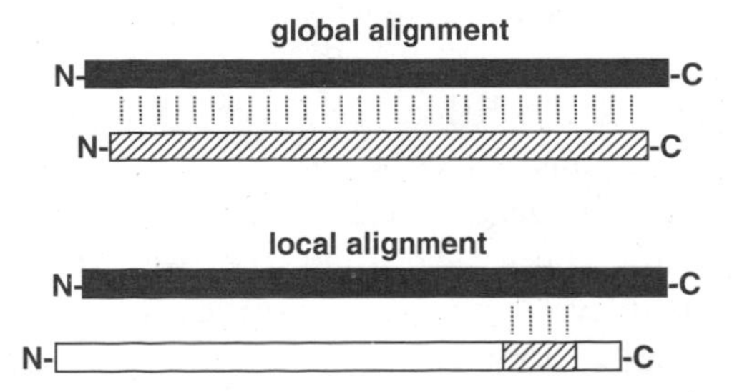

FIGURE 9.31. The two main strategies for nucleic acid and protein sequence alignment are depicted as comparisons (dotted lines) between four hypothetical proteins (rectangles) that share sequence identity (hatched lines) across the entire sequence (top) or a small region (bottom).

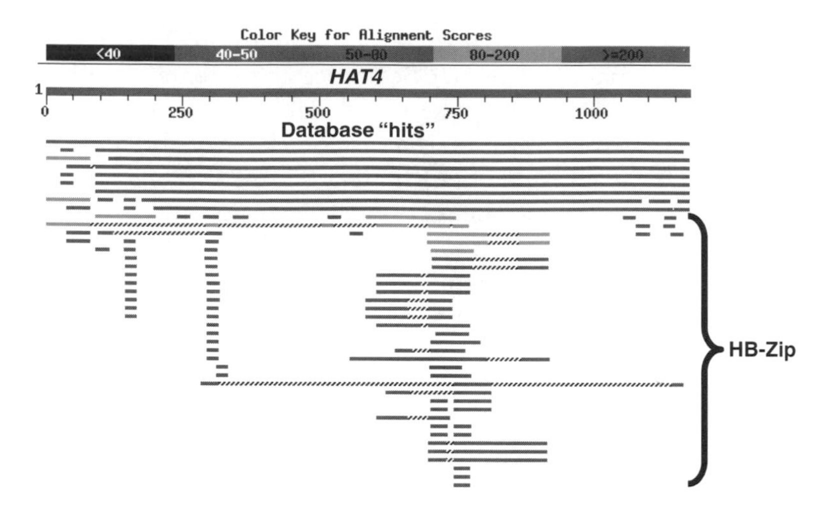

FIGURE 9.32. The results of a BLAST sequence alignment search using the ***homeobox*** *cDNA from* Arabidopsis thaliana *as the query sequence* (thick line) *against the entire GenBank database. Several dozen positive scores (thin lines), arranged according to percent identity (top), were obtained across the entire region (thin lines). Related genes (orthologs) from different plants share homology to the local homeobox-leucine zipper (HB-Zip) region.*

Built on the concept of local alignments, *FASTA* (Lipman and Pearson, 1985) and BLAST (Altschul et al., 1990) provide increasingly rapid sequence alignments strategies, and both can be executed via a personal computer. Sequence alignment needs for most microarray applications can be performed using one of the BLAST programs in conjunction with the NCBI databases (see above).

The value of local alignment strategies is seen in database searches with sequences that contain discrete functional domains, such as ***homeobox-leucine zipper*** regions that encode a DNA-binding motif conserved throughout the plant kingdom (Ruberti et al. 1991; Schena and Davis, 1992). A homology search with the 1176-nucleotide *HAT4* homeobox-leucine zipper sequence from *Arabidopsis thaliana* produces hits that share homology across the entire *HAT4* sequence, as well as more than a dozen sequences that share homology confined to the homeobox-leucine zipper region (Fig. 9.32). Sequences that align with *HAT4* across the homeobox-leucine zipper region include paralogs from *Arabidopsis* as well as orthologs from other plants, including soybean, *Craterostigma, Pimpinella brachycarpa* (related to anise), *Capsella rubella,* and sunflower.

Homeobox. *Highly conserved DNA binding motif found in transcription factors from yeast, plants, and animals.*

Homeobox-leucine Zipper. *Novel functional domain in higher plants containing a DNA binding domain and a dimerization motif.*

The amount of similarity between two sequences is expressed as the ***percent identity,*** given as the quotient of the number of identical positions (e.g., nucleotides, amino acids), divided by the total number of positions, times 100:

Percent identity. *Similarity between two sequences, expressed as the number of identical positions divided by the total number of positions compared times 100.*

$$\text{Percent identity} = \left(\frac{\text{Identical positions}}{\text{Total positions}}\right) \times 100 \tag{9.2}$$

If two nucleic acid sequences share identical nucleotides at 75 out of 100 total positions, the two sequences would have 75% percent identity. In a slightly more complicated comparison, two sequences that share 194 out of 248 nucleotides would be 78% (194/248 × 100) identical (Fig. 9.33). Percent identity is useful for microarray analysis because it provides a quantitative measure of sequence similarity and, therefore, quantitative evolutionary information. Percent identity is also helpful in selecting target sequences for microarray hybridization experiments (see below).

The degeneracy of the genetic code allows multiple codons to encode the same amino acid, such that two proteins that are highly similar at the amino acid level can possess significantly less identity at the nucleotide level. The human and rat polyubiquitin coding sequences are 100% identical at the amino

```
gi|1149534|emb|X94449.1|PBPHZ4GEN
P.brachycarpa mRNA for homeobox-leucine zipper protein (PHZ4)Length = 1401

Score = 63.9 bits (32), Expect = 3e-07
Identities = 194/248 (78%)
Strand = Plus / Plus

Query: 556 agtgacgatgaagatggtgataactccaggaaaaagcttagactttccaaagatcaatct 615
           ||||| || |||||||||||||| || || || || || || ||||| ||||| || ||
Sbjct: 505 agtgatgaagaagatggtgataattctagaaagaaactcaggctttctaaagaccagtcc 564

Query: 616 gctattcttgaagagaccttcaaagatcacagtactctcaatccgaagcagaagcaagca 675
           |||||||||||||| |  || ||||| ||||  ||||| ||||| ||||| ||||  ||
Sbjct: 565 gctattcttgaagatagttttaaagaacacaacactcttaatccaaagcaaaagctggcc 624

Query: 676 ttggctaaacaattagggttacgagcaagacaagtggaagtttggtttcagaacagacga 735
           ||||| |||  | |||||||  || |  | || ||||| |||||||| ||||| ||| |
Sbjct: 625 ttggcaaaaagactagggttgagacctcgtcaggtggaggtttggttccagaatagaagg 684

Query: 736 gcaagaacaaagctgaagcaaacggaggtagactgcgagttcttacggagatgctgcgag 795
           ||||| || ||| |||||||||| ||||| || || ||||||||    |||||||| |||
Sbjct: 685 gcaaggaccaagttgaagcaaaccgaggttgattgtgagttcttgaaaagatgctgtgag 744

Query: 796 aatctaac 803
           ||||||||
Sbjct: 745 aatctaac 752
```

FIGURE 9.33. *The* HAT4 *sequence from* Arabidopsis thaliana, *spanning nucleotides 556–803* (Query) *aligns optimally with the PHZ4 cDNA sequence from* Pimpinella brachycarpa *along nucleotides 505–752* (Sbjct). *The two sequences have a percent identity value of 78%* (arrow), *and identical nucleotides are designated (bars).*

acid level, but only 88.9% identical at the nucleotide level (Fig. 9.34). The presence of wobble positions at the third codon position causes degeneracy, making it essential to compare nucleic acid sequences rather than protein sequences when assessing stringency and potential cross-hybridization. A good metric for cDNA microarrays is that sequences will cross-hybridize at high stringency if the percent identity exceeds 75–80% at the nucleotide level, a value observed for members of some gene families and highly conserved orthologs. Sequences with < 75% identity will exhibit little or no cross-hybridization unless the stringency of the assay is reduced intentionally to allow, for example, comparative studies between different organisms.

Oligonucleotide Targets

With the rapid expansion of gene databases and the increasing availability of affordable oligonucleotides, there is a trend toward the use of long oligonucleotides (50–70 nt) for gene expression applications. Long oligonucleotides provide some key advantages over cDNAs as gene expression targets, including the fact that oligonucleotides can be prepared directly from sequence database information rather than by PCR, making it easier to synthesize target material and reducing "sample tracking" errors. Unlike cDNAs, the sequence of long oligonucleotides can be verified using mass spectrometry, expediting target verification that in the case of cDNAs requires time-consuming and costly DNA sequencing. Long oligonucleotides are also single stranded rather than double stranded, avoiding the requirement to denature the target sequences before hybridization and preventing snap back during the hybridization reaction, both of which are issues with cDNAs. Long synthetic oligonucleotides can also be made in much larger molar quantities than cDNA targets, reducing the cost of microarray manufacturing in high-throughput settings. Cross-hybridization can also be minimized or reduced more easily with long oligonucleotides than with cDNAs, owing to the fact that it is easier to design around

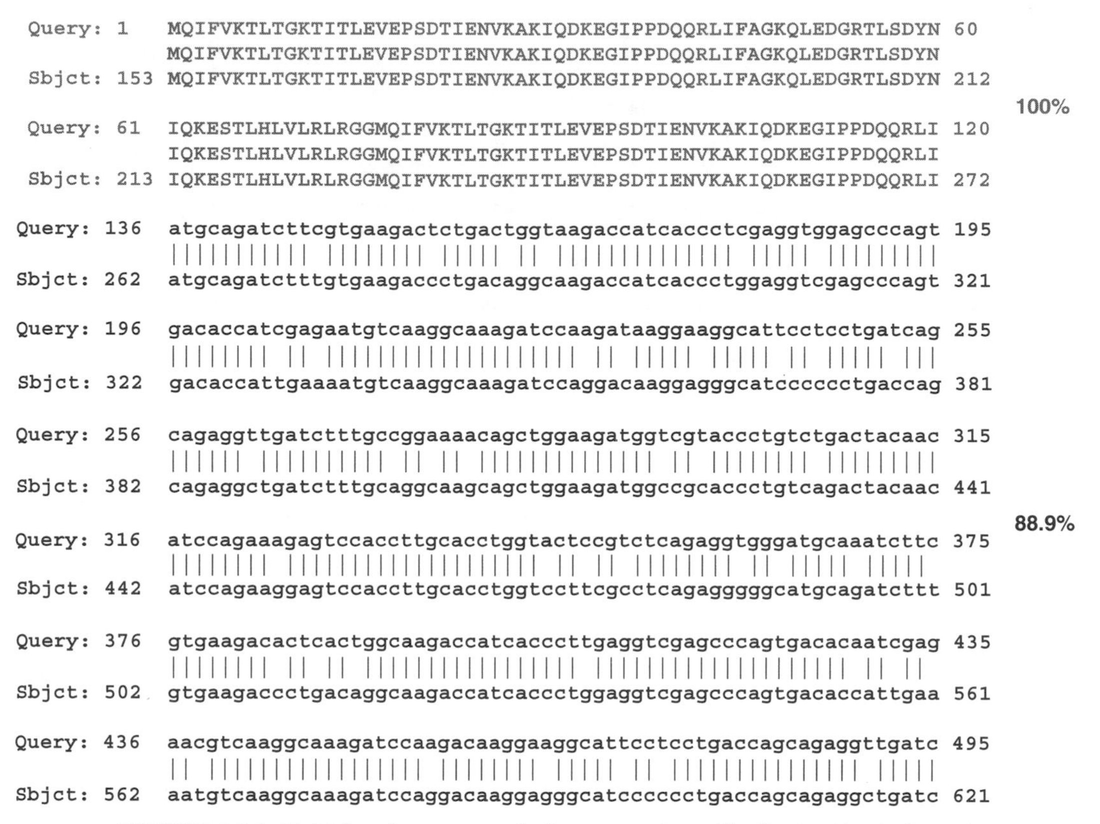

FIGURE 9.34. Multiple codons can encode the same amino acid, allowing identical protein sequences to show significantly less identity at the nucleic acid level. The human (query) and rat (subject) ubiquitin coding sequences were compared using BLAST, revealing 100% amino acid identity and 88.9% nucleotide identity.

regions of high sequence identity with a 50mer than it is with a longer PCR product.

Long oligonucleotides also possess some advantages over short oligonucleotides (25 nt), such as those that are made using photolithography, micromirrors, and other combinatorial synthesis techniques. Owing to the higher melting temperature of long oligonucleotides compared to short oligonucleotides, elevated hybridization temperatures (e.g., 60–65°C) can be used, thereby melting secondary structure out of the target sequences and enabling more efficient hybridization. Long oligonucleotides also increase the chances of productive collisions with probe molecules, which increases the rate of hybridization and the resultant signal intensities.

The trend of using long oligonucleotides for gene expression applications has prompted the formulation of a "design directive," which provides a set of criteria to expedite the selection of optimal oligonucleotide targets from coding sequences and genomic DNA. Targets should be 50–70 nucleotides in length and include sequences from the organism of interest (e.g., human), as well as additional sequences from a heterologous organism (e.g., *Arabidopsis*) to provide positive and negative hybridization controls. Because most mRNA purification and labeling procedures enrich for sequences corresponding to the 3′ end of transcripts (see Chapter 6), oligonucleotide targets should be designed to the 3′ end of each cDNA to achieve maximal hybridization signals (Fig. 9.35). Except in cases of sense strand labeling such as probes derived from amplified RNA (aRNA), long oligonucleotides should be designed using sequences

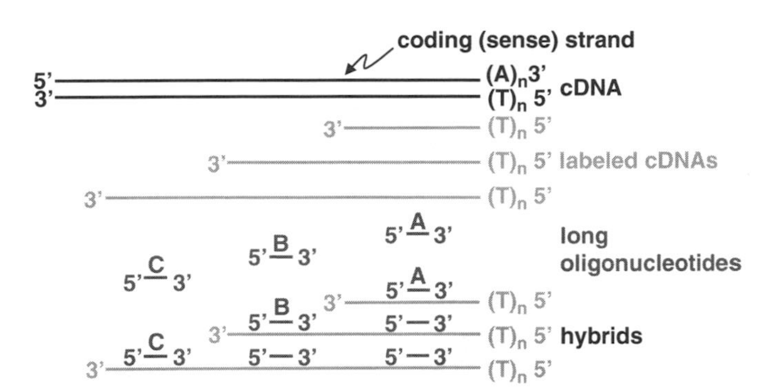

FIGURE 9.35. mRNA synthesis and labeling procedures based on oligo-dT priming produce labeled cDNAs enriched in 3′ sequences (green lines), *relative to the full length cDNA* (black double lines). *The 3′ bias requires the use of 3′ oligonucleotide targets* (short lines) *to obtain optimal hybridization signals. As the location of a long oligonucleotide moves toward the 5′ end of the coding sequence (A to C), hybridization signals diminish because targets form productive hybrids with a smaller population of labeled cDNA molecules. Target A hybridizes with all three labeled cDNAs, producing a stronger signal than targets B and C.*

from the sense (top) strand of the mRNA or cDNA. To achieve matched melting temperatures for a set of oligonucleotides, designs should avoid extremes in sequence composition (e.g., GC bias) and repeated homonucleotide stretches (e.g., polyA). Every target sequence should be checked against a large sequence database (e.g., GenBank) using a powerful local search tool (e.g., BLAST) to assess the uniqueness of a given sequence with respect to the genome.

The concept of sense and antisense orientation deserves some additional discussion in the context of oligonucleotide microarrays because choosing the incorrect strand can lead to catastrophic errors (e.g., no signal) in chip design. The sense–antisense consideration is not an issue with double-stranded cDNA microarrays because both strands are present on microarray. Oligonucleotide microarrays containing sense strand targets will hybridize to probe molecules made from the antisense strand such as single-stranded cDNAs made by oligo-dT priming of mRNA. Potential confusion about sense and antisense arises from two possible sources, including uncertainties in sequence databases and the availability of myriad labeling procedures, some of which produce sense probe molecules and some of which produce antisense probes.

To avoid potential catastrophic design errors, microarray researchers should make an effort to verify that each sequence in the sequence database represents the coding (sense) strand of the mRNA. This can be achieved using a number of different procedures, one of which involves the verification of an open reading frame electronically, using informatics tools that allow the virtual translation of coding sequences into protein. Putative coding sequences that do not create a virtual protein in one of the three frames should be examined more closely to make sure that such sequences are not antisense strands. Researchers should also examine all labeling procedures to verify the strand that is produced in the labeling reaction. Enzymatic labeling of mRNA primed with oligo-dT and extended with reverse transcriptase produces labeled antisense cDNA strands. Labeling aRNA with fluorescent nucleotides and klenow will produce labeled sense molecules if prepared using the standard T7-based Eberwine procedure (see Chapter 6). It is important to understand the sense–antisense distinction before embarking on target and probe design, as many a researcher has stumbled over the fundamental and often overlooked issue of strand orientation.

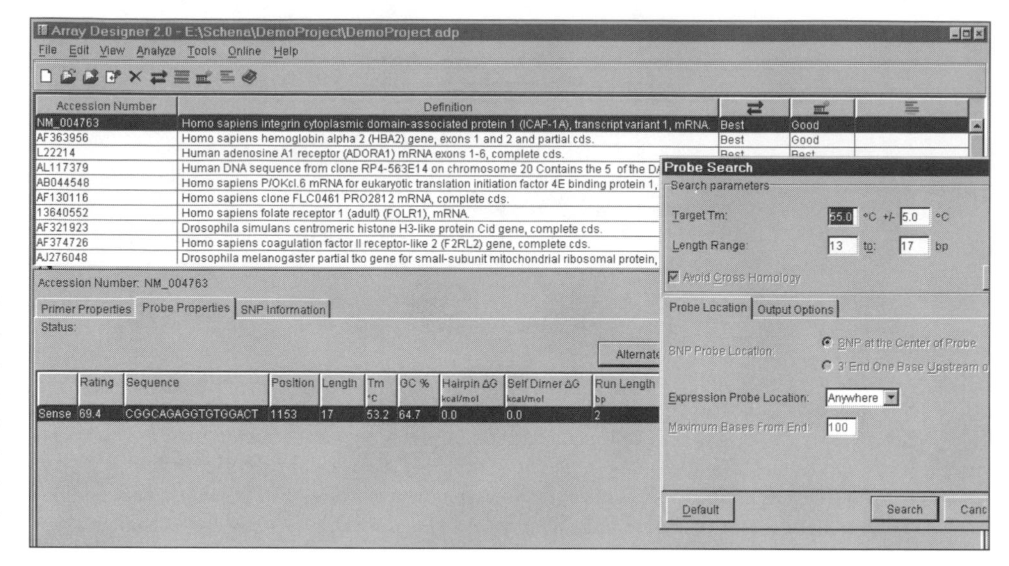

FIGURE 9.36. *A screen in* Array Designer 2 *allows microarray researchers to design oligonucleotide targets and probes with specified design criteria, including melting temperature, oligonucleotide length, GC content, and position within the query sequence. (Courtesy of K. Brown, Premier Biosoft, Palo Alto, CA.)*

Microarray researchers benefit from the availability of automated oligonucleotide target and probe selection tools, which provide optimal target sequences from a larger query sequence string (Fig. 9.36). *Array Designer* 2 from Premier Biosoft International (Palo Alto, CA) allows the user to specify many important parameters for oligonucleotide selection, including a range of melting temperatures, target length, GC content, and position within the query sequence. Automated target selection software greatly expedites the speed and precision of oligonucleotide design relative to manual procedures; and as sequence databases expand and assays become more and more sophisticated, informatics tools for target and probe selection will play an increasingly central role in microarray design.

PCR Primers

PCR is used widely in microarray analysis to generate target sequences for microarray manufacture. PCR-generated amplicons possess a number of advantages over oligonucleotides, including the fact that sense and antisense strands are both present on the chip, ensuring that microarrays containing these denatured double-stranded targets will hybridize efficiently to probe molecules of either strand. PCR-generated targets (100–5000 base pairs) are much longer than oligonucleotides (25–70 nt), providing greater probe complementarity and minimizing signal loss that can occur if labeled probe mixtures are complementary to a restricted portion of the target sequence. Amplified targets can be prepared from cDNA libraries from any organism (plant, animal, bacteria, etc.) without any clone sequence information. PCR amplification and target purification can be automated completely, allowing researchers to generate cDNA targets for entire genomes rapidly and economically.

The widespread use of PCR amplification in microarray manufacture places much importance on the availability of informatics tools for automated primer selection. A *primer* is a synthetic oligonucleotide that hybridizes to a complementary nucleic acid template, and expedites (primes) enzymatic synthesis by providing a starting point for the enzyme. PCR primers bind to proximal and distal sites on cDNA templates, allowing target sequences to be amplified with DNA polymerase in the PCR process (Fig. 9.37). Microarray researchers can

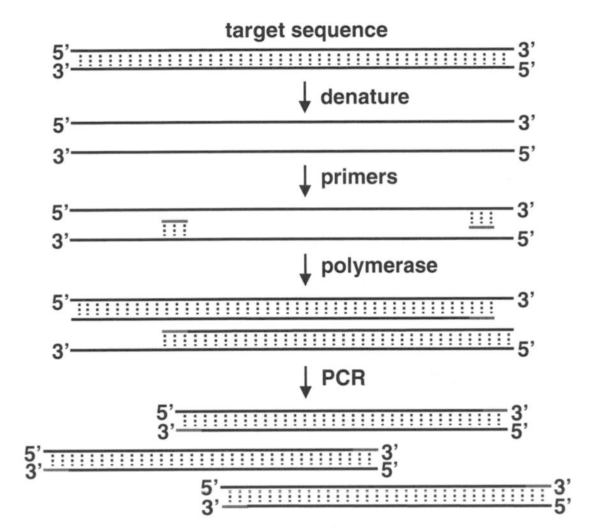

FIGURE 9.37. Target sequences such as cDNA clones (top) can be amplified using PCR by denaturing the template with heat to melt the hydrogen bonds (dashed lines). *Annealing a pair of primers* (short lines) *to the denatured strands* (black lines) *provide starting points for polymerase, which performs DNA synthesis in a 5′ to 3′ direction with respect to the synthesized strand. Multiple rounds of PCR create a large number of amplicons (bottom) whose borders are defined precisely by the sequence of the primers used for amplification, enabling researchers to synthesize custom cDNA target elements.*

specify the size and location of the amplified target (amplicon) by choosing primer sequences that produce targets with the desired properties.

As with long oligonucleotides, there are a number of design considerations when choosing PCR primers, including length, melting temperature, sequence composition, secondary structure, and position within the DNA template. Primer design is facilitated by the availability of informatics software that allow automated primer selection. *Array Designer* 2 allows microarray researchers to specify all of the important parameters and generate primers for any template of interest (Fig. 9.38). As a rule, primers designed by computational means perform much better in PCR reactions than primers chosen manually, placing

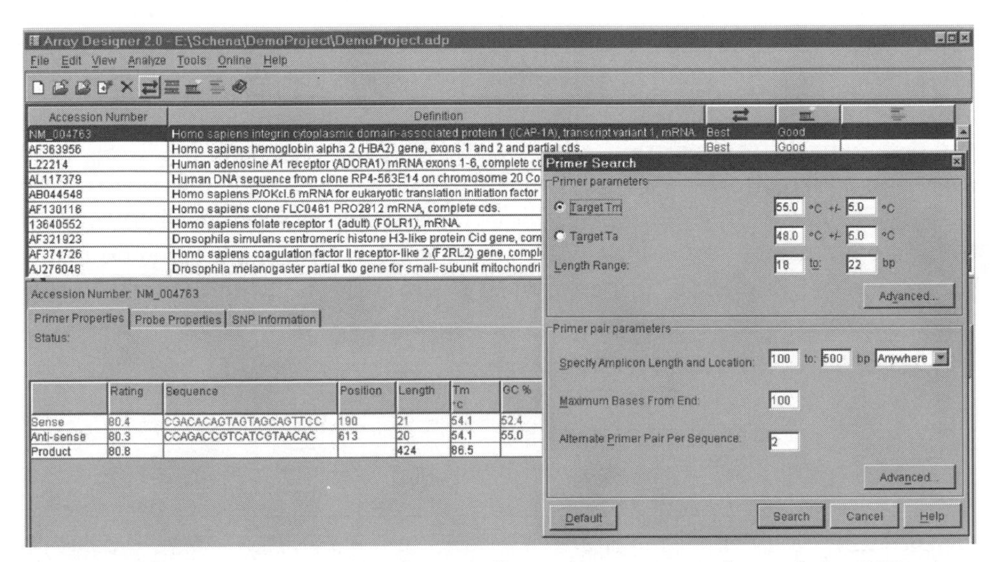

FIGURE 9.38. A screen in Array Designer 2 *allows microarray researchers to design PCR primers with user-specified design criteria, including melting temperature, length range, location within the target sequence, sequence composition, and amplicon size. (Courtesy of K. Brown, Premier Biosoft, Palo Alto, CA.)*

considerable value on bioinformatics tools that expedite primer selection for microarray applications.

DESIGN INFORMATICS

Computers play a central role in many aspects of microarray analysis, including design informatics, the information science that provides the underpinnings for microarray manufacture. Microarray printing robots can be directed through software to manufacture microarrays of myriad physical characteristics, and this design flexibility is critical for the study of different organisms, the implementation of different methods, and the development of new assays. Though this section is most pertinent to researchers who make their own arrays, those who purchase off-the-shelf microarrays are also served well by understanding design informatics. The location of the microarray on the printing substrate, the number of features and genes, genetic complexity, the number of pins and samples, and content maps are among the concepts addressed in this section.

Microarray Location

The location of a microarray on the printing substrates must be specified by the user before manufacture by a contact or noncontact method. Microarray location can be specified automatically using the controller software provided with most commercial arrayers. Location is designated as an offset in millimeters from the top and left edges of the substrate. For robots in which the substrates are configured in a horizontal manner on the platen, the distances from the top and left edges are measured as positive offsets from the x and y origins, so that a microarray located 15 mm from the top and 5 mm from the left is given as $x = 15$ and $y = 5$ (Fig. 9.39).

Specifying microarray location is important because it affects downstream steps, including hybridization, detection, and data analysis. Microarrays printed too close to the edge of the substrate can be problematic because samples applied to the microarray under a cover slip are prone to spilling over the edge and wicking under the substrate, depleting the sample under the cover slip. Microarrays printed too far from the top of the substrate (> 40–50 mm) can impair detection by scanners and imagers that have restricted imaging areas. Microarray users should print large microarrays using a small offset from the top edge to prevent overlap with bar code labels on the bottom end of the substrate. A convenient location for most microarrays is approximately 10 mm

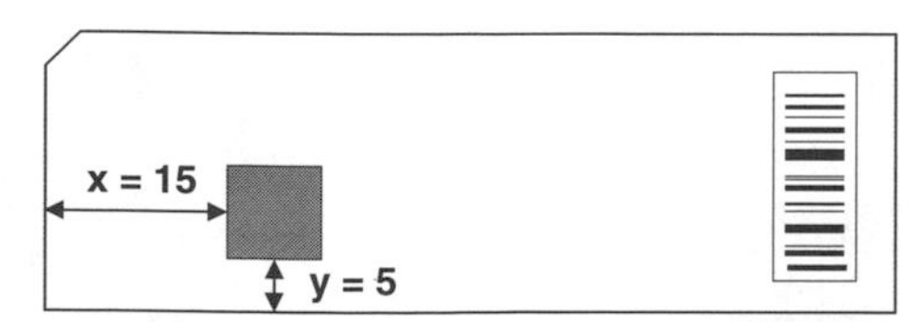

FIGURE 9.39. *The location of a microarray (stipled square) on a printing substrate (rectangle) is designated by the distance of the microarray from the top (x axis) and left edge of the substrate (y axis). Some microarray robots hold printing the substrates in a horizontal orientation (shown here); thus the user must rotate the printed substrates clockwise by 90° to obtain the correct orientation such that the barcode is at the bottom.*

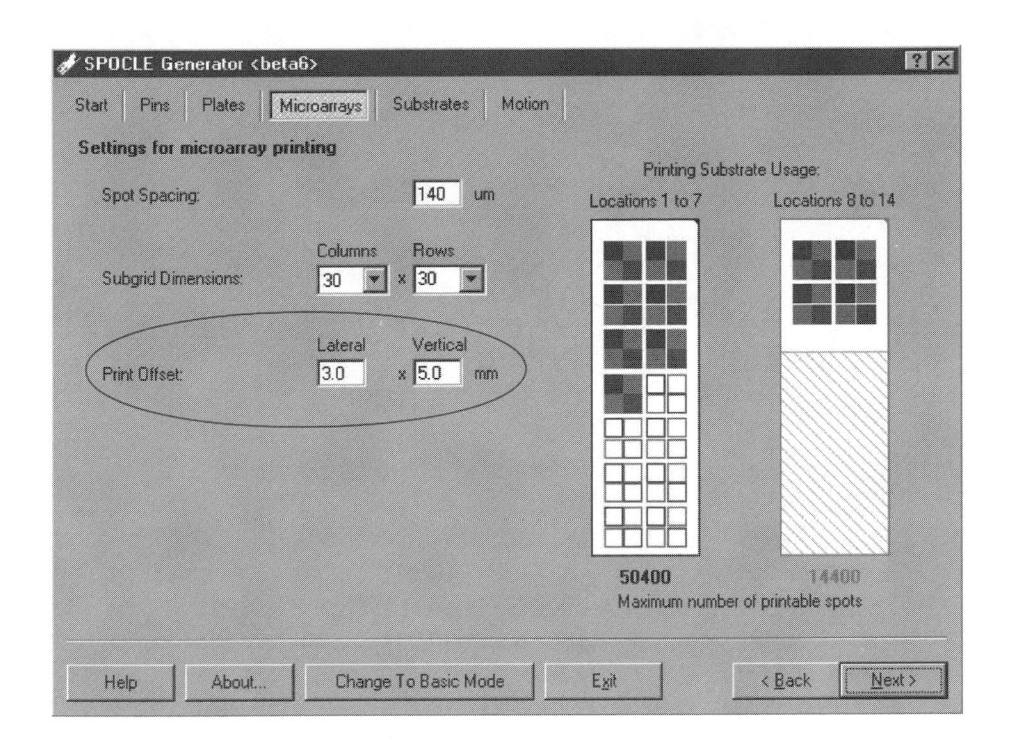

FIGURE 9.40. Most software packages for printing robots (e.g., SpotBot*) allow the user to specify offset values for the* x *(vertical) and* y *(lateral) axes using text windows* (oval). *(Courtesy of H. Bird, TeleChem/arrayit.com, Sunnyvale, CA.)*

from the top edge and at least 2 mm from the left or right edge, with a center location from left to right being optimal. The left to right (y axis) offset required to print a microarray in the center of a standard 25-mm-wide substrate is calculated by subtracting the width of the microarray from 25 and dividing by 2:

$$y \text{ axis offset} = \frac{25 - \text{Microarray width}}{2} \tag{9.3}$$

A printhead that contains four printing implements (e.g., pins, capillaries, ink jets) at 4.5 mm center-to-center spacing will produce an 18-mm-wide microarray, requiring a 3.5-mm offset [(25 − 18)/2] for center alignment. A microarray manufactured with two pins at 4.5 mm spacing will produce a 9-mm-wide microarray and require a y axis offset of 8 mm. Most graphical user interfaces allow the researcher to specify the exact x and y offsets via software (Fig. 9.40), ensuring precise microarray centering.

Features, Genes, and Complexity

The total number of features or spots in a printed microarray is important because it allows the user to calculate the volume of data yielded by a given set of experiments. Knowing the data volume helps one assess the computational and storage requirements needed to complete a project. Microarray scientists often use the number of spots as a shorthand to convey the type of microarray used for a particular application. A researcher might say that "17K arrays" were used to profile gene expression in the worm, meaning that microarrays containing 17,000 spots were employed in the study of *C. elegans*. The total number of features is assessed by first determining the total number of spots in a subgrid. A ***subgrid*** (also *subarray* or *block*) is defined as the unit microarray produced on a substrate by a single printing implement. A printhead containing 16 pins would produce a printed microarray containing 16 subgrids (Fig. 9.41). The total number of spots in a subgrid is the product of the number of spots in a row

Subgrid. *Unit microarray produced on a substrate by a single printing implement, also known as a subarray or block.*

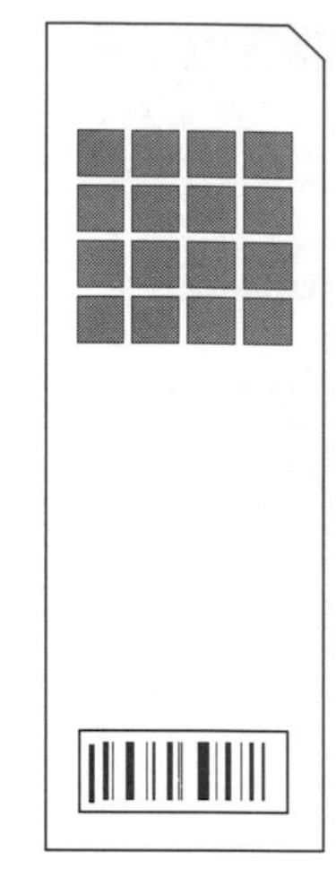

FIGURE 9.41. A printhead with a 4 × 4-configuration of printing implements set at 4.5 mm center to center produces an 18- × 18-mm microarray containing 16 subgrids. Printed microarrays are often labeled with a barcode for identification purposes, and the cropped corner provides unambiguous orientation.

times the number of rows in a subgrid:

(9.4) Number of spots (subgrid) = Spots per row × Number of rows

A subgrid containing 30 spots per row and 24 rows of spots would contain 720 (30 × 24) spots. The total number of features or spots in a microarray is the number of spots per subgrid, times the number of subgrids in a microarray:

(9.5) Total spots (microarray) = Spots per subgrid × Number of subgrids

The total number of spots in a microarray containing 16 subgrids and 720 spots per subgrid is 11,520 (720 × 16). In the common parlance, a microarray with 11,520 spots is known as an 11K array. If a substrate contains multiple microarrays, the total number of spots per substrate is the product of the number of spots per microarray, times the number of microarrays:

$$\text{Number of spots (substrate)} = \text{Spots per microarray} \times \text{Number of microarrays} \tag{9.6}$$

One common misconception is that the total number of spots present on a microarray is equivalent to the total genes or gene products present. In fact, the total number of genes examined is equal to the total number of spots on the microarray divided by the number of replicates per unique gene:

$$\text{Total number of genes} = \frac{\text{Total number of spots}}{\text{Number of replicates per unique gene}} \tag{9.7}$$

The total number of genes assayable on a 21,000-spot microarray with each unique gene printed in triplicate is 7,000 (21,000/3). This is not to say that replicates should be avoided, but simply that replicates reduce the total number of genes that can be examined on a microarray containing a given number of spots. Replicate printing provides an excellent means of improving the precision of microarray data, but users should be aware that the number of genes does not always equal the total number of spots. So-called 30K microarrays might actually correspond to 15,000 cDNAs printed in duplicate, and researches should be cognizant of the distinction between total genes and total spots. Most commercial controller packages calculate the number of spots automatically once the user specifies the configuration of each subgrid and the number of replicates per sample (Fig. 9.42).

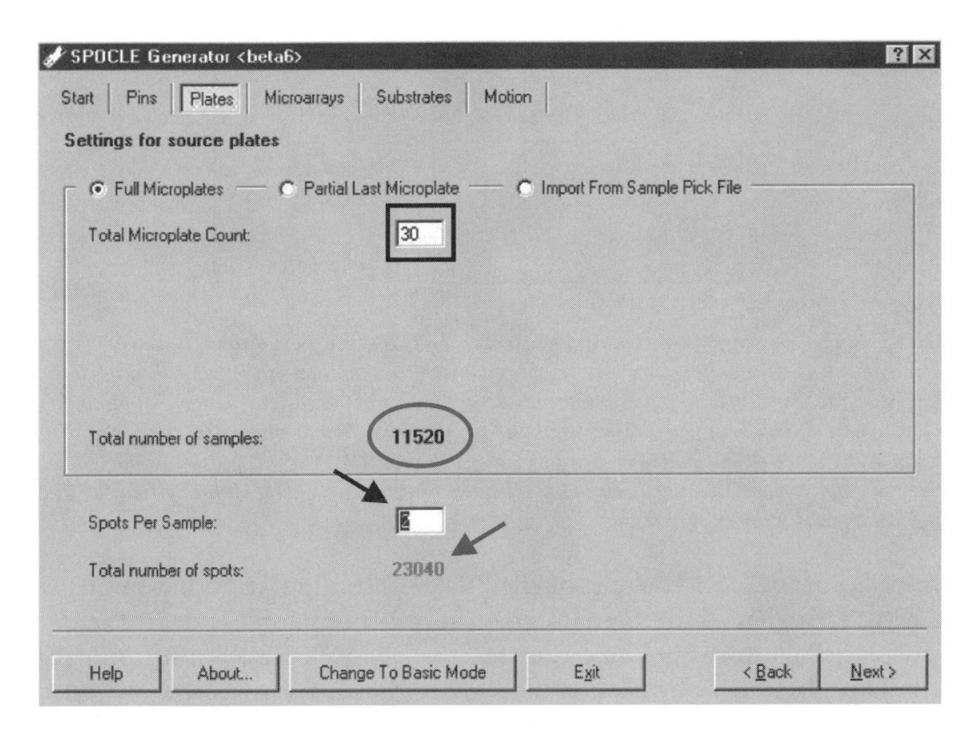

FIGURE 9.42. The total number of spots or features in a printed microarray can be calculated automatically by the software packages for most microarray robots (e.g., SpotBot*). The user enters the number of microplates* (square) *and replicates* (top arrow) *into text windows, and the computer calculates the total number of samples* (oval) *and spots* (bottom arrow). *(Courtesy of H. Bird, TeleChem/arrayit.com, Sunnyvale, CA.)*

The amount of genomic information on a microarray is expressed as microarray ***complexity,*** calculated as the product of the number of unique sequences, times their average length in nucleotides:

Complexity. *Amount of genomic information on a microarray, calculated as the product of the number of unique sequences times their average length in nucleotides.*

$$\text{Microarray complexity} = \text{Number of unique sequences} \times \text{Average length} \quad (9.8)$$

A microarray containing 10,000 unique cDNAs with an average length of 1,000 nt would have a complexity of 10,000,000 nucleotides (10 megabases), or approximately 1/300 (0.0033) of the human genome. The complexity of a microarray is often expressed as a function of the size of the genome represented in terms of a ***genome equivalent***. The number of genome equivalents in nucleotides is measured as the complexity of the microarray divided by the size of the genome:

Genome equivalent. *Sequence complexity corresponding to all of the nucleotides present in the genome of an organism.*

$$\text{Genome equivalent} = \frac{\text{Microarray complexity}}{\text{Genome size}} \quad (9.9)$$

A human microarray with a complexity of 10 megabases, divided by the size of the human genome (3,000 megabases) would have a genome equivalent of 1/300. It is noteworthy that a printed microarray containing 30,000 bacterial artificial chromosome (BAC) clones with an average length of 100,000 nt would have a genome equivalent of 1.0, meaning that each printed microarray would contain all 3,000,000,000 bases in the human genome. It is remarkable that microarray technology has already advanced to the point where a single laboratory could manufacture microarrays containing the entire human genome on a single chip!

Samples and Content Maps

Microarrays are printed from microplates or source plates having 384 wells, though 96-well plates and to a much lesser extent 1536-well plates are used by

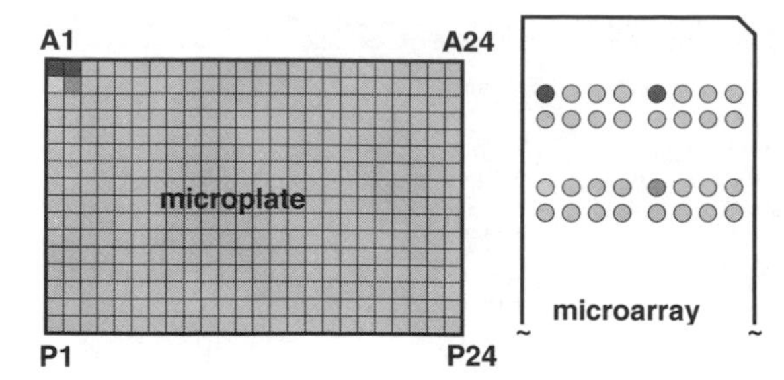

FIGURE 9.43. The density of a printed microarray (right) is much greater than the density of a 384-well microplate (left), resulting in a change of position or transposition of the printed samples on the chip relative to the microplate. A four-pin printhead with the pins arranged in a two by two configuration (shown here) will load four adjacent samples (upper left) *from a microplate and print them at different addresses on the microarray substrate* (dark spheres). *The location of printed samples relative to their microplate position depends on the number of printing implements, printing routine, subgrid configuration, and number of replicates.*

some laboratories. Microplates of the 384-well configuration have wells that are small enough to accommodate a large number of samples but large enough to enable straightforward liquid transfer. The 384-well plate is also compatible with many glass fiber purification kits that find wide use in microarray sample preparation. The total number of samples used in a microarray printing session is simply the product of the number of wells in the microplate, times the number of microplates:

(9.10) Number of samples = Number of wells × Number of microplates

A printing session that uses 27 microplates of the 384-well configuration would comprise 10,368 samples (384 × 27), equivalent approximately to the total number of genes expressed in most human tissues. The use of 75 microplates of the 384-well design would provide 28,800 samples (384 × 75), a number sufficient to represent all of the genes in the human genome. Laboratories equipped with basic automation can easily prepare 75 microplates of samples.

The standard printing convention uptakes samples from the upper left corner of the microplate and proceeds from left to right to the end of the row. The robot then performs a "carriage return and line feed," moving the printhead leftward and downward to uptake samples from a new set of microplate rows, and proceeds in this manner until the entire microplate is printed. The first and last wells sampled from a 384-well plate are A1 and P24, respectively (Fig. 9.43).

Transposition. *Computational process that converts the order and location of microarray samples in microplates into a printed microarray.*

Because the density of a printed microarray is much greater than the density of samples in the microplate, the order and location of microarray samples in the printed microarray are different from their corresponding locations in the microplate (Fig. 9.43). The ***transposition*** of samples from microplate into microarray is dictated by the number and configuration of printing implements in the printhead, the printing routine, the density of the printed spots, and the number of replicates. Most commercial software packages allow the user to design the microarray by specifying all of the important printing parameters via the graphical user interface (Fig. 9.44), and icon-based user interfaces speed the scripting of custom-printing routines (Fig. 9.45). Many commercial packages also allow the user to customize motion-control parameters, such as velocity and acceleration, and specify the use of microplate stackers, ultrasonic baths, and other modular accessories (Fig. 9.46).

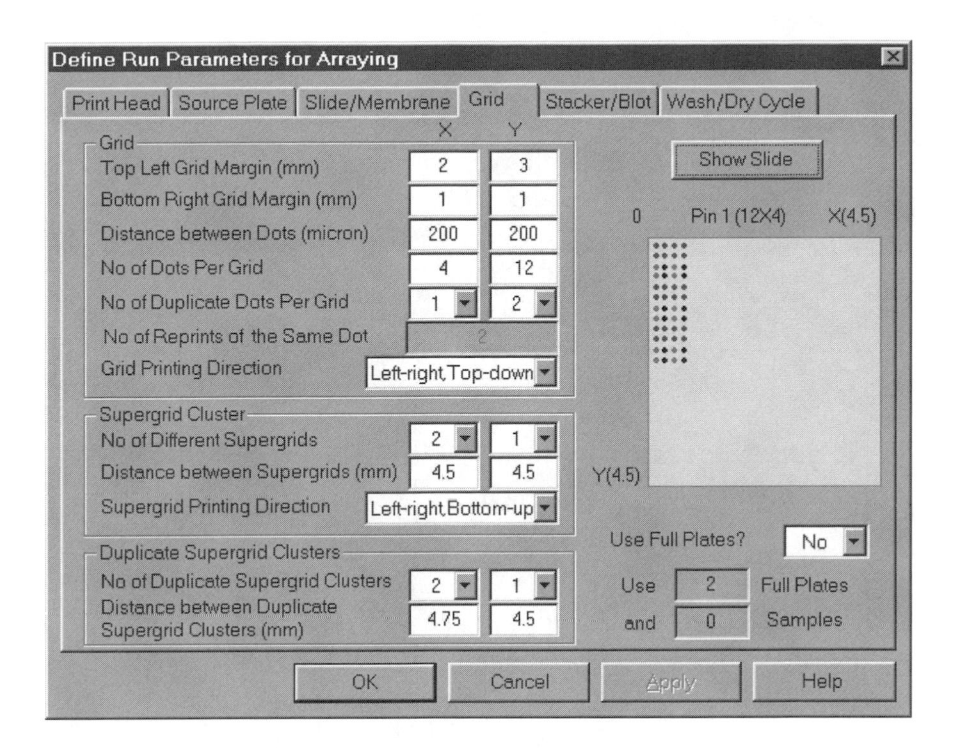

FIGURE 9.44. ChipWriter *software allows the user to specify myriad printing parameters, including specifications for the printhead and pins, source plates, substrate, subgrid, automatic microplate stacker, and the wash and dry cycles. (Courtesy of Dr. T. Bond, Dr. M. Kircanski and colleagues, Virtek Vision International, Waterloo Ontario.)*

The transposition of samples from the microplate into a printed microarray and the storage and retrieval of all of the relevant sample information are enormously complex tasks to perform manually, but they are accomplished readily with the proper database management software. The process of ***sample tracking,*** whereby sample identity is maintained between microplates and printed microarrays, along with all associated biological information, can be performed computationally using sample tracking software available from a number of commercial providers (Fig. 9.47). *CloneTracker* from BioDiscovery (Marina del

Sample tracking. *Computational process in which sequence identity, clone size, organism, and other associated biological information is correlated and maintained between liquid samples in microplates and printed spots in a microarray.*

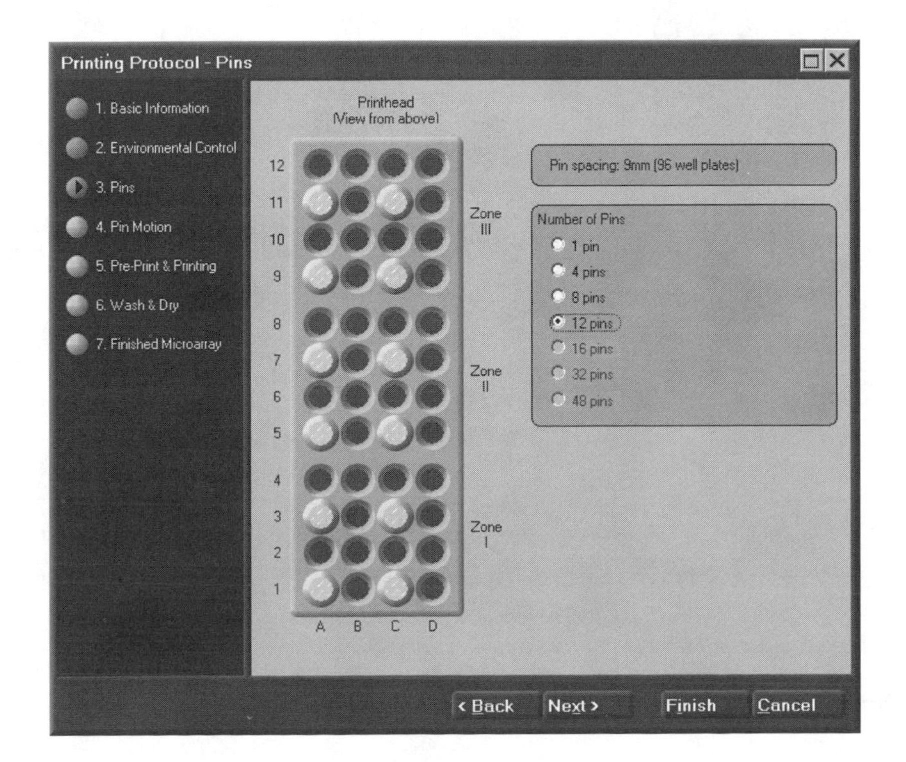

FIGURE 9.45. The graphical user interface for SpotArray *allows the user to specify printing parameters as a linear sequence of commands, using screen icons to guide the entire gamut of robotic functions. (Courtesy of Dr. D. Holton and colleagues, Packard BioScience, Billerica, MA.)*

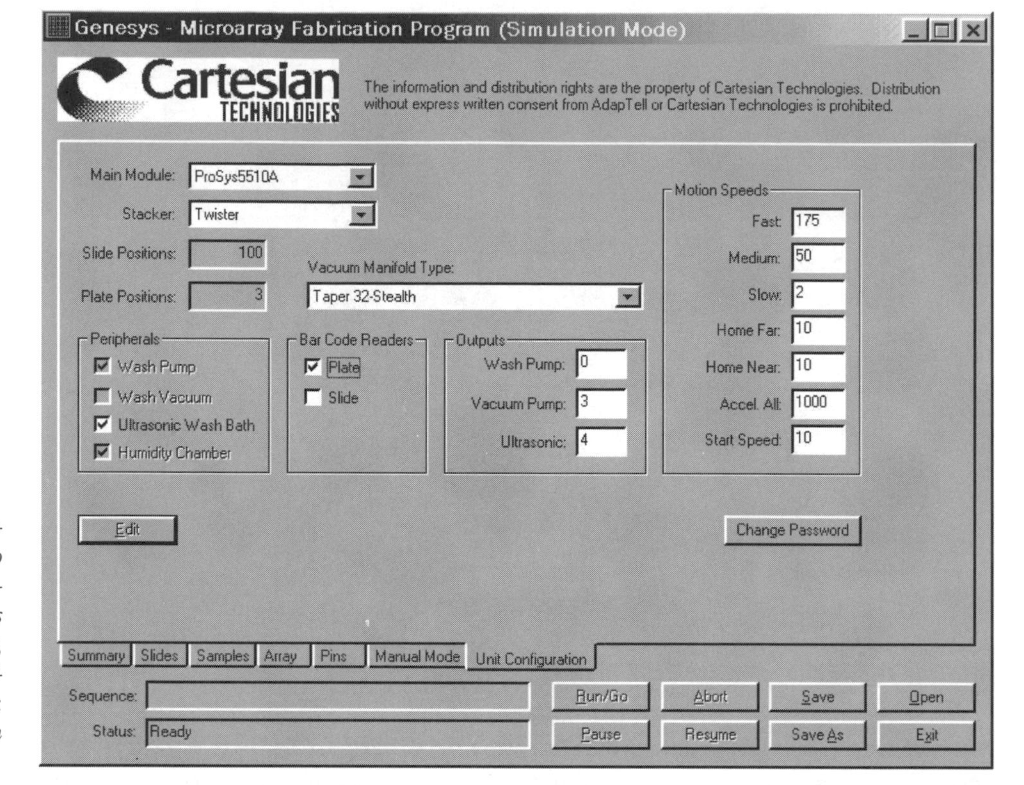

FIGURE 9.46. The Genesys *motion-control software allows the user to set detailed printing parameters, including velocity and acceleration, as well as to specify hardware accessories, such as a microplate stacker and ultrasonic wash bath. (Courtesy of Dr. D. Rose and Dr. M. Faugh, Cartesian Technologies, Durham, NC.)*

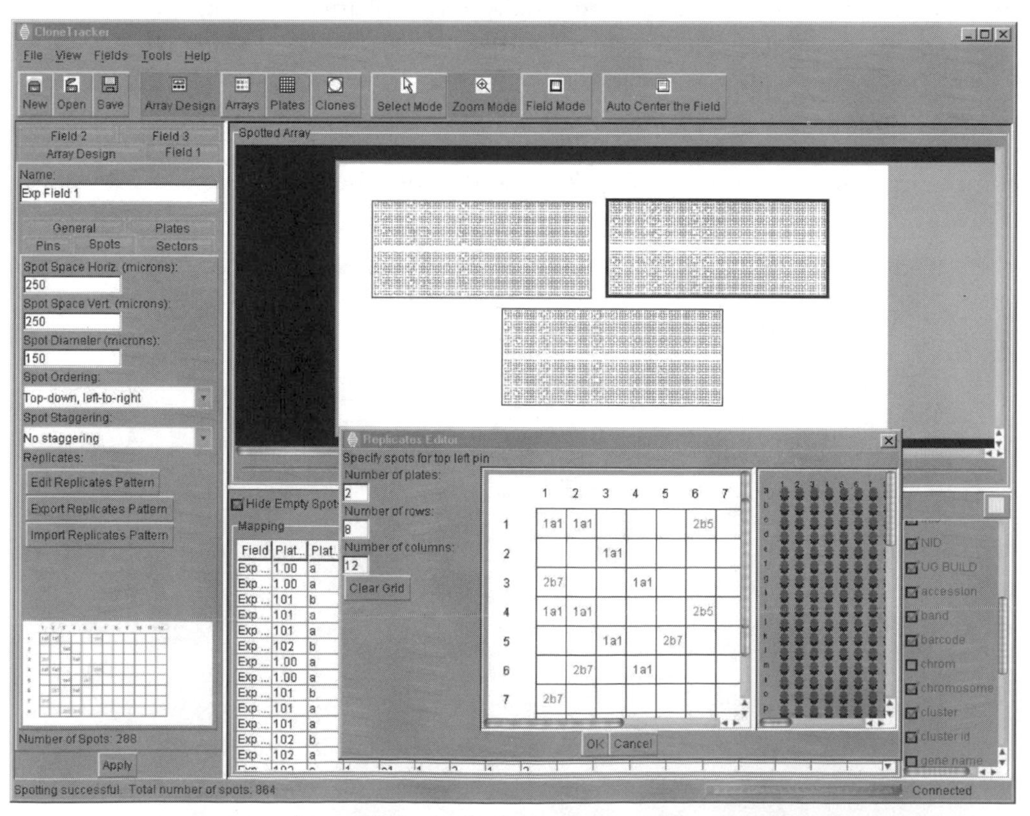

FIGURE 9.47. Using Clontracker, *The researcher specifies the details of the microarray design, and powerful SQL-based informatics provides graphical representations and electronic files of the printed microarray, including the location of each sample in the source plates. (Courtesy of Dr. S. Shams, BioDiscovery, Marina del Rey, CA.)*

TABLE 9.5. Content Map Parameters

Parameter	Description	Importance
Microarray column	column of spot (e.g., 1–36) in printed microarray	exact spot location essential for data analysis
Microarray row	row of spot (e.g., 1–36) in printed microarray	exact spot location essential for data analysis
Microarray subgrid	subgrid of spot (e.g., 1–48) in printed microarray	exact spot location essential for data analysis
Microplate well	sample location in microplate (e.g., A1–P24)	exact sample location essential for data analysis
Microplate number	number of microplates (e.g., 1–120) that contain a given sample	exact sample location essential for data analysis
Sequence information (oligonucleotides, peptides)	sequence string (e.g., nucleotides, amino acids) of target sample	target sequence determines reactivity of molecules in each spot
Primer information (cDNAs)	nucleotide sequence of primers determines amplicon obtained by PCR	sequence of amplified target determines reactivity of molecules in each spot
GenBank accession number	unique identifier (e.g., M73060) for each sequence in the GenBank database	electronic sequence identifier provides unambiguous and universal access to microarray content
Gene name	scientific name (RNA polymerase II subunit RPB9) of a gene or gene product represented in the microarray	assists the researcher in understanding the biological function of the target sequence at a given location
Gene source	name of the organism (e.g., *Drosophila melanogaster*) represented in the microarray	allows the researcher to understand the organism under investigation and tailor samples accordingly

Rey, CA) uses a relational database system based on ***structured query language*** (SQL; pronounced "sequel") to store and retrieve sample tracking information. A number of popular relational database management systems including *Oracle,* Microsoft *SQL Server,* and *Sybase* all use SQL, which is the international standard for databases. The electronic file that contains all of the information pertinent to each microarray spot is known as a ***content map***. Content maps include microplate location, microarray number, sequence information, GenBank accession number, gene names, and other important data (Table 9.5), as well as normalized intensity values obtained through quantitation (see below).

Structured query language. *Computing language used in relational database systems.*

Content map. *Electronic file that contains all of the information pertinent to each microarray spot or feature.*

DATA QUANTITATION

The speed and ease of quantitation is one of the many attributes of microarray experimentation, allowing users to obtain quantitative biological information for thousands of genes in less than a second. But the facility of the process should not be taken to mean that data quantitation informatics is based on trivial mathematics. The theory behind absolute quantitation, ratio calculations, normalization, and other aspects of microarray data quantitation is quite sophisticated; and microarray researchers are served well by acquiring a working knowledge of the concepts and formulae. Treating quantitation as a software black box yields meaningful and accurate results in most cases, but experimental precision can be improved by understanding quantitation at the computational and statistical levels and by making minor adjustments in software to obtain optimal results.

Absolute Quantitation

Quantitation. *Computational process by which numerical values are obtained from microarray data files.*

Quantification. *Synonym of quantitation.*

Microarray scanners and imagers acquire fluorescent microarray data as tagged image file format (TIFF) files. The TIFF file (.tif) is a two-dimensional intensity map of the microarray surface, with fluorescent signals stored in square bins known as pixels (see Chapter 8). Each microarray spot, often representing an individual gene, contains multiple pixels each of which contains quantitative information about the gene or gene product present at a microarray location. The process by which numerical values are obtained from flat microarray data files is known as ***quantitation*** or ***quantification***. Microarray quantitation allows the researcher to extract numerical values for microarray spots from TIFF files. The numerical values provide information that can be used to determine the concentration of mRNAs, proteins and other biomolecules of interest in complex samples.

Absolute quantitation. *Computational process of obtaining numerical values for each microarray spot based on the signal intensity at each location.*

Relative quantitation. *Computational process of measuring the ratio of absolute signals for each microarray spot in two samples on a microarray.*

Absolute quantitation is process of obtaining numerical values for each microarray spot based on the signal intensity at each location (e.g., 17,451 counts). Absolute numbers are used to measure of the abundance or molar amount of a given species in a probe mixture and to calculate ratios between multiple samples and different experiments (see below). ***Relative quantitation*** refers to the process of measuring the ratio of absolute signals in two samples on a microarray and is reported as the quotient of one channel divided by the other channel (e.g., 6.1-fold). Relative quantitation is used in conjunction with multicolor labeling and detection schemes to gain comparative information concerning gene expression levels, genotypes, and other important data.

Manual quantitation. *Computational process in which signal intensities are generated for microarray spots one at a time.*

All forms of microarray quantitation are performed by computer. The two main types, distinguished on the basis of whether single or multiple spots are quantified in a given step, are known as manual quantitation and automated quantitation. ***Manual quantitation*** is the process by which signal intensities are generated for single spots one at a time, and researchers employ manual methods in cases for which a microarray contains a small number of spots or to gain rapid, intuitive information about a microarray image without constructing a quantitation template (see below). Most commercial scanners and imagers enable manual quantitation in the controller software via icon-based commands in the graphical user interface.

Each microarray spot contains two numerical components known as signal and background. Signal corresponds to the intensity values associated with true microarray data, and these numbers provide quantitative information concerning gene expression patterns, genotypes, protein–protein interactions, and other essential microarray data. Background, or noise, corresponds to the unwanted intensity values associated with spurious biochemical events, such as nonspecific binding of probe, substrate reflection or fluorescence, and other nonsignal sources that are registered as counts by the detection instrument but do not contain genuine microarray data. Distinguishing signal from background is an essential step in microarray quantitation.

The three steps in manual quantitation include opening a microarray image (.tif file), defining a quantitation region, and acquiring the data. Most detection

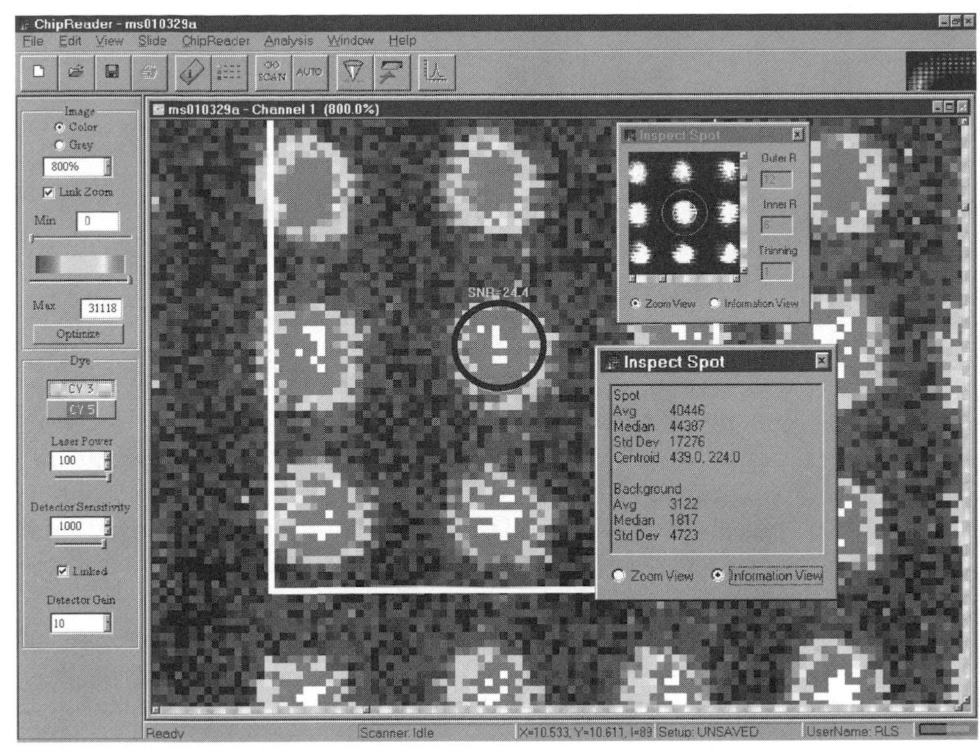

FIGURE 9.48. *With* ChipReader, *a spot of interest is demarcated by a boundary* (circle) *within a larger quantitation area* (circle, *inset) that defines the background. The computer then calculates the average signal intensity of the pixels inside the border (40,446), the median signal intensity (44,387), the standard deviation (17,276), the average (3,122) and median (1,817) background values, and the signal-to-noise ratio (24.4).*

instruments allow users to open a stored microarray image using the "File open" command in the "File" menu. Once the data file is opened, a boundary is placed around a spot of interest using a software tool to demarcate the signal region (Fig. 9.48). The user also typically defines a slightly larger area outside of the spot to allow the computer to calculate a background value (Fig. 9.49). Once the signal and background boundaries are specified by the user, the computer calculates the signal and background intensities automatically and provides the values in fluorescent counts.

Signal and background values can be computed in several different ways, with the mean (average) and median being the most common. The mean or average signal intensity is the sum of the intensity values of each pixel inside

FIGURE 9.49. *Enlarged view of the microarray spot shown in Figure 9.48, with the boundaries for signal* (inner circle) *and background* (outer circle) *demarcated. Some pixels that contain noise reside within the signal region* (white arrow) *and some pixels that contain signal end up in the background region* (dark arrow).

the designated signal area divided by the total number of pixels in the signal region.

$$\text{Mean signal intensity} = \frac{\text{Sum of intensity values}}{\text{Total number of pixels}} \tag{9.11}$$

The median signal intensity is the intensity value corresponding to the middle score in the range of pixel intensities specified in the signal region. The mean and median values for background are calculated using the same procedure used for signal, except that the pixels used to calculate background values lie outside of the specified signal region (Figs. 9.48 and 9.49).

The mean signal and background values for the intensely fluorescent spot shown in Figure 9.48 are 40,446 and 3,122 counts, respectively, compared to the median values of 44,387 and 1,817. It is important to note that in this example, the average signal intensity is lower than the median signal intensity and the average background value is higher than the median background value. The discrepancy between average and median scores indicates that some background pixels have "contaminated" the signal region and some signal pixels reside within the area specified as background (Fig. 9.49). The inclusion of background pixels in the signal region and signal pixels in the background region reduce the signal reading and inflate the actual background value, which explains the discrepancy between the mean and median.

Signal segmentation. *Computational process in which microarray signals are demarcated from background in a microarray image.*

Spatial segmentation. *Method of signal segmentation that relies on careful positioning of the quantitation boundaries for signal and noise, using physical location to segment signal from background.*

Intensity segmentation. *Method of signal segmentation that uses pixel values and mathematical and statistical approaches to exclude anomalous pixel intensities and segment signal from background.*

The process by which microarray signals are demarcated from background in a microarray image is known as ***signal segmentation***. The user can increase the precision of signal segmentation by adjusting the physical and statistical parameters used to demarcate signal and background. The two main forms of signal segmentation are spatial segmentation and intensity segmentation. Methods that use ***spatial segmentation*** rely on careful positioning of the quantitation boundaries for signal and noise, using physical location to segment signal from background. ***Intensity segmentation*** uses pixel intensity values to separate signal from background, relying on mathematical and statistical approaches to exclude anomalous pixel intensities. Some sophisticated commercial quantitation packages, such as *ImaGene* (BioDiscovery, Marina del Ray, CA), incorporate both spatial and intensity segmentation to yield highly precise microarray data. The capacity to adjust the demarcation regions is helpful when microarray images contain slightly irregular spot spacing or morphology. Microarrays that have highly regular and circular features with uniform signal intensity across the entire spot are extremely easy to quantitate, placing added value on surfaces, buffers, and printing technologies that afford high-quality microarray spots.

Thinning. *Method of spatial segmentation that maintains the outer border of the signal region but restricts the number of pixels that are used to obtain the average and median signal intensities.*

Methods of spatial segmentation use a number of adjustments to the physical dimensions that define signal and noise, including a process known as ***thinning,*** which maintains the outer border of the signal region but restricts the number of pixels that are used to obtain the average and median signal intensities (Fig. 9.50). By specifying different boundary radii and thinning values, the user can adjust the physical size of signal and background regions and minimize the presence of signal pixels in the background region, and vice versa. One common representation of signal and background is the quotient of the two values, which yields the signal-to-noise ratio (SNR):

$$\text{SNR} = \frac{\text{Median signal intensity}}{\text{Median noise value}} \tag{9.12}$$

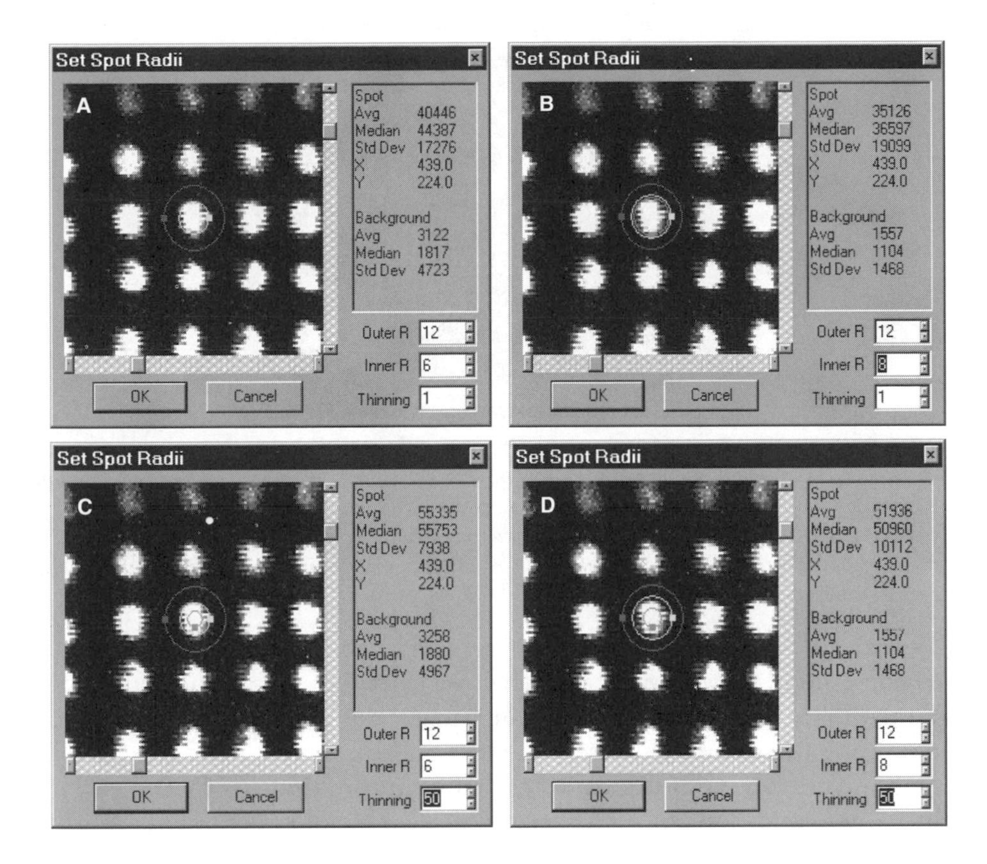

FIGURE 9.50. *Spatial segmentation, using* ChipReader, *adjusts physical quantitation parameters to improve quantitation precision: standard settings* (**A**), *signal radius increased from 6 to 8 pixels* (**B**), *signal region thinning increased from 1 to 50%* (**C**), *and signal radius increased from 6 to 8 pixels and signal region thinning increased from 1 to 50%* (**D**).

For the example shown in Figure 9.48, a microarray spot with median signal and background intensities of 44,387 and 1,817, respectively, has a 24.4 SNR value.

Nearly all microarray detection instruments acquire data in 16-bit format, producing intensity values ranging from 1 to 65,536 (see Chapter 8). As a general rule, signals in excess of ~50,000 counts define the upper limit of what can be detected accurately, above this the signals exceed the detection limit of the photomultiplier tube (PMT) or charge-coupled device (CCD). Signals that exceed the upper detection limit are known as saturated signals (Fig. 9.51). All detectors used in microarray scanners and imagers are capable of saturation if exposed to intensely fluorescent spots or to moderately fluorescent microarrays for excessive periods of time.

Representing microarray data in a rainbow palette allows the user to choose detection instrument settings that provide the greatest intensity values without saturating the detector. Many microarray palettes use white to depict saturated signals, and users should avoid deriving quantitative information from the white spots. Fluorescent spots that are saturated appear nearly identical in scanned images even though the true fluorescent intensity of the spots can differ by, for example, 1000-fold. Figure 9.51 also emphasizes the lack of spreading of intensely fluorescent signals beyond the physical spot boundaries, allowing the capture of enormously bright signals next to much weaker signals on a single microarray. Signals from radioactive labels, by contrast, spread aggressively in two dimensions, preventing radioisotope-based labeling and detection methods from being used in a microarray format.

Images that contain saturated signals prevent the user from obtaining accurate intensity values at the high end of the intensity scale, though saturated

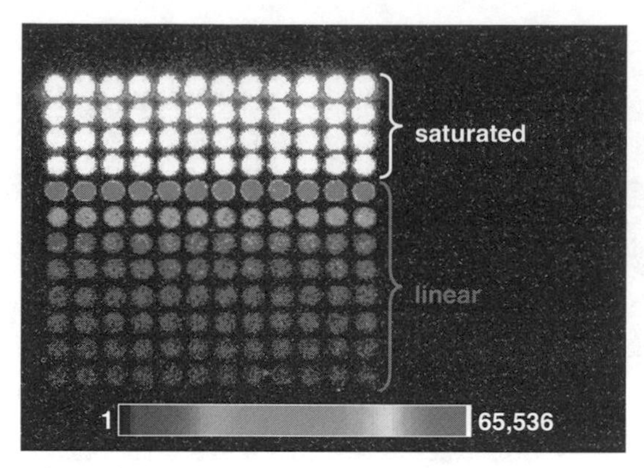

FIGURE 9.51. Saturated signals. Microarray data from a fluorescent scan are represented in a rainbow palette coded to signal intensities from 1 to 65,536 counts as shown in the color bar (bottom). Rows correspond to 10-fold dilutions of a fluorescent oligonucleotide, ranging from 100 pmol/μL to 0.1 attomol/μL (rows 1–10) with 12 replicates per dilution. Rows 11 and 12 (bottom two rows) contain buffer only. (Data courtesy of Dr. R. Stears, TeleChem/arrayit.com Sunnyvale, CA.) Figure also appears in Color Figure section.

images can be useful for eking out weak signals on the microarray. Weak signals often correspond to rare transcripts that are extremely important biologically (e.g., membrane receptors, transcription factors, kinases), placing a premium on obtaining these data. The researcher can capture weak signals by increasing the gain on the photomultiplier tube or the capture time on CCD-based systems. Under conditions of high sensitivity, weakly fluorescent spots can produce signals that are above the background and therefore can provide useful data even though the all of the moderate and intensely fluorescent spots are saturated at high sensitivity settings.

Subsaturating signals provide graphical and numerical data that correspond to true fluorescence intensity and are said to fall within the linear range of the assay. Because fluorescence detection allows such enormous detectivity, trace quantities of fluorescent or reflective reagents in microarray printing buffers can contribute to fluorescence measurements as illustrated by the fact that spots corresponding to buffer alone (Fig. 9.51, rows 11 and 12) produce detectable signals. These values can be excluded from signal reading using a process known as ***background subtraction***. Background subtraction is the process by which the fluorescent counts corresponding to background (or buffer) are subtracted from the true signals to obtain more accurate quantitation.

Background subtraction. *Computational process in which the fluorescent counts corresponding to the noise are subtracted from true signals to obtain more accurate quantitation.*

The data in Figure 9.51 also reveal the fact that fluorescent signals are not always directly proportional to the number of fluorescent molecules present at a given microarray location. Rows 5 and 6 contain 10 fmol/μl and 1 fmol/μl fluorescent oligonucleotide, respectively, though the fluorescent signals (33,870 counts vs. 4,971 counts) differ by a factor of 6.8 instead of the 10-fold ratio expected for spots that contain a 10-fold difference in dye molecules. Signal compression is the process by which fluorescent spots produce less than a directly proportional number of counts relative to their fluorescent dye content, a phenomenon that arises from energy transfer between dye molecules and other detection hardware factors (see Chapter 8). Signal compression does *not* mean that microarray assays are nonquantitative, but rather that the data must be normalized to control spots of known concentration before deriving quantitative information (see below). A gene that registers a 6.8-fold ratio in two samples may actually differ by 10-fold concentration, making it imperative for users to understand the phenomenon of signal compression.

Weak microarray signals present a challenge to accurate quantitation; and as a general rule, signals lower than 1.5-fold above the median background are generally considered unreliable. If the median background value is 1,817 counts (Fig. 9.48), users should be wary of signals that measure less than about 2,700 counts (1,817 × 1.5) because such signals could correspond either to weak signals (e.g., rare transcripts) or to slightly elevated background in a local region. Weak signals also present a problem when calculating ratios of rare transcripts in comparative analyses (see below).

A median background value of 1,817 counts (Fig. 9.48) is rather high for a microarray assay, but such values can be observed when using dilute probe mixtures or high sensitivity settings. One obvious outcome of Equation 9.12 is that assays that reduce median noise values increase SNR proportionately, so that a median signal intensity of 44,387 against a median noise value of 1,817 would produce an SNR of 24.4, compared to a background value of 18, which would yield an SNR of nearly 2500. The dynamic range and detectivity of microarray assays is almost always limited by the background noise, placing tremendous value on surfaces, labeling and purification procedures, hybridization techniques, and other approaches that reduce background. Reducing the background intensity increases the dynamic range and detectivity of an assay, enabling fluorescence detection of transcripts approaching one mRNA molecule per 10 human cells.

Microarrays routinely contain thousands or even tens of thousands of features, requiring software tools to quantitation the microarray spots. The process by which many microarray features are quantified simultaneously is know as ***automated quantitation***. Though automated quantitation produces a dizzying volume of data—indeed a single researcher can produce a million data points in a few weeks—the basic principles established for manual quantitation guide the automated methods.

Automated quantitation. *Computational process in which a large number of microarray features are quantified simultaneously without manual intervention.*

Automated quantitation requires the user to specify a graphical pattern known as a ***quantitation template*** or ***quantitation grid*** across the microarray image (Fig. 9.52). The template is positioned by defining the four corners of the microarray, specifying the number of rows and columns of spots, and stretching the template into place using drag icons located on the boundaries of the template. Once a quantitation template is positioned correctly over a microarray image, the computer calculates the signal intensities of each spot located within a quantitation element. The speed of modern quantitation packages is such that precise signal intensities can be obtained for a thousand spots in a few seconds. Automated quantitation procedures are so robust that a single research laboratory can generate gigabyte and even terabyte (1,000 gigabyte) quantities of microarray data using high-density chips.

Quantitation template. *Graphical checkerboard pattern placed over a microarray image to facilitation data extraction.*

Quantitation grid. *Synonym for quantitation template.*

The data outputs from most automated quantitation packages are formatted as text (.txt) files. The most common format of text files separates each field or parameter with a recognizable keystroke (e.g., comma or tab) to create a ***delimited file*** (Fig. 9.53). A file in which each record is delimited with a tab is known as a ***tab-delimited file***. Each record (microarray spot) is further distinguished by another keystroke such as a carriage return. Delimiting every field and every record in a microarray text file allows large datasets to be saved and imported systematically and precisely into spreadsheet programs (Fig. 9.54) or into data visualization software (see below) for further analysis. Microarray datasets that are not saved in a delimited format are unwieldy to the point of being difficult or impossible to analyze.

Delimited file. *File format used in microarray quantitation in which each field or parameter is separated with a recognizable keystroke.*

Tab-delimited file. *File format used in microarray quantitation in which each field or parameter is separated with a tab keystroke.*

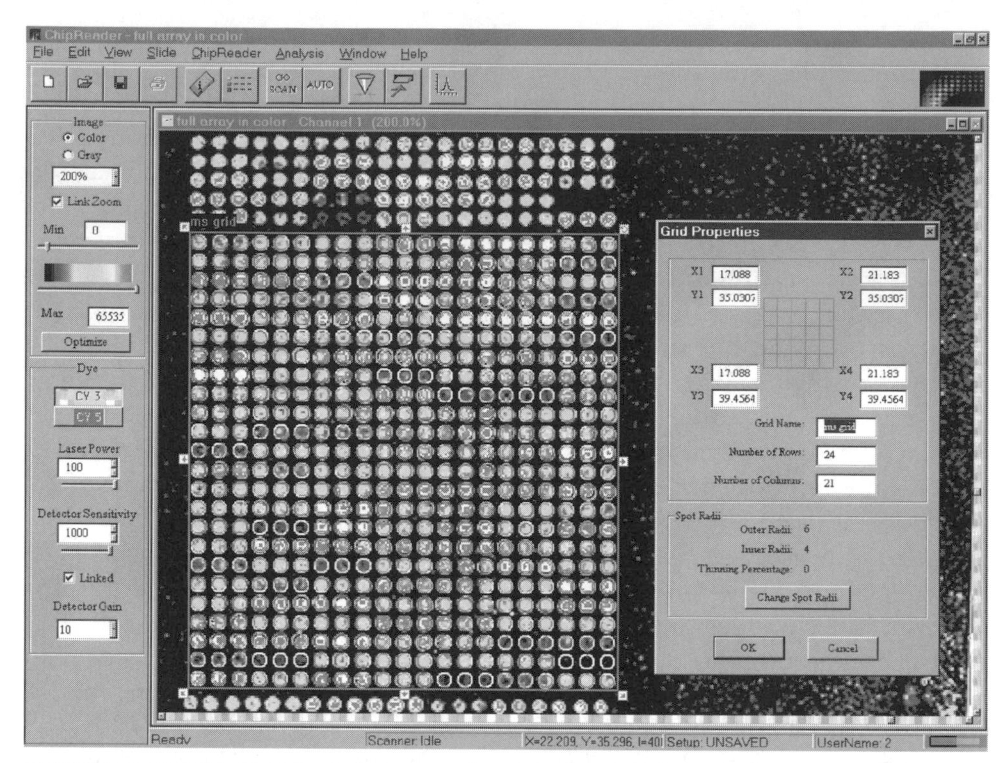

FIGURE 9.52. With ChipReader, *automated quantitation is performed by superimposing a quantitation template* (square) *over a microarray image. The user specifies the number of rows, columns, and other quantitation parameters and positions the template precisely over the image. The computer calculates the signal intensity for each spot located inside the template. Figure also appears in Color Figure section.*

full array in color_1.txt - Notepad

File Edit Search Help

```
Slide#,User Name,Sample#,Experiment#,Unit,X1,Y1,X2,Y2,Resolution,Scan
Speed(LPS),Channel,Laser Intensity,Detector Sensitivity,Gain,Number of Scans,Scan Date,Notes
2,2,2,2,mm,1.5,0.0,23.5,59.6,18.00,30,0,100,1000,10,1,2001-09-11 16:55:37,2
GridName,Column#,Row#,CentroidX,CentroidY,SNR,Signal Average,Signal Median,Signal
Std,Background Average,Background Median,Background Std
ms grid,1,1,973.0,1952.0,0.0,0.0,0.0,0.0,31783.0,22584.0,29419.8
ms grid,2,1,985.0,1952.0,0.0,0.0,0.0,0.0,33091.6,30186.0,28785.0
ms grid,3,1,997.0,1952.0,0.0,0.0,0.0,0.0,27796.6,25074.0,25500.5
ms grid,4,1,1009.4,1951.6,2.4,50167.7,53369.0,13472.4,20740.8,6016.0,24816.4
ms grid,5,1,1021.1,1950.9,4.7,58165.5,63641.0,10884.4,12483.7,5014.0,16552.8
ms grid,6,1,1032.9,1951.8,4.3,56067.3,63159.0,13589.5,13049.1,3896.0,17739.6
ms grid,7,1,1044.6,1951.8,2.8,55026.4,58403.0,10679.5,19448.9,6848.0,22140.5
ms grid,8,1,1055.6,1951.6,3.3,51390.6,52087.0,11172.5,15557.1,6012.0,19766.8
ms grid,9,1,1069.0,1952.0,0.0,0.0,0.0,0.0,9014.4,8110.0,7415.2
ms grid,10,1,1081.0,1952.0,0.0,0.0,0.0,0.0,9146.4,6428.0,8381.3
ms grid,11,1,1088.2,1952.0,2.0,15117.5,13944.0,3272.2,7635.9,3922.0,9340.5
ms grid,12,1,1108.8,1953.6,2.8,6713.0,5780.0,1212.6,2367.2,1846.0,1613.5
ms grid,13,1,1118.0,1952.0,0.0,0.0,0.0,0.0,2619.6,2472.0,1757.4
ms grid,14,1,1130.5,1951.8,0.0,0.0,0.0,0.0,2317.3,2456.0,1277.0
ms grid,15,1,1142.0,1952.0,0.0,0.0,0.0,0.0,3085.7,3102.0,1134.5
ms grid,16,1,1152.2,1955.1,3.3,12795.3,12806.0,2016.0,3883.9,2988.0,3076.4
ms grid,17,1,1166.0,1952.0,0.0,0.0,0.0,0.0,3871.8,3164.0,2554.6
ms grid,18,1,1178.0,1952.0,0.0,0.0,0.0,0.0,4121.3,3868.0,2177.1
ms grid,19,1,1192.0,1953.0,2.7,10994.0,0.0,0.0,4073.5,3432.0,2884.4
ms grid,20,1,1202.0,1952.0,0.0,0.0,0.0,0.0,2626.4,2098.0,2190.0
ms grid,21,1,1214.0,1952.0,2.8,12830.0,0.0,0.0,4586.8,3854.0,3270.3
ms grid,1,2,973.0,1963.0,0.0,0.0,0.0,0.0,20846.6,19652.0,20960.7
ms grid,2,2,988.1,1961.0,4.1,38542.5,41947.0,13260.5,9442.6,1992.0,13293.8
ms grid,3,2,999.2,1961.8,3.5,54032.8,65087.0,16824.0,15411.0,2662.0,20962.4
ms grid,4,2,1010.1,1961.9,3.4,55477.7,64997.0,15619.4,16304.0,5308.0,20547.3
ms grid,5,2,1021.2,1961.7,3.1,61588.8,65535.0,8173.2,19634.8,11702.0,21477.0
ms grid,6,2,1033.0,1961.9,4.6,57526.8,65535.0,13767.1,12557.2,2912.0,16845.1
ms grid,7,2,1043.9,1961.8,4.0,56964.3,65535.0,13990.0,14307.0,6524.0,18284.0
```

FIGURE 9.53. Output data from microarray quantitation packages are saved as text files in which the each field (parameter) is separated by a comma, and each record (microarray spot) ends with a carriage return.

linear signals.xl

	A	B	C	D	E	F	G	H	I	J	K	L	M	
1	Slide#	User Name	Sample#	Experiment	Unit	X1	Y1	X2	Y2	Resolution	Scan Speed(LPS)	Channel	Laser Inten	Detec
2	3	3	3	3	mm	7.4	8.7	19.3	20	12	25	0	100	
3	GridName	Column#	Row#	CentroidX	CentroidY	SNR	Signal Aver	Signal Medi	Signal Std	Background Avera	Background Medi	Background Std		
4	ms grid	1	5	163.8	189.6	2	29560	31088	10023.5	14907.8	2154	17905.2		
5	ms grid	2	5	180.8	189.6	2.3	28866	28684	12299.9	12540.2	1748	15441.2		
6	ms grid	3	5	197.8	189.6	2.2	32275.9	32604	14149	14980.4	1188	18245.7		
7	ms grid	4	5	213.5	188.6	2.6	36965.1	35725	6471.9	14048.3	8980	15326		
8	ms grid	5	5	230.7	189.5	2.5	38224.4	38621	10142.9	15333.9	3654	17836.7		
9	ms grid	6	5	247.8	189.6	2.2	34724.3	37105	16581.6	15958	532	19099.2		
10	ms grid	7	5	264.6	189.5	2.2	35084.5	37505	13383.1	16052.5	2120	19641.6		
11	ms grid	8	5	281.6	189.6	2	32867.3	33117	14357.3	16751.8	816	20154.1		
12	ms grid	9	5	297.6	188.8	2.2	33716.6	36701	12652	15598.9	646	19624.4		
13	ms grid	10	5	313.7	189.7	2.2	34392.4	38109	13846.8	15351.7	1822	18408.1		
14	ms grid	11	5	331.6	189.8	2.1	31108.8	35799	15429.2	15097.8	456	19410.2		
15	ms grid	12	5	348.6	189.7	2.1	27088.7	30776	13614.4	12925	760	16852.2		
16	ms grid	1	6	163.8	205.6	2.3	5477.8	5732	1530.5	2429.4	400	3062.4		
17	ms grid	2	6	181.1	205.7	2.6	6198.8	6050	1425.7	2399.5	1318	2626.4		
18	ms grid	3	6	197.8	205.6	2	5342.2	5042	1886.6	2653.4	680	3087.6		
19	ms grid	4	6	214.7	205.5	2.6	5870.7	6332	1868	2279.6	582	2492		
20	ms grid	5	6	230.8	205.7	2.1	6323.1	6402	1305.2	2960.2	1244	3213.3		
21	ms grid	6	6	247.7	205.6	2.2	6681.6	6096	2150	3079.1	1210	3375.3		
22	ms grid	7	6	264.6	205.7	1.9	4823	4546	2102.5	2517.9	742	2861.7		
23	ms grid	8	6	281.6	205.8	1.9	5673	4882	3753.7	3001.7	430	4300		
24	ms grid	9	6	298.6	205.8	2	5374.9	5440	2118.4	2730	256	3263.2		
25	ms grid	10	6	314.6	205.7	2.1	6270.5	6622	2405.8	3029.1	140	3781.4		
26	ms grid	11	6	331.7	205.7	1.8	5352.4	5866	2657.9	2916.4	254	3722		
27	ms grid	12	6	348.5	205.7	2.2	6081.3	6040	2384.9	2745.9	434	3411.6		
28	ms grid	1	7	163.8	221.6	2.8	2033.4	1756	854.4	713.9	252	799		
29	ms grid	2	7	180.5	221.8	2.9	2387.6	2384	706.5	817.4	422	858.1		
30	ms grid	3	7	197.7	221.7	1.7	1768.1	1668	939.6	1020.2	188	1249.3		
31	ms grid	4	7	213	218.6	3.9	3766	3200	847.2	953.7	966	761.9		
32	ms grid	5	7	230.5	221.6	2.9	2692.3	2752	455.3	933.3	474	1051.7		
33	ms grid	6	7	247.2	221.7	3.7	3879.1	3628	521.1	1039	810	1033.8		
34	ms grid	7	7	263.6	220.4	4.1	3956	3796	164.1	966.1	808	906.2		
35	ms grid	8	7	279.7	220.2	2.8	3164.3	3010	505.0	1147.5	884	1104.2		
36	ms grid	9	7	297.6	221.6	1.9	2418	2420	756.1	1245.6	304	1398.9		
37	ms grid	10	7	313.9	219.8	0	0	0	0	1167.9	966	874.9		
38	ms grid	11	7	329.8	220.8	3.1	3212.3	3092	418.1	1024.7	998	826		
39	ms grid	12	7	346.1	221.4	3.7	3214	3416	353.8	861.9	818	543.1		
40	ms grid	1	8	163.8	237.7	2.8	1613.3	1570	419.1	569.8	336	551.5		
41	ms grid	2	8	179.8	237.6	3.4	2047.2	1908	415.5	597.8	420	576.8		
42	ms grid	3	8	197.7	237.8	2.9	1793.3	1768	511.8	627.3	320	670.4		
43	ms grid	4	8	215	237.6	2.4	1636.9	1568	430.6	673	320	728.4		
44	ms grid	5	0	230.9	236.3	3.5	2393.9	2240	436.2	684.9	520	595.6		
45	ms grid	6	8	247.6	238.6	3.1	1722.7	1556	857.4	559.4	230	571.7		
46	ms grid	7	8	264.8	237.8	2.4	1434.6	1364	477.3	609.5	256	647.2		
47	ms grid	8	8	278.9	236.5	3.4	2490.8	2430	401.7	725.8	682	373.5		
48	ms grid	9	8	298.6	237.6	2.7	992.4	968	583.2	373.5	118	443.5		
49	ms grid	10	8	315	236.1	3	1512.9	1544	286.2	504.6	380	521.6		
50	ms grid	11	8	331.6	237.8	1.7	1396.6	1186	850.2	803.3	152	1023.7		
51	ms grid	12	8	347.8	238.5	3	1978.7	1904	395.7	663.6	408	713.1		
52	ms grid	1	9	163.8	254.7	2.3	1088.8	1084	194.5	464.8	254	470.5		

linear signals.xl

Ready NUM

FIGURE 9.54. *Microarray data delimited by commas or tabs can be imported into spreadsheet software such as Microsoft* Excel *(shown here) or into data visualization software.*

Calculating Ratios

The process by which signal intensities from two microarray images are divided to obtain a quotient for each datum point is known as a ***ratio calculation***. Ratio calculations are used widely in microarray analysis to obtain quantitative information concerning genotypes, gene expression patterns, and other biological processes. Absolute signal intensities based on comparisons of two labeled samples provide a quantitative measure of the amount of a gene or gene product present in two labeled mixtures and the extent of the change in concentration of a given molecule in two samples, which provides a measure of gene activation or repression.

Ratio calculation. *Computational process by which signal intensities from two microarray images are divided to obtain a quotient for each data point.*

A popular ratio measurement for gene expression applications uses a test sample and a reference sample to investigate a biological phenomenon of interest. A test sample is a labeled mixture derived from a biological source subjected to a specific experimental manipulation (e.g., heat shock, drug treatment), and a reference sample is a labeled mixture prepared from the same biological source but without the experimental stimulus. By comparing the two samples on a gene-by-gene basis, the ratios identify genes that are activated or repressed by the experimental stimulus. In this manner, it is possible to explore biology in spectacular detail, discovering new genes and identifying regulatory relationships between genes on a genomic scale that is virtually impossible using any other approach. Comparisons based on ratio calculations have become a gold standard in microarray analysis and published reports from hundreds of laboratories demonstrate the usefulness of this approach (see arrayit.com/e-library).

Ratios can also be obtained for a large number of test samples (or different chips), using a universal reference sample as a control in each experiment. A ***universal reference sample*** is a mixture of molecules (e.g., mRNA) used as a common control sample to generate ratio measurements for many different test samples. The use of a single, common source as a control sample allows comparisons of

Universal reference sample. *Specialized reference sample used to generate ratios for a complete set of test samples.*

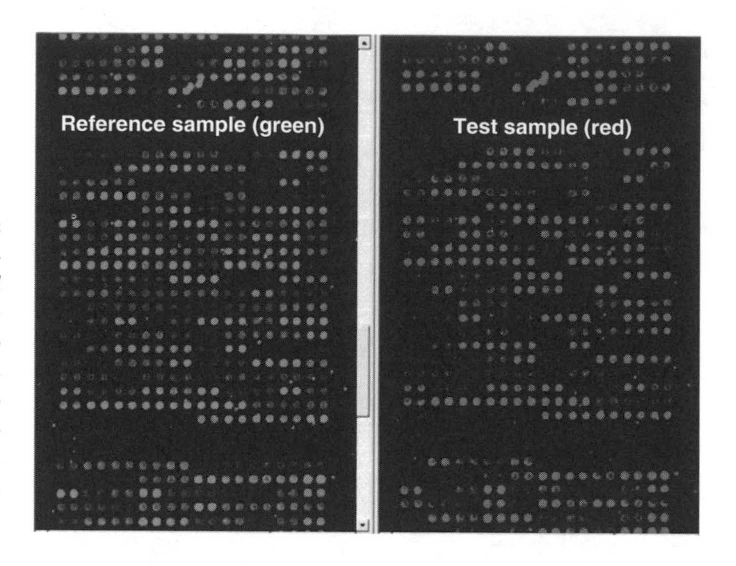

FIGURE 9.55. Scanned images of a yeast genome microarray co-hybridized with a test sample (right; red channel) and a reference sample (left; green channel), correspond to cyanine 5 and cyanine 3 emission channels, respectively, with the fluorescent signals indexed to red and green color palettes to distinguish the data. (Data courtesy of Dr. M. McGovern, Virtek Vision International, Waterloo Ontario.)

many different ratio calculations, and the subsequent assembly of databases of information obtained from myriad experiments.

Ratio experiments often make use of multicolor-color labeling and detection strategies described first by Pinkel and colleagues at the University of California at San Francisco (San Francisco, CA), wherein each sample is labeled separately with a fluor that absorbs and emits light at a different wavelength, allowing data to be collected uniquely for each sample (Weier et al., 1991). Scanners and imagers often index data from separate channels using a red and green color palette, which allows the researcher to visualize signals from test and reference samples (Fig. 9.55). Calculating the ratios of signals from the red and green channels on a yeast whole genome microarray provides a quantitative measure of the abundance of each transcript in the two mixtures across the entire yeast genome (Fig. 9.56). Red and green channels can also be displayed as a single

FIGURE 9.56. A composite image of two microarray images are superimposed into a single image to allow visualization of a test and reference sample simultaneously. (Data courtesy of Dr. M. McGovern, Virtek Vision International, Waterloo Ontario.)

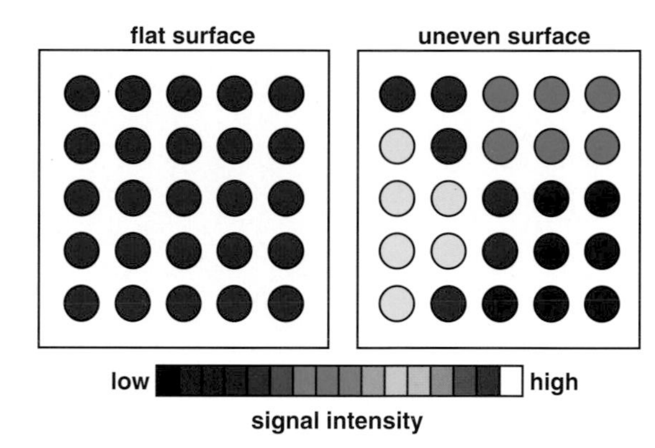

FIGURE 5.2. Because microarray detectors have a limited depth of focus, a hypothetical microarray of identical samples formed on a flat surface will produce a uniform image, whereas the same microarray formed on an uneven surface will produce an image that has variations in signal intensity.

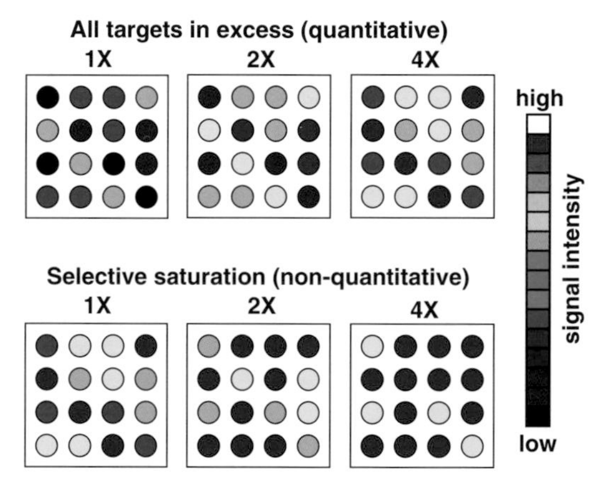

FIGURE 5.11. ***Top,*** *When all the targets are present in excess of probe, successive twofold increases in probe concentration produce successive twofold increases in signal intensity at each microarray location.* ***Bottom,*** *Under conditions of probe excess, there is a selective saturation of targets corresponding to abundant species, so that successive twofold increases in probe concentration produce twofold increases in signal intensity at only some microarray locations.*

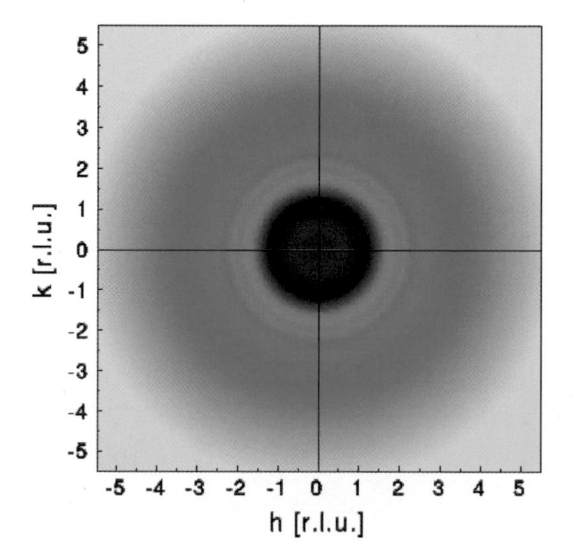

FIGURE 5.13. An x-ray diffraction pattern of a silicon atom, one of the two main atomic components of glass. Electron densities are represented in a gray scale. (Data courtesy of T. Proffen and R. B. Neder, University of Wurzburg, Germany.)

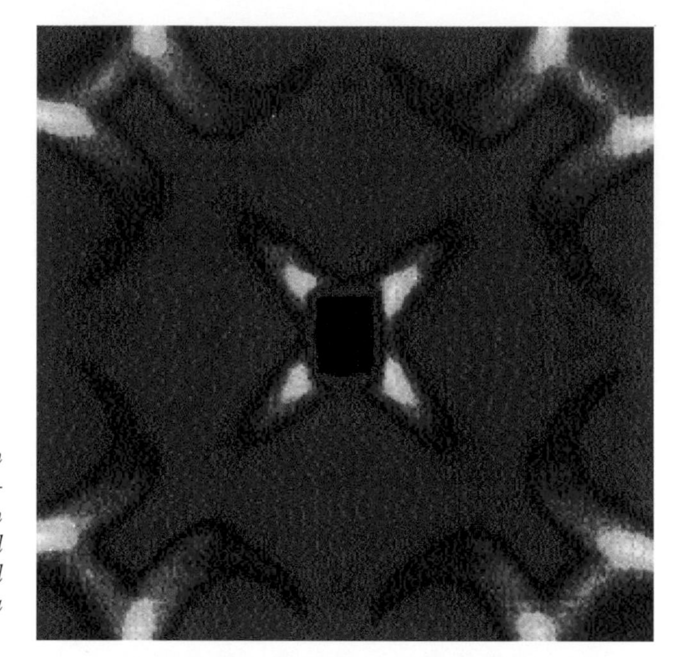

FIGURE 5.14. *SANS of silicon dioxide (SiO_2), the principle component of glass. A single crystal of silicon was heated for 500 h at 600°C, and trace oxygen (~ 30 ppm) diffused to form SiO_2 precipitates. (Data courtesy of ILL, France.)*

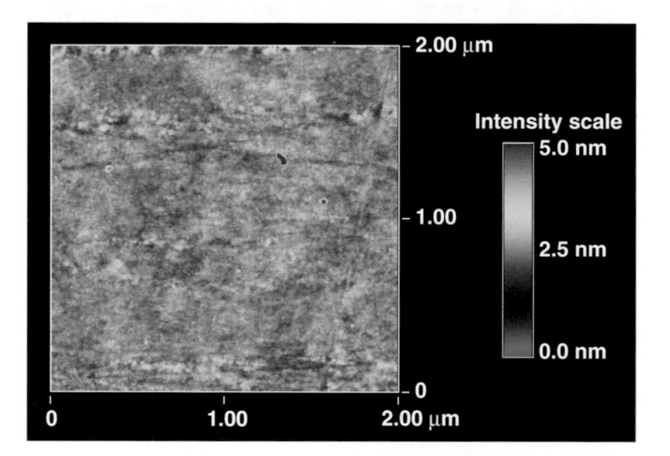

FIGURE 5.15. *An optically flat glass substrate was examined by AFM over an area of 4 μm^2. The roughness image is displayed in a rainbow intensity scale, and the mean flatness is ~2.1 nm. (Data courtesy of TeleChem/ArrayIt.com, Sunnyvale, CA.)*

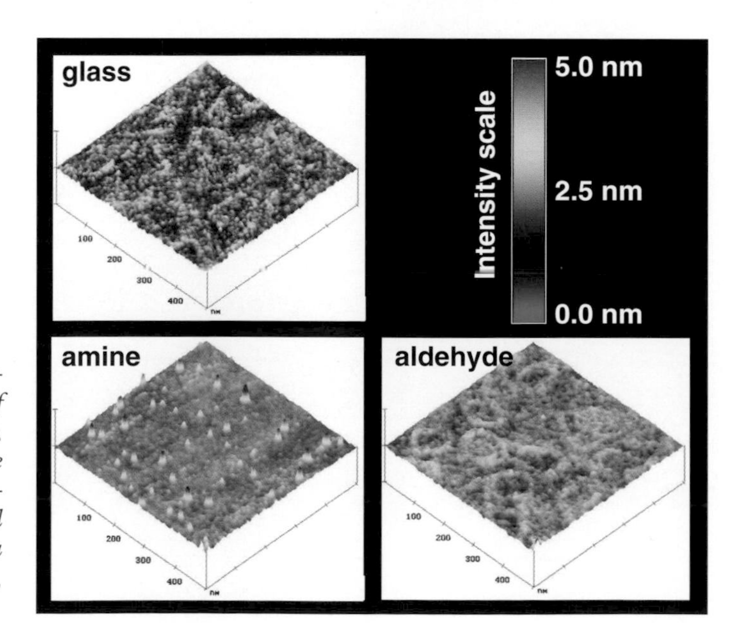

FIGURE 5.18. *AFM was used to examine three surfaces over an area of 0.25 μm^2. Data for optically flat glass, optically flat glass with reactive amine groups, and optically flat glass with reactive aldehyde groups are represented in a rainbow intensity scale. (Data courtesy of TeleChem/ArrayIt.com, Sunnyvale, CA.)*

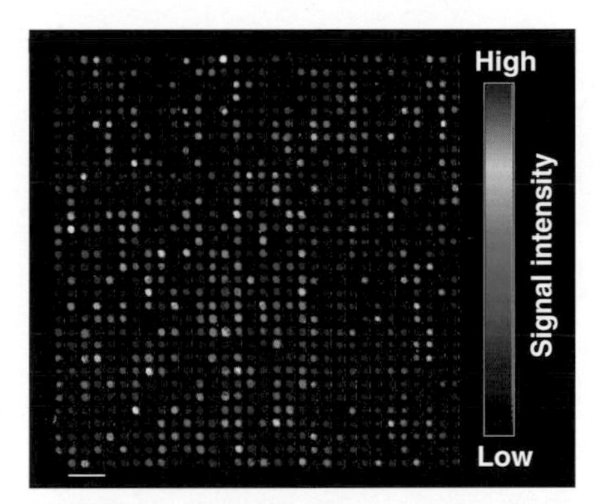

FIGURE 5.21. *Target cDNAs corresponding to 1024 human genes were printed on a reactive amine surface and hybridized to fluorescent probes prepared from human liver RNA. The scanned data are represented in a rainbow color palette. Space bar = 500μm. (Data courtesy of Y. Li, Fudan University, Shanghai, People's Republic of China.)*

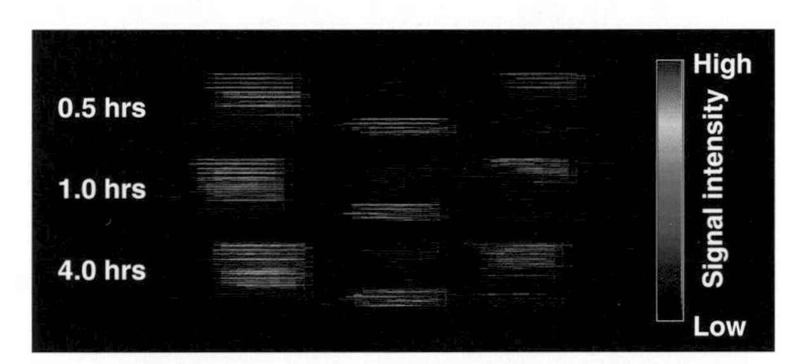

FIGURE 5.25. *Oligonucleotides (15mers) containing amino linkers were attached to an aldehyde surface and hybridized for different lengths of time (0.5–4 h) with a mixture of fluorescent oligonucleotides (15mers) at a concentration of 2 μM for each microarray. After hybridization, the chips were washed to remove unbound probe and scanned for Cy3 emission. Fluorescence intensities are represented in a rainbow scale. Because the hybridization was carried out under conditions of target excess, signal intensities increased as a function of time.*

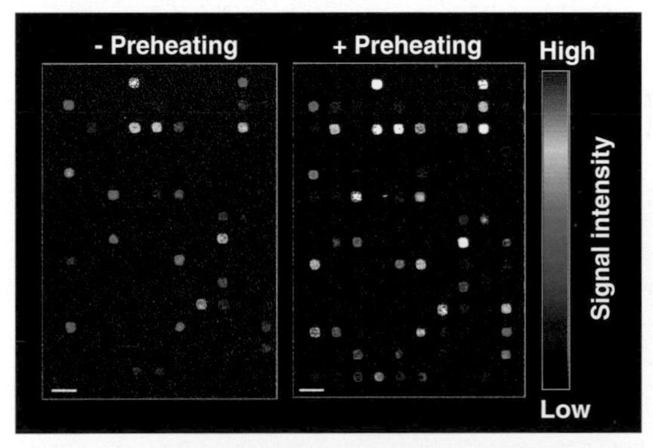

FIGURE 5.26. *Two microarrays were hybridized with fluorescent probes prepared from 3 μg of total* Arabidopsis *RNA and scanned for Cy3 emission. Image data are displayed in rainbow palette. Hybridizations were initiated by adding room-temperature probe to a room-temperature aldehyde microarray (−Preheating) or a 55°C probe to a 55°C aldehyde microarray (+Preheating). Space bar = 250 μm. The materials used include a Submicro Expression Array detection kit (Genisphere, Montvale, NJ), GlassHyb buffer (Clontech, Palo Alto, CA), and SuperAldehyde substrates (TeleChem/ArrayIt.com, Sunnyvale, CA). (Data courtesy of S. Ruuska, Michigan State University, East Lansing, MI.)*

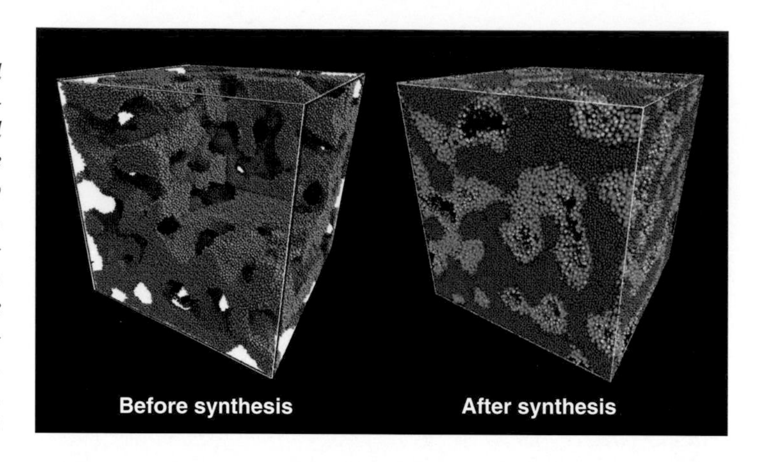

FIGURE 6.12. *Computer-generated models of CPG simulating the oligonucleotide synthesis matrix before and after oligonucleotide synthesis. The first DNA base is bound directly to silicon dioxide atoms* (gray matrix). *During the synthesis process, phosphoramidite reagents* (white spheres) *enter the porous CPG and become incorporated into the growing oligonucleotide chains.(Courtesy of Dr. L. Gelb, Department of Chemistry, Florida State University, Tallahassee.)*

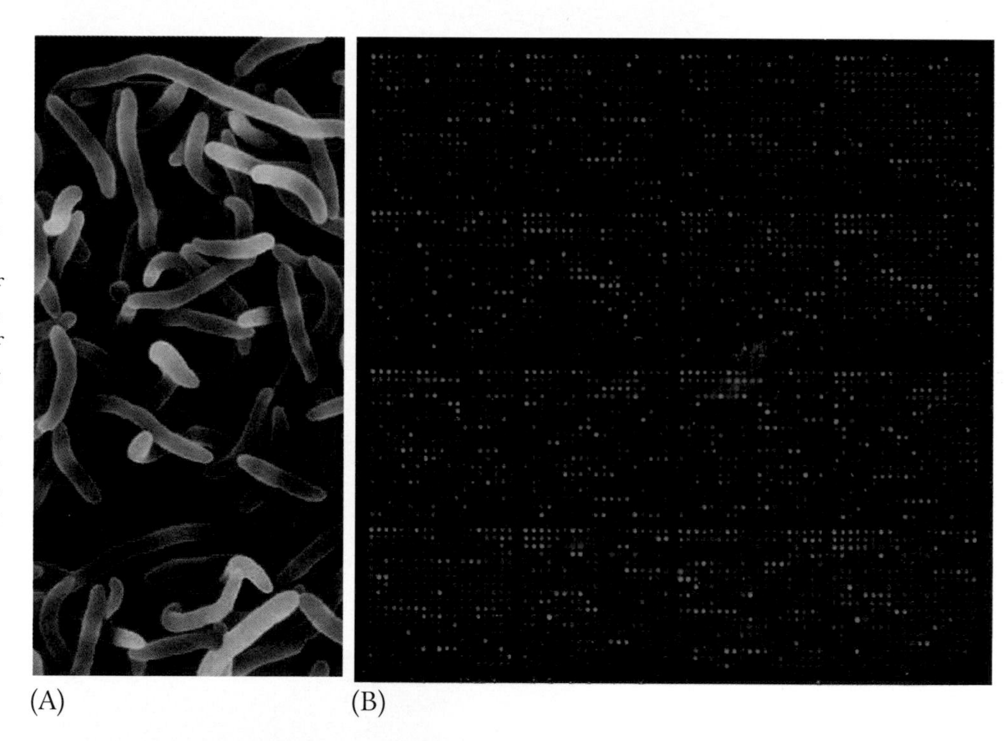

FIGURE 6.21. **A,** *Photomicrograph of the bacterium* V. cholera. *(Courtesy of Dennis Kunkel Microscopy, Kailua, HI.)* **B,** *Fluorescent image of a microarray containing the entire genome of* V. cholera. *(Courtesy of K. Chong, Schoolnik Laboratory at Stanford University, Stanford, CA.) Total RNA from* V. cholera *was labeled with reverse transcriptase in the presence of Alexa546-dUTP, and the direct-labeled probe was hybridized to the microarray and scanned with a ScanArray 5000 (Packard Biochip Technologies, Billerica, MA). The fluorescent TIF data are coded to a monochromatic green look-up table. (Photomicrograph was provided)*

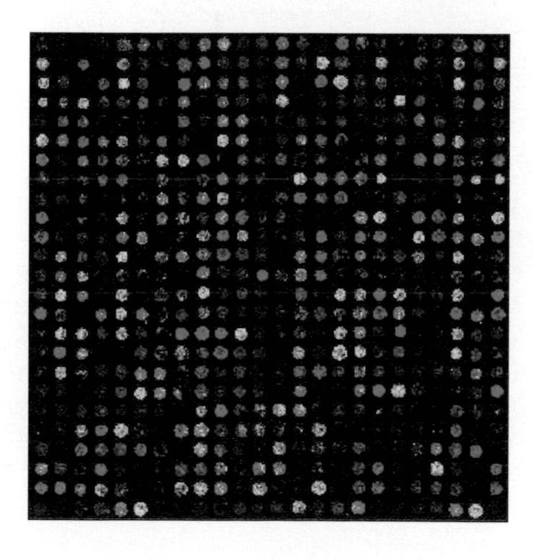

FIGURE 6.24. *Human total RNA (4 μg) was labeled and hybridized to a microarray containing full-length human cDNAs prepared using Full-Length Expressed Gene (FLEX) technology (AlphaGene, Woburn, MA). The hybridized microarray was then stained with TSA reagents from a MICROMAX system (NEN Life Science Products, Boston), and fluorescent gene expression signals were displayed in a rainbow palette.*

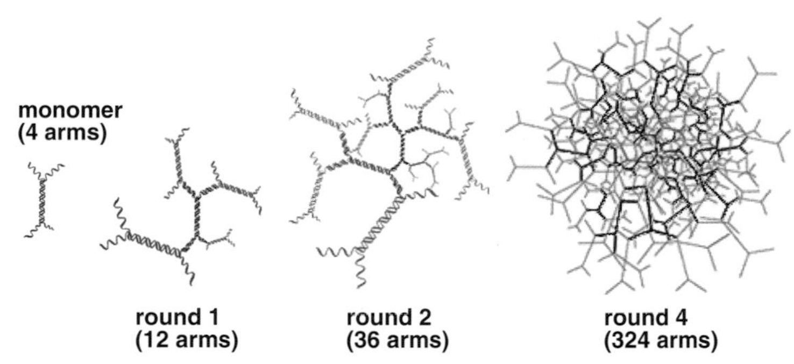

FIGURE 6.25. *For the dendrimer synthesis process, two oligonucleotides are annealed to produce a molecule (monomer) that has a double-stranded central portion and four single-stranded arms. Monomers are then annealed and cross-linked in a stepwise manner; four successive rounds of annealing and cross-linking produce dendrimers with 324 arms. If the synthesis process is performed with fluorescent oligonucleotides, each dendrimer contains as many as 324 fluorescent dye molecules. (Courtesy of Dr. R. C. Getts, Genisphere, Montvale, NJ.)*

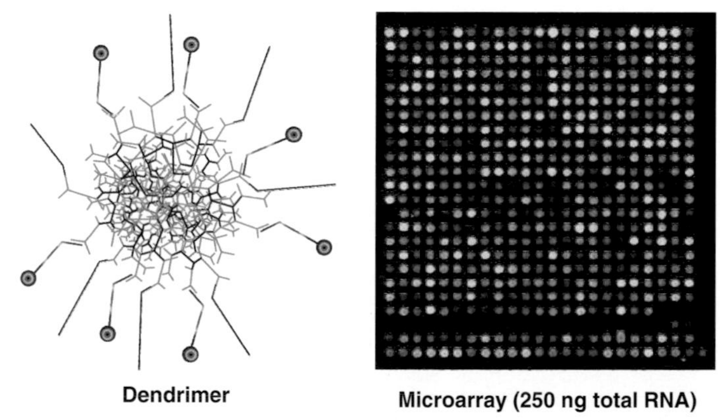

FIGURE 6.26. *A dendrimer reagent was used to label a microarray indirectly by hybridizing the single-stranded capture sequences* (black lines) *on the dendrimers to a microarray hybridized with a probe mixture prepared from 250 ng mouse total RNA. The intense fluorescent signals are due to passive signal enhancement by the dye molecules* (circles) *attached to the dendrimers. (Dendrimer courtesy of Dr. R. C. Getts, Genisphere, Montvale, NJ; microarray data courtesy of University of California at San Francisco Cancer Center.)*

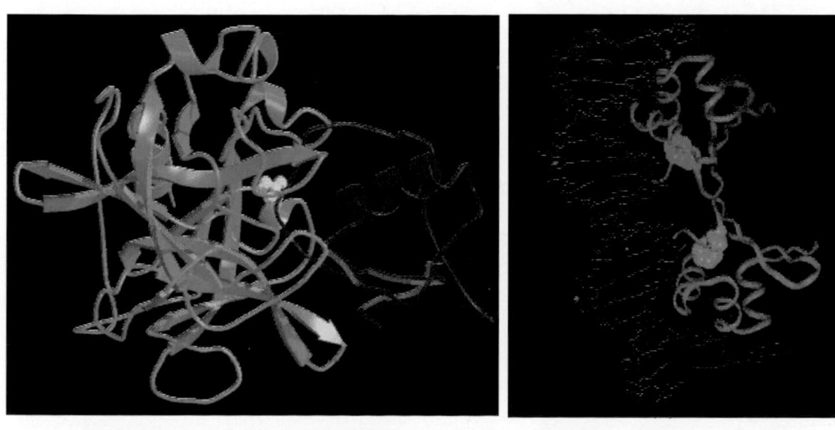

FIGURE 6.31. *Crystal structures of a protein–protein binding reaction between the elastase enzyme* (light blue) *and the ovomucoid substrate* (red), *and a protein–DNA binding reaction between the Cro-repressor protein* (green) *and double-stranded DNA* (blue and magenta). *(Protein-protein complex courtesy of Dr. K. P. Murphy, Department of Biochemistry, University of Iowa at Iowa City; protein–DNA complex courtesy of Dr. M. C. Mossing, University of Notre Dame, Notre Dame, IN.)*

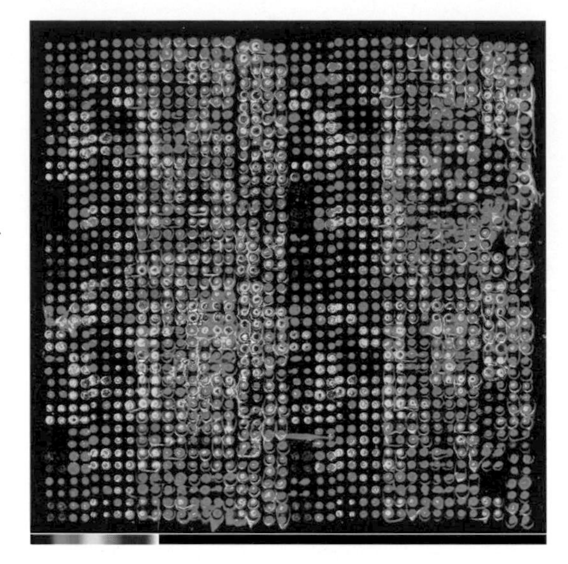

FIGURE 8.18. *Fluorescent image of a cDNA microarray printed with a prototype ink-jet dispensor and stained with the OliGreen dye. Detection was made with a confocal scanning device using an Argon laser for excitation; data are represented in a rainbow palette. (Courtesy of M. Schena and R. W. Davis, Stanford University, Stanford, CA, and Combion, Pasadena, CA.)*

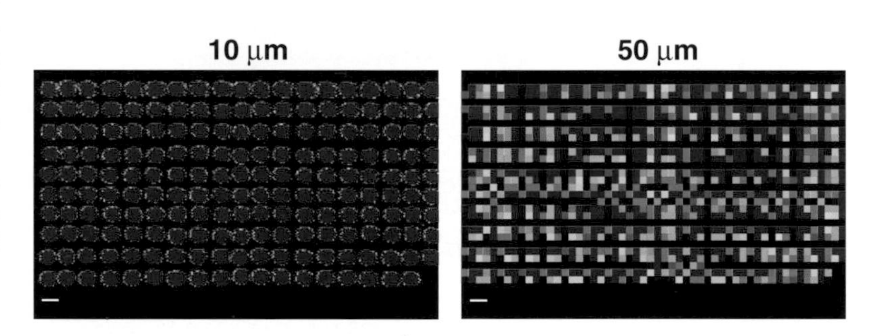

FIGURE 8.36. *Two microarray images obtained from the same fluorescent microarray, scanned at a pixel size of 10 μm and 50 μm. The TIFF data were obtained using a ScanArray 3000 (Packard Biochip, Billerica, MA) and are presented in a rainbow palette from* NIH Image 1.62 *software developed by Rosband (National Institutes of Health, Bethesda, MD). Space bars* = 150 *μm.*

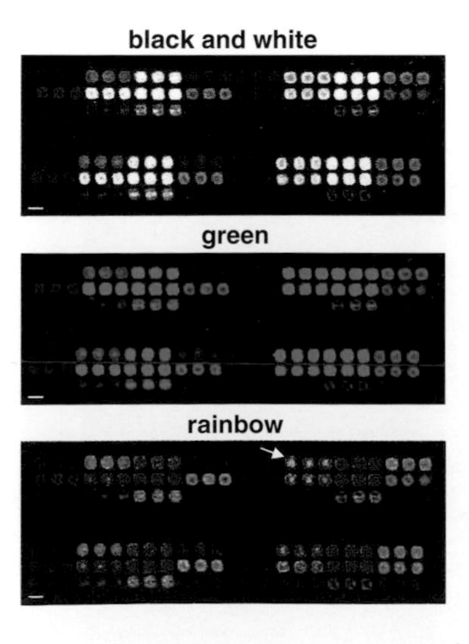

FIGURE 8.37. *A single microarray image with the values from the TIFF file used to index a LUT composed of a black and white, green, or rainbow scale. The brightest fluorescent signals are represented as white, bright green, and red in the three respective palettes. Space bars* = 150 *μm.* Arrow, *the portion of the image shown in Figure 8.38. (Data generated with a ScanArray 3000 instrument set to the Cy3 channel.)*

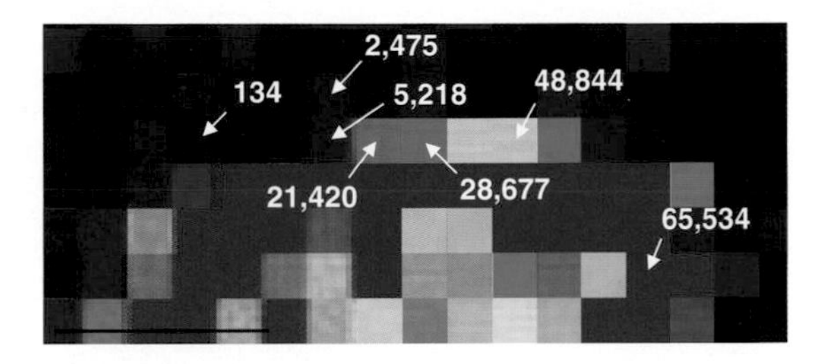

FIGURE 8.38. *An enlarged image of a microarray (from Fig. 8.37) with intensity values given for seven different pixels. The custom rainbow palette was indexed with a 16-bit TIFF file. Space bar = 50 μm.*

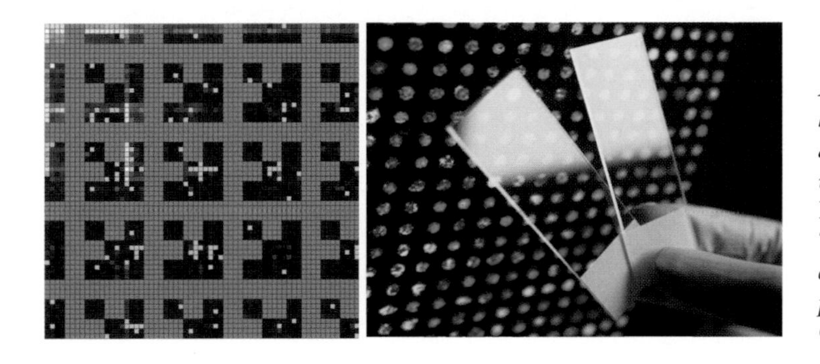

FIGURE 8.39. A, *Oligonucleotide hybridization data displayed via a custom color palette. (Reprinted with permission from* Nucleic Acids Research.*)* ***B,*** *Corning Microarray Technology slides and gene expression data displayed in a custom color palette. (Data courtesy of Corning, Corning NY.)*

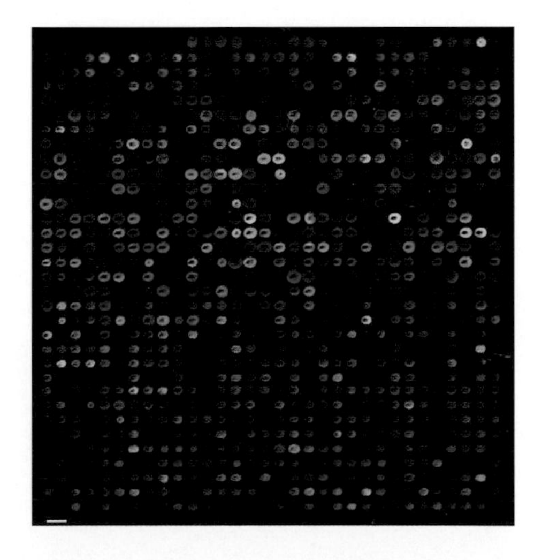

FIGURE 8.40. *A green–red overlay is created by subtracting image values obtained from separate channels of a microarray scanner and superimposing the images to create a composite. The samples were prepared by labeling mRNA from human cells grown at 37 and 42°C, respectively. (Courtesy of Dr. M. Schena and Dr. R. W. Davis, Stanford University, Stanford, CA, and by Dr. D. Shalon, Synteni, Palo Alto, CA.)*

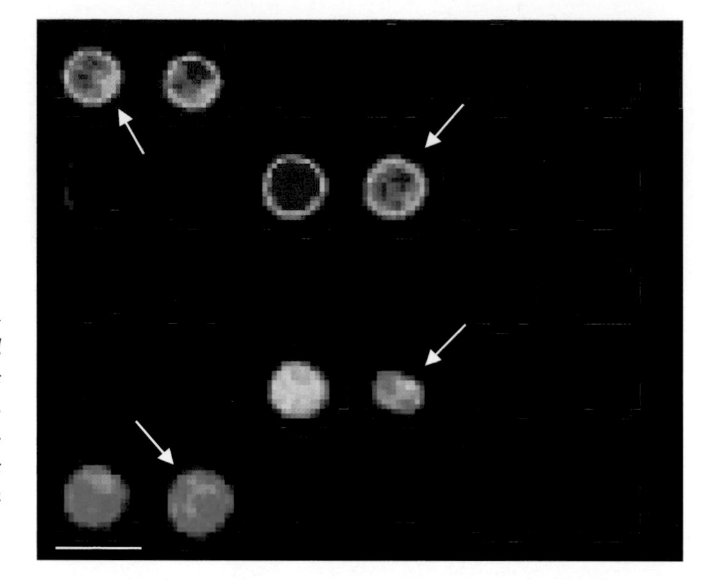

FIGURE 8.48. *Fluorescent microarray spots captured with a CCD-based camera, with high signal intensities manifesting a blooming phenomenon* (arrows). *Fully charged pixels transfer excess charge to adjacent pixels that are empty or filled partially. Data are represented in a rainbow palette. Space bar = 150* μm.

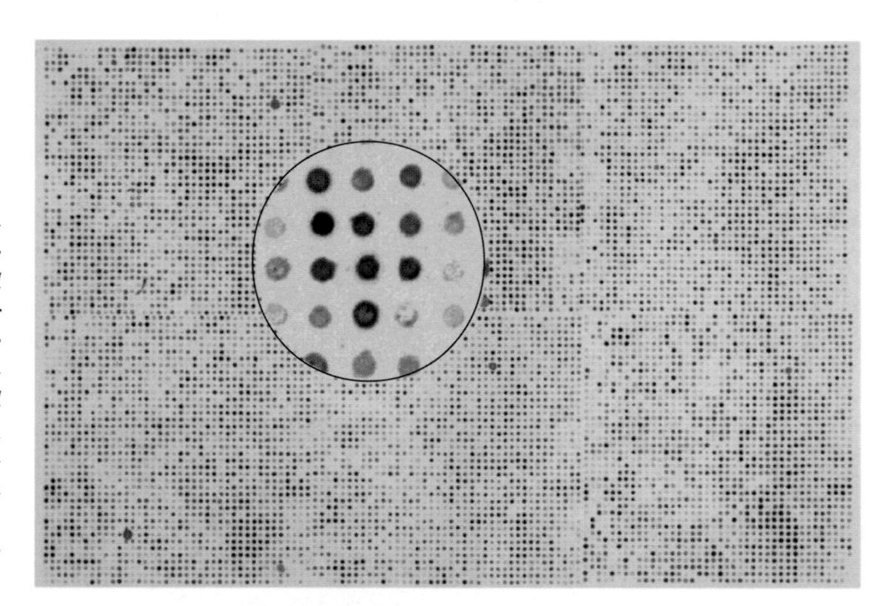

FIGURE 8.50. *Human oral carcinoma cells were treated with the chemotherapy drug camptothecin and compared to untreated cells. Two-color data are represented in a composite image so that dark and light circles correspond to genes that are activated and repressed, respectively, by the drug. A portion of the array* (circle) *is enlarged for ease of viewing and the data were obtained by Dr. Konan Peck. (Reprinted with permission from Academic Press, San Diego, CA.)*

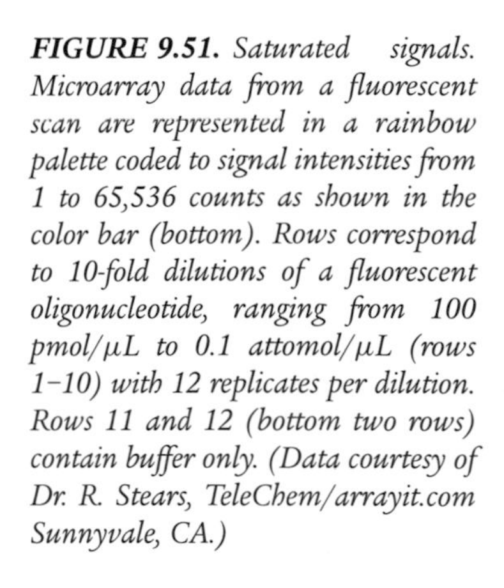

FIGURE 9.51. *Saturated signals. Microarray data from a fluorescent scan are represented in a rainbow palette coded to signal intensities from 1 to 65,536 counts as shown in the color bar (bottom). Rows correspond to 10-fold dilutions of a fluorescent oligonucleotide, ranging from 100 pmol/*μL *to 0.1 attomol/*μL *(rows 1–10) with 12 replicates per dilution. Rows 11 and 12 (bottom two rows) contain buffer only. (Data courtesy of Dr. R. Stears, TeleChem/arrayit.com Sunnyvale, CA.)*

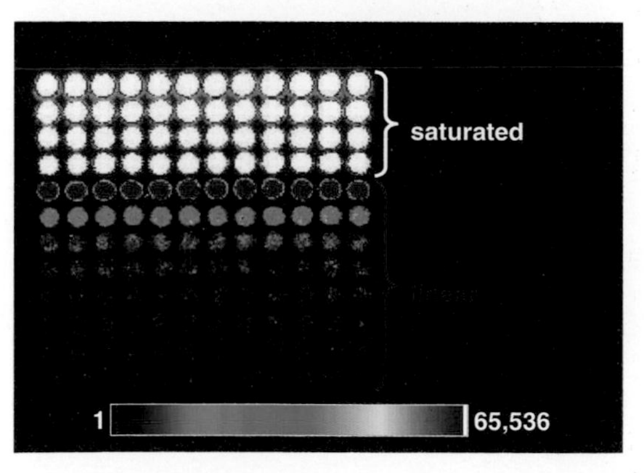

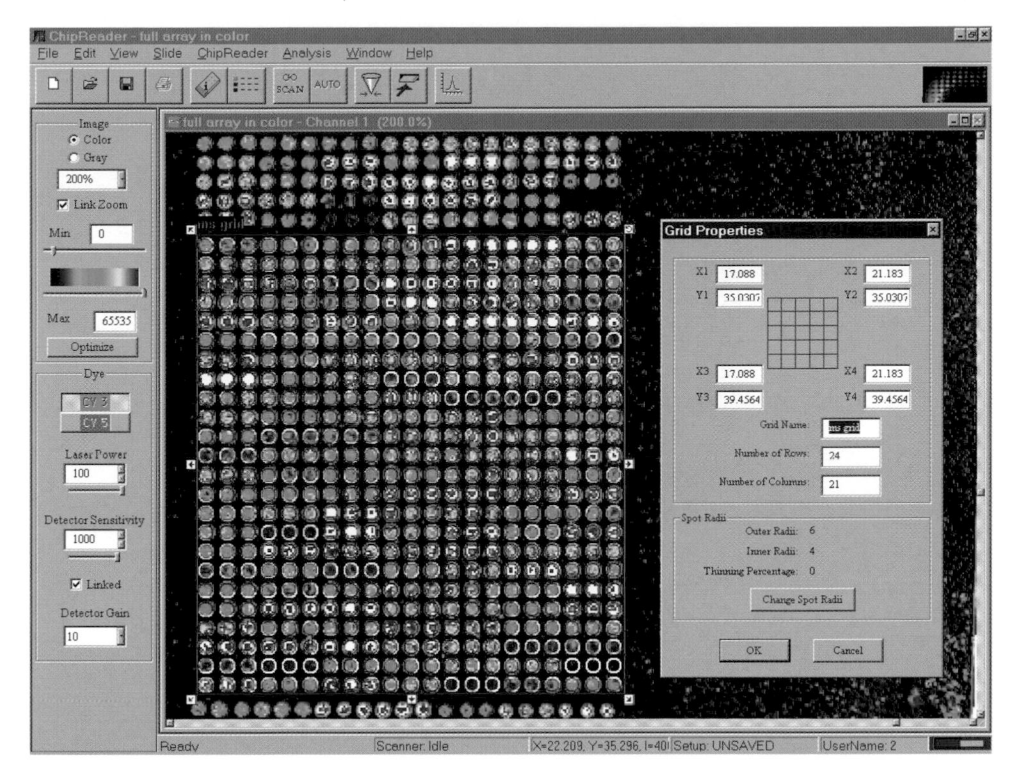

FIGURE 9.52. *With* ChipReader, *automated quantitation is performed by superimposing a quantitation template* (square) *over a microarray image. The user specifies the number of rows, columns, and other quantitation parameters and positions the template precisely over the image. The computer calculates the signal intensity for each spot located inside the template.*

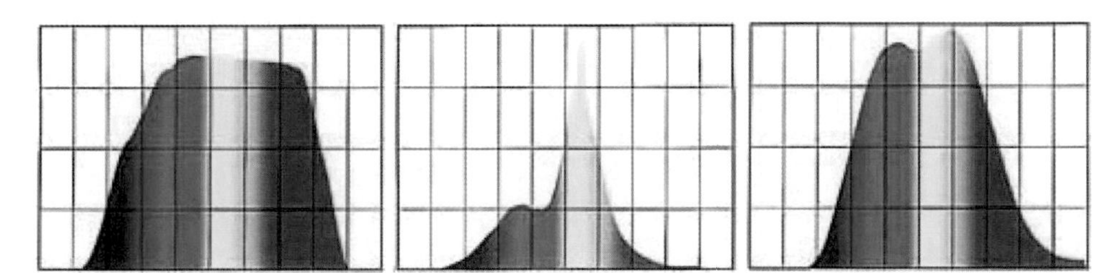

FIGURE 11.2. *Emission spectra for natural sunlight (left), cool white fluorescent lights (center), and full-spectrum fluorescent lights (right). The intensity* (I) *of each source across the visible spectrum of wavelengths is depicted in a rainbow palette according to actual color of each portion of the spectrum. (Data adapted courtesy of Litelids.com, Euclid, OH.)*

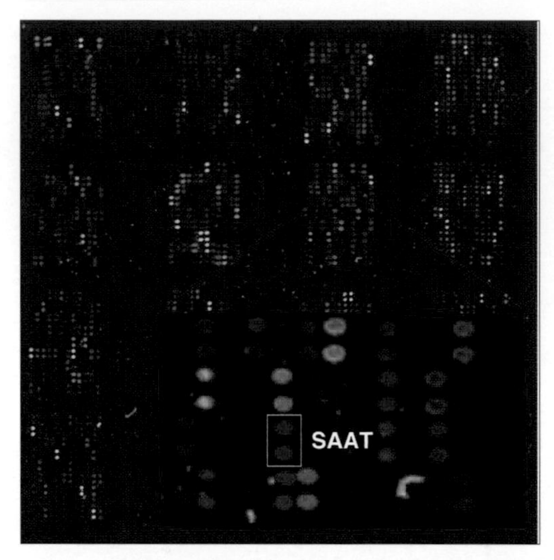

FIGURE 12.7. *To examine gene expression levels in ripe and immature strawberry fruits, total mRNA was isolated, labeled separately, mixed, hybridized to a single microarray, and scanned for cyanine-3 (Cy3) and cyanine-5 (Cy5) emission. The composite fluorescent image is a two-color overlay, with fluorescent signals from ripe fruits (red channel) and immature fruits (green channel). One portion of the microarray (red box) contains targets corresponding to SAAT (white box), a gene whose expression is induced in ripe fruits. (Data Reprinted with permission from the American Society of Plant Physiologists.)*

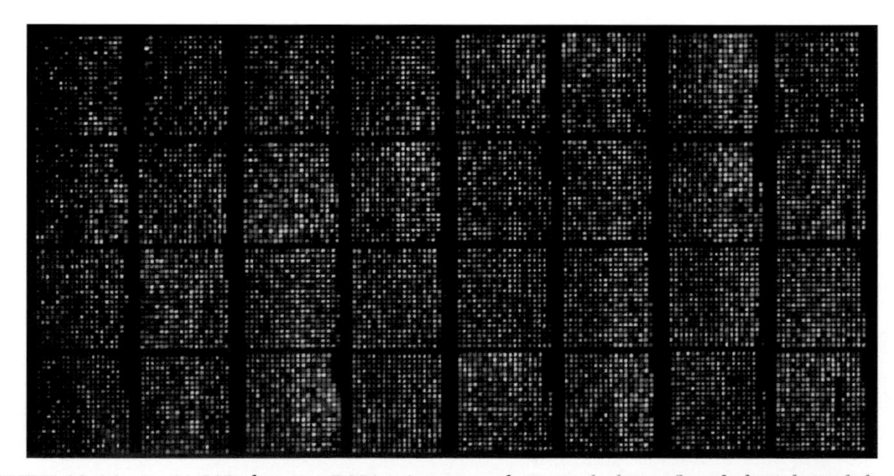

FIGURE 12.14. *An 11,500-element cDNA microarray that reveals the profound physiological changes that occur when plants are exposed to light. Flourescent probes were prepared from RNA isolated from 7-day-old dark-grown (etiolated) seedlings (Cy3) or from adult leaves (Cy5), mixed together, hybridized to a single microarray, and scanned for emission of Cy3 and Cy5. Red and green spots in the two-color overlay correspond to genes that are light-induced or light-repressed, respectively. (Data courtesy of Dr. S.-H. Wu and Dr. S. Somerville, AFGC, Carnegie Institute of Washington, Stanford, CA.)*

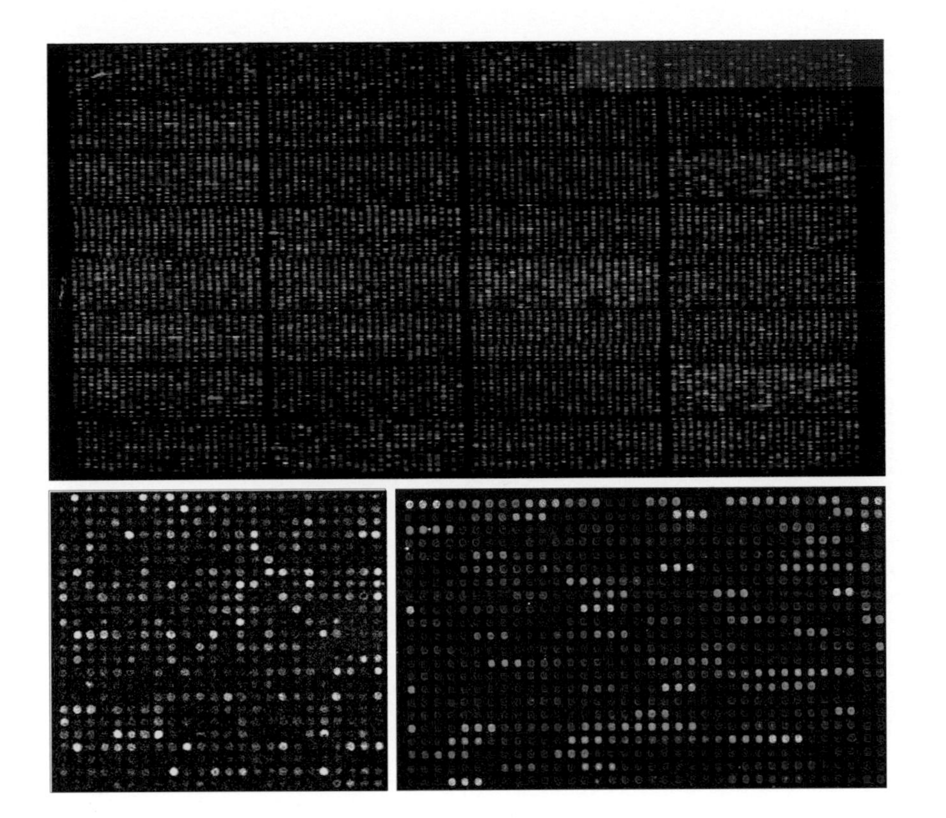

FIGURE 12.15. **A**, *Arabidopsis microarray containing 18,432 elements with one scanned channel shown in a rainbow palette.* **B**, *A two-color representation of Arabidopsis genes whose expression changes up or down upon salt treatment.* **C**, *A rice microarray showing salt-induced and salt-repressed genes with one scanned channel shown in a rainbow palette. (Data courtesy of Dr. Deyholos, Dr. Kawasaki, and Dr. Galbraith, University of Arizona, Tucson).*

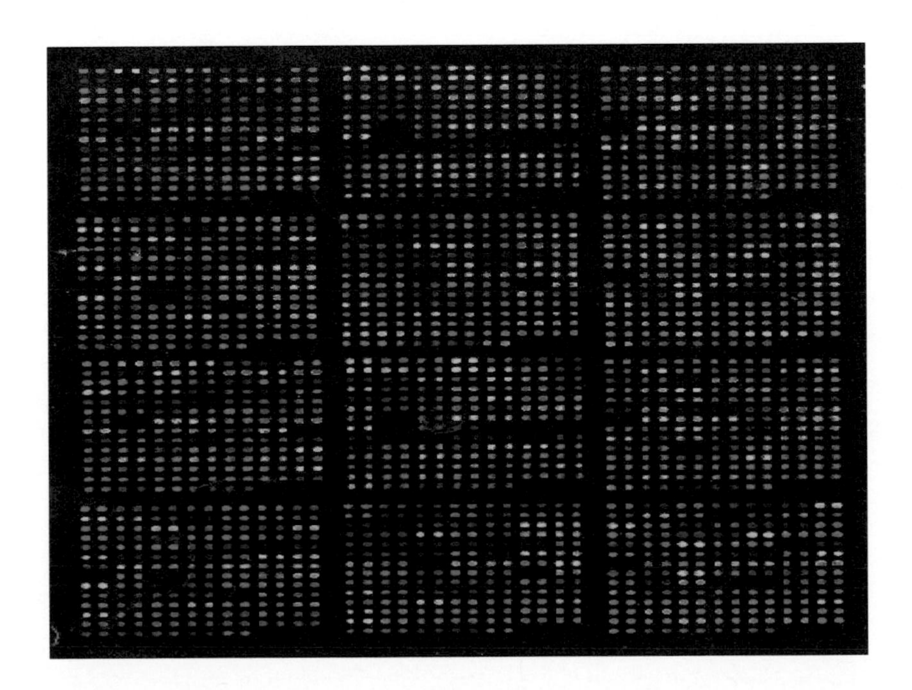

FIGURE 12.16. *One channel of a toxicology experiment comparing a canine treated with a test compound versus an untreated canine. The cDNAs were printed in duplicate, and one channel of the scanned fluorescent image is shown in a rainbow scale. (Data courtesy of Dr. B. Whitford, Phase 1 Molecular Toxicology, Santa Fe, NM.)*

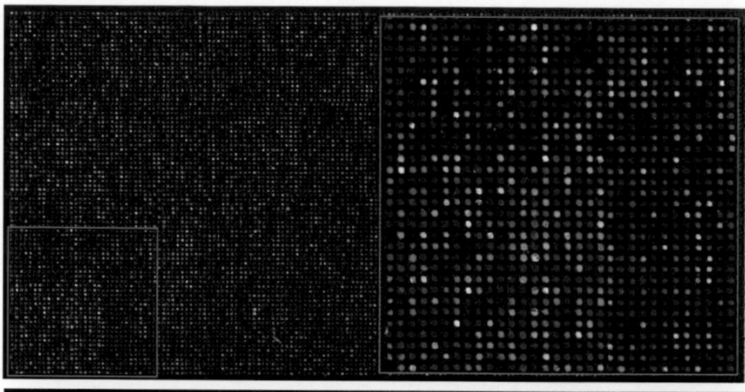

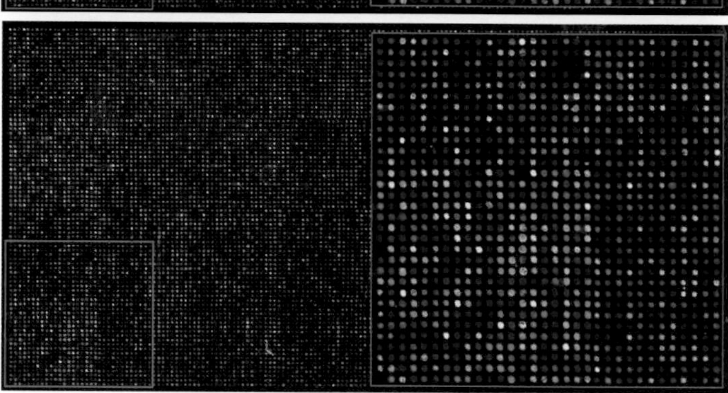

FIGURE 12.17. *Scanned images of microarrays used to study liver cancer in human patients. A total of 12,800 genes were printed as cDNA elements and hybridized with fluorescent probes prepared from control liver samples (upper panel) or from liver samples taken from patients diagnosed postmortem with hepatocellular carcinoma (lower panel). (Data courtesy of Y. Li et al., Fudan University, Shanghai, China.)*

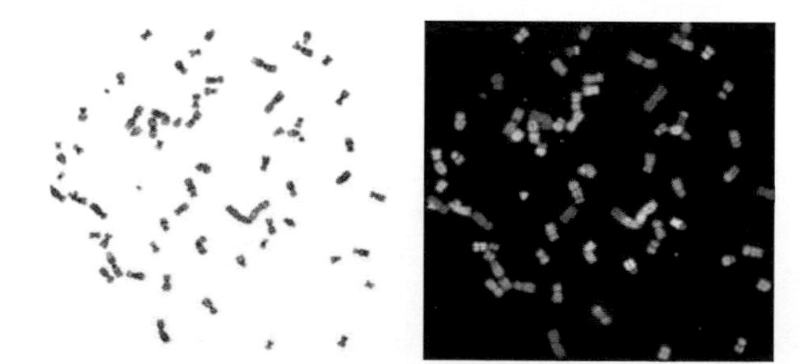

FIGURE 13.2. *Two methods for examining human chromosomes. Traditional Geimsa staining reveals the familiar G-banding pattern, and spectral karyotyping by hybridization with fluorescent SKY probes provides a unique color for each chromosome, allowing translocations to be detected more easily. (Data courtesy of Dr. M. Shuster, Gollin laboratory, University of Pittsburgh Cancer Institute, Pittsburgh, PA.)*

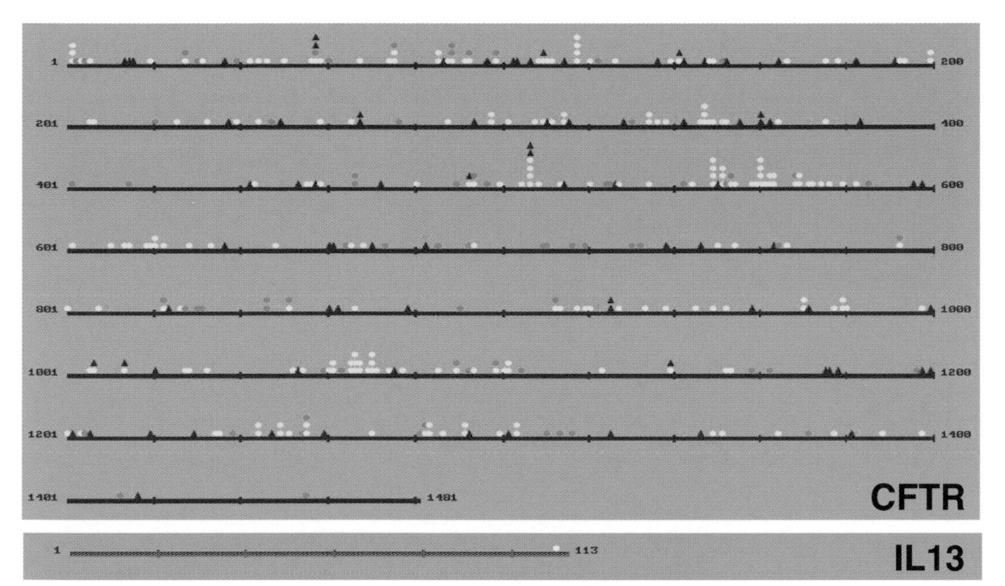

FIGURE 13.3. *The CFTR and* IL13 *mutations known to cause cystic fibrosis and to be associated with asthma, respectively. Missense mutations (*light circles*), nonsense mutations (*gray circles*), insertion mutations (*light triangles*), and deletion mutations (*dark triangles*) are shown. (Data adapted from the Human Gene Mutation Database, University of Wales College of Medicine, Cardiff, Wales, UK, in conjunction with Celera Genomics, Rockville, MD.)*

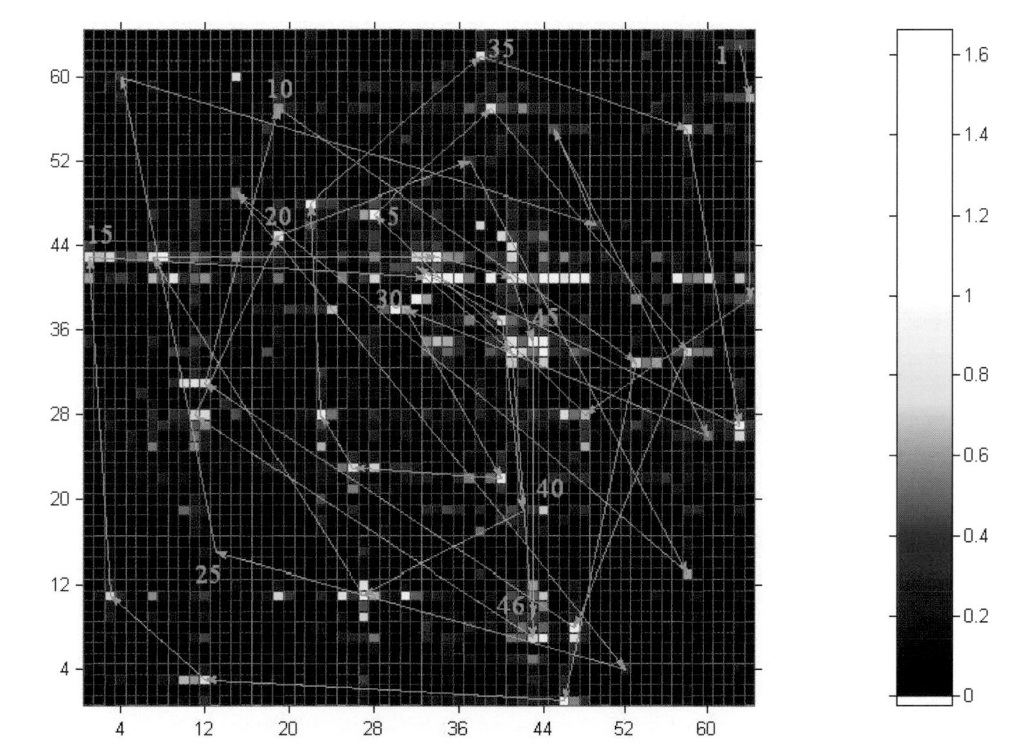

```
GAAAAGCGAGTCAGTTTGGGTCGATATTAGACGCCCGAACGGCTGGGGTGT
1------    11------    21------    31------    41------
 2------    12------    22------    32------    42------
  3------    13------    23------    33------    43------
   4------    14------    24------    34------    44------
    5------    15------    25------    35------    45------
     6------    16------    26------    36------    46------
      7------    17------    27------    37------
       8------    18------    28------    38------
        9------    19------    29------    39------
         10------    20------    30------    40------
```

FIGURE 13.7. **A,** *Fluorescent image of the hybridized 6mer (hexamer) array, with the intensity values indexed to a fire look-up table (right). Interconnected arrows (light green) indicate the array locations that are perfectly complementary to the fluorescent 51mer.* **B,** *Sequence of the 51mer and the 46 perfectly complementary hexamers that span the polymer. (Reprinted with permission from* Journal of Biomolecular Structure and Dynamics.*)*

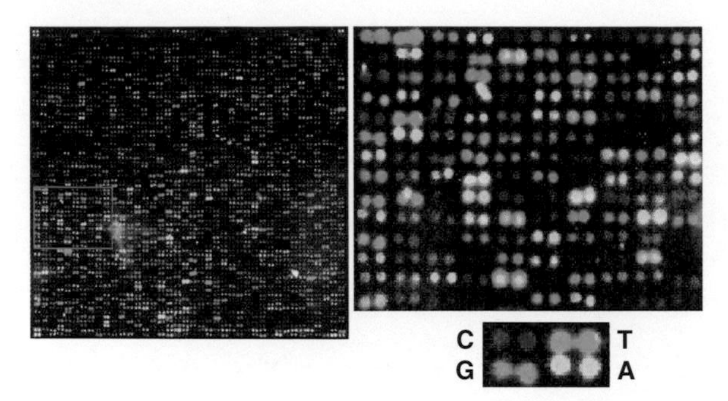

*FIGURE 13.10. Microarray image from APEX showing 1334 SNP markers from human chromosome 22 (*left*), with a small portion of the array enlarged (*right*). Genotypes of both DNA strands are determined (*bottom*). (Data courtesy of Dr. A. Kurg and A. Metspalu, Institute of Molecular and Cell Biology, University of Tartu, Tartu, Estonia.)*

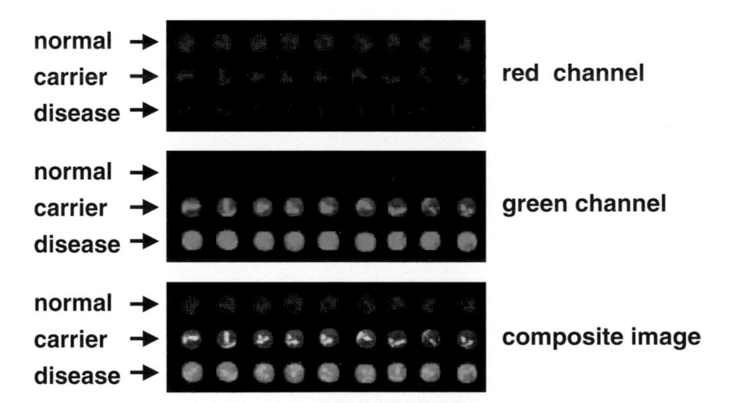

*FIGURE 13.14. A two-color NGS™ method was used to screen for sickle cell anemia. A Cyanine-3 labeled synthetic oligonucleotide probe (*red*) perfectly complementary to a wild typebeta-globin allele (*normal*), and a Cyanine-5 labeled synthetic oligonucleotide probe (*green*) perfectly complementary to a mutant beta-globin allele (*disease*) were hybridized to a single microarray containing neonatal amplicons; the chip was scanned in the red and green channels. A composite image (*bottom*) reveals the ease by which the normal, carrier, and disease genotypes are distinguished by color. (Data courtesy of NGS-ArrayIt, Inc., Sunnyvale, CA.)*

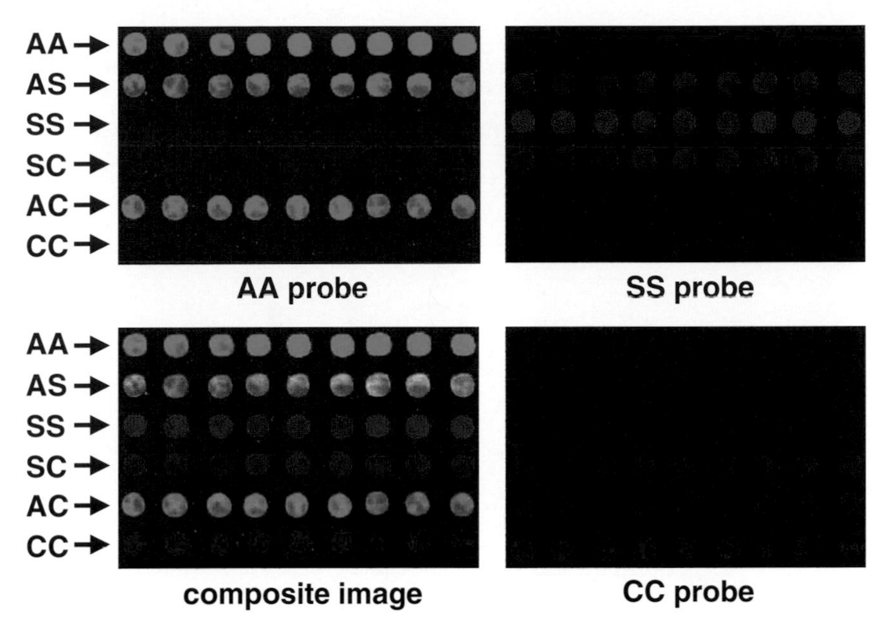

FIGURE 13.15. *The three-color. The NGS™ method was used to screen for six sickle cell anemia genotypes. Three oligonucleotide probes perfectly complementary to three genotypes (AA, SS and CC) were co-hybridized and scanned to reveal hybridization signals for the AA, SS, and CC probes. A composite image reveals the ease by which the six genotypes are distinguished by color. (Data courtesy of NGS-ArrayIt, Inc., Sunnyvale, CA.)*

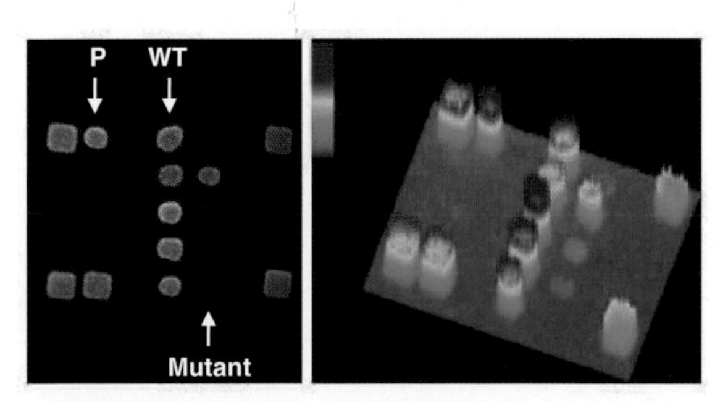

FIGURE 14.12. **A,** *DNA probes labeled with RLS particles produce much stronger signals when hybridized to wild type sequences (WT) and positive controls (P) than to mutant sequences (mutant).* **B,** *Microarray data are represented in a three-dimensional graphic with signal intensities indexed to a rainbow palette. (Data courtesy of Dr. J. P. Linton, Genicon Sciences, San Diego, CA.)*

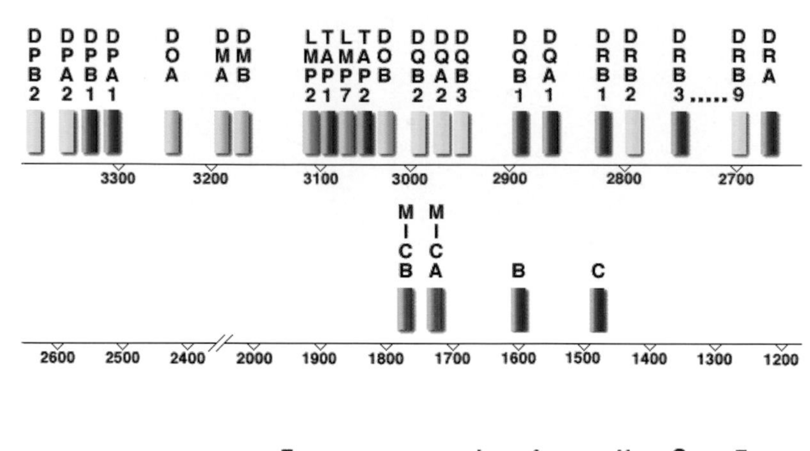

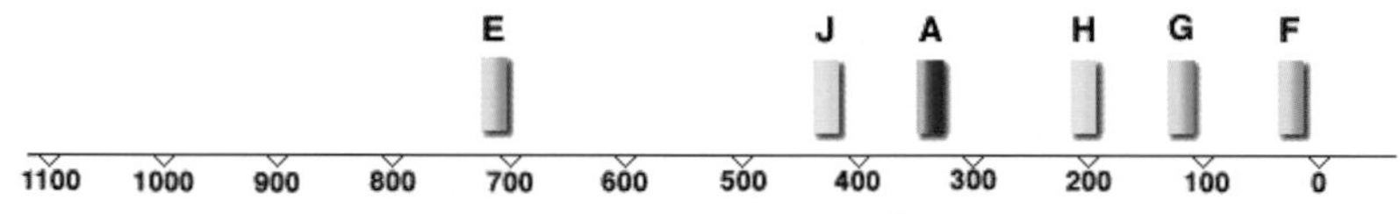

FIGURE 15.5. *Some of the HLA alleles* (shaded rectangles), *along with their names. (Courtesy of the Anthony Nolan Research Institute, London.)*

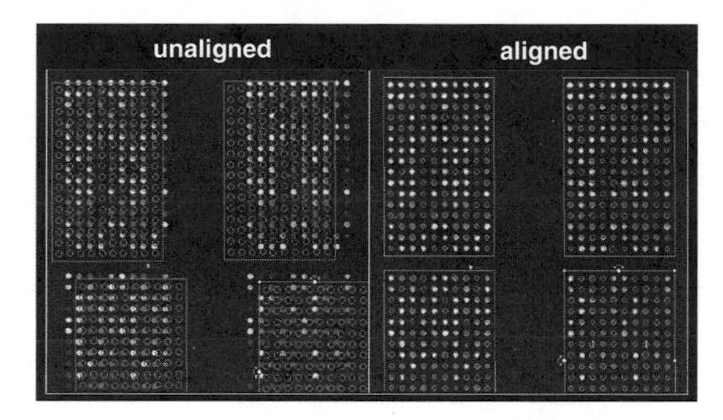

FIGURE 9.57. Ratio calculations are facilitated by software such as GenePix Pro *3.0, which allows automated placement of a predefined quantitation template (white circles) over a composite (red–green) microarray image. (Data courtesy of Dr. D. Wellis, Dr. S. Elletson, and colleagues, Axon Instruments, Union City, CA.)*

image known as a two-color overlay, in which activated and repressed genes appear as red and green spots against a background of yellow spots, which correspond to transcripts present at equal concentrations in the two samples. Most commercial scanners and imagers provide software that enables the representation of data in red, green, and two-color overlay formats.

A more general type of ratio calculation is used for genotyping applications involving microarrays of patient amplicons that are hybridized with mixtures of fluorescent oligonucleotides (see Chapter 13). In this approach, it is often convenient to represent the wild-type allele in one channel (e.g., green) and the mutant allele in a second channel (e.g. red), such that the normal, carrier and disease genotypes are distinguished readily (see Chapter 13). Because information is sought for all of the alleles with equal interest, this more generalized ratio experiment involves two different fluorescent oligonucleotide mixtures rather than a test sample and a reference sample. Genotyping assays also make use of three- and four-color strategies and ratio calculations derived from the ratios based on two, three, or four different images (see Chapter 13).

The physical location of microarrays can vary from chip to chip, owing to small differences in the printing origin on each of the different substrates, producing microarray images in which the location of the spots varies relative to the image border. Images generated from different chips often require slight alignment of quantitation templates before extracting the data, a process that can be rather time consuming and tedious if performed manually. Most scanners and imagers and stand-alone software packages provide automatic spot-finding algorithms that allow the alignment of preformed quantitation templates on microarray images that share a common spot configuration (Fig. 9.57). Automatic template alignment compensates for small differences in the location of microarrays on different substrates, speeding ratio calculations by obviating manual template placement.

Normalization

The process by which data from different channels or different chips are equalized before analysis is known as ***normalization,*** and the value (e.g., 1.3) that is used to normalize different datasets is known as a ***normalization factor***. If performed properly, the normalization process does not alter the content of the data but rather corrects for minor imbalances that arise during the imaging process owing to differences in labeling and hybridization efficiency, washes, differential quantum yield of dyes, variations in laser power and detector sensitivity, and many other causes of slight disparities between different chips or

Normalization. *Computational process in which data from different channels or different chips are equalized before analysis.*

Normalization factor. *Value used to equalize data from different channels or chips before data analysis.*

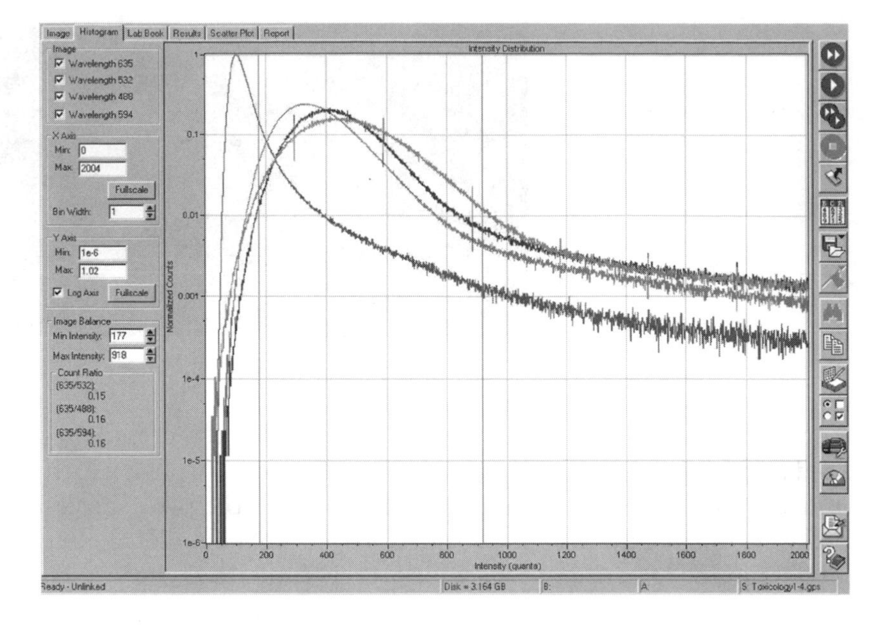

FIGURE 9.58. The GenePix *4000-B microarray scanner depicts a series of histograms* (lines) *corresponding to the pixel intensities acquired from four separate scanner channels (488, 532, 594, and 635 nm). (Courtesy of Dr. D. Wellis, Dr. S. Elletson, and colleagues, Axon Instruments, Union City, CA.)*

channels. Normalization enables precise ratio calculations and other valuable comparisons that would be ill-advised with nonnormalized images.

Balance. *Propensity of two or more channels in a microarray detection instrument to produce images with matched signal intensities.*

Matched intensities. *When two or more microarray images possess similar signal readings across the full gamut of signal values.*

As a prerequisite to normalization, hardware settings on the detection instrument should be adjusted before imaging to produce images that are as closely matched as possible. Checking the ***balance*** of the channels before imaging improves the ease and reliability of normalization. Attempting to normalize microarray images that differ widely in overall signal intensity is difficult and likely to produce many more artifacts than images with ***matched intensities***.

Some scanners, such as the *GenePix* 4000B from Axon Instruments (Union City, CA), allow the user to display the complete spectrum of pixel intensities in the form of histograms, enabling to researcher to balance each channel before or during image acquisition (Fig. 9.58). The user can adjust the gain on the photomultiplier tube dynamically to superimpose the intensity histograms in each channel, thereby achieving scanner settings that produce precisely matched images. Other detection instruments allow the user to control the intensity of the light source, the detector gain, or both to achieve balanced images upstream of the normalization process. Some scanners also provide a set of performance diagnostics that monitor and record changes in the instrument over time, ensuring the reliability of data gathered in different channels over an extended period of operation (Fig. 9.59).

Global intensity. *Sum of all of the signal readings on a microarray, particularly for use as a normalization factor.*

With matched images in hand, data normalization is a relatively straightforward process. Normalization can be accomplished using a variety of different criteria, including global intensities, housekeeping genes, and internal standards. ***Global intensity*** normalization uses the sum of signals in multiple images to provide equalized signals. Images that have slightly different signal ratios can be adjusted computationally using a normalization factor (e.g., 1.2) to balance the data globally. Global intensity matching works extremely well for samples that share many common signal intensities, such as differential gene expression experiments that compare a test and reference sample prepared from a common biological source (see above).

Normalizing data from different biological tissues or from many different microarray experiments presents a greater normalization challenge, owing mainly to the fact that global intensities and median signals may differ

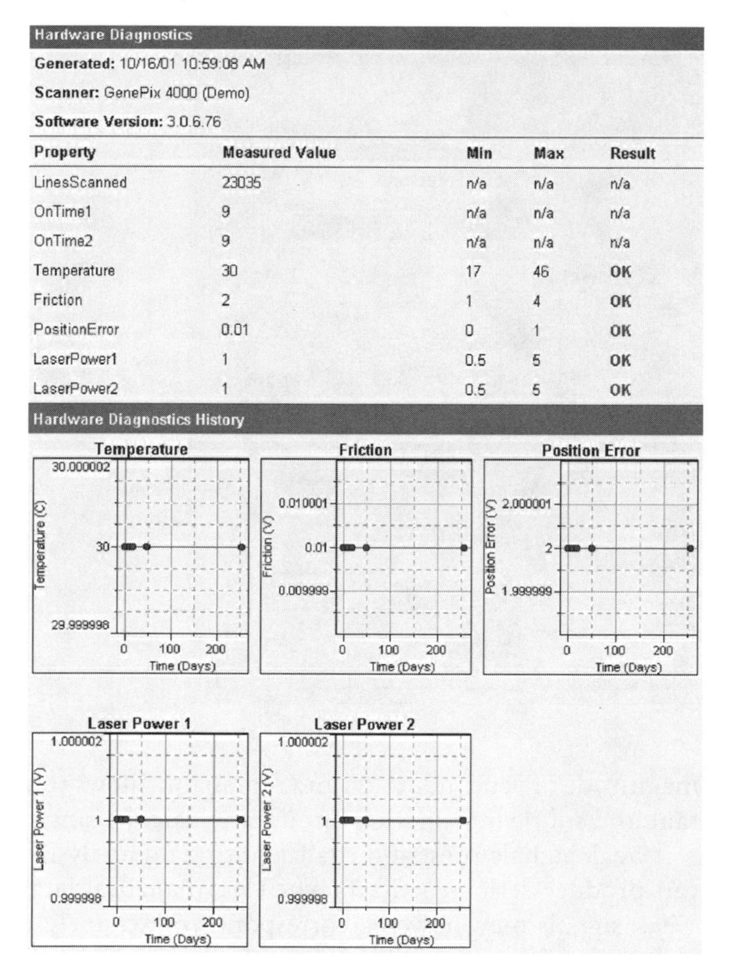

Hardware Diagnostics

Generated: 10/16/01 10:59:08 AM

Scanner: GenePix 4000 (Demo)

Software Version: 3.0.6.76

Property	Measured Value	Min	Max	Result
LinesScanned	23035	n/a	n/a	n/a
OnTime1	9	n/a	n/a	n/a
OnTime2	9	n/a	n/a	n/a
Temperature	30	17	46	OK
Friction	2	1	4	OK
PositionError	0.01	0	1	OK
LaserPower1	1	0.5	5	OK
LaserPower2	1	0.5	5	OK

Hardware Diagnostics History

FIGURE 9.59. *The* GenePix *4000-B microarray scanner provides performance diagnostics, allowing the user to monitor critical instrument parameters over time, including motion control, lasers, and other components. The diagnostics history output (bottom) ensures the reliability of data acquired in different instrument channels over the course of successive experiments and improves data quality before normalization. (Courtesy of Dr. D. Wellis, Dr. S. Elletson, and colleagues, Axon Instruments, Union City, CA.)*

substantially for a large number of datum points on each microarray. Normalizing or equalizing signals to a set of cellular genes known as ***housekeeping genes,*** which are expressed at approximately the same level in many tissues, is one approach to normalizing data derived from different tissues or chips. Another approach is to dope in small quantities of control samples (e.g., mRNAs) from a foreign source into each labeling reaction, and normalize the images using the signal intensities obtained from the control spots. Yet another approach is to dedicate a universal control sample to the same detection channel in each experiment and normalize each test sample to the signal intensities derived from the universal controls. Most commercial scanners and imagers allow the user to compute normalized values for multiple channels automatically (Fig. 9.60).

Housekeeping gene. *Cellular gene that plays a central role in all cells and correspondingly is expressed nearly equally in all cells and tissues.*

One advantage of doping a series of control mRNAs of known quantities into a gene expression labeling reactions is that the intensities of the resultant control spots allow the construction of a ***standard curve,*** which is a ladder of signals based on known quantities of mRNA. The standard curve provides a means of quantifying absolute and ratio measurements, deriving true estimates of levels and ratios rather than relying on raw fluorescent counts. Because fluorescent signal intensities are often compressed with respect to true quantities and ratios (see above), extrapolations from standard curves usually provide more accurate measurements than raw intensity values. A gene that is calculated as a 12-fold ratio based on signal intensities, might actually be induced by >25-fold, depending to the technical details of the measurement. Knowing the true

Standard curve. *Graph of signal intensity versus concentration constructed using known concentrations of messenger RNA or some other material and used to assign quantitative values to genes or gene products of unknown concentration.*

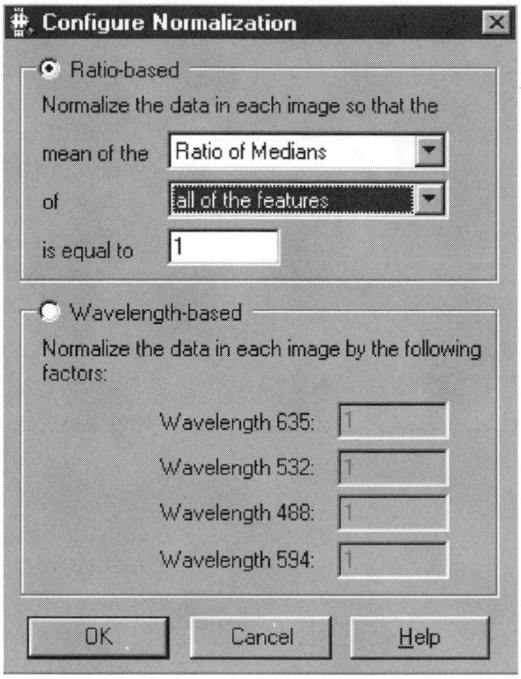

FIGURE 9.60. *The* GenePix *4000-B microarray scanner allows automated image normalization, providing equalized datasets for ratio calculations. (Courtesy of Dr. D. Wellis, Dr. S. Elletson, and colleagues, Axon Instruments, Union City, CA.)*

magnitude of gene induction and repression allows researchers to glean important and subtle information out of ratio calculations.

Weak signals present a challenge to data analysis because small numbers can produce serious artifacts when normalizing data and calculating ratios. Weak signals may, however, correspond to extremely important cellular transcripts that are expressed at low levels, making it imperative to treat these datum points cautiously but not with irreverence. Missing features, fluorescent blemishes, and other artifacts in microarray images present additional normalization challenges, making it imperative to identify or ***flag*** suspect features before normalization and ratio calculation.

Flag. *Manual or automated procedure in which missing or anomalous microarray features are identified before quantitation.*

Boolean. *Computational process in which logical operators (e.g., AND, OR) are employed in microarray spot analysis or some other search function.*

Most sophisticated software packages allow the user to identify unreliable microarray spots based on a series of user-specified parameters (Fig. 9.61). The flagging process often involves the use of a ***Boolean*** procedure, which is a combinatorial process that employs a set of operators (e.g., AND, NOT, OR) to

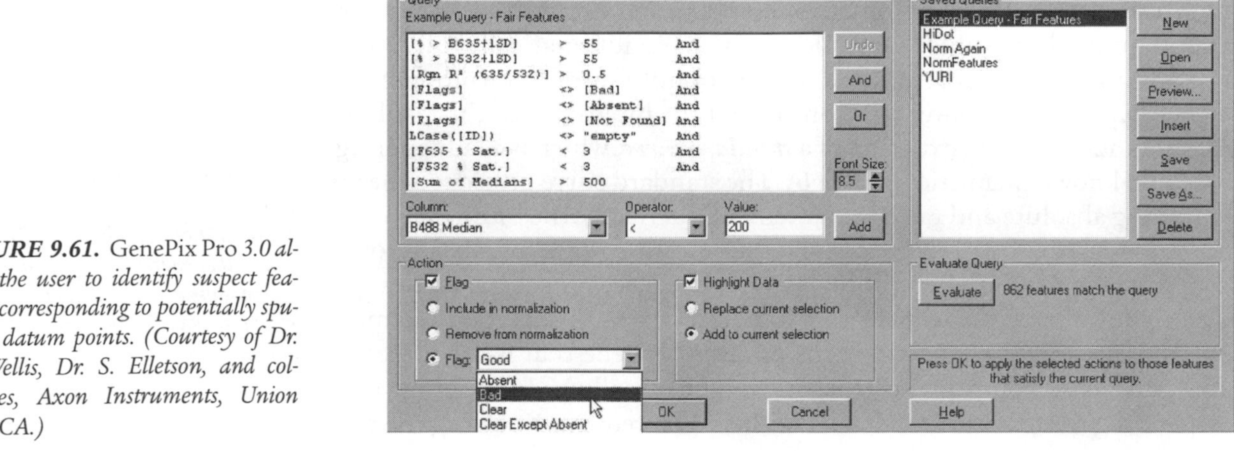

FIGURE 9.61. GenePix Pro *3.0 allows the user to identify suspect features corresponding to potentially spurious datum points. (Courtesy of Dr. D. Wellis, Dr. S. Elletson, and colleagues, Axon Instruments, Union City, CA.)*

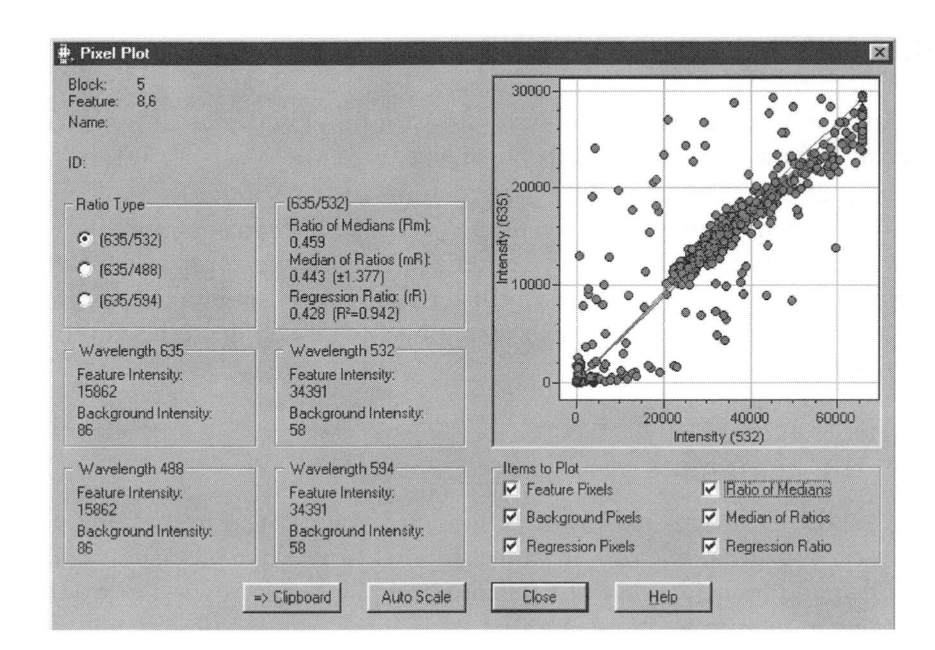

FIGURE 9.62. GenePix Pro *3.0 allows users to select a microarray spot of interest and generate an intensity distribution for both channels on a pixel-by-pixel basis, enabling quality assessment. (Courtesy of Dr. D. Wellis, Dr. S. Elletson, and colleagues, Axon Instruments, Union City, CA.)*

identify potentially bogus spots. A Boolean query that is designed to flag weak signals OR saturated signals would identify every microarray spot that satisfies either criterion. Flagged datum points should be examined carefully before being included in a microarray database. Some data analysis packages allow the user to examine each spot on a pixel-by-pixel basis using a ***pixel plot,*** which graphs each pixel intensity in both channels in a two-dimensional manner, and allows the user to assess whether the pixel intensities distribute normally as would be expected for a bona fide printed spot (Fig. 9.62). Other sophisticated data analysis informatics allow the researcher to make fine adjustments to the quantitation template to ensure the most accurate fit of the template over the image, thereby ensuring optimal normalization and quantitation (Fig. 9.63).

Pixel plot. *Two-dimensional representation of a microarray spot in which all of the pixel intensities are plotted at two emission wavelengths to allow the user to assess whether the pixel intensities distribute normally as would be expected for a legitimate printed spot.*

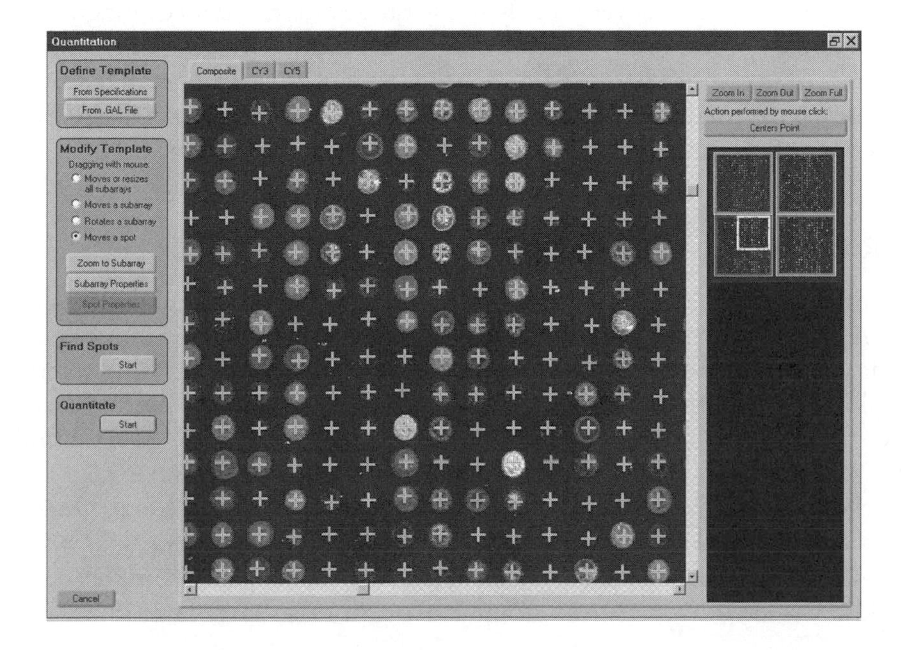

FIGURE 9.63. ScanArray Express *1.0 enables custom template placement, which ensures optimal spot finding. (Data courtesy of Dr. D. Holton and colleagues, Packard BioScience, Meriden, CT.)*

DATA MINING AND MODELING

The final step in microarray analysis is data mining and modeling. Once microarray data are quantified and normalized, it is necessary to interpret the meaning of the data vis-à-vis the vast amount of scientific information available on the genes, gene products, tissues, cells, and organisms used in the experiments. Data transformation, scatter plots, principle component analysis, expression maps, pathway analysis, cluster analysis, self-organizing maps, and workflow management are among the mining and modeling methods and tools described in this section.

Data Transformation

Transform. *Computational process of converting microarray data from raw counts into a logarithmic or some other scale to improve the statistical soundness of an analysis procedure.*

Logarithm. *Power to which 10 must be raised to produce a given microarray value.*

Logarithmic scale. *Series of microarray data values expressed as logarithms rather than raw counts.*

Data obtained from microarray detection instruments are expressed in units of fluorescent "counts," with 16-bit values ranging from 1 to 65,536. Raw data can be mined and modeled, though it is common practice to convert or ***transform*** the raw counts into a different scale that is more convenient and statistically sound. Transformation of microarray data into a logarithmic scale is the most common procedure. A ***logarithm,*** or log, is defined as the power to which 10 must be raised to produce a given microarray value. Using a ***logarithmic scale,*** 16-bit microarray data are expressed in values ranging from 0 (log 1) to 4.8 (log 65,536), so that quantified spots yielding 1,000 and 10,000 counts, respectively, would have a log values of 3.0 and 4.0.

Microarray data transformed into a log scale (0–4.8) exhibit a much more uniform distribution than raw signal intensities (1–65,536), enabling straightforward application of parametric procedures for statistical analyses that leverage data that exhibit a normal distribution. Log scale data also facilitate the analysis of ratios, allowing dynamic changes in gene expression to be represented as simple positive and negative integers, rather than unwieldy values. Two genes that are induced and repressed by 20-fold and 127-fold, respectively, would have logarithm values of 1.3 and −2.1. A log scale of −5.0 to +5.0 allows the representation of genes that are repressed by 100,000-fold and activated by 100,000-fold, covering a dynamic range of ten orders of magnitude or 10,000,000,000-fold. Most researchers find an expression value of −2.6 more palatable than trying to visualize a gene that is expressed at a level of 0.0025 relative to a control sample. Transformed microarray data are used widely in scatter plots, clustering, and other forms of data analysis and visualization (see below).

Scatter Plots

Scatter plot. *Graphical representation of microarray data in which the signal intensities of two samples are plotted along the x and y axes, and the ratio values are plotted as a distance from the diagonal.*

Identity line. *Diagonal in a scatter plot that defines the signal intensities and gene expression values that are equivalent in two samples.*

One of the most useful representations of gene expression data is the ***scatter plot.*** A scatter plot is a graphical two-dimensional representation of microarray data in which the signal intensities of two samples (e.g., test and reference) are plotted along the x- and y-axes, and the ratio values are plotted as a distance from the diagonal (Fig. 9.64). Genes expressed at a high level (abundant mRNAs) reside at a greater distance from the origin than genes expressed at a low level (rare mRNAs), and genes that are activated or repressed fall above and below the diagonal or ***identity line,*** respectively. Scatter plots provide an intuitive visualization tool that assists the researcher in analyzing complex datasets quickly. The key advantage over numerical outputs is that scatter plot data can

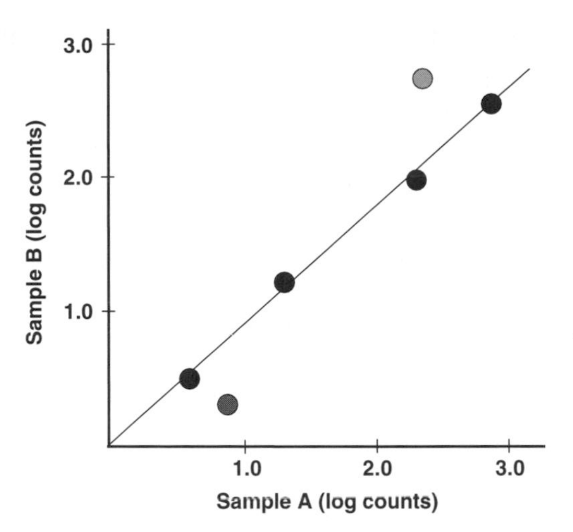

FIGURE 9.64. *A scatter plot of six hypothetical genes* (circles) *expressed in two different samples (A and B). Signal intensities are represented in a log scale (0–3.0) corresponding to raw counts of 1–1000. Four of the six genes* (black spheres) *are expressed at equal levels in the two samples and fall along the diagonal identity line.* One gene (light gray circle) *is an abundant transcript induced in sample B relative to sample A, and one gene* (dark gray circle) *is a rare transcript repressed in sample B relative to sample A.*

be assimilated more efficiently than vast collections of signal intensities and ratios in spreadsheets.

One of the most useful applications of scatter plots is precision assessment in ratio calculations. The precision of ratio measurements can be assessed by labeling two identical samples with two different fluors and visualizing the data in a scatter plot. Assuming identical labeling efficiencies for the two fluors, balanced detection channels, accurate normalization, and the like, two identical samples should yield a ratio of 1.0 for every gene, and all of the data should fall precisely along the identity line (Fig. 9.65). Comparing two identical samples of mRNA from cultured human cells labeled separately with Cy3 and Cy5, produces ratio data that fall tightly (but not precisely) along the identity line, as expected for two identical samples that differ with respect to the fluorescent tag used for labeling. Comparing samples from

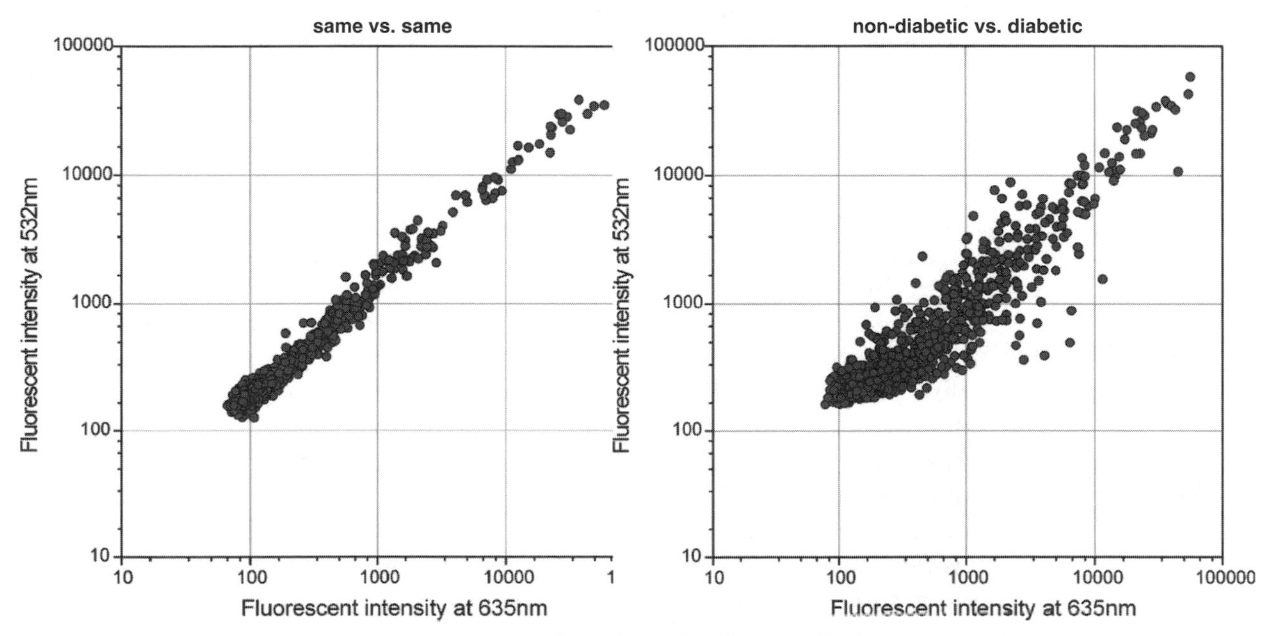

FIGURE 9.65. ***A,*** *For the same–same comparison, two identical samples of human cell culture mRNA were labeled with cyanine 3 and cyanine 5; note the expected tight concordance of the data along the diagonal, indicating ratios of ~1.0 for all of the genes in the two samples.* ***B,*** *A comparison of skeletal muscle mRNA from a patient with diabetes and one without, reveals significant scattering of the data, as expected for differentially expressed genes.*

two identical samples is known as a same–same comparison. Same–same comparisons should be performed before commencing ratio studies, particular those that may involve novel fluorescent tags, labeling procedures, or detection equipment. The concordance of a same–same scatter plot data provides a good assessment of ratio measurement precision.

Another application of scatter plots is the visualization of ratios obtained from two different samples, such as human skeletal muscle mRNA isolated from nondiabetic and diabetic patients (Fig. 9.65). The scatter plot data reveal a significant scattering above and below the identity line, indicating differences in mRNA levels in the two patient samples. The activated and repressed genes identify physiological differences between the patients as well as changes due to diabetes. By comparing a spectrum of patients afflicted with a common medical condition (e.g., diabetes), it is possible to identify a set of stereotyped genes whose expression is altered by a specific disease (see Chapter 12). If a common control sample is used as a reference sample in each measurement, scatter plots enable direct visual comparisons of many different samples, such as a group of individuals afflicted with diabetes.

Principle Component Analysis

Scatter plots facilitate visualization of two samples in two-dimensional space, and a three-dimensional scatter plot could be used to compare three samples. But in many microarray applications, researchers wish to compare expression profiles in many different samples, as illustrated by the pioneering work of Kim and colleagues (2001) at Stanford Medical School (Stanford, CA), in which microarrays were used to examine and compare hundreds of samples from the nematode *C. elegans*. Comparisons between 481 different *C. elegans* samples by scatter plot would require a 481-dimensional plot, which is impossible to display graphically. But graphical visualization of many comparisons is possible using techniques that allow ***dimensionality reduction***. Reducing complex comparisons into three-dimensional space allows representation of the data from hundreds of different samples on a conventional computer screen.

Dimensionality reduction. *Computational process used in principle component analysis and other data-modeling procedures in which complex datasets are compressed into x, y, and z coordinates to allow their representation on a conventional computer screen.*

Principle component analysis. *Computational method that reduces relationships that exist in high-dimensional space into three dimensions, enabling complex data to be visualized in standard graphical form.*

Multivariate. *Any statistical technique, including principle component analysis, that allows the comparison of many different variables.*

Outlier. *Statistical term corresponding to a gene or datum point that behaves differently from the majority of data when examined across multiple experiments.*

One of the most common methods of dimensionality reduction is ***principle component analysis*** (PCA). PCA is a method that reduces relationships that exist in high-dimensional space into three dimensions, enabling complex data to be visualized in standard graphical form (Fig. 9.66). PCA is known as a ***multivariate*** technique, because it is a statistical approach that allows the comparison of many different variables (genes). The PCA method preserves closeness relationships between genes in multidimensional space, so that genes residing in close proximity in many dimensions are configured close to each other in three dimensions. Principle component analysis is used widely in microarray analysis to identify groups of genes that are regulated in a similar manner across many experiments, providing functional clues to genes of unknown sequence or biological function based on their proximity to known genes. PCA also finds ***outlier*** genes that behave differently from most genes across many experiments, suggesting novel functionality.

Expression Maps

The genes of every organism are configured on contiguous pieces of DNA known as chromosomes, which together make up the genome (see Chapter 4).

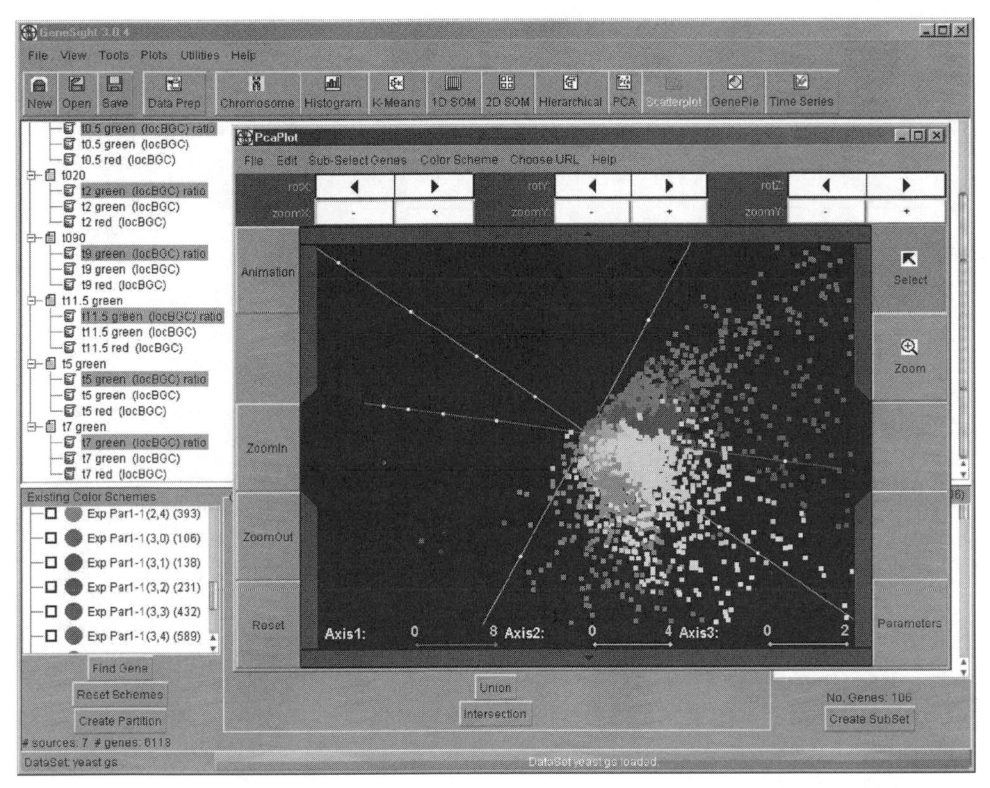

FIGURE 9.66. GeneSight *3.0 enables comparisons of many different samples using dimensionality reduction by the method of PCA. (Data courtesy of Dr. S. Shams, BioDiscovery, Marina del Rey, CA.)*

The complete genomic sequence of a eukaryotic organism such as yeast provides a linear map of the genes along each chromosome, and the map can be depicted graphically using a computer. A microarray containing targets for every yeast gene allows expression monitoring of the entire yeast genome in a single experiment (Fig. 9.67). If the position of each yeast gene is known on the microarray and in the genome, microarray data can be superimposed on the genome to provide an ***expression map***. An expression map allows the researcher to visualize transcript abundance and regulation with respect to their physical location on the chromosomes, enabling extremely sophisticated analyses, including co-regulation, chromosome structure, and evolution. Expression maps are finding greater and greater use as complete genomic sequences are obtained. Genotyping information such as ploidy and the location of SNPs can also be superimposed over genomic sequences.

Expression map. *Graphical representation in which gene expression values are superimposed onto their cognate genes along the chromosomes.*

Pathway Analysis

Gene expression information can be superimposed onto the genome to provide a view of how the genes are regulated vis-à-vis the physical structure of chromosomes. Expression data can also be superimposed onto biochemical and genetic pathways to correlate gene expression with metabolic activity (Derisi et al., 1997) and developmental programming (Heller et al, 1997). The practice of correlating gene expression information with biochemical, genetic, and development processes is known as ***pathway analysis***.

Pathway analysis. *Computational procedure in which gene expression data are correlated with a biochemical, genetic, or developmental process.*

One tour de force in microarray-based pathway analysis was carried out by Miki and colleagues (2001) at the RIKEN Genomic Sciences Center (Yokohama Japan). Microarrays containing >18,000 mouse cDNAs were used to profile

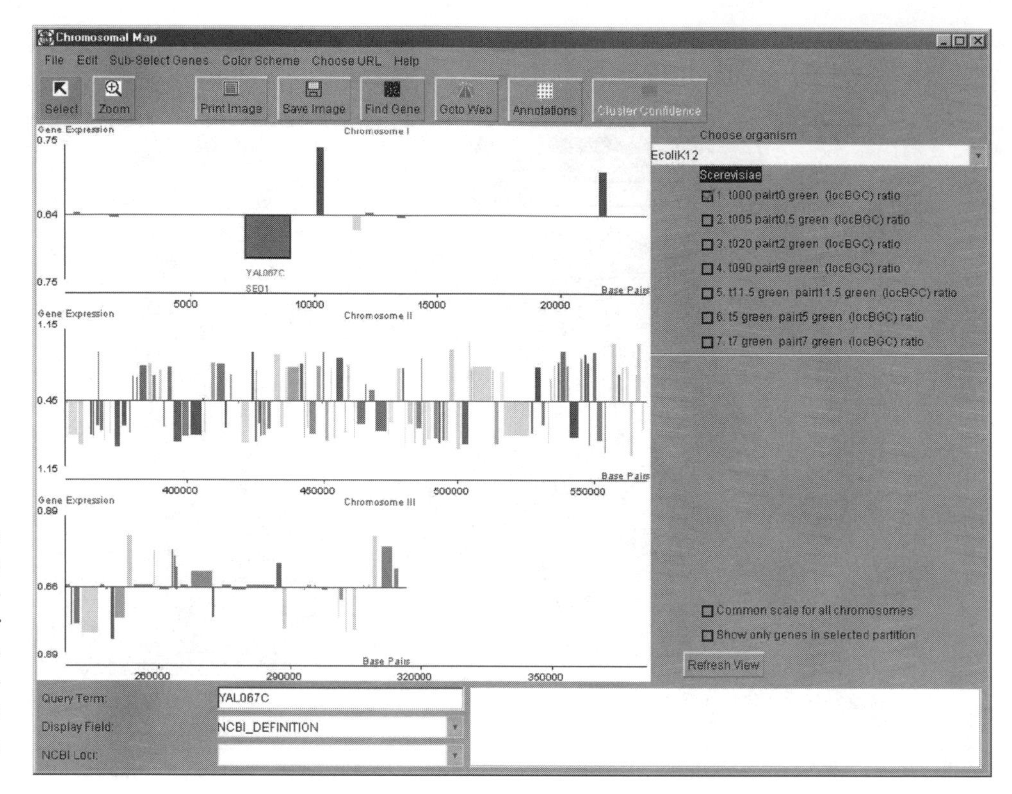

FIGURE 9.67. GeneSight *3.0 can depict the yeast genome sequence in a linear manner as a series of chromosomes. Microarray expression data for each gene* (vertical bars) *can be superimposed on each physical gene location to provide a genome expression map. (Data courtesy of Dr. S. Shams, BioDiscovery, Marina del Rey, CA.)*

Dendrogram. *Graphical representation of a set of relationships based on the closeness of data or datasets to one another.*

gene expression patterns in 49 embryonic and adult mouse tissues. Comparing expression levels in each tissue using cluster analysis (see below), the authors were able to construct a dendrogram (Fig. 9.68). A ***dendrogram*** is a graphical representation of a set of relationships based on the closeness of datasets to one another, and gene expression information acquired from different mouse tissues allowed the classification of mouse tissues according to their developmental closeness. Experiments like this demonstrate how expression analysis can be used to delineate the intricate pathways underlying mouse body plan programming. Similar experiments could be used to establish body plan relationships in human, other animals, and higher plants.

Cluster Analysis

Cluster analysis. *Multivariate classification method that uses dendrograms as a general means of categorizing data based on the similarity of the datum points to one another.*

Clustering. *Process of performing cluster analysis.*

Another multivariate classification method, developed in the 1930s, is known as ***cluster analysis,*** which is a general means of categorizing data based on the similarity of the datum points to one another. Similar to the strategies used to construct conventional dendrograms (see above), clustering places the similar data into a cluster and then organizes the clusters with respect to one another, so that similar clusters are close to one another and dissimilar groups are far apart. ***Clustering*** was first applied to microarray analysis in the late 1990s (Eisen et al. 1998), and has become an increasingly important tool for gene expression analysis and other applications.

Clustering algorithms are similar to those employed for sequence comparisons, but cluster diagrams display the "terminal branches" as a list of genes that share similar behavior across multiple experiments. In a typical gene expression clustering diagram, the genes and gene clusters are tabulated vertically, and each gene expression experiment is configured across the horizontal axis (Fig. 9.69).

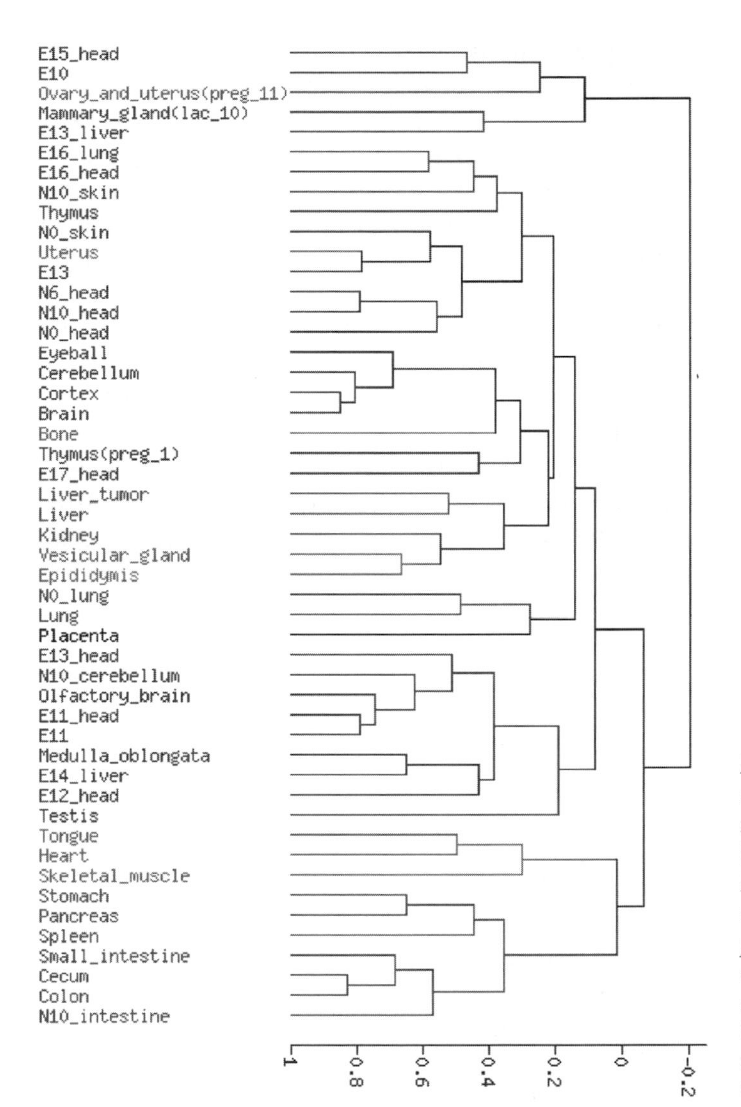

FIGURE 9.68. A tissue dendrogram reveals the developmental relationships of 49 different mouse tissues based on gene expression profiles. Branch lengths, based on correlation coefficients (scale bar, bottom), *are proportional to gene expression similarity. Terminal branches are color coded to reflect the presumptive germ layer tissues. (Data courtesy of Dr. Y. Okazaki, RIKEN Genomic Sciences Center, Ibaraki, Japan.)*

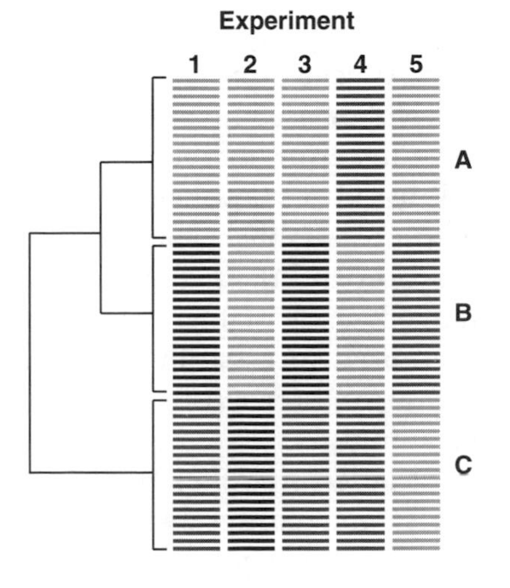

FIGURE 9.69. A gene expression cluster diagram depicting hypothetical expression ratios of 60 genes (horizontal bars) *across five different experiments (1–5), color coded to indicate up-regulation* (green bars), *down-regulation* (red bars), *and genes whose expression is the same in the two samples* (black bars). *Genes are grouped into clusters* (horizontal branches, left) *according to gene expression similarity, providing three sets of 20 cellular genes (A–C).*

Gene expression ratios are often color coded (e.g., red and green), so that activated genes are represented by green bars, genes that do not change are depicted in black, and repressed genes are shown in red. Patterns of expression for each gene are analyzed across all of the experiments, and genes that share the most common pattern of expression are placed into a single cluster. The clustering process is repeated until every gene has been placed into a cluster and until each cluster is configured into a single gene expression dendrogram.

The tight correlation between gene expression and function (see Chapter 12) means that genes that cluster together often share a common function. Clustering therefore provides a rapid and intuitive means of identifying and visualizing genes that share a similar function. Clustering also provides a way to ascribe putative function to novel sequences, if those sequences fall into a cluster that contains genes of well-known function (e.g., cell cycle genes). Clustering provides a rapid means of establishing a putative function for genes that have no known function, such as the >10,000 human genes for which functional roles have yet to be established. Cluster analysis further provides a functional entry point into organisms that have intractable genetics or unwieldy genomes.

Pairwise. *Mathematic procedure used in clustering and other forms of data analysis in which genes or gene products are compared in groups of two.*

Vector. *Mathematical quantity that has both a magnitude and a direction and is thus a good descriptor of gene expression. Alternatively, a plasmid or other molecule that carries a DNA insert.*

Correlation coefficient. *Mathematical quantity that corresponds to the dot product of two normalized vectors.*

Supervised. *Clustering method that exploits known reference vectors to classify and organize gene expression and other types of data.*

One prerequisite for cluster analysis is to choose a mathematic description that allows genes to be compared in a ***pairwise*** (two at a time) manner. Changes in gene expression embody both a quantity (mRNA concentration) and a direction (up or down), making a mathematical ***vector*** a well-suited approximation for gene expression. A vector is a mathematical quantity that has both a magnitude (e.g., fold change) and a direction (e.g., activation or repression) and is thus a good descriptor of gene expression. Pairwise comparisons between two gene expression ratios can be assessed using a number of different mathematical quantities, including the dot product of two normalized vectors, known as the ***correlation coefficient***. Correlation coefficients provide a standardized means of comparing large quantities of gene expression data.

Unsupervised. *Clustering method that does not use reference vectors to classify and organize gene expression and other types of data.*

Hierarchical. *Clustering method that identifies a small group of genes that share a common pattern of expression and then constructs the dendrogram in a sequential manner using a ranked series of clusters.*

There are two broad categories of clustering: supervised and unsupervised methods. ***Supervised*** clustering exploits known reference vectors (e.g., known gene expression ratios) to classify and organize gene expression data, whereas ***unsupervised*** techniques do not use reference vectors. Supervised clustering methods are somewhat better suited to gene expression analysis at present, because the current state of knowledge concerning expression patterns is limited enough to prevent the establishment of systematic and reliable reference vectors. As gene expression information accumulates, supervised clustering methods will be used more frequently.

Nonhierachical. *Clustering method such as k-means analysis that assigns each gene expression datum to a cluster based on its expression profile, and then repeats the process until every datum point has been placed in a cluster.*

K-means. *Nonhierarchical clustering method in which the user specifies the number of clusters (k) into which the genes are to be assigned.*

Clustering methods are also classified as hierarchical and nonhierarchical, based on how the dendrograms are constructed. ***Hierarchical*** clustering methods establish a small group of genes that share a common pattern of expression and then construct the dendrogram in a sequential manner using a ranked series or hierarchy of clusters. Hierarchical clustering methods have been used successfully in the context of evolutionary studies for decades (see above), and the clustering data provided by hierarchical method is reasonably meaningful.

Nonhierachical methods, such as ***k-means*** clustering, are also gaining wide use in microarray analysis (Fig. 9.70). In the nonhierarchical k-means approaches, each gene expression datum is assigned to a cluster based on its expression profile, and the process is repeated until every datum point has been

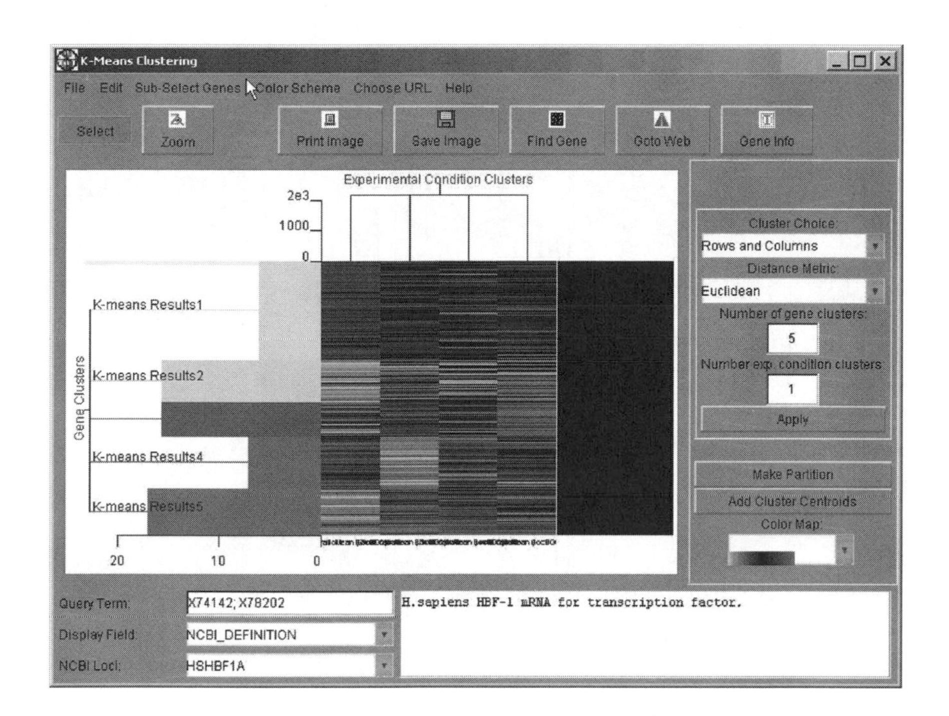

FIGURE 9.70. Genes were clustered according to an expression profile, using the nonhierarchical k-means clustering method in GeneSight *3.0. (Data courtesy of Dr. S. Shams, BioDiscovery, Marina del Rey, CA.)*

placed into a cluster. The user initiates the k-means process by specifying the number of clusters (k) into which the genes are to be assigned, and the computer constructs a dendrogram of the clusters. The k-means approaches have the advantages of speed and somewhat greater precision relative to the hierarchical approaches, depending on the details of the sameness algorithms used. A repetitive and cyclical computational process is known as an ***iterative*** procedure. All forms of clustering are iterative processes.

Iterative. *Any computational procedure, including clustering, that employs a repetitive and cyclical computational process.*

In addition to viewing clustered information in dendrogram and cluster formats, it is possible to display cluster information in a two-dimensional manner using ***self-organizing map*** (SOM) algorithms. The SOM is a powerful visualization tool that is particularly useful for charting genes that exhibit temporal (over time) patterns of regulation. A typical two-dimensional SOM displays each cluster in a separate software window, allowing the user to specify the physical layout (Fig. 9.71). Self-organizing maps enable the identification of regulatory relationships that are difficult or impossible to observe using conventional clustering. By charting the linear sequence of gene expression as a ***time series*** (Fig. 9.72), regulatory patterns, such as cell cycle control, emerge within clusters of genes. Software packages from more than a dozen commercial sources provide an extensive assortment of sophisticated data mining and modeling tools for the microarray community (Table 9.6).

Self-organizing map. *Algorithm that displays cluster information in a two-dimensional manner.*

Time series. *Graphical representation in which data from related biological samples are displayed sequentially as a function of time.*

Workflow Management

The process of controlling and recording the physical steps by which microarray data are generated and stored is known as ***workflow management***. The management of microarray workflow is a complex task that is best performed using computer tools that provide the user a means of coordinating clone sets, sample tracking information, image quantitation, other data outputs such as substrate processing and probe preparation protocols, and other workflow

Workflow management. *Process of controlling and recording the physical steps used to generate and store microarray data.*

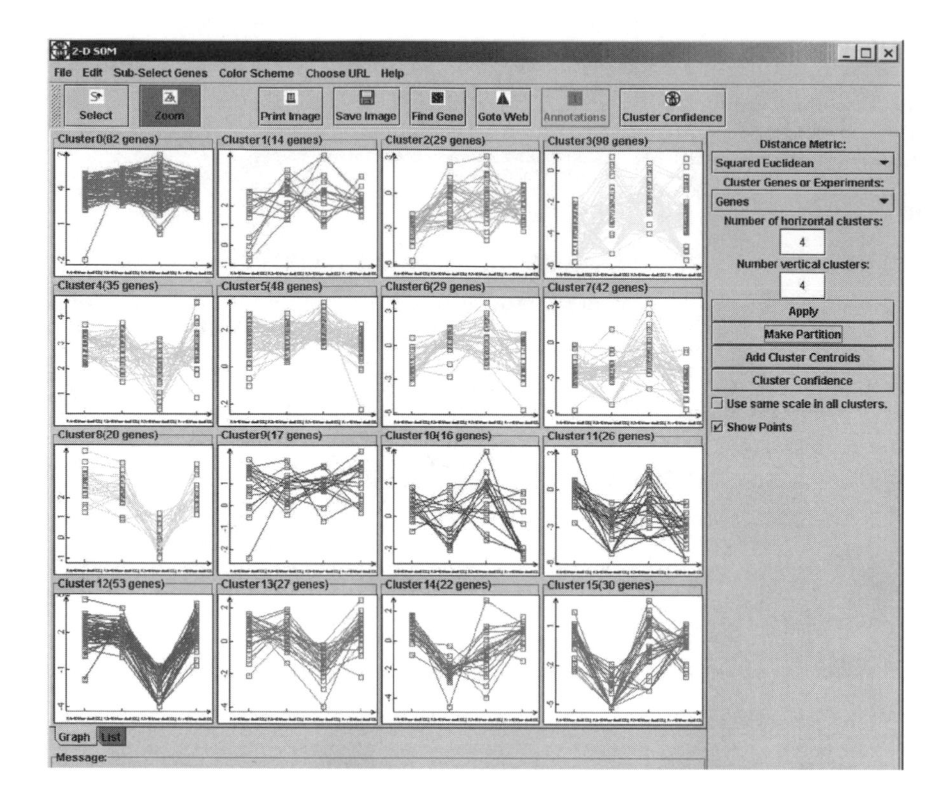

FIGURE 9.71. *Clustered gene sets can be viewed in a two-dimensional manner using an SOM algorithm. (Data courtesy of Dr. S. Shams, BioDiscovery, Marina del Rey, CA.)*

details that bear on the quality and reliability of the data (Fig. 9.73). Workflow information is stored in a specialized database known as a ***relational database,*** which stores microarray data as values in tables and allows every datum to be queried and retrieved in a systematic manner. Careful workflow management enhances the downstream steps of data mining and visualization.

Relational database. *Specialized information storage and retrieval system in which microarray data are represented in tables, allowing each and every datum to be queried and retrieved in a systematic manner.*

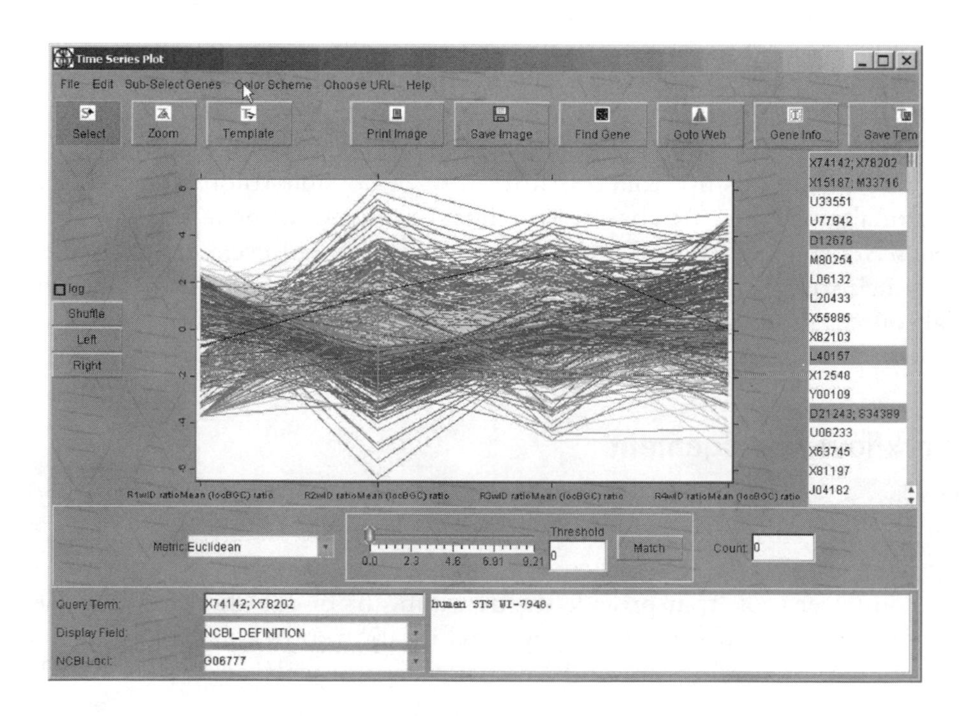

FIGURE 9.72. *Temporal gene expression data can be represented as a time series, allowing groups of genes to be examined at high resolution. Gene expression ratios are plotted on a log scale along the vertical (y) axis and time is displayed along the horizontal (x) axis. Subsets of genes are color coded for enhanced visualization. (Data courtesy of Dr. S. Shams, BioDiscovery, Marina del Rey, CA.)*

TABLE 9.6. Data Mining and Modeling Software

Company	Product	Cluster Analysis	PCA	SOM	Custom Algorithms
Affymetrix (affymetrix.com)	*Data Mining Tool* 2.0	yes	no	yes	yes
BioDiscovery (biodiscovery.com)	*GeneSight* 3.0	yes	yes	yes	yes
BioMax Informatics AG (www.biomax.de)	*Gene Expression Analysis Suite*	yes	no	yes	yes
Gene Network Sciences (genenetworksciences.com)	*BioMine* 1.0	yes	yes	yes	yes
InforMax (www.informaxinc.com)	*XpressionNTI*	yes	yes	yes	yes
Molecular Mining (molecularmining.com)	*GeneLinker Gold*	yes	yes	yes	yes
OmniViz (omniviz.com)	*OmniViz Pro*	yes	yes	no	yes
Packard BioScience (www.packardbioscience. com)	*ArrayInformatics*	yes	yes	yes	yes
Partek (partek.com)	Partek Pro 5.0	yes	yes	yes	yes
Rosetta Biosoftware (www.rosettabio.com)	*Resolver* 3.0	yes	yes	yes	yes
Silicon Genetics (sigenetics.com)	*GeneSpring* 4.1	yes	yes	yes	yes
Spotfire (spotfire.com)	*DecisionSite*	yes	yes	no	yes
VizXLabs (vizxlabs.com)	GeneSifter.net	yes	yes	yes	yes

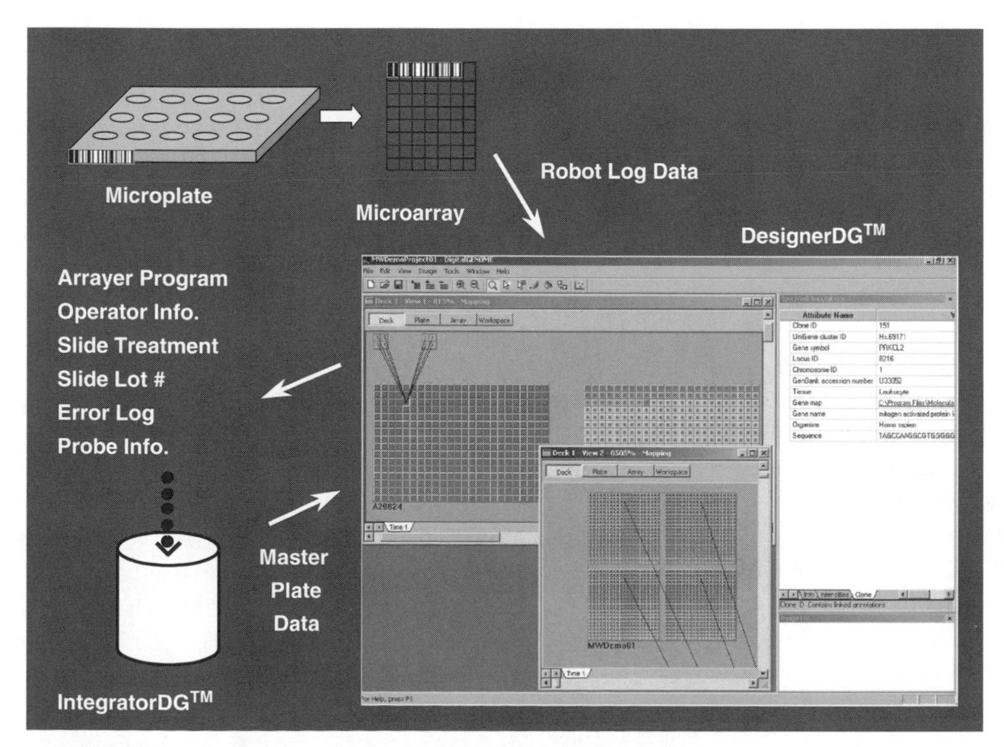

FIGURE 9.73. Microarray manufacture and other physical processes in microarray analysis require software tools to manage the massive information flow. (Courtesy of Dr. S. Thomas, MolecularWare, Cambridge, MA.)

SUMMARY

Computers play a central role in virtually every aspect of microarray research. The rapid proliferation of personal computers and the Internet enable the use of extremely powerful communication tools including the World Wide Web and electronic mail. The availability of electronic libraries provide universal access to the microarray literature, and biological databases containing gene, protein, structure, and genomic information enable homology searches, identification of sequence variants, and evolutionary studies. Sequence informatics facilitate melting temperature calculations, sequence alignments, and the design of oligonucleotide targets, probes, and PCR primers. Computers also enable microarray design informatics, including microarray location and feature spacing, gene number and complexity measurements, and sample and content maps.

Informatics is also used for data quantitation, allowing researchers to calculate absolute expression levels and ratios and to normalize data. Data mining and modeling are a rapidly advancing, speeding data transformation, and enabling data visualization via scatter plots, PCA, expression maps, and pathway and cluster analysis. Enterprise-level workflow management software using relational database architecture allows researchers to integrate the physical aspects of data collection and archiving, with the computation tools used for data analysis and visualization.

SELECTED READING

Abarbanel, R. M., Wieneke, P. R., Mansfield, E., Jaffe, D. A., and Brutlag, D. L. Rapid searches for complex patterns in biological molecules. *Nucleic Acids Res* 12:263–280, 1984.

Altschul, S. F., Gish, W., Miller, W., Myers, E. W., and Lipman, D. J. Basic local alignment search tool. *J Mol Biol* 215: 403–410, 1990.

Baxevanis, A. D., and Francis Ouellette, B. F. F., eds. *Bioinformatics: A Practical Guide to the Analysis of Genes and Proteins,* 2nd ed. Wiley-Interscience, New York, 2001.

Berson, A., and Smith, S. J. *Data Warehousing, Data Mining, and OLAP.* McGraw-Hill, New York, 1997.

Cormen, T. H., Leiserson, C. E., Rivest, R. L., and Stein, C. *Introduction to Algorithms.* MIT Press, Cambridge, MA, 2001.

DeRisi, J. L., Iyer, V. R., and Brown, P. O. Exploring the metabolic and genetic control of gene expression on a genomic scale. *Science* 278:680–686, 1997.

Diamantaras, K. I., and Kung, S. Y. *Principal Component Neural Networks: Theory and Applications.* Wiley, New York, 1996.

Heller, R. A., Schena, M., Chai, A., Shalon, D., Bedilion, T., Gilmore, J., Woolley, D. E., and Davis, R.W. Discovery and analysis of inflammatory disease-related genes using cDNA microarrays. *Proc Natl Acad Sci U S A* 94:2150–2155, 1997.

Higgins, D., and Taylor, W., eds. *Bioinformatics: Sequence, Structure, and Databanks: A Practical Approach.* Oxford University Press, Oxford, UK, 2000.

Kaufman, L., and Rousseeuw, P. J. *Finding Groups in Data: An Introduction to Cluster Analysis.* Wiley, New York, 1990.

Kim, S. K., Lund, J., Kiraly, M., Duke, K., Jiang, M., Stuart, J. M., Eizinger, A., Wylie, B. N., and Davidson, G. S. A gene expression map for *Caenorhabditis elegans*. *Science* 293:2087–2092, 2001.

Lehninger, A. L., Nelson, D. C., and Cox, M. M. *Principles of Biochemistry,* 3rd ed. Freeman, New York, 2000.

Lewin, B. *Genes VII.* Oxford University Press, Oxford, UK, 2000.

Lipman, D. J., and Pearson, W. R. Rapid and sensitive protein similarity searches. *Science* 227:1435–1441, 1985.

Miki, R., Kadota, K., Bono, H., Mizuno, Y., Tomaru, Y., Carninci, P., Itoh, M., Shibata, K., Kawai, J., Konno, H., Watanabe, S., Sato, K., Tokusumi, Y., Kikuchi, N., Ishii, Y., Hamaguchi, Y., Nishizuka, I. .I, Goto, H., Nitanda, H., Satomi, S., Yoshiki, A., Kusakabe, M., DeRisi, J. L., Eisen, M. B., Iyer, V. R., Brown,

P. O., Muramatsu, M., Shimada, H., Okazaki, Y., and Hayashizaki, Y. Delineating developmental and metabolic pathways in vivo by expression profiling using the RIKEN set of 18,816 full-length enriched mouse cDNA arrays. *Proc Natl Acad Sci U S A* 98:2199–2204, 2001.

Needleman, S. B., and Wunsch, C. D. A general method applicable to the search for similarities in the amino acid sequence of two proteins. *J Mol Biol* 48:443–453, 1970.

Obermayer, K., and Sejnowski, T. J., eds. *Self-Organizing Map Formation*. MIT Press, Cambridge, MA, 2001.

Ruberti, I, Sessa, G., Lucchetti, S., and Morelli, G. A novel class of plant proteins containing a homeodomain with a closely linked leucine zipper motif. *EMBO J* 10:1787–1791, 1991.

Schena, M., ed. *DNA Microarrays: A Practical Approach,* 2nd ed. Oxford University Press, Oxford, UK, 2000.

Schena, M., ed. *Microarray Biochip Technology*. Eaton, Natick, MA, 2000.

Schena, M., and Davis, R. W. HD-Zip proteins: Members of an *Arabidopsis* homeodomain protein superfamily. *Proc Natl Acad Sci U S A* 89:3894–3898, 1992.

Schena, M., and Davis, R. W. Parallel analysis with biological chips. In M. Innis, D. Gelfand, and J. Sninsky, eds. *PCR Methods Manual*. Academic Press, San Diego, 1999:445–456.

Schena, M., Shalon, D., Davis, R. W., and Brown, P. O. Quantitative monitoring of gene expression patterns with a complementary DNA microarray. *Science* 270:467–470, 1995.

Schena, M., Shalon, D., Heller, R., Chai, A., Brown, P. O., and Davis, R. W. Parallel human genome analysis: Microarray-based expression monitoring of 1,000 genes. *Proc Natl Acad Sci U S A* 93:10614–10619, 1996.

Singer, M., and Berg, P. *Genes and Genomes*. University Science Books, Herdon, VA, 1991.

Skiena, S. S. *The Algorithm Design Manual*. Springer-Verlag, New York, 1998.

Smith T. F., and Waterman, M. S. Identification of common molecular subsequences. *J Mol Biol* 147:195–197, 1981.

Strachan, T., and Read, A. P. *Human Molecular Genetics,* 2nd ed. Wiley, New York, 1999.

Stryer, L. *Biochemistry,* 4th ed. Freeman, New York, 1995.

Thomsen, E. *OLAP Solutions: Building Multidimensional Information Systems*. Wiley, New York, 1997.

Wallace, R. B., Shaffer, J., Murphy R. F., Bonner, J., Hirose, T., and Itakura, K. Hybridization of synthetic oligodeoxyribonucleotides to phi chi 174 DNA: The effect of single base pair mismatch. *Nucleic Acid Res* 6:3543–3557, 1979.

Weier, H. U., Lucas, J. N., Poggensee, M., Segraves, R., Pinkel, D., and Gray, J. W. Two-color hybridization with high complexity chromosome-specific probes and a degenerate alpha satellite probe DNA allows unambiguous discrimination between symmetrical and asymmetrical translocations. *Chromosoma* 100:371–376, 1991.

REVIEW QUESTIONS

1. The international computer network that supports the World Wide Web (WWW) and electronic mail is known as what? (*Hint:* Al Gore was an early supporter of this.)
2. The 168.17.14.126 string of numbers has the appearance of a: (a) file transfer protocol, (b) transmission control protocol, (c) domain name, (d) Internet protocol address, or (e) post office protocol?
3. Which of the following are available in the Microarray Electronic Library (arrayit.com/e-library): (a) microarray citation information, (b) author contact information, (c) citation abstracts and summaries, (d) electronic links to full manuscripts, and (e) keyword search capabilities?
4. Using the Entrez search and retrieval system at the NCBI (www.ncbi.nlm.nih.gov/), describe the gene with GenBank accession number U14680. How many nucleotides are contained in the U14680 sequence?
5. Use the protein basic local alignment search tool (BLAST) at the NCBI (www.ncbi.nlm.nih.gov/) to find the best match to the 20 amino acid sequence MQKSPLEKASFISKLFFSWT. What protein from which organism most closely matches this 20 amino acid query sequence?

6. Use Equation 9.1 to calculate the melting temperature (T_m) of the 20-nucleotide sequence AAGACTAAAGGATCGATTAG.
7. A researcher hybridizes a cDNA target with a fluorescent 50mer containing a single nucleotide mismatch at position 49, compares the signal obtained with a perfectly complementary 50mer and finds that the two 50mer probes produce nearly identical signals on the microarray. Provide an explanation for these results.
8. Calculate the percent identity of two homologous genes from human and chimpanzee that share sequence identity at 746 out of 798 nucleotide positions.
9. A researcher prints 1000 human 70mer targets into a microarray, hybridizes the microarray with a fluorescent probe mixture of unknown origin, and obtains extremely weak fluorescent signals at every microarray location. With respect to the sequences of the targets and probes, provide two possible explanations for the results.
10. Name one type of information that is required to synthesize oligonucleotide targets but not cDNA targets.
11. Calculate the total number of features in a microarray with 48 subgrids of 900 spots each.
12. A researcher uses a manual quantitation procedure and calculates a median signal and median noise value of 45,897 and 124 counts, respectively, in a 16-bit microarray image. What is the signal-to-noise ratio (SNR) of these data?
13. An automated quantitation procedure is used to generate delimited data files from a single microarray scanned at two different sensitivity settings. Dividing the data from the high-sensitivity scan by the data from the low-sensitivity scan produces signal ratios of approximately 3.5 for every microarray element except those with high-intensity values. Provide an explanation for these data.
14. A microarray image from mRNA sample A is coded to a green color palette and a microarray image from mRNA sample B is coded to a red color palette. Spots that appear red in a two-color overlay of the sample A and B images correspond to genes that are (a) expressed equally in samples A and B, (b) activated in sample A relative to sample B, (c) repressed in sample A relative to sample B, or (d) activated in both samples A and B?
15. Data normalization in a two-color microarray experiment can be used to correct for minor differences in which of the following: (a) mRNA labeling, (b) dye quantum yield, (c) PMT performance, (d) laser power, and (e) hybridization efficiency?
16. A 10-K gene expression microarray is used to examine a space alien that is being held in a secret government facility, and scatter plot analysis of the alien expression profile relative to the human control reveals 147 genes lying above the scatter plot identity line. If the alien data are plotted along the y axis, what can be said about the 147 genes?
17. Principle component analysis (PCA) enables which of the following: (a) visualization on a computer screen, (b) dimensionality reduction, (c) comparisons of many different microarray data sets, and (d) preservation of closeness relationships?
18. A researcher constructs an expression map of gene expression data on a human chromosome and finds that the expression of two genes expressed in

opposite directions is regulated coordinately. In the context of transcriptional regulation, provide an explanation for the expression map data.

19. Which of the following could be true for cellular genes that group together tightly in a cluster diagram: (a) share a common cellular function, (b) are activated coordinately, (c) are repressed coordinately, (d) are not expressed in any of the experiments examined, and (e) share no sequence similarity.

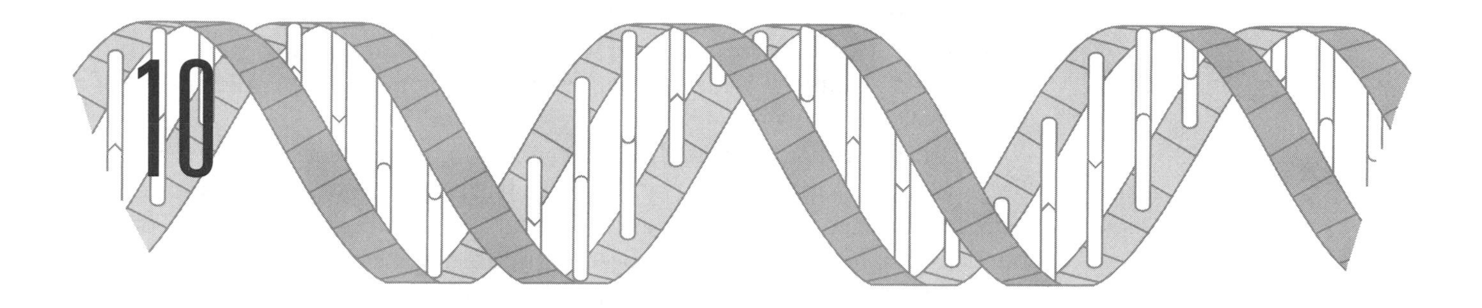

Methodological Architecture

> Therefore O students study mathematics and do not build without foundations.
>
> —*Leonardo Da Vinci*

Microarray analysis is built on an elaborate foundation of theoretical and practical knowledge. This sophisticated methodological architecture contains all of the components required for successful microarray experimentation including worldview, theory, rules, methods, techniques, tools, and implementation. A methodology provides a long-term road map to guide the advances in genetics, biochemistry, chemistry, engineering, and computer science that are required to optimally leverage gene expression microarray analysis. A set of structured and rigorous principles also streamline the development of new products in the commercial sector. This chapter provides a brief glimpse into the methodological architecture of microarray analysis, describing how an engineering approach to biology can be used to build a comprehensive framework on which to understand biological systems and the microarray assays that are used to explore them.

WHAT IS A METHODOLOGY?

The term ***methodology*** refers to the complete body of theories, concepts, principles, rules, methods, protocols, techniques, formulae, and the like that guide a particular process or undertaking. A microarray analysis methodology would therefore refer to the complete set of information that governs microarray research. The completeness of a methodology implies inherently an enormous

Methodology. *Complete body of theories, concepts, principles, rules, methods, protocols, techniques, formulas, and the like that guide microarray analysis or some other process or procedure.*

compilation of material. Currently, one could imagine easily assembling tens or even hundreds of volumes of information to embody all that is used in the day-to-day course of microarray analysis. Currently, a microarray analysis methodology does not exist, except in scattered pieces in the form of scientific publications, books, Web sites, lectures, textbooks, and so forth. Assembling this complete body of information would be a difficult task but well worth the effort. This chapter provides a brief insight into some of the components of a methodology.

Method. *Specific process, such as the polymerase chain reaction, used to perform a given task.*

A defining feature of a methodology is that it tells the scientist *what* needs to be done and *why*, rather than simply how. The "how" part of science is contained generally in scientific methods and protocols, which are part of a methodology but would not constitute the highest methodology levels. It is important to emphasize that the terms methodology and ***method*** are often used interchangeably in the scientific literature, but the two are vastly different and should not be confused. A method is simply a specific process that is used to perform a task, and hundreds of methods for gene amplification, nucleic acid precipitation, RNA purification, microarray hybridization, and the like are used every day in microarray research. A methodology contains each of these methods, but a single method would never contain the contents of a methodology. Scientists often attempt to use methodology as a pretentious synonym for method, but this usage is linguistically incorrect and should be avoided.

The distinction between methodology (what and why) and method (how) is illustrated by considering any process in microarray research. Unincorporated fluorescent nucleotides should be removed from fluorescent probes (what) because failure to do so will lead to elevated background (why). Probes can be purified using any number of methods, including ethanol precipitation (how) and column purification (how). Reactive groups on a microarray substrate should be spaced uniformly (what) because even spacing leads to an equal distribution of bound target molecules across the surface (why). Uniformity of reactive groups on a surface is accomplished by any number of advanced manufacturing procedures (how). Photobleaching of fluorescent dyes in the scanning procedure (what) is problematic because it can lead to erroneous readings in scanned images (why). Photobleaching can be avoided by taking several different measures, including the use of moderate laser power (how) and short dwell times (how). More extensive collections of "what" and "why" can be found as part of the previous chapters on surfaces, targets and probes, manufacturing, detection, and data analysis and modeling (see Chapters 5–9).

Understanding the distinction between methodology and method is much more than an intellectual exercise. Being able to draw the distinction allows scientists to develop new methods, techniques, and protocols to better expedite a given task in microarray research. It is also indispensable in the commercial world, because it enables companies to develop innovative products that meet all of the criteria required to perform a procedure of interest. Most scientists think about methodology and method on a daily basis without drawing the distinction formally. A methodology can be formulated for all of microarray research (e.g., a microarray analysis methodology) or for a subset of microarray science (e.g., gene expression, see below). In all cases, a more complete and accurate methodology enables better methods to be developed and more robust commercial products. Scientists with a knack for formalistic thinking are encouraged to dissect a given process and formulate a simple

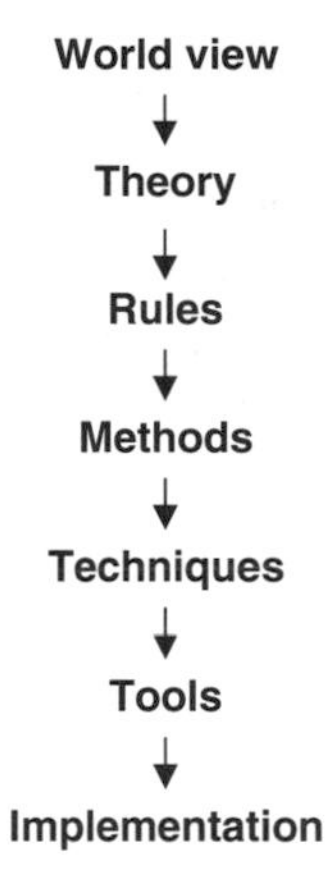

FIGURE 10.1. The epistemological components or tiers of a methodology, in order of most to least abstract.

methodology for a process of interest. Such thinking is of enormous scientific and commercial value.

STRUCTURE OF A METHODOLOGY

A methodology provides a comprehensive and systematic architecture for a given task or process that can structured epistemologically, according to different levels of abstraction (Fig. 10.1). These epistomological tiers are worldview, theory, rules, methods, techniques, tools, and experimental implementation, in the order of most to least abstract. A key concept here is that a methodology is distinct fundamentally from the methods and tools that are used in its implementation, as emphasized in the following equation:

$$\text{Methodology} = \text{World view} + \text{Theory} + \text{Rules} + \text{Methods} + \text{Techniques} + \text{Tools} + \text{Implementation} \quad (10.1)$$

A brief discussion of each of these components is useful in understanding how they differ and their relationships to one another within the larger context of a methodology.

WORLDVIEW AND THEORY

Worldview and theory represent the most abstract components of a methodology, but their consideration is key to understanding microarray analysis at the highest level. If one were to formulate a methodology for gene expression analysis, it would include the worldview and theory of genes and proteins, cells, organisms, cultures, and the like. A closer look at these components, as well as the foundation, approach, and "readout" that might be used to formulate a gene expression analysis methodology, illustrates the concepts encompassed in worldview and theory.

Genomics as the Foundation

The foundation of a microarray-based gene expression methodology could, in principle, be built on any number of academic disciplines including theology, sociology, chemistry, and physics. The theologian might view the results of microarray assays in such a way as to perceive God as a central force in determining the results of the assays. A sociological view is likely to evoke the role of social pressures in affecting the outcome of the gene expression results. The particle physicist might endeavor to model or view biological systems and biomolecules in terms of neutrinos, quarks, and other subatomic particles. The chemist is likely to take a mechanistic, kinetic, and thermodynamic approach to microarray analysis.

A microarray methodology could be built on the social sciences, but these disciplines lack the technical rigor required for a robust science methodology. Physics and chemistry provide key information for many processes in microarray analysis, but these fields are too specific to serve as the foundational discipline. The proper level of "resolution" falls soundly on the biological sciences, and any microarray methodology for gene expression analysis would use biology as the best-suited academic discipline.

Biology and the biological sciences encompass a wide range of subdisciplines, including more than a dozen fields from ecology to structural biology. Though each field contributes to the successful deployment of microarray-based assays, biochemistry seems the best suited as the foundational subdiscipline on which to construct a microarray methodology. Biochemistry places a premium on chemical principles, technology development, whole genome analysis, bioinformatics, advanced technologies, and other tenets and is the ideal subdiscipline for these reasons.

Gene Expression as the Core Assay

Another main tenet of the methodology is that gene expression is the most powerful and informative readout of biological function. Complex biological processes are best studied by the systematic and global monitoring of gene expression patterns. Because gene expression correlates so strongly with function (see Chapter 12), it makes sense to use gene expression data as the best source of information on biological systems. The primary sequence of a genome tells us what the genome *is,* and genome expression data tell us what the genome *does.*

Engineering Approach

Biological systems have been studied historically in a rather descriptive and nonsystematic manner. Proof of this abounds, and a good example is found in the fact that no systematic nomenclature exists for the naming of genes in most organisms. Some of the most intensively studied genes in fruit fly development include *gooseberry, tinman, bride of sevenless, hunchback, aceate-scute,* and *krupple.* Though these gene names have meaning to the practitioners of *Drosophila* development and provide a user-friendly system for an organism that has ~13,000 genes, they are difficult to remember and understand for scientists in other areas of the biological sciences and in other fields. The nonsystematic and anecdotal naming of genes in *Drosophila* or any other organism has some additional negative consequences, including that the fact that it uncouples the gene name with the biochemical activity, overemphasizes the importance of single genes, and undervalues the importance of genome-wide interactions.

The lack of a systematic naming system for genes makes sense with respect to the tools that were available in previous decades. Recombinant DNA and the polymerase chain reaction (PCR) excel at facilitating the isolation, manipulation, and amplification of single genes and were developed primarily to expedite single gene studies. These technologies play a key role modern microarray science, though any methodological approach must go well beyond single genes in terms of the approach taken to study gene expression or any other process on a genomic scale. A methodology for microarray-based gene expression analysis would benefit from taking a quantitative and comprehensive look at genes, proteins, cells, and the processes that govern them. The capacity to generate and manipulate huge amounts of biological information using microarray methods makes a global view tractable.

One useful idea is to take an engineering approach to biology, whereby genes and organisms are viewed as components or parts of larger entities (see Chapter 12). An engineering perspective is helpful because it allows for the systematic and quantitative analysis of every gene in an organism and because it paints a global and interactive picture of genes functioning in the larger context of cells, organisms, and beyond. A brief discussion of the components of an engineering view of biology vis-á-vis a methodology is useful in illustrating the essence of such an approach.

System Variable

In an engineering approach to biology, genes can be viewed as ***system variables,*** and this concept is helpful because it emphasizes the notion that the expression of genes is highly variable and that gene products function in the larger context of cells and organisms (Fig. 10.2). Because each location on a gene expression microarray measures a single gene, it makes sense to use the system variable as the fundamental unit of function in a methodology. System variables (genes) differ from each other by a unique ***biological potential*** (see below). The process of expression of a system variable leads to the genesis of different states of a gene (see below). System variables are the functional units of ***system modules***. The expression of each system variable is affected by the system module, culture, biosphere, and universe (Fig. 10.2).

System variable. *Methodological term corresponding to a gene or gene product and the unit of function in the methodology.*

Biological potential. *Full gamut of biological activities that are possible for a system variable, system module, system, culture, or biosphere in the methodological framework.*

System module. *Methodological term corresponding to a cell, the unit of life in the methodology.*

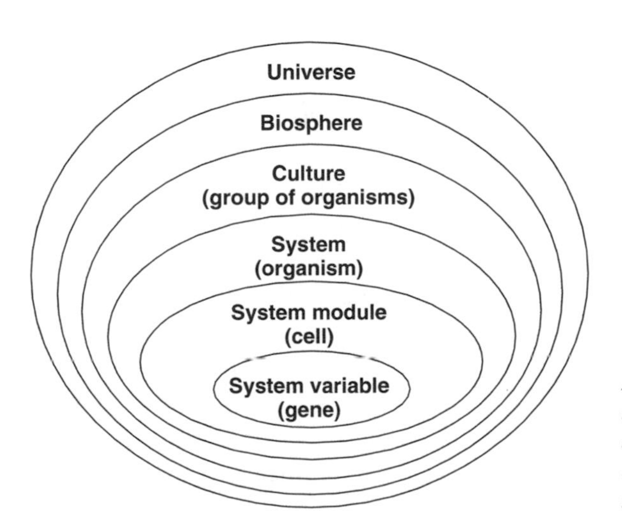

FIGURE 10.2. The levels of organization that could be used to structure a gene expression analysis methodology, in order from most general to most specific.

System Module

The cell or system module is the fundamental unit of life in the methodology. The term is introduced to emphasize the fact that cells are not isolated compartments but rather highly interactive and integrated components of complex biological systems, responding to myriad internal and external genetic and biochemical cues and integrating these into cellular function. Similar to the case with system variables, each system module or cell differs from every other in having a unique biological potential (see below). The differentiation of a system module causes changes in form of the cell (see below). System modules are viewed as the building blocks of systems, the next highest level of biological organization in the methodology (Fig. 10.2).

System

System. *Methodological term corresponding to an intact organism.*

An intact biological organism or ***system*** is viewed as the fundamental life form in the methodology. In the case of single-celled organisms (e.g., bacteria and yeast), the system and the system module are identical. Most biological systems (e.g., flies, worms, humans) are composed of many cells or system modules, which, in turn, are composed of many of system variables (genes). The function of a biological system is determined by the combined effects of intrinsic properties (i.e., system modules and system variables) and external forces such as the interaction of the system with other biological systems, and with the environment. Systems are the building blocks of cultures.

Culture

Culture. *Methodological term corresponding to a group of systems or organisms in the methodology.*

A group of organisms or biological systems is termed a ***culture***. The word culture was chosen purposely because of its broad usage. Culture is used to describe life in a broad range of contexts, including human beings (e.g., French culture) and microorganisms (e.g., bacterial culture). The use of culture assists in appreciating the unity of living things and it emphasizes the fact that biological systems interact on a daily basis in the biosphere (Fig. 10.2). These interactions influence gene function. Biological function in humans, yeast, bacteria, and other organisms is determined by additive effects of intrinsic and extrinsic factors (see below).

Biosphere

Biosphere. *Methodological term corresponding to the complete collection of all systems and cultures on the planet.*

The term ***biosphere*** in the methodology describes the complete collection of all organisms on the planet. Cultures and systems are the two main components of the biosphere (Fig. 10.2). The function of the biosphere is determined mainly by intrinsic forces, though external factors from the universe (e.g., sunlight, meteors) can also play a role. The fact that we can see the stars and travel to nearby points in space affect human thought and function.

Universe

Universe. *Methodological term corresponding to the highest level of biological organization, encompassing everything living and nonliving in existence.*

A gene expression analysis methodology would see the ***universe*** as the highest order of biological organization, encompassing everything living and nonliving in existence (Fig. 10.2). The universe, by definition, is affected by intrinsic forces only and encompasses every lower level of organization, including all of the biospheres, cultures, systems, system modules, and system variables in existence. Life forms (systems) on other planets, if they exist, would be included in the universe.

State (of a System Variable)

System variables (genes) can be modeled as occurring in three ***states,*** corresponding to DNA, RNA, and protein; and these states obviously need to be consistent with the fundamental dogma of molecular biology. The term state is used in the methodology to draw a parallel to the physical state of a substance (e.g., water in the solid state), to reinforce the notion that the expression of a gene results in the genesis of products that possess strikingly different chemical and biochemical properties. Gene expression can be viewed as the process by which a system variable in the DNA state is converted into RNA and subsequently into protein. The analogy of a "change of states" is useful in that it suggests reversibility, as in the case of an RNA being converted into DNA by reverse transcriptase (and water returning to the solid state).

State. *Any of three biochemical manifestations of a system variable or gene in the methodology, corresponding to DNA, RNA, and protein.*

Form (of a System Module)

System modules (cells) can be viewed in a methodology as existing in numerous morphologies or ***forms***. System module forms correspond to differences in both intracellular and extracellular features, including differences in the morphology of organelles and other subcellular structures, cell shape, and cell size. They are determined partly by the repertoire of genes they express. The term is introduced to underscore the fact that most cell types in a given biological system share an identical set of system variables (genes) but differ in appearance, or form, due to the combined effects of intrinsic and extrinsic factors.

Form. *Methodological term corresponding to the myriad physiological and morphological manifestations of a system module or cell.*

Stage (of a System)

The overall physical appearance of a biological system is viewed as a ***stage***. The term is useful in a methodology to reinforce the idea that the life cycle of an organism is a continuum of stages, each of which is amenable to quantitative analysis with respect to system variables and modules.

Stage. *Corresponds to the number of dynodes in a photomultiplier tube. Alternatively, the methodological term corresponding to the overall physical appearance of a system or organism in the methodology.*

Biological Potential

The biological potential of a system variable, system module, system, culture, or biosphere refers to the full gamut of biological activities that are possible for each component of biological organization. The biological potential of a system module (cell) or system (organism) depends on dozens of variables, including the expression of system variables (genes), the presence of external cues (hormones), and the proximity of neighboring system modules (cells). Biological potential is an appropriate term because it emphasizes the fact that most attributes in biology are not absolute but rather are malleable and contingent on complex interactions with other system variables, system modules, systems, cultures, and biospheres.

External force. *Methodological term corresponding to an extrinsic factor, such as phosphorylation, that has an effect on a system variable, system module, system, culture, or biosphere.*

Biological potential is determined by the additive contributions of intrinsic properties and ***external forces*** (see below). Each system (organism) has a different spectrum of biological potential, owing to the fact that each system contains a unique repertoire of system variables (genes) and interacts with a unique set of system modules (cells). Biological potential is an important component of worldview and theory in microarray analysis because it emphasizes the wide range of gene expression outcomes that are possible in the course of microarray experimentation.

Intrinsic Property

Intrinsic property. *Methodological term corresponding to an attribute, such as primary nucleotide sequence, that is inherent to a system variable, system module, system, culture, or biosphere.*

An attribute that is inherent to a system variable, system module, system, culture, or biosphere is an ***intrinsic property***. At present, no term in the scientific parlance has an equivalent meaning. The primary nucleotide sequence of a gene and the primary amino acid sequence of a gene product provide good examples of intrinsic properties (Table 10.1), because these attributes are contained within the DNA and protein molecules, respectively. But primary nucleotide and amino acid sequence can be influenced by the external forces of mutation

TABLE 10.1. Biological Parameters and the Intrinsic Properties and External Forces That Act on Them

Parameter	Intrinsic Property	External Force
Gene sequence	primary nucleotide sequence	recombination, transposition, mutation, repair, amplification, methylation, restriction
mRNA sequence	primary ribonucleotide sequence	splicing, capping, polyadenylation,
Protein sequence	primary amino acid sequence	protein splicing, posttranslational modification
DNA replication	replication origin	replication proteins
Gene expression	enhancers, promoters, and other *cis*-acting sequences	enhancer binding proteins, promoter factors, and other *trans*-acting components
Protein folding	primary amino acid sequence	intracellular milieu, chaperonins
Protein regulatory properties	protein functional domains	regulatory elements, hormones,
Protein half-life	primary amino acid sequence	proteolysis, ubiquitination
Enzymatic capacity	sequence and structure of active site	covalent modification, substrate availability, cofactors
Subcellular localization	signal sequences	cellular organelles, trafficking machinery
Transcript profile	identity and level of cellular transcripts	hormone signals, cell–cell interactions, environmental stimuli
Cell morphology	repertoire of expressed genes	proximity of neighboring cells, hormone signals, morphogens
Behavior of an organism	genomic sequence	prenatal environment, diet, social pressure

and posttranslational modification, respectively, underscoring the fact that every intrinsic property can be acted up by external forces (see below). The physical shape of a cell, metastatic capacity, the size of a system (organism), and the morals of a culture are additional examples of intrinsic properties, as are expression level, regulatory properties, catalytic capacity, half-life, subcellular localization, mutability, and diffusability.

External Force

An extrinsic factor that has an effect on a system variable, system module, system, culture, or biosphere is termed an external force. No term in the current scientific parlance has an equivalent meaning. The phosphorylation of a protein, the influence of one cell on another, human contact, and the interactions of societies constitute external forces acting on a system variable, system module, system, and culture, respectively. DNA mismatch repair of a nucleotide sequence, phosphorylation and myristilation, gene induction and repression, proteolysis, protein trafficking, and the like provide additional examples of external forces (Table 10.1). Sunlight, wind, rain, peer pressure, and other natural and social factors are also external forces. The effect of any external force on biological potential is determined by the intrinsic properties of system variables and their products, the system module, system and so forth. Every external force—natural, social, or molecular—requires collaboration with one or more intrinsic properties. It is therefore straightforward to see that some systems (organisms) are effected by some external forces and not by others, and that not all systems are created the same but rather are distinct based on the sequence and function of their genes (system variables).

Order from Chaos

One concept that is helpful in formulating a gene expression analysis methodology is the order from chaos idea, which is simply the realization that microarray samples are random (chaotic) mixtures of molecules that are deconvoluted by microarray analysis to yield the identity and amount of each species in the mixture on a gene-by-gene basis (Fig. 10.3). This simple concept helps us appreciate the importance of sample preparation and microarray manufacture. It also provides a solidifying theme by which to understand the 12 rules of parallel gene analysis (see below).

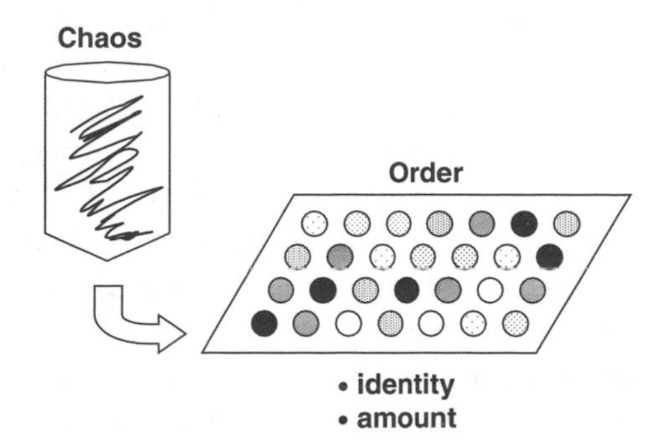

FIGURE 10.3. Samples derived from biological sources are chaotic mixtures (wavy lines) *that can be ordered by microarray analysis to yield the identity and amount* (shaded circles) *of each gene product present in the mixture.*

RULES

Twelve rules of parallel gene analysis. *Methodological set of regulations and procedures used to guide gene expression analysis using microarrays.*

Parallel. *Methodological description of a microarray substrate referring to the fact that microarray spots lie in the same plane and never converge.*

Rules are a set of regulations or procedures used to guide a process defined by a methodology. In order of epistemological level, rules occupy the third tier, below worldview and theory (Fig. 10.1). In a gene expression analysis methodology, the ***twelve rules of parallel gene analysis*** (Schena and Davis, 1999) provide a road map by which to obtain optimal experimental results with microarrays. Each of the rules is presented below, along with a brief explanation.

1. *All gene analyses must be performed in parallel.* Rule 1 states that each and every microarray experiment must be performed on a parallel or planar surface, such as glass or another solid material. ***Parallel*** is defined as "lying in the same plane" and "never converging," both of which are accurate descriptions of treated and coated glass but not of flexible filters made of nylon, nitrocellulose, and other materials. The requirement for parallelism in the methodology ensures the simultaneous evaluation of all of the genes represented in the parallel assay. Using a solid support, such as nylon or nitrocellulose, without a planar surface is not sufficient and does not constitute a microarray assay. Nonparallel assays for gene expression, including northern blotting, S1 nuclease analysis, ribonuclease protection, primer extension, plaque hybridization, differential display, serial analysis of gene expression (SAGE), transcript imaging, and the PCR-based protocols, are serial methods that do not meet the requirement. Parallelism, miniaturization, and automation (see below) are three foundational tenets employed by the microelectronics industry and by the microarray field.

2. *Technologies for parallel format preparation must be amenable to miniaturization and automation.* Rule 2 states that any and all methods of microarray manufacture must be amenable to miniaturization and automation. The combinatorial methods that use photolithography and micromirrors, ink-jetting procedures, and contact printing technologies all meet the requirements of this rule. Rule 2 is useful in that it encourages current microarray manufacturers to employ miniaturization and automation as key principles, and guides the developers of new manufacture technologies. Methods of manufacture that do not provide for relatively high density and throughput are unlikely to meet the needs of the microarray community satisfactorily and should be avoided for this reason.

3. *Each round of parallel gene analysis has five steps: biological question, sample preparation, biochemical reaction, detection, and data analysis and modeling.* The third rule of parallel gene analysis defines five distinct steps that should be used in each and every gene expression microarray experiment. The steps can be viewed as an experimental cycle, with each round of experimentation yielding a more accurate view of the biological system until the question posed initially is answered (Fig. 10.4). The stepwise and repetitive nature of the experimental cycle has the appearance of birth, development, growth, and death; for this reason, the basic components of parallel gene analysis can be viewed as a life cycle.

Biological questions can be broad or narrow in the context of microarray-based gene expression analysis. How does treatment of a mouse with dexamethasone alter the expression of liver genes? What is the temporal order of induction of genes in a kinase signaling cascade? Is autism diagnosable by examining patterns of gene expression in patients? These are the types of questions that can and should be put forth before embarking on a microarray experiment. Simply using microarrays because they are "sexy" or powerful will

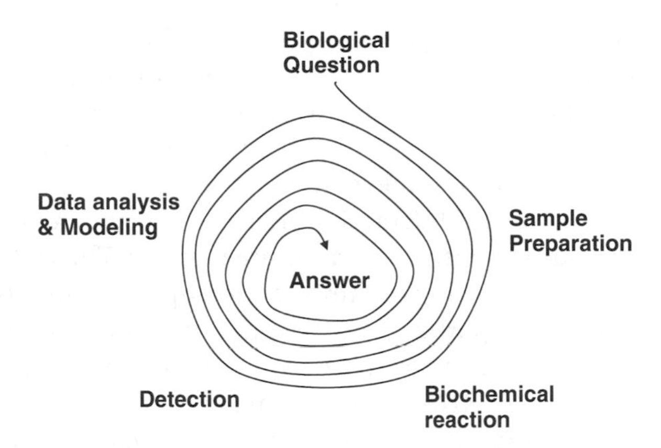

FIGURE 10.4. The five steps of the experimental life cycle in a gene expression microarray methodology starts with the researcher posing a biological question and ends with data analysis and modeling.

do little but generate an enormous amount of data that could take weeks or months to analyze, and analysis is not worth the time if the experiments are not based on a properly structured question.

The soundness of a biological question can be checked by asking questions of the question. Is the question appropriate for the experimental system? Is the question too broad or narrow? What are the possible artifacts? Is the experiment reproducible? Once a sound question is formulated and evaluated, the researcher can then proceed to the wet lab to initiate sample preparation and the other downstream components of the experimental life cycle.

Sample preparation may include preparing targets for microarray manufacture and fluorescent probes for hybridization. The biochemical reaction most commonly involves hybridization of a sample to the microarray, but protein microarrays could also be used for gene expression analysis, in which case the biochemical reaction would involve protein–protein interactions. Detection pertains to scanning or imaging the microarray, and data analysis and modeling uses the tools of bioinformatics to mine and evaluate the results. Each round of experimentation produces a clearer picture of the question posed at the beginning, and the cyclic process should be used in an iterative manner until an answer to the question is obtained (Fig. 10.4).

4. *Manipulations of biological systems must reflect precisely the biological question.* The fourth rule of the methodology means that bacterial and yeast cells, transgenic organisms, plants, animals, and other biological systems used to address biological questions must be handled with care to prevent experimental artifact. If a biological question involves asking how a specific chemical compound affects gene expression in yeast, the cells must be manipulated in such a way that controls for other potential modulators of gene expression, including changes in temperature, centrifugation, ethanol addition, and other experimental manipulations. If a biological question hopes to address how light alters gene expression in plants, the control plants must be treated in exactly the same manner as the experimental plants. Differences in temperature, carbon dioxide, mechanical vibration (Braam and Davis, 1990), and other stimuli can and will alter gene expression in plants, producing unexpected changes in expression.

Growth media, rates of agitation, the shape and size of culture flasks and dishes, strain backgrounds, environmental and dietary differences in transgenic colonies, and many other factors can affect gene expression and should be considered in detail to make sure that the question that is being answered through the course of microarray experimentation is the same as the question posed at the beginning of the experimental process. Poor planning and sloppy technique with respect to the cells and organisms can lead to serious artifact and the contamination of gene expression databases.

5. *Biochemical samples must reflect precisely the biological specimen.* Rule 5 states that care must be taken in sample preparation to ensure that the probe mixtures derived from RNA (and protein) reflect the starting material present in the biological specimen in an accurate manner. The isolation, purification, amplification, and labeling of samples encompass hundreds of methods and techniques, and it is important to realize that deviations at any point in the sample preparation procedure can cause inaccuracy and artifact in microarray analysis (see Chapter 6). RNA molecules are susceptible to rapid and differential turnover and degradation by ribonucleases, placing a premium on robust RNA isolation procedures and careful technique. Failure to isolate the RNA quickly and employ the necessary precautions against ribonucleases can doom a microarray experiment. The differential stability of proteins poses unique challenges to gene expression experiments designed to monitor protein levels.

Methods of RNA amplification and labeling must also be considered carefully to ensure that skewing of the molecular population with respect to the specimen is minimized. Linear amplification methods such as those that use T7 polymerase are preferable in most cases to the exponential PCR-based procedures, because the amplified sample more closely reflects the starting material in the former case. Most direct and indirect labeling procedures manifest minimal sequence bias, but care should be taken when choosing a labeling procedure. An ideal set of sample preparation procedures yields a probe mixture that provides an exact measure of the types and amounts of each molecule present in the specimen.

6. *Parallel formats must provide a precise and ordered reflection of the biochemical sample.* The sixth rule of parallel gene analysis states simply that the microarrays used in the course of experimentation must provide an orderly and accurate measurement of the molecules present in the probe mixture. Substrates with poor surface treatments and coatings cause inaccuracies in gene expression readings and should not be used for microarray experiments (see Chapter 5). Insufficient target density can lead to a saturation of target molecules and a clipping of the readout at feature locations corresponding to abundant species (see Chapter 6). Microarrays that contain uneven spot spacing can cause serious sample tracking and quantification problems and should also be avoided (see Chapter 7). Most current providers of commercial microarrays provide parallel assay formats that allow precise sample analysis, and researchers preparing their own microarrays can adhere successfully to rule 6, as long as the methods of manufacture yield orderly microarrays of high quality.

7. *Detection systems must allow the precise acquisition of data from the parallel format.* Rule 7 states that the detection system (e.g., scanner or imager) must capture the microarray data in a precise manner. Most parallel formats (microarrays) emit fluorescent signals, and so accurate signal acquisition in most cases involves efficient fluorescence detection. The careful choice of optics, lasers, photomultiplier tubes, analog-to-digital converters and other detection components ensure that the detection system faithfully captures the signals emitted from microarrays. Photobleaching, quantum yield, channel-to-channel cross-talk, dark current, shot noise, light scatter and reflection, and a host of other theoretical and practical considerations bear on the issue of detection quality (see Chapter 8). Most commercial scanners and imagers provide

extremely high precision in data acquisition, with lesser and more variable quality seen in home-built systems. The key here is to conceptualize the importance of maintaining the linearity, detectivity, and dynamic range of the signals on the microarray and converting those data into numbers that provide an accurate snapshot of the microarray.

8. Data from the detection system must be manipulated and modeled in a precise manner. The eighth rule of parallel gene analysis addresses the computational issues involved in microarray analysis, stating that data quantification must be performed in a manner that accurately reflects the images produced by the scanner or imager and that modeling of the quantified data must be carried out in a way that mirrors the biological system. Nearly all scanners and imagers provide graphical images (e.g., TIFF files) of microarray data, but extracting the meaning from the data requires accurate and robust bioinformatics tools (see Chapter 9). The most basic challenge is that the quantification software must provide a numerical output (e.g., 16-bit data) that closely matches the data contained in the graphical image. Spot selection, background subtraction, artifact rejection, and many other considerations bear on the issue of accurate spot quantification (see Chapter 9).

Another fundamental issue is that each spot in the image must correlate precisely with the content map, which contains the identity of the target sequences deposited at each microarray location. Assigning signal intensities or gene expression ratios to the wrong genes would be catastrophic with respect to the modeling of the results. Most commercial quantification packages provide highly accurate sample tracking and content map capabilities, virtually eliminating concerns about spot identity. Another requirement in data quantification pertains to the problems associated with small numbers, which are microarray spots with weak signal intensities. Ratiometric measurements with small denominators can produce wildly erroneous expression ratios if care is not taken to account for weak signal intensities (see Chapter 9). Statistically insignificant numbers is a matter that must be addressed for proper data modeling.

Regulatory relationships, signaling hierarchies, temporal control, and a host of other fundamental issues in gene expression can be addressed by microarray experimentation, but it is imperative to have well-characterized biochemical properties for all of the gene products that are used in model building (see rule 11). Improper information concerning DNA binding and substrate specificity, protein–protein interactions, subcellular localization, and other functional properties open the possibility of improper data modeling. The issue of whether a microarray contains a complete or partial gene set also affects data modeling and the conclusions that can be drawn from a given experiment (see rule 10).

9. Comparisons of two or more parallel datasets shall be subject to the limitations inherent in comparing separate experiments. Rule 9 of parallel gene analysis reinforces the important concept that data collected from a single microarray is inherently more accurate than data from two or more separate microarrays; in the latter case, great care must be taken when comparing expression profiles derived from two or more chips. Gene expression ratios derived from one microarray using two-color fluorescence controls for minor differences in chip manufacturing, sample preparation, hybridization conditions, and other potential sources of variation that can be problematic when comparing different samples reacted with different chips. Gene expression databases built using ratiometric information derived by comparing separate chips are subject to the shortcomings that limit such comparisons.

10. Conclusions concerning gene relationships can be made only from parallel experiments that singularly interrogate a complete set of system variables. The tenth rule reminds the researcher that global conclusions concerning gene function can be drawn only if the microarrays that are used for the experiments contain a complete set of genes (system variables) for that organism (system). Microarrays with partial gene sets can be extremely useful and economical when examining a relatively well understood process, but it is essential to understand that global conclusions concerning biological processes require that all of the genes in the genome be represented to avoid overlooking key interactions or relationships. Microarrays containing 10,000 genes often yield an enormous volume of useful information, but these parallel platforms are methodologically incomplete if the organism (system) that is being examined contains 15,000 genes. Disease pathways and the targets of drug action may involve a single gene of diagnostic and therapeutic value, and such a gene may be overlooked if the microarrays used in the analysis contain a partial set of system variables.

11. A universal parallel format is one that contains analytical elements for complete set of system variables for which all of the intrinsic and extrinsic properties have been delineated. Rule 11 defines the *ultimate microarray* as one that contains every gene in the genome, along with a complete functional understanding of each gene and gene product represented in the microarray. A complete set of system variables (genes) ensures that every possible gene expression relationship is examined in a gene expression experiment. A companion database that contains comprehensive information on every gene and gene product, including a complete understanding of internal potential and external force for each gene (see above), enables powerful modeling of the gene expression data.

Complete knowledge of each system variable includes information concerning the genetic, molecular, biochemical, chemical, biophysical, histological, enzymatic, and structural properties of the genes and their products. The characterization of a system variable is complete only when such detailed information is available for each of these parameters. Rule 11 places tremendous value and importance on detailed studies of single genes and proteins, undermining the common misconception that genomics and microarray analysis minimizes the importance of more traditional research endeavors. It is just the opposite. Microarray data are interpreted best in the context of a large body of detailed information, the majority of which is assembled only through exhaustive gene-by-gene and protein-by-protein analysis of each organism (system) of interest.

12. Parallel gene analysis for an organism is said to be complete when a four-dimensional dataset has been assembled for all of the system variables in each system module. The twelfth rule outlines what is required for a complete gene expression analysis of an organism, namely quantitative information for every gene (system variable) in every cell (system module) over the complete lifetime of the organism (system). A three-dimensional dataset is one that contains gene expression information for every gene in every cell at a single moment in time, providing a temporal gene expression snapshot. A four-dimensional dataset includes the time dimension and thus provides a "motion picture" of gene expression data. If this sort of movie were available, it would allow the researcher to understand how gene expression events are coordinated in real time in the context of the whole organism.

A comprehensive, four-dimensional gene expression dataset is reminiscent of the fate map that is available for the nematode *C. elegans,* and the worm might be a good place to start in terms of generating a genome-wide temporal gene expression database for a whole organism. One challenging requirement of such a database is the mandated single-cell resolution, a capability that may be possible using currently available RNA amplification strategies (see Chapter 6), providing that each cell could be separated and identified. Most current microarray analyses involve mixtures and populations of cells, and samples of this type yield extremely useful data; but it is important for the researcher to appreciate the methodological fact that microarray experiments with mixed-cell populations yield average values at the level of the system module.

METHODS, TECHNIQUES, AND TOOLS

Microarray analysis employs myriad methods, techniques, and tools (Fig. 10.1), and their use is guided by the methodological principles embodied in the more abstract components of the methodology (worldview, theory, and rules). The key distinction is that methodology defines "what" and "why," and methods, techniques, and tools provide "how." Short lists of examples for microarray surfaces (Table 10.2), targets and probes (Table 10.3), manufacturing (Table 10.4), detection (Table 10.5), and data analysis and modeling (Table 10.6) underscore

TABLE 10.2. Methodological Underpinnings for Microarray Surfaces

Methodology		Method, Technique, or Tool
What	Why	How
Planar substrate	enables parallel analysis of genes	glass substrate
Dimensional substrate	uniform physical size facilitates manufacture and detection	precision glass cutting
Flat substrate	provides added precision in printing and detection	precision polishing to optical flatness
Uniform reactive groups	enables uniform density of target molecules	treatment with organosilane reagents
Durable surface chemistry	prevents loss of target molecules during the assay	schiff base attachment chemistry
Low background fluorescence	substrates that elevate fluorescent background reduce the detectivity of the assay	background minimization during substrate manufacture
Efficient binding of probe molecules to targets	efficient binding produces intense and uniform signals	surface treatments and coatings that allow facile probe accessibility
Hydration during the hybridization reaction	prevents drying of the sample and elevated background	hybridization cassette
Uniform layer of probe solution	increases assay precision by providing the same concentration of probe molecules across the entire microarray	glass cover slip

TABLE 10.3. Methodological Underpinnings for Microarray Targets and Probes

Methodology		Method, Technique, or Tool
What	Why	How
Efficient oligonucleotide synthesis	pure populations of target molecules improve hybridization specificity	photolithography, micromirrors, phosphoramidite synthesis
Amplified cDNAs	amplified cDNA targets enable gene expression monitoring	PCR
Purified cDNA targets	purified targets attach more efficiently and produce less background	membrane purification kits
Denatured cDNA targets	single-stranded cDNAs are essential for hybridization to probe molecules	treatment in boiling water
Covalent modification of oligonucleotides	covalent modifiers allow directional coupling to the surface	synthetic addition using modified phosphoramidites
High quantum yield dyes	bright dyes produce stronger fluorescent signals	cyanine 3 and cyanine 5
Purify probes to remove unincorporated fluorescent nucleotides	free fluorescent nucleotides can elevate fluorescent background	spin column purification
Signal amplification and enhancement	amplification and enhancement of fluorescent signal improves assay detectivity	tyramide signal amplification, dendrimer staining
Rapid and thorough washing after hybridization	removes unbound fluorescent probe molecules and reduces background	wash station

TABLE 10.4. Methodological Underpinnings for Microarray Manufacture

Methodology		Method, Technique, or Tool
What	Why	How
Controlled environment	environments free of particulate and biological contamination produce superior microarrays	class 100 microarray cleanroom
Microarrays with highly ordered elements	regular spot spacing enables rapid and accurate quantification	photolithography, ink-jet printing, contact printing
Microarrays containing a complete set of genes	whole genome representation provides complete biological information	genome sequence databases, clone sets, oligonucleotide synthesis
Microarrays containing partial gene sets	gene subsets allow targeted studies and reduce cost	genome sequence databases, clone sets, oligonucleotide synthesis
Microarrays containing protein targets	enable the determination of protein levels and detection of posttranslational modification	protein and antibody microarrays
Contact printing devices that deliver small volumes of sample	contact printing implements enable microarray manufacture when combined with suitable motion control	solid pins, tweezers, microspotting pins, pin and ring
Positional accuracy in three dimensions	highly accurate motion control is required for many different methods of microarray manufacture	stepper motor, servomotor, linear actuator, linear encoder
Attaching DNA to the surface following microarray printing	stable attachment improves assay performance by providing stable target sequences	dehydration, baking, cross-linking with ultraviolet light

TABLE 10.5. Methodological Underpinnings for Microarray Detection

Methodology		Method, Technique, or Tool
What	Why	How
Rapid data acquisition	high-speed data gathering improves the robustness of microarray analysis	confocal scanning, nonconfocal scanning, imaging
High throughput	detecting many substrates or slides in a short period speeds microarray analysis	autoloading devices for slides or substrates
High resolution	large number of pixels per unit area improves image quality when reading high-density microarrays	PMT, CCD, or CMOS detector[a]
High detectivity	high detectivity allows detection of low abundance transcripts and other rare species	Pin hole in optical path, PMT with low dark current, nonreflective substrate holder
Detection at multiple wavelengths	multicolor detection allows the analysis of two or more samples in a single experiment	Separate lasers for each wavelength, continuous light source with proper filtering
Low amount of optical cross-talk	reducing channel-to-channel spill over improves the precision of ratiometric analysis	beam splitter, excitation filter, emission filter
Low amount of photobleaching	little or no photobleaching improves detection precision by minimizing fluorescent dye degradation	low output lasers, neutral density filters, short dwell times
Digital intensity values from analog detector sources	digital data are required for computer-based analysis	analog-to-digital converter

[a] *PMT,* photomultiplier; *CCD,* charge-coupled device; *CMOS,* complementary metal oxide semiconductor.

TABLE 10.6. Methodological Underpinnings for Microarray Data Analysis and Modeling

Methodology		Method, Technique, or Tool
What	Why	How
Precise knowledge of the target sequence present at each microarray location	identity of the target sequence is essential to meaningful data analysis	algorithms that generate content files based on target sequences
Quantification of each microarray location	intensity values at each location are required to derive information about each gene represented in the microarray	manual, semiautomatic, and automatic spot-finding programs
Depict changes in expression for all of the genes in two experiments	allows researcher to visualize activation and repression and absolute expression on a global scale	scatter plot
Functional classification of genes used in expression experiments	gene classification based on function allows the assembly of regulatory relationships and the assignment of function to unknown sequences	hierarchical, self-organizing map, support vector machines and related clustering approaches
Automated image processing	allows multiple microarray images to be processed simultaneously, speeding up data analysis	batch-mode procedures for quantification
Depict changes in expression for many genes across many experiments	provides researcher with a global view of gene expression for complex datasets	principle component analysis
Regulatory models based on accurate protein function data	regulatory interactions implied by microarray data must occur in vivo	two hybrid analysis, filter binding, column chromatography
Correct for minor differences in intensity between two channels	allows the generation of more accurate expression ratios	multiply one of the two channels using a correction factor obtained by taking the quotient of the sums of the two datasets

the need to distinguish methodology and method and highlight the value embodied in this distinction with respect to developing powerful new methods, techniques, and tools for microarray analysis.

IMPLEMENTATION

The least abstract component of methodology is implementation (Fig. 10.1), which is the literal task of doing microarray experiments. Implementation is carried out using the methods, techniques, and tools of microarray science vis-à-vis the worldview, theory, and rules, but implementation can be achieved by anyone who is trained to perform microarray experiments. One evaluative criterion for methods, techniques, and tools is the extent to which they can be implemented by users with minimal advanced training, a characteristic that allows scientists of many different levels of expertise to participate in microarray science. Procedures that can be performed by microarray technicians reduce the cost associated with implementing microarray technology in basic research and commercial laboratories.

SUMMARY

The complete body of theories, concepts, principles, rules, methods, protocols, techniques, formulae, and the like that guide microarray research is known as the methodological architecture. The methodological framework for microarray analysis uses the principles of engineering to configure biology into a set of informational tiers that include the system variable (gene), system module (cell), system, culture, biosphere, and universe. Different states, forms, and stages describe system variables, system modules, and systems, respectively. Proteins and cells are described as having biological potential that is determined by the intrinsic properties and external forces. The twelve rules of parallel gene analysis govern microarray-based gene expression monitoring in a formalistic manner. Methods, techniques, and tools are used in conjunction with the methodological underpinnings to implement microarray analysis.

SELECTED READING

Baxevanis, A. D., and Ouellette, B. F. F., eds. *Bioinformatics: A Practical Guide to the Analysis of Genes and Proteins,* 2nd ed. Wiley-Interscience, New York, 2001.

Berson, A., and Smith, S. J. *Data Warehousing, Data Mining, and OLAP.* McGraw-Hill, New York, 1997.

Braam J., and Davis R. W. Rain-, wind-, and touch-induced expression of calmodulin and calmodulin-related genes in Arabidopsis. *Cell* 60:357–364, 1990.

Fersht, A. *Structure and Mechanism in Protein Science: A Guide to Enzyme Catalysis and Protein Folding.* Freeman, New York, 1999.

Higgins, D., and Taylor, W., eds. *Bioinformatics: Sequence, Structure, and Databanks: A Practical Approach.* Oxford University Press, Oxford, UK, 2000.

Lehninger, A. L., Nelson, D. C., and Cox, M. M. *Principles of Biochemistry,* 3rd ed. Freeman, New York, 2000.

Lewin, B. *Genes VII.* Oxford University Press, Oxford, UK, 2000.

Schena, M., ed. *DNA Microarrays: A Practical Approach,* 2nd ed. Oxford University Press, Oxford, UK, 2000.

Schena, M., ed. *Microarray Biochip Technology.* Eaton Natick, MA, 2000.

Schena, M., and Davis, R. W. Parallel analysis with biological chips. In M. Innis, D. Gelfand, and J. Sninsky, eds.

PCR Methods Manual. Academic Press, San Diego, CA, 1999:445–456.

Singer, M., and Berg, P. *Genes and Genomes.* University Science Books, Herdon, VA, 1991.

Strachan, T., and Read, A. P. *Human Molecular Genetics,* 2nd ed. Wiley, New York, 1999.

Thomsen, E. *OLAP Solutions: Building Multidimensional Information Systems.* Wiley, New York, 1997.

REVIEW QUESTIONS

1. Which of the following in *not* a tier in the structure of a methodology: (a) theory, (b) worldview, (c) rules, (d) methods, (e) techniques, (f) Ten Commandments, and (g) implementation?
2. A methodology focuses *primarily* on telling a scientist which one of the following: (a) what and how, (b) what and why, (c) what and when, or (d) how and where?
3. Match the what, why, and how in a methodology with the following steps in a microarray procedure: (a) use distilled water at 100°C, (b) denature the two strands of the PCR products in a cDNA microarray, and (c) allow hybridization of the cDNA strands with fluorescent probe molecules.
4. With respect to the engineering aspect of the methodology, assign one of the following to the terms *system module, system, system variable,* and *culture:* (a) nerve cell, (b) human population of New York City, (c) fruit fly, and (d) *p53* gene.
5. With respect to the biological samples employed, microarray analysis creates informational order out of biochemical: (a) chaos, (b) unity, (c) data, or (d) feedback loops?
6. Explain the significance of each aspect of the circular arrow in Figure 10.4: (a) free end of the arrow, (b) arrowhead, (c) concentric pattern, and (d) clockwise revolutions.
7. With respect to the catalytic activity of an enzyme, which of the following are external forces: (a) phosphorylation, (b) myristylation, (c) primary amino acid sequence, and (d) glycerol concentration.
8. Which of the following procedures might make up rule 5 of the Twelve Rules of Parallel Gene Analysis: (a) Eberwine amplification, (b) tyramide signal amplification, (c) reverse transcription, (d) replacing a linear actuator, (e) dendrimer staining, (f) changing an mRNA purification protocol, and (g) increasing the gain on a photomultiplier tube?
9. Which of the following hardware components might make up rule 7 of the Twelve Rules of Parallel Gene Analysis: (a) microspotting pin, (b) laser, (c) imaging stage, (d) photomultiplier tube, (e) lens, and (f) excitation filter?
10. A researcher uses a 10,000-gene microarray to explore the mechanism of action of insulin in human cells, but his supervisor reminds him that he would be compromising one of the twelve rules by his choice of microarray. Which rule would be compromised in this experimental approach? Explain your answer.
11. A genomics researcher exclaims that single-gene studies are dead, but one of the twelve rules suggests a critical role for detailed biochemical, structural, and genetic studies of single genes and proteins. Which rule?

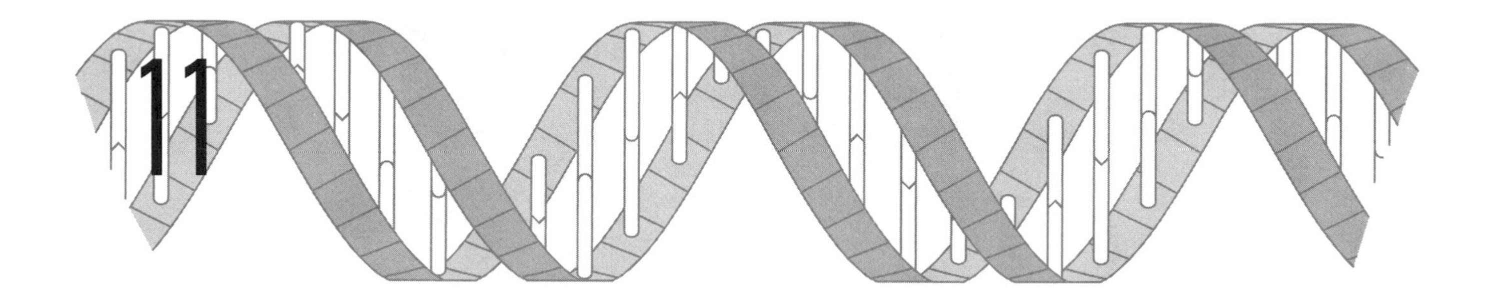

11 Cleanroom Technology

Never go to a doctor whose office plants have died. —*Erma Bombeck*

A cleanroom is an advanced, controlled enclosure in which the levels of particles, biological agents, and other forms of contamination are reduced greatly relative to the ambient environment. Cleanroom technology has its roots in the semiconductor industry dating back to the mid-1970s, during which time commercial computer chip manufacturers such as Intel (Santa Clara, CA) identified the need to create cleaner working environments to minimize the contamination of their silicon wafers (Fig. 11.1). The microarray field has expanded on the principles established for semiconductor cleanrooms, developing specialized technologies to meet the unique requirements of microarray manufacture and use. This chapter provides a glimpse into state-of-the art microarray cleanroom technology, providing discussions of the criteria that are used to benchmark cleanrooms and the different forms of contamination that must be controlled on a daily basis to maintain the exacting specifications of these sophisticated environments.

EVALUATIVE CRITERIA

Microarray cleanrooms are evaluated using five specific criteria: *class, environmental control, functionality, durability,* and *safety*. This section provides a brief explanation of each of these important criteria.

FIGURE 11.1. **A,** *An early semiconductor cleanroom built in 1975.* **B,** *A modern facility in 2000. (Courtesy of Intel, Santa Clara, CA.)*

Class

Class. *Cleanroom evaluative criterion corresponding to the number of particles ≥ 0.5 μm per cubic foot of cleanroom air.*

The ***class*** of a microarray cleanroom refers to the quality of the air inside the room in terms of its particle content. Because dust and other airborne particles elevate background fluorescence and cause punctate blemishes in scanned images, particle contamination should be reduced as much as possible in microarray cleanrooms, particularly in commercial facilities that manufacture substrates and printed chips. The standards used for microarray cleanrooms are identical to those established by the Institute of Environmental Sciences and Technology (IEST; Mount Prospect, IL), developed principally for the microelectronics industry. Federal Standard 209E of the IEST specifies six cleanroom classes ranging from 1 to 100,000 (Table 11.1), with smaller numbers corresponding to successively cleaner environments. A class 1 facility has an allowable limit of 1 particle of 0.5 μm diameter per cubic foot (ft^3) of cleanroom air, whereas a class 100,000 facility has a limit of 100,000 particles/ft^3 of 0.5 μm diameter.

Most commercial microarray cleanrooms have ratings of class 100 or class 1,000, providing air that has >1,000 times fewer particles than a traditional laboratory environment, which typically contains 20,000–200,000 particles/ft^3 (Table 11.1). Unfiltered outdoor air in urban areas contains as many as 5,000,000–500,000,000 particles/ft^3, or more than 1 million-fold the particle count of a typical microarray cleanroom. Particulates complicate microarray analysis at multiple steps, and reducing their count is essential for

TABLE 11.1. Particle Limits for Various Classes of Cleanrooms[a]

Type	Number of Particles[b]
Cleanroom	
Class 1	1
Class 10	10
Class 100	100
Class 1,000	1,000
Class 10,000	10,000
Class 100,000	100,000
Research laboratory	20,000–200,000
Urban air	5,000,000–500,000,000

[a] Per Federal Standard 209E.
[b] Number of particles with a diameter ≥ 0.5 $\mu m/ft^3$ of cleanroom air.

obtaining high-quality microarray data. Class 100 or better is recommended for substrate manufacturing, whereas class 1,000 generally suffices for research and development facilities, particularly if the manufacturing robots provide additional filtration as part of their environmental enclosures.

Environmental Control

In addition to reducing and controlling the levels of airborne particles (class rating), cleanrooms are evaluated based on their capacity to control other aspects of the environment. ***Environmental control*** includes ***temperature, humidity, lighting,*** and ***uniformity***.

Because a cleanroom is essentially an airtight enclosure, temperature control presents a surprisingly difficult challenge (see below). Though there are no formal standards for temperature control, most cleanrooms endeavor to maintain a temperature of 20°C (68°F), with a range of ± 3°C (± 5°F). The conversion from Celsius (T_c) to Farenheit (T_f) is achieved by multiplying the Celsius temperature by 1.8 and then adding 32:

$$T_f = 1.8 \times T_c + 32 \quad (11.1)$$

A 20°C (68°F) temperature provides a comfortable working environment for personnel and prevents the overheating of motion control systems and other robotic devices, some of which have thermal protection circuitry that will cease operation if instrument temperatures become excessive. Maintaining a relatively constant temperature range of ± 3°C (± 5°F), ensures the day-to-day consistency of important chemical reactions that depend on the ambient temperature to dictate their reaction kinetics. Large fluctuations in cleanroom temperature can lead to variability in chemical coupling reactions and other manufacturing processes and should be avoided, particularly in a commercial setting where premium microarray quality is essential.

Another component of environmental control is humidity, or more precisely, ***relative humidity,*** which is the ratio of the ambient moisture (M_{amb}), at a defined temperature and pressure, divided by the maximum moisture capacity (M_{max}) of the air. Relative humidity (H_{rel}) is expressed as a percent saturation:

$$H_{rel} = \frac{M_{amb}}{M_{max}} \times 100 \quad (11.2)$$

A relative humidity of 40 ± 5% provides a convenient specification for most microarray cleanrooms. If the relative humidity exceeds 50%, moisture can accumulate on surfaces, trapping particles and preventing efficient air filtration. Extended periods of high humidity followed by a sudden drop into the normal range, can cause a spike in the particle account in the cleanroom because particles trapped on moist surfaces become dislodged when the relative humidity drops. Wide humidity swings up and down can cause fluctuations in particle count and variability in the quality of manufactured substrates and printed microarrays from day to day. For premium-quality microarrays, it is essential to maintain the humidity within the relatively narrow range of 35–45%.

Excessive humidity can also reduce the efficiency of important chemical processes, such as those that involve dehydration reactions. Dehydration is required to couple DNA to amine and aldehyde surfaces, though these reactions

Environmental control. *Cleanroom evaluative criterion that includes temperature, relative humidity, lighting, and uniformity.*

Temperature. *Cleanroom environmental control parameter corresponding to the Celcius reading of the air inside a cleanroom.*

Humidity. *Cleanroom environmental control parameter corresponding to the relative humidity reading inside a cleanroom.*

Lighting. *Cleanroom environmental control parameter corresponding to the intensity and quality of the illumination inside a cleanroom.*

Uniformity. *Microarray detection system criterion that refers to the consistency of the recorded fluorescent signal at different locations across a perfect substrate, typically measured as a percentage. Alternatively, a microarray cleanroom environmental control parameter that refers to the extent to which the temperature, humidity, and lighting are maintained evenly in the cleanroom space.*

Relative humidity. *Cleanroom environmental control parameter defined as the ratio of the ambient moisture at a defined temperature and pressure, divided by the maximum moisture capacity of the air inside a cleanroom.*

can be driven by the use of drying ovens, which can be used to obtain low relative humidity <20% if the cleanroom environment is more humid. Another hazard of excessive humidity is the accumulation of moisture on robotic devices, leading to corrosion and reduced efficiency and longevity of electrical and mechanical components.

Contact printing with pin technologies generally requires 55% relative humidity, which is beyond the upper limit suggested for microarray cleanrooms (45%). To achieve elevated relative humidity for microarray manufacture without humidifying the entire room, environmental enclosures are used to house microarray robots and other manufacturing devices inside the cleanroom. These humidity chambers enable elevated humidity for microarray manufacture, while maintaining a 35–45% reading in the main cleanroom space. Most commercial arrayers include some form of environmental control as standard equipment to allow the user to vary the temperature and humidity of the printing environment, which is almost always different from the ambient conditions of the cleanroom.

Efficient microarray cleanroom operation can also be complicated by excessively low relative humidity. The main risk of low humidity (<25%) is the accumulation of static charge (see below) on surfaces, which can bind particles and reduce the efficiency of air filtration. Extended periods of low humidity followed by a period of normal humidity (35–45%) can produce an sudden increase in the particle count in the cleanroom, due to the release of particles that have accumulated on surfaces because of static charge. Proper selection of cleanroom construction materials prevents static buildup, though the relative humidity should be kept in the 35–45% range to dissipate any static charge that might accumulate (see below).

Lighting in a cleanroom sounds like a trivial matter, but it is nothing of the sort. Proper selection of cleanroom lighting minimizes thermal contamination (see below) and maximizes the quality of the illumination source. Fluorescent lights are preferable to incandescent lights because they emit much less heat. Thermal contamination in cleanrooms is a major consideration due to the fact that extraneous sources of heat lead to elevated temperatures, which presents cooling problems in an enclosed environment.

Color temperature. *Correlation between the thermal state and the color of light emitted by a hot substance, typically expressed in degrees Kelvin.*

Color rendition index. *Measure of how closely a given light source approximates the wavelengths of light emitted by the sun, expressed as a percent.*

There are a wide variety of different fluorescent bulbs available commercially, and illumination sources differ considerably in terms of their ***color temperature*** and ***color rendition index***. The color temperature of a light source is a measure of the whiteness of the light source expressed in degrees Kelvin (K), with greater color temperatures correlating with a whiter light appearance. Natural midday sunlight has a Kelvin rating of approximately 5250 K and a white-yellow appearance. Standard fluorescent lights (Table 11.2) emit light in the

TABLE 11.2. Lighting Specifications

Source	Type	Color Temperature (K)	CRI (%)[a]
Wax candle	candlelight	1000–2000	N.A.
Warm white	fluorescent	3000–3500	50–60
Cool white	fluorescent	4000–5000	60–70
Midday sun	sunlight	5000–5500	100
Full spectrum	fluorescent	5000–6500	> 94

[a] N.A., not applicable.

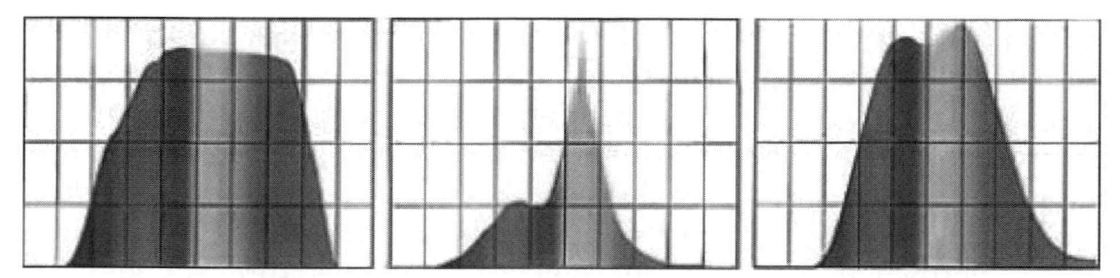

FIGURE 11.2. *Emission spectra for natural sunlight (left), cool white fluorescent lights (center), and full-spectrum fluorescent lights (right). The intensity* (I) *of each source across the visible spectrum of wavelengths is depicted in a rainbow palette according to actual color of each portion of the spectrum. (Data adapted courtesy of Litelids.com, Euclid, OH.) Figure also appears in Color Figure section.*

range of 3000–4250 K, which provides a more yellow appearance than natural sunlight. Full-spectrum fluorescent bulbs are available, with color temperatures in the 5000–6500 K range and should be used in microarray cleanrooms to provide a much whiter light source than standard fluorescent bulbs. Cleanrooms illuminated with 5000–6500 K light offer an optical environment that facilitates the visual inspection of manufactured substrates and printed microarrays. Blemishes that are virtually impossible to detect with standard fluorescent lighting are easily observed if full-spectrum bulbs are used.

The color rendition index (CRI) is a measure of how closely a given light source approximates the wavelengths emitted by the sun. Natural midday sunlight has, by definition, a color rendition index of 100, and standard fluorescent lights provide a CRI in the range of 50–70% (Table 11.2). Full-spectrum fluorescent bulbs offer a color rendition index of > 94% and should be used in microarray cleanrooms to provide a set of wavelengths that covers the visible spectrum in a uniform manner (Fig. 11.2). Standard cool white fluorescent lights have an uneven distribution of emission wavelengths, which impairs illumination quality and can cause eyestrain among cleanroom personnel.

The intensity required to illuminate a microarray cleanroom is approximately 4 W of full-spectrum fluorescent light per square foot of cleanroom space. The 4 W/ft^2 (0.4 W/m^2) illumination density corresponds to approximately 10 full-spectrum bulbs (48-in. size) for a cleanroom space that measures 10 ft^2. Fluorescent lighting fixtures should be sealed completely to prevent cleanroom contamination (Fig. 11.3). Illumination intensity should be uniform across the two-dimensional cleanroom space (see below), to prevent the occurrence of bright and dim areas. Mirrored walls and other surfaces can be used to enhance the uniformity of the illumination by reflecting the full-spectrum wavelengths that are emitted from the cleanroom lights. The light density can

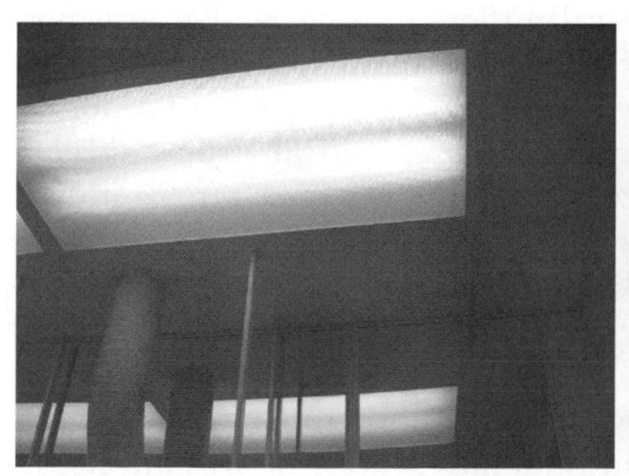

FIGURE 11.3. **A,** *Full-spectrum fluorescent lights in a sealed fixture allow the light to pass but prevent particulate contamination.* **B,** *Mirrored walls can be used to enhance the uniformity of cleanroom illumination by reflecting the full-spectrum fluorescent light. (Courtesy of TeleChem/arrayit.com, Sunnyvale, CA.)*

be increased or reduced by twofold for specific applications, but the 4 W/ft^2 specification provides a good measure in most cases.

Temperature, humidity, and lighting are key criteria by which the environmental control capacities of microarray cleanrooms are evaluated. But all of these criteria must be enabled evenly throughout the two- and three-dimensional space of the room. Uniformity, the fourth component of environmental control, refers to the extent to which the temperature, humidity, and lighting are maintained evenly in the cleanroom space. The standard specification is that each region in a given cleanroom should fall within the allowable range for each of the three specifications. A temperature of 19–25°C, a relative humidity of 35–45% and a lighting intensity that is 4 W/ft^2 $\pm$ 30% should be maintained in all cleanroom areas. The entire room should also be consistent in terms of particle count, with every portion maintaining a single class rating (100, 1000, etc.). Uniformity of environmental control and class rating ensures that every portion of the cleanroom space will have a defined temperature, humidity, lighting, and particle count. Cleanroom uniformity guarantees the consistency of the manufacturing processes throughout the entire microarray facility. A large number of design considerations determine uniformity (see below).

Functionality

Functionality. *Cleanroom evaluative criterion that includes the sum of all of the physical attributes of a cleanroom, including overall design, airflow, environmental control, chemical containment, utilities, surface materials, shelving, bench design, and accessibility.*

The sum total of all of the physical attributes of a cleanroom is referred to as ***functionality***. Microarray cleanroom functionality includes overall design, airflow, environmental control, chemical containment, utilities, surface materials, shelving, bench design, and accessibility.

With respect to the overall design of a microarray cleanroom, the most important concept is that the room should be designed around the process that it endeavors to enable. A substrate manufacturing cleanroom might be different from one destined for use in research and development. Surface chemistry considerations are likely to be different from the issues that surround genetic screening and diagnostics. A cleanroom intended for use as an educational and demonstration facility has different considerations from one that might find use in forensics. Specific applications should be considered in detail before a cleanroom is constructed. A secondary design consideration is that the cleanroom should be adapted to the existing exterior facility. If the overall design can be tailored for compatibility with standard cleanroom construction materials, the cost of building is reduced considerably compared to custom designs.

Laminarity. *Cleanroom parameter that refers to the direction of the airflow inside a cleanroom.*

The direction of the airflow or ***laminarity*** of a cleanroom is another aspect of functionality. The two types of laminarity are vertical and horizontal, corresponding to ceiling-to-floor and wall-to-wall airflow, respectively. Because most commercial buildings contain air ducting in the ceiling space, vertical laminarity is the most common microarray cleanroom design, as air-conditioning feeds can be plumbed into the top of the structure. Vertical airflow also has the advantage of using the ceiling space for the supply of fresh (makeup) air, which preserves the wall space for shelving and storage. Horizontal laminarity provides the requisite cleanroom specifications, though this design is less common than vertical flow because it is generally more expensive and somewhat more difficult to implement.

Environmental control is an important aspect of functionality in the sense that the location of filtration and humidity control devices affects the physical space of the cleanroom. The environmental control systems of most microarray cleanrooms contain main components and ancillary components. The main components are nonportable and supply the majority of air at defined filtration, temperature, and humidity parameters. The ancillary components are portable devices that provide additional environmental control. Freestanding air filtration devices are used commonly in microarray cleanrooms to enable further reductions is particle count (see below).

Substrate manufacturing and other cleanroom processes that require the use of organic solvents and other volatile compounds pose a serious challenge to cleanroom functionality, in that enclosed environments do not dissipate chemical fumes nearly as readily as open laboratory space. Chemical containment is achieved in cleanrooms by the construction of subchambers and through the use of fume hoods that remove volatile substances and prevent their accumulation in the air supply. Minimizing the accumulation of chemical fumes inside the cleanroom is critical with respect to worker safety and because solvents can react with microarray surfaces, leading to changes in surface chemistry and elevated background fluorescence. Chemical containment is readily achieved with proper microarray cleanroom design.

Functionality is also determined by the efficiency and design of cleanroom utilities, including electrical outlets, distilled water supplies, and drains. Depending on the activities inside the room, electrical loads can be extremely large, particularly in manufacturing environments that may require extensive lighting as well as a large number of robots, vacuum pumps, freestanding air-conditioners, and other electrical devices that draw considerable current. Care should be taken to ensure that the electrical outlets are sufficient to handle the load of a given facility. With respect to the water supply, it is imperative that the water quality be compatible with each cleanroom application. The use of solder, soldering agents, solvents, and other traditional plumbing materials may negatively impact the cleanroom water quality, and a professional should be consulted to make sure that the proper materials are used. To minimize fire hazard, the installation of natural gas lines and the use of open flames in microarray cleanrooms is strongly discouraged and should be avoided if at all possible to maximize worker safety.

Cleanroom functionality is also assessed on the basis of the construction materials that are used for the ceiling, walls, benches, shelves, and floor. All surface materials should be particle free, vapor free, smooth, resistant to moisture and chemicals, and white or light colored. Surfaces that release particles or chemical vapors into the cleanroom space will diminish the class rating and reduce the quality of the air. Such materials should be avoided at all cost. Smooth surfaces prevent particles from being trapped and expedite airflow in the space. Surfaces that are resistant to water and mild chemicals provide durability during cleanroom use and periodic cleaning. Materials that are white or light colored reflect light and therefore provide much more uniform illumination than pigmented surfaces.

Shelving and bench construction should be sufficient to accommodate the weight load required in a given cleanroom environment. Microarray manufacturing robots can be quite heavy (>200 kg) and benches designed to hold multiple robots deserve special attention in terms of design and construction. Wire shelving units should be used when possible, to expedite airflow and air filtration (see below). Accessibility of the cleanroom is determined by the design of the room itself and by the components inside it. Walkways should be constructed to a minimum width of 36 in. (91 cm) to allow easy access during daily use and rapid passage in the case of emergency. Portable carts and chairs provide

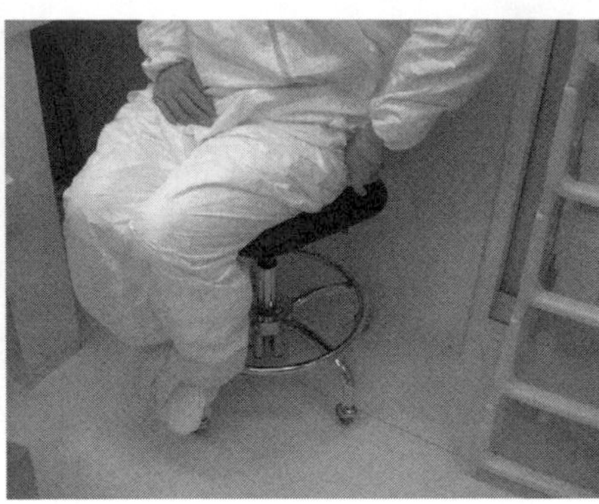

FIGURE 11.4. Portable cleanroom accessories, including a stainless-steel cart (A) and chair (B), enhance functionality while maintaining the flexibility of the floor space. (Courtesy of TeleChem/arrayit.com, Sunnyvale, CA.)

efficient transport and seating accessories, while maintaining the openness of the cleanroom interior (Fig. 11.4).

Durability

Durability. *Cleanroom evaluative criterion that refers to the extent to which the physical structure, environmental control, and other components of a microarray cleanroom maintain their specifications over time. Alternatively, the total length of time that an electric motor or microarray robot can be used without visible deterioration in performance, expressed typically in machine hours.*

The ***durability*** of a microarray cleanroom refers to the extent to which the physical structure, environmental control, and other components maintain their specifications over time. High-quality cleanrooms should provide near maintenance-free operation for 2–3 years. The use of advanced building materials and high-performance environmental control components ensures durability. Caution should be exercised when considering prefabricated or portable designs, both with respect to durability and to whether class ratings can be maintained in the context of heavy use. High-quality microarray cleanrooms are currently available from several commercial sources.

Safety

Safety. *Cleanroom evaluative criterion that pertains to the extent to which a cleanroom maximizes the physical comfort and well-being of cleanroom personnel and minimizes or eliminates any chance of injury.*

A most important criterion for microarray cleanrooms is user ***safety***. The room should be designed to eliminate any chance of personnel discomfort or injury. The use of heavy-duty construction materials ensures the stability of the cleanroom structure, as well as that of the benches and shelves. Proper airflow and chemical containment maintain an abundant supply of fresh air into the facility at all times. Advanced electrical, plumbing, heating, and cooling practices guarantee the safe functioning of the utilities. The use of portable accessories and sliding doors allow easy access and enable rapid egress in the case of an emergency (Fig. 11.5**A**). Readily accessible fire extinguishers should be available

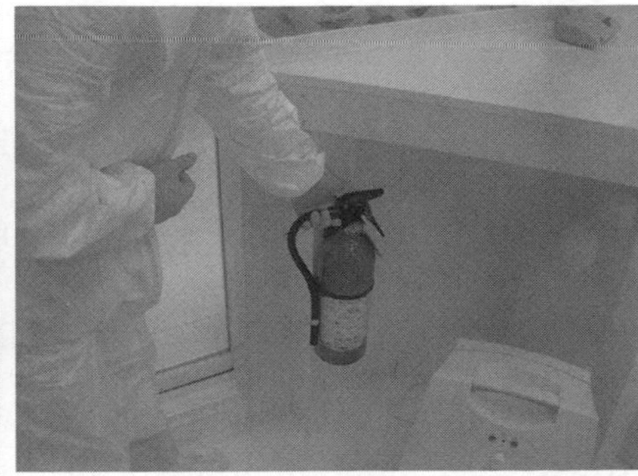

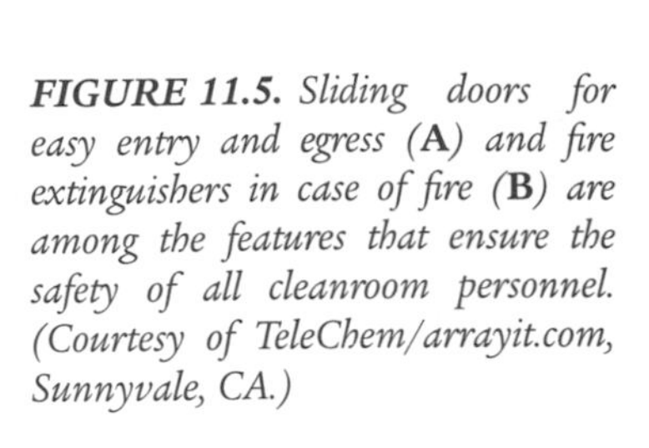

FIGURE 11.5. Sliding doors for easy entry and egress (A) and fire extinguishers in case of fire (B) are among the features that ensure the safety of all cleanroom personnel. (Courtesy of TeleChem/arrayit.com, Sunnyvale, CA.)

in case of fire (Fig. 11.5**B**). Commercially available microarray cleanrooms offer comprehensive safety features.

CONTAMINATION

Contamination presents the single biggest challenge to effective day-to-day use of a microarray cleanroom. Momentary lapses in cleanroom protocol can lead to devastating increases in particle count and sudden and dramatic decreases in performance. There are five main forms of contamination: particle, biological, chemical, thermal, and electrostatic. This section explores each of these sources of contamination and provides guidelines for minimizing or eliminating them from the cleanroom environment. Proper use of the facilities by trained personnel ensures the success of microarray cleanrooms as the environment of choice for microarray research and guarantees consistent manufacturing quality in commercial settings.

Contamination. *Any of five different forms of impurities including particle, biological, chemical, thermal, and electrostatic that compromise the integrity of a microarray cleanroom.*

Particle

A ***particle*** is defined as any form of airborne debris that contaminates a microarray cleanroom facility and has a width and length ratio not exceeding 10:1. Common particle types include dust, dander, lint, spores, pollen, fungi, and bacteria. Fibers are specialized forms of airborne debris that have width and length dimensions exceeding the 10:1 limit established for particles. Airborne particle and fiber contamination derive from many different sources, including leaks in ventilation systems as well as from equipment (computers, robotic devices, and supplies) brought into the cleanroom enclosure from the ambient environment. In most cleanrooms, human-borne particle and fiber contamination represents the largest single source of contamination during day-to-day use. And in many cleanrooms, human contributions exceed all of the other forms of particle and fiber contamination combined, placing a premium on minimizing human impact.

Airborne contamination can settle into solutions and onto microarray substrates, printed microarrays, robots, scanners, and other devices used for microarray analysis. Particles interfere with microarray printing by disrupting the uniform spreading of printed samples, leading to spot merging and other printing artifacts. Particles can also interfere with scanning and imaging by reflecting laser light, causing spurious background that usually manifests as punctate artifacts in microarray images. The miniaturized microarray format and the increasing requirement for precision in microarray analysis in genetic screening and forensics contexts demand low particle count environments. A brief discussion of the subject will assist researchers in minimizing particle contamination of microarray cleanrooms.

Cleanroom particle count is often viewed as a fixed value, but the reality for most cleanrooms is that the particle count varies considerably, depending on when it is measured. The three main categories of cleanroom status are known as ***setup, operational,*** and ***at rest***. The setup phase of a microarray cleanroom refers to the period immediately following cleanroom construction, during which time the environmental control systems are fully operational and equipment, supplies, robots, and accessories are being introduced into the facility from the ambient environment. Due to the movement of a large number

Particle. *One of five main forms of cleanroom contamination corresponding to airborne debris, including dust, dander, lint, spores, pollen, fungi, bacteria, and any other form of debris that contaminates the air space in a microarray cleanroom and has a width and length ratio not exceeding 10:1.*

Setup. *One of three main categories of microarray cleanroom status corresponding to the period after cleanroom construction in which the environmental-control systems are fully operational and equipment, supplies, robots, and accessories are being introduced into the facility from the ambient environment.*

Operational. *One of three main categories of microarray cleanroom status, corresponding to the period in which a cleanroom is actively engaged in a process; contains cleanroom personnel; and involves the normal transfer of supplies, reagents, and equipment from the ambient environment.*

At rest. *One of three main categories of microarray cleanroom status corresponding to the period in which a cleanroom contains all of the necessary equipment and accessories required for operation but is not actively engaged in a process and does not contain any cleanroom staff.*

of items into the cleanroom and the excessive human traffic, the setup phase typically corresponds to the period of highest particle count. This anomalous period therefore does not provide good measure of particle count. Microarray cleanrooms should be allowed to equilibrate at least 1 week after the setup phase. During the equilibration period, the main and ancillary filtration systems remove particles and fibers from the air and reduce their levels to what is expected for a given class of cleanroom. After the equilibration phase, the cleanroom is ready to be used on a daily basis.

An operational cleanroom is one that is actively engaged in a process; contains personnel; and involves the normal transfer of supplies, reagents, and equipment from the ambient environment. A microarray cleanroom is said to be at rest when it contains all of the necessary equipment and accessories required for operation but is not actively engaged in a process and does not contain any staff. The operational phase of most microarray cleanrooms corresponds to the normal workday and provides a conservative measure of cleanroom particle count, whereas the at rest period usually falls on nights and weekends and provides an optimistic measure of particle count. Though there is some debate about when a microarray cleanroom class rating should be established, a reasonable measure involves taking the average of the particle count during the operational and at rest periods. If sound protocol is adhered to (see below), particle count should be similar for these two phases, though the count at rest will usually be lower than the operational period, mainly due to human contamination during the operational period.

During the setup phase, new equipment rather than used equipment should be used in the cleanroom if at all possible, because particles accumulate inside computers, robots, and other devices over time, particularly those that use ambient air and motorized fans to cool their internal components. Before introducing any piece of equipment into a microarray cleanroom, the device must be decontaminated completely using an oil-free and low particle source of pressurized air. Inexpensive air compressors using purified ambient air suffice for decontamination. Casings, skins, and housings should be removed from every piece of equipment, and the internal components–drive axes, motors, circuit boards, and other mechanical and electrical components–should be subjected to a stream of forced air for several minutes to remove all visible particle contamination. A 1-year-old computer used in an ambient environment can accumulate several grams of dust internally, enough to contaminate a microarray cleanroom the size of a football field for years! Contaminated equipment should not be introduced into a microarray cleanroom *under any circumstances*!

The setup phase often involves the introduction of a large number of supplies and reagents into the cleanroom, and all of these items should be decontaminated with forced air, to remove topical particle contamination before introduction into the cleanroom. Consumable items should be removed from cardboard shipping boxes and decontaminated with forced air. Cardboard, paper, and other types of packaging materials should not be taken into the cleanroom under any circumstances. Petri dishes, microplates, beakers, pipettes, conical tubes, and the like should be purchased in bulk or obtained from commercial sources that use low particle count or particle-free packaging materials (e.g., plastic foam) to avoid contamination by fibrous paper materials. Traditional notebooks and notepads, pipettes with cotton plugs, cardboard storage containers, and other items used in traditional laboratories should not be used in a microarray cleanroom. Graphite pencils, paper laboratory wipes, and paper towels generate an enormous number of particles and fibers and should not be used in a cleanroom under any circumstance. A wide assortment of cleanroom supplies is available at an affordable price from many commercial vendors, and these items should be used at all times in microarray cleanrooms. Polyester cleanroom wipes with laser-cut edges are a particularly useful substitute for traditional paper towels.

The continuous introduction of particle contamination by personnel represents a major source of contamination. The human body can generate 1 million particles/h, a number that is significant in the sense that a standard 10 × 10 ft class 100 cleanroom (1,000 ft^3 or 28 m^3) has an upper limit of 100,000 particles. Thus the human body can generate 10 times more particles per hour than are present in all of the air in an entire standard cleanroom! Reducing particle contamination by personnel is extremely important and can be accomplished by following some simple guidelines.

All cleanroom personnel should shower daily to reduce topical particle count. Personnel exposed to anomalously high concentrations of particles (e.g., a freeway construction zone) should shower before entering a microarray cleanroom. Personal effects, including watches, jewelry, and cosmetics, should not be taken into the room. Residual cigarette smoke from the lungs of a smoker generates an enormous number of particles, and smokers should not be allowed in a cleanroom under any circumstances. Dust, dander, and other particle sources are shed in a continuous manner from the human body and must be contained by proper cleanroom clothing at all times.

Full cleanroom suits, including tight-fitting hoods, elastic sleeves, and booties, must be worn under all circumstances (Fig. 11.6). One-piece designs are generally more effective than modular suits and preferable for use in microarray cleanrooms for this reason. Cleansuits made of nonwoven polyethylene materials (e.g., Tyvek), which are chemically inert and low particle and fiber emitters (Fig. 11.7), provide an excellent barrier against human contamination. Face shields and respirators can be somewhat uncomfortable, but are effective in reducing particle counts, especially when the highest levels of class rating (e.g., 1 and 10) are required. Face coverings can, however, lead to elevated biological contamination (see below) and this must be balanced against their effectiveness in reducing particle count.

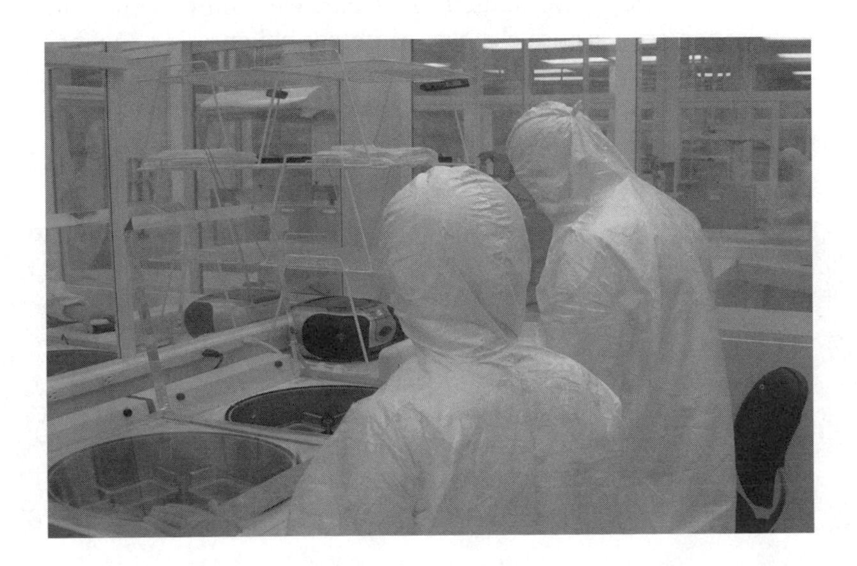

FIGURE 11.6. *Researchers T. Costa and J. Chung wearing protective cleansuits to reduce particle contamination derived from the human body. (Courtesy of TeleChem/arrayit.com, Sunnyvale, CA.)*

FIGURE 11.7. *Magnified view of Tyvek, a nonwoven polyethylene fabric used to make cleansuits for use in microarray cleanrooms. (Courtesy of Dupont, Wilmington, DE.)*

Personnel should remove their shoes before entering a microarray cleanroom and cover their hands with gloves (see below). Adhesive mats containing sticky sheets should be used at the base of all entryways to reduce the introduction of particles into the room from the outside and to prevent particle transfer between different cleanroom chambers. Sticky sheets should be changed at least once a week to provide fresh adhesive surfaces at all times. Cleansuits that contain pigmented booties tend to slough particles and are less effective than white booties made of Tyvek or similar material.

Most microarray cleanrooms use tandem entry chambers to prevent direct exposure of the main cleanroom facility to the outside environment. Each entry chamber is equipped with one or more portable air filters to reduce the particle count (Fig. 11.8) as the cleanroom technician or researcher passes through each pre-cleanroom chamber. The disposable filters should be changed every 60 days. Active reduction of particle count in the entry space is extremely important because cleanroom personnel entering the facility don regular street clothes, which contain extensive particle contamination. Before putting on a cleansuit, both hands should be covered with gloves (see below) to prevent biological and particle contamination of external cleansuit surfaces.

FIGURE 11.8. *Portable air filtration units containing ultra-high-efficiency particulate air filters. (Courtesy of TeleChem/arrayit.com, Sunnyvale, CA.)*

Powder-containing gloves are a major source of particulate contamination, so the powder-free variety should be used at all times (see below).

Outer garments such as sweaters and heavy shirts should be removed before putting on a cleansuit to reduce particle contamination and to prevent overheating while working in the cleanroom. The latter is a concern, particularly because cleansuits pass body heat slowly, and elevated body temperature and dehydration can occur if cleanroom personnel wear excessive clothing under a cleansuit. The cleansuit should be zipped up completely and the elastic sleeves should be drawn tightly around the wrists so that no skin is exposed. All cleanroom personnel should don cleansuits before entering the cleanroom and the cleansuits should never be outside of the cleanroom facility.

Particles are removed from cleanroom air by both the main and the auxiliary purification systems. The auxiliary purification system often includes one or more portable air filters positioned at various locations in the room to reduce particles that may enter the room through the main purification system. These freestanding air filters are available from a variety of commercial providers. Early cleanroom technology made use of high-efficiency particulate air (HEPA) filters, though the newer ultra high efficiency particulate air (ULPA) filters are preferred in microarray cleanrooms because they provide better filtration efficiency and retain smaller particles. The filtration efficiency of ULPA filters is 99.999% compared to the 99.97% efficiency of HEPA filters (Fig. 11.8), which means that on average only 1 particle out of 100,000 escapes filtration by the ULPA technology compared to 30 particles out of 100,000 that pass through a HEPA filter (Fig. 11.9). Furthermore, ULPA filters trap 0.1-μm-diameter particles, where as HEPA filters trap 0.3-μm particles. ULPA filters are disposable and should be changed at regular 30- to 60-day intervals to maintain optimal filtration performance.

The efficiency of particle removal by the air filtration systems also depends on the airflow dynamics in a microarray cleanroom. To maintain optimal airflow, wire shelves with an open geometry (Fig. 11.10) are preferable to solid shelving units. Wire shelves minimize the disruption of cleanroom laminarity and maximize the filtration capabilities of the air filtration systems. Open shelving also minimizes the accumulation of particles and expedites routine cleanroom maintenance. The floors of a microarray cleanroom should be mopped at every 30 to 60 days to remove particles that are too large to be removed by the main and auxiliary filtration systems. Cleaning should be carried out with particle-free mops that are designed exclusively for cleanroom use. All cleaning and maintenance equipment should be dedicated to cleanroom use. The use of conventional vacuums and hand-held vacuums with unfiltered air outlets are strictly forbidden. General-purpose cleaning supplies and solutions used in an ambient setting should not be used in a microarray cleanroom *under any circumstances*!

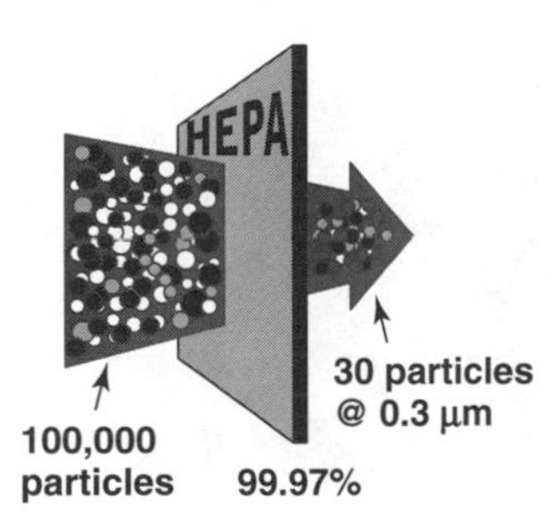

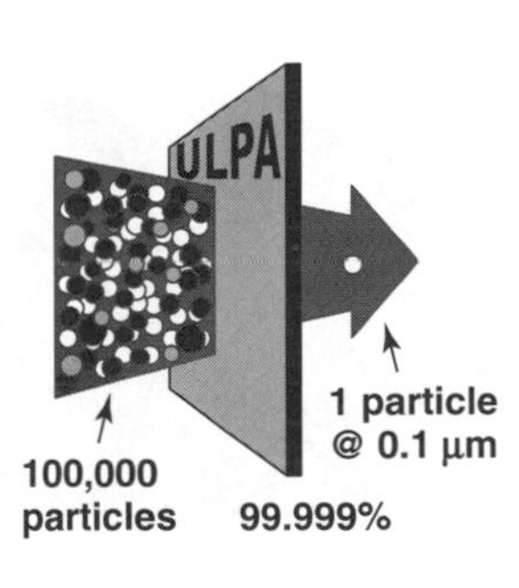

FIGURE 11.9. **A,** *The HEPA filter provides a 99.97% filtration efficiency, allowing 30 particles out of 100,000 to pass through on average. The ULPA filter is 99.999% efficient and allows, on average, only 1 particle out of 100,000 to pass through. (Adapted courtesy of InTech Marketing, Oakville Ontario, Canada.)*

FIGURE 11.10. Wire shelves, with their open configuration, are used to maintain optimal airflow dynamics in a microarray cleanroom. (Courtesy of TeleChem/arrayit.com, Sunnyvale, CA.)

Biological

Biological. *One of five main forms of cleanroom contamination corresponding to human hand oils and fingerprints, hair, saliva, tears, oral condensation, perspiration, sloughed skin cells, and the like.*

Second only to particle contamination, ***biological*** contaminants constitute the most serious form of microarray cleanroom contamination. The main sources of biological contamination include human hand oils and fingerprints, hair, saliva, tears, oral condensation, perspiration, and sloughing skin cells. All of these contaminants contain biological macromolecules that can compromise microarray cleanroom processes.

Hand oils and fingerprints contain fats, lipids, DNA, ribonucleases, and other biological molecules that can impair many aspects of cleanroom experimentation and production. Oily deposits can alter chemical reactivity in substrate manufacturing and degrade the quality of printed microarrays. Skin cells present in human hand oils contain nucleic acid, which can produce anomalies in polymerase chain reaction studies. Microarray cleanrooms used for forensics and other diagnostic applications must be completely DNA free. Hand oils and fingerprints also contain ribonucleases, which will degrade RNA molecules on microarrays that contain ribonucleic acid targets and probes. Both hands should be covered with gloves before one enters a microarray cleanroom, and the gloves should be worn at all times while inside any of the cleanroom chambers (Fig. 11.11). In cleanrooms where personnel wear open-faced cleansuits, workers should never touch their faces with gloved hands. Oils, perspiration,

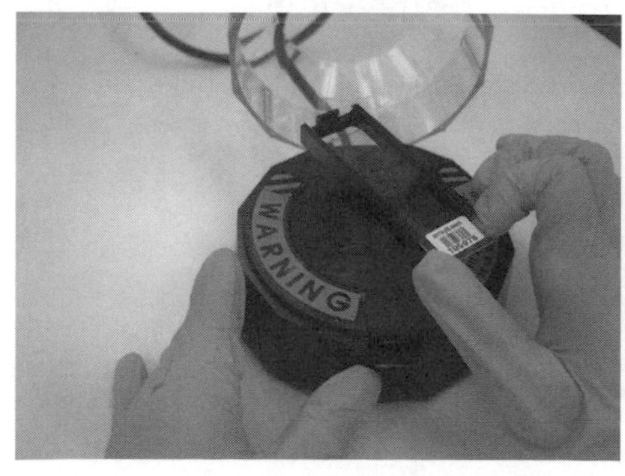

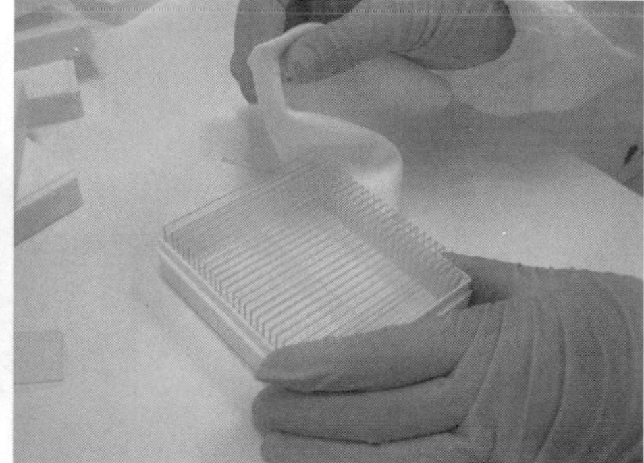

FIGURE 11.11. **A,** *Protective gloves should be worn in a cleanroom at all times.* **B,** *Particle-free polyester wipes should be used remove trace liquids from processed microarrays. (Courtesy of TeleChem/arrayit.com (Sunnyvale, CA.)*

and other facial secretions can be transferred from contaminated gloves onto cleanroom surfaces, analytical equipment, and other exacting devices. Personnel that need to scratch an itch should move into the changing area and use a polyester cleanroom wipe. Gloved hands should then be decontaminated completely with 100% ethanol before returning to the main room. When glove contamination is suspected, fresh gloves should be donned.

Gloves made of synthetic rubber are preferable to those made of latex and other natural materials because of their high tensile strength and protein-free composition (Fig. 11.11). Natural rubber latex, derived from the milky fluid of tropical rubber trees (*H. basiliensis*), contains high concentrations of plant proteins that can contaminate glass substrates and printed microarrays. Protein contamination, though generally less problematic than nucleic acid contamination, is a growing concern with the expanding interest in protein microarrays. Some personnel may also exhibit allergic reactions to latex gloves, and latex allergies are an increasing issue as the number of workers donning gloves on a regular basis increases. Synthetic rubber is vastly superior to natural rubber gloves and should be used in microarray cleanrooms whenever possible.

Two layers of gloves can be worn when tearing is a concern. Double gloves also allow fresh gloves to be donned without exposing the cleanroom changing area to bare hands. Two layers of gloves can be useful when handling corrosive chemicals and for other times when fresh gloves are needed at regular intervals.

Human hair poses another source of biological contamination in microarray cleanrooms. Personnel with moustaches and beards should wear face coverings whenever possible. Human hair contains DNA, ribonucleases, and oils, all of which can impair various aspects of microarray analysis in a negative way. The same is true for human saliva, tears, and oral vapors. Gloved hands should never make direct contact with saliva or tears. Polyester wipes should be used to blot these secretions when necessary, and gloves should be decontaminated with ethanol after any such procedure. Fresh gloves should be put on if there is any concern whatsoever of contamination.

Face coverings reduce particle count by preventing sloughing skin cells from contaminating cleanroom air. Depending on the design, however, face coverings may pose a greater threat of biological contamination owing to the fact that they tend to cause accumulation of oral vapors, which can transfer or sputter onto cleanroom surfaces. Exhaled air contains a wide spectrum of biological compounds and is an unwanted substance in any microarray cleanroom. Open-faced cleansuits may thus be preferable to respirators and other face coverings, depending on the specific applications of a given cleanroom.

Coughing and sneezing is forbidden in a cleanroom setting under all circumstances. Rapid discharge of air from the human nose and mouth releases a large number of droplets than contain saliva and mucus, both of which pose a serious threat to a microarray cleanroom. Coughing and sneezing should be done outside the cleanroom if possible. In cases of an emergency, personnel should excite the main cleanroom quickly and cough or sneeze into a waste container in the changing facility.

Human perspiration contains biological contaminants, and cleanroom personnel should take care to wear light clothing under their cleansuits to prevent contamination of the cleanroom environment. Running, lifting, and other forms of excessive movement in the cleanroom should also be avoided at all times to prevent excessive perspiration. Cleansuits should be changed at regularly weekly intervals to avoid transfer of perspiration-based contaminants from soiled cleansuits onto cleanroom chairs and other surfaces.

Chemical

Chemical. *One of five main forms of cleanroom contamination, corresponding to the vapors emitted from paints, glues, adhesives, packaging materials, organic solvents, and other substances that emit volatile chemicals.*

Outgassing. *Process in which paints, glues, adhesives, packaging materials, and other cleanroom surfaces emit vapors into the cleanroom air space.*

Because most microarray surfaces are composed of glass (instead of silicon) and glass (silicon dioxide) is less reactive than pure silicon, chemical vapors and other ***chemical*** contaminants are somewhat less of a concern in microarray cleanrooms than they are in semiconductor fabrication facilities. Nonetheless, chemical contamination is an important issue in microarray cleanrooms and should be minimized or eliminated whenever possible. The most common sources of chemical contaminants originate from paints, glue, adhesives, and packaging materials that emit vapors in a process known as ***outgassing***. Substances that exhibit outgassing can emit vapors for weeks or months after being installed and should be avoided at all times.

Volatile chemicals such as organic solvents used in cleanroom manufacturing processes can also contribute to chemical contamination, particularly if the chemical containment facilities are inadequate for the solvents used. Solvents and other organic compounds that contaminate the cleanroom air can form chemical bonds with reactive groups on microarray substrates. Most organic compounds are only weakly fluorescent, but the extraordinary detectivity required for certain microarray applications such as human gene expression monitoring is negatively affected by any increase in background. Elevated background fluorescence can mask important data, such as those corresponding to rare gene transcripts, making it imperative to reduce or eliminate chemical contamination in a microarray cleanroom.

The advanced construction materials used in most commercial cleanrooms are essentially vapor free, and thus contribute little if any chemical contamination to microarray cleanrooms. Shakers, centrifuges, and other pieces of equipment vary considerably and should be checked for the presence of volatile emissions before being moved into a cleanroom. Routine cleaning and sterilization of cleanroom surfaces should be performed with sterile water, dilute ethanol, or similar solvents that are known to be free of volatile chemical contaminants. Cleanroom personnel should avoid the use of cologne, perfume, mouthwash, and other personal products that have a high vapor pressure. Chewing gum, soda pop, food, and other substances that are likely to release fumes into the cleanroom air are strictly forbidden at all times.

Thermal

Thermal. *One of five main forms of cleanroom contamination corresponding to any source that produces an artificial increase in cleanroom temperature, including microarray robots, refrigerators, freezers, lights, and cleanroom personnel.*

Heat load. *Amount of thermal contamination present in a microarray cleanroom.*

Any artificial increase in cleanroom temperature is known as ***thermal*** contamination. Because of the essentially airtight nature of the cleanroom environment, seemingly innocuous sources such as refrigerators, freezers, lights, and humans can be important contributors to thermal contamination. Robots and other systems that employ stepper motors, shakers, centrifuges, and air-conditioning units also contribute to thermal contamination. Elevated temperatures can alter chemical reactivity and create an environment that is uncomfortable for personnel. Thermal contamination can be minimized using a few simple guidelines.

Before construction, the amount of thermal contamination or ***heat load*** for a given facility should be evaluated and used to determine the number of air changes per unit time and the cooling system required to maintain a constant temperature. Electrical equipment that is not essential for a given cleanroom process should be moved into ancillary chambers to prevent thermal contamination of the main cleanroom space. Auxiliary cooling devices such as portable air-conditioners can be used to supplement the main cooling system when extra heat load requires amelioration.

Electrostatic

The presence of static electricity in a cleanroom is known as ***electrostatic*** contamination. Static electricity is generated by the imbalance and negative and positive charges caused by the movement of electrons between two substances. When static electricity accumulates in an environment, particles from the air are attracted to the charged surfaces and can be trapped by the forces of static electricity. Electrostatic contamination in a microarray cleanroom degrades the environment by trapping particles on surfaces, such as microarray substrates that are supposed to be particle free. Electrostatic contamination can also inhibit processes, such as ink-jet printing, that involve the ejection of tiny sample droplets, causing the deflection of droplets from their downward path and leading to irregular arrays. Particles trapped on microarray substrates due to static electricity reduces the quality of the surfaces by scattering excitation light and causing punctate background in microarray images. Particles trapped by static electricity can also interfere with contact printing, leading to spot merging and other blemishes because the printed droplets spread unevenly on the surface. Electrostatic contamination can be minimized or eliminated in microarray cleanrooms by following a number of basic guidelines.

Electrostatic. *One of five main forms of cleanroom contamination corresponding to any source, including plastics, rubber, and cellophane, that produces an imbalance of negative and positive charges inside a cleanroom environment.*

Proper selection of cleanroom construction materials and surfaces minimizes the generation of static electricity inside the room. All of the commercial microarray cleanroom builders use such materials. Careful control of cleanroom humidity is also an effective means of reducing electrostatic contamination. If the humidity is maintained in the range of 35–45%, moisture in the air absorbs the static charge and prevents its accumulation. Cleanroom personnel should take care never to rub two objects together while working in the room. Manufactured substrates and microarrays should be packaged in antistatic bags to prevent static transfer onto the glass surfaces (Fig. 11.12).

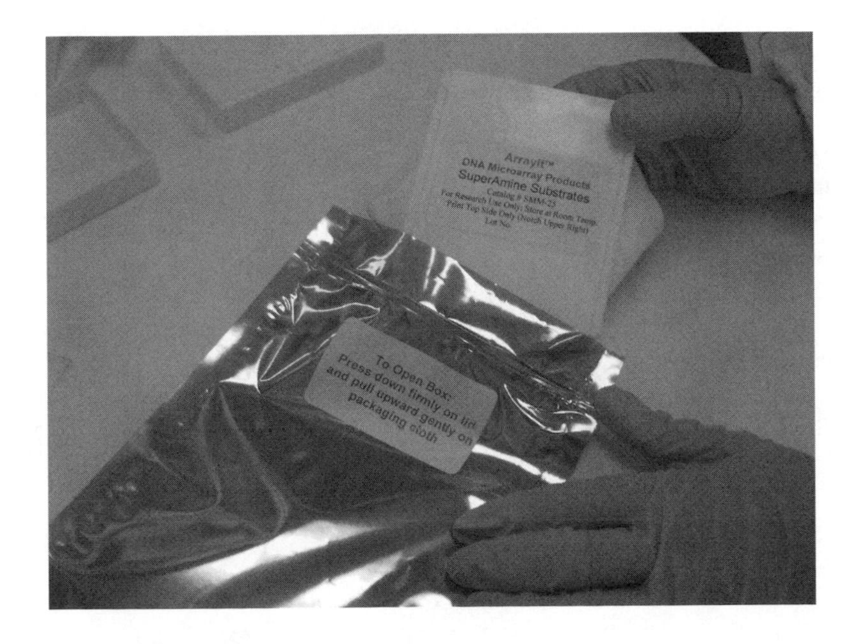

FIGURE 11.12. Silver antistatic bags should be used at all times in microarray cleanrooms to prevent electrostatic contamination of manufactured substrates and microarrays. (Courtesy of TeleChem/arrayit.com, Sunnyvale, CA.)

Sealed bags and other packaging materials should be opened slowly with scissors, rather than by tearing. Chronic generators of static charge, such as natural rubber and cellophane, should not be introduced into a microarray cleanroom under any circumstances.

SUMMARY

Microarray cleanrooms are sophisticated laboratories that reduce the levels of contaminants found in the ambient environment and increase the consistency and quality of all processes related to microarray manufacture and use. Microarray cleanrooms are evaluated according to a number of important criteria, including class, environmental control, functionality, durability, and safety. A class 100 cleanroom has fewer than 100 particles/ft^3 of air, which is approximately 10,000 times cleaner than a traditional research laboratory. Environmental control includes temperature, humidity, lighting, and uniformity, all of which are monitored closely in a cleanroom setting. Functionality pertains to overall cleanroom design, and durability refers to the extent to which the physical parameters can be maintained over time. User safety is a key issue, and appropriate design and use ensure the safety of personnel at all times.

The main forms of cleanroom contamination include particle, biological, chemical, thermal, and electrostatic. Dust, dander, fiber, lint, spores, pollen, fungi, bacteria, and many other forms of debris are removed from the cleanroom air by sophisticated filtration systems. Traditional research environments are contaminated extensively by hand oils, fingerprints, hair, saliva, tears, oral condensation, perspiration, and sloughing skin cells; but these biological contaminants can be virtually eliminated by the use of a cleanroom. Chemical, thermal, and electrostatic contaminants are also minimized greatly in a cleanroom setting, providing a pristine environment in which to conduct microarray research. Cleanrooms are seeing increasing use given the expanding interest in microarray-based genetic screening, diagnostics, and forensics.

SELECTED READING

Benson, H. *University Physics,* rev. ed. Wiley, New York, 1996.

Blackburn, G. M., and Gait, M. J., eds. *Nucleic Acids in Chemistry and Biology,* 2nd ed. Oxford University Press, Oxford, UK, 1996.

Bloomfield, V. A., Crothers, D. M., and Tinoco, I. *Nucleic Acids: Structure.* University Science Books, Herndon, VA, 2000.

Brook, M. A. *Silicon in Organic, Organometallic, and Polymer Chemistry.* Wiley, New York, 2000.

Ehrlich, R., Tuszynski, J., Roelofs, L., and Stoner, R. *Electricity and Magnetism Simulations.* Wiley, New York, 1995.

Giancoli, D. C. *Physics for Scientists and Engineers,* 3rd ed. Prentice Hall, Upper Saddle River, NJ, 2000.

Innis, M. A., Gelfand, D. H., and Sninsky, J. J., eds. *PCR Applications: Protocols for Functional Genomics.* Academic Press, San Diego, CA, 1999.

Lehninger, A. L., Nelson, D. C., and Cox, M. M. *Principles of Biochemistry,* 3rd ed. Freeman, New York, 2000.

Martin, F. G. *Robotic Explorations: A Hands-on Introduction to Engineering.* Prentice Hall, Upper Saddle River, NJ, 2001.

Nishi, Y., and Doering, R., eds. *Handbook of Semiconductor Manufacturing Technology.* Marcel-Dekker, New York, 2000.

Ramstorp, M. *Introduction to Contamination Control and Cleanroom Technology.* Wiley-VCH, Weinheim, Germany, 2000.

Schena, M., ed. *DNA Microarrays: A Practical Approach,* 2nd ed. Oxford University Press, Oxford, UK, 2000.

Schena, M., ed. *Microarray Biochip Technology.* Eaton, Natick, MA, 2000.

Singer, M., and Berg, P. *Genes and Genomes.* University Science Books, Herdon, VA, 1991.

Streitweiser, A., and Heathcock, C. H. *Introduction to Organic Chemistry,* 2nd ed. Macmillan, New York, 1981.

Sun, S. F. *Physical Chemistry of Macromolecules: Basic Principles and Issues.* Wiley, New York, 1994.

Wenckebach, W. T. *Essentials of Semiconductor Physics.* Wiley, New York, 1999.

Whyte, W. *Cleanroom Design,* 2nd ed. Wiley, Chichester, UK, 1999.

Whyte, W. *Cleanroom Technology: Fundamentals of Design, Testing and Operation.* Wiley, New York, 2001.

REVIEW QUESTIONS

1. A controlled laboratory facility that reduces particle, biological, and other forms of contamination is known as a what?
2. According to Federal Standard 209E, a cleanroom containing 350 particles of ≥ 0.5-μm diameter/ft^3 of air would receive what class rating: (a) class 100, (b) class 1000, (c) class 10, or (d) class 10,000?
3. An unsophisticated researcher exclaims that "I don't need no darn cleanroom for my microarray experiments because dust particles are so darn small they're hardly visible." In the context microarray printing, explain the flaw in this argument.
4. Which of the following are cleanroom environmental control parameters: (a) lighting, (b) humidity, and (c) temperature?
5. A researcher notices that a batch of microarrays printed with an ink-jet dispenser are unusually irregular in terms of spot spacing, at about the same time that the cleanroom humidity was measured at 17%. Provide an explanation for the correlation between humidity and variable spot spacing.
6. A technician replacing the full-spectrum fluorescent bulbs in a microarray cleanroom is provided two separate sets of unlabeled bulbs, along with their color temperatures of 3500 K and 6000 K. Which color temperature would correspond to the full-spectrum bulbs?
7. Explain why each of the following materials would not be permitted in a microarray cleanroom: (a) paper towels, (b) bare hands, (c) ether, and (d) a computer from a flea market.
8. At the 99.999% filtration rating for an ULPA filter, how many particles would be trapped by an ULPA filter exposed to 55,000 particles of 1 μm diameter: (a) 55, (b) 55,000, (c) 54,999, or (d) 0?
9. One drawback of stepper motors compared to servomotors is that stepper motors require a constant supply of electrical current to maintain robot position, a quality that challenges predominantly which of the five main categories of cleanroom contamination: (a) biological, (b) electrostatic, (c) thermal, (d) chemical, or (e) particle?

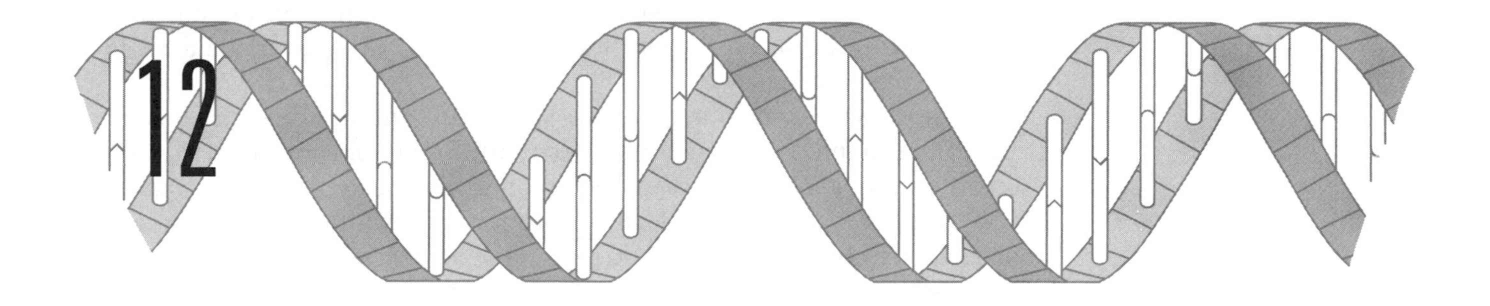

Gene Expression Profiling

I am following nature without being able to grasp her. —*Claude Monet*

Gene expression profiling with microarrays is facilitating the exploration of the natural world with a precision and breadth that are unprecedented in the history of biology. Chips containing hundreds, thousands, or even tens of thousands of genes arrayed in parallel are revealing the function of genes and relationships between genetic and biochemical pathways that are virtually impossible to grasp by any other means. Entire genomic sequences are available for yeast, worm, fruit fly, mustard plant, and human, allowing microarrays containing a complete set of genes to be constructed. Mouse biology, strawberry fruit development, the effects of ionizing radiation, acute renal failure, light responses and osmotic stress in plants, drug toxicology, and liver cancer are but a few examples of how basic science, medicine, and agriculture are being revolutionized by microarray-based gene expression profiling. A comprehensive collection of scientific manuscripts is available electronically (arrayit.com/e-library) to assist the reader in identifying additional biological problems that are under current investigation with microarrays.

BASICS OF GENE EXPRESSION

Why does gene expression profiling tell us so much about biological processes? The answer is simple. Organisms express their genes at a relatively constant rate until the products encoded by those genes are needed for a specific function.

When needed, genes are activated or repressed rapidly and in dramatic fashion, changing by 10-, 100- or even 1000-fold or more, depending on the particular gene and the strength of the regulatory cue. The expression of genes changes in response to a wide spectrum of signals, including hormones, chemicals, nutrients, stress, changes in cell division and development, light stimulation and the like, providing a gene expression "fingerprint" that is characteristic for a given physiological state.

Because gene expression correlates specifically and tightly with function, it is possible to infer the function of genes and the interaction of pathways by documenting which genes are turned up or down in a given physiological state. Gene regulation provides a selective evolutionary advantage by conserving cellular building blocks (e.g., nucleotides, amino acids) and enzymatic machinery (e.g., transcription factors, polymerases) until they are needed, and by allowing the organism to adapt to a plethora of different environmental conditions to which it is exposed during its lifetime.

When cells are subjected to elevated temperature, for example, heat shock genes are activated to protect cellular proteins from thermal damage. Disease states, drug treatment, different developmental stages, and many other processes can be examined by cataloging gene expression profiles. The logical extension of this concept is to build comprehensive gene expression databases for each organism that contain expression profiles for each gene across thousands of different conditions, thereby allowing biological exploration to take place predominately by means of a computer. The concept of "in silico biology," a term dubbed by Scott and co-workers at Incyte Genomics (Palo Alto, CA), is being brought to fruition through data acquisition via gene expression microarrays and through the mining of these data with the appropriate tools of computer science.

MOUSE BIOLOGY

The laboratory mouse (*Mus musculus*) is one of the most powerful experimental organisms available to modern researchers. The usefulness of the mouse derives from its extensive genomic similarity to human, plurality of human disease gene counterparts, advanced genetics, and rapid generation time. The mouse is probably the most useful model for human biology available, and it is important to consider this experimental system in greater detail.

The mouse genome contains approximately 3 billion (3×10^9) bases of DNA, very similar in size to the human genome. The ~30,000 mouse genes are distributed over 19 autosonal and 2 sex chromosomes, 3 fewer than the 22 autosomal and two sex chromosomes that make up the human genome. Comparisons between mouse and human reveal extensive homology on every chromosome (Fig. 12.1). At present only about 1% of the mouse genome is available in finished form, but comparison of the known sequences from mouse against the complete human genome sequence reveals a fairly uniform distribution of homologous regions across all of the mouse chromosomes (Fig. 12.2).

Synteny. *Extended region of similarity in the chromosomal or genomic sequences of two organisms, resulting in a conservation in the order of some or most of the genes along a genomic segment.*

The extended regions of similarity or ***synteny*** between the mouse and human genomes, in which the order of some or most of the genes is conserved along the length of a genomic segment, are such that each mouse chromosome shares extensive homology with approximately a half dozen human chromosomes. It appears that these two mammalian genomes contain roughly the same genetic content in terms of total gene number and gene function, but

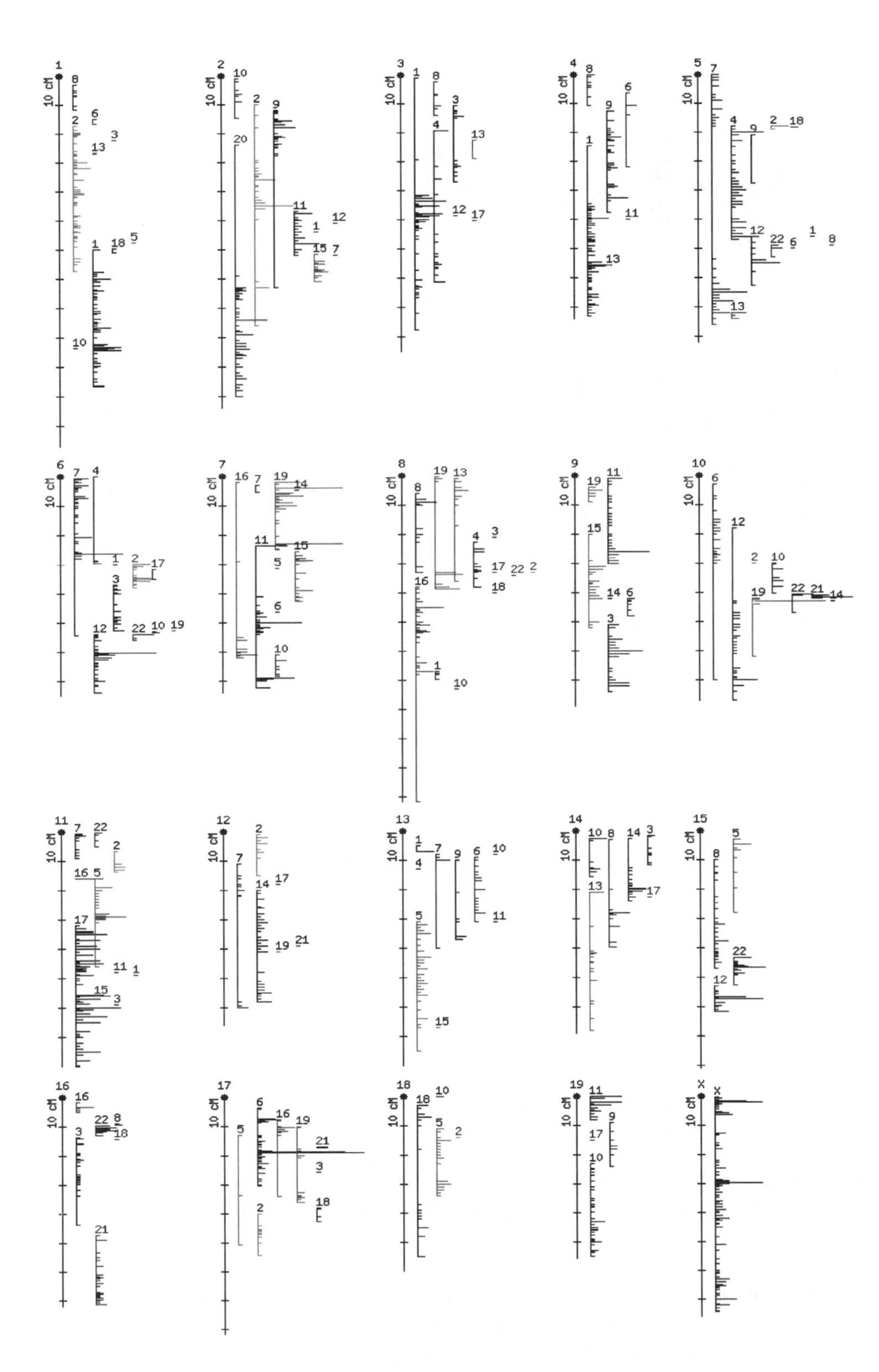

FIGURE 12.1. *Regions of similarity between 20 mouse (black) and human (gray) chromosomes. Horizontal hash marks along each mouse chromosome denote a recombination distance of 10 centimorgans (cM). (Data courtesy of Mouse Genome Informatics, The Jackson Laboratory, Bar Harbor, ME.)*

the two genomes contain chunks of homology if compared to each other (Fig. 12.2). The gene content appears to have been shuffled through the course of evolution, and the 100–120 million years that separate the mouse and human provided sufficient time for the gradual reorganization of the genomes to meet the unique biological requirements of rodents and humans. Mouse and rat are much closer evolutionarily (35–45 million years) than mouse and human, and correspondingly the amount of synteny between mouse and rat is vastly greater than that between mouse and human.

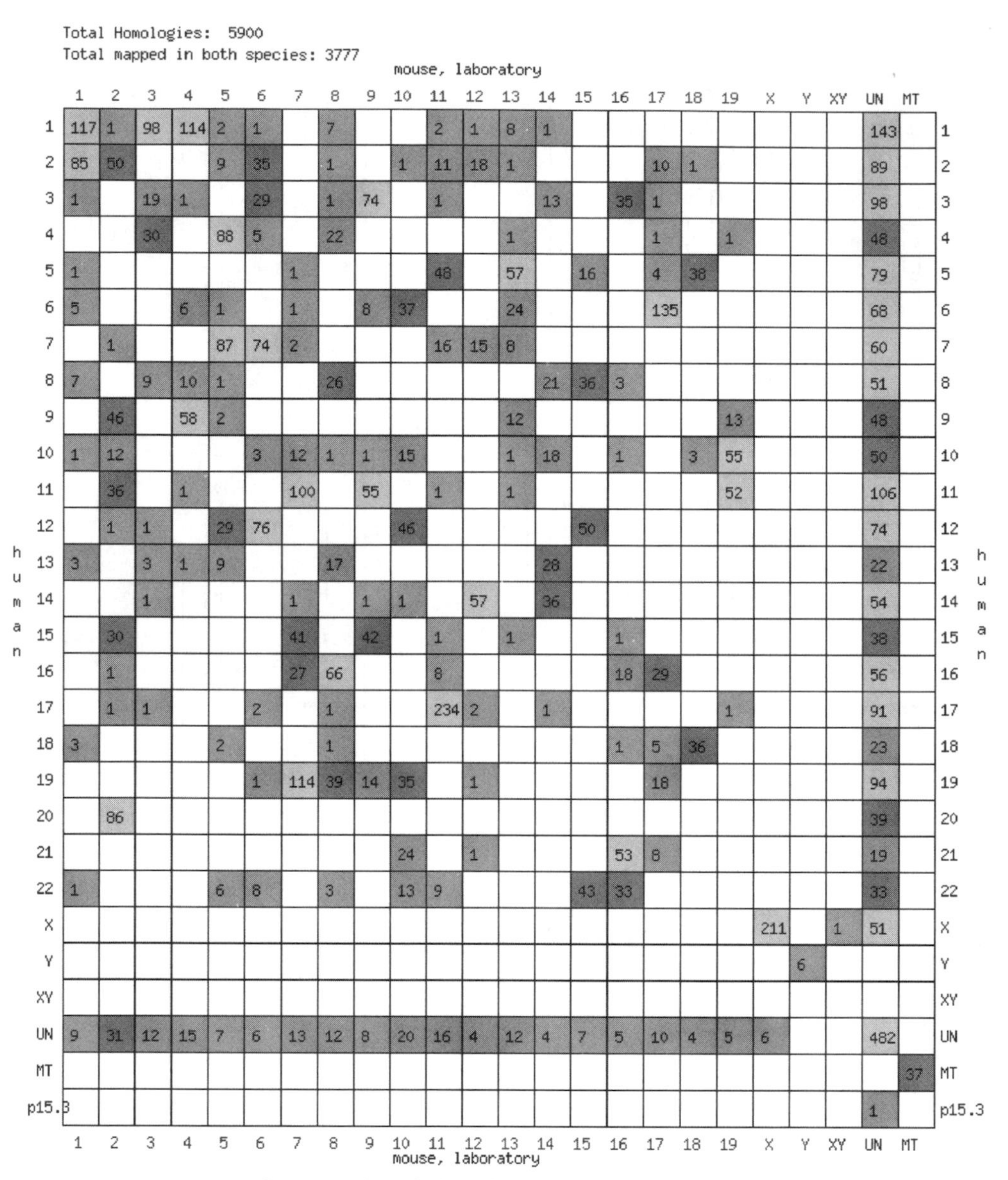

Total Homologies: 5900
Total mapped in both species: 3777

mouse, laboratory

human	1	2	3	4	5	6	7	8	9	10	11	12	13	14	15	16	17	18	19	X	Y	XY	UN	MT	human
1	117	1	98	114	2	1		7			2	1	8	1									143		1
2	85	50			9	35		1		1	11	18	1				10	1					89		2
3	1		19	1		29		1	74		1			13		35	1						98		3
4			30		88	5		22					1				1		1				48		4
5	1						1				48		57		16		4	38					79		5
6	5			6	1		1		8	37			24				135						68		6
7		1			87	74	2				16	15	8										60		7
8	7		9	10	1			26						21	36	3							51		8
9		46		58	2								12						13				48		9
10	1	12				3	12	1	1	15			1	18		1		3	55				50		10
11		36		1			100		55		1		1						52				106		11
12		1	1		29	76				46					50								74		12
13	3		3	1	9			17						28									22		13
14			1				1		1	1		57		36									54		14
15		30					41		42		1		1			1							38		15
16		1					27	66			8					18	29						56		16
17		1	1			2		1			234	2		1					1				91		17
18	3				2			1								1	5	36					23		18
19						1	114	39	14	35		1					18						94		19
20		86																					39		20
21										24		1				53	8						19		21
22	1				6	8		3		13	9				43	33							33		22
X																				211		1	51		X
Y																					6				Y
XY																									XY
UN	9	31	12	15	7	6	13	12	8	20	16	4	12	4	7	5	10	4	5	6			482		UN
MT																								37	MT
p15.3																							1		p15.3

mouse, laboratory

FIGURE 12.2. *A two-dimensional Oxford Grid comparison of the mouse and human genomes. The mouse chromosomes are arranged along the top and bottom and the human chromosomes are arranged along the left and right. A total of 5900 homologies were examined and 3777 mapped to discrete chromosomal positions in both species. Each cell in the grid is color coded according to the number of homologous regions: 0 (white), 1 (gray), 2–10 (blue), 11–25 (green), 26–50 (orange), and >50 (yellow). (Data courtesy of Mouse Genome Informatics, The Jackson Laboratory, Bar Harbor, ME.)*

A comparative analysis of a large number of genes in mouse and human reveals extensive conservation at both the DNA and the protein level (Fig. 12.3). A statistical analysis of nearly 1196 orthologous coding sequences, which are mRNAs that have a clear counterpart in the two species, shows ~85% sequence identity on average at both the nucleotide and the amino acid level for the 1.7 megabases of complete coding sequence analyzed. A similar comparison of >30 human disease genes with their mouse orthologs reveals extensive similarity (Table 12.1), underscoring the likely usefulness of mouse as an accurate model for many human diseases.

Mouse as a human disease model is straightforward to explore at the molecular level, because of the powerful genetics afforded by this model system. One

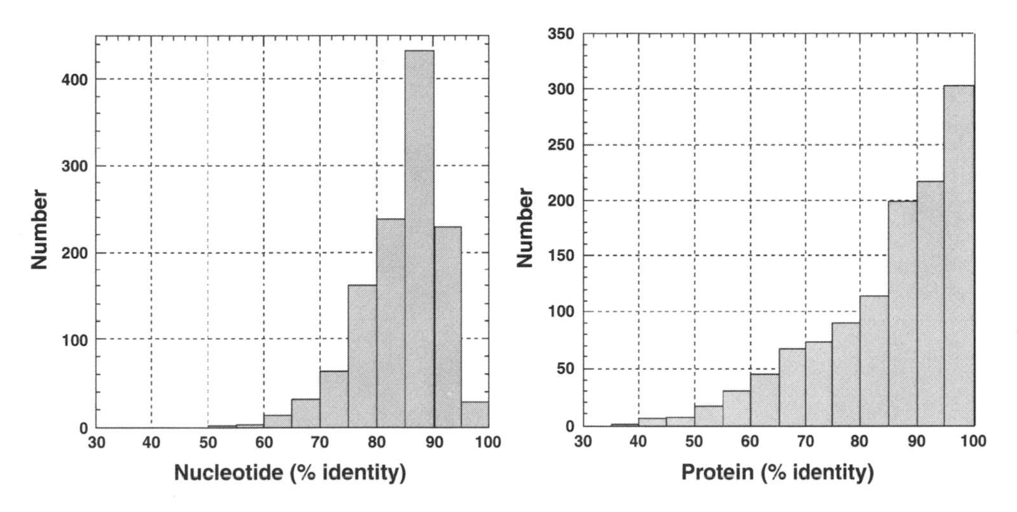

FIGURE 12.3. Sequence alignment data for 1196 cognate (orthologous) mouse and human coding regions. (Data reprinted with permission from Genome Research, *Cold Spring Harbor, NY.)*

particularly powerful approach involves the use of ***embryonic stem cells*** to generate strains of mice that contain mutated copies of a disease gene of interest. In the stem cell approach, the presumptive disease gene is replaced with a mutated version of the gene using a process known ***homologous recombination,*** which targets the mutated gene copy to the correct chromosomal position in the manipulated stem cells. Embryonic stem cells carrying the modified disease gene are introduced into a host mouse embryo, resulting in the development of adult mice that contain cells bearing the mutated disease gene. Mating experiments between the transgenic mice yield progeny that are homozygous for the mutated disease gene, allowing the researcher to examine the phenotypic consequences of loss of function mutations in live mice. By comparing and contrasting the phenotypes in the transgenic mice and in human patients afflicted with a disease caused by mutations in the orthologous gene, provides clues to the function of any human disease gene of interest.

Embryonic stem cell. *Any cell from an embryo not confined to a particular developmental pathway that can give rise to many types of differentiated cells.*

Homologous recombination. *Experimental approach used to examine gene function in intact cells, animals, and plants in which the normal gene is replaced with a defective gene copy at the chromosomal position of the normal gene.*

Because of the high degree of similarity between mouse and human proteins, mice also provide useful models for studying the action of human drugs. Most drugs work by binding to cellular proteins and altering their activity. In cases in which drug-binding sites are conserved evolutionarily, mice can provide extremely useful models for the action of pharmaceutical compounds of interest.

Gene expression monitoring with microarrays is useful for a wide spectrum of different applications in mouse genetics, including studies of transgenic lines for human disease analysis, mechanism of drug action, tissue specificity of gene expression, and many other fundamental problems in basic and applied research. The extensive database of mouse gene sequences has expedited the use of oligonucleotide microarrays to study mouse biology. The Affymetrix GeneChip technology uses up to 400,000 different oligonucleotides of 20–25 nucleotides in length per microarray to study approximately 10,000 mouse genes. Each gene is represented by a series of 20 oligonucleotides that match a given mouse mRNA perfectly (perfect match; PM) and 20 oligonucleotides that are identical to the PM sequences but contain a single nucleotide mismatch (single mismatch; SM) at the center position (Fig. 12.4). The average hybridization intensity at each set of 20 PM sequences provides a quantitative measure of expression of the cognate gene, and the use of a tiled set of 20 PM targets across the mRNA enables the detection of splice variants, splicing defects, and the like. The reduced signals at each of the 20 SM locations validate the specificity of the hybridization.

TABLE 12.1. Sequence Conservation of Human Disease Genes in the Mouse[a,b]

Human Disease	Gene Function	Nucleotide Identity (%)	Protein Identity (%)
Aarskog-Scott syndrome	rho/rac GEF	92	94
Achondroplasia	fibroblast growth factor receptor-3	86	93
Adenomatous polyposis coli	cell adhesion	88	91
Amyotrophic lateral sclerosis	Cu-Zn superoxide dismutase	84	84
Aniridia	oculorhombin	93	99
Breast cancer type 1	tumor-suppressor gene	75	57
Chronic granulomatous disease	neutrophil cytosol factor 1	81	82
Congenital adrenal hyperplasia	steroid 21-monooxygenase	77	72
Cystic fibrosis	transmembrane conductance regulator	81	79
Diastrophic dysplasia	sulfate transporter	83	81
Duchenne muscular dystrophy	dystrophin	88	87
Fragile-X syndrome	unknown	96	97
Glycerol kinase deficiency	glycerol kinase	92	97
Gonadal dysgenesis	testis-determining factor	65	42
Hereditary nonpolyposis colon cancer	DNA mismatch repair enzyme	81	73
Hereditary nonpolyposis colon cancer	DNA mismatch repair enzyme	87	93
Huntington disease	huntingtin	86	91
Hyperexplexia	inhibitory glycine receptor	91	98
Menkes disease	heavy-metal binding protein	86	84
Miller-Dieker lissencephaly	acetylhydrolase subunit	95	>99
Multiple endocrine neoplasia 2A	receptor tyrosine kinase	84	85
Myotonic dystrophy	mutonin protein kinase	83	84
Neurofibromatosis type 1	neurofibromin	92	>99
Neurofibromatosis type 2	merlin	90	99
Norrie disease[b]	norrin	92	95
Retinoblastoma[b]	tumor-suppressor gene	89	90
Thomsen disease	skeletal muscle chloride channel	88	92
vonHippel-Lindau syndrome[b]	tumor suppressor gene	84	90
Waardenburg syndrome	paired box homeodomain protein	92	98
Wilms tumor	zinc-finger protein	91	97
Wilson disease	copper-transporting ATPase	82	81
X-linked adrenoleukodystrophy[b]	ABC protein transporter	86	91

[a] Data reprinted with permission from *Genome Research,* (Cold Spring Harbor, NY.)
[b] Data added since original publication.

FIGURE 12.4. *Gene expression data acquired using Affymetrix GeneChip technology. Fluorescent hybridization data are indexed to a rainbow palette for ease of viewing.* **Inset:** *A total of 40 oligonucleotides are used to monitor the expression of each gene. (Data courtesy of Dr. K. Tillman, Affymetrix, Santa Clara, CA.)*

Tissue rejection, the allergic response, and autoimmune diseases in human all depend on understanding the mechanisms that regulate the proliferation of antibody-producing B cells. The immunosuppressive drug FK506 is used in organ transplantation because it blocks the development of B cells and thus reduces the incidence of rejection of the new organ by the immune system. FK506 also increases the likelihood of blood clot formation, and this serious side effect can complicate the health of heart transplant patients by increasing the chance of stroke, heart attack, or death. Gene expression profiles of mice treated with FK506-provide a detailed molecular portrait of the gene expression events that accompany FK506-dependent immunosuppression (Fig. 12.5). The availability of a molecular fingerprint of FK506 action may allow the development of more efficacious immunosuppressive drugs that circumvent the side effects associated with their use.

STRAWBERRY FRUIT DEVELOPMENT

The developmental processes that convert precursor plant structures into fruits and vegetables underlie worldwide agriculture. Corn, wheat, rice, citrus fruits, and berries are among the many agricultural products that are staples in the human diet. Traditional crop breeding has been used for centuries to select traits of interest, including fruit size, shape, color, flavor, and aroma. In some cases,

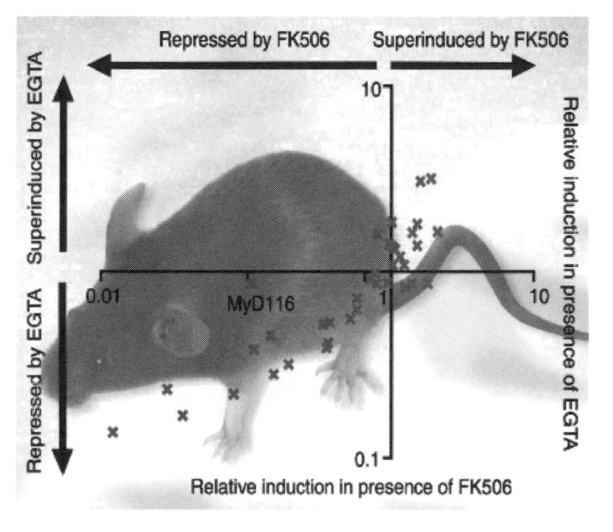

FIGURE 12.5. *Gene expression profiles (scatter plot) obtained from Affymetrix GeneChip technology. Mice were treated with either the immunosuppressive drug FK506 or the chelating agent ethylene glycoltetraccetic acid (EGTA) or were left untreated. Mouse genes that are induced or repressed by either compound (x) span a range of 10-fold induction (10) to 10-fold repression (0.1). The expression of* MyD116, *a gene thought to be involved in apoptosis, is repressed 10-fold by FK506 and is unaffected by EGTA. (Data courtesy of Dr. K. Tillman, Affymetrix, Santa Clara, CA.)*

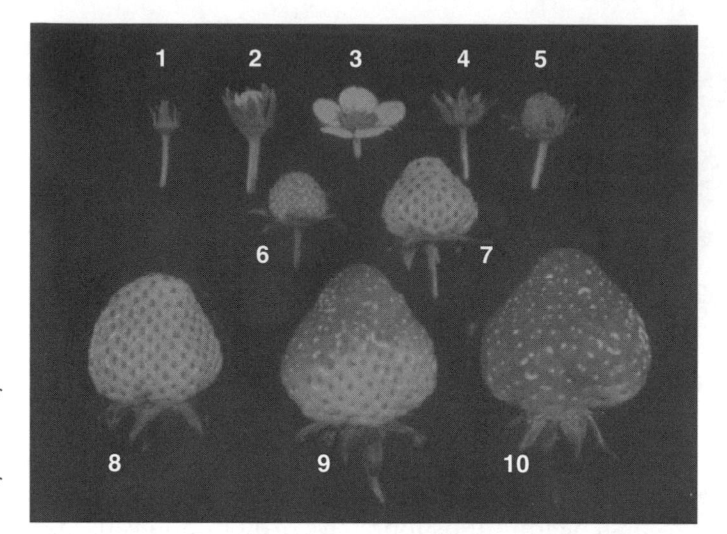

FIGURE 12.6. *The key stages of the ripening process of the strawberry. (Data reprinted with permission from the American Society of Plant Physiologists.)*

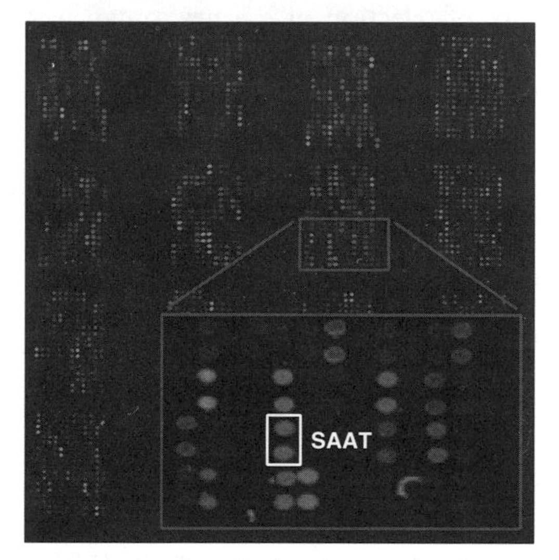

FIGURE 12.7. *To examine gene expression levels in ripe and immature strawberry fruits, total mRNA was isolated, labeled separately, mixed, hybridized to a single microarray, and scanned for cyanine-3 (Cy3) and cyanine-5 (Cy5) emission. The composite fluorescent image is a two-color overlay, with fluorescent signals from ripe fruits (red channel) and immature fruits (green channel). One portion of the microarray (red box) contains targets corresponding to SAAT (white box), a gene whose expression is induced in ripe fruits. (Data Reprinted with permission from the American Society of Plant Physiologists.) Figure also appears in Color Figure section.*

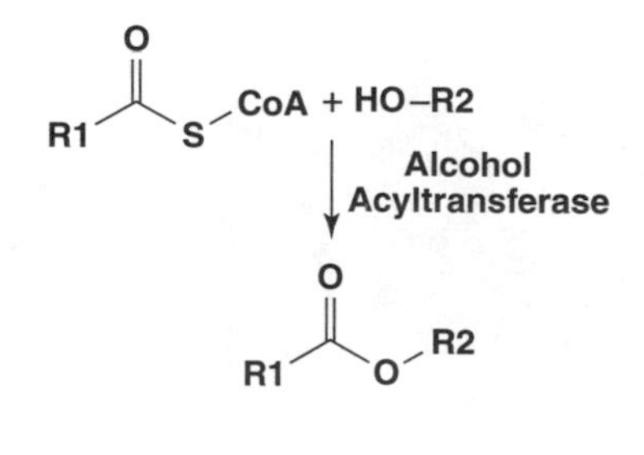

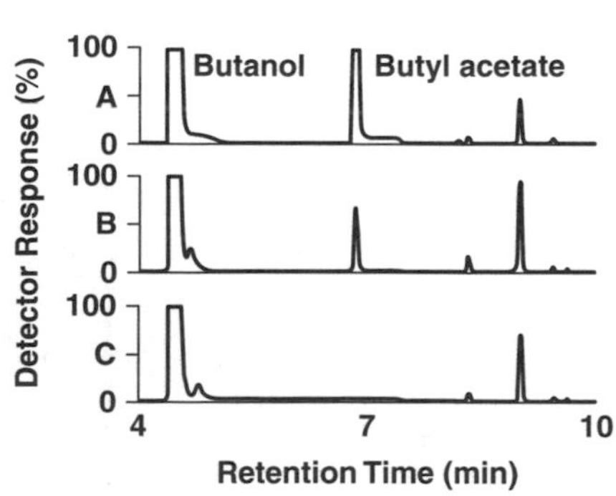

FIGURE 12.8. **A,** *SAAT encodes an enzyme that catalyzes the acetylation of alcohols to produce fruity-smelling esters.* **B,** *Studies with the strawberry enzyme purified from bacteria reveal the production of esters (butyl acetate)* in vitro. *(Data reprinted with permission from the American Society of Plant Physiologists.)*

molecular and biochemical studies have succeeded in identifying key genes involved in these processes. With genes in hand, it is possible modify crop plants further by elevating or reducing the levels of specific gene products and modulating the biochemical and developmental pathways that are controlled by the genes. ***Transgenic crops*** or ***genetically modified organisms*** (GMO) offer great promise in terms of providing fruits and vegetables with improved yield, durability, disease resistance, and nutritional composition. GMOs are likely to reduce the need for fertilizers, pesticides, and herbicides, enabling an era of commercial agriculture that is environmentally friendlier. Key to the success of ***molecular agriculture*** is the comprehensive understanding of the regulatory events that underlie crop plant development. Microarray analysis offers great promise for understanding fruits and vegetables of economic interest.

Transgenic crop. *Any agricultural plant that contains a genetic modification introduced using the tools of recombinant DNA or some other molecular technology.*

Genetically modified organism. *An agricultural plant that contains a genetic modification introduced using recombinant DNA, transgenic technology, or some other molecular method.*

Molecular agriculture. *Agronomic approach that relies on transgenic and other molecular technologies to improve the taste, nutrition, appearance, yield, and other traits in fruits and vegetables.*

Strawberries constitute a multibillion-dollar industry annually, with the United States providing about one-quarter of the world's supply of this important berry. Improving the flavor, aroma, and other qualities of the strawberry fruit are thus of considerable commercial interest. Strawberry fruits undergo a stereotyped developmental process, whereby the fruits transition from white to red, with increases in size, pigmentation, and aroma (Fig. 12.6). To study the changes in gene expression accompanying strawberry ripening, Aharoni and co-workers at Plant Research International (Wageningen, The Netherlands) employed cDNA microarray analysis to examine mRNA populations in mature and immature strawberry fruits (Fig. 12.7). This analysis identified dozens of known and new genes whose expression changes during the ripening process, including a novel strawberry alcohol acetyltransferase (SAAT) that encodes an enzyme involved in the production of strawberry esters. Biochemical studies indicate that the strawberry enzyme is active if expressed in bacteria (Fig. 12.8), supporting the sequence data that suggest its role in acetylation. The novel SAAT gene is almost certainly responsible for producing the ester compounds that give ripe strawberries their characteristic aroma, and transgenic studies with this novel aroma gene are under way to determine whether enhanced aroma can be achieved in strawberries and other fruits by elevating the level of this key enzyme identified by microarray analysis.

IONIZING RADIATION

Intense forms of electromagnetic radiation that have the capacity to dislodge electrons from the outer shells of atoms are known as ***ionizing radiation***. The most common forms of ionizing electromagnetic radiation include X rays and gamma rays. Because of their short wavelength (0.1–10 Å) and high energy, X rays and gamma rays can penetrate the skin and reach internal organs.

Ionizing radiation. *Intense form of electromagnetic radiation that has the capacity to dislodge electrons from the outer shells of atoms.*

Mammals are exposed to many different sources of ionizing radiation, including natural radiation from cosmic rays and radioactive elements found in the earth, as well as a wide spectrum of human sources owing to airline travel, nuclear weapons testing, atomic energy, food processing, and medical diagnostics. Exposure to X rays and gamma rays for medical procedures including dental X rays, positron emission tomography (PET) scans, and radiation-based cancer therapy represent the largest single source of exposure to ionizing radiation in humans, with the total exposure per person being larger than all of the other natural and human-made sources combined. Human exposure to extreme

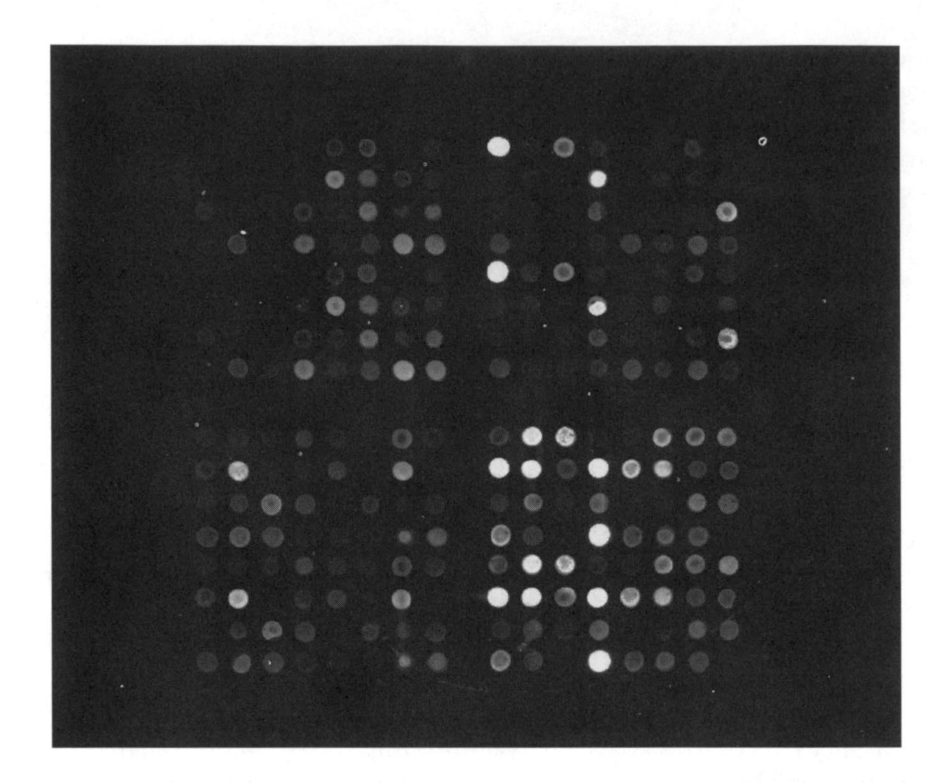

FIGURE 12.9. *To examine the changes in gene expression that accompany exposure to ionizing radiation, mice were exposed to 5 J/kg of radiation. Spleen mRNA was isolated 2 h after exposure, labeled with Alexa 594 dye, and hybridized to a microarray together with a labeled control sample. The microarray was detected in two channels with an ArrayWoRx charge-coupled device imager, and the data were displayed as a two-color overlay, with up- and down-regulated genes appearing as red and green spots, respectively. (Data courtesy of Dr. R. Mitra, Baylor College of Medicine, Houston, TX, and Dr. M. Kapoor, M.D. Anderson Cancer Center, Houston, TX.)*

levels of ionizing radiation can lead to point mutations, deletions, chromosomal rearrangements, and cell death owing to the ionization of DNA, though the modest levels used for most medical procedures pose little risk to patients.

As a first step toward a detailed understanding of the physiological effects of ionizing radiation, Mitra at the Baylor College of Medicine (Houston, TX) and Kapoor at the M. D. Anderson Cancer Center (Houston, TX) used DNA microarrays to profile radiation-induced changes in gene expression in the mouse. Mice were exposed to 5 Grays (5 J/kg) of radiation, and gene expression patterns in the spleen were compared to control mice. The microarray data reveal both up- and down-regulated genes in mice (Fig. 12.9), providing clues to the molecular mechanisms underlying radiation exposure in whole animals. The genetic and biochemical similarities between mouse and human suggest that genes identified in the mouse experiments will provide clues to the physiological consequences of radiation exposure in humans. Such information will be extremely valuable, given the importance of radiation therapy as a cancer treatment in humans.

ACUTE RENAL FAILURE

Acute renal failure. *Medical condition characterized by a sudden decline in kidney function.*

The human kidneys, two bean-shaped organs located in the lower back, remove toxins, excess fluids and salts, and waste products from the bloodstream. The kidneys filter vast quantities of blood each day, producing urine containing the unwanted waste components, which are excreted out of the body during urination. A sudden decline in kidney function, known as ***acute renal failure*** (ARF), can be triggered by a number of different medical insults, including bacterial infection, kidney stones, exposure to drugs or poisons, shock, and trauma. The sudden loss of kidney function results in the rapid accumulation of waste

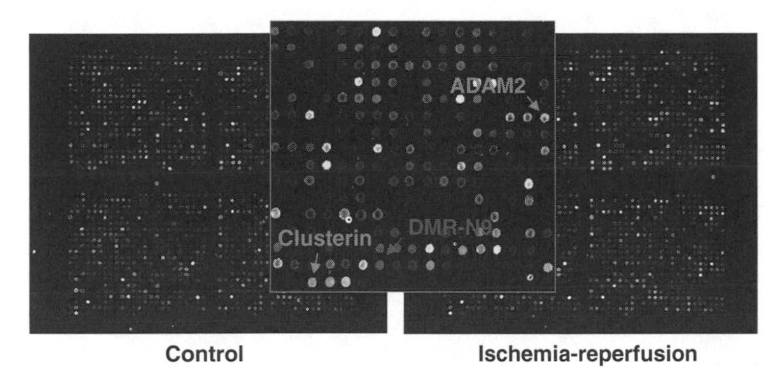

FIGURE 12.10. *Microarray analysis was used to examine the changes in gene expression that accompany ARF in the rat as ischemia followed by reperfusion. Fluorescent probes prepared from kidney samples were co-hybridized to a mouse cDNA microarray, and the Cy3* (left) *and Cy5* (right) *scans correspond to control and surgically treated rats, respectively. The two-color overlay* (center) *shows the identity of two activated genes (Clusterin and* ADAM2) *and one repressed gene* (DMR-N9). *(Data courtesy of Dr. T. Yoshida and Dr. S. Gullans, Brigham and Women's Hospital, Harvard Institutes of Medicine, (Boston, MA.)*

products in the bloodstream, leading to the serious condition known as ***uremia***. ARF can be fatal if left untreated and little is known about the molecular basis of this relatively common medical threat.

Uremia. *Medical condition characterized by the rapid accumulation of waste products in the bloodstream.*

To explore the molecular basis of ARF, Yoshida and co-workers at the Brigham and Women's Hospital (Boston, MA) employed microarray analysis in conjunction with a well-characterized rat model for the condition. Kidney failure was induced surgically in male rats by blocking blood flow to the kidney for 30 min, causing oxygen shortage or ***ischemia,*** followed by the reintroduction or ***reperfusion*** of blood and fluids into the kidneys to restore function. Then 4 days after the ischemia–reperfusion procedure, the kidneys were harvested from the ARF and control rats. Fluorescent probes were prepared from the experimental and control kidney samples and co-hybridized to a microarray containing 2100 mouse genes. Mouse cDNAs were chosen as the microarray elements because much more sequence information is available at present for the mouse than for the rat, and the two organisms share extensive sequence conservation, which allows mouse microarrays to be used to study rat gene expression.

Ischemia. *Medical condition of oxygen shortage caused by a blockage in blood flow.*

Reperfusion. *Medical procedure involving the reintroduction of blood or fluids into a tissue.*

Microarray analysis of the ischemia–reperfusion process in rat identified key genes whose expression is altered during acute renal failure (Fig. 12.10). Genes that were up-regulated during ARF included the acidic glycoprotein clusterin and the cell adhesion molecule ADAM2. *DMR-N9* was among the repressed genes. Two-dimensional analysis of expression ratios confirmed the precision of the data (Fig. 12.11) and identified many more genes whose expression is altered by ischemia–reperfusion. These data suggest that microarrays can be used for detailed analysis of the physiological changes that occur upon renal failure and may suggest new diagnostic markers and drug therapies for treating this acute medical condition.

LIGHT RESPONSES IN PLANTS

Light exerts a profound effect on plant development, triggering organelle biogenesis, the formation of chlorophyll pigments, photosynthesis, and the growth and proliferation of myriad vegetative and reproductive plant tissues,

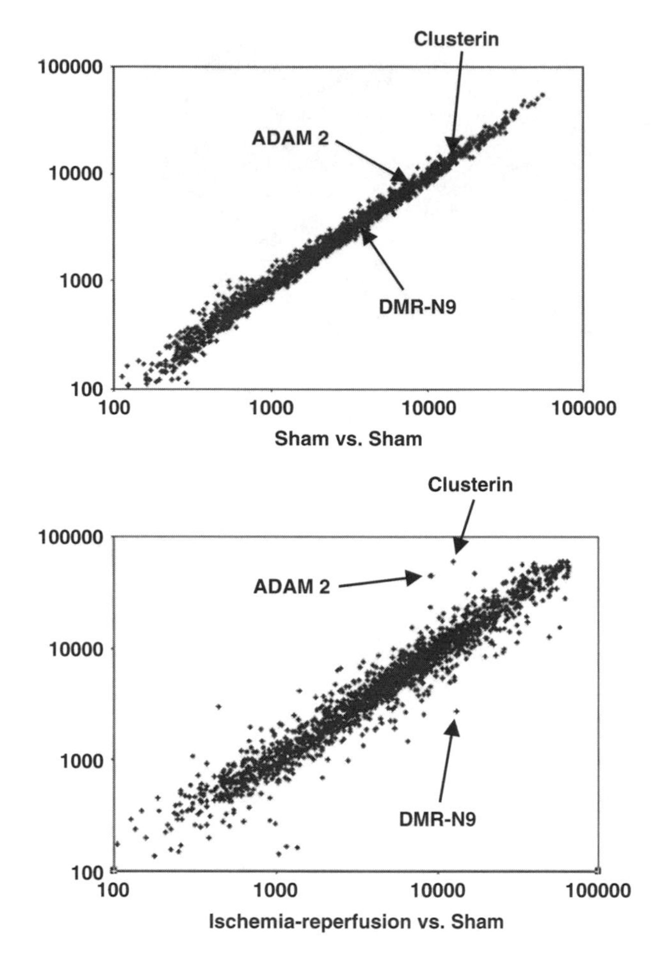

FIGURE 12.11. *Ratios of signal intensities for control (sham) vs. control (sham) rats, and ischemia–reperfusion vs. control rats. Data from Figure 12.10. (Courtesy of Dr. T. Yoshida and Dr. S. Gullans, Brigham and Women's Hospital, Harvard Institutes of Medicine, Boston, MA.)*

Phytochrome. *Light receptor protein in plants that undergoes a conformational change upon absorption of specific wavelengths of electromagnetic radiation, rendering the protein active to execute cellular signals, including changes in expression of specific cellular genes.*

including leaves, roots, stems, and flowers. Light receptor proteins or ***phytochromes*** absorb specific wavelengths of electromagnetic radiation, which trigger conformational changes in the protein and render the protein active to execute cellular signals, such as changes in the expression of specific cellular genes. Light signaling has been studied in many species, but a favorite model system for such studies is the small mustard plant *Arabidopsis thaliana* (Fig. 12.12). This famous weed has been used extensively in molecular studies because of its small genome (125 Mb), manageable gene number (~25,000), rapid generation time (4–6 weeks), diploid genome, and facile genetics. The recent completion of the complete genomic sequence of *Arabidopsis thaliana* (Fig. 12.13) is causing an explosion in experimentation with this organism.

Etiolated. *Morphology and appearance of plants, including extreme elongation and the absence of chlorophyll, observed when plants are grown in the dark.*

Wu and co-workers at the *Arabidopsis* Functional Genomic Consortium (AFGC) at the Carnegie Institute of Washington (Stanford, CA) performed a genomewide survey to identify changes in plant gene expression that occur when plants are shifted from darkness to light. Microarrays containing cDNAs derived from more than 7500 unique *Arabidopsis* genes were used to compare dark-grown or ***etiolated*** seedlings, with plants grown in the light (Fig. 12.14). The experiments reveal that changes in expression occur for approximately 25% of the genes in the genome, supporting the classical literature, which provides examples of dozens of light-regulated plant genes. Microarray-based analysis of light signaling will provide a paradigm for understanding how sunlight exerts its profound effect on plant development. Such studies in *Arabidopsis* will inform parallel processes in all flowering plants, including the crop plants.

FIGURE 12.12. *Three lines of A. thaliana. (Reprinted with permission from Cold Spring Harbor Laboratory Press, Cold Spring Harbor, NY.)*

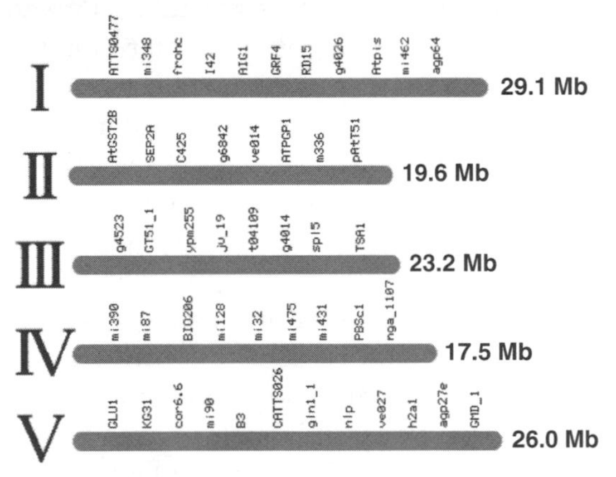

FIGURE 12.13. *The five chromosomes of A. thaliana, with the locations of key genetic and molecular markers. The approximate length of each chromosome is estimated from the recently completed genomic sequence. (Data courtesy of the Kazusa DNA Research Institute, Chiba, Japan, and The Institute for Genomic Research, Rockville, MD.)*

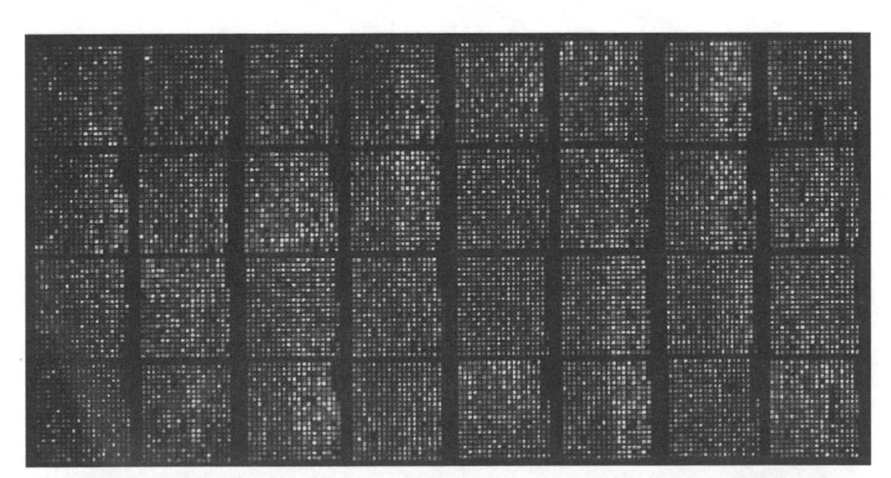

FIGURE 12.14. *An 11,500-element cDNA microarray that reveals the profound physiological changes that occur when plants are exposed to light. Flourescent probes were prepared from RNA isolated from 7-day-old dark-grown (etiolated) seedlings (Cy3) or from adult leaves (Cy5), mixed together, hybridized to a single microarray, and scanned for emission of Cy3 and Cy5. Red and green spots in the two-color overlay correspond to genes that are light-induced or light-repressed, respectively. (Data courtesy of Dr. S.-H. Wu and Dr. S. Somerville, AFGC, Carnegie Institute of Washington, Stanford, CA.) Figure also appears in Color Figure section.*

OSMOTIC STRESS

Salinity. *Salt content of irrigation water.*

Abiotic stress. *Category of environmental stress in plants attributable to nonliving sources.*

Osmotic stress. *Form of stress in plants and other organisms and cells due to an imbalance in salt concentration between the interior and exterior of cells.*

Most plants grow best in water containing low concentrations of sodium chloride and other salts, and such is the case with plants growing in natural settings. But nearly all crop plants grow by commercial agriculture are hydrated using some form of irrigation. During the irrigation process, sunlight causes evaporation, leading to an increase in the salt content or ***salinity*** of the irrigation water. Irrigation, deicing, and other forms of unnatural elevation in salts are contributors to a broad category of plant stresses known as ***abiotic stress*** Abiotic stress involving elevated salt concentrations causes what is known as ***osmotic stress***. Plant growth and development are affected negatively by osmotic stress, because elevated salt concentration on the exterior of the plant causes an influx of sodium ions, which perturbs biochemical reactions inside the cells. Osmostic stress threatens major crop plants in nearly 30% of the world's irrigated land and is thus a major ecological challenge, particularly as an expanding global population places greater and greater pressure on commercial agriculture. There is thus significant interest in understanding the physiological basis of osmotic stress, with the hopes of engineering variants that can better cope with elevated salts in irrigation water.

As a comprehensive step toward studying osmotic stress in plants at the genomic level, Deyholos and co-workers at the University of Arizona (Tucson, AZ) prepared microarrays of *Arabidopsis* and rice cDNAs and used those microarrays to examine salt effects on gene expression (Fig. 12.15). A microarray containing 18,432 elements from *Arabidopsis* allowed identification of dozens of genes whose expression changes upon treatment with sodium chloride. Analogous studies with rice are also yielding promising results. The prospects of using microarrays to study and develop salt-resistant rice plants is likely to receive much attention, given that this crop plant provides the main source of nutrition for approximately 50% of the world's population.

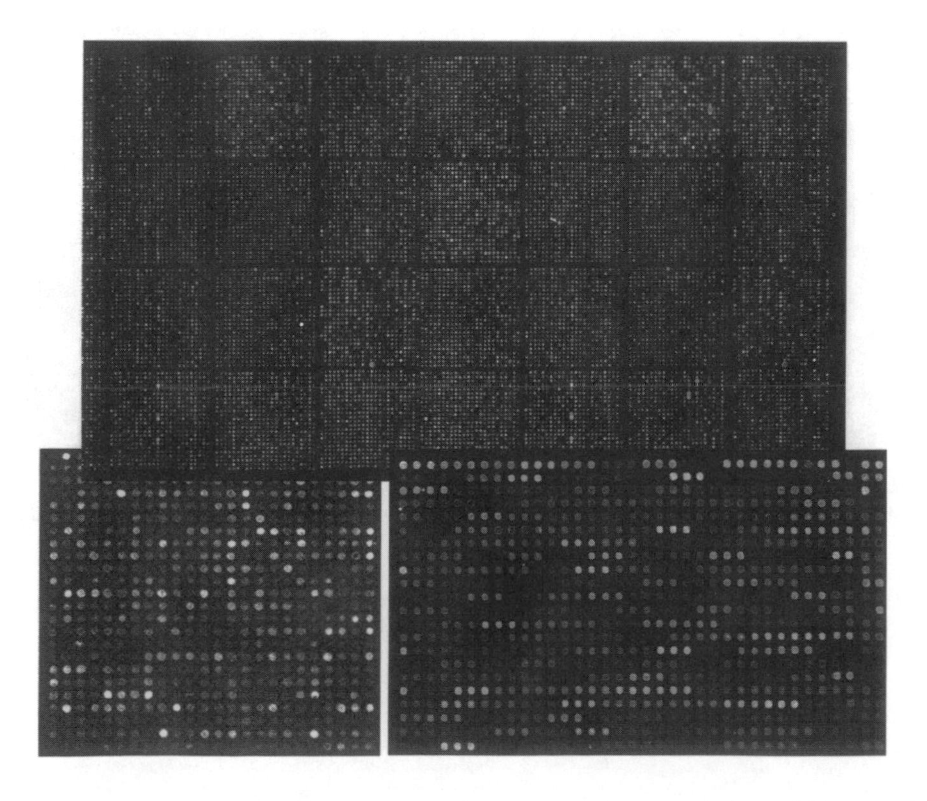

FIGURE 12.15. **A**, Arabidopsis *microarray containing 18,432 elements with one scanned channel shown in a rainbow palette.* **B**, *A two-color representation of Arabidopsis genes whose expression changes up or down upon salt treatment.* **C**, *A rice microarray showing salt-induced and salt-repressed genes with one scanned channel shown in a rainbow palette. (Data courtesy of Dr. Deyholos, Dr. Kawasaki, and Dr. Galbraith, University of Arizona, Tucson).* Figure also appears in Color Figure section.

TOXICOLOGY

The branch of science that focuses on the effects of poisonous substances is known as ***toxicology***. Pharmaceuticals, herbicides, cosmetics, and many other types of chemical compounds are potential toxins in humans by virtue of their ability to alter normal physiology in an undesired manner. The so-called side effects of many drugs result from inadvertent alterations in biochemical pathways that are distinct from the pathways of interest. Before the advent of genomics and microarray technology, it was not possible to test a pharmaceutical substance in a comprehensive, gene-by-gene manner before its commercial sale. Most drugs on the market today possess one or more side effects in some fraction of the human population; consequently there is tremendous medical interest in identifying molecular pathways that may serve as diagnostic markers for side effects with the hopes of developing safer drugs. Global gene expression monitoring with microarrays provides a comprehensive way to examine the physiological consequences of drug treatment in a quantitative manner.

Toxicology. *Branch of science that focuses on the biological effects of poisonous substances.*

Whitford and co-workers at Phase 1 Molecular Toxicology (Santa Fe, NM) have pioneered the use of microarrays in toxicology screening. One test system that has been used involves the administration of a compound of interest into a canine model, followed by gene expression analysis with microarrays of various organs that are affected by the drug (Fig. 12.16). The gene expression readout provides a highly sensitive measure of the physiological effects of a chemical compound at various times after administration. Test compounds that alter the expression of genes other than the desired target genes may exhibit toxic side effects when introduced into humans. Systematic use of microarrays may streamline the drug discovery process by allowing researchers to eliminate toxic substances before their entry into the enormously expensive phase known as ***clinical trials***. Microarray-based toxicology profiling may also allow the optimization of existing drugs, thereby eliminating or minimizing the side effects of drugs that are already known to work well for treating specific medical conditions.

Clinical trial. *Medical study conducted in human patients to assess the safety and efficacy of new drugs, vaccines, and other types of therapies.*

FIGURE 12.16. *One channel of a toxicology experiment comparing a canine treated with a test compound versus an untreated canine. The cDNAs were printed in duplicate, and one channel of the scanned fluorescent image is shown in a rainbow scale. (Data courtesy of Dr. B. Whitford, Phase 1 Molecular Toxicology, Santa Fe, NM.) Figure also appears in Color Figure section.*

LIVER CANCER

Hepatocellular carcinoma. *Malignant form of liver cancer thought to be associated with hepatitis B. It has a poor prognosis and is prevalent in parts of Southeast Asia and Africa.*

Co-factor. *Environmental, genetic, or infectious contributor that raises the incidence of a specific disease.*

The clinical condition of liver cancer or ***hepatocellular carcinoma*** is a worldwide health problem, with >1 million new cases being diagnosed each year. It is the seventh and ninth most common malignant tumor in males and females, respectively, and most patients with the disease die within weeks or months after diagnosis. The disease exhibits uneven geographical distribution, and populations in Southeast Asia and Africa exhibit elevated incidence of the disease. Though the reasons for the uneven geographical distribution are unknown, chronic hepatitis B infection is often suspected as an environmental contributor or ***co-factor*** in the disease. One model is that the hepatitis B virus integrates into the human genome and alters the expression of one or more cellular genes, triggering the cancerous transformation of liver cells and the growth of liver tumors. Altered expression of various cellular genes has been identified in hepatocellular carcinoma, though there has yet to be a systematic study of the changes that occur at the level of genomic expression in individuals afflicted with this disease.

To better understand, diagnose and treat liver cancer, Li and co-workers at Fudan University (Shanghai, China) used microarrays to identify genes that show elevated or reduced expression in cancerous liver tissue isolated from patients afflicted with the disease. Six patients were screened using microarrays containing cDNAs derived from 12,800 human genes (Fig. 12.17), and scatter plot analysis of the data indicated changes in the expression of many human genes when normal and cancerous tissues were compared (Fig. 12.18). Of the genes examined, 199 showed elevated expression in liver cancer tissue (Table 12.2) and 1,633 exhibited reduced expression (Table 12.3). Altered expression was observed for genes involved in the function of peroxisomes, serum control, polycyclic aromatic hydrocarbon (PAH) carcinogenesis, cell growth and differentiation, metastasis, the function of the immune system, apoptosis, and

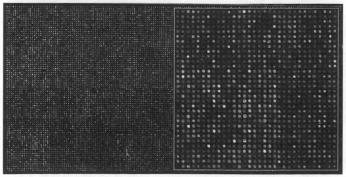

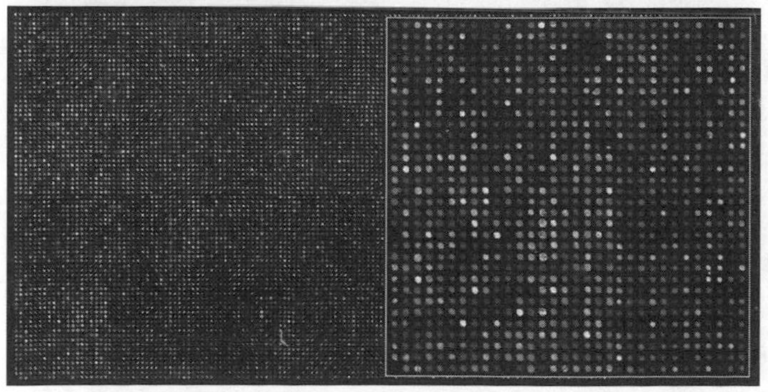

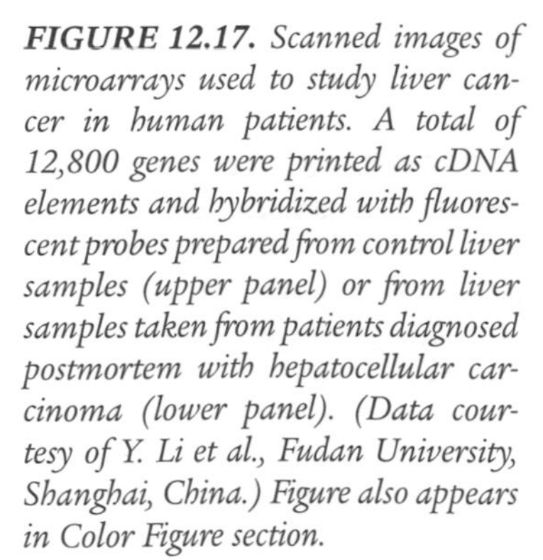

FIGURE 12.17. *Scanned images of microarrays used to study liver cancer in human patients. A total of 12,800 genes were printed as cDNA elements and hybridized with fluorescent probes prepared from control liver samples (upper panel) or from liver samples taken from patients diagnosed postmortem with hepatocellular carcinoma (lower panel). (Data courtesy of Y. Li et al., Fudan University, Shanghai, China.) Figure also appears in Color Figure section.*

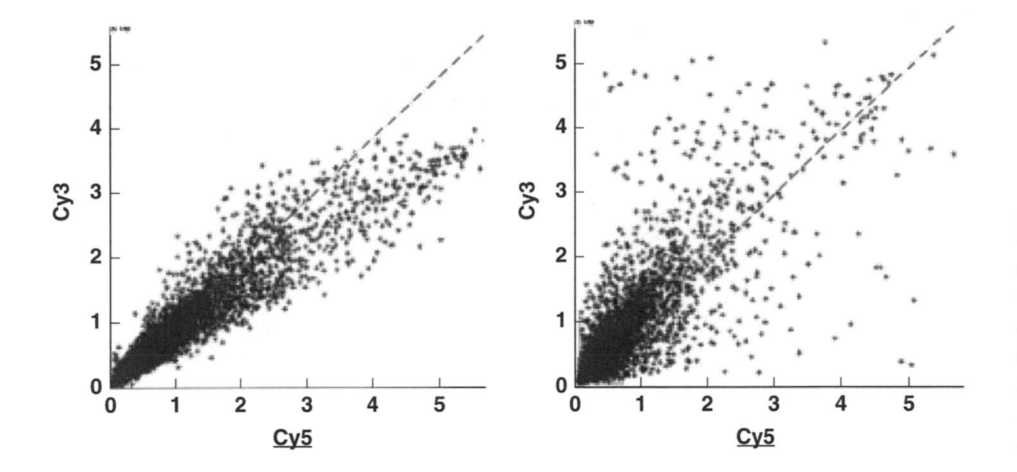

FIGURE 12.18. *Cy3 and Cy5 ratios in log scale are represented as two-dimensional scatter plots for control versus control (left) and control versus liver cancer (right), indicating up- and down-regulation for a large number of human genes. Data from Figure 12.17. (Data courtesy of Y. Li et al., Fudan University, Shanghai, China.)*

TABLE 12.2. Genes Up-Regulated in Liver Cancer Patients[a]

Accession Number	Gene Product	Patient Concordance	Ratio
U05684	dihydrodiol dehydrogenase	6	62
X07173	inter-alpha-trypsin inhibitor complex, second protein	6	46
K02091	prealbumin	6	37
X02761	ficronectin (FN precursor)	6	25
4557312	angiogenin, ribonuclease, RNase A family, 5 precursor	5	23
X13694	osteopontin	6	17
X86401	L-arginine:glycine amidinotransferase	6	16
U50078	guanine nucleotide exchange factor p532	3	14
M14219	chondroitin/dermatan sulfate proteoglycan (PG40) core protein	3	14
Y00716	complement factor H	6	13

[a] Data from Figure 12.17; courtesy of Y. Li et al., Fudan University, Shanghai, China.

TABLE 12.3. Genes Down-Regulated in Liver Cancer Patients[a]

Accession Number	Gene Product	Patient Concordance	Ratio
V00497	beta-globin	6	12.5
L02326	hu lambda-17 lambda-like protein	5	10
M27749	immunoglobulin-related 14.1 protein	5	10
L54057	CLP	6	8.3
NM_001558	interleukin 10 receptor, alpha	6	7.7
X16663	hS1 gene for heamatopoietic lineage cell-specific protein	6	7.7
X54469	beta-preprotachykinin	6	7.7
NM_003079	SWI/SNF related, matrix associated, actin-dependent regulator of chromatin	5	7.7
AF127670	hyaluronic acid receptor	6	7.1
L13463	Helix-loop-helix basic phosphoprotein (G0S8)	4	6.7

[a] Data from Figure 12.17; courtesy of Y. Li. et al., Fudan University, Shanghai, China.

TABLE 12.4. Concordance of Altered Genes in Six Liver Cancer Patients[a]

Number of Patients	Concordant Genes
6	213
5	227
4	496
3	846
≤ 2	50

[a] Data courtesy of Y. Li et al., Fudan University, Shanghai, China.

remodeling of the cytoskeleton; all of these functions make physiological sense in view of the cellular changes that take place during hepatocellular carcinoma. More than 97% of the 1,832 genes with altered expression were identified in three or more of the patients examined (Table 12.4), and this high patient ***concordance*** indicates that one or more of the genes may find use as a diagnostic marker.

Concordance. *Agreement or coincidence of data or datasets between a group of patients.*

SUMMARY

Gene expression profiling with microarrays enables researches to explore the bacterial, plant, and animal worlds with a depth and breadth that are unprecedented in the history of science. Gene expression reveals much about biological processes because organisms express their genes only when they are needed to fulfill a specific biological function. In response to environmental, genetic, or biochemical signals, genes are activated or repressed by as much as 1000-fold or more, providing a characteristic gene expression fingerprint, which can be used to understand gene function and the interplay between regulatory pathways. Microarrays containing target elements for all of the genes in a genome reveal comprehensive genetic and biochemical relationships that are virtually impossible to obtain by any other means. Mouse biology, strawberry fruit development, the effects of ionizing radiation, acute renal failure, light responses and osmotic stress in plants, drug toxicology, and liver cancer are some of the areas of basic science, medicine, and agriculture being revolutionized by microarray analysis. Gene expression information assembled into databases allows researchers to carry out biology in silico, in which the secrets of genomes are explored via computer.

SELECTED READING

Aharoni, A., Keizer, L. C., Bouwmeester, H. J., Sun, Z., Alvarez-Huerta, M., Verhoeven, H. A., Blaas, J., van Houwelingen, A. M., et al. Identification of the SAAT gene involved in strawberry flavor biogenesis by use of DNA microarrays. *Plant Cell* 12:647–662, 2000.

Baxevanis, A. D., and Ouellette, B. F. F., eds. *Bioinformatics: A Practical Guide to the Analysis of Genes and Proteins*, 2nd ed. Wiley-Interscience, New York, 2001.

Blackburn, G. M., and Gait, M. J., eds. *Nucleic Acids in Chemistry and Biology*, 2nd ed. Oxford University Press, Oxford, UK, 1996.

Bloomfield, V. A., Crothers, D. M., and Tinoco, I. *Nucleic Acids: Structure*. University Science Books, Herndon, VA, 2000.

Brown, T. A. Genomes. Wiley, New York, 1999.

Higgins, D., and Taylor, W., eds. *Bioinformatics: Sequence, Structure, and Databanks: A Practical Approach*. Oxford University Press, Oxford, UK, 2000.

Holter, N. S., Mitra, M., Maritan, A., Cieplak, M., Banavar, J. R., and Fedoroff, N. V. Fundamental patterns underlying gene expression profiles: simplicity from complexity. *Proc Natl Acad Sci USA* 97:8409–8414, 2000.

Hsiao, L. L., Stears, R. L., Hong, R. L., and Gullans, S. R. Prospective use of DNA microarrays for evaluating renal function and disease. *Curr Opin Nephrol Hypertens* 9:253–258, 2000.

Innis, M. A., Gelfand, D. H., and Sninsky, J. J., eds. *PCR Applications: Protocols for Functional Genomics*. Academic Press, San Diego, CA, 1999.

Kawasaki, S., Borchert, C., Deyholos, M., Wang, H., Brazille, S., Kawai, K., Galbraith, D., and Bohnert, H. J. Gene expression profiles during the initial phase of salt stress in rice. *Plant Cell* 13:889–906, 2001.

Kehoe, D. M., Villand, P., and Somerville. S. DNA microarrays for studies of higher plants and other photosynthetic organisms. *Trends Plant Sci* 4:38–41, 1999.

Lehninger, A. L., Nelson, D. C., and Cox, M. M. *Principles of Biochemistry*, 3rd ed. Freeman, New York, 2000.

Lewin, B. *Genes VII*. Oxford University Press, Oxford, UK, 2000.

Li, Y., Qiu, M. Y., Wu, C. Q., Cao, Y. Q., Tang, R., Chen, Q., Shi, X. Y., Hu, Z. Q., et al. Detection of differentially expressed genes in hepatocellular carcinoma using DNA microarray. *Yi Chuan Xue Bao* 27:1042–1048, 2000.

Richmond, T., and Somerville, S. Chasing the dream: Plant EST microarrays. *Curr Opin Plant Biol* 3:108–116, 2000.

Robbins, S. L., Cotran, R. S., Kumar, V., and Collins, T., eds. *Pocket Companion to Robbins Pathologic Basis of Disease*, 6th ed. Saunders, Philadelphia, 1999.

Schena, M., ed. *DNA Microarrays: A Practical Approach*, 2nd ed. Oxford University Press, Oxford, UK, 2000.

Schena, M., ed. *Microarray Biochip Technology*. Eaton Publishing, Natick, MA, 2000.

Schena, M., Shalon, D., Davis, R. W., and Brown, P. O. Quantitative monitoring of gene expression patterns with a complementary DNA microarray. *Science* 270:467–470, 1995.

Scriver, C. R., Sly, W. S., Childs, B., Beaudet, A., Valle, D., Kinzler, K., and Vogelstein, B. *The Metabolic and Molecular Bases of Inherited Disease*, 8th ed. McGraw-Hill Professional, Hightstown, NJ, 2000.

Singer, M., and Berg, P. *Genes and Genomes*. University Science Books, Herdon, VA, 1991.

Strachan, T., and Read, A. P. *Human Molecular Genetics*, 2nd ed. Wiley, New York, 1999.

Stryer, L. *Biochemistry*, 4th ed. Freeman, New York, 1995.

REVIEW QUESTIONS

1. The cellular process whereby mRNA is synthesized from DNA templates is known as what?
2. Which of the following stimuli might alter patterns of gene expression in a human being: (a) cup of coffee, (b) falling stock prices, (c) sleep deprivation, (d) vodka tonic, and (e) Viagra?
3. Patterns of gene expression are informative biologically because the expression of cellular genes is linked tightly to what? One word will suffice.
4. A researcher wants to study human gene expression with chimpanzee cDNA microarrays. An adjustment in which experimental parameter would be most effective in permitting the use of chimp cDNA microarrays for human studies: (a) size of the cover slip, (b) target length, (c) hybridization temperature, or (d) probe volume?
5. A human cDNA microarray would yield the greatest fluorescent signals with a probe prepared by labeling total mRNA from which organism: (a) yeast, (b) *Arabidopsis,* (c) mouse, (d) *Drosophila,* or (e) *E. coli*?
6. The single-mismatch (SM) oligonucleotides in Affymetrix gene expression microarrays are used to assess primarily which aspect of hybridization: (a) specificity, (b) time, (c) temperature, or (d) strength?
7. A genetically modified organism (GMO) is compared to an unmodified parent plant using a gene expression microarray. Analysis reveals elevated expression of five genes in a metabolic pathway even though only a single cellular gene was modified in the GMO. Anti-GMO activists use the

microarray data to demonstrate the dangers of genetically modified plants. Explain the flaw in their position.

8. *Arabidopsis thaliana* plants were grown for 4 weeks in total darkness; groups of plants were shifted to growth in bright light for 2 hr, 24 hr, or 7 days; and mRNA was isolated from the three groups of plants and analyzed on 20-K microarrays versus mRNA from a dark-grown control. The differential expression patterns revealed 17 altered genes in the 2-hr sample, 653 altered genes in the 24-hr sample and 4457 altered genes in the 7-day sample. Explain these results vis-à-vis gene regulation.
9. Canine microarrays are used to profile a protein-binding drug before clinical trials, and a minor side effect is observed in canine, along with specific changes in gene expression. A subsequent small-scale clinical trial in humans reveals the same minor side effect and a similar gene expression profile. What do these data suggest about the protein involved in the side effect in dogs and humans? One sentence will suffice.
10. Five sets of concordant genes (213, 227, 496, 846, and 50) are listed in Table 12.4. Which gene set is most likely to yield the most effective protein target for a liver cancer drug?

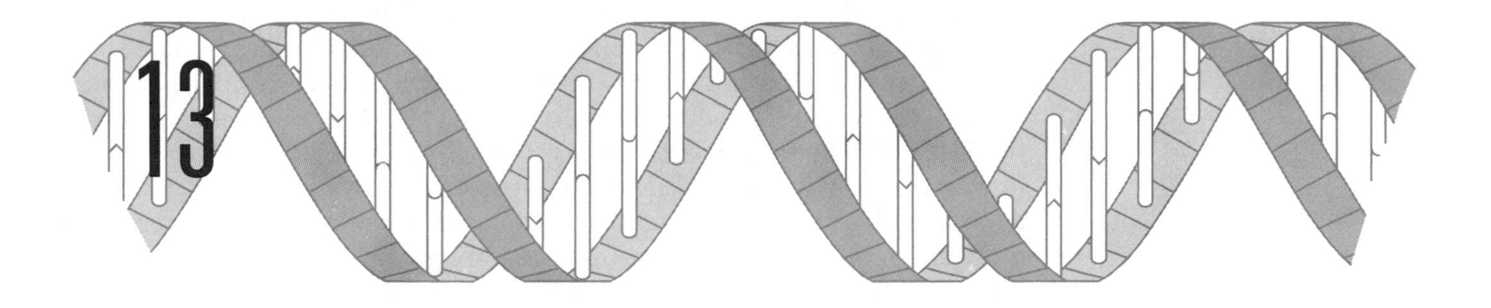

13 Genetic Screening and Diagnostics

Without virtue, happiness cannot be.
—*Thomas Jefferson*

With the completion of the sequence of the human genome and a massive collection of carefully compiled clinical data from decades of work, we are poised to revolutionize health care by ushering in the era of molecular medicine. One component of molecular medicine will include widespread genetic screening and diagnostics, to identify treatable and curable genetic diseases either before or at an early stage of onset and minimize the suffering and loss of life associated with phenotypic diagnosis at later times. Microarray analysis offers a set of analytical platforms that provide rapid, affordable and extremely accurate genetic information at the DNA level. Microarray-based neonatal and adult screening promises to provide comprehensive testing capabilities to identify mutations and polymorphisms that underlie inherited and infectious diseases. Working hand in hand with ethicists and clergy ensures the virtuous application of microarray tests for the benefit of everyone.

HUMANS AS A SYMPHONY ORCHESTRA

Many analogies have been put forth over the years to explain the genetic functionality of biological systems, but none is perhaps as instructive as the human being as a symphony orchestra. One can think of each player in the orchestra as representing a single gene or protein in the human body. Similar to the group of musicians, the human genome encodes families of genes that perform similar cellular functions, including DNA replication (strings), ion transport (brass), gene expression (woodwinds), and cell division (percussion). Within a given gene family, such as the transcription factors that control gene expression, there are specific motifs including zinc finger (flute), homeodomain (oboe), and helix-loop-helix (clarinet). Each gene in the human body encodes a unique biochemical entity, in much the same way that each member of the orchestra produces a distinct musical sound.

The ~30,000 genes in the human genome work in concert to carry out the intricate instructions of cells, using a series of complex regulatory pathways and hierarchies akin to the tightly coordinated events required to produce a musical symphony. Specific proteins (master regulators) control the activity of many different downstream genes by functioning as "conductors" of biological function, whereas a given downstream gene (third violin) mediates its effect once the proper instructions are received from the upstream regulator (conductor). Though all human beings share > 99.9% DNA sequence identity, small differences in genomic sequence (e.g., single nucleotide polymorphisms) give rise to significant individual differences, including physical appearance, disease susceptibility, longevity, and so forth. Analogously, the identical score of Beethovan's fifth performed by the San Francisco Symphony and the New York Philharmonic would likely yield very similar and pleasing but nonidentical performances of the musical masterpiece.

Optimal musical performance demands precise timing and execution by each member of the orchestra. An occasional missed note by one performer may produce an error (human freckle) that goes unnoticed by all but the most highly trained musical expert. On the other hand, a major mistake in timing by multiple members of the orchestra would produce a cacophony, in much that same way that global miscues in cell division can lead to cancer.

Essentially every cell in the human body contains an identical genome consisting of ~3 billion bases of DNA, but different cell types express a different repertoire of the ~30,000 genes as required to perform their specialized cellular functions. Nerve cells and skin cells each express approximately 5,000 genes, but a different subset of genes is expressed in the two cell types, allowing each to carry out respective neural and epidermal functions. An evening at the symphony often includes the presence of all of the members of the orchestra, but each musician is used in accordance to the demands of piece that is being performed (e.g. Mozart's thirty-fifth or thirty-ninth).

Human health is determined by the combined factors of genes and environment. Individuals with identical genomes (identical twins) can experience different frequencies of lung cancer, stroke, and infectious disease based on their different lifestyles (e.g., smoking, diet, sexual conduct). We know from common experience that more subtle day-to-day environmental factors such as stress and sleeplessness can alter how we feel (our phenotype), even though our genomic sequence (genotype) is invariant on this timescale. The environment can have a subtle but noticeable impact on the performance of the symphony orchestra; an off night can be associated with the state of mind and body of the conductor and key performers in the symphony. Excessive use of alcohol has been known to diminish the performance of conductors, and make no mistake that an alcohol-related hangover is entirely biochemical.

OVERVIEW OF HUMAN DISEASE

A human being can be modeled as a symphony orchestra, though an equally meaningful analogy is to think of humans as machines wherein each gene represents one part of the mechanical device. By this analogy, healthy humans would be those who contain a complete set of fully functional parts (genes), whereas humans with diseases contain one or more defective parts (genes) or are infected with foreign bacterial or viral genes or gene products that disrupt the normal function of the human cellular genes. Human diseases are commonly categorized according to the phenotype associated with the disease (e.g., tumor or blood cell count). The main categories of human disease include cancer, and disorders that impair the nervous system, immune system, muscle and bone formation, and cellular metabolism. Serious illnesses are also associated with defects in cell signaling and cellular transport. To place human disease in a broader clinical framework, it is instructive to consider the main categories of disease in more detail.

In the Nobel Prize–winning work of Bishop and Varmus (Varmus, 1984), malignancy was linked directly to the inappropriate function of cellular genes, establishing the fact that cancer is intrinsically a genetic phenomenon. Cancer arises when the genetic mechanisms that regulate cell division are perturbed or damaged (Fig. 13.1). In the human body, cell division is controlled tightly in both a ***temporal*** and a ***spatial*** manner, such that the rate and location of mitosis is under strict regulation. Alterations in timing or spatial signaling can lead to excessive cell division in improper locations, causing the formation of tumors. The growth of tumors can impair the organs in which the tumor mass is growing and disrupt the normal physiology of the afflicted individual. Tumor cells can also break away from the tumor mass, move into the bloodstream, and migrate to a secondary location in the body in a process known as ***metastasis.*** Tumor cells that undergo metastasis can themselves form tumors, compounding the seriousness of a cancer. Early diagnosis of cancer improves the prognosis in many cases, in part because little or no metastasis occurs at the early stages of tumor formation. Microarray tests for tumor typing are showing great promise for cancer diagnosis, prognosis and therapy and a complete list of cancer-related microarray publications is available electronically (www.arrayit.com/e-library/).

Temporal. *Gene expression, developmental, or other biochemical event that is controlled or restricted in time.*

Spatial. *Gene expression, developmental or other biochemical event that is controlled or restricted according to biological location or region.*

Metastasis. *Cancerous process in which tumor cells dislodge from a tumor mass, move into the bloodstream, and migrate to a secondary location in the body.*

Several hundred different types of cancer have been characterized and extensive information is available electronically (cancernet.nci.nih.gov/alphalist.html#The%20As). Cancer of the lung, colon, breast, prostate, and pancreas claim more than 300,000 American lives annually (Table 13.1). Nearly 90% of all lung cancer deaths are attributable to cigarette smoking, underscoring that fact that environmental exposure to chemical ***carcinogens*** can present a serious health risk. In the case of colon, breast, and pancreatic cancers, a few

Carcinogen. *Chemical, biological, radiological or environmental agent that causes cancer.*

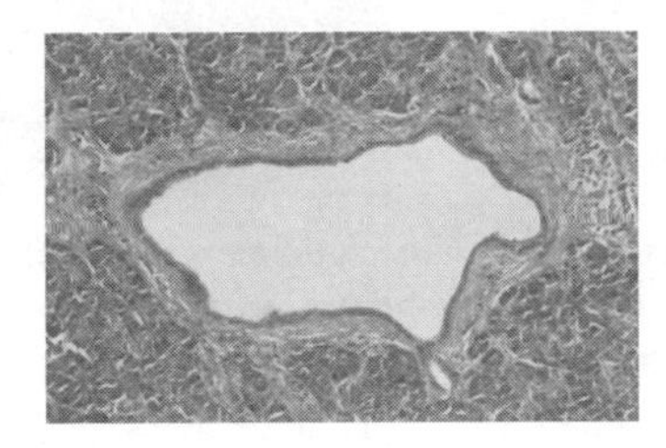

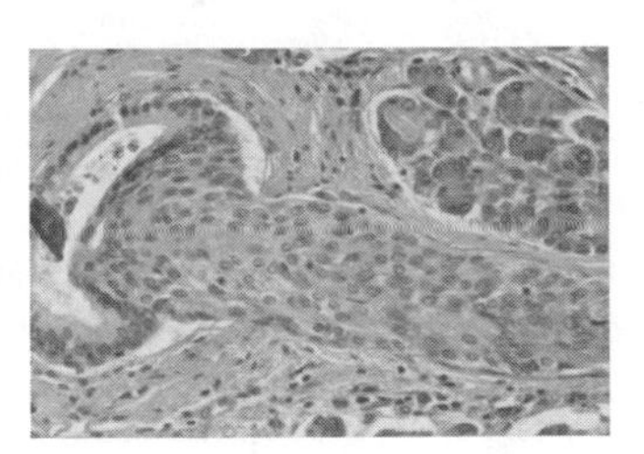

FIGURE 13.1. Healthy pancreatic tissue with normal ductal intraepithelial cells and cancerous, or malignant, tissue with squamous cells. (Courtesy of the Johns Hopkins Oncology Center, Baltimore, MD.)

TABLE 13.1. Common Cancers in the United States[a]

Type	Number of Deaths (per year)	Susceptibility Genes	Comments
Lung	160,000		leading cause of cancer death in men and women; 80–90% of cases attributable to cigarette smoking
Colon	48,000	*MLH1, MSH2, MSH6*	
Breast	44,000	*BRCA1, BRCA2*	> 99% of deaths occur in women
Prostate	39,000	*HPC1*	second leading cause of cancer death in males
Pancreas	28,000	*DPC4* (*Smad4*)	

[a]Data courtesy of the National Center for Biotechnology Information (Bethesda, MD).

disease susceptibility genes have been identified. Cancers are known to arise both from mutations in single genes and from chromosomal breaks and rearrangements, the latter phenomenon being observable by fluorescence analysis of human chromosomes (Fig. 13.2).

Neurodegenerative. *Category of human disease including Parkinson, Alzheimer, epilepsy, and Huntington, that degrade neural function and impair the human brain and nervous system.*

Inclusion body. *Foreign tissue deposit that forms in the brain and other human organs, and is associated with Parkinson, Alzheimer, and other diseases.*

Immune system. *Highly complex cellular defense mechanism in humans and other mammals that combats bacterial, fungal, and viral infection.*

Antigen. *Any non-self-protein or other biomolecule that illicits an immune response.*

Diseases that affect the human brain and nervous system encompass a family of serious disorders, including Parkinson disease, Alzheimer disease, epilepsy, Huntington disease, and amyotrophic lateral sclerosis. These diseases result from perturbations in the electrical signals that coordinate memory, movement, speech, and comprehension. Because there is a common degeneration of neural function, illnesses of this type are often known collectively as ***neurodegenerative*** diseases. In the case of Parkinson and Alzheimer diseases, incidence rises sharply with age, and both diseases result in the formation of foreign tissue deposits known as ***inclusion bodies*** in the brain. For many neurodegenerative diseases, one or more genes implicated in the disorders have been identified, and their biochemical mechanisms are being delineated.

To defend against bacterial, fungal and viral invaders, humans and other mammals possess a highly complex cellular defense mechanism known as the ***immune system.*** The immune system is capable of distinguishing between normal cellular components (self) and foreign agents such as infectious organisms (non-self). The immune system produces millions of different types of antibodies, each with a highly specific reactivity to non-self proteins known as ***antigens.*** Antibodies work in conjunction with lymphoid cells to react with foreign

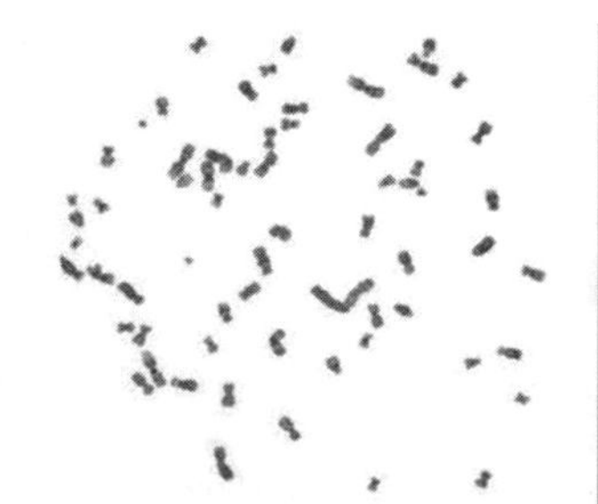
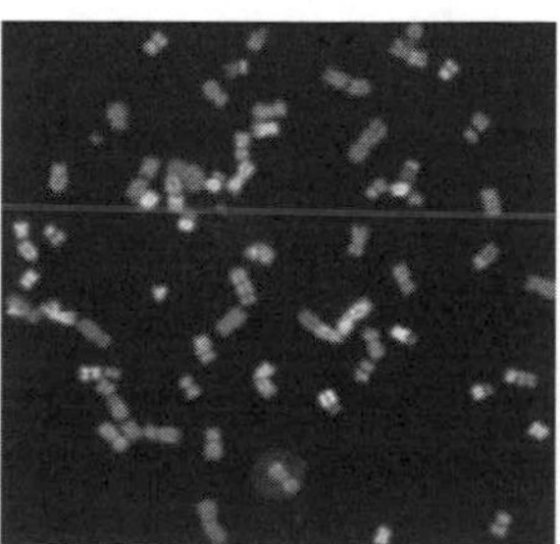

FIGURE 13.2. Two methods for examining human chromosomes. Traditional Geimsa staining reveals the familiar G-banding pattern, and spectral karyotyping by hybridization with fluorescent SKY probes provides a unique color for each chromosome, allowing translocations to be detected more easily. (Data courtesy of Dr. M. Shuster, Gollin laboratory, University of Pittsburgh Cancer Institute, Pittsburgh, PA.) Figure also appears in Color Figure section.

antigens and clear the body of the infectious organisms that are often associated with them. Antibodies and lymphoid cells function in an interconnected system known as the lymphatic system, which provides immune defensiveness throughout the human body.

Defects in the immune system are attributable to inherited disorders, such as severe combined immunodeficiency (SCID), and to damage to the immune system through viral infection, such as with AIDS. A large fraction of SCID cases result from the inheritance of either a defective interleukin 2 receptor gamma gene or a defective *JAK3* gene, both of which are required for proper development of T lyphocytes, key cellular players in the cellular immune response. AIDS patients also have impaired T cell function, but not because of an inherited disorder but rather because HIV infects T cells and kills them. AIDS patients have a reduced number of $CD4^+$ T cells and are thus susceptible to infection by microorganisms, fungi, and viruses. It is opportunistic infection that kills most patients infected with HIV. There is also a family of serious disorders known as autoimmune diseases, in which the immune system attacks itself and causes tissues and organs to malfunction. Rheumatoid arthritis, certain forms of diabetes, and multiple sclerosis are among the diseases in this category.

The human skeleton and musculature provide the means for our locomotion. Bones, muscle, cartilage, and tendons work in concert to execute signals from the brain that carry instructions for walking, running, and other forms of movement. A number of serious disorders, including muscular dystrophy, spinal muscular atrophy, myotonic dystrophy, and Marfan syndrome, impair the proper functioning of bone and muscle. Mutations in the gene encoding dystrophin underlie the malfunction of muscle cells in duchenne muscular dystrophy, and defects in the gene encoding the connective tissue protein fibrillin lead to Marfan syndrome. Both of these diseases are readily amenable to DNA-based analysis.

Cellular metabolism is the process by which enzymes in the body convert the food we eat into essential sources of energy and nutrients required for growth, development, and maintenance. Mutations in genes encoding key metabolic enzymes can lead to serious disease, either due to a shortage of one or more key nutrients or due to the accumulation of toxic by-products resulting from blockage in a metabolic pathway. Because of their biochemical nature, metabolic diseases often manifest at an early stage of development, and these so-called inborn errors of metabolism pose a serious health risk to newborn children. Defects in the metabolism of amino acids, fatty acids, organic acids, sugars, and other small molecules afflict > 1 infant in 5000, leading to mental retardation and death in serious cases (Table 13.2). The genes and enzymes underlying many metabolic disorders have been identified, serving as a basis for newborn screening by mass spectrometry or microarray analysis (see below).

Defects in genes that transport gases, ions, and other cellular materials are responsible for a class of serious conditions that include sickle cell anemia, cystic fibrosis, and other important diseases. Patients afflicted with sickle cell anemia have defects in the genes encoding the hemoglobin (Hb) proteins, which transport oxygen in the body. Sickle cell patients suffer from chronic anemia and periodic episodes of pain. Cystic fibrosis patients have one or more mutations in the cystic fibrosis transmembrane conductance regulator (CFTR), a gene encoding a protein that transports sodium and chloride ions in and out of cells. CFTR mutations result in the accumulation of a sticky DNA mucus in the lungs, owing to the lysis of defective lung cells and a release of their

TABLE 13.2. Neonatal Diseases

Class	Diseases	Frequency[a]
Amino acid	phenylketonuria, maple syrup urine disease	1:7, 728
Fatty acid	medium-chain acyl-CoA dehydrogenase, multiple acyl-CoA dehydrogenase deficiency	1:15, 675
Organic acid	glutaric acidemia, methylmalonic acidemia, propionic acidemia	1:16, 388
Ion transport	cystic fibrosis	1:5,979
Endocrine	congenital hypothyroidism, congenital adrenal hyperplasia	1:14, 630
Hemoglobin	sickle cell anemia type SS, sickle cell anemia type SC	1:1, 777
Vitamin and sugar	biotinidase deficiency, galactosemia	1:33, 867

[a]Combined rates of the class as measured by Neo Gen Screening (Bridgeville, PA).

genomic DNA into the lungs. Cystic fibrosis patients suffer from a clogging of the bronchial passages, leading to shortness of breath, wheezing, and infection as well as other serious conditions. Cystic fibrosis is the most common fatal genetic disorder in the United States, afflicting > 1 out of every 6000 newborns (Table 13.2).

Cystic fibrosis provides an extreme example of a disease that has a large number of different mutations associated with the disease phenotype, with more than 500 gene variants or alleles (Table 13.3) distributed across the coding sequence of the 1481-amino acid CFTR protein (Fig. 13.3). This is in contrast with the interleukin 13 (*IL13*) gene found in association with asthma, which contains only a single allele. The typical human disease gene contains one or two common alleles and a dozen or so rare alleles that cause the disease phenotype.

In addition to the commonly occurring human diseases, such as cystic fibrosis and asthma, there are literally thousands of less common illnesses encompassing a wide spectrum of categories, and hundreds of these have one or more genes that have been shown to cause or be associated with the disease (Scriver et al., 2001). There are dozens of infectious diseases such as AIDS, hepatitis, typhoid, Lyme disease, malaria, yellow fever, and tuberculosis; and

TABLE 13.3. Mutations in CFTR and IL13[a]

Mutation Type	CFTR	*IL13*
Single nucleotide (missense and nonsense)	306	1
Single nucleotide (splicing)	68	0
Single nucleotide (regulatory)	3	0
Small deletions	87	0
Small insertions	26	0
Small indels	5	0
Gross deletions	11	0
Gross insertions and duplications	1	0
Complex rearrangements (including inversions)	4	0
Repeat variations	0	0
Total	511	1

[a]Data courtesy of the Human Gene Mutation Database, University of Wales College of Medicine (Wales, UK) in conjunction with Celera Genomics (Rockville, MD).

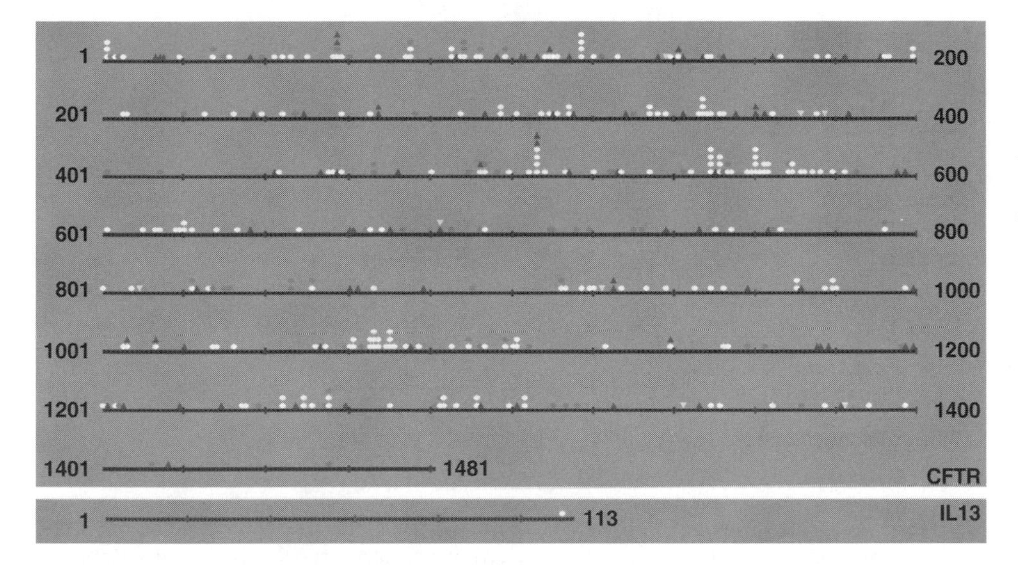

FIGURE 13.3. The CFTR and IL13 mutations known to cause cystic fibrosis and to be associated with asthma, respectively. Missense mutations (light circles), nonsense mutations (gray circles), insertion mutations (light triangles), and deletion mutations (dark triangles) are shown. (Data adapted from the Human Gene Mutation Database, University of Wales College of Medicine, Cardiff, Wales, UK, in conjunction with Celera Genomics, Rockville, MD.) Figure also appears in Color Figure section.

their modes of transfer are well understood. Nearly every infectious disease is attributable to the presence of bacterial or viral DNA in the infected individual, rendering infectious disease amenable to analysis at the DNA level.

Microarrays provide a broad and accurate technology for examining DNA sequences in a massively parallel and highly automated fashion. Microarray methods are poised to change health care in a fundamental way by allowing mutations, chromosomal translocations, and the presence of foreign DNA to be identified in an economical manner in patients afflicted with inherited, environmentally triggered, or infectious disease. By leveraging clinical information and the complete sequence of the human genome, microarrays are enabling populationwide screening and diagnosis of human diseases that have an unequivocal genetic basis, including those that involve point mutations, deletions, translocations, or the acquisition of foreign DNA. Some of the most promising microarray methods for genetic screening and disease diagnostics are described below.

SEQUENCING BY HYBRIDIZATION

Sequencing by hybridization (SBH) was developed by Drmanac, Crkvenjakov and their co-workers at the University of Belgrade (Yugoslavia) in the late 1980s. The SBH approaches use hybridization with oligonucleotides to determine the sequence of an unknown target DNA. In one form of the method (format 1), clones of unknown sequence are arrayed onto filters or glass, the clones are hybridized with labeled oligonucleotides of known sequence, and the oligonucleotides that produce positive hybridization signals are scored (Fig. 13.4). By assembling and aligning the overlapping oligonucleotide sequences via the computer, it is possible to gain detailed information about the sequence of the unknown target.

Sequencing by hybridization. *Molecular method that exploits nucleic acid hybridization for primary sequence determination.*

Because there are 4 bases in the genetic code (A, G, C, and T), the number of possible oligonucleotides that can be created from the four bases increases by a factor of 4 for each additional nucleotide present in the oligonucleotide probe (Table 13.4). There are 64 possible 3mers (4^3), 256 possible 4mers (4^4), and 65,536 possible 8mers (4^8). As the number of possible oligonucleotide

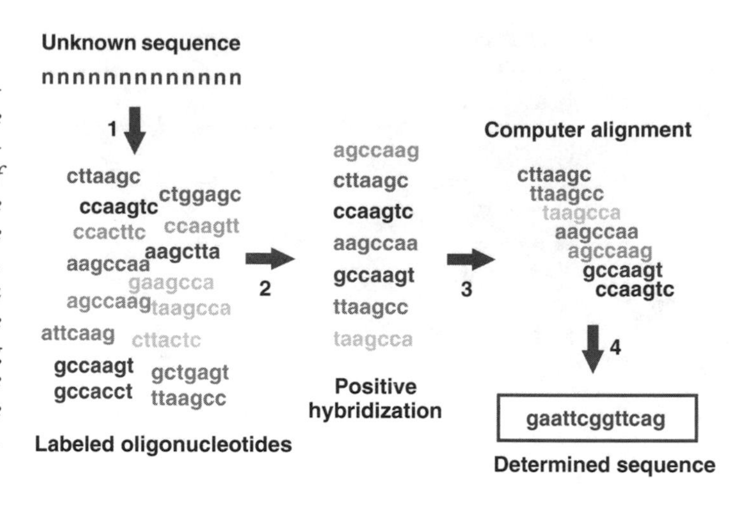

FIGURE 13.4. In one SBH approach, an unknown target sequence is attached to a filter or glass substrate and hybridized with a set of labeled oligonucleotides (1). The oligonucleotides that give positive hybridization signals are detected (2), and their sequences are aligned via a computer (3). By assembling the positively hybridizing, overlapping oligonucleotides the DNA sequence can be determined with a high degree of accuracy (4). (Courtesy of Dr. R. Drmanac, Hyseq, Sunnyvale, CA.)

probes increases, so does the theoretical read length. With a complete set of more than 68 million 18mers, it is possible, in theory, to determine the sequence of a 7-gigabase target sequence, which is larger than the human genome (3 gigabases). In practice, SBH has not been applied widely to de novo genomic sequencing due to several practical limitations inherent in the biochemistry of hybridization and in genome structure. Different oligonucleotides hybridize with different efficiencies, making it difficult to unambiguously distinguish positive and negative hybridization signals for a large number of probes. In addition, genomic DNA contains many repetitive regions, which generate branch points in the sequence assembly step and make it difficult to assemble long stretches of DNA as would be required to sequence the human genome.

The more practical applications of the SBH methods include the identification of novel clones (e.g., expressed sequence tags), sequence verification, mutation detection, and single nucleotide polymorphism analysis. For these applications, oligonucleotide arrays have been used in a number of extremely powerful ways. One approach has been to employ a form of SBH (format 3) that uses arrays containing all possible 5mer oligonucleotides (4^5 or 1024) in a

TABLE 13.4. Probe Sets for SBH[a]

Probe Length (n)	Number of Possible Probes (4^n)	Theoretical Read Length (bases)
3	64	6
4	256	25
5	1,024	100
6	4,096	500
7	16,384	1,500
8	65,536	6,500
9	262,144	25,000
10	1,048,576	100,000
12	16,777,216	1,500,000
14	268,435,456	25,000,000
16	4,294,967,296	450,000,000
18	68,719,476,736	7,000,000,000
20	1,099,511,627,780	100,000,000,000

[a]Data courtesy of Dr. R. Drmanac, Hyseq (Sunnyvale, CA).

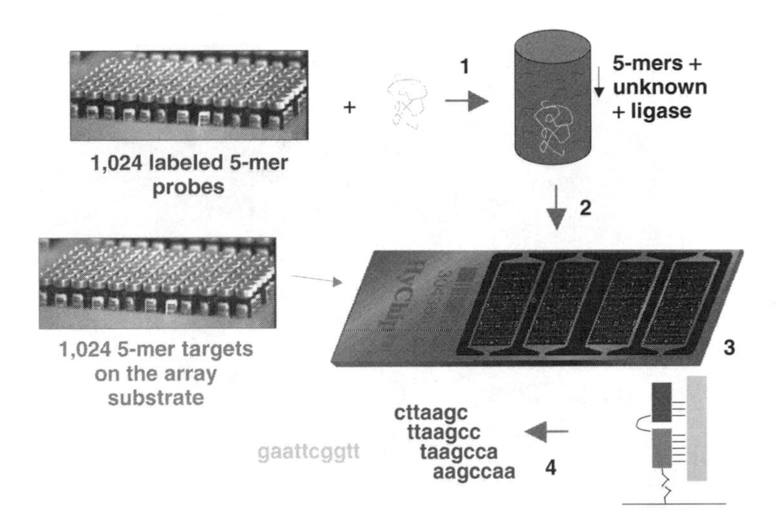

FIGURE 13.5. For format 3 SBH, labeled oligonucleotide 5mer probes (red) are mixed with a unknown sequence (yellow) and DNA ligase (1), and the mixture is hybridized (2) to an oligonucleotide array containing all possible (1024) 5mer targets (green). Positive hybridization events are scored as fluorescent signals on the array (3), and the overlapping probes are aligned (4) and assembled by computer to obtain the sequence (gaattcggtt) of the unknown molecule. (Courtesy of Dr. R. Drmanac, Hyseq, Sunnyvale, CA.)

hybridization–ligation reaction, with fluorescent probes consisting of an unknown sequence and 5mers. The hybridization–ligation reaction involving 5mer targets and 5mer probes affords the effective specificity of a 10mer oligonucleotide (Fig. 13.5), enabling relatively accurate sequence determination by SBH. The format 3 SBH approach have been used for a wide spectrum of applications, including the identification of heterozygous *ApoB* locus in a cardiovascular patient and mutation detection in the p53 tumor-suppressor gene (Fig. 13.6). Array-based sequencing by hybridization offers a massively parallel format that provides an inexpensive alterative to traditional DNA sequencing methods.

Hybridization to arrays can also be used to resequence an oligonucleotide or other molecule that contains a sequence that is already known. The resequencing approaches employ degenerate arrays, which are chips that contain every possible combination of DNA sequence for a given length of oligonucleotide. Mirzabekov and co-workers (Chechetkin et al. 2000) used degenerate 6mer oligonucleotide arrays containing all possible 4096 hexamers (4^6) to examine a fluorescent 51mer oligonucleotide. By identifying the strongly hybridizing locations on the array, the 51mer can be resequenced by aligning

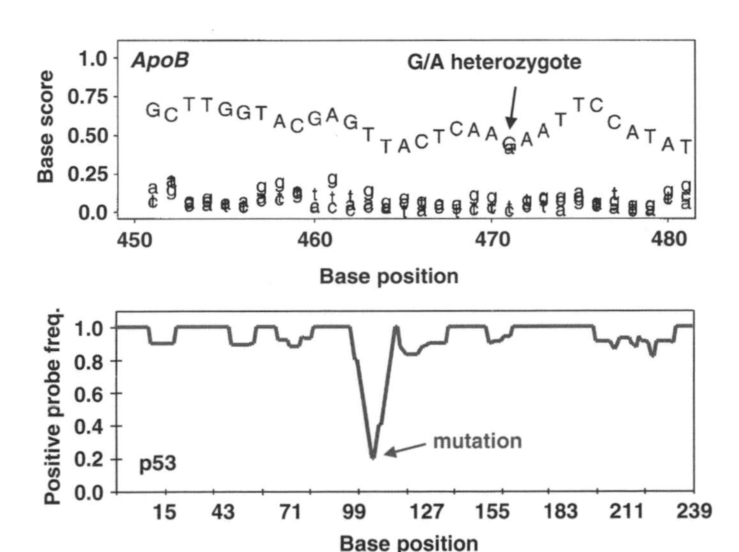

FIGURE 13.6. Format 3 SBH was used to analyze an ApoB *gene from a cardiovascular patient (top) and a human* p53 *gene (bottom). A G-A heterozygote (arrow) in the* ApoB *gene was noted, as was a mutation (arrow) in the p53 sequence. (Data courtesy of Dr. R. Drmanac, Hyseq, Sunnyvale, CA.)*

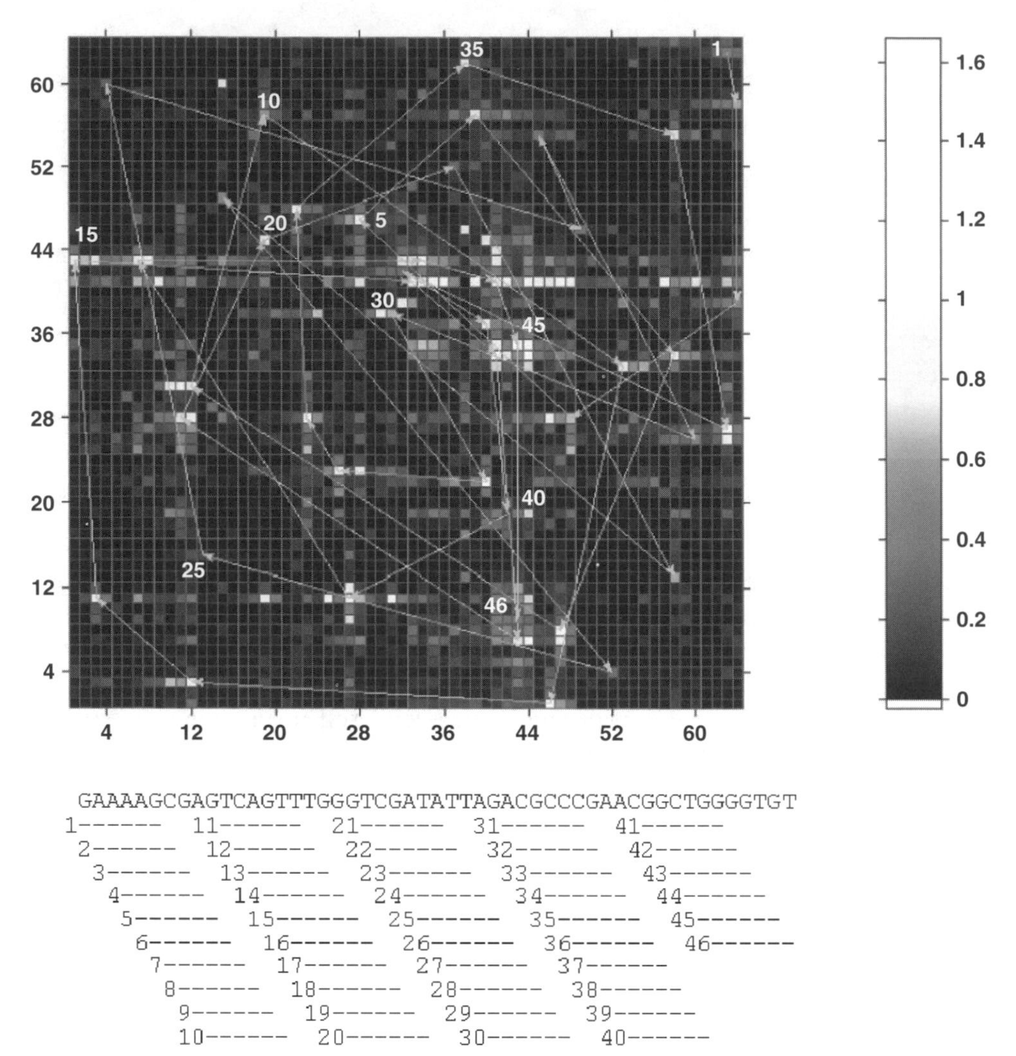

FIGURE 13.7. **A,** *Fluorescent image of the hybridized 6mer (hexamer) array, with the intensity values indexed to a fire look-up table (right). Interconnected arrows (light green) indicate the array locations that are perfectly complementary to the fluorescent 51mer.* **B,** *Sequence of the 51mer and the 46 perfectly complementary hexamers that span the polymer. (Reprinted with permission from* Journal of Biomolecular Structure and Dynamics.*) Figure also appears in Color Figure section.*

the overlapping hexamers (Fig. 13.7). Resequencing by hybridization provides a rapid means of verifying the sequence of synthetic molecules.

OLIGONUCLEOTIDE ARRAYS

Oligonucleotide array. *Ordered array of oligonucleotides on a solid surface.*

Large collections of oligonucleotides provide a rapid means of genotyping, by hybridizing patient samples to the ***oligonucleotide arrays*** and reading the fluorescence intensities at each location on the chip. The GeneChip technology (Affymetrix, Santa Clara, CA) provides an extremely powerful platform for examining small variations in DNA sequence, including mutations and polymorphisms (see Chapter 7). One class of sequence variant that is of great interest is the single nucleotide polymorphism (SNP), which consists of a single nucleotide change at a particular location in the genome. By examining a large number of SNPs in many patients using one chip per patient, it is possible to gain unambiguous genotyping information for an entire population.

SNP analysis with GeneChip technology uses chips containing up to 400,000 different oligonucleotide sequences, representing > 10,000 SNPs. Each

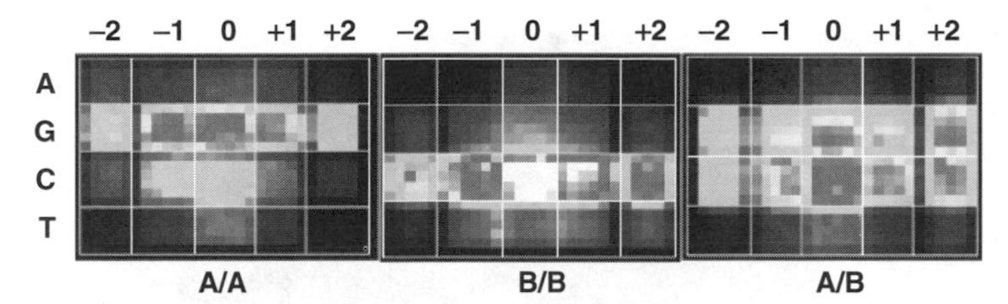

FIGURE 13.8. The variable central position of the oligonucleotide microarray corresponds to location of the SNP (0) and to positions one or two nucleotides upstream (−1, −2) and downstream (+1, +2) of the polymorphism. The fluorescence intensity at each oligonucleotide location allows the users to call the base sequence. Hybridization signals for homozygous (AA and BB), and heterozygous (AB) patients are distinguished easily. (Data courtesy of Dr. K. Tillman, Affymetrix, Santa Clara, CA.)

location on the chip contains a different oligonucleotide sequence, but the sequences are grouped in sets of four, with the central position of each set of four oligonucleotides containing one of the four possible bases (A, G, C, T) in the genetic code. By comparing the fluorescence intensity at the four locations in each cell on the chip, the presence of an SNP can be detected with extremely high accuracy (Fig. 13.8). The bordering oligonucleotides that tile across the SNP provide additional confirmation of the SNP. Homozygotes and heterozygotes can also be distinguished with ease using oligonucleotide arrays, the latter being detected as giving strong signals at two of the four locations in a cell. Because so many oligonucleotides can be synthesized on a single chip, it is possible to genotype a single patient in detail with a single hybridization. The massively parallel format for SNP analysis can provide extremely detailed information about human migration, descent, and the genesis of human disease. SNP analysis with oligonucleotide arrays can also provide comprehensive information for forensic and paternity applications.

MINISEQUENCING

The method of genotyping by ***minisequencing*** was developed by Jalanko and co-workers (Helsinki, Finland) in the early 1990s. Arrays of amplified target sequences, such as fragments of disease genes, are prepared and hybridized with marker-specific primers that form duplexes immediately upstream of a mutation or SNP of interest. The last nucleotide in the primer corresponds to the presumptive mutation or SNP. The primed templates are then incubated with DNA polymerase in the presence of a labeled dideoxynucleotide, resulting in the incorporation of a labeled nucleotide on the templates that are primed perfectly. Four rounds of synthesis, one for each of the four DNA bases, allow genotyping of all of the templates on the array. This solid-phase approach has been used successfully to identify patients suffering from cystic fibrosis, aspartylglucosaminuria (AGU), and other serious illnesses.

Minisequencing. *Genotyping method in which templates are hybridized with primers so that the last base overlaps a mutation of interest and the primed templates are extended with polymerase and fluorescent dideoxynucleotides to determine the sequence.*

A reverse minisequencing format was developed recently for microarray analysis. In the reverse minisequencing approach, marker-specific primers are printed onto glass and used to prime amplified patient samples of interest corresponding to gene regions containing mutations or SNPs. A set of four primers is used for each allele, and the primers are identical except at the last (3′-most) position, which is either A, G, C, or T. The last primer position corresponds to the presumptive mutation or polymorphism, such that only one of

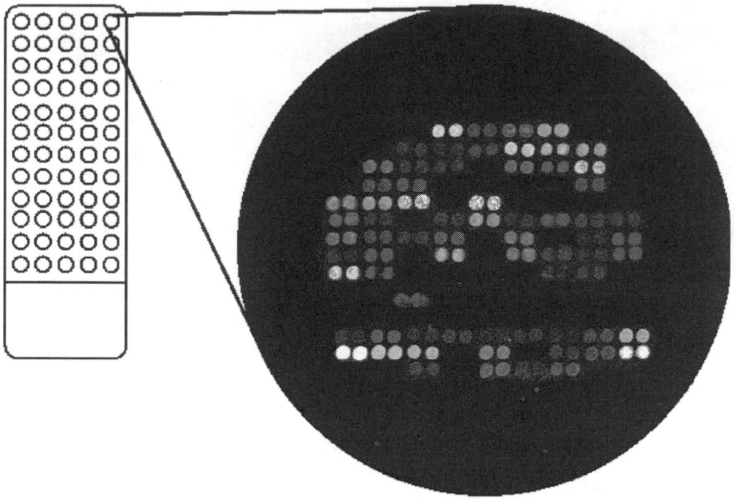

FIGURE 13.9. Minisequencing allows the analysis of 118 SNPs from 60 different patients on a single chip. The fluorescent image depicts one of the patients. (Data courtesy of U. Liljedahl and Dr. A.-C. Syvänen, Molecular Medicine, Department of Medical Sciences, Uppsala University, Uppsala, Sweden.)

the four templates will form a perfect hybrid. The primed templates are incubated with polymerase in the presence of a fluorescent dideoxynucleotide, resulting in efficient enzymatic incorporation on the perfectly primed templates and lesser incorporation on the templates that contain a mismatch at the site of the allele. Dideoxynucleotides are used to prevent the incorporation of more than one base, which would degrade the specificity of the assay. Genotypes are easily derived by identifying the primer in each set of four that produces the strongest fluorescent signal on the microarray (Fig. 13.9). Glass substrates containing 60 identical primer microarrays allow genotyping of 60 different patients for > 100 SNPs on a single chip. This novel microarray method was used to generate $> 10{,}000$ genotypes for the human Y chromosome, providing detailed insight into human migration and genetic diversity in a Finnish population

AMPLIFIED PRIMER EXTENSION

Amplified primer extension. *Genotyping method by which templates are hybridized with primers one nucleotide upstream of a mutation of interest. The primed templates are extended with polymerase and fluorescent dideoxynucleotides to determine the sequence.*

The microarray method of ***amplified primer extension*** (APEX) uses marker-specific arrays of primers to hybridize amplified templates of interest. The primers hybridize at a position one nucleotide upstream of mutations or SNPs present in the amplified templates. The templates are pooled so that dozens or even hundreds of loci can be genotyped in a single hybridization. Once the templates are hybridized to the microarrays, the primed templates are incubated with polymerase in the presence of four fluorescent dideoxynucleotides, each modified with a different fluorescent label. The nucleotide corresponding to the mutation or SNP determines which of the four fluorescent nucleotides is incorporated, producing an allele-dependent color bias at each genotyping location on the microarray. As in the case of minisequencing, the use of a dideoxynucleotide in APEX prevents incorporation of more than one nucleotide, enabling a high degree of specificity in the assay. Genotypes are easily determined by detecting fluorescent signals across the entire microarray and indexing the signals to a palette that assigns a different color to each of the four bases (Fig. 13.10). Mutation detection in the human breast cancer gene *BRCA1* is one of many

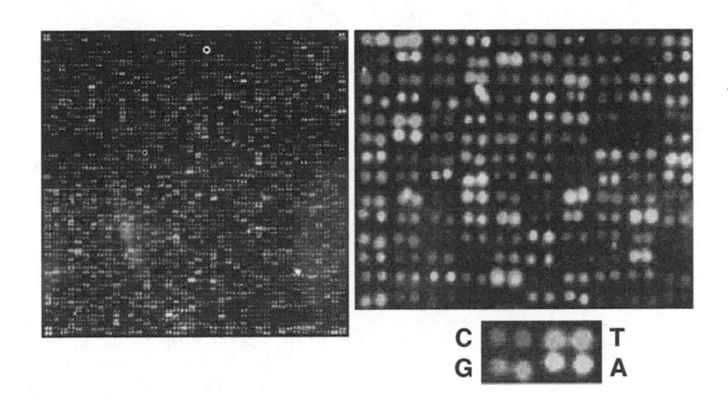

FIGURE 13.10. *Microarray image from APEX showing 1334 SNP markers from human chromosome 22 (*left*), with a small portion of the array enlarged (*right*). Genotypes of both DNA strands are determined (*bottom*). (Data courtesy of Dr. A. Kurg and A. Metspalu, Institute of Molecular and Cell Biology, University of Tartu, Tartu, Estonia.) Figure also appears in Color Figure section.*

examples of how microarray-based APEX can be applied to important research questions.

NEXT GENERATION SCREENING

Neonatal screening and other genetic testing and diagnostic procedures require affordable platforms that are rapid and highly accurate. Current neonatal screening procedures are based largely on a technology known as ***tandem mass spectrometry,*** an analytical technique that allows measurement of small quantities of metabolites and other compounds in neonatal blood samples. Mass spectrometry provides a quantitative readout of dozens of small molecules, identifying babies that have inherited the serious genetic disorders that perturb these key compounds. Mass spectrometry is used routinely to screen for phenylketonuria (PKU), maple syrup urine disease (MSUD), glutaric acidemia (GA1), medium-chain acyl-CoA dehydrogenase (MCAD), and many other diseases (Table 13.2). Mass spectrometry is rapid, affordable, and extremely accurate and has been used with great success for many years in neonatal screening laboratories.

Tandem mass spectrometry. *Analytical method used in traditional neonatal screening that allows the identification of patients with metabolic disorders by measuring the levels of small molecules.*

There are a great many genetic disorders, however, that do not perturb the levels of small molecules in a readily measurable manner and thus are not easily amenable to screening by mass spectrometry, including most cancers, disorders of the nervous and immune system, defects in muscle and bone formation, and many of the cell signaling and transport diseases. The relatively common sickle cell diseases, which result from mutations in the oxygen-carrying protein hemoglobin, are currently screened with rather cumbersome gel-based techniques rather than by mass spectrometry. All of the above disease categories are amenable to DNA-based screening, underscoring the need for widespread use of microarray analysis for neonates and adults. The recent development of a new microarray method meets the criteria of speed, accuracy, and affordability required for populationwide clinical use.

Next Generation Screening (NGS), developed at TeleChem International (Sunnyvale, CA), uses high-density microarrays of amplified patient samples coupled with allele-specific hybridization with fluorescent synthetic oligonucleotides to examine disease loci of interest (Fig. 13.11). Each location on the chip essentially corresponds to a single locus from a single patient, and thus each spot on the chip provides genotyping information for one allele from one patient. At the current microarray printing density, 10,000 patients can be screened for 12 disease loci in a single hybridization experiment using the

Next Generation Screening. *Genotyping method by which a large number of amplified patient samples are printed into a microarray and hybridized with fluorescent oligonucleotides to determine the genotypes of multiple patients for multiple diseases in a single test.*

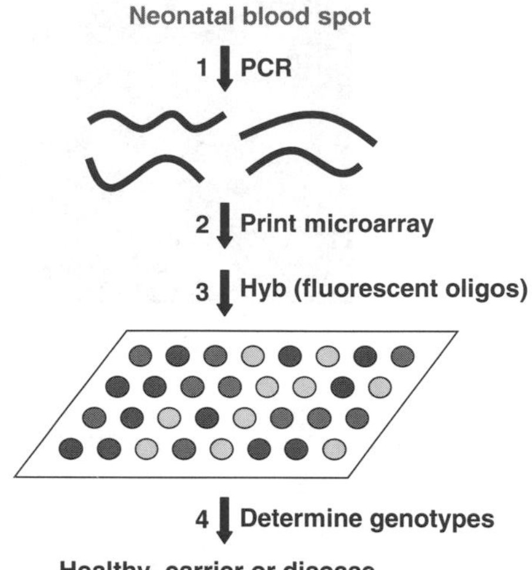

FIGURE 13.11. *A simplified view of the Next Generation Screening (NGS) approach to genetic screening with microarrays. Genomic DNA from neonatal blood spots is amplified with the polymerase chain reaction* (PCR, 1) *to generate a series of amplicons essentially corresponding to a single locus or gene fragment from a single patient* (wavy lines). *The amplicons are printed into a microarray* (2) *and hybridized* (Hyb) *with fluorescent synthetic oligonucleotides* (oligos) *complementary to the disease alleles of interest* (3). *Fluorescent signals, which differ because the oligonucleotides hybridize with different efficiencies to the different loci, provide the genotypes* (4). *(Courtesy of TeleChem International, Sunnyvale, CA.)*

NGS method. Three signal intensities are observed for a given locus, identifying healthy, carrier, and disease genotypes (Fig. 13.12 and Fig. 13.13). Multiple fluorescent labels can be used to provide reciprocal signals for wild type and disease loci, with carriers giving equivalent signals in the two channels (Fig. 13.14 and Fig. 13.15).

The assay differs from the typical microarray format that uses oligonucleotide capture probes (e.g., 15mers) to screen one patient per microarray (Dobrowolski et al., 1999). The massively parallel NGS format, with the patient samples attached to the substrate, allows the cost of a single test to be amortized over hundreds or thousands of patients, providing extremely cost-effective genetic screening information. Efforts are under way to examine hereditary hearing loss, blood typing, HLA analysis, forensics, and infectious diseases as applications. Researchers at NGS-ArrayIt, Inc. (Sunnyvale, CA) are exploring ways to optimize the NGS method, including the use of dendrimers (see Chapter 6), advances in surface chemistry, and sophisticated cleanroom manufacturing procedures.

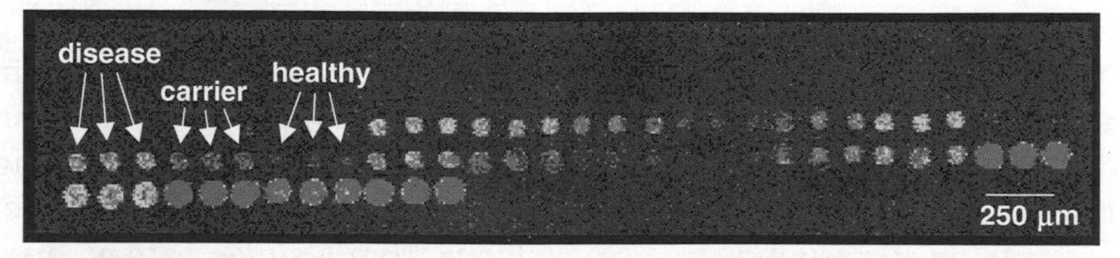

FIGURE 13.12. *Fluorescent microarray neonatal screening data indexed to a rainbow palette. Amplified DNA segments from neonatal blood spots, corresponding to different alleles of sickle cell anemia and galactosemia, were printed onto glass substrates and hybridized with fluorescent synthetic oligonucleotides. The genotypes for healthy, carrier, and sickle cell disease are easily observed (arrows). (Data courtesy of TeleChem International, Sunnyvale, CA.)*

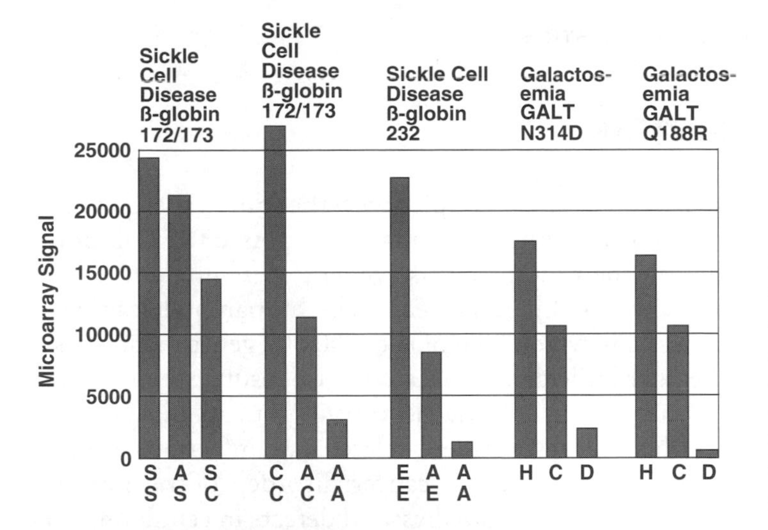

*FIGURE 13.13. Fluorescent signals obtained from the NGS™ microarray assay for neonatal screening. Different alleles of sickle cell disease (S, A, C and E) and healthy (H), carrier (C), and disease (D) genotypes for galactosemia (*N314D *and* Q188R*) were examined. (Data courtesy of TeleChem International, Sunnyvale, CA.)*

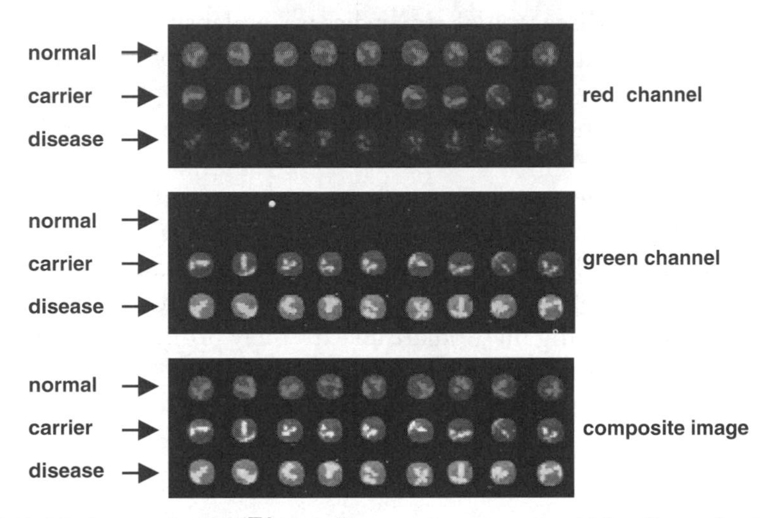

FIGURE 13.14. *A two-color NGS™ method was used to screen for sickle cell anemia. A Cyanine-3 labeled synthetic oligonucleotide probe* (red) *perfectly complementary to a wild typebeta-globin allele* (normal)*, and a Cyanine-5 labeled synthetic oligonucleotide probe* (green) *perfectly complementary to a mutant beta-globin allele* (disease) *were hybridized to a single microarray containing neonatal amplicons; the chip was scanned in the red and green channels. A composite image* (bottom) *reveals the ease by which the normal, carrier, and disease genotypes are distinguished by color. (Data courtesy of NGS-ArrayIt, Inc., Sunnyvale, CA.) Figure also appears in Color Figure section.*

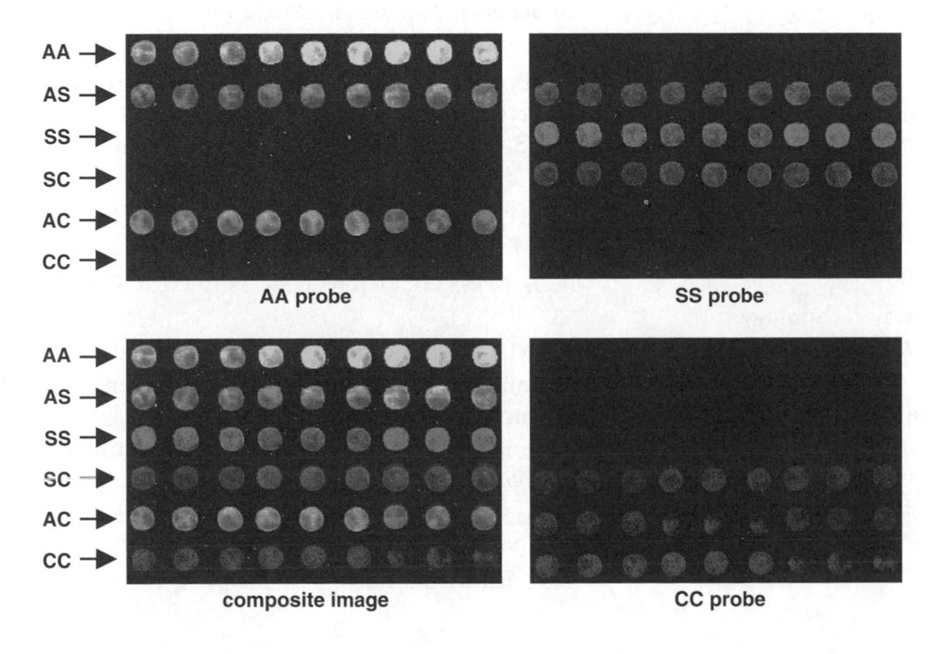

FIGURE 13.15. *The three-color. The NGS™method was used to screen for six sickle cell anemia genotypes. Three oligonucleotide probes perfectly complementary to three genotypes (AA, SS and CC) were co-hybridized and scanned to reveal hybridization signals for the AA, SS, and CC probes. A composite image reveals the ease by which the six genotypes are distinguished by color. (Data courtesy of NGS-ArrayIt, Inc., Sunnyvale, CA.) Figure also appears in Color Figure section.*

SUMMARY

Microarray analysis, coupled with the complete sequence of the human genome and extensive clinical information, is poised to revolutionize health care by ushering in the widespread use genetic screening and diagnostic information to treat and cure human disease. The human body can be viewed as a symphony orchestra in which each of the ~30,000 genes must work in concert to produce a healthy individual. Human disease results when one or more genetic players are impaired or defective in carrying out its biochemical function. Nearly every inherited, environment-related, and infectious disease is amenable to DNA-based analysis, including cancer; disorders of the immune and nervous system; muscle and bone anomalies; and defects in cellular metabolism, signaling, and transport. Parkinson disease, Alzheimer disease, epilepsy, Huntington disease, ALS, SCID, rheumatoid arthritis, diabetes, multiple sclerosis, muscular dystrophy, sickle cell anemia, cystic fibrosis, AIDS, hepatitis, typhoid, Lyme disease, malaria, yellow fever, and tuberculosis are among the diseases that have a clear DNA basis.

Microarrays provide an accurate technology for examining DNA sequences in a massively parallel and highly automated fashion and are thus well suited for genetic screening and diagnostic applications in infants and adults. Sequencing by hybridization, oligonucleotide arrays, minisequencing, APEX, and the Next Generation Screening method are all extremely promising approaches for microarray-based analysis of human disease.

SELECTED READING

Blackburn, G. M., and Gait, M. J., eds. *Nucleic Acids in Chemistry and Biology,* 2nd ed. Oxford University Press, Oxford, UK, 1996.

Bloomfield, V. A., Crothers, D. M., and Tinoco, I. *Nucleic Acids: Structure*. University Science Books, Herndon, VA, 2000.

Chechetkin, V. R., Turygin, A. Y., Proudnikov, D. Y., Prokopenko, D. V., Kirillov, E. V., and Mirzabekov, A. D. Sequencing by hybridization with the generic 6-mer oligonucleotide microarray: an advanced scheme for data processing. *J Biomol Struct Dynamics* 18:83–101, 2000.

Clarke, A. J., ed. *The Genetic Testing of Childen*. Bios Scientific, Oxford, UK, 1998.

Dobrowolski, S. F., Banas, R. A., Naylor, E. W., Powdrill, T., and Thakkar, D. DNA microarray technology for neonatal screening. *Acta Paediatr Suppl* 88:61–64, 1999.

Drmanac, R., Labat, I., Brukner, I., and Crkvenjakov, R. Sequencing of megabase plus DNA by hybridization: Theory of the method. *Genomics* 4:114–128, 1989.

Innis, M. A., Gelfand, D. H., and Sninsky, J. J., eds. *PCR Applications: Protocols for Functional Genomics*. Academic Press, San Diego, CA, 1999.

Jalanko, A., Kere, J., Savilahti, E., Schwartz, M., Syvanen, A. C., Ranki, M., and Soderlund, H. Screening for defined cystic fibrosis mutations by solid-phase minisequencing. *Clin Chem* 38:39–43, 1992.

Khoury, M. J., Burke, W., and Thomson, E., eds. *Genetics and Public Health in the 21st Century: Using Genetic Information to Improve Health and Prevent Disease*. Oxford University Press, Oxford, UK, 2000.

Kurg, A., Tonisson, N., Georgiou, I., Shumaker, J., Tollett, J., and Metspalu, A. Arrayed primer extension: Solid-phase four-color DNA resequencing and mutation detection technology. *Genet Test* 4:1–7, 2000.

Lehninger, A. L., Nelson, D. C., and Cox, M. M. *Principles of Biochemistry,* 3rd ed. Freeman, New York, 2000.

Lewin, B. *Genes VII*. Oxford University Press, Oxford, UK, 2000.

Raitio, M., Lindroos, K., Laukkanen, M., Pastinen, T., Sistonen, P., Sajantila, A., and Syvanen, A. C. Y-chromosomal SNPs in Finno-Ugric-speaking populations analyzed by minisequencing on microarrays. *Genome Res* 11:471–482, 2001.

Robbins, S. L., Cotran, R. S., Kumar, V., and Collins, T., eds. *Pocket Companion to Robbins Pathologic Basis of Disease,* 6th ed. Saunders, Philadelphia, 1999.

Schena, M., ed. *DNA Microarrays: A Practical Approach,* 2nd ed. Oxford University Press, Oxford, UK, 2000.

Schena, M., ed. *Microarray Biochip Technology*. Eaton, Natick, MA, 2000.

Schena, M., Shalon, D., Davis, R. W., and Brown, P. O. Quantitative monitoring of gene expression patterns with a complementary DNA microarray. *Science* 270:467–470, 1995.

Scriver, C. R., Sly, W. S., Childs, B., Beaudet, A., Valle, D., Kinzler, K., and Vogelstein, B. *The Metabolic and Molecular Bases of Inherited Disease,* 8th ed. McGraw-Hill, Hightstown, NJ, 2000.

Singer, M., and Berg, P. *Genes and Genomes*. University Science Books, Herdon, VA, 1991.

Strachan, T., and Read, A. P. *Human Molecular Genetics,* 2nd ed. Wiley, New York, 1999.

Streitweiser, A., and Heathcock, C. H. *Introduction to Organic Chemistry,* 2nd ed. Macmillan, New York, 1981.

Syvanen, A. C., Ikonen, E., Manninen, T., Bengtstrom, M., Soderlund, H., Aula, P., and Peltonen, L. Convenient and quantitative determination of the frequency of a mutant allele using solid-phase minisequencing: Application to aspartylglucosaminuria in Finland. *Genomics* 12:590–595, 1992.

Tonisson, N., Kurg, A., Kaasik, K., Lohmussaar, E., and Metspalu, A. Unravelling genetic data by arrayed primer extension. *Clin Chem Lab Med* 38:165–170, 2000.

Varmus, HE. The molecular genetics of cellular oncogenes. Ann. Rev. Genetics 18, 553–612, 1884.

REVIEW QUESTIONS

1. A genetic screening assay designed to detect a single nucleotide polymorphism would most likely examine which biochemical entity: (a) mRNA, (b) DNA, (c) protein, or (d) carbohydrate?
2. A human cell has been compared to a symphony orchestra, but other types of analogies are conceivable. If a human cell where race car, the four wheels of the car might correspond to which two of these: (a) human body, (b) human mind, (c) human genes, or (d) human proteins?
3. Parkinson, Alzheimer, and many other genetic disorders are said to run in families. What does this mean at the DNA level?
4. Patients with AIDS show impaired T cell function. T cells are part of the human: (a) nervous system, (b) immune system, (c) visual system, or (d) auditory system.
5. Metabolic diseases result primarily in changes in the levels of: (a) proteins, (b) nucleic acids, (c) proteases, or (d) small molecules?
6. A 2116-base-pair gene contains a total of 12 sequence variants known to cause a human disease. The 12 sequence variants of the gene are known as: (a) alleles, (b) lipids, (c) genomes, or (d) polyploid?
7. Which of the following are examples of mutations: (a) missense, (b) nonsense, (c) insertion, (d) deletion, and (e) phosphorylation?
8. Lung cancer due to smoking claims the lives of approximately what fraction of the U.S. population annually: (a) 0.005%, (b) 5%, (c) 0.05%, or (d) 0.5%?
9. Microarrays containing human gene sequences were hybridized with fluorescent oligonucleotides of known sequence, scored for positive signals, and a computer was used to align the positively hybridizing oligonucleotide sequences. This procedure describes what microarray method: (a) minisequencing, (b) next generation screening, (c) APEX, or (d) SBH?
10. The genotyping approach that uses the hybridization of primers one nucleotide upstream of a sequence variant, followed by polymerase incorporation of fluorescent dideoxynucleotides is known as: (a) minisequencing, (b) next generation screening, (c) APEX, or (d) SBH?

11. The microarray method that allows thousands of patients to be genotyped on a single microarray using a single hybridization reaction is known as: (a) minisequencing, (b) next generation screening, (c) APEX, or (d) SBH?
12. The genotyping methods that use primers that hybridize with the last nucleotide overlapping the sequence variant, followed by polymerase incorporation of fluorescent dideoxynucleotides is known as: (a) minisequencing, (b) next generation screening, (c) APEX, or (d) sequencing by hybridization?

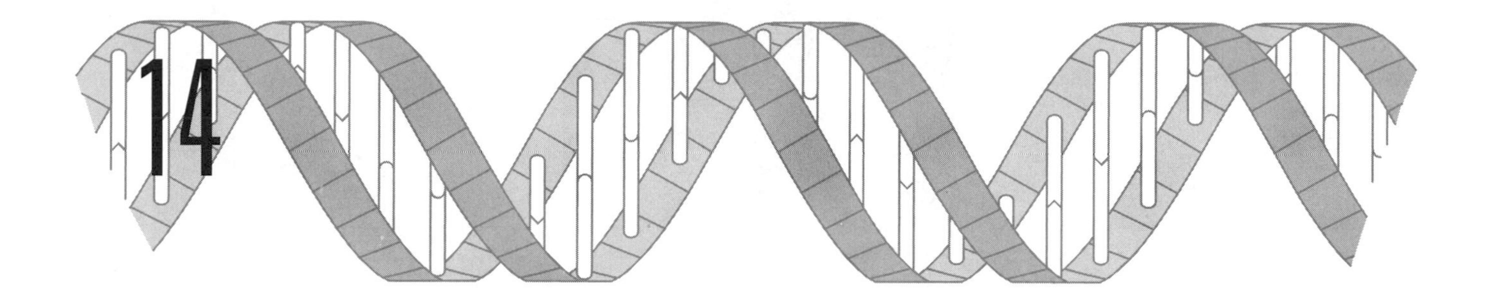

Novel Microarray Technologies

Every great advance in science has issued from a new audacity of the imagination —*John Dewey*

The microarray field has been anointed with many assets, but one of the main ones is the curiosity and ingenuity of its scientists. This collective cleverness has given rise to many novel microarray formats and technologies and will ensure a steady supply of new ideas for future innovation in the field. This chapter describes some of the novel microarray tools that have recently been developed, including printing technologies, biosensors, new surfaces, protein microarrays, and nonfluorescent detection methods. Many more new and promising technologies are on the horizon.

PRINTING TECHNOLOGY

Because microarray manufacture plays such a important role in determining microarray quality, continued improvements in printing technologies are key to advancing the field. By leveraging their understanding of surfaces and high-end fabrication techniques, scientists at Corning have developed a new printing technology that contains a large cylindrical bundle of sample intakes that are consolidated into a small printhead, effectively allowing many samples to be printed in a small area with one loading of the printing tool (Fig. 14.1). This new device appears to use an ink-stamping-type printing mechanism, whereby droplets at the end of each printing channel are deposited onto a solid surface by direct contact with the surface, followed by adhesive removal

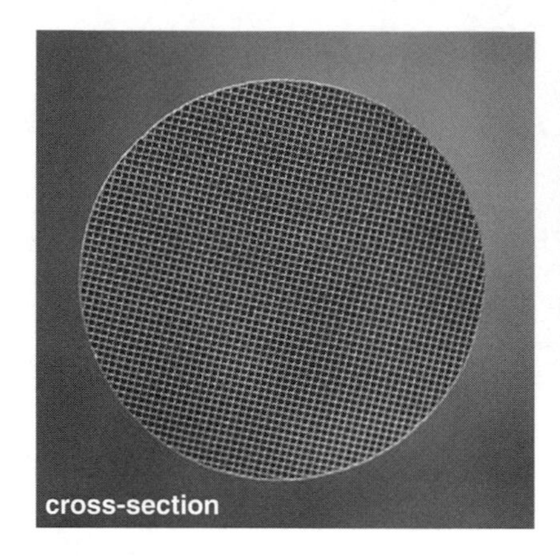

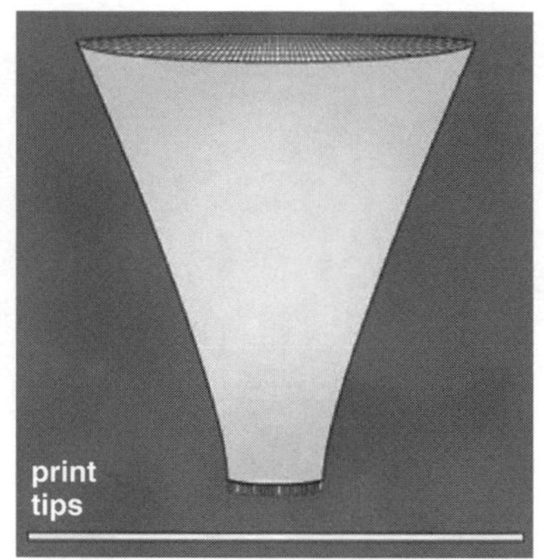

FIGURE 14.1. A new microarray printing device from Corning Microarray Technology. **A,** *cross-section of the sample intake bundle.* **B,** *Side view of the device and print tips. (Courtesy of Dr. A. Bhatia, Corning, Corning, NY.)*

of the droplet from the printing device by the substrate. Because the channels hold several microliters of sample, the device prints many spots with a single loading.

This new printing technology offers the potential advantage of allowing hundreds or even thousands of samples to be printed simultaneously. Based on the closed design of the printing channels, the device might be effective at reducing evaporation during printing. Provided that it can be fabricated at a reasonable cost, this new printing technology might find use in commercial manufacturing setting or as a tool to allow researchers to make their own microarrays. One potential challenge of the device will be to implement efficient sample loading and cleaning procedures. The new printing technology integrates nicely into a larger suite of tools developed at Corning that also includes advanced surface chemistry and printed microarrays for gene expression applications. The quality of the hybridization data obtained on Corning slides (Fig. 14.2) illustrates the quality of these tools.

BIOSENSORS

Biosensors make up family of advanced technologies that are akin to microarrays in that they enable molecular analyses to be performed in a parallel and miniaturized format. Biosensors are distinct from microarrays in a number

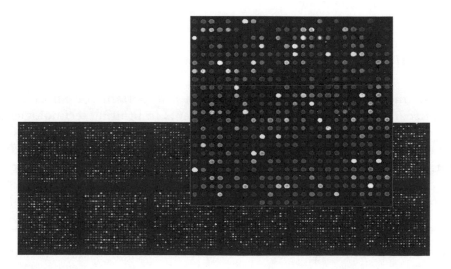

FIGURE 14.2. Two-color overlay showing differential gene expression from comparative analysis of RNA samples from wild type and mutant strains of yeast. (Data courtesy of Dr. A. Bhatia, Corning, Corning, NY.)

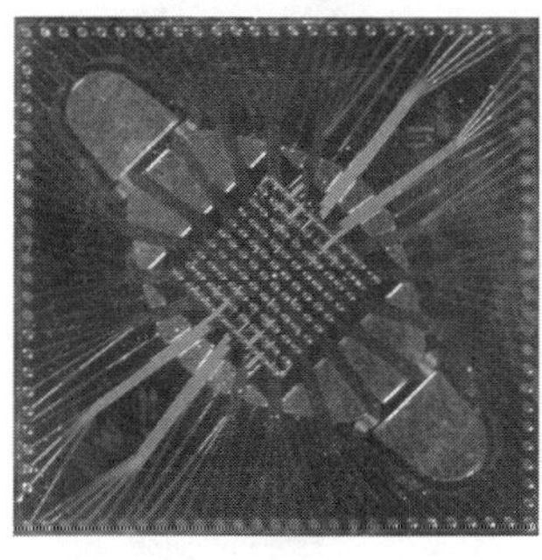

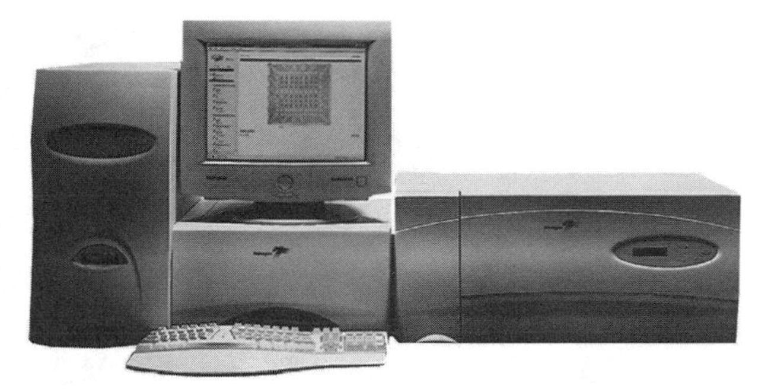

*FIGURE 14.3. A microelectronic array (**A**) and an integrated workstation (**B**) from Nanogen. NanoChip technology uses electronic addressing of individual microelectrodes on the biosensor surface to mediate rapid biochemical reactions such as hybridization. Electronic stringency at each site enables detection of single nucleotide mismatches, providing a robust platform for genetic analysis. (Courtesy of Dr. M. J. Heller, Nanogen, San Diego, CA.)*

of key ways, however, with an important technological distinction being that biosensors use an active process (e.g., electrical charge) to mediate various aspects of biosensor manufacture, whereas microarrays generally rely on passive (e.g., diffusion) reaction conditions.

One type of biosensor that has received much attention is the ***active microelectronic array*** developed by scientists at Nanogen (San Diego, CA). This ingenious technology combines microelectronics and molecular biology to create active analytical devices for genetic analysis. Nanogen's microelectronic arrays are made on silicon substrates using sophisticated photolithographic techniques. Each microarray element or ***microlocation*** is a built on top of a platinum microelectrode, fitted with a platinum wire that is connected to an electrical contact on the chip periphery (Fig. 14.3). The platinum microelectrodes are coated with a 1- to 2-μm-thick agarose ***permeation layer,*** which serves to embed target and probe molecules during operation and protect the biomolecules from electrochemical damage during microelectrode biasing.

Active microelectronic array. *Biosensor technology that combines microelectronics and molecular biology to create active analytical devices for genetic analysis.*

Microlocation. *Analytical element on an active microelectronic array.*

Permeation layer. *Thin coating on an active microelectronic array substrate into which target molecules are embedded and probe molecules diffuse.*

Microelectronic arrays are used in a two-step process. The first step involves array manufacture, which is accomplished by applying an electrical bias (e.g., positive charge) at each array location and attracting target molecules to them. Applying a positive charge to a microelectrode in the presence of a target oligonucleotide, for example, results in the rapid migration of the negatively charged DNA to that location and an embedding of the targets into the permeation layer over the microelectrode surface (Fig. 14.4). The capacity to control the current, charge, polarity, and voltage of each microelectronic site, as well as the chemistry of the permeation layer, allows biosensor fabrication of virtually any charged molecule of interest. Once fabricated, the biosensors can be used to mediate specific biochemical reactions, such as hybridization, when the user simply applies a positive charge to the electrode, which attracts fluorescent probe molecules present in the solution to one or more sites. Other types of reactions such as protein–protein interactions can be mediated using a similar scheme.

Biosensors offer several advantages over traditional microarrays. One advantage is accelerated reaction time, which derives from the use of active electronic biasing instead of passive diffusion. Hybridization kinetics for microelectronic arrays are routinely 10- to 100-fold more rapid than for traditional microarrays, enabling hybridization reactions to be performed in <5 min. Another advantage of biosensors is that they enable the use of ***electronic stringency,***

Electronic stringency. *Attribute of an active microelectronic array that allows the efficiency of a hybridization reaction to be controlled by adjusting the bias at each microelectrode.*

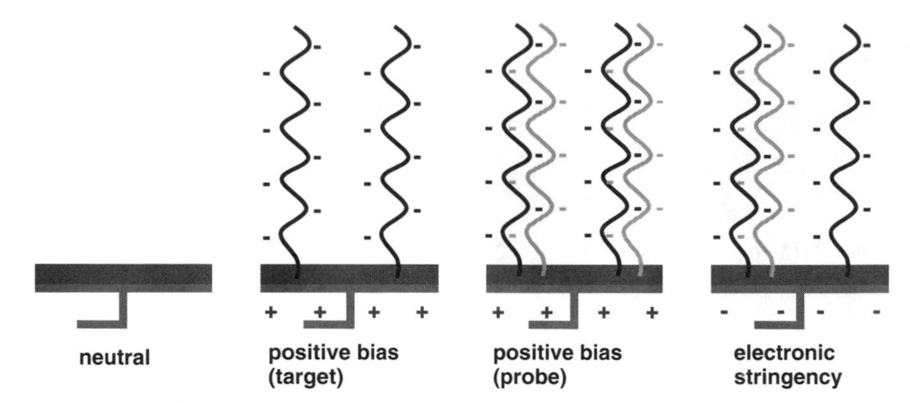

*FIGURE 14.4. Each element in a microelectronic array corresponds to a microelectrode (*gray lines*) coated with a thin layer of agarose (*rectangle*), which acts as a permeation layer. Applying a positive bias to the microelectrode (+) attracts negatively charged molecules, such as oligonucleotides (*black wavy lines*), to the site, embedding the molecules in the permeation layer. Once the target molecules are embedded, they can be hybridized to probe molecules (*gray wavy lines*) if a positive bias is maintained at the electrode. After hybridization, the bias can be reversed (−), driving off imperfect hybrids and allowing detection of single nucleotide substitutions through the use of electronic stringency.*

which allows the efficiency of a hybridization reaction to be controlled by adjusting the bias at each microelectrode. Electronic stringency is extremely useful in single nucleotide genotyping and diagnostics applications because it allows the formation each heteroduplex to be fine tuned, thereby maximizing the capacity to distinguish single base changes irrespective of sequence composition. One disadvantage of microelectronic arrays is that their manufacture requires independent addressing of each array site, and this serial process can be arduous and time consuming, particularly for arrays that contain many microlocations. There is also some elevated cost associated with biosensor technology, though Nanogen has been able to deliver a fully integrated package (Fig. 14.3) at an affordable price.

NITROCELLULOSE SURFACE CHEMISTRY

Nitrocellulose. *Novel microarray surface coating consisting of a nitrated glucose polymer.*

Surface chemistry is one of the most rapidly advancing forefronts in microarray science. Intensive studies are being conducted in dozens of laboratories worldwide to characterize and improve the substrates, surface treatments, and surface coatings. Through efforts of scientists at Schiecher and Schuel (Keene, NH), one recent development has been the adaptation of ***nitrocellulose*** as a novel microarray surface coating.

Nitrocellulose is a flexible polymeric material that has been used widely for 30 years in traditional molecular biology assays such as Southern and northern blots (see Chapter 4). Nitrocellulose is manufactured from raw materials by treating cotton with a concentrated mixture of sulfuric acid and nitric acid in a chemical reaction known as nitration, which replaces hydroxyl groups in the glucose polymer with nitro groups (Fig. 14.5). The nitrated cotton is then rolled into thin (e.g., 100 μm) sheets, and the resultant nitrocellulose membranes are used in myriad methods to bind DNA, RNA, and proteins, which allows subsequent reaction of labeled probe solutions with the bound target molecules. Though the mechanistic details by which charged molecules bind to nitrocellulose remain unclear, DNA, RNA, and proteins are thought to bind the

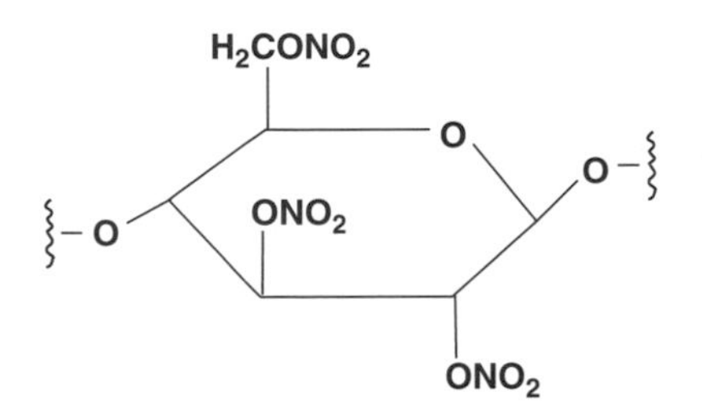

FIGURE 14.5. *The molecular structure of a single unit (monomer) of the glucose-based polymer nitrocellulose. The thin nitrocellulose membranes used as microarray coatings contain a nitrocellulose matrix consisting of nitrocellulose monomers joined at the polymer junctions (wavy lines).*

membranes in a noncovalent manner. Binding of radioactive probe molecules to nitrocellulose filters is assessed in traditional methods by autoradiography.

For microarray assays, nitrocellulose is applied as a 20- to 30-μm coating to glass substrates, to provide a coated substrate that is compatible with existing detection hardware. The novel nitrocellulose-coated microarray slide contains of a transparent periphery and a white, opaque central region corresponding to a 22 × 40-mm nitrocellulose coating. Microarray target molecules can be printed onto the nitrocellulose portion of the substrate using contact or noncontact printing methods. Because the coating is extremely thin (e.g., 20 μm) and contains an underlying glass substrate, the printing surface is much more durable than the flexible nitrocellulose filters used in traditional blotting assays. Once the target molecules are printed onto the nitrocellulose, the slides are processed and reacted with fluorescent probe solutions derived from DNA, RNA, or a variety of biomolecular sources. Hybridization experiments with cyanine-labeled DNA molecules reveal the detectivity and linearity of the signals observed using a nitrocellulose surface (Fig. 14.6).

Two advantages of nitrocellulose-based microarray slides are binding capacity and familiarity. Nitrocellulose-coated slides contain a 20- to 30-μm thick membrane and thus provide a three-dimensional matrix into which target molecules can attach. The three-dimensional nature of the coating provides a much greater binding capacity than two-dimensional surfaces, such as glass, that contain reactive amine and aldehyde groups; thus the nitrocellulose offers improved detectivity in certain cases. The extensive history of nitrocellulose in traditional assays provides a wealth of practical information in the form of reagents, protocols, and methods that can be adapted for use in a microarray setting.

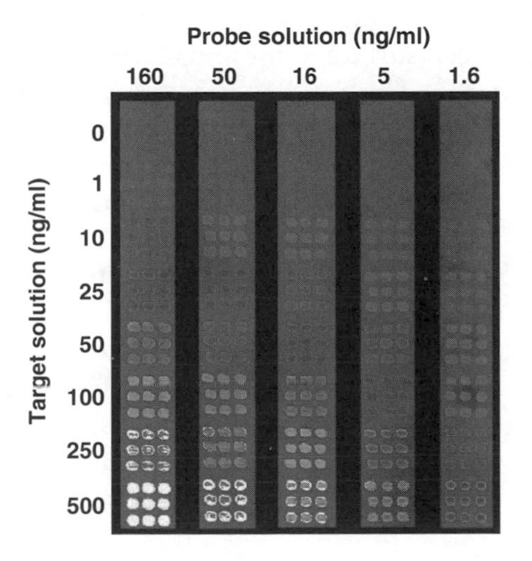

FIGURE 14.6. *Data from a microarray hybridization experiment using five nitrocellulose-coated FAST slides (Schleicher & Schuell) that were printed with 0–500 ng/mL of DNA target solutions. Each was hybridized with a different concentration (1.6–600 ng/mL) of Cyanine-5 labeled probe solution. The slides were scanned with a ScanArray 4000 (Packard Biochip Technologies, Billerica, MA), and TIFF data were indexed to a rainbow palette for ease of viewing. (Data courtesy of Dr. B. O. Parker, Schleicher & Schuell, Keene, NH).*

A disadvantage of nitrocellulose surfaces compared to organic treatments on glass is that the white nitrocellulose membrane reflects excitation light, which typically elevates background noise in microarray experiments. Nitrocellulose is not intrinsically fluorescent; it reflects light because white surfaces reflect all of the colors in the visible spectrum, including laser light from microarray scanners and white excitation light from microarray imagers. Reflection can lead to a loss of assay detectivity, though background noise can be compensated for by the increased signals that are observed due to the elevated binding capacity of nitrocellulose compared to chemically treated glass. Owing to the thickness of the nitrocellulose coating, minor adjustments to scanners, imagers and arrayers might also be necessary when using this surface rather than treated glass.

PROTEIN MICROARRAYS

Protein microarrays promise to replace many traditional assays in protein biochemistry. The capacity to analyze protein–protein, antibody–antigen, protein–drug, and many other types of protein reactions in a fluorescent, massively parallel format offers a level of sensitivity, precision, and safety that older formats do not provide. Dozens of laboratories are working to develop novel protein microarray platforms, and several examples are provided here. Protein microarrays appear to be poised to revolutionize protein biochemistry in a fundamental sense.

Joos and co-workers at the Natural and Medical Sciences Institute (Tuebingen Germany) have developed novel microarray assays to examine autoimmune diseases in afflicted patients. In autoimmune disease, individuals mount an immune response to their own cellular proteins, and the so-called autoantibodies disrupt cellular physiology by impairing the normal function of endogenous proteins. Understanding the mechanisms that trigger autoimmune disorders, identifying the cellular targets of autoantibodies, and developing suitable diagnostics tests are important challenges in clinical medicine.

To begin to develop microarray tests for autoimmune disease, Joos and co-workers prepared microarrays of antigens corresponding to some of the known cellular proteins of autoantibodies. The antigens were printed onto glass substrates containing reactive aldehyde groups, which allowed covalent binding of the antigen targets to the microarray surface via Schiff base formation between surface lysine and arginine residues and the reactive aldehydes (see Chapter 6). The antigen microarrays were then reacted with antisera isolated from patients with immune disorders (Fig. 14.7). Antibodies present in the patient sera bound with high affinity to antigens attached to the microarray substrate, providing a quantitative readout of antibody reactivity and titer. Positive signals to the topoisomerase I (Scl70) and centromere component (CENP-B) antigens, reveal the high degree of specificity, detectivity, and linearity of the microarray assay.

Antigen microarrays provide an advantage over traditional immunological assays in that a single patient sample can be analyzed simultaneously for many or possibly all autoimmune disorders in a single microarray test. Microarray assays use fluorescence instead of radioactivity, providing a greater level of safety than traditional assays that use radioactive probes. Antigen microarrays can be

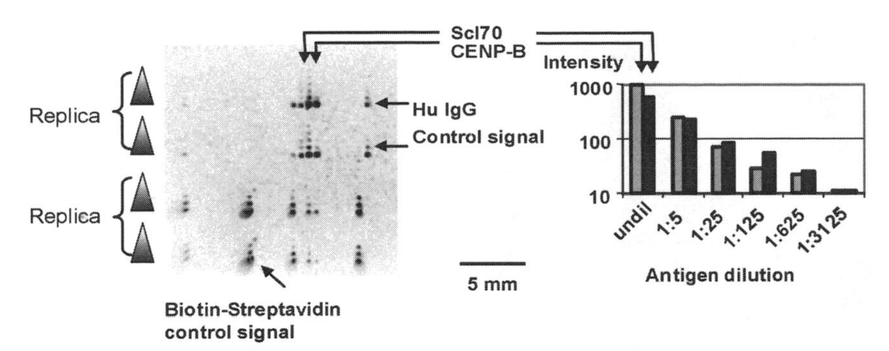

FIGURE 14.7. *Cellular proteins (antigens) were printed into microarrays (*left*) and probed with sera from a patient afflicted with an autoimmune disorder. The targets included antigenic regions from human topoisomerase I (Scl70), and the centromere component protein (CENP-B). Human immunoglobulin G (Hu IgG) and a biotin-streptavidin conjugate were included as control targets. A graph of the microarray signal intensities (*right*) reveals the linear response observed with antigen concentration (antigen dilution). (Data courtesy of Dr. T. Joos, Natural and Medical Sciences Institute, Tuebingen Germany.)*

mass produced, providing the promise of affordable and effective diagnostic tools for patient screening. In the future, protein microarrays representing the entire human genome would allow patient samples to be screened against every human protein in a single reaction. The capacity to use multicolor fluorescence allows comparisons among patients to be performed and enhances the reliability of the diagnostic tests.

Protein microarray technology development is also being carried out intensively at ZYOMYX (Hayward, CA). Nock and co-workers (Wilson and Nock, 2002) developed a platform that combines biosensor technology and protein microarrays as tools to study a broad spectrum of problems in protein biochemistry (Fig. 14.8). The silicon-based biosensors provide a 50-μm feature size and allow binding of 10,000 proteins in an area of 1.0 cm^2. Such tools, in principle, allow exploration of protein–protein interactions, protein–drug binding, and the entire gamut of protein reactions under conditions of accelerated kinetics, owing to the fact that proteins can be drawn to their cognate locations on the chip by applying an electrical charge to each electrode. The capacity to use electronic stringency may allow small differences in binding constants to be distinguished.

FIGURE 14.8. *A protein biochip prepared on a silicon substrate allows immobilization of up to 10,000 different proteins. A scanned image of a 40 × 40 element sensor containing bound fluorescent proteins (*large inset*) reveals the regularity of the 50-μm features. The space bar in the enlarged 8 × 8 portion of the microarray (*white outlined inset*) corresponds to 200 μm. (Data courtesy of Dr. S. Nock, ZYOMYX, Hayward, CA.)*

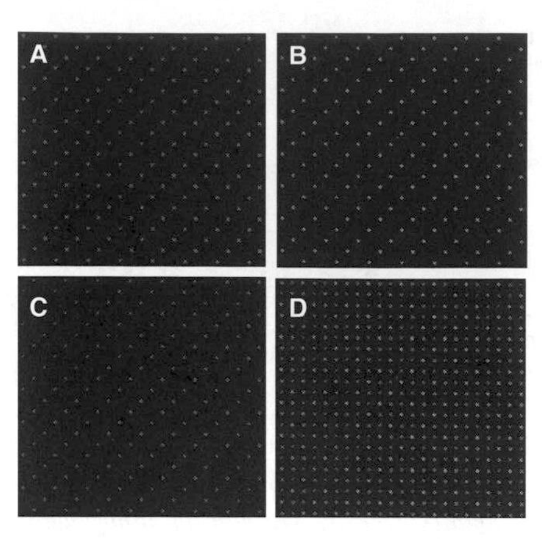

*FIGURE 14.9. Proteins were labeled with three different fluors, printed into a microarray, and detected using three different excitation and emission wavelengths: excitation at 490 nm and emission at 520 nm (**A**), excitation at 555 nm and emission at 570 nm (**B**) and excitation at 649 nm and emission at 670 nm (**C**). **D**, A composite image. (Data courtesy of Dr. S. Nock, ZYOMYX, Hayward, CA.)*

Multicolor protein microarray formats have also been developed. Nock and co-workers were able to demonstrate multicolor fluorescence detection by printing microarrays with proteins labeled with three different fluors and performing the detecting step using three separate excitation and emission wavelengths (Fig. 14.9). This technical advance allows exploration of many new areas of protein biochemistry, including studies of protein complex formation. Because many important enzymes and molecular motors contain multiple protein subunits, multicolor protein microarrays allow examination of protein assembly and enzymatic activity, all in a massively parallel microarray format.

Mirzabekov and co-workers (Khrapko et al., 1991) at the Russian Academy of Sciences (Moscow) developed a novel 20-μm-thick polyacrylamide microarray coating that allows the immobilization of proteins of up to 400 kDal in molecular weight. These gel pads enable square microarray features of 100 × 100 μm. Proteins of interest are printed onto the gel pads and enter the polyacrylamide layer by diffusion or by the assistance of an electrophoretic current. Once immobilized in the polyacrylamide layer, the proteins are available for binding interactions with target proteins of interest. Test studies with printed monoclonal antibodies prepared against laminin, fibronectin, c-reactive protein, alpha-fetoprotein, and human immunoglobulin G reveal highly specific reactivity with cognate antigen targets labeled with the fluorescent dye fluorescein isothiocyanate (Fig. 14.10).

RESONANCE LIGHT SCATTERING

Fluorescence is the standard detection platform for microarray analysis, and there are many advantages to fluorescence-based detection schemes, though fluorescent molecules are susceptible to photobleaching and have limited brightness owing to the intrinsic properties of the fluors and detection schemes. Fluorescent scanning and imaging technologies are also rather expensive and complicated, because of the presence of moving stages, lasers, photomultiplier tubes, and other sophisticated components. For these reasons, several groups have sought to explore alternatives to fluorescence detection for microarray-based studies.

Researchers at Genicon Sciences (San Diego, CA) developed an extraordinarily sensitive microarray detection technology based on ***resonance light***

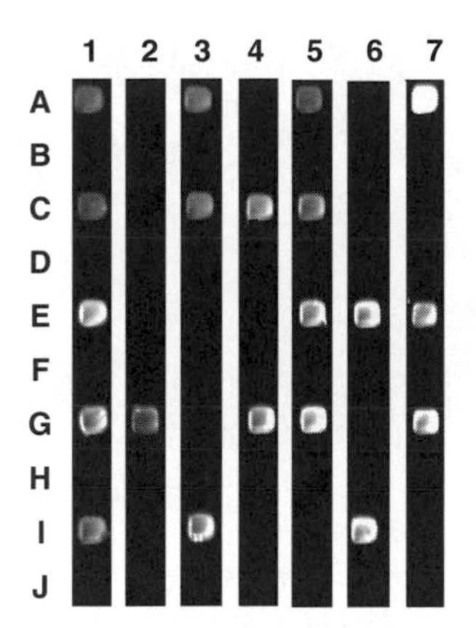

FIGURE 14.10. Polyacrylamide gel pads coated on a glass substrate were used to immobilize several different monoclonal antibody preparations raised against: laminin (A), fibronectin (C), c-reactive protein (E), alpha-fetoprotein (G), and human immunoglobulin G (I). The control proteins include bovine immunoglobulin G (B), monoclonal antibody to bovine serum albumin (D), and bovine serum albumin. (F, H, J)). The protein microchips were reacted with fluorescent probe mixtures containing the following proteins: laminin, fibronectin, c-reactive protein, alpha-fetoprotein, and human immunoglobulin G (1); laminin, fibronectin, alpha-fetoprotein (2); human immunoglobulin G (3); fibronectin and alpha-fetoprotein (4); laminin, fibronectin, c-reactive protein, and alpha-fetoprotein (5); c-reactive protein and human immunoglobulin G (6); and hIgG; laminin, c-reactive protein, and alpha-fetoprotein (7). Fluorescent signals were recorded and displayed in a black and white color palette. (Data courtesy of Dr. A. Mirzabekov and co-workers, Russian Academy of Sciences, Moscow.)

scattering (RLS). RLS technology exploits the fact that gold and silver particles of small diameter (40–120 nm) generate intense monochromatic light when exposed to a white excitation source. The intense monochromatic signal is produced by the oscillation of electrons in the gold and silver atoms located on the surface of the nanoparticles, which vibrate in phase in the presence of white excitation light, producing an emission wavelength that depends on particle diameter and composition and on the intensity of the excitation source. By controlling the diameter (e.g., ±5–10 nm) and composition of the particles during manufacturing, it is possible to generate a series of labels with different diameters that have discrete emission wavelengths spanning the entire visible spectrum (Fig. 14.11).

Resonance light scattering. *Microarray detection technology that exploits the fact that small gold and silver particles generate intense monochromatic light when exposed to a white excitation source.*

RLS particles can be attached in a covalent manner to oligonucleotides, cDNAs, proteins, antibodies, and many other biomolecules of interest and used as microarray labeling reagents in a manner analogous to fluorescent dyes. The successful application of RLS technology to mutation detection studies (Fig. 14.12) illustrates the specificity of this novel nonfluorescent labeling approach.

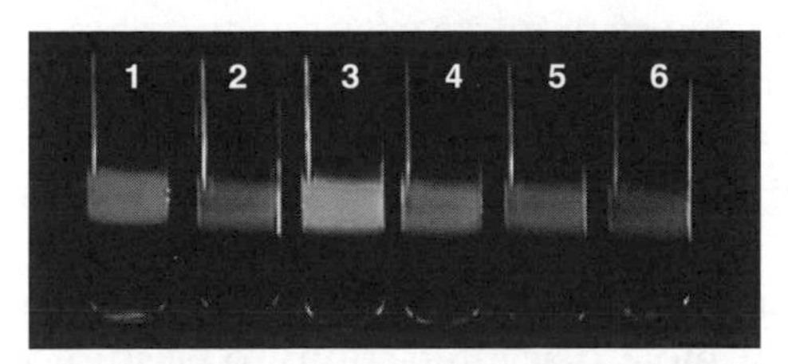

FIGURE 14.11. Gold (Au) and silver (Ag) nanoparticles of a defined diameter emit intense monochromatic light in the presence of a white light source, as can be seen from these preparations of different composition and diameter: 40 nm Ag at 2 pM (1), 40 nm Au at 13 pM (2), 78 nm Au at 1.7 pM (3), 118 nm Au at 0.5 pM (4), 140 nm Au at 0.3 pM (5), and fluorescein dye at 2000 pM (6). (Data courtesy of Dr. J. P. Linton, Genicon Sciences, San Diego, CA.)

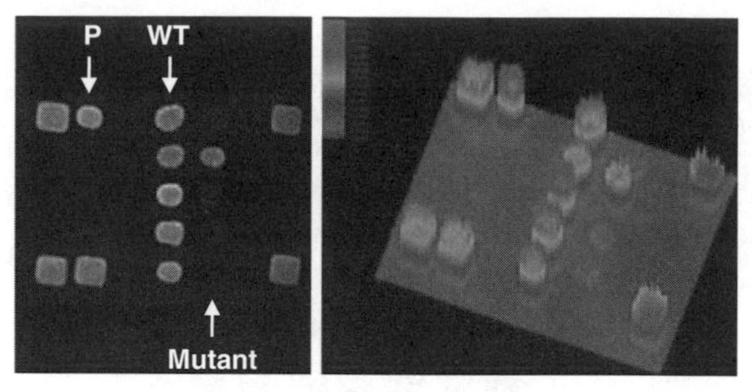

FIGURE 14.12. **A,** *DNA probes labeled with RLS particles produce much stronger signals when hybridized to wild type sequences (WT) and positive controls (P) than to mutant sequences (mutant).* **B,** *Microarray data are represented in a three-dimensional graphic with signal intensities indexed to a rainbow palette. (Data courtesy of Dr. J. P. Linton, Genicon Sciences, San Diego, CA.) Figure also appears in Color Figure section.*

There are several advantages of RLS technology over fluorescence detection for microarray applications, including little or no quenching or photobleaching, and a 1000-fold greater signal on a molar basis. The fact that RLS is based on light scattering using durable metal nanoparticles rather than fluorescent organic molecules obviates signal losses due to energy transfer (e.g., quenching) and the degradation of the fluorescent dyes (e.g., photobleaching). The 1000-fold greater signal should not be overstated, because the gold and silver particles are much larger (~100 nm) than fluorescent dyes (~2 nm); but the greater specific activity of RLS particles is highly significant because many microarray labeling and detection schemes use a 1:1 ratio of biomolecule (e.g., oligonucleotide) to label. A single oligonucleotide probe molecule labeled with an RLS particle is orders of magnitude brighter than the same oligonucleotide labeled with a fluorescent dye.

There are several disadvantages of RLS technology, including mixing and detection platform compatibility issues. The size and density of the gold and silver nanoparticles can lead to a settling of the particles out of solution if labeled samples are stored for prolonged periods in the liquid state, though the particles appear to diffuse rapidly in microarray hybridization experiments. The requirement for white excitation light of a defined angle of incidence means that microarrays stained with RLS particles require a dedicated detection platform, though several extremely affordable systems have already been developed at Genicon.

SUMMARY

The curiosity and ingenuity of microarray scientists are giving rise to a steady flow of novel microarray formats and technologies that are advancing all of the forefronts in microarray analysis. A new printing technology that contains a large cylindrical bundle of sample intakes may allow hundreds or even thousands of samples to be loaded and printed simultaneously in a small substrate area. Biosensors, a fusion of microelectronics and molecular biology, are analytical devices that allow molecular interactions to be studied in a miniature format. The capacity to apply electrical charge to each biosensor electrode enables extremely rapid hybridizations for mutation detection applications. A nitrocellulose surface coating has been developed for microarray applications, promising increased binding capacity and a wealth of protocols from traditional blotting methods.

Newly developed protein, antigen, and monoclonal antibody microarrays allow entry into **proteomics**, including studies of the genetic basis of autoimmune disease. Proteome microarrays may ultimately allow patients to be screened against all of the proteins expressed in the human genome on a single chip. Protein biosensors coupled with three-color protein detection should allow protein–protein interactions and enzyme assembly to be studied with a precision not afforded by traditional methods. RLS, which exploits the fact that gold and silver nanoparticles generate intense monochromatic light when exposed to a white excitation source, provides a nonfluorescent detection platform that obviates the limitations of photobleaching and offers stronger signals than traditional fluorescent schemes.

Proteomics. *Study of cellular proteins on a genomic scale.*

SELECTED READING

Arenkov, P., Kukhtin, A., Gemmell, A., Chupeeva, V., Voloschuk, S., and Mirzabekov, A. Protein microchips: Use for immunoassay and enzymatic reactions. *Anal Biochem* 278:123–131, 2000.

Benson, H. *University Physics,* rev. ed. Wiley, New York, 1996.

Bloomfield, V. A., Crothers, D. M., and Tinoco, I. *Nucleic Acids: Structure*. University Science Books, Herndon, VA, 2000.

Fersht, A. *Structure and Mechanism in Protein Science: A Guide to Enzyme Catalysis and Protein Folding.* Freeman New York, 1999.

Giancoli, D. C. *Physics for Scientists and Engineers,* 3rd ed. Prentice Hall, Upper Saddle River, NJ, 2000.

Joos T. O., Schrenk, M., Hopfl, P., Kroger, K., Chowdhury, U., Stoll, D., Schorner, D., Durr, M., et al. A microarray enzyme-linked immunosorbent assay for autoimmune diagnostics. *Electrophoresis* 21:2641–2650, 2000.

Khrapko, K. R., Lysov, Yu. P., Khorlin, A. A., Ivanov, I. B., Yershov, F. M., Vasilenko, S. K., Florentiev, V. L., Mirzabekov, A. D. A method for DNA sequencing by hybridization with oligonucleotide matrix. DNA seq. 1, 375–388, 1991.

Lakowicz, J. R. *Principles of Fluorescence Spectroscopy,* 2nd ed. Kluwer Academic/Plenum, New York, 1999.

Lehninger, A. L., Nelson, D. C., and Cox, M. M. *Principles of Biochemistry,* 3rd ed. Freeman, New York, 2000.

Lewin, B. *Genes VII.* Oxford University Press, Oxford, UK, 2000.

Macbeath, G., and Schrieber, S. L. Printing proteins as microarrays for high-throughput function determination. *Science* 289:1760–1763, 2000.

Schena, M., ed. *DNA Microarrays: A Practical Approach,* 2nd ed. Oxford University Press, Oxford, UK, 2000.

Schena, M., ed. *Microarray Biochip Technology*. Eaton, Natick, MA, 2000.

Schultz, S., Smith, D. R., Mock, J. J., and Schultz, D. A. Single-target molecule detection with nonbleaching multicolor optical immunolabels. *Proc Natl Acad Sci USA* 97:996–1001, 2000.

Singer, M., and Berg, P. *Genes and Genomes*. University Science, Herdon, VA, 1991.

Sosnowski, R. G., Tu, E., Butler, W. F., O'Connell, J. P., and Heller, M. J. Rapid determination of single base mismatch mutations in DNA hybrids by direct electric field control. *Proc Natl Acad Sci U S A* 94:1119–1123, 1997.

Streitweiser, A., and Heathcock, C. H. *Introduction to Organic Chemistry,* 2nd ed. Macmillan, New York, 1981.

Stryer, L. *Biochemistry,* 4th ed. Freeman, New York, 1995.

Wilson, D. S., Nock, S. Functional protein microarrays. Curr. Opin. Chem. Biol. 6, 81–85, 2002.

REVIEW QUESTIONS

1. Which of the following would be advantageous for a novel microarray printing technology: (a) 100-μm spot size, (b) 1.5-mm spot size, (c) 0.2-μl loading volume, (d) 300-pl delivery volume, (e) 300-μl delivery volume, and (f) regular spot spacing?
2. Most microarray printing devices achieve high-density printing by moving the printhead in small increments with a robot during successive printing

cycles. How does the device shown in Figure 14.1 achieve high-density microarray printing?

3. Which of the following biomolecules would be expected to bind *rapidly* to a biosensor array with a positive bias: (a) DNA, (b) RNA, (c) protein, and (d) carbohydrate?
4. Schiecher and Schuel sell a glass microarray substrate containing a thin nitrocellulose coating that allows attachment of DNA, RNA and protein. Chapter 1 insists that microarray assays are distinct from earlier assays that use flexible nylon and nitrocellulose filters. Reconcile this apparent paradox.
5. Traditional protein assays are being replaced quickly by protein microarrays. Which of the following advantages are afforded by the protein microarray format: (a) parallel analysis of protein binding, (b) low reaction volume, (c) rapid kinetics, (d) high density, and (e) multicolor analysis?
6. Which of the following, if any, are components of resonance light scattering (RLS) technology: (a) fluorescent dyes, (b) laser excitation, (c) photosensitive labels, or (d) monochromatic emission?

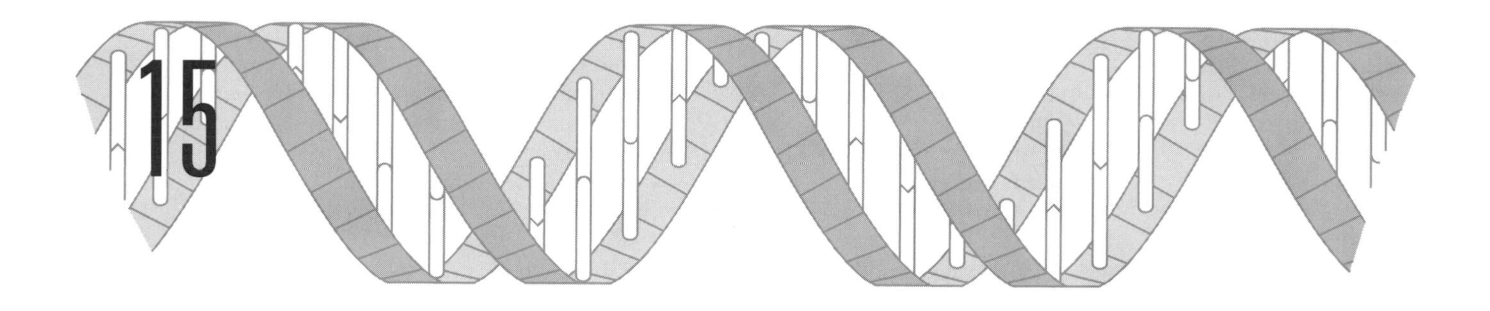

15

Commercial Opportunities A to Z

If you can count your money, you don't have a billion dollars. —*J. Paul Getty*

Microarray technology has created scores of interesting commercial opportunities, including motion control devices, detection equipment, business-to-business relationships, microarray manufacture, genetic screening services, kits and other consumables, oligonucleotides, slides, substrates, and software tools. The microarray field is expected to include > 400,000 researchers worldwide and sales of > 2 billion dollars by about 2006. The close alliance of scientists, investors, legal experts, and business people is driving this robust economic and technological expansion. This chapter provides an A to Z overview of commercial opportunities in microarray technology.

ARRAYERS

The sale of automated systems for microarray manufacture (arrayers) has grown into a cottage industry, with > 15 commercial organizations offering robotic microarray devices at present (Table 15.1). The availability of the first commercial arrayers and scanners in 1997 triggered an explosion in basic research, yielding a rapid increase in the number of scientific publications on microarrays (Fig. 15.1). Commercial arrayers offer a number of key advantages over home-built devices, including improved resolution and accuracy, greater durability, better software options, technical support, and the immediate availability of replacement parts if repairs are needed. Taken together, the advantages

TABLE 15.1. Arrayers

Company	Products	Web Site	Location
Affymetrix	417 arrayer	www.affymetrix.com	Santa Clara, CA
Beecher Instruments	microarray technology	www.beecherinstruments.com	Silver Spring, MD
Bio-Rad Laboratories	VersArray chipwriter	www.bio-rad.com	Hercules, CA
*Bio*Robotics	*Micro*Grid II	www.biorobotics.com	Cambridge, UK
Cartesian Technologies	MicroSys, PixSys, ProSys	www.cartesiantech.com	Irvine, CA
GeneMachines	OmniGrid	www.genemachines.com	San Carlos, CA
Genetix	QArray, genpakARRAY 21	www.genetix.com	Hampshire, UK
Genomic Solutions	GeneTAC	www.genomicsolutions.com	Ann Arbor, MI
Intelligent Bio-Instruments	microarray technology	www.intelligentbio.com	Cambridge, MA
Labman Automation	microarray technology	www.labman.co.uk	North Yorkshire, UK
MiraiBio	SPBIO	www.miraibio.com	Alameda, CA
Molecular Dynamics	Generation III Array Spotter	www.mdyn.com	Sunnyvale, CA
NimbleGen Systems,	Maskless Array Synthesizer	www.nimblegen.com	Madison, WI
Packard BioScience	SpotArray	www.packardbioscience.com	Meriden, CT
TeleChem/Arrayit.com	SpotBot	www.arrayit.com	Sunnyvale, CA

offered by commercial systems render home-made robots obsolete, though makeshift research instruments played an important role in the early stages of the field.

Commercial arrayers run the gamut from affordable desktop devices that hold 10–12 glass substrates, to expensive industrial-scale systems with substrate capacities of 100–300. Microarray robots embody a fusion of technologies, including electric motors, linear actuators, linear encoders, printing pins, ink-jet nozzles, and environmental control chambers as well as all of the expertise that draws on mechanical and software engineering, micromachining, and material science. As the use of microarray technology expands, the number of commercial arrayers is expected to increase dramatically. In the not too distant future, it is likely that nearly every laboratory in the world, studying problems spanning biology to material science, will use a commercial microarray device. The expected sales volume of this industry is expected to be ~100,000 units, offering a total commercial opportunity of about 5 billion dollars.

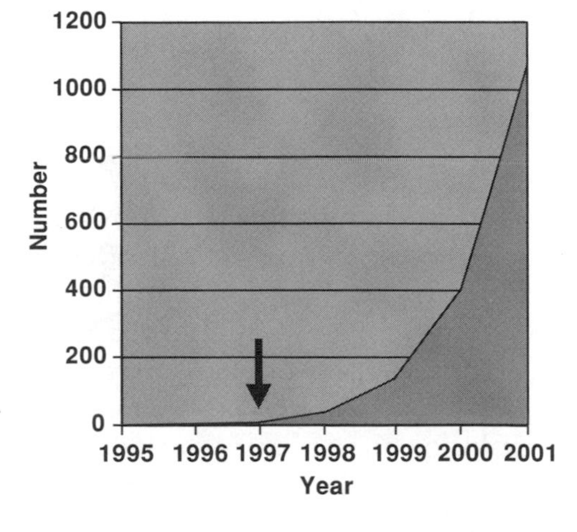

FIGURE 15.1. *The total number of microarray scientific publications as a function of calendar year, beginning with the first publication in* Science *in 1995. The first commercial microarray robots and scanners were introduced in 1997* (arrow). *(Data from the e-library of TeleChem/arrayit.com, Sunnyvale, CA.)*

BUSINESS TO BUSINESS

One of the promises of the Internet era is the potential for seamless interaction between companies. Business-to-business (B2B) models, which leverage the Internet to link companies with related business activities, have been applied extensively to the financial and consumer markets, and the same basic principles can be used to unite microarray technology companies. The Internet is well suited as an information transfer mechanism for microarray technology, in that microarray design and manufacture draw extensively from private and public sequence databases, and all of these data are readily mined and sent electronically. Microarray design and redesign are enhanced greatly by rapid iteration, and electronic mail (e-mail) works well to send iterative data sets quickly and economically.

Microarray manufacture often requires coordinated efforts among three or more companies, placing a premium on integrated commerce, and the B2B models make sense for this reason (Fig. 15.2). One obvious set of interactions is among companies that provide genomic sequence information, oligonucleotide synthesis services, and microarray manufacture capabilities, respectively. The three companies and the customer can be linked via the Internet, which streamlines the design process and minimizes manufacturing costs. The B2B model for custom microarray manufacture allows a high degree of technological specialization at the three companies, which provides the customer with a level of product quality and affordability that cannot be achieved by any other means. Many other B2B configurations can be envisioned for microarray-based services.

COMPUTER SCIENCE

Computers play a central role in many different aspects of microarray analysis, and computer-related opportunities will continue to be abundant. Motion control software and graphical user interface (GUI) programs for arrayers and scanners, data visualization and mining tools for gene expression and genotyping information, and sequence search engines are some of the exciting commercial areas likely to see major growth. The use of computers for microarray

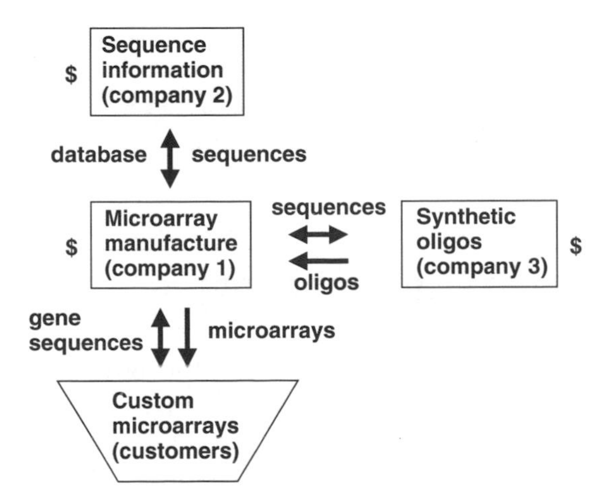

FIGURE 15.2. A model of how companies are linked in a B2B configuration via the Internet to provide custom microarrays at an affordable price.

TABLE 15.2. Microarray Software Companies

Company	Products	Web Site	Location
Applied Maths	*GeneMaths*	www.applied-maths.com/	Sint-Martens-Latem, Belgium
Applied Precision	*DataInspector*	www.api.com/	Issaquah, WA
Array Genetics	*SeqIt, NucIt*	www.arraygenetics.com	Newtown, CT
Axon	*GenePix Pro* 3.0	www.axon.com/	Union City, CA
BioDiscovery	*ImaGene* 4.1, *GeneSight* 2.1, *CloneTracker*	www.biodiscovery.com/	Marina Del Rey, CA
Compugen	DNA chip design and analysis	cgen.com/	Jamesburg, NJ
Imaging Research	*ArrayStat*	imaging.brocku.ca/	Ontario, Canada
Lion bioscience	*arraySCOUT, arrayTAG, arrayBASE*	www.lionbioscience.com/	Heidelberg, Germany
MediaCybernetics	*Array-Pro*	www.mediacy.com/	Silver Spring, MD
MiraiBio	*DNASpace*	www.miraibio.com/	Alameda, CA
Packard BioScience	*QuantArray*	www.packardbioscience.com/	Meriden, CT
Premier Biosoft International	*Array Designer* 2	www.premierbiosoft.com	Palo Alto, CA
Rosetta Inpharmatics	*Resolver*	www.rii.com/	Kirkland,WA
Silicon Genetics	*GeneSpring* 4.0	www.sigenetics.com/	Redwood City, CA

applications falls into the specialized category of computer science known as ***bioinformatics*** or ***computational biology,*** which focus on biological applications (see Chapter 9). More than a dozen companies have commercial software products that expedite various aspects of microarray analysis, and this number is likely to grow as the field expands (Table 15.2).

Bioinformatics. *Specialized field of computer science focused on the analysis of biological data.*

Computational biology. *Specialized field of computer science focused on the analysis of biological data.*

DIAGNOSTICS

Diagnostics. *Branch of medicine devoted to identifying or determining the nature and cause of disease by combining patient information and laboratory testing data.*

Diagnostics is the branch of medicine devoted to identifying or determining the nature and cause of disease by combining patient information and laboratory testing data. With the recent completion of the human genome sequence and the rapid proliferation of microarray technology, the twenty-first century is poised to see an explosion in high-quality diagnostic tests based on DNA and protein microarrays. Many if not all of the current assays can be supplemented or replaced by microarray tests, providing more accurate and rapid information, greater sensitivity, and reduced patient cost. Important diagnostic areas include tests for infectious diseases, substance abuse, cancer, cardiac illnesses, blood disorders, and diabetes. Microarray-based diagnostic kits can also be applied to cattle, sheep, and other livestock to diagnose bacterial, viral, and prion-based diseases.

EXPRESSION MONITORING

Global gene expression monitoring with microarrays provides an amount of biological information that is unprecedented in biology. Because gene expression and biological function are linked (see Chapter 12), expression databases are virtual treasure troves of important information. Gene expression profiles can be used to elucidate the molecular basis of diseases, identify new drugs, and chart the efficacy of clinical trials. Given the massive medical implications of gene expression information (see Chapter 16), expression monitoring tools and services represent a dynamic commercial opportunity. Microarrays for gene expression analysis, kits, software, and database information are some of the commercial areas likely to see expansion.

FORENSICS

The branch of medicine that gathers, examines, and analyzes specimens found at crime scenes is known as ***forensics***. Because of the chemical and biochemical stability of DNA, modern molecular forensics focuses almost exclusively on nucleic acid analysis as the primary forensic tool. The polymerase chain reaction (PCR) is used to obtain analyzable material from the DNA present in samples as minute as the follicle cells of a single human hair. The amplified material is then examined for small differences in DNA sequences known as ***polymorphisms***, which allow unambiguous identity to be established for individuals present at a crime scene. Establishing the identity of a person by examining DNA polymorphisms is known as ***DNA typing***. Two specific types of polymorphisms—***tandem repeats*** and ***restriction fragment length polymorphisms*** (RFLPs)—are used most often in DNA typing (Fig. 15.3).

Forensics. *Branch of medicine that gathers, examines, and analyzes specimens found at crime scenes.*

Polymorphism. *Any of a family of minor DNA sequence variants, including single nucleotide polymorphisms and restriction fragment length polymorphisms.*

DNA typing. *Experimental approach in which the identity of a person is established by examining DNA polymorphisms.*

Tandem repeats are short 4- to 25-base pair sequences that are joined in a head-to-tail configuration in the human genome. Though the function of tandem repeats is unknown, they provide useful polymorphisms, because they occur in different numbers in the population; these polymorphic differences among people can be observed by examining the products by gel electrophoresis (Fig. 15.4). By determining the number of tandem repeats in each individual, researchers can obtain identity information.

Tandem repeat. *Short nucleotide sequences that are joined in a head-to-tail configuration in the human genome, vary among individuals, and find use in genotyping applications.*

RFLPs, the second common type of polymorphism used for DNA typing, are single nucleotide changes in the genome that result in the creation or loss of restriction sites (Fig. 15.4). Digestion of amplified genomic DNA from different

Restriction fragment length polymorphisms. *Single nucleotide change in the genome that results in the creation or loss of a restriction site.*

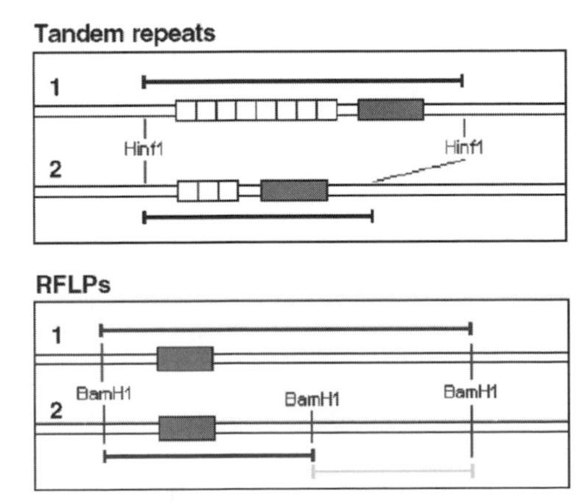

FIGURE 15.3. **A,** *Tandem repeats* (squares) *are short sequences that differ in number among individuals. Digestion of genomic DNA* (double lines) *with the HinfI enzyme produces different size fragments* (solid line) *for a person* (1) *Who has eight tandem repeats* (1) *compared to a person with three tandem repeats* (2). **B,** *RFLPs are polymorphisms that result in the presence or absence of a restriction site in the DNA, producing different size fragments* (solid line) *for a person with 2* BamHI *sites* (1) *than for a person with 3* BamHI *sites* (2). *(Courtesy of Dr. R. A. Bowen and co-workers, Animal Reproduction and Biotechnology Laboratory, Colorado State University, Fort Collins.)*

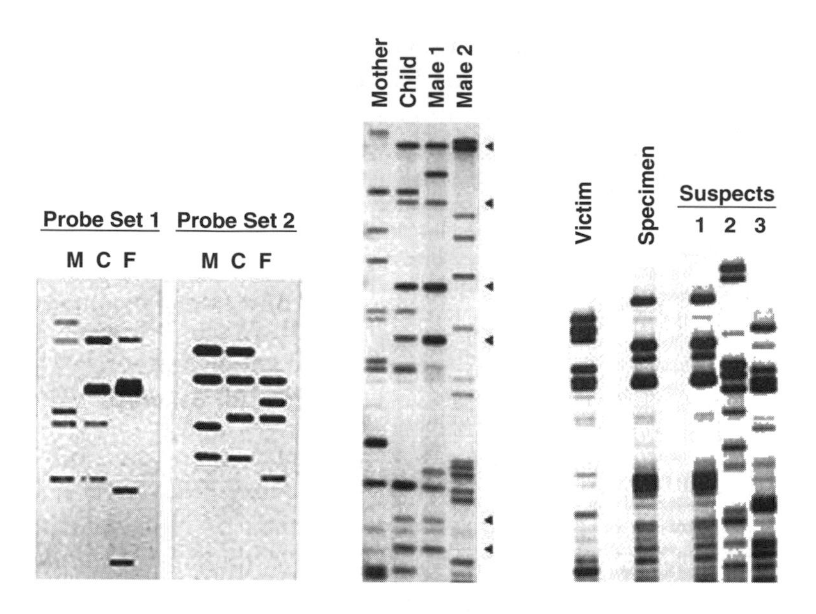

FIGURE 15.4. *DNA regions containing polymorphisms are amplified by PCR, size-separated by gel electrophoresis, blotted onto filter paper, and hybridized with radioactive probes to determine the identity and size of a given set of DNA fragments. Identical fragments observed in a child* (C), *mother* (M) *and father* (F), *allow kinship to be established. Identical fragments can also be used to distinguish between three potential suspects (1–3) in forensic analyses. (Courtesy of Dr. R. A. Bowen and co-workers, Animal Reproduction and Biotechnology Laboratory, Colorado State University, Fort Collins.)*

individuals, followed by fragment size examination using gel electrophoresis, provides useful identity information. By combining multiple tandem repeat and RFLP markers, it is possible to type an individual to greater than 1 in a 100,000,000 in the population. Combining forensic DNA data with other investigative information can establish a strong case for guilt or innocence in a criminal case. DNA typing analysis is also used widely to establish kinship in paternity cases, for immigration, child support, adoption, and other forms of identity testing.

There are several shortcomings to the current techniques used for DNA typing. Most techniques use restriction enzymes, which add cost and time and can lead to ambiguous readings owing to incomplete DNA digestion. Current techniques also employ gel electrophoresis to separate the DNA fragments on the basis of size, and gels can be laborious to run and difficult to read. Many modern DNA typing techniques use radioactive probes, which can reduce researcher safety and present radioisotope storage and disposal issues. Though current DNA typing methods for forensics and other forms of identity testing are highly accurate, additional accuracy, speed, affordability, and precision could be obtained by converting the present tests into a DNA microarray format.

One could envision a universal DNA typing microarray that uses gene-specific primers and PCR to amplify a hundred or so polymorphic loci from the human genome. A set of unique amplicons for each person could be printed into a microarray and hybridized with fluorescent oligonucleotides, allowing hundreds or even thousands of persons to be typed in a single test. By typing all of the amplicons in parallel as described for the NGS format (see Chapter 13), an unambiguous identity could be established for each person in the human population in simple and affordable manner. The crime chip would be easy to use, safe to perform, and yield unambiguous DNA typing information within 24 h. The same set of primers could be used to manufacture father and mother chips in paternity and parentage cases. Forensics and other applications of DNA typing by microarray represent a significant commercial opportunity that will likely yield a quality and reliability of data vastly superior to the current state-of-the-art.

GENETIC SCREENING

The systematic examination of a population for genetic information is known as ***genetic screening***. The targeted examination of a specific subset of the population for a specific disease or condition is known as ***genetic testing***. Genetic screening and genetic testing often use the same assays, but are generally distinguished by the scope of the population tested. Screening is generally performed in a comprehensive manner without any a priori motivation for conducting the assay, whereas in testing cases, individuals are often examined as a result of being at high risk for a given condition. Microarrays offer enormous commercial promise for both genetic screening and testing (see Chapter 13). Neonatal and adult screens and tests for metabolic and other inherited disorders, infectious diseases, and any other condition that is amenable to DNA or protein analysis can be performed in a microarray format. The affordability and reliability of microarray screens and tests renders these assays applicable to the entire human population.

Genetic screening. *Systematic examination of a population for genetic information by microarray analysis or some other means, without any a priori motivation for conducting a specific screen.*

Genetic testing. *Targeted examination of a specific subset of the population for a specific disease or condition using microarray analysis of some other testing procedure.*

HUMAN LEUKOCYTE ANTIGEN TYPING

The major histocompatibility complex (MHC) in the human genome is a gene-dense, 8-Mb region on chromosome 6 that contains a large number of genes involved in immune system function. One very important region of the MHC is known as the ***human leukocyte antigen*** (HLA) locus (Fig. 15.5). The HLA locus consists of two principle classes of genes (I and II), all of which encode cell surface glycoproteins. Class I proteins– HLA-A, HLA-B, and HLA-C– are abundant proteins expressed in nearly all cells and are involved in antigen presentation to cytotoxic T cells. The HLA-D protein, a class II HLA antigen, is expressed in a restricted set of cell types and mediates interactions with helper T cells. The HLA locus is one of the most highly polymorphic coding regions in the human genome, with hundreds of alleles. HLA variability appears to have been selected for during evolution, probably as a defense against lethal infection by pathogenic bacteria.

Human leukocyte antigen locus. *Important region of the major histocompatibility complex in the human genome that encodes a family of surface glycoproteins, and is highly polymorphic.*

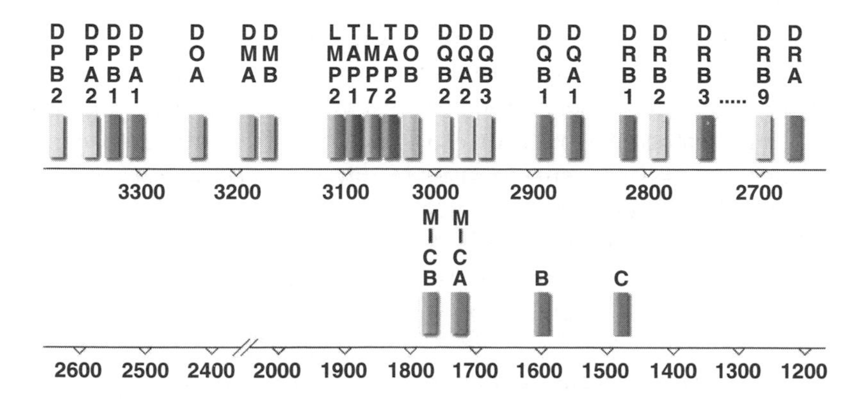

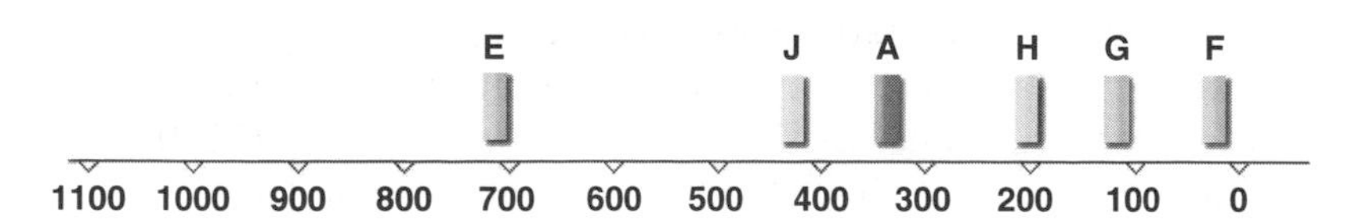

FIGURE 15.5. Some of the HLA alleles (shaded rectangles), *along with their names. (Courtesy of the Anthony Nolan Research Institute, London.) Figure also appears in Color Figure section.*

Foreign HLAs are highly antigenic, and T cells mount a vigorous attack against non-self HLA proteins. HLA incompatibility is a major cause of organ and bone marrow rejection, placing enormous medical importance on HLA typing before transplantation procedures. Once HLA compatibility is established between donor and acceptor, the chances of successful transplantation of heart, kidney, lung, and other solid organs increase dramatically. Other factors being equal, a highly matched HLA genotype improves the chance of successful transplantation.

PCR-based HLA typing has replaced seriological tests, but the current state-of-the-art still relies on data from a limited set of alleles and the rather cumbersome and sometimes ambiguous analysis of the PCR amplicons by gel electrophoresis. There is a substantial commercial opportunity in introducing a rapid, economical, and highly precise microarray assay for HLA typing; and the positive medical consequences of this new technology would be considerable. The availability of a complete genomic sequence of the HLA locus and hundreds of characterized HLA alleles will expedite the development of a microarray-based HLA test.

Intellectual property. *Any product of the human mind that meets the criteria required for patentability.*

Utility. *One of the three formal types of patents issued by the U.S. Patent and Trademark Office.*

Design. *One of the three formal types of patents issued by the U.S. Patent and Trademark Office.*

Plant. *One of the three formal types of patents issued by the U.S. Patent and Trademark Office.*

INTELLECTUAL PROPERTY

Provisional patent. *Abbreviated invention disclosure to the U.S. Patent and Trademark Office that establishes a priority date and requires a complete application to be filed within 12 months of the provisional filing.*

Disclose. *Process of revealing an invention by written or oral communication.*

Priority date. *First day that the patent office was informed of an invention, given by day, month and year, and used to establish legal ownership in patent disputes.*

Useful. *One of the three formal requirements for patentability, pertaining to whether a given invention has utility by improving on the state of the art in a technology area.*

Novel. *One of the three formal requirements for patentability, pertaining to whether a given invention is new with respect to the state of the art in a technology area.*

Unobvious. *One of the three formal requirements for patentability, pertaining to whether a given invention could not be ascertained by a simple reformulation of the state of the art in a technology area.*

Any product of the human mind that meets the criteria required for patentability is known as ***intellectual property***. A patent provides a legal framework within which to receive credit for contributing a piece of intellectual property and a means of protecting that property from commercial misuse by others. In the context of microarray technology, intellectual property can take the form of a method, analytical device, manufacturing process, computer algorithm, gene sequence, chemical compound, transgenic crop, and so forth. Each of these forms would fall into one of the three formal types of patents that are issued: ***utility, design,*** and ***plant***.

In addition to patents of utility, design, and plant, there is a fourth type of patent known as a ***provisional patent,*** granted not on the basis of the type of invention but rather to denote an invention that has been ***disclosed*** to the patent office but has not been filed as a complete application. Provisional patents can be written quickly and submitted for a nominal fee, allowing the inventor 12 months of breathing room to develop the invention further, acquire additional funding, or seek supplemental marketing and manufacturing expertise before filing a complete application. Provisional patents provide a means of establishing a ***priority date,*** which corresponds to the first day that the patent office was informed of the invention, given by day, month, and year. Priority dates are important legally because they help clarify who invented what and when, if disputes arise between holders of related patents. If a provisional patent has been filed, it also allows the inventor and the future assignee to use the term *patent pending* on manufactured items, marketing and financial reports, the Internet, and so forth.

To meet the basic threshold of patentability, a piece of intellectual property must be ***useful, novel,*** and ***unobvious***. The usefulness or utility of a patent can be established simply by demonstrating or explaining how the proposed invention would improve the state-of-the-art in a particular area or stimulate an important commercial sector. The novelty requirement addresses the "Is this new?" question and is met by reviewing what has been invented or described in the past and showing that the proposed invention has not been described in any previous patent or publication. The comprehensive body of information, including patents, scientific publications, academic theses, and so forth, is known as collectively as the ***prior art***. Examination of the prior art assists the patent office in evaluating a patent application and affects the validity and scope of a patent once it has issued.

The third criterion for patentability is that an invention must be unobvious. An invention is deemed unobvious if it cannot be derived in a trivial way from the prior art and cannot be obtained by combining two pieces of related art in a straightforward and expected manner. A key job of the patent attorney is to convince the U.S. ***Patent and Trademark Office*** (PTO), the agency of the Department of Commerce that issues patents, that the proposed invention represents a significant piece of intellectual property that cannot be derived simply by common sense. The process by which applications, documents, and arguments are exchanged between the patent attorney and the PTO examiner is known as ***patent prosecution,*** a "tango" that usually takes 2–3 years from the time an application is received to the time that a patent is issued. Once issued, a patent provides either 20 years (utility patent) or 14 years (design patent) of protection to the patent holder. Most patents in the microarray field are utility patents, allowing the patent holder to commercialize the invention and prevent others from doing so for 20 years from the priority date.

A patent is, in essence, a contract between the U.S. government and the patent holder that maintains that if the patent holder describes or ***teaches*** how to practice the invention fully, the federal government will protect the patent holder against others from doing so. Each issued U.S. patent is given a unique identifying number (e.g., 6,101,946), derived from the consecutive list of patents that have been issued since the inception of the PTO. To date, > 6.2 million patents have been issued through the U.S. government, and nearly half of these were issued in the last quarter of the twentieth century.

Once a patent issues, it prevents another party from ***making, using, selling,*** or ***importing*** the invention anywhere within the United States. Worldwide patent protection can be achieved by following the guidelines of the ***Patent Cooperation Treaty*** (PCT). A single international application under the PCT allows an inventor to file for patent protection in > 100 countries simultaneously, which streamlines the process of obtaining worldwide protection. It should be noted that PCT applications become public information 18 months after the U.S. priority date, which may be a consideration commercially, because it allows competitors to get a glimpse of the invention before the issuance of the patent in the United States. PCT applications are therefore a mixed bag, particularly for small companies or when an extended prosecution process is anticipated domestically.

The coverage or ***scope*** of a patent refers to the breadth of the invention and what exactly it prevents others from making, using, selling, or importing. The patent scope is defined by the ***claims,*** a numbered list of sentences or paragraphs found at the end of a patent document. ***Claim construction,*** the precise language used to draft the claims, is extremely important because it determines the eventual extent of what is protected by the patent. Goods claims are very difficult to draft, require a high level of training and knowledge of patent law, and are best drafted and prosecuted by a highly trained (and highly paid) patent attorney. Most patents contain approximately a dozen claims, and the content of the claims is explained or clarified by the ***specification***. The specification is the main

Prior art. *Term in patent law encompassing any publication or previous invention that can be used to reject or invalidate a patent or patent claim.*

Patent and Trademark Office. *Agency of the U.S. Department of Commerce that issues patents.*

Patent prosecution. *Process by which applications and supporting documents are exchanged between a patent attorney and the Patent and Trademark Office en route to obtaining a patent.*

Teach. *Term in patent law that refers to the extent to which a patent instructs and informs the reader as to the content of the invention.*

Making. *One of the four main activities that is legally prohibited with respect to an invention once a patent issues.*

Using. *One of the four main activities that is legally prohibited with respect to an invention once a patent issues.*

Selling. *One of the four main activities that is legally prohibited with respect to an invention once a patent issues.*

Importing. *One of the four main activities that is legally prohibited with respect to an invention once a patent issues.*

Patent Cooperation Treaty. *Agreement signed by 115 countries that expedites patent filing and prosecution worldwide.*

Scope. *Term in patent law that refers to the breadth of the invention.*

Claim. *Term in patent law corresponding to a short paragraph that defines the scope of an invention.*

Claim construction. *Term in patent law that refers to the precise language used in the claims at the end of a patent application.*

Specification. *Term in patent law that refers to the main body of a patent application that teaches, explains, and clarifies an invention.*

body of the patent and explains how to make and use the invention, provides examples, and defines key terms.

With respect to microarray technology, the commercial implications and opportunities of intellectual property and patents are profound. Innovative ideas, coupled with strong patents and sound marketing, can be worth hundreds of millions of dollars. One compelling reason to file for patent protection is that it expedites the commercialization of microarray technology and, therefore, enables many scientists to use this powerful set of molecular tools. Microarray methods, chips, analytical devices, instruments, and the like that are not protected by patents are less likely to be commercialized because there is no legal framework for a company to prevent other commercial entities from manufacturing the invention independently. Attorneys, expert witnesses, and consultants are needed on a regular basis to draft patent applications, prosecute patents, and litigate disputes, all of which provide generous compensation to participants who understand microarray technology.

JOURNALS

The number of scientific publications using microarray technology has exploded in the past few years (Fig. 15.1), offering exciting opportunities for existing and new journals to publish the findings of microarray science. Microarray papers have appeared in > 100 scholarly journals to date, and the journals have been Herculean in their capacity of proliferating microarray science in an accurate and efficient manner. Many journals offer publications in an electronic format, which facilitates access, particularly in foreign countries that may not have the infrastructure required to maintain hard copies of every exciting journal. Microarray books have also been quite successful commercially, with $> 20,000$ copies sold worldwide to date. Newsletters distributed weekly or monthly provide a means of keeping track of the latest developments in commercial products, legal activities, mergers, and acquisitions. Commercial opportunities for laboratory manuals and specialty bulletins are plentiful. A microarray electronic library (arrayit.com/e-library), provided free of charge, contains abstract, summary information, and electronic links for all of the papers published on microarrays since the first paper appeared in *Science* in 1995.

KITS

The availability of commercial kits has revolutionized scientific research in a way unimaginable 25 years ago. There are great stories about the early Genentech (S San Francisco, CA) pioneers going door to door asking whether researchers might they be interested in a career in commercial science, and most scientists wondered how biology could *ever* be worth anything beyond the value of the basic knowledge acquired through the course of doing experiments. Today, the biotechnology industry has grown to include > 10 billion dollars in sales annually, and companies selling microarray products represent an expanding component of this commercial opportunity. Sales for microarray biochip hardware, kits, software, services, and other consumable items are expected to exceed 2 billion dollars by the year 2006 (Fig. 15.6), and this figure may grow

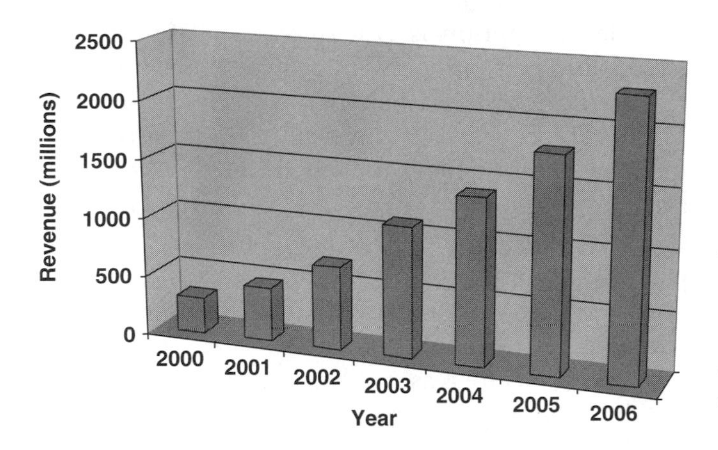

FIGURE 15.6. Conservative estimates of sales figures in millions of dollars (U.S.) for microarrays and biochips, hardware, consumables, and software. (Data courtesy of Drug and Market Development Publications, Westborough, MA.)

by 10-fold if microarrays continue to move steadily into the genetic screening and diagnostics markets.

Dozens of kits for microarray analysis are available currently, falling into five main categories: premade microarrays, DNA and RNA isolation and purification, amplification and labeling, cDNA clones and biological reagents, and chemical reagents and buffers. A rapid proliferation of new kits is expected as the microarray field expands, and kits offer a number of distinct advantages for microarray analysis. Kits containing premade microarrays and supporting buffers and reagents provide standardization, which allows scientists using the same product to exchange technical information and compare their results. Kits provide a level of quality control and consistency that does not exist if each researcher prepares a separate set of buffers and solutions for his or her own use. The availability of kits minimizes trivial labor, freeing up more time for experimental design, data analysis, and creative thinking.

Because of the miniature format of microarray assays, sample and reagent purity are much more important than in traditional molecular biology research. Many microarray kits are manufactured in controlled environments and some in class 100 cleanrooms. If prepared properly, kits provide a consistent level of purity and quality that is not readily obtained with homemade reagents prepared in an ambient environment. As microarray analysis moves rapidly into the areas of genetic screening, genetic testing and diagnostics, and forensics, kits will play an increasingly important role in ensuring the validity of the clinical data. Commercial kits that allow patients to withdraw small samples of their own blood in the privacy of their own homes are likely to be big sellers, particularly if the results from microarray-based genetic screens and tests were provided via the Internet. At-home microarray tests would provide the public with advanced genetic health care, without requiring time-consuming visits to hospitals or clinical laboratories, and without compromising privacy.

LICENSE

Patents prevent another party from making, using, selling, or importing an invention, unless the patent holder grants specific permission to do so. The legal document that defines the terms and conditions by which another party may use patented technology is known as a ***license***. Patents and patents pending may be licensed to all interested parties, and microarray patent holders should

License. *Term in patent law that refers to the legal document that defines the terms and conditions by which another party may use a patented or patent-pending technology.*

consider this option as a exciting means of generating revenue, stimulating research, and preventing invent-around innovation.

Contract. *Term in patent law that refers to a document signed by two parties in which both parties agree that the laws of the land will be used to enforce the terms of the agreement.*

Tort. *Term in patent law that refers to branch of law that enforces the conduct of the licensing parties and provides for remedies in case of breach or failed performance.*

Assignee. *Term in patent law that refers to the holder of a patent, also known as the licensor.*

Licensee. *Term in patent law that refers to the licensing party of a patent.*

Nonexclusive. *Term in patent law that refers to a licensing arrangement in which the assignee reserves the right to license the same technology to multiple licensees.*

Exclusive. *Term in patent law that refers to a licensing arrangement in which the assignee guarantees unique access to an invention for a single licensee.*

Licensor. *Term in patent law that refers to the holder of a patent, also known as the assignee.*

Rights. *Term in patent law that refers to the legal privileges, including ownership and licensing, afforded to a patent holder.*

Patent licenses are based on two specific bodies of law known as ***contracts*** and ***torts***. A contract is essentially a promise that the laws of the land will enforce the terms of the agreement and comes into existence when both the patent holder and the party that wishes to license the technology mutually agree to the terms of the license. The law of torts enforces the conduct of the licensing parties and provides for remedies in case of breach or failed performance. The patent holder and licensing party are known as ***assignee*** and ***licensee,*** respectively, and fair licenses for access to patented and patent pending microarray technology ensure reasonable compensation for the assignee and a nice upside for the licensee in having access to a technology that is otherwise precluded legally.

A general guideline is that licenses are granted for a modest annual fee (e.g., \$10,000–\$100,000) and a small percentage (e.g., 5%) of revenues that result from use of the technology. Moderate licensing terms are the best means of successful licensing, and assignees should negotiate in good faith and avoid the temptation of overcharging the licensee. The terms shown here obviously apply to ***nonexclusive*** licenses, which means that the assignee reserves the right to license the same technology to as many parties as are interested in it. An ***exclusive*** license to patented or patent pending technology, which would legally guarantee unique access by a single licensee, is expected to be much more expensive than a nonexclusive license. It should be noted that in cases of both nonexclusive and exclusive licenses, the assignee or ***licensor*** must provide access to the technology but retains legal title.

A more expensive arrangement than both nonexclusive and exclusive licenses are situations in which the holder of patents or patents pending assigns the ***rights*** to an interested party. Assignment of rights would mean that the new assignee would possess all of the legal rights of the initial assignee, including the power to grant licenses to the technology. Assigning patent rights often occurs when one microarray company acquires another company in a merger or acquisition. Compensation for patent rights may exceed tens or hundreds of millions of dollars for valuable microarray technologies.

MICROARRAYS

Premanufactured microarrays represent a growing commercial opportunity that is likely to exceed 1 billion dollars within the decade, and probably sooner. Revenues on Affymetrix GeneChip systems from Affymetrix (Santa Clara, CA), the most successful commercial microarrays to date, have grown rapidly since the late 1990s (Fig. 15.7) and are expected continue to do so, particularly as microarray analysis becomes the platform of choice for functional genomics. More than 20 commercial organizations currently provide manufactured microarrays (Table 15.3), and this number is expected to grow as the demand for microarray products expands and the applications diversify.

The two predominant drivers for commercial microarrays are increasing accessibility of complete genomic sequences and the increasing affordability of synthetic oligonucleotides. Complete genomic sequences provide a complete blueprint or database of all of the genes in an organism and thus a comprehensive set of microarray target sequences representing all of the genes can be selected computationally. Once the target set is established, microarrays can be

TABLE 15.3. Commercial Microarrays

Company	Products	Web Site	Location
Affymetrix	GeneChip	www.affymetrix.com	Santa Clara, CA
Agilent Technologies	microarray technology	www.agilent.com	Palo Alto, CA
AlphaGene	FLEX microarrays	www.alphagene.com	Woburn, MA
Capital Biochip	microarray technology	www.capitalbiochip.com	Beijing, China
Clincal Microsensors	eSensor	www.microsensor.com	Pasadena, CA
Clontech	Atlas	www.clontech.com	Palo Alto, CA
Corning	CMT	www.corning.com	Corning, NY
DNAmicroarray.com	microarray technology	www.dnamicroarray.com	
GeneScan Europe	microarray technology	www.biochip.com	Freiburg, Germany
Hyseq	HyChip	www.hyseq.com	Sunnyvale, CA
Incyte Genomics	GEM	www.incyte.com	Palo Alto, CA
Mergen	*Express*Chip	www.mergen-ltd.com	San Leandro, CA
Nanogen	NanoChip	www.nanogen.com	San Diego, CA
Operon Technologies	OpArrays	www.operon.com	Alameda, CA
OriGene Technologies	SmartArray	www.origene.com	Rockville, MD
PerkinElmer Life Sciences	MICROMAX	lifesciences.perkinelmer.com	Boston, MA
ProtoGene Laboratories	microarray technology	www.protogene.com	Menlo Park, CA
Sigma-Genosys	Panorama	www.genosys.com	Woodlands, TX
Stratagene	GeneConnection	www.stratagene.com	La Jolla, CA
TeleChem/Arrayit.com	custom microarray services	arrayit.com	Sunnyvale, CA
Vysis	GenoSensor	www.vysis.com	Downers Grove, IL

prepared either by in situ synthesis methods, such as photolithography, or with one of the delivery techniques, such as contact printing (see Chapter 7). Affordable oligonucleotides allow a complete set of gene targets to be synthesized off line and then printed into microarrays, affording whole genome chips to all researchers. Commercial genome chips allow researchers to address a nearly unlimited set of interesting biological questions with a level of ease and sophistication that would have been unimaginable a few years ago.

Microarrays sold by most vendors can be used with little or no preparation, allowing the customer to obtain data immediately. Chips made in a cleanroom setting, with sophisticated motion control technology, sample tracking, and the like, offer a level of quality that is difficult to achieve in most laboratory settings. Cost-effectiveness can be enjoyed at both low and high use volumes, because there is significant cost associated with making microarrays. Customers that use a few dozen chips a month would probably spend 10–100 times as much to make their own chips. Companies that provide custom manufacturing services alleviate some of the serious technical and logistical challenges posed by

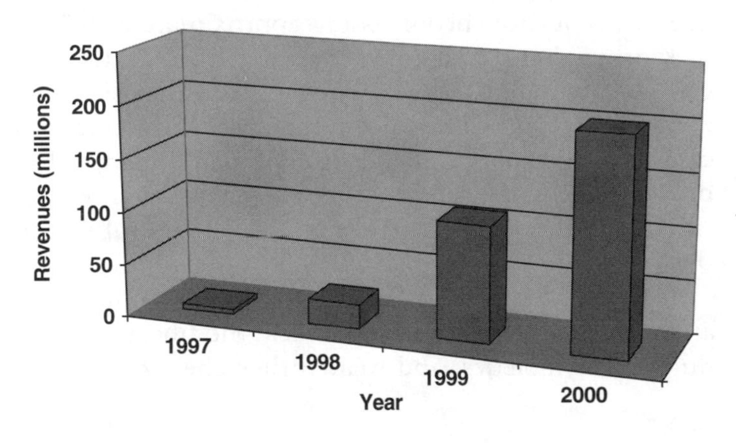

FIGURE 15.7. Approximate sales figures for Affymetrix, provider of GeneChip technology, the world's best-selling microarray. Revenues represent the totals from chips, instruments, and supporting consumables, and are shown in dollars (U.S.). (Data courtesy of Affymetrix, Santa Clara, CA.)

high-volume microarray manufacturing, allowing laboratories engaged in applications such as drug screening and toxicology to subcontract or outsource this component of their high-throughput activities.

NEW TECHNOLOGY

One of the more prudent mottos of Silicon Valley is innovate instead of litigate. Though litigation is sometimes a necessary part of business, it is an arduous, costly, disruptive, and cynical process that pits colleagues and companies against one another and drains valuable resources. The outcome of a lawsuit is difficult to predict; and verdicts, financial awards, and damages are decided, in part, by a jury that has little detailed knowledge of the technical subject at issue in most legal disputes. Innovation, on the other hand, creates tremendous technological value and thus a large and sustainable financial opportunity. New microarray technologies (see Chapter 14) and products are essential to maintaining cutting-edge microarray science and are necessary for meeting the stringent requirements of genetic screening and diagnostic agencies. Improvements in manufacturing, detection, surface chemistry, amplification, and labeling are some of the many areas that could benefit from new technology.

Innovation in microarray science has taken place at commercial organizations, running the full gamut of sizes, from small businesses of < 10 employees to large multinational pharmaceutical companies with $> 100{,}000$ workers. This trend will undoubtedly continue, and microarray scientists should scratch the itch to innovate whenever possible. Entrepreneurs entering the microarray field should take solace in the fact that American small businesses—defined as firms with < 500 employees—generate nearly half of all the nation's revenues each year and more than half of the private gross domestic product annually. Small, high-tech microarray companies can and will make key contributions in the form of new technology. A recent search of the Internet using the search term *microarray* produced 145,000 matches, providing a staggering indication of the amount of interest that exists for microarray technology and the potential for rapid acceptance of innovative microarray products.

OLIGONUCLEOTIDES

Synthetic oligonucleotides are playing an increasingly important role in microarray analysis, both as target sequences for microarray manufacture and as fluorescent probes for hybridization. The availability of a complete human genome sequence encompassing approximately 30,000 genes opens the door to synthesizing a complete set of oligonucleotides representing each human gene. Genome chips allow gene expression analysis at the level of the entire human genome, providing a comprehensive view of the physiological effects of disease progression, aging, drug action, and many other important biological processes. Oligonucleotides that represent small sequence variants, including single nucleotide polymorphisms, provide useful reagents for genotyping applications.

Oligonucleotide synthesis enables the manufacture of native DNA chains and sequences that contain chemical modifications, including amino linkers, fluorescent moieties, and many other chemical substituents. Amino linkers

TABLE 15.4. Oligonucleotide Providers

Company	Web Site	Location
Alpha DNA	www.alphadna.com	Montreal, Canada
Annovis	www.cybersyn.com/	Aston, PA
Biosearch Technologies	www.biosearchtech.com	Novato, CA
BioServe Biotechnologies	www.bioserve.com/	Laurel, MD
Biosource International	www.keydna.com/	Foster City, CA
Bio-Synthesis	www.biosyn.com/	Lewisville, TX
Commonwealth Biotechnologies	www.cbi-biotech.com/	Richmond, VA
Cruachem	www.cruachem.com/	Glasgow, Scotland
Dharmacon Research	www.dharmacon.com/	Lafayette, CO
DNAgency	www.dnagency.com/	Malvern, PA
Eurogentec	www.eurogentec.com	Seraing, Belgium
GenoMechanix	geno-mechanix.com/	Alachua, FL
Gene Link	www.genelink.com	Hawthorne, NY
Genset	www.gensetoligos.com	Paris, France
Great American Gene Company	www.geneco.com	Ramona, CA
Illumina,	www.illumina.com	San Diego, CA
Integrated DNA Technologies	www.idtdna.com	Coralville, IA
Invitrogen	www.invitrogen.com	Carlsbad, CA
Metabion	www.metabion.com	Planegg-Martinsried, Germany
Midland Certified Reagent	www.mcrc.com	Midland, TX
MWG-BIOTECH	www.mwg-biotech.com	Ebersberg, Germany
New England Biolabs	www.neb.com	Beverly, MA
Operon Technologies	www.operon.com	Alameda, CA
Peninsula Laboratories	www.penlabs.com	San Carlos, CA
Promega	www.promega.com	Madison, WI
Ransom Hill Bioscience	www.ransomhill.com	Ramona, CA
ResGen	www.resgen.com	Huntsville, AL
Scandinavian Gene Synthesis	www.sgsdna.com	Köping, Sweden
Sigma-Genosys	www.genosys.com	The Woodlands, TX
Synthegen	www.synthegen.com	Houston, TX

allow covalent attachment of the oligonucleotides to microarray substrates containing reactive aldehyde groups, providing stably attached target elements of exceptional quality. Dozens of different types of fluorescent tags allow direct detection of hybridized oligonucleotide probes, a capacity that lends itself well to NGS genotyping and other assays (see Chapter 13). Other types of synthetic groups–including biotin, dinitrophenol, thiol, cholesterol, and acridine–enable myriad attachment chemistries and detection schemes. Oligonucleotide manufacture, purification quality control, and verification represent significant commercial opportunities. Dozens of synthetic oligonucleotide providers (Table 15.4) offer a wide range of synthetic products for microarray applications.

PROTEIN MICROARRAYS

Protein biochemistry has a glorious past and boasts numerous successes in characterizing the catalytic, structural, and mechanistic basis of protein function. Our knowledge of enzymes, the specialized proteins that catalyze

TABLE 15.5. Protein Drug Sales in 2000

Therapeutic Area	Products	Company	Revenues (millions of U.S. dollars)
Cardiovascular	Activase	Genentech (South San Francisco, CA)	206
Cystic fibrosis	Pulmozyme	Genentech (South San Francisco, CA)	122
Human growth	Nutropin Depot, Nutropin AQ, Nutropin, Protropin, Pulmozyme	Genentech (South San Francisco, CA)	227
Multiple sclerosis	AVONEX	Biogen (Cambridge, MA)	624
Oncology	Herceptin, Rituxan	Genentech (South San Francisco, CA)	720
Red blood cell	Epogen	Amgen (Thousand Oaks, CA)	1960
White blood cell	Neupogen	Amgen (Thousand Oaks, CA)	1220

biochemical reactions, led to a detailed understanding of how cells synthesize their DNA and to the revolutionary technologies of recombinant DNA and polymerase chain reaction. Studies of enzymes, protein–protein interactions, and cellular signaling also made possible a series of blockbuster protein drugs, including Activase, Avonex, Epogen, Herceptin, Neupogen, Nutropin, Protropin, Pulmozyme, Rituxan, and TNKase. Combined sales for these products exceeded 5 billion dollars in the year 2000 (Table 15.5). Protein drugs have a bright future, particularly if improvements can be made on existing products and new products can be developed.

Protein microarrays, analytical devices that allow tens of thousands of protein reactions to be studied in a single experiment (see Chapter 14), promise to accelerate our understanding of protein biochemistry in a manner unprecedented in history. Microarray-based protein assays allow the enzymatic specificity, catalytic rate and protein–protein affinities of native proteins to be studied in a massively parallel format. The availability of a complete human genome sequence, coupled with protein microarrays assays and a solid infrastructure for drug development, will enable the development of safer and more efficacious derivatives of current protein therapeutics and many new drugs in the near future. The human genome is likely to contain hundreds of protein therapeutics that could be brought to market if the cost of development could be reduced. Protein microarrays, by virtue of their parallel analysis capabilities, offer the promises of reduced development time and expense, both of which would have a positive impact on the protein drug pipeline.

QUALITY CONTROL

Quality control. *Commercial manufacturing procedure in which the characteristics of a product are defined and regulated to guarantee conformity of the product to within specified parameters.*

One of the keys to commercial success in the microarray field is ***quality control*** (QC), the process by which the features and characteristics of a product are defined and regulated to guarantee the uniformity of the product within specified parameters. Quality control is important in the manufacture of microarrays, substrates, kits, and other products. Uniform products enable greater precision during use and improved customer satisfaction. QC parameters are one aspect that is emphasized heavily in a commercial setting, but is often lacking or nonexistent in basic research laboratories.

Quality control encompasses a broad scope of activities, including process design, statistical control, downstream analysis, and follow-up work. With respect to the design of a microarray manufacturing process, it is often useful to start with the end in mind, meaning that a focus on product consistency should be engineered into the manufacturing process. If the goal is to make highly consistent substrates or printed microarrays, idiosyncratic steps and baroque protocols should be eliminated from the process before that process is brought on-line. A high level of quality control is much easier to implement if the entire process is examined and fine-tuned before it is implemented on a major scale. Poor quality control will frustrate staff and customers, damage product image, and elevate production costs, all of which can be avoided by considering QC issues before introducing a product.

The use of statistical methods such as quantitative analytical devices and measuring tools are very useful in the context of quality control. Fluorescent scanners, dissecting microscopes, and other instruments are used as regular components of microarray substrate quality control, ensuring the day-to-day consistency of products. Statistical methods have the advantage of providing numerical assessments of quality and permanent records, obviating human error and allowing quality to be charted over time. Analytical devices can be applied during a manufacturing process and after the product has been made; their combined use during these separate phases usually generates a superior and more consistent product.

In cases of customer complaints, follow-up QC work is often essential and includes the review of statistical records and QC data, product reexamination, and customer contacts. Lot numbers, manufacturing dates, and employee identification numbers are useful in expediting follow-up work related to quality control. It is important to emphasize that a customer complaint does not necessarily equate with a defective product. Most microarray products are used in the context of complicated protocols, methods, samples, and instruments and may appear to be defective when, in fact, the fault lies with the user or one of the other components. In this respect, contacting other customers that have used the same product lot is an effective way of assigning the cause of customer dissatisfaction.

Another important aspect of QC is defining the acceptable range variability for a given product. Variability exists in every manufacturing process and includes both intrinsic and assignable sources. Intrinsic variation refers to the noise associated with a given process that is not detectable or assignable within the scope of the QC process and may include small differences in reagent purity, subtle temperature variation, instrument scatter, and other imperceptible or uncontrollable differences that occur from lot to lot. Assignable sources of variability are those that can be traced to a given step in a process. Quality control mandates that the sum of the intrinsic and assignable variability fall within an acceptable and predetermined range. The establishment of ***standard operating procedure*** (SOP) guidelines is useful for minimizing product variability. The International Organization for Standardization (ISO) is a worldwide agency dedicated to promoting and ensuring standardization in manufacturing

Standard operating procedure. *Commercial manufacturing protocol, employed to minimize product variability.*

processes. Microarray manufacturing facilities can achieve ISO 9000 certification by following an eight-step procedure.

RAMPING UP

A key to commercial success is the ability to expand production capacity or ramp up to meet customer demands. One strategy useful in this regard, also recommended with respect to quality control (see above), is to start with the end in mind. If the expectation or goal is to manufacture a certain number of chips per day, for example, the production strategy must be tailored to facilitate a scaling of the process in an efficient and economical manner. A production scheme that works for low volumes but not for high volumes has much less value than a process that can be increased by orders of magnitude to meet the explosive growth often associated with microarray products. Millions of manufactured chips and substrates are used every year in the microarray community, and these numbers will undoubtedly grow in the coming years. Personnel training, instrument and facilities costs, reagent availability, waste disposal, and integration and distribution are some of the key considerations required to ramp up.

SURFACE CHEMISTRY

The pressure to perform, generate results, acquire data, and make discoveries can create a frenetic environment in the research laboratory, and most scientists depend heavily on reliable microarray surfaces and sound attachment chemistries. But surface chemistry, substrate manufacturing, and quality control are difficult to implement in most research laboratories, placing slides and substrates at ground zero in microarray science. The tremendous importance of high-quality surfaces and the relative difficulty in making them has created a large commercial opportunity for organizations with the know-how and infrastructure to mass produce slides and substrates at an affordable price. More than a half dozen commercial entities (Table 15.6) provide slides and substrates at present, offering a spectrum of different treatments, coatings, and attachment schemes. Many more opportunities in surface chemistry will be available in the near future as the breadth of microarray assays increases and the proliferation

TABLE 15.6. Microarray Substrates and Slides

Company	Products	Web Site	Location
CEL Associates	silanated, silylated	www.cel-1.com	Houston, TX
Corning	CMT-GAPS	www.corning.com	Acton, MA
Erie Scientific	UltraArray, UltraClean, BioGold,	www.eriesci.com	Portsmouth, NH
Microarray Shop	slides, substrates	www.microarrayshop.com	
Schleicher & Schuell	FAST slides	www.s-und-s.de	Keene, NH
Sigma-Aldrich	SigmaScreen	www.sigma-aldrich.com	St. Louis, MO
TeleChem/arrayit.com	ArrayIt, SuperClean, SuperAmine, SuperAldehyde, SuperEpoxy	arrayit.com	Sunnyvale, CA
Xenopore	Xenobind, Xenoprobe	www.xenopore.com	Hawthorne, NJ

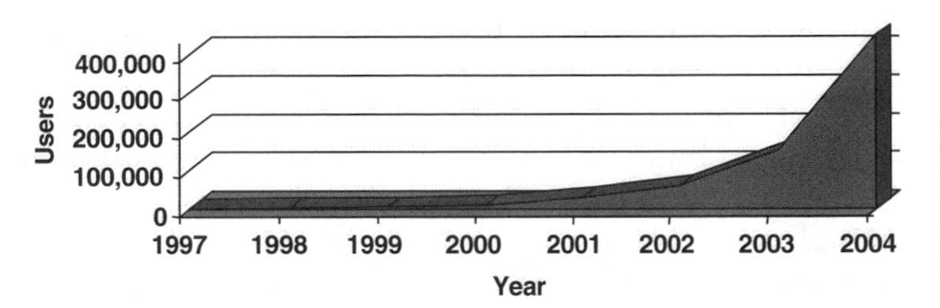

FIGURE 15.8. The number of current and projected users of microarray technology. (Data courtesy of K. Smith and co-workers, Front Line Strategic Management Consulting, Foster City, CA.)

of the technology continues. As noted above, the number of researchers in the microarray industry is growing nearly exponentially (Fig. 15.8).

TRADE SECRETS

A ***trade secret*** is defined as any piece of confidential information, including inventions, prototypes, codes, formulas, manufacturing methods, protocols, reagents and suppliers, customer information, purchasing practices, marketing reports, and business alliances, that are used to develop, conduct, and promote business. To be considered a trade secret under recent federal legislation (see below), the information must provide a competitive advantage in the marketplace, and the owner of the information must have taken reasonable measures to maintain its secrecy. Trade secrets must also not be part of what is known generally outside the organization.

Trade secret. *Any piece of confidential information, including inventions, prototypes, codes, formulas, manufacturing methods, protocols, reagents, suppliers, customer information, purchasing practices, marketing reports, and business alliances, that are used to develop, conduct, and promote business.*

The massive commercial importance of trade secrets in the United States and the increasing concern over trade secret theft, prompted President Clinton to pass The Economic Espionage Act of 1996, making stealing trade secrets a federal criminal offense. One consequence of the Espionage Act is that it empowers the Justice Department to charge and prosecute cases of trade secret theft both inside and outside of the United States, and over the Internet. The sternness of the legislation is underscored by the fact that the act provides for prison sentences of up to 15 years to individuals and corporate fines of up to 10 million dollars for trade secret violations.

Trade secrets in the microarray industry are numerous and extremely valuable. In some cases, a trade secret is worth more than a patent in the sense that trade secret information, by definition, is not divulged to the outside community. Patents are also valuable, but there are two sides to the patent coin. In exchange for receiving patent protection from the federal government, detailed plans of how to make and use the invention must be disclosed in the patent application, and those plans become public information. Public accessibility of an invention may result in a loss of strategic commercial advantage, particularly if the specification of the patent teaches an important concept or process that is not covered in the patent claims. In the process of obtaining patent protection, a company may actually unwittingly lose some valuable information to a competitor. Trade secrets, on the other hand, are maintained tightly within a company; and if they remain confidential, considerable strategic advantage is afforded. It is noteworthy that the recipe for Coca-Cola is protected by trade secret laws and not by patents.

Trade secrets in the microarray field include microarray manufacture methods, cleanroom construction materials, designs and practices, suppliers of chemical and biochemical reagents, recipes for buffers and solutions, substrate printing and processing techniques and protocols, engineering strategies, drawings and designs, software algorithms, graphical user interfaces and architecture, subcontractors, joint venture participants, financial projections, sources and reports, trade show and promotions strategies, and myriad other important sources of confidential information. Every employee in the microarray field should work together to ensure the confidentiality of trade secret information, both for the sake of the commercial success of the field and for the legal safety of each of the practitioners.

USING HUMAN RESOURCES

The microarray field is a young field, both with respect to the newness of the technology and in the sense that many microarray scientists are at an early stage in their careers. Many senior-level people also participate and have also made key discoveries, inventions, and advances and will continue to do so in the future. Another attribute of the microarray field is the extensive collaboration between academic and commercial organizations. A third defining feature is the highly interdisciplinary aspect of microarray science, drawing on the expertise of scientists, engineers, computational biologists, business people, lawyers, and investors. The genesis of microarray technology and its present prosperity are very much attributable to the fact that the field has always embraced a diversity of opinions, tools, technologies, approaches, and strategies producing a high-tech melting pot as robust as any in recent memory. The future of the field depends, in part, on the extent to which these abundant human resources can be used to their fullest.

Using human resources in a commercial setting is key to profitability and expansion. With respect to hiring practices, there are two schools of thought representing the philosophical extremes. One philosophy dictates that highly specialized tasks require highly specialized skills, and every professional billet in a microarray company requires someone with extensive specialized training. The other extreme suggests that nearly any talented person, irrespective of their technical background, can learn a specialized skill and so jobs should be filled on the basis of talent and not specialized training. Both philosophies have merit and both suffice to build the human resources required to run a successful microarray company. There are pros and cons to the two approaches.

Highly trained personnel with advanced degrees, distinguished publication records, and the like offer the advantage of having an immediate and self-sufficient impact on the company, attributes that are extremely advantageous in a highly competitive commercial environment. One disadvantage of experts is that they often carry along strong opinions and views, some of which are useful and others that can be disruptive to a smooth-running organization. In a crude managerial sense, the challenge is to maximize the upside of the expertise and minimize the downside of the baggage that is almost always part and parcel of a high level of training.

Talented employees who lack highly specialized skills offer the key asset of being much more malleable than technical experts and often integrate much more easily into an existing organization. One disadvantage of such personnel is that they often require a much greater investment in training than pretrained employees and are generally less adept at identifying problems in processes and in suggesting remedies. A mixture of personnel is each organization provides sufficient structure while maintaining the flexibility required for the smooth, day-to-day operations.

Another aspect of using human resources pertains to the narrowness or amount of flexibility that is allowed with respect to job descriptions and duties. There seem to be two philosophies in this respect, one that dictates extreme narrowness and the other that allows complete freedom. Employment duties that are defined too narrowly often lead to drudgery and a loss of enthusiasm, and those that lack any definition often result in a loss of productivity and direction. One solution is to require all personnel to perform a set of required duties but to have time open for creative exploration, discovery, and self-motivated direction each day. This middle-ground approach ensures that day-to-day operations run smoothly in a fun and interesting environment. A happy employee is a productive employee, and microarray executives should go to great lengths to ensure that everyone in their organization has a sense of well-being and prosperity.

VENTURE CAPITAL

Professional money made available to companies in exchange for an equity stake is known as ***venture capital,*** and the investors involved in structuring deals of this type are known as ***venture capitalists***. Because venture investments are much less certain than traditional investments, venture capital is also known as risk capital. At present there are > 500 firms specializing in venture capital financing in the United States, and the amount of venture money available has increased dramatically since 1996. In 1999, > 3500 companies in the United States received venture capital investments totaling nearly 40 billion dollars. In 2000, the total venture capital investment worldwide was approximately 140 billion dollars (Fig. 15.9), with major investments in the United States, Europe, and Asia.

Venture capital. *Professional money made available to companies in exchange for an equity stake, also known as risk capital.*

Venture capitalists. *An investor who structures deals involving an equity stake in exchange for risk capital.*

The predominant view is that the biotechnology industry, which accounts for approximately 5% of all venture financing, would be greatly diminished or even nonexistent if there were not venture capital available. Support for this argument is easy to garner in the drug-discovery areas, where product development is relatively slow and research is extremely expensive. Venture financing contributes much to the current vigor and prosperity observed in the traditional biotechnology sector, and close alliances among and scientists, businesspeople, and investors are likely to be key to the commercial success of microarray technology. Microarray companies contemplating venture financing should

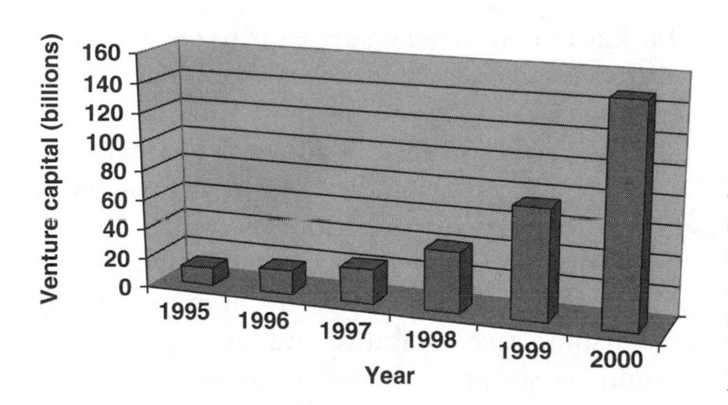

FIGURE 15.9. The approximate amount of venture capital investment in billions of dollars (U.S.) by venture capital firms worldwide. (Data courtesy of VentureEconomics.com, Thomson Financial Securities Data, New York.)

Due diligence. *Scrutinizing process of a company by a venture capitalist in anticipation of making a venture capital investment.*

Stage of development. *Term used in venture capital financing that refers to the maturity of a business with respect to products and sales.*

Startup. *Term used in venture-capital financing that refers to a business that has important ideas and technologies but has yet to develop products or generate revenues.*

Seed capital. *Term used in venture-capital financing that refers to the initial investment of venture capital in a startup business.*

Early stage. *Term used in venture-capital financing that refers to a business that is developing one or products but has yet to ship the products or generate sales.*

Expansion. *Term used in venture-capital financing that refers to the third stage of business development, after the startup and early stage phases; characterized by an increasing product line and sales.*

Profitability. *Term used in venture-capital financing that refers to the fourth stage of business development (after the startup, early stage, and expansion phases); characterized by a development stage in which company sales exceed company expenses.*

Initial public offering. *Term used in venture-capital financing that refers to the first sale of company shares to the public.*

Mezzanine round. *Term used in venture-capital financing that refers to last round of venture capital before an initial public offering.*

Exit strategy. *Term used in venture-capital financing that refers to a plan for beneficially terminating an investment relationship through the sale of the business, company shares, or through some other revenue-generating means.*

be aware that the venture capital company will scrutinize their organization closely as part of their ***due diligence*** before investing. Companies should make sure to have their ducks in a row before this due diligence process.

The venture capital community classifies companies according to their ***stage of development***, and the different stages correspond to the maturity of the business. A ***startup*** is a business that has important ideas and technologies but has yet to develop products or generate revenues. Venture firms use ***seed capital*** to stimulate startup companies, and seed rounds of financing usually range from 1–5 million dollars. An ***early stage*** company is one that is developing products but is not yet shipping them and has yet to generate sales. During the early stage of a business, venture capital investment rounds are usually numbered (first round, second round, etc.), and the investment amounts are usually 5–100 million dollars per round. ***Expansion*** and ***profitability*** are generally recognized as the third and fourth stages of development, and additional rounds of venture financing are used to promote these stages of a business.

Some companies may offer to sell shares to the public for the first time, in a move known as an ***initial public offering*** (IPO). The last round of venture financing before an IPO is known as the ***mezzanine round.*** Going public has appeal to business owners and venture capitalists alike. To the business owner, an IPO is a way to raise a large amount of capital in a short period of time, without the hassles of bank financing, which invariably involve more stringent criteria and conservative benchmarks than those governing IPOs. The venture capital community embraces the IPO because it provides rapid means of realizing a gain in the investments that were made en route to taking a company public.

Venture capital firms often sell some of their shares shortly after the IPO, and this ***exit strategy*** has appeal in that it allows the venture capitalist to recoup their initial investment and realize a profit. Venture capitalists, owners, company employees, and major shareholders may sell their shares on the open market after a designated 3- to 24-month period known as the ***lockup period***. The process by which insiders sell their shares is known as ***insider trading***, and strict rules of conduct, overseen by the U.S. Securities and Exchange Commission (SEC), govern these activities.

Venture capital investment and IPOs are powerful strategies for raising large amounts of operating capital, and microarray companies should consider these strategies as they progress along the path of business development. But these issues are complex and highly technical and should be considered only under advice of legal counsel, financial officers, and other experts.

WORKING ENVIRONMENT

One key to commercial success is to create a company atmosphere that fosters productivity and creativity. How is this accomplished? Minimizing hierarchies, limiting the number of meetings and other bureaucratic distractions, creating intensity without tension, allowing independence and freedom of expression, and organizing extracurricular events are a few suggestions.

Some level of organizational hierarchy is required in every organization, and the amount tends to increase as the size of the company increases. Having a chain of command is important in maintaining smooth company operations, establishing accountability, and settling disputes. But executives and managers should be sensitive to the needs of those they supervise and should avoid heavy-handed tactics. Employees should always feel free to engage their supervisors, and open communication is extremely important for managers and workers alike because it is the main vehicle by which successes are shared, problems are identified, and new opportunities created.

As the size of a company grows so does the complexity, and one symptom is an increase in the number of meetings. Regular meetings are important for the success of any company, but there is a clear limit of return on this investment. On a scale of 1 to 10, most meetings receive about a 3.5 in terms of their impact on the company, and it is important to remember that meetings represent down time for everyone involved. When meetings are in session, many other operations suffer, leading to a drain on company resources and diminishing productivity. Other bureaucratic requirements including the writing of reports and memos, unnecessary and redundant bookkeeping, and other excessive distractions and time sinks should be minimized whenever possible.

Intensity and tension are sometimes used interchangeably, but the two are very different. The pace at most microarray companies is fast and frenetic, and this hard-driving style is required to excel in a competitive field. But intensity embodies vigor and enthusiasm, not the uneasiness that comes with tension. A tense work environment, one in which employees are walking on pins and needles, is a detriment to any organization and does little to create the sense of harmony and well-being required for a productive workplace. Fostering intensity and working to eliminate tension should be top priorities for executives and managers.

Reasonable independence and freedom of expression enhance the workplace and should be fostered, within limits, by all executives and managers. Every employee has different family and lifestyle considerations, and these differences should be respected and embraced. Flexible work schedules do much to allow for childcare, commuting, and the pursuit of hobbies and other personal matters. Working nights, weekends, and holidays, should be encouraged as a means of making up lost hours and as a mechanism for supplementing income in the form of overtime. Relaxed attire and casual Fridays promote a sense of individuality and congeniality. It is not clear that the hypercorporate environment does much to increase productivity, and it may be a detriment to companies that expect longer hours than the traditional 40-h workweek.

Most employees operate in the midst of considerable pressure, and meeting an extensive list of expectations—including shipping orders, collecting payments, manufacturing products, inventing new methods, answering customer support, and managing Web sites—requires great skill and energy. One potential downside of this pressure-filled environment is the tendency for colleagues to drift apart in the sense of personal friendship. One way to restore collegiality and ensure functional personal ties is to organize extracurricular events at periodic intervals. Corporate retreats, concerts, sporting events, and parties do much to rehumanize the workplace, and executives and managers should make certain to schedule such events regularly.

Lockup period. *Term used in venture-capital financing that refers to a 3- to 24-month period during which venture capitalists, owners, company employees, and major shareholders are prevented from selling their shares on the open market.*

Insider trading. *Term used in venture-capital financing that refers to the selling of shares by a major company share holder.*

X-RAY CRYSTALLOGRAPHY

The method whereby an X-ray beam is used to solve macromolecular structures has played a central role in understanding protein function. But the current methods for X-ray studies have changed little in decades, still relying heavily on the rate at which protein crystals can be obtained from pure proteins in solution. Stories abound of scientists spending months or even years crystallizing a single protein; and the crystallization process is poorly understood and idiosyncratic. In stochastic events, there is value in being able to parallelize the process. In this regard, microarray printing as means of creating crystals for X-ray crystallography may be of great scientific and commercial value.

Supersaturated. *Solution that contains an amount of protein or some other solute in excess of what can be maintained in solution by a given solution at a defined temperature.*

Protein crystals form in ***supersaturated*** solutions as a movement of proteins from the liquid phase into the solid phase. A supersaturated solution is one that contains an amount of protein in excess of what can be maintained in solution by a given buffer at a defined temperature. Conditions for supersaturation depend on a complex set of variables, including temperature, the type of solvents used, and the particulars of a given protein. Many macroscopic methods are used to grow protein crystals, including batch crystallization, dialysis, microseeding, and vapor diffusion. The vapor diffusion method, commonly known as the hanging drop method, is probably the most frequent approach used to crystallize proteins. All of the approaches are serial processes requiring relatively large amounts of protein and considerable time and labor.

Contact and noncontact printing are potentially amenable to massive parallelization of the protein crystallization process. Tens of thousands of setups could be implemented on a single solid surface in a few hours, allowing literally millions of different crystallization trials to be performed in a few weeks. Experiments with different buffers, solvents, protein concentrations, and crystallographic conditions could be implemented in parallel, in the hopes of generating usable crystals at one or more locations on the microarray. Once grown, the microarray-generated crystals could be subjected to X-ray analysis.

As complete genomic sequences for many organisms are compiled, it is important for structural genomics to keep pace. In this respect, the prospects of parallelizing and automating X-ray crystallography have great scientific and commercial implications. Most small molecule drugs exert their biological activity by binding to discrete sites on the surface of proteins. Having complete structural libraries of every human protein would greatly facilitate the drug development process. Global information on bacterial and fungal proteins would also be of great assistance in discovering better drugs to combat microbial infections. Structural information for all of the proteins in agricultural crops and insects would speed the development of new and improved pesticides and insecticides. The prospects of combining microarray-based crystallization and X-ray analysis may be a way to speed up the otherwise arduous process of structure determination. Microarray companies with expertise in physical biochemistry and structural biology may consider exploring this interesting avenue.

YOUTHFUL EXUBERANCE

Young people are an asset to every microarray company, owing to the energy and enthusiasm they are able to bring to the workplace. What is lacking in experience and technological expertise can be acquired and compensated by a willingness to learn and commitment to handling the large workload that is common in most high-tech settings. Executives and managers should give young people a chance to prove themselves; and if guided properly, youthful exuberance is an asset to any microarray company.

ZOOLOGY

The study of the behavior, reproduction, structure, diet, evolution, development, classification, and distribution of animals is known as ***zoology***. This complex and important discipline encompasses a broad category of subdisciplines including entomology, herpetology, ichthyology, mammalogy, marine biology, oceanography, ornithology, paleontology, and primatology. Zoology interfaces with many other areas of biology, such as conservation, ecology, evolution, sociobiology, and veterinary science. A commercial opportunity exists to offer zoo chips to zoologists, as a means of bringing microarray technology to bear on the full spectrum of questions and challenges that face zoologists on a daily basis. Microarray-based zoological services might be useful in the context of aquariums, marine parks, wildlife refuges, and zoos. Every animal in the biosphere is amenable to microarray analysis as a means of studying physiology, identity, health, heredity, and other important areas, and it makes sense to move microarrays into zoology as quickly as possible.

Zoology. *Study of the behavior, reproduction, structure, diet, evolution, development, classification, and distribution of animals.*

SUMMARY

Since the 1980s, there has been a rapid expansion of commercial science in the United States and abroad, and microarray technology offers many exciting opportunities for those with a flair for entrepreneurial activities. This chapter explores a variety of hot commercial areas. The number of researchers using microarrays is expected to grow to half a million within the next decade, with worldwide sales approaching 10 billion dollars.

SELECTED READING

Baxevanis, A. D., and Ouellette, B. F. F., eds. *Bioinformatics: A Practical Guide to the Analysis of Genes and Proteins,* 2nd ed., Wiley-Interscience, New York, 2001.

Brook, M. A. *Silicon in Organic, Organometallic, and Polymer Chemistry*. Wiley, New York, 2000.

Clarke, A. J., ed. *The Genetic Testing of Childen*. Bios Scientific, Oxford, UK, 1998.

Ferraris, J. D., and Palumbi, S. R., eds. *Molecular Zoology: Advances, Strategies and Protocols*. Wiley, New York, 1996.

Foster, F. H., and Shook, R. L. *Patents, Copyrights, & Trademarks,* 2nd ed. Wiley, New York, 1993.

Higgins, D., and Taylor, W., eds. *Bioinformatics: Sequence, Structure, and Databanks: A Practical Approach*. Oxford University Press, Oxford, UK, 2000.

Innis, M. A., Gelfand, D. H., and Sninsky, J. J., eds. *PCR Applications: Protocols for Functional Genomics*. Academic Press, San Diego, CA, 1999.

Lehninger, A. L., Nelson, D. C., and Cox, M. M. *Principles of Biochemistry,* 3rd ed. Freeman, New York, 2000.

Lerner, J. *Venture Capital and Private Equity: A Casebook*. Wiley New York, 1999.

Lewin, B. Genes VII. Oxford University Press, Oxford, UK, 2000.

Macbeath, G., and Schrieber, S. L. Printing proteins as microarrays for high-throughput function determination. *Science* 289:1760–1763, 2000.

Miller, A. R., and Davis, M. H. *Intellectual Property: Patents, Trademarks, and Copyright,* 3rd ed. West Wadsworth, 2000.

Peppers, D., and Rogers, M. *One to One B2B : Customer Development Strategies for the Business-to-Business World*. Random House, New York, 2001.

Robbins, S. L., Cotran, R. S., Kumar, V., and Collins, T., eds. *Pocket Companion to Robbins Pathologic Basis of Disease,* 6th ed. Saunders, Philadelphia, 1999.

Rousseau, J. *Basic Crystallography*. Wiley, New York, 1998.

Schena, M., ed. *DNA Microarrays: A Practical Approach,* 2nd ed. Oxford University Press, Oxford, UK, 2000.

Schena, M., ed. *Microarray Biochip Technology*. Eaton, Natick, MA, 2000.

Schena, M., Shalon, D., Davis, R. W., and Brown, P. O. Quantitative monitoring of gene expression patterns with a complementary DNA microarray. *Science* 270:467–470, 1995.

Scriver, C. R., Sly, W. S., Childs, B., Beaudet, A., Valle, D., Kinzler, K., and Vogelstein, B. *The Metabolic and Molecular Bases of Inherited Disease,* 8th ed. McGraw-Hill, Hightstown, NJ, 2000.

Singer, M., and Berg, P. *Genes and Genomes.* University Science Books, Herdon, VA, 1991.

Strachan, T., and Read, A. P. *Human Molecular Genetics,* 2nd ed. Wiley, New York, 1999.

Stryer, L. *Biochemistry,* 4th ed. Freeman, New York, 1995.

REVIEW QUESTIONS

1. Describe the trend in microarray publications per year in the scientific literature: (a) increasing, (b) decreasing, or (c) constant?
2. A B2B model involves which type of interaction: (a) university to business, (b) university to university, or (c) business to business?
3. A $10 microarray test is used to determine the genotypes of the entire U.S. population of 280 million. What is the total commercial opportunity afforded by this microarray application: (a) $2.8 billion, (b) $280 million, (c) $28 million, or $28 billion?
4. Which of the following show an increasing trend: (a) Affymetrix sales, (b) projected microarray sales, (c) number of microarray users, and (d) venture capital investment?
5. Which of the following commercial organizations provide automated microarray robots: (a) TeleChem, (b) Ford, (c) Cartesian, (d) GeneMachines, (e) Virtek, and (f) Genetix?
6. Which of the following commercial organizations provide microarray software tools: (a) Silicon Genetics, (b) Rosetta, (c) PREMIER Biosoft, (d) BioDiscovery, (e) Oracle, and (f) Microsoft?
7. Which of the following infectious agents could *not* be detected with a nucleic acid microarray: (a) HIV, (b) prion, (c) pathogenic *E. coli,* or (d) cytomegalovirus?
8. Which of the following organizations might be interested in a microarray-based genotype database: (a) military, (b) law enforcement, (c) insurance companies, (d) drug companies, and (e) health departments?
9. A patient who received a liver transplant regrettably rejects the organ. Microarray-based genotyping of which genomic locus might have prevented the organ rejection: (a) p53, (b) insulin receptor, (c) HLA, or (d) BRCA?
10. The three criteria for patentability include (a) cleaver, (b) useful, (c) revolutionary, (d) novel, (e) sublime, and (f) unobvious?
11. Once a patent issues, other parties are prohibited from doing which of the following with the patented invention: (a) making, (b) thinking about, (c) selling, (d) importing, and (e) using.
12. Which of the following are a positive outcome of commercial technology development: (a) job creation, (b) tax base expansion, (c) international strategic advantage, and (d) cloned aliens?
13. Which of the following could constitute a trade secret: (a) market information, (b) business relationship, (c) microarray cleanroom, (d) method, (e) protocol, (f) reagent, and (g) supplier?
14. Which of the following may be provided by a venture capitalist: (a) operating capital, (b) financial advice, (c) management assistance, and (d) IPO strategy?

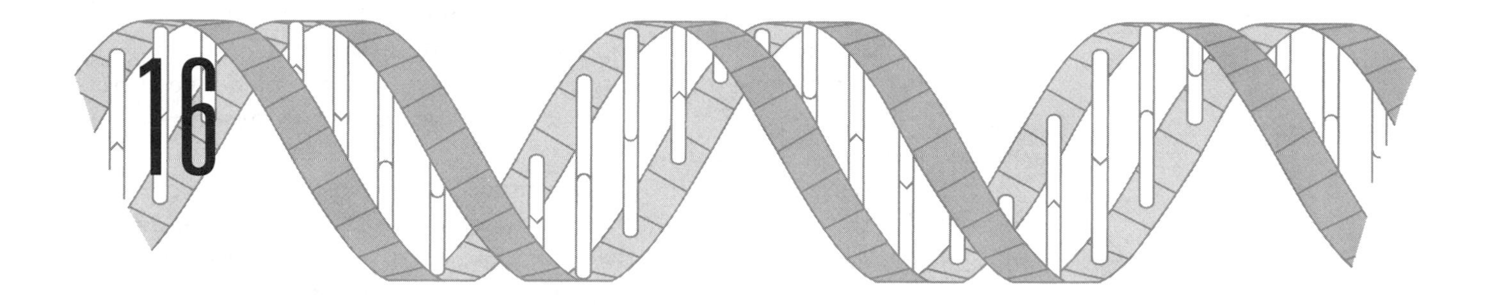

Future Trends: Chips in the Clinics?

You are forgiven for your happiness and your successes only if you generously consent to share them. —*Albert Camus*

Western medicine has a glorious history, but much of medical practice is based on observation and description. With the recent completion of the human genome sequence and the rise of microarray analysis as a formidable research tool, there is an opportunity to revolutionize medical practice in a way that has not occurred since the genesis of the field. Genetic and biochemical data from microarrays, combined with the full spectrum of clinical tools, will enable precise and profound insights into the molecular basis of mental and physical disorders. Chips could be used as a regular component of physical examinations, enabling the physician to view the molecular landscape of a patient by correlating anomalous patterns of gene expression with inactivity, poor diet, drug and alcohol abuse, tobacco use, and a host of mental disorders. The integration of microarray data into clinical medicine would enable physicians to make more informed decisions concerning suggested changes in patient lifestyle, drug treatment, therapy, surgery, and other medical procedures. Doctors and scientists have training and expertise that can be combined to better inform clinical matters, and it makes sense to share these synergistic talents to create an improved health care system.

PHYSICAL EXAMINATION

Physical examination. *Medical practice used to assess the general health of a patient and includes first-tier diagnostics.*

The ***physical examination*** is used widely by the medical community as a means of assessing the general health of a patient and for first-tier diagnostics. The tools of the trade include a stethoscope, reflex hammer, tongue depressor, blood pressure cuff, and tuning fork. A carefully administered physical can tell much about a patient's health, including information on the pulmonary, circulatory, respiratory, auditory, visual, and nervous systems. But the current tools and protocols used in physical examinations are premolecular, providing little or no detailed information about the genetic, biochemical, and physiological health of a patient. Microarray technology could be used to supplement the current spectrum of tests used in regular physical checkups and therefore enhance the quality of the data obtained.

It makes sense to consider a whole genome gene expression test as one component of future physical examinations. Because gene expression and human physiology are coordinated tightly (see Chapter 12), whole genome scans can provide a molecular snapshot of a patient's health, providing useful information concerning his or her mental and physical state. Small blood samples are the likely specimen of choice because blood is easily obtained and because the genes expressed in blood cells can probably be used as sensors that respond to genetic and biochemical perturbations that accompany smoking, alcohol abuse, drug abuse, eating disorders, depression, schizophrenia, and many other conditions of medical concern (see below). Altered patterns of blood cell gene expression, defected as altered mRNA or protein, may also provide early warning signs for breast cancer, prostate cancer, Alzheimer disease, Parkinson disease, epilepsy, dysfunction of the immune and skeletal systems, and metabolic and transport disorders. As a companion to gene expression microarray tests, DNA-based analysis of patient gene sequences would provide important information on inherited and infectious diseases.

It is tempting to speculate a significant role of microarray technology and genomics in the physical examination of the future, and their gradual and expanded role in health care over time (Fig. 16.1). One could envision microarray screens, tests, and diagnostics as a core component of medical information and, together with traditional physicals and other clinical data, could be used by physicians to deliver more comprehensive care, guiding lifestyle changes, drug therapy, counseling, surgery, and other medical treatments. Doctors of the future will likely require a detailed knowledge of genetics and

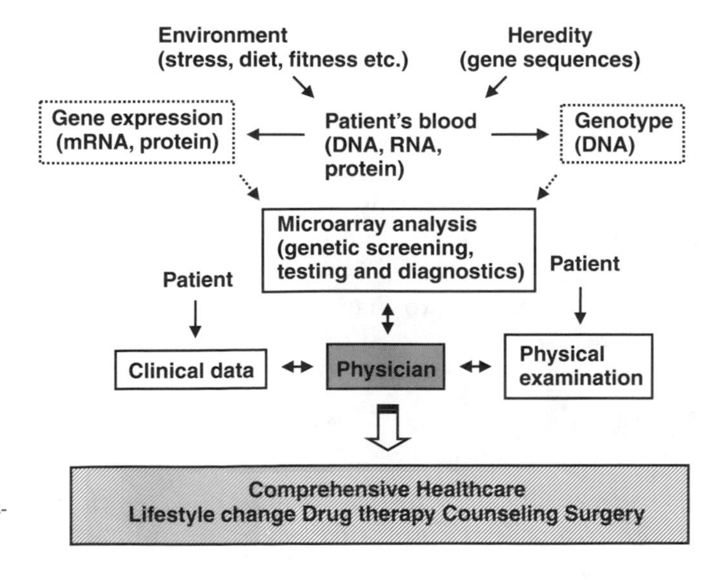

FIGURE 16.1. *Microarrays in health care.*

biochemistry and the associated technologies of genomics and microarray analysis.

PHYSICAL FITNESS

According to the American Heart Association, the positive health benefits of regular physical exercise are striking. Physical fitness achieved through walking, jogging, dancing, cycling, swimming, hiking, weight training, and aerobics has been shown to correlate with a reduced risk of cardiovascular disease and diabetes, lower levels of blood cholesterol and blood pressure, reduced body weight, and improved mental health. Regular exercise also promotes a sense of well-being, confidence, and vigor and should be practiced as a regular component of health maintenance.

Microarray-based analyses are well suited to provide comprehensive genetic and biochemical information on physical fitness, and key questions can be addressed immediately. What are the changes in gene expression that occur during and after moderate physical exercise? Do physically fit individuals possess physiological differences that correlate with the known health benefits of exercise? Do different forms of physical fitness produce different physiological changes? How does physical fitness work in concert with genetic and environmental factors to mediate healthy physiology?

Global gene expression analysis with samples prepared from individuals before and after physical activity are likely to yield stereotyped expression patterns, providing a human gene subset that could be used to manufacture fitness chips. Microarrays containing exercise-response genes could be used to screen the population for physical fitness at the molecular level. Fitness chips could also be used to monitor patients who have recently begun fitness regimens as a means of assessing the effectiveness of their workouts. Understanding the molecular basis of physical fitness with microarrays is likely to shed light on why regular exercise exerts its many positive effects on human health.

SMOKING

The main bioactive substance in tobacco smoke is nicotine, a plant alkaloid that is structurally similar to caffeine. Nicotine dissolves in mucus in the lungs, resulting in its absorption into the bloodstream. Nicotine, like other alkaloids, exerts its psychoactive effect by perturbing the activity of brain chemicals known as ***neurotransmitters,*** elevating the levels of ***acetylcholine*** and ***endorphins*** and producing a stimulating and slightly euphoric sensation. The short half-life of nicotine in the bloodstream requires regular reintroduction of nicotine to maintain the stimulation and euphoria associated with smoking, rendering tobacco use highly addictive.

According to the National Center for Health Statistics, nearly 50 million Americans, or just over 20% of the population, smoke tobacco products on a regular basis, causing 400,000 preventable deaths annually and making smoking the single largest source of preventable mortality in the country. There is enormous medical interest in understanding the full scope of questions pertaining to smoking at the molecular level. What is the spectrum of gene expression changes that take place on exposure to nicotine? Do some patients

Neurotransmitter. *Any of a family of signaling molecules including acetylcholine and endorphin that serve to communicate chemical messages in the brain and peripheral nervous system.*

Acetylcholine. *Neurotransmitter that communicates chemical messages in the brain and peripheral nervous system.*

Endorphin. *Endogenous morphine-like signaling molecule that functions by binding to cell surface receptors.*

respond differently from others to nicotine? Do specific single nucleotide polymorphisms (SNPs) and other minor and detectable differences in gene sequence predispose a certain subset of the population to smoking-related lung cancer, atherosclerosis, and other diseases? Do smoking, poor diet, lack of physical fitness, and stress co-conspire to create a set of biochemical conditions that encourage tobacco use and exacerbate addiction? Do the global gene expression data suggest new drug therapies to combat nicotine dependence?

These and other questions are soon to be answered using the human genome sequence and microarray analysis. Nicotine can be administered to test animals in model studies or samples can be examined from smokers. Microarrays constructed to examine gene variants might provide useful information in the context of genetic screening and testing. RNA and protein microarrays could be used to explore the global changes in gene expression that occur on exposure to nicotine and the other toxic substances present in cigarette smoke. It may eventually be possible, using patient blood samples, to identify patients at high risk for lung cancer, atheriosclerosis, and other diseases. A smoking chip may eventually become a regular component of a physical examination for patients who use tobacco products.

ALCOHOL ABUSE

Alcohol is the most widely used drug in the United States, with two thirds of adults consuming alcoholic beverages on a regular basis. Beer, wine, and other liquors used in moderation contribute in a positive way to the society by increasing relaxation, encouraging socializing, and improving fine dining. But excessive alcohol use and alcoholism inflict a terrible toll on the United States; alcohol is correlated with approximately 50% of all murders, suicides, fatal car accidents, and accidental deaths. According to McGinnis and Foege (1993), the 100,000 deaths in 1990 caused by alcohol abuse make it the third leading cause of preventable mortality behind tobacco (400,000) and poor diet and inactivity (300,000). Prolonged alcohol use can reduce life expectancy by more than 10 years, contribute to liver damage, and cause fetal alcohol syndrome in mothers who drink during pregnancy.

The interest from a public health perspective in understanding the genetic and biochemical bases of alcoholism are considerable. Family, adoption, and twin studies suggest a hereditary component to alcoholism, but the disease is complex and undoubtedly involves the interplay of genes and environment. Interesting questions concerning the mechanism of alcohol action and the basis for alcohol abuse beckon comprehensive molecular analysis. What is the biochemical basis of intoxication and how and why are neurological functions impaired? Why are some individuals predisposed to alcohol abuse and not others? How do genes and the environment co-conspire to cause alcoholism? Do the molecular data suggest new drugs, lifestyle changes, or other therapies to reduce the incidence of alcohol abuse?

Preliminary microarray studies show some encouraging data with respect to defining the changes in gene expression that occur upon alcohol abuse. The expression of > 4000 human genes have been compared for alcoholics and nonalcoholics, and 163 genes with cell cycle, neuronal, and myelin-related functions were identified as differentially expressed in frontal cortex samples (Lewohl et al., 2000, 2001). These and more detailed studies will undoubtedly provide a comprehensive molecular portrait of alcohol abuse, yielding markers for genetic screening, testing, and diagnostics. Microarrays are poised to assist in the clinics by allowing physicians to identify patients with a genetic predisposition to alcoholism, identify current alcohol abusers, and monitor the effectiveness of treatments. Small samples of blood, yielding genomic DNA and mRNA, would provide the sources for gene and gene expression analyses. SNP analysis would probably be most effective for hereditary studies, whereas the gene expression data would provide the phenotypic assessment required to integrate genetic and environmental factors.

DRUG ABUSE

Though the social cost of illicit drug abuse is less than that of alcohol and smoking, illegal drugs—including marijuana, heroin, cocaine, methamphetamine, inhalants, and hallucinogens—extract a considerable toll on the American health care system. Studies by the National Institute on Drug Abuse indicate that millions of Americans use illegal drugs on a regular basis, and approximately 50% of all adults have used illegal substances at least once in their lives. The estimated cost of illegal drug use owing to medical costs, treatment and prevention, crime, lost earnings, and the like exceed 100 billion dollars annually. Intravenous drug use elevates the risk of AIDS, hepatitis, and tuberculosis; more than one third of the diagnosed cases correlate with the use of injected drugs. A more detailed understanding of the factors that lead to substance abuse would have wide sweeping social benefit.

Data from the Holman Group (Canoga Park, CA) reveal that nearly a half million Americans are treated at drug abuse treatment centers each year; of these, men account for $>70\%$ of the patients and $>71\%$ of the patients are between the ages of 21 and 44. The elevated incidence in men suggests possible genetic factors, but environmental components are likely to work together with genes to produce the addictive phenotype.

Cocaine, heroin, LSD, and other illicit drugs are members of a large family of soluble, nitrogen-containing small molecules known as ***alkaloids*** that exert their psychoactive effects by altering biochemical function in the brain. Cocaine and related stimulants appear to work by increasing the length of time that neurotransmitters such as serotonin, dopamine, and norepinephrine reside in the synapse. That illegal drugs modulate neurotransmitter function and signaling cascades suggests that these compounds exert profound and complex changes on gene expression and are thus prime candidates for detailed study by microarray analysis.

Alkaloid. *Any member of a family of small molecules including cocaine and heroin that exert their psychoactive effects by altering biochemical functions in the brain.*

There are myriad interesting basic and clinical questions that deserve immediate attention with respect to the microarray studies. What are the molecular targets (i.e., receptors) of the important illicit drugs? What is the spectrum of changes in gene expression observed when illicit substances are administered to test animals, such as mice and rats, and how do such changes correlate in humans? Is there a genetic and biochemical basis of the addictive phenotype, and if so, which genes are involved? How does the environment contribute to illegal drug abuse at the biochemical level? What is the mechanistic basis of drug tolerance and what changes occur upon prolonged drug use?

Molecular studies using microarrays to study the mechanism of action of illegal drugs have been proposed or performed by several groups (Kuhar et al., 2001; Lichtermann et al., 2000). The cocaine and amphetamine-regulated transcript (CART) corresponds to one gene that appears to be associated with drug abuse (Kuhar et al., 2001). Microarray analysis holds the potential to define dozens and probably hundreds of genes involved in drug use and abuse, some of which may have important diagnostic uses. Patient blood samples as sources of genetic material would allow genomewide expression profiling and genotyping, potentially enabling clinicians to identify drug abusers, chart the course of therapies, and obtain information on patients predisposed to drug abuse.

EATING DISORDERS

Anorexia nervosa. *Serious eating disorder, mostly affecting adolescent and young adult females, characterized by an emaciated appearance, excessively low body weight, and malnutrition caused by intentional starvation.*

Bulimia nervosa. *Serious eating disorder that mostly afflicts adolescent and young adult females in which sufferers consume large quantities of food and then rid themselves of the extra calories by vomiting, the use of laxatives, and other purging methods.*

The National Institutes of Health report that up to 5 million people in the United States, mostly adolescents and young adult females, are affected by serious eating disorders. The two most common are ***anorexia nervosa*** and ***bulimia nervosa.*** Young women afflicted with anorexia starve themselves intentionally and without medical treatment, some may actually starve to death. Individuals with anorexia have exhibit an emaciated appearance, excessively low body weight, and suffer from malnutrition, the combined effects of which can elevate the risk of infection, cardiac arrest, and other serious medical complications. Bulimia sufferers consume large quantities of food and rid themselves of the extra calories by vomiting and using laxatives and other purging methods. Bulimia patients often maintain a normal body weight, but extended cycles of binging and purging can lead to tooth decay and other forms of dental damage and well as acid-induced ulceration of the esophagus, increasing the chances of internal bleeding.

Nearly 90% of all anorexia and bulimia patients are young females, and eating disorders run in families, both of which suggest possible genetic components to these serious conditions. Like alcoholism, drug abuse, and many other behavioral disorders, however, environmental conditions appear to play a central role in the onset and progression of eating disorder diseases. Studies show that many women with anorexia and bulimia live in overbearing type A household environments, where a tremendous pressure on being slim appears to promote eating disorders. Parents should be sensitive to the needs of their daughters and should avoid under all circumstances making insensitive comments about weight and appearance, which may result in the development of life-threatening eating disorders.

Genomics and microarray analysis hold the promise of improving our understanding of anorexia and bulimia by answering some basic questions about these diseases. What are the global physiological changes that occur upon severe dieting or starvation? How do anorexia and bulimia patients and healthy women compare in terms of their gene expression patterns? Are there genes that may serve provide early warning signs of eating disorders and can these be incorporated into a microarray test? Are there inherited traits that predispose young women to anorexia and bulimia? Do the biochemical changes that occur during the disease progression suggest new therapies in the form of drugs or changes in lifestyle and environment?

Microarray technology and the Human Genome Project have yet to be brought to bear on eating disorders, but a major push is expected in this direction in the near future. Careful clinical studies of anorexia and bulimia patients may identify diagnostic genes that could serve to warn physicians of potential eating disorders before they become serious. Global expression monitoring with genome chips as part of a regular physical examination might be means of surveying the physiological landscape to cull out the molecular culprits responsible for eating disorders.

DEPRESSION

According to the National Institute of Mental Health, clinical depression effects >10 million American adults annually, with symptoms that include whole body changes in feelings, thinking, and behavior as well as loss of appetite and energy, fatigue, headaches, stomachaches, and thoughts of suicide in severe cases. Women are twice as likely to suffer from depression than men, and depression-induced suicide is the third leading cause of death among teenagers. Major depression, double depression, seasonal affective disorder (SAD), and bipolar disorder are some of the major clinically recognized classes. Bipolar disorder, which affects more than 1 million adults each year, is characterized by dramatic high to low mood swings, with manias and depression defining the extremes. Imbalances in the levels and transmission of neurotransmitters are believed to play a role in the depression illnesses, but a precise molecular picture has not been forthcoming. Bipolar disorder and other forms of depression run in families, suggesting a hereditary basis for the diseases, but the culprit genes have yet to be identified in most cases.

Psychiatry uses a large number of drug therapies to treat depressive disorders including tricyclic antidepressants, serotonin selective reuptake inhibitors, monoamine oxidase inhibitors, and lithium. Antidepressants are extremely effective medications, but toxic side effects, efficacy in some patients but not others, and poorly understood mechanisms of action dog current treatments. Pressing questions concerning all forms of depression remain and could be addressed by advances in genomics and microarray technology. Are there stereotyped gene expression patterns that accompany the major forms of clinical depression? Which genes confer a genetic predisposition to clinical depression? What are the molecular targets (i.e., receptors) for antidepressants? What are the comprehensive changes in gene expression that occur on administration of the different classes of antidepressants? Can current antidepressant drugs be modified to avoid toxic side effects? What is the biochemical basis of nonresponsiveness in certain patients?

Microarray analysis in model systems and humans should allow the identification of a complete set of human genes involved in depression, including neurotransmitters, receptors, and G proteins. Genomewide expression profiling should allow diagnosis of the illness and may provide a fingerprint for each form of depression. Chip-based analyses might also allow antidepressant responders and nonresponders to be distinguished before drug administration. Genetic screening and testing at the DNA level may allow at risk patients to be identified before disease onset.

SCHIZOPHRENIA

A mental illness characterized by delusions, disorganized thinking, abnormal speech, hallucinations, catatonia, and other behavioral symptoms is known as ***schizophrenia.*** Some patients suffering from schizophrenia may believe that spies or aliens are controlling their lives or that bugs are crawling on them,

Schizophrenia. *Mental illness characterized by delusions, disorganized thinking, abnormal speech, hallucinations, catatonia, and other behavioral symptoms.*

while others have delusions of grandeur, equating themselves with major political or religious leaders. Rigid posture, disjointed speech, poor hygiene, lethargy, and other negative forms of movement and behavior accompany the disease. Schizophrenia affects nearly 1% of the world's population, cutting across racial, cultural, socioeconomic, and gender boundaries. According to the Society for Neuroscience (Washington, DC), schizophrenia takes a serious economic toll on American society, resulting in >30 billion dollars lost annually due to hospitalization, treatment, loss of productivity, and the like. Families, friends, colleagues, and co-workers also suffer at the hands of this serious mental illness.

Schizophrenia runs in families, suggesting a genetic component to the illness and the existence of one or more disease genes. Though the biochemical basis for schizophrenia is poorly understood, several lines of evidence point to excessive dopamine transmission as one cause of the illness, including the observation that dopamine blockers such as clozapine are extremely effective is treating serious forms of the disease. Myriad additional ***antipsychotic*** medications are used to treat schizophrenia, including fluphenazine, chlorprothixene, chlorpromazine, molindone, loxapine, haloperidol, olanzapine, and risperidone. The existing medications are quite effective in treating the disease but show variable efficacy in the population and often elicit serious side effects, including tremors, cardiac abnormalities, seizures, blood cell depletion, and abnormal movements of the face and mouth.

Antipsychotic. *Any of a family of medications used to treat schizophrenia, including fluphenazine, chlorprothixene, chlorpromazine, molindone, loxapine, haloperidol, olanzapine, and risperidone.*

The availability of a complete sequence of the human genome, coupled with microarray analysis, is likely to render psychiatry a DNA science in the near future. There are myriad interesting questions that beckon comprehensive molecular analysis. What are the physiological differences between healthy individuals and schizophrenic patients at the gene expression level? What is the biochemical basis of schizophrenia? What are the biochemical receptors of the antipsychotic drugs? What is the genetic and biochemical basis of antipsychotic drug responsiveness and nonresponsiveness? Why do certain drugs elicit side effects and can the microarray data inform and guide the development of new and improved medications?

Once a comprehensive genetic and biochemical understanding of schizophrenia is obtained, microarrays containing the diagnostic genes can be constructed and used in a clinical setting to screen and test for this form of mental illness at the DNA, RNA, and protein levels. Microarray data, combined with other forms of clinical information, will assist physicians, nurses, and other health care officials in prescribing counseling, drug treatment, and other forms of therapy to combat schizophrenia.

SUMMARY

Human beings are gene machines, responsive to myriad genetic and environmental signals, and these signals are integrated by each person to produce resultant phenotypes that manifest as mental and physical wellness, normal and aberrant behavior, disease predisposition, longevity, and so forth. Medical practice has succeeded in identifying and curing disease using a host of clinical procedures, but medicine is lacking in the sense that many illnesses, therapies, and drugs are poorly understood at the genomic level.

The use of microarray technology in a clinical setting holds the promise of providing detailed molecular information to the physician, enabling more informed decision making and better health care. Chips could be used to profile patients for predispositions to drug and alcohol abuse, eating disorders, depression, and schizophrenia. The biochemical and physiological changes that accompany physical exercise, poor diet, substance abuse, antidepressant and antipsychotic medications, and many other environmental and chemical stimuli that alter body chemistry are probably detectable by microarray. As microarray analysis is used more extensively in clinical research and genetic screening, testing and diagnostics, it makes sense to consider whether future visits to the doctor's office might include one or more microarray tests as components of the physical examination.

SELECTED READING

Baxevanis, A. D., and Ouellette, B. F. F., eds. *Bioinformatics: A Practical Guide to the Analysis of Genes and Proteins,* 2nd ed. Wiley-Interscience, New York, 2001.

Bickley, L. S., and Hoekelman, R. A. *Bates' Guide to Physical Examination and History Taking,* 7th ed. Lippincott Williams & Wilkins, Philadelphia, 1998.

Brown, P. O., and Botstein, D. Exploring the new world of the genome with DNA microarrays. *Nat Genet* 21 (1 Suppl):33–37, 1999.

Clarke, A. J., ed. *The Genetic Testing of Childen.* Bios Scientific, Oxford, UK, 1998.

Galanter, M., and Kleber, H. D. *The American Psychiatric Press Textbook of Substance Abuse,* 2nd ed. American Psychiatric Press, Washington, DC, 1999.

Gordon, R. A. A. *Anorexia and Bulimia: Anatomy of a Social Epidemic.* Blackwell, Oxford, UK, 1992.

Innis, M. A., Gelfand, D. H., and Sninsky, J. J., eds. *PCR Applications: Protocols for Functional Genomics.* Academic Press, San Diego, CA, 1999.

Khoury, M. J., Burke, W., and Thomson, E., eds. *Genetics and Public Health in the 21st Century: Using Genetic Information to Improve Health and Prevent Disease.* Oxford University Press, Oxford, UK, 2000.

Kuhar, M. J., Joyce, A., and Dominguez, G. Genes in drug abuse. *Drug Alcohol Depend* 62:157–162, 2001.

Lehninger, A. L., Nelson, D. C., Cox, M. M. *Principles of Biochemistry,* 3rd ed. Freeman, New York, 2000.

Lewin, B. *Genes VII.* Oxford University Press, Oxford, UK, 2000.

Lewohl, J. M., Dodd, P. R., Mayfield, R. D., and Harris, R. A. Application of DNA microarrays to study human alcoholism. *J Biomed Sci* 8:28–36, 2001.

Lewohl, J. M., Wang, L., Miles, M. F., Zhang, L., Dodd, P. R., and Harris, R. A. Gene expression in human alcoholism: microarray analysis of frontal cortex. *Alcohol Clin Exp Res* 24:1873–1882, 2000.

Lichtermann, D., Franke, P., Maier, W., and Rao, M. L. Pharmacogenomics and addiction to opiates. *Eur J Pharmacol* 410:269–279, 2000.

McGinnis, J., and Foege, W. Actual causes of death in the United States. *JAMA* 270:2208–, 1993.

Robbins, S. L., Cotran, R. S., Kumar, V., and Collins, T., eds. *Pocket Companion to Robbins Pathologic Basis of Disease,* 6th ed. Saunders, Philadelphia, 1999.

Schena, M., ed. *DNA Microarrays: A Practical Approach,* 2nd ed. Oxford University Press, Oxford, UK, 2000.

Schena, M., ed. *Microarray Biochip Technology.* Eaton, Natick, MA, 2000.

Scriver, C. R., Sly, W. S., Childs, B., Beaudet, A., Valle, D., Kinzler, K., and Vogelstein, B. *The Metabolic and Molecular Bases of Inherited Disease,* 8th ed. McGraw-Hill, Hightstown, NJ, 2000.

Seidel, H. M., Ball, J. W., Dains, J. E., and Benedict, G. W. *Mosby's Physical Examination Handbook,* 4th ed. Mosby, St Louis, MO, 1998.

Singer, M., and Berg, P. *Genes and Genomes.* University Science Books, Herdon, VA, 1991.

Strachan, T., and Read, A. P. *Human Molecular Genetics,* 2nd ed. Wiley, New York, 1999.

Streitweiser, A., and Heathcock, C. H. *Introduction to Organic Chemistry,* 2nd ed. Macmillan, New York, 1981.

Stryer, L. *Biochemistry,* 4th ed. Freeman, New York, 1995.

REVIEW QUESTIONS

1. A whole-genome gene expression profile for a patient might provide important information regarding which of the following: (a) schizophrenia, (b) depression, (c) anorexia, (d) drug abuse, and (e) smoking?

2. A traditional physical examination provides specific information regarding which of the following: (a) gene sequences, (b) transcript levels, (c) protein structure, and (d) hormone levels?
3. Identifying carriers of human genetic diseases would be expected to improve health care by: (a) reducing health-care costs, (b) expediting family planning, (c) reducing insurance costs, and (d) improving mental health?

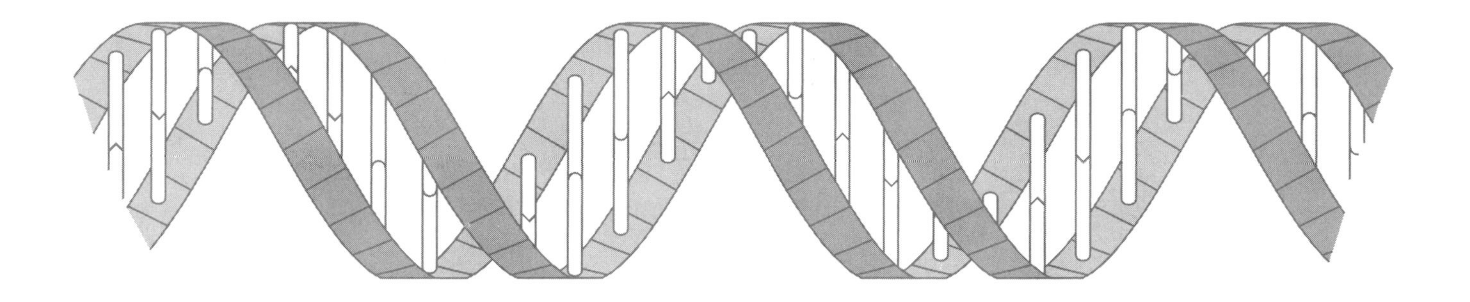

Appendix A
Prefixes

Prefix (Abbreviation)	Factor	Example
tera (T)	multiply by 10^{12}	terabyte (Tb) of microarray data
giga (G)	multiply by 10^{9}	gigabyte (Gb) of microarray data
mega (M)	multiply by 10^{6}	megabase (Mb) of human DNA
kilo (k)	multiply by 10^{3}	kilogram (kg) of instrument weight
centi (c)	multiply by 10^{-2}	centimeter (cm) of microarray substrate
milli (m)	multiply by 10^{-3}	millimeter (mm) of microarray subgrid
micro (μ)	multiply by 10^{-6}	micrometer (μm) of microarray spot diameter
nano (n)	multiply by 10^{-9}	nanometer (nm) of light wavelength
pico (p)	multiply by 10^{-12}	picoliter (pl) of deposited DNA sample
fempto (f)	multiply by 10^{-15}	femptomoles (fmole) of oligonucleotide
atto (a)	multiply by 10^{-18}	attomoles (amole) of oligonucleotide

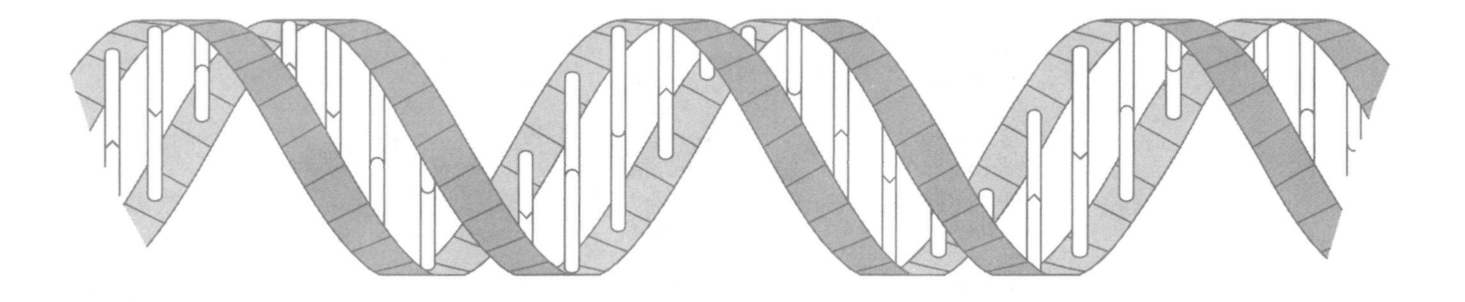

Appendix B
Unit Abbreviations

Unit	Type	Abbreviation	Symbol
ampere	electricity	a	A
angstrom	length	ang	Å
atomic mass unit	atomic mass	amu	u
bar	pressure	bar	bar
base pair	DNA length	bp	bp
Celsius	temperature	cel	°C
centiMorgan	recombination distance	cM	cM
degree	temperature	deg	°
Fahrenheit	temperature	degF	°F
gigabase	DNA length	Gb	Gb
gigabyte	computer file size	GB	GB
gram	mass	g	g
Gray	energy	Gy	Gy
Hertz	frequency	hz	Hz
hour	time	hr	h
kilobase	DNA length	kb	kb

(*Continued*)

Unit	Type	Abbreviation	Symbol
kilobyte	computer file size	KB	KB
kilodalton	atomic mass	kDa	kDa
kilogram	mass	kg	kg
liter	volume	L	L
megabase	DNA length	Mb	Mb
megabyte	computer file size	MB	MB
meter	length	m	m
microgram	mass	μg	μg
microliter	volume	μl	μl
micromolar	concentration	μM	μM
milliliter	volume	ml	ml
milliwatt	power	mW	mW
minute	time	min	min
molar	concentration	M	M
mole	constant	mol	mol
molecular weight	mass	M.W.	M.W.
parts per million	concentration	ppm	ppm
percent (chemistry)	concentration	%	%
pH	concentration	pH	pH
second	time	sec	s
volt	electricity	v	V
watt	energy	w	W

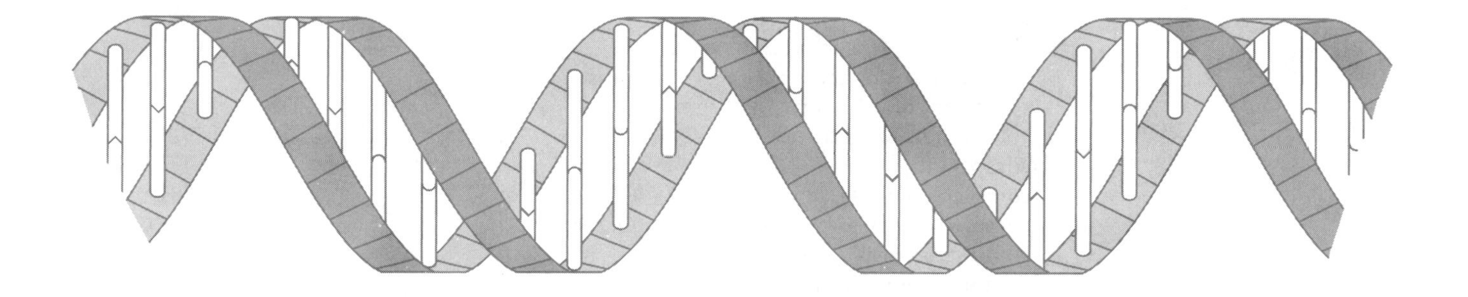

Appendix C
Technical Abbreviations

Technical Term	Abbreviation
acquired immunodeficiency syndrome	AIDS
acute renal failure	ARF
adenosine triphospate	ATP
α-fetoprotein	AFP
amplified primer extension	APEX
amplified RNA	aRNA
amyotrophic lateral sclerosis	ALS
analog-to-digital	A/D
aspartylglucosaminuria	AGU
atomic force microscopy	AFM
bacterial artificial chromosome	BAC
basic local alignment search tool	BLAST
because it's time network	BITNET
binary digit map	bit map
bovine serum albumin	BSA
breast and ovarian cancer susceptibility gene	*BRCA*

(*Continued*)

Technical Term	Abbreviation
business-to-business	B2B
capillary electrophoresis	CE
center-to-center	CTC
charge-coupled device	CCD
chloramphenicol acetyl transferase	CAT
cocaine and amphetamine-regulated transcript	CART
coefficient of variation	CV
color rendition index	CRI
compact disk	CD
complementary DNA	cDNA
complementary metal oxide semiconductor	CMOS
computer-assisted design	CAD
computer numerical control	CNC
congenital adrenal hyperplasia	CAH
congenital hypothyroidism	CH
controlled pore glass	CPG
cyanine	Cy
cystic fibrosis	CF
cystic fibrosis transmembrane conductance regulator	CFTR
cytidine triphosphate	CTP
deoxyadenosine triphosphate	dATP
deoxycytidine triphosphate	dCTP
deoxyguanosine triphosphate	dGTP
deoxythymidine triphosphate	dTTP
deoxynucleotide triphosphate	dNTP
deoxyribonucleic acid	DNA
digital light processing	DLP
digital micromirror device	DMD
dimethoxytrityl	DMT
dimethylformamide	DMF
dimethylsulfoxide	DMSO
dinitrophenol	DNP
diode-pumped solid-state	DPSS
dithiothreitol	DTT
domain name system	DNS
electrical discharge machining	EDM
electronic mail	e-mail
expressed sequence tag	EST
file transfer protocol	FTP
fluorescein isothiocyanate	FITC
Full-Length Expressed Gene	FLEX
galactosemia	GAL
genetically modified organism	GMO
glucocorticoid receptor	GR
glutaric acidemia	GA1
glutathione-*S*-transferase	GST
graphical interchange format	GIF
graphical user interface	GUI
green fluorescent protein	GFP
guanosine triphosphate	GTP
heat-shock protein	hsp
helium and neon gas	HeNe
hemagglutinin	HA
hemoglobin	Hb
high-efficiency particulate air	HEPA
high-pressure liquid chromatography	HPLC

(*Continued*)

homozygous for hemoglobin S	HbSS
horseradish peroxidase	HRP
human immunodeficiency virus	HIV
human leukocyte antigen	HLA
immunoglobulin	Ig
initial public offering	IPO
interleukin 13	*IL13*
intermediate range order	IRO
Internet protocol	IP
Internet service provider	ISP
joint photographic experts group	JPEG
laser capture microdissection	LCM
light amplification by stimulated emission of radiation	laser
light-emitting diode	LED
local area network	LAN
long range order	LRO
look-up table	LUT
lysergic acid diethylamide	LSD
major histocompatibility complex	MHC
maple syrup urine disease	MSUD
maskless array synthesizer	MAS
mass spectrometry	MS
medium-chain acyl-CoA dehydrogenase	MCAD
melting temperature	T_m
messenger RNA	mRNA
metal oxide semiconductor	MOS
methylmalonic acidemia	MMA
methylnitropiperonyloxycarbonyl	MeNPOC
monoclonal antibody	Mab
multiple acyl-CoAdehydrogenase deficiency	MAAD
Next Generation Screening	NGS
nucleotide	nt
numerical aperture	NA
oligonucleotide	oligo
open reading frame	ORF
operating system	OS
parallel gene analysis	PGA
perfect match	PM
personal computer	PC
phage artificial chromosome	PAC
phenylketonuria	PKU
photomultiplier tube	PMT
phycoerythrin	PE
pin and ring	PAR
polyacrylamide gel electrophoresis	PAGE
polyadenylation	polyA
polycyclic aromatic hydrocarbon	PAH
polymerase chain reaction	PCR
positron emission tomography	PET
post office protocol	POP
principle component analysis	PCA
Pro/ENGINEER	ProE
propionic acidemia	PA
quality control	QC
quantum efficiency	QE
resonance light scattering	RLS
restriction fragment length polymorphism	RFLP
reverse transcriptase	RT

(*Continued*)

Technical Term	Abbreviation
ribonucleic acid	RNA
ribonucleotide triphosphate	rNTP
ribosomal RNA	rRNA
scanning electron micrograph	SEM
seasonal affective disorder	SAD
self-organizing map	SOM
sequencing by hybridization	SBH
serial analysis of gene expression	SAGE
serotonin selective reuptake inhibitor	SSRI
severe combined immunodeficiency	SCID
short range order	SRO
sickle-cell hemoglobin C	HbSC
signal-to-noise ratio	SNR
simple mail transfer protocol	SMTP
single mismatch	SM
single nucleotide polymorphism	SNP
small-angle neutron scattering	SANS
sodium dodecylsulfate	SDS
Spectral Karyotyping	SKY
standard operating procedure	SOP
strawberry alcohol acetyltransferase	SAAT
structured query language	SQL
tagged image file format	TIFF
tetramethylammonium chloride	TMAC
transfer RNA	tRNA
transmission control protocol	TCP
tyramide signal amplification	TSA
ultra-high-efficiency particulate air	ULPA
uridine triphosphate	UTP
very large scale immobilized polymer synthesis	VLSIPS
World Wide Web	WWW
yeast artificial chromosome	YAC
yttrium argon garnet	YAG

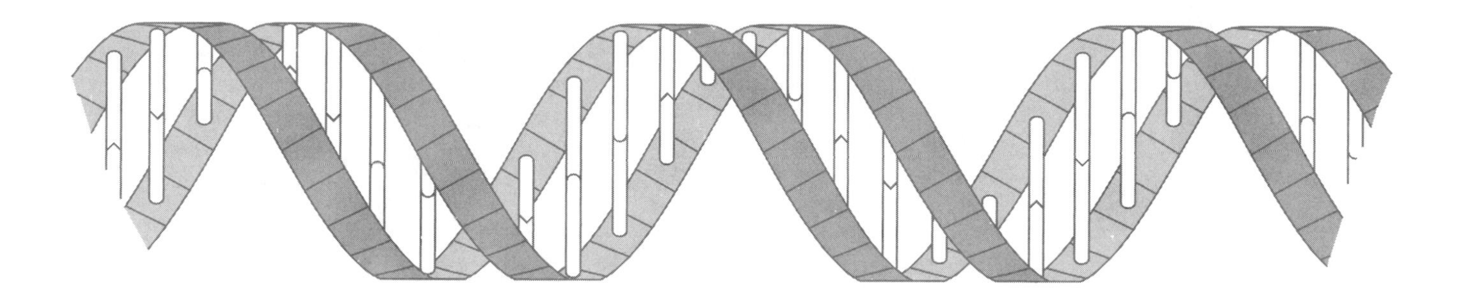

Appendix D Conversions

Number of molecules: 1.0 mole = 10^3 mmole = 10^6 μmole = 10^9 nmole = 10^{12} pmole = 10^{15} fmole = 10^{18} amole
Concentration: 1.0 molar (M) = 10^3 mM = 10^6 μM = 10^9 nM = 10^{12} pM = 10^{15} fM = 10^{18} aM
Length: 1.0 meter (m) = 10^2 cm = 10^3 mm = 10^6 μm = 10^9 nm = 10^{10} Å
Weight: 1.0 kilogram (kg) = 10^3 mg = 10^6 μg = 10^9 ng = 10^{12} pg = 10^{15} fg
Volume: 1.0 liter (L) = 10^3 ml = 10^6 μl = 10^9 nl = 10^{12} pl

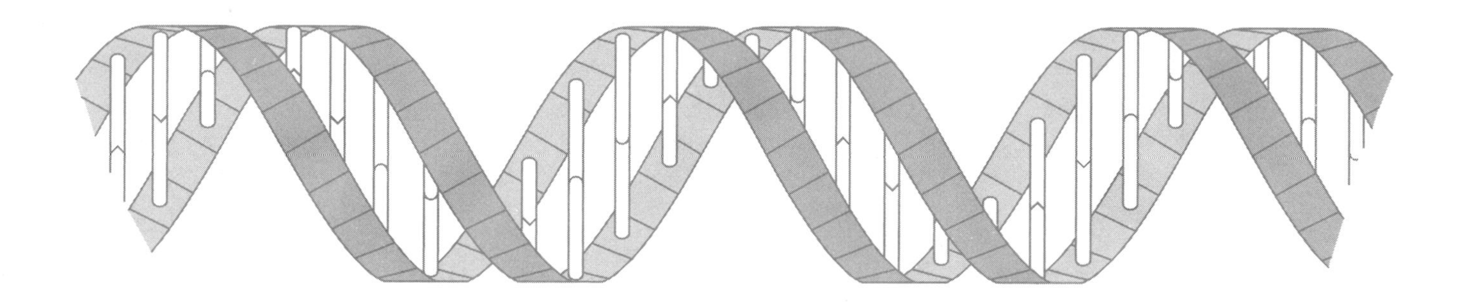

Appendix E
Nucleic Acid Molecular Weight Calculations

Oligonucleotides (average for a mixed sequence): 1 nucleotide = 330 g/mole
Messenger RNA (average for a mixed sequence): 1 nucleotide = 330 g/mole
Single-stranded DNA (average for a mixed sequence): 1 nucleotide = 330 g/mole
Double-stranded DNA (average for a mixed sequence): 1 base pair = 660 g/mole
Average molecular weight (oligonucleotides , messenger RNA, single-stranded DNA): M.W. = 330 g/mole × number of bases in sequence

Calculate the molecular weight of a 55-nt oligonucleotide: M.W. = 330 g/mole × 55 nt = 18,150 g/molc

Calculate the molecular weight of a 1.0 kb mRNA molecule: M.W. = 330 g/mole × 1,000 nt = 330,000 g/mole

Average molecular weight (double-stranded DNA): M.W. = 660 g/mole × number of base pairs in sequence

Calculate the molecular weight of a 12,557 bp gene: M.W. = 660 g/mole × 12,557 bp = 8,287,620 g/mole

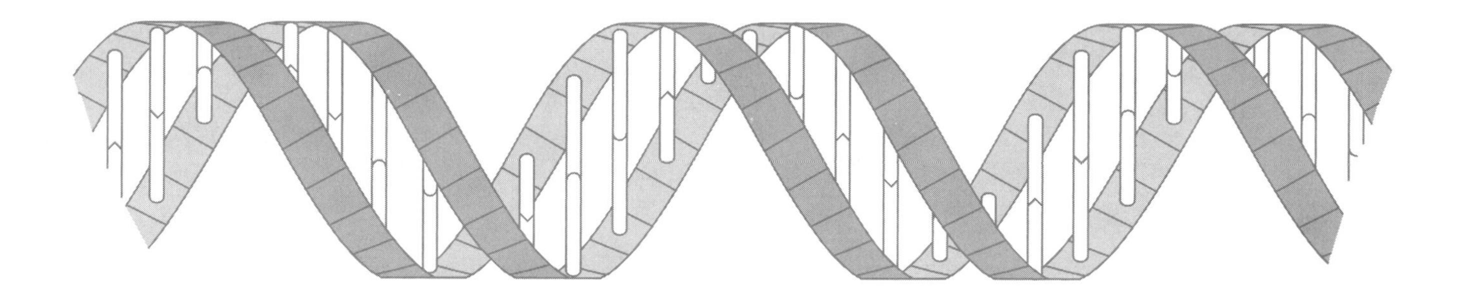

Appendix F
Microarray Spot Area Calculations

Microarray spot radius = 1/2 × spot diameter
Microarray spot area = πr^2 (r = spot radius)
  90 μm diameter spot
  $r = 45\ \mu m$
  area = $\pi r^2 = 3.1 \times (45\ \mu m)^2 = 6278\ \mu m^2$

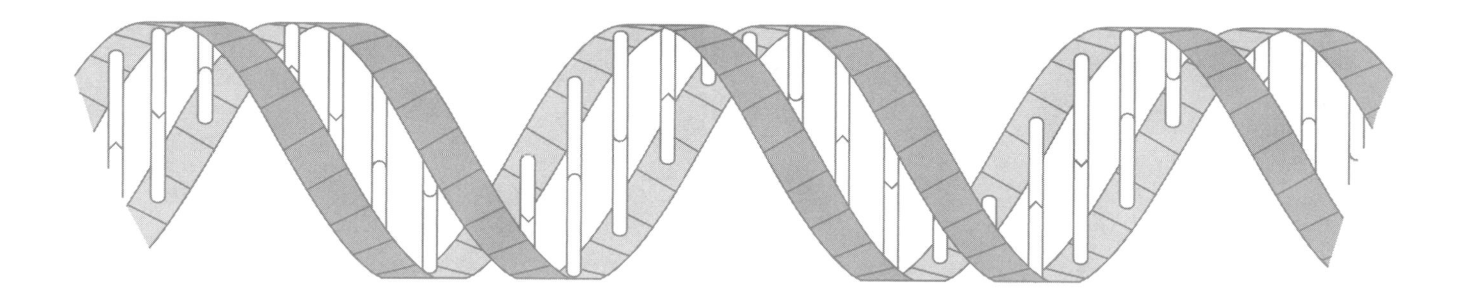

Appendix G
Microarray Fluor Density Calculations

Number of fluors/μm^2 (example for a stock solution of Cy3-dCTP)

Amersham stock dye concentration = 1.0 mM = 1.0×10^{-3} M = 1.0×10^{-3} moles/L

Stock dye = 6.0×10^{23} molecules/mole $\times$ 1.0×10^{-3} moles/L = 6×10^{20} molecules/L

The stock dye was diluted 100,000-fold for printing

Each molecule of dye has one fluor, so the diluted dye = 6×10^{15} fluors/L

The microarray spot sample volume = 300 pL = 3×10^{-10} L

Each microarray spot contains 3×10^{-10} L $\times$ 6×10^{15} fluors/L = 1.8×10^{6} fluors

Area of the 90 μm diameter microarray spot = 6,278 μm^2

Microarray dye concentration = 1.8×10^{6} fluors $\div$ 6,278 μm^2 = 286 $\approx$ 300 fluors/μm^2

A 100,000-fold diluted stock of Cy3-dCTP gives $\approx$ 300 Cy3 fluors/μm^2

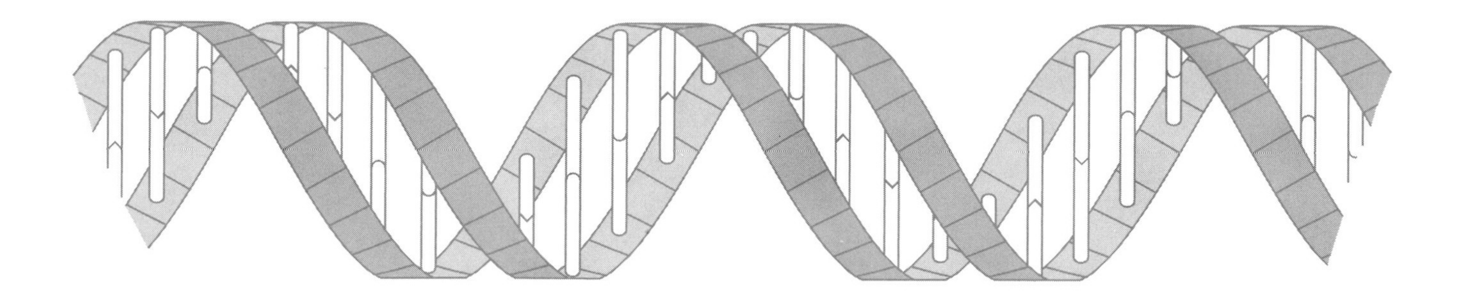

Appendix H
Microarray Target Density Calculations

Target density = number of target molecules/μm^2 of microarray substrate
  $= 1.6 \times 10^9$molecules $\div$ 6,278 $\mu m^2 = 2.6 \times 10^5$ molecules/μm^2
  $= 2.6 \times 10^5$ molecules/$\mu m^2 \div 10^8 Å^2/\mu m^2 = 2.6 \times 10^{-3}$ molecules/$Å^2$

A target solution of 30 μM = 30 pmoles/μl = 30×10^{-12} moles/μl = 3×10^{-11} moles/μl

A target concentration of 3×10^{-11} moles/μl $\times$ 6.02×10^{23} molecules/mole = 1.8×10^{13} molecules/μl

A microarray spot droplet = 1.8×10^{13} molecules/μl $\times$ 3×10^{-4} μl = 5.4×10^9 molecules

The substrate binding efficiency = 30% = $0.3 \times 5.4 \times 10^9$ molecules = 1.6×10^9 molecules

The number of targets in 400 $Å^2 = 2.6 \times 10^{-3}$ molecules/$Å^2 \times$ 400 $Å^2$= 1 molecule

The number of targets in 20 Å of linear microarray substrate = $\sqrt{1} = 1$ molecule

For a 100 μM solution
  the target density = $1 \times 100/30 = 3.3$ molecules/400 $Å^2$
  there are $\sqrt{3.3} = 1.8$ molecules in 20 Å of linear microarray substate
  there is 1 target molecule every 10.8 Å of linear microarray substate

For a 3 μM solution

the target density = $1 \times 3/30 = 0.1$ molecule/400 Å^2

there are $\sqrt{0.1} = 0.32$ molecules in 20 Å of linear microarray substate

there is 1 target molecule every 63 Å of linear microarray substate

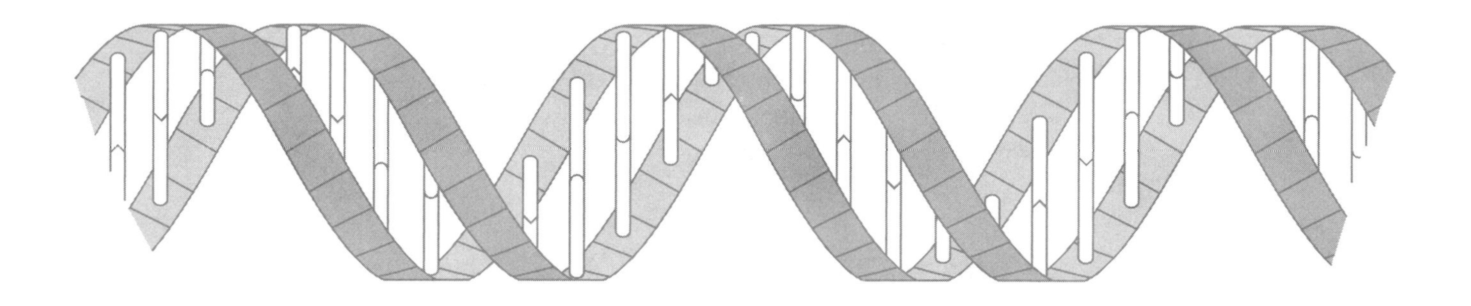

Appendix I
Microarray Web Sites

ACADEMIC SITES

abs.cit.nih.gov/fscan/
afgc.stanford.edu/afgc_html/site2.htm
array.sdsc.edu/
arrays.rockefeller.edu/xenopus/
bioinf.man.ac.uk/microarray/
bioinformatics.duke.edu/camda/
cmgm.stanford.edu/pbrown/mguide.index
discover.nci.nih.gov/nature2000/
ee.tamu.edu/~camdi/subpages/cdna.html
genome-www.stanford.edu/nci60/
info.med.yale.edu/microarray/
mach1.nci.nih.gov/
oz.berkeley.edu/users/terry/zarray/Html/
quantgen.med.yale.edu/
workhorse.stanford.edu/utils/Primer_search.html
www.bio.davidson.edu/biology/courses/genomics/chip/chip.html
www.cse.ucsc.edu/research/compbio/genex/genex.html

www.ebi.ac.uk/microarray/
www.epa.gov/nheerl/epamac/links.htm
www.mcb.arizona.edu/wardlab/microarray.html
www.mged.org/
www.microarrays.org/
www.nhgri.nih.gov/DIR/LCG/15K/HTML/
www.nigms.nih.gov/funding/microarray.html
www.sghms.ac.uk/depts/medmicro/bugs/
www.tigr.org/tdb/microarray/
www-stat.stanford.edu/~tibs/lab/

COMMERCIAL SITES

arctur.com/
arrayit.com/
ngs-arrayit.com/
www.aat-array.com/
www.affymetrix.com/
www.axon.com/GN_Genomics.html
biodiscovery.com/
www.cartesiantech.com/
www.corning.com/cmt/
www.d-trends.com/
www.genemachines.com
www.incyte.com/
www.mediacy.com/arraypro.htm
www.microarrays.com/
www.perkinelmer.com/
www.premierbiosoft.com/dnamicroarray/dnamicroarray.html
www.scanalytics.com/product/hts/microarray.html
stratagene.com/
www.bio-rad.com/

MICROARRAY CORE FACILITIES

www.emory.edu/WHSC/YERKES/VRC/genomicsstats.html
sgio2.biotec.psu.edu/
www.array.saci.org/
www.bcm.tmc.edu/microarray/
www.ccgpm.org
www.uhnres.utoronto.ca/services/microarray/
www.umich.edu/~caparray/

OTHER

linkage.rockefeller.edu/wli/microarray/
www.biologie.ens.fr/en/genetiqu/puces/links.html
www.bsi.vt.edu/ralscher/gridit/
www.gene-chips.com/
www.sciencemag.org/feature/e-market/benchtop/micro.shl

References

Aach, J., Rindone, W., and Church, G. M. Systematic management and analysis of yeast gene expression data. *Genome Res* 10:431–445, 2000.

Abdellatif, M. Leading the way using microarray: A more comprehensive approach for discovery of gene expression patterns. *Circ Res* 86:919–920, 2000.

Achary, M. P., Jaggernauth, W., Gross, E., Alfieri, A., Klinger, H. P., and Vikram, B. Cell lines from the same cervical carcinoma but with different radiosensitivities exhibit different cDNA microarray patterns of gene expression. *Cytogenet Cell Genet* 91:39–43, 2000.

Adler, K., Broadbent, J., Garlick, R., Joseph, R., Khimani, A., Mikulskis, A., Rapiejko, P., and Killian, J. MICROMAX^{TM}: A highly sensitive system for differential gene expression on microarrays. In: M. Schena, ed. *Microarray Biochip Technology,* M. Schena (editor), Eaton Publishing, Natick, MA, 2000:221–230.

Afshari, C. A., Nuwaysir, E. F., and Barrett, J. C. Application of complementary DNA microarray technology to carcinogen identification, toxicology, and drug safety evaluation. *Cancer Res* 59:4759–4760, 1999.

Aharoni, A., Keizer, L. C., Bouwmeester, H. J., Sun, Z., Alvarez-Huerta, M., Verhoeven, H. A., Blaas, J., van Houwelingen, A. M., and co-workers. Identification of the SAAT gene involved in strawberry flavor biogenesis by use of DNA microarrays. *Plant Cell* 12:647–662, 2000.

Ahram, M., Best, C. J., Flaig, M. J., Gillespie, J. W., Leiva, I. M., Chuaqui, R. F., Zhou, G., Shu, H., and co-workers. Proteomic analysis of human prostate cancer. Mol Carcinog 33:9–15, 2002.

Aitman, T. J. DNA microarrays in medical practice. *Brit Med J* 323:611–615, 2001.

Aitman, T. J., Glazier, A. M., Wallace, C. A., Cooper, L. D., Norsworthy, P. J., Wahid, F. N., Al-Majali, K. M., Trembling, P. M., and co-workers. Identification of Cd36 (Fat) as an insulin-resistance gene causing defective fatty acid and glucose metabolism in hypertensive rats. *Nat Genet* 21:76–83, 1999.

Akagi, K., Kanai, M., Saya, H., Kozu, T., and Berns, A. A novel tetracycline-dependent transactivator with E2F4 transcriptional activation domain. *Nucleic Acids Res* 29:E23, 2001.

Akopyants, N. S., Clifton, S. W., Martin, J., Pape, D., Wylie, T., Li, L., Kissinger, J. C., Roos, D. S., and Beverley, S. M. A survey of the *Leishmania major* Friedlin strain V1 genome by shotgun sequencing: A resource for DNA microarrays and expression profiling. *Mol Biochem Parasitol* 113:337–340, 2001.

Akutsu, N., Lin, R., Bastien, Y., Bestawros, A., Enepekides, D. J., Black, M. J., and White, J. H. Regulation of gene Expression by 1alpha,25-dihydroxyvitamin D3 and its analog EB1089 under growth-inhibitory conditions in squamous carcinoma cells. *Mol Endocrinol* 15:1127–1139, 2001.

Akutsu, T., Miyano, S., and Kuhara, S. Algorithms for identifying Boolean networks and related biological networks based on matrix multiplication and fingerprint function. *J Comput Biol* 7:331–343, 2000.

Alaiya, A. A., Franzen, B., Hagman, A., Silfversward, C., Moberger, B., Linder, S., and Auer, G. Classification of human ovarian tumors using multivariate data analysis of polypeptide expression patterns. *Int J Cancer.* 86:731–736, 2000.

Alam, R., and Gorska, M. Genomic microarrays. Arraying order in biological chaos? *Am J Respir Cell Mol Biol* 25:405–408, 2001.

Alcantara, O., Kalidas, M., Baltathakis, I., and Boldt, D. H. Expression of multiple genes regulating cell cycle and apoptosis in differentiating hematopoietic cells is dependent on iron. *Exp Hematol* 29:1060–1069, 2001.

Alevizos, I., Mahadevappa, M., Zhang, X., Ohyama, H., Kohno, Y., Posner, M., Gallagher, G. T., Varvares, M., and co-workers. Oral cancer in vivo gene expression profiling assisted by laser capture microdissection and microarray analysis. *Oncogene* 20:6196–6204, 2001.

Alexandre, H., Ansanay-Galeote, V., Dequin, S., and Blondin, B. Global gene expression during short-term ethanol stress in *Saccharomyces cerevisiae.* FEBS Lett 498:98–103.

Alizadeh, A., Eisen, M., Botstein, D., Brown, P. O., and Staudt, L. M. Probing lymphocyte biology by genomic-scale gene expression analysis. *J Clin Immunol* 18:373–379, 1998.

Alizadeh, A., Eisen, M., Davis, R. E., Ma, C., Sabet, H., Tran, T., Powell, J. I., Yang, L., and co-workers. The lymphochip: A specialized cDNA microarray for the genomic-scale analysis of gene expression in normal and malignant lymphocytes. *Cold Spring Harb Symp Quant Biol.* 64:71–78, 1999.

Alizadeh, A. A., Eisen, M. B., Davis, R. E., Ma, C., Lossos, I. S., Rosenwald, A., Boldrick, J. C., Sabet, H., and co-workers. Distinct types of diffuse large B-cell lymphoma identified by gene expression profiling. *Nature* 403:503–511, 2000.

Alizadeh, A. A., and Staudt, L. M. Genomic-scale gene expression profiling of normal and malignant immune cells. Curr Opin Immunol 12:219–225, 2000.

Alizadeh, A. A., Ross, D. T., Perou, C. M., and van de Rijn, M. Towards a novel classification of human malignancies based on gene expression patterns. *J Pathol* 195:41–52, 2001.

Allen, J. F. Bioinformatics and discovery: induction beckons again. *Bioessays* 23:104–107, 2001.

Allander, S. V., Nupponen, N. N., Ringner, M., Hostetter, G., Maher, G. W., Goldberger, N., Chen, Y., Carpten, J., and co-workers. Gastrointestinal stromal tumors with KIT mutations exhibit a remarkably homogeneous gene expression profile. *Cancer Res* 61:8624–8628, 2001.

Almeida, R., Norrish, A., Levick, M., Vetrie, D., Freeman, T., Vilo, J., Ivens, A., Lange, U., and co-workers. From genomes to vaccines: Leishmania as a model. *Philos Trans R Soc Lond B Biol Sci* 357:5–11, 2002.

Altmann, C. R., Bell, E., Sczyrba, A., Pun, J., Bekiranov, S., Gaasterland, T., and Brivanlou, A. H. Microarray-based analysis of early development in *Xenopus laevis. Dev Biol* 236:64–75, 2001.

Altman, R. B., and Raychaudhuri, S. Whole-genome expression analysis: challenges beyond clustering. *Curr Opin Struct Biol* 11:340–347, 2001.

Ameen, M., Smith, C. H., and Barker, J. N. Pharmacogenetics in clinical dermatology. *Br J Dermatol* 146:2–6, 2002.

Amick, J. D., and Brun, Y. V. Anatomy of a bacterial cell cycle [Reviews]. *Genome Biol* 2:1020, 2001.

Amler, L. C., Agus, D. B., LeDuc, C., Sapinoso, M. L., Fox, W. D., Kern, S., Lee, D., Wang, V., and co-workers. Dysregulated expression of androgen-responsive and nonresponsive genes in the androgen-independent prostate cancer xenograft model CWR22-R1. *Cancer Res* 60:6134–6141, 2000.

Amundson, S. A., Bittner, M., Chen, Y., Trent, J., Meltzer, P., and Fornace, A. J. Jr. Fluorescent cDNA microarray hybridization reveals complexity and heterogeneity of cellular genotoxic stress responses. *Oncogene* 18:3666–3672, 1999.

Amundson, S. A., Bittner, M., Meltzer, P., Trent, J., and Fornace, A. J. Induction of gene expression as a monitor of exposure to ionizing radiation. *Radiat Res* 156:657–661, 2001.

Amundson, S. A., Bittner, M., Meltzer, P., Trent, J., and Fornace, A. J. Jr. Physiological function as regulation of large transcriptional programs: The cellular response to genotoxic stress. *Comp Biochem Physiol B Biochem Mol Biol* 129:703–710, 2001.

Amundson, S. A., Do, K. T., Shahab, S., Bittner, M., Meltzer, P., Trent, J., and Fornace, A. J. Jr. Identification of potential mRNA biomarkers in peripheral blood lymphocytes for human exposure to ionizing radiation. *Radiat Res* 154:342–346, 2000.

Amundson, S. A., and Fornace, A. J. Jr. Gene expression profiles for monitoring radiation exposure. *Radiat Prot Dosimetry* 97:11–16, 2001.

Andersen, C. L., Hostetter, G., Grigoryan, A., Sauter, G., and Kallioniemi, A. Improved procedure for fluorescence in situ hybridization on tissue microarrays. *Cytometry* 45:83–86, 2001.

Andrew, R. Clinical measurement of steroid metabolism. *Best Pract Res Clin Endocrinol Metab* 15:1–16, 2001.

Andrews, J., Bouffard, G. G., Cheadle, C., Lu, J., Becker, K. G., and Oliver, B. Gene discovery using computational and microarray analysis of transcription in the *Drosophila melanogaster* testis. *Genome Res* 10:2030–2043, 2000.

Ang, E., Chen, J., Zagouras, P., Magna, H., Holland, J., Schaeffer, E., and Nestler, E. J. Induction of nuclear factor-kappaB in nucleus accumbens by chronic cocaine administration. *J Neurochem* 79:221–224, 2001.

Ang, S., Lee, C. Z., Peck, K., Sindici, M., Matrubutham, U., Gleeson, M. A., and Wang, J. T. Acid-induced gene expression in Helicobacter pylori: study in genomic scale by microarray. *Infect Immun* 69:1679–1686, 2001.

Angelastro, J. M., Klimaschewski, L. P., and Vitolo, O. V. Improved NlaIII digestion of PAGE-purified 102 bp ditags by addition of a single purification step in both the SAGE and microSAGE protocols. *Nucleic Acids Res* 28:E62, 2000.

Angus-Hill, M. L., Schlichter, A., Roberts, D., Erdjument-Bromage, H., Tempst, P., and Cairns, B. R. A rsc3/rsc30 zinc cluster dimer reveals novel roles for the chromatin remodeler rsc in gene expression and cell cycle control. *Mol Cell* 7:741–51, 2001.

Aprelikova, O., Pace, A. J., Fang, B., Koller, B. H., and Liu, E. T. BRCA1 is a selective co-activator of 14-3-3 sigma gene transcription in mouse embryonic stem cells. *J Biol Chem* 276:25647–25650, 2001.

Arai, S., Osawa, T., Ohigashi, H., Yoshikawa, M., Kaminogawa, S., Watanabe, M., Ogawa, T., Okubo, K., and co-workers. A mainstay of functional food science in Japan–history, present status, and future outlook. *Biosci Biotechnol Biochem* 65:1–13, 2001.

Arcellana-Panlilio, M., and Robbins, S. M. I. Global gene expression profiling using DNA microarrays. *Am J Physiol Gastrointest Liver Physiol* 282:G397–402, 2002.

Arendt, C. W., and Littman, D. R. HIV: Master of the host cell [Reviews]. *Genome Biol* 2:1030, 2001.

Arimura, G., Tashiro, K., Kuhara, S., Nishioka, T., Ozawa, R., and Takabayashi, J. Gene responses in bean leaves induced by herbivory and by herbivore-induced volatiles. *Biochem Biophys Res Commun* 277:305–310, 2000.

Aronow, B. J., Richardson, B. D., and Handwerger, S. Microarray analysis of trophoblast differentiation: Gene expression reprogramming in key gene function categories. *Physiol Genomics* 6:105–116, 2001.

Aronow, B. J., Toyokawa, T., Canning, A., Haghighi, K., Delling, U., Kranias, E., Molkentin, J. D., and Dorn, G. W. II. Divergent transcriptional responses to independent genetic causes of cardiac hypertrophy. *Physiol Genomics* 6:19–28, 2001.

Arvand, A., Welford, S. M., Teitell, M. A., and Denny, C. T. The COOH-terminal domain of FLI-1 is necessary for full tumorigenesis and transcriptional modulation by EWS/FLI-1. *Cancer Res* 61:5311–5317, 2001.

Ashby, J. Increasing the sensitivity of the rodent uterotrophic assay to estrogens, with particular reference to bisphenol a. *Environ Health Perspect* 109:1091–1094, 2001.

Ashman, K., Moran, M. F., Sicheri, F., Pawson, T., and Tyers, M. Cell signalling—The proteomics of it all. *Sci STKE* 103:PE33, 2001.

Aubele, M., and Werner, M. Heterogeneity in breast cancer and the problem of relevance of findings. *Anal Cell Pathol* 19:53–58, 1999.

Augenlicht, L. H., Bordonaro, M., Heerdt, B. G., Mariadason, J., and Velcich, A. Cellular mechanisms of risk and transformation. *Ann N Y Acad Sci* 889:20–31, 1999.

Avseenko, N. V., Morozova, Tya., Ataullakhanov, F. I., and Morozov, V. N. Immobilization of proteins in immunochemical microarrays fabricated by electrospray deposition. *Anal Chem* 73:6047–6052, 2001.

Azuaje, F. A computational neural approach to support the discovery of gene function and classes of cancer. *IEEE Trans Biomed Eng* 48:332–339, 2001.

Baba, Y. Development of novel biomedicine based on genome science. *Eur J Pharm Sci* 13:3–4, 2001.

Badarinarayana, V., Estep, P. W. III, Shendure, J., Edwards, J., Tavazoie, S., Lam, F., and Church, G. M. Selection analyses of insertional mutants using subgenic-resolution arrays. *Nat Biotechnol* 19:1060–1065, 2001.

Baelde, H. J., Cleton-Jansen, A. M., van Beerendonk, H., Namba, M., Bovee, J. V., and Hogendoorn, P. C. High quality RNA isolation from tumours with low cellularity and high extracellular matrix component for cDNA microarrays: Application to chondrosarcoma. *J Clin Pathol* 54:778–782, 2001.

Baggerly, K. A., Coombes, K. R., Hess, K. R., Stivers, D. N., Abruzzo, L. V., and Zhang, W. Identifying differentially expressed genes in cDNA microarray experiments. *J Comput Biol* 8:639–659, 2001.

Bahl, A., Brunk, B., Coppel, R. L., Crabtree, J., Diskin, S. J., Fraunholz, M. J., Grant, G. R., Gupta, D., and co-workers. PlasmoDB: The Plasmodium genome resource. An integrated database providing tools for accessing, analyzing and mapping expression and sequence data (both finished and unfinished). *Nucleic Acids Res* 30:87–90, 2002.

Bahn, S., Augood, S. J., Ryan, M., Standaert, D. G., Starkey, M., and Emson, P. C. Gene expression profiling in the post-mortem human brain - no cause for dismay. *J Chem Neuroanat* 22:79–94, 2001.

Balasubramaniam, J., and Del Bigio, M. R. Analysis of age-dependant alteration in the brain gene expression profile following induction of hydrocephalus in rats. *Exp Neurol* 173:105–113, 2002.

Baldi, P., and Long, A. D. A Bayesian framework for the analysis of microarray expression data: Regularized t-test and statistical inferences of gene changes. *Bioinformatics* 17:509–519, 2001.

Baldwin, D., Crane, V., and Rice, D. A comparison of gel-based, nylon filter and microarray techniques to detect differential RNA expression in plants. *Curr Opin Plant Biol* 2:96–103, 1999.

Baldwin, K. M. Research in the exercise sciences: Where do we go from here? *J Appl Physiol* 88:332–336, 2000.

Baetz, K., Moffat, J., Haynes, J., Chang, M., and Andrews, B. Transcriptional coregulation by the cell integrity mitogen-activated protein kinase Slt2 and the cell cycle regulator Swi4. *Mol Cell Biol* 21:6515–6528, 2001.

Baker, T. K., Carfagna, M. A., Gao, H., Dow, E. R., Li, Q., Searfoss, G. H., and Ryan, T. P. Temporal gene expression analysis of monolayer cultured rat hepatocytes. *Chem Res Toxicol* 14:1218–12131, 2001.

Ball, C. A., Jin, H., Sherlock, G., Weng, S., Matese, J. C., Andrada, R., Binkley, G., Dolinski, K., and co-workers. Saccharomyces genome database provides tools to survey gene expression and functional analysis data. *Nucleic Acids Res* 29:80–81, 2001.

Baltathakis, I., Alcantara, O., and Boldt, D. H. Expression of different NF-kappaB pathway genes in dendritic cells (DCs) or macrophages assessed by gene expression profiling. *J Cell Biochem* 83:281–290, 2001.

Bammert, G. F., and Fostel, J. M. Genome-wide expression patterns in *Saccharomyces cerevisiae:* Comparison of drug treatments and genetic alterations affecting biosynthesis of ergosterol. *Antimicrob Agents Chemother* 44:1255–1265, 2000.

Bard, J. B. A bioinformatics approach to investigating developmental pathways in the kidney and other tissues. *Int J Dev Biol* 43(5):397–403, 1999.

Bard, J., and Winter, R. Ontologies of developmental anatomy: Their current and future roles. *Brief Bioinform* 2:289–299, 2001.

Barlund, M., Forozan, F., Kononen, J., Bubendorf, L., Chen, Y., Bittner, M. L., Torhorst, J., Haas, P., and co-workers. Detecting activation of ribosomal protein S6 kinase by complementary DNA and tissue microarray analysis. *J Natl Cancer Inst* 92:1252–1259, 2000.

Barlund, M., Monni, O., Kononen, J., Cornelison, R., Torhorst, J., Sauter, G., Kallioniemi, O. L. L. I.-P., and Kallioniemi, A. Multiple genes at 17q23 undergo amplification and overexpression in breast cancer. *Cancer Res* 60:5340–5344, 2000.

Barrans, J. D., Stamatiou, D., and Liew, C. Construction of a human cardiovascular cDNA microarray: Portrait of the failing heart. *Biochem Biophys Res Commun* 280:964–969, 2001.

Barrett, T., Xie, T., Piao, Y., Dillon-Carter, O., Kargul, G. J., Lim, M. K., Chrest, F. J., Wersto, R., and co-workers. A murine dopamine neuron-specific cDNA library and microarray: Increased COXI expression during methamphetamine neurotoxicity. *Neurobiol Dis* 8:822–833, 2001.

Barry, C. E. III, and Schroeder, B. G. DNA microarrays: Translational tools for understanding the biology of *Mycobacterium tuberculosis*. *Trends Microbiol* 8:209–210, 2000.

Barry, C. E. III, Slayden, R. A., Sampson, A. E., and Lee, R. E. Use of genomics and combinatorial chemistry in the development of new antimycobacterial drugs. *Biochem Pharmacol* 59:221–231, 2000.

Barry, C. E. III, Wilson, M., Lee, R., and Schoolnik, G. K. DNA microarrays and combinatorial chemical libraries: Tools for the drug discovery pipeline. *Int J Tuberc Lung Dis* 4(12 Suppl 2):S189–S193, 2000.

Bartlett, J. Technology evaluation: SAGE, genzyme molecular oncology. *Curr Opin Mol Ther* 3:85–96, 2001.

Bartosiewicz, M., Penn, S., and Buckpitt, A. Applications of gene arrays in environmental toxicology: Fingerprints of gene regulation associated with cadmium chloride, benzo(a)pyrene, and trichloroethylene. *Environ Health Perspect* 109:71–74, 2001.

Bartosiewicz, M., Trounstine, M., Barker, D., Johnston, R., and Buckpitt, A. Development of a toxicological gene array and quantitative assessment of this technology. *Arch Biochem Biophys* 376:66–73, 2000.

Basarsky, T., Verdnik, D., Zhai, J. Y., and Wellis, D. Overview of a microarray scanner: Design essentials for an integrated acquisition and analysis platform. In: M. Schena, ed. *Microarray Biochip Technology.* Eaton Publishing, Natick, MA, 2000: 265–284.

Battaglia, C., Salani, G., Consolandi, C., Bernardi, L. R., and De Bellis, G. Analysis of DNA microarrays by non-destructive fluorescent staining using SYBR green II. *Biotechniques* 29:78–81, 2000.

Bavykin, S. G., Akowski, J. P., Zakhariev, V. M., Barsky, V. E., Perov, A. N., and Mirzabekov, A. D. Portable system for microbial sample preparation and oligonucleotide microarray analysis. *Appl Environ Microbiol* 67:922–928, 2001.

Beaucage, S. L. Strategies in the preparation of DNA oligonucleotide arrays for diagnostic applications. *Curr Med Chem* 8:1213–1244, 2001.

Becich, M. J. The role of the pathologist as tissue refiner and data miner: The impact of functional genomics on the modern pathology laboratory and the critical roles of pathology informatics and bioinformatics. *Mol Diagn* 5:287–299, 2000.

Beck, M. T., Holle, L., and Chen, W. Y. Combination of PCR subtraction and cDNA microarray for differential gene expression profiling. *Biotechniques* 31:782–784, 786, 2001.

Becker, K. G. The sharing of cDNA microarray data. *Nat Rev Neurosci* 2:438–440, 2001.

Bednar, M. DNA microarray technology and application. *Med Sci Monit* 6:796–800, 2000.

Bedalov, A., Gatbonton, T., Irvine, W. P., Gottschling, D. E., and Simon, J. A. Identification of a small molecule inhibitor of Sir2p. *Proc Natl Acad Sci U S A* 98:15113–15118, 2001.

Behr, M. A., Wilson, M. A., Gill, W. P., Salamon, H., Schoolnik, G. K., Rane, S., and Small, P. M. Comparative genomics of BCG vaccines by whole-genome DNA microarray. *Science* 284:1520–1523, 1999.

Beier, M., and Hoheisel, J. D. Analysis of DNA-microarrays produced by inverse in situ oligonucleotide synthesis. *J Biotechnol* 94:15–22, 2002.

Beier, M., and Hoheisel, J. D. Production by quantitative photolithographic synthesis of individually quality checked DNA microarrays. *Nucleic Acids Res* 28:E11, 2000.

Beitner-Johnson, D., Seta, K., Yuan, Y., Kim, H., Rust, R. T., Conrad, P. W., Kobayashi, S., and Millhorn, D. E. Identification of hypoxia-responsive genes in a dopaminergic cell line by subtractive cDNA libraries and microarray analysis. *Parkinsonism Relat Disord* 7:273–281, 2001.

Belcher, C. E., Drenkow, J., Kehoe, B., Gingeras, T. R., McNamara, N., Lemjabbar, H., Basbaum, C., and Relman, D. A. The transcriptional responses of respiratory epithelial cells to *Bordetella pertussis* reveal host defensive and pathogen counter-defensive strategies. *Proc Natl Acad Sci U S A* 97:13847–13852, 2000.

Belenkiy, R., Haefele, A., Eisen, M. B., and Wohlrab, H. The yeast mitochondrial transport proteins: New sequences and consensus residues, lack of direct relation between consensus residues and transmembrane helices, expression patterns of the transport protein genes, and protein-protein interactions with other proteins. *Biochim Biophys Acta* 1467:207–218, 2000.

Bell, S. E., Mavila, A., Salazar, R., Bayless, K. J., Kanagala, S., Maxwell, S. A., and Davis, G. E. Differential gene expression during capillary morphogenesis in 3D collagen matrices: Regulated expression of genes involved in basement membrane matrix assembly, cell cycle progression, cellular differentiation and G-protein signaling. *J Cell Sci* 114(Pt 15):2755–2773, 2001.

Bellenson, J. L. Expression data and the bioinformatics challenges. In: M. Schena, ed. *DNA Microarrays: A Practical Approach,* 2nd ed. Oxford University Press, Oxford, UK, 2000:139–165.

Belosludtsev, Y., Belosludtsev, I., Iverson, B., Lemeshko, S., Wiese, R., Hogan, M., and Powdrill, T. Nearly instantaneous, cation-independent, high selectivity nucleic acid hybridization to dna microarrays. *Biochem Biophys Res Commun* 282:1263–1267, 2001.

Belov, L., de la Vega, O., dos Remedios, C. G., Mulligan, S. P., and Christopherson, R. I. Immunophenotyping of leukemias using a cluster of differentiation antibody microarray. *Cancer Res* 61:4483–4489, 2001.

Ben, M. C., Gluzman, I. Y., Hott, C., MacMillan, S. K., Amarakone, A. S., Anderson, D. L., Carlton, J. M., Dame, J. B., and co-workers. Co-ordinated programme of gene expression during asexual intraerythrocytic development of the human malaria parasite *Plasmodium falciparum* revealed by microarray analysis. *Mol Microbiol* 39:26–36, 2001.

Bennicelli, J. L., and Barr, F. G. Genetics and the biologic basis of sarcomas. *Curr Opin Oncol* 11:267–274, 1999.

Benoit, G. R., Tong, J. H., Balajthy, Z., and Lanotte, M. Exploring (novel) gene expression during retinoid-induced maturation and cell death of acute promyelocytic leukemia. *Semin Hematol* 38:71–85, 2001.

Benters, R., Niemeyer, C. M., Drutschmann, D., Blohm, D., and Wohrle, D. DNA microarrays with PAMAM dendritic linker systems. *Nucleic Acids Res* 30:E10, 2002.

Benters, R., Niemeyer, C. M., and Wohrle, D. Dendrimer-activated solid supports for nucleic Acid and protein microarrays. *Chembiochem Europ J Chem Biol* 2:686–694, 2001.

Bertucci, F., Bernard, K., Loriod, B., Chang, Y. C., Granjeaud, S., Birnbaum, D., Nguyen, C., Peck, K., and Jordan, B. R. Sensitivity issues in DNA array-based expression measurements and performance of nylon microarrays for small samples. *Hum Mol Genet* 8:1715–1722, 1999.

Bertucci, F., Houlgatte, R., Nguyen, C., Benziane, A., Nasser, V., Granjeaud, S., Tagett, B., Loriod, B., and co-workers. Molecular typing of breast cancer: transcriptomics and DNA microarrays. *Bull Cancer* 88:277–286, 2001.

Bertucci, F., Loriod, B., Tagett, R., Granjeaud, S., Birnbaum, D., Nguyen, C., and Houlgatte, R. DNA arrays: Technological aspects and applications. *Bull Cancer* 88:243–252, 2001.

Beuzen, N. D., Stear, M. J., and Chang, K. C. Molecular markers and their use in animal breeding. *Vet J* 160:42–52, 2000.

Beverley, S. M., Akopyants, N. S., Goyard, S., Matlib, R. S., Gordon, J. L., Brownstein, B. H., Stormo, G. D., Bukanova, E. N., and co-workers. Putting the *Leishmania* genome to work: Functional genomics by transposon trapping and expression profiling. *Philos Trans R Soc Lond B Biol Sci* 357:47–53, 2002.

Bhattacharjee, A., Richards, W. G., Staunton, J., Li, C., Monti, S., Vasa, P., Ladd, C., Beheshti, J., and co-workers. Classification of human lung carcinomas by mRNA expression profiling reveals distinct adenocarcinoma subclasses. *Proc Natl Acad Sci U S A* 98:13790–13795, 2001.

Bier, F. F., and Kleinjung, F. Feature-size limitations of microarray technology—A critical review. *Fresenius J Anal Chem* 371:151–156, 2001.

Bigger, C. B., Brasky, K. M., and Lanford, R. E. DNA microarray analysis of chimpanzee liver during acute resolving hepatitis C virus infection. J Virol 75:7059–7066, 2001.

Bilban, M., Head, S., Desoye, G., and Quaranta, V. DNA microarrays: A novel approach to investigate genomics in trophoblast invasion–a review. Placenta 21(Suppl A):S99–S105, 2000.

Bjorkholm, B., Lundin, A., Sillen, A., Guillemin, K., Salama, N., Rubio, C., Gordon, J. I., Falk, P., and Engstrand, L. Comparison of genetic divergence and fitness between two subclones of *Helicobacter pylori*. *Infect Immun* 69:7832–7838, 2001.

Blader, I. J., Manger, I. D., and Boothroyd, J. C. Microarray analysis reveals previously unknown changes in toxoplasma gondii infected human cells. *J Biol Chem* 276:24223–24231, 2001.

Blagoev, B., and Pandey, A. Microarrays go live—New prospects for proteomics. *Trends Biochem Sci* 26:639–641, 2001.

Blanchard, R. K., Moore, J. B., Green, C. L., and Cousins, R. J. Modulation of intestinal gene expression by dietary zinc status: Effectiveness of cDNA arrays for expression profiling of a single nutrient deficiency. *Proc Natl Acad Sci U S A* 98:13507–13513, 2001.

Bleul, C. C., and Boehm, T. Laser capture microdissection-based expression profiling identifies PD1-ligand as a target of the nude locus gene product. *Eur J Immunol* 31:2497–2503, 2001.

Block-Alper, L., Webster, P., Zhou, X., Supekova, L., Wong, W. H., Schultz, P. G., and Meyer, D. I. IN02, a positive regulator of lipid biosynthesis, is essential for the formation of inducible membranes in yeast. *Mol Biol Cell* 13:40–51, 2002.

Blohm, D. H., and Guiseppi-Elie, A. New developments in microarray technology. *Curr Opin Biotechnol* 12:41–47, 2001.

Blumcke, I., Becker, A. J., Normann, S., Hans, V., Riederer, B. M., Krajewski, S., Wiestler, O. D., and Reifenberger, G. Distinct expression pattern of microtubule-associated protein-2 in human oligodendrogliomas and glial precursor cells. *J Neuropathol Exp Neurol* 60:984–993, 2001.

Bochner, B. R., Gadzinski, P., and Panomitros, E. Phenotype microarrays for high-throughput phenotypic testing and assay of gene function. *Genome Res* 11:1246–1255, 2001.

Boel, E., Albrektsen, T., Fleckner, J., and Selmer, J. Modulation of metabolism through transcriptional control has created new treatment opportunities for type 2 diabetes. *Curr Pharm Biotechnol* 1:63–71, 2000.

Boeuf, S., Klingenspor, M., Van Hal, N. L., Schneider, T., Keijer, J., and Klaus, S. Differential gene expression in white and brown preadipocytes. *Physiol Genomics* 7:15–25, 2001.

Bonilla, I. E., Tanabe, K., and Strittmatter, S. M. Small proline-rich repeat protein 1A is expressed by axotomized neurons and promotes axonal outgrowth. *J Neurosci* 22:1303–1315, 2002.

Bono, H., Kasukawa, T., Furuno, M., and Hayashizaki, Y., Okazaki, Y. FANTOM DB: Database of functional annotation of RIKEN mouse cDNA clones. *Nucleic Acids Res* 30:116–118, 2002.

Bono, H., Kasukawa, T., Hayashizaki, Y., Okazaki, Y. READ: RIKEN expression array database. *Nucleic Acids Res* 30:211–213, 2002.

Borneman, J., Chrobak, M., Della Vedova, G., Figueroa, A., and Jiang, T. Probe selection algorithms with applications in the analysis of microbial communities. *Bioinformatics* 17(Suppl 1):S39–S48, 2001.

Bornholdt, S. Modeling genetic networks and their evolution: A complex dynamical systems perspective. *Biol Chem* 382:1289–1299, 2001.

Borrebaeck, C. A. Antibodies in diagnostics–From immunoassays to protein chips. *Immunol Today* 21:379–382, 2000.

Bortoluzzi, S., d'Alessi, F., Romualdi, C., and Danieli, G. A. The human adult skeletal muscle transcriptional profile reconstructed by a novel computational approach. *Genome Res* 10:344–349, 2000.

Bosetti, F., Seemann, R., Bell, J. M., Zahorchak, R., Friedman, E., Rapoport, S. I., and Manickam, P. Analysis of gene expression with cDNA microarrays in rat brain after 7 and 42 days of oral lithium administration. *Brain Res Bull* 57:205–209, 2002.

Bouchie, A. Shift anticipated in DNA microarray market. *Nat Biotechnol* 20:8, 2002.

Bouton, C. M., Hossain, M. A., Frelin, L. P., Laterra, J., and Pevsner, J. Microarray analysis of differential gene expression in lead-exposed astrocytes. *Toxicol Appl Pharmacol* 176(1):34–53, 2001.

Bouton, C. M., and Pevsner, J. DRAGON view: Information visualization for annotated microarray data. *Bioinformatics* 18:323–324, 2002.

Bouton, C. M., and Pevsner, J. Effects of lead on gene expression. *Neurotoxicology* 21:1045–1055, 2000.

Bova, G. S., Parmigiani, G., Epstein, J. I., Wheeler, T., Mucci, N. R., and Rubin, M. A. Web-based tissue microarray image data analysis: Initial validation testing through prostate cancer Gleason grading. *Hum Pathol* 32:417–427, 2001.

Bowen, C., Bubendorf, L., Voeller, H. J., Slack, R., Willi, N., Sauter, G., Gasser, T. C., Koivisto, P., and co-workers. Loss of NKX3.1 expression in human prostate cancers correlates with tumor progression. *Cancer Res* 60:6111–6115, 2000.

Bowtell, D. D. Options available—from start to finish—for obtaining expression data by microarray. *Nat Genet* 21(Suppl 1):25–32, 1999.

Braasch, D. A., and Corey, D. R. Locked nucleic acid (LNA): Fine-tuning the recognition of DNA and RNA. *Chem Biol* 8:1–7, 2001.

Brachat, A., Pierrat, B., Brungger, A., and Heim, J. Comparative microarray analysis of gene expression during apoptosis-induction by growth factor deprivation or protein kinase C inhibition. *Oncogene* 19:5073–5082, 2000.

Brady, G. Expression profiling of single mammalian cells—Small is beautiful. *Yeast* 17:211–217, 2000.

Brail, L. H., Jang, A., Billia, F., Iscove, N. N., Klamut, H. J., and Hill, R. P. Gene expression in individual cells: Analysis using global single cell reverse transcription polymerase chain reaction (GSC RT-PCR). *Mutat Res* 406:45–54, 1999.

Brar, A. K., Handwerger, S., Kessler, C. A., and Aronow, B. J. Gene induction and categorical reprogramming during in vitro human endometrial fibroblast decidualization. *Physiol Genomics* 7:135–148, 2001.

Braxton, S., and Bedilion, T. The integration of microarray information in the drug development process. *Curr Opin Biotechnol* 9:643–649, 1998.

Brazma, A., Robinson, A., Cameron, G., and Ashburner, M. One-stop shop for microarray data. *Nature* 403:699–700, 2000.

Brazma, A., Hingamp, P., Quackenbush, J., Sherlock, G., Spellman, P., Stoeckert, C., Aach, J., Ansorge, W., and co-workers. Minimum information about a microarray experiment (MIAME)—toward standards for microarray data. *Nat Genet* 29:365–371, 2001.

Brazma, A., and Vilo, J. Gene expression data analysis. *FEBS Lett* 480:17–24, 2000.

Brazma, A., and Vilo, J. Gene expression data analysis. *Microbes Infect* 3:823–829, 2001.

Breitkreutz, B. J., Jorgensen, P., Breitkreutz, A., and Tyers, M. AFM 4.0: A toolbox for DNA microarray analysis. *Genome Biol* 2: 2001.

Brem, R., Certa, U., Neeb, M., Nair, A. P., and Moroni, C. Global analysis of differential gene expression after transformation with the v-H-ras oncogene in a murine tumor model. *Oncogene* 20:2854–2858, 2001.

Brem, R., Hildebrandt, T., Jarsch, M., Van Muijen, G. N., and Weidle, U. H. Identification of metastasis-associated genes by transcriptional profiling of a metastasizing versus a non-metastasizing human melanoma cell line. *Anticancer Res* 21:1731–1740, 2001.

Brenman, J. E., Gao, F. B., Jan, L. Y., and Jan, Y. N. Sequoia, a tramtrack-related zinc finger protein, functions as a pan-neural regulator for dendrite and axon morphogenesis in *Drosophila*. *Dev Cell* 1:667–677, 2001.

Brenner, V., Lindauer, K., Parkar, A., Fordham, J., Hayes, I., Stow, M., Gama, R., Pollock, K., and Jupp, R. Analysis of cellular adhesion by microarray expression profiling. *J Immunol Methods* 250:15–28, 2001.

Brenner, S., Williams, S. R., Vermaas, E. H., Storck, T., Moon, K., McCollum, C., Mao, J. I., Luo, S., and co-workers. In vitro cloning of complex mixtures of DNA on microbeads: physical separation of differentially expressed cDNAs. *Proc Natl Acad Sci U S A* 97:1665–1670, 2000.

Brenton, J. D., Aparicio, S. A., and Caldas, C. Molecular profiling of breast cancer: portraits but not physiognomy. *Breast Cancer Res* 3:77–80, 2001.

Breyne, P., and Zabeau, M. Genome-wide expression analysis of plant cell cycle modulated genes. *Curr Opin Plant Biol* 4:136–142, 2001.

Brignac, S. J. Jr., Gangadharan, R., McMahon, M., Denman, J., Gonzales, R., Mendoza, L. G., and Eggers, M. A proximal CCD imaging system for high-throughput detection of microarray-based assays. *IEEE Eng Med Biol Mag* 18:120–122, 1999.

Brockman, J. A., and Tamminga, C. A. The human genome: Microarray expression analysis. *Am J Psychiatry* 158:1199, 2001.

Brodersen, P., Petersen, M., Pike, H. M., Olszak, B., Skov, S., Odum, N., Jorgensen, L. B., Brown, R. E., and Mundy, J. Knockout of *Arabidopsis* ACCELERATED-CELL-DEATH11 encoding a sphingosine transfer protein causes activation of programmed cell death and defense. *Genes Dev* 16:490–502, 2002.

Broggini, M. Prospects of molecular pharmacology in oncology. *Recenti Prog Med* 92:93–97, 2001.

Broude, N. E., Woodward, K., Cavallo, R., Cantor, C. R., and Englert, D. DNA microarrays with stem-loop DNA probes: Preparation and applications. *Nucleic Acids Res* 29:E92, 2001.

Brown, A. J., Planta, R. J., Restuhadi, F., Bailey, D. A., Butler, P. R., Cadahia, J. L., Cerdan, M. E., De Jonge, M., and co-workers. Transcript analysis of 1003 novel yeast genes using high-throughput northern hybridizations. *EMBO J* 20:3177–3186, 2001.

Brown, C. S., Goodwin, P. C., and Sorger, P. K. Image metrics in the statistical analysis of DNA microarray data. *Proc Natl Acad Sci U S A* 98:8944–8949, 2001.

Brown, M. P., Grundy, W. N., Lin, D., Cristianini, N., Sugnet, C. W., Furey, T. S., Ares, M. Jr., and Haussler, D. Knowledge-based analysis of microarray gene expression data by using support vector machines. *Proc Natl Acad Sci U S A* 97:262–267, 2000.

Brown, P. O., and Botstein, D. Exploring the new world of the genome with DNA microarrays. *Nat Genet.* 21(Suppl 1):33–37, 1999.

Brown, V., Jin, P., Ceman, S., Darnell, J. C., O'Donnell, W. T., Tenenbaum, S. A., Jin, X., Feng, Y., and co-workers. Microarray identification of FMRP-associated brain mRNAs and altered mRNA translational profiles in fragile X syndrome. *Cell* 107:477–487, 2001.

Brown, V. M., Ossadtchi, A., Khan, A. H., Cherry, S. R., Leahy, R. M., and Smith, D. J. High-throughput imaging of brain gene expression. *Genome Res* 12:244–254, 2002.

Bruder, C. E., Hirvela, C., Tapia-Paez, I., Fransson, I., Segraves, R., Hamilton, G., Zhang, X. X., Evans, D. G., and co-workers. High resolution deletion analysis of constitutional DNA from neurofibromatosis type 2 (NF2) patients using microarray-CGH. *Hum Mol Genet* 10:271–282, 2001.

Brun, Y. V. Global analysis of a bacterial cell cycle: Tracking down necessary functions and their regulators. *Trends Microbiol* 9:405–407, 2001.

Brunskill, E. W., Witte, D. P., Yutzey, K. E., and Potter, S. S. Novel cell lines promote the discovery of genes involved in early heart development. *Dev Biol* 235:507–520, 2001.

Bryant, Z., Subrahmanyan, L., Tworoger, M., LaTray, L., Liu, C. R., Li, M. J., van den Engh, G., and Ruohola-Baker, H. Characterization of differentially expressed genes in purified *Drosophila* follicle cells: Toward a general strategy for cell type-specific developmental analysis. *Proc Natl Acad Sci U S A* 96:5559–5564, 1999.

Bubendorf, L. High-throughput microarray technologies: from genomics to clinics. *Eur Urol* 40:231–238, 2001.

Bubendorf, L., Kolmer, M., Kononen, J., Koivisto, P., Mousses, S., Chen, Y., Mahlamaki, E., Schraml, P., and co-workers. Hormone therapy failure in human prostate cancer:

Analysis by complementary DNA and tissue microarrays. *J Natl Cancer Inst* 91:1758–1764, 1999.

Bubendorf, L., Kononen, J., Koivisto, P., Schraml, P., Moch, H., Gasser, T. C., Willi, N., Mihatsch, M. J., and co-workers. Survey of gene amplifications during prostate cancer progression by high-throughout fluorescence in situ hybridization on tissue microarrays. *Cancer Res* 59:803–806, 1999.

Bubendorf, L., Nocito, A., Moch, H., and Sauter, G. Tissue microarray (TMA) technology: Miniaturized pathology archives for high-throughput in situ studies. *J Pathol* 195:72–79, 2001.

Bucher, P. Regulatory elements and expression profiles. *Curr Opin Struct Biol* 9:400–407, 1999.

Buechler, C., Ullrich, H., Ritter, M., Porsch-Oezcueruemez, M., Lackner, K. J., Barlage, S., Friedrich, S. O., Kostner, G. M., and Schmitz, G. Lipoprotein (a) up-regulates the expression of the plasminogen activator inhibitor 2 in human blood monocytes. *Blood* 97:981–986, 2001.

Bulera, S. J., Eddy, S. M., Ferguson, E., Jatkoe, T. A., Reindel, J. F., Bleavins, M. R., and De La Iglesia, F. A. RNA expression in the early characterization of hepatotoxicants in Wistar rats by high-density DNA microarrays. *Hepatology* 33:1239–1258, 2001.

Bull, J. H., Ellison, G., Patel, A., Muir, G., Walker, M., Underwood, M., Khan, F., and Paskins, L. Identification of potential diagnostic markers of prostate cancer and prostatic intraepithelial neoplasia using cDNA microarray. *Br J Cancer* 84:1512–1519, 2001.

Bulyk, M. L., Huang, X., Choo, Y., and Church, G. M. Exploring the DNA-binding specificities of zinc fingers with DNA microarrays. *Proc Natl Acad Sci U S A* 98:7158–7163, 2001.

Bumm, K., Zheng, M., Bailey, C., Zhan, F., Chiriva-Internati, M., Eddlemon, P., Terry, J., Barlogie, B., and Shaughnessy, J. D. Jr. CGO: Utilizing and integrating gene expression microarray data in clinical research and data management. *Bioinformatics* 18:327–328, 2002.

Burczynski, M. E., McMillian, M., Ciervo, J., Li, L., Parker, J. B., Dunn, R. T. II, Hicken, S., Farr, S., and Johnson, M. D. Toxicogenomics-based discrimination of toxic mechanism in HepG2 human hepatoma cells. *Toxicol Sci* 58:399–415, 2000.

Burgess, J. K. Gene expression studies using microarrays. *Clin Exp Pharmacol Physiol* 28:321–328, 2001.

Burgess, J. K., and Hazelton, R. H. New developments in the analysis of gene expression. *Redox Rep* 5:63–73, 2000.

Burke, H. B. Discovering patterns in microarray data. *Mol Diagn* 5:349–357, 2000.

Burmeister, J. J., Moxon, K., and Gerhardt, G. A. Ceramic-based multisite microelectrodes for electrochemical recordings. *Anal Chem* 72:187–192, 2000.

Burns, C. G., Ohi, R., Mehta, S., O'Toole, E. T., Winey, M., Clark, T. A., Sugnet, C. W., Ares, M. Jr., and Gould, K. L. Removal of a single alpha-tubulin gene intron suppresses cell cycle arrest phenotypes of splicing factor mutations in Saccharomyces cerevisiae. *Mol Cell Biol* 22:801–815, 2002.

Bushel, P. R., Hamadeh, H., Bennett, L., Sieber, S., Martin, K., Nuwaysir, E. F., Johnson, K., Reynolds, K., and co-workers. MAPS: A microarray project system for gene expression experiment information and data validation. *Bioinformatics* 17:564–565, 2001.

Butte, A. J., Tamayo, P., Slonim, D., Golub, T. R., and Kohane, I. S. Discovering functional relationships between RNA expression and chemotherapeutic susceptibility using relevance networks. *Proc Natl Acad Sci U S A* 97:12182–12186, 2000.

Butte, A. J., Ye, J., Haring, H. U., Stumvoll, M., White, M. F., and Kohane, I. S. Determining significant fold differences in gene expression analysis. *Pac Symp Biocomput*:6–17, 2001.

Calaluce, R., Kunkel, M. W., Watts, G. S., Schmelz, M., Hao, J., Barrera, J., Gleason-Guzman, M., Isett, R., and co-workers. Laminin-5-mediated gene expression in human prostate carcinoma cells. *Mol Carcinog* 30:119–129, 2001.

Calame, K. L. Plasma cells: Finding new light at the end of B cell development. *Nat Immunol* 2:1103–1108, 2001.

Caldwell, R., Sapolsky, R., Weyler, W., Maile, R. R., Causey, S. C., and Ferrari, E. Correlation between *Bacillus subtilis* scoC phenotype and gene expression determined using microarrays for transcriptome analysis. *J Bacteriol* 183:7329–7340, 2001.

Califano, A. Advances in sequence analysis. *Curr Opin Struct Biol* 11:330–333, 2001.

Califano, A., Stolovitzky, G., and Tu, Y. Analysis of gene expression microarrays for phenotype classification. *Proc Int Conf Intell Syst Mol Biol* 8:75–85, 2000.

Call, D. R., Chandler, D. P., and Brockman, F. Fabrication of DNA microarrays using unmodified oligonucleotide probes. *Biotechniques* 30:368–372, 374, 376, 2001.

Call, D. R., Brockman, F. J., and Chandler, D. P. Detecting and genotyping *Escherichia coli* O157:H7 using multiplexed PCR and nucleic acid microarrays. *Int J Food Microbiol* 67:71–80, 2001.

Callow, M. J., Dudoit, S., Gong, E. L., Speed, T. P., and Rubin, E. M. Microarray expression profiling identifies genes with altered expression in HDL-deficient mice. *Genome Res* 10:2022–2029, 2000.

Camp, R. L., Charette, L. A., and Rimm, D. L. Validation of tissue microarray technology in breast carcinoma. *Lab Invest* 80:1943–1949, 2000.

Campbell, W. G., Gordon, S. E., Carlson, C. J., Pattison, J. S., Hamilton, M. T., and Booth, F. W. Differential global gene expression in red and white skeletal muscle. *Am J Physiol Cell Physiol* 280:C763–C768, 2001.

Cao, S. X., Dhahbi, J. M., Mote, P. L., and Spindler, S. R. Genomic profiling of short- and long-term caloric restriction effects in the liver of aging mice. *Proc Natl Acad Sci U S A* 98:10630–10635, 2001.

Cao, Y., and Dulac, C. Profiling brain transcription: neurons learn a lesson from yeast. *Curr Opin Neurobiol* 11:615–620, 2001.

Cardozo, A. K., Heimberg, H., Heremans, Y., Leeman, R., Kutlu, B., Kruhoffer, M., Orntoft, T., and Eizirik, D. L. A comprehensive analysis of cytokine-induced and NF-kB dependent genes in primary rat pancreatic β-cells. *J Biol Chem* 2001.

Carles, M., Lee, T., Moganti, S., Lenigk, R., Tsim, K. W., Ip, N. Y., Hsing, I. M., and Sucher, N. J. Chips and Qi: Microcomponent-based analysis in traditional Chinese medicine. *Fresenius J Anal Chem* 371:190–194, 2001.

Carlson, M. Regulation of glucose utilization in yeast. *Curr Opin Genet Dev* 8:560–564, 1998.

Carmel, J. B., Galante, A., Soteropoulos, P., Tolias, P., Recce, M., Young, W., and Hart, R. P. Gene expression profiling of acute spinal cord injury reveals spreading inflammatory signals and neuron loss. *Physiol Genomics* 7:201–213, 2001.

Carson, J. A., Nettleton, D., and Reecy, J. M. Differential gene expression in the rat soleus muscle during early work overload-induced hypertrophy. *FASEB J* 2001.

Carucci, D. J. Technologies for the study of gene and protein expression in *Plasmodium*. *Philos Trans R Soc Lond B Biol Sci* 357:13–16, 2002.

Carulli, J. P., Artinger, M., Swain, P. M., Root, C. D., Chee, L., Tulig, C., Guerin, J., Osborne, M., and co-workers. High throughput analysis of differential gene expression. *J Cell Biochem Suppl* 30–31:286–296, 1998.

Case-Green, S, Pritchard, C., and Southern, E. Use of oligonucleotide arrays in enzymatic assays: Assay optimisation. In: M. Schena, ed. *DNA Microarrays: A Practical Approach,* 2nd ed. Oxford University Press, Oxford, UK, 2000:61–76.

Catani, M. V., Rossi A., Costanzo, A., Sabatini, S., Levrero, M., Melino, G., and Avigliano, L. Induction of gene expression via activator protein-1 in the ascorbate protection against UV-induced damage. *Biochem J* 356:77–85, 2001.

Cavalieri, D., Townsend, J. P., and Hartl, D. L. Manifold anomalies in gene expression in a vineyard isolate of *Saccharomyces cerevisiae* revealed by DNA microarray analysis. *Proc Natl Acad Sci U S A* 97:12369–12374, 2000.

Celis, J. E., Kruhoffer, M., Gromova, I., Frederiksen, C., Ostergaard, M., Thykjaer, T., Gromov, P., Yu, J., and co-workers. Gene expression profiling: monitoring transcription and translation products using DNA microarrays and proteomics. *FEBS Lett* 480: 2–16, 2000.

Celniker, S. E. *The Drosophila genome. Curr Opin Genet Dev* 10:612–616, 2000.

Chabas, D., Baranzini, S. E., Mitchell, D., Bernard, C. C., Rittling, S. R., Denhardt, D. T., Sobel, R. A., Lock, C., and co-workers. The influence of the proinflammatory cytokine, osteopontin, on autoimmune demyelinating disease. *Science* 294:1731–1735, 2001.

Chaib, H., Cockrell, E. K., Rubin, M. A., and Macoska, J. A. Profiling and verification of gene expression patterns in normal and malignant human prostate tissues by cDNA microarray analysis. *Neoplasia* 3:43–52, 2001.

Chaib, H., Rubin, M. A., Mucci, N. R., Li, L., Taylor, J. M. G., Day, M. L., Rhim, J. S., and Macoska, J. A. Protein nucleotide activated in prostate cancer: A PDZ domain-containing protein highly expressed in human primary prostate tumors. *Cancer Res* 61:2390–2394, 2001.

Chambergo, F. S., Bonaccorsi, E. D., Ferreira, A. J., Ramos, A. S., Ferreira, J. R. Jr., Abrahao-Neto, J., Simon Farah, J. P., and El-Dorry, H. Elucidation of the metabolic fate of glucose in the filamentous fungus *Trichoderma reesei* using EST analysis and cDNA microarrays. *J Biol Chem* 2002.

Chambers, J., Angulo, A., Amaratunga, D., Guo, H., Jiang, Y., Wan, J. S., Bittner, A., Frueh, K., and co-workers. DNA microarrays of the complex human cytomegalovirus genome: profiling kinetic class with drug sensitivity of viral gene expression. *J Virol* 73:5757–5766, 1999.

Chan, J. K. The new World Health Organization classification of lymphomas: The past, the present and the future. *Hematol Oncol* 19:129–150, 2001.

Chan, W. C., and Huang, J. Z. Gene expression analysis in aggressive NHL. *Ann Hematol* 80(Suppl 3):B38–41, 2001.

Chandler, D. P., Brown, J., Call, D. R., Wunschel, S., Grate, J. W., Holman, D. A., Olson, L., Stottlemyre, M. S., and Bruckner-Lea, C. J. Automated immunomagnetic separation and microarray detection of *E. coli* O157:H7 from poultry carcass rinse. *Int J Food Microbiol* 70:143–154, 2001.

Chang, B. D., Swift, M. E., Shen, M., Fang, J., Broude, E. V., and Roninson, I. B. Molecular determinants of terminal growth arrest induced in tumor cells by a chemotherapeutic agent. *Proc Natl Acad Sci U S A* 2001.

Chang, K. C., Yeh, Y. C., Lin, T. L., and Wang, J. T. Identification of genes associated with natural competence in *Helicobacter pylori* by transposon shuttle random mutagenesis. *Biochem Biophys Res Commun* 288:961–968, 2001.

Chang, Y. E., and Laimins, L. A. Interferon-inducible genes are major targets of human papillomavirus type 31: Insights from microarray analysis. *Dis Markers* 17:139–142, 2001.

Chang, Y. E., and Laimins, L. A. Microarray analysis identifies interferon-inducible genes and Stat-1 as major transcriptional targets of human papillomavirus type 31. *J Virol* 74:4174–4182, 2000.

Chapman, S., Schenk, P., Kazan, K., and Manners, J. Using biplots to interpret gene expression patterns in plants. *Bioinformatics* 181:202–204, 2002.

Chatterjee, A., and Roux, S. J. *Ceratopteris richardii:* A productive model for revealing secrets of signaling and development. *J Plant Growth Regul* 19:284–289, 2000.

Chattopadhyay, S., Muzaffar, N. E., Sherman, F., and Pearce, D. A. The yeast model for batten disease: Mutations in BTN1, BTN2, and HSP30 alter pH homeostasis. *J Bacteriol* 182:6418–6423, 2000.

Chauhan, B. K., Reed, N. A., Zhang, W., Duncan, M. K., Kilimann, M., and Cvekl, A. Identification of genes downstream of Pax6 in the mouse lens using cDNA microarrays. *J Biol Chem* 2002.

Chaussee, M. S., Sylva, G. L., Sturdevant, D. E., Smoot, L. M., Graham, M. R., Watson, R. O., and Musser, J. M. Rgg influences the expression of multiple regulatory loci to coregulate virulence factor expression in *Streptococcus pyogenes. Infect Immun* 70:762–770, 2002.

Che, D., Bao, Y., and Muller, U. R. Novel surface and multicolor charge coupled device-based fluorescent imaging system for DNA microarrays. *J Biomed Opt* 6:450–456, 2001.

Chechetkin, V. R., Turygin, A. Y., Proudnikov, D. Y., Prokopenko, D. V., Kirillov, E. V., and Mirzabekov, A. D. Sequencing by hybridization with the generic 6-mer oligonucleotide microarray: An advanced scheme for data processing. *J Biomol Struct Dyn* 18:83–101, 2000.

Cheek, B. J., Steel, A. B., Torres, M. P., Yu, Y. Y., and Yang, H. Chemiluminescence detection for hybridization assays on the flow-thru chip, a three-dimensional microchannel biochip. *Anal Chem* 73:5777–5783, 2001.

Chen, B. P., Li, Y. S., Zhao, Y., Chen, K. D., Li, S., Lao, J., Yuan, S., Shyy, J. Y., and Chien, S. DNA microarray analysis of gene expression in endothelial cells in response to 24-h shear stress. *Physiol Genomics* 7:55–63, 2001.

Chen, C. C., Shieh, B., Jin, Y. T., Liau, Y. E., Huang, C. H., Liou, J. T., Wu, L. W., Huang, W., and co-workers. Microarray profiling of gene expression patterns in bladder tumor cells treated with genistein. *J Biomed Sci* 8:214–222, 2001.

Chen, F., Bower, J., Leonard, S. S., Ding, M., Lu, Y., Rojanasakul, Y., and co-workers. Protective roles of NF-kB for chromium(VI)-induced cytotoxicity is revealed by expression of IkB kinase-b mutant. *J Biol Chem* 2001.

Chen, F., Demers, L. M., Vallyathan, V., Lu, Y., Castranova, V., and Shi, X. Impairment of NF-kappaB activation and modulation of gene expression by calpastatin. *Am J Physiol Cell Physiol* 279:C709–C716, 2000.

Chen, H., Liu, J., Merrick, B. A., and Waalkes, M. P. Genetic events associated with arsenic-induced malignant transformation: Applications of cDNA microarray technology. *Mol Carcinog* 30:79–87, 2001.

Chen, H., Liu, J., Zhao, C. Q., Diwan, B. A., Merrick, B. A., and Waalkes, M. P. Association of c-myc overexpression and hyperproliferation with arsenite-induced malignant transformation. *Toxicol Appl Pharmacol* 175:260–268, 2001.

Chen, J. N., Li, Y., Li, Y. L., Qin, H. Y., Li, R. Y., Cao, H. M., Xie, Y., and Mao, Y. M. Investigation of genotyping HLA DRB1 gene using oligoneucleotide arrays. *Yi Chuan Xue Bao* 28(10):887–894, 2001.

Chen, J., Zhong, Q., Wang, J., Cameron, R. S., Borke, J. L., Isales, C. M., and Bollag, R. J. Microarray analysis of Tbx2-directed gene expression: A possible role in osteogenesis. *Mol Cell Endocrinol* 177:43–54, 2001.

Cheran, L. E., McGovern, M. E., and Thompson, M. Surface immobilized biochemical macromolecules studied by scanning Kelvin microprobe. *Faraday Discuss* 116:23–34, 2000.

Cherkaoui-Malki, M., Meyer, K., Cao, W. Q., Latruffe, N., Yeldandi, A. V., Rao, M. S., Bradfield, C. A., and Reddy, J. K. Identification of novel peroxisome proliferator-activated receptor alpha (PPARalpha) target genes in mouse liver using cDNA microarray analysis. *Gene Expr* 9:291–304, 2001.

Cheung, V. G., Gregg, J. P., Gogolin-Ewens, K. J., Bandong, J., Stanley, C. A., Baker, L., Higgins, M. J., Nowak, N. J., and co-workers. Linkage-disequilibrium mapping without genotyping. *Nat Genet* 18:225–230, 1998.

Cheung, V. G., Morley, M., Aguilar, F., Massimi, A., Kucherlapati, R., and Childs, G. Making and reading microarrays. *Nat Genet* 21(Suppl 1):15–91, 1999.

Cheung, V. G., and Nelson, S. F. Genomic mismatch scanning identifies human genomic DNA shared identical by descent. *Genomics* 47:1–6, 1998.

Chiang, D. Y., Brown, P. O., and Eisen, M. B. Visualizing associations between genome sequences and gene expression data using genome-mean expression profiles. *Bioinformatics* 17(Suppl 1):S49–S55, 2001.

Chinnaiyan, A. M., Huber-Lang, M., Kumar-Sinha, C., Barrette, T. R., Shankar-Sinha, S., Sarma, V. J., Padgaonkar, V. A., and Ward, P. A. Molecular signatures of sepsis: Multiorgan gene expression profiles of systemic inflammation. *Am J Pathol* 159:1199–1209, 2001.

Chinnaiyan, A. M., and Rubin, M. A. Gene-expression profiles in hereditary breast cancer: On gene-expression profiles in hereditary breast cancer. *Adv Anat Pathol* 9:1–6, 2002.

Chizhikov, V., Rasooly, A., Chumakov, K., and Levy, D. D. Microarray analysis of microbial virulence factors. *Appl Environ Microbiol* 67:3258–3263, 2001.

Cho, J. C., and Tiedje, J. M. Bacterial species determination from DNA-DNA hybridization by using genome fragments and dna microarrays. *Appl Environ Microbiol* 67:3677–3682, 2001.

Cho, S. H., Tam, S. W., Demissie-Sanders, S., Filler, S. A., and Oh, C. K. Production of plasminogen activator inhibitor-1 by human mast cells and its possible role in asthma. *J Immunol* 165:3154–3161, 2000.

Cho, Y. S., Kim, M. K., Cheadle, C., Neary, C., Becker, K. G., and Cho-Chung, Y. S. Antisense DNAs as multisite genomic modulators identified by DNA microarray. *Proc Natl Acad Sci U S A* 98:9819–9823, 2001.

Cho, Y. J., Meade, J. D., Walden, J. C., Chen, X., Guo, Z., and Liang, P. Multicolor fluorescent differential display. *Biotechniques* 30:562–568, 570, 572, 2001.

Choi, D., Whittier, P. S., Oshima, J., and Funk, W. D. Telomerase expression prevents replicative senescence but does not fully reset mRNA expression patterns in Werner syndrome cell strains. *FASEB J* 15:1014–1020, 2001.

Chtanova, T., Kemp, R. A., Sutherland, A. P., Ronchese, F., and Mackay, C. R. Gene microarrays reveal extensive differential gene expression in both cd4(+) and cd8(+) type 1 and type 2 t cells. *J Immunol* 167:3057–3063, 2001.

Chu, S., DeRisi, J., Eisen, M., Mulholland, J., Botstein, D., Brown, P. O., and Herskowitz, I. The transcriptional program of sporulation in budding yeast. *Science* 282:699–705, 1998.

Chuaqui, R., Cole, K., Cuello, M., Silva, M., Quintana, M. E., and Emmert-Buck, M. R. Analysis of mRNA quality in freshly prepared and archival Papanicolaou samples. *Acta Cytol* 43:831–836, 1999.

Chuaqui, R. F., Cole, K. A., Emmert-Buck, M. R., and Merino, M. J. Histopathology and molecular biology of ovarian epithelial tumors. *Ann Diagn Pathol* 2:195–207, 1998.

Chudin, E., Walker, R., Kosaka, A., Wu, S. X., Rabert, D., Chang, T. K., and Kreder, D. E. Assessment of the relationship between signal intensities and transcript concentration for Affymetrix GeneChip(R) arrays. *Genome Biol* 3: 2002.

Chun, H. S., Gibson, G. E., DeGiorgio, L. A., Zhang, H., Kidd, V. J., and Son, J. H. Dopaminergic cell death induced by MPP(+), oxidant and specific neurotoxicants shares the common molecular mechanism. *J Neurochem* 76:1010–1021, 2001.

Chung, G. G., Provost, E., Kielhorn, E. P., Charette, L. A., Smith, B. L., and Rimm, D. L. Tissue microarray analysis of beta-catenin in colorectal cancer shows nuclear phospho-beta-catenin is associated with a better prognosis. *Clin Cancer Res* 7:4013–4020, 2001.

Churchill, G. A., and Oliver, B. Sex, flies and microarrays. *Nat Genet* 29:355–356, 2001.

Cirelli, C., and Tononi, G. Differences in brain gene expression between sleep and waking as revealed by mRNA differential display and cDNA microarray technology. *J Sleep Res* 8(Suppl 1):44–52, 1999.

Cirelli, C., and Tononi, G. Gene expression in the brain across the sleep-waking cycle. *Brain Res* 885:303–321, 2000.

Clark, M. D., Hennig, S., Herwig, R., Clifton, S. W., Marra, M. A., Lehrach, H., Johnson, S. L., and WU-GSCnEST Group. An oligonucleotide fingerprint normalized and expressed sequence tag characterized zebrafish cDNA library. *Genome Res* 11:1594–1602, 2001.

Clarke, P. A., te Poele, R., Wooster, R., and Workman, P. Gene expression microarray analysis in cancer biology, pharmacology, and drug development: progress and potential. *Biochem Pharmacol* 62:1311–1336, 2001.

Claverie, J. M. Computational methods for the identification of differential and coordinated gene expression. *Hum Mol Genet* 8:1821–1832, 1999.

Cleary, J. D., Rogers, P. D., and Chapman, S. W. Differential transcription factor expression in human mononuclear cells in response to amphotericin B: Identification with complementary DNA microarray technology. *Pharmacotherapy* 21:1046–1054, 2001.

Clement, K., Viguerie, N., Diehn, M., Alizadeh, A., Barbe, P., Thalamas, C., Storey, J. D., Brown, P. O., and co-workers. In-vivo regulation of human skeletal muscle gene expression by thyroid hormone. *Genome Res* 12:281–291, 2002.

Clewley, J. P. DNA microarrays. *Commun Dis Public Health* 3:71–72, 2000.

Cobb, J. P., Brownstein, B. H., Watson, M. A., Shannon, W. D., Laramie, J. M., Qiu, Y., Stormo, G. D., Morrissey, J. J., and co-workers. Injury in the era of genomics. *Shock* 15:165–170, 2001.

Cobb, J. P., and Buchman, T. G. microArRAY of hope. *Shock* 16:264–265, 2001.

Coffey, R. J., and Threadgill, D. Microarray foray. *Breast Cancer Res* 2:8–9, 2000.

Cohen, N., Rozenfeld-Granot, G., Hardan, I., Brok-Simoni, F., Amariglio, N., Rechavi, G., and Trakhtenbrot, L. Subgroup of patients with Philadelphia-positive chronic myelogenous leukemia characterized by a deletion of 9q proximal to ABL gene: Expression profiling, resistance to interferon therapy, and poor prognosis. *Cancer Genet Cytogenet* 128:114–119, 2001.

Cohen, P., Bouaboula, M., Bellis, M., Baron, V., Jbilo, O., Poinot-Chazel, C., Galiegue, S., Hadibi, E. H., and Casellas, P. Monitoring cellular responses to *Listeria* monocytogenes with oligonucleotide arrays. *J Biol Chem* 275:11181–11190, 2000.

Cojocaru, G. S., Rechavi, G., and Kaminski, N. The use of microarrays in medicine. *Isr Med Assoc J* 3:292–296, 2001.

Colantuoni, C., Jeon, O.H., Hyder, K., Chenchik, A., Khimani, A. H., Narayanan, V., Hoffman, E. P., Kaufmann, W. E., and co-workers. Gene expression profiling in postmortem Rett syndrome brain: Differential gene expression and patient classification. *Neurobiol Dis* 8:847–865, 2001.

Colantuoni, C., Purcell, A. E., Bouton, C. M., and Pevsner, J. High throughput analysis of gene expression in the human brain. *J Neurosci Res* 59:1–10, 2000.

Cole, K. A., Krizman, D. B., and Emmert-Buck, M. R. The genetics of cancer—A 3D model. *Nat Genet* 21(Suppl 1):38–41, 1999.

Coller, H. A., Grandori, C., Tamayo, P., Colbert, T., Lander, E. S., Eisenman, R. N., and Golub, T. R. Expression analysis with oligonucleotide microarrays reveals that MYC regulates genes involved in growth, cell cycle, signaling, and adhesion. *Proc Natl Acad Sci U S A* 97:3260–3265, 2000.

Comander, J., Weber, G. M., Gimbrone, M. A. Jr., and Garcia-Cardena, G. Argus—A new database system for Web-based analysis of multiple microarray data sets. *Genome Res* 11:1603–1610, 2001.

Connor, J. R., Kumar, S., Sathe, G., Mooney, J., O'Brien, S. P., Mui, P., Murdock, P. R., Gowen, M., and Lark, M. W. Clusterin expression in adult human normal and osteoarthritic articular cartilage. *Osteoarthritis Cartilage* 9:727–737, 2001.

Conway, T., Kraus, B., Tucker, D. L., Smalley, D. J., Dorman, A. F., and McKibben, L. DNA array analysis in a Microsoft *Windows* environment. *Biotechniques* 32:110, 112–114, 116, 118–119, 2002.

Coombes, B. K., and Mahony, J. B. cDNA array analysis of altered gene expression in human endothelial cells in response to *Chlamydia pneumoniae* infection. *Infect Immun* 69:1420–1427, 2001.

Cooper, C. S. Applications of microarray technology in breast cancer research. *Breast Cancer Res* 3:158–175, 2001.

Cossman, J. Gene expression analysis of single neoplastic cells and the pathogenesis of Hodgkin's lymphoma. *J Histochem Cytochem* 49:799–800, 2001.

Courcelle, J., Khodursky, A., Peter, B., Brown, P. O., and Hanawalt, P. C. Comparative gene expression profiles following UV exposure in wild-type and SOS-deficient *Escherichia coli*. *Genetics* 158:41–64, 2001.

Cowman, A. F. Functional analysis of drug resistance in *Plasmodium falciparum* in the post-genomic era. *Int J Parasitol* 31:871–878, 2001.

Crescenzi, M., and Giuliani, A. The main biological determinants of tumor line taxonomy elucidated by a principal component analysis of microarray data. *FEBS Lett* 507:114–118, 2001.

Crews, F. T. Summary report of a symposium: genes and gene delivery for diseases of alcoholism. *Alcohol Clin Exp Res* 25:1778–1800, 2001.

Cristillo, A. D., and Bierer, B. E. Identification of novel targets of immunosuppressive agents by cDNA-based microarray analysis. *J Biol Chem* 2001.

Cronin, M. T., Pho, M., Dutta, D., Frueh, F., Schwarcz, L., and Brennan, T. Utilization of new technologies in drug trials and discovery. *Drug Metab Dispos* 29(4):586–590, 2001.

Cummings, C. A., and Relman, D. A. Using DNA microarrays to study host-microbe interactions. *Emerg Infect Dis* 6:513–525, 2000.

Cunningham, M. J. Genomics and proteomics. The new millennium of drug discovery and development. *J Pharmacol Toxicol Methods* 44:291–300, 2000.

Cunningham, M. J., Liang, S., Fuhrman, S., Seilhamer, J. J., and Somogyi, R. Gene expression microarray data analysis for toxicology profiling. *Ann N Y Acad Sci* 919:52–67, 2000.

Cutler, D. J., Zwick, M. E., Carrasquillo, M. M., Yohn, C. T., Tobin, K. P., Kashuk, C., Mathews, D. J., Shah, N. A., and co-workers. High-throughput variation detection and genotyping using microarrays. *Genome Res* 11:1913–1925, 2001.

Custodia, N., Won, S. J., Novillo, A., Wieland, M., Li, C., and Callard, I. P. *Caenorhabditis elegans* as an environmental monitor using DNA microarray analysis. *Ann N Y Acad Sci* 948:32–42, 2001.

Dadgostar, H., Zarnegar, B., Hoffmann, A., Qin, X. F., Truong, U., Rao, G., Baltimore, D., and Cheng, G. Cooperation of multiple signaling pathways in CD40-regulated gene expression in B lymphocytes. *Proc Natl Acad Sci U S A* 99:1497–1502, 2002.

Daigo, Y., Chin, S. F., Gorringe, K. L., Bobrow, L. G., Ponder, B. A., Pharoah, P. D., and Caldas, C. Degenerate oligonucleotide primed-polymerase chain reaction-based array comparative genomic hybridization for extensive amplicon profiling of breast cancers: A new approach for the molecular analysis of paraffin-embedded cancer tissue. *Am J Pathol* 158:1623–1631, 2001.

Dales, J. P., Plumas, J., Palmerini, F., Devilard, E., Defrance, T., Lajmanovich, A., Pradel, V., Birg, F., and Xerri, L. Correlation between apoptosis microarray gene expression profiling and histopathological lymph node lesions. *Mol Pathol* 54:17–23, 2001.

Damrauer, S. M., DeFina, R., He, H., Haley, K. J., and Perkins, D. L. Molecular profiles of allograft rejection following inhibition of CD40 ligand costimulation differentiated by cluster analysis. *J Leukoc Biol* 71:348–358, 2002.

Dangond, F. Chips around the world: Proceedings from the *Nature Genetics* Microarray Meeting. *Physiol Genomics* 2:53–58, 2000.

Datta, S. Exploring relationships in gene expressions: A partial least squares approach. *Gene Expr* 9:249–255, 2001.

Davenport, R. J. Microarrays. Data standards on the horizon. *Science* 292:414–415, 2001.

Davies, M. J. Microarrays in disarray? *Trends Biotechnol* 19:163, 2001.

Davis, S. J., and Millar, A. J. Watching the hands of the *Arabidopsis* biological clock. *Genome Biol* 2: 2001.

Day, W. A. Jr., Fernandez, R. E., and Maurelli, A. T. Pathoadaptive mutations that enhance virulence: Genetic organization of the cadA regions of *Shigella* spp. *Infect Immun* 69:7471–7480, 2001.

De Bellis, G., Salani, G., Battaglia, C., Pietta, P., Rosti, E., and Mauri, P. Electrospray ionization mass spectrometry of synthetic oligonucleotides using 2-propanol and spermidine. *Rapid Commun Mass Spectrom* 14:243–249, 2000.

Debinski, W., Slagle-Webb, B., Achen, M. G., Stacker, S. A., Tulchinsky, E., Gillespie, G. Y., and Gibo, D. M. VEGF-D is an X-linked/AP-1 regulated putative onco-angiogen in human glioblastoma multiforme. *Mol Med* 7:598–608, 2001.

Debouck, C., and Goodfellow, P. N. DNA microarrays in drug discovery and development. DNA microarrays in drug discovery and development. *Nat Genet* 21(Suppl 1): 48–50, 1999.

Debouck, C., and Metcalf, B. The impact of genomics on drug discovery. *Annu Rev Pharmacol Toxicol* 40:193–207, 2000.

Degrave, W. M., Melville, S., Ivens, A., and Aslett, M. Parasite genome initiatives. *Int J Parasitol* 31:531–535, 2001.

DeLisa, M. P., Wu, C. F., Wang, L., Valdes, J. J., and Bentley, W. E. DNA microarray-based identification of genes controlled by autoinducer 2-stimulated quorum sensing in *Escherichia coli*. *J Bacteriol* 183:5239–5247, 2001.

del Rey Calero, J. Infectious disease trends. *Ann R Acad Nac Med (Madr)* 116:41–68; 69–72 (discussion), 1999.

del Rio, G., Bartley, T. F., del-Rio, H., Rao, R., Jin, K., Greenberg, D. A., Eshoo, M., and Bredesen, D. E. Mining DNA microarray data using a novel approach based on graph theory. *FEBS Lett* 509:230–234, 2001.

Dempsey, A. A., Dzau, V. J., and Liew, C. C. Cardiovascular genomics: Estimating the total number of genes expressed in the human cardiovascular system. *J Mol Cell Cardiol* 33:1879–1886, 2001.

De Nigris, F., Lerman, L. O., Condorelli, M., Lerman, A., and Napoli, C. Oxidation-sensitive transcription factors and molecular mechanisms in the arterial wall. *Antioxid Redox Signal* 3:1119–1130, 2001.

Dent, G. W., O'Dell, D. M., and Eberwine, J. H. Gene expression profiling in the amygdala: An approach to examine the molecular substrates of mammalian behavior. *Physiol Behav* 73:841–847, 2001.

DeRisi, J. L., Iyer, V. R., and Brown, P. O. Exploring the metabolic and genetic control of gene expression on a genomic scale. *Science* 278:680–686, 1997.

DeRisi, J., Penland, L., Brown, P. O., Bittner, M. L., Meltzer, P. S., Ray, M., Chen, Y., Su, Y. A., and Trent, J. M. Use of a cDNA microarray to analyse gene expression patterns in human cancer. *Nat Genet* 14:457–460, 2000.

de Saizieu, A., Gardes, C., Flint, N., Wagner, C., Kamber, M., Mitchell, T. J., Keck, W., Amrein, K. E., and Lange, R. Microarray-based identification of a novel *Streptococcus pneumoniae* regulon controlled by an autoinduced peptide. *J Bacteriol* 182:4696–4703, 2000.

De Sanctis, V., Bertozzi, C., Costanzo, G., Di Mauro, E., and Negri, R. Cell cycle arrest determines the intensity of the global transcriptional response of *Saccharomyces cerevisiae* to ionizing radiation. *Radiat Res* 156:379–387, 2001.

Desikan, R., A -H -Mackerness, S., Hancock, J. T., and Neill, S. J. Regulation of the *Arabidopsis* transcriptome by oxidative stress. *Plant Physiol* 127:159–172, 2001.

Dessens, J. T., Margos, G., Rodriguez, M. C., and Sinden, R. E. Identification of differentially regulated genes of *Plasmodium* by suppression subtractive hybridization. *Parasitol Today* 16:354–356, 2000.

Dessus-Babus, S., Knight, S. T., and Wyrick, P. B. *Chlamydial* infection of polarized HeLa cells induces PMN chemotaxis but the cytokine profile varies between disseminating and non-disseminating strains. *Cell Microbiol* 2:317–327, 2000.

Detera-Wadleigh, S. D. Lithium-related genetics of bipolar disorder. *Ann Med* 33:272–285, 2001.

Detweiler, C. S., Cunanan, D. B., and Falkow, S. Host microarray analysis reveals a role for the salmonella response regulator phoP in human macrophage cell death. *Proc Natl Acad Sci U S A* 98:5850–5855, 2001.

Devaux, F., Marc, P., Bouchoux, C., Delaveau, T., Hikkel, I., Potier, M. C., and Jacq, C. An artificial transcription activator mimics the genome-wide properties of the yeast Pdr1 transcription factor. *EMBO Rep* 2:493–498, 2001.

Devaux, F., Marc, P., and Jacq, C. Transcriptomes, transcription activators and microarrays. *FEBS Lett* 498:140–144, 2001.

de Veer, M. J., Holko, M., Frevel, M., Walker, E., Der, S., Paranjape, J. M., Silverman, R. H., and Williams, B. R. Functional classification of interferon-stimulated genes identified using microarrays. *J Leukoc Biol* 69:912–920, 2001.

Devireddy, L. R., Teodoro, J. G., Richard, F. A., and Green, M. R. Induction of apoptosis by a secreted lipocalin that is transcriptionally regulated by IL-3 deprivation. *Science* 293:829–834, 2001.

DeVita, V. T. Jr., and Bleickardt, E. W. National oncology forum: Perspectives for the year 2000. *Cancer J* 1 (Suppl 7):S2–13, 2001.

Deyholos, M. K., and Galbraith, D. W. High-density microarrays for gene expression analysis. *Cytometry* 43:229–238, 2001.

Dhanasekaran, S. M., Barrette, T. R., Ghosh, D., Shah, R., Varambally, S., Kurachi, K., Pienta, K. J., Rubin, M. A., and Chinnaiyan, A. M. Delineation of prognostic biomarkers in prostate cancer. *Nature* 412:822–826, 2001.

Dhiman, N., Bonilla, R., O'Kane, J. D., and Poland, G. A. Gene expression microarrays: A 21st century tool for directed vaccine design. *Vaccine* 20:22–30, 2001.

Diamandis, E. P. Sequencing with microarray technology—A powerful new tool for molecular diagnostics. *Clin Chem* 46:1523–1525, 2000.

Diehn, M., Alizadeh, A. A., and Brown, P. O. Examining the living genome in health and disease with DNA microarrays. *JAMA* 283:2298–2299, 2000.

Diehn, M., Eisen, M. B., Botstein, D., and Brown, P. O. Large-scale identification of secreted and membrane-associated gene products using DNA microarrays. *Nat Genet* 25:58–62, 2000.

Diehl, F., Grahlmann, S., Beier, M., and Hoheisel, J. D. Manufacturing DNA microarrays of high spot homogeneity and reduced background signal. *Nucleic Acids Res* 29:E38, 2001.

Diehn, M., and Relman, D. A. Comparing functional genomic datasets: Lessons from DNA microarray analyses of host-pathogen interactions. *Curr Opin Microbiol* 4: 95–101, 2001.

Dietz, A. B., Bulur, P. A., Knutson, G. J., Matasic, R., and Vuk-Pavlovic, S. Maturation of human monocyte-derived dendritic cells studied by microarray hybridization. *Biochem Biophys Res Commun* 275:731–738, 2000.

Dobrowolski, S. F., Banas, R. A., Naylor, E. W., Powdrill, T., and Thakkar, D. DNA microarray technology for neonatal screening. *Acta Paediatr Suppl* 88:61–64, 1999.

Doerge, R. W. Mapping and analysis of quantitative trait loci in experimental populations. *Nat Rev Genet* 3:43–52, 2002.

Doi, M., Nagano, A., and Nakamura, Y. Genome-wide Screening by cDNA microarray of genes associated with matrix mineralization by human mesenchymal stem cells in vitro. *Biochem Biophys Res Commun* 290:381–390, 2002.

Dolan, P. L., Wu, Y., Ista, L. K., Metzenberg, R. L., Nelson, M. A., and Lopez, G. P. Robust and efficient synthetic method for forming DNA microarrays. *Nucleic Acids Res* 29:E107–7, 2001.

Dolferus, R., Klok, E. J., Ismond, K., Delessert, C., Wilson, S., Good, A., Peacock, J., and Dennis, L. Molecular basis of the anaerobic response in plants. *IUBMB Life* 51:79–82, 2001.

Dolganov, G. M., Woodruff, P. G., Novikov, A. A., Zhang, Y., Ferrando, R. E., Szubin, R., and Fahy, J. V. A novel method of gene transcript profiling in airway biopsy homogenates reveals increased expression of a Na^{+}-K^{+}-Cl^{-} cotransporter (NKCC1) in asthmatic subjects. *Genome Res* 11:1473–1483, 2001.

Dominiczak, A. F., Negrin, D. C., Clark, J. S., Brosnan, M. J., McBride, M. W., and Alexander, M. Y. Genes and hypertension: From gene mapping in experimental models to vascular gene transfer strategies. *Hypertension* 35(1):164–172, 2000.

Dong, X. Genetic dissection of systemic acquired resistance. *Curr Opin Plant Biol* 4:309–314, 2001.

Dong, Y., Ganther, H. E., Stewart, C., and Ip, C. Identification of molecular targets associated with selenium-induced growth inhibition in human breast cells using cDNA microarrays. *Cancer Res* 62:708–714, 2002.

Dong, Y., Glasner, J. D., Blattner, F. R., and Triplett, E. W. Genomic interspecies microarray hybridization: Rapid discovery of three thousand genes in the maize endophyte, *Klebsiella pneumoniae* 342, by microarray hybridization with *Escherichia coli* K-12 open reading frames. *Appl Environ Microbiol* 67:1911–1921, 2001.

Dong, G., Loukinova, E., Chen, Z., Gangi, L., Chanturita, T. I., Liu, E. T., and Van Waes, C. Molecular profiling of transformed and metastatic murine squamous carcinoma cells by differential display and cDNA microarray reveals altered expression of multiple genes related to growth, apoptosis, angiogenesis, and the NF-kappaB signal pathway. *Cancer Res* 61:4797–4808, 2001.

Dorrell, N., Mangan, J. A., Laing, K. G., Hinds, J., Linton, D., Al-Ghusein, H., Barrell, B. G., Parkhill, J., and co-workers. Whole Genome Comparison of *Campylobacter jejuni* human isolates using a low-cost microarray reveals extensive genetic diversity. *Genome Res* 11:1706–1715, 2001.

Draghici, S., Kuklin, A., Hoff, B., and Shams, S. Experimental design, analysis of variance and slide quality assessment in gene expression arrays. *Curr Opin Drug Discov Devel* 4:332–337, 2001.

Dugas, M., Schoch, C., Schnittger, S., Haferlach, T., Danhauser-Riedl, S., Hiddemann, W., Messerer, D., and Uberla, K. A comprehensive leukemia database: integration of cytogenetics, molecular genetics and microarray data with clinical information, cytomorphology and immunophenotyping. *Leukemia* 15:1805–1810, 2001.

Duggan, D. J., Bittner, M., Chen, Y., Meltzer, P., and Trent, J. M. Expression profiling using cDNA microarrays. *Nat Genet* 21(Suppl 1):10–14, 1999.

Dunman, P. M., Murphy, E., Haney, S., Palacios, D., Tucker-Kellogg, G., Wu, S., Brown, E. L., Zagursky, R. J., and co-workers. Transcription profiling-based identification of *Staphylococcus aureus* genes regulated by the agr and/or sarA loci. *J Bacteriol* 183:7341–7353, 2001.

Dupont, J., Khan, J., Qu, B. H., Metzler, P., Helman, L., and LeRoith, D. Insulin and IGF-1 induce different patterns of gene expression in mouse fibroblast NIH-3T3 cells: Identification by cDNA microarray analysis. *Endocrinology* 142:4969–4975, 2001.

Dusetti, N. J., Tomasini, R., Azizi, A., Barthet, M., Vaccaro, M. I., Fiedler, F., Dagorn, J. C., and Iovanna, J. L. Expression profiling in pancreas during the acute phase of pancreatitis using cDNA microarrays. *Biochem Biophys Res Commun* 277:660–667, 2000.

Dutt, M. J., and Lee, K. H. Proteomic analysis. *Curr Opin Biotechnol* 11:176–179, 2000.

Dysvik, B., and Jonassen, I. J-Express: exploring gene expression data using Java. *Bioinformatics* 17:369–370, 2001.

Dziejman, M., Balon, E., Boyd, D., Fraser, C. M., Heidelberg, J. F., and Mekalanos, J. J. Comparative genomic analysis of *Vibrio cholerae:* Genes that correlate with cholera endemic and pandemic disease. *Proc Natl Acad Sci U S A* 99:1556–1561, 2002.

Eaves, I. A., Wicker, L. S., Ghandour, G., Lyons, P. A., Peterson, L. B., Todd, J. A., and Glynne, R. J. Combining mouse congenic strains and microarray gene expression analyses to study a complex trait: the NOD model of type 1 diabetes. *Genome Res* 12:232–243, 2002.

Edman, C. F., Mehta, P., Press, R., Spargo, C. A., Walker, G. T., and Nerenberg, M. Pathogen analysis and genetic predisposition testing using microelectronic arrays and isothermal amplification. *J Investig Med* 48:93–101, 2000.

Ehrlich, M., Buchanan, K. L., Tsien, F., Jiang, G., Sun, B., Uicker, W., Weemaes, C. M., Smeets, D., and co-workers. DNA methyltransferase 3B mutations linked to the ICF syndrome cause dysregulation of lymphogenesis genes. *Hum Mol Genet* 10:2917–2931, 2001.

Ehrt, S., Schnappinger, D., Bekiranov, S., Drenkow, J., Shi, S., Gingeras, T. R., Gaasterland, T., Schoolnik, G., and Nathan, C. Reprogramming of the macrophage transcriptome in response to interferon-gamma and *Mycobacterium tuberculosis.* Signaling roles of nitric oxide synthase-2 and phagocyte oxidase. *J Exp Med* 194:1123–1140, 2001.

Eidelman, O., Srivastava, M., Zhang, J., Leighton, X., Murtie, J., Jozwik, C., Jacobson, K., Weinstein, D. L., and co-workers. Control of the proinflammatory state in cystic fibrosis lung epithelial cells by genes from the TNF-alphaR/NFkappaB pathway. *Mol Med* 7:523–534, 2001.

Eisen, M. B., Spellman, P. T., Brown, P. O., and Botstein, D. Cluster analysis and display of genome-wide expression patterns. *Proc Natl Acad Sci U S A* 95:14863–14868, 1998.

Ekins, R. P. Ligand assays: From electrophoresis to miniaturized microarrays. *Clin Chem* 44:2015–2030, 1998.

Ekins, R., and Chu, F. Immunoassay and other ligand assays: Present status and future trends. *J Int Fed Clin Chem* 9:100–109, 1997.

Ekins, R., and Chu, F. W. Microarrays: Their origins and applications. *Trends Biotechnol* 17:217–218, 1999.

Elek, J., Park, K. H., and Narayanan, R. Microarray-based expression profiling in prostate tumors. *In Vivo* 14:173–182, 2000.

Elder, J. K, Johnson, M., Milner, N., Mir, K. U., Sohail, M., and Southern, E. M. Antisense oligonucleotide scanning arrays. In: M. Schena, ed. *DNA Microarrays: A Practical Approach,* 2nd ed. Oxford University Press, Oxford, UK, 2000: 77–99.

Ellisen, L. W., Palmer, R. E., Maki, R. G., Truong, V. B., Tamayo, P., Oliner, J. D., and Haber, D. A. Cascades of transcriptional induction during human lymphocyte activation. *Eur J Cell Biol* 80:321–328, 2001.

El-Sayed, N. M., Hegde, P., Quackenbush, J., Melville, S. E., and Donelson, J. E. The African trypanosome genome. *Int J Parasitol* 30:329–345, 2000.

Emi, M. cDNA microarray and SNP analysis. *J Nippon Med Sch* 68:411–412, 2001.

Engellau, J., Akerman, M., Anderson, H., Domanski, H. A., Rambech, E., Alvegard, T. A., and Nilbert, M. Tissue microarray technique in soft tissue sarcoma: immunohistochemical Ki-67 expression in malignant fibrous histiocytoma. *Appl Immunohistochem Mol Morphol* 9:358–363, 2001.

Englert, D. Production of microarrays on porous substrates using non-contact piezoelectric dispensing. In: M. Schena, ed. *Microarray Biochip Technology.* Eaton Publishing, Natick, MA, 2000:231–246.

Epstein, C. B., and Butow, R. A. Microarray technology—Enhanced versatility, persistent challenge. *Curr Opin Biotechnol* 11:36–41, 2000.

Epstein, C. B., Hale, W. IV, and Butow, R. A. Numerical methods for handling uncertainty in microarray data: An example analyzing perturbed mitochondrial function in yeast. *Methods Cell Biol* 65:439–452, 2001.

Epstein, C. B., Waddle, J. A., Hale, W. IV, Dave, V., Thornton, J., Macatee, T. L., Garner, H. R., and Butow, R. A. Genome-wide responses to mitochondrial dysfunction. *Mol Biol Cell* 12:297–308, 2001.

Erdogan, F., Kirchner, R., Mann, W., Ropers, H. H., and Nuber, U. A. Detection of mitochondrial single nucleotide polymorphisms using a primer elongation reaction on oligonucleotide microarrays. *Nucleic Acids Res* 29:E36, 2001.

Ermolaeva, O., Rastogi, M., Pruitt, K. D., Schuler, G. D., Bittner, M. L., Chen, Y., Simon, R., Meltzer, P., and co-workers. Data management and analysis for gene expression arrays. *Nat Genet* 20:19–23, 1998.

Erwin, C. R., Falcone, R. A., Stern, L. E., Kemp, C. J., and Warner, B. W. Analysis of intestinal adaptation gene expression by cDNA expression arrays. *JPEN J Parenter Enteral Nutr* 24:311–316, 2000.

Escalante, R., Vicente, J. J. Dictyostelium discoideum: A model system for differentiation and patterning. *Int J Dev Biol* 44:819–835, 2000.

Evans, C. O., Young, A. N., Brown, M. R., Brat, D. J., Parks, J. S., Neish, A. S., and Oyesiku, N. M. Novel patterns of gene expression in pituitary adenomas identified by complementary deoxyribonucleic acid microarrays and quantitative reverse transcription-polymerase chain reaction. *J Clin Endocrinol Metab* 86:3097–3107, 2001.

Evertsz, E., Gupta, R., Starink, P., and Watson, D. Technology and applications of gene expression microarrays. In: M. Schena, ed. *Microarray Biochip Technology.* Eaton Publishing, Natick, MA, 2000:149–166.

Evertsz, E. M., Au-Young, J., Ruvolo, M. V., Lim, A. C., and Reynolds, M. A. Hybridization cross-reactivity within homologous gene families on glass cDNA microarrays. *Biotechniques* 31:1182, 1184, 1186, 2001.

Ewing, R. M., and Claverie, J. M. EST databases as multi-conditional gene expression datasets. EST databases as multi-conditional gene expression datasets. *Pac Symp Biocomput* 430–442, 2000.

Eyster, K. M., Boles, A. L., Brannian, J. D., and Hansen, K. A. DNA microarray analysis of gene expression markers of endometriosis. *Fertil Steril* 77:38–42, 2002.

Eyster, K. M., and Lindahl, R. Molecular medicine: A primer for clinicians. Part XII: DNA microarrays and their application to clinical medicine. *S D J Med* 54:57–61, 2001.

Fahrer, A. M., Bazan, J. F., Papathanasiou, P., Nelms, K. A., and Goodnow, C. C. A genomic view of immunology. *Nature* 409:836–838, 2001.

Fahrer, A. M., Konigshofer, Y., Kerr, E. M., Ghandour, G., Mack, D. H., Davis, M. M., and Chien, Y. H. Attributes of gammadelta intraepithelial lymphocytes as suggested by their transcriptional profile. *Proc Natl Acad Sci U S A* 98:10261–10266, 2001.

Fan, M., Goodwin, M. E., Birrer, M. J., and Chambers, T. C. The c-Jun NH(2)-terminal protein kinase/AP-1 pathway is required for efficient apoptosis induced by vinblastine. *Cancer Res* 61:4450–4458, 2001.

Fan, J. B., Surti, U., Taillon-Miller, P., Hsie, L., Kennedy, G. C., Hoffner, L., Ryder, T., Mutch, D. G., and Kwok, P. Y. Paternal origins of complete hydatidiform moles proven by whole genome single-nucleotide polymorphism haplotyping. *Genomics* 79:58–62, 2002.

Farjo, R., Yu, J., Othman, M. I., Yoshida, S., Sheth, S., Glaser, T., Baehr, W., and Swaroop, A. Mouse eye gene microarrays for investigating ocular development and disease. *Vision Res* 42:463–470, 2002.

Farlow, D. N., Vansant, G., Cameron, A. A., Chang, J., Khoh-Reiter, S., Pham, N. L., Wu, W., Sagara, Y., and co-workers. Gene expression monitoring for gene discovery in models of peripheral and central nervous system differentiation, regeneration, and trauma. *J Cell Biochem* 80:171–180, 2000.

Farook, J. M., Zhu, Y. Z., Wang, H., Moochhala, S., Lee, L., and Wong, P. T. Strain differences in freezing behavior of PVG hooded and Sprague-Dawley rats: Differential cortical expression of cholecystokinin2 receptors. *Neuroreport* 12:2717–2720, 2001.

Favis, R., and Barany, F. Mutation detection in K-ras, BRCA1, BRCA2, and p53 using PCR/LDR and a universal DNA microarray. *Ann N Y Acad Sci* 906:39–43, 2000.

Favis, R., Day, J. P., Gerry, N. P., Phelan, C., Narod, S., and Barany, F. Universal DNA array detection of small insertions and deletions in BRCA1 and BRCA2. *Nat Biotechnol* 18:561–564, 2000.

Fazzio, T. G., Kooperberg, C., Goldmark, J. P., Neal, C., Basom, R., Delrow, J., and Tsukiyama, T. Widespread collaboration of Isw2 and Sin3-Rpd3 chromatin remodeling complexes in transcriptional repression. *Mol Cell Biol* 21:6450–6460, 2001.

Felix, C. A., Lange, B. J., and Chessells, J. M. Pediatric acute lymphoblastic leukemia: Challenges and controversies in 2000. *Hematology* 285–302, 2000.

Fellenberg, K., Hauser, N. C., Brors, B., Neutzner, A., Hoheisel, J. D., and Vingron, M. Correspondence analysis applied to microarray data. *Proc Natl Acad Sci U S A* 98:10781–10786, 2001.

Feng, H. P. Picture story. A protein microarray. *Nat Struct Biol* 7:829, 2000.

Feng, X., Jiang, Y., Meltzer, P., and Yen, P. M. Thyroid hormone regulation of hepatic genes in vivo detected by complementary DNA microarray. *Mol Endocrinol* 14:947–955, 2000.

Feng, X., Jiang, Y., Meltzer, P., and Yen, P. M. Transgenic targeting of a dominant negative corepressor to liver blocks basal repression by thyroid hormone receptor and increases cell proliferation. *J Biol Chem* 276:15066–15072, 2001.

Feng, Y., Yang, J. H., Huang, H., Kennedy, S. P., Turi, T. G., Thompson, J. F., Libby, P., and Lee, R. T. Transcriptional profile of mechanically induced genes in human vascular smooth muscle cells. *Circ Res* 85:1118–1123, 1999.

Feriotto, G., Corradini, R., Sforza, S., Bianchi, N., Mischiati, C., Marchelli, R., and Gambari, R. Peptide nucleic acids and biosensor technology for real-time detection of the cystic fibrosis w1282x mutation by surface plasmon resonance. *Lab Invest* 81:1415–1427, 2001.

Ferrante, A. W. Jr., Thearle, M., Liao, T., and Leibel, R. L. Effects of leptin deficiency and short-term repletion on hepatic gene expression in genetically obese mice. *Diabetes* 50:2268–2278, 2001.

Ferea, T. L., Botstein, D., Brown, P. O., and Rosenzweig, R. F. Systematic changes in gene expression patterns following adaptive evolution in yeast. *Proc Natl Acad Sci U S A* 96:9721–9726, 1999.

Ferea, T. L., and Brown, P. O. Observing the living genome. *Curr Opin Genet Dev* 9:715–722, 1999.

Ferguson, J. A., Boles, T. C., Adams, C. P., and Walt, D. R. A fiber-optic DNA biosensor microarray for the analysis of gene expression. *Nat Biotechnol* 14:1681–1684, 1996.

Ferguson, J. A., Steemers, F. J., and Walt, D. R. High-density fiber-optic DNA random microsphere array. *Anal Chem* 72:5618–5624, 2000.

Fernandez Zapico, M. E., Ahmad, U. S., and Urrutia, R. DNA microarrays: revolutionary insight into the living genome. *Surgery* 130:403–407, 2001.

Feske, S., Giltnane, J., Dolmetsch, R., Staudt, L. M., and Rao, A. Gene regulation mediated by calcium signals in T lymphocytes. *Nat Immunol* 2:316–324, 2001.

Finlin, B. S., Gau, C. L., Murphy, G. A., Shao, H., Kimel, T., Seitz, R. S., Chiu, Y. F., Botstein, D., and co-workers. Rerg is a novel ras-related, estrogen-regulated and growth-inhibitory gene in breast cancer. *J Biol Chem* 2001.

Fitzgerald, J. R., and Musser, J. M. Evolutionary genomics of pathogenic bacteria. *Trends Microbiol* 9:547–553, 2001.

Fitzgerald, J. R., Sturdevant, D. E., Mackie, S. M., Gill, S. R., and Musser, J. M. Evolutionary genomics of *Staphylococcus aureus:* Insights into the origin of methicillin-resistant strains and the toxic shock syndrome epidemic. *Proc Natl Acad Sci U S A* 98:8821–8826, 2001.

Fletcher, S. T., Baker, V. A., Fentem, J. H., Basketter, D. A., and Kelsell, D. P. Gene expression analysis of EpiDerm() following exposure to SLS using cDNA microarrays. *Toxicol In Vitro* 15:393–398, 2001.

Florell, S. R., Coffin, C. M., Holden, J. A., Zimmermann, J. W., Gerwels, J. W., Summers, B. K., Jones, D. A., and Leachman, S. A. Preservation of RNA for functional genomic studies: a multidisciplinary tumor bank protocol. *Mod Pathol* 14:116–128, 2001.

Flores-Morales, A., Stahlberg, N., Tollet-Egnell, P., Lundeberg, J., Malek, R. L., Quackenbush, J., Lee, N. H., and Norstedt, G. Microarray analysis of the in vivo effects of hypophysectomy and growth hormone treatment on gene expression in the rat. *Endocrinology* 142:3163–3176, 2001.

Fogolari, F., Tessari, S., and Molinari, H. Singular value decomposition analysis of protein sequence alignment score data. *Proteins* 46:161–170, 2002.

Ford, B. N., Wilkinson, D., Thorleifson, E. M., and Tracy, B. L. Gene expression responses in lymphoblastoid cells after radiation exposure. *Radiat Res* 156(5):668–671, 2001.

Fornace, A. J. Jr., Amundson, S. A., Bittner, M., Myers, T. G., Meltzer, P., Weinsten, J. N., and Trent, J. The complexity of radiation stress responses: analysis by informatics and functional genomics approaches. *Gene Expr* 7:387–400, 1999.

Forozan, F., Karhu, R., Kononen, J., Kallioniemi, A., and Kallioniemi, O. P. Genome screening by comparative genomic hybridization. *Trends Genet* 13:405–409, 1997.

Forozan, F., Mahlamaki, E. H., Monni, O., Chen, Y., Veldman, R., Jiang, Y., Gooden, G. C., Ethier, S. P., and co-workers. Comparative genomic hybridization analysis of 38 breast cancer cell lines: A basis for interpreting complementary DNA microarray data. *Cancer Res* 60:4519–4525, 2000.

Forrest, K. A. The first hope for the "big picture" of infection—DNA microarrays. *Fertil Steril* 72:185–187, 1999.

Fortin, A., Cregan, S. P., MacLaurin, J. G., Kushwaha, N., Hickman, E. S., Thompson, C. S., Hakim, A., Albert, P. R., and co-workers. APAF1 is a key transcriptional target for p53 in the regulation of neuronal cell death. *J Cell Biol* 155:207–216, 2001.

Foster, W. R., and Huber, R. M. Current themes in microarray experimental design and analysis. *Drug Discov Today* 7:290–292, 2002.

Fox, J. L. Complaints raised over restricted microarray access. *Nat Biotechnol* 17:325–326, 1999.

Frantz, D. J., Hughes, B. G., Nelson, D. R., Murray, B. K., and Christensen, M. J. Cell cycle arrest and differential gene expression in HT-29 cells exposed to an aqueous garlic extract. *Nutr Cancer* 38:255–264, 2000.

Frantz, G. D., Pham, T. Q., Peale, F. V. Jr., and Hillan, K. J. Detection of novel gene expression in paraffin-embedded tissues by isotopic in situ hybridization in tissue microarrays. *J Pathol* 195:87–96, 2001.

Freed, W. J., and Vawter, M. P. Microarrays: Applications in neuroscience to disease, development, and repair. *Restor Neurol Neurosci* 18:53–56, 2001.

Freeman, T. C., Lee, K., and Richardson, P. J. Analysis of gene expression in single cells. *Curr Opin Biotechnol* 10:579–582, 1999.

Freeman, W. M., Robertson, D. J., and Vrana, K. E. Fundamentals of DNA hybridization arrays for gene expression analysis. *Biotechniques* 29:1042–1046, 1048–1055, 2000.

Friedman, N., Linial, M., Nachman, I., and Pe'er, D. Using Bayesian networks to analyze expression data. *J Comput Biol* 7:601–620, 2000.

Friend, S. H. How DNA microarrays and expression profiling will affect clinical practice. *Brit Med J* 319:1306–1307, 1999.

Friend, S. H., and Stoughton, R. B. The magic of microarrays. *Sci Am* 286:44–49, 53, 2002.

Frueh, F. W., Hayashibara, K. C., Brown, P. O., and Whitlock, J. P. Jr. Use of cDNA microarrays to analyze dioxin-induced changes in human liver gene expression. *Toxicol Lett* 122:189–203, 2001.

Fruman, D. A., Ferl, G. Z., An, S. S., Donahue, A. C., Satterthwaite, A. B., and Witte, O. N. Phosphoinositide 3-kinase and Bruton's tyrosine kinase regulate overlapping sets of genes in B lymphocytes. *Proc Natl Acad Sci U S A* 99:359–364, 2002.

Fuhrman, S., Cunningham, M. J., Wen, X., Zweiger, G., Seilhamer, J. J., and Somogyi, R. The application of shannon entropy in the identification of putative drug targets. *Biosystems* 55:5–14, 2000.

Fujibuchi, W., Anderson, J. S., and Landsman, D. PROSPECT improves *cis*-acting regulatory element prediction by integrating expression profile data with consensus pattern searches. *Nucleic Acids Res* 29:3988–3996, 2001.

Fujimori, F., Gunji, W., Kikuchi, J., Mogi, T., Ohashi, Y., Makino, T., Oyama, A., Okuhara, K., and co-workers. Crosstalk of prolyl isomerases, pin1/ess1, and cyclophilin a. *Biochem Biophys Res Commun* 289:181–190, 2001.

Fujita, M., Furukawa, Y., Tsunoda, T., Tanaka, T., Ogawa, M., and Nakamura, Y. Up-regulation of the ectodermal-neural cortex 1 (ENC1) gene, a downstream target of the beta-catenin/T-cell factor complex, in colorectal carcinomas. *Cancer Res* 2001 61:7722–7726, 2001.

Fukusaki, E., Oishi, T., Tanaka, H., Kajiyama, S., and Kobayashi, A. Identification of genes induced by taxol application using a combination of differential display RT-PCR and DNA microarray analysis. *Z Naturforsch [C]* 56:814–819, 2001.

Fukushima, H. Forensic DNA analysis–past and future. *Nippon Hoigaku Zasshi* 53:276–284, 1999.

Fuller, C. E., and Perry, A. Fluorescence in situ hybridization (FISH) in diagnostic and investigative neuropathology. *Brain Pathol* 12:67–86, 2002.

Fung, E. T., Thulasiraman, V., Weinberger, S. R., and Dalmasso, E. A. Protein biochips for differential profiling. *Curr Opin Biotechnol* 12:65–69, 2001.

Funk, W. D., Wang, C. K., Shelton, D. N., Harley, C. B., Pagon, G. D., and Hoeffler, W. K. Telomerase expression restores dermal integrity to in vitro-aged fibroblasts in a reconstituted skin model. *Exp Cell Res* 258:270–278, 2000.

Furey, T. S., Cristianini, N., Duffy, N., Bednarski, D. W., Schummer, M., and Haussler, D. Support vector machine classification and validation of cancer tissue samples using microarray expression data. *Bioinformatics* 16:906–914, 2000.

Furlong, E. E., Andersen, E. C., Null, B., White, K. P., and Scott, M. P. Patterns of gene expression during Drosophila mesoderm development. *Science* 293:1629–1633, 2001.

Furlong, E. E., Profitt, D., and Scott, M. P. Automated sorting of live transgenic embryos. *Nat Biotechnol* 19:153–156, 2001.

Furushima, K., Shimo-Onoda, K., Maeda, S., Nobukuni, T., Ikari, K., Koga, H., Komiya, S., Nakajima, T., and co-workers. Large-scale screening for candidate genes of ossification of the posterior longitudinal ligament of the spine. *J Bone Miner Res* 17:128–137, 2002.

Futcher, B. Microarrays and cell cycle transcription in yeast. *Curr Opin Cell Biol* 12:710–715, 2000.

Gaasterland, T., and Bekiranov, S. Making the most of microarray data. *Nat Genet* 24:204–206, 2000.

Gabig, M., and Wegrzyn, G. An introduction to DNA chips: Principles, technology, applications and analysis. *Acta Biochim Pol* 48:615–622, 2001.

Gachotte, D., Eckstein, J., Barbuch, R., Hughes, T., Roberts, C., and Bard, M. A novel gene conserved from yeast to humans is involved in sterol biosynthesis. *J Lipid Res* 42:150–154, 2001.

Gai, X., Lal, S., Xing, L., Brendel, V., and Walbot, V. Gene discovery using the maize genome database ZmDB. *Nucleic Acids Res* 28:94–96, 2000.

Galbraith, D. W., Macas, J., Pierson, E. A., Xu, W., and Nouzova, M. Printing DNA microarrays using the Biomek 2000 laboratory automation workstation. *Methods Mol Biol* 2001;170:131–140.

Galitski, T., Saldanha, A. J., Styles, C. A., Lander, E. S., and Fink, G. R. Ploidy regulation of gene expression. *Science* 285:251–254, 1999.

Gallinat, S., Busche, S., Yang, H., Raizada, M. K., and Sumners, C. Gene expression profiling of rat brain neurons reveals angiotensin II-induced regulation of calmodulin and synapsin I: possible role in neuromodulation. *Endocrinology* 142:1009–1016, 2001.

Galon, J., Franchimont, D., Hiroi, N., Frey, G., Boettner, A., Ehrhart-Bornstein, M., O'Shea, J. J., Chrousos, G. P., and Bornstein, S. R. Gene profiling reveals unknown enhancing and suppressive actions of glucocorticoids on immune cells. *FASEB J* 16:61–71, 2002.

Gao, X., LeProust, E., Zhang, H., Srivannavit, O., Gulari, E., Yu, P., Nishiguchi, C., Xiang, Q., and Zhou, X. A flexible light-directed DNA chip synthesis gated by deprotection using solution photogenerated acids. *Nucleic Acids Res* 29:4744–4750, 2001.

Garber, M. E., Troyanskaya, O. G., Schluens, K., Petersen, S., Thaesler, Z., Pacyna-Gengelbach, M., van de Rijn, M., Rosen, G. D., and co-workers. Diversity of gene expression in adenocarcinoma of the lung. *Proc Natl Acad Sci U S A* 98:13784–13789, 2001.

Gardiner-Garden, M., and Littlejohn, T. G. A comparison of microarray databases. *Brief Bioinform* 2:143–158, 2001.

Gasch, A. P., Huang, M., Metzner, S., Botstein, D., Elledge, S. J., and Brown, P. O. Genomic expression responses to dna-damaging agents and the regulatory role of the yeast atr homolog mec1p. *Mol Biol Cell* 12:2987–3003, 2001.

Gasch, A. P., Spellman, P. T., Kao, C. M., Carmel-Harel, O., Eisen, M. B., Storz, G., Botstein, D., and Brown, P. O. Genomic expression programs in the response of yeast cells to environmental changes. *Mol Biol Cell* 11:4241–4257, 2000.

Gaudet, J., and Mango, S. E. Regulation of organogenesis by the *Caenorhabditis elegans* FoxA protein PHA-4. *Science* 295:821–825, 2002.

Ge, H. UPA, a universal protein array system for quantitative detection of protein-protein, protein-DNA, protein-RNA and protein-ligand interactions. *Nucleic Acids Res* 28:E3, 2000.

Geiss, G. K., Bumgarner, R. E., An, M. C., Agy, M. B., van 't Wout, A. B., Hammersmark, E., Carter, V. S., Upchurch, D., and co-workers. Large-scale monitoring of host cell gene expression during HIV-1 infection using cDNA microarrays. *Virology* 266:8–16, 2000.

Geiss, G., Jin, G., Guo, J., Bumgarner, R., Katze, M. G., and Sen, G. C. A comprehensive view of regulation of gene expression by double-stranded RNA-mediated cell signaling. *J Biol Chem* 276:30178–30182, 2001.

Genoud, T., and Metraux, J. P. Crosstalk in plant cell signaling: structure and function of the genetic network. *Trends Plant Sci* 4:503–507, 1999.

Genoud, T., Trevino Santa Cruz, M. B., and Metraux, J. P. Numeric simulation of plant signaling networks. *Plant Physiol* 126:1430–1437, 2001.

George, R. A., Woolley, J. P., and Spellman, P. T. Ceramic capillaries for use in microarray fabrication. *Genome Res* 11:1780–1783, 2001.

Geraci, M. W., Gao, B., Hoshikawa, Y., Yeager, M. E., Tuder, R. M., and Voelkel, N. F. Genomic approaches to research in pulmonary hypertension. *Respir Res* 2:210–215, 2001.

Geraci, M. W., Moore, M., Gesell, T., Yeager, M. E., Alger, L., Golpon, H., Gao, B., Loyd, J. E., and co-workers. Gene expression patterns in the lungs of patients with primary pulmonary hypertension: a gene microarray analysis. *Circ Res* 88:555–562, 2001.

Gerhold, D., Lu, M., Xu, J., Austin, C., Caskey, C. T., and Rushmore, T. Monitoring expression of genes involved in drug metabolism and toxicology using DNA microarrays. *Physiol Genomics* 5:161–170, 2001.

Gerry, N. P., Witowski, N. E., Day, J., Hammer, R. P., Barany, G., and Barany, F. Universal DNA microarray method for multiplex detection of low abundance point mutations. *J Mol Biol* 292:251–262, 1999.

Gerton, J. L., DeRisi, J., Shroff, R., Lichten, M., Brown, P. O., and Petes, T. D. Inaugural article: global mapping of meiotic recombination hotspots and coldspots in the yeast Saccharomyces cerevisiae. *Proc Natl Acad Sci U S A* 97:11383–11390, 2000.

Geschwind D. H. Mice, microarrays, and the genetic diversity of the brain. *Proc Natl Acad Sci U S A* 97:10676–10678, 2000.

Geschwind, D. H. Sharing gene expression data: An array of options. *Nat Rev Neurosci* 2:435–438, 2001.

Geschwind, D. H., Gregg, J., Boone, K., Karrim, J., Pawlikowska-Haddal, A., Rao, E., Ellison, J., Ciccodicola, A., and co-workers. Klinefelter's syndrome as a model of anomalous cerebral laterality: Testing gene dosage in the X chromosome pseudoautosomal region using a DNA microarray. *Dev Genet* 23:215–229, 1998.

Geschwind, D. H., Ou, J., Easterday, M. C., Dougherty, J. D., Jackson, R. L., Chen, Z., Antoine, H., Terskikh, A., and co-workers. A genetic analysis of neural progenitor differentiation. *Neuron* 29:325–339, 2001.

Getz, G., Levine, E., and Domany, E. Coupled two-way clustering analysis of gene microarray data. *Proc Natl Acad Sci U S A* 97:12079–12084, 2000.

Ghassemian, M., Waner, D., Tchieu, J., Gribskov, M., and Schroeder, J. I. An integrated Arabidopsis annotation database for Affymetrix Genechip(R) data analysis, and tools for regulatory motif searches. *Trends Plant Sci* 6:448–449, 2001.

Ghazal, P. DNA microarray technology and integrative viral genomics. *Drug Discov Today* 6:1046, 2001.

Ghazizadeh, M., Kawanami, O., and Araki, T. Assessment of gene expression profile by cDNA microarray analysis. *J Nippon Med Sch* 68:460–461, 2001.

Ghosh, D. High throughput and global approaches to gene expression. *Comb Chem High Throughput Screen* 3:411–420, 2000.

Ghosh, D., and Chinnaiyan, A. M. Mixture modelling of gene expression data from microarray experiments. *Bioinformatics* 18:275–286, 2002.

Gilbert, D. R., Schroeder, M., and van Helden, J. Interactive visualization and exploration of relationships between biological objects. *Trends Biotechnol* 18:487–494, 2000.

Gill, R. T., DeLisa, M. P., Valdes, J. J., and Bentley, W. E. Genomic analysis of high-cell-density recombinant Escherichia coli fermentation and "cell conditioning" for improved recombinant protein yield. *Biotechnol Bioeng* 72:85–95, 2001.

Gim, B. S., Park, J. M., Yoon, J. H., Kang, C., and Kim, Y. J. *Drosophila* Med6 is required for elevated expression of a large but distinct set of developmentally regulated genes. *Mol Cell Biol* 21:5242–5255, 2001.

Gingeras, T. R., Ghandour, G., Wang, E., Berno, A., Small, P. M., Drobniewski, F., Alland, D., Desmond, E., and co-workers. Simultaneous genotyping and species identification using hybridization pattern recognition analysis of generic *Mycobacterium* DNA arrays. *Genome Res* 8:435–448, 1998.

Ginsberg, S. D., Hemby, S. E., Lee, V.M., Eberwine, J. H., and Trojanowski, J. Q. Expression profile of transcripts in Alzheimer's disease tangle-bearing CA1 neurons. *Ann Neurol* 48:77–87, 2000.

Girke, T., Todd, J., Ruuska, S., White, J., Benning, C., and Ohlrogge, J. Microarray analysis of developing *Arabidopsis* seeds. *Plant Physiol* 124:1570–1581, 2000.

Gitan, R. S., Shi, H., Chen, C. M., Yan, P. S., and Huang, T. H. Methylation-specific oligonucleotide microarray: a new potential for high-throughput methylation analysis. *Genome Res* 12:158–164, 2002.

Glynne, R., Ghandour, G., Rayner, J., Mack, D. H., and Goodnow, C. C. B-lymphocyte quiescence, tolerance and activation as viewed by global gene expression profiling on microarrays. *Immunol Rev* 176:216–246, 2000.

Glynne, R. J., and Watson, S. R. The immune system and gene expression microarrays—New answers to old questions. *J Pathol* 195:20–30, 2001.

Gmuender, H. Perspectives and challenges for DNA microarrays in drug discovery and development. *Biotechniques* 32:152–154, 156, 158, 2002.

Gmuender, H., Kuratli, K., Di Padova, K., Gray, C. P., Keck, W., and Evers, S. Gene expression changes triggered by exposure of Haemophilus influenzae to novobiocin or ciprofloxacin: combined transcription and translation analysis. *Genome Res* 11:28–42, 2001.

Golpon, H. A., Geraci, M. W., Moore, M. D., Miller, H. L., Miller, G. J., Tuder, R. M., and Voelkel, N. F. HOX genes in human lung: altered expression in primary pulmonary hypertension and emphysema. *Am J Pathol* 158:955–966, 2001.

Golub, T. R. Genomic approaches to the pathogenesis of hematologic malignancy. *Curr Opin Hematol* 8:252–261, 2001.

Golub, T. R., Slonim, D. K., Tamayo, P., Huard, C., Gaasenbeek, M., Mesirov, J. P., Coller, H., Loh, M. L., and co-workers. Molecular classification of cancer: class discovery and class prediction by gene expression monitoring. *Science* 286:531–537, 1999.

Gomes, M. D., Lecker, S. H., Jagoe, R. T., Navon, A., and Goldberg, A. L. Atrogin-1, a muscle-specific F-box protein highly expressed during muscle atrophy. *Proc Natl Acad Sci U S A* 98:14440–14445, 2001.

Gonzalez, A., and Cantu, J. M. The thread and the labyrinth. Use and applications of DNA microarrays in the genomic sciences. *Rev Invest Clin* 53:298–301, 2001.

Goodwin, L. O., Mason, J. M., and Hajdu, S. I. Gene expression patterns of paired bronchioloalveolar carcinoma and benign lung tissue. *Ann Clin Lab Sci* 31:369–375, 2001.

Goryachev, A. B., Macgregor, P. F., and Edwards, A. M. Unfolding of microarray data. *J Comput Biol* 8:443–461, 2001.

Goto, T., Holding, C., Daniels, R., Salpekar, A., and Monk, M. Gene expression studies on human primordial germ cells and preimplantation embryos. *Ital J Anat Embryol* 106:119–127, 2001.

Gracey, A. Y., Troll, J. V., and Somero, G. N. Hypoxia-induced gene expression profiling in the euryoxic fish Gillichthys mirabilis. *Proc Natl Acad Sci U S A* 98:1993–1998, 2001.

Grandi, G. Antibacterial vaccine design using genomics and proteomics. *Trends Biotechnol* 19:181–188, 2001.

Granger, J. P., Alexander, B. T., Bennett, W. A., and Khalil, R. A. Pathophysiology of pregnancy-induced hypertension. *Am J Hypertens* 14:178S–185S, 2001.

Granjeaud, S., Bertucci, F., and Jordan, B. R. Expression profiling: DNA arrays in many guises. *Bioessays* 21:781–790,1999.

Grant, S. G., and Blackstock, W. P. Proteomics in neuroscience: from protein to network. *J Neurosci* 21:8315–8318, 2001.

Grant, W. N., and Viney, M. E. Post-genomic nematode parasitology. *Int J Parasitol* 31:879–888, 2001.

Granucci, F., Castagnoli, P. R., Rogge, L., and Sinigaglia, F. Gene expression profiling in immune cells using microarray. *Int Arch Allergy Immunol* 126:257–266, 2001.

Granucci, F., Vizzardelli, C., Pavelka, N., Feau, S., Persico, M., Virzi, E., Rescigno, M., Moro, G., and Ricciardi-Castagnoli, P. Inducible IL-2 production by dendritic cells revealed by global gene expression analysis. *Nat Immunol* 2:882–888, 2001.

Granucci, F., Vizzardelli, C., Virzi, E., Rescigno, M., and Ricciardi-Castagnoli, P. Transcriptional reprogramming of dendritic cells by differentiation stimuli. *Eur J Immunol* 31:2539–2546, 2001.

Graveel, C. R., Jatkoe, T., Madore, S. J., Holt, A. L., and Farnham, P. J. Expression profiling and identification of novel genes in hepatocellular carcinomas. *Oncogene* 20:2704–2712, 2001.

Graves, D. J., Su, H. J., Addya, S., Surrey, S., and Fortina, P. Four-laser scanning confocal system for microarray analysis. *Biotechniques* 32:346–348, 350, 352, 354, 2002.

Graves, D. J., Su, H. J., McKenzie, S. E., Surrey, S., and Fortina, P. System for preparing microhybridization arrays on glass slides. *Anal Chem* 70:5085–5092, 1998.

Gray, S. G. Immunophenotyping leukemia by microarray. *Trends Genet* 17:442, 2001.

Gray, J. W., and Collins, C. Genome changes and gene expression in human solid tumors. *Carcinogenesis* 21:443–452, 2000.

Grayhack, E. J., and Phizicky, E. M. Genomic analysis of biochemical function. *Curr Opin Chem Biol* 5:34–39, 2001.

Greenberg, D. A. Microarrays, markers of disease, and the myth of "nonhypothesis-driven" research. *Ann Neurol* 50:695, 2001.

Greenberg, S. A. DNA microarray gene expression analysis technology and its application to neurological disorders. *Neurology* 57:755–761, 2001.

Greenfield, A. Applications of DNA microarrays to the transcriptional analysis of mammalian genomes. *Mamm Genome* 11:609–613, 2000.

Grimmond, S., Van Hateren, N., Siggers, P., Arkell, R., Larder, R., Soares, M.B., de Fatima Bonaldo, M., and co-workers. Sexually dimorphic expression of protease nexin-1 and vanin-1 in the developing mouse gonad prior to overt differentiation suggests a role in mammalian sexual development. *Hum Mol Genet* 9:1553–1560, 2000.

Groen, A. K. The pros and cons of gene expression analysis by microarrays. *J Hepatol* 35:295–296, 2001.

Gross, C., Kelleher, M., Iyer, V. R., Brown, P. O., and Winge, D. R. Identification of the copper regulon in *Saccharomyces cerevisiae* by DNA microarrays. *J Biol Chem* 275:32310–32316, 2000.

Grouse, L. H., Munson, P. J., and Nelson, P. S. Sequence databases and microarrays as tools for identifying prostate cancer biomarkers. Urology 57(4 Suppl 1):154–159, 2001.

Grundschober, C., Delaunay, F., Puhlhofer, A., Triquenaux, G., Laudet, V., Bartfai, T., and Nef, P. Circadian regulation of diverse gene products revealed by mRNA expression profiling of synchronized fibroblasts. *J Biol Chem* 2001.

Gruvberger, S., Ringner, M., Chen, Y., Panavally, S., Saal, L. H., Borg, A., Ferno, M., Peterson, C., and Meltzer, P. S. Estrogen receptor status in breast cancer is associated with remarkably distinct gene expression patterns. *Cancer Res* 61:5979–5984, 2001.

Gullans, S. R. Of microarrays and meandering data points. *Nat Genet* 26:4–5, 2000.

Guo, X., and Liao, K. Analysis of gene expression profile during 3T3-L1 preadipocyte differentiation. *Gene* 251:45–53, 2000.

Guo, Q. M., Malek, R. L., Kim, S., Chiao, C., He, M., Ruffy, M., Sanka, K., Lee, N. H., and co-workers. Identification of c-myc responsive genes using rat cDNA microarray. *Cancer Res* 60:5922–5928, 2000.

Gupta, R. A., Brockman, J. A., Sarraf, P., Willson, T. M., and DuBois, R. N. Target genes of peroxisome proliferator-activated receptor gamma in colorectal cancer cells. *J Biol Chem* 276:29681–29687, 2001.

Gurova, K. V., Roklin, O. W., Krivokrysenko, V. I., Chumakov, P. M., Cohen, M. B., Feinstein, E., and Gudkov, A. V. Expression of prostate specific antigen (PSA) is negatively regulated by p53. *Oncogene* 21:153–157, 2002.

Haab, B. B. Advances in protein microarray technology for protein expression and interaction profiling. *Curr Opin Drug Discov Devel* 4:116–123, 2001.

Haab, B. B., Dunham, M. J., and Brown, P. O. Protein microarrays for highly parallel detection and quantitation of specific proteins and antibodies in complex solutions. *Genome Biol* 2: 2001.

Haase, D., Lehmann, M. H., Ko-rner, M. M., Ko-rfer, R., Sigusch, H. H., and Figulla, H. R. Identification and validation of selective upregulation of ventricular myosin light chain type 2 mRNA in idiopathic dilated cardiomyopathy. *Eur J Heart Fail* 4:23–31, 2002.

Haber, L. T., Maier, A., Zhao, Q., Dollarhide, J. S., Savage, R. E., and Dourson, M. L. Applications of mechanistic data in risk assessment: The past, present, and future. *Toxicol Sci* 61:32–39, 2001.

Hacia, J. G. Resequencing and mutational analysis using oligonucleotide microarrays. *Nat Genet* 21(Suppl 1):42–47, 1999.

Hacia, J. G., and Collins, F. S. Mutational analysis using oligonucleotide microarrays. *J Med Genet* 36:730–736, 1999.

Hacia, J. G., Fan, J. B., Ryder, O., Jin, L., Edgemon, K., Ghandour, G., Mayer, R. A., Sun, B., and co-workers. Determination of ancestral alleles for human single-nucleotide polymorphisms using high-density oligonucleotide arrays. *Nat Genet* 22:164–167, 1999.

Hacia, J. G., Edgemon, K., Fang, N., Mayer, R. A., Sudano, D., Hunt, N., and Collins, F. S. Oligonucleotide microarray based detection of repetitive sequence changes. *Hum Mutat* 16:354–363, 2000.

Haddock, S. H., Quartararo, C., Cooley, P., and Dao, D. D. Low-resolution typing of HLA-DQA1 using DNA mmicroarray. *Methods Mol Biol* 170:201–210, 2001.

Hakak, Y., Walker, J. R., Li, C., Wong, W. H., Davis, K. L., Buxbaum, J. D., Haroutunian, V., and Fienberg, A. A. Genome-wide expression analysis reveals dysregulation of myelination-related genes in chronic schizophrenia. *Proc Natl Acad Sci U S A* 98:4746–4751, 2001.

Hakenbeck, R., Balmelle, N., Weber, B., Gardes, C., Keck, W., and de Saizieu, A. Mosaic genes and mosaic chromosomes: Intra- and interspecies genomic variation of *Streptococcus pneumoniae*. *Infect Immun* 69:2477–2486, 2001.

Haley, K. J., Lilly, C. M., Yang, J. H., Feng, Y., Kennedy, S. P., Turi, T. G., Thompson, J. F., Sukhova, G. H., and co-workers. Overexpression of eotaxin and the CCR3 receptor in human atherosclerosis: Using genomic technology to identify a potential novel pathway of vascular inflammation. *Circulation* 102:2185–2190, 2000.

Halgren, R. G., Fielden, M. R., Fong, C. J., and Zacharewski, T. R. Assessment of clone identity and sequence fidelity for 1189 IMAGE cDNA clones. *Nucleic Acids Res* 29:582–588, 2001.

Hamadeh, H. K., Bushel, P., Tucker, C. J., Martin, K., Paules, R., and Afshari, C. A. Detection of diluted gene expression alterations using cDNA microarrays. *Biotechniques* 32:322, 324, 326–329, 2002.

Hamalainen, H., Zhou, H., Chou, W., Hashizume, H., Heller, R., and Lahesmaa, R. Distinct gene expression profiles of human type 1 and type 2 T helper cells. *Genome Biol* 2: 2001.

Hamel, B. C., and Poppelaars, F. A. Sex-linked mental retardation. *Ned Tijdschr Geneeskd* 144:1713–1716, 2000.

Hamels, S., Gala, J. L., Dufour, S., Vannuffel, P., Zammatteo, N., and Remacle, J. Consensus PCR and microarray for diagnosis of the genus *Staphylococcus,* species, and methicillin resistance. *Biotechniques* 31:1364–1366, 1368, 1370–1372, 2001.

Hampson, S., Baldi, P., Kibler, D., and Sandmeyer, S. B. Analysis of yeast's ORF upstream regions by parallel processing, microarrays, and computational methods. *Proc Int Conf Intell Syst Mol Biol* 8:190–201, 2000.

Han, J., Yoo, H. Y., Choi, B. H., and Rho, H. M. Selective transcriptional regulations in the human liver cell by hepatitis B viral X protein. *Biochem Biophys Res Commun* 272:525–530, 2000.

Hanash, S. M. Global profiling of gene expression in cancer using genomics and proteomics. *Curr Opin Mol Ther* 3(6):538–545, 2001.

Hanash, S. M., Bobek, M. P., Rickman, D. S., Williams, T., Rouillard, J. M., Kuick, R., and Puravs, E. Integrating cancer genomics and proteomics in the post-genome era. *Proteomics* 2:69–75, 2002.

Hansen-Hagge, T. E., Trefzer, U., zu Reventlow, A. S., Kaltoft, K., and Sterry, W. Identification of sample-specific sequences in mammalian cDNA and genomic DNA by the novel ligation-mediated subtraction (LIMES). *Nucleic Acids Res* 29:E20, 2001.

Hanzel, D. K., Trojanowski, J. Q., Johnston, R. F., and Loring, J. F. High-throughput quantitative histological analysis of Alzheimer's disease pathology using a confocal digital microscanner. *Nat Biotechnol* 17:53–57, 1999.

Harkin, D. P. Uncovering functionally relevant signaling pathways using microarray-based expression profiling. *Oncologist* 5:501–507, 2000.

Harmer, S. L., Hogenesch, J. B., Straume, M., Chang, H. S., Han, B., Zhu, T., Wang, X., Kreps, J. A., and Kay, S. A. Orchestrated transcription of key pathways in *Arabidopsis* by the circadian clock. *Science* 290:2110–2113, 2000.

Harmer, S. L., and Kay, S. A. Microarrays: determining the balance of cellular transcription. *Plant Cell* 12:613–616, 2000.

Harper, J. C., and Wells, D. Recent advances and future developments in PGD. *Prenat Diagn* 19:1193–1199, 2000.

Harrington, C. A., Rosenow, C., and Retief, J. Monitoring gene expression using DNA microarrays. *Curr Opin Microbiol* 3:285–291, 2000.

Harris, E. D. Differential PCR and DNA microarrays: The modern era of nutritional investigations. *Nutrition* 16:714–715, 2000.

Harris, N. L., Stein, H., Coupland, S. E., Hummel, M., Favera, R. D., Pasqualucci, L., and Chan, W. C. New approaches to lymphoma diagnosis. *Hematology* 194–220, 2001.

Harris, T. M., Massimi, A., and Childs, G. Injecting new ideas into microarray printing. *Nat Biotechnol* 18:384–385, 2000.

Hassmann, J., Misch, A., Schulein, J., Krause, J., Grassl, B., Muller, P., and Bertling, W. M. Development of a molecular diagnosis assay based on electrohybridization at plastic electrodes and subsequent PCR. *Biosens Bioelectron* 16:857–863, 2001.

Hata, R., Masumura, M., Akatsu, H., Li, F., Fujita, H., Nagai, Y., Yamamoto, T., Okada, H., Kosaka, K., and co-workers. Up-regulation of calcineurin Abeta mRNA in the Alzheimer's disease brain: assessment by cDNA microarray. *Biochem Biophys Res Commun* 284:310–316, 2001.

Haugen, A. Progress and potential of genetic susceptibility to environmental toxicants. *Scand J Work Environ Health* 25:537–540, 1999.

Hautefort, I., and Hinton, J. C. Measurement of bacterial gene expression in vivo. *Philos Trans R Soc Lond B Biol Sci* 355:601–611, 2000.

Hayes, A. The second international meeting on microarray data standards, annotations, ontologies and databases. *Yeast* 17:238–240, 2000.

Hayward, R. E. *Plasmodium falciparum* phosphoenolpyruvate carboxykinase is developmentally regulated in gametocytes. *Mol Biochem Parasitol* 107:227–240, 2000.

Hayward, R. E., Derisi, J. L., Alfadhli, S., Kaslow, D. C., Brown, P. O., and Rathod, P. K. Shotgun DNA microarrays and stage-specific gene expression in *Plasmodium falciparum* malaria. *Mol Microbiol* 35:6–14, 2000.

Hayashi, I., and Nishiyama, M. Genome research and anticancer chemotherapy. *Gan To Kagaku Ryoho* 28:1183–1189, 2001.

He, L. Z., Tolentino, T., Grayson, P., Zhong, S., Warrell, R. P. Jr., Rifkind, R. A., Marks, P. A., Richon, V. M., and Pandolfi, P. P. Histone deacetylase inhibitors induce remission in transgenic models of therapy-resistant acute promyelocytic leukemia. *J Clin Invest* 108:1321–1330, 2001.

He, Y. D., and Friend, S. H. Microarrays—The 21st century divining rod? *Nat Med* 7:673–679, 2001.

Heck, D. E., Roy, A., and Laskin, J. D. Nucleic acid microarray technology fortoxicology: Promise and practicalities. *Adv Exp Med Biol* 500:709–714, 2001.

Hedberg, J. J., Grafstrom, R. C., Vondracek, M., Sarang, Z., Warngard, L., and Hoog, J. O. Micro-array chip analysis of carbonyl-metabolising enzymes in normal, immortalised and malignant human oral keratinocytes. *Cell Mol Life Sci* 58:1719–1726, 2001.

Hedenfalk, I., Duggan, D., Chen, Y., Radmacher, M., Bittner, M., Simon, R., Meltzer, P., Gusterson, B., and co-workers. Gene-expression profiles in hereditary breast cancer. *N Engl J Med* 344:539–548, 2001.

Hedvat, C. V., Jaffe, E. S., Qin, J., Filippa, D. A., Cordon-Cardo, C., Tosato, G., Nimer, S. D., and Teruya-Feldstein, J. Macrophage-derived chemokine expression in classical Hodgkin's lymphoma: Application of tissue microarrays. *Mod Pathol* 14:1270–1276, 2001.

Hegde, P., Qi, R., Abernathy, K., Gay, C., Dharap, S., Gaspard, R., Hughes, J. E., Snesrud, E., and co-workers. A concise guide to cDNA microarray analysis. *Biotechniques* 29:548–550, 2000.

Hegde, P., Qi, R., Gaspard, R., Abernathy, K., Dharap, S., Earle-Hughes, J., Gay, C., Nwokekeh, N. U., and co-workers. Identification of tumor markers in models of human colorectal cancer using a 19,200-element complementary DNA microarray. *Cancer Res* 61:7792–7797, 2001.

Heinrichs, A. Microarray consortium. *Trends Mol Med* 7:200, 2001.

Heiskanen, M. A., Bittner, M. L., Chen, Y., Khan, J., Adler, K. E., Trent, J. M., and Meltzer, P. S. Detection of gene amplification by genomic hybridization to cDNA microarrays. *Cancer Res* 60:799–802, 2000.

Heiskanen, M., Kononen, J., Barlund, M., Torhorst, J., Sauter, G., Kallioniemi, A., and Kallioniemi, O. CGH, cDNA and tissue microarray analyses implicate FGFR2 amplification in a small subset of breast tumors. *Anal Cell Pathol* 22:229–234, 2001.

Hellauer, K., Sirard, E., and Turcotte, B. Decreased expression of specific genes in yeast cells lacking histone h1. *J Biol Chem* 276:13587–13592, 2001.

Heller, R. A., Allard, J., Zuo, F., Lock, C., Wilson, S., Klonowski, P., Gmuender, H., Van Wart, H., and Booth, R. Gene-chips and Micro-arrays: Applications in disease profiles, drug target discovery, drug action and toxicity. In: M. Schena, ed. *DNA Microarrays: A Practical Approach,* 2nd ed. Oxford University Press, Oxford, UK, 2000:187–202.

Heller, M. J., Holmsen, A., Sosnowski, R. G., and O'Connell, J. Active microelectronic arrays for DNA hybridization analysis. In: M. Schena, ed. *DNA Microarrays: A Practical Approach,* 2nd ed. Oxford University Press, Oxford, UK, 2000:167–185.

Heller RA, Schena M, Chai A, Shalon D, Bedilion T, Gilmore J, Woolley DE, and Davis RW. Discovery and analysis of inflammatory disease-related genes using cDNA microarrays. *Proc Natl Acad Sci U S A* 94:2150–2155, 1997.

Helliwell, C. A., Chin-Atkins, A. N., Wilson, I. W., Chapple, R., Dennis, E. S., and Chaudhury, A. The *Arabidopsis* amp1 gene encodes a putative glutamate carboxypeptidase. *Plant Cell* 13:2115–2125, 2001.

Helmann, J. D., Wu, M. F., Kobel, P. A., Gamo, F. J., Wilson, M., Morshedi, M. M., Navre, M., and Paddon, C. Global transcriptional response of *Bacillus subtilis* to heat shock. *J Bacteriol* 183:7318–7328, 2001.

Helmberg, A. DNA-microarrays: novel techniques to study aging and guide gerontologic medicine. *Exp Gerontol* 36:1189–1198, 2001.

Hendrix, M. J., Seftor, E. A., Meltzer, P. S., Gardner, L. M., Hess, A. R., Kirschmann, D. A., Schatteman, G. C., and Seftor, R. E. Expression and functional significance of VE-cadherin in aggressive human melanoma cells: Role in vasculogenic mimicry. *Proc Natl Acad Sci U S A* 98:8018–8023, 2001.

Henn, W. Genetic screening with the DNA chip: A new Pandora's box? *J Med Ethics* 25:200–203, 1999.

Henson, M., Damm, D., Lam, A., Garrard, L. J., White, T., Abraham, J. A., Schreiner, G. F., Stanton, L. W., and Joly, A. H. Insulin-like growth factor-binding protein-3 induces fetalization in neonatal rat cardiomyocytes. *DNA Cell Biol* 19:757–763, 2000.

Herrler, M. Use of SMART-generated cDNA for differential gene expression studies. *J Mol Med* 78:B23, 2000.

Hertzberg, M., Aspeborg, H., Schrader, J., Andersson, A., Erlandsson, R., Blomqvist, K., Bhalerao, R., Uhlen, M., and co-workers. A transcriptional roadmap to wood formation. *Proc Natl Acad Sci U S A* 98:14732–14737, 2001.

Hertzberg, M., Sievertzon, M., Aspeborg, H., Nilsson, P., Sandberg, G., and Lundeberg, J. cDNA microarray analysis of small plant tissue samples using a cDNA tag target amplification protocol. *Plant J* 25:585–591, 2001.

Hess, A. R., Seftor, E. A., Gardner, L. M., Carles-Kinch, K., Schneider, G. B., Seftor, R. E., Kinch, M. S., and Hendrix, M. J. Molecular regulation of tumor cell vasculogenic mimicry by tyrosine phosphorylation: Role of epithelial cell kinase (Eck/EphA2). *Cancer Res* 61:3250–3255, 2001.

Hess, K. R., Zhang, W., Baggerly, K. A., Stivers, D. N., Coombes, K. R., and Zhang, W. Microarrays: Handling the deluge of data and extracting reliable information. *Trends Biotechnol* 19:463–468, 2001.

Hicks, J. S., Harker, B. W., Beattie, K. L., and Doktycz, M. J. Modification of an automated liquid-handling system for reagent-jet, nanoliter-level dispensing. *Biotechniques* 30:878–885, 2001.

Hidalgo, A., and Salcedo, M. Global analysis strategies: Toward the genetic management of neoplasias. *Rev Invest Clin* 53:430–443, 2001.

Hihara, Y., Kamei, A., Kanehisa, M., Kaplan, A., and Ikeuchi, M. DNA microarray analysis of cyanobacterial gene expression during acclimation to high light. *Plant Cell* 13:793–806, 2001.

Hiller, R., Laffer, S., Harwanegg, C., Huber, M., Schmidt, W. M., Twardosz, A., Barletta, B., Becker, W. M., and co-workers. Microarrayed allergen molecules: diagnostic gatekeepers for allergy treatment. *FASEB J* 2002.

Hippo, Y., Taniguchi, H., Tsutsumi, S., Machida, N., Chong, J. M., Fukayama, M., Kodama, T., and Aburatani, H. Global gene expression analysis of gastric cancer by oligonucleotide microarrays. *Cancer Res* 62:233–240, 2002.

Hishikawa, K., Oemar, B. S., and Nakaki, T. Static pressure regulates connective tissue growth factor expression in human mesangial cells. *J Biol Chem* 2001.

Hodgson, G., Hager, J. H., Volik, S., Hariono, S., Wernick, M., Moore, D., Albertson, D. G., Pinkel, D., and co-workers. Genome scanning with array CGH delineates regional alterations in mouse islet carcinomas. *Nat Genet* 29:459–464, 2001.

Hoffman, E. P., and Dressman, D. Molecular pathophysiology and targeted therapeutics for muscular dystrophy. *Trends Pharmacol Sci* 22:465–470, 2001.

Hofmann, W. K., de Vos, S., Elashoff, D., Gschaidmeier, H., Hoelzer, D., Koeffler, H. P., and Ottmann, O. G. Relation between resistance of Philadelphia-chromosome-positive acute lymphoblastic leukaemia to the tyrosine kinase inhibitor STI571 and gene-expression profiles: A gene-expression study. *Lancet* 359:481–486, 2002.

Hofmann, W. K., de Vos, S., Tsukasaki, K., Wachsman, W., Pinkus, G. S., Said, J. W., and Koeffler, H. P. Altered apoptosis pathways in mantle cell lymphoma detected by oligonucleotide microarray. *Blood* 98:787–794, 2001.

Ho, L., Guo, Y., Spielman, L., Petrescu, O., Haroutunian, V., Purohit, D., Czernik, A., Yemul, S., and co-workers. Altered expression of a-type but not b-type synapsin isoform in the brain of patients at high risk for Alzheimer's disease assessed by DNA microarray technique. *Neurosci Lett* 298:191–194, 2001.

Holden, P. R., James, N. H., Brooks, A. N., Roberts, R. A., Kimber, I., and Pennie, W. D. Identification of a possible association between carbon tetrachloride-induced hepatotoxicity and interleukin-8 expression. *J Biochem Mol Toxicol* 14:283–290, 2000.

Hollon, T. Comparing microarray data: What technology is needed? *J Natl Cancer Inst* 93:1126–1127, 2001.

Holter, N. S., Maritan, A., Cieplak, M., Fedoroff, N. V., and Banavar, J. R. Dynamic modeling of gene expression data. *Proc Natl Acad Sci U S A* 98:1693–1698, 2001.

Holter, N. S., Mitra, M., Maritan, A., Cieplak, M., Banavar, J. R., and Fedoroff, N. V. Fundamental patterns underlying gene expression profiles: simplicity from complexity. *Proc Natl Acad Sci U S A* 97:8409–8414, 2000.

Holsboer, F. Prospects for antidepressant drug discovery. *Biol Psychol* 57:47–65, 2001.

Honda, M., Kaneko, S., Kawai, H., Shirota, Y., and Kobayashi, K. Differential gene expression between chronic hepatitis B and C hepatic lesion. *Gastroenterology* 120:955–966, 2001.

Honda, S., Farboud, B., Hjelmeland, L. M., and Handa, J. T. Induction of an aging mRNA retinal pigment epithelial cell phenotype by matrix-containing advanced glycation end products in vitro. *Invest Ophthalmol Vis Sci* 42:2419–2425, 2001.

Hong, T. M., Yang, P. C., Peck, K., Chen, J. J., Yang, S. C., Chen, Y. C., and Wu, C. W. Profiling the downstream genes of tumor suppressor PTEN in lung cancer cells by complementary DNA microarray. *Am J Respir Cell Mol Biol* 23:355–363, 2000.

Honma, K., Ochiya, T., Nagahara, S., Sano, A., Yamamoto, H., Hirai, K., Aso, Y., and Terada, M. Atelocollagen-based gene transfer in cells allows high-throughput screening of gene functions. *Biochem Biophys Res Commun* 289:1075–1081, 2001.

Hooper, L. V., Wong, M. H., Thelin, A., Hansson, L., Falk, P. G., and Gordon, J. I. Molecular analysis of commensal host-microbial relationships in the intestine. *Science* 291:881–884, 2001.

Hoos, A., and Cordon-Cardo, C. Tissue microarray profiling of cancer specimens and cell lines: opportunities and limitations. *Lab Invest* 81:1331–1338, 2001.

Hoos, A., Stojadinovic, A., Mastorides, S., Urist, M. J., Polsky, D., Di Como, C. J., Brennan, M. F., and Cordon-Cardo, C. High Ki-67 proliferative index predicts disease specific survival in patients with high-risk soft tissue sarcomas. *Cancer* 92:869–874, 2001.

Hoos, A., Stojadinovic, A., Singh, B., Dudas, M. E., Leung, D. H., Shaha, A. R., Shah, J. P., Brennan, M. F., and co-workers. Clinical significance of molecular expression profiles of Hurthle cell tumors of the thyroid gland analyzed via tissue microarrays. *Am J Pathol* 160:175–183, 2002.

Hoos, A., Urist, M. J., Stojadinovic, A., Mastorides, S., Dudas, M. E., Leung, D. H., Kuo, D., Brennan, M. F., and co-workers. Validation of tissue microarrays for immunohistochemical profiling of cancer specimens using the example of human fibroblastic tumors. *Am J Pathol* 158:1245–1251, 2001.

Hoovers, J. M., Mellink, C. H., and Leschot, N. J. Fluorescence in situ hybridization in the study of chromosomal abnormalities. *Ned Tijdschr Geneeskd* 143:2265–2268, 1999.

Horimoto, K., and Toh, H. Statistical estimation of cluster boundaries in gene expression profile data. *Bioinformatics* 17:1143–1151, 2001.

Horvath, L., and Henshall, S. The application of tissue microarrays to cancer research. *Pathology* 33:125–129, 2001.

Horvath, S., and Baur, M. P. Future directions of research in statistical genetics. *Stat Med* 19:3337–3343, 2000.

Hossain, M. A., Bouton, C. M., Pevsner, J., and Laterra, J. Induction of vascular endothelial growth factor in human astrocytes by lead. Involvement of a protein kinase C/activator protein-1 complex-dependent and hypoxia-inducible factor 1-independent signaling pathway. *J Biol Chem* 275:27874–27882, 2000.

Hou, Z., Bailey, J. P., Vomachka, A. J., Matsuda, M., Lockefeer, J. A., and Horseman, N. D. Glycosylation-dependent cell adhesion molecule 1 (GlyCAM 1) is induced by prolactin and suppressed by progesterone in mammary epithelium. *Endocrinology* 141:4278–4283, 2000.

Houghton, R. L., Dillon, D. C., Molesh, D. A., Zehentner, B. K., Xu, J., Jiang, J., Schmidt, C., Frudakis, A., and co-workers. Transcriptional complementarity in breast cancer: application to detection of circulating tumor cells. *Mol Diagn* 6:79–91, 2001.

Howell, S. B. DNA microarrays for analysis of gene expression. *Mol Urol* 3:295–300, 1999.

Hsiao, L. L., Dangond, F., Yoshida, T., Hong, R., Jensen, R. V., Misra, J., Dillon, W., Lee, K. F., and co-workers. A compendium of gene expression in normal human tissues. *Physiol Genomics* 7:97–104, 2001.

Hsiao, L. L., Jensen, R. V., Yoshida, T., Clark, K. E., Blumenstock, J. E., and Gullans, S. R. Correcting for signal saturation errors in the analysis of microarray data. *Biotechniques* 32:330–332, 334, 336, 2002.

Hsiao, L. L., Stears, R. L., Hong, R. L., and Gullans, S. R. Prospective use of DNA microarrays for evaluating renal function and disease. *Curr Opin Nephrol Hypertens* 9:253–258, 2000.

Hsieh, H. B., Lersch, R. A., Callahan, D. E., Hayward, S., Wong, M., Clark, O. H., and Weier, H. U. Monitoring signal transduction in cancer: cDNA microarray for semiquantitative analysis. *J Histochem Cytochem* 49:1057–1058, 2001.

Hua, S., and Sun, Z. A novel method of protein secondary structure prediction with high segment overlap measure: Support vector machine approach. *J Mol Biol* 308:397–407, 2001.

Huang, G. S., Yang, S. M., Hong, M. Y., Yang, P. C., and Liu, Y. C. Differential gene expression of livers from ApoE deficient mice. *Life Sci* 68:19–28, 2000.

Huang, J., Lih, C. J., Pan, K. H., and Cohen, S. N. Global analysis of growth phase responsive gene expression and regulation of antibiotic biosynthetic pathways in *Streptomyces coelicolor* using DNA microarrays. *Genes Dev* 15:3183–3192, 2001.

Huang, J. X., Mehrens, D., Wiese, R., Lee, S., Tam, S. W., Daniel, S., Gilmore, J., Shi, M., and Lashkari, D. High-throughput genomic and proteomic analysis using microarray technology. *Clin Chem* 47:1912–1916, 2001.

Huang, J., Qi, R., Quackenbush, J., Dauway, E., Lazaridis, E., and Yeatman, T. Effects of ischemia on gene expression. *J Surg Res* 99:222–227, 2001.

Huang, P., Feng, L., Oldham, E. A., Keating, M. J., and Plunkett, W. Superoxide dismutase as a target for the selective killing of cancer cells. *Nature* 407:390–395, 2000.

Huang, Q., Dunn, R. T. 2nd , Jayadev, S., DiSorbo, O., Pack, F. D., Farr, S. B., Stoll, R. E., and Blanchard, K. T. Assessment of cisplatin-induced nephrotoxicity by microarray technology. *Toxicol Sci* 63:196–207, 2001.

Huang, Q., Liu, N., Majewski, P., Schulte, AC., Korn, J. M., Young, R. A., Lander, E. S., and Hacohen, N. The plasticity of dendritic cell responses to pathogens and their components. *Science* 294:870–875, 2001.

Huang, W., Carlsen, B., Rudkin, G. H., Shah, N., Chung, C., Ishida, K., Yamaguchi, D. T., and Miller, T. A. Effect of serial passage on gene expression in MC3T3-E1 preosteoblastic cells: a microarray study. *Biochem Biophys Res Commun* 281:1120–1126, 2001.

Huang, W., Sher, Y. P., Delgado-West, D., Wu, J. T., Peck, K., and Fung, Y. C. Tissue remodeling of rat pulmonary artery in hypoxic breathing. I. Changes of morphology, zero-stress state, and gene expression. *Ann Biomed Eng* 29:535–551, 2001.

Huang, Y., Uchiyama, Y., Fujimura, T., Kanamori, H., Doi, T., Takamizawa, A., Hamakubo, T., and Kodama, T. A human hepatoma cell line expressing hepatitis c virus nonstructural proteins tightly regulated by tetracycline. *Biochem Biophys Res Commun* 281:732–740, 2001.

Huber, M., Losert, D., Hiller, R., Harwanegg, C., Mueller, M. W., and Schmidt, W. M. Detection of single base alterations in genomic DNA by solid phase polymerase chain reaction on oligonucleotide microarrays. *Anal Biochem* 299:24–30, 2001.

Hui, A. B., Lo, K. W., Teo, P. M., To, K. F., and Huang, D. P. Genome wide detection of oncogene amplifications in nasopharyngeal carcinoma by array based comparative genomic hybridization. *Int J Oncol* 20:467–473, 2002.

Hughes, T. R., Mao, M., Jones, A. R., Burchard, J., Marton, M. J., Shannon, K. W., Lefkowitz, S. M., Ziman, M., and co-workers. Expression profiling using microarrays fabricated by an ink-jet oligonucleotide synthesizer. *Nat Biotechnol* 19:342–347, 2001.

Hughes, T. R., Roberts, C. J., Dai, H., Jones, A. R., Meyer, M. R., Slade, D., Burchard, J., Dow, S., and co-workers. Widespread aneuploidy revealed by DNA microarray expression profiling. *Nat Genet* 25:333–337, 2000.

Hughes, T. R., and Shoemaker, D. D. DNA microarrays for expression profiling. *Curr Opin Chem Biol* 5:21–25, 2001.

Hunt, J. S., Petroff, M. G., Morales, P., Sedlmayr, P., Geraghty, D. E., and Ober, C. HLA-G in reproduction: Studies on the maternal-fetal interface. *Hum Immunol* 61:1113–1117, 2000.

Hurt, R. A., Qiu, X., Wu, L., Roh, Y., Palumbo, A. V., Tiedje, J. M., and Zhou, J. Simultaneous recovery of RNA and DNA from soils and sediments. *Appl Environ Microbiol* 67:4495–4503, 2001.

Husbeck, B., Berggren, M. I., and Powis, G. DNA microarray reveals increased expression of thioredoxin peroxidase in thioredoxin-1 transfected cells and its functional consequences. *Adv Exp Med Biol* 500:157–168, 2001.

Hvidsten, T. R., Komorowski, J., Sandvik, A. K., and Laegreid, A. Predicting gene function from gene expressions and ontologies. *Pac Symp Biocomput* 299–310, 2001.

Hwang, J. J., Dzau, V. J., and Liew, C. C. Genomics and the pathophysiology of heart failure. *Curr Cardiol Rep* 3:198–207, 2001.

Ibarrola, N., and Pandey, A. Integrating DNA and tissue microarrays for cancer profiling. *Trends Biochem Sci* 26:589, 2001.

Ibrahim, S. M., Mix, E., Bottcher, T., Koczan, D., Gold, R., Rolfs, A., and Thiesen, H. J. Gene expression profiling of the nervous system in murine experimental autoimmune encephalomyelitis. *Brain* 124:1927–1938, 2001.

Ichikawa, J. K., Norris, A., Bangera, M. G., Geiss, G. K., van 't Wout, A. B., Bumgarner, R. E., and Lory, S. Interaction of *Pseudomonas aeruginosa* with epithelial cells: Identification of differentially regulated genes by expression microarray analysis of human cDNAs. *Proc Natl Acad Sci U S A* 97:9659–9664, 2000.

Ideker, T., Thorsson, V., Ranish, J. A., Christmas, R., Buhler, J., Eng, J. K., Bumgarner, R., Goodlett, D. R., and co-workers. Integrated genomic and proteomic analyses of a systematically perturbed metabolic network. Science 292:929–934, 2001.

Ideker, T., Thorsson, V., Siegel, A. F., and Hood, L. E. Testing for differentially-expressed genes by maximum-likelihood analysis of microarray data. *J Comput Biol* 7:805–817, 2000.

Imai, E., Takenaka, M., Nagasawa, Y., Kaimori, J., and Hori, M. Application of microarray assay to nephrology. *Nephrol Dial Transplant* 15 Suppl 6:78–80, 2000.

Iranfar, N., Fuller, D., Sasik, R., Hwa, T., Laub, M., and Loomis, W. F. Expression patterns of cell-type-specific genes in dictyostelium. *Mol Biol Cell* 12:2590–2600, 2001.

Irving, P., Troxler, L., Heuer, T. S., Belvin, M., Kopczynski, C., Reichhart, J. M., Hoffmann, J. A., and Hetru, C. A genome-wide analysis of immune responses in *Drosophila. Proc Natl Acad Sci U S A* 2001.

Irwin, L. N. Gene expression in the hippocampus of behaviorally stimulated rats: Analysis by DNA microarray. *Brain Res Mol Brain Res* 96:163–169, 2001.

Isaksson, A., and Landegren, U. Accessing genomic information: alternatives to PCR. *Curr Opin Biotechnol* 10:11–15, 1999.

Isambert, H., and Siggia, E. D. Modeling RNA folding paths with pseudoknots: Application to hepatitis delta virus ribozyme. *Proc Natl Acad Sci U S A* 97:6515–6520, 2000.

Ishida, S., Huang, E., Zuzan, H., Spang, R., Leone, G., West, M., and Nevins, J. R. Role for E2F in control of both DNA replication and mitotic functions as revealed from DNA microarray analysis. *Mol Cell Biol* 21:4684–4699, 2001.

Ishiguro, H., Tsunoda, T., Tanaka, T., Fujii, Y., Nakamura, Y., and Furukawa, Y. Identification of AXUD1, a novel human gene induced by AXIN1 and its reduced expression in human carcinomas of the lung, liver, colon and kidney. *Oncogene* 20:5062–5066, 2001.

Ishikawa, K., and Tsujimoto, G. New strategy on medical research after completion of genome sequencing. *Nippon Yakurigaku Zasshi* 118:170–176, 2001.

Israel, D. A., Salama, N., Arnold, C. N., Moss, S. F., Ando, T., Wirth, H. P., Tham, K. T., Camorlinga, M., and co-workers. *Helicobacter pylori* strain-specific differences in genetic content, identified by microarray, influence host inflammatory responses. *J Clin Invest* 107:611–620, 2001.

Israel, D. A., Salama, N., Krishna, U., Rieger, U. M., Atherton, J. C., Falkow, S., and Peek, R. M. Jr. *Helicobacter pylori* genetic diversity within the gastric niche of a single human host. *Proc Natl Acad Sci U S A* 98:14625–14630, 2001.

Ivanov, I., Schaab, C., Planitzer, S., Teichmann, U., Machl, A., Theml, S., Meier-Ewert, S., Seizinger, B., and Loferer, H. DNA microarray technology and antimicrobial drug discovery. *Pharmacogenomics* 1:169–178, 2000.

Iyer, V. R., Eisen, M. B., Ross, D. T., Schuler, G., Moore, T., Lee, J. C., Trent, J. M., Staudt, L. M., and co-workers. The transcriptional program in the response of human fibroblasts to serum. *Science* 283:83–87, 1999.

Iyer, V. R., Horak, C. E., Scafe, C. S., Botstein, D., Snyder, M., and Brown, P. O. Genomic binding sites of the yeast cell-cycle transcription factors SBF and MBF. *Nature* 409:533–538, 2001.

Jaccoud, D., Peng, K., Feinstein, D., and Kilian, A. Diversity arrays: A solid state technology for sequence information independent genotyping. *Nucleic Acids Res* 29:E25, 2001.

Jackson-Grusby, L., Beard, C., Possemato, R., Tudor, M., Fambrough, D., Csankovszki, G., Dausman, J., Lee, P., and co-workers. Loss of genomic methylation causes p53-dependent apoptosis and epigenetic deregulation. *Nat Genet* 27:31–39, 2001.

Jain, A. N., Tokuyasu, T. A., Snijders, A. M., Segraves, R., Albertson, D. G., and Pinkel, D. Fully automatic quantification of microarray image data. *Genome Res* 12:325–332, 2002.

Jain, K. K. Applications of biochip and microarray systems in pharmacogenomics. *Pharmacogenomics* 1:289–307, 2000.

Jain, K. K. Biotechnological applications of lab-chips and microarrays. *Trends Biotechnol* 18:278–280, 2000.

Jain, K. K. Cambridge Healthtech Institute's third annual conference on lab-on-a-chip and microarrays. *Pharmacogenomics* 2:73–77, 2001.

Janowski, B. A., Shan, B., and Russell, D. W. The hypocholesterolemic agent LY295427 reverses suppression of SREBP processing mediated by oxysterols. *J Biol Chem* 2001.

Janssen, P. J., Audit, B., and Ouzounis, C. A. Strain-specific genes of *Helicobacter pylori:* Distribution, function and dynamics. *Nucleic Acids Res* 29:4395–4404, 2001.

Jarmer, H., Berka, R., Knudsen, S., and Saxild, H. H. Transcriptome analysis documents induced competence of *Bacillus subtilis* during nitrogen limiting conditions. *FEMS Microbiol Lett* 206:197–200, 2002.

Jbilo, O., Derocq, J. M., Segui, M., Le Fur, G., and Casellas, P. Stimulation of peripheral cannabinoid receptor CB2 induces MCP-1 and IL-8 gene expression in human promyelocytic cell line HL60. *FEBS Lett* 448:273–277, 1999.

Jenssen, T. K., Laegreid, A., Komorowski, J., and Hovig, E. A literature network of human genes for high-throughput analysis of gene expression. *Nat Genet* 28:21–28, 2001.

Jenssen, T. K., and Vinterbo, S. A set-covering approach to specific search for literature about human genes. *Proc AMIA Symp* 384–8, 2000.

Jervis, K. M., and Robaire, B. Dynamic changes in gene expression along the rat epididymis. *Biol Reprod* 65:696–703, 2001.

Jhaveri, M. S., Wagner, C., and Trepel, J. B. Impact of extracellular folate levels on global gene expression. *Mol Pharmacol* 60:1288–1295, 2001.

Jia, M. H., Larossa, R. A., Lee, J. M., Rafalski, A., Derose, E., Gonye, G., and Xue, Z. Global expression profiling of yeast treated with an inhibitor of amino acid biosynthesis, sulfometuron methyl. *Physiol Genomics* 3:83–92, 2000.

Jiang, H., Kang, D. C., Alexandre, D., and Fisher, P.B. RaSH, a rapid subtraction hybridization approach for identifying and cloning differentially expressed genes. *Proc Natl Acad Sci U S A* 97:12684–12690, 2000.

Jiang, L., Tsubakihara, M., Heinke, M. Y., Yao, M., Dunn, M. J., Phillips, W., Remedios, C. G., and Nosworthy, N. J. Heart failure and apoptosis: Electrophoretic methods support data from micro- and macro-arrays. A critical review of genomics and proteomics. *Proteomics* 1:1481–1488, 2001.

Jiang, M., Ryu, J., Kiraly, M., Duke, K., Reinke, V., and Kim, S. K. Genome-wide analysis of developmental and sex-regulated gene expression profiles in *Caenorhabditis elegans. Proc Natl Acad Sci U S A* 98:218–223, 2001.

Jiang, Z., Woda, B. A., Rock, K. L., Xu, Y., Savas, L., Khan, A., Pihan, G., Cai, F., and co-workers. P504S: A new molecular marker for the detection of prostate carcinoma. *Am J Surg Pathol* 25:1397–1404, 2001.

Jin, H., Yang, R., Awad, T. A., Wang, F., Li, W., Williams, S. P., Ogasawara, A., Shimada, B., and co-workers. Effects of early angiotensin-converting enzyme inhibition on cardiac gene expression after acute myocardial infarction. *Circulation* 103:736–742, 2001.

Jin, K., Mao, X. O., Eshoo, M. W., Nagayama, T., Minami, M., Simon, R. P., and Greenberg, D. A. Microarray analysis of hippocampal gene expression in global cerebral ischemia. *Ann Neurol* 50:93–103, 2001.

Jin, W., Riley, R. M., Wolfinger, R. D., White, K. P., Passador-Gurgel, G., and Gibson, G. The contributions of sex, genotype and age to transcriptional variance in *Drosophila melanogaster. Nat Genet* 29:389–395, 2001.

Jin-Lee, H., Goodrich, T. T., and Corn, R. M. SPR imaging measurements of 1-D and 2-D DNA microarrays created from microfluidic channels on gold thin films. *Anal Chem* 73:5525–5531, 2001.

Jo, H., Cho, Y. J., Zhang, H., and Liang, P. Differential display analysis of gene expression altered by *ras* oncogene. *Methods Enzymol* 332:233–244, 2001.

Jobs, M., Fredriksson, S., Brookes, A. J., and Landegren, U. Effect of oligonucleotide truncation on single-nucleotide distinction by solid-phase hybridization. *Anal Chem* 74:199–202, 2002.

Johannes, G., Carter, M. S., Eisen, M. B., Brown, P. O., and Sarnow, P. Identification of eukaryotic mRNAs that are translated at reduced cap binding complex eIF4F concentrations using a cDNA microarray. *Proc Natl Acad Sci U S A* 96:13118–13123, 1999.

Johnson, K., and Lin, S. Call to work together on microarray data analysis. *Nature* 411:885, 2001.

Johnson, K. F., and Lin, S. M. Critical assessment of microarray data analysis: The 2001 challenge. *Bioinformatics* 17:857–858, 2001.

Johnston, C. J., Williams, J. P., Okunieff, P., and Finkelstein, J. N. Radiation-induced pulmonary fibrosis: Examination of chemokine and chemokine receptor families. *Radiat Res* 157:256–265, 2002.

Johnston, M. V., Jeon, O. H., Pevsner, J., Blue, M. E., and Naidu, S. Neurobiology of Rett syndrome: A genetic disorder of synapse development. *Brain Dev* 23:S206–S213, 2001.

Johnston-Wilson, N. L., Bouton, C. M., Pevsner, J., Breen, J. J., Torrey, E. F., and Yolken, R. H. Emerging technologies for large-scale screening of human tissues and fluids in the study of severe psychiatric disease. *Int J Neuropsychopharmacol* 4:83–92, 2001.

Joki, T., Carroll, R. S., Dunn, I. F., Zhang, J., Abe, T., and Black, P. M. Assessment of alterations in gene expression in recurrent malignant glioma after radiotherapy using complementary deoxyribonucleic acid microarrays. *Neurosurgery* 48:195–201; 201–202 (discussion), 2001.

Jones, D., Amin, M., Ordonez, N. G., Glassman, A. B., Hayes, K. J., and Jeffrey Medeiros, L. Reticulum cell sarcoma of lymph node with mixed dendritic and fibroblastic features. *Mod Pathol* 14:1059–1067, 2001.

Jones, D. A., and Fitzpatrick, F. A. Genomics and the discovery of new drug targets. Genomics and the discovery of new drug targets. *Curr Opin Chem Biol* 3:71–76, 1999.

Joos, T. O., Schrenk, M., Hopfl, P., Kroger, K., Chowdhury, U., Stoll, D., Schorner, D., Durr, M., and co-workers. A microarray enzyme-linked immunosorbent assay for autoimmune diagnostics. *Electrophoresis* 21:2641–2650, 2000.

Joos, T. O., Stoll, D., and Templin, M. F. Miniaturised multiplexed immunoassays. *Curr Opin Chem Biol* 6:76–80, 2002.

Jordan, B. R. Large-scale expression measurement by hybridization methods: from high-density membranes to "DNA chips." *J Biochem [Tokyo]* 124:251–258, 1998.

Joseph, P., Muchnok, T., and Ong, T. M. Gene expression profile in BALB/c-3T3 cells transformed with beryllium sulfate. *Mol Carcinog* 32:28–35, 2001.

Jost, J. P., Oakeley, E. J., Zhu, B., Benjamin, D., Thiry, S., Siegmann, M., and Jost, Y. C. 5-Methylcytosine DNA glycosylase participates in the genome-wide loss of DNA methylation occurring during mouse myoblast differentiation. *Nucleic Acids Res* 29:4452–4461, 2001.

Joussen, A. M., and Huang, S. Possibilities of broad spectrum analysis of gene expression patterns with cDNA arrays. *Ophthalmologe* 98:568–573, 2001.

Jun, A. S., Liu, S. H., Koo, E. H., Do, D. V., Stark, W. J., and Gottsch, J. D. Microarray analysis of gene expression in human donor corneas. *Arch Ophthalmol* 119:1629–1634, 2001.

Jurecic, R., and Belmont, J. W. Long-distance DD-PCR and cDNA microarrays. *Curr Opin Microbiol* 3:316–321, 2000.

Jurecic, R., Nachtman, R. G., Colicos, S. M., and Belmont, J. W. Identification and cloning of differentially expressed genes by long-distance differential display. *Anal Biochem* 259:235–244, 1998.

Kacharmina, J. E., Crino, P. B., and Eberwine, J. Preparation of cDNA from single cells and subcellular regions. *Methods Enzymol* 303:3–18, 1999.

Kadota, K., Miki, R., Bono, H., Shimizu, K., Okazaki, Y., and Hayashizaki, Y. Reprocessing implementation for microarray (PRIM): An efficient method for processing cDNA microarray data. *Physiol Genomics* 4:183–188, 2001.

Kagnoff, M. F., and Eckmann, L. Analysis of host responses to microbial infection using gene expression profiling. *Curr Opin Microbiol* 4:246–250, 2001.

Kahmann, R., and Basse, C. Fungal gene expression during pathogenesis-related development and host plant colonization. *Curr Opin Microbiol* 4:374–380, 2001.

Kallioniemi, O. P. Biochip technologies in cancer research. *Ann Med* 33:142–147, 2001.

Kallioniemi, O. P., Wagner, U., Kononen, J., and Sauter, G. Tissue microarray technology for high-throughput molecular profiling of cancer. *Hum Mol Genet* 10:657–662, 2001.

Kalma, Y., Marash, L., Lamed, Y., and Ginsberg, D. Expression analysis using DNA microarrays demonstrates that E2F-1 up-regulates expression of DNA replication genes including replication protein A2. *Oncogene* 20:1379–1387, 2001.

Kamb, A., and Ramaswami, M. A simple method for statistical analysis of intensity differences in microarray-derived gene expression data. *BMC Biotechnol* 1:8, 2001.

Kampke, T., Kieninger, M., and Mecklenburg, M. Efficient primer design algorithms. *Bioinformatics* 17:214–225, 2001.

Kan, T., Shimada, Y., Sato, F., Maeda, M., Kawabe, A., Kaganoi, J., Itami, A., Yamasaki, S., and Imamura, M. Gene expression profiling in human esophageal cancers using cDNA microarray. *Biochem Biophys Res Commun* 286:792–801, 2001.

Kanamaru, R. Cancer diagnosis using microarray technologies. *Gan To Kagaku Ryoho* 27:949–953, 2000.

Kane, M. D., Jatkoe, T. A., Stumpf, C. R., Lu, J., Thomas, J. D., and Madore, S. J. Assessment of the sensitivity and specificity of oligonucleotide (50mer) microarrays. *Nucleic Acids Res* 28:4552–4557, 2000.

Kanehisa, M., Goto, S., Kawashima, S., and Nakaya, A. The KEGG databases at GenomeNet. *Nucleic Acids Res* 30:42–46, 2002.

Kanesaki, Y., Suzuki, I., Allakhverdiev, S. I., Mikami, K., and Murata, N. Salt stress and hyperosmotic stress regulate the expression of different sets of genes in *Synechocystis* sp. PCC 6803. *Biochem Biophys Res Commun* 290:339–348, 2002.

Kang, D., Jiang, H., Wu, Q., Pestka, S., and Fisher, P. B. Cloning and characterization of human ubiquitin-processing protease-43 from terminally differentiated human melanoma cells using a rapid subtraction hybridization protocol RaSH. *Gene* 267:233–242, 2001.

Kang, J. J., Watson, R. M., Fisher, M. E., Higuchi, R., Gelfand, D. H., and Holland, M. J. Transcript quantitation in total yeast cellular RNA using kinetic PCR. *Nucleic Acids Res* 28:E2, 2000.

Kannan, K., Amariglio, N., Rechavi, G., Jakob-Hirsch, J., Kela, I., Kaminski, N., Getz, G., Domany, E., and Givol, D. DNA microarrays identification of primary and secondary target genes regulated by p53. *Oncogene* 20:2225–2234, 2001.

Kannan, K., Kaminski, N., Rechavi, G., Jakob-Hirsch, J., Amariglio, N., and Givol, D. DNA microarray analysis of genes involved in p53 mediated apoptosis: Activation of Apaf-1. *Oncogene* 20:3449–3455, 2001.

Kapp, U., Yeh, W. C., Patterson, B., Elia, A. J., Kagi, D., Ho, A., Hessel, A., Tipsword, M., and co-workers. Interleukin 13 is secreted by and stimulates the growth of Hodgkin and Reed-Sternberg cells. *J Exp Med* 189:1939–1946, 1999.

Kapteyn, J. C., ter Riet, B., Vink, E., Blad, S., De Nobel, H., Van Den Ende, H., and Klis, F. M. Low external pH induces HOG1-dependent changes in the organization of the *Saccharomyces cerevisiae* cell wall. *Mol Microbiol* 39:469–479, 2001.

Karp, C. L., Grupe, A., Schadt, E., Ewart, S. L., Keane-Moore, M., Cuomo, P. J., Kohl, J., Wahl, L., and co-workers. Identification of complement factor 5 as a susceptibility locus for experimental allergic asthma. *Nat Immunol* 1:221–226, 2201.

Karpf, A. R., Peterson, P. W., Rawlins, J. T., Dalley, B. K., Yang, Q., Albertsen, H., and Jones, D. A. Inhibition of DNA methyltransferase stimulates the expression of signal transducer and activator of transcription 1, 2, and 3 genes in colon tumor cells. *Proc Natl Acad Sci U S A* 96:14007–14012, 1999.

Karsten, S. L., Van Deerlin, V. M., Sabatti, C., Gill, L. H., and Geschwind, D. H. An evaluation of tyramide signal amplification and archived fixed and frozen tissue in microarray gene expression analysis. *Nucleic Acids Res* 30:E4, 2002.

Kashiwagi, H., and Uchida, K. Genome-wide profiling of gene amplification and deletion in cancer. *Hum Cell* 13:135–141, 2000.

Kato, T., Satoh, S., Okabe, H., Kitahara, O., Ono, K., Kihara, C., Tanaka, T., Tsunoda, T., and co-workers. Protein nucleotide isolation of a novel human gene, markl1, homologous to mark3 and its involvement in hepatocellular carcinogenesis. *Neoplasia* 3:4–9, 2001.

Kato, M., Seki, N., Sugano, S., Hashimoto, K., Masuho, Y., Muramatsu, Ma. MA., Kaibuchi, K., and Nakafuku, M. Identification of sonic hedgehog-responsive genes using cDNA microarray. *Biochem Biophys Res Commun* 289:472–478, 2001.

Kato, M., Tsunoda, T., and Takagi, T. Inferring genetic networks from DNA microarray data by multiple regression analysis. *Genome Inform Ser Workshop Genome Inform* 11:118–128, 2000.

Kato-Maeda, M., Gao, Q., and Small, P. M. Microarray analysis of pathogens and their interaction with hosts. *Cell Microbiol* 3:713–719, 2001.

Kato-Maeda, M., Rhee, J. T., Gingeras, T. R., Salamon, H., Drenkow, J., Smittipat, N., and Small, P. M. Comparing genomes within the species *Mycobacterium tuberculosis*. *Genome Res* 11:547–554, 2001.

Katsuma, S., Nishi, K., Tanigawara, K., Ikawa, H., Shiojima, S., Takagaki, K., Kaminishi, Y., Suzuki, Y., and co-workers. Molecular monitoring of bleomycin-induced pulmonary fibrosis by cDNA microarray-based gene expression profiling. *Biochem Biophys Res Commun* 288:747–751, 2001.

Kauraniemi, P., Barlund, M., Monni, O., and Kallioniemi, A. New amplified and highly expressed genes discovered in the ERBB2 amplicon in breast cancer by cDNA microarrays. *Cancer Res* 61:8235–8240, 2001.

Kauraniemi, P., Hedenfalk, I., Persson, K., Duggan, D. J., Tanner, M., Johannsson, O., Olsson, H., Trent, J. M., and co-workers. MYB oncogene amplification in hereditary BRCA1 breast cancer. *Cancer Res* 60:5323–5328, 2000.

Kausch, A. P., Owen, T. P. Jr., Narayanswami, S., and Bruce, B. D. Organelle isolation by magnetic immunoabsorption. *Biotechniques* 26:336–343, 1999.

Kawai, H. F., Kaneko, S., Honda, M., Shirota, Y., and Kobayashi, K. Alpha-fetoprotein-producing hepatoma cell lines share common expression profiles of genes in various categories demonstrated by cDNA microarray analysis. *Hepatology* 33:676–691, 2001.

Kawamoto, S., Ohnishi, T., Kita, H., Chisaka, O., and Okubo, K. Expression profiling by iAFLP: A PCR-based method for genome-wide gene expression profiling. *Genome Res* 9:1305–1312, 1999.

Kawasaki, S., Borchert, C., Deyholos, M., Wang, H., Brazille, S., Kawai, K., Galbraith, D., and Bohnert, H. J. Gene expression profiles during the initial phase of salt stress in rice. *Plant Cell* 13:889–906, 2001.

Kehoe, D. M., Villand, P., and Somerville, S. DNA microarrays for studies of higher plants and other photosynthetic organisms. *Trends Plant Sci* 4:38–41, 1999.

Kellam, P. Host-pathogen studies in the post-genomic era. *Genome Biol* 1: 2000.

Kellam, P. Microarray gene expression database: Progress towards an international repository of gene expression data. *Genome Biol* 2: 2001.

Kellam, P. Post-genomic virology: The impact of bioinformatics, microarrays and proteomics on investigating host and pathogen interactions. *Rev Med Virol* 11:313–329, 2001.

Kelly, D. L., and Rizzino, A. DNA microarray analyses of genes regulated during the differentiation of embryonic stem cells. *Mol Reprod Dev* 56:113–123, 2000.

Kendall, L. V., Myles, M., and Riley, L. K. DNA microarray. *Contemp Top Lab Anim Sci* 40:60, 2001.

Kennedy, G. C. The impact of genomics on therapeutic drug development. *EXS* 89:1–10, 2000.

Kerley, J. S., Olsen, S. L., Freemantle, S. J., and Spinella, M. J. Transcriptional activation of the nuclear receptor corepressor RIP140 by retinoic acid: A potential negative-feedback regulatory mechanism. *Biochem Biophys Res Commun* 285:969–975, 2001.

Kerr, M. K., and Churchill, G. A. Bootstrapping cluster analysis: assessing the reliability of conclusions from microarray experiments. *Proc Natl Acad Sci U S A* 98:8961–8965, 2001.

Kerr, M. K., Martin, M., and Churchill, G. A. Analysis of variance for gene expression microarray data. *J Comput Biol* 7:819–837, 2000.

Kersten, S., Mandard, S., Escher, P., Gonzalez, F. J., Tafuri, S., Desvergne, B., and Wahli, W. The peroxisome proliferator-activated receptor alpha regulates amino acid metabolism. *FASEB J* 15:1971–1978, 2001.

Keshelava, N., Zuo, J. J., Chen, P., Waidyaratne, S. N., Luna, M. C., Gomer, C. J., Triche, T. J., and Reynolds, C. P. Loss of p53 function confers high-level multidrug resistance in neuroblastoma cell lines. *Cancer Res* 61:6185–6193, 2001.

Kessler, D., Dethlefsen, S., Haase, I., Plomann, M., Hirche, F., Krieg, T., and Eckes, B. Fibroblasts in mechanically stressed collagen lattices assume a "synthetic" phenotype. *J Biol Chem* 276:36575–36585, 2001.

Khan, J., Bittner, M. L., Chen, Y., Meltzer, P. S., and Trent, J. M. DNA microarray technology: The anticipated impact on the study of human disease. *Biochim Biophys Acta* 1423:M17–M28, 1999.

Khan, J., Bittner, M. L., Saal, L. H., Teichmann, U., Azorsa, D. O., Gooden, G. C., Pavan, W. J., Trent, J. M., and Meltzer, P. S. cDNA microarrays detect activation of a myogenic transcription program by the PAX3-FKHR fusion oncogene. *Proc Natl Acad Sci U S A* 96:13264–13269, 1999.

Khan, J., Saal, L. H., Bittner, M. L., Chen, Y., Trent, J. M., and Meltzer, P. S. Expression profiling in cancer using cDNA microarrays. *Electrophoresis* 20:223–229, 1999.

Khan, J., Simon, R., Bittner, M., Chen, Y., Leighton, S. B., Pohida, T., Smith, P. D., Jiang, Y., and co-workers. Gene expression profiling of alveolar rhabdomyosarcoma with cDNA microarrays. *Cancer Res* 58:5009–5013, 1998.

Khanna, C., Khan, J., Nguyen, P., Prehn, J., Caylor, J., Yeung, C., Trepel, J., Meltzer, P., and Helman, L. Metastasis-associated differences in gene expression in a murine model of osteosarcoma. *Cancer Res* 61:3750–3759, 2001.

Khaoustov, V. I., Risin, D., Pellis, N. R., and Yoffe, B. Microarray analysis of genes differentially expressed in HepG2 cells cultured in simulated microgravity: preliminary report. *In Vitro Cell Dev Biol Anim* 37:84–88, 2001.

Khatri, P., Draghici, S., Ostermeier, G. C., and Krawetz, S. A. Profiling gene expression using onto-express. *Genomics* 79:266–270, 2002.

Khitrov, G. Use of inexpensive dyes to calibrate and adjust your microarray printer. *Biotechniques* 30:748, 2001.

Khodursky, A. B., Peter, B. J., Cozzarelli, N. R., Botstein, D., Brown, P. O., and Yanofsky, C. DNA microarray analysis of gene expression in response to physiological and genetic changes that affect tryptophan metabolism in *Escherichia coli*. *Proc Natl Acad Sci U S A* 97:12170–12175, 2000.

Khodursky, A. B., Peter, B. J., Schmid, M. B., DeRisi, J., Botstein, D., Brown, P. O., and Cozzarelli, N. R. Analysis of topoisomerase function in bacterial replication fork movement: Use of DNA microarrays. *Proc Natl Acad Sci U S A* 97:9419–9424, 2000.

Kiguchi, T., Niiya, K., Shibakura, M., Miyazono, T., Shinagawa, K., Ishimaru, F., Kiura, K., Ikeda, K., and co-workers. Induction of urokinase-type plasminogen activator by the anthracycline antibiotic in human RC-K8 lymphoma and H69 lung-carcinoma cells. *Int J Cancer* 93:792–797, 2001.

Kihara, C., Tsunoda, T., Tanaka, T., Yamana, H., Furukawa, Y., Ono, K., Kitahara, O., Zembutsu, H., and co-workers. Prediction of sensitivity of esophageal tumors to adjuvant chemotherapy by cDNA microarray analysis of gene-expression profiles. *Cancer Res* 61:6474–6479, 2001.

Kim, I. J., Kang, H. C., Park, J. H., Ku, J. L., Lee, J. S., Kwon, H. J., Yoon, K. A., Heo, S. C., and co-workers. RET oligonucleotide microarray for the detection of RET mutations in multiple endocrine neoplasia type 2 syndromes. *Clin Cancer Res* 8:457–463, 2002.

Kim, J. H., Kim, H. Y., and Lee, Y. S. A novel method using edge detection for signal extraction from cDNA microarray image analysis. *Exp Mol Med* 33:83–88, 2001.

Kim, J. H., Ohno-Machado, L., and Kohane, I. S. Unsupervised learning from complex data: The matrix incision tree algorithm. *Pac Symp Biocomput* 30–41, 2001.

Kim, J. S., Pirnia, F., Choi, Y. H., Nguyen, P. M., Knepper, B., Tsokos, M., Schulte, T. W., Birrer, M. J., and co-workers. Lovastatin induces apoptosis in a primitive neuroectodermal tumor cell line in association with RB down-regulation and loss of the G1 checkpoint. *Oncogene* 19:6082–6090, 2000.

Kim, S., Dougherty, E. R., Bittner, M. L., Chen, Y., Sivakumar, K., Meltzer, P., and Trent, J. M. General nonlinear framework for the analysis of gene interaction via multivariate expression arrays. *J Biomed Opt* 5:411–424, 2000.

Kim, S., Dougherty, E. R., Chen, Y., Sivakumar, K., Meltzer, P., Trent, J. M., and Bittner, M. Multivariate measurement of gene expression relationships. *Genomics* 67:201–209, 2000.

Kim, S. K. Http://*C. elegans:* Mining the functional genomic landscape. *Nat Rev Genet* 2:681–689, 2001.

Kim, S. K., Lund, J., Kiraly, M., Duke, K., Jiang, M., Stuart, J. M., Eizinger, A., Wylie, B. N., and Davidson, G. S. A gene expression map for *Caenorhabditis elegans*. *Science* 293:2087–2092, 2001.

Kim, W. H. Tissue array technology for translational research. From gene discovery to application. *Exp Mol Med* 33(Suppl 1):135–148, 2001.

King, H. C., and Sinha, A. A. Gene expression profile analysis by DNA microarrays: Promise and pitfalls. *JAMA* 286:2280–2288, 2001.

Kipps, T. J. Advances in classification and therapy of indolent B-cell malignancies. *Semin Oncol* 29:98–104, 2002.

Kishino, H., and Waddell, P. J. Correspondence analysis of genes and tissue types and finding genetic links from microarray data. *Genome Inform Ser Workshop Genome Inform* 11:83–95, 2000.

Kita, Y., Shiozawa, M., Jin, W., Majewski, R. R., Besharse, J. C., Greene, A. S., and Jacob, H. J. Implications of circadian gene expression in kidney, liver and the effects of fasting on pharmacogenomic studies. *Pharmacogenetics* 12:55–65, 2002.

Kitahara, O., Furukawa, Y., Tanaka, T., Kihara, C., Ono, K., Yanagawa, R., Nita, M. E., Takagi, T., and co-workers. Alterations of gene expression during colorectal carcinogenesis revealed by cDNA microarrays after laser-capture microdissection of tumor tissues and normal epithelia. *Cancer Res* 3544–3549, 2001.

Kitchens, W. H., Byrne, M. C., Strominger, J. L., and Wilson, S. B. Using DNA chips to unravel the genetics of type 1 diabetes. *Diabetes Technol Ther* 2:249–258, 2000.

Klebl, B., Kozian, D., Leberer, E., and Kukuruzinska, M. A. A comprehensive analysis of gene expression profiles in a yeast N-glycosylation mutant. *Biochem Biophys Res Commun* 286:714–720, 2001.

Klein, U., Tu, Y., Stolovitzky, G. A., Mattioli, M., Cattoretti, G., Husson, H., Freedman, A., Inghirami, G., and co-workers. Gene expression profiling of B cell chronic lymphocytic leukemia reveals a homogeneous phenotype related to memory B cells. *J Exp Med* 194:1625–1638, 2001.

Klevecz, R. R. Dynamic architecture of the yeast cell cycle uncovered by wavelet decomposition of expression microarray data. *Funct Integr Genomics* 1:186–192, 2000.

Klysik, J. Concept of immunomics: a new frontier in the battle for gene function? *Acta Biotheor* 49:191–202, 2001.

Knapp, D. R., and Kim, J. S. Inexpensive cleanroom for microfabrication, microarray work, and other activities requiring protection from airborne contamination. *Biotechniques* 31:842, 844, 846, 2001.

Knezevic, V., Leethanakul, C., Bichsel, V. E., Worth, J. M., Prabhu, V. V., Gutkind, J. S., Liotta, L. A., Munson, P. J., and co-workers. Proteomic profiling of the cancer microenvironment by antibody arrays. *Proteomics* 1:1271–1278, 2001.

Knowles, M. A. What we could do now: molecular pathology of bladder cancer. *Mol Pathol* 54:215–221, 2001.

Knox, D. P., Redmond, D. L., Skuce, P. J., and Newlands, G. F. The contribution of molecular biology to the development of vaccines against nematode and trematode parasites of domestic ruminants. *Vet Parasitol* 101:311–335, 2001.

Kobayashi, K., Ogura, M., Yamaguchi, H., Yoshida, K., Ogasawara, N., Tanaka, T., and Fujita, Y. Comprehensive DNA microarray analysis of *Bacillus subtilis* two-component regulatory systems. *J Bacteriol* 183:7365–7370, 2001.

Kodadek, T. Protein microarrays: prospects and problems. *Chem Biol* 8:105–115, 2001.

Kohro, T., Nakajima, T., Wada, Y., Sugiyama, A., Ishii, M., Tsutsumi, S., Aburatani, H., Imoto, I., and co-workers. Genomic structure and mapping of human orphan receptor LXR alpha: upregulation of LXRa mRNA during monocyte to macrophage differentiation. *J Atheroscler Thromb* 7:145–151, 2001.

Komatani, H., Kotani, H., Hara, Y., Nakagawa, R., Matsumoto, M., Arakawa, H., and Nishimura, S. Identification of breast cancer resistant protein/mitoxantrone resistance/placenta-specific, ATP-binding cassette transporter as a transporter of NB-506 and J-107088, topoisomerase I inhibitors with an indolocarbazole structure. *Cancer Res* 61:2827–2832, 2001.

Kononen, J., Bubendorf, L., Kallioniemi, A., Barlund, M., Schraml, P., Leighton, S., Torhorst, J., Mihatsch, M. J., and co-workers. Tissue microarrays for high-throughput molecular profiling of tumor specimens. *Nat Med* 4:844–847, 1998.

Kontoyiannis, D. P., and May, G. S. Identification of azole-responsive genes by microarray technology: Why are we missing the efflux transporter genes? *Antimicrob Agents Chemother* 45:3674–3676, 2001.

Konu, O., Kane, J. K., Barrett, T., Vawter, M. P., Chang, R., Ma, J. Z., Donovan, D. M., Sharp, B., Becker, K. G., and Li, M. D. Region-specific transcriptional response to chronic nicotine in rat brain. *Brain Res* 909:194–203, 2001.

Konu, O. O., and Li, M. D. Correlations between mRNA expression levels and GC contents of coding and untranslated regions of genes in rodents. *J Mol Evol* 54:35–41, 2002.

Korade-Mirnics, Z., Burnside, J., and Corey, S. Dna microarray analysis of g-csf dependent, lyn-dependent genes. *Exp Hematol* 28:1497, 2000.

Korbel, G. A., Lalic, G., and Shair, M. D. Reaction microarrays: A method for rapidly determining the enantiomeric excess of thousands of samples. *J Am Chem Soc* 123:361–362, 2001.

Kornblum, H., and Geschwind, D. The use of representational difference analysis and cDNA microarrays in neural repair research. *Restor Neurol Neurosci* 18:89–94, 2001.

Kricka, L. J. Microchips, microarrays, biochips and nanochips: Personal laboratories for the 21st century. *Clin Chim Acta* 307:219–223, 2001.

Kricka, L. J., and Fortina, P. Microarray technology and applications: an all-language literature survey including books and patents. *Clin Chem* 47:1479–1482, 2001.

Krogan, N. J., and Greenblatt, J. F. Characterization of a six-subunit holo-elongator complex required for the regulated expression of a group of genes in *Saccharomyces cerevisiae. Mol Cell Biol* 21:8203–8212, 2001.

Kudoh, K., Ramanna, M., Ravatn, R., Elkahloun, A. G., Bittner, M. L., Meltzer, P. S., Trent, J. M., Dalton, W. S., and Chin, K. V. Monitoring the expression profiles of doxorubicin-induced and doxorubicin-resistant cancer cells by cDNA microarray. *Cancer Res* 60:4161–4166, 2000.

Kuhar, M. J., Joyce, A., and Dominguez, G. Genes in drug abuse. *Drug Alcohol Depend* 62:157–162, 2001.

Kuhn, K. M., DeRisi, J. L., Brown, P. O., and Sarnow, P. Global and specific translational regulation in the genomic response of Saccharomyces cerevisiae to a rapid transfer from a fermentable to a nonfermentable carbon source. *Mol Cell Biol* 21:916–927, 2001.

Kuiper, H. A., Kleter, G. A., Noteborn, H. P., and Kok, E. J. Assessment of the food safety issues related to genetically modified foods. *Plant J* 27:503–528, 2001.

Kuipers, O. P. Genomics for food biotechnology: Prospects of the use of high-throughput technologies for the improvement of food microorganisms. *Curr Opin Biotechnol* 10:511–516, 1999.

Kuipers, O. P., de Jong, A., Holsappel, S., Bron, S., Kok, J., and Hamoen, L. W. DNA-microarrays and food-biotechnology. *Antonie Van Leeuwenhoek* 76:353–355, 1999.

Kuklin, A., Shams, S., and Shah, S. High throughput screening of gene expression signatures. *Genetica* 108:41–46, 2000.

Kumar, A., Harrison, P. M., Cheung, K. H., Lan, N., Echols, N., Bertone, P., Miller, P., Gerstein, M. B., and Snyder, M. An integrated approach for finding overlooked genes in yeast. *Nat Biotechnol* 20:58–63, 2002.

Kumar, A., Soprano, D. R., and Parekh, H. K. Cross-resistance to the synthetic retinoid cd437 in a paclitaxel-resistant human ovarian carcinoma cell line is independent of the overexpression of retinoic acid receptor-gamma. *Cancer Res* 61:7552–7555, 2001.

Kuno, N., and Furuya, M. Phytochrome regulation of nuclear gene expression in plants. *Semin Cell Dev Biol* 11:485–493, 2000.

Kurella, M., Hsiao, L. L., Yoshida, T., Randall, J. D., Chow, G., Sarang, S. S., Jensen, R. V., and Gullans, S. R. DNA microarray analysis of complex biologic processes. *J Am Soc Nephrol* 12:1072–1078, 2001.

Kurian, K. M., Watson, C. J., and Wyllie, A. H. DNA chip technology. *J Pathol* 187:267–271, 1999.

Kuwabara, I., Kuwabara, Y., Yang, R. Y., Schuler, M., Green, D. R., Zuraw, B. L., Hsu, D. K., and Liu, F. T. Galectin-7 (PIG1:p53-induced gene 1) exhibits pro-apoptotic function through JNK activation and mitochondrial cytochrome c release. *J Biol Chem* 2001.

Kwiatkowski, M., Fredriksson, S., Isaksson, A., Nilsson, M., and Landegren, U. Inversion of in situ synthesized oligonucleotides: improved reagents for hybridization and primer extension in DNA microarrays. *Nucleic Acids Res* 27:4710–4714, 1999.

Kwitek-Black, A. E., and Jacob, H. J. The use of designer rats in the genetic dissection of hypertension. *Curr Hypertens Rep* 3:12–18, 2001.

Labourier, E., Blanchette, M., Feiger, J. W., Adams, M. D., and Rio, D. C. The KH-type RNA-binding protein PSI is required for *Drosophila* viability, male fertility, and cellular mRNA processing. *Genes Dev* 16:72–84, 2002.

Ladner, D. P., Leamon, J. H., Hamann, S., Tarafa, G., Strugnell, T., Dillon, D., Lizardi, P., and Costa, J. Multiplex detection of hotspot mutations by rolling circle-enabled universal microarrays. *Lab Invest* 81:1079–1086, 2001.

LaForge, K. S., Shick, V., Spangler, R., Proudnikov, D., Yuferov, V., Lysov, Y., Mirzabekov, A., and Kreek, M. J. Detection of single nucleotide polymorphisms of the human mu opioid receptor gene by hybridization or single nucleotide extension on custom oligonucleotide gelpad microchips: Potential in studies of addiction. *Am J Med Genet* 96:604–615, 2000.

Lage, J. M., Hamann, S., Gribanov, O., Leamon, J. H., Pejovic, T., and Lizardi, P. M. Microgel assessment of nucleic acid integrity and labeling quality in microarray experiments. *Biotechniques* 32:312–314, 2002.

Lai, R., Hirsch-Ginsberg, C. F., and Bueso-Ramos, C. Pathologic diagnosis of acute lymphocytic leukemia. *Hematol Oncol Clin North Am* 14:1209–1235, 2000.

Lam, L. T., Pickeral, O. K., Peng, A. C., Rosenwald, A., Hurt, E. M., Giltnane, J. M., Averett, L. M., Zhao, H., and co-workers. Genomic-scale measurement of mRNA turnover and the mechanisms of action of the anti-cancer drug flavopiridol. *Genome Biol* 2: 2001.

Lam, L. T., Ronchini, C., Norton, J., Capobianco, A. J., and Bresnick, E. H. Suppression of erythroid but not megakaryocytic differentiation of human K562 erythroleukemic cells by notch-1. *J Biol Chem* 275:19676–19684, 2000.

Lapteva, N., Nieda, M., Ando, Y., Ide, K., Hatta-Ohashi, Y., Dymshits, G., Ishikawa, Y., Juji, T., and Tokunaga, K. Expression of renin-angiotensin system genes in immature and mature dendritic cells identified using human cDNA microarray. *Biochem Biophys Res Commun* 285:1059–1065, 2001.

Larsen, L. A., Christiansen, M., Vuust, J., and Andersen, P. S. Recent developments in high-throughput mutation screening. *Pharmacogenomics* 2:387–399, 2001.

Lashkari, D. A., DeRisi, J. L., McCusker, J. H., Namath, A. F., Gentile, C., Hwang, S. Y., Brown, P. O., and Davis, R. W. Yeast microarrays for genome wide parallel genetic and gene expression analysis. *Proc Natl Acad Sci U S A* 94:13057–13062, 1997.

Lassus, H., Laitinen, M. P., Anttonen, M., Heikinheimo, M., Aaltonen, L. A., Ritvos, O., and Butzow, R. Comparison of serous and mucinous ovarian carcinomas: distinct pattern of allelic loss at distal 8p and expression of transcription factor GATA-4. *Lab Invest* 81:517–526, 2001.

Lau, W. Y., Lai, P. B., Leung, M. F., Leung, B. C., Wong, N., Chen, G., Leung, T. W., and Liew, C. T. Differential gene expression of hepatocellular carcinoma using cDNA microarray analysis. *Oncol Res* 12:59–69, 2000.

Law, A. S., and Archibald, A. L. Farm animal genome databases. *Brief Bioinform* 1:151–160, 2000.

Lawn, R. M., Wade, D. P., Garvin, M. R., Wang, X., Schwartz, K., Porter, J. G., Seilhamer, J. J., Vaughan, A. M., and Oram, J. F. The Tangier disease gene product ABC1 controls the cellular apolipoprotein-mediated lipid removal pathway. *J Clin Invest* 104:R25–R31, 1999.

Lawrance, I., Fiocchi, C., and Chakravarti, S. Ulcerative colitis and Crohn's Disease: Distinctive gene expression profiles and novel susceptibility candidate genes. *Hum Mol Genet* 10:445–456, 2001.

Lawrence, D. S. Functional proteomics: Large-scale analysis of protein kinase activity. Functional proteomics: Large-scale analysis of protein kinase activity. *Genome Biol* 2:REVIEWS 1007, 2001.

Lebkowski, J. S., Gold, J., Xu, C., Funk, W., Chiu, C. P., and Carpenter, M. K. Human embryonic stem cells: Culture, differentiation, and genetic modification for regenerative medicine applications. *Cancer J* 2(Suppl 7):S83–S93, 2001.

Le Crom, S., Devaux, F., Jacq, C., and Marc, P. yMGV: Helping biologists with yeast microarray data mining. *Nucleic Acids Res* 30:76–79, 2002.

Lee, H., Greeley, G. H., and Englander, E. W. Age-associated changes in gene expression patterns in the duodenum and colon of rats. *Mech Ageing Dev* 122:355–371, 2001.

Lee, J. M., Zhang, S., Saha, S., Santa Anna, S., Jiang, C., and Perkins, J. RNA expression analysis using an antisense *Bacillus subtilis* genome array. *J Bacteriol* 183:7371–7380, 2001.

Lee, M., and Walt, D. R. A fiber-optic microarray biosensor using aptamers as receptors. *Anal Biochem* 282:142–146, 2000.

Lee, M. L., Kuo, F. C., Whitmore, G. A., and Sklar, J. Importance of replication in microarray gene expression studies: Statistical methods and evidence from repetitive cDNA hybridizations. *Proc Natl Acad Sci U S A* 97:9834–9839, 2000.

Lee, P. D., Sladek, R., Greenwood, C. M., and Hudson, T. J. Control genes and variability: Absence of ubiquitous reference transcripts in diverse mammalian expression studies. *Genome Res* 12:292–297, 2002.

Lee, P. H., Sawan, S. P., Modrusan, Z., Arnold, L. J. Jr., and Reynolds, M. A. An efficient binding chemistry for glass polynucleotide microarrays. *Bioconjug Chem* 13:97–103, 2002.

Lee, P. S., and Lee, K. H. Genomic analysis. *Curr Opin Biotechnol* 11:171–175, 2000.

Lee, R. T., Yamamoto, C., Feng, Y., Potter-Perigo, S., Briggs, W. H., Landschulz, K. T., Turi, T. G., Thompson, J. F., and co-workers. Mechanical strain induces specific changes in the synthesis and organization of proteoglycans by vascular smooth muscle cells. *J Biol Chem* 276:13847–13851, 2001.

Lee, S. Y., Madan, A., Furuta, G. T., Colgan, S. P., and Sibley, E. Lactase gene transcription is activated in response to hypoxia in intestinal epithelial cells. *Mol Genet Metab* 75:65–69, 2002.

Lee, W. K., Padmanabhan, S., and Dominiczak, A. F. Genetics of hypertension: from experimental models to clinical applications. *J Hum Hypertens* 14:631–647, 2000.

Lehman, T. A., Haffty, B. G., Carbone, C. J., Bishop, L. R., Gumbs, A. A., Krishnan, S., Shields, P. G., Modali, R., and Turner, B. C. Elevated frequency and functional activity of a specific germ-line p53 intron mutation in familial breast cancer. *Cancer Res* 60:1062–1069, 2000.

Leikauf, G. D., McDowell, S. A., Bachurski, C. J., Aronow, B. J., Gammon, K., Wesselkamper, S. C., Hardie, W., Wiest, J. S., and co-workers. Functional genomics of oxidant-induced lung injury. *Adv Exp Med Biol* 500:479–487, 2001.

Le Menuet, D., Isnard, R., Bichara, M., Viengchareun, S., Muffat-Joly, M., Walker, F., Zennaro, M. C., and Lombes, M. Alteration of cardiac and renal functions in transgenic mice overexpressing human mineralocorticoid receptor. *J Biol Chem* 276:38911–38920, 2001.

Lemeshko, S. V., Powdrill, T., Belosludtsev, Y. Y., and Hogan, M. Oligonucleotides form a duplex with non-helical properties on a positively charged surface. *Nucleic Acids Res* 29:3051–3058, 2001.

Lemieux, B., Aharoni, A., and M. Schena. Overview of DNA chip technology. *Molecular Breeding* 4:277–289, 1998.

Lemkin, P. F., Thornwall, G. C., Walton, K. D., and Hennighausen, L. The microarray explorer tool for data mining of cDNA microarrays: application for the mammary gland. *Nucleic Acids Res* 28:4452–4459, 2000.

Le Naour, F. Contribution of proteomics to tumor immunology. *Proteomics* 1:1295–1302, 2001.

Le Naour, F., Hohenkirk, L., Grolleau, A., Misek, D. E., Lescure, P., Geiger, J. D., Hanash, S. M., and Beretta, L. Profiling changes in gene expression during differentiation and maturation of monocyte-derived dendritic cells using both oligonucleotide microarrays and proteomics. *J Biol Chem* 276:17920–17931, 2001.

Lennon, G. G. High-throughput gene expression analysis for drug discovery. *Drug Discov Today* 5:59–66, 2000.

LeProust, E., Pellois, J. P., Yu, P., Zhang, H., Gao, X., Srivannavit, O., Gulari, E., and Zhou, X. Digital light-directed synthesis. A microarray platform that permits rapid reaction optimization on a combinatorial basis. *J Comb Chem* 2:349–354, 2000.

Lersch, R. A., Fung, J., Hsieh, H. B., Smida, J., and Weier, H. U. Monitoring signal transduction in cancer: from chips to fish. *J Histochem Cytochem* 49:925–926, 2001.

Leung, Y. F., Lam, D. S., and Pang, C. P. The miracle of microarray data analysis. *Genome Biol* 2: 2001.

Lewis, F., Maughan, N. J., Smith, V., Hillan, K., and Quirke, P. Unlocking the archive—Gene expression in paraffin-embedded tissue. *J Pathol* 195:66–71, 2001.

Lewis, M. L., Cubano, L. A., Zhao, B., Dinh, H. K., Pabalan, J. G., Piepmeier, E. H., and Bowman, P. D. cDNA microarray reveals altered cytoskeletal gene expression in space-flown leukemic T lymphocytes (Jurkat). *FASEB J* 15:1783–1785, 2001.

Lewohl, J. M., Dodd, P. R., Mayfield, R. D., and Harris, R. A. Application of DNA microarrays to study human alcoholism. *J Biomed Sci* 8:28–36, 2001.

Lewohl, J. M., Wang, L., Miles, M. F., Zhang, L., Dodd, P. R., and Harris, R. A. Gene expression in human alcoholism: Microarray analysis of frontal cortex. *Alcohol Clin Exp Res* 24:1873–1882, 2000.

Li, F., and Stormo, G. D. Selection of optimal DNA oligos for gene expression arrays. *Bioinformatics* 17:1067–1076, 2001.

Li, H., and Hong, F. Cluster-Rasch models for microarray gene expression data. *Genome Biol* 2: 2001.

Li, J., Chen, S., and Evans, D. H. Typing and subtyping influenza virus using DNA microarrays and multiplex reverse transcriptase PCR. *J Clin Microbiol* 39:696–704, 2001.

Li, J., Lee, J. M., and Johnson, J. A. Microarray analysis reveals an antioxidant responsive element-driven gene set involved in conferring protection from an oxidative stress-induced apoptosis in IMR-32 cells. *J Biol Chem* Oct 30, 2001.

Li, J., Peet, G. W., Balzarano, D., Li, X., Massa, P., Barton, R. W., and Marcu, K. B. Novel NEMO/IKKγ and NF-κB target genes at the pre-B to immature B cell transition. *J Biol Chem* 276:18579–18590, 2001.

Li, J. Y., Boado, R. J., and Pardridge, W. M. Blood-brain barrier genomics. *J Cereb Blood Flow Metab* 21:61–68, 2001.

Li, J. Y., Lescure, P. A., Misek, D. E., Lai, Y. M., Chai, B. X., Kuick, R., Thompson, R. C., Demo, R. M., and co-workers. Food deprivation induced expression of Minoxidil sulfotransferase in the hypothalamus uncovered by microarray analysis. *J Biol Chem* 2002.

Li, L., Weinberg, C. R., Darden, T. A., and Pedersen, L. G. Gene selection for sample classification based on gene expression data: study of sensitivity to choice of parameters of the GA/KNN method. *Bioinformatics* 17:1131–1142, 2001.

Li, Q. Z., Eckenrode, S., Ruan, Q. G., Wang, C. Y., Shi, J. D., McIndoe, R. A., and She, J. X. Rapid decrease of RNA level of a novel mouse mitochondria solute carrier protein (Mscp) gene at 4-5 weeks of age. *Mamm Genome* 12(11):830–836, 2001.

Li, S., Ross, D. T., Kadin, M. E., Brown, P. O., and Wasik, M. A. Comparative genome-scale analysis of gene expression profiles in T cell lymphoma cells during malignant progression using a complementary DNA microarray. *Am J Pathol* 158:1231–1237, 2001.

Li, Y., Qiu, M. Y., Wu, C. Q., Cao, Y. Q., Tang, R., Chen, Q., Shi, X. Y., Hu, Z. Q., and co-workers. Detection of differentially expressed genes in hepatocellular carcinoma using DNA microarray. *Yi Chuan Xue Bao* 27:1042–1048, 2000.

Li, Z., He, L., Wilson, K., and Roberts, D. Thrombospondin-1 inhibits TCR-mediated T lymphocyte early activation. *J Immunol* 166:2427–2436, 2001.

Li, Z., Khaletskiy, A., Wang, J., Wong, J. Y., Oberley, L. W., and Li, J. J. Genes regulated in human breast cancer cells overexpressing manganese-containing superoxide dismutase. *Free Radic Biol Med.* 30:260–267, 2001.

Liang, C. P., and Tall, A. R. Transcriptional profiling reveals global defects in energy metabolism, lipoprotein and bile acid synthesis and transport with reversal by leptin treatment in ob/ob mouse liver. *J Biol Chem* 2001.

Liao, B., Hale, W., Epstein, C. B., Butow, R. A., and Garner, H. R. MAD: A suite of tools for microarray data management and processing. *Bioinformatics* 16:946–947, 2000.

Liau, L. M., Lallone, R. L., Seitz, R. S., Buznikov, A., Gregg, J. P., Kornblum, H. I., Nelson, S. F., and Bronstein, J. M. Identification of a human glioma-associated growth factor gene, granulin, using differential immuno-absorption. *Cancer Res* 60:1353–1360, 2000.

Lichter, P., Joos, S., Bentz, M., and Lampel, S. Comparative genomic hybridization: Uses and limitations. *Semin Hematol* 37:348–357, 2000.

Lichtermann, D., Franke, P., Maier, W., and Rao, M. L. Pharmacogenomics and addiction to opiates. *Eur J Pharmacol* 410:269–279, 2000.

Lichtlen, P., Wang, Y., Belser, T., Georgiev, O., Certa, U., Sack, R., and Schaffner, W. Target gene search for the metal-responsive transcription factor MTF-1. *Nucleic Acids Res* 29:1514–1523, 2001.

Lieberman, R. Androgen deprivation therapy for prostate cancer chemoprevention: current status and future directions for agent development. *Urology* 58(Suppl 1):83–90, 2001.

Lieberman, R., Crowell, J. A., Hawk, E. T., Boone, C. W., Sigman, C. C., and Kelloff, G. J. Development of new cancer chemoprevention agents: Role of pharmacokinetic/pharmacodynamic and intermediate endpoint biomarker monitoring. *Clin Chem* 44:420–427, 1998.

Lieuallen, K., Pennacchio, L. A., Park, M., Myers, R. M., and Lennon, G. G. Cystatin B-deficient mice have increased expression of apoptosis and glial activation genes. *Hum Mol Genet* 10:1867–1871, 2001.

Lievens, S., Goormachtig, S., and Holsters, M. A critical evaluation of differential display as a tool to identify genes involved in legume nodulation: Looking back and looking forward. *Nucleic Acids Res* 29:3459–3468, 2001.

Lin, B., Ferguson, C., White, J. T., Wang, S., Vessella, R., True, L. D., Hood, L., and Nelson, P. S. Prostate-localized and androgen-regulated expression of the membrane-bound serine protease TMPRSS2. *Cancer Res* 59:4180–4184, 1999.

Lin, B., White, J. T., Ferguson, C., Bumgarner, R., Friedman, C., Trask, B., Ellis, W., Lange, and co-workers . PART-1: A novel human prostate-specific, androgen-regulated gene that maps to chromosome 5q12. *Cancer Res* 60:858–863, 2000.

Lin, C. S., Ho, H. C., Gholami, S., Chen, K. C., Jad, A., and Lue, T. F. Gene expression profiling of an arteriogenic impotence model. *Biochem Biophys Res Commun* 285:565–569, 2001.

Lin, B., White, J. T., Ferguson, C., Wang, S., Vessella, R., Bumgarner, R., True, L. D., Hood, L., and Nelson, P. S. Prostate short-chain dehydrogenase reductase 1 (PSDR1): A new member of the short-chain steroid dehydrogenase/reductase family highly expressed in normal and neoplastic prostate epithelium. *Cancer Res* 61:1611–1618, 2001.

Lin, S. S., Manchester, J. K., and Gordon, J. I. Enhanced gluconeogenesis and increased energy storage as hallmarks of aging in *Saccharomyces cerevisiae*. *J Biol Chem* 276:36000–36007, 2001.

Lin, Y. M., Ono, K., Satoh, S., Ishiguro, H., Fujita, M., Miwa, N., Tanaka, T., Tsunoda, T., and co-workers. Identification of AF17 as a downstream gene of the beta-catenin/T-cell factor pathway and its involvement in colorectal carcinogenesis. *Cancer Res* 61:6345–6349, 2001.

Lindroos, K., Liljedahl, U., Raitio, M., and Syvanen, A. C. Minisequencing on oligonucleotide microarrays: comparison of immobilisation chemistries. *Nucleic Acids Res* 29:E69–9, 2001.

Ling, V., Wu, P. W., Finnerty, H. F., Agostino, M. J., Graham, J. R., Chen, S., Jussiff, J. M., Fisk, G. J., and co-workers. Assembly and annotation of human chromosome 2q33 sequence containing the CD28, CTLA4, and ICOS gene cluster: Analysis by computational, comparative, and microarray approaches. *Genomics* 78:155–168, 2001.

Linnarsson, S., Mikaels, A., Baudet, C., and Ernfors, P. Activation by GDNF of a transcriptional program repressing neurite growth in dorsal root ganglia. *Proc Natl Acad Sci U S A* 98:14681–14686, 2001.

Liu, H., He, Z., and Rosenwaks, Z. Application of complementary DNA microarray (DNA chip) technology in the study of gene expression profiles during folliculogenesis. *Fertil Steril* 75:947–955, 2001.

Liu, H. C., Cheng, H. H., Tirunagaru, V., Sofer, L., and Burnside, J. A strategy to identify positional candidate genes conferring Marek's disease resistance by integrating DNA microarrays and genetic mapping. *Anim Genet* 32:351–359, 2001.

Liu, J., Chen, H., Miller, D. S., Saavedra, J. E., Keefer, L. K., Johnson, D. R., Klaassen, C. D., and Waalkes, M. P. Overexpression of glutathione S-transferase II and multidrug resistance transport proteins is associated with acquired tolerance to inorganic arsenic. *Mol Pharmacol* 60:302–309, 2001.

Liu, J., Corton, C., Dix, D. J., Liu, Y., Waalkes, M. P., and Klaassen, C. D. Genetic background but not metallothionein phenotype dictates sensitivity to cadmium-induced testicular injury in mice. *Toxicol Appl Pharmacol* 176:1–9, 2001.

Liu, J., Saavedra, J. E., Lu, T., Song, J. G., Clark, J., Waalkes, M. P., and Keefer, L. K. O(2)-Vinyl 1-(pyrrolidin-1-yl)diazen-1-ium-1,2-diolate protection against D-galactosamine/endotoxin-induced hepatotoxicity in mice: Genomic analysis using microarrays. *J Pharmacol Exp Ther* 300:18–25, 2002.

Liu, T. J., Lai, H. C., Wu, W., Chinn, S., and Wang, P. H. Developing a strategy to define the effects of insulin-like growth factor-1 on gene expression profile in cardiomyocytes. *Circ Res* 88:1231–1238, 2001.

Liu, W. T., Mirzabekov, A. D., and Stahl, D. A. Optimization of an oligonucleotide microchip for microbial identification studies: A non-equilibrium dissociation approach. *Environ Microbiol* 3:619–629, 2001.

Liu, X., Brutlag, D. L., and Liu, J. S. BioProspector: Discovering conserved DNA motifs in upstream regulatory regions of co-expressed genes. *Pac Symp Biocomput*:127–138, 2001.

Livesey, F. J., Furukawa, T., Steffen, M. A., Church, G. M., and Cepko, C. L. Microarray analysis of the transcriptional network controlled by the photoreceptor homeobox gene Crx. *Curr Biol* 10:301–310, 2000.

Ljubimova, J. Y., Lakhter, A. J., Loksh, A., Yong, W. H., Riedinger, M. S., Miner, J. H., Sorokin, L. M., Ljubimov, A. V., and Black, KL. Overexpression of alpha4 chain-containing laminins in human glial tumors identified by gene microarray analysis. *Cancer Res* 61:5601–5610, 2001.

Locklin, R. M., Riggs, B. L., Hicok, K. C., Horton, H. F., Byrne, M. C., and Khosla, S. Assessment of gene regulation by bone morphogenetic protein 2 in human marrow stromal cells using gene array technology. *J Bone Miner Res* 16:2192–2204, 2001.

Loferer, H. Mining bacterial genomes for antimicrobial targets. *Mol Med Today* 6: 470–474, 2000.

Loftus, S. K., Chen, Y., Gooden, G., Ryan, J. F., Birznieks, G., Hilliard, M., Baxevanis, A. D., Bittner, M., and co-workers. Informatic selection of a neural crest-melanocyte cDNA set for microarray analysis. *Proc Natl Acad Sci U S A* 96:9277–9280, 1999.

Loftus, S. K., and Pavan, W. J. The use of expression profiling to study pigment cell biology and dysfunction. *Pigment Cell Res* 13:141–146, 2000.

Loguinov, A. V., Anderson, L. M., Crosby, G. J., and Yukhananov, R. Y. Gene expression following acute morphine administration. *Physiol Genomics* 6:169–181, 2001.

Lomax, M. I., Huang, L., Cho, Y., Gong, T. L., and Altschuler, R. A. Differential display and gene arrays to examine auditory plasticity. *Hear Res* 147:293–302, 2000.

Lomri, A., Lemonnier, J., Delannoy, P., and Marie, P. J. Increased expression of protein kinase calpha, interleukin-1alpha, and RhoA guanosine 5′-triphosphatase in osteoblasts expressing the Ser252Trp fibroblast growth factor 2 apert mutation: Identification by analysis of complementary DNA microarray. *J Bone Miner Res* 16:705–712, 2001.

Long, A. D., Mangalam, H. J., Chan, B. Y., Tolleri, L., Hatfield, G. W., and Baldi, P. Gene expression profiling in escherichia coli K12: Improved statistical inference from DNA microarray data using analysis of variance and a Bayesian statistical framework. *J Biol Chem* 2001.

Lonning, P. E., Sorlie, T., Perou, C. M., Brown, P. O., Botstein, D., and Borresen-Dale, A. L. Microarrays in primary breast cancer—Lessons from chemotherapy studies. *Endocr Relat Cancer* 8:259–263, 2001.

Loos, A., Glanemann, C., Willis, L. B., O'Brien, X. M., Lessard, P. A., Gerstmeir, R., Guillouet, S., and Sinskey, A. J. Development and validation of corynebacterium DNA microarrays. *Appl Environ Microbiol* 67:2310–2318, 2001.

Lopez, M. C., and Baker, H. V. Understanding the growth phenotype of the yeast gcr1 mutant in terms of global genomic expression patterns. *J Bacteriol* 182:4970–4978, 2000.

Lopez, P. J., and Seraphin, B. YIDB: The yeast intron database. *Nucleic Acids Res* 28:85–86, 2000.

Lopez Figueroa, A. L., Watson, S. J., and Lopez Figueroa, M. O. Use of the technology of the genetic chip to study of the nitric oxide synthase. *Rev Neurol* 33:555–560, 2001.

Loring, J. F., Porter, J. G., Seilhammer, J., Kaser, M. R., and Wesselschmidt, R. A gene expression profile of embryonic stem cells and embryonic stem cell-derived neurons. *Restor Neurol Neurosci* 18:81–88, 2001.

Loring, J. F., Wen, X., Lee, J. M., Seilhamer, J., and Somogyi, R. A gene expression profile of Alzheimer's disease. *DNA Cell Biol* 20:683–695, 2001.

Lossos, I. S., Alizadeh, A. A., Eisen, M. B., Chan, W. C., Brown, P. O., Botstein, D., Staudt, L. M., and Levy, R. Ongoing immunoglobulin somatic mutation in germinal center B cell-like but not in activated B cell-like diffuse large cell lymphomas. *Proc Natl Acad Sci U S A* 97:10209–10213, 2000.

Lou, X. J., Schena, M., Horrigan, F. T., Lawn, R. M., and Davis, R. W. Expression monitoring using cDNA microarrays. A general protocol. *Methods Mol Biol* 175:323–340, 2001.

Lu, C., Kasik, J., Stephan, D. A., Yang, S., Sperling, M. A., and Menon, R. K. Grtp1, a novel gene regulated by growth hormone. *Endocrinology* 142:4568–4571, 2001.

Lu, J., Liu, Z., Xiong, M., Wang, Q., Wang, X., Yang, G., Zhao, L., Qiu, Z., and co-workers. Gene expression profile changes in initiation and progression of squamous cell carcinoma of esophagus. *Int J Cancer* 91:288–294, 2001.

Lu, T., Liu, J., LeCluyse, E. L., Zhou, Y. S., Cheng, M. L., and Waalkes, M. P. Application of cDNA microarray to the study of arsenic-induced liver diseases in the population of Guizhou, China. *Toxicol Sci* 59:185–192, 2001.

Lu, Y. J., Williamson, D., Clark, J., Wang, R., Tiffin, N., Skelton, L., Gordon, T., Williams, R., and co-workers. Comparative expressed sequence hybridization to chromosomes for tumor classification and identification of genomic regions of differential gene expression. *Proc Natl Acad Sci U S A* 98:9197–9202, 2001.

Lucchini, S., Thompson, A., and Hinton, J. C. Microarrays for microbiologists. *Microbiology* 147:1403–1414, 2001.

Lucito, R., West, J., Reiner, A., Alexander, J., Esposito, D., Mishra, B., Powers, S., Norton, L., and Wigler, M. Detecting gene copy number fluctuations in tumor cells by microarray analysis of genomic representations. *Genome Res* 10:1726–1736, 2000.

Lueking, A., Horn, M., Eickhoff, H., Bussow, K., Lehrach, H., and Walter, G. Protein microarrays for gene expression and antibody screening. *Anal Biochem* 270:103–111, 1999.

Lukas, J., Gao, D. Q., Keshmeshian, M., Wen, W. H., Tsao-Wei, D., and Rosenberg, S. Alternative and aberrant messenger RNA splicing of the mdm2 oncogene in invasive breast cancer. *Cancer Res* 61:3212–3219, 2001.

Lukashin, A. V., and Fuchs, R. Analysis of temporal gene expression profiles: clustering by simulated annealing and determining the optimal number of clusters. *Bioinformatics* 17:405–414, 2001.

Luo, J., Duggan, D. J., Chen, Y., Sauvageot, J., Ewing, C. M., Bittner, M. L., Trent, J. M., and Isaacs, W. B. Human prostate cancer and benign prostatic hyperplasia: Molecular dissection by gene expression profiling. *Cancer Res* 61:4683–4688, 2001.

Luo, L. Y., and Diamandis, E. P. Preliminary examination of time-resolved fluorometry for protein array applications. *Luminescence* 7:409–413, 2000.

Luo, Z., and Geschwind, D. H. Microarray applications in neuroscience. *Neurobiol Dis* 8:183–193, 2001.

Lyakhovich, A., Aksenov, N., Pennanen, P., Miettinen, S., Ahonen, M. H., Syvala, H., Ylikomi, T., and Tuohimaa, P. Vitamin D induced up-regulation of keratinocyte growth factor (FGF-7/KGF) in MCF-7 human breast cancer cells. *Biochem Biophys Res Commun* 273:675–680, 2000.

Lyons, T. J., Gasch, A. P., Gaither, L. A., Botstein, D., Brown, P. O., and Eide, D. J. Genome-wide characterization of the Zap1p zinc-responsive regulon in yeast. *Proc Natl Acad Sci U S A* 97:7957–7962, 2000.

Ma, C., and Staudt, L. M. Molecular definition of the germinal centre stage of B-cell differentiation. *Philos Trans R Soc Lond B Biol Sci* 356:83–89, 2001.

Ma, L., Li, J., Qu, L., Hager, J., Chen, Z., Zhao, H., and Deng, X. W. Light control of *Arabidopsis* development entails coordinated regulation of genome expression and cellular pathways. *Plant Cell* 13:2589–2607, 2001.

Macas, J., Nouzova, M., and Galbraith, D. W. Adapting the Biomek 2000 Laboratory Automation Workstation for printing DNA microarrays. *Biotechniques* 25:106–110, 1998.

MacBeath, G., and Schreiber, S. L. Printing proteins as microarrays for high-throughput function determination. *Science* 289:1760–1763, 2000.

Mace, M. L., Jr., Montagu, J., Rose, S. D., and McGuinness, G. Novel microarray printing and detection technologies. In: M. Schena, ed. *Microarray Biochip Technology*. Eaton Publishing, Natick, MA, 2000:39–64.

Macreadie, I., Ginsburg, H., Sirawaraporn, W., and Tilley, L. Antimalarial drug development and new targets. *Parasitol Today* 16:438–444, 2000.

Macoska, J. A. The progressing clinical utility of DNA microarrays. *CA Cancer J Clin* 52:50–59, 2002.

Madden, S. L., Wang, C. J., and Landes, G. Serial analysis of gene expression: From gene discovery to target identification. *Drug Discov Today* 5:415–425, 2000.

Madoz-Gurpide, J., Wang, H., Misek, D. E., Brichory, F., and Hanash, S. M. Protein based microarrays: A tool for probing the proteome of cancer cells and tissues. *Proteomics* 1:1279–1287, 2001.

Maeda, S., Otsuka, M., Hirata, Y., Mitsuno, Y., Yoshida, H., Shiratori, Y., Masuho, Y., Muramatsu, M., and co-workers. cDNA microarray analysis of *Helicobacter pylori*-mediated alteration of gene expression in gastric cancer cells. *Biochem Biophys Res Commun* 284:443–449, 2001.

Maekawa, T., Bernier, F., Sato, M., Nomura, S., Singh, M., Inoue, Y., Tokunaga, T., Imai, H., and co-workers. Mouse ATF-2 null mutants display features of a severe type of meconium aspiration syndrome. *J Biol Chem* 274:17813–17819, 1999.

Mahalingam, R., and Fedoroff, N. Screening insertion libraries for mutations in many genes simultaneously using DNA microarrays. *Proc Natl Acad Sci U S A* 98:7420–7425, 2001.

Maleck, K., Levine, A., Eulgem, T., Morgan, A., Schmid, J., Lawton, K. A., Dangl, J. L., and Dietrich, R. A. The transcriptome of *Arabidopsis thaliana* during systemic acquired resistance. Nat Genet 26:403–410, 2000.

Maldonado-Rodriguez, R., and Beattie, K. L. Analysis of nucleic acids by tandem hybridization on oligonucleotide microarrays. *Methods Mol Biol* 170:157–171, 2001.

Malloff, C. A., Fernandez, R. C., and Lam, W. L. Bacterial comparative genomic hybridization: A method for directly identifying lateral gene transfer. *J Mol Biol* 312:1–5, 2001.

Mandel, S., Grunblatt, E., and Youdim, M. cDNA microarray to study gene expression of dopaminergic neurodegeneration and neuroprotection in MPTP and 6-hydroxydopamine models: Implications for idiopathic Parkinson's disease. *J Neural Transm Suppl* 60:117–124, 2000.

Manderson, E. N., Mes-Masson, A. M., Novak, J., Lee, P. D., Provencher, D., Hudson, T. J., and Tonin, P. N. Expression profiles of 290 ESTs mapped to chromosome 3 in human epithelial ovarian cancer cell lines using DNA expression oligonucleotide microarrays. *Genome Res* 12:112–121, 2002.

Manduchi, E., Grant, G. R., McKenzie, S. E., Overton, G. C., Surrey, S., and Stoeckert, C. J. Jr. Generation of patterns from gene expression data by assigning confidence to differentially expressed genes. *Bioinformatics* 16:685–698, 2000.

Manganelli, R., Voskuil, M. I., Schoolnik, G. K., and Smith, I. The *Mycobacterium tuberculosis* ECF sigma factor sigmaE: Role in global gene expression and survival in macrophages. *Mol Microbiol* 41:423–437, 2001.

Manger, I. D., and Relman, D. A. How the host 'sees' pathogens: Global gene expression responses to infection. *Curr Opin Immunol* 12:215–218, 2000.

Maniotis, A. J., Folberg, R., Hess, A., Seftor, E. A., Gardner, L. M., Pe'er, J., Trent, J. M., Meltzer, P. S., and Hendrix, M. J. Vascular channel formation by human melanoma cells in vivo and in vitro: Vasculogenic mimicry. *Am J Pathol* 155:739–752, 1999.

Manley, S., Mucci, N. R., De Marzo, A. M., and Rubin, M. A. Relational database structure to manage high-density tissue microarray data and images for pathology studies focusing on clinical outcome: The prostate specialized program of research excellence model. *Am J Pathol* 159:837–843, 2001.

Manos, E. J., and Jones, D. A. Assessment of tumor necrosis factor receptor and Fas signaling pathways by transcriptional profiling. *Cancer Res* 61:433–438, 2001.

Manos, E. J., Kim, M. L., Kassis, J., Chang, P. Y., Wells, A., and Jones, DA. Dolichol-phosphate-mannose-3 (DPM3)/prostin-1 is a novel phospholipase C-gamma regulated gene negatively associated with prostate tumor invasion. *Oncogene* 20:2781–2790, 2001.

Mansergh, F. C., Wride, M. A., and Rancourt, D. E. Neurons from stem cells: Implications for understanding nervous system development and repair. *Biochem Cell Biol* 78:613–628, 2000.

Marc, P., Devaux, F., and Jacq, C. yMGV: A database for visualization and data mining of published genome-wide yeast expression data. *Nucleic Acids Res* 29:E63–3, 2001.

Marc, P., Margeot, A., Devaux, F., Blugeon, C., Corral-Debrinski, M., and Jacq, C. Genome-wide analysis of mRNAs targeted to yeast mitochondria. *EMBO Rep* 3:159–164, 2002.

Marciano, P., Eberwine, J. H., Raghupathi, R., and McIntosh, T. K. The assessment of genomic alterations using DNA arrays following traumatic brain injury: A review. *Restor Neurol Neurosci* 18:105–113, 2001.

Marcotte, E. R., Srivastava, L. K., and Quirion, R. DNA microarrays in neuropsychopharmacology. *Trends Pharmacol Sci* 22:426–436, 2001.

Mariadason, J. M., Corner, G. A., and Augenlicht, L. H. Genetic reprogramming in pathways of colonic cell maturation induced by short chain fatty acids: comparison with trichostatin A, sulindac, and curcumin and implications for chemoprevention of colon cancer. *Cancer Res* 60:4561–4572, 2000.

Mariani, L., Beaudry, C., McDonough, W. S., Hoelzinger, D. B., Demuth, T., Ross, K. R., Berens, T., Coons, S. W., and co-workers. Glioma cell motility is associated with reduced transcription of proapoptotic and proliferation genes: A cDNA microarray analysis. *J Neurooncol* 53:161–176, 2001.

Mariani, L., Beaudry, C., McDonough, W. S., Hoelzinger, D. B., Kaczmarek, E., Ponce, F., Coons, S. W., Giese, A., and co-workers. Death-associated protein 3 (Dap-3) is overexpressed in invasive glioblastoma cells in vivo and in glioma cell lines with induced motility phenotype in vitro. *Clin Cancer Res* 7:2480–2489, 2001.

Markert, J. M., Fuller, C. M., Gillespie, G. Y., Bubien, J. K., McLean, L. A., Hong, R. L., Lee, K., Gullans, S. R., and co-workers. Differential gene expression profiling in human brain tumors. *Physiol Genomics* 5:21–33, 2001.

Marra, M., Hillier, L., Kucaba, T., Allen, M., Barstead, R., Beck, C., Blistain, A., Bonaldo, M., and co-workers. An encyclopedia of mouse genes. *Nat Genet* 21:191–194, 1999.

Marti, G. E., Gaigalas, A., and Vogt, R.F. Jr. Recent developments in quantitative fluorescence calibration for analyzing cells and microarrays. *Cytometry* 42:263, 2000.

Martinez, E. J., Corey, E. J., and Owa, T. Antitumor activity- and gene expression-based profiling of ecteinascidin Et 743 and phthalascidin Pt 650. *Chem Biol* 8:1151–1160, 2001.

Martinsky, T., and Haje, P. Microarray tools, kits, reagents and services. In: M. Schena, ed. *Microarray Biochip Technology*. Eaton Publishing, Natick, MA, 2000:201–220.

Marton, M. J., DeRisi, J. L., Bennett, H. A., Iyer, V. R., Meyer, M. R., Roberts, C. J., Stoughton, R., Burchard, J., and co-workers. Drug target validation and identification of secondary drug target effects using DNA microarrays. *Nat Med* 4:1293–1301, 1998.

Masters, J. R., and Lakhani, S. R. How diagnosis with microarrays can help cancer patients. Nature 404:921, 2000.

Masys, D. R. Linking microarray data to the literature. *Nat Genet* 28:9–10, 2001.

Masys, D. R., Welsh, J. B., Lynn-Fink, J., Gribskov, M., Klacansky, I., and Corbeil, J. Use of keyword hierarchies to interpret gene expression patterns. *Bioinformatics* 17:319–326, 2001.

Mataki, C., Murakami, T., Umetani, M., Wada, Y., Ishii, M., Tsutsumi, S., Aburatani, H., Hamakubo, T., and Kodama, T. A novel zinc finger protein mRNA in human umbilical vein endothelial cells is profoundly induced by tumor necrosis factor alpha. *J Atheroscler Thromb* 7:97–103, 2000.

Matejuk, A., Dwyer, J., Zamora, A., Vandenbark, A. A., and Offner, H. Evaluation of the effects of 17beta-estradiol (17beta-e2) on gene expression in experimental autoimmune encephalomyelitis using DNA microarray. *Endocrinology* 143:313–319, 2002.

Mathiassen, S., Lauemoller, S. L., Ruhwald, M., Claesson, M. H., and Buus, S. Tumor-associated antigens identified by mRNA expression profiling induce protective anti-tumor immunity. *Eur J Immunol* 31:1239–1246, 2001.

Matsunaga, T., Nakayama, H., Okochi, M., and Takeyama, H. Fluorescent detection of cyanobacterial DNA using bacterial magnetic particles on a MAG-microarray. *Biotechnol Bioeng* 73:400–405, 2001.

Matsushima-Nishiu, M., Unoki, M., Ono, K., Tsunoda, T., Minaguchi, T., Kuramoto, H., Nishida, M., Satoh, T., and co-workers. Growth and gene expression profile analyses of endometrial cancer cells expressing exogenous pten. *Cancer Res* 61:3741–3749, 2001.

Maughan, N. J., Lewis, F. A., and Smith, V. An introduction to arrays. *J Pathol* 195:3–6, 2001.

Maurer, M., Trajanoski, Z., Frey, G., Hiroi, N., Galon, J., Willenberg, H. S., Gold, P. W., Chrousos, G. P., and co-workers. Differential gene expression profile of glucocorticoids, testosterone, and dehydroepiandrosterone in human cells. *Horm Metab Res* 33:691–695, 2001.

Maxwell, S. A., and Davis, G. E. Biological and molecular characterization of an ECV-304-derived cell line resistant to p53-mediated apoptosis. *Apoptosis* 5:277–290, 2000.

Maxwell, S. A., and Davis, G. E. Differential gene expression in p53-mediated apoptosis-resistant vs. apoptosis-sensitive tumor cell lines. *Proc Natl Acad Sci U S A* 97:13009–13014, 2000.

May, B. J., Zhang, Q., Li, L. L., Paustian, M. L., Whittam, T. S., and Kapur, V. Complete genomic sequence of *Pasteurella multocida,* Pm70. *Proc Natl Acad Sci U S A* 98:3460–3465, 2001.

Mayanil, C. S., George, D., Freilich, L., Miljan, E. J., Mania-Farnell, B., McLone, D. G., and Bremer, E. G. Microarray analysis detects novel Pax3 downstream target genes. *J Biol Chem*, 2001.

McCracken, A. A., and Brodsky, J. L. A molecular portrait of the response to unfolded proteins. *Genome Biol* 1: 2000.

McCormack, G. W. Subcellular micromolecular pharmacotherapy using computer-assisted microarray analysis. *J Natl Cancer Inst* 93:1350–1351, 2001.

McDonald, M. J., and Rosbash, M. Microarray analysis and organization of circadian gene expression in *Drosophila*. *Cell* 107:567–578, 2001.

McDonald, W. H., and Yates, J. R. III. Proteomic tools for cell biology. *Traffic* 1:747–754, 2000.

McDonough, P. G. The first hope for the "big picture" of infection—DNA microarrays. *Fertil Steril* 72:187, 1999.

McDowell, S. A., Mallakin, A., Bachurski, C. J., Toney-Earley, K., Prows, D. R., Bruno, T., Kaestner, K. H., Witte, D. P., and co-workers. The role of the receptor tyrosine kinase ron in nickel-induced acute lung injury. *Am J Respir Cell Mol Biol* 26:99–104, 2002.

McGonigle, B., Keeler, S. J., Lau, S. M., Koeppe, M. K., and O'Keefe, D. P. A genomics approach to the comprehensive analysis of the glutathione S-transferase gene family in soybean and maize. *Plant Physiol* 124:1105–1120, 2000.

McKenzie, S. E., Mansfield, E., Rappaport, E., Surrey, S., and Fortina, P. Parallel molecular genetic analysis. *Eur J Hum Genet* 6:417–429, 1998.

Meadows, L. Microarrays with a twist: Identification of mesoderm development genes. *Trends Genet* 17:693–694, 2001.

Medhora, M., Bousamra, M. II, Zhu, D., Somberg, L., and Jacobs, E. R. Uppregulation of collagens detected by gene array in a model of flow-induced pulmonary vascular remodeling. *Am J Physiol Heart Circ Physiol* 282:H414–H422, 2002.

Medico, E., Gentile, A., Lo Celso, C., Williams, T. A., Gambarotta, G., Trusolino, L., and Comoglio, P. M. Osteopontin is an autocrine mediator of hepatocyte growth factor-induced invasive growth. *Cancer Res* 61:5861–5868, 2001.

Medlin, J. Array of hope for gene technology. *Environ Health Perspect* 109:A34–A37, 2001.

Medlin, J. F. Timely toxicology. *Environ Health Perspect* 107:A256–A258, 1999.

Mehrotra, J., and Bishai, W. R. Regulation of virulence genes in *Mycobacterium tuberculosis*. *Int J Med Microbiol* 291:171–182, 2001.

Meinhold-Heerlein, I., Stenner-Liewen, F., Liewen, H., Kitada, S., Krajewska, M., Krajewski, S., Zapata, J. M., Monks, A., and co-workers. Expression and potential role of Fas-associated phosphatase-1 in ovarian cancer. *Am J Pathol* 158:1335–1344, 2001.

Meiyanto, E., Mineno, J., Ishida, N., and Takeya, T. Application of fluorescently labeled poly(dU) for gene expression profiling on cDNA microarrays. *Biotechniques* 31:406–410, 412–413, 2001.

Meldrum, D. Automation for genomics, part two: sequencers, microarrays, and future trends. *Genome Res* 10:1288–1303, 2000.

Meltzer, P. S. Spotting the target: Microarrays for disease gene discovery. *Curr Opin Genet Dev* 11:258–263, 2001.

Mendoza, L. G., McQuary, P., Mongan, A., Gangadharan, R., Brignac, S., and Eggers, M. High-throughput microarray-based enzyme-linked immunosorbent assay (ELISA). *Biotechniques* 27:778–780, 1999.

Mercier, G., Denis, Y., Marc, P., Picard, L., and Dutreix, M. Transcriptional induction of repair genes during slowing of replication in irradiated *Saccharomyces cerevisiae*. *Mutat Res* 487:157–172, 2001.

Mezzasoma, L., Bacarese-Hamilton, T., Di Cristina, M., Rossi, R., Bistoni, F., and Crisanti, A. Antigen microarrays for serodiagnosis of infectious diseases. *Clin Chem* 48:121–130, 2002.

Mhlanga, M. M., and Malmberg, L. Using molecular beacons to detect single-nucleotide polymorphisms with real-time PCR. *Methods* 25:463–471, 2001.

Miettinen, H. E., Jarvinen, T. A., Kellner, U., Kauraniemi, P., Parwaresch, R., Rantala, I., Kalimo, H., Paljarvi, L., and co-workers. High topoisomerase IIalpha expression associates with high proliferation rate and and poor prognosis in oligodendrogliomas. *Neuropathol Appl Neurobiol* 26:504–512, 2000.

Miettinen, H. E., Paunu, N., Rantala, I., Kalimo, H., Paljarvi, L., Helin, H., and Haapasalo, H. Cell cycle regulators (p21, p53, pRb) in oligodendrocytic tumors: a study by novel tumor microarray technique. *J Neurooncol* 55(1):29–37, 2001.

Mifflin, T.E. DNA Microarrays: A Practical Approach. Mark Schena, ed. Oxford: Oxford University Press, 1999, 209 pp., $55.00, paperback. ISBN 0-19-963777-8. Microarray Biochip Technology. Mark Schena, ed. Natick, MA: Eaton Publishing, 2000, 298 pp., $45.95. ISBN 1-881299-37-6. DNA Arrays: Methods and Protocols. Jang B. Rampal, ed. Totowa, NJ: Humana Press, 2001, 264 pp., $89.50. ISBN 0-89603-822-X. Clin Chem 48:211–213, 2002.

Miki, R., Kadota, K., Bono, H., Mizuno, Y., Tomaru, Y., Carninci, P., Itoh, M., Shibata, K., and co-workers. Delineating developmental and metabolic pathways in vivo by expression profiling using the RIKEN set of 18,816 full-length enriched mouse cDNA arrays. *Proc Natl Acad Sci U S A* 98:2199–2204, 2001.

Mikita, T., Porter, G., Lawn, R. M., and Shiffman, D. Oxidized LDL exposure alters the transcriptional response of macrophages to inflammatory stimulus. *J Biol Chem* 27, 2001.

Miklos, G. L., and Maleszka, R. Protein functions and biological contexts. *Proteomics* 1:169–178, 2001.

Mikovits, J., Ruscetti, F., Zhu, W., Bagni, R., Dorjsuren, D., and Shoemaker, R. Potential cellular signatures of viral infections in human hematopoietic cells. *Dis Markers* 17:173–178, 2001.

Miles, M. F. Microarrays: Lost in a storm of data? *Nat Rev Neurosci* 2:441–443, 2001.

Miller, J. C., Butler, E. B., The, B. S., and Haab, B. B. The application of protein microarrays to serum diagnostics: Prostate cancer as a test case. *Dis Markers* 17:225–234, 2001.

Miller, L. D., Park, K. S., Guo, Q. M., Alkharouf, N. W., Malek, R. L., Lee, N. H., Liu, E. T., and Cheng, S. Y. Silencing of WNT signaling and activation of multiple metabolic pathways in response to thyroid hormone-stimulated cell proliferation. *Mol Cell Biol* 21:6626–6639, 2001.

Miller, P. L. Opportunities at the intersection of bioinformatics and health informatics: A case study. *J Am Med Inform Assoc* 7:431–438, 2000.

Mills, J. C., and Gordon, J. I. A new approach for filtering noise from high-density oligonucleotide microarray datasets. *Nucleic Acids Res* 29:E72–E82, 2001.

Mills, J. C., Roth, K. A., Cagan, R. L., and Gordon, J. I. DNA microarrays and beyond: completing the journey from tissue to cell. *Nat Cell Biol* 3:E175–E178, 2001.

Mills, J. C., Syder, A. J., Hong, C. V., Guruge, J. L., Raaii, F., and Gordon, J. I. A molecular profile of the mouse gastric parietal cell with and without exposure to *Helicobacter pylori*. *Proc Natl Acad Sci U S A* 98:13687–13692, 2001.

Mir, K. U. The hypothesis is there is no hypothesis. *Trends Genet* 16:63–64, 2000.

Mir, K. U., and Southern, E. M. Determining the influence of structure on hybridization using oligonucleotide arrays. *Nat Biotechnol* 17:788–792, 1999.

Mirnics, K. Microarrays in brain research: The good, the bad and the ugly. *Nat Rev Neurosci* 2:444–447, 2001.

Mirnics, K., Middleton, F. A., Lewis, D. A., and Levitt, P. Analysis of complex brain disorders with gene expression microarrays: schizophrenia as a disease of the synapse. *Trends Neurosci* 24:479–486, 2001.

Mirnics, K., Middleton, F. A., Marquez, A., Lewis, D. A., and Levitt, P. Molecular characterization of schizophrenia viewed by microarray analysis of gene expression in prefrontal cortex. *Neuron* 28:53–67, 2000.

Mirnics, K., Middleton, F. A., Stanwood, G. D., Lewis, D. A., and Levitt, P. Disease-specific changes in regulator of G-protein signaling 4 (RGS4) expression in schizophrenia. *Mol Psychiatry* 6:293–301, 2001.

Mirowski, M., and Bartkowiak, J. DNA microarrays in biomedical studies. *Postepy Biochem* 46:272–281, 2000.

Mirzabekov, A., and Kolchinsky, A. Emerging array-based technologies in proteomics. *Curr Opin Chem Biol* 6:70–75, 2002.

Mitchell, T. C., Hildeman, D., Kedl, R. M., Teague, T. K., Schaefer, B. C., White, J., Zhu, Y., Kappler, J., and Marrack, P. Immunological adjuvants promote activated T cell survival via induction of Bcl-3. *Nat Immunol* 2:397–402, 2001.

Miyachi, H. The present status and future prospect of the molecular diagnostic tests. *Rinsho Byori* 49:139–149, 2001.

Miyazato, A., Ueno, S., Ohmine, K., Ueda, M., Yoshida, K., Yamashita, Y., Kaneko, T., Mori, M., and co-workers. Identification of myelodysplastic syndrome-specific genes by DNA microarray analysis with purified hematopoietic stem cell fraction. *Blood* 98:422–427, 2001.

Miyazaki, Y. J., Hamada, J., Tada, M., Furuuchi, K., Takahashi, Y., Kondo, S., Katoh, H., and Moriuchi, T. HOXD3 enhances motility and invasiveness through the TGF-beta-dependent and -independent pathways in A549 cells. *Oncogene* 21:798–808, 2002.

Mizuno, Y., Sotomaru, Y., Katsuzawa, Y., Kono, T., Meguro, M., Oshimura, M., Kawai, J., Tomaru, Y., and co-workers. Asb4, ata3, and dcn are novel imprinted genes identified by high-throughput screening using RIKEN cDNA microarray. *Biochem Biophys Res Commun* 290:1499–1505, 2002.

Moch, H., Kononen, T., Kallioniemi, O. P., and Sauter, G. Tissue microarrays: What will they bring to molecular and anatomic pathology? *Adv Anat Pathol* 8:14–20, 2001.

Moch, H., Schraml, P., Bubendorf, L., Mirlacher, M., Kononen, J., Gasser, T., Mihatsch, M. J., Kallioniemi, O. P., and Sauter, G. High-throughput tissue microarray analysis to evaluate genes uncovered by cDNA microarray screening in renal cell carcinoma. *Am J Pathol* 154:981–986, 1999.

Moch, H., Schraml, P., Bubendorf, L., Mirlacher, M., Kononen, J., Gasser, T., Mihatsch, M. J., Kallioniemi, O. P., and Sauter, G. Identification of prognostic parameters for renal cell carcinoma by cDNA arrays and cell chips. *Verh Dtsch Ges Pathol* 83:225–232, 1999.

Modrusan, Z., Marlowe, C., Wheeler, D., Pirseyedi, M., and Bryan, R. N. CPT-EIA assays for the detection of vancomycin resistant vanA and vanB genes in *Enterococci. Diagn Microbiol Infect Dis* 37:45–50, 2000.

Mody, M., Cao, Y., Cui, Z., Tay, K. Y., Shyong, A., Shimizu, E., Pham, K., Schultz, P., and co-workers. Genome-wide gene expression profiles of the developing mouse hippocampus. *Proc Natl Acad Sci U S A* 98:8862–8867, 2001.

Moerman, R., Frank, J., Marijnissen, J. C., Schalkhammer, T. G., and van Dedem, G. W. Miniaturized electrospraying as a technique for the production of microarrays of reproducible micrometer-sized protein spots. *Anal Chem* 73:2183–2189, 2001.

Mohrm S., and Rihn, B. Profiling gene expression in human mesothelioma cells using DNA microarray and high-density filter array technologies. *Bull Cancer* 88:305–313, 2001.

Mok, S. C., Chao, J., Skates, S., Wong, K. K., Yiu, G. K., Muto, M. G., Berkowitz, R. S., and Cramer, D. W. Prostasin, a potential serum marker for ovarian cancer: Identification through microarray technology. *J Natl Cancer Inst* 93:1458–1464, 2001.

Mokbel, K. The Twenty-third Annual San Antonio Breast Cancer Symposium. *Curr Med Res Opin* 16:276–284, 2001.

Momose, Y., and Iwahashi, H. Bioassay of cadmium using a DNA microarray: Genome-wide expression patterns of *Saccharomyces cerevisiae* response to cadmium. *Environ Toxicol Chem* 20:2353–2360, 2001.

Monni, O., Barlund, M., Mousses, S., Kononen, J., Sauter, G., Heiskanen, M., Paavola, P., Avela, K., and co-workers. Comprehensive copy number and gene expression profiling of the 17q23 amplicon in human breast cancer. *Proc Natl Acad Sci U S A* 98:5711–5716, 2001.

Monni, O., Hyman, E., Mousses, S., Barlund, M., Kallioniemi, A., and Kallioniemi, O. P. From chromosomal alterations to target genes for therapy: Integrating cytogenetic and functional genomic views of the breast cancer genome. *Semin Cancer Biol* 11:395–401, 2001.

Monti, J., Gross, V., Luft, F. C., Franca, Milia A., Schulz, H., Dietz, R., Sharma, A. M., and Hubner, N. Expression analysis using oligonucleotide microarrays in mice lacking bradykinin type 2 receptors. *Hypertension* 38:E1–E3, 2001.

Morgan, R. W., Sofer, L., Anderson, A. S., Bernberg, E. L., Cui, J., and Burnside, J. Induction of host gene expression following infection of chicken embryo fibroblasts with oncogenic Marek's disease virus. *J Virol* 75:533–539, 2001.

Mori, M., Mimori, K., Yoshikawa, Y., Shibuta, K., Utsunomiya, T., Sadanaga, N., Tanaka, F., Matsuyama, A., and co-workers. Analysis of the gene-expression profile regarding the progression of human gastric carcinoma. *Surgery* 131(1):S39–47, 2002.

Morozov, V. N., and Morozova, T. Y. Electrospray deposition as a method for mass fabrication of mono- and multicomponent microarrays of biological and biologically active substances. *Anal Chem* 71:3110–3117, 1999.

Mossman, K. L., Macgregor, P. F., Rozmus, J. J., Goryachev, A. B., Edwards, A. M., and Smiley, J. R. Herpes simplex virus triggers and then disarms a host antiviral response. *J Virol* 75:750–758, 2001.

Mousses, S., Kallioniemi, A., Kauraniemi, P., Elkahloun, A., and Kallioniemi, O. P. Clinical and functional target validation using tissue and cell microarrays. *Curr Opin Chem Biol* 6:97–101, 2002.

Mousses, S., Wagner, U., Chen, Y., Kim, J. W., Bubendorf, L., Bittner, M., Pretlow, T., Elkahloun, A. G., and co-workers. Failure of hormone therapy in prostate cancer involves systematic restoration of androgen responsive genes and activation of rapamycin sensitive signaling. *Oncogene* 20:6718–6723, 2001.

Moxon, R., and Tang, C. Challenge of investigating biologically relevant functions of virulence factors in bacterial pathogens. *Philos Trans R Soc Lond B Biol Sci* 355:643–656, 2000.

Mross, K., and Marz, W. Clinical trials: prerequisite of evidence-based oncology: Reality, perspectives and a new tool recruited—The Internet. *Onkologie* 24(Suppl 1):24–34, 2001.

Mu, X., Zhao, S., Pershad, R., Hsieh, T. F., Scarpa, A., Wang, S. W., White, R. A., Beremand, P. D., and co-workers. Gene expression in the developing mouse retina by EST sequencing and microarray analysis. *Nucleic Acids Res* 29:4983–4993, 2001.

Mucci, N. R., Akdas, G., Manely, S., and Rubin, M. A. Neuroendocrine expression in metastatic prostate cancer: Evaluation of high throughput tissue microarrays to detect heterogeneous protein expression. *Hum Pathol* 31:406–414, 2000.

Muhle, R. A., Pavlidis, P., Grundy, W. N., and Hirsch, E. A high-throughput study of gene expression in preterm labor with a subtractive microarray approach. *Am J Obstet Gynecol* 185:716–724, 2001.

Mullan, P. B., McWilliams, S., Quinn, J., Andrews, H., Gilmore, P., McCabe, N., McKenna, S., and Harkin, D. P. Uncovering BRCA1-regulated signalling pathways by microarray-based expression profiling. *Biochem Soc Trans* 29:678–683, 2001.

Murakami, T., Fujimoto, M., Ohtsuki, M., and Nakagawa, H. Expression profiling of cancer-related genes in human keratinocytes following non-lethal ultraviolet B irradiation. *J Dermatol Sci* 27:121–129, 2001.

Murakami, T., Fukasawa, T., Fukayama, M., Usui, K., Ohtsuki, M., and Nakagawa, H. Gene expression profile in a case of primary cutaneous CD30-negative large T-cell lymphoma with a blastic phenotype. *Clin Exp Dermatol* 26:201–204, 2001.

Murakami, T., Mataki, C., Nagao, C., Umetani, M., Wada, Y., Ishii, M., Tsutsumi, S., Kohro, T., and co-workers. The gene expression profile of human umbilical vein endothelial cells stimulated by tumor necrosis factor alpha using DNA microarray analysis. *J Atheroscler Thromb* 7:39–44, 2000.

Murakami, T., Ohtsuki, M., and Nakagawa, H. Angioimmunoblastic lymphadenopathy-type peripheral T-cell lymphoma with cutaneous infiltration: Report of a case and its gene expression profile. *Br J Dermatol* 144:878–884, 2001.

Murakami, Y., and Tamiya, E. Department of microbiosensing system using micromachine techniques. *Rinsho Byori* 47:1105–1112, 1999.

Murphy, G. M., Pollock, B. G., Kirshner, M. A., Pascoe, N., Cheuk, W., Mulsant, B. H., and Reynolds, C. F. CYP2D6 genotyping with oligonucleotide microarrays and nortriptyline concentrations in geriatric depression. *Neuropsychopharmacology* 25:737–743, 2001.

Murray, A. E., Lies, D., Li, G., Nealson, K., Zhou, J., and Tiedje, J. M. DNA/DNA hybridization to microarrays reveals gene-specific differences between closely related microbial genomes. *Proc Natl Acad Sci U S A* 98:9853–9858, 2001.

Muta, H., Boise, L. H., Fang, L., and Podack, E. R. CD30 signals integrate expression of cytotoxic effector molecules, lymphocyte trafficking signals, and signals for proliferation and apoptosis. *J Immunol* 165:5105–5111, 2000.

Mysorekar, I. U., Mulvey, M. A., Hultgren, S. J., and Gordon, J. I. Molecular regulation of urothelial renewal and host defenses during infection with uropathogenic *E. coli. J Biol Chem* 2001.

Nadler, S. T., and Attie, A. D. Please pass the chips: Genomic insights into obesity and diabetes. *J Nutr* 131:2078–2081, 2001.

Nadder, T. S., and Langley, M. R. The new millennium laboratory: molecular diagnostics goes clinical. Clin Lab Sci 14:252–259; quiz 260, 2001.

Nadler, S. T., Stoehr, J. P., Schueler, K. L., Tanimoto, G., Yandell, B. S., and Attie, A. D. The expression of adipogenic genes is decreased in obesity and diabetes mellitus. *Proc Natl Acad Sci U S A* 97:11371–11376, 2000.

Naef, F., Hacker, C. R., Patil, N., and Magnasco, M. Characterization of the expression ratio noise structure in high-density oligonucleotide arrays. *Genome Biol* 3: 2002.

Nagai, T., and Chapman, W. H. Jr. Analysis of microliter volumes of dye-labeled nucleic acids. *Biotechniques* 32:356–358, 360, 362, 364, 2002.

Nagai, M., Tanaka, S., Tsuda, M., Endo, S., Kato, H., Sonobe, H., Minami, A., Hiraga, H., and co-workers. Analysis of transforming activity of human synovial sarcoma-associated chimeric protein SYT-SSX1 bound to chromatin remodeling factor hBRM/hSNF2 alpha. *Proc Natl Acad Sci U S A* 98:3843–3848, 2001.

Nagamine, K., Kuzuhara, Y., and Notomi, T. Isolation of single-stranded DNA from loop-mediated isothermal amplification products. *Biochem Biophys Res Commun* 290:1195–1198, 2002.

Nagan, N., and O'Kane, D.J. Validation of a single nucleotide polymorphism genotyping assay for the human serum paraoxonase gene using electronically active customized microarrays. *Clin Biochem* 34:589–592, 2001.

Nagarajan, R., Svaren, J., Le, N., Araki, T., Watson, M., and Milbrandt, J. EGR2 mutations in inherited neuropathies dominant-negatively inhibit myelin gene expression. *Neuron* 30:355–368, 2001.

Nagarajan, U. M., Lochamy, J., Chen, X., Beresford, G. W., Nilsen, R., Jensen, P. E., and Boss, J. M. Class II transactivator is required for maximal expression of HLA-DOB in B cells. *J Immunol* 168:1780–1786, 2002.

Nagasawa, Y., Takenaka, M., Kaimori, J., Matsuoka, Y., Akagi, Y., Tsujie, M., Imai, E., and Hori, M. Rapid and diverse changes of gene expression in the kidneys of protein-overload proteinuria mice detected by microarray analysis. *Nephrol Dial Transplant* 16:923–931, 2001.

Nagle, R. B. New molecular approaches to tissue analysis. *J Histochem Cytochem* 49:1063–1064, 2001.

Nahm, O., Woo, S. K., Handler, J. S., and Kwon, H. M. Involvement of multiple kinase pathways in stimulation of gene transcription by hypertonicity. *Am J Physiol Cell Physiol* 282:C49–C58, 2002.

Naiki, T., Nagaki, M., Shidoji, Y., Kojima, H., Imose, M., Kato, T., Ohishi, N., Yagi, K., and Moriwaki, H. Analysis of gene expression profile induced by hepatocyte nuclear factor 4a in hepatoma cells using an oligonucleotide microarray. *J Biol Chem* 2002.

Nakahigashi, K., Kubo, N., Narita, Si. S., Shimaoka, T., Goto, S., Oshima, T., Mori, H., Maeda, M., and co-workers. HemK, a class of protein methyl transferase with similarity to DNA methyl transferases, methylates polypeptide chain release factors, and hemK knockout induces defects in translational termination. *Proc Natl Acad Sci U S A* 99:1473–1478, 2002.

Nakaigawa, N., Yao, M., Kishida, T., and Kubota, Y. Molecular genetic mechanism of the kidney cancer. *Nippon Rinsho* 59:104–109, 2001.

Nakanishi, T., Oka, T., and Akagi, T. Recent advances in DNA microarrays. *Acta Med Okayama* 55:319–328, 2001.

Nakao, M., Bono, H., Kawashima, S., Kamiya, T., Sato, K., Goto, S., and Kanehisa, M. Genome-scale gene expression analysis and pathway reconstruction in KEGG. *Genome Inform Ser Workshop Genome Inform* 10:94–103, 1999.

Nakaya, A., and Goto, S. Extraction of correlated gene clusters by multiple graph comparison. *Genome Inform Ser Workshop Genome Inform* 12:44–53, 2001.

Nakeff, A., Sahay, N., Pisano, M., and Subramanian, B. Painting with a molecular brush: Genomic/proteomic interfacing to define the drug action profile of novel solid-tumor selective anticancer agents. *Cytometry* 47:72–79, 2002.

Nal, B., Mohr, E., and Ferrier, P. Location analysis of DNA-bound proteins at the whole-genome level: untangling transcriptional regulatory networks. *Bioessays* 23:473–476, 2001.

Nallur, G., Luo, C., Fang, L., Cooley, S., Dave, V., Lambert, J., Kukanskis, K., Kingsmore, S., and co-workers. Signal amplification by rolling circle amplification on DNA microarrays. *Nucleic Acids Res* 29:E118, 2001.

Napoli, C., de Nigris, F., and Palinski, W. Multiple role of reactive oxygen species in the arterial wall. *J Cell Biochem* 82:674–682, 2001.

Narayanan, B. A., Narayanan, N. K., and Reddy, B. S. Docosahexaenoic acid regulated genes and transcription factors inducing apoptosis in human colon cancer cells. *Int J Oncol* 19:1255–1262, 2001.

Narravula, S., and Colgan, S. P. Hypoxia-inducible factor 1-mediated inhibition of peroxisome proliferator-activated receptor alpha expression during hypoxia. *J Immunol* 166:7543–7548, 2001.

Natarajan, K., Meyer, M. R., Jackson, B. M., Slade, D., Roberts, C., Hinnebusch, A. G., and Marton, M. J. Transcriptional profiling shows that Gcn4p is a master regulator of gene expression during amino acid starvation in yeast. *Mol Cell Biol* 21:4347–4368, 2001.

Natkunam, Y., Warnke, R. A., Montgomery, K., Falini, B., and van De Rijn, M. Analysis of MUM1/IRF4 protein expression using tissue microarrays and immunohistochemistry. *Mod Pathol* 14:686–694, 2001.

Nau, G. J., Richmond, J. F., Schlesinger, A., Jennings, E. G., Lander, E. S., and Young, R. A. Human macrophage activation programs induced by bacterial pathogens. *Proc Natl Acad Sci U S A* 99:1503–1508, 2002.

Nees, M., Geoghegan, J. M., Hyman, T., Frank, S., Miller, L., and Woodworth, C. D. Papillomavirus type 16 oncogenes downregulate expression of interferon-responsive genes and upregulate proliferation-associated and nf-kappab-responsive genes in cervical keratinocytes. *J Virol* 75:4283–4296, 2001.

Nelson, B. P., Grimsrud, T. E., Liles, M. R., Goodman, R. M., and Corn, R. M. Surface plasmon resonance imaging measurements of DNA and RNA hybridization adsorption onto DNA microarrays. *Anal Chem* 73:1–7, 2001.

Nelson, N. Microarrays pave the way to 21st century medicine. *J Natl Cancer Inst* 88:1803–1805, 1996.

Nelson N. J. Microarrays have arrived: Gene expression tool matures. *J Natl Cancer Inst* 93:492–494, 2001.

Nelson, P. S., Han, D., Rochon, Y., Corthals, G. L., Lin, B., Monson, A., Nguyen, V., Franza, B. R., and co-workers. Comprehensive analyses of prostate gene expression: Convergence of expressed sequence tag databases, transcript profiling and proteomics. *Electrophoresis* 21:1823–1831, 2000.

Nelson, S. F., and Denny, C. T. Representational differences analysis and microarray hybridization for efficient cloning and screening of differentially expressed genes. In: M. Schena, ed. *DNA Microarrays: A Practical Approach*, 2nd ed. Oxford University Press, Oxford, UK, 2000:43–59.

Nene, V., Bishop, R., Morzaria, S., Gardner, M. J., Sugimoto, C., Ole-MoiYoi, O. K., Fraser, C. M., and Irvin, A. Theileria parva genomics reveals an atypical apicomplexan genome. *Int J Parasitol* 30:465–474, 2000.

Nesbit, C. E., Tersak, J. M., Grove, L. E., Drzal, A., Choi, H., and Prochownik, E.V. Genetic dissection of c-myc apoptotic pathways. *Oncogene* 19:3200–3212, 2000.

Newton, M. A., Kendziorski, C. M., Richmond, C. S., Blattner, F. R., and Tsui, K. W. On differential variability of expression ratios: Improving statistical inference about gene expression changes from microarray Data. *J Comput Biol* 8:37–52, 2001.

Ng, P. W., Iha, H., Iwanaga, Y., Bittner, M., Chen, Y., Jiang, Y., Gooden, G., Trent, J. M., and co-workers. Genome-wide expression changes induced by HTLV-1 Tax: Evidence for MLK-3 mixed lineage kinase involvement in Tax-mediated NF-kappaB activation. *Oncogene* 20:4484–4496, 2001.

Nguyen, D. V., and Rocke, D. M. Tumor classification by partial least squares using microarray gene expression data. *Bioinformatics* 18:39–50, 2002.

Niculescu, A. B. III, Segal, D. S., Kuczenski, R., Barrett, T., Hauger, R. L., and Kelsoe, J. R. Identifying a series of candidate genes for mania and psychosis: A convergent functional genomics approach. *Physiol Genomics* 4:83–91, 2000.

Niculescu, A. B. III, and Kelsoe, J. R. Convergent functional genomics: application to bipolar disorder. *Ann Med* 33:263–271, 2001.

Niemeyer, C. M., and Blohm, D. DNA Microarrays. *Angew Chem Int Ed Engl* 38:2865–2869, 1999.

Nierman, W. C., Eisen, J. A., Fleischmann, R. D., and Fraser, C. M. Genome data: What do we learn? *Curr Opin Struct Biol* 10:343–348, 2000.

Nilsson, M., Barbany, G., Antson, D. O., Gertow, K., and Landegren, U. Enhanced detection and distinction of RNA by enzymatic probe ligation. *Nat Biotechnol* 18:791–793, 2000.

Nishizuka, S., Winokur, S. T., Simon, M., Martin, J., Tsujimoto, H., and Stanbridge, E. J. Oligonucleotide microarray expression analysis of genes whose expression is correlated with tumorigenic and non-tumorigenic phenotype of HeLaxhuman fibroblast hybrid cells. *Cancer Lett* 165:201–209, 2001.

Nocito, A., Bubendorf, L., Maria, Tinner, E., Suess, K., Wagner, U., Forster, T., Kononen, J., and co-workers. Microarrays of bladder cancer tissue are highly representative of proliferation index and histological grade. *J Pathol* 194:349–357, 2001.

Noensie, E. N., and Dietz, H. C. A strategy for disease gene identification through nonsense-mediated mRNA decay inhibition. *Nat Biotechnol* 19:434–439, 2001.

Nolan, J. P., and Sklar, L. A. Suspension array technology: Evolution of the flat-array paradigm. *Trends Biotechnol* 20:9–12, 2002.

Nolan, P. M. Generation of mouse mutants as a tool for functional genomics. *Pharmacogenomics* 1:243–255, 2000.

Noordewier, M. O., and Warren, P. V. Gene expression microarrays and the integration of biological knowledge. *Trends Biotechnol* 19:412–415, 2001.

Nouzova, M., Neumann, P., Navratilova, A., Galbraith, D.W., and Macas, J. Microarray-based survey of repetitive genomic sequences in *Vicia* spp. *Plant Mol Biol* 45:229–244, 2001.

Novak, J. P., Sladek, R., and Hudson, T. J. Characterization of variability in large-scale gene expression data: Implications for study design. *Genomics* 79:104–113, 2002.

Nuwaysir, E. F., Bittner, M., Trent, J., Barrett, J. C., and Afshari, C. A. Microarrays and toxicology: The advent of toxicogenomics. *Mol Carcinog* 24:153–159, 1999.

Ogawa, N., DeRisi, J., and Brown, P. O. New components of a system for phosphate accumulation and polyphosphate metabolism in *Saccharomyces cerevisiae* revealed by genomic expression analysis. *Mol Biol Cell* 11:4309–4321, 2000.

Ogura, M., Yamaguchi, H., Yoshida, K-i., Fujita, Y., and Tanaka, T. DNA microarray analysis of *Bacillus subtilis* DegU, ComA and PhoP regulons: An approach to comprehensive analysis of *B. subtilis* two-component regulatory systems. *Nucleic Acids Res* 29:3804–3813, 2001.

Oh, J. M., Brichory, F., Puravs, E., Kuick, R., Wood, C., Rouillard, J. M., Tra, J., Kardia, S., and co-workers. A database of protein expression in lung cancer. *Proteomics* 1:1303–1319, 2001.

Oh, M. K., and Liao, J. C. DNA microarray detection of metabolic responses to protein overproduction in *Escherichia coli. Metab Eng* 2:201–209, 2000.

Oh, M. K., and Liao, J. C. Gene expression profiling by DNA microarrays and metabolic fluxes in *Escherichia coli. Biotechnol* Prog 16:278–286, 2000.

Oh, M. K., Rohlin, L., Kao, K. C., and Liao, J. C. Global expression profiling of acetate-grown *Escherichia coli. J Biol Chem* 2002.

Ohlrogge, J., Pollard, M., Bao, X., Focke, M., Girke, T., Ruuska, S., Mekhedov, S., and Benning, C. Fatty acid synthesis: From CO_2 to functional genomics. *Biochem Soc Trans* 28:567–573, 2000.

Ohmine, K., Ota, J., Ueda, M., Ueno, S., Yoshida, K., Yamashita, Y., Kirito, K., Imagawa, S., and co-workers. Characterization of stage progression in chronic myeloid leukemia by DNA microarray with purified hematopoietic stem cells. *Oncogene* 20:8249–8257, 2001.

Ohyama, H., Zhang, X., Kohno, Y., Alevizos, I., Posner, M., Wong, D. T., and Todd, R. Laser capture microdissection-generated target sample for high-density oligonucleotide array hybridization. *Biotechniques* 29:530–536, 2000.

Oka, T., Yoshino, T., Hayashi, K., Ohara, N., Nakanishi, T., Yamaai, Y., Hiraki, A., Sogawa, C. A., and co-workers. Reduction of hematopoietic cell-specific tyrosine phosphatase SHP-1 gene expression in natural killer cell lymphoma and various types of lymphomas/leukemias : Combination analysis with cDNA expression array and tissue microarray. *Am J Pathol* 159:1495–1505, 2001.

Okabe, H., Satoh, S., Kato, T., Kitahara, O., Yanagawa, R., Yamaoka, Y., Tsunoda, T., Furukawa, Y., and Nakamura, Y. Genome-wide analysis of gene expression in human hepatocellular carcinomas using cDNA microarray: Identification of genes involved in viral carcinogenesis and tumor progression. *Cancer Res* 61:2129–2137, 2001.

Okamoto, T., Suzuki, T., and Yamamoto, N. Microarray fabrication with covalent attachment of DNA using bubble jet technology. *Nat Biotechnol* 18:438–441, 2000.

Oliver, D. J., Nikolau, B., and Wurtele, E. S. Functional genomics: High-throughput mRNA, protein, and metabolite analyses. *Metab Eng* 4:98–106, 2002.

Ollila, J., and Vihinen, M. Stimulation of B and T cells activates expression of transcription and differentiation factors. *Biochem Biophys Res Commun* 249:475–480, 1998.

Ono, K., Tanaka, T., Tsunoda, T., Kitahara, O., Kihara, C., Okamoto, A., Ochiai, K., Takagi, T., and Nakamura, Y. Identification by cDNA microarray of genes involved in ovarian carcinogenesis. *Cancer Res* 60:5007–5011, 2000.

Ooi, S. L., Shoemaker, D. D., and Boeke, J. D. A DNA microarray-based genetic screen for nonhomologous end-joining mutants in *Saccharomyces cerevisiae. Science* 2001.

Ortaldo, J. R., Bere, E. W., Hodge, D., and Young, H. A. Activating ly-49 nk receptors: Central role in cytokine and chemokine production. *J Immunol* 166:4994–4999, 2001.

Osin, P., Shipley, J., Lu, Y. J., Crook, T., and Gusterson, B. A. Experimental pathology and breast cancer genetics: New technologies. *Recent Results Cancer Res* 152:35–48, 1998.

Otsuka, M., Kato, M., Yoshikawa, T., Chen, H., Brown, E. J., Masuho, Y., Omata, M., and Seki, N. Differential expression of the L-plastin gene in human colorectal cancer progression and metastasis. *Biochem Biophys Res Commun* 289:876–881, 2001.

Outinen, P. A., Sood, S. K., Liaw, P. C., Sarge, K. D., Maeda, N., Hirsh, J., Ribau, J., Podor, T. J., and co-workers. Characterization of the stress-inducing effects of homocysteine. *Biochem J* 332:213–221, 1998.

Outinen, P. A., Sood, S. K., Pfeifer, S. I., Pamidi, S., Podor, T. J., Li, J., Weitz, J. I., and Austin, R. C. Homocysteine-induced endoplasmic reticulum stress and growth arrest leads to specific changes in gene expression in human vascular endothelial cells. *Blood* 94:959–967, 1999.

Ozawa, K. Perspectives on postgenome medicine: hematological diseases. *Nippon Rinsho* 59:59–64, 2001.

Pabon, C., Modrusan, Z., Ruvolo, M. V., Coleman, I. M., Daniel, S., Yue, H., and Arnold, L. J. Jr. Optimized T7 amplification system for microarray analysis. *Biotechniques* 31:874–879, 2001.

Pan, W., Zhang, Q., Xi, Q. S., Gan, R. B., and Li, T. P. FUP1, a gene associated with hepatocellular carcinoma, stimulates NIH3T3 cell proliferation and tumor formation in nude mice. *Biochem Biophys Res Commun* 286:1033–1038, 2001.

Pandita, A., Zielenska, M., Thorner, P., Bayani, J., Godbout, R., Greenberg, M., and Squire, J. A. Application of comparative genomic hybridization, spectral karyotyping, and microarray analysis in the identification of subtype-specific patterns of genomic changes in rhabdomyosarcoma. *Neoplasia* 1:262–275, 1999.

Panisko, E. A., Conrads, T. P., Goshe, M. B., and Veenstra, T. D. The postgenomic age. Characterization of proteomes. *Exp Hematol* 30:97–107, 2002.

Park, I. K., He, Y., Lin, F., Laerum, O. D., Tian, Q., Bumgarner, R., Klug, C. A., Li, K., and co-workers. Differential gene expression profiling of adult murine hematopoietic stem cells. *Blood* 99:488–498, 2002.

Park, P. J., Pagano, M., and Bonetti, M. A nonparametric scoring algorithm for identifying informative genes from microarray data. *Pac Symp Biocomput:* 52–63, 2001.

Park, W. Y. Application of DNA chip technology to biomedical research. *Exp Mol Med* 33(Suppl 1):113–124, 2001.

Pasinetti, G. M. Use of cDNA microarray in the search for molecular markers involved in the onset of Alzheimer's disease dementia. *J Neurosci Res* 65:471–476, 2001.

Pasinetti, G. M., and Ho, L. From cDNA microarrays to high-throughput proteomics. Implications in the search for preventive initiatives to slow the clinical progression of Alzheimer's disease dementia. *Restor Neurol Neurosci* 18:137–142, 2001.

Pastinen, T., Raitio, M., Lindroos, K., Tainola, P., Peltonen, L., and Syvanen, A. C. A system for specific, high-throughput genotyping by allele-specific primer extension on microarrays. *Genome Res* 10:1031–1042, 2000.

Paulose-Murphy, M., Ha, N. K., Xiang, C., Chen, Y., Gillim, L., Yarchoan, R., Meltzer, P., Bittner, M., and co-workers. Transcription program of human herpesvirus 8 (Kaposi's sarcoma-associated herpesvirus). *J Virol* 75:4843–4853, 2001.

Paulsen, I. T., Chen, J., Nelson, K. E., and Saier, M. H, Jr. Comparative genomics of microbial drug efflux systems. *J Mol Microbiol Biotechnol* 3:145–150, 2001.

Pavelic, K., and Gall-Troselj, K. Recent advances in molecular genetics of breast cancer. *J Mol Med* 79:566–573, 2001.

Pavlidis, P., and Noble, W. S. Analysis of strain and regional variation in gene expression in mouse brain. *Genome Biol* 2: 2001.

Paweletz, C. P., Charboneau, L., Bichsel, V. E., Simone, N. L., Chen, T., Gillespie, J. W., Emmert-Buck, M. R., Roth, M. J., and co-workers. Reverse phase protein microarrays which capture disease progression show activation of pro-survival pathways at the cancer invasion front. *Oncogene* 20:1981–1989, 2001.

Pearce, D. A., Ferea, T., Nosel, S. A., Das, B., and Sherman, F. Action of BTN1, the yeast orthologue of the gene mutated in Batten disease. *Nat Genet* 22:55–58, 1999.

Pecheniuk, N. M., Walsh, T. P., and Marsh, N. A. DNA technology for the detection of common genetic variants that predispose to thrombophilia. *Blood Coagul Fibrinolysis* 11:683–700, 2000.

Pe'er, I., and Shamir, R. Spectrum alignment: efficient resequencing by hybridization. *Proc Int Conf Intell Syst Mol Biol* 8:260–268, 2000.

Pellois, J. P., Wang, W., and Gao, X. Peptide synthesis based on t-Boc chemistry and solution photogenerated acids. *J Comb Chem* 2:355–360, 2000.

Pendurthi, U. R., Allen, K. E., Ezban, M., and Rao, L. V. Factor VIIa and thrombin induce the expression of Cyr61 and connective tissue growth factor, extracellular matrix signaling proteins that could act as possible downstream mediators in factor VIIa x tissue factor-induced signal transduction. *J Biol Chem* 275:14632–14641, 2000.

Penn, S. G., Rank, D. R., Hanzel, D. K., and Barker, D. L. Mining the human genome using microarrays of open reading frames. *Nat Genet* 26:315–318, 2000.

Purcell, A. E., Rocco, M. M., Lenhart, J. A., Hyder, K., Zimmerman, A. W., and Pevsner, J. Assessment of neural cell adhesion molecule (NCAM) in autistic serum and postmortem brain. *J Autism Dev Disord* 31:183–194, 2001.

Perez-Amador, M. A., Lidder, P., Johnson, M. A., Landgraf, J., Wisman, E., and Green, P. J. New Molecular phenotypes in the dst mutants of arabidopsis revealed by DNA microarray analysis.*Plant Cell* 13:2703–2717, 2001.

Perego, P., Jimenez, G. S., Gatti, L., Howell, S. B., and Zunino, F. Yeast mutants as a model system for identification of determinants of chemosensitivity. *Pharmacol Rev* 52:477–492, 2000.

Perou, C. M., Jeffrey, S. S., van de Rijn, M., Rees, C. A., Eisen, M. B., Ross, D. T., Pergamenschikov, A., Williams, C. F., and co-workers. Distinctive gene expression patterns in human mammary epithelial cells and breast cancers. *Proc Natl Acad Sci U S A* 96:9212–9217, 1999.

Perou, C. M., Sorlie, T., Eisen, M. B., van de Rijn, M., Jeffrey, S. S., Rees, C. A., Pollack, J. R., Ross, D. T., and co-workers. Molecular portraits of human breast tumours. *Nature* 406:747–752, 2000.

Perrone, E. E., Theoharis, C., Mucci, N. R., Hayasaka, S., Taylor, J. M., Cooney, K. A., and Rubin, M. A. Tissue microarray assessment of prostate cancer tumor proliferation in African-American and white men. *J Natl Cancer Inst* 92:937–939, 2000.

Peterson, A. W., Heaton, R. J., and Georgiadis, R. M. The effect of surface probe density on DNA hybridization. *Nucleic Acids Res* 29:5163–5168, 2001.

Petersen, M., Brodersen, P., Naested, H., Andreasson, E., Lindhart, U., Johansen, B., Nielsen, H. B., Lacy, M., and co-workers. Arabidopsis map kinase 4 negatively regulates systemic acquired resistance. *Cell* 103:1111–1120, 2000.

Peterson, S., Cline, R. T., Tettelin, H., Sharov, V., and Morrison, D. A. Gene expression analysis of the *Streptococcus pneumoniae* competence regulons by use of DNA microarrays. *J Bacteriol* 182:6192–6202, 2000.

Petrik, J. Microarray technology: The future of blood testing? *Vox Sang* 80:1–11, 2001.

Pfister, T., Feng, H., Wimmer, E., and Jones, K. W. Synchrotron radiation-induced x-ray emission to identify metal ions in preparations of purified protein. *Biotechniques* 32:134–136, 138, 140–141, 2002.

Phan, J., Pesaran, T., Davis, R. C., and Reue, K. The diet1 locus confers protection against hypercholesterolemia through enhanced bile acid metabolism. *J Biol Chem* 277:469–477, 2002.

Phelps, T. J., Palumbo, A. V., and Beliaev, A. S. Metabolomics and microarrays for improved understanding of phenotypic characteristics controlled by both genomics and environmental constraints. *Curr Opin Biotechnol* 13:20–24, 2002.

Phillips, J. L., Hayward, S. W., Wang, Y., Vasselli, J., Pavlovich, C., Padilla-Nash, H., Pezullo, J. R., Ghadimi, B. M., and co-workers. The consequences of chromosomal aneuploidy on gene expression profiles in a cell line model for prostate carcinogenesis. *Cancer Res* 61:8143–8149, 2001.

Pilpel, Y., Sudarsanam, P., and Church, G. M. Identifying regulatory networks by combinatorial analysis of promoter elements. *Nat Genet* 29:153–159, 2001.

Pinkel, D., Segraves, R., Sudar, D., Clark, S., Poole, I., Kowbel, D., Collins, C., Kuo, W. L., and co-workers. High resolution analysis of DNA copy number variation using comparative genomic hybridization to microarrays. *Nat Genet* 20:207–211, 1998.

Pinheiro, N. A., Caballero, O. L., Soares, F., Reis, L. F., and Simpson, A. J. Significant overexpression of oligophrenin-1 in colorectal tumors detected by cDNA microarray analysis. *Cancer Lett* 172:67–73, 2001.

Planet, P. J., DeSalle, R., Siddall, M., Bael, T., Sarkar, I. N., and Stanley, S. E. Systematic analysis of DNA microarray data: Ordering and interpreting patterns of gene expression. *Genome Res* 11:1149–1155, 2001.

Plate, J. M., Petersen, K. S., Buckingham, L., Shahidi, H., and Schofield, C. M. Gene expression in chronic lymphocytic leukemia B cells and changes during induction of apoptosis. *Exp Hematol* 28:1214–1224, 2000.

Podyminogin, M. A., Lukhtanov, E. A., and Reed, M. W. Attachment of benzaldehyde modified oligodeoxynucleotide probes to semicarbazide-coated glass. *Nucleic Acids Res* 29:5090–5098, 2001.

Polager, S., Kalma, Y., Berkovich, E., and Ginsberg, D. E2Fs up-regulate expression of genes involved in DNA replication, DNA repair and mitosis. *Oncogene* 21:437–446, 2002.

Pollak, E. S., Feng, L., Ahadian, H., and Fortina, P. Microarray-based genetic analyses for studying susceptibility to arterial and venous thrombotic disorders. *Ital Heart* J 2:568–572, 2001.

Pollack, J. R., Perou, C. M., Alizadeh, A. A., Eisen, M. B., Pergamenschikov, A., Williams, C. F., Jeffrey, S. S., Botstein, D., and Brown, P. O. Genome-wide analysis of DNA copy-number changes using cDNA microarrays. *Nat Genet* 23:41–46, 1999.

Pollock, P. M., and Trent, J. M. The genetics of cutaneous melanoma. *Clin Lab Med* 20:667–690, 2000.

Polyak, K. On the birth of breast cancer. *Biochim Biophys Acta* 1552:1–13, 2001.

Polyak, K., and Riggins, G. J. Gene discovery using the serial analysis of gene expression technique: implications for cancer research. *J Clin Oncol* 19:2948–2958, 2001.

Pomeroy, S. L., Tamayo, P., Gaasenbeek, M., Sturla, L. M., Angelo, M., McLaughlin, M. E., Kim, J. Y., Goumnerova, L. C., and co-workers. Prediction of central nervous system embryonal tumour outcome based on gene expression. *Nature* 415:436–442, 2002.

Popovici, R. M., Kao, L. C., and Giudice, L. C. Discovery of new inducible genes in in vitro decidualized human endometrial stromal cells using microarray technology. *Endocrinology* 141:3510–3513, 2000.

Porter, J. D., Khanna, S., Kaminski, H. J., Rao, J. S., Merriam, A. P., Richmonds, C. R., Leahy, P., Li, J., and Andrade, F. H. Extraocular muscle is defined by a fundamentally distinct gene expression profile. *Proc Natl Acad Sci U S A* 98:12062–12067, 2001.

Porter, J. D., Khanna, S., Kaminski, H. J., Rao, J. S., Merriam, A. P., Richmonds, C. R., Leahy, P., Li, J., and co-workers. A chronic inflammatory response dominates the skeletal muscle molecular signature in dystrophin-deficient mdx mice. *Hum Mol Genet* 11:263–272, 2002.

Porwollik, S., Wong, R. M., Sims, S. H., Schaaper, R. M., DeMarini, D. M., and McClelland, M. The DeltauvrB mutations in the Ames strains of *Salmonella* span 15 to 119 genes. *Mutat Res* 483:1–11, 2001.

Postier, R. G. Past, present, and future of pancreatic surgery. *Am J Surg* 182:547–551, 2001.

Presti, R. M., Popkin, D. L., Connick, M., Paetzold, S., and Virgin, H. W. Novel cell type-specific antiviral mechanism of interferon gamma action in macrophages. *J Exp Med* 193:483–496, 2001.

Price, C. P. Microarrays: The reincarnation of multiplexing in laboratory medicine, but now more relevant? *Clin Chem* 47:1345–1346, 2001.

Prichard, R., and Tait, A. The role of molecular biology in veterinary parasitology. *Vet Parasitol* 98:169–194, 2001.

Primdahl, H., Wikman, F. P., von der Maase, H., Zhou, X. G., Wolf, H., and Orntoft, T. F. Allelic imbalances in human bladder cancer: Genome-wide detection with high-density single-nucleotide polymorphism arrays. *J Natl Cancer Inst* 94:216–223, 2002.

Primig, M., Williams, R. M., Winzeler, E. A., Tevzadze, G. G., Conway, A. R., Hwang, S. Y., Davis, R. W., and Esposito, R. E. The core meiotic transcriptome in budding yeasts. *Nat Genet* 26:415–423, 2000.

Prince, L. S., Karp, P. H., Moninger, T. O., and Welsh, M. J. KGF alters gene expression in human airway epithelia: Potential regulation of the inflammatory response. *Physiol Genomics* 6:81–89, 2001.

Pritchard, C. C., Hsu, L., Delrow, J., and Nelson, P. S. Project normal: Defining normal variance in mouse gene expression. *Proc Natl Acad Sci U S A* 98:13266–13271, 2001.

Proudnikov, D., Timofeev, E., and Mirzabekov, A. Immobilization of DNA in polyacrylamide gel for the manufacture of DNA and DNA-oligonucleotide microchips. *Anal Biochem* 259:34–41, 1998.

Puga, A., Maier, A., and Medvedovic, M. The transcriptional signature of dioxin in human hepatoma HepG2 cells. *Biochem Pharmacol* 60:1129–1142, 2000.

Pugh, B. F., and Gilmour, D. S. Genome-wide analysis of protein-DNA interactions in living cells. *Genome Biol* 2: 2001.

Purcell, A. E., Jeon, O. H., and Pevsner, J. The abnormal regulation of gene expression in autistic brain tissue. *J Autism Dev Disord* 31:545–549, 2001.

Qi, L., and Sit, K. H. Housekeeping genes commanded to commit suicide in CpG-cleavage commitment upstream of Bcl-2 inhibition in caspase-dependent and -independent pathways. *Mol Cell Biol Res Commun* 3:319–327, 2000.

Qi, L., and Sit, K. H. Suicidal differential housekeeping gene activity in apoptosis induced by DCNP. *Apoptosis* 5:379–388, 2000.

Qi, Z., Hui, G., and Li, Y. Isolation and study of one novel full-length gene related to human glioma. *Zhonghua Yi Xue Za Zhi* 81:1124–1127, 2001.

Qian, J., Ramroop, K., McLeod, A., Bandari, P., Livingston, D. H., Harrison, J. S., and Rameshwar, P. Induction of hypoxia-inducible factor-1alpha and activation of caspase-3 in hypoxia-reoxygenated bone marrow stroma is negatively regulated by the delayed production of substance p. *J Immunol* 167:4600–4608, 2001.

Quackenbush, J. Computational analysis of microarray data. *Nat Rev Genet* 2:418–427, 2001.

Quarmby, S., West, C., Magee, B., Stewart, A., Hunter, R., and Kumar, S. Differential expression of cytokine genes in fibroblasts derived from skin biopsies of patients who developed minimal or severe normal tissue damage after radiotherapy. *Radiat Res* 157:243–248, 2002.

Rabitsch, K. P., Toth, A., Galova, M., Schleiffer, A., Schaffner, G., Aigner, E., Rupp, C., Penkner, A. M., and co-workers. A screen for genes required for meiosis and spore formation based on whole-genome expression. *Curr Biol* 11:1001–1009, 2001.

Raddatz, G., Dehio, M., Meyer, T. F., and Dehio, C. PrimeArray: Genome-scale primer design for DNA-microarray construction. *Bioinformatics* 17:98–99, 2001.

Radich, J., and Sievers, E. New developments in the treatment of acute myeloid leukemia. *Oncology* 14:125–31, 2000.

Radtkey, R., Feng, L., Muralhidar, M., Duhon, M., Canter, D., DiPierro, D., Fallon, S., and Tu, E., Rapid, high fidelity analysis of simple sequence repeats on an electronically active DNA microchip. *Nucleic Acids Res* 28:E17, 2000.

Rae, J. M., Johnson, M. D., Lippman, M. E., and Flockhart, D. A. Rifampin is a selective, pleiotropic inducer of drug metabolism genes in human hepatocytes: Studies with cDNA and oligonucleotide expression arrays. *J Pharmacol Exp Ther* 299:849–857, 2001.

Raetz, E. A., Moos, P. J., Szabo, A., and Carroll, W. L. Gene expression profiling. Methods and clinical applications in oncology. *Hematol Oncol Clin North Am* 15:911–930, 2001.

Raghuraman, M. K., Winzeler, E. A., Collingwood, D., Hunt, S., Wodicka, L., Conway, A., Lockhart, D. J., Davis, R. W., and co-workers. Replication dynamics of the yeast genome. Science 294:115–121, 2001.

Ragno, S., Romano, M., Howell, S., Pappin, D. J., Jenner, P. J., and Colston, M. J. Changes in gene expression in macrophages infected with *Mycobacterium tuberculosis*: A combined transcriptomic and proteomic approach. *Immunology* 104:99–108, 2001.

Raguenez, G., Douc-Rasy, S., Blanc, E., Goldschneider, D., Barrois, M., Valteau-Couanet, D., and Benard, J. A functional gene map is required to adapt therapy of metastatic neuroblastoma. *Bull Cancer* 88:295–304, 2001.

Rainey, W. E., Carr, B. R., Wang, Z. N., and Parker, C. R. Jr. Gene profiling of human fetal and adult adrenals. *J Endocrinol* 171:209–215, 2001.

Raitio, M., Lindroos, K., Laukkanen, M., Pastinen, T., Sistonen, P., Sajantila, A., and Syvanen, A. C. Y-chromosomal snps in finno-ugric-speaking populations analyzed by minisequencing on microarrays. *Genome Res* 11:471–482, 2001.

Rajeevan, M. S., Dimulescu, I. M., Unger, E. R., and Vernon, S. D. Chemiluminescent analysis of gene expression on high-density filter arrays. *J Histochem Cytochem* 47:337–342, 1999.

Rajeevan, M. S., Ranamukhaarachchi, D. G., Vernon, S. D., and Unger, E. R. Use of real-time quantitative PCR to validate the results of cDNA array and differential display PCR technologies. *Methods* 25:443–451, 2001.

Ramarathnam, R., and Subramaniam, S. A novel microarray strategy for detecting genes and pathways in microbes with unsequenced genomes. *Microb Comp Genomics* 5:153–161, 2000.

Ramaswamy, S., Tamayo, P., Rifkin, R., Mukherjee, S., Yeang, C. H., Angelo, M., Ladd, C., Reich, M., and co-workers. Multiclass cancer diagnosis using tumor gene expression signatures. *Proc Natl Acad Sci U S A* 2001.

Ramdas, L., Coombes, K. R., Baggerly, K., Abruzzo, L., Highsmith, W. E., Krogmann, T., Hamilton, S. R., and Zhang, W. Sources of nonlinearity in cDNA microarray expression measurements. *Genome Biol* 2: 2001.

Ramdas, L., Wang, J., Hu, L., Cogdell, D., Taylor, E., and Zhang, W. Comparative evaluation of laser-based microarray scanners. *Biotechniques* 31:546, 548, 550, 2001.

Rampon, C., Jiang, C. H., Dong, H., Tang, Y. P., Lockhart, D. J., Schultz, P. G., Tsien, J. Z., and Hu, Y. Effects of environmental enrichment on gene expression in the brain. *Proc Natl Acad Sci U S A* 97:12880–12884, 2000.

Ramsay, G. DNA chips: state-of-the art. *Nat Biotechnol* 16:40–44, 1998.

Rao, J. S., and Bond, M. Microarrays: Managing the data deluge. *Circ Res* 88:1226–1227, 2001.

Rappuoli, R. Pushing the limits of cellular microbiology: microarrays to study bacteria-host cell intimate contacts. *Proc Natl Acad Sci U S A* 97:13467–13469, 2000.

Rathod, P. K., Ganesan, K., Hayward, R. E., Bozdech, Z., and DeRisi, J. L. DNA microarrays for malaria. *Trends Parasitol* 18:39–45, 2002.

Ravasi, T., Wells, C., Forest, A., Underhill, D. M., Wainwright, B. J., Aderem, A., Grimmond, S., and Hume, D. A. Generation of diversity in the innate immune system: Macrophage heterogeneity arises from gene-autonomous transcriptional probability of individual inducible genes. *J Immunol* 168:44–50, 2002.

Ravine, D. Automated mutation analysis. *J Inherit Metab Dis* 22:503–518, 1999.

Raychaudhuri, S., Stuart, J. M., and Altman, R. B. Principal components analysis to summarize microarray experiments: Application to sporulation time series. *Pac Symp Biocomput* :455–466, 2000.

Raychaudhuri, S., Stuart, J. M., Liu, X., Small, P. M., and Altman, R. B. Pattern recognition of genomic features with microarrays: Site typing of *Mycobacterium tuberculosis* strains. *Proc Int Conf Intell Syst Mol Biol* 8:286–295, 2000.

Raychaudhuri, S., Sutphin, P. D., Chang, J. T., and Altman, R. B. Basic microarray analysis: Grouping and feature reduction. *Trends Biotechnol* 19:189–193, 2001.

Reese, J., Das, S. K., Paria, B. C., Lim, H., Song, H., Matsumoto, H. M., Knudtson, K. L., DuBois, R. N., and Dey, S. K. Global gene expression analysis to identify molecular markers of uterine receptivity and embryo implantation. *J Biol Chem* 2001.

Reddy, A. S. Calcium: Silver bullet in signaling. *Plant Sci* 160:381–404, 2001.

Reick, M., Garcia, J. A., Dudley, C., and McKnight, S. L. NPAS2: An analog of clock operative in the mammalian forebrain. *Science* 293:506–509, 2001.

Reid, R., Dix, D. J., Miller, D., and Krawetz, S. A. Recovering filter-based microarray data for pathways analysis using a multipoint alignment strategy. *Biotechniques* 30:762–768, 2001.

Reilly, T. P., Bourdi, M., Brady, J. N., Pise-Masison, C. A., Radonovich, M. F., George, J. W., and Pohl, L. R. Expression profiling of acetaminophen liver toxicity in mice using microarray technology. *Biochem Biophys Res Commun* 282:321–328, 2001.

Reilly, T. P., Brady, J. N., Marchick, M. R., Bourdi, M., George, J. W., Radonovich, M. F., Pise-Masison, C. A., and Pohl, L. R. A protective role for cyclooxygenase-2 in drug-induced liver injury in mice. *Chem Res Toxicol* 14:1620–1628, 2001.

Reinke, V., Smith, H. E., Nance, J., Wang, J., Van Doren, C., Begley, R., Jones, S. J., Davis, E. B., and co-workers. A global profile of germline gene expression in *C. elegans*. *Mol Cell* 6:605–616, 2000.

Reitzer, L., and Schneider, B. L. Metabolic context and possible physiological themes of sigma(54)-dependent genes in *Escherichia coli. Microbiol Mol Biol Rev* 65:422–444, 2001.

Ren, B., Robert, F., Wyrick, J. J., Aparicio, O., Jennings, E. G., Simon, I., Zeitlinger, J., Schreiber, J., and co-workers. Genome-wide location and function of DNA binding proteins. *Science* 290:2306–2309, 2000.

Resor, L., Bowen, T. J., and Wynshaw-Boris, A. Unraveling human cancer in the mouse: Recent refinements to modeling and analysis. *Hum Mol Genet* 10:669–675, 2001.

Reusch, J. E., and Klemm, D. J. Inhibition of CREB activity decreases protein kinase B/Akt expression in 3T3-L1 adipocytes and induces apoptosis. *J Biol Chem* 2001.

Revel, A. T., Talaat, A. M., and Norgard, M. V. DNA microarray analysis of differential gene expression in *Borrelia burgdorferi,* the Lyme disease spirochete. *Proc Natl Acad Sci U S A* 99:1562–1567, 2002.

Rew, D. A. DNA microarray technology in cancer research. *Eur J Surg Oncol* 27:504–508, 2001.

Reymond, P., Weber, H., Damond, M., and Farmer, E. E. Differential gene expression in response to mechanical wounding and insect feeding in *Arabidopsis. Plant Cell* 12:707–720, 2000.

Reynolds, H. Y. Use of bronchoalveolar lavage in humans—Past necessity and future imperative. *Lung* 178:271–293, 2000.

Rhind, S. M. Veterinary oncological pathology—Current and future perspectives. *Vet J* 163:7–18, 2002.

Rhodes, D. R., Miller, J. C., Haab, B. B., and Furge, K. A. CIT: Identification of differentially expressed clusters of genes from microarray data. *Bioinformatics* 18:205–206, 2002.

Richer, J. K., Jacobsen, B. M., Manning, N. G., Abel, M. G., Wolf, D. M., and Horwitz, K. B. Differential gene regulation by the two progesterone receptor isoforms in human breast cancer cells. *J Biol Chem* 2001.

Richmond, C. S., Glasner, J. D., Mau, R., Jin, H., and Blattner, F. R. Genome-wide expression profiling in *Escherichia coli* K-12. *Nucleic Acids Res* 27:3821–3835, 1999.

Richmond, T., and Somerville, S. Chasing the dream: Plant EST microarrays. *Curr Opin Plant Biol* 3:108–116, 2000.

Richmond, T. A., and Somerville, C. R. Integrative approaches to determining CSl function. *Plant Mol Biol* 47:131–143, 2001.

Richter, J., Wagner, U., Kononen, J., Fijan, A., Bruderer, J., Schmid, U., Ackermann, D., Maurer, R., and co-workers. High-throughput tissue microarray analysis of cyclin E gene amplification and overexpression in urinary bladder cancer. *Am J Pathol* 157:787–794, 2000.

Rickman, D. S., Bobek, M. P., Misek, D. E., Kuick, R., Blaivas, M., Kurnit, D. M., Taylor, J., and Hanash, S. M. Distinctive molecular profiles of high-grade and low-grade gliomas based on oligonucleotide microarray analysis. *Cancer Res* 61:6885–6891, 2001.

Riewald, M., Kravchenko, V. V., Petrovan, R. J., O'Brien, P. J., Brass, L. F., Ulevitch, R. J., and Ruf, W. Gene induction by coagulation factor Xa is mediated by activation of protease-activated receptor 1. *Blood* 97:3109–3116, 2001.

Rihn, B. H., Mohr, S., McDowell, S. A., Binet, S., Loubinoux, J., Galateau, F., Keith, G., and Leikauf, G. D. Differential gene expression in mesothelioma. *FEBS Lett* 480:95–100, 2000.

Rimm, D. L. Impact of microarray technologies on cytopathology. Overview of technologies and commentary on current and future implications for pathologists and cytopathologists. *Acta Cytol* 45:111–114, 2001.

Rimm, D. L. Molecular biology in cytopathology: current applications and future directions. *Cancer* 90:1–9, 2000.

Rimm, D. L., Camp, R. L., Charette, L. A., Olsen, D. A., and Provost, E. Amplification of tissue by construction of tissue microarrays. *Exp Mol Pathol* 70:255–264, 2001.

Rimm, D. L., Camp, R. L., Charette, L. A., Costa, J., Olsen, D. A., and Reiss, M. Tissue microarray: A new technology for amplification of tissue resources. *Cancer J* 7:24–31, 2001.

Rininger, J. A., DiPippo, V. A., and Gould-Rothberg, B. E. Differential gene expression technologies for identifying surrogate markers of drug efficacy and toxicity. *Drug Discov Today* 5:560–568, 2000.

Robert, J. New concepts for the study of anticancer drug resistance. *Bull Cancer* 89:17–22, 2002.

Robert, J. Resistance to cytotoxic agents. *Curr Opin Pharmacol* 1(4):353–357, 2001.

Robles, A. I., Bemmels, N. A., Foraker, A. B., and Harris, C. C. APAF-1 is a transcriptional target of p53 in DNA damage-induced apoptosis. *Cancer Res* 61:6660–6664, 2001.

Rockett, J. C., Christopher, Luft J., Brian, Garges J., Krawetz, S. A., Hughes, M. R., Hee Kirn, K., and co-workers. Development of a 950-gene DNA array for examining gene expression patterns in mouse testis. Genome Biol 2: 2001.

Rockett, J. C., and Dix, D. J. Application of DNA arrays to toxicology. *Environ Health Perspect* 107:681–685, 1999.

Rockett, J. C., Mapp, F. L., Garges, J. B., Luft, J. C., Mori, C., and Dix, D. J. Effects of hyperthermia on spermatogenesis, apoptosis, gene expression, and fertility in adult male mice. *Biol Reprod* 65:229–239, 2001.

Roda, A., Guardigli, M., Russo, C., Pasini, P., and Baraldini, M. Protein microdeposition using a conventional ink-jet printer. *Biotechniques* 28:492–496, 2000.

Roda, A., Pasini, P., Guardigli, M., Baraldini, M., Musiani, and M., Mirasoli, M. Bio- and chemiluminescence in bioanalysis. *Fresenius J Anal Chem* 366:752–759, 2000.

Rodi, C. P., Bunch, R. T., Curtiss, S. W., Kier, L. D., Cabonce, M. A., Davila, J. C., Mitchell, M. D., and co-workers. Revolution through genomics in investigative and discovery toxicology. *Toxicol Pathol* 27:107–110, 1999.

Rose, D. Microfluidic technologies and instrumentation for printing DNA microarrays. In: M. Schena, ed. *Microarray Biochip Technology*. Eaton Publishing, Natick, MA, 2000:19–38.

Rosenfeld, C. S., Wagner, J. S., Roberts, R. M., and Lubahn, D. B. Intraovarian actions of oestrogen. *Reproduction* 122:215–226, 2001.

Ross, D. T., Scherf, U., Eisen, M. B., Perou, C. M., Rees, C., Spellman, P., Iyer, V., Jeffrey, S. S., and co-workers. Systematic variation in gene expression patterns in human cancer cell lines. *Nat Genet* 24:227–235, 2000.

Rossman, T. G. Cloning genes whose levels of expression are altered by metals: implications for human health research. *Am J Ind Med* 38:335–339, 2000.

Roth, C. M., and Yarmush, M. L. Nucleic acid biotechnology. *Annu Rev Biomed Eng* 1:265–297, 1999.

Rowland, O., and Jones, J. D. Unraveling regulatory networks in plant defense using microarrays. *Genome Biol.* 2: 2001.

Rozovskaia, T., Feinstein, E., Mor, O., Foa, R., Blechman, J., Nakamura, T., Croce, C. M., Cimino, G., and Canaani, E. Upregulation of Meis1 and HoxA9 in acute lymphocytic leukemias with the t(4 : 11) abnormality. *Oncogene* 20:874–8, 2001.

Rozzo, S. J., Allard, J. D., Choubey, D., Vyse, T. J., Izui, S., Peltz, G., and Kotzin, B. L. Evidence for an interferon-inducible gene, IFI202, in the susceptibility to systemic lupus. *Immunity* 15:435–443, 2001.

Ruan, Y., Gilmore, J., and Conner, T. Towards *Arabidopsis* genome analysis: Monitoring expression profiles of 1400 genes using cDNA microarrays. *Plant J* 15:821–833, 1998.

Rubin, M. A. Use of laser capture microdissection, cDNA microarrays, and tissue microarrays in advancing our understanding of prostate cancer. *J Pathol* 195:80–86, 2001.

Rubin, M. A., Mucci, N. R., Figurski, J., Fecko, A., Pienta, K. J., and Day, M. L. E-cadherin expression in prostate cancer: A broad survey using high-density tissue microarray technology. *Hum Pathol* 32:690–697, 2001.

Rudolph, A. S., and Reasor, J. Cell and tissue based technologies for environmental detection and medical diagnostics. *Biosens Bioelectron* 16:429–431, 2001.

Rusiniak, M. E., Yu, M., Ross, D. T., Tolhurst, E. C., and Slack, J. L. Identification of B94 (TNFAIP2) as a potential retinoic acid target gene in acute promyelocytic leukemia. *Cancer Res* 60:1824–1829, 2000.

Rutka, J. T., Taylor, M., Mainprize, T., Langlois, A., Ivanchuk, S., Mondal, S., and Dirks, P. Molecular biology and neurosurgery in the third millennium. *Neurosurgery* 46:1034–1051, 2000.

Rutherford, J. C., Jaron, S., Ray, E., Brown, P. O., and Winge, D. R. A second iron-regulatory system in yeast independent of Aft1p. *Proc Natl Acad Sci U S A* 98:14322–14327, 2001.

Ryu, D. D., and Nam, D. H. Recent progress in biomolecular engineering. *Biotechnol Prog* 16:2–16, 2000.

Saban, M. R., Hellmich, H., Nguyen, N. B., Winston, J., Hammond, T. G., and Saban, R. Time course of LPS-induced gene expression in a mouse model of genitourinary inflammation. *Physiol Genomics* 5:147–160, 2001.

Sakai, K., Higuchi, H., Matsubara, K., and Kato, K. Microarray hybridization with fractionated cDNA: Enhanced identification of differentially expressed genes. *Anal Biochem* 287:32–37, 2000.

Sakaki, Y. Genomics and atherosclerosis. *Ann N Y Acad Sci* 947:254–258, 2001.

Salama, N., Guillemin, K., McDaniel, T. K., Sherlock, G., Tompkins, L., and Falkow, S. A whole-genome microarray reveals genetic diversity among Helicobacter pylori strains. *Proc Natl Acad Sci U S A* 97:14668–14673, 2000.

Salin, H., Vujasinovic, T., Mazurie, A., Maitrejean, S., Menini, C., Mallet, J., and Dumas, S. A novel sensitive microarray approach for differential screening using probes labelled with two different radioelements. *Nucleic Acids Res* 30:17, 2002.

Sallinen, S. L., Sallinen, P. K., Haapasalo, H. K., Helin, H. J., Helen, P. T., Schraml, P., Kallioniemi, O. P., and Kononen, J. Identification of differentially expressed genes in human gliomas by DNA microarray and tissue chip techniques. *Cancer Res* 60:6617–6622, 2000.

Salunga, R. C., Guo, H., Luo, L., Bittner, A., Joy, K. C., Chambers, J. R., Wan, J. S., Jackson, M. R., and Erlander, M. G. Gene expression analysis via cDNA microarrays of laser capture microdissected cells from fixed tissue. In: M. Schena, ed. *DNA Microarrays: A Practical Approach,* 2nd ed. Oxford University Press, Oxford, UK, 2000:121–137.

Samir, A. A., Ropolo, A., Grasso, D., Tomasini, R., Dagorn, J. C., Dusetti, N., Iovanna, J. L., and Vaccaro, M. I. Cloning and expression of the mouse PIP49 (pancreatitis induced protein 49) mRNA which encodes a new putative transmembrane protein activated in the pancreas with acute pancreatitis. *Mol Cell Biol Res Commun* 4:188–193, 2001.

Samuel, C. E. Antiviral actions of interferons. *Clin Microbiol Rev* 14:778–809, 2001.

Samura, O., Sohda, S., Johnson, K. L., Pertl, B., Ralston, S., Delli-Bovi, L. C., and Bianchi, D. W. Diagnosis of trisomy 21 in fetal nucleated erythrocytes from maternal blood by use of short tandem repeat sequences. *Clin Chem* 47:1622–1626, 2001.

Sanchez-Ramos, J. R., Song, S., Kamath, S. G., Zigova, T., Willing, A., Cardozo-Pelaez, F., Stedeford, T., Chopp, M., and Sanberg, P. R. Expression of neural markers in human umbilical cord blood. *Exp Neurol* 171:109–115, 2001.

Sanson, M., Marie, Y., Huguet, S., He, J., and Hoang-Xuan, K. Genes implicated in glial tumors. *Morphologie* 84:51–55, 2000.

Saramaki, O., Willi, N., Bratt, O., Gasser, T. C., Koivisto, P., Nupponen, N. N., Bubendorf, L., and Visakorpi, T. Amplification of EIF3S3 gene is associated with advanced stage in prostate cancer. *Am J Pathol* 159:2089–2094, 2001.

Sarwal, M., Chang, S., Barry, C., Chen, X., Alizadeh, A., Salvatierra, O., and Brown, P. Genomic analysis of renal allograft dysfunction using cDNA microarrays. *Transplant Proc* 33:297–298, 2001.

Sato, H., Sagai, M., Suzuki, K. T., and Aoki, Y. Identification, by cDNA microarray, of A-raf and proliferating cell nuclear antigen as genes induced in rat lung by exposure to diesel exhaust. *Res Commun Mol Pathol Pharmacol* 105:77–86, 1999.

Sato, N. Genomic medicine. *Nippon Rinsho* 59:2445–2450, 2001.

Saiura, A., Mataki, C., Murakami, T., Umetani, M., Wada, Y., Kohro, T., Aburatani, H., Harihara, Y., and co-workers. A comparison of gene expression in murine cardiac allografts and isografts by means DNA microarray analysis. *Transplantation* 72:320–329, 2001.

Sasik, R., Iranfar, N., Hwa, T., and Loomis, W. F. Extracting transcriptional events from temporal gene expression patterns during *Dictyostelium* development. *Bioinformatics* 18:61–66, 2002.

Schaffer, R., Landgraf, J., Accerbi, M., Simon, V. V., Larson, M., and Wisman, E. Microarray analysis of diurnal and circadian-regulated genes in arabidopsis. *Plant Cell* 13:113–123, 2001.

Schaffer, R., Landgraf, J., Perez-Amador, M., and Wisman, E. Monitoring genome-wide expression in plants. *Curr Opin Biotechnol* 11:162–167, 2000.

Schageman, J. J., Basit, M., Gallardo, T. D., Garner, H. R., and Shohet, R. V. MarC-V: A spreadsheet-based tool for analysis, normalization, and visualization of single cDNA microarray experiments. *Biotechniques* 32:338–340, 342, 344, 2002.

Schena, M. Genome analysis with gene expression microarrays. *Bioessays* 18:427–431, 1996.

Schena, M. Paper presented at the Fourth International Congress of Plant Molecular Biology, Amsterdam, June 20, 1994.

Schena, M. Paper presented at the Stanford Sierra Camp, October 5, 1994.

Schena, M. Paper presented at the Stanford Sierra Retreat, October 12, 1995.

Schena, M., and Davis, R. W. Genes, genomes and chips. In: M. Schena, ed. *DNA Microarrays: A Practical Approach,* 2nd ed. Oxford University Press, Oxford, UK, 2000:1–16.

Schena, M., and R. W. Davis. Parallel analysis with biological chips. In: M. Innis, D. Gelfand, and J. Sninsky, eds. *PCR Methods Manual*. Academic Press, San Diego, CA, 1999:445–456.

Schena, M., and Davis, R. W. Technology standards for microarray research. In: M. Schena, ed. Microarray *Biochip Technology*. Eaton Publishing, Natick, MA, 2000: 1–18.

Schena, M., Heller, R.A., Theriault, T. P., Konrad, K., Lachenmeier, E., and Davis, R. W. Microarrays: Biotechnology's discovery platform for functional genomics. *Trends Biotechnol* 16:301–306, 1998.

Schena, M., Shalon, D., Davis, R. W., and Brown, P. O. Quantitative monitoring of gene expression patterns with a complementary DNA microarray. *Science* 270:467–470, 1995.

Schena, M., and Shalon, D. Paper presented at the Sixth International Meeting or *Arabidopsis* Research, Madison, WI, June 8, 1995.

Schena, M., Shalon, D., Heller, R., Chai, A., Brown, P. O., and Davis, R. W. Parallel human genome analysis: Microarray-based expression monitoring of 1000 genes. *Proc Natl Acad Sci U S A* 93:10614–10619, 1996.

Schenk, P. M., Kazan, K., Wilson, I., Anderson, J. P., Richmond, T., Somerville, S. C., and Manners, J. M. Coordinated plant defense responses in *Arabidopsis* revealed by microarray analysis. *Proc Natl Acad Sci U S A* 97:11655–11660, 2000.

Scherf, U., Ross, D. T., Waltham, M., Smith, L. H., Lee, J. K., Tanabe, L., Kohn, K. W., Reinhold, W. C., and co-workers. A gene expression database for the molecular pharmacology of cancer. *Nat Genet* 24:236–244, 2000.

Schermer, M. J. Confocal scanning microscopy in microarray detection. In: M. Schena, ed. *DNA Microarrays: A Practical Approach*, 2nd ed. Oxford University Press, Oxford, UK, 2000:17–42.

Schibler, U., Rifat, D., and Lavery, D. J. The isolation of differentially expressed mRNA sequences by selective amplification via biotin and restriction-mediated enrichment. *Methods* 24:3–14, 2001.

Schlake, T., and Boehm, T. Expression domains in the skin of genes affected by the nude mutation and identified by gene expression profiling. *Mech Dev* 109:419–422, 2001.

Schmidt, C. W. Charting the map of life. *Environ Health Perspect* 109:A24–A29, 2001.

Schneider, M. D., and Schwartz, R. J. Chips ahoy: Gene expression in failing hearts surveyed by high-density microarrays. *Circulation* 102:3026–3027, 2000.

Schoenberg Fejzo, M., and Slamon, D. J. Frozen tumor tissue microarray technology for analysis of tumor RNA, DNA, and proteins. *Am J Pathol* 159:1645–1650, 2001.

Scholl, F. A., Betts, D. R., Niggli, F. K., and Schafer, B. W. Molecular features of a human rhabdomyosarcoma cell line with spontaneous metastatic progression. *Br J Cancer* 82:1239–1245, 2000.

Schoolnik, G. K. Functional and comparative genomics of pathogenic bacteria. *Curr Opin Microbiol* 5:20–26, 2002.

Schoolnik, G. K., Voskuil, M. I., Schnappinger, D., Yildiz, F. H., Meibom, K., Dolganov, N. A., Wilson, M. A., and Chong, K. H. Whole genome DNA microarray expression analysis of biofilm development by *Vibrio cholerae* O1 E1 Tor. *Methods Enzymol* 336: 3–18, 2001.

Schraml, P., Kononen, J., Bubendorf, L., Moch, H., Bissig, H., Nocito, A., Mihatsch, M. J., Kallioniemi, O. P., and Sauter, G. Tissue microarrays for gene amplification surveys in many different tumor types. *Clin Cancer Res* 5:1966–1975, 1999.

Schraml, P., Struckmann, K., Bednar, R., Fu, W., Gasser, T., Wilber, K., Kononen, J., Sauter, and co-workers. CDKNA2A mutation analysis, protein expression, and deletion mapping of chromosome 9p in conventional clear-cell renal carcinomas: Evidence for a second tumor suppressor gene proximal to CDKN2A. *Am J Pathol* 158:593–601, 2001.

Schuchhardt, J., Beule, D., Malik, A., Wolski, E., Eickhoff, H., Lehrach, H., and Herzel, H. Normalization strategies for cDNA microarrays. *Nucleic Acids Res* 28:E47, 2000.

Schuhmacher, M., Kohlhuber, F., Holzel, M., Kaiser, C., Burtscher, H., Jarsch, M., Bornkamm, G. W., Laux, G., and co-workers. The transcriptional program of a human B cell line in response to Myc. *Nucleic Acids Res* 29:397–406, 2001.

Schuldiner, O., Benvenisty, N. A DNA microarray screen for genes involved in c-MYC and N-MYC oncogenesis in human tumors. *Oncogene* 20:4984–4994, 2001.

Schultz, R. A., Nielsen, T., Zavaleta, J. R., Ruch, R., Wyatt, R., and Garner, H. R. Hyperspectral imaging: A novel approach for microscopic analysis. *Cytometry* 43:239–247, 2001.

Schulze, A., and Downward, J. Analysis of gene expression by microarrays: cell biologist's gold mine or minefield? *J Cell Sci* 113 Pt 23:4151–4156, 2000.

Schulze, A., and Downward, J. Navigating gene expression using microarrays—A technology review. *Nat Cell Biol* 3:E190–E195, 2001.

Schut, G. J., Zhou, J., and Adams, M. W. DNA Microarray analysis of the hyperthermophilic archaeon *Pyrococcus furiosus:* Evidence for a new type of sulfur-reducing enzyme complex. *J Bacteriol* 183:7027–7036, 2001.

Schwab, W., Aharoni, A., Raab, T., Perez, A. G., and Sanz, C. Cytosolic aldolase is a ripening related enzyme in strawberry fruits (*Fragaria* × *ananassa*). *Phytochemistry* 56:407–415, 2001.

Schwarze, S. R., DePrimo, S. E., Grabert, L. M., Fu, V. X., Brooks, J. D., and Jarrard, D. F. Novel pathways associated with bypassing cellular senescence in human prostate epithelial cells. *J Biol Chem* 2002.

Schwartz, A., and Fernandez-Repollet, E. Quantitative flow cytometry. *Clin Lab Med* 21:743–761, 2001.

Schweitzer, B., and Kingsmore, S. Combining nucleic acid amplification and detection. *Curr Opin Biotechnol* 12:21–27, 2001.

Schweitzer, B., and Kingsmore, S. F. Measuring proteins on microarrays. *Curr Opin Biotechnol* 13:14–19, 2002.

Schweitzer, B., Wiltshire, S., Lambert, J., O'Malley, S., Kukanskis, K., Zhu, Z., Kingsmore, S. F., Lizardi, P. M., and Ward, D. C. Immunoassays with rolling circle DNA amplification: A versatile platform for ultrasensitive antigen detection. *Proc Natl Acad Sci U S A* 97:10113–10119, 2000.

Scorilas, A., Bjartell, A., Lilja, H., Moller, C., and Diamandis, E. P. Streptavidin-polyvinylamine conjugates labeled with a europium chelate: Applications in immunoassay, immunohistochemistry, and microarrays. *Clin Chem* 46:1450–1455, 2000.

Seftor, R. E., Seftor, E. A., Koshikawa, N., Meltzer, P. S., Gardner, L. M., Bilban, M., Stetler-Stevenson, W. G., Quaranta, V., and Hendrix, M. J. Cooperative interactions of laminin 5 gamma2 chain, matrix metalloproteinase-2, and membrane type-1-matrix/metalloproteinase are required for mimicry of embryonic vasculogenesis by aggressive melanoma. *Cancer Res* 61:6322–6327, 2001.

Sehl, P. D., Tai, J. T., Hillan, K. J., Brown, L. A., Goddard, A., Yang, R., Jin, H., and Lowe, D. G. Application of cDNA microarrays in determining molecular phenotype in cardiac growth, development, and response to injury. *Circulation* 101(16):1990–1999, 2000.

Seki, M., Narusaka, M., Abe, H., Kasuga, M., Yamaguchi-Shinozaki, K., Carninci, P., Hayashizaki, Y., and Shinozaki, K. Monitoring the expression pattern of 1300 *Arabidopsis* genes under drought and cold stresses by using a full-length cDNA microarray. *Plant Cell* 13:61–72, 2001.

Selaru, F. M., Zou, T., Xu, Y., Shustova, V., Yin, J., Mori, Y., Sato, F., Wang, S., and coworkers. Global gene expression profiling in Barrett's esophagus and esophageal cancer: A comparative analysis using cDNA microarrays. *Oncogene* 21:475–478, 2002.

Semov, A., Marcotte, R., Semova, N., Ye, X., and Wang, E. Microarray analysis of E-Box binding-related gene expression in young and replicatively senescent human fibroblasts. *Anal Biochem* 302:38–51, 2002.

Sengupta, R., and Tompa, M. Quality control in manufacturing oligo arrays: A combinatorial design approach. *Pac Symp Biocomput* :348–359, 2001.

Seo, J., Kim, M., and Kim, J. Identification of novel genes differentially expressed in PMA-induced HL-60 cells using cDNA microarrays. *Mol Cells* 10:733–739, 2001.

Seta, K. A., Kim, R., Kim, H. W., Millhorn, D. E., and Beitner-Johnson, D. Hypoxia-induced regulation of MKP-1: Identification by subtractive suppression hybridization and cDNA microarrays. *J Biol Chem* 2001.

Sgroi, D. C., Teng, S., Robinson, G., LeVangie, R., Hudson, J. R. Jr. and Elkahloun, A. G. In vivo gene expression profile analysis of human breast cancer progression. *Cancer Res* 59:5656–5661, 1999.

Shaffer, A. L., Rosenwald, A., Hurt, E. M., Giltnane, J. M., Lam, L. T., Pickeral, O. K., and Staudt, L. M. Signatures of the immune response. *Immunity* 15:375–385, 2001.

Shaffer, A. L., Yu, X, He, Y., Boldrick, J., Chan, E. P., and Staudt, L. M. BCL-6 represses genes that function in lymphocyte differentiation, inflammation, and cell cycle control. *Immunity* 13:199–212, 2000.

Shah, R., Mucci, N. R., Amin, A., Macoska, J. A., and Rubin, M. A. Postatrophic hyperplasia of the prostate gland : Neoplastic precursor or innocent bystander? *Am J Pathol* 158:1767–1773, 2001.

Shaheduzzaman, S., Krishnan, V., Petrovic, A., Bittner, M., Meltzer, P., Trent, J., Venkatesan, S., and Zeichner, S. Effects of HIV-1 nef on cellular gene expression profiles. *J Biomed Sci* 9:82–96, 2002.

Shalev, A., Valasek, L., Pise-Masison, C. A., Radonovich, M., Phan, L., Clayton, J., He, H., Brady, J. N., and co-workers. *Saccharomyces cerevisiae* protein Pci8p and human protein eIF3e/Int-6 Interact with the eIF3 core complex by binding to cognate eIF3b subunits. *J Biol Chem* 276:34948–34957, 2001.

Shalon, D., Smith, S. J., and Brown, P. O. A DNA microarray system for analyzing complex DNA samples using two-color fluorescent probe hybridization. *Genome Res* 6:639–645, 1996.

Sharan, R., and Shamir, R. CLICK: A clustering algorithm with applications to gene expression analysis. *Proc Int Conf Intell Syst Mol Biol* 8:307–316, 2000.

Shatkay, H., Edwards, S., Wilbur, W. J., and Boguski, M. Genes, themes and microarrays: Using information retrieval for large-scale gene analysis. *Proc Int Conf Intell Syst Mol Biol* 8:317–328, 2000.

Shaughnessy, J. Jr., Gabrea, A., Qi, Y., Brents, L., Zhan, F., Tian, E., Sawyer, J., Barlogie, B., and co-workers. Cyclin D3 at 6p21 is dysregulated by recurrent chromosomal translocations to immunoglobulin loci in multiple myeloma. *Blood* 98:217–223, 2001.

Shcherbakova, P. V., Hall, M. C., Lewis, M. S., Bennett, S. E., Martin, K. J., Bushel, P. R., Afshari, C. A., and Kunkel, T. A. Inactivation of DNA mismatch repair by increased expression of yeast MLH1. *Mol Cell Biol* 21:940–951, 2001.

Shelling, A. N., and Foulkes, W. Molecular genetics of ovarian cancer. *Mol Biotechnol* 19:13–28, 2001.

Shelton, D. N., Chang, E., Whittier, P. S., Choi, D., and Funk, W. D. Microarray analysis of replicative senescence. *Curr Biol* 9:939–945, 1999.

Shen, X., Li, J., Hu, P. P., Waddell, D., Zhang, J., and Wang, X. F. The activity of guanine exchange factor net1 is essential for transforming growth factor-beta -mediated stress fiber formation. *J Biol Chem* 276:15362–13568, 2001.

Sherlock, G. Analysis of large-scale gene expression data. *Curr Opin Immunol* 12:201–205, 2000.

Sherlock, G. Analysis of large-scale gene expression data. *Brief Bioinform* 2:350–362, 2001.

Sherlock, G., Hernandez-Boussard, T., Kasarskis, A., Binkley, G., Matese, J. C., Dwight, S. S., Kaloper, M., Weng, S., and co-workers. The Stanford Microarray Database. *Nucleic Acids Res* 29:152–155, 2001.

Sherman, D. R., Voskuil, M., Schnappinger, D., Liao, R., Harrell, M. I., and Schoolnik, G. K. Regulation of the *Mycobacterium tuberculosis* hypoxic response gene encoding alpha-crystallin. *Proc Natl Acad Sci U S A* 98:7534–7539, 2001.

Shi, M. M. Enabling large-scale pharmacogenetic studies by high-throughput mutation detection and genotyping technologies. *Clin Chem* 47:164–172, 2001.

Shi, S. J., Scheffer, A., Bjeldanes, E., Reynolds, M. A., and Arnold, L. J. DNA exhibits multi-stranded binding recognition on glass microarrays. *Nucleic Acids Res* 29:4251–4256, 2001.

Shiffman, D., Mikita, T., Tai, J. T., Wade, D. P., Porter, J. G., Seilhamer, J. J., Somogyi, R., Liang, S., and Lawn, R. M. Large scale gene expression analysis of cholesterol-loaded macrophages. *J Biol Chem* 275:37324–37332, 2000.

Shih, J. Y., Yang, S. C., Hong, T. M., Yuan, A., Chen, J. J., Yu, C. J., Chang, Y. L., Lee, Y. C., and co-workers. Collapsin response mediator protein-1 and the invasion and metastasis of cancer cells. *J Natl Cancer Inst* 93:1392–1400, 2001.

Shirota, Y., Kaneko, S., Honda, M., Kawai, H. F., and Kobayashi, K. Identification of differentially expressed genes in hepatocellular carcinoma with cDNA microarrays. *Hepatology* 33:832–840, 2001.

Shoemaker, D. D., Schadt, E. E., Armour, C. D., He, Y. D., Garrett-Engele, P., McDonagh, P. D., Loerch, P. M., and Leonardson, A. Experimental annotation of the human genome using microarray technology. *Nature* 409:922–927, 2001.

Shridhar, V., Lee, J., Pandita, A., Iturria, S., Avula, R., Staub, J., Morrissey, M., Calhoun, E., and co-workers. Genetic analysis of early- versus late-stage ovarian tumors. *Cancer Res* 61:5895–5904, 2001.

Shultz, V. D., Phillips, S., Sar, M., Foster, P. M., and Gaido, K. W. Altered gene profiles in fetal rat testes after in utero exposure to di(*n*-butyl) phthalate. *Toxicol Sci* 64:233–242, 2001.

Sibille, E., and Hen, R. Combining genetic and genomic approaches to study mood disorders. *Eur Neuropsychopharmacol* 11:413–421, 2001.

Sibley, L. D., Mordue, D. G., Su, C., Robben, P. M., and Howe, D. K. Genetic approaches to studying virulence and pathogenesis in *Toxoplasma gondii. Philos Trans R Soc Lond B Biol Sci* 357:81–88, 2002.

Siebert, R., Rosenwald, A., Staudt, L. M., and Morris, S. W. Molecular features of B-cell lymphoma. *Curr Opin Oncol* 13:316–324, 2001.

Siedow, J. N. Making sense of microarrays. *Genome Biol* 2: 2001.

Silani, V., Braga, M., Botturi, A., Cardin, V., Bez, A., Pizzuti, A., and Scarlato, G. Human developing motor neurons as a tool to study ALS. *Amyotroph Lateral Scler Other Motor Neuron Disord* 2(Suppl 1):S69–S76, 2001.

Silzel, J. W., Cercek, B., Dodson, C., Tsay, T., and Obremski, R. J. Mass-sensing, multianalyte microarray immunoassay with imaging detection. *Clin Chem* 44:2036–2043, 1998.

Simbulan-Rosenthal, C. M., Ly, D. H., Rosenthal, D. S., Konopka, G., Luo, R., Wang, Z. Q., Schultz, P. G., and Smulson, M. E. Misregulation of gene expression in primary fibroblasts lacking poly(ADP-ribose) polymerase. *Proc Natl Acad Sci U S A* 97:11274–11279, 2000.

Simeonova, P. P., Wang, S., Toriuma, W., Kommineni, V., Matheson, J., Unimye, N., Kayama, F., Harki, D., and co-workers. Arsenic mediates cell proliferation and gene expression in the bladder epithelium: Association with activating protein-1 transactivation. *Cancer Res* 60:3445–3453, 2000.

Simmen, K. A., Singh, J., Luukkonen, B. G., Lopper, M., Bittner, A., Miller, N. E., Jackson, M. R., Compton, T., and Fruh, K. Global modulation of cellular transcription by human cytomegalovirus is initiated by viral glycoprotein B. *Proc Natl Acad Sci U S A* 98:7140–7145, 2001.

Simon, R., Nocito, A., Hubscher, T., Bucher, C., Torhorst, J., Schraml, P., Bubendorf, L., Mihatsch, M. M., and co-workers. Patterns of her-2/neu amplification and overexpression in primary and metastatic breast cancer. *J Natl Cancer Inst* 93:1141–1146, 2001.

Simon, R., Richter, J., Wagner, U., Fijan, A., Bruderer, J., Schmid, U., Ackermann, D., Maurer, R., and co-workers. High-throughput tissue microarray analysis of 3p25

(RAF1) and 8p12 (FGFR1) copy number alterations in urinary bladder cancer. *Cancer Res* 61:4514–4519, 2001.

Singh-Gasson, S., Green, R. D., Yue, Y., Nelson, C., Blattner, F., Sussman, M. R., and Cerrina, F. Maskless fabrication of light-directed oligonucleotide microarrays using a digital micromirror array. *Nat Biotechnol* 17:974–978, 1999.

Sinha, S., and Tompa, M. A statistical method for finding transcription factor binding sites. *Proc Int Conf Intell Syst Mol Biol* 8:344–354, 2000.

Skacel, M., Ormsby, A. H., Pettay, J. D., Tsiftsakis, E. K., Liou, L. S., Klein, E. A., Levin, H. S., Zippe, C. D., and Tubbs, R. R. Aneusomy of chromosomes 7, 8, and 17 and amplification of HER-2/neu and epidermal growth factor receptor in Gleason score 7 prostate carcinoma: A differential fluorescent in situ hybridization study of Gleason pattern 3 and 4 using tissue microarray. *Hum Pathol* 32:1392–1397, 2001.

Skrabanek, L., and Campagne, F. TissueInfo: High-throughput identification of tissue expression profiles and specificity. *Nucleic Acids Res* 29(21):E102–102, 2001.

Small, J., Call, D. R., Brockman, F. J., Straub, T. M., and Chandler, D. P. Direct detection of 16S rRNA in soil extracts by using oligonucleotide microarrays. *Appl Environ Microbiol* 67:4708–4716, 2001.

Smoot, L. M., Smoot, J. C., Graham, M. R., Somerville, G. A., Sturdevant, D. E., Migliaccio, C. A., Sylva, G. L., and Musser, J. M. Global differential gene expression in response to growth temperature alteration in group A *Streptococcus*. *Proc Natl Acad Sci U S A* 98:10416–10421, 2001.

Smothers, J. F., and Henikoff, S. Predicting in vivo protein peptide interactions with random phage display. *Comb Chem High Throughput Screen* 4:585–591, 2001.

Smulski, D. R., Huang, L. L., McCluskey, M. P., Reeve, M. J., Vollmer, A. C., Van Dyk, T. K., and LaRossa, R. A. Combined, functional genomic-biochemical approach to intermediary metabolism: Interaction of acivicin, a glutamine amidotransferase inhibitor, with *Escherichia coli* K-12. *J Bacteriol* 183:3353–3364, 2001.

Sniegowski, P. The genomics of adaptation in yeast. *Curr Biol* 9:R897-R898, 1999.

Snijders, A. M., Meijer, G. A., Brakenhoff, R. H., van den Brule, A. J., and van Diest, P. J. Microarray techniques in pathology: Tool or toy? *Mol Pathol* 53:289–294, 2000.

Snijders, A. M., Nowak, N., Segraves, R., Blackwood, S., Brown, N., Conroy, J., Hamilton, G., and Hindle, A. K., Assembly of microarrays for genome-wide measurement of DNA copy number. *Nat Genet* 29:263–264, 2001.

Soares, M. B. Identification and cloning of differentially expressed genes. *Curr Opin Biotechnol* 8:542–546, 1997.

Sobue, G. Molecular pathogenesis of motor neuron diseases. *Nihon Shinkei Seishin Yakurigaku Zasshi* 21:21–25, 2001.

Soeth, E., Wirth, T., List, H. J., Kumbhani, S., Petersen, A., Neumaier, M., Czubayko, F., and Juhl, H. Controlled ribozyme targeting demonstrates an antiapoptotic effect of carcinoembryonic antigen in HT29 colon cancer cells. *Clin Cancer Res* 7:2022–2030, 2001.

Soini, H., and Musser, J. M. Molecular diagnosis of mycobacteria. *Clin Chem* 47:809–814, 2001.

Sone, H., Shimano, H., Sakakura, Y., Inoue, N., Amemiya-Kudo, M., Yahagi, N., Osawa, M., and Suzuki, H., Acetyl-coenzyme A synthetase is a lipogenic enzyme controlled by SREBP-1 and energy status. *Am J Physiol Endocrinol Metab* 282:E222–E230, 2002.

Soref, C. M., Di, Y. P., Hayden, L., Zhao, Y. H., Satre, M. A., and Wu, R. Characterization of a novel airway epithelial cell-specific short chain alcohol dehydrogenase/reductase gene whose expression is up-regulated by retinoids and is involved in the metabolism of retinol. *J Biol Chem,* 2001.

Sorlie, T., Perou, C. M., Tibshirani, R., Aas, T., Geisler, S., Johnsen, H., Hastie, T., Eisen, M. B., and co-workers. Gene expression patterns of breast carcinomas distinguish tumor subclasses with clinical implications. *Proc Natl Acad Sci U S A* 98:10869–10874, 2001.

Sotiriou, C., Khanna, C., Jazaeri, A. A., Petersen, D., and Liu, E. T. Core biopsies can be used to distinguish differences in expression profiling by cDNA microarrays. *J Mol Diagn* 4:30–36, 2002.

Soukas, A., Cohen, P., Socci, N. D., and Friedman, J. M. Leptin-specific patterns of gene expression in white adipose tissue. *Genes Dev* 14:963–980, 2000.

Soukas, A., Socci, N. D., Saatkamp, B. D., Novelli, S., and Friedman, J. M. Distinct transcriptional profiles of adipogenesis in vivo and in vitro. *J Biol Chem* 276:34167–34174, 2001.

Soulez, M., and Parker, M. G. Identification of novel oestrogen receptor target genes in human ZR75-1 breast cancer cells by expression profiling. *J Mol Endocrinol* 27:259–274, 2001.

Southern, E., Mir, K., and Shchepinov, M. Molecular interactions on microarrays. *Nat Genet* 21(Suppl 1):5–9, 1999.

Spellman, P. T. The future of publishing microarray data. *Brief Bioinform* 2:316–318, 2001.

Spellman, P. T., Sherlock, G., Zhang, M. Q., Iyer, V. R., Anders, K., Eisen, M. B., Brown, P. O., Botstein, D., and Futcher, B. Comprehensive identification of cell cycle-regulated genes of the yeast *Saccharomyces cerevisiae* by microarray hybridization. *Mol Biol Cell* 9:3273–3297, 1998.

Spiegelman, J. I., Mindrinos, M. N., and Oefner, P. J. High-accuracy DNA sequence variation screening by DHPLC. *Biotechniques* 29:1084–1090, 1092, 2000.

Spiro, A., Lowe, M., and Brown, D. A bead-based method for multiplexed identification and quantitation of DNA sequences using flow cytometry. *Appl Environ Microbiol* 66:4258–4265, 2000.

Sreekumar, A., Nyati, M. K., Varambally, S., Barrette, T. R., Ghosh, D., Lawrence, T. S., and Chinnaiyan, A. M. Profiling of cancer cells using protein microarrays: Discovery of novel radiation-regulated proteins. *Cancer Res* 61:7585–7593, 2001.

Srivastava, M., Bubendorf, L., Srikantan, V., Fossom, L., Nolan, L., Glasman, M., Leighton, X., Fehrle, W., and co-workers. ANX7, a candidate tumor suppressor gene for prostate cancer. *Proc Natl Acad Sci U S A* 98:4575–4580, 2001.

Srivastava, M., Eidelman, O., and Pollard, H. B. Pharmacogenomics of the cystic fibrosis transmembrane conductance regulator (CFTR) and the cystic fibrosis drug CPX using genome microarray analysis. *Mol Med* 5:753–767, 1999.

Stafford, D. E., and Stephanopoulos, G. Metabolic engineering as an integrating platform for strain development. *Curr Opin Microbiol* 4:336–340, 2001.

Stagliano, N. E., Carpino, A. J., Ross, J. S., and Donovan, M. Vascular gene discovery using laser capture microdissection of human blood vessels and quantitative PCR. *Ann N Y Acad Sci* 947:344–349, 2001.

Stanton, J. L., and Green, D. P. Meta-analysis of gene expression in mouse preimplantation embryo development. *Mol Hum Reprod* 7:545–552, 2001.

Stanton, L. W., Garrard, L. J., Damm, D., Garrick, B. L., Lam, A., Kapoun, A. M., Zheng, Q., Protter, A. A., and co-workers. Altered patterns of gene expression in response to myocardial infarction. *Circ Res* 86:939–945, 2000.

Staudt, L. M. Gene expression profiling of lymphoid malignancies. *Annu Rev Med* 53:303–318, 2002.

Staudt, L. M., and Brown, P. O. Genomic views of the immune system. *Annu Rev Immunol* 18:829–859, 2000.

Staunton, J. E., Slonim, D. K., Coller, H. A., Tamayo, P., Angelo, M. J., Park, J., Scherf, U., Lee, J. K., and co-workers. Chemosensitivity prediction by transcriptional profiling. *Proc Natl Acad Sci U S A* 98:10787–10792, 2001.

Stears, R. L., Getts, R. C., and Gullans, S. R. A novel, sensitive detection system for high-density microarrays using dendrimer technology. *Physiol Genomics* 3:93–99, 2000.

Steckler, T. The molecular neurobiology of stress—Evidence from genetic and epigenetic models. *Behav Pharmacol* 12:381–427, 2001.

Steel, A. B., Levicky, R. L., Herne, T. M., and Tarlov, M. J. Immobilization of nucleic acids at solid surfaces: Effect of oligonucleotide length on layer assembly. *Biophys J* 79:975–981, 2000.

Steel, A., Torres, M., Hartwell, J., Yu, Y-Y., Ting, N., Hoke G., and Yang, H. The Flow-Thru Chip™: A three-dimensional biochip platform. In: M. Schena, ed. *Microarray Biochip Technology*. Eaton Publishing, Natick, MA, 2000:87–117.

Steemers, F. J., Ferguson, J. A., and Walt, D. R. Screening unlabeled DNA targets with randomly ordered fiber-optic gene arrays. *Nat Biotechnol* 18:91–94, 2000.

Steinman, L. Gene microarrays and experimental demyelinating disease: A tool to enhance serendipity. *Brain* 124:1897–1899, 2001.

Stephan, D. A., Chen, Y., Jiang, Y., Malechek, L., Gu, J. Z., Robbins, C. M., Bittner, M. L., Morris, J. A., and co-workers. Positional cloning utilizing genomic DNA microarrays: The Niemann-Pick type C gene as a model system. *Mol Genet Metab* 70:10–18, 2000.

Stephens, C. Bacterial cell cycle: Seeing the big picture with microarrays. *Curr Biol* 11:R222–R225, 2001.

Stern, L. E., Erwin, C. R., Falcone, R. A., Huang, F. S., Kemp, C. J., Williams, J. L., and Warner, B. W. cDNA microarray analysis of adapting bowel after intestinal resection. *J Pediatr Surg* 36:190–195, 2001.

Stewart, D. J. Making and using DNA microarrays: a short course at Cold Spring Harbor Laboratory. *Genome Res* 10:1–3, 2000.

Stillman, B. A., and Tonkinson, J. L. Expression microarray hybridization kinetics depend on length of the immobilized DNA but are independent of immobilization substrate. *Anal Biochem* 295:149–157, 2001.

Stillman, B. A., and Tonkinson, J. L. FAST slides: A novel surface for microarrays. *Biotechniques* 29:630–635, 2000.

Stingley, S. W., Ramirez, J. J., Aguilar, S. A., Simmen, K, Sandri-Goldin, R. M., Ghazal, P., and Wagner, E. K. Global analysis of herpes simplex virus type 1 transcription using an oligonucleotide-based DNA microarray. *J Virol* 74:9916–9927, 2000.

Stokes, D. G., Liu, G., Coimbra, I. B., Piera-Velazquez, S., Crowl, R. M., and Jimenez, S. A. Assessment of the gene expression profile of differentiated and dedifferentiated human fetal chondrocytes by microarray analysis. *Arthritis Rheum* 46:404–419, 2002.

Stolarov, J., Chang, K., Reiner, A., Rodgers, L., Hannon, G. J., Wigler, M. H., and Mittal, V. Design of a retroviral-mediated ecdysone-inducible system and its application to the expression profiling of the PTEN tumor suppressor. *Proc Natl Acad Sci U S A* 98:13043–13048, 2001.

Stoll, D., Templin, M. F., Schrenk, M., Traub, P. C., Vohringer, C. F., and Joos, T. O. Protein microarray technology. *Front Biosci* 7:C13–C32, 2002.

Stomakhin, A. A., Vasiliskov, V. A., Timofeev, E., Schulga, D., Cotter, R. J., and Mirzabekov, A. D. DNA sequence analysis by hybridization with oligonucleotide microchips: MALDI mass spectrometry identification of 5mers contiguously stacked to microchip oligonucleotides. *Nucleic Acids Res* 28:1193–1198, 2000.

Stone, D. H., Sivamurthy, N., Contreras, M. A., Fitzgerald, L., LoGerfo, F. W., and Quist, W. C. Altered ubiquitin/proteasome expression in anastomotic intimal hyperplasia. *J Vasc Surg* 34:1016–1022, 2001.

Storz, M., Zepter, K., Kamarashev, J., Dummer, R., Burg, G., and Haffner, A. C. Coexpression of CD40 and CD40 ligand in cutaneous T-cell lymphoma (mycosis fungoides). *Cancer Res.* 61:452–454, 2001.

Strakhova, M. I., and Skolnick, P. Can 'differential display' methodologies make an impact on biological psychiatry? *Int J Neuropsychopharmacol* 4:75–82, 2001.

Stratowa, C., Loffler, G., Lichter, P., Stilgenbauer, S., Haberl, P., Schweifer, N., Dohner, H., and Wilgenbus, K. K. CDNA microarray gene expression analysis of B-cell chronic lymphocytic leukemia proposes potential new prognostic markers involved in lymphocyte trafficking. *Int J Cancer* 91:474–480, 2001.

Strehlow, D. R. The promise of transcription profiling for understanding the pathogenesis of scleroderma. *Curr Rheumatol Rep* 2:506–511, 2000.

Strizhkov, B. N., Drobyshev, A. L., Mikhailovich, V. M., and Mirzabekov, A. D. PCR amplification on a microarray of gel-immobilized oligonucleotides: Detection of bacterial toxin- and drug-resistant genes and their mutations. *Biotechniques* 29:844–848, 2000.

Sturn, A., Quackenbush, J., and Trajanoski, Z. Genesis: Cluster analysis of microarray data. *Bioinformatics* 18:207–208, 2002.

Sturniolo, T., Bono, E., Ding, J., Raddrizzani, L., Tuereci, O., Sahin, U., Braxenthaler, M., Gallazzi, F., and co-workers. Generation of tissue-specific and promiscuous HLA ligand databases using DNA microarrays and virtual HLA class II matrices. *Nat Biotechnol* 17:555–561, 1999.

Su, Y. A., Bittner, M. L., Chen, Y., Tao, L., Jiang, Y., Zhang, Y., Stephan, D. A., and Trent, J. M. Identification of tumor-suppressor genes using human melanoma cell lines UACC903, UACC903(+6), and SRS3 by comparison of expression profiles. *Mol Carcinog* 28:119–127, 2000.

Suciu-Foca Cortesini, N., Piazza, F., Ho, E., Ciubotariu, R., LeMaoult, J., Dalla-Favera, R., and Cortesini, R. Distinct mRNA microarray profiles of tolerogenic dendritic cells. *Hum Immunol* 62:1065–1072, 2001.

Sudarsanam, P., Iyer, V. R., Brown, P. O., and Winston, F. Whole-genome expression analysis of snf/swi mutants of *Saccharomyces cerevisiae*. *Proc Natl Acad Sci U S A* 97:3364–3369, 2000.

Sudbrak, R., Wieczorek, G., Nuber, U. A., Mann, W., Kirchner, R., Erdogan, F., Brown, C. J., Wohrle, D., and co-workers. X chromosome-specific cDNA arrays: Identification of genes that escape from X-inactivation and other applications. *Hum Mol Genet* 10:77–83, 2001.

Sun, Y., Huang, P. L., Li, J. J., Huang, Y. Q., Zhang, L., Huang, P. L., and Lee-Huang, S. Anti-HIV agent map30 modulates the expression profile of viral and cellular genes for proliferation and apoptosis in aids-related lymphoma cells infected with Kaposi's sarcoma-associated virus. *Biochem Biophys Res Commun* 287:983–994, 2001.

Sundberg, S. A., Chow, A., Nikiforov, T., and Wada, H. G. Microchip-based systems for target validation and HTS. *Drug Discov Today* 5(Suppl 1):92–103, 2000.

Suzuki, I., Kanesaki, Y., Mikami, K., Kanehisa, M., and Murata, N. Cold-regulated genes under control of the cold sensor Hik33 in synechocystis. *Mol Microbiol* 40:235–244, 2001.

Suzuki, J., Shen, W. J., Nelson, B. D., Patel, S., Veerkamp, J. H., Selwood, S. P., Murphy, G. M. Jr., Reaven, E., and Kraemer, F. B. Absence of cardiac lipid accumulation in transgenic mice with heart-specific HSL overexpression. *Am J Physiol Endocrinol Metab* 281:E857–E866, 2001.

Suzuki, T., and Kanamaru, R. Genetic background in carcinogenesis. *Gan To Kagaku Ryoho* 26:1971–1979, 1999.

Suzuki, Y., Rahman, M., and Mitsuya, H. Diverse transcriptional response of CD4(+) T cells to stromal cell-derived factor (SDF)-1: Cell survival promotion and priming effects of SDF-1 on CD4(+) T cells. *J Immunol* 167:3064–3073, 2001.

Svaren, J., Ehrig, T., Abdulkadir, S. A., Ehrengruber, M. U., Watson, M. A., and Milbrandt, J. EGR1 target genes in prostate carcinoma cells identified by microarray analysis. *J Biol Chem* 275:38524–38531, 2000.

Swerdlow, S. H., and Williams, M. E. From centrocytic to mantle cell lymphoma: A clinicopathologic and molecular review of 3 decades. *Hum Pathol* 33:7–20, 2002.

Syvanen, A. C. From gels to chips: "Minisequencing" primer extension for analysis of point mutations and single nucleotide polymorphisms. *Hum Mutat* 13:1–10, 1999.

Szallasi, Z. Genetic network analysis in light of massively parallel biological data acquisition. *Pac Symp Biocomput* 5–16, 1999.

Tabuchi, Y., Kondo, T., Ogawa, R., and Mori, H. DNA microarray analyses of genes elicited by ultrasound in human U937 cells. *Biochem Biophys Res Commun* 290: 498–503, 2002.

Tackels-Horne, D., Goodman, M. D., Williams, A. J., Wilson, D. J., Eskandari, T., Vogt, L. M., Boland, J. F., Scherf, U., and Vockley, J. G. Identification of differentially expressed genes in hepatocellular carcinoma and metastatic liver tumors by oligonucleotide expression profiling. *Cancer* 92:395–405, 2001.

Taetle, R., Aickin, M., Panda, L., Emerson, J., Roe, D., Thompson, F., Davis, J., Trent, J., and Alberts, D. Chromosome abnormalities in ovarian adenocarcinoma: II. Prognostic impact of nonrandom chromosome abnormalities in 244 cases. *Genes Chromosomes Cancer* 25:46–52, 1999.

Takahashi, M., Rhodes, D. R., Furge, K. A., Kanayama, H., Kagawa, S., Haab, B. B., and The, B. T. Gene expression profiling of clear cell renal cell carcinoma: Gene identification and prognostic classification. *Proc Natl Acad Sci U S A* 98:9754–9759, 2001.

Takahashi, M., Nishihira, J., Shimpo, M., Mizue, Y., Ueno, S., Mano, H., Kobayashi, E., Ikeda, U., and Shimada, K. Macrophage migration inhibitory factor as a redox-sensitive cytokine in cardiac myocytes. *Cardiovasc Res* 52:438–445, 2001.

Takahashi, Y., Nagata, T., Ishii, Y., Ikarashi, M., Ishikawa, K., and Asai, S. Up-regulation of vitamin D3 up-regulated protein 1 gene in response to 5-fluorouracil in colon carcinoma SW620. *Oncol Rep* 9:75–79, 2002.

Takemasa, I., Higuchi, H., Yamamoto, H., Sekimoto, M., Tomita, N., Nakamori, S., Matoba, R., Monden, M., and Matsubara, K. Construction of preferential cDNA microarray specialized for human colorectal carcinoma: molecular sketch of colorectal cancer. *Biochem Biophys Res Commun* 285:1244–1249, 2001.

Takenaka, M., Imai, E., Nagasawa, Y., Matsuoka, Y., Moriyama, T., Kaneko, T., Hori, M., Kawamoto, S., and Okubo, K. Gene expression profiles of the collecting duct in the mouse renal inner medulla. *Kidney Int* 57:19–24, 2000.

Takizawa, P. A., DeRisi, J. L., Wilhelm, J. E., and Vale, R. D. Plasma membrane compartmentalization in yeast by messenger RNA transport and a septin diffusion barrier. *Science* 290:341–344, 2000.

Talaat, A. M., Hunter, P., and Johnston, S. A. Genome-directed primers for selective labeling of bacterial transcripts for DNA microarray analysis. *Nat Biotechnol* 18:679–682, 2000.

Tanabe, L., Scherf, U., Smith, L. H., Lee, J. K., Hunter, L., and Weinstein, J. N. MedMiner: An Internet text-mining tool for biomedical information, with application to gene expression profiling. *Biotechniques* 27:1210–1214, 1216–1217, 1999.

Tanaka, F., Hori, N., and Sato, K. Identification of differentially expressed genes in rat hepatoma cell lines using subtraction and microarray. *J Biochem [Tokyo]* 131:39–44, 2002.

Tanaka, T. Transcriptome analysis and pharmacogenomics. *Nippon Yakurigaku Zasshi* 116:241–246, 2000.

Tanaka, T., Nishimura, Y., Tsunoda, H., Kitaoka, Y., and Naka, M. Pharmacogenomics and pharmainformatics. *Nippon Rinsho* 601:39–50, 2002.

Tanaka, T. S., Jaradat, S. A., Lim, M. K., Kargul, G. J., Wang, X., Grahovac, M. J., Pantano, S., Sano, Y., and co-workers. Genome-wide expression profiling of mid-gestation placenta and embryo using a 15,000 mouse developmental cDNA microarray. *Proc Natl Acad Sci U S A* 97:9127–9132, 2000.

Tang, Y., Lu, A., Aronow, B. J., and Sharp, F. R. Blood genomic responses differ after stroke, seizures, hypoglycemia, and hypoxia: blood genomic fingerprints of disease. *Ann Neurol* 50:699–707, 2001.

Tanghe, A., Teunissen, A., Van Dijck, P., and Thevelein, J. M. Identification of genes responsible for improved cryoresistance in fermenting yeast cells. *Int J Food Microbiol* 55(1–3):259–262, 2000.

Taniguchi, M., Miura, K., Iwao, H., and Yamanaka, S. Quantitative assessment of DNA microarrays—Comparison with Northern blot analyses. *Genomics* 71:34–39, 2001.

Tannapfel, A., Geissler, F., Witzigmann, H., Hauss, J., and Wittekind, C. Analysis of liver allograft rejection related genes using cDNA-microarrays in liver allograft specimen. *Transplant Proc* 33:3283–3284, 2001.

Tawata, M., Aida, K., and Onaya, T. Screening for genetic mutations. A review. *Comb Chem High Throughput Screen* 3:1–9, 2000.

Taylor, E., Cogdell, D., Coombes, K., Hu, L., Ramdas, L., Tabor, A., Hamilton, S., and Zhang, W. Sequence verification as quality-control step for production of cDNA microarrays. *Biotechniques* 31:62–65, 2001.

Taylor, L. A., Carthy, C. M., Yang, D., Saad, K., Wong, D., Schreiner, G., Stanton, L. W., and McManus, B. M. Host gene regulation during coxsackievirus B3 infection in mice: Assessment by microarrays. *Circ Res* 87:328–334, 2000.

Taylor, M. F., Wiederholt, K., and Sverdrup, F. Antisense oligonucleotides: A systematic high-throughput approach to target validation and gene function determination. *Drug Discov Today* 4:562–567, 1999.

Temple, R., Allen, E., Fordham, J., Phipps, S., Schneider, H. C., Lindauer, K., Hayes, I., Lockey, J., and co-workers. Microarray analysis of eosinophils reveals a number of candidate survival and apoptosis genes. *Am J Respir Cell Mol Biol* 25:425–433, 2001.

Tepperman, J. M., Zhu, T., Chang, H. S., Wang, X., and Quail, P. H. Multiple transcription-factor genes are early targets of phytochrome A signaling. *Proc Natl Acad Sci U S A* 98:9437–9442, 2001.

ter Linde, J. J., Liang, H., Davis, R.W., Steensma, H. Y., van Dijken, J. P., and Pronk, J. T. Genome-wide transcriptional analysis of aerobic and anaerobic chemostat cultures of *Saccharomyces cerevisiae*. *J Bacteriol* 181:7409–7413, 1999.

Terskikh, A. V., Easterday, M. C., Li, L., Hood, L., Kornblum, H. I., Geschwind, D. H., and Weissman, I. L. From hematopoiesis to neuropoiesis: evidence of overlapping genetic programs. *Proc Natl Acad Sci U S A* 98:7934–7939, 2001.

Teruyama, K., Abe, M., Nakano, T., Takahashi, S., Yamada, S., and Sato, Y. Neurophilin-1 is a downstream target of transcription factor Ets-1 in human umbilical vein endothelial cells. *FEBS Lett* 504:1–4, 2001.

Theilhaber, J., Bushnell, S., Jackson, A., and Fuchs, R. Bayesian estimation of fold-changes in the analysis of gene expression: The PFOLD algorithm. *J Comput Biol* 8:585–614, 2001.

Theilhaber, J., Connolly, T., Roman-Roman, S., Bushnell, S., Jackson, A., Call, K., Garcia, T., and Baron, R. Finding genes in the C2C12 osteogenic pathway by k-nearest-neighbor classification of expression data. *Genome Res* 12:165–176, 2002.

Theriault, T. P., Winder, S. C., and Gamble, R. C. Application of ink-jet printing technology to the manufacture of molecular arrays. In: M. Schena, ed. *DNA Microarrays: A Practical Approach,* 2nd ed. Oxford University Press, Oxford, UK, 2000:101–120.

Thieffry, D. From global expression data to gene networks. *Bioessays* 21:895–899, 1999.

Thielecke, H., Mack, A., and Robitzki, A. Biohybrid microarrays—Impedimetric biosensors with 3D in vitro tissues for toxicological and biomedical screening. *Fresenius J Anal Chem* 369:23–29, 2001.

Thijs, G., Moreau, Y., De Smet, F., Mathys, J., Lescot, M., Rombauts, S., Rouze, P., De Moor, B., and Marchal, K. INCLUSive: Integrated clustering, upstream sequence retrieval and motif sampling. *Bioinformatics* 18:331–332, 2002.

Thimm, O., Essigmann, B., Kloska, S., Altmann, T., and Buckhout, T. J. Response of *Arabidopsis* to iron deficiency stress as revealed by microarray analysis. *Plant Physiol* 127:1030–1043, 2001.

Thomas, J. G., Olson, J. M., Tapscott, S. J., and Zhao, LP. An efficient and robust statistical modeling approach to discover differentially expressed genes using genomic expression profiles. *Genome Res* 11:1227–1236, 2001.

Thomas, J. T., Oh, S. T., Terhune, S. S., and Laimins, L. A. Cellular changes induced by low-risk human papillomavirus type 11 in keratinocytes that stably maintain viral episomes. *J Virol* 75:7564–7571, 2001.

Thomas, R. S., Rank, D. R., Penn, S. G., Zastrow, G. M., Hayes, K. R., Pande, K., Glover, E., Silander, T., and co-workers. Identification of toxicologically predictive gene sets using cDNA microarrays. *Mol Pharmacol* 60:1189–1194, 2001.

Tomiuk, S., and Hofmann, K. Microarray probe selection strategies. *Brief Bioinform* 2:329–340, 2001.

Thompson, A., Lucchini, S., and Hinton, J. C. It's easy to build your own microarrayer! *Trends Microbiol* 9:154–156, 2001.

Thompson, D. K., Beliaev, A. S., Giometti, C. S., Tollaksen, S. L., Khare, T., Lies, D. P., Nealson, K. H., Lim, H., and co-workers. Transcriptional and proteomic analysis of a ferric uptake regulator (Fur) mutant of Shewanella oneidensis: possible involvement of fur in energy metabolism, transcriptional regulation, and oxidative stress. *Appl Environ Microbiol* 68:881–892, 2002.

Thompson, M., and Furtado, L. M. High density oligonucleotide and DNA probe arrays for the analysis of target DNA. *Analyst* 124:1133–1136, 1999.

Tice, D. A., Soloviev, I., and Polakis, P. Activation of the wnt pathway interferes with SRE-driven transcription of immediate early genes. *J Biol Chem* 2001.

Tillib, S. V., and Mirzabekov, A. D. Advances in the analysis of DNA sequence variations using oligonucleotide microchip technology. *Curr Opin Biotechnol* 12:53–58, 2001.

Timofeev, E., and Mirzabekov, A. Binding specificity and stability of duplexes formed by modified oligonucleotides with a 4096-hexanucleotide microarray. *Nucleic Acids Res* 29:2626–2634, 2001.

Tobe, K., Suzuki, R., Aoyama, M., Yamauchi, T., Kamon, J., Kubota, N., Terauchi, Y., Matsui, J., and co-workers. Increased expression of the sterol regulatory element-binding protein-1 gene in insulin receptor substrate-2(−/−) mouse liver. *J Biol Chem* 276:38337–38340, 2001.

Toh, H., and Horimoto, K. Inference of a genetic network by a combined approach of cluster analysis and graphical Gaussian modeling. *Bioinformatics* 18:287–297, 2002.

Tohda, S., and Nara, N. Molecular diagnostic tests in hematologic diseases. *Rinsho Byori* 49:205–209, 2001.

Tollet-Egnell, P., Flores-Morales, A., Stahlberg, N., Malek, R. L., Lee, N., and Norstedt, G. Gene expression profile of the aging process in rat liver: Normalizing effects of growth hormone replacement. *Mol Endocrinol* 15:308–318, 2001.

Tomasini, R., Azizi Samir, A., Vaccaro, M. I., Pebusque, M. J., Dagorn, J. C., Iovanna, J. L., and Dusetti, N. J. Molecular and functional characterization of the SIP gene and its two transcripts generated by alternative splicing. SIP is induced by stress and promotes cell death. *J Biol Chem* 2001.

Tomlinson, I. M., and Holt, L. J. Protein profiling comes of age. Genome Biol 2: 2001.

Tompa, R., McCallum, C. M., Delrow, J., Henikoff, J. G., van Steensel, B., and Henikoff, S. Genome-wide profiling of DNA methylation reveals transposon targets of CHROMOMETHYLASE3. *Curr Biol* 12:65–68, 2002.

Tong, L., Shen, H., Perreau, V. M., Balazs, R., and Cotman, C. W. Effects of exercise on gene-expression profile in the rat hippocampus. *Neurobiol Dis* 8:1046–1056, 2001.

Tõnisson, N., Kurg, A., Lohmussaar, E., and Metspalu, A. Arrayed primer extension on the dNA chip—Method and applications. In: M. Schena, ed. *Microarray Biochip Technology*. Eaton Publishing, Natick, MA, 2000:247–263.

Torhorst, J., Bucher, C., Kononen, J., Haas, P., Zuber, M., Kochli, O. R., Mross, F., Dieterich, H., and coo-workers. Tissue microarrays for rapid linking of molecular changes to clinical endpoints. *Am J Pathol* 159:2249–2256, 2001.

Toronen, P., Kolehmainen, M., Wong, G., and Castren, E. Analysis of gene expression data using self-organizing maps. *FEBS Lett* 451:142–146, 1999.

Traven, A., Wong, J. M., Xu, D., Sopta, M., and Ingles, C. J. Inter-organellar communication: Altered nuclear gene expression profiles in a yeast mitochondrial DNA mutant. *J Biol Chem* 2000.

Travers, K. J., Patil, C. K., Wodicka, L., Lockhart, D. J., Weissman, J. S., and Walter, P. Functional and genomic analyses reveal an essential coordination between the unfolded protein response and ER-associated degradation. *Cell* 101:249–258, 2000.

Triche, T. J., Schofield, D., and Buckley, J. DNA microarrays in pediatric cancer. *Cancer J* 7:2–15, 2001.

Triendl, R., and Yoon, R. Singapore—From microprocessors to microarrays? *Nat Biotechnol* 19:521–522, 2001.

Troyanskaya, O., Cantor, M., Sherlock, G., Brown, P., Hastie, T., Tibshirani, R., Botstein, D., and Altman, R. B. Missing value estimation methods for DNA microarrays. *Bioinformatics* 17:520–525, 2001.

Tseng, G. C., Oh, M. K., Rohlin, L., Liao, J. C., and Wong, W. H. Issues in cDNA microarray analysis: quality filtering, channel normalization, models of variations and assessment of gene effects. *Nucleic Acids Res* 29:2549–2557, 2001.

Tsirulnikov, N., and Shohat, M. Microarray technology—Clinical aspects. *Harefuah* 140(11):1072–1077, 2001.

Tsodikov, A., Szabo, A., and Jones, D. Adjustments and measures of differential expression for microarray data. *Bioinformatics* 18:251–260, 2002.

Tsou, R., Cole, J. K., Nathens, A. B., Isik, F. F., Heimbach, D. M., Engrav, L. H., and Gibran, N. S. Analysis of hypertrophic and normal scar gene expression with cDNA microarrays. *J Burn Care Rehabil* 21:541–550, 2000.

Tusher, V. G., Tibshirani, R., and Chu, G. Significance analysis of microarrays applied to the ionizing radiation response. *Proc Natl Acad Sci U S A* 98:5116–5121, 2001.

Tzou, P., De Gregorio, E., and Lemaitre, B. How *Drosophila* combats microbial infection: A model to study innate immunity and host-pathogen interactions. *Curr Opin Microbiol* 5:102–110, 2002.

Ueda, A., Hamadeh, H. K., Webb, H. K., Yamamoto, Y., Sueyoshi, T., Afshari, C. A., Lehmann, J. M., and Negishi, M. Diverse roles of the nuclear orphan receptor CAR in regulating hepatic genes in response to phenobarbital. *Mol Pharmacol* 61:1–6, 2002.

Ueda, K. Detection of the retinoic acid-regulated genes in a RTBM1 neuroblastoma cell line using cDNA microarray. *Kurume Med J* 48:159–164, 2001.

Uhl, G. R., Liu, Q. R., Walther, D., Hess, J., and Naiman, D. Polysubstance abuse-vulnerability genes: Genome scans for association, using 1,004 subjects and 1,494 single-nucleotide polymorphisms. *Am J Hum Genet* 69:1290–1300, 2001.

Uhlmann, E. Oligonucleotide technologies: synthesis, production, regulations and applications. *Expert Opin Biol Ther* 1:319–328, 2001.

Umek, R. M., Lin, S. W., Vielmetter, J., Terbrueggen, R. H., Irvine, B., Yu, C. J., Kayyem, J. F., Yowanto, H., and co-workers. Electronic detection of nucleic acids: a versatile platform for molecular diagnostics. *J Mol Diagn* 3:74–84, 2001.

Unger, M. A., Rishi, M., Clemmer, V. B., Hartman, J. L., Keiper, E. A., Greshock, J. D., Chodosh, L. A., Liebman, M. N., and Weber, B. L. Characterization of adjacent breast tumors using oligonucleotide microarrays. *Breast Cancer Res* 3:336–341, 2001.

Urakawa, H., Noble, P. A., El Fantroussi, S., Kelly, J. J., and Stahl, D. A. Single-base-pair discrimination of terminal mismatches by using oligonucleotide microarrays and neural network analyses. *Appl Environ Microbiol* 68:235–244, 2002.

Vaarala, M. H., Porvari, K., Kyllonen, A., and Vihko, P. Differentially expressed genes in two LNCaP prostate cancer cell lines reflecting changes during prostate cancer progression. *Lab Invest* 80:1259–1268, 2000.

Valencia, M., Bentele, M., Vaze, M. B., Herrmann, G., Kraus, E., Lee, S. E., Schar, P., and Haber, J. E. NEJ1 controls non-homologous end joining in Saccharomyces cerevisiae. *Nature* 414:666–669, 2001.

Valerius, M.T., Patterson, L.T., Witte, D.P., Potter, S.S. Microarray analysis of novel cell lines representing two stages of metanephric mesenchyme differentiation. Mech Dev 112:219–232, 2002.

van Berkum, N. L., and Holstege, F. C. DNA microarrays: Raising the profile. *Curr Opin Biotechnol* 12:48–52, 2001.

van Dam, R. M., and Quake, S. R. Gene expression analysis with universal *n*-mer arrays. *Genome Res* 12:145–152, 2002.

van De Rijke, F., Zijlmans, H., Li, S., Vail, T., Raap, A. K., Niedbala, R. S., and Tanke, H. J. Up-converting phosphor reporters for nucleic acid microarrays. *Nat Biotechnol* 19:273–276, 2001.

Van Dyk, T. K., Wei, Y., Hanafey, M. K., Dolan, M., Reeve, M. J., Rafalski, J. A., Rothman-Denes, L. B., and LaRossa, R. A. A genomic approach to gene fusion technology. *Proc Natl Acad Sci U S A* 98:2555–2560, 2001.

van Hal, N. L., Vorst, O., van Houwelingen, A. M., Kok, E. J., Peijnenburg, A., Aharoni, A., van Tunen, A. J., and Keijer, J. The application of DNA microarrays in gene expression analysis. *J Biotechnol* 78:271–280, 2000.

van Helden, J., Rios, A. F., and Collado-Vides, J. Discovering regulatory elements in non-coding sequences by analysis of spaced dyads. *Nucleic Acids Res* 28:1808–1818, 2000.

van Steensel, B., Delrow, J., and Henikoff, S. Chromatin profiling using targeted DNA adenine methyltransferase. *Nat Genet* 27:304–308, 2001.

van Tilburg, J., van Haeften, T. W., Pearson, P., and Wijmenga, C. Defining the genetic contribution of type 2 diabetes mellitus. *J Med Genet* 38:569–578, 2001.

van 't Veer, L. J., Dai, H., van de Vijver, M. J., He, Y. D., Hart, A. A., Mao, M., Peterse, H. L., van der Kooy, K., and co-workers. Gene expression profiling predicts clinical outcome of breast cancer. *Nature* 415:530–536, 2002.

Van't Veer, L. J., and De Jong, D. The microarray way to tailored cancer treatment. *Nat Med* 8:13–14, 2002.

Varedi, M., Lee, H. M., Greeley, G. H. Jr., Herndon, D. N., and Englander, E. W. Gene expression in intestinal epithelial cells, IEC-6, is altered by burn injury-induced circulating factors. *Shock* 16:259–263, 2001.

Vasiliskov, A. V., Timofeev, E. N., Surzhikov, S. A., Drobyshev, A. L., Shick, V. V., and Mirzabekov, A. D. Fabrication of microarray of gel-immobilized compounds on a chip by copolymerization. *Biotechniques* 27:592–594, 596–598, 600, 1999.

Vawter, M. P., Barrett, T., Cheadle, C., Sokolov, B. P., Wood, W. H., Donovan, D. M., Webster, M., Freed, W. J., and Becker, K. G. Application of cDNA microarrays to examine gene expression differences in schizophrenia. *Brain Res Bull* 55:641–650, 2001.

Vedoy, C. G., Bengtson, M. H., and Sogayar, M. C. Hunting for differentially expressed genes. *Braz J Med Biol Res* 32:877–884, 1999.

Venkatasubramanian, R., Siivola, E., Colpitts, T., and O'Quinn, B. Thin-film thermoelectric devices with high room-temperature figures of merit. *Nature* 413:597–602, 2001.

Vente, A., Korn, B., Zehetner, G., Poustka, A., and Lehrach, H. Distribution and early development of microarray technology in Europe. *Nat Genet* 22:22, 1999.

Vernon, S. D., Unger, E. R., Rajeevan, M., Dimulescu, I. M., Nisenbaum, R., and Campbell, C. E. Reproducibility of alternative probe synthesis approaches for gene expression profiling with arrays. *J Mol Diagn* 2:124–127, 2000.

Verrecchia, F., Chu, M. L., and Mauviel, A. Identification of novel TGF-beta/Smad gene targets in dermal fibroblasts using a combined cDNA microarray/promoter transactivation approach. *J Biol Chem* 2001.

Vershon, A. K., and Pierce, M. Transcriptional regulation of meiosis in yeast. *Curr Opin Cell Biol* 12:334–339, 2000.

Vikhanskaya, F., Marchini, S., Marabese, M., Galliera, E., and Broggini, M. P73a overexpression is associated with resistance to treatment with DNA-damaging agents in a human ovarian cancer cell line. *Cancer Res* 61:935–938, 2001.

Villaret, D. B., Wang, T., Dillon, D., Xu, J., Sivam, D., Cheever, M. A., and Reed, S. G. Identification of genes overexpressed in head and neck squamous cell carcinoma using a combination of complementary DNA subtraction and microarray analysis. *Laryngoscope* 110(3):374–381, 2000.

Vilo, J., and Kivinen, K. Regulatory sequence analysis: application to the interpretation of gene expression. *Eur Neuropsychopharmacol* 11:399–411, 2001.

Vincenti, M. P., and Brinckerhoff, C. E. Early response genes induced in chondrocytes stimulated with the inflammatory cytokine interleukin-1beta. *Arthritis Res* 3:381–388, 2001.

Virtaneva, K., Wright, F. A., Tanner, S. M., Yuan, B., Lemon, W. J., Caligiuri, M. A., Bloomfield, C. D., de La Chapelle, A., and Krahe, R. Expression profiling reveals fundamental biological differences in acute myeloid leukemia with isolated trisomy 8 and normal cytogenetics. *Proc Natl Acad Sci U S A* 98:1124–1129, 2001.

Virasch, N., and Kruse, C. A. Strategies using the immune system for therapy of brain tumors. *Hematol Oncol Clin North Am* 15:1053–1071, 2001.

Vo-Dinh, T., Alarie, J. P., Isola, N., Landis, D., Wintenberg, A. L., and Ericson, M. N. DNA biochip using a phototransistor integrated circuit. *Anal Chem* 71:358–363, 1999.

Voehringer, D. W., Hirschberg, D. L., Xiao, J., Lu, Q., Roederer, M., Lock, C. B., Herzenberg, L. A., Steinman, L., and Herzenberg, L. A. Gene microarray identification of redox and mitochondrial elements that control resistance or sensitivity to apoptosis. *Proc Natl Acad Sci U S A* 97:2680–2685, 2000.

Vogt, P. K. Jun, the oncoprotein. *Oncogene* 20:2365–2367, 2001.

Voit, E. O., and Radivoyevitch, T. Biochemical systems analysis of genome-wide expression data. *Bioinformatics* 16:1023–1037, 2000.

von Heydebreck, A., Huber, W., Poustka, A., and Vingron, M. Identifying splits with clear separation: a new class discovery method for gene expression data. *Bioinformatics* 17(Suppl 1):S107–S114, 2001.

Wada, R., Tifft, C. J., and Proia, R. L. Microglial activation precedes acute neurodegeneration in Sandhoff disease and is suppressed by bone marrow transplantation. *Proc Natl Acad Sci USA* 97:10954–10959, 2000.

Waddell, P. J., and Kishino, H. Cluster inference methods and graphical models evaluated on NCI60 microarray gene expression data. *Genome Inform Ser Workshop Genome Inform* 11:129–140, 2000.

Waddell, E., Wang, Y., Stryjewski, W., McWhorter, S., Henry, A. C., Evans, D., McCarley, R. L., and Soper, S. A. High-resolution near-infrared imaging of DNA microarrays with time-resolved acquisition of fluorescence lifetimes. *Anal Chem* 72:5907–5917, 2000.

Wagner, A. Decoupled evolution of coding region and mRNA expression patterns after gene duplication: implications for the neutralist-selectionist debate. *Proc Natl Acad Sci USA* 97:6579–6584, 2000.

Wai, D. H., Schaefer, K. L., Schramm, A., Korsching, E., Van Valen, F., Ozaki, T., Boecker, W., Schweigerer, L., and co-workers. Expression analysis of pediatric solid tumor cell lines using oligonucleotide microarrays. *Int J Oncol* 20:441–451, 2002.

Walker, J., Flower, D., and Rigley, K. Microarrays in hematology. *Curr Opin Hematol* 9:23–29, 2002.

Wall, M. E., Dyck, P. A., and Brettin, T. S. SVDMAN—Singular value decomposition analysis of microarray data. *Bioinformatics* 17:566–568, 2001.

Wandinger, K. P., Sturzebecher, C. S., Bielekova, B., Detore, G., Rosenwald, A., Staudt, L. M., McFarland, H. F., and Martin, R. Complex immunomodulatory effects of interferon-beta in multiple sclerosis include the upregulation of T helper 1-associated marker genes. *Ann Neurol* 50:349–357, 2001.

Wang, A., Pierce, A., Judson-Kremer, K., Gaddis, S., Aldaz, C. M., Johnson, D. G., and MacLeod, M. C. Rapid analysis of gene expression (RAGE) facilitates universal expression profiling. *Nucleic Acids Res* 27:4609–4618, 1999.

Wang, C., Francis, R., Harirchian, S., Batlle, D., Mayhew, B., Bassett, M., Rainey, W. E., and Pestell, R. G. The application of high density microarray for analysis of mitogenic signaling and cell-cycle in the adrenal. *Endocr Res* 26:807–823, 2000.

Wang, E., Miller, L. D., Ohnmacht, G. A., Liu, E. T., and Marincola, F. M. High-fidelity mRNA amplification for gene profiling. *Nat Biotechnol* 18:457–459, 2000.

Wang, H., Wang, H., Zhang, W., and Fuller, G. N. Tissue microarrays: Applications in neuropathology research, diagnosis, and education. *Brain Pathol* 12:95–107, 2002.

Wang, J., and Zhang, C. T. Identification of protein-coding genes in the genome of *Vibrio cholerae* with more than 98% accuracy using occurrence frequencies of single nucleotides. *Eur J Biochem* 268:4261–4268, 2001.

Wang, K., Gan, L., Jeffery, E., Gayle, M., Gown, A. M., Skelly, M., Nelson, P. S., Ng, W. V., and co-workers. Monitoring gene expression profile changes in ovarian carcinomas using cDNA microarray. *Gene* 229:101–108, 1999.

Wang, L., Wu, Q., Qiu, P., Mirza, A., McGuirk, M., Kirschmeier, P., Greene, J. R., Wang, Y., and co-workers. Analyses of P53 target genes in the human genome by bioinformatic and microarray approaches. *J Biol Chem* 2001.

Wang, R., Guegler, K., LaBrie, S. T., and Crawford, N. M. Genomic analysis of a nutrient response in *Arabidopsis* reveals diverse expression patterns and novel metabolic and potential regulatory genes induced by nitrate. *Plant Cell* 12:1491–1509, 2000.

Wang, T., Fan, L., Watanabe, Y., McNeill, P., Fanger, G. R., Persing, D. H., and Reed, S. G. L552S, an alternatively spliced isoform of XAGE-1, is over-expressed in lung adenocarcinoma. *Oncogene* 20:7699–7709, 2001.

Wang, T., Hopkins, D. A., Fan, L., Fanger, G. R., Houghton, R., Vedvick, T. S., Repasky, E., and Reed, S. G. A p53 homologue and a novel serine proteinase inhibitor are over-expressed in lung squamous cell carcinoma. *Lung Cancer* 34:363–374, 2001.

Wang, T., Hopkins, D., Schmidt, C., Silva, S., Houghton, R., Takita, H., Repasky, E., and Reed, S. G. Identification of genes differentially over-expressed in lung squamous cell carcinoma using combination of cDNA subtraction and microarray analysis. *Oncogene* 19:1519–1528, 2000.

Wang, W., Lee, S. B., Palmer, R., Ellisen, L. W., and Haber, D. A. A functional interaction with CBP contributes to transcriptional activation by the Wilms tumor suppressor WT1. *J Biol Chem* 276:16810–16816, 2001.

Wang, X., Ghosh, S., and Guo, S. W. Quantitative quality control in microarray image processing and data acquisition. *Nucleic Acids Res* 29:E75–E85, 2001.

Wang, X., Quail, E., Hung, N. J., Tan, Y., Ye, H., and Costa, R. H. Increased levels of forkhead box M1B transcription factor in transgenic mouse hepatocytes prevent age-related proliferation defects in regenerating liver. *Proc Natl Acad Sci U S A* 98:11468–11473, 2001.

Wang, Y., Hu, L., Yao, R., Wang, M., Crist, K. A., Grubbs, C. J., Johanning, G. L., Lubet, R. A., and You, M. Altered gene expression profile in chemically induced rat mammary adenocarcinomas and its modulation by an aromatase inhibitor. *Oncogene* 20:7710–7721, 2001.

Wardrop, S. L., Wells, C., Ravasi, T., Hume, D. A., and Richardson, D. R. Induction of Nramp2 in activated mouse macrophages is dissociated from regulation of the Nramp1, classical inflammatory genes, and genes involved in iron metabolism. *J Leukoc Biol* 71:99–106, 2002.

Waring, J. F., Ciurlionis, R., Jolly, R. A., Heindel, M., and Ulrich, R. G. Microarray analysis of hepatotoxins in vitro reveals a correlation between gene expression profiles and mechanisms of toxicity. *Toxicol Lett* 120:359–368, 2001.

Waring, J. F., and Ulrich, R. G. The impact of genomics-based technologies on drug safety evaluation. *Annu Rev Pharmacol Toxicol* 40:335–352, 2000.

Warrington, J. A., Dee, S., and Trulson, M. large scale genomic analysis using affymetrix GeneChip(R) probe arrays. In: M. Schena, ed. *Microarray Biochip Technology*. Eaton Publishing, Natick, MA, 2000:119–148.

Wassarman, K. M., Repoila, F., Rosenow, C., Storz, G., and Gottesman, S. Identification of novel small RNAs using comparative genomics and microarrays. *Genes Dev* 15:1637–1651, 2001.

Watakabe, A., Sugai, T., Nakaya, N., Wakabayashi, K., Takahashi, H., Yamamori, T., and Nawa, H. Similarity and variation in gene expression among human cerebral cortical subregions revealed by DNA macroarrays: Technical consideration of RNA expression profiling from postmortem samples. *Brain Res Mol Brain Res* 88:74–82, 2001.

Watanabe, C. M., Wolffram, S., Ader, P., Rimbach, G., Packer, L., Maguire, J. J., Schultz, P. G., and Gohil, K. The in vivo neuromodulatory effects of the herbal medicine ginkgo biloba. *Proc Natl Acad Sci U S A* 98:6577–6580, 2001.

Waters, K. M., Safe, S., and Gaido, K. W. Differential gene expression in response to methoxychlor and estradiol through ERalpha, ERbeta, and AR in reproductive tissues of female mice. *Toxicol Sci* 63:47–56, 2001.

Watson, A., Mazumder, A., Stewart, M., and Balasubramanian, S. Technology for microarray analysis of gene expression. *Curr Opin Biotechnol* 9:609–614, 1998.

Watson, M. A., Perry, A., Budhjara, V., Hicks, C., Shannon, W. D., and Rich, K. M. Gene expression profiling with oligonucleotide microarrays distinguishes World Health Organization grade of oligodendrogliomas. *Cancer Res* 61:1825–1829, 2001.

Watson, S. J., Meng, F., Thompson, R. C., and Akil, H. The "chip" as a specific genetic tool. *Biol Psychiatry* 48:1147–1156, 2000.

Watson, T. M., Reynolds, S. D., Mango, G. W., Boe, I. M., Lund, J., and Stripp, B. R. Altered lung gene expression in CCSP-null mice suggests immunoregulatory roles for Clara cells. *Am J Physiol Lung Cell Mol Physiol* 281:L1523–L1530, 2001.

Watterson, J. H., Piunno, P. A., Wust, C. C., Raha, S., and Krull, U. J. Influences of non-selective interactions of nucleic acids on response rates of nucleic acid fiber optic biosensors. *Fresenius J Anal Chem* 369(7–8):601–608, 2001.

Watts, G. S., Futscher, B. W., Isett, R., Gleason-Guzman, M., Kunkel, M. W., and Salmon, S. E. cDna microarray analysis of multidrug resistance: Doxorubicin selection produces multiple defects in apoptosis signaling pathways. *J Pharmacol Exp Ther* 299:434–441, 2001.

Webb, G. C., Akbar, M. S., Zhao, C., and Steiner, D. F. Expression profiling of pancreatic beta cells: Glucose regulation of secretory and metabolic pathway genes. *Proc Natl Acad Sci U S A* 97:5773–5778, 2000.

Wei, Y., Lee, J. M., Richmond, C., Blattner, F. R., Rafalski, J. A., and LaRossa, R. A. High-density microarray-mediated gene expression profiling of *Escherichia coli. J Bacteriol* 183:545–556, 2001.

Wei, Y., Lee, J. M., Smulski, D. R., and LaRossa, R. A. Global impact of sdiA amplification revealed by comprehensive gene expression profiling of *Escherichia coli. J Bacteriol* 183:2265–2272, 2001.

Weier, H. U., Zitzelsberger, H. F., Hsieh, H. B., Sun, M. V., Wong, M., Lersch, R. A., Yaswen, P., Smida, J., and co-workers. Monitoring signal transduction in cancer. Tyrosine kinase gene expression profiling. *J Histochem Cytochem* 49:673–674, 2001.

Weindruch, R., Kayo, T., Lee, C. K., and Prolla, T. A. Gene expression profiling of aging using DNA microarrays. *Mech Ageing Dev* 123:177–193, 2002.

Weinmann, A. S., Yan, P. S., Oberley, M. J., Huang, T. H., and Farnham, P. J. Isolating human transcription factor targets by coupling chromatin immunoprecipitation and CpG island microarray analysis. *Genes Dev* 16:235–244, 2002.

Weinstein, J. N., Scherf, U., Lee, J. K., Nishizuka, S., Gwadry, F., Bussey, A. K., Kim, S., Smith, L. H., and co-workers. The bioinformatics of microarray gene expression profiling. *Cytometry* 47:46–49, 2002.

Welford, S. M., Gregg, J., Chen, E., Garrison, D., Sorensen, P. H., Denny, C. T., and Nelson, S. F. Detection of differentially expressed genes in primary tumor tissues using representational differences analysis coupled to microarray hybridization. *Nucleic Acids Res* 26:3059–3065, 1998.

Welle, S., Brooks, A., and Thornton, C. A. Senescence-related changes in gene expression in muscle: similarities and differences between mice and men. *Physiol Genomics* 5:67–73, 2001.

Wells, D. B., Tighe, P. J., Wooldridge, K. G., Robinson, K., and Ala' Aldeen, D. A. Differential gene expression during meningeal-meningococcal interaction: evidence for self-defense and early release of cytokines and chemokines. *Infect Immun* 69:2718–2722, 2001.

Welsh, J. B., Zarrinkar, P. P., Sapinoso, L. M., Kern, S. G., Behling, C. A., Monk, B. J., Lockhart, D. J., Burger, R. A., and Hampton, G. M. Analysis of gene expression profiles in normal and neoplastic ovarian tissue samples identifies candidate molecular markers of epithelial ovarian cancer. *Proc Natl Acad Sci U S A* 98:1176–1181, 2001.

Wen, W. H., Bernstein, L., Lescallett, J., Beazer-Barclay, Y., Sullivan-Halley, J., White, M., and Press, M. F. Comparison of TP53 mutations identified by oligonucleotide microarray and conventional DNA sequence analysis. *Cancer Res* 60:2716–2722, 2000.

Wendisch, V. F., Zimmer, D. P., Khodursky, A., Peter, B., Cozzarelli, N., and Kustu, S. Isolation of *Escherichia coli* mRNA and comparison of expression using mRNA and total RNA on DNA microarrays. *Anal Biochem* 290:205–213, 2001.

Weng, S., Gu, K., Hammond, P. W., Lohse, P., Rise, C., Wagner, R. W., Wright, M. C., and Kuimelis, R. G. Generating addressable protein microarrays with PROfusiontrade mark covalent mRNA-protein fusion technology. *Proteomics* 2:48–57, 2002.

Weng, Z., and DeLisi, C. Protein therapeutics: promises and challenges for the 21st century. *Trends Biotechnol* 20:29–35, 2002.

Wessels, L. F., van Someren, E. P., and Reinders, M. J. A comparison of genetic network models. *Pac Symp Biocomput* :508–519, 2001.

Wessendorf, S., Fritz, B., Wrobel, G., Nessling, M., Lampel, S., Goettel, D., Kuepper, M., Joos, S., and co-workers. Automated screening for genomic imbalances using matrix-based comparative genomic hybridization. *Lab Invest* 82:47–60, 2002.

West, M., Blanchette, C., Dressman, H., Huang, E., Ishida, S., Spang, R., Zuzan, H., Olson, J. A. Jr., and co-workers. Predicting the clinical status of human breast cancer by using gene expression profiles. *Proc Natl Acad Sci U S A* 98:11462–11467, 2001.

Whetten, R., Sun, Y. H., Zhang, Y., and Sederoff, R. Functional genomics and cell wall biosynthesis in loblolly pine. *Plant Mol Biol* 47:275–291, 2001.

White, K. P., Rifkin, S. A., Hurban, P., and Hogness, D. S. Microarray analysis of *Drosophila* development during metamorphosis. *Science* 286:2179–2184, 1999.

Whitney, L. W., and Becker, K. G. Radioactive 33-P probes in hybridization to glass cDNA microarrays using neural tissues. *J Neurosci Methods* 106:9–13, 2001.

Whitney, L. W., Becker, K. G., Tresser, N. J., Caballero-Ramos, C. I., Munson, P. J., Prabhu, V. V., Trent, J. M., McFarland, H. F., and Biddison, W. E. Analysis of gene expression in mutiple sclerosis lesions using cDNA microarrays. *Ann Neurol* 46:425–428, 1999.

Whitney, L. W., Ludwin, S. K., McFarland, H. F., Biddison, and W. E. Microarray analysis of gene expression in multiple sclerosis and EAE identifies 5-lipoxygenase as a component of inflammatory lesions. *J Neuroimmunol* 121:40–48, 2001.

Wiechen, K., Diatchenko, L., Agoulnik, A., Scharff, K. M., Schober, H., Arlt, K., Zhumabayeva, B., Siebert, P. D., and co-workers. Caveolin-1 is down-regulated in human ovarian carcinoma and acts as a candidate tumor suppressor gene. *Am J Pathol* 159:1635–1643, 2001.

Wiese, R., Belosludtsev, Y., Powdrill, T., Thompson, P., and Hogan, M. Simultaneous multianalyte ELISA performed on a microarray platform. *Clin Chem* 47:1451–1457, 2001.

Wigle, D. A., Rossant, J., and Jurisica, I. Mining mouse microarray data. *Genome Biol* 2: 2001.

Wikman, F. P., Lu, M. L., Thykjaer, T., Olesen, S. H., Andersen, L. D., Cordon-Cardo, C., and Orntoft, T. F. Evaluation of the performance of a p53 sequencing microarray chip using 140 previously sequenced bladder tumor samples. *Clin Chem* 46:1555–1561, 2000.

Wildsmith, S. E., Archer, G. E., Winkley, A. J., Lane, P. W., and Bugelski, P. J. Maximization of signal derived from cDNA microarrays. *Biotechniques* 30:202–206, 208, 2001.

Wildsmith, S. E., and Elcock, F. J. Microarrays under the microscope. *Mol Pathol* 54:8–16, 2001.

Wilkins Stevens, P., Hall, J. G., Lyamichev, V., Neri, B. P., Lu, M., Wang, L., Smith, L. M., and Kelso, D. M. Analysis of single nucleotide polymorphisms with solid phase invasive cleavage reactions. *Nucleic Acids Res* 29:E77, 2001.

Williams, E. D., and Brooks, J. D. New molecular approaches for identifying novel targets, mechanisms, and biomarkers for prostate cancer chemopreventive agents. *Urology* 57:100–102, 2001.

Wilson, D. S., and Nock, S. Functional protein microarrays. *Curr Opin Chem Biol* 6:81–85, 2002.

Wilson, M., DeRisi, J., Kristensen, H. H., Imboden, P., Rane, S., Brown, P. O., and Schoolnik, G. K. Exploring drug-induced alterations in gene expression in *Mycobacterium tuberculosis* by microarray hybridization. *Proc Natl Acad Sci U S A* 96:12833–12838, 1999.

Wilson, S. B., and Byrne, M. C. Gene expression in NKT cells: Defining a functionally distinct CD1d-restricted T cell subset. *Curr Opin Immunol* 13:555–561, 2001.

Wilson, S. B., Kent, S. C., Horton, H. F., Hill, A. A., Bollyky, P. L., Hafler, D. A., Strominger, J. L., and Byrne, M. C. Multiple differences in gene expression in regulatory Valpha 24Jalpha Q T cells from identical twins discordant for type I diabetes. *Proc Natl Acad Sci U S A* 97:7411–7416, 2000.

Wiltshire, S., O'Malley, S., Lambert, J., Kukanskis, K., Edgar, D., Kingsmore, S. F., and Schweitzer, B. Detection of multiple allergen-specific IgEs on microarrays by immunoassay with rolling circle amplification. *Clin Chem* 46:1990–1993, 2000.

Winzeler, E. A., Schena, M., and Davis, R. W. Fluorescence-based expression monitoring using microarrays. *Methods Enzymol* 306:3–18, 1999.

Wisman, E., and Ohlrogge, J. Arabidopsis microarray service facilities. *Plant Physiol* 124:1468–1471, 2000.

Witowski, N. E., Leiendecker-Foster, C., Gerry, N. P., McGlennen, R. C., and Barany, G. Microarray-based detection of select cardiovascular disease markers. *Biotechniques* 29:936–938, 940, 942, 2000.

Wittes, J., and Friedman, H. P. Searching for evidence of altered gene expression: a comment on statistical analysis of microarray data. *J Natl Cancer Inst* 91:400–401, 1999.

Wixon, J., Blaxter, M., Hope, I., Barstead, R., and Kim, S. *Caenorhabditis elegans. Yeast* 17:37–42, 2000.

Wolf, D., Gray, C. P., and de Saizieu, A. Visualising gene expression in its metabolic context. *Brief Bioinform* 1:297–304, 2000.

Wolf, M., El-Rifai, W., Tarkkanen, M., Kononen, J., Serra, M., Eriksen, E. F., Elomaa, I., Kallioniemi, A., and co-workers. Novel findings in gene expression detected in human osteosarcoma by cDNA microarray. *Cancer Genet Cytogenet* 123:128–132, 2000.

Wolfgang, C. D., Essand, M., Lee, B., and Pastan, I. T-cell receptor gamma chain alternate reading frame protein (TARP) expression in prostate cancer cells leads to an increased growth rate and induction of caveolins and amphiregulin. *Cancer Res* 61:8122–8126, 2001.

Wolfinger, R. D., Gibson, G., Wolfinger, E. D., Bennett, L., Hamadeh, H., Bushel, P., Afshari, C., and Paules, R. S. Assessing gene significance from cDNA microarray expression data via mixed models. *J Comput Biol* 8:625–637, 2001.

Wolfsberg, T. G., Gabrielian, A. E., Campbell, M. J., Cho, R. J., Spouge, J. L., and Landsman, D. Candidate regulatory sequence elements for cell cycle-dependent transcription in Saccharomyces cerevisiae. *Genome Res* 9:775–792, 1999.

Wolter, S., Mushinski, J. F., Saboori, A. M., Resch, K., and Kracht, M. Inducible expression of a constitutively active mutant of MAP kinase kinase (MKK) 7 specifically activates JUN N-terminal protein kinase (JNK), alters expression of at least nine genes, and inhibits cell proliferation. *J Biol Chem* 2001.

Wong, K. K., Cheng, R. S., and Mok, S. C. Identification of differentially expressed genes from ovarian cancer cells by MICROMAX cDNA microarray system. *Biotechniques* 30:670–675, 2001.

Wood, W. M., Sarapura, V. D., Dowding, J. M., Woodmansee, W. W., Haakinson, D. J., Gordon, D. F., and Ridgway, E. C. Early gene expression changes preceding thyroid hormone-induced involution of a thyrotrope tumor. *Endocrinology* 143:347–359, 2002.

Woods, T. C., Blystone, C. R., Yoo, J., and Edelman, E. R. Activation of EphB2 and its ligands promote vascular smooth muscle cell proliferation. *J Biol Chem Nov* 12, 2001.

Wooster, R. Cancer classification with DNA microarrays is less more? *Trends Genet* 16:327–329, 2000.

Worley, J., Bechtol, K., Penn, S., Roach, D., Hanzel, D., Trounstine, M., and Barker, D. A Systems approach to fabricating and analyzing DNA microarrays. In: M. Schena, ed. *Microarray Biochip Technology*. Eaton Publishing, Natick, MA, 2000:65–85.

Worsham, M. J., Pals, G., Raju, U., and Wolman, S. R. Establishing a molecular continuum in breast cancer DNA microarrays and benign breast disease. *Cytometry* 47: 56–59, 2002.

Wren, B. W., Linton, D., Dorrell, N., and Karlyshev, A. V. Post genome analysis of *Campylobacter jejuni. Symp Ser Soc Appl Microbiol* 30:36S–44S, 2001.

Wu, C. G., Salvay, D. M., Forgues, M., Valerie, K., Farnsworth, J., Markin, R. S., and Wang, X. W. Distinctive gene expression profiles associated with hepatitis B virus x protein. *Oncogene* 20:3674–3682, 2001.

Wu, L., Thompson, D. K., Li, G., Hurt, R. A., Tiedje, J. M., and Zhou, J. Development and evaluation of functional gene arrays for detection of selected genes in the environment. *Appl Environ Microbiol* 67:5780–5790, 2001.

Wu, T. D. Analysing gene expression data from DNA microarrays to identify candidate genes. *J Pathol* 195:53–65, 2001.

Wurmbach, E., Yuen, T., Ebersole, B. J., and Sealfon, S. C. Gonadotropin releasing hormone receptor-coupled gene gene network organization. *J Biol Chem* 2001.

Xerri, L. Recent advance in the classification of tumors using microarrays. *Ann Pathol* 20:396–397, 2000.

Xerri, L., Dales, J. P., Devilard, E., Hassoun, J., and Birg, F. DNA-array analysis of the expression profile of apoptosis gene regulators in malignant lymphoma. *Bull Acad Natl Med* 185:963–74; 974–975 (discussion), 2001.

Xia, X., and Xie, Z. AMADA: Analysis of microarray data. *Bioinformatics* 17:569–570, 2001.

Xiao, J., Gregersen, S., Kruhoffer, M., Pedersen, S. B., Orntoft, T. F., and Hermansen, K. The effect of chronic exposure to fatty acids on gene expression in clonal insulin-producing cells: Studies using high density oligonucleotide microarray. *Endocrinology* 142:4777–4784, 2001.

Xiao, J., Xu, M., Li, J., Chang Chan, H., Lin, M., Zhu, H., Zhang, W., Zhou, Z., and co-workers. NYD-SP6, a novel gene potentially involved in regulating testicular development/spermatogenesis. *Biochem Biophys Res Commun* 291:101–110, 2002.

Xie, T., Tong, L., Barrett, T., Yuan, J., Hatzidimitriou, G., McCann, U. D., Becker, K. G., Donovan, D. M., and Ricaurte, G. A. Changes in gene expression linked to methamphetamine-induced dopaminergic neurotoxicity. *J Neurosci* 22:274–283, 2002.

Xie, W., Mertens, J. C., Reiss, D. J., Rimm, D. L., Camp, R. L., Haffty, B. G., and Reiss, M. Alterations of Smad signaling in human breast carcinoma are associated with poor outcome: A tissue microarray study. *Cancer Res* 62:497–505, 2002.

Xing, E. P., and Karp, R. M. CLIFF: Clustering of high-dimensional microarray data via iterative feature filtering using normalized cuts. *Bioinformatics* 17(Suppl 1):S306–S315, 2001.

Xiong, M., Li, W., Zhao, J., Jin, L., and Boerwinkle, E. Feature (gene) selection in gene expression-based tumor classification. *Mol Genet Metab* 73:239–247, 2001.

Xu, B., Sakkas, L.I., Slachta, C. A., Goldman, B. I., Jeevanandam, V., Oleszak, E. L., and Platsoucas, C. D. Apoptosis in chronic rejection of human cardiac allografts. *Transplantation* 71:1137–1146, 2001.

Xu, H., and Raafat el-Gewely, M. P53-responsive genes and the potential for cancer diagnostics and therapeutics development. *Biotechnol Annu Rev* 7:131–164, 2001.

Xu, J., Kalos, M., Stolk, J. A., Zasloff, E. J., Zhang, X., Houghton, R. L., Filho, A. M., Nolasco, M., and co-workers. Identification and characterization of prostein, a novel prostate-specific protein. *Cancer Res* 61:1563–1568, 2001.

Xu, J., Stolk, J. A., Zhang, X., Silva, S. J., Houghton, R. L., Matsumura, M., Vedvick, T. S., Leslie, K. B., and co-workers. Identification of differentially expressed genes in human prostate cancer using subtraction and microarray. *Cancer Res* 60: 1677–1682, 2000.

Xu, W., Bak, S., Decker, A., Paquette, S. M., Feyereisen, R., and Galbraith, D. W. Microarray-based analysis of gene expression in very large gene families: the cytochrome P450 gene superfamily of Arabidopsis thaliana. *Gene* 272(1–2):61–74, 2001.

Xu, X. R., Huang, J., Xu, Z. G., Qian, B. Z., Zhu, Z. D., Yan, Q., Cai, T., Zhang, X., and co-workers. Insight into hepatocellular carcinogenesis at transcriptome level by comparing gene expression profiles of hepatocellular carcinoma with those of corresponding noncancerous liver. *Proc Natl Acad Sci U S A* 98:15089–15094, 2001.

Xu, Y., and Olman, V. Minimum spanning trees for gene expression data clustering. *Genome Inform Ser Workshop Genome Inform* 12:24–33, 2001.

Xynos, I. D., Edgar, A. J., Buttery, L. D., Hench, L. L., and Polak, J. M. Gene-expression profiling of human osteoblasts following treatment with the ionic products of Bioglass(R) 45S5 dissolution. *J Biomed Mater Res* 55:151–157, 2001.

Xynos, I. D., Edgar, A. J., Buttery, L. D., Hench, L. L., and Polak, J. M. Ionic products of bioactive glass dissolution increase proliferation of human osteoblasts and induce insulin-like growth factor II mRNA expression and protein synthesis. *Biochem Biophys Res Commun* 276:461–465, 2000.

Yale, J., and Bohnert, H. J. Transcript expression in saccharomyces cerevisiae at high salinity. *J Biol Chem* 276:15996–6007, 2001.

Yamada, M., Yamada, M., Yamazaki, S., Takahashi, K., Nara, K., Ozawa, H., Yamada, S., Kiuchi, Y., and co-workers. Induction of cysteine string protein after chronic antidepressant treatment in rat frontal cortex. *Neurosci Lett* 301:183–186, 2001.

Yamada, M., Yamada, M., Yamazaki, S., Takahashi, K., Nishioka, G., Kudo, K., Ozawa, H., Yamada, S., and co-workers. Identification of a novel gene with RING-H2 finger motif induced after chronic antidepressant treatment in rat brain. *Biochem Biophys Res Commun* 278:150–157, 2000.

Yamada, T., Takaoka, A. S., Naishiro, Y., Hayashi, R., Maruyama, K., Maesawa, C., Ochiai, A., and Hirohashi, S. Transactivation of the multidrug resistance 1 gene by T-cell factor 4/beta-catenin complex in early colorectal carcinogenesis. *Cancer Res* 60:4761–4766, 2000.

Yamamoto, F., Yamamoto, M., Soto, J. L., Kojima, E., Wang, E. N., Perucho, M., Sekiya, T., and Yamanaka, H. Notl-Msell methylation-sensitive amplied fragment length polymorhism for DNA methylation analysis of human cancers. *Electrophoresis* 22:1946–1956, 2001.

Yamamoto, N. The practice of the genetical diagnosis and its problems. *Hinyokika Kiyo* 47:825–828, 2001.

Yamanaka, Y., Tamari, M., Nakahata, T., and Nakamura, Y. Gene expression profiles of human small airway epithelial cells treated with low doses of 14- and 16-membered macrolides. *Biochem Biophys Res Commun* 287:198–203, 2001.

Yamashita, K., Takagi, M., Uchida, K., Kondo, H., and Shigeori, T. Visualization of DNA microarrays by scanning electrochemical microscopy (SECM). *Analyst* 126:1210–1211, 2001.

Yamazaki, K., Kuromitsu, J., and Tanaka, I. Microarray analysis of gene expression changes in mouse liver induced by peroxisome proliferator—Activated receptor alpha agonists. *Biochem Biophys Res Commun* 290:1114–1122, 2002.

Yan, P. S., Chen, C. M., Shi, H., Rahmatpanah, F., Wei, S. H., Caldwell, C. W., and Huang, T. H. Dissecting complex epigenetic alterations in breast cancer using CpG island microarrays. *Cancer Res* 61:8375–8380, 2001.

Yan, Q., and Sadee, W. Human membrane transporter database: A web-accessible relational database for drug transport studies and pharmacogenomics. *AAPS PharmSci* 2:E20, 2000.

Yanagawa, R., Furukawa, Y., Tsunoda, T., Kitahara, O., Kameyama, M., Murata, K., Ishikawa, O., and Nakamura, Y. Genome-wide screening of genes showing altered expression in liver metastases of human colorectal cancers by cDNA microarray. *Neoplasia* 3:395–401, 2001.

Yang, G. P., Ross, D. T., Kuang, W. W., Brown, P. O., and Weigel, R. J. Combining SSH and cDNA microarrays for rapid identification of differentially expressed genes. *Nucleic Acids Res* 27:1517–1523, 1999.

Yang, J., Fizazi, K., Peleg, S., Sikes, C. R., Raymond, A. K., Jamal, N., Hu, M., Olive, M., and co-workers. Prostate cancer cells induce osteoblast differentiation through a Cbfa1-dependent pathway. *Cancer Res* 61:5652–5659, 2001.

Yang, J., Moravec, C. S., Sussman, M. A., DiPaola, N. R., Fu, D., Hawthorn, L., Mitchell, C. A., Young, J. B., and co-workers. Decreased SLIM1 expression and increased gelsolin expression in failing human hearts measured by high-density oligonucleotide arrays. *Circulation* 102:3046–3052, 2000.

Yang, M. C., Ruan, Q. G., Yang, J. J., Eckenrode, S., Wu, S., McIndoe, R. A., and She, J. X. A statistical method for flagging weak spots improves normalization and ratio estimates in microarrays. *Physiol Genomics* 7:45–53, 2001.

Yang, Y. H., Buckley, M. J., and Speed, T. P. Analysis of cDNA microarray images. *Brief Bioinform* 2:341–349, 2001.

Yang, Y. H., Dudoit, S., Luu, P., Lin, D. M., Peng, V., Ngai, J., and Speed, T. P. Normalization for cDNA microarray data: A robust composite method addressing single and multiple slide systematic variation. *Nucleic Acids Res* 30:e15, 2002.

Yano, K., and Miki, Y. Perspectives on postgenome medicine: Cancer. *Nippon Rinsho* 59:31–37, 2001.

Yarwood, J. M., McCormick, J. K., Paustian, M. L., Kapur, V., and Schlievert, P. M. Repression of the *Staphylococcus aureus* accessory gene regulator in serum and in vivo. *J Bacteriol* 184:1095–1101, 2002.

Yarwood, S. J., and Woodgett, J. R. Extracellular matrix composition determines the transcriptional response to epidermal growth factor receptor activation. *Proc Natl Acad Sci U S A* 98:4472–4477, 2001.

Yasui, W., Oue, N., Kuniyasu, H., Ito, R., Tahara, E., and Yokozaki, H. Molecular diagnosis of gastric cancer: present and future. *Gastric Cancer* 4:113–121, 2001.

Yazaki, J., Kishimoto, N., Nakamura, K., Fujii, F., Shimbo, K., Otsuka, Y., Wu, J., Yamamoto, K., and co-workers. Embarking on rice functional genomics via cDNA microarray: use of 3′ UTR probes for specific gene expression analysis. *DNA Res* 7:367–370, 2000.

Ye, R. W., Tao, W., Bedzyk, L., Young, T., Chen, M., and Li, L. Global gene expression profiles of *Bacillus subtilis* grown under anaerobic conditions. *J Bacteriol* 182:4458–4465, 2000.

Ye, R. W., Wang, T., Bedzyk, L., and Croker, K. M. Applications of DNA microarrays in microbial systems. *J Microbiol Methods* 47:257–272, 2001.

Yogi, O., Kawakami, T., Yamauchi, M., Ye, J. Y., and Ishikawa, M. On-demand droplet spotter for preparing pico- to femtoliter droplets on surfaces. *Anal Chem* 73:1896–1902, 2001.

Yoon, Y. R., Cha, I. J., Shon, J. H., Kim, K. A., Cha, Y. N., Jang, I. J., Park, C. W., Shin, S. G., and co-workers. Relationship of paroxetine disposition to metoprolol metabolic ratio and CYP2D6*10 genotype of Korean subjects. *Clin Pharmacol Ther* 67:567–576, 2000.

Yoshida, K., Kobayashi, K., Miwa, Y., Kang, C. M., Matsunaga, M., Yamaguchi, H., Tojo, S., Yamamoto, M., and co-workers. Combined transcriptome and proteome analysis as a powerful approach to study genes under glucose repression in *Bacillus subtilis*. *Nucleic Acids Res* 29:683–692, 2001.

Yoshihara, T., Ishigaki, S., Yamamoto, M., Liang, Y., Niwa, J., Takeuchi, H., Doyu, M., and Sobue, G. Differential expression of inflammation- and apoptosis-related genes in spinal cords of a mutant SOD1 transgenic mouse model of familial amyotrophic lateral sclerosis. *J Neurochem* 80:158–167, 2002.

Yoshikawa, T., Nagasugi, Y., Azuma, T., Kato, M., Sugano, S., Hashimoto, K., Masuho, Y., Muramatsu, M., and Seki, N. Isolation of novel mouse genes differentially expressed in brain using cDNA microarray. *Biochem Biophys Res Commun* 275:532–537, 2000.

Yoshimura, N. Retinal neuronal cell death: molecular mechanism and neuroprotection. *Nippon Ganka Gakkai Zasshi* 105:884–902, 2001.

Young, A. N., Amin, M. B., Moreno, C. S., Lim, S. D., Cohen, C., Petros, J. A., Marshall, F. F., and Neish, A. S. Expression profiling of renal epithelial neoplasms: A method for tumor classification and discovery of diagnostic molecular markers. *Am J Pathol* 158:1639–1651, 2001.

Yowe, D., Cook, W. J., and Gutierrez-Ramos, J. Microarrays for studying the host transcriptional response to microbial infection and for the identification of host drug targets. *Microbes Infect* 3:813–821, 2001.

Yu, Q., He, M., Lee, N. H., and Liu, E. T. Identification of Myc-mediated death response pathways by microarray analysis. *J Biol Chem* 2002.

Yuan, R. Q., Fan, S., Achary, M., Stewart, D. M., Goldberg, I. D., and Rosen, E. M. Altered gene expression pattern in cultured human breast cancer cells treated with hepatocyte growth factor/scatter factor in the setting of DNA damage. *Cancer Res* 61:8022–8031, 2001.

Yue, H., Eastman, P. S., Wang, B. B., Minor, J., Doctolero, M. H., Nuttall, R. L., Stack, R., Becker, J. W., and co-workers. An evaluation of the performance of cDNA microarrays for detecting changes in global mRNA expression. *Nucleic Acids Res.* 29:E41–1, 2001.

Yun, C. W., Ferea, T., Rashford, J., Ardon, O., Brown, P. O., Botstein, D., Kaplan, J., and Philpott, C. C. Desferrioxamine-mediated iron uptake in *Saccharomyces cerevisiae*. Evidence for two pathways of iron uptake. *J Biol Chem* 275:10709–10715, 2000.

Zammatteo, N., Jeanmart, L., Hamels, S., Courtois, S., Louette, P., Hevesi, L., and Remacle, J. Comparison between different strategies of covalent attachment of DNA to glass surfaces to build DNA microarrays. *Anal Biochem* 280:143–150, 2000.

Zammatteo, N., Lockman, L., Brasseur, F., De Plaen, E., Lurquin, C., Lobert, P. E., Hamels, S., Boon, T., and Remacle, J. DNA microarray to monitor the expression of MAGE-A genes. *Clin Chem* 48:25–34, 2002.

Zanders, E. D. Gene expression analysis as an aid to the identification of drug targets. *Pharmacogenomics* 1:375–384, 2000.

Zarrinkar, P. P., Mainquist, J. K., Zamora, M., Stern, D., Welsh, J. B., Sapinoso, L. M., Hampton, G. M., and Lockhart, D. J. Arrays of arrays for high-throughput gene expression profiling. *Genome Res* 11:1256–1261, 2001.

Zeller, K. I., Haggerty, T., Barrett, J. F., Guo, Q., Wonsey, D. R., and Dang, C. V. Characterization of nucleophosmin (B23) as a Myc target by scanning chromatin immunoprecipitation (SChIP). *J Biol Chem* 2001.

Zembutsu, H., Ohnishi, Y., Tsunoda, T., Furukawa, Y., Katagiri, T., Ueyama, Y., Tamaoki, N., Nomura, T., and co-workers. Genome-wide cDNA microarray screening to correlate gene expression profiles with sensitivity of 85 human cancer xenografts to anticancer drugs. *Cancer Res* 62:518–527, 2002.

Zeng, G., Gao, L., Suetake, K., Joshi, R. M., and Yu, R. K. Variations in gene expression patterns correlated with phenotype of F-11 tumor cells whose expression of GD3-synthase is suppressed. *Cancer Lett* 178:91–98, 2002.

Zhang, H., Yu, C. Y., Singer, B., and Xiong, M. Recursive partitioning for tumor classification with gene expression microarray data. *Proc Natl Acad Sci U S A* 98:6730–6735, 2001.

Zhang, K., and Zhao, H. Assessing reliability of gene clusters from gene expression data. *Funct Integr Genomics* 1:156–173, 2000.

Zhang, M. Q. Large-scale gene expression data analysis: a new challenge to computational biologists. *Genome Res* 9:681–688, 1999.

Zhang, M. Q. Promoter analysis of co-regulated genes in the yeast genome. *Comput Chem* 23:233–250, 1999.

Zhang, P., Briones, N., Liu, C. G., Brush, C. K., Powdrill, T., Belosludtsev, Y., and Hogan, M. Acceleration of nucleic acid hybridization on DNA microarrays driven by pH tunable modifications. *Nucleosides Nucleotides Nucleic Acids* 20:1251–1254, 2001.

Zhang, W., Bardwell, P. D., Woo, C. J., Poltoratsky, V., Scharff, M. D., and Martin, A. Clonal instability of V region hypermutation in the Ramos Burkitt's lymphoma cell line. *Int Immunol* 13:1175–1184, 2001.

Zhang, W., Wang, H., Song, S. W., and Fuller, G. N. Insulin-like growth factor binding protein 2: Gene expression microarrays and the hypothesis-generation paradigm. *Brain Pathol* 12:87–94, 2002.

Zhang, Y., Luxon, B. A., Casola, A., Garofalo, R. P., Jamaluddin, M., and Brasier, A. R. Expression of respiratory syncytial virus-induced chemokine gene networks in lower airway epithelial cells revealed by cDNA microarrays. *J Virol* 75:9044–9058, 2001.

Zhang, Y., Price, B. D., Tetradis, S., Chakrabarti, S., Maulik, G., and Makrigiorgos, G. M. Reproducible and inexpensive probe preparation for oligonucleotide arrays. *Nucleic Acids Res* 29:E66–6, 2001.

Zhao, L. P., Aragaki, C., Hsu, L., and Quiaoit, F. Mapping of complex traits by single-nucleotide polymorphisms. *Am J Hum Genet* 63:225–240, 1998.

Zhao, L. P., Prentice, R., and Breeden, L. Statistical modeling of large microarray data sets to identify stimulus-response profiles. *Proc Natl Acad Sci U S A* 98:5631–5636, 2001.

Zhao, R., Gish, K., Murphy, M., Yin, Y., Notterman, D., Hoffman, W. H., Tom, E., Mack, D. H., and Levine, A. J. Analysis of p53-regulated gene expression patterns using oligonucleotide arrays. *Genes Dev* 14:981–993, 2000.

Zhao, X., Demary, K., Wong, L., Vaziri, C., McKenzie, A. B., Eberlein, T. J., and Spanjaard, R. A. Retinoic acid receptor-independent mechanism of apoptosis of melanoma cells by the retinoid CD437 (AHPN). *Cell Death Differ* 8:878–886, 2001.

Zhao, X., Lein, E. S., He, A., Smith, S. C., Aston, C., and Gage, F. H. Transcriptional profiling reveals strict boundaries between hippocampal subregions. *J Comp Neurol* 441:187–196, 2001.

Zhao, Y., Gran, B., Pinilla, C., Markovic-Plese, S., Hemmer, B., Tzou, A., Whitney, L. W., and Biddison, W. E., and co-workers. Combinatorial peptide libraries and biometric score matrices permit the quantitative analysis of specific and degenerate interactions between clonotypic TCR and MHC peptide ligands. *J Immunol* 167:2130–2141, 2001.

Zheng, M., Wang, X., Templeton, L. J., Smulski, D. R., LaRossa, R. A., and Storz, G. DNA microarray-mediated transcriptional profiling of the *Escherichia coli* response to hydrogen peroxide. *J Bacteriol* 183:4562–4570, 2001.

Zhou, F. C., Duguid, J. R., Edenberg, H. J., McClintick, J., Young, P., and Nelson, P. DNA microarray analysis of differential gene expression of 6-year-old rat neural striatal progenitor cells during early differentiation. *Restor Neurol Neurosci* 18:95–104, 2001.

Zhou, X., Tan, F. K., Xiong, M., Milewicz, D. M., Feghali, C. A., Fritzler, M. J., Reveille, J. D., and Arnett, F. C. Systemic sclerosis (scleroderma): specific autoantigen genes are selectively overexpressed in scleroderma fibroblasts. *J Immunol* 167:7126–7133, 2001.

Zhou, Y., and Abagyan, R. Match-only integral distribution (MOID) algorithm for high-density oligonucleotide array analysis. *BMC Bioinformatics* 3:3, 2002.

Zhou, Y-X, Kalocsai, P., Chen, J-Y, and Shams, S. Information processing issues and solutions associated with microarray technology. In: M. Schena, ed. *Microarray Biochip Technology*. Eaton Publishing, Natick, MA, 2000:167–200.

Zhu, H., Bilgin, M., Bangham, R., Hall, D., Casamayor, A., Bertone, P., Lan, N., and Jansen, R., and co-workers. Global analysis of protein activities using proteome chips. *Science* 293:2101–2105, 2001.

Zhu, H., and Snyder, M. Protein arrays and microarrays. *Curr Opin Chem Biol* 5:40–45, 2001.

Ziauddin, J., and Sabatini, D. M. Microarrays of cells expressing defined cDNAs. *Nature* 411:107–110, 2001.

Zien, A., Aigner, T., Zimmer, R., and Lengauer, T. Centralization: a new method for the normalization of gene expression data. *Bioinformatics* 17(Suppl 1):S323–S331, 2001.

Zimmer, D. P., Soupene, E., Lee, H. L., Wendisch, V. F., Khodursky, A. B., Peter, B. J., Bender, R. A., and Kustu, S. Nitrogen regulatory protein C-controlled genes of *Escherichia coli:* Scavenging as a defense against nitrogen limitation. *Proc Natl Acad Sci U S A* 97:14674–14679, 2000.

Zimmermann, J., Erdmann, D., Lalande, I., Grossenbacher, R., Noorani, M., and Furst, P. Proteasome inhibitor induced gene expression profiles reveal overexpression of transcriptional regulators ATF3, GADD153 and MAD1. *Oncogene* 19:2913–2920, 2000.

Zirlinger, M., Kreiman, G., and Anderson, D. J. Amygdala-enriched genes identified by microarray technology are restricted to specific amygdaloid subnuclei. *Proc Natl Acad Sci U S A* 98:5270–5275, 2001.

Zlatanova, J., and Mirzabekov, A. Gel-immobilized microarrays of nucleic acids and proteins. Production and application for macromolecular research. *Methods Mol Biol* 170:17–38, 2001.

Zou, Z. L., Wang, S. Q., and Wang, Z. Q. Preparation optimization and properties of the aldehyde microscopic slides for oligonucleotide microarray fabrication. *Sheng Wu Gong Cheng Xue Bao* 17:498–502, 2001.

Zubritsky, E. Spotting a microarray system. *Anal Chem* 72:761A–767A, 2000.

Zweiger, G. Knowledge discovery in gene-expression-microarray data: Mining the information output of the genome. *Trends Biotechnol* 17:429–436, 1999.

Glossary

abiotic stress Category of environmental stress in plants attributable to nonliving sources.

absolute quantitation Computational process of obtaining numerical values for each microarray spot based on the signal intensity at each location.

absorption Process by which light is captured by a molecule.

absorption maximum Wavelength of light that is captured most efficiently by a fluorescent molecule.

accessible Evaluative criterion for a microarray substrate that refers to the efficacy by which the surface enables productive interactions between target and probe molecules.

accession number Unique identification number attached to each sequence submitted to GenBank.

accuracy Maximal observed discrepancy between the desired and observed movement of an electric motor or microarray robot, expressed typically in microns.

acetylcholine Neurotransmitter that communicates chemical messages in the brain and peripheral nervous system.

acid A functional group or molecule that donates a proton in a chemical reaction.

activation An increase in the rate of transcription of a cellular gene.

activation energy Barrier in a reaction profile that must be overcome for a chemical or biochemical reaction to occur.

activator Cellular protein that binds to a regulatory element and increases the rate of transcription of a gene.

active microelectronic array Biosensor technology that combines microelectronics and molecular biology to create active analytical devices for genetic analysis.

Acute renal failure (ARF) Medical condition characterized by a sudden decline in kidney function.

adenine (A) One of the four nitrogen-containing bases of DNA and RNA.

adhesion Attractive force exerted by a solid surface such as a printing pin on a liquid sample, contributing to meniscus formation and other phenomena.

adsorption Nonspecific attachment of a molecule to a surface.

affordability Criterion of microarray manufacture that pertains to cost of a given technology.

algorithm Any sequence of computational steps performed by computer.

aliphatic A functional group or molecule containing a straight chemical chain.

aliphatic amine Chemical compound in which the amino group is attached to the end of a straight-chain hydrocarbon.

alkaloid Any member of a family of small molecules including cocaine and heroin that exert their psychoactive effects by altering biochemical functions in the brain.

allele Any sequence variant of a gene.

alternating current (AC) Electricity that flows in a bidirectional manner and at a defined frequency.

amino acid Any of a family of the 20 common biochemical building blocks that make up cellular proteins.

amino linker Chemical group consisting of a primary amine that is used to attach DNA to an amine-reactive microarray substrate.

aminoacyl-tRNA Cellular molecule that binds a specific amino acid and facilitates protein synthesis by codon recognition.

aminoallyl Aliphatic primary amine containing a double bond that allows direct labeling of microarray probe molecules by nucleophilic attack of fluorescent reagents by the modified bases.

amorphous solid A substance such as glass having a defined shape and volume but lacking the regular pattern of molecules found in true solids.

amplified primer extension (APEX) Genotyping method by which templates are hybridized with primers one nucleotide upstream of a mutation of interest. The primed templates are extended with polymerase and fluorescent dideoxynucleotides to determine the sequence.

analog Electrical signal from the anode of a photomultiplier tube or other source that consists of a continuously varying flow of electrons.

analog-to-digital converter (A/D) Solid-state device that converts a continuous flow of electrons into binary signals that can be read and manipulated with a computer.

angle of incidence (θ_1) Arc formed between a mirror and the incoming light beam in a microarray detection instrument.

angle of reflection (θ_2) Arc formed between a mirror and the outgoing light beam in a microarray detection instrument.

anode Component in a photomultiplier tube that collects the amplified electron stream from the dynodes.

anorexia nervosa Serious eating disorder, mostly affecting adolescent and young adult females, characterized by an emaciated appearance, excessively low body weight, and malnutrition caused by intentional starvation.

antiblooming gate Charge-coupled device chip design feature that increases the spacing between detector pixels to reduce pixel-to-pixel charge transfer and blooming.

anticodon loop The portion of a transfer RNA molecule that interacts with cognate codons in the messenger RNA.

antigen Any non-self-protein or other biomolecule that illicits an immune response.

antiparallel The configuration of the two strands in double-stranded DNA, wherein one strand runs in the 5′ to 3′ direction and the other strand runs in the 3′ to 5′ direction.

antipsychotic Any of a family of medications used to treat schizophrenia, including fluphenazine, chlorprothixene, chlorpromazine, molindone, loxapine, haloperidol, olanzapine, and risperidone.

aqueous A buffer that is composed mainly of H_2O. Alternatively, a chemical reaction that occurs in H_2O.

arc Discharge of current produced by applying a large voltage across two electrodes.

architecture Microarray detection system criterion that refers to the overall layout of the filters, lasers, light source, detectors, optics, stage, and other hardware components.

area detector Microarray light-sensing implement, such as a charge-coupled device, that captures data from a large area of a microarray substrate in a single detection step.

aromatic A functional group or molecule containing a ring structure and conjugated double bonds.

artificial intelligence Capacity of a computer to think and learn in a manner analogous to the human mind.

assignee Term in patent law that refers to the holder of a patent, also known as the licensor.

asynchronous Aspect of e-mail that provides tremendous flexibility by allowing senders and recipients to compose, read, and respond to e-mail messages without direct communication between the sender and recipient.

at rest One of three main categories of microarray cleanroom status corresponding to the period in which a cleanroom contains all of the necessary equipment and accessories required for operation but is not actively engaged in a process and does not contain any cleanroom staff.

atom The indivisible building block of all molecules, including solids, liquids, and gases.

atomic force microscopy (AFM) Analytical technique used to generate high-resolution, three-dimensional plots of microarray surfaces.

atomic mass The weight of an atom given in atomic mass units.

atomic mass unit One-twelfth (1/12) the mass of a carbon atom.

atomic number Corresponds to the number of protons in the nucleus of the atom and provides the basis by which the 118 known elements are arranged in the periodic table.

attenuate Process of reducing the intensity of a laser beam.

automated quantitation Computational process in which a large number of microarray features are quantified simultaneously without manual intervention.

Avagadro's number The constant 6.022×10^{23}, which is equivalent to the number of atoms in 12 grams of carbon.

avalanche photodiode Semiconductor-based light-sensing device used in microarray detection instruments.

background Unwanted contribution to detection instrument readings arising from nonspecific interactions between probe molecules and the microarray surface.

background subtraction Computational process in which the fluorescent counts corresponding to the noise are subtracted from true signals to obtain more accurate quantitation.

backlash Magnitude of the input that produces no movement of an electric motor or microarray robot when the actuator is moved in the opposite direction along a given axis, expressed typically in microns.

bacterial artificial chromosome (BAC) Recombinant plasmid that replicates autonomously in a prokaryotic host and contains 50–250 kilobase segments of inserted DNA.

bacterium Single-celled prokaryotic organism, including *Esherichia coli,* that can be propogated quickly in the laboratory, contains no cellular nucleus, and has a relatively simple circular genome.

balance Propensity of two or more channels in a microarray detection instrument to produce images with matched signal intensities.

balanced Refers to a chemical equation written so that the same number of atoms occur on both sides of the equation.

ball screw Linear actuator that uses ball bearings and a corkscrew-style track to convert circular motion to linear motion.

ball spline Linear actuator component that uses a bearing and a stainless-steel shaft with uniformly spaced ridges as the guide mechanism.

band gap Energy separating the valence band and the conduction band in a conductor, semiconductor, or insulator.

BankIt Graphical form used to submit sequences to GenBank via the World Wide Web.

basal expression Low level of gene transcription that occurs in the absence of activation.

base A functional group or molecule that acquires a proton in a chemical reaction. Alternatively, any one of five nitrogenous molecules (adenine, cytosine, guanine, thymine, and uracil) contained in the nucleotides that make up cellular DNA and RNA.

base pair Unit of measure of double-stranded DNA making up one set of complementary nucleotides.

beam splitter Optical filter in a microarray detection instrument that divides excitation or emission light into reflected and transmitted components.

Beer's law Mathematical equation that represents light absorption as the product of the molar extinction coefficient times the path length times the concentration of the fluorescent molecule: $A = \varepsilon \times b \times C$.

belt drive Linear actuator that uses flat rubber or plastic cords and pulleys to convert circular motion to linear motion.

binary code Nomenclature system used in computers that represents all specified values in terms of 0 or 1.

biochemistry The field of study that endeavors to understand the chemical basis of life by focusing on the study of DNA, RNA, proteins, and other biomolecules.

bioinformatics Specialized field of computer science focused on the analysis of biological data.

biological One of five main forms of cleanroom contamination corresponding to human hand oils and fingerprints, hair, saliva, tears, oral condensation, perspiration, sloughed skin cells, and the like.

biological potential Full gamut of biological activities that are possible for a system variable, system module, system, culture, or biosphere in the methodological framework.

biopolymer Biological molecules, including DNA, RNA, protein, and carbohydrate, characterized by many small chemical building blocks linked together into a highly repeated structure.

biosphere Methodological term corresponding to the complete collection of all systems and cultures on the planet.

bit Shortened form of binary digit that refers to the number of values assigned to the full spectrum of analog values.

bit map Digital file corresponding to a microarray surface.

blocking agent Chemical or biochemical agent such as borohydride or bovine serum albumin used to inactivate reactive groups on a microarray substrate to prevent nonspecific reactivity.

blooming Unwanted phenomenon produced by a charge-coupled device in which charge from a completely filled pixel transfers to an adjacent, partially filled pixel to produce a blurry image.

Boolean Computational process in which logical operators (e.g., AND, OR) are employed in microarray spot analysis or some other search function.

bulimia nervosa Serious eating disorder that mostly afflicts adolescent and young adult females in which sufferers consume large quantities of food and then rid themselves of the extra calories by vomiting, the use of laxatives, and other purging methods.

byte Amount of memory on a personal computer required to store eight bits of information.

capacitor Charge-coupled device pixel or other device that stores electrical charge in a temporary manner.

capillary action The spontaneous loading of a liquid into an elongated tube or channel.

capillary electrophoresis Method of chromatography used in DNA sequencing in which the DNA molecules are separated by movement through thin glass tubes.

capillary tube Enclosed, circular microarray printing implement that loads a sample by capillary action and delivers the sample onto the microarray substrate by direct contact.

capped Inactivated functional group prevented from undergoing chemical reactivity by the addition of a blocking group.

capping Enzymatic processing of messenger RNA whereby a modified G nucleotide is added to the 5′ end of the transcript. Alternatively, the step in the chemical synthesis of oligonucleotides and other synthetic molecules in which an unreacted group such as the 5′ hydroxyl is blocked by acetylation or some other means to prevent incorrect additions to the growing polymer chain.
carbohydrate Any of a large family of cellular biomolecules composed of repeating sugar monomers.
carcinogen Chemical, biological, radiological or environmental agent that causes cancer.
carrier An individual with a normal phenotype that is heterozygous for a disease gene and capable of passing on the disease gene to its progeny.
Cartesian coordinate Mapping system used for most microarray printing robots and scanners consisting of *x*, *y*, and *z* values, each of which are perpendicular to each other.
Cartesian image Microarray data file that can be described in terms of *x* and *y* values.
catalyst Enzyme, chemical, or other agent that increases the rate of a biochemical or chemical reaction but does not change the energy of the reactants and products.
cell cycle Biochemical process that divides mitosis into four discrete phases known as presynthetic gap, synthesis, postsynthetic gap, and mitosis.
cellulose Plant carbohydrate that is the main structural component of the plant cell wall.
center-to-center spacing Physical distance between the centers of two printed microarray spots, expressed typically in microns.
change of state A chemical transition characterized by the conversion of one physical state to another.
chemical One of five main forms of cleanroom contamination, corresponding to the vapors emitted from paints, glues, adhesives, packaging materials, organic solvents, and other substances that emit volatile chemicals.
chemical bond An attractive force between two atoms.
chemical equation A shorthand notation used by chemists and biochemists to describe a chemical reaction.
chemical formula A shorthand notation that describes the type and number of atoms present in a molecule.
chemical polarity The unequal sharing of electrons in a chemical bond.
chemical reaction The process whereby chemical bonds are formed and broken.
chromosome Large segment of genomic DNA that replicates autonomously in the cell and segregates during cell division.
claim Term in patent law corresponding to a short paragraph that defines the scope of an invention.
claim construction Term in patent law that refers in the precise language used in the claims at the end of a patent application.
class Cleanroom evaluative criterion corresponding to the number of particles $\geq$ 0.5 μm per cubic foot of cleanroom air.
clinical trial Medical study conducted in human patients to assess the safety and efficacy of new drugs, vaccines, and other types of therapies.
cloning Generation of a fertile, adult animal by nuclear transfer of somatic DNA; alternatively, recombinant DNA procedures.
cluster Small group of adjacent pixels that are partially or fully defective in a charged-coupled device.
cluster analysis Multivariate classification method that uses dendrograms as a general means of categorizing data based on the similarity of the datum points to one another.
clustering Process of performing cluster analysis.
codon Any one of 64 three-nucleotide sequences or triplets in messenger RNA that specify one of the 20 amino acids used for protein synthesis.

co-factor Environmental, genetic, or infectious contributor that raises the incidence of a specific disease.

coherent Property of laser light in which all of the photons in the beam share the same wavelength.

co-linear Any case in which two molecules share genetic information along the primary sequence, such as the codons in messenger RNA that specify a cognate series of amino acids.

collimated Property of laser light in which all of the photons in the beam are parallel to one another.

color palette A look-up table used to index a graphical microarray image, producing red, green, and rainbow representations of microarray data.

color rendition index (CRI) Measure of how closely a given light source approximates the wavelengths of light emitted by the sun, expressed as a percent.

color temperature Correlation between the thermal state and the color of light emitted by a hot substance, typically expressed in degrees Kelvin.

color temperature range Kelvin scale for color emission derived from the correlation between temperature and the color of light emitted by a hot substance.

combinatorial method Chemical synthesis approach that allows the generation of enormous chemical diversity by stepwise coupling of synthetic building blocks in a spatially addressable manner.

common primers Two oligonucleotides used for polymerase chain reaction amplification that bind to vector sequences bordering each target sequence and allow amplification of all of the targets using a single oligonucleotide pair.

complementary DNA (cDNA) DNA version of messenger RNA.

complementary metal oxide semiconductor Relatively new type of microarray detector that has an opposite polarity charge relative to the metal oxide semiconductor switches used in charge-coupled device cameras.

complementary strand A DNA strand that has the opposite nucleotide sequence and chemical polarity of a second DNA strand.

complexity Amount of genomic information on a microarray, calculated as the product of the number of unique sequences times their average length in nucleotides.

compress Act of reducing the size of a computer file.

computational biology Specialized field of computer science focused on the analysis of biological data.

computer assisted design (CAD) Sophisticated electronic approach to generating two- and three-dimensional representations of objects used in micromachining and other manufacturing approaches.

computer numerical control (CNC) Sophisticated machining approach that uses computer drawings, stepper motors, and other technologies to prepare high-precision parts and implements used in microarray analysis.

concordance Agreement or coincidence of data or datasets between a group of patients.

conduction band Location of electrons that participate in electrical flow in a conductor, semiconductor, or insulator.

conductor Material that supports the flow of electrons.

conjugated bond Double bond that occurs on every other carbon atom containing π electrons; often associated with fluorescence.

conjugated system A series of alternating single and double bonds that contains electrons residing above and below the plane of the molecule.

contact angle Tangent formed between a printed droplet and a microarray substrate, measured typically 1 s after the droplet is deposited onto the surface.

contact distance The atomic separation between two atoms that maximizes the strength of a van der Waals bond.

contact printing Microarray manufacturing technology that involves direct interaction between the printing implement or the sample and the microarray substrate.

contamination Any of five different forms of impurities including particle, biological, chemical, thermal, and electrostatic that compromise the integrity of a microarray cleanroom.

content Criterion of microarray manufacture that pertains to the amount of biological or chemical information on a microarray.

content map Electronic file that contains all of the information pertinent to each microarray spot or feature.

continuous spectrum Light source produced by an arc lamp or a related device characterized by the presence of many different wavelengths owing to the fact that the excited electrons occupy poorly defined orbits around the atom.

contract Term in patent law that refers to a document signed by two parties in which both parties agree that the laws of the land will be used to enforce the terms of the agreement.

controlled pore glass (CPG) Porous silicon dioxide matrix used in the manufacture of synthetic oligonucleotides.

converge Optical process of focusing a beam of light into a point source, such as is achieved with a lens.

coordinate regulation Process by which the expression of multiple cellular genes is controlled by a common mechanism.

correlation coefficient Mathematical quantity that corresponds to the dot product of two normalized vectors.

corrosion The oxidation of iron and other metals, producing rust and other oxidative surface changes.

coupling Step in the chemical synthesis of oligonucleotides and other synthetic molecules in which a monomer is added to the growing polymer chain.

covalent bond An attractive chemical force characterized by the sharing of electrons between two atoms.

covalent coupling Attachment scheme that involves electron sharing between target molecules and the microarray substrate.

coverage Term from DNA sequencing that refers to the number of times a given genomic segment or genome is represented in the sequencing data, represented typically as fold coverage.

cross-hybridization Undesired hybridization between target molecules and probe molecules that share sequence similarity.

cross-talk Microarray detection system criterion that refers to the unwanted phenomenon in which signal from one channel is detected in another channel, measured typically as a percentage.

culture Methodological term corresponding to a group of systems or organisms in the methodology.

current (I) Flow of electrons per unit time.

cytoplasm Portion of a cell located outside the nucleus.

cytosine (C) One of the four nitrogen-containing bases of DNA and RNA.

dark count Numerical value in a microarray detector produced by dark current.

dark current Instrument noise that originates in the absence of light, deriving mainly from thermal emissions and from current leaks, and measured typically in electrons per pixel per second at a given temperature.

dark pixel Defect in a charge-coupled device detector corresponding to a picture element that produces too little signal relative to a normally functioning picture element.

dead pixel Defect in a charge-coupled device detector corresponding to a picture element that produces no signal.

degenerate Term used to describe the fact that many of the 20 common cellular amino acids are encoded by more than one codon.

dehydration reaction A chemical reaction involving the loss of water.

deletion Mutation that results in the removal of one or more nucleotides from a DNA sequence.

delimited file File format used in microarray quantitation in which each field or parameter is separated with a recognizable keystroke.

delivery Method of microarray manufacture in which the target molecules are made offline and delivered to the microarray surface by contact printing or some other means.

delocalized The quantum mechanical state of electrons characterized by a freedom to move within and maintain indefinite positions within a conjugated electron system.

denaturation Process of denaturing.

denature Process of converting DNA into single-stranded form or, more generally, any process that reduces the structure of a molecule.

dendrimer Large, highly fluorescent, branchlike microarray labeling reagent built by the sequential hybridization and cross-linking of fluorescent oligonucleotides.

dendrogram Graphical representation of a set of relationships based on the closeness of data or datasets to one another.

density Criterion of microarray manufacture equal to the number of elements or features per unit area of microarray substrate, expressed typically as spots per square centimeter.

deoxyribonucleic acid (DNA) The biopolymeric molecule that constitutes the genetic blueprint of virtually every organism in the biosphere.

deoxyribose The five-carbon DNA sugar that contains a hydrogen atom at the 2′ position instead of the hydroxyl group, which is found in the ribose RNA sugar.

deprotection Step in oligonucleotide synthesis in which a protecting group such as dimethoxytrityl is removed from the 5′ hydroxyl group to allow a subsequent round of coupling.

depth of focus Distance between the closest and farthest images that remain in focus during microarray detection.

depurination Chemical cleavage and loss of the purine bases adenine and guanine that occurs by treatment of DNA with acids and other low pH agents.

design One of the three formal types of patents issued by the U.S. Patent and Trademark Office.

detectivity Microarray detection system criterion that refers to the dimmest signal that can be detected over the total noise of the system, typically expressed in fluors per square micron.

detector Photomultiplier tube, charge-coupled device, or other component of a microarray detection instrument that converts photons into electrical current.

detector lens Optical component in a microarray detection instrument that focuses the emitted light onto the photomultiplier tube, charge-coupled device, or other light-sensing device.

detritylation Step in oligonucleotide synthesis in which the dimethytrityl group is removed from the 5′ hydroxyl group to allow a subsequent round of coupling.

diagnostics Branch of medicine devoted to identifying or determining the nature and cause of disease by combining patient information and laboratory testing data.

dideoxy sequencing Method of DNA sequencing that relies on chain termination by modified nucleotides that lack 2′ and 3′ hydroxyl groups.

dielectric coating Special multiplayer material containing aluminum and other materials, used in the preparation of mirrors and optical components in microarray detection instruments.

digital Electrical signal or other data source represented as a series of binary (0 or 1) values that can be read with a computer.

digital subscriber line (DSL) New technology that provides accelerated access and data transfer through the Internet.

dimensional Evaluative criterion for a microarray substrate that refers to the physical size of the substrate.

dimensionality reduction Computational process used in principle component analysis and other data-modeling procedures in which complex datasets are compressed into x, y, and z coordinates to allow their representation on a conventional computer screen.

dinucleotide A molecule consisting of two nucleotides joined in succession.

diode laser Compact laser used in many microarray scanners that contains semiconductor materials as the gain medium.

diode-pumped solid-state laser (DPSS laser) Type of laser that uses the beam of a diode laser to excite the semiconductor gain medium in the second laser.

diploid Cell that possesses two copies of each chromosome.

dipole A transient or permanent asymmetry in electron density producing a positive or negative bias around an atom.

direct current (DC) Electricity that flows in a unidirectional manner and at a constant rate.

direct labeling Probe preparation scheme in which the fluorescent tags are attached in a covalent manner directly to the probe molecule using an enzymatic or chemical means.

disclose Process of revealing an invention by written or oral communication.

disaccharide Any of a family of sugar molecules containing two monomer building blocks.

dissociation energy The amount of heat required to break the chemical bonds in 1 mole of a substance, expressed typically in kilocalories per mole.

disulfide bond Covalent chemical interaction between two sulfur atoms.

diverge To accumulate changes in gene and protein sequences over evolutionary time.

DNA ligase Enzyme used in gene splicing that creates phosphodiester bonds between free ends of double-stranded DNA.

DNA polymerase Cellular enzyme that synthesizes an exact DNA copy from a single-stranded DNA template.

DNA sequencing Experimental process of determining the primary nucleotide sequence of a DNA molecule.

DNA synthesizer An automated machine used to manufacture synthetic oligonucleotides.

DNA typing Experimental approach in which the identity of a person is established by examining DNA polymorphisms.

domain name Intuitive Internet identifier (e.g., arrayit.com) that corresponds to the type of institution or business that holds the domain.

domain name system (DNS) Internet function that interconverts Internet protocol addresses and domain names.

dominant An allele that manifests a phenotype in the presence of a normal copy of the gene.

doping Process of adding trace quantities of atoms that contain a lower or higher number of valence electrons to a pure substance, such as silicon, to increase or decrease its electrical conductivity.

double helix The two-stranded configuration of native DNA, wherein the complementary strands are interwoven around a center axis like a spiral staircase.

double-stranded The form of naturally occurring DNA consisting of two complementary strands held together by hydrogen bonding.

due diligence Scrutinizing process of a company by a venture capitalist in anticipation of making a venture capital investment.

durability Cleanroom evaluative criterion that refers to the extent to which the physical structure, environmental control, and other components of a microarray cleanroom maintain their specifications over time. Alternatively, the total length of time that an electric motor or microarray robot can be used without visible deterioration in performance, expressed typically in machine hours.

durable Evaluative criterion for a microarray substrate that refers to the stability of the chemical groups, treatment, or coating on the surface.

dwell time Duration of a pause in microarray printing, detection, or some other process, measured typically in milliseconds.

dynamic programming Mathematical strategy that tackles the alignment problem backwards to find the best alignment by making decisions with a series of subalignments scored one after another.

dynamic range Microarray detection system criterion that refers to the full gamet of signal intensities that can be discerned with a given detection instrument, typically 100,000-fold or more.

dynode Charged electrode in a photomultiplier tube that amplifies the flow of electrons produced by the photocathode.

early stage Term used in venture-capital financing that refers to a business that is developing one or more products but has yet to ship the products or generate sales.

ease of implementation Criterion of microarray manufacture that pertains to the simplicity by which a given technology can be used in a laboratory setting.

efficiency Light-capturing ability of a lens in a microarray detection instrument.

efficient Evaluative criterion for a microarray substrate that refers to the efficacy by which the surface attaches printed target molecules.

electrical discharge machining (EDM) Manufacturing procedure that uses a solid electric conductor to cut metal by generating sparks between the two surfaces.

electron A subatomic particle that resides in an orbit around the nucleus of an atom, and carries a charge of -1 and a mass of approximately 0.0005 atomic mass units.

electron affinity The propensity of an atom to acquire or lose electrons.

electronegative The chemical property of an atom or functional group, such as a nucleophile, characterized by a propensity for negative charge.

electronic mail (e-mail) Messages containing text, graphics, images, and other forms of data sent back and forth between computers.

electronic noise Unwanted contribution to detection instrument readings arising from the nondetector electrical components of the detection system notably the amplifiers, circuitry and analog-to-digital converter.

electronic stringency Attribute of an active microelectronic array that allows the efficiency of a hybridization reaction to be controlled by adjusting the bias at each microelectrode.

electrophile A molecule or functional group that has strongly positive charge bias and an affinity for a nucleophile.

electropositive The chemical property of an atom or functional group, such as an electrophile, characterized by a propensity for positive charge.

electrostatic One of five main forms of cleanroom contamination corresponding to any source, including plastics, rubber, and cellophane, that produces an imbalance of negative and positive charges inside a cleanroom environment.

electrostatic bond A noncovalent chemical interaction formed by the association of two molecules of opposite charge.

electrostatic interaction Noncovalent attractive force between positive and negative charges such as an amine surface and the phosphate groups on DNA.

electrostatic repulsion Repellant force that occurs when molecules with the same electrostatic charge are brought into close proximity.

element Any of a set of the 118 fundamental buildings block of all matter. Alternatively, a printed spot or feature in a microarray.

embryonic stem cell Any cell from an embryo not confined to a particular developmental pathway that can give rise to many types of differentiated cells.

emission maximum Wavelength of light that is emitted most efficiently by a fluorescent molecule.

emitter Optical filter in a microarray detection instrument that allows passage of the fluorescence emission.

endorphin Endogenous morphine-like signaling molecule that functions by binding to cell surface receptors.

enhancer Regulatory element that alters promoter efficiency by increasing or decreasing the rate of transcription.

Entrez Search and retrieval system used to query the contents of GenBank via a standard World Wide Web browser.

environmental control Cleanroom evaluative criterion that includes temperature, relative humidity, lighting, and uniformity.

enzyme A protein that carries out a biochemical reaction in the cell.

epifluorescence Confocal microarray scanning architecture in which the excitation and emission beams pass through a single objective lens.

epitope Protein moiety used in antibody binding or to stimulate an immune response.

etching Chemical process used to score glass surfaces for the purpose of labeling or identification.

etiolated Morphology and appearance of plants, including extreme elongation and the absence of chlorophyll, observed when plants are grown in the dark.

evaporation The chemical transition from a liquid to a gas.

excitation Light absorption process resulting in the conversion of electrons in a fluorophore from the ground state into a higher energy state.

exciter Optical filter in a microarray detection instrument that transmits the excitation light.

exclusive Term in patent law that refers to a licensing arrangement in which the assignee guarantees unique access to an invention for a single licensee.

exit strategy Term used in venture-capital financing that refers to a plan for beneficially terminating an investment relationship through the sale of the business, company shares, or through some other revenue-generating means.

exon Segment of a gene retained in the messenger RNA after processing, and often containing the codons represented as amino acids in proteins.

exon shuffling Evolutionary model that postulates that novel genes are formed by the mixing and matching of coding sequences through genetic recombination over time.

expansion Term used in venture-capital financing that refers to the third stage of business development, after the startup and early stage phases; characterized by an increasing product line and sales.

expressed sequence tag (EST) Specialized complementary DNA molecule that has been subjected to a single pass of DNA sequencing.

expression map Graphical representation in which gene expression values are superimposed onto their cognate genes along the chromosomes.

extension Text identifier in a microarray computer file that allows file recognition by the operating system, which in DOS corresponds to the three letters to the right of the decimal point.

external force Methodological term corresponding to an extrinsic factor, such as phosphorylation, that has an effect on a system variable, system module, system, culture, or biosphere.

extrinsic One of the three signal determinants in the microarray detection process, corresponding to the light source, excitation wavelength, pH, and other instrument and environmental contributions to signal.

feature reactivity Criterion of microarray manufacture that pertains to the chemical or biochemical activity of the target molecules in a given feature, expressed typically as a percentage of the total target molecules present.

feature purity Criterion of microarray manufacture that pertains to the chemical or biochemical homogeneity of the target molecules in a given feature, expressed typically as a percentage of the desired sequence or of the total target molecules present.

feature size Criterion of microarray manufacture that pertains to the physical size of a printed microarray element or feature, expressed typically as the diameter in microns.

fertilization The cellular union of egg and sperm to create an embryo.

fidelity Measure of the precision of DNA replication, represented typically as the number of errors per nucleotide copied.

field size Microarray detection system criterion that refers to the size of the physical area that can be read in a given detection step or the overall area that can be read on a standard microarray substrate using a given detection instrument, measured typically in millimeters.

file Collection of digital microarray data stored on a computer.

file name Text identifier for a microarray computer file, which in the DOS system corresponds to the letters to the left of the decimal point.

filter Round optical component of a microarray detection instrument that excludes specific wavelengths of light.

filter set Pair of filters in a microarray detection instrument, typically consisting of an exciter and emitter.

filter wheel Rotary device in a microarray detection instrument that holds a plurality of optical filters.

finishing Late phase of the DNA sequencing process in which small gaps are filled and sequences aligned.

5′ end End of a DNA or RNA chain that contains the free 5′ phosphate group.

flag Manual or automated procedure in which missing or anomalous microarray features are identified before quantitation.

flat Evaluative criterion for a microarray substrate that refers to the smoothness of the surface over a small area.

fluorescence Light emission process in which a fluorophore in the excited state releases a photon of slightly longer wavelength than the excitation light.

fluorophore Specialized class of organic molecules capable of light absorption and fluorescence emission owing to their conjugated configuration of π electrons.

flying objective Moving optics microarray scanner design in which the objective, rather than the laser beam, is the dynamic component.

footprint Physical width, height, and length of a microarray instrument, typically measured in centimeters.

forensics Branch of medicine that gathers, examines, and analyzes specimens found at crime scenes.

form Methodological term corresponding to the myriad physiological and morphological manifestations of a system module or cell.

format Microarray detection system criterion that refers to the physical dimensions of the microarray substrate that can read with a given detection instrument.

free radical An atom with a single, unpaired electron.

functional group A group of atoms that possesses a characteristic chemical property.

functionality Cleanroom evaluative criterion that includes the sum of all of the physical attributes of a cleanroom, including overall design, airflow, environmental control, chemical containment, utilities, surface materials, shelving, bench design, and accessibility.

fundamental dogma of molecular biology Doctrine that specifies that genetic information flows from DNA into RNA into protein.

fuzzy logic Programming system that allows logical conclusions to drawn from inexact, complex, or nonbinary relationships.

gain Amplification factor produced by a photomultiplier tube or other detector in a microaray detection instrument.

galvonometer Position transducer in a microarray scanner that translates rapidly across the x axis and enables microarray image capture.

gamete Cell such as an egg or sperm that carries a single copy of each chromosome.

gas The physical state of matter characterized by an indefinite shape and an indefinite volume.

GenBank Public sequence database maintained by the National Center for Biotechnology Information, consisting of an annotated collection of all publicly available sequences from the United States, Japan, and Europe; includes >13 billion bases of DNA from approximately 12 million submissions.

gene Segment of genomic DNA that encodes a specific cellular mRNA and protein.

gene expression The cellular process by which genetic information flows from gene to messenger RNA to protein.

gene-specific primers Two oligonucleotides used for polymerase chain reaction amplification that bind uniquely to a given target sequence and allow amplification of a unique target from a complex mixture of complementary or genomic DNA.

genetic code The cellular alphabet that specifies one of the 20 common cellular amino acids or stop codons from the 64 triplets in messenger RNA.

genetic recombination Genetic crossing over between two genomic segments that share sequence similarity.

genetic screening Systematic examination of a population for genetic information by microarray analysis or some other means, without any a priori motivation for conducting a specific screen.

genetic testing Targeted examination of a specific subset of the population for a specific disease or condition using microarray analysis of some other testing procedure.

genetically modified organism (GMO) An agricultural plant that contains a genetic modification introduced using recombinant DNA, transgenic technology, or some other molecular method.

genome Entire DNA content of a cell, including the nucleotides, genes, and chromosomes.

genome equivalent Sequence complexity corresponding to all of the nucleotides present in the genome of an organism.

genomics Field of study focused on the analysis of genomes.

genotype The genetic makeup of an organism.

germ cell Cell such as an egg or sperm that carries a single copy of each chromosome.

global alignment Computational strategy in which the entire sequence of the genes or proteins is compared.

global intensity Sum of all of the signal readings on a microarray, particularly for use as a normalization factor.

glycogen Highly branched glucose-based carbohydrate used to maintain blood sugar levels in animal cells.

graphical user interface (GUI) Intuitive software package, that allows a user to efficiently control a robotic device such as a microarray robot or scanner via computer.

graphics file Any of several different file types consisting of ordered rows and columns of digital values that provides a 1:1 numerical representation of the fluorescent values acquired from a microarray by a detection instrument.

green–red overlay Two-color composite microarray image generated by superimposing a green image and a red image.

guanine (G) One of the four nitrogen-containing bases of DNA and RNA.

haploid Cell that possesses a single copy of each chromosome.

heat load Amount of thermal contamination present in a microarray cleanroom.

hepatocellular carcinoma Malignant form of liver cancer thought to be associated with hepatitis B. It has a poor prognosis and is prevalent in parts of Southeast Asia and Africa.

heredity Transmission of genetic traits from one generation to the next.

heterozygous Diploid cell or organism that contains two different variants of a given gene.

heuristic Algorithm that provides an approximate solution to an extremely complex problem.

hierarchical Clustering method that identifies a small group of genes that share a common pattern of expression and then constructs the dendrogram in a sequential manner using a ranked series of clusters.

higher eukaryote Category of biological organisms that includes primates, rodents, insects, plants, and other multicellular life forms.

homeobox Highly conserved DNA binding motif found in transcription factors from yeast, plants, and animals.

homeobox-leucine zipper Novel functional domain in higher plants containing a DNA binding domain and a dimerization motif.

homolog Gene or protein that shares sequence similarity with another gene or protein.

homologous Two or more biological sequences that display sequence similarity.

homologous recombination Experimental approach used to examine gene function in intact cells, animals, and plants in which the normal gene is replaced with a defective gene copy at the chromosomal position of the normal gene.

homology search Computational process in which a query sequence is aligned with every subject sequence in a gene or protein database.

homozygous Diploid cell or organism that contains two identical copies of a given gene.

horizontal shift register Device in a charge-coupled device camera that reads the charge accumulated at each pixel by transferring the charge from left to right.

hot pixel Defect in a charge-coupled device detector corresponding to a picture element that produces too much signal relative to a normally functioning picture element.

housekeeping gene Cellular gene that plays a central role in all cells and correspondingly is expressed nearly equally in all cells and tissues.

human leukocyte antigen locus (HLA locus) Important region of the major histocompatibility complex in the human genome that encodes a family of surface glycoproteins, and is highly polymorphic.

humidity Cleanroom environmental control parameter corresponding to the relative humidity reading inside the cleanroom.

hybridization The chemical process by which two complementary DNA or RNA strands zipper up to form a double-stranded molecule.

hydrogen acceptor An electron-rich atom that has an electron lone pair that interacts with the partially positively charged hydrogen atom in a hydrogen bond.

hydrogen bond A noncovalent chemical interaction formed by the sharing of a hydrogen atom between two molecules.

hydrogen donor An atom bonded to hydrogen that creates a partial positive charge on the hydrogen atom and facilitates the formation of a hydrogen bond.

hydrophilic Measure of the extent to which a microarray surface or some other substance or molecule dissolves water.

hydrophobic Measure of the extent to which a microarray surface or some other substance or molecule repels water.

identity line Diagonal in a scatter plot that defines the signal intensities and gene expression values that are equivalent in two samples.

imager One of the two categories of microarray detection instrument that relies on stationary optics and substrate to capture large microarray images in single imaging steps.

immune system Highly complex cellular defense mechanism in humans and other mammals that combats bacterial, fungal, and viral infection.

importing One of the four main activities that is legally prohibited with respect to an invention once a patent issues.

incident light Electromagnetic radiation that impinges on a microarray substrate, mirror, or other component during microarray detection.

inclusion body Foreign tissue deposit that forms in the brain and other human organs, and is associated with Parkinson, Alzheimer, and other diseases.

indirect labeling Probe preparation scheme in which the fluorescent tags are attached in a noncovalent and indirect manner to the probe molecules using dendrimers, antibodies, or some other reagent.

inert Evaluative criterion for a microarray substrate that refers to the non-contribution of the substrate to the assay signal.
initial public offering (IPO) Term used in venture-capital financing that refers to the first sale of company shares to the public.
ink jet Noncontact liquid dispensing device that uses piezoelectric, microsolenoid, and other types of actuators to expel droplets in a controlled manner for microarray printing.
inorganic compound A molecule that does not contain carbon-to-hydrogen bonds.
insertion Mutation that results in the addition of one or more nucleotides to a DNA sequence.
insider trading Term used in venture-capital financing that refers to the selling of shares by a major company share holder.
insulator Material that inhibits the flow of electrons.
intellectual property Any product of the human mind that meets the criteria required for patentability.
intensity (I) Amount of light used for microarray excitation equivalent to the power of the light source divided by the illumination area.
intensity segmentation Method of signal segmentation that uses pixel values and mathematical and statistical approaches to exclude anomalous pixel intensities and segment signal from background.
intermolecular bond An attractive chemical force between two atoms in different molecules.
Internet International network of computers that allows global communication in the form of the World Wide Web, electronic mail, and other applications.
Internet service provider (ISP) Company or institution that provides Internet access to users.
intramolecular Biochemical interactions that occur between functional groups within a single molecule of DNA, RNA, protein, and the like.
intramolecular bond An attractive chemical force between two atoms in the same molecule.
intrinsic One of the three signal determinants in the microarray detection process, corresponding to the signal inherent to the labels attached to the probe molecules.
intrinsic property Methodological term corresponding to an attribute, such as primary nucleotide sequence, that is inherent to a system variable, system module, system, culture, or biosphere.
intron Segment of a gene removed from the messenger RNA during processing and not represented in proteins.
ionization The chemical process whereby a molecule acquires a positive or negative charge.
ionizing radiation Intense form of electromagnetic radiation that has the capacity to dislodge electrons from the outer shells of atoms.
Internet protocol address (IP address) Four sets of eight-bit numbers separated by periods (e.g., 171.65.21.118) that provide a unique identifier, allowing computers to communicate globally over the Internet.
ischemia Medical condition of oxygen shortage caused by a blockage in blood flow.
iterative Any computational procedure, including clustering, that employs a repetitive and cyclical computational process.
Jablonski diagram Schematic representation of the fluorescence process in which energy is plotted as a function of time, including excitation, emission, and quenching.
kilobase Segment of DNA equal to 1000 base pairs.
k-means Nonhierarchical clustering method in which the user specifies the number of clusters (k) into which the genes are to be assigned.
laminarity Cleanroom parameter that refers to the direction of the airflow inside a cleanroom.

laser Any of a family of devices that emits an intense beam of collimated monochromatic light.

leucine zipper Protein dimerization motif found in many eukaryotic proteins; includes the specialized family of plant transcription factors known as the homeodomain-leucine zipper proteins.

license Term in patent law that refers to the legal document that defines the terms and conditions by which another party may use a patented or patent-pending technology.

licensee Term in patent law that refers to the licensing party of a patent.

licensor Term in patent law that refers to the holder of a patent, also known as the assignee.

life cycle Experimental cycle of microarray analysis that contains the five components of biological question, sample preparation, biochemical reaction, detection, and data analysis and modeling.

lighting Cleanroom environmental control parameter corresponding to the intensity and quality of the illumination inside a cleanroom.

linear actuator Device used in microarray manufacturing robots that converts circular motion into linear motion.

linear amplification Experimental procedure in which multiple copies of nucleic acids are obtained by a nonexponential process.

linear drive system Microarray robot component consisting of a motor, linear actuator, and other components required to convert circular motion into precise linear motion.

linear encoder Measuring device on the linear actuator of a microarray robot consisting of a bar-code-like tape with microscopic hash marks and a reading device that allows increased precision in motion control.

linear range Portion of microarray data in which an increase in probe concentration produces a concomitant increase in assay signal.

liquid The physical state of matter characterized by an indefinite shape and a defined volume.

load Maximal weight that can be moved without a significant deterioration in the performance specifications of an electric motor or microarray robot, expressed typically in grams or kilograms.

local alignment Computational strategy in which small regions of sequence similarity between genes or proteins is compared.

local area network (LAN) Group of interconnected computers that provide speedy and economical computer connectivity for a set of users that share common interests.

lockup period Term used in venture-capital financing that refers to a 3- to 24-month period during which venture capitalists, owners, company employees, and major shareholders are prevented from selling their shares on the open market.

logarithm Power to which 10 must be raised to produce a given microarray value.

logarithmic scale Series of microarray data values expressed as logarithms rather than raw counts.

lone pair Two electrons that occupy the outer energy level of an atom but do not participate in covalent bond formation.

lowest total energy Minimum energy state sought by a liquid spreading across a solid surface.

macro Higher-level routine in a scripting language used to achieve sophisticated motion control without requiring the user to code in the primary programming language.

magnetic field Force created by the flow of electrons through a conductive material that is perpendicular to the direction of the current.

making One of the four main activities that is legally prohibited with respect to an invention once a patent issues.

manual quantitation Computational process in which signal intensities are generated for microarray spots one at a time.
matched intensities When two or more microarray images possess similar signal readings across the full gamut of signal values.
matter Any substance that has mass and occupies space.
mechanical shearing Experimental procedure that uses high pressure, sonic waves, or some other means to break nucleic acid molecules into smaller fragments.
MEDLINE Main source of content for PubMed containing a comprehensive electronic library of medically oriented scientific publications.
megabase Length of DNA containing 1 million base pairs.
meiosis Process of cell division resulting in the production of gametes.
melting The chemical transition from a solid to a liquid.
melting temperature (T_m) Thermal point at which 50% of the complementary molecules hybridize and 50% dissociate from their cognate strands.
meniscus The curved surface of a liquid produced by the interaction between the liquid and the walls of a capillary device or container.
messenger RNA (mRNA) The class of cellular RNA that undergoes extensive editing, contains the protein-coding sequences of genes, and functions as an informational intermediate between DNA and protein.
messenger RNA processing Cellular editing steps that include capping, polyadenylation, and splicing.
metastasis Cancerous process in which tumor cells dislodge from a tumor mass, move into the bloodstream, and migrate to a secondary location in the body.
method Specific process, such as the polymerase chain reaction, used to perform a given task.
methodology Complete body of theories, concepts, principles, rules, methods, protocols, techniques, formulas, and the like that guide microarray analysis or some other process or procedure.
mezzanine round Term used in venture-capital financing that refers to last round of venture capital before an initial public offering.
micro spotting Contact printing method that uses printing pins that contain a defined uptake channel and a flat or horizontally level tip; allows microarray printing without tapping the printing implement on the surface.
microarray An ordered array of microscopic elements on a planar substrate that allows the specific binding of genes or gene products.
microarray analysis Five-step experimental procedure that uses microarrays to explore the biological, chemical, and physical world.
microarray noise Unwanted contribution to detection instrument readings arising from the substrate, cross-reactivity of the sample, and any other chip-based source.
microlocation Analytical element on an active microelectronic array.
micromachining High-precision machining technology used to manufacture printing pins and other fine implements used in microarray analysis.
micromirror Microscopic reflective device configured in a large array that is used to selectively direct light onto a microarray substrate during the digital light processing manufacturing approach.
microscopic A object that cannot be seen clearly without the use of a microscope and is measured typically in microns (1000 microns = 1 mm).
microstepping Stepper motor technique that allows increased motion control precise by moving in fractions of steps.
minisequencing Genotyping method in which templates are hybridized with primers so that the last base overlaps a mutation of interest and the primed templates are extended with polymerase and fluorescent dideoxynucleotides to determine the sequence.
mirror Optical component used in microarray detection instruments that reflects light in a highly precise manner.

mismatch A location in a DNA double helix wherein the bases on opposite strands do not share complementary sequences and therefore do not form productive adenine–thymine or guanine–cytosine base pairs.

mispairing Juxtaposition of a noncomplementary base or bases in a hybridized nucleic molecule.

mitosis Cellular process whereby two identical cells are produced from a single dividing cell.

mixed gas Type of laser, such as a krypton–argon laser, used in some microarray scanners that contains two or more gases as the gain medium.

mixed sequence Nucleic acid containing a heterogenous composition of adenine, cytosine, guanine, and thymine bases.

molarity The number of moles of a substance present per liter of liquid.

molecular agriculture Agronomic approach that relies on transgenic and other molecular technologies to improve the taste, nutrition, appearance, yield, and other traits in fruits and vegetables.

molecular weight The sum of the atomic masses of all the atoms in a molecule.

molecule A substance that contains two or more atoms.

monochromatic light Electromagnetic radiation of a single or narrow wavelength, such as that emitted by a laser.

monomer A nucleotide, amino acid, sugar, or other biochemical building block that makes up a biopolymer.

monosaccharide A highly soluble chemical building block of carbohydrates containing carbon, hydrogen, and oxygen in a ratio of about 1:2:1.

moving optics Microarray detection instrument design that employs physical translocation of the laser path, lens, filters, or other optical components to acquire a microarray image.

moving platen Microarray robot design that achieves movement along the *x* or *y* axis by movement of the platform that holds the printing substrates.

moving substrate Microarray detection instrument design that employs physical translocation of the microarray substrate or slide to acquire an image.

multivariate Any statistical technique, including principle component analysis, that allows the comparison of many different variables.

mutagen Chemical agent that alters the primary nucleotide sequence of DNA.

mutation Any change in a DNA sequence, but typically acquired during the life span of an organism.

negative control A microarray element or substrate that provides little or no readable signal, irrespective of the results obtained from the experimental component of the assay.

neurodegenerative Category of human disease including Parkinson, Alzheimer, epilepsy, and Huntington, that degrade neural function and impair the human brain and nervous system.

neurotransmitter Any of a family of signaling molecules including acetylcholine and endorphin that serve to communicate chemical messages in the brain and peripheral nervous system.

neutral density filter Optical component in a microarray detection instrument that reduces the power of a laser beam.

neutron A subatomic particle that resides in the nucleus of an atom, carries no charge, and has a mass of approximately 1 atomic mass unit.

next generation screening (NGS) Genotyping method by which a large number of amplified patient samples are printed into a microarray and hybridized with fluorescent oligonucleotides to determine the genotypes of multiple patients for multiple diseases in a single test.

nitrocellulose Novel microarray surface coating consisting of a nitrated glucose polymer.

nitrogenous base A heterocyclic chemical structure that contains multiple nitrogen atoms, and includes the five bases (adenine, guanine, cytosine, thymine, and uracil) that make up DNA and RNA.

noise Numerical output acquired by the detection instrument, corresponding to background fluorescence, dark current, shot noise and other nondata components of the assay.

noncoding Any gene sequence that does not contain protein coding information.

noncontact printing Microarray manufacturing technology, including ink-jet printing, that does not require direct interaction between the printing implement or the sample and the microarray substrate.

noncovalent Any of several types of attractive chemical forces between atoms possessing opposite charge character.

nonexclusive Term in patent law that refers to a licensing arrangement in which the assignee reserves the right to license the same technology to multiple licensees.

nonhierarchical Clustering method such as *k*-means analysis that assigns each gene expression datum to a cluster based on its expression profile, and then repeats the process until every datum point has been placed in a cluster.

nonpolar solvent A solvent that contains nonpolar bonds or a uniform distribution of charge.

normalization Computational process in which data from different channels or different chips are equalized before analysis.

normalization factor Value used to equalize data from different channels or chips before data analysis.

novel One of the three formal requirements for patentability, pertaining to whether a given invention is new with respect to the state of the art in a technology area.

***n*-type** Negatively charged silicon made by substituting a trace amount of phorphorus to provide a doped material that has slightly greater conductivity than pure silicon.

nuclear transfer Technology used in cloning that allows the introduction of a somatic cell nucleus into an egg lacking a nucleus.

nucleic acid Any of a family of negatively charged polymeric biomolecules, including DNA and RNA, that contain a repeating series of nucleotide building blocks.

nucleic acid microarray A broad category of microarray that includes any microarray that contains DNA or RNA target elements.

nucleophile A molecule or functional group that contains a free pair of electrons or a strongly negative charge bias, and has an affinity for an electrophile.

nucleoside A biomolecule consisting of a base and a sugar.

nucleosome Higher order nuclear structure in which DNA is wound up in three dimensions.

nucleotide A biochemical building block that makes up cellular DNA and RNA, consisting of a base, sugar, and phosphate group.

nucleotide monophosphate A nucleotide in DNA and RNA that contains a single phosphate group.

nucleotide triphosphate A nucleotide in DNA and RNA that contains three phosphate groups.

nucleus The dense, positively charged core of an atom. Alternatively, the organelle inside a cell that contains the genetic material.

number of channels Microarray detection system criterion that refers to the number of different wavelengths of light that can be detected with a given instrument.

numerical aperture (NA) Measure of the angle or cone of light that is captured by an objective lens during microarray scanning or imaging; the larger the NA value the light more that is efficiently captured and the smaller the depth of focus.

objective lens Optical component in a microarray detection instrument that focuses the laser light on the microarray.

octet rule The chemical propensity of an atom to obtain a complete set of eight valence electrons.

Ohm's law Mathematical relationship between current, voltage and resistance, given as $I = V/R$.

oligonucleotide Short chain of single-stranded DNA or RNA.

oligonucleotide array Ordered array of oligonucleotides on a solid surface.

operational One of three main categories of microarray cleanroom status, corresponding to the period in which a cleanroom is actively engaged in a process; contains cleanroom personnel; and involves the normal transfer of supplies, reagents, and equipment from the ambient environment.

optical encoder Measuring device on the linear actuator of a microarray robot consisting of a bar-code-like tape with microscopic hash marks and an optical reading device, such as a laser, that allows increased precision in motion control.

optical filter Optical component in a microarray detection instrument that transmits light over a very narrow bandwidth.

optical laser Type of laser, such as a ruby laser, that uses a light source for pumping.

optical noise Unwanted contribution to detection instrument readings, arising from reflected light from the substrate holder, spurious reflections from instrument enclosures, light leaks impinging on the detectors, cosmic rays, and other light-dependent sources.

optimal probe concentration Number of probe molecules per unit volume of sample that provides the strongest linear signal in a microarray assay.

optimal target concentration Number of target molecules per unit volume of printed sample that provides the strongest signal in a microarray assay.

optimal target density Number of target molecules per unit area on a microarray substrate that provides the strongest signal in a microarray assay.

orbital diagram An illustration of the energy levels, occupancy, and spin states of the electrons in atom.

ordered A collection of microarray elements configured in rows and columns.

organic compound A molecule that contains one or more carbon-to-hydrogen bonds.

organosilane Any member of the specialized class of chemical compounds that contains silicon-to-carbon bonds.

ortholog Related gene or protein that occurs in two different organisms.

osmotic stress Form of stress in plants and other organisms and cells due to an imbalance in salt concentration between the interior and exterior of cells.

outgassing Process in which paints, glues, adhesives, packaging materials, and other cleanroom surfaces emit vapors into the cleanroom air space.

outlier Statistical term corresponding to a gene or datum point that behaves differently from the majority of data when examined across multiple experiments.

overhead gantry Microarray robot design that achieves movement along all three axes, using a suspended motion-control system that moves over a stationary platen.

oxidation The chemical process involving the loss of electrons.

oxidizing agent A chemical compound that removes electrons in a chemical reaction.

pairwise Mathematic procedure used in clustering and other forms of data analysis in which genes or gene products are compared in groups of two.

parallel Methodological description of a microarray substrate referring to the fact that microarray spots lie in the same plane and never converge.

paralog Related gene or protein present in the same organism.

particle One of five main forms of cleanroom contamination corresponding to airborne debris, including dust, dander, lint, spores, pollen, fungi, bacteria, and any other form of debris that contaminates the air space in a microarray cleanroom and has a width and length ratio not exceeding 10:1.

Patent and Trademark Office (PTO) Agency of the U.S. Department of Commerce that issues patents.

Patent Cooperation Treaty (PCT) Agreement signed by 115 countries that expedites patent filing and prosecution worldwide.
patent prosecution Process by which applications and supporting documents are exchanged between a patent attorney and the Patent and Trademark Office en route to obtaining a patent.
pathway analysis Computational procedure in which gene expression data are correlated with a biochemical, genetic, or developmental process.
pentose Any five-carbon sugar molecule, including deoxyribose and ribose.
peptide bond An amide linkage between two amino acids formed between the carboxyl group of the first amino acid to the amino group of the second.
percent concentration The number of grams of solid per 100 mL liquid.
percent identity Similarity between two sequences, expressed as the number of identical positions divided by the total number of positions compared times 100.
perfect heteroduplex Target and probe hybrid that shares 100% complementarity.
periodic table An organizational chart that contains the names and atomic properties of the 118 known elements.
permeation layer Thin coating on an active microelectronic array substrate into which target molecules are embedded and probe molecules diffuse.
personal computer (PC) Any of a wide spectrum of binary hardware devices used to control the robots employed in microarray manufacture and detection and to analyze and model microarray data.
phenotype The physical manifestation of genotype.
phosphate (PO_4^{3-}) The negatively charged component of DNA and RNA.
phosphoramidite Any of the set of modified oligonucleotides used in the manufacture of synthetic oligonucleotides.
photobleaching Specialized form of sample degradation in which fluorescent molecules are damaged by exposure to excessive excitation light during the detection process, typically measured as a percentage.
photocathode Light-sensitive device in a photomultiplier tube that captures the incident fluorescent photons and converts them into electrons.
photochemistry A chemical reaction involving the interaction of organic molecules with light.
photoelectric effect Interplay of photons and electron flow in which light impinging on a metal surface can produce an electrical current.
photolithography Method of microarray manufacture that uses a combination of modified phosphoramidite nucleotides, ultraviolet light, and photomasks to achieve solid phase chemical synthesis.
photolysis Chemical bond breakage caused by the absorption of light.
photomultiplier tube Detector in a microarray scanner that converts photons into an amplified electrical signal.
photon Particle form of light.
photoprotecting group Chemical protecting group, such as methylnitropiperonyloxycarbonyl, used in microarray manufacture by photolithography that requires exposure to ultraviolet light for removal.
physical examination Medical practice used to assess the general health of a patient and includes first-tier diagnostics.
phytochrome Light receptor protein in plants that undergoes a conformational change upon absorption of specific wavelengths of electromagnetic radiation, rendering the protein active to execute cellular signals, including changes in expression of specific cellular genes.
piezoelectric Noncontact liquid-dispensing device that uses electrical pulses to deform a transducer material, such as ceramic and expel droplets, in a controlled manner to achieve microarray printing.
pin and ring (PAR) Contact printing device that uses a circular loop to load sample by capillary action and a solid pin that moves up and down through the

liquid in the loop to enable printing by direct contact with the microarray substrate.

pixel Numerical bin or picture element used to store data in a graphics computer file.

pixel plot Two-dimensional representation of a microarray spot in which all of the pixel intensities are plotted at two emission wavelengths to allow the user to assess whether the pixel intensities distribute normally as would be expected for a legitimate printed spot.

planar Evaluative criterion for a microarray substrate that refers to the parallelism of the surface over the entire substrate.

plant One of the three formal types of patents issued by the U.S. Patent and Trademark Office.

plasmid Any circular DNA molecule that has the capacity to replicate independently in the cell.

platen Bed or base of a microarray robot that holds the microarray substrates, microplates, and other components and accessories.

point defect Single defective pixel in a charge-coupled device detector.

polar A chemical bond or molecule that exhibits an unequal distribution of electron density.

polar coordinate Mapping system used in rotary flying objective scanners, consisting of a radius and a θ angle instead of the traditional x and y Cartesian coordinates.

polar solvent A solvent that contains polar bonds or uneven distribution of charge.

polarity A term that refers to the chemical nonsymmetry of DNA and RNA chains, specified as $5'$ and $3'$ ends.

polyadenylation Enzymatic processing of mRNA whereby a series of adenine nucleotides are added to the $3'$ end of the transcript.

poly A sequence Sequence of 10–100 adenine residues added to the $3'$ end of eukaryotic messenger RNAs.

polychromatic Any light source, such as a xenon lamp, that produces electromagnetic radiation of many different wavelengths.

polymerase chain reaction (PCR) Revolutionary technology developed during the 1980s that allows massive amplification of any gene sequence of interest.

polymorphism Any of a family of minor DNA sequence variants, including single nucleotide polymorphisms and restriction fragment length polymorphisms.

polypeptide A naturally occurring chain of amino acids.

population inversion Phenomenon in which the majority of electrons in an atomic mixture exist in the excited state, providing a highly efficient source of electrons for stimulated emission.

position transducer Scanner component such as a galvonometer that translates rapidly along the x axis and enables the capture of a complete microarray image from the substrate.

positive control A microarray element or substrate that provides a readable signal, irrespective of the results obtained from the experimental component of the assay.

post office protocol (POP) Internet procedure that allows messages to be received from remote mail servers.

potential difference Discrepancy in the capacity to support the flow of electricity between two terminals of a conducting substance, measured in volts.

power density Intensity of a light source used for microarray excitation.

precision Range of deviations that occur for 95% of the movements of an electric motor or microarray robot for a given input, typically expressed in microns. Alternatively, a microarray detection system criterion that refers to the consistency of microarray images acquired during different detection sessions.

preobjective Moving optics microarray scanner design in which the laser beam, rather than the lens, is the dynamic component.

primary amine Amino group in which the nitrogen atom is bonded to a single organic group, usually carbon.

primary response gene Cellular gene under direct and immediate regulation in a regulatory cascade.

primer Oligonucleotide that hybridizes to a complementary nucleic acid template and expedites enzymatic synthesis by providing a starting point for polymerase.

principle component analysis (PCA) Computational method that reduces relationships that exist in high-dimensional space into three dimensions, enabling complex data to be visualized in standard graphical form.

printhead Printing device that guides the movement of pins, ink jets, and other printing implements with high precision.

printing buffer Chemical mixture used in microarray manufacture to stabilize target molecules and improve sample spreading and attachment.

prior art Term in patent law encompassing any publication or previous invention that can be used to reject or invalidate a patent or patent claim.

priority date First day that the patent office was informed of an invention, given by day, month and year, and used to establish legal ownership in patent disputes.

probe Labeled molecule in solution that reacts with a complementary target molecule on the substrate.

probe excess Microarray assay condition in which the probe concentration results in target saturation and a loss of quantiation.

product A compound or chemical produced a chemical reaction.

profitability Term used in venture-capital financing that refers to the fourth stage of business development (after the startup, early stage, and expansion phases); characterized by a development stage in which company sales exceed company expenses.

promoter Genomic location upstream of a cellular gene that determines the start site for RNA polymerase.

protein Any member of the major family of cellular biomolecules encoded by a unique cellular gene and consisting of a repeating series of amino acids linked together by peptide bonds.

proteomics Study of cellular proteins on a genomic scale.

proton A subatomic particle that resides in the nucleus of an atom and carries a charge of $+1$ and a mass of approximately 1 atomic mass unit.

provisional patent Abbreviated invention disclosure to the U.S. Patent and Trademark Office that establishes a priority date and requires a complete application to be filed within 12 months of the provisional filing.

pseudo first order Second-order kinetic reaction that can be approximated as a first-order reaction because the concentration of one of the two reactants remains nearly constant over the duration of the reaction.

p-type Positively charged silicon made by substituting a trace amount of boron to provide a doped material that has slightly less conductivity than pure silicon.

PubMed Electronic resource from the National Library of Medicine that provides electronic access to >10 million scientific papers from nearly 5000 journals published in 70 countries since the mid-1960s.

pumping Process in the operation of a laser in which an energy source is used to excite the atoms in the gain medium.

purine A two-ringed chemical structure that includes adenine and guanine.

pyrimidine A single-ringed chemical structure that includes cytosine, thymine, and uracil.

quality control (QC) Commercial manufacturing procedure in which the characteristics of a product are defined and regulated to guarantee conformity of the product to within specified parameters.

quantification Synonym of quantitation.

quantitation Computational process by which numerical values are obtained from microarray data files.

quantitation grid Synonym for quantitation template.

quantitation template Graphical checkerboard pattern placed over a microarray image to facilitation data extraction.

quantitative Any microarray assay that provides a precise measure of the number, amount, or concentration of the molecules present in a sample.

quantity One of the three signal determinants in the microarray detection process, corresponding to the number of probe molecules and labels present at a given microarray location.

quantum yield *(Q)* Efficiency of photon release by a fluorescent molecule, defined as the ratio of emitted photons divided by absorbed photons.

raster pattern Rapid translation of a microarray scanner back and forth across the x axis of a substrate.

rate constant Number in a kinetic equation that has a fixed value.

ratio calculation Computational process by which signal intensities from two microarray images are divided to obtain a quotient for each data point.

reactant A substance consumed in a chemical reaction.

read time Number of seconds or minutes required to capture a microarray image with a microarray imaging instrument.

recessive An allele that does not manifest a phenotype in the presence of a normal copy of the gene.

recognition sequence Nucleotide target site at which a restriction enzyme cleaves the DNA.

recombinant DNA Revolutionary technology developed in the 1970s that allows genes from different organisms to be spliced together.

rectilinear Microarray scanning profile characterized by a series of straight lines achieved by a 90° alignment of the optics with the substrate.

reducing agent A chemical compound that contributes electrons to a chemical reaction.

reduction The chemical process involving the acquisition of electrons.

redundant Two or more target sequences in a microarray that share perfect or near-perfect sequence identity.

reference sample Population of molecules, such as messenger RNA and complementary DNA, used as a control to facilitate comparison of one or more test samples.

regularity Criterion of microarray manufacture that pertains to the evenness of the spacing of rows and columns in a printed microarray, expressed typically as a coefficient of variation percentage of the theoretical center-to-center distance.

regulatory cascade Group of cellular genes configured in a complex gene expression hierarchy.

regulatory element Short segment of DNA near a promoter that binds one or more cellular proteins and modulates gene expression.

relational database Specialized information storage and retrieval system in which microarray data are represented in tables, allowing each and every datum to be queried and retrieved in a systematic manner.

relative humidity Cleanroom environmental control parameter defined as the ratio of the ambient moisture at a defined temperature and pressure, divided by the maximum moisture capacity of the air inside a cleanroom.

relative quantitation Computational process of measuring the ratio of absolute signals for each microarray spot in two samples on a microarray.

repeatability Reliability of achieving a specified position of an electric motor or microarray robot over repeated attempts, typically expressed in microns. Alternatively, a microarray detection system criterion that refers to the line-to-line accuracy of a scanning device.

reperfusion Medical procedure involving the reintroduction of blood or fluids into a tissue.

replacement reaction A chemical reaction in which a more reactive molecule takes the place of a less reactive molecule during the formation of a new substance.
replication Cellular process by which DNA is copied from a DNA template to produce an exact copy of the genome.
replication origin Discrete chromosomal location at which DNA replication occurs.
repression A decrease in the rate of transcription of a cellular gene.
repressor Cellular protein that binds to a regulatory element and decreases the rate of transcription of a gene.
resistance (*R*) Propensity of a material to impede the flow of electricity.
resolution Smallest detectable movement increment of an electric motor or microarray robot, typically expressed in microns. Alternatively, a microarray detection system criterion that refers to the pixel size in a microarray image, expressed in microns.
resonance light scattering (RLS) Microarray detection technology that exploits the fact that small gold and silver particles generate intense monochromatic light when exposed to a white excitation source.
restriction enzyme Any of a family specialized enzymes used in genetic engineering that cleave DNA at discrete locations.
restriction fragment length polymorphism (RFLP) Single nucleotide change in the genome that results in the creation or loss of a restriction site.
reverse bias Electrical configuration maintained by applying an external voltage to a charge-coupled device pixel so that the p-side carries a negative charge and the n-side carries a positive charge; thus allows charge to accumulate in the pixel equivalent to the number of photons emitted at a given position in the microarray.
reverse transcriptase Enzyme that synthesizes complementary DNA from a messenger RNA template.
reverse transcription Enzymatic process in which genetic information in cellular RNA is converted into DNA.
ribonucleic acid Any of a broad family of single stranded, negatively charged nucleic acids composed of ribonucleotides.
ribose The five-carbon RNA sugar that contains an hydroxyl group at the 2′ position, instead of the hydrogen atom found in the doexyribose DNA sugar.
ribosomal RNA (rRNA) Specialized class of cellular RNA, located in ribosomes, that plays structural and catalytic roles during protein synthesis.
ribosome Large cytoplasmic structure that facilitates protein synthesis.
rights Term in patent law that refers to the legal privileges, including ownership and licensing, afforded to a patent holder.
rigid Physically inflexible or unbending, such as glass and other materials used for microarray manufacture.
RNA polymerase Enzyme that synthesizes RNA molecules from a DNA template.
RNA-dependent DNA polymerase Specialized family of enzymes, including reverse transcriptase, that synthesize DNA from a ribonucleic acid template.
rotary flying objective Moving optics microarray scanner design in which the objective moves in a circular arc relative to the microarray substrate.
rotor Permanent magnet that rotates in a circular manner in an electric motor.
round rail Linear actuator component that uses linear bushings and a round stainless-steel shaft as the guide mechanism.
S phase Second step in the cell cycle, during which DNA synthesis occurs.
safety Cleanroom evaluative criterion that pertains to the extent to which a cleanroom maximizes the physical comfort and well-being of cleanroom personnel and minimizes or eliminates any chance of injury.
salinity Salt content of irrigation water.

sample degradation Unwanted phenomenon in which probe molecules are damaged by photobleaching or some other means during the detection process, typically measured as a percentage of the total molecules.

sample noise Unwanted contribution to detection instrument readings arising from nonspecific attachment of probe molecules, probe reflection, salt crystals, or any other probe- or buffer-dependent source.

sample tracking Computational process in which sequence identity, clone size, organism, and other associated biological information is correlated and maintained between liquid samples in microplates and printed spots in a microarray.

saturated Microarray target element in which most or all of the target molecules contain bound probe molecules.

scanner One of the two categories of microarray detection instrument that relies on rapid movement of the optics, substrate, or both to capture a microarray image.

scatter plot Graphical representation of microarray data in which the signal intensities of two samples are plotted along the x and y axes, and the ratio values are plotted as a distance from the diagonal.

Schiff base A specialized imine consisting of a primary amine that forms a double bond between carbon and nitrogen.

schizophrenia Mental illness characterized by delusions, disorganized thinking, abnormal speech, hallucinations, catatonia, and other behavioral symptoms.

scope Term in patent law that refers to the breadth of the invention.

scratch/dig Glass polishing specification used as a benchmark for microarray substrates.

scripting language Software routines used in microarray motion control programs that run on top of the underlying programming language and simplify coding.

second order Biochemical reaction, such as hybridization, in which the reaction rate depends on the concentration of two reactants.

secondary response gene Cellular gene under indirect regulation in a regulatory cascade.

secondary structure Two- and three-dimensional configurations of proteins and nucleic acids that originate from intramolecular interactions of primary linear sequences.

seed capital Term used in venture-capital financing that refers to the initial investment of venture capital in a startup business.

selective target saturation Microarray assay condition in which a subset of the target elements become largely or fully bound, leading to a loss of quantiation.

self-organizing map (SOM) Algorithm that displays cluster information in a two-dimensional manner.

self-quenching Unwanted detection phenomenon in which emission signal is lost by energy transfer between adjacent fluorescent molecules, resulting in a diminished signal.

selling One of the four main activities that is legally prohibited with respect to an invention once a patent issues.

semiconductor technology Any of the microarray manufacturing approaches, including photolithography, that uses technology developed in the computer chip industry for microarray manufacture.

sensitivity Microarray detection system criterion that refers to the efficiency by which a microarray detection device, such as a photomultiplier tube, converts photons from a fluorescent microarray into an electronic signal.

sequence alignment Computational juxtaposition of two or more linear strings of nucleotides or amino acids.

sequence trace Data from a DNA-sequencing machine corresponding to the nucleotide sequence of a DNA sample.

sequence variant Any minor change in the primary nucleotide sequence of a gene.

sequencing by hybridization (SBH) Molecular method that exploits nucleic acid hybridization for primary sequence determination.

sequential Microarray scanner design that captures multiple channels one at a time.

servomotor Electric motor that achieves precise mechanical motion by the use of control circuitry.

setup One of three main categories of microarray cleanroom status corresponding to the period after cleanroom construction in which the environmental-control systems are fully operational and equipment, supplies, robots, and accessories are being introduced into the facility from the ambient environment.

shaft coupling Linking device in a microarray robot that connects a motor to a linear actuator.

shot noise Unwanted contribution to detection instrument readings arising from the fundamental process of current flow, which corresponds to the discrete movement of electrons rather than a continuous flow process.

shotgun sequencing Method of determining an unknown sequence by breaking the sequence into small pieces, analyzing the fragments at random, and compiling the overlapping fragments to generate a complete sequence.

side chain A chemical group on an amino acid that determines its chemical and biochemical properties.

signal Numerical output from a microarray assay acquired by the detection instrument, and corresponding to the true experimental data.

signal compression Microarray assay condition in which the fluorescent readings underestimate the number of molecules present on the target element or in the probe mixture, leading to a loss of assay quantitation.

signal segmentation Computational process in which microarray signals are demarcated from background in a microarray image.

signal-to-noise ratio Quotient of signal and total noise in the microarray detection process.

silane reagent Any member of the specialized class of chemical compounds containing a silicon atom that is used to add organic groups to glass microarray surfaces.

silicon dioxide (SiO_2) Main component of glass.

silicon ester Any member of the specialized class of chemical compounds that contains an oxygen atom between the silicon-to-carbon bond.

simple mail transfer protocol (SMTP) Internet procedure that allows messages to be sent from remote mail servers.

simultaneous Microarray scanner design that captures multiple channels at the same time.

single nucleotide polymorphism (SNP) Common sequence variant containing a one-base-pair change relative to the normal gene.

software Programs, routines, and commands that a computer uses to control hardware devices.

solid The physical state of matter characterized by a defined shape and volume.

solid modeling Computer-assisted design approach that allows the rendering of three-dimensional objects on a computer screen.

solid pin Microarray printing implement containing a solid shaft and flat tip that holds sample on the end of the pin by adsorption and delivers sample onto the microarray substrate by direct contact.

solubility Refers to the amount of solid that can be dissolved in a defined volume of liquid.

somatic cell Diploid cell of an organism.

spacer arm Chemical group such as an aliphatic hydrocarbon used to create distance between a functional group and a microarray substrate.

spatial Gene expression, developmental or other biochemical event that is controlled or restricted according to biological location or region.

spatial segmentation Method of signal segmentation that relies on careful positioning of the quantitation boundaries for signal and noise, using physical location to segment signal from background.

specific activity Number of labels or fluorescent tags attached per probe molecule.

specific coupling Any covalent attachment scheme including a Schiff base, in which target attachment to the microarray surface occurs with defined chemistry.

specification Term in patent law that refers to the main body of a patent application that teaches, explains, and clarifies an invention.

specificity Uniqueness of the biochemical reaction between a target and probe molecule.

spectra Plural of spectrum.

spectrum Plot of light absorption or emission as a function of wavelength for a fluorescent molecule or light source.

speed Magnitude of travel per unit time of an electric motor or microarray robot, expressed typically in centimeters per second. Alternatively, a microarray detection system criterion that refers to the area of microarray substrate that can be captured per unit time at a given resolution, expressed typically as square centimeters per minute.

spin state The intrinsic angular momentum of an electron characterized by the quantum number $\frac{1}{2}$; depicted with an arrow in orbital diagrams.

splicing Enzymatic processing of messenger RNA whereby the introns are removed from the nascent messenger RNA.

split pin Tweezerlike printing implement that loads a volume of sample between two uneven points and enables microarray printing by tapping the implement on the substrate to expel the sample.

spot A microarray element or feature that contains target molecules.

square rail Linear actuator component that uses linear bushings and a square stainless-steel shaft as the guide mechanism.

stage Corresponds to the number of dynodes in a photomultiplier tube. Alternatively, the methodological term corresponding to the overall physical appearance of a system or organism in the methodology.

stage of development Term used in venture capital financing that refers to the maturity of a business with respect to products and sales.

standard curve Graph of signal intensity versus concentration constructed using known concentrations of messenger RNA or some other material and used to assign quantitative values to genes or gene products of unknown concentration.

standard operating procedure (SOP) Commercial manufacturing protocol, employed to minimize product variability.

starch Plant and animal carbohydrate that provides a cellular reservoir of stored energy.

startup Term used in venture-capital financing that refers to a business that has important ideas and technologies but has yet to develop products or generate revenues.

state Any of three biochemical manifestations of a system variable or gene in the methodology, corresponding to DNA, RNA, and protein.

stator Fixed electromagnetic that deflects a rotor in an electric motor.

step Precise clockwise or counterclosewise movement of an electric motor.

stepper motor Electric motor that uses rotors and stators to achieve stepwise mechanical movement.

steric availability Desirable spatial configuration, such as end attachment, that maximizes the physical accessibility of target molecules on a microarray substrate to incoming probe molecules.

stimulated emission Phenomenon in which excited electrons exposed to a light source emit two photons of the same energy and direction, providing the physical basis of a laser.

Stokes shift Difference in wavelength between the excitation light and the emitted light in a fluorescent assay, typically expressed in nanometers.

stringency Sum total of the external factors that affect hybridization efficiency, including temperature, salt, and pH, with greater stringency corresponding to excessive temperature, low salt, high pH, and other conditions that reduce hybridization efficiency.

structured query language (SQL) Computing language used in relational database systems.

subatomic particle Any of a family of quantum mechanical entities, including protons, neutrons, and electrons, that make up atoms.

subgrid Unit microarray produced on a substrate by a single printing implement, also known as a subarray or block.

sublimation The chemical transition from a solid to a gas.

substituted imine Chemical function that contains a carbon atom double bonded to the nitrogen atom of a primary, secondary, or tertiary amine.

substrate noise Unwanted contribution to detection instrument readings arising from surface reflection, organic treatments, surface coatings, or any other surface-dependent source.

sugar Any of a family of biomolecules, including deoxyribose and ribose in DNA and RNA, that contains carbon, hydrogen, and oxygen in a ratio of approximately 1:2:1.

supersaturated Solution that contains an amount of protein or some other solute in excess of what can be maintained in solution by a given solution at a defined temperature.

supervised Clustering method that exploits known reference vectors to classify and organize gene expression and other types of data.

surface coating Thin film placed in a noncovalent manner on a microarray substrate.

surface free energy Thermodynamic description of a liquid spreading across a solid surface.

surface tension Cohesive force, such as hydrogen bonding exerted between liquid molecules in a droplet, contributing to meniscus formation and other properties of liquids.

surface treatment Chemically reactive monolayer formed by the covalent linkage of reactive groups to the microarray substrate.

surfactant A chemical agent that reduces the surface tension of a liquid, typically by disrupting hydrogen bonding interactions between water molecules.

synteny Extended region of similarity in the chromosomal or genomic sequences of two organisms, resulting in a conservation in the order of some or most of the genes along a genomic segment.

synthesis Method of microarray manufacture in which the target molecules are manufactured directly on the microarray surface using photolithography or some other chemical means.

synthesis reaction A chemical reaction in which two or more compounds are combined into a more complex compound.

synthetic oligonucleotide A short chain of synthetic DNA manufactured using a DNA synthesizer.

system Methodological term corresponding to an intact organism.

system module Methodological term corresponding to a cell, the unit of life in the methodology.

system variable Methodological term corresponding to a gene or gene product, and the unit of function in the methodology.

tab-delimited file File format used in microarray quantitation in which each field or parameter is separated with a tab keystroke.

tandem mass spectrometry Analytical method used in traditional neonatal screening that allows the identification of patients with metabolic disorders by measuring the levels of small molecules.

tandem repeat Short nucleotide sequences that are joined in a head-to-tail configuration in the human genome, vary among individuals, and find use in genotyping applications.

target Molecule tethered to a microarray substrate that reacts with a complementary probe molecule in solution.

target excess Kinetic conditions in a microarray assay in which the concentration of the target molecules on the surface exceeds the concentration of the probe molecules in solution.

taxonomy Grouping of organisms according to their natural relationship.

teach Term in patent law that refers to the extent to which a patent instructs and informs the reader as to the content of the invention.

temperature Cleanroom environmental control parameter corresponding to the Celcius reading of the air inside a cleanroom.

temporal Gene expression, developmental, or other biochemical event that is controlled or restricted in time.

test sample Population of molecules such as messenger RNA and complementary DNA from which experimental data are derived.

thermal One of five main forms of cleanroom contamination corresponding to any source that produces an artificial increase in cleanroom temperature, including microarray robots, refrigerators, freezers, lights, and cleanroom personnel.

thinning Method of spatial segmentation that maintains the outer border of the signal region but restricts the number of pixels that are used to obtain the average and median signal intensities.

3′ end End of a DNA or RNA chain that contains the free 3′ hydroxyl group.

throughput Criterion of microarray manufacture that pertains to the number of microarrays that can be manufactured per unit time, typically expressed as the number of chips per day. Alternatively, the microarray detection system criterion that refers to the number of microarrays of a defined size that can be scanned or imaged at a defined resolution per unit time, typically expressed as microarrays per hour.

thymine (T) One of the four nitrogen-containing bases of DNA.

time illumination Quantity in microarray detection equal to the product of the light intensity times the dwell time.

time series Graphical representation in which data from related biological samples are displayed sequentially as a function of time.

tolerance Amount of error that can be tolerated in the manufacture of a component by micromachining, typically expressed as $\pm$ 0.0001″.

tort Term in patent law that refers to branch of law that enforces the conduct of the licensing parties and provides for remedies in case of breach or failed performance.

total noise Sum of all of the nondata numerical outputs acquired by the detection instrument, including background fluorescence, dark current, shot noise, and other unwanted sources.

toxicology Branch of science that focuses on the biological effects of poisonous substances.

trade secret Any piece of confidential information, including inventions, prototypes, codes, formulas, manufacturing methods, protocols, reagents, suppliers, customer information, purchasing practices, marketing reports, and business alliances, that are used to develop, conduct, and promote business.

trait Phenotypic quality produced by a gene.

transcript Synonym for cellular messenger RNA.

transcription First step in gene expression in which messenger RNA is synthesized from a DNA template.

transcription factor Cellular protein that binds to a promoter, enhancer, or other regulatory element and modulates expression of a cellular gene.

transducer An element such as a ceramic coupling that confers the piezoelectric effect.

transfer RNA (tRNA) Specialized class of cellular RNA that binds specific amino acids and facilitates protein synthesis by mediating codon recognition.

transform Computational process of converting microarray data from raw counts into a logarithmic or some other scale to improve the statistical soundness of an analysis procedure.

transgenic crop Any agricultural plant that contains a genetic modification introduced using the tools of recombinant DNA or some other molecular technology.

translation Synthesis of a polypeptide chain from a processed messenger RNA molecule.

transposition Computational process that converts the order and location of microarray samples in microplates into a printed microarray.

triplet Any one of 64 three-nucleotide sequences or triplets in messenger RNA that specify one of the 20 common cellular amino acids or stop codons during protein synthesis.

tweezer Early microarray printing implement that forms a meniscus on loading and requires a tapping force to break the meniscus and expel the sample onto the microarray printing surface.

twelve rules of parallel gene analysis Methodological set of regulations and procedures used to guide gene expression analysis using microarrays.

two-color overlay Graphical combination of two separate microarray images created by subtracting the values in the two images to obtain a composite image and used to depict gene expression and genotyping data from two samples in a single image.

tyramide signal amplification (TSA) Microarray labeling procedure in which fluorescent signals are enhanced by the enzymatic deposition of fluorescent reagents.

uniform Evaluative criterion for a microarray substrate that refers to the regularity of the spacing of the chemical groups on the surface.

uniformity Microarray detection system criterion that refers to the consistency of the recorded fluorescent signal at different locations across a perfect substrate, typically measured as a percentage. Alternatively, a microarray cleanroom environmental control parameter that refers to the extent to which the temperature, humidity, and lighting are maintained evenly in the cleanroom space.

universal reference sample Specialized reference sample used to generate ratios for a complete set of test samples.

universal solvent Water, so called because of its capacity to dissolve a wide range of different solids.

universe Methodological term corresponding to the highest order of biological organization, encompassing everything living and nonliving in existence.

unobvious One of the three formal requirements for patentability, pertaining to whether a given invention could not be ascertained by a simple reformulation of the state of the art in a technology area.

unsupervised Clustering method that does not use reference vectors to classify and organize gene expression and other types of data.

uracil (U) The nitrogen-containing base found in RNA but not in DNA.

uremia Medical condition characterized by the rapid accumulation of waste products in the bloodstream.

useful One of the three formal requirements for patentability, pertaining to whether a given invention has utility by improving on the state of the art in a technology area.

using One of the four main activities that is legally prohibited with respect to an invention once a patent issues.

utility One of the three formal types of patents issued by the U.S. Patent and Trademark Office.

valence band Location of electrons that participate in chemical bonding in a conductor, semiconductor or insulator.

valence electron An electron that occupies the outer energy shell of an atom and plays a major role in chemical reactivity and bonding.

van der Waals bond A noncovalent interaction between two molecules, created by transient fluctuations in electron density.

vapor pressure The propensity of a liquid to enter the gas phase.

vector Mathematical quantity that has both a magnitude and a direction and is thus a good descriptor of gene expression. Alternatively, a plasmid or other molecule that carries a DNA insert.

venture capital Professional money made available to companies in exchange for an equity stake, also known as risk capital.

venture capitalist An investor who structures deals involving an equity stake in exchange for risk capital.

vertical shift register Device in a charge-coupled device camera that reads the charge accumulated at each pixel by transferring the charge from top to bottom.

virtual instrument (VI) Software module in *LabVIEW* software that can be constructed and used to control robots and other instruments.

virus Simple and often pathogenic organism, typically containing a protein coat and a small DNA or RNA genome.

viscosity Property of a liquid corresponding to its resistance to flowing or spreading.

voice coil Scanner component consisting of a magnet and a current carrying wire that allows the rapid translation of the scanner objective along the x axis.

voltage (V) Difference in electric potential between two terminals in an electrical source.

wettable Extent to which a solid surface allows spreading of a liquid.

wild type Genetics term that refers to a gene or organism that represents the most commonly found allele or strain.

workflow management Process of controlling and recording the physical steps used to generate and store microarray data.

working distance Measure of the physical distance between the end of the objective in a microarray detection instrument and the microarray surface when the lens is in focus.

xenon arc lamp Polychromatic light source used for excitation in microarray imaging devices.

xenon ion Negatively charged atom produced when electrons are lost from xenon in an arc lamp or other device.

zoology Study of the behavior, reproduction, structure, diet, evolution, development, classification, and distribution of animals.

zwitterion Molecule that bears a both positive and a negative charge.

Answers

CHAPTER 1

1. Microscopic array.
2. Microarray analysis uses microarrays, which are characterized by the four basic criteria of ordered, microscopic, planar, and specific. The filter blot experiment probably meets the specificity requirement but not the first three criteria. Phosphorus-32 labels spread signals in two dimensions and fail the ordered array requirement. A filter is a flexible nylon membrane that is not a planar substrate. Spots of 5 mm in diameter are not microscopic.
3. An ordered array is one in which the spots are configured in rows and columns, allowing the use of standard motion-control technology for manufacture and detection. Microscopic spots are measured in microns, allowing tens of thousands of spots and entire genomes to be examined on a single microarray. A planar substrate is flat and rigid, allowing the use of high-precision printing and detection technologies. Specific binding between target and probe molecules provides quantitative assay data.
4. Answers (A) and (B) because both are planar, flat, and rigid. A 50-μm plastic slide (C) is too thin to be planar and rigid, and nitrocellulose (D) is a flexible membrane.
5. Quantitation.

6. A microarray scanner is technologically complex, the discipline of computer science is used for data analysis and mining, and gene expression can be examined on a genomic scale.
7. DNA polymerase (1950s), reverse transcriptase (1960s), recombinant DNA (1970s), and polymerase chain reaction (1980s).
8. Parallelism, miniaturization, and automation.
9. Human.
10. Positive control, negative control, and experimental component.
11. (1) Biological question–Which genes are expressed in this cell type? (2) Sample preparation–mRNA isolation. (3) Biochemical reaction–hybridization. (4) Detection–microarray scanning. (5) Data analysis and modeling–quantify signal intensities.
12. Here are answers for the first three. (1) Follow the protocol. Most protocols provide a good starting point for experimentation. Avoid changing the protocol arbitrarily until functional results have been obtained. (2) Read the manual. Microarray instruments are expensive and can be damaged through misuse. Read the instructions to avoid damaging your equipment. (3) Think small. Microarray assays are quite different from traditional methods. Small volumes have profound implications for concentrations and kinetics.
13. The activated and repressed genes common to all three patients are probably related to the unexplained illness, whereas the unique changes in expression are probably explained by patient differences (gender, race, age, diet, etc.).
14. 997 nondisease, 1 carrier, and 2 disease.
15. Yellow.

CHAPTER 2

1. Periodic table.
2. Sulfur, phosphorus, oxygen, and nitrogen.
3. Electron, proton, nucleus, atom, golf ball, and baseball stadium.
4. HCl: 1 hydrogen and 1 chlorine; CH_3OH: 4 hydrogens, 1 carbon, and 1 oxygen; HCN: 1 hydrogen, 1 carbon, and 1 nitrogen; and C_6H_6: 6 hydrogens and 6 carbons.
5. Six. Electron pair with opposite spin states.
6. 1×10^{22} molecules.
7. 2.0 g.
8. Van der Waals, hydrogen, electrostatic, and covalent.
9. Hydrogen bonds.
10. Ether, ester, and aldehyde.
11. Oxygen, oxygen, oxygen, carbon, and sulfur.
12. Reactants and products.
13. Polar.
14. Nucleophile. Ring strain.
15. Conjugated system of π electrons.

CHAPTER 3

1. Biochemistry.
2. All four are polymers.
3. Base, sugar, and phosphate.
4. Adenine, guanine, cytosine, and thymine.
5. DNA is double-stranded and RNA is single-stranded. DNA contains thymine and RNA contains uracil. DNA contains deoxyribose and RNA contains ribose.
6. A: alanine, E: glutamate, G: glycine, F: phenylalanine, and K: lysine.
7. Carbohydrate.
8. Phe-Tyr-Ile, because it is highly hydrophobic.
9. 5′ CCTGAATTACGT 3′.
10. 160,000.
11. DNA is much more stable to chemical and enzymatic degradation than RNA owing to the absence of an OH group in the 2′ position of deoxyribose compared to presence of a 2′ OH group on ribose. DNA microarrays are more stable than RNA microarrays because the target molecules are more stable.
12. Met-Pro-Stop.
13. Degeneracy. Pool containing 18,432 different oligonucleotides.
14. DNase, RNase, and protease. Small portions of the three unlabeled microarrays could be treated with each of the three enzymes and then reacted with fluorescent nucleic acid or protein probe mixtures. DNase, RNase, and protease would degrade DNA, RNA, and protein target molecules, respectively, producing localized loss of fluorescent signal on the corresponding microarray.
15. A hybrid double helix containing a DNA strand and an RNA strand.

CHAPTER 4

1. Gene.
2. Exons.
3. SNP, insertion, and deletion.
4. Recombinant DNA, gene splicing, or genetic engineering allows researchers to cut and paste any gene sequences of interest to create novel molecules. PCR allows large quantities of any gene sequence to be obtained by primer-directed amplification of a small quantity of starting material.
5. A cDNA molecule is a DNA copy of an mRNA molecule. An EST is a cDNA molecule for which single-pass DNA sequence information is available.
6. Transcription is DNA into mRNA and translation is mRNA into protein.
7. The 10 cellular genes activated by the steroid hormone probably share a common 8-base pair enhancer sequence upstream of the TATA box, which binds a steroid-dependent activator and activates expression of the 10 genes in a coordinate manner.
8. Splicing removes introns and joins exons together, polyadenylation adds a string of A nucleotides on the 3′ end, and capping adds a modified G nucleotide on the 5′ end.

9. mRNA and tRNA.
10. PolyA sequence.
11. Differential gene expression allows different cell types to express unique repertoires of cellular genes, corresponding to subsets of the genes in the genome. The protein subset produced by differential gene expression imparts different functionality to each cell type.
12. No. A patient with a 1:1 ratio of the normal and mutated allele of a recessive disorder is a genetic carrier, who contains one normal chromosome and one mutant chromosome. The protein encoded by the wild-type locus provides sufficient function to mask the illness that is observed in patients with two mutated alleles.
13. Errors in DNA replication (attributed to DNA polymerase) and exposure to environmental mutagens (e.g., ultraviolet light and dietary carcinogens) that damage the DNA and produce sequence changes.
14. Hydroxyl group.
15. Virus, bacterium, yeast, fruit fly, and mouse.
16. No. Expressed genes can interact in a combinatorial manner to produce a nearly infinite number of functional states. If each expressed gene contributes to function and the expression of a gene is modeled in a binary fashion with off and on states as 0 and 1, the number of functional states (f) is calculated as $f = 2^n$, where n is the number of genes in an organism. An organism with 50 more genes than another organism would be capable of 1.1×10^{15} or 1 million trillion additional functional states. It should be clear from this calculation that a mammal with 2.5 times as many genes as an insect (e.g., 25,000 vs. 10,000), or 15,000 additional genes, would have more than enough genes to impart its additional complexity.

CHAPTER 5

1. Dimensional, flat, and planar are three of the eight criteria. Dimensional refers to the physical size of a microarray substrate, and these dimensions should be held to tolerances measured in microns. Dimensionality is important because arrayers, scanners, and other hardware components use substrate size to register the substrate during various phases of the experimental process. Flat refers to the evenness of the surface over small regions of the substrate and is measured in microns. Flat surfaces produce printed spots of improved morphology and more a uniform hybridization layer, increasing the precision of hybridization signals. Planar refers to the rigid, two-dimensional quality of the entire substrate and is measured in microns. Printing pins, ink-jet nozzles, confocal optics, and other hardware components excel if the surface is planar. Planar surfaces (e.g., glass) also tend to be impermeable, which allows a high degree of miniaturization.
2. There are several different possible explanations. An uneven surface that deviates from flatness would produce spots of different diameters, producing a spectrum of signal intensities due to variable target density (measured as fluors/μm^2). A surface with a nonuniform density of functional groups would also produce different signal intensities from spot to spot because different target densities would be present after substrate processing owing to variable target coupling.

3. The highest oligonucleotide sample concentration produces excessive or prohibitive target density in which the coupled oligos are spaced too closely to allow productive interactions with probe molecules. Steric interference on the surface produces a weaker hybridization signal because the efficiency of target–probe interactions is reduced at the highest oligonucleotide concentration.
4. If target is in excess to probe at every probe concentration, the greatest fluorescent signal will be observed with the highest probe concentration (10 μM).
5. Silicon dioxide (SiO_2). No, glass is actually an amorphous solid because of the variable bond angles between SiO_4 tetrahedra in the solid lattice.
6. 2.3 nm (23 Å or 0.0023 μm) corresponds to the smoothest surface.
7. A surface with reactive amine groups carries a positive charge that binds nucleic acids and proteins primarily through electrostatic interactions. The negative charge on DNA and proteins is conferred by the phosphate backbone and negatively charged amino acids (Asp and Glu), respectively. A surface with reactive aldehyde groups is neutral and binds nucleic acids and proteins through covalent interactions. Primary amine groups act as nucleophiles, attacking the aldehyde groups and forming Schiff base linkages upon dehydration. Primary amines can be added to DNA using synthetic linkers, and they are also present in the less reactive aromatic amines in three of the four DNA bases (A, G, and C). Proteins contain primary amines as side chains of lysine and arginine.
8. The density of reactive groups is nonuniform (uneven) on the substrate.
9. The researcher probably forgot to denature the DNA to form single-stranded targets. Double-stranded PCR products will not hybridize with single-stranded fluorescent probe molecules, and the signals will be weak or undetectable at every microarray location.
10. Efficient and stable coupling to an aldehyde surface requires a dehydration reaction, which is prevented from occurring at high humidity. An easy solution is to turn off the humidity control and allow the humidity to drop to ambient (e.g., 40%) to favor the dehydration reaction and stable coupling.
11. Melting point.
12. Greater discrimination on aldehyde compared to amine is the result of end attachment of the target molecules, which increases the hybridization specificity and improves SNP analysis. Molecules on an amine surface couple nonspecifically, which reduces hybridization specificity.

CHAPTER 6

1. Target.
2. Synthesis.
3. The PCR products are longer targets that have greater complementarity to the fluorescent probe molecules and, therefore, produce more intense fluorescent signals, but the specificity is less than the 70mers because the long PCR targets cross-hybridize with probe molecules from multiple members of the gene family.
4. Mouse is the closest organism to human with respect to evolutionary distance and is likely to have the greatest homology across the 500-nucleotide query sequence.

5. The student could grow the moss, isolate mRNA, prepare a cDNA library using a common cloning vector, and amplify moss cDNA inserts using a pair of common primers that hybridize to vector sequences bordering each moss cDNA. This would provide the target elements required for microarray manufacture, without requiring any knowledge of moss DNA sequence.
6. Cyanoethyl group on the 3′ phosphate, a dimethoxytrityl group on the 5′ hydroxyl, and a benzoyl group (A and C) on the primary amine position in the base.
7. A porous glass surface will attach DNA efficiently because it can be treated with silane reagents to produce a high density of chemical coupling groups, but the porous glass surface may trap probe molecules during hybridization and prevent their removal during the wash steps, leading to elevated background.
8. Proteins contain primary amines as side chains of the surface lysine and arginine residues, which function as nucleophiles and attack the electrophilic carbon atoms in the epoxide ring, which are highly reactive because of ring strain. Simply, nucleophilic attack produces a secondary amine and a hydroxyl group on the adjacent carbon atom.
9. A. The addition of 0.1% SDS will reduce the surface tension of the droplet and produce larger spots than a lesser concentration of 0.01% SDS (D). Addition of sodium chloride (B and C) will increase the ionic strength of the printing buffers and *decrease* spot diameter.
10. C, B, A, D.
11. 36% (0.96^{25}).
12. 125. The reduced incorporation of Cy3-dCTP relative to the natural nucleotide is due to the presence of the bulky Cy3 moiety, which interferes with binding of Cy3-dCTP to RT and reduces its incorporation efficiency.
13. B.
14. All.
15. The microarray format is comprehensive, allowing the researcher to examine all of the potential cellular kinase targets in a single experiment and to identify 117 novel targets. The parallel microarray format also allows the simultaneous assessment of kinase efficiency for each target and the classification of cellular proteins, according to the extent of phosphorylation.

CHAPTER 7

1. All.
2. D.
3. 2,727 features/cm^2.
4. 4.6×10^5 oligonucleotides/μm^2.
5. All.
6. Magnetism.
7. Linear.
8. A, B, C.
9. C.
10. All.
11. B, C, D.
12. B, D, E.

13. The MeNPOC group contains only four double bonds and this conjugated system of π electrons is too small to allow absorption of visible range photons, which have less energy than those of ultraviolet light.
14. All.
15. A, C, B, D.

CHAPTER 8

1. Architecture.
2. D.
3. A.
4. B.
5. 3.9.
6. A, B, C, D.
7. The 500-ng/mL probe solution may be saturating the target molecules on the surface, or the greater density of dye molecules may be causing quenching by energy transfer during detection.
8. B, F, E, C, D, A.
9. Larger.
10. B.
11. The sulfate groups add negative charge to the hydrophobic cyanine dyes, improving their solubility in aqueous buffers.
12. The simplest interpretation is that the two dyes have slightly different incorporation efficiencies, which depend on the primary nucleotide sequence and which would produce slightly different emission patterns even though the mRNA samples are identical in both cases.
13. A, B, C, D, F.
14. D.
15. A large NA lens has improved light gathering efficiency and, therefore, increases signal; but it also has a reduced lens working distance and requires greater positional precision for optimal performance.
16. A, B, C, D.
17. B.
18. C.
19. C.
20. Scanning resolution.
21. A.
22. A, B, C.

CHAPTER 9

1. Internet.
2. D.
3. All.
4. Accession number U14680 is the *Homo sapiens* breast and ovarian cancer susceptibility (BRCA1) mRNA, and it contains 5711 nucleotides.
5. The 20 amino acid query sequence most closely matches CFTR from mouse.
6. 54°C.

7. A single mismatch at the end of a relatively long oligonucleotide (50mer) will have little destabilizing effect relative to a perfectly complementary 50mer, and the hybridization efficiencies will be nearly identical for the two oligonucletides, producing similar fluorescent signals.
8. 93.5%.
9. The human target sequences may have little homology with the probe mixture, as would be the case if a human microarray were hybridized at high stringency with a yeast cDNA probe mixture. Alternatively, the single-stranded 70mer human targets may share extensive homology with the *opposite strand* of the probe mixture but would produce very weak signals if the wrong strand were labeled.
10. Primary nucleotide sequence information.
11. 43,200.
12. 370.
13. The high-intensity values exceed the linear range of the 16-bit detector in both data sets, and so the ratio will be <3.5.
14. C.
15. All.
16. The 147 genes are activated in the space alien relative to the human.
17. All.
18. The most likely explanation is that an enhancer element possessing bidirectionality and spacing independence controls the expression of the two genes that face in opposite directions.
19. All.

CHAPTER 10

1. F.
2. B.
3. What is: B, why is: C, and how is: A.
4. System module: A, system: C, system variable: D, and culture: B.
5. A.
6. A: represents the first step in experimental cycle; B: represents the outcome of the experimental cycle, C: represents the fact that each cycle of experimentation brings the researcher closer to the answer, and D: represents the fact that multiple cycles are used to answer a biological question.
7. A, B, D.
8. A, B, C, E, F.
9. B, C, D, E, F.
10. Rule 10, because the human genome contains >25,000 genes and a 10,000-gene microarray would possess only a subset of the system variables.
11. Rule 11 specifies that a universal parallel format is one that contains complete knowledge of each system variable, attainable to a large extent through detailed studies of single genes and proteins.

CHAPTER 11

1. Cleanroom.
2. B.

3. Microarray spots are also extremely small and, therefore, dust particles present a *huge* problem for microarray printing.
4. All.
5. Low humidity can increase electrostatic contamination (i.e., static electricity), causing electrostatic deflection of droplets dispensed by ink-jet devices and producing irregular spot spacing.
6. 6000 K.
7. Paper towels shed tremendous numbers of particles, bare hands contain ribonucleases and other biological contaminants, ether emits volatile chemical vapor contaminants, and used computers contain large quantities of dust drawn in by the cooling fan.
8. B.
9. C.

CHAPTER 12

1. Transcription.
2. All.
3. Function.
4. C.
5. C.
6. A.
7. The flaw in this position is that many metabolic pathways are subject to coordinate regulation by a single gene product. Modification of a single cellular gene could alter the expression of five genes in a metabolic pathway in an entirely safe manner, and such an effect may be essential to achieve the phenotype desired in the GMO. Every crop plant in the world has undergone extensive genetic modification by traditional crop breeding, and GMOs are actually likely to be *safer* than most crop plants because their modifications are precise and known.
8. Genes are regulated in hierarchies or cascades consisting of primary, secondary, and tertiary response genes that function in a temporal manner such that primary response genes regulate secondary response genes, secondary response genes regulate tertiary response genes, and so forth. The increasing number of regulated genes in the 2-h, 24-h and 7-day samples reflects the cumulative effect of a light-signaling cascade in plants.
9. The protein involved in the side effect shares sufficient homology to bind the drug in both species and produce the same side effect.
10. The 213-gene set, which is concordant in all six patients, because concordance suggests a common functional role in liver cancer and, therefore, possible candidacy for an anticancer drug.

CHAPTER 13

1. B.
2. C, D.
3. The disease genes are passed onto the progeny in the course of sexual reproduction.
4. B.

5. D.
6. A.
7. A, B, C, D.
8. C.
9. D.
10. C.
11. B.
12. A.

CHAPTER 14

1. A, C, D, F.
2. The device in Figure 14.1 achieves high-density printing by consolidating the samples physically in a funnel-like apparatus.
3. A, B.
4. The paradox is reconciled by the fact that the flexible nitrocellulose coating is placed on a planar and rigid glass substrate and, therefore, satisfies the planarity requirement of a microarray.
5. All.
6. D.

CHAPTER 15

1. A.
2. C.
3. A.
4. All.
5. A, C, D, E, F.
6. A, B, C, D.
7. B.
8. All.
9. C.
10. B, D, F.
11. A, C, D, E.
12. A, B, C.
13. All.
14. All.

CHAPTER 16

1. All.
2. None.
3. All.

Index